Alloy Physics

Edited by
Wolfgang Pfeiler

1807–2007 Knowledge for Generations

Each generation has its unique needs and aspirations. When Charles Wiley first opened his small printing shop in lower Manhattan in 1807, it was a generation of boundless potential searching for an identity. And we were there, helping to define a new American literary tradition. Over half a century later, in the midst of the Second Industrial Revolution, it was a generation focused on building the future. Once again, we were there, supplying the critical scientific, technical, and engineering knowledge that helped frame the world. Throughout the 20th Century, and into the new millennium, nations began to reach out beyond their own borders and a new international community was born. Wiley was there, expanding its operations around the world to enable a global exchange of ideas, opinions, and know-how.

For 200 years, Wiley has been an integral part of each generation's journey, enabling the flow of information and understanding necessary to meet their needs and fulfill their aspirations. Today, bold new technologies are changing the way we live and learn. Wiley will be there, providing you the must-have knowledge you need to imagine new worlds, new possibilities, and new opportunities.

Generations come and go, but you can always count on Wiley to provide you the knowledge you need, when and where you need it!

William J. Pesce
President and Chief Executive Officer

Peter Booth Wiley
Chairman of the Board

Alloy Physics

A Comprehensive Reference

Edited by
Wolfgang Pfeiler

WILEY-VCH Verlag GmbH & Co. KGaA

The Editor

Professor Wolfgang Pfeiler
Universität Wien
Fakultät für Physik
Dynamik Kondensierter Systeme
Strudlhofgasse 4
1090 Wien
Österreich

Cover

Domain structure in an $L1_2$-ordered single crystal as obtained by Monte-Carlo simulation. Colours show the four possible variants of ordered domains.

All books published by Wiley-VCH are carefully produced. Nevertheless, authors, editors, and publisher do not warrant the information contained in these books, including this book, to be free of errors. Readers are advised to keep in mind that statements, data, illustrations, procedural details or other items may inadvertently be inaccurate.

Library of Congress Card No.: applied for
British Library Cataloguing-in-Publication Data
A catalogue record for this book is available from the British Library.

Bibliographic information published by the Deutsche Nationalbibliothek
The Deutsche Nationalbibliothek lists this publication in the Deutsche Nationalbibliografie; detailed bibliographic data are available in the Internet at ⟨http://dnb.d-nb.de⟩.

© 2007 WILEY-VCH Verlag GmbH & Co. KGaA, Weinheim

All rights reserved (including those of translation into other languages). No part of this book may be reproduced in any form – by photoprinting, microfilm, or any other means – nor transmitted or translated into a machine language without written permission from the publishers. Registered names, trademarks, etc. used in this book, even when not specifically marked as such, are not to be considered unprotected by law.

Typesetting Asco Typesetters, Hong Kong
Printing betz-Druck GmbH, Darmstadt
Binding Litges & Dopf Buchbinderei GmbH, Heppenheim
Wiley Bicentennial Logo Richard J. Pacifico

Printed in the Federal Republic of Germany
Printed on acid-free paper

ISBN 978-3-527-31321-1

Contents

Preface *XIX*

Foreword *XXI*
by Robert W. Cahn

Motto *XXIII*

List of Contributors *XXV*

1	**Introduction** *1*	
	Wolfgang Pfeiler	
1.1	The Importance of Alloys at the Beginning of the Third Millennium *1*	
1.2	Historical Development *5*	
1.2.1	Historical Perspective *5*	
1.2.2	The Development of Modern Alloy Science *9*	
1.3	Atom Kinetics *12*	
1.4	The Structure of this Book *13*	
	References 18	
2	**Crystal Structure and Chemical Bonding** *19*	
	Yuri Grin, Ulrich Schwarz, and Walter Steurer	
2.1	Introduction *19*	
2.2	Factors Governing Formation, Composition and Crystal Structure of Intermetallic Phases *20*	
2.2.1	Mappings of Crystal Structure Types *21*	
2.3	Models of Chemical Bonding in Intermetallic Phases *25*	
2.3.1	Models Based on the Valence (or Total) Electron Numbers *25*	
2.3.2	Quantum Mechanical Models for Metallic Structures *29*	
2.3.3	Electronic Closed-Shell Configurations and Two-Center Two-Electron Bonds in Intermetallic Compounds *31*	
2.3.3.1	Zintl–Klemm Approach *32*	

Alloy Physics: A Comprehensive Reference. Edited by Wolfgang Pfeiler
Copyright © 2007 WILEY-VCH Verlag GmbH & Co. KGaA, Weinheim
ISBN: 978-3-527-31321-1

2.3.3.2	Extended 8 − N Rule	33
2.3.3.3	Bonding Models in Direct Space	34
2.4	Structure Types of Intermetallic Compounds	36
2.4.1	Classification of the Crystal Structures of Intermetallic Compounds	37
2.4.2	Crystal Structures Derived from the Closest Packings of Equal Spheres	37
2.4.3	Crystal Structures Derived from the Close Packings of Equal Spheres	40
2.4.4	Crystal Structures Derived from the Packings of the Spheres of Different Sizes	43
2.4.5	Selected Crystal Structures with Complex Structural Patterns	44
2.5	Quasicrystals	48
2.5.1	Introduction	48
2.5.2	Quasiperiodic Structures in Direct and Reciprocal Space	50
2.5.3	Formation and Stability	52
2.5.4	Structures of Decagonal Quasicrystals (DQCs)	53
2.5.5	Structures of Icosahedral Quasicrystals	55
2.6	Outlook	59
	References	60

3 Solidification and Grown-in Defects 63
Thierry Duffar

3.1	Introduction: the Solid–Liquid Interface	63
3.1.1	Structure of the Solid–Liquid Interface	63
3.1.2	Kinetics of the Solid–Liquid Interface	65
3.1.3	Chemistry of the Solid–Liquid Interface: the Segregation Problem	67
3.1.4	Temperature of the Solid–Liquid Interface	69
3.2	Solidification Structures	70
3.2.1	The Interface Stability and Cell Periodicity	71
3.2.2	Dendrites	74
3.2.2.1	Different Types of Dendrites	75
3.2.2.2	Kinetics of Columnar Dendrites	78
3.2.2.3	Kinetics of Equiaxed Dendrites	81
3.2.2.4	Characteristic Dimensions of the Dendrite	83
3.2.2.5	Microsegregation	85
3.2.3	Rapid Solidification	86
3.2.3.1	Absolute Stability and Diffusionless Solidification	86
3.2.3.2	Nonequilibrium Phase Diagrams	87
3.2.3.3	Structure of the Rapidly Solidified Phase	87
3.2.4	Eutectic Structures	90
3.2.4.1	Size of the Eutectic Structure	90
3.3	Defects in Single and Polycrystals	93
3.3.1	Defects in Single Crystals	94

3.3.1.1	Point Defects	94
3.3.1.2	Twins	97
3.3.1.3	Grains	98
3.3.2	Grain Structure of an Alloy	101
3.3.2.1	Equiaxed Growth in Presence of Refining Particles	103
3.3.2.2	Columnar to Equiaxed Transition	107
3.3.3	Macro- and Mesosegregation	110
3.4	Outlook	114
	References	117
4	**Lattice Statics and Lattice Dynamics**	**119**
	Véronique Pierron-Bohnes and Tarik Mehaddene	
4.1	Introduction: The Binding and Atomic Interaction Energies	119
4.2	Elasticity of Crystalline Lattices	124
4.2.1	Linear Elasticity	125
4.2.2	Elastic Constants	125
4.2.3	Cases of Cubic and Tetragonal Lattices	127
4.2.4	Usual Elastic Moduli	128
4.2.5	Link with Sound Propagation	130
4.3	Lattice Dynamics and Thermal Properties of Alloys	132
4.3.1	Normal Modes of Vibration in the Harmonic Approximation	133
4.3.1.1	Classical Theory	133
4.3.1.2	Diatomic Linear Chain	136
4.3.1.3	Quantum Theory	138
4.3.1.4	Phonon Density of States	141
4.3.1.5	Lattice Specific Heat	143
4.3.1.6	Debye's Model	144
4.3.1.7	Elastic Waves in Cubic Crystals	146
4.3.1.8	Vibrational Entropy	147
4.4	Beyond the Harmonic Approximation	149
4.4.1	Thermal Expansion	150
4.4.2	Thermal Conductivity	151
4.4.3	Soft Phonon Modes and Structural Phase Transition	153
4.5	Experimental Investigation of the Normal Modes of Vibration	156
4.5.1	Raman Spectroscopy	156
4.5.2	Inelastic Neutron Scattering	157
4.6	Phonon Spectra and Migration Energy	160
4.7	Outlook	165
	References	168
5	**Point Defects, Atom Jumps, and Diffusion**	**173**
	Wolfgang Püschl, Hiroshi Numakura, and Wolfgang Pfeiler	
5.1	Point Defects	173
5.1.1	A Brief Overview	173
5.1.1.1	Types of Point Defects	173

5.1.1.2	Formation of Equilibrium and Nonequilibrium Defects *175*
5.1.1.3	Mobility *178*
5.1.1.4	Experimental Techniques *179*
5.1.2	Point Defects in Pure Metals and Dilute Alloys *187*
5.1.2.1	Vacancies *187*
5.1.2.2	Self-Interstitial Atoms *193*
5.1.2.3	Solute Atoms *195*
5.1.3	Point Defects in Ordered Alloys *197*
5.1.3.1	Point Defects and Properties of the Material *197*
5.1.3.2	Statistical Thermodynamics *199*
5.1.3.3	Equilibrium Concentrations – Examples *208*
5.1.3.4	Abundant Vacancies in some Intermetallic Compounds *213*
5.2	Defect Migration: Microscopic Diffusion *217*
5.2.1	The Single Atom Jump *217*
5.2.1.1	Transition State Theory *217*
5.2.1.2	Alternative Methods *221*
5.2.2	Solid Solutions *222*
5.2.2.1	Random Walk *222*
5.2.2.2	Correlated Walk – the Interaction of Defect and Atom *228*
5.2.2.3	Diffusion Walk with Chemical Driving Force *234*
5.2.2.4	Diffusion Walk in an Inhomogeneous Crystal *237*
5.2.3	Atom Migration in Ordered Alloys *238*
5.2.3.1	Experimental Approach to Atom Kinetics in Ordered Alloys *238*
5.2.3.2	Jumps Within and Between Sublattices *239*
5.2.3.3	Jump Cycles and Cooperative Atom Jumps *246*
5.3	Statistical Methods: from Single Jump to Configuration Changes *252*
5.3.1	Master Equation Method *253*
5.3.2	Continuum Approaches to Microscopic Diffusion and their Interrelationship with Atom Jump Statistics *253*
5.3.3	Path Probability Method *255*
5.3.4	Monte Carlo Simulation Method *255*
5.4	Macroscopic Diffusion *256*
5.4.1	Formal Description *256*
5.4.1.1	Fick's Laws *256*
5.4.1.2	Nonreciprocal Diffusion, the Kirkendall Effect *259*
5.4.1.3	Nonideal Solutions *261*
5.4.2	Phase Transformations as Diffusion Phenomena *263*
5.4.2.1	Spinodal Decomposition *263*
5.4.2.2	Nucleation, Growth, Coarsening *264*
5.4.3	Enhanced Diffusion Paths *265*
5.4.3.1	Dislocation-Core Diffusion *266*
5.4.3.2	Grain-Boundary Diffusion *268*
5.4.3.3	Diffusion along Interfaces and Surfaces *270*
5.5	Outlook *272*
	References *274*

6 Dislocations and Mechanical Properties *281*
Daniel Caillard

6.1 Introduction *281*
6.2 Thermally Activated Mechanisms *283*
6.2.1 Introduction to Thermal Activation *283*
6.2.2 Interactions with Solute Atoms *285*
6.2.2.1 General Aspects *285*
6.2.2.2 Low Temperatures (Domain 2, Interaction with Fixed Solute Atoms) *286*
6.2.2.3 Intermediate Temperatures (Domain 3, Stress Instabilities) *289*
6.2.2.4 High Temperatures (Domain 4, Diffusion-Controlled Glide) *291*
6.2.3 Forest Mechanism *292*
6.2.4 Peierls-Type Friction Forces *293*
6.2.4.1 The Kink-Pair Mechanism *293*
6.2.4.2 Locking–Unlocking Mechanism *295*
6.2.4.3 Transition between Kink-Pair and Locking–Unlocking Mechanisms *297*
6.2.4.4 Observations of Peierls-Type Mechanisms *298*
6.2.5 Cross-Slip in fcc Metals and Alloys *305*
6.2.5.1 Elastic Calculations *305*
6.2.5.2 Atomistic Calculations *307*
6.2.5.3 Experimental Results *307*
6.2.6 Dislocation Climb *309*
6.2.6.1 Emission of Vacancies at Jogs *309*
6.2.6.2 Diffusion of Vacancies from Jogs *310*
6.2.6.3 Jog Density and Jog-Pair Mechanism *311*
6.2.6.4 Effect of Over- (Under-) Saturations of Vacancies: Chemical Force *313*
6.2.6.5 Stress Dependence of the Dislocation Climb Velocity *314*
6.2.6.6 Experimental Results *314*
6.2.7 Conclusions on Thermally Activated Mechanisms *316*
6.3 Hardening and Recovery *316*
6.3.1 Dislocation Multiplication versus Exhaustion *317*
6.3.1.1 Dislocation Sources *318*
6.3.1.2 Dislocation Exhaustion and Annihilation *320*
6.3.2 Dislocation–Dislocation Interaction and Internal Stress: the Taylor Law *321*
6.3.3 Hardening Stages in fcc Metals and Alloys *323*
6.3.3.1 Stage II (Linear Hardening) *324*
6.3.3.2 Stage III *329*
6.3.3.3 Stage IV *330*
6.3.3.4 Strain-Hardening in Intermetallic Alloys *330*
6.4 Complex Behavior *330*
6.4.1 Yield Stress Anomalies *330*
6.4.1.1 Dynamic Strain Aging *331*

6.4.1.2	Cross-Slip Locking 332
6.4.2	Fatigue 333
6.4.2.1	Microstructure of Fatigued Metals and Alloys 334
6.4.2.2	Comparison with Stages II and III of Monotonic Strain Hardening 335
6.4.2.3	Intrusions, Extrusions and Fracture 335
6.4.2.4	Conclusions 336
6.4.3	Strength of Nanocrystalline Alloys and Thin Layers 336
6.4.3.1	The Hall–Petch Law (Grain Size $D \geq 20$ nm) 337
6.4.3.2	Hall–Petch Law Breakdown (Grain Size $D \leq 20$ nm) 337
6.4.4	Fracture 338
6.4.5	Quasicrystals 339
6.5	Outlook 342
	References 342

7 Phase Equilibria and Phase Transformations 347
Brent Fultz and Jeffrey J. Hoyt

7.1	Alloy Phase Diagrams 347
7.1.1	Solid Solutions 347
7.1.2	Free Energy and the Lever Rule 351
7.1.3	Common Tangent Construction 353
7.1.4	Unmixing and Continuous Solid Solubility Phase Diagrams 354
7.1.5	Eutectic and Peritectic Phase Diagrams 356
7.1.6	More Complex Phase Diagrams 357
7.1.7	Atomic Ordering 359
7.1.8	Beyond Simple Models 362
7.1.9	Entropy of Configurations 363
7.1.10	Principles of Phonon Entropy 365
7.1.11	Trends of Phonon Entropy 367
7.1.12	Phonon Entropy at Elevated Temperatures 369
7.2	Kinetics and the Approach to Equilibrium 371
7.2.1	Suppressed Diffusion in the Solid (Nonequilibrium Compositions) 371
7.2.2	Nucleation Kinetics 373
7.2.3	Suppressed Diffusion in the Liquid (Glasses) 374
7.2.4	Suppressed Diffusion in a Solid Phase (Solid-State Amorphization) 375
7.2.5	Combined Reactions 376
7.2.6	Statistical Kinetics of Phase Transformations 377
7.2.7	Kinetic Pair Approximation 378
7.2.8	Equilibrium State of Order 380
7.2.9	Kinetic Paths 380
7.3	Nucleation and Growth Transformations 382
7.3.1	Definitions 382

7.3.2	Fluctuations and the Critical Nucleus	384
7.3.3	The Nucleation Rate	387
7.3.4	Time-Dependent Nucleation	391
7.3.5	Effect of Elastic Strain	393
7.3.6	Heterogeneous Nucleation	395
7.3.7	The Kolmogorov–Johnson–Mehl–Avrami Growth Equation	397
7.4	Spinodal Decomposition	399
7.4.1	Concentration Fluctuations and the Free Energy of Solution	400
7.4.2	The Diffusion Equation	402
7.4.3	Effects of Elastic Strain Energy	404
7.5	Martensitic Transformations	406
7.5.1	Characteristics of Martensite	406
7.5.2	Massive and Displacive Transformations	411
7.5.3	Bain Strain Mid-Lattice Invariant Shear	412
7.5.4	Martensite Crystallography	413
7.5.5	Nucleation and Dislocation Models of Martensite	415
7.5.6	Soft Mode Transitions, the Clapp Lattice Instability Model	417
7.6	Outlook	418
	References	420

8 Kinetics in Nonequilibrium Alloys 423
Pascal Bellon and Georges Martin

8.1	Relaxation of Nonequilibrium Alloys	424
8.1.1	Coherent Precipitation: Nothing but Solid-State Diffusion	425
8.1.2	Cluster Dynamics, Nucleation Theory, Diffusion Equations: Three Tools for Describing Kinetic Pathways	426
8.1.3	Cluster Dynamics	427
8.1.3.1	Dilute Alloy at Equilibrium	427
8.1.3.2	Fluctuations in the Gas of Clusters at Equilibrium	429
8.1.3.3	Relaxation of a Nonequilibrium Cluster Gas	429
8.1.4	Classical Nucleation Theory	432
8.1.4.1	Summary of CNT	432
8.1.4.2	Source of Fluctuations Consistent with CNT	433
8.1.4.3	A First Application	435
8.1.5	Kinetics of Concentration Fields	436
8.1.6	Conclusion	438
8.2	Driven Alloys	438
8.2.1	Examples of Driven Alloys	439
8.2.1.1	Alloys Subjected to Sustained Irradiation	439
8.2.1.2	Alloys Subjected to Sustained Plastic Deformation	447
8.2.1.3	Alloys Subjected to Sustained Electrochemical Exchanges	449
8.2.2	Identification of the Relevant Control Parameters: Toward a Dynamical Equilibrium Phase Diagram	450
8.2.3	Theoretical Approaches and Simulation Techniques	454

8.2.3.1	Molecular Dynamics Simulations 455
8.2.3.2	Microscopic Master Equation 456
8.2.3.3	Kinetic Monte Carlo Simulations 458
8.2.3.4	Kinetics of Concentration Fields under Irradiation 460
8.2.3.5	Nucleation Theory under Irradiation 466
8.2.4	Self-Organization in Driven Alloys: Role of Length Scales of the External Forcing 468
8.2.4.1	Compositional Patterning under Irradiation 469
8.2.4.2	Patterning of Chemical Order under Irradiation 478
8.2.4.3	Compositional Patterning under Plastic Deformation 480
8.2.5	Practical Applications and Extensions 481
8.2.5.1	Tribochemical Reactions 481
8.2.5.2	Pharmaceutical Compounds Synthesized by Mechanical Activation 483
8.3	Outlook 484
	References 484

9 Change of Alloy Properties under Dimensional Restrictions 491
Hirotaro Mori and Jung-Goo Lee

9.1	Introduction 491
9.2	Instrumentation for in-situ Observation of Phase Transformation of Nanometer-Sized Alloy Particles 492
9.3	Depression of the Eutectic Temperature and its Relevant Phenomena 494
9.3.1	Atomic Diffusivity in Nanometer-Sized Particles 494
9.3.2	Eutectic Temperature in Nanometer-Sized Alloy Particles 496
9.3.3	Structural Instability 500
9.3.4	Thermodynamic Discussion 503
9.3.4.1	Gibbs Free Energy in Nanometer-Sized Alloy Systems 503
9.3.4.2	Result of Calculations 505
9.4	Solid/Liquid Two-Phase Microstructure 508
9.4.1	Solid–Liquid Phase Transition 508
9.4.2	Two-Phase Microstructure 514
9.5	Solid Solubility in Nanometer-Sized Alloy Particles 518
9.6	Summary and Future Perspectives 521
	References 522

10 Statistical Thermodynamics and Model Calculations 525
Tetsuo Mohri

10.1	Introduction 525
10.2	Statistical Thermodynamics on a Discrete Lattice 527
10.2.1	Description of Atomic Configuration 527
10.2.2	Internal Energy 534
10.2.3	Entropy and Cluster Variation Method 536

10.2.4	Free Energy *542*	
10.2.5	Relative Stability and Intrinsic Stability *544*	
10.2.6	Atomistic Kinetics by the Path Probability Method *549*	
10.3	Statistical Thermodynamics on Continuous Media *552*	
10.3.1	Ginzburg–Landau Free Energy *552*	
10.3.2	Diffusion Equation and Time-Dependent Ginzburg–Landau Equation *554*	
10.3.3	Width of an Interface *557*	
10.3.4	Interface Velocity *559*	
10.4	Model Calculations *560*	
10.4.1	Calculation of a Phase Diagram *561*	
10.4.1.1	Ground-State Analysis *561*	
10.4.1.2	Effective Cluster Interaction Energy *564*	
10.4.1.3	Phase Diagram *568*	
10.4.2	Microstructural Evolution Calculated by the Phase Field Method *572*	
10.4.2.1	Hybrid Model *572*	
10.4.2.2	Toward the First-Principles Phase Field Calculation *576*	
10.5	Future Scope and Outlook *580*	
	Appendix: CALPHAD Free Energy *582*	
	References *585*	
11	**Ab-Initio Methods and Applications** *589*	
	Stefan Müller, Walter Wolf, and Raimund Podloucky	
11.1	Introduction *589*	
11.2	Theoretical Background *590*	
11.2.1	Density Functional Theory *590*	
11.2.2	Computational Methods *594*	
11.2.3	Elastic Properties *598*	
11.2.4	Vibrational Properties *601*	
11.3	Applications *606*	
11.3.1	Structural and Phase Stability *606*	
11.3.2	Point Defects *612*	
11.3.3	Diffusion Processes *616*	
11.3.4	Impurity Effects on Grain Boundary Cohesion *622*	
11.3.5	Toward Multiscale Modeling: Cluster Expansion *625*	
11.3.6	Search for Ground-State Structures *639*	
11.3.7	Ordering and Decomposition Phenomena in Binary Alloys *641*	
11.4	Outlook *648*	
	References *649*	
12	**Simulation Techniques** *653*	
	Ferdinand Haider, Rafal Kozubski, and T.A. Abinandanan	
12.1	Introduction *653*	
12.2	Molecular Dynamics Simulations *654*	

12.2.1	Basic Ideas	*654*
12.2.2	Atomic Interaction, Potential Models	*656*
12.2.2.1	Pairwise Interaction	*656*
12.2.2.2	Many-Body Potentials, the EAM Method	*657*
12.2.3	Practical Considerations	*659*
12.2.4	Different Thermodynamic Ensembles: Thermostats, Barostats	*659*
12.2.5	Implementation of MD Algorithms	*661*
12.2.6	Practical Aspects: Time Steps	*662*
12.2.7	Evaluation of Data: Use of Correlation Functions	*662*
12.2.8	Applications to Alloys, Alloy Dynamics, and Alloy Kinetics	*664*
12.3	Monte Carlo Simulations	*667*
12.3.1	Foundations of Stochastic Processes – Markov Chains and the Master Equation	*667*
12.3.2	The Idea of Sampling	*668*
12.3.3	Markov Chains as a Tool for Importance Sampling	*670*
12.3.4	General Applicability	*671*
12.3.4.1	Simulation and Characterization of System Properties in Thermodynamic Equilibrium	*671*
12.3.4.2	Simulation of Relaxation Processes Toward Equilibrium	*673*
12.3.4.3	Simulation of Nonequilibrium Processes and Transport Phenomena	*673*
12.3.5	Limitations: Finite-Size Effects and Boundary Conditions	*674*
12.3.6	Numerical Implementation of MC	*675*
12.3.6.1	Classical Realization of Markov Chains	*675*
12.3.6.2	"Residence Time" Algorithm	*676*
12.3.6.3	The Problem of Time Scales	*677*
12.3.7	Applications to Alloys	*678*
12.3.7.1	General Assumptions	*678*
12.3.7.2	Physical Model of an Alloy	*679*
12.3.8	Practical Aspects	*681*
12.3.9	Review of Current Applications in Studies of Alloys	*682*
12.3.9.1	Computation of Phase Diagrams using Grandcanonical Ensemble	*683*
12.3.9.2	Reverse and Inverse Monte Carlo Methods: from Experimental SRO Parameters to Atomic Interaction Energies	*683*
12.3.10	Going beyond the Ising Model and Rigid-Lattice Simulations	*685*
12.3.11	Monte Carlo Simulations in View of other Techniques of Alloy Modeling	*686*
12.4	Phase Field Models	*686*
12.4.1	Introduction	*686*
12.4.2	Cahn–Hilliard Model	*687*
12.4.2.1	Energetics	*687*
12.4.2.2	Interfacial Energy and Width	*689*
12.4.2.3	Dynamics	*691*
12.4.3	Numerical Implementation	*691*

12.4.4	Application: Spinodal Decomposition	693
12.4.5	Cahn–Allen Model	694
12.4.5.1	Kinetics	695
12.4.6	Generalized Phase Field Models	696
12.4.6.1	Key Features of Phase Field Models	696
12.4.6.2	Precipitation of an Ordered Phase	697
12.4.6.3	Grain Growth in Polycrystals	698
12.4.6.4	Solidification	700
12.4.7	Other Topics	700
12.4.7.1	Anisotropy in Interfacial Energy	700
12.4.7.2	Elastic Strain Energy	701
12.5	Outlook	702
	Appendix	702
	References	703

13 High-Resolution Experimental Methods 707

13.1 High-Resolution Scattering Methods and Time-Resolved Diffraction 707
Bogdan Sepiol and Karl F. Ludwig

13.1.1	Introduction: Theoretical Concepts, X-Ray, and Neutron Scattering Methods	707
13.1.2	Magnetic Scattering	710
13.1.2.1	Magnetic Neutron Scattering	710
13.1.2.2	Magnetic X-Ray Scattering	715
13.1.3	Spectroscopy	721
13.1.3.1	Coherent Time-Resolved X-Ray Scattering	722
13.1.3.1.1	Homodyne X-Ray Studies of Equilibrium Fluctuation Dynamics	723
13.1.3.1.2	Heterodyne X-Ray Studies of Equilibrium Fluctuation Dynamics	725
13.1.3.1.3	Studies of Critical Fluctuations with Microbeams	726
13.1.3.1.4	Coherent X-Ray Studies of the Kinetics of Nonequilibrium Systems	726
13.1.3.1.5	Coherent X-Ray Studies of Microscopic Reversibility	729
13.1.3.2	Phonon Excitations	729
13.1.3.2.1	Inelastic X-Ray Scattering	730
13.1.3.2.2	Nuclear Inelastic Scattering	732
13.1.3.3	Quasielastic Scattering: Diffusion	733
13.1.3.3.1	Quasielastic Methods: Mössbauer Spectroscopy and Neutron Scattering	738
13.1.3.3.2	Nuclear Resonant Scattering of Synchrotron Radiation	741
13.1.3.3.3	Pure Metals and Dilute Alloys	743
13.1.3.3.4	Ordered Alloys	744
13.1.3.3.5	Amorphous Materials	745
13.1.4	Time-Resolved Scattering	749

13.1.4.1 Technical Capabilities 750
13.1.4.2 Time-Resolved Studies – Examples 751
13.1.5 Diffuse Scattering from Disordered Alloys 756
13.1.5.1 Metallic Glasses and Liquids 757
13.1.5.2 Diffuse Scattering from Disordered Crystalline Alloys 759
13.1.6 Surface Scattering – Atomic Segregation and Ordering near Surfaces 762
13.1.7 Scattering from Quasicrystals 763
13.1.8 Outlook 764
References 765

13.2 High-Resolution Microscopy 774
Guido Schmitz and James M. Howe

13.2.1 Surface Analysis by Scanning Probe Microscopy 775
13.2.1.1 Functional Principle of Scanning Tunneling and Atomic Force Microscopy 776
13.2.1.2 Modes of Measurement in AFM 779
13.2.1.3 Cantilever Design for the AFM 781
13.2.1.4 Exemplary Studies by Scanning Probe Microscopy 783
13.2.1.4.1 Chemical Contrast by STM and Surface Ordering 783
13.2.1.4.2 Microstructure Characterization and Surface Topology by AFM 785
13.2.1.4.3 Imaging of Nanomagnets by Magnetic Force Microscopy 789
13.2.2 High-Resolution Transmission Electron Microscopy and Related Techniques 791
13.2.2.1 Principles of Image Formation in and Practical Aspects of High-Resolution Transmission Electron Microscopy 793
13.2.2.1.1 Principles of Image Formation 793
13.2.2.1.2 Practical Aspects of HRTEM 796
13.2.2.2 In-Situ Hot-Stage High-Resolution Transmission Electron Microscopy 797
13.2.2.3 Examples of HRTEM Studies of Dislocation and Interphase Boundaries 799
13.2.2.3.1 Disclinations in Mechanically Milled Fe Powder 799
13.2.2.3.2 Interphase Boundaries in Metal Alloys 802
13.2.2.3.3 Diffuse Interface in Cu–Au 802
13.2.2.3.4 Partly Coherent Interfaces in Al–Cu 807
13.2.2.3.5 Incoherent Interfaces in Ti–Al 811
13.2.3 Local Analysis by Atom Probe Tomography 817
13.2.3.1 The Functional Principle of Atom Probe Tomography 819
13.2.3.2 Two-Dimensional Single-Ion Detector Systems 823
13.2.3.3 Ion Trajectories and Image Magnification 827
13.2.3.4 Tomographic Reconstruction 830
13.2.3.5 Accuracy of the Reconstruction 833
13.2.3.6 Specimen Preparation 836
13.2.3.7 Examples of Studies by Atom Probe Tomography 837

13.2.3.7.1	Decomposition in Supersaturated Alloys	837
13.2.3.7.2	Nucleation of the First Product Phase	843
13.2.3.7.3	Diffusion in Nanocrystalline Thin Films	847
13.2.3.7.4	Thermal Stability of GMR Sensor Layers	850
13.2.4	Future Development and Outlook	853
	References	857

14	**Materials and Process Design**	**861**
14.1	**Soft and Hard Magnets**	**861**
	Roland Grössinger	
14.1.1	What do "Soft" and "Hard" Magnetic Mean?	861
14.1.1.1	Intrinsic Properties Determining the Hysteresis Loop (Anisotropy, Magnetostriction)	863
14.1.1.2	Extrinsic Properties – Microstructure	864
14.1.2	Soft Magnetic Materials	865
14.1.2.1	Pure Fe and Fe–Si	867
14.1.2.2	Ni–Fe Alloys	868
14.1.2.3	Soft Magnetic Ferrites	869
14.1.2.4	Amorphous Materials	871
14.1.2.5	Nanocrystalline Materials	872
14.1.3	Hard Magnetic Materials	873
14.1.3.1	AlNiCo	876
14.1.3.2	Ferrites	877
14.1.3.3	Sm–Co	878
14.1.3.4	Nd–Fe–B	879
14.1.3.5	Nanocrystalline Materials	880
14.1.3.6	Industrial Nanocrystalline Hard Magnetic Materials	882
14.1.4	Outlook	883
	References	883

14.2	**Invar Alloys**	**885**
	Peter Mohn	
14.2.1	Introduction and General Remarks	885
14.2.2	Spontaneous Volume Magnetostriction	888
14.2.3	The Modeling of Invar Properties	889
14.2.4	A Microscopic Model	893
14.2.5	Outlook	894
	References	895

14.3	**Magnetic Media**	**895**
	Laurent Ranno	
14.3.1	Data Storage	895
14.3.1.1	Information Storage	895

14.3.1.2	Competing Physical Effects	896
14.3.1.3	Magnetic Storage	897
14.3.2	Magnetic Recording Media	905
14.3.2.1	Particulate Media	905
14.3.2.2	Continuous Media – Film Media	906
14.3.2.3	Perpendicular Recording	907
14.3.3	Outlook	909
	Further Reading	910

14.4 Spin Electronics (Spintronics) 911
Laurent Ranno

14.4.1	Electrical Transport in Conductors	911
14.4.1.1	Conventional Transport	911
14.4.1.2	Role of Disorder	913
14.4.1.3	Transport in Magnetic Conductors	914
14.4.2	Magnetoresistance	915
14.4.2.1	Cyclotron Magnetoresistance	916
14.4.2.2	Anisotropic Magnetoresistance (AMR)	916
14.4.2.3	Giant MR (GMR) and Tunnel MR (TMR)	916
14.4.2.4	Magnetic Field Sensors	918
14.4.2.5	Magnetic RAM	920
14.4.3	Outlook	921
	Further Reading	921

14.5 Phase-Change Media 921
Takeo Ohta

14.5.1	Electrically and Optically Induced Writing and Erasing Processes	921
14.5.2	Phase-Change Dynamic Model	925
14.5.3	Alternative Functions	933
14.5.4	Outlook	938
	References	938

14.6 Superconductors 939
Harald W. Weber

14.6.1	Fundamentals	939
14.6.2	Superconducting Materials	944
14.6.3	Technical Superconductors	946
14.6.4	Applications	952
	Further Reading	953

Index 955

Preface

Just after Wiley-VCH's consulting editor, Ed Immergut, invited me to edit a book on "alloy physics", I hesitated greatly before agreeing, as I was a bit shocked by the breadth of the topic to be covered in such a book. I therefore first suggested a volume on *Atom Jumps in Intermetallics* to reduce the area correspondingly, but finally agreed to *Alloy Physics* because it would appeal to a much broader audience including physicists, materials scientists, physical chemists, and metallurgists in academic as well as in industrial laboratories.

Writing "alloy physics" into an Internet search engine, I suddenly realized that indeed there seems to be a need for such a reference book. I therefore started to prepare a tentative outline, and found that this could be done in a straightforward way and in a comparatively short time. Another surprise was the great number of very positive referee reports on my outline – four of them arrived within just two days after I sent out my outline!

The next step was to find and motivate chapter authors to take part in the project. The resulting book now contains 14 chapters, written by an international team of authors, some of whom have written their text by correspondence across the Atlantic.

An essential point is that the book does not avoid overlaps between chapters: on the contrary, I hope that looking at fundamental problems in alloy physics from different aspects will make the book even more interesting and useful.

I am very grateful to my colleague Wolfgang Püschl for continual discussions, comments, and suggestions, to our student David Reith for the use of one of his simulation pictures for the cover, and last but not least to my dear wife Heidi for her great patience and encouragement.

Vienna, May 2007 *Wolfgang Pfeiler*

Foreword*

The word 'alloy', from the old French, originally carried overtones of base metals contaminating noble ones, and the word is still sometimes used in this sense in literary metaphor. Today, however, we know that the true practical value of an alloy derives from the knowledge and ingenuity that has gone into the choice of constituents, proportions and heat-treatment, and an aristocratic alloy often emerges from a mix of base metals. So, 'nobility' in alloys is a function of understanding, not of scarcity in the earth's crust.

This important book sets out the sources of the modern understanding of alloys, in great depth. The book represents the intimate mingling of physics with metallurgy – whether that is called alloy physics or physical metallurgy is of no consequence. It is some time since an overview of this whole field has been published and the time is very ripe for this new venture. The book not only mixes theory and experiment in a very helpful way but it also comprehensively covers the world community of alloy physicists.

Sitting at the centre of the spider's web is a notable Austrian physicist: Wolfgang Pfeiler's office is in the Boltzmanngasse, named after the genial theorist who first properly understood the entropy of mixtures. All of us in the profession owe Wolfgang a great debt for his skill and persuasiveness in drawing together such a distinguished array of contributors.

Cambridge, March 2007 *Robert W. Cahn, FRS*

---

* Robert Wolfgang Cahn (1924–2007) passed away on April 9th, 2007.

Professor Robert Cahn was one of the first promoters of 'Materials Science' (see his recent book 'The Coming of Materials Science', Pergamon 2001) and soon became an international leader in this field. Aside of his own intense scientific research he was editor of many compendia and book series and founded four journals. His memoirs have recently been published ('The Art of Belonging', Book Guild Publishing 2005).

Alloy Physics: A Comprehensive Reference. Edited by Wolfgang Pfeiler
Copyright © 2007 WILEY-VCH Verlag GmbH & Co. KGaA, Weinheim
ISBN: 978-3-527-31321-1

Motto

Georgii Agricolae, *De re metallica libri XII, quibus Officia, Instrumenta, Machinae, ac omnia denique ad Metallicam spectantia, non modo luculentissime describuntur, sed & per effigies, suis locis insertas, adiunctis Latinis, Germanisque appellationibus ita ob oculos ponuntur, ut clarius tradi non possint. Eiusdem de animantibus subterraneis Liber, ab Autore recognitus: cum Indicibus diversis, quicquid in opere tractatum est, pulchre demonstrantibus.* Basileae MDLVI:

Metallicus praeterea sit oportet multarum artium & disciplinarum non ignarus: Primo Philosophiae, ut subterraneorum ortus & causas, naturasque noscat: Nam ad fodiendas venas faciliore & commodiore uia perveniet, & ex effossis uberiores capiet fructus. Secundo Medicinae, ut fossoribus & aliis operariis providere possit, ne in morbos, quibus prae caeteris urgentur, incidant: aut si inciderint, vel ipse eis curationes adhibere, vel ut medici adhibeant curare. Tertio Astronomiae, ut cognoscat coeli partes, atque ex eis venarum extensiones iudicet. Quarto Mensurarum disciplinae, ut & metiri queat, quam alte fodiendus sit puteus, ut pertineat ad cuniculum usque qui eo agitur, & certos cuique fodinae, praesertim in profundo, constituere fines terminosque. Tum numerorum disciplinae sit intelligens, ut sumptus, qui in machinas & fossiones habendi sunt, ad calculos revocare possit. Deinde Architecturae, ut diversas machinas substructionesque ipse fabricari, vel magis fabricandi rationem aliis explicare queat. Postea Picturae, ut machinarum exempla deformare possit. Postremo Iuris, maxime metallici sit peritus, ut & alteri nihil surripiat, & sibi petat non iniquum, munusque aliis de iure respondendi sustineat. Itaque necesse est ut is, cui placent certae rationes & praecepta rei metallicae hos aliosque nostros libros studiose diligenterque legat, aut de quaque re consulat experientes metallicos, sed paucos inveniet gnaros totius artis.

"Furthermore, there are many arts and sciences of which a miner should not be ignorant. First there is Philosophy, that he may discern the origin, cause, and nature of subterranean things; for then he will be able to dig out the veins easily and advantageously, and to obtain more abundant results from his mining. Secondly, there is Medicine, that he may be able to look after his diggers and other workmen, that they do not meet with those diseases to which they are more liable than workmen in other occupations, or if they do meet with them, that he himself may be able to heal them or may see that the doctors do so. Thirdly follows Astronomy, that he may know the divisions of the heavens and from them judge the direction of the veins. Fourthly, there is the science of Surveying that he may be able to estimate how deep a shaft should be sunk to reach the tunnel which is being driven to it, and to determine the limits and boundaries in these workings, especially in depth. Fifthly, his knowledge in Arithmetical Science should be such that he may calculate the cost to be incurred in the machinery and the working of the mine. Sixthly, his learning must comprise Architecture, that he himself may construct various machines and timber work required underground, or that he may be able to explain the method of the construction to others. Next, he must have knowledge of Drawing, that he can draw plans of his machinery. Lastly, there is the Law, especially that dealing with metals, that he may claim his own rights, that he may undertake the duty of giving others his opinion on legal matters, that he may not take another man's property and so make trouble for himself, and that he may fulfill his obligations according to the law.

It is therefore necessary that those who take interest in the methods and precepts of mining and metallurgy should read these and others of our books studiously and diligently; or on every point they should consult expert mining people, though they will discover few who are skilled in the whole art."

This translation follows Hoover and Hoover (1950).

Thus, here in the introduction to his famous work, Agricola gives us a plethora of skills and arts which are indispensable for anyone extracting metal ores from the bowels of the earth and processing them. This enumeration is more than symbolic for the conditions the modern alloy physicist has to fulfill. Although the development of science has carried his tasks to other levels of sophistication, as a result they are no less numerous and diverse than those of the medieval metalworker. It is the intended spirit of this book to provide, by presenting solid knowledge and different viewpoints, the wide horizon which favors significant scientific progress.

Reference

Hoover, H.C. and Hoover, L.H. (1950) Georgius Agricola: De Re Metallica, *Translation of the first Latin Edition of 1556*, Dover Publications, Inc., New York, p. 3.

List of Contributors

T. A. Abinandanan
Indian Institute of Science
Department of Materials
Engineering
Bangalore 560 012
India

Pascal Bellon
University of Illinois at Urbana-
Champaign
Department of Materials Science
and Engineering
1304 W. Green St.
Urbana, IL 61801
USA

Daniel Caillard
CEMES-CNRS
29 rue Jeanne Marvig, BP4347
31055 Toulouse Cedex 4
France

Thierry Duffar
INPG Grenoble, SIMAP-EPM
ENSEEG – BP 75
38402 St. Martin d'Hères
France

Brent Fultz
California Institute of Technology,
mail 138-78
Pasadena, CA 91125
USA

Yuri Grin
Max-Planck-Institut für Chemische
Physik fester Stoffe
Nöthnitzer Straße 40
01187 Dresden
Germany

Roland Grössinger
Vienna University of Technology
Institut für Festkörperphysik
Wiedner Hptstr. 8-10/138
1040 Vienna
Austria

Ferdinand Haider
Universität Augsburg
Institut für Physik
Universitätsstr. 1
86135 Augsburg
Germany

James M. Howe
University of Virginia
School of Engineering and Applied
Science
Department of Materials Science and
Engineering
P.O. Box 400745
Charlottesville, VA 22904-4745
USA

Alloy Physics: A Comprehensive Reference. Edited by Wolfgang Pfeiler
Copyright © 2007 WILEY-VCH Verlag GmbH & Co. KGaA, Weinheim
ISBN: 978-3-527-31321-1

Jeffrey J. Hoyt
Mc Master University and
Brockhouse
Institute for Materials Research
Department of Materials Science
and Engineering
1280 Main Street West
Hamilton, Ontario, L8S 4L7
Canada

Rafal Kozubski
Jagellonian University
M. Smoluchowski Institute of
Physics
ul. Reymonta 4
30-059 Cracow
Poland

Jung-Goo Lee
Korea Institute of Machinery &
Materials
Advanced Materials Research
Division
66 Sangnam-dong, Changwon-
City
Kyeongnam 641-831
Korea

Karl F. Ludwig, Jr.
Boston University
Physics Department,
590 Commonwealth Ave
Boston, MA 02215
USA

Georges Martin
CEA-Siège, Cab. H.C.
91191 Gif sur Yvette Cedex
France

Tarik Mehaddene
Technische Universität München
Physik Department E13/FRMII
James Franck Str. 1
85747 Garching
Germany

Peter Mohn
Vienna University of Technology
Center for Computational Materials
Science
Institut für Allgemeine Physik
Gumpendorferstraße 1a/134
1060 Vienna
Austria

Tetsuo Mohri
Hokkaido University
Graduate School of Engineering
Division of Materials Science and
Engineering
Sapporo 060-8628
Japan

Hirotaro Mori
Osaka University
Research Center for Ultra-High Voltage
Electron Microscopy
7-1 Mihogaoka, Ibaraki
Osaka 567-0047
Japan

Stefan Müller
Universität Erlangen-Nürnberg
Lehrstuhl für Festkörperphysik
Staudtstraße 7
91058 Erlangen
Germany

Hiroshi Numakura
Kyoto University
Department of Materials Science and
Engineering
Yoshida Hon-machi, Sakyo-ku
Kyoto 606-8501
Japan

now with:

Osaka Prefecture University
Department of Materials Science
Gakuen-cho 1-1, Naka-ku
Sakai 599-8531
Japan

Takeo Ohta
Ovonic Phase Change Institute
Yamato Sogyo Incubation Center
102 Takabatake-cho, Nara-city,
Nara 630-8301
Japan

and

Energy Conversion Devices, Inc.
2956 Waterview Drive
Rochester Hills, MI 48309
USA

Wolfgang Pfeiler
University of Vienna
Faculty of Physics
Dynamics of Condensed Systems
Strudlhofgasse 4
1090 Vienna
Austria

Véronique Pierron-Bohnes
Institut de Physique et Chimie
des Matériaux de Strasbourg
UMR 7504 CNRS-ULP
23 rue du Loess, BP 43
67034 Strasbourg Cedex 2
France

Raimund Podloucky
University of Vienna
Faculty of Chemistry
Department of Physical
Chemistry
Sensengasse 8
1090 Vienna
Austria

Wolfgang Püschl
University of Vienna
Faculty of Physics
Dynamics of Condensed Systems
Strudlhofgasse 4
1090 Vienna
Austria

Laurent Ranno
Institut Néel, CNRS/UJF
25 avenue des Martyrs
BP 166
38042 Grenoble Cedex 9
France

Guido Schmitz
Westfälische Wilhelms-Universität
Münster
Institut für Materialphysik
Wilhelm-Klemm-Straße 10
48149 Münster
Germany

Ulrich Schwarz
Max-Planck-Institut für Chemische
Physik fester Stoffe
Nöthnitzer Straße 40
01187 Dresden
Germany

Bogdan Sepiol
University of Vienna
Faculty of Physics
Dynamics of Condensed Systems
Strudlhofgasse 4
1090 Vienna
Austria

Walter Steurer
ETH Zurich
Department of Materials
Laboratory of Crystallography
Wolfgang-Pauli-Str. 10
8093 Zurich
Switzerland

Harald W. Weber
Vienna University of Technology
Atominstitut der Österreichischen
Universitäten
Stadionallee 2
1020 Vienna
Austria

Walter Wolf
Materials Design s.a.r.l.
44 avenue F.-A. Bartholdi
72000 Le Mans
France

1
Introduction

Wolfgang Pfeiler

1.1
The Importance of Alloys at the Beginning of the Third Millennium

An alloy is a "mixture" of two or more chemical elements, one of which at least is a metal. The alloying element can be distributed over the crystal lattice sites of the host element and yield a solid solution, or it can form different phases showing up as particles in a "matrix." Whereas the physical properties of a solid solution are essentially determined by the chemical composition of the constituents, the properties of a multiphase alloy are determined largely by the spatial distribution of the second-phase particles. This possibility of "designing" physical and technical properties of a material by a careful selection of alloying elements and alloying concentrations put alloys in the forefront of materials from early human history up to our time at the beginning of the third millennium.

We are facing tremendous progress in materials development and design of advanced materials, driven by technical needs in all fields of modern production. A real revolution in materials science, however, could be observed during very recent years, a great leap from understanding bulk materials to the study, development, and application of nanostructured materials. In the present book we give accounts of the state-of-the-art of alloy physics and the challenges of future research in the field together with the basic knowledge necessary to understand the radical changes currently happening.

Materials today, aside from metals and alloys, include such different substances as ceramics, high-T_c (HT_c) superconductors, liquid crystals, polymers, foams, biomimetic materials, nanotubes, nanocomposites ... and it is justified to ask: "What is the importance of *alloys* in the field of modern materials today? What role do *alloys* play in the fascinating advance toward nanostructurization?"

We will later cast a glance at the historical perspectives together with a short description of the development of modern alloy science. However, let us get started by affirming the importance of alloys at the onset of the third millennium.

Alloy Physics: A Comprehensive Reference. Edited by Wolfgang Pfeiler
Copyright © 2007 WILEY-VCH Verlag GmbH & Co. KGaA, Weinheim
ISBN: 978-3-527-31321-1

Conductors and superconductors In July 1996 overloaded transmission lines sagging low enough to touch trees caused a blackout of electric power which affected millions of people in the western part of the United States, Canada, and parts of Mexico. The blackout on August 14 2003 was the biggest in US history. Roughly 50 million people lost power due to a failure in the "Lake Erie Loop." A similar event occurred in Europe on September 29 of the same year: some 57 million people lost their electrical power in Italy at the weekend when the national grid crashed. The minor event that a single tree fell across a line in the Alps during a storm started a domino effect which soon knocked out the entire Italian grid.

A key step in the process of improving the distribution of electric power lies in reducing the weight/conductivity ratio of utility transmission and distribution lines.

Aside from conventional conductors used for overhead transmission and distribution lines, special systems making use of superconductivity may be applied. The capacity of superconducting materials to handle large currents with no resistance and extremely low energy losses can be applied to electric devices such as motors and generators as well as to transmission in electric power lines. A superconducting power system would meet the growing demand for electricity with fewer power plants and transmission lines than was otherwise needed. High-temperature superconductors which need cooling to only about 80 K seem especially promising and are just about to enter economical use for utility applications. Possible applications are cables which carry up to five times more power than conventional utility cables, smaller and more efficient motors, smaller and lighter generators, compact transformers without oil (but with liquid nitrogen) as the cooling medium, fault-current limiters, and superconducting magnetic energy storage (SMES) systems which store energy in a magnetic field created by the flow of direct current in a coil of superconducting material.

Soft and hard magnets Tailoring the ferromagnetic hysteresis curves is an ongoing demand in the field of soft and hard magnets. At one end of the spectrum magnetically soft materials reduce hysteresis losses, where necessary. Amorphous alloys, for example, show a high permeability and very low losses (especially low losses at elevated frequencies), but also a high magnetization at room temperature. At the other end, using rare earth alloys we get very high values of magnetization which have been increasing exponentially during recent years. Magnetically hard materials with high coercivity are needed for magnetic storage. To increase information storage density, nanosized magnetic domain structures are being developed. Nanocrystalline soft magnetic materials are already commercially available. Nanocrystalline hard magnetic materials exhibit interesting properties; an industrial breakthrough has not yet been achieved, however.

Intermetallics and superalloys Intermetallic compounds, because of their long-range ordered structure, show a temperature range where their mechanical strength increases with temperature (yield stress anomaly). Together with their advantageous corrosion properties this selects them as potential high-temperature structural materials. The present challenge is to combine extreme strength and hardness with sufficient ductility and surface stability, a problem which is not yet really solved.

At present, turbo superchargers and aircraft turbine engines are made of two-phase nickel-based superalloys, usually strengthened by a coherent precipitation of $L1_2$ ordered γ' particles in the disordered face-centered cubic γ matrix. The operating surface temperatures of turbine blades are close to 1150 °C, and the average bulk metal temperature approaches 1000 °C. Ni-based superalloys are also used in load-bearing structures at up to 90% of their melting temperature.

Another possible use of intermetallics when they are ferromagnetic and nanostructured is that for high-density magnetic recording, a field in which much effort is invested at the moment, especially in order to overcome the superparamagnetic limit.

Shape memory alloys These are alloys which exhibit thermoelastic effects and revert to the original shape by a phase transformation when they are heated after a plastic deformation. The applications of shape memory alloys are essentially in electrical/mechanical junctions, actuators which are the most promising alternatives to hydraulic systems, actuators for microelectromechanical systems (MEMS), surgical tools such as NiTi bone plates which apply a steady pressure to assist the healing process and thus reduce recovery time, and robotic muscles (shape memory alloys mimic human muscles and tendons very well). In addition, magnetic shape memory alloys (MSM) based on NiMnGa exist in which the martensitic transition can be triggered by an external magnetic field. These are alloys for which many applications can be expected.

Nanocrystalline and amorphous alloys Glassy forms of metallic alloys have quite unique mechanical, electrical, and magnetic properties. It has been found in the recent past that for certain alloy compositions appropriate cooling rates can be applied to form "bulk amorphous alloys." Many of these alloys have a very high strength to weight ratio with excellent elastic energy storage. Other applications are in ferromagnets ranging from the lowest known coercivities to extremely high coercivities with high saturation induction values, as well as in composites made of ferromagnetic and antiferromagnetic components (giant magneto resistance), and in magnetic reading heads.

Dimensionally restricted alloys Dimensionally restricted systems are currently very fashionable in physics and materials science and research in this area easily attracts funding. In electronics, for example, the ever greater complexity in microprocessor and memory chips means an exponentially increasing chip density (with a doubling rate about two years at the moment: Moore's law). Individual electronic components are therefore smaller than 100 nm, bringing these conducting elements into the low-dimensional regime. Another field where nanostructurization commends itself is high-density data storage and magnetic or magneto-optic recording. Further materials enhancement is achieved by incorporating nanocrystalline particles to increase toughness or to improve catalyst properties.

In addition, there is a steadily growing general interest in the basic physics of low-dimensional alloys. Significant differences are obtained in the physical properties of nanosized structures as compared to the bulk, for instance a change in melting temperature or modification of microstructures. The changes in diffusion processes seem to be essential but are not known in detail. Other fundamental questions are which structural defects are stable and what their role is in phase transformations. It is therefore important to study all the changes which occur when going from bulk to thin films, nanowires, or small atom clusters.

Friction and wear The reduction of friction between moving parts becomes more and more important for future mechanical applications, especially in modern combustion engines. On the one hand there are efforts to minimize the energy loss during operation; on the other hand there is the trend to simultaneously maximize the output and minimize the engine volume, in this way increasing the efficiency of engines. Friction properties and the resistance to high thermal stresses therefore become main design factors. Besides low friction and low wear rate, the materials must also have high fatigue strength and good ductility to sustain the deformation of the different parts caused by the firing pressure and by the inertial forces. To close the gap between the performance of conventional materials and the needs of the engines of the next generation, there is a very acute demand in alloy design for the design of new tribological alloys.

Medical and biological applications Special alloy products are frequently used for medical applications, e.g. in accident surgery. Typically, these include orthopedic and dental implants, fracture fixation devices, suture needles and staples, surgical blades and saw cutting tools, dental burrs and reamers, catheter guide wire systems and wire for diagnostic and sensor electrical leads. Another area of current interest is biomimetic engineering. Inspired by Nature, biomimetic engineering copies natural systems, utilizing molecular self-assembly as the key link between physics, chemistry, and biology, and thereby creating novel advanced structures. The driving force stems from the recognition that there are a number of areas where biological methods are more efficient and environmentally friendly than current technology, and superior to it overall. These areas are, for example, energy storage and utilization, low-temperature fabrication of complex materials, linear motors and actuators, dirt-repellent surfaces, neural computation, sensors, and many other possible applications. Alloys play a major role in this field.

In conclusion, alloy physics is at the hub of materials science and materials physics. It turns out that virtually all high-technology developments are involving modern alloys, and alloy design is a driving force of scientific research in materials physics. This means that within the field of technological applications specialized alloys are involved in all cutting-edge technologies. In addition, alloys have essential repercussions in other fields as well, such as environmental technology to limit the growing ecocide, modern surgery and health care, and even the social and economic development of underdeveloped regions.

1.2 Historical Development

1.2.1 Historical Perspective

Modern anthropology dates the rise of the genus "human" at more than three million years ago (*Homo rudolfensis* and *Homo habilis*, both being potential ancestors of *Homo ergaster*, who thrived about 1.5 million years ago). Looking into the early history of man, certain development "steps" can be distinguished, which are correlated with a "revolutionary" advance of skills. One example is the first handling of fire in the middle of the Pleistocene (the Ice Ages) together with the manufacture of stone tools. This age is therefore also called the Old Stone Age (the Paleolithic period) and marks the first appearance of human species closely resembling ourselves (*Homo ergaster* → *Homo heidelbergensis* → *Homo sapiens*). It is certainly at that moment that "matter" changed into "material" – in the sense that a material is matter with some special use. Another "revolutionary" change led directly to the class of materials which is the focus of this book: The development of metallic alloys had probably started already in the Far East, in China and Indochina, Thailand, Vietnam, and East India before 5000 BC. Among the oldest known tin bronzes are the dagger of Veľke Raškovce in central Europe (Slovakia), containing 4.5 wt.% tin from the second half of the fifth millennium BC and pieces from Mundigak, Afghanistan, from the second half of the fourth millennium BC (Trnka 2006). The reason why those processes of human development look fast and "revolutionary" to us, while they are in fact slow and evolutionary, is to be found in the large timescale for counting these developments in the far past. Actually, a drastic reduction of the periods in which such far-reaching changes took place can already be recognized by comparing the Stone Age with the Bronze Age. Whereas the human development during the Stone Age took roughly a million years, the Bronze Age in Middle Europe was superseded by the Iron Age after no more than about 2000 years. A corresponding acceleration in technical development continues into present times, the replacement of electron tubes by transistors after only 50 years being such an example.

It is known that in early history the extended use of tools and weapons was essential for the survival of the human species, in the permanent search for food and in the fight against animals as well as other human species. Tools and weapons were first found accidentally in the near surroundings, then later made artificially from specific materials. This documents the importance of materials for mankind, which has continued right up to the 21st century. Very early in human history, alloys started to play a central role among the materials used, because it was quickly recognized that many of the desired properties can be created by deliberately "mixing" together a metal with other components, especially other metals. Metal alloys, therefore, immediately started a new era of human life when tools of copper bronze (70–95% copper and 30–5% tin) prevailed over earlier ones made from stone or pure copper. The evolution of hardening a metal by alloying was accompanied by the possibility of casting complicated shapes in special molds made of wax embedded in clay.

We will now give an overview of the historical perspectives of alloys, followed by a short description of the development of modern alloy science.

During the final period of the Stone Age men became settled and started to live together in small tribes and communities. Lumps of native copper and later of gold collected from ice-free mountains and out of rivers, respectively, were probably the first metals to be discovered and both were immediately recognized as being excellently suited to substitute wood and bone in artwork. The first use of copper dates back more than 10 000 years (Ergani Maden and Catal Hüyük in the highlands of Anatolia). An intensified search then probably brought to light the bright green malachite and the blue azurite. It was learned that copper got hard when hammered and could be softened again by heating it. Early smelting techniques were already being applied around 6500 BC, first in East Asia (China) and West Asia (Near East), and from the first half of the fifth millennium BC onward in Europe also.

The very first bronze pieces used were made from "arsenic bronze," arsenic accidentally contained in the copper ore. It turned out to be easier to cast, more ductile, and harder than pure copper. It is now known that arsenic deoxidizes the harmful oxygen and thereby reduces brittleness. Later, by intentional addition of tin to copper, true "bronze" (tin bronze) was made and the production became independent of deposits with special ore. Its use spread very rapidly out of Mesopotamia to the Mediterranean countries, where bronze lasted for over 2000 years and even longer in north-western Europe (see Fig. 1.1).

It is assumed that the knowledge about iron smelting and how to obtain steel was first gained in the highlands of Anatolia where the Hittites flourished from about 1450 to 1200 BC, but several objects of smelted iron date back further than 2000 BC (Cowen 2005). When Tutankhamen was buried in about 1400 BC he had with him an iron dagger with a hilt and sheath of gold decorated with rock crystal. Surprisingly enough, the dagger blade had not rusted within the 3000 years and more until its discovery, whereas many other ancient iron objects probably perished as they were transformed into shapeless rust. Since the element iron practically always occurs as a compound, usually siderite ($FeCO_3$) and often together with malachite ($CuCO_3$), most of these very old iron weapons have been made from meteoritic iron or from the black sands on the south coast of the Black Sea containing the iron-rich magnetite (Fe_3O_4). One of the oldest known furnaces for "cooking" iron from iron ore are from Tell Hammeh, on the north bank of the Zarqa River in Jordan, and date to the eighth century BC (Veldhuijzen and Van der Steen, 2000). In Europe iron came into use with the development of iron smelting and working which began around 800 BC during the

Fig. 1.1 (a) "Strettweger Kultwagen," a beautifully made artefact showing a sacrificial procession with a cart, bearing a vessel containing some liquid for offerings. It was produced about 600 BC and found in 1851 near Judenburg, Austria (Egg 1996). (b) "Kultwagen von Acholshausen," made about 1000 BC, found near Acholshausen, Germany (Worschech 1977/78). Both objects show the importance of coaches with spoke wheels as prestigious objects used for religious ceremonies.

Hallstatt culture, named after the famous cemetery near a salt mine at Hallstatt, Austria.

The usual way to obtain useful products from the unattractive, unmelted iron which is left in the furnace (a spongy mass, the "bloom", containing slag) is to hammer the liquid slag out of the hot lump, yielding fairly pure wrought iron. The result may not be better than bronze. Only in a forging process with a long procedure of hammering and reheating were superior steels obtained. In charcoal-fired forges the fresh, hot surface of the wrought iron comes into contact with carbon and carbon monoxide which thus "carburize" the iron surface to form steel as an iron–carbon alloy.

Alloying iron with more than 2% carbon considerably reduces the melting temperature with respect to iron and makes it possible to cast in molds. Whereas in the western hemisphere this development was discouraged by the knowledge that more than 1% carbon makes the iron alloy brittle, the superior furnace technology in China already allowed iron casting in the first half of the first millennium BC, and the technique of decarburization by reheating to 800–900 °C in air between 400 and 300 BC (Cowen 2005).

When they came to the Near East during the Crusades, European knights learned to fear the extreme quality of razor-sharp Damascene blades (see Fig. 1.2). Two production methods have been applied, both dating back to before 500 AD. One method was to forge two ingots of high- and low-carbon steels, respectively, into flat sheets and after cutting them to pieces, intimately bonding the

(a) (b)

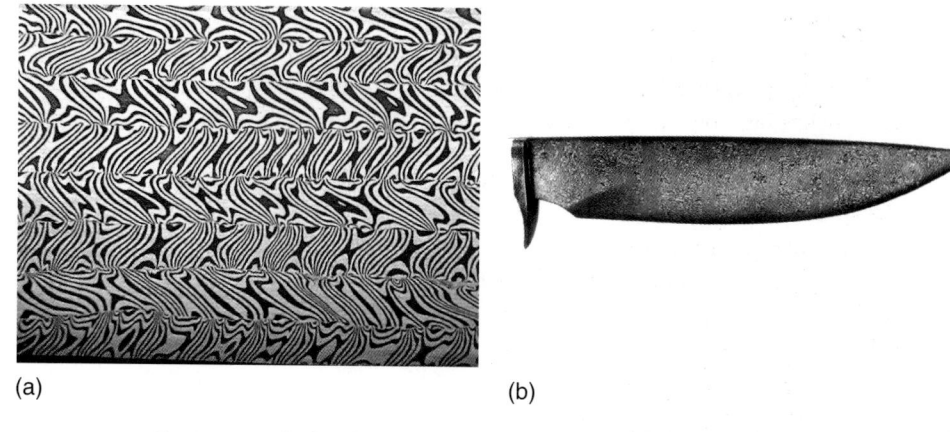

Fig. 1.2 (a) Etched surface of a Damascus blade showing in detail the Damascene surface pattern generated by continued folding and twisting during forge welding of alternating sheets of high- and low-carbon steels. (b) A reconstructed Damascus blade showing the Damascene surface pattern ("Mohammed's ladders" and "rose patterns") produced from a single wootz steel ingot (Verhoeven et al. 1998).

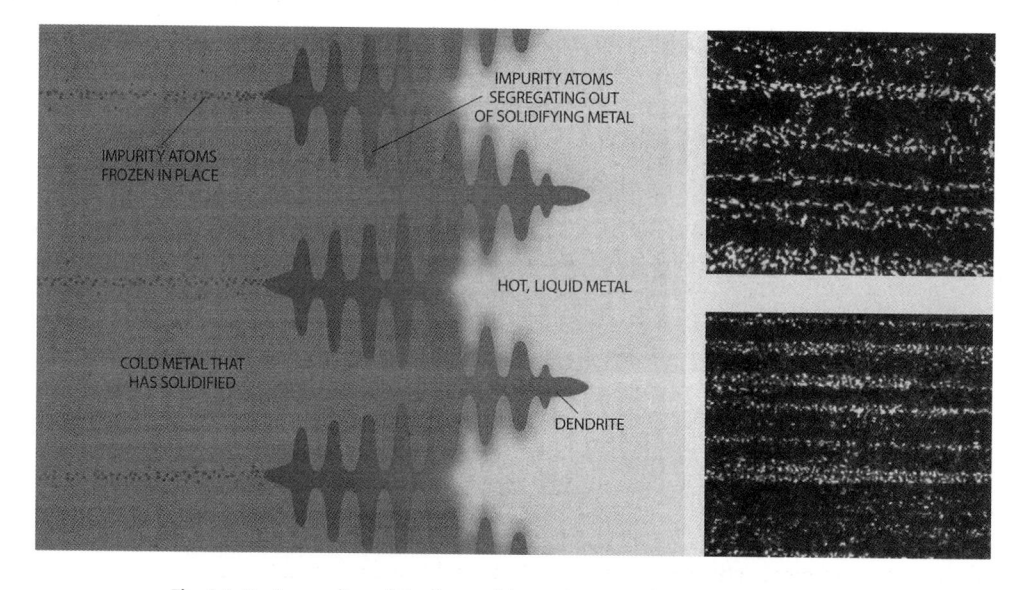

Fig. 1.3 During cooling of the "wootz" ingot, impurity elements (e.g. vanadium) segregate out of the solid iron and freeze in an aligned way. Subsequent cycles of heating and cooling lead to a growth of iron carbide particles (Fe_3C, cementite) along the lines, forming the light-colored bands in the Damascene blade. Right: Micrograph showing light and dark bands in an original Damascene blade (top) and a modern reconstruction (bottom) (Verhoeven 2001).

pieces together by continued folding, twisting, and hammering the work piece (Fig. 1.2a). The other method started from a single high-carbon steel ingot, called "wootz", originally produced from special ore in India. The wootz was then forged in many heating–cooling cycles (Fig. 1.2b).

The secret of the production of Damascene blades from a wootz steel ingot was unfortunately lost during the 18th century, but has recently been rediscovered in a co-operation between a scientist and a blacksmith (Verhoeven et al. 1998; Verhoeven 2001). It turned out that the surface patterns on the blades resulted from the formation of bands of iron carbide (Fe_3C) particles initiated by the microsegregation of carbide-forming impurity elements being present in the wootz ingots (Fig. 1.3).

1.2.2
The Development of Modern Alloy Science

Whereas, as we have seen, the roots of alloy production go back into early human history, what we can call "scientific alloy research" started in the 19th century. It is well known that the 19th century was very productive in several fields; the time of the industrial revolution was accompanied by a deep belief in progress and in universal feasibility, and several new branches of science (social, medical, biological, technical) were developed. The progress in science and technology began to be tightly connected. The foundations were laid for more sophisticated studies of alloys by the development of crystallography, elasticity, electrochemistry, and metallography, together with a parallel advance in experimental methods (e.g., thermometry and measurement of electrical resistivity and hardness).

In his two articles "On the Equilibrium of Heterogeneous Substances" (in 1876 and 1878, respectively), J. W. Gibbs applied thermodynamics to the phase equilibria of alloys, introduced the phase diagram as a basis of modern alloy science, and derived his famous phase rule. We can indeed connect the rise of modern alloy physics with this pivotal work of Gibbs. An important first step toward an understanding of alloy kinetics was the investigation of solid-state diffusion by W.C. Roberts-Austen, showing in 1896 that the diffusion coefficient of gold in lead was surprisingly far greater than supposed. The systematic study of binary phase diagrams is closely connected with G. Tammann. Generations of students used Tammann's famous book *Lehrbuch der Metallographie* (first published in 1914) for getting started with binary phase diagrams. The idea of the crystallinity of metals and alloys was already growing during the 18th and 19th centuries, but conclusive evidence of their crystalline structure was furnished after the discovery of X-rays as electromagnetic waves which could be diffracted by crystals (M. von Laue in 1912). Immediately afterward, W. H. Bragg and his son W. L. Bragg used X-rays systematically for the study of crystal structures (in 1913–1914). The determination of crystal unit cells of many primary and intermediate phases brought about a real revolution. This was also the onset of the study of chemical long-range order (LRO), i.e., the condensation of the different sorts of atom in an alloy on specific sublattices of the crystal structure. The three papers by W. L. Bragg and E. J. Williams discussed for the first time the degree of LRO and LRO kinetics with respect to atomic interaction energies (Bragg and Williams 1934, 1935; Williams 1935). These papers are still great reading for anybody entering the field of ordering kinetics.

In 1930, F. Laves tried a classification of crystal structures from a topological point of view and found special compounds which are stabilized by their high packing density. These "Laves phases" have a packing density higher than a close-packed structure of identical hard spheres.

Besides phase diagrams, Tammann investigated extensively the mechanisms of phase transformations during freezing, which led to the development of a theory of nucleation by M. Volmer and A. Weber in 1926, later applied to solid-state phase transformations by R. Becker and W. Döring (in 1935).

In 1926 W. Hume-Rothery showed that the electron/atom ratio plays a decisive role in the determination of the solid solubility of copper, silver, and gold alloys with B-subgroup metals and also for the so-called "magic numbers" of concentrations involved in the intermediate phases of these alloys. The reason lies in variations of the band structure for different crystallographic systems which comes into play when metals of different electron/atom ratio are alloyed together.

Much effort within physical metallurgy and alloy physics has of course been put into the study of plastic deformation. Great progress in this respect was the possibility of growing single crystals (J. Czochralski in 1917; P.W. Bridgman in 1925), allowing the investigation of plasticity without any influence of grain boundaries and with a well-defined orientation. The fundamental studies of M. Polanyi, E. Schmid, and W. M. Boas were summarized in the famous book by Schmid and Boas in 1935 establishing the critical resolved shear stress as a fundamental criterion for slip. In this macroscopic description of plasticity, a fundamental problem remained unsolved: The experimentally observed yield stress of single crystals is several orders of magnitude lower than calculated from the theoretical strength determined for the rigid glide of lattice planes. This led to the postulation of dislocations by M. Polanyi, E. Orowan, and G. I. Taylor (separately, in 1934), which later, after a further development of the electron microscope by E. Ruska (in 1931), was brilliantly confirmed by transmission electron microscopy (TEM) of thin crystals by Heidenreich (in 1949) and P. B. Hirsch (in 1954).

J. Frenkel (in 1926) and C. Wagner, as well as W. Schottky (in 1930), postulated point defects, vacancies, and interstitials as very important for the physical properties of metals and alloys. Vacancies turned out to be essential for diffusion, as was shown theoretically by H. B. Huntington and F. Seitz in 1942, later confirmed experimentally as fact by A. D. Smigelskas and E. O. Kirkendall (in 1947), and critically analyzed by L. S. Darken (in 1948). The Kirkendall effect of marker movement during chemical diffusion of two components with different diffusion coefficients is generally accepted as evidence for the vacancy diffusion mechanism.

The theory of phase transformations in alloys, originally dominated mainly by the nucleation and growth concept, was greatly developed by efforts to explain superconductivity (H. Kamerlingh-Onnes in 1911) and superfluidity (P. L. Kapitsa, J. F. Allen, and D. Misener in 1937). First a concept was developed on a phenom-

1.2 Historical Development

enological level by L. D. Landau in 1938 to explain second-order phase transitions, but it applies to more general physical transition problems (see, e.g., Melbourne 2000). Another essential development reaches back to Gibbs and Volmer: spinodal decomposition, with the consequence of "uphill" diffusion. The problem was described theoretically by J. W. Cahn and J. E. Hilliard in 1958 in a continuum approach and by M. Hillert (in 1961; and also in his 1956 thesis) in an atomistic approach. Corroboration came from small-angle scattering experiments in the early stages of decomposing systems (see, e.g., Gerold and Kostorz (1978), for a first review on this matter). An atomistic approach to obtain theoretically phase diagrams and information on the kinetics of phase transformations from statistical thermodynamics was developed in 1950 by R. Kikuchi, called cluster variation method (CVM), and was later extended to the path probability method (PPM) for describing transformation kinetics (in 1966). In 1949 N. Metropolis and S. Ulam first described the Monte Carlo method (MC) of random variations of the microstate; today it is one of the leading methods in alloy physics for studying structure and kinetics.

After C. J. Davisson and L. H. Germer in 1927 had observed the wave character of electrons in diffraction experiments on Ni crystals, it was only a small step to scattering experiments with neutrons, yielding phonon dispersion curves, diffuse scattering, and magnetic scattering with polarized neutrons.

The density functional theory (DFT) by P. Hohenberg and W. Kohn (in 1964) and W. Kohn and L. J. Sham (in 1965) solved the fundamental problem of many-body interactions in a solid. It was therefore a starting point for a quantum mechanical ab-initio approach to realistic phase diagrams and materials parameters. The theory has been very successful in recent years at describing the ground-state properties of metals and alloys. Very recently the DFT was combined with statistical thermodynamics within the cluster expansion method (CEM) by J. M. Sanchez, F. Ducastelle, and D. Gratias (in 1984). When this is applied together with MC simulations, a *quantitative* study of alloy properties is also made possible with respect to the temperature dependence.

The work of Cahn and Hilliard, and later S. M. Allen and J. W. Cahn (in 1979), laid the basis for a general modeling of multiphase systems, called phase field modeling (PFM), which has been used successfully to study solidification and phase transformation in alloys.

Since the mid-1990s, electron microscopy as well as field ion microscopy, originally invented by E. W. Müller in 1951, made a big advance. Methods for electron microscopy were developed to obtain atomic resolution (high-resolution TEM, or HRTEM) by special observation techniques and technical correction of aberrations. The original field ion microscope and its analytical version, the atom probe, have recently been upgraded to 3D tomography atom probe (A. Cerezo in 1988; D. Blavette in 1993): The field-evaporated part of a very fine tip is simultaneously imaged on a position-sensitive detector and chemically analyzed by a time-of-flight (TOF) measurement. If all data are carefully stored as a function of time, the sample can be computer-reconstructed on an atomic scale. The development of scanning probes for atomic-scale microscopy led to further extremely fruitful experimental tools: the scanning tunneling microscope (STM) invented by G.

Binning and H. Rohrer in 1982, and the atomic force microscope (AFM) by C. F. Quate, G. Binning, and C. Gerber in 1985. Using a cantilever with magnetic coating enables the AFM to be extended to the study of magnetic structures as the magnetic force microscope (MFM).

Since their use in historical times, alloys have never left the front row of advanced materials. Among the greatest challenges of alloy physics today are the following topics:

- ductile, high-temperature, high-strength, structural materials
- high-density, magnetic and magneto-optic, recording media
- physical behavior with restricted dimensions
- alloy behavior under extreme conditions
- increasingly accurate computation of physical parameters, especially by ab-initio calculations
- access to alloy kinetics by PPM, PFM, molecular dynamics (MD), and MC simulations supported by CEM.

Having at hand elaborate experimental and theoretical methods together with the possibility of computer simulations, further developments in the field of advanced alloys and alloy design will continue to contribute to the overall scientific and technological progress.

1.3
Atom Kinetics

As we have seen already in the historical perspective, producing an alloy with specific properties has always needed specific microstructural changes, such as take place in the carburizing process when forging iron in the flame to get steel or in the special segregation processes for the famous Damascene blades. It is absolutely clear that all diffusive structural changes in a crystal are ultimately brought about by atoms jumping between different lattice sites. If atoms are not able to move, no change at all is possible and the alloy remains frozen in the current state. For physical properties of an alloy to be meaningful they have to be measured in a thermodynamic equilibrium state. The knowledge of atom jump processes is an essential prerequisite to decide if the system is in a true equilibrium state. Furthermore, detailed information on alloy kinetics is required for the design and development of advanced alloys.

Some examples will underline the importance of atom jumps for alloy physics. During the solidification process the homogeneity/heterogeneity/crystallinity of the solid alloy is almost completely determined by atomic mobility. Fast cooling of the melt, for instance, enables the production of metallic glasses with alloy properties far different from crystalline materials.

Dislocation reactions essentially determining the mechanical properties are widely influenced by atom jump processes bringing about the motion of vacancies and solute atoms.

Atom jumps, mainly into neighboring vacancies, are the underlying mechanism of nondisplacive phase transformations. Therefore, what really happens depends critically on the local atomic motion in an alloy (see, e.g., Soffa et al. 2003).

Phase changes of the order–disorder type are of special interest due to their reversibility. In this case atom jumps occur between different sublattices of the ordered or at least partially ordered alloy. A fundamental understanding of the role of defects in intermetallic compounds can then be attempted, which will allow open questions on configurational kinetics of alloys to be answered.

We have further seen that size reduction to obtain nanostructures is a current challenge in alloy physics. Essential changes in physical properties of alloys accompany dimensional restrictions. These are due to the increase in surface/volume ratio so that the atom jump processes run in a different manner than those in bulk alloys.

The necessary information on atom jump processes may be obtained by a combination of experimental and theoretical methods, together with computer simulations (see, e.g., Pfeiler et al. 2004; Kozubski et al. 2004, 2006).

Because of their importance and ubiquity in all microstructural changes, both desired and unwanted, it was decided to choose "atom jump processes" as a guiding principle and leitmotiv through this book.

1.4
The Structure of this Book

When one looks at the Contents pages, the structure of this book (or let's say the order parameter) will become apparent: from statics to dynamics and kinetics, from fundamental to complex structures, from phenomenology via theory and experiment to applications.

This section gives a short overview of the chapters and their interconnections.

Chapter 2, "Crystal Structure and Chemical Bonding," is dedicated to a classification of structure types and the relationship between structures and structure-determining parameters. A modern view of Hume-Rothery alloys and intermetallic compounds is given as well as of Laves phases and structurally more complex alloy systems such as magnetic materials of the $CaCu_5$ type or quasicrystals.

Alloy structure and structural defects depend of course on the solidification process. In Chapter 3, "Solidification and Grown-in Defects," it is shown how diffusion, mainly on the liquid side, and convection processes close to the solid–liquid interface determine details of the resulting alloy structure. The generation of defects, formation of grain structure, twinning, and chemical segregation effects are other topics in this chapter.

Starting from a discussion on atomic binding and atomic interaction energies, a review is given in Chapter 4, "Lattice Statics and Lattice Dynamics," on the elasticity of crystalline lattices, lattice vibrations, and thermal properties of alloys. Examples are shown of soft phonon modes and their influence on structural phase transitions, and special emphasis is put on the connection of phonon spectra with atomic migration, allowing a determination of migration energies from lattice-dynamical properties.

The importance of lattice defects for atomic migration plays a central role in Chapter 5, "Point Defects, Atom Jumps, and Diffusion." The discussion starts with point defects and their energetics in alloys and continues with the different descriptions of single and multiple atom jump processes leading to atomic diffusion under various conditions. There is a fairly detailed discussion of diffusion in ordered intermetallics and of various theoretical approaches to obtain the macroscopic diffusion behavior from atom kinetics.

Plastic deformation of alloys due to dislocation processes is reviewed in detail in Chapter 6, "Dislocations and Mechanical Properties." The mechanisms discussed include thermally activated processes such as dislocation glide and climb, as well as hardening and recovery, complex deformation behavior like the famous yield strength anomaly and fatigue, the strength of nanocrystalline alloys and thin layers, and the mechanical properties of quasi-crystals.

The mechanisms of phase transformations in alloys are reviewed in Chapter 7, "Phase Equilibria and Phase Transformation." After a solid introduction on phase diagrams, phase transformation kinetics is described, and how it can be suppressed to get amorphous alloys is dicussed. The difference between nucleation and growth on the one hand, and spinodal decomposition on the other, is discussed for the cases of phase separations as well as for order–disorder transformations. Martensitic transformations, which are massive and displacive, are presented at the end of this important chapter together with some applications.

Chapter 8, "Relaxation of Nonequilibrium Alloys," first concerns description of alloy relaxation to an equilibrium structure by diffusion-controlled, thermally activated processes, starting from a nonequilibrium state. It is shown that a realistic kinetic pathway can be found by linking the classical nucleation theory to the details of atomic jumps. Secondly driven alloys are considered, for which an external forcing maintains the alloy in nonequilibrium states. The theory and modeling of such alloys rests on the combination of kinetics: that resulting from the external forcing and that of natural relaxation to equilibrium.

In Chapter 9, "Change of Alloy Properties under Dimensional Restrictions," detailed studies of phase equilibrium and phase transformations of nanometer-sized alloy particles are presented and a discussion on atom movement in nanoparticles is given. It is shown that a stable amorphous phase can form in a eutectic system for a small enough particle size (below 10 nm), which promises an insight into the liquid–glass transition.

The theoretical basis of Chapter 10, "Statistical Thermodynamics and Model Calculations," is the cluster variation method (CVM). The statistical thermodynamics on a discrete lattice and for continuous media are reviewed and it is shown that by combining the CVM with calculations of the total energy of the electronic structure, self-consistent first-principles calculations of phase equilibria are made possible. Results of calculations using CVM and PPM (the path probability method) are presented and a possible approach toward first-principles phase field calculations is discussed.

Chapter 11, "Ab-initio Methods and Applications," is dedicated to a review of the present state of density functional theory. After an account of the theoretical background, the different computational methods and approaches are discussed critically and several examples of applications are given, such as elastic and vibrational properties, point defects and diffusion, the search for ground-state structures, and ordering phenomena. It is shown that the very successful cluster expansion method yields access to multiscale modeling.

In Chapter 12, "Simulation Techniques," advantages and disadvantages of molecular dynamics simulations are discussed first and some interesting results are given. Then various Monte Carlo (MC) simulation techniques, including kinetic MC, are described, and their possibilities and limits are shown in several examples. In the final part of the chapter, phase field models are explained, starting from their principles. These models are used to describe, e.g., the precipitation of a new phase, grain growth in polycrystals, and solidification.

Two chapters on experimental methods focus on high-resolution techniques: Chapters 13.1, "High-Resolution Scattering Methods and Time-Resolved Diffraction," and 13.2, "High-Resolution Microscopy." To study nonequilibrium alloys or equilibrium dynamics a high temporal experimental resolution is of great interest, which can be met under certain conditions, especially using the third generation of synchrotron sources. In Chapter 13.1, examples are given of high-resolution X-ray and neutron scattering methods, which include magnetic scattering, coherent time-resolved scattering, elastic, quasielastic, and inelastic scattering as well as diffuse scattering and scattering from surfaces.

In Chapter 13.2, after a short introduction to the field of surface analysis by scanning probe microscopy, the basic principles and practical aspects of HRTEM and related techniques are provided. This is followed by several examples, particularly emphasizing the use of in-situ HRTEM. The second part of this chapter focuses on up-to-date atom probe tomography, giving a detailed description of the method. Exemplary studies are described on decomposition, diffusion in nanocrystalline thin films, and grain boundary diffusion.

Chapter 14, "Materials and Process Design," is dedicated to various applications of alloys, and tries to convey a sound understanding of the corresponding effects and their up-to-date technical uses. These include soft and hard magnetic materials, invar alloys, magnetic media for the use in hard disk memories, spintronic materials showing giant magnetoresistance, and phase change media which use the phase transformation from crystalline to amorphous state for data storage. Last but not least the story of superconductors is related.

It is clear that the chapters summarized above have multiple connections; some important concepts are discussed in several chapters from different aspects. An example is the formation of a new phase from a supersaturated matrix. This process, which is fundamental for alloys, is discussed from the viewpoint of atom jumps mediating the transformation process in Chapter 5, a description starting from the basics is presented in Chapter 7, a prospective theoretical modeling of the kinetic process is formulated in Chapter 8, changes in the transformation behavior under the dimensional restrictions of nanometer-sized particles are illus-

Table 1.1 Key to the topics in this book.

Chapter no.:	1	2	3	4	5	6	7	8	9	10	11	12	13.1	13.2	14
	Introduction	Structure	Solidification	Lattice dynamics	Defects & diffusion	Mechanical properties	Phase transformations	Relaxation & driven alloys	Dimensional restrictions	Statistics & modelling	Ab-initio applications	Simulations	Scattering methods	HR microscopy	Design & applications
Topic															
Ab-initio	○				○	○				○	●	○			○
Alloy design	○		○		○	○	○	○			○				●
Amorphous alloys	○	●			○		○		○				○		○
Anisotropy		●		○		○	○				○				○
Atom jump	○		○	○	●	○	○	○	○	○	○	○			
Cahn–Hilliard	○				○		○	○		●		○			
Cluster expansion	○				○					○	●				
Cluster variation	○				○		○			●					
Complex alloys		●				○				○		○			○
Crystal structure		●	○	○	○	○	○		○	○	○	○	○	○	○
Diffusion	○		○	○	●	○	○	○	○	○	○	○	○	○	○
Dislocations			○		○	●	○		○		○		○	○	
Elasticity				●		○	○				○				
Grain structure			●		○	○	○		○		○	○	○	○	○
HR microscopy	○					○	○	○	○					●	
Inelastic scattering				○								○	●		
Interaction energies	○	○	○	●	○	○	○	○	○	○	○	○			
Interfaces			○		○	○	○		○	○	○		○	●	○
Intermetallics	○	●	○	○	○	○	○	○	○	○	○	○	○	○	○
Jump barriers				○	●			○		○	○	○			
Lattice dynamics				●			○				○		○		
Liquid alloys		●							○				○		
Long-range order	○	●	○	○	○	○	○	○	○	○	○	○	○	○	○
Low-dimension	○				○	○			●		○	○	○	○	○

Table 1.1 (continued)

Topic \ Chapter no.:	1 Introduction	2 Structure	3 Solidification	4 Lattice dynamics	5 Defects & diffusion	6 Mechanical properties	7 Phase transformations	8 Relaxation & driven alloys	9 Dimensional restrictions	10 Statistics & modelling	11 Ab-initio applications	12 Simulations	13.1 Scattering methods	13.2 HR microscopy	14 Design & applications
Magnetism	○										○		○		●
Martensitic trafo				○		○	●							○	
Master equation					●		○	○				○			
MC simulation	○			○	○		○	○	○	○	○	●	○	○	
Metastability	○	○	○		○		●	○	○	○					
Molecular dynamics				○	○						○	●			
Multi-scaling			○							○	●				
Phase field method	○		○							○		●			
Phase transformation	○	○	○	○	○	○	●	○	○	○	○	○	○	○	○
Phonons				●		○	○				○		○		
Plastic deformation			○			●	○	○						○	
Point defects			○	○	●	○	○	○	○		○	○	○		
Precipitation				○	○	○	●	○	○	○	○	○	○	○	○
Quasicrystals		●				○							○		
Recovery & recryst						●	○							○	
Scattering methods				○	○		○	○					●		
Short-range order					○		●			○	○	○	○		
Solute–defect interface					●	○	○	○	○			○			
Spinodal					○		○	○		●		○	○	○	

● principal treatment of this topic, ○ minor treatment.

trated in Chapter 9, statistical models of phase transitions are presented in Chapter 10, the advantage of DFT together with the cluster expansion method and kinetic MC simulations is documented in Chapter 11, the application of phase field models is discussed in Chapter 12, the experimental access to details of phase transformations by high-resolution scattering techniques as well as by TEM and atom probe tomography is presented in Chapters 13.1 and 13.2, respectively, and finally, possibilities of technical applications, e.g., of the phase change media for data storage in CDs and DVDs, are described in Chapter 14. Table 1.1 shows how important topics figure in different chapters. It is our explicit intention to present these various aspects, instead of conscientiously avoiding any overlap. We hope that this many-sided approach makes this book even more attractive.

In the course of this introductory chapter, we have tried to demonstrate that alloys still have to be counted among the most important materials of the 21st century, especially when considering the ongoing shift in interest from bulk materials to nanostructures. It goes without saying that a thorough understanding of the physics of alloys is a precondition for anyone who wants to do serious research or to engage in technical application of alloys. The present book covers in much detail and considerable depth nearly all fields of alloy physics necessary to know when entering one of these areas either as a graduate student or post-doc. It should be valuable for scientists of neighboring fields or somebody already working in an area of alloy physics who wants more information in one of the specific directions covered by this compendium.

References

Bragg, W.L. and Williams, E.J. (1934), *Proc. Roy. Soc. (London)* **A145**, 699.

Bragg, W.L. and Williams, E.J. (1935), *Proc. Roy. Soc. (London)* **A151**, 540.

Cowen, R. (2005), http://www.geology.ucdavis.edu/~cowen/~GEL115/115CH5.html

Egg, M. (1996), *Monogr. RGZM – Römisch-Germanisches Zentralmuseum Forschungsinstitut für Vor- und Frühgeschichte*, **37**.

Gerold, V. and Kostorz, G. (1978), *J. Appl. Crystallogr.* **11**, 376.

Kozubski, R., Kozlowski, M., Pierron-Bohnes, V., and Pfeiler, W. (2004), *Z. Metallkde.* **95**, 880.

Kozubski, R., Issro, Ch., Zapała, K., Kozłowski, M., Rennhofer, M., Partyka, E., Pierron-Bohnes, V., and Pfeiler, W. (2006), *Z. Metallkde.* **97**, 273.

Melbourne, I., in: *Nonlinear Instability, Chaos and Turbulence*, Vol. 2 (Debnath, L. and Riahi, D.N., Eds.), WIT Press, Southampton, 2000, p. 79.

Pfeiler, W., Püschl, W., and Podloucky, R. (2004), *J. Mater. Sci.* **39**, 3877.

Soffa, W.A., Püschl, W., and Pfeiler, W. (2003), *Intermetallics* **11**, 161.

Trnka, G., Institut für Ur- und Frühgeschichte, University of Vienna (2006), personal communication.

Veldhuijzen, H. and Van der Steen, E. (2000), *Archeology* **53**(1), 21.

Verhoeven, J.D., Pendray, A.H., and Dauksch, W.E. (1998), *JOM* **50**, 58.

Verhoeven, J.D. (2001), *Scientific American* **284**, 62.

Williams, E.J. (1935), *Proc. Roy. Soc. (London)* **A152**, 231.

Worschech, R. (1977/78), *Geschichte am Obermain* **11**, 45.

2
Crystal Structure and Chemical Bonding

Yuri Grin, Ulrich Schwarz, and Walter Steurer

2.1
Introduction

For the following systematic considerations we define alloys as a class of materials which may be multiphase systems. The components of these systems are binary or multinary phases, e.g., or intermetallic compounds. Hence, the relevant chemical and physical properties of alloys can be described in a slightly simplified manner as being dependent on the properties of each of the constituents and on the interactions between these constituents at the phase boundaries.

Intermetallic compounds (phases) show physical and chemical properties which are often interesting for applications, so they frequently serve as important components for materials design. Due to the strongly application-governed development of knowledge on intermetallic compounds, understanding of their chemical nature is still incomplete. Important questions can not be answered definitely: Why do some intermetallic compounds only form at distinct compositions? Why do others form homogeneity ranges? Why are certain structural motifs stable and can be found frequently among intermetallic phases? Why do others appear only occasionally? One of the reasons for this situation is the insufficient knowledge about chemical bonding in intermetallic compounds. Additionally, attempts to find a direct causal link between properties and chemical bonding for this group of inorganic substances often leads to the conclusion that there is a lack of reliable information on both bonding and properties, and, consequently, on their relationship. The nature of chemical bonding in intermetallic compounds, in particular, is still under debate (Nesper 1991, Grin 2000).

In this chapter, we discuss the most common crystal structure motifs of intermetallic compounds, factors which are suggested to govern formation and properties, and different approaches to understanding the atomic interactions in intermetallic phases.

Alloy Physics: A Comprehensive Reference. Edited by Wolfgang Pfeiler
Copyright © 2007 WILEY-VCH Verlag GmbH & Co. KGaA, Weinheim
ISBN: 978-3-527-31321-1

2.2
Factors Governing Formation, Composition and Crystal Structure of Intermetallic Phases

Compound formation, spatial arrangement of atoms, and bond constitution are influenced by factors which can be traced back to parameters such as the average number of valence electrons per atom (valence electron concentration), the electronegativity of the constituents (electrochemical factor), and atomic size. With the additional condition that the two metals realize the same crystal structure, the same factors have been used as parameters controlling solid solubility (Hume-Rothery and Raynor 1938). As additional parameters for compound formation, measures of the cohesive energy such as boiling point, compressibility, or lattice energy are employed, but these parameters have no apparent relevance to the realization of certain atomic arrangements.

For crystal chemistry considerations, sets of self-consistent atomic radii were introduced which are dependent on the realized coordination number due to repulsive interactions between adjacent ions (Goldschmidt 1926). Moreover, in order to take into account different types of chemical bonding there are sets of metallic, covalent, and ionic radii (Teatum et al. 1960; Emsley 1991). Despite these drawbacks, the correlation of radius ratios and coordination number results in a crude separation of different structure types.

Figure 2.1 shows an example of a mapping using the radius ratio as the only structure-determining factor for the Laves phases AB_2 (Stein et al. 2004). From

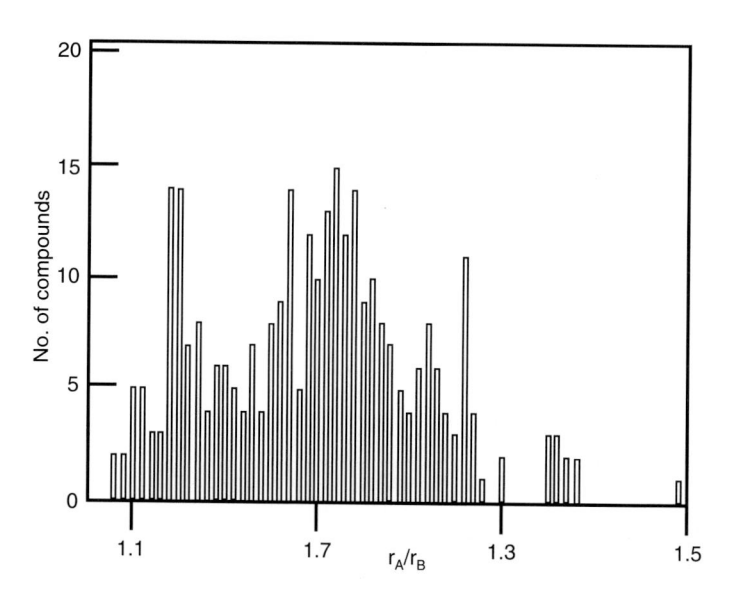

Fig. 2.1 Number of known Laves phases as a function of the radius ratio of the constituting metals.

the geometrical analysis (cf. Section 3.2.4), it was expected that the ideal size ratio for this structural motif is 1.225:1. In reality, only a few compounds form with this size ratio. Moreover, maximal numbers of the phases are obtained at $r_A/r_B = 1.15$ and 1.33. Obviously, an individual determining factor is not sufficient to predict the realization of a distinct crystal structure. Taking this shortcoming into account, a later approach analyzed the dependence of the stabilization of Laves phases on the heteronuclear A–B and homonuclear B–B next-neighbor distances (Simon 1983). Similarly, later classifications of atomic arrangements employed several factors, a procedure which is known as structure mapping.

2.2.1
Mappings of Crystal Structure Types

The influence of a combination of two or more factors on the stability of certain atomic patterns can be studied in a systematic way by employing so-called structure maps. Here, normally factors such as valence electron concentration, electronegativity, or atomic size are used as coordinates for multidimensional diagrams of crystal structure types.

An early mapping of the atomic arrangements of valence compounds (Mooser and Pearson 1959) employed the difference in electronegativity and the averaged value of the principle quantum number of the constituents as coordinates (Fig. 2.2).

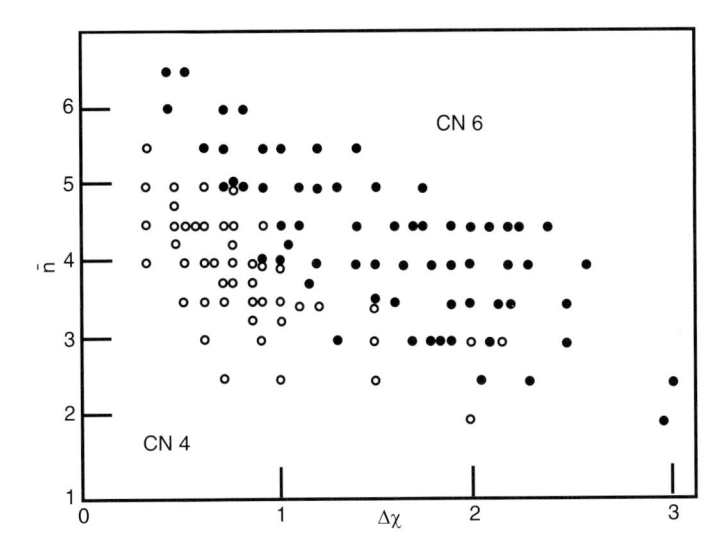

Fig. 2.2 Mooser and Pearson representation of binary valence compounds with composition AB and eight valence electrons, as a function of average quantum number $\bar{n}$ and electronegativity difference $\Delta\chi$. Open symbols indicate ZnS or ZnO types of crystal structures (CN 4), filled symbols NaCl arrangements (CN 6).

Small values of the parameters favor covalent bonding and a low coordination number (e.g., four in zincblende and wurtzite-type arrangements), whereas large values correspond to ionic bonding and higher coordination numbers (e.g., six in sodium chloride-type arrangements or eight in CsCl-type structures). Consequently, the graphical representation revealed a satisfactory separation of the stability fields of binary *AB* compounds with eight valence electrons and coordination numbers of four and six (Fig. 2.2). However, different atomic patterns of intermetallic non-octet compounds are often located within narrow parameter domains and are not well separated into different domains by these coordinates (Zunger 1980).

On the basis of pseudopotential calculations, the average band gap E_h and the ionicity C of group 14 elements and binary semiconductors with four valence electrons per atom was calculated (Phillips 1968; van Vechten 1969a,b; Phillips 1970). A plot using the square roots of the homopolar and ionic fractions of the band gap, respectively, results in a clear separation of fourfold and sixfold coordination. The positions of the stability fields reveal that CN 4 is more stable for dominantly covalent compounds and CN 6 for more ionic binaries (Fig. 2.3).

In another approach, two-dimensional structure maps of *AB* compounds were constructed on the basis of calculated pseudopotential radii and two coordinates [Eqs. (1, 2)] derived thereof (St. John and Bloch 1974; Zunger 1980).

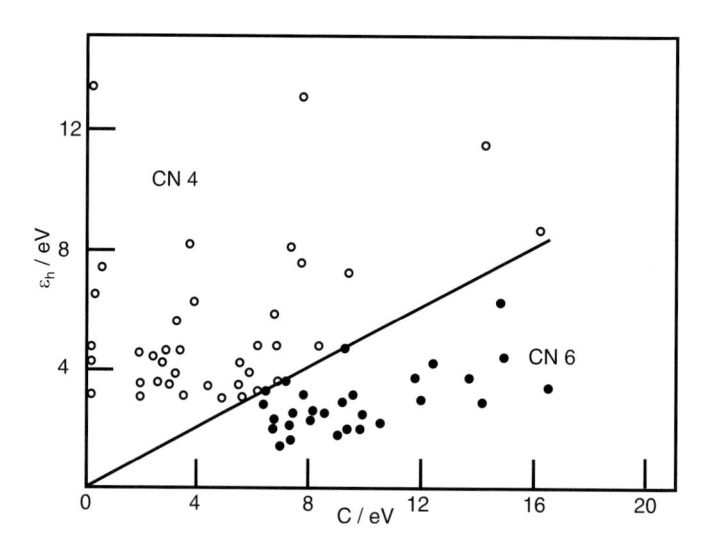

Fig. 2.3 Stability fields of tetrahedral and octahedral coordination in binary valence compounds as a function of the homopolar band gap ε_h and the ionicity C. Open symbols indicate CN 4, filled symbols CN 6.

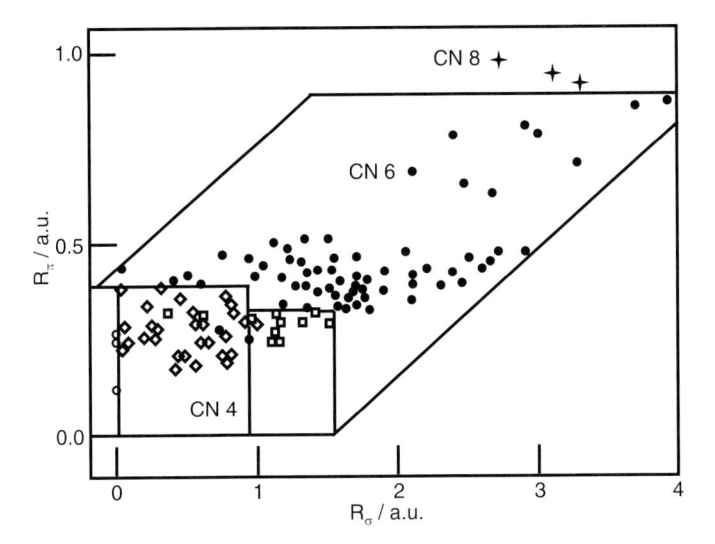

Fig. 2.4 Stability fields of selected crystal structures of valence compounds as a function of calculated pseudopotential radii (see text). Open circles: diamond-type structure (CN 4); open diamonds: ZnS type (CN 4); open squares: ZnO type (wurtzite, CN 4); filled dots: NaCl type (CN 6); crosses: CsCl type (CN 8).

$$R_\sigma^{AB} = (r_p^A + r_s^A) - (r_p^B + r_s^B) \tag{1}$$

$$R_\pi^{AB} = (r_p^A - r_s^A) + (r_p^B - r_s^B) \tag{2}$$

The sums are measures of the atomic radii of the constituents and the differences are a measure for the promotion energy from s to p states. The coordinate R_σ^{AB} represents a size difference and the value of R_π^{AB} the tendency for sp hybridization. The resulting separation of stability fields of structure types (Fig. 2.4) works for 437 non-octet compounds adopting 27 different crystal structures, but other arrangements with similar atomic coordination are not resolved, e.g., the NiAs and MnP.

For a more efficient separation, three-dimensional mappings of crystal structure types have been accomplished (Villars 1983, 1984a,b). For different compositions stability fields are shown in cross-sections sorted for certain ranges of the number of valence electrons (defined as electron per atom ratio) and the crystal structures of compounds are displayed using the differences of orbital electronegativities $\Delta\chi_{AB}$ and orbital radii $\Delta(r_s + r_p)_{AB}$ as coordinates (Fig. 2.5). Although a reasonable separation is achieved in general, a shortcoming of the mapping is that the NiAs type had to be excluded and that some stability fields contain more than one structure type. The incomplete separation is taken as evidence for the importance of the angular character of the valence orbitals which is not included

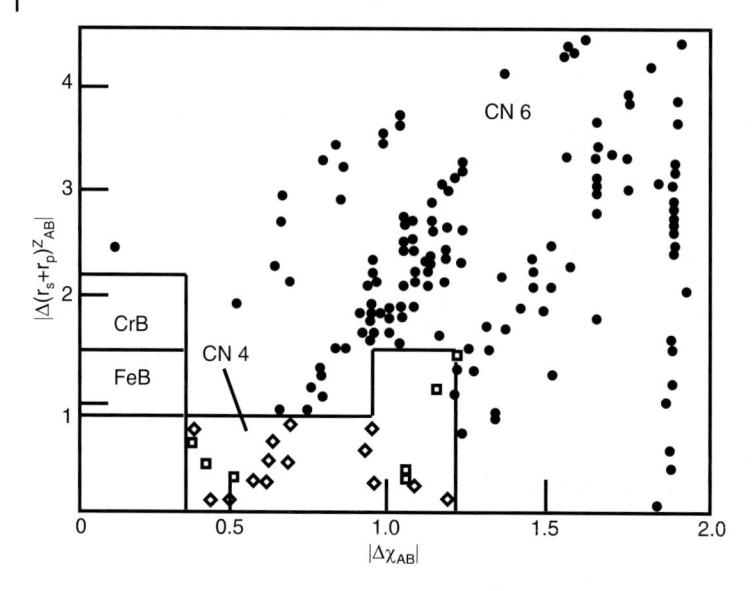

Fig. 2.5 Stability fields of selected structure types of compounds with composition AB and a valence electron concentration of four, as a function of orbital radii and electronegativity difference $\Delta\chi$. Open diamonds: ZnS type (CN 4); open squares: ZnO type (wurtzite, CN 4); filled dots: NaCl type (CN 6). Stability fields of FeB- and CrB-type arrangements are indicated.

in Villars' maps. Moreover, the sorting does not separate elemental structures ($\Delta\chi_{AB} = 0$ and $\Delta(r_s + r_p)_{AB} = 0$) with the same number of electrons but different crystal structures, e.g., those of sulfur, selenium, and chromium with six valence electrons.

In an attempt to overcome the shortcomings of structure mapping which are due to the neglect of the angular dependence, i.e., the s, p or d character of the involved electrons, Pettifor introduced a phenomenological coordinate in order to group structural data within a simple and coherent framework. By running a one-dimensional string through the two-dimensional periodic system of elements, a so-called Mendeleev number (a combination of the factors atomic number and valence electron configuration) is defined (Pettifor 1990). Plotting the stable atomic arrangements at a given composition as a function of the Mendeleev number of the constituents provides a clear separation of structure types (Fig. 2.6). Moreover, the implicit use of the electronic configuration as a coordinate also carries predictive power for targeted synthesis and modification of properties by partial substitution. With regard to regions where different arrangements compete, the diagrams provide optimized test structures for total energy calculations. Although the Pettifor' mapping is widely accepted as optimal, even this type of structure plot does not reveal a clear and easy separation of some structure types.

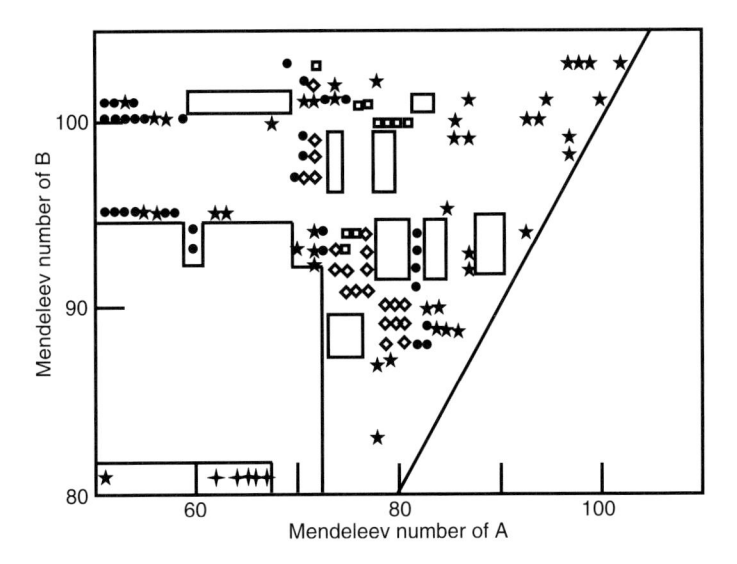

Fig. 2.6 Stability fields of the AB crystal structure types as a function of the Mendeleev number. Open diamonds: ZnS type (CN 4); open squares: ZnO type (wurtzite, CN 4); filled dots: NaCl type (CN 6); crosses: CsCl type (CN 8). Stars and open fields indicate other structure types.

2.3
Models of Chemical Bonding in Intermetallic Phases

One of the main reasons for problems associated with definitions of structural fields in different mapping systems originates from the fact that the parameters involved are describing particular or even singular aspects of atomic interactions. The complete picture of chemical bonding in intermetallic phases is not developed to the same extent as it is for other classes of inorganic or organic compounds. Nevertheless, several successful models were proposed for the description of particular groups of intermetallic compounds which are the subject of this chapter.

2.3.1
Models Based on the Valence (or Total) Electron Numbers

An early, qualitative systematization of bond types (the so-called van Arkel–Ketelaar triangle; van Arkel 1956; Ketelaar 1958) included ionic, covalent, and metallic interactions. The van der Waals force was not considered as a physical interaction although a large number of crystal structures comprising low-dimensional building units provide evidence of the importance of this type of interaction for the cohesion of condensed phases.

Based on early concepts of chemical bonding and the octet-rule (Abegg 1904; Drude 1904), the number of valence electrons turned out to be the essential quantity in elements and compounds with covalent bonding (Lewis 1916) and in the configuration of ionic solids (Kossel 1916). A simple and semi-quantitative description of homodesmic ionic bonding in strongly polar crystalline solids assumes a full charge transfer from the electropositive component, usually a main group metal, to the electronegative constituent, typically a halogen or chalcogen. A model of the different coordination in compounds constituted by spherically symmetric closed-shell ions was elaborated (Pauling 1960). The compositions are balanced so that the resulting ions can adopt a noble gas configuration. The resulting cohesion is based on nondirected Coulomb (electrostatic) interactions between charged particles which give rise to attractive and repulsive forces in a three-dimensionally periodic arrangement of ions. The dimensionless Madelung factor sums up attractive and repulsive forces of the specific spatial arrangement of ions. The potential energy in equilibrium conditions is a function of the ionic charge, the equilibrium distance, and the spatial arrangement. An additional characteristic repulsion due to the interaction of the electron shells (Pauli exclusion principle) is taken into account by a correction term which depends on the electron configuration of the ions. Typically, the calculated full lattice energy is about 10% smaller than the Coulomb energy alone. The concept is very useful for an understanding of the atomic coordination on a basis of ionic interactions (Hoppe 1970, 1979). The approximation of this approach is, first, an assignment of certain radii to atoms or ions although the extension of their wavefunction is infinite. Moreover, in accordance with earlier experimental data, elaborated quantum mechanical treatments reveal additional cohesion contributions besides the ionic ones, e.g., covalent bonding and static dipole interaction.

In contrast to the undirected forces between ions in a solid, covalent bonds are directional between distinct neighboring atoms. The case of the two-electron-two-centre bonds can easily be visualized by Lewis formulae. In the majority of compounds formed by main group elements, atoms in molecules or solids fulfill the octet rule.

A severe conceptual problem of discussing metallic bonding is that the terms "covalent" or "ionic" only describe types of chemical interaction, while the term "metallic" can characterize either a type of bonding or a kind of electronic transport behavior, i.e., a physical property. Thus, metallic conductivity is generated by the presence of partially filled bands and is not strictly correlated with a certain bonding mechanism. An example of this situation in chemical elements is the carbon modification graphite, which combines covalent bonding with a 2D metal-type conductivity. A different aspect of the problem is highlighted by the binary ionic compound CsI, which is characterized by filled p and empty d states and, thus, semiconducting properties at ambient pressure. Upon compression the p and d states start to overlap and partly filled bands result. As a consequence, a pressure-induced insulator–metal transition is observed (Asaumi and Kondo 1981; Reichlin et al. 1986).

In a metal, the diffuse wavefunctions of the constituting atoms in some cases enable a description of the chemical bonding by the model of a free-electron-gas. In contrast to the local organization concepts for valence compounds, this model for metallic solids describes valence electrons as negatively charged particles interacting in a field constituted by a regular arrangement of positively charged atom cores. The resulting band dispersions (widths) are large in comparison with the band splittings (gaps) at special points of the Brillouin zone and partially filled bands result. Experimental and theoretical investigations reveal that, to a good approximation, a nearly unstructured electron gas with a square-root shaped density of states (DOS) is found for elemental sodium (Fig. 2.7). With increasing atomic weight, a growing contribution of d states brings about a structuring of the DOS. Similarly, enhancing the occupation of p states by proceeding from sodium via magnesium to aluminum gives rise to a subtle structuring and a clearly increasing dispersion of the bands evidences the intensification of next-neighbor interactions (Demchyna et al. 2006).

With regard to electron count, the description of the bonding situation becomes even more complicated when we analyze band structures of transition metals and their compounds which are normally characterized by broad s and narrow d states (Burdett 1995). Here, typically one electron per metal atom is located in a band (or bands) constituted by predominantly s orbitals. As a consequence, transition metals exhibit a maximum of the cohesion energy in the region of six valence electrons because one s and five d electrons are located in bonding states. In late transition metals like Fe, Ni or Co, the interaction of completely filled, non- or anti-bonding d orbitals influences the structural arrangement critically.

Several binary intermetallic systems exhibit a similar sequence (cf. Section 2.4) of crystal structures as observed in the binary system Cu–Zn. The similarity of behavior is attributed to the systematically varying valence electron concentration, which changes proportionally to the composition and stabilizes certain atomic arrangements under the condition that the constituting atoms have only small size differences. Mixing of metals with similar radii and electronegativity, such as copper and zinc, does not require a significant amount of elastic energy, and the phase exhibits a crystal structure which would be realized by a hypothetical pure element with the same (average) electron configuration. For example, in the γ-brass phase, the excess charge of zinc with respect to copper is screened and the electron densities of the different types of atoms are adjusted. A similar process takes place for the core states so that the electron wavefunction is only slightly modified by the substitution. In these so-called electron compounds or Hume-Rothery phases (frequently known as alloys components), the dominant factor for the phase stability is the number of electrons in the Brillouin zone so that certain magic electron numbers correspond to maxima in stability. Small deviations of the optimal filling can be tolerated, with the consequence that a certain range of slightly varying compositions also fulfill the stability criteria. Well-known examples of electronic compounds are the α phases (fcc crystal structures with approximately one electron per atom), the β phases (bcc-type crystal structures with 1.5 electrons per atom), the γ phases (a defect variety of bcc phases with 52 atoms in the unit cell and an electron concentration of $\frac{21}{13}$ electrons per atom), and the ε phases (hcp arrangements with $\frac{7}{4}$ electrons per atom).

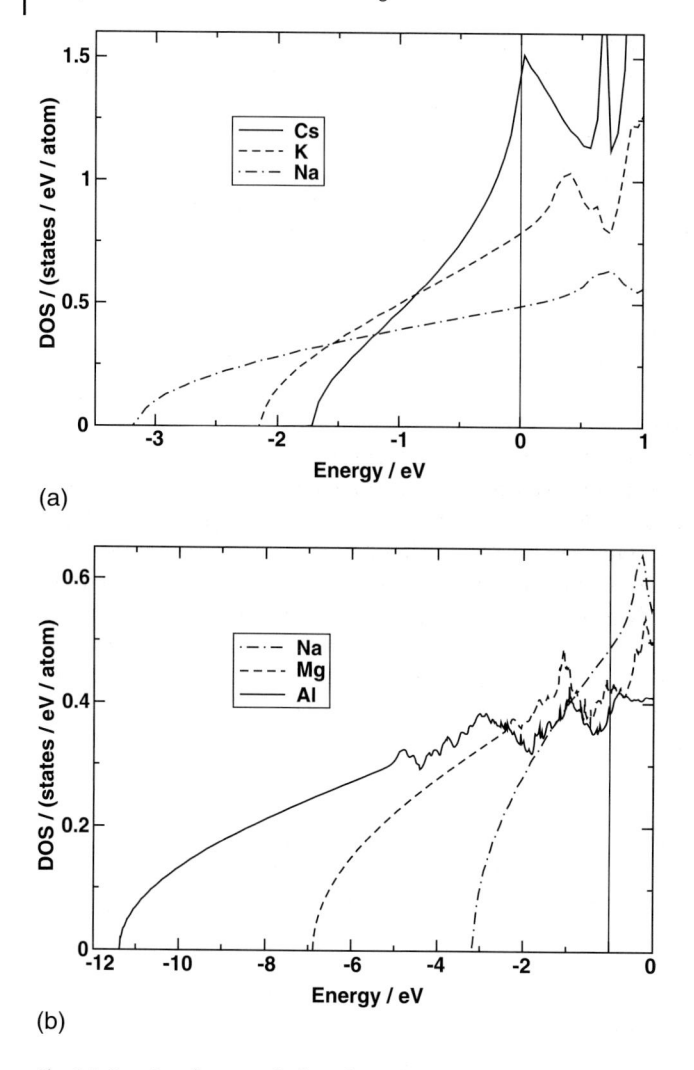

Fig. 2.7 Density of states of selected (a) alkali metals and (b) third row elements. For the heavy alkali metals, the DOS close to the Fermi level is structured by the presence of d states; subtle deviations for Mg and Al are attributed to the occupation of p states.

Assuming that in these electron phases the energy difference is given entirely by the difference in band structure energies and that the electron–electron interactions can be neglected, the effect of substitution can be described in the rigid-band approximation as a change in the number of occupied orbitals but not the band structure. Thus, a determination of the stability of atomic patterns requires a calculation of reasonable densities of states and band structure energies at certain electron concentrations.

Early attempts to explain the concurrence of certain structural patterns at specific electron concentrations (Mott and Jones 1936) attribute the stabilization to a drop in the density of states after a pronounced maximum. According to this idea, this abrupt decrease is located at an energy where the free-electron sphere expands beyond the boundary of the Brillouin zone. Addition of electrons after this point causes an energy cost which is higher than that for arrangements where the density of states continues to rise. An advantage of this model is the accurate prediction of phase changes, but it has the shortcoming that the Fermi surface of the metal is approximated as a sphere.

In a later attempt (Jones 1937), the energy bands of the intermetallic compounds were assumed to be free-electron-like except when the wave vector approaches Bragg reflecting planes (nearly a free-electron model). This scenario has the severe disadvantage that the drop in the density of states is located at concentrations far below the observed structure changes.

More elaborate quantum mechanical calculations reveal that the free-electron model (Mott and Jones 1936) works surprisingly well. It was shown in detail that peaks in the DOS tend to stabilize a phase, but the maximum stability is reached at a significantly higher electron concentration (Paxton et al. 1997). The density of states does not directly give the band structure energy, but its second derivative. Integration of the function (calculating the energy by filling up electronic levels) shifts its extrema to higher values, and integrating two times changes arguments (i.e., it exchanges the positions of minima and maxima). Thus, a Fermi energy which is located in a dip in a density of states often leads to a maximum in stability. It was shown that band structure calculations taking into account sp bands exclusively reproduce the results of Mott and Jones, but those including d bands also correctly predict the structure of copper metal (one valence electron per atom) and confirm roughly the correctness of the earlier models. Thus, structural stability is largely determined by the density of states, e.g., the stability of γ-brass-type crystal structures is due to a lowering of the DOS around an e/a of 1.7:1. Additionally, the structural distortion of the atomic planes from bcc gives rise to extra scattering of the electrons.

Recent theoretical investigation on the γ-brass phases Cu_5Zn_8, Cu_9Al_4 (Asahi et al. 2005a) and T_2Zn_{11} (T = Ni, Pd, Co, and Fe; Asahi et al. 2005b) revealed a stabilization of these electron phases by a pseudogap formation across the Fermi level caused by a combination of orbital hybridization and Fermi level-Brillouin zone interaction.

2.3.2
Quantum Mechanical Models for Metallic Structures

The structural systematics of elemental metals as a function of the atomic number stimulated investigations on the stability of atomic patterns by quantum mechanical calculations (Skriver 1985). The study of close-packed (metallic) crystal structures was performed by taking advantage of the frozen potential approximation. Here, the potentials were not self-consistently relaxed, with the consequence

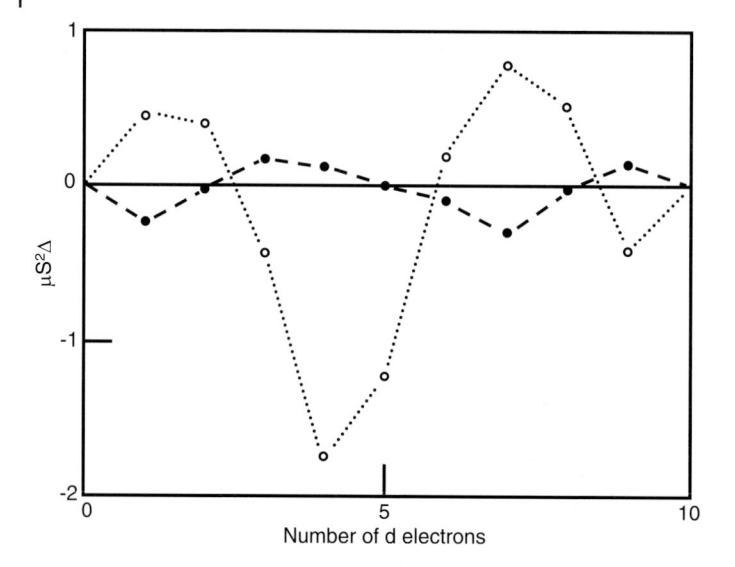

Fig. 2.8 Differences in the structural energy as calculated using canonical d bands. Reference line $\mu S^2 \Delta = 0$ corresponds to fcc, open dots to bcc, and filled dots to hcp. Broken and dotted lines are guides to the eye. Δ stands for energy difference between the fcc and hcp or bcc structure, μ means d-band mass, S is the atomic Wigner-Seitz radius.

that both core level contributions and double-counting terms of different atomic patterns cancel in the calculation of the structural energy differences. As a consequence, the only remaining quantities are the one-electron energies (which in general underestimate electron–electron correlations) and the electrostatic term. Since cohesion of elemental metals has no electrostatic contribution, the energy difference of the fcc and hcp or bcc structural patterns can be obtained alone by integrating the one-electron state densities (Fig. 2.8). The calculation procedure reproduced the stability of the empirically observed crystal structures for Na, and all transition metals except gold, ytterbium, and the three 3d metals exhibiting ferromagnetic behavior. However, the calculated energy differences are approximately five times larger than the experimental values.

Systematics of interatomic distances within the transition metal series can be understood in the framework of the Friedel model (Friedel 1969; Söderlind et al. 1994). For the nonmagnetic 4d transition metal series (Fig. 2.9, broken curve) the successive filling of the bonding, nonbonding, and anti-bonding states results in a parabolic dependence of the binding energies and the interatomic distances on the number of valence electrons. In the 4d series the magnetic interaction is relatively weak and cannot compete with the elastic energy. In contrast, magnetic order occurs within the 3d series. Here, the metals with four to eight d electrons show pronounced deviations from the expected parabolic behavior (Fig. 2.9, full curve) toward longer distances. The weakening of the bonds is driven by a gain

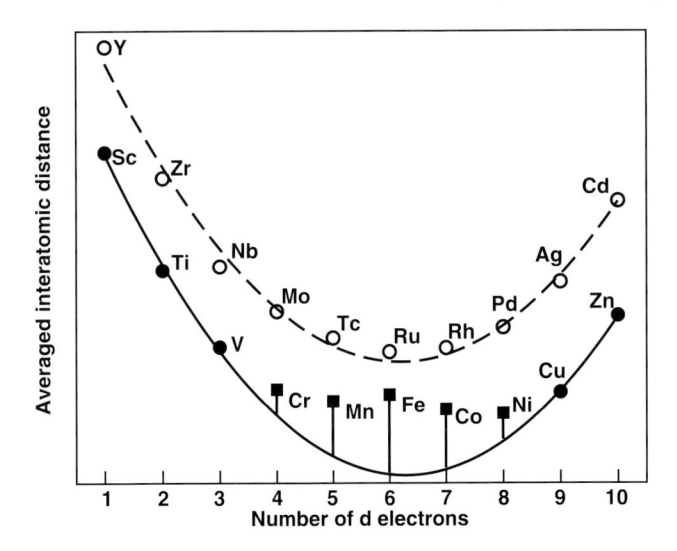

Fig. 2.9 Averaged interatomic distance in 3d and 4d metals as a function of the number of d electrons. Full circles indicat nonmagnetic 3d metals, full squares magnetically ordered 3d metals, and open circles the nonmagnetic 4d metals.

in magnetic energy. The magnetic order shifts the majority spin states to lower energy and, in the case of the magnetic 3d elements, increases the occupation of the anti-bonding majority spin states. The corresponding depopulation of the minority spins has less influence since it moves the nonbonding states to the Fermi level. Increasing the band dispersion, e.g., by application of pressure can suppress the magnetic ordering and cancel the anomaly (Rosner et al. 2006).

For elemental rare-earth metals, band structure calculations (Duthie and Pettifor 1977) revealed that the relative volume of the ion (metal) core influences the d band occupancy quite critically. Within the series of metals, the d band occupancy changes from about 1.5 to 2.5 and it is this fraction of the total energy which drives the crystal structure sequence in the trivalent 4f metals.

2.3.3
Electronic Closed-Shell Configurations and Two-Center Two-Electron Bonds in Intermetallic Compounds

While valence compounds usually comprise fully occupied valence bands and empty conduction bands, so-called semimetals such as the compound LiIn with a diamond-type In^- partial structure have an indirect overlap of bands (Schwarz et al. 1998). The band structure reveals strong similarities to those of group 14 semiconductors with the energy of the conduction band dropping below the top of the valence band. This electronic situation induces the formation of an equivalent number of electrons and holes, which causes a metal-like conductivity. Despite the presence of partially filled bands, counting rules are valid since the number of orbitals is still in accordance with the simple empirical rules for valence compounds.

2.3.3.1 Zintl–Klemm Approach

With respect to both experimental and theoretical work, the so-called Zintl phases (Zintl and Kaiser 1933; Zintl 1939; Schäfer et al. 1973) are of special interest since chemical bonding in this group of intermetallic compounds can be described in most cases in full accordance with the counting rules for ionic and covalent compounds, i.e., the correlation between composition and number of covalent bonds per atom can be predicted by means of the "$8 - N$" rule (cf. Section 2.3.3.2). Thus, in compounds like K_4Si_4 the valence electron of sodium is transferred to silicon. The resulting three-bonded Si^- ions with five valence electrons form tetrahedra $(Si^{4-})_4$ similar to those in white phosphorus. Analogously, in the Zintl phase NaTl a complete transfer of the valence electron of the electropositive metal sodium to the more electronegative thallium is assumed. The Tl^- anions have the same number of valence electrons as group 14 elements and consequently realize a diamond-type network of four-bonded thallium species. However, the isotypic compound LiCd comprises four-bonded Cd^- with only three valence electrons per Cd^- thus providing evidence that, even within this subset of intermetallic compounds, violations of the simple valence rules occur.

Syntheses and characterization of compounds like CsAu (Sommer 1943; Kienast et al. 1961) and Cs_2Pt (Karpov et al. 2003) which are constituted by a main group and a noble metal have completed the group of intermetallic compounds, in which the constituting metals exhibit large differences in electronegativity, resulting in a full charge transfer from cations to anions. Experimental data reveal that, e.g., the auride anion Au^- with a filled d shell plus a filled s shell is a surprisingly stable species, and that CsAu exhibits semiconducting behavior (Spicer et al. 1959) like a valence compound. Heavy noble metals such as platinum and gold are characterized by high electron affinities originating from their special electron configuration comprising 4f and 5d shells which are just filled. Additional stabilization of heavy metal anions by relativistic effects is discussed in theoretical treatments within the local density approximation (LDA) (Karpov et al. 2003). However, the stability of the CsCl-type compound LiAg (Freeth and Ragnor 1953/54) demonstrates that s^2d^{10} configurations are also adopted by lighter elements with a less pronounced relativistic stabilization of the s states. Moreover, the compound $Na_{16}Rb_7Sb_7$, which may be more precisely written as $Na_{16}{}^+Rb_6{}^+Rb^-Sb_7{}^{3-}$, indicates that even a filled s subshell can be realized (von Schnering et al. 1995).

The given examples expand the list of compounds where filled valence electron subshells are stable, and oxidation states can be attributed in a number of strongly polar intermetallic compounds comprising discrete building units which can be described with known counting rules such as $8 - N$ (Hume-Rothery 1930; see below). Thus, the identification of covalent fragments in intermetallic compounds appears to be an especially promising approach to advance the counting rules into the field of intermetallic compounds.

2.3.3.2 Extended 8 − N Rule

Elemental semi-metals like arsenic or tellurium are elements with partly filled p bands in a hypothetical primitive cubic arrangement. Symmetry breaking by specific distortions opens band gaps at the zone bondary, with the consequence of metal–insulator transitions. The number of covalent bonds in the resulting distorted patterns, e.g., $8 − 5 = 3$ for arsenic and $8 − 6 = 2$ for tellurium, is in accordance with the $8 − N$ rule (Hume-Rothery 1930), which correlates the number of valence electrons N with the number of atomic bonds for post-transition main group elements. The basic idea is that the atoms complete their octet by forming $8 − N$ two-center two-electron bonds. Later, the idea was extended to binary semiconductors and generalized to take into account homonuclear bonding. In the so-called general valence rule [Eq. (3)], where n_e is the total number of valence electrons, b_a and b_c represent the number of bonds between anions and cations, respectively, and n_a stands for the number of anions (Pearson 1964).

$$(n_e + b_a − b_c)/n_a = 8 \tag{3}$$

In b_c, the number of unshared valence electron pairs has to be included. All values relate to one formula unit of the compound. Thus, for GaTe with $n_e = 9$, $n_a = 1$ and $b_a = 0$, we find one cation–cation bond per formula unit; for PbSe with $n_e = 10$, $n_a = 1$, $b_a = 0$, we find a stereochemically active electron pair.

A special case of the $8 − N$ rule is realized by the weakly polar binary 1:1 Grimm–Sommerfeld compounds. These are composed of elements of groups $N − k$ and $N + k$ ($k = 1, 2, 3$) and exhibit properties similar to those of the elements in group N. Examples are Ge and the binaries GaAs, ZnS, and CuBr, or As and binary SnTe (Grimm and Sommerfeld 1926).

There is no direct way to predict immediately the structural effect of the electron deficiency with respect to the $(8 − N)$ electron count. Of crucial importance seems to be the relation between skeletal bonding, i.e, within the anionic part of the crystal structure, and exohedral bonding, i.e., on the periphery of the anionic part of the crystal structure. For example, Wade's rules (Wade 1976) for boranes imply counts lower than two for both skeletal and exohedral bonds, with a preference for the exohedral bonds compared with the skeletal bonds, since in the series from *closo*- to *arachno*-pentaborane $B_5H_5^{2-}$ (22 valence electrons), B_5H_9 (24 valence electrons), B_5H_{11} (26 valence electrons), the number of skeletal B–B bonding electrons increases from 12 via 14 to 16 while the number of exohedral bonding electrons is constantly 10. For the electron-deficient intermetallic compound $Eu_3Ge_5 = (Eu^{2+})_3(Ge_5)^{6-}$ it was shown by means of the electron localization function/electron density (ELF/ED) approach (cf. Section 2.3.2.3) that the skeletal Ge–Ge bonds have electron counts lower than two, whereas in the lone-pair regions the expected numbers of electrons are observed (Budnyk et al. 2006). For systems like silicides of alkali and alkaline earth metals (Nesper 2003) or $(Eu^{2+})_5(Ga_9)^{10-}$ (Grin et al. 2003) it was demonstrated that the skeletal bonds remain as two-electron–two-center bonds, whereas the electron counts for the exohedral part, i.e., lone pairs, are less then two.

2.3.3.3 Bonding Models in Direct Space

The local environment of atoms in the crystal structures of intermetallic compounds is difficult to describe by means of traditional MO pictures as well as by means of the free-electron gas model. The atomic coordination is usually more complex, as in salt-like inorganic compounds, and cannot necessarily be derived from the motifs corresponding to closest packing of spheres, which are characteristic of crystal structures of, e.g., elemental metals.

Bonding analysis in real space is a promising way to obtain deeper insight into this field. Recently, some quantum chemical functions such as the electron localization function (ELF) or the localized orbital locator (LOL; Schmider and Becke 2000) have been developed as quantum-chemical tools for analyzing and investigating bonding problems of chemical compounds in direct space. The ELF is especially useful for investigating bond formation in compounds with typically metallic physical properties and it is widely used for describing and visualizing chemical bonds in solids. Originally defined within the framework of Hartree–Fock theory (Becke and Edgecome 1990) and later developed for density functional theory (Savin et al. 1992), the ELF belongs to the so-called bonding indicators in real space (i.e., position space). As was shown later (Kohout 2004), the functions and functionals of this group trace the correlation of electronic motion of same-spin electrons, and a more general function of this kind is called the electron localizability indicator (ELI). The ELI represents, in a general sense, a charge distribution of electron pairs and, thus, are suitable tools for the analysis of chemical bonding in the sense of the Lewis theory (Lewis 1916), where pair formation plays the central role. In the vicinity of the nuclei, maxima indicate closed atomic shells. More detailed investigations show that not only the number of ELI (ELF) maxima for a given atom but also the integrated electronic population per shell is very close to the values expected from the *Aufbau* principle (Kohout and Savin 1996). The maxima of ELI (ELF) in the valence region (valence shells) or/and structurization of the penultimate (outer core) shell (Kohout et al. 2002) provide signatures for directed (covalent) bonding in position space. The ELI (ELF) tools are especially suitable to detect directed (covalent) bonding in materials with bands which are not fully occupied or are strongly overlapping, a situation which is typical for intermetallic compounds. The analysis of the topology of ELI (ELF) can be combined with the consecutive integration of the electron density (ED) in basins, which are bounded by zero-flux surfaces in the ELI (ELF) gradient field. This procedure, similar to the one proposed for ED (Bader 1999), makes it possible to assign an electron count for each basin, revealing the basic information about the chemical bonding. This combined application of ELI (ELF) together with ED offers the possibility of Zintl-like electron counts for a large group of intermetallic phases and of getting access to a bond definition in real space. The application of ELF in different kinds of bonding situations has been reviewed (Gatti 2005; for more details see http://www.cpfs.mpg.de/ELF).

In metals, the atoms adopt high coordination numbers but have relatively few valence electrons. Therefore, it seems natural to expect multicenter bonding. There is also an alternative view (Pauling 1960) which regards metallic bonding

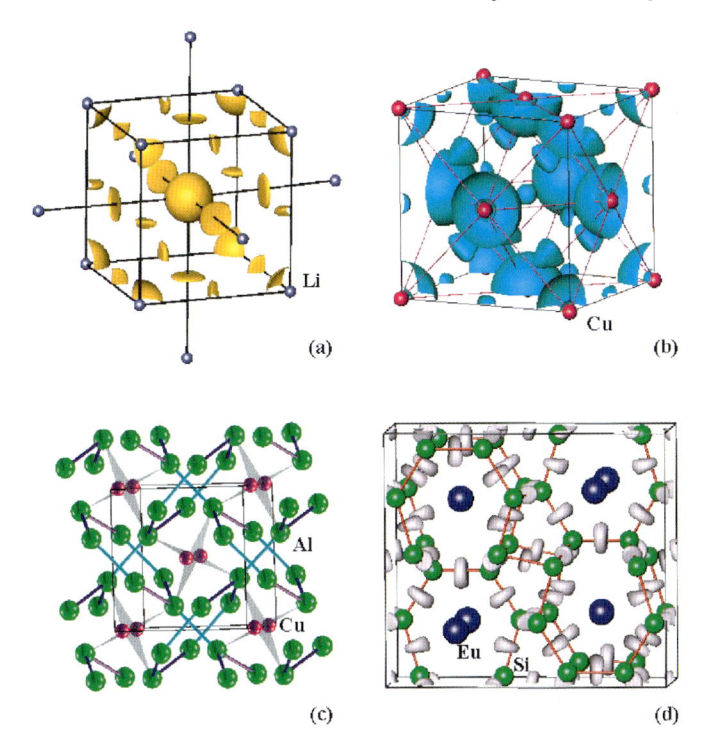

Fig. 2.10 Chemical bonding in metals and intermetallic compounds as evidenced by the ELF (ELI) approach. (a) Location of ELF maxima in the octahedral holes (six-center bonds visualized by the appropriate isosurface of ELF, orange) in elemental Li from TB-LMTO calculation. (b) Four-center bonds in the fcc structure of copper from FPLO calculations, with ELF maxima in the tetrahedral holes shown by the isosurface (turquoise). (c) Interpenetrating graphite-like nets of aluminum (blue) with copper atoms embedded in the framework by three-center bonds (dark gray) in the crystal structure of $Cu_{1-x}Al_2$. (d) Two-center bonds in the silicon framework (red) of $EuSi_6$ visualized by the ELF isosurface (gray).

as a partial (or unsaturated) bonding between nearest neighbors. For bcc Li (Fig. 2.10a), ELF (ELI) bonding attractors are observed in the centers of octahedral holes, thus indicating six-center bonds for calculations with the Hartree-Fock or the tight binding–linear muffin tin orbital (TB-LMTO) method. In the full-potential calculation, attractors occur at the tetrahedral voids, and imply four-center bonds. Despite the differences, these calculations show consistently that bonding is multicentered. ELF (ELI) for fcc Cu (Fig. 2.10b), as computed by the full potential local orbital (FPLO) and TB-LMTO-ASA (atomic spheres approximation) methods, yields a single set of identical attractor positions. Each bonding basin has a count of 1.05 electrons and represents a four-center bond. This implies that each Cu atom participates with 2.1 electrons in chemical bonding (Ormeci et al. 2006).

The atomic coordination in the crystal structure of the binary phase $Cu_{1-x}Al_2$ is pronouncedly different from that of the constituting elements in having copper in a tetragonal-antiprismatic coordination of aluminum. Despite this, the electronic density of states is metal-like since the valence and conduction bands overlap. The analysis of ELF reveals a larger amount of directed bonding than expected from the constituting elements. The covalently three-bonded aluminum atoms form interpenetrating graphite-like nets by two-center (and nearly two-electron) bonds. The copper atoms are embedded in the aluminum framework by means of three-center Cu–Al–Cu bonds (Fig. 2.10c; Grin et al. 2006).

Four-bonded silicon atoms in the crystal structure of the high-pressure phase $EuSi_6$ form a 3D network with a reduced density which is similar to that of diamond. The europium atoms are embedded in large cavities of the framework. Covalent bonding within the network is confirmed by ELF calculation (Fig. 2.10d; Wosylus et al. 2006) and, as a consequence, the calculated electronic density of states shows a clear (pseudo-) gap. From the ELF/ED approach, the valence electrons of Eu are transferred to the anionic network. In the electronic density of states (DOS) they fill bands above the pseudo-gap, causing metal-like electronic transport properties. In total, this compound is an example of so-called polar intermetallic compounds (Nesper 1990) and can be described by the charge-transfer balance $Eu^{2+}(Si^0)_6 \times 2e^-$, which visualizes the organization of bonding electrons causing metallic behavior.

2.4
Structure Types of Intermetallic Compounds

The crystal structures of the approximately 20 000 intermetallic compounds known in total belong to about 1200 structure types. The structure type is defined crystallographically as an atomic arrangement characterized by:
- a symmetry of the same space group
- a lattice with the same or a similar ratio of lattice parameters $a:b:c$ and angles α,β,γ
- a set of occupied Wyckoff positions (Wondratschek 1983) (sites) within a given space group
- a distinct distribution of the sites among different components
- similar values of the free positional parameters for according sites.

A more detailed analysis (Parthé et al. 1993/94) revealed the presence of subtypes (branches) within the structure types. Nevertheless, the definition above allows a classification of the whole variety of crystal structures of intermetallic compounds into a manageable number of basic motifs. Some of the structure types have hundreds of representatives (e.g., $BaAl_4$, AlB_2), others are found for only a few compounds, and some of them are unique (Villars and Calvert 1996).

2.4.1
Classification of the Crystal Structures of Intermetallic Compounds

The first classification of the crystal structures of the intermetallic compounds was introduced in the *Strukturbericht* and continued in *Structure Reports* (Anon. 1931–1943, 1940–1990; see References); it was based on the composition of the materials: A = elements, B = two-component compounds of 1:1 composition, C = two-component compounds with 1:2 stoichiometry, ... This system offers the possibility of only very restricted and formal descriptions. Thus currently it is not used very much; however, descriptors like B2 (CsCl type) and A15 (Cr_3Si type) can still be found in the literature for ordered bcc phases or for conventional superconductors. For ordered alloys, labels such as $L1_a$ ($CuPt_3$ type), $L1_0$ (AuCu type), $L1_1$ (CuPt type), and $L1_2$ ($AuCu_3$) are still in use.

Early attempts to classify the crystal structures of intermetallic compounds according to chemical bonding were not successful because of a lack of efficient tools for bonding investigations (cf. Section 2.3.1). Thus, the combined application of structural and chemical principles was considered as an appropriate route for classification of the crystal structures of intermetallic compounds.

Pauling (1960) pointed out the following groups: crystal structures of elemental metals (Cu, Mg, α-Fe) and their derivatives based on simple atomic packings, crystal structures with packings of atoms with different sizes (Laves phases, $BaAl_4$), electronic phases (β- and γ-brass), structures of metals with half- and non-metals (Fe_3C, FeB, AlB_2).

The structural principle of classification can be superimposed on the chemical aspects (Schubert 1964). According to the structural pattern, the derivatives of the closest- and closed-sphere packings are grouped together with the brass-like structures. The remaining crystal structures are classified according to their chemical composition (structures formed by main group elements, by a main group element and a transition metal, and by different transition metals).

Another way of taking chemical principles into account for the classification of crystal structures of intermetallic phases is to focus on the atomic environment. Krypiakevich (1977) classified the structure types according to the coordination number and the shape of the coordination polyhedra of the atoms with the smallest CN in the investigated structure. Seventeen groups were defined with CN values from 2 to 14 and with a coordination ranging from linear to rhombododecahedral. This classification offers the possibility of recognizing relationships between structure types which are not easily detectable on the level of structural patterns.

The following description of basic structures of intermetallic compounds is performed by combining the structural approach and atomic coordination analysis.

2.4.2
Crystal Structures Derived from the Closest Packings of Equal Spheres

From a chemical point of view, the easiest way to describe intermetallic compounds is to start with the crystal structures of elemental metals. Most crystal

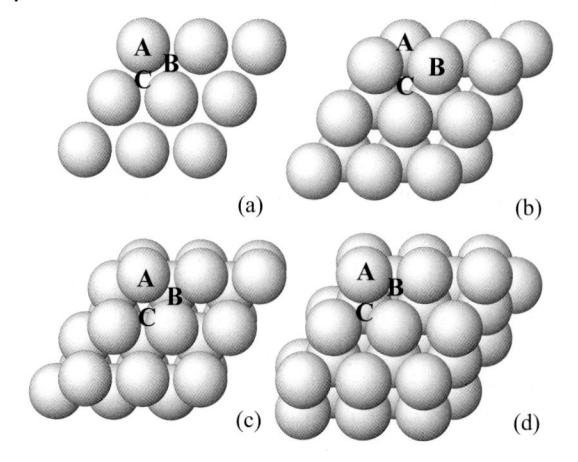

Fig. 2.11 Formation of basic closest-packed structure patterns. (a) Closest-packed atomic layer. (b) Stacking of the second layer in position B. (c) Stacking of the third layer in position A (...AB... packing sequence). (d) Stacking of the third and fourth layers in positions C and A (...ABC... packing sequence).

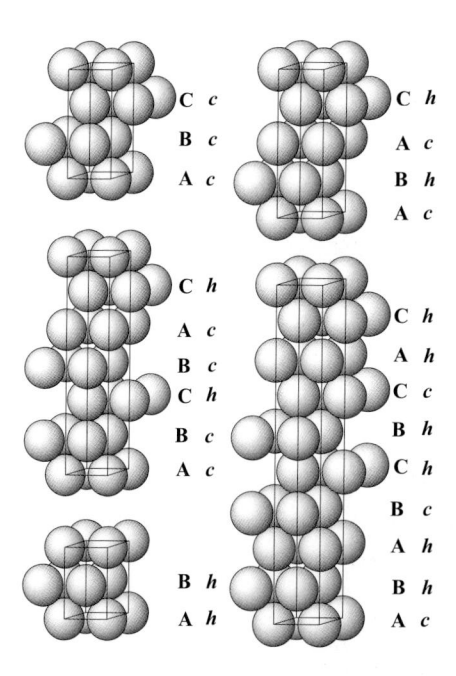

Fig. 2.12 Basic stacking patterns for closest-sphere packings and their notation.

structures of main group and transition metals are based on the closest and close packings of spheres of the same size. The closest packing of the spheres in the plane (2D) is defined by the coordination number 6. Each atom has six ligands forming a six-membered ring (Fig. 2.11).

The 3D closest packing is formed when the atoms of adjacent layers are in contact with three atoms of each neighboring layer. This can be realized in two different positions B and C when the arrangement of the starting layer is labeled A (Fig. 2.11). Two different positions, A and C, are available for the stacking of the third atomic plane when, e.g., the second one is in the position B. The resulting stacking sequences are ...ABC... and ...AB..., respectively. This defines the endless variety of the closest packings according to their stacking pattern. Only two different coordination environments are present in the closest packings: cuboctahedral in case of the ...ABC... stacking and its hexagonal analogue in

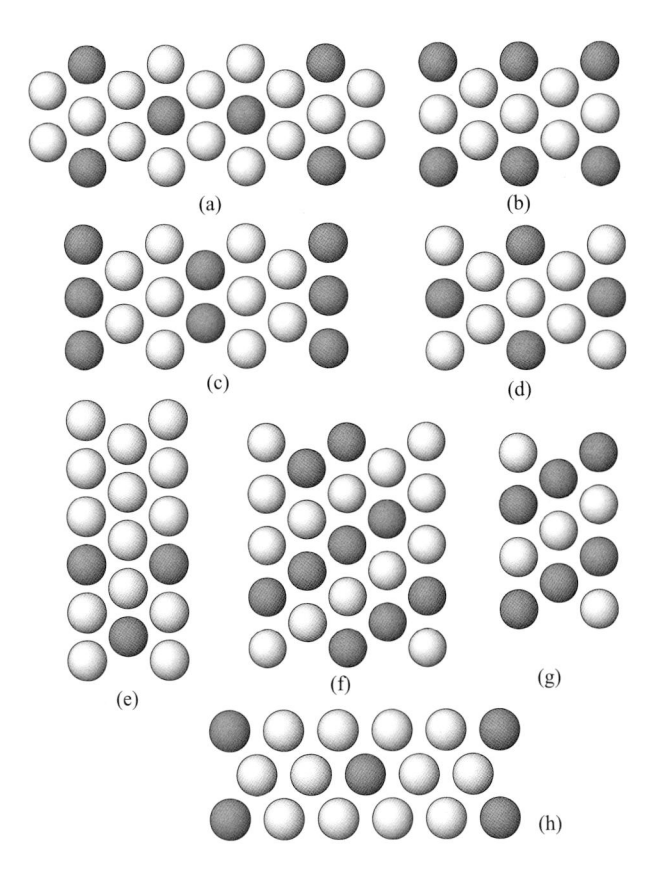

Fig. 2.13 Ordering pattern in the closest-packed layers in the crystal structures of (a) $(Cd_{0.5}In_{0.5})Au_3$; (b) $NbPd_3$; (c) $CdMg_3$, $AuCu_3$, $TiNi_3$, $PuAl_3$; (d) $MoPt_2$, $TaPt_2$; (e) $ZrAu_4$; (f) Ti_2Ga_3; (g) TiAl, AuCu, $(Nb_{0.75}Rh_{0.25})Rh$, α-TaRh, β'-AuCd; (h) $MoNi_4$.

case of the ...AB... stacking. Alternatively to the stacking sequences (positional symbols), the so-called coordinational symbols are used for the notation of the layer sequence: c for the layers with the cuoboctahedral coordination and h for the layers with the coordination in the shape of the hexagonal analogue of the cuboctahedron. In this so-called Jagodzinski–Wyckoff notation (Jagodzinski 1954), the ...ABC..., ...AB..., and ...ABAC... stackings are labeled c_3, h_2, and $(ch)_2$, respectively (Fig. 2.12).

When all the atoms in the crystal structure are chemically equal, the positional or Jagodzinki symbol notates uniquely the symmetry, axis ratio, and atomic positions in the structure. In cases where the atoms in the crystal structure are different, they can be ordered in different ways within the layer, depending on the composition. Typical ordering patterns are shown in Fig. 2.13. In this way, the whole variety of the ordered variants of closest packing structures can be classified and notated with stacking sequence, composition, and ordering pattern in the layer.

2.4.3
Crystal Structures Derived from the Close Packings of Equal Spheres

The bcc structural pattern can be obtained from an fcc arrangement by compression of the latter either along the four-fold or along the three-fold axis. In case of a distortion along a four-fold axis, the intermediate structure is tetragonal (Fig. 2.14). After reducing the compressed axis length to $\sqrt{2}/2$ of the initial value, the structure becomes bcc. The unit cell contains only two atoms; thus, the bcc structure is one of the simplest among intermetallic compounds. Several metals and intermetallic phases adopt this type of crystal structure and the relatively simple pattern allows for a large number of derivatives.

The first route to obtain new arrangements is an ordered occupation of the sites by different atoms, as already shown for the closest packings. Ordered re-

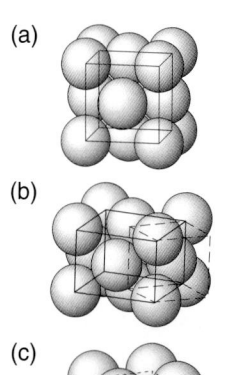

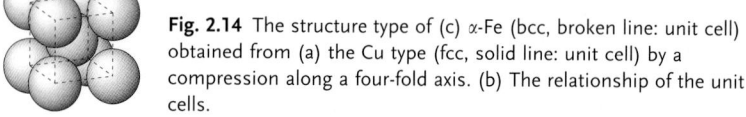

Fig. 2.14 The structure type of (c) α-Fe (bcc, broken line: unit cell) obtained from (a) the Cu type (fcc, solid line: unit cell) by a compression along a four-fold axis. (b) The relationship of the unit cells.

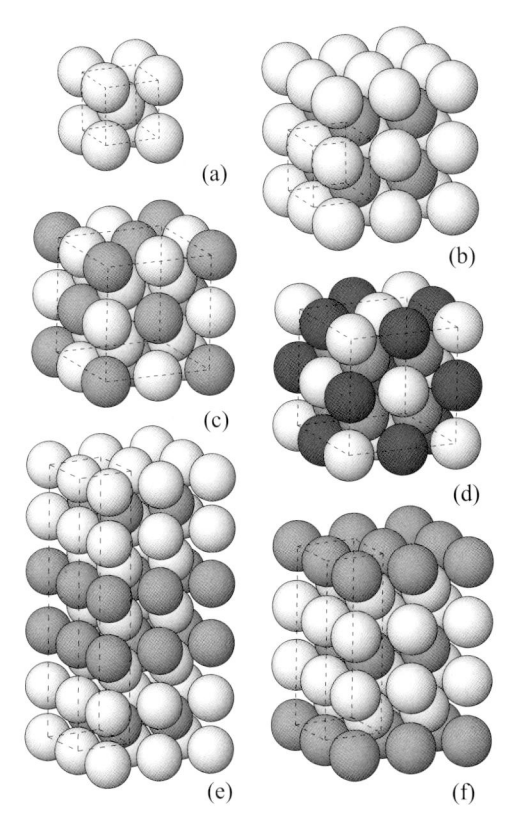

Fig. 2.15 Ordering patterns for (a) the basic bcc structure: structure types (b) CsCl; (c) Fe_3Si; (d) $MnCu_2Al$; (e) Os_2Al_3; (f) Cr_2Al.

placement of the positions in the ratio 1:1 without increasing the unit cell size leads to the formation of the cubic CsCl (or B2) type of structure (Fig. 2.15), which is often observed in metallic construction materials. Ordered occupation of the sites in the ratio 1:3 leads to the cubic structure of the BiF_3 (or Fe_3Si, $D0_3$) type with a unit cell which is eight times larger than that of the CsCl type (16 atoms per unit cell). This crystal structure is characteristic of, e.g., the Ni_3Al phase (superalloy). If the positions in the bcc motif are occupied in an ordered manner by three different kinds of atoms in the ratio 1:2:1, the cubic structure of the $AlCu_2Mn$ type ($L2_1$) forms. The unit cell volume is also eight times that of the CsCl arrangement. The phases showing this type of crystal structure are called Heusler phases. These materials are currently becoming the focus of research because of their promising electronic properties (spintronics). Numerous derivatives of the bcc motif with ordered occupation (superstructures) are known, and some have lower symmetry than cubic. As examples, we show the tetragonal structures of the Cr_2Al and Os_2Al_3 types with three and five times the unit cell volume of the initial bcc structure, respectively (Fig. 2.15).

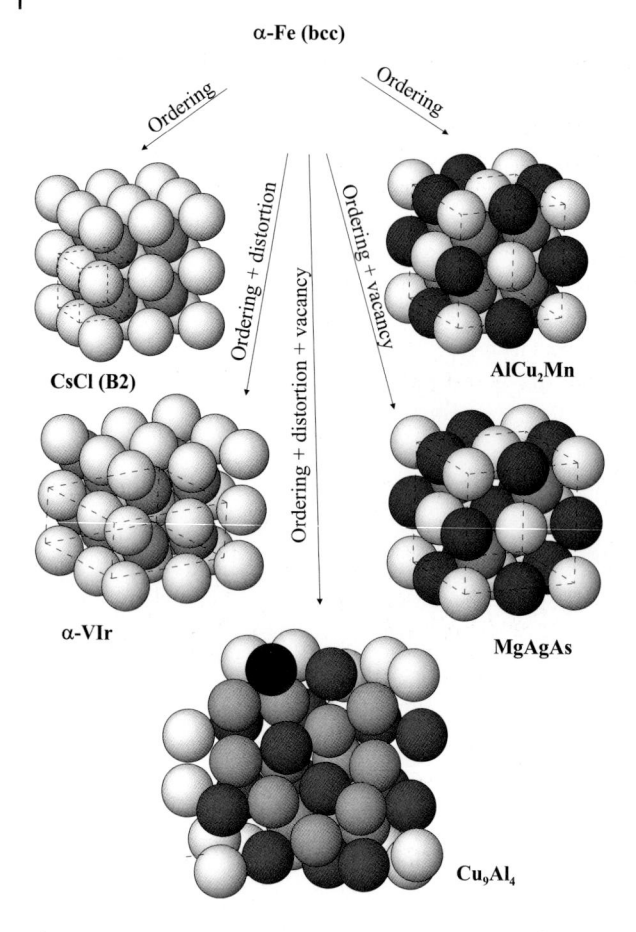

Fig. 2.16 Formation of structural patterns derived from the bcc packing.

Formation of superstructures is not a unique way to obtain structurally new materials derived from the bcc pattern (Fig. 2.16). Ordered occupation may lead to a local distortion resulting in a symmetry reduction to orthorhombic in the case of α-VIr. However, the formation of a superstructure can be combined with the occurrence of vacancies: bcc → $AlCu_2Mn$ → MgAgAs. The phases with the crystal structure of the latter type can be found in the literature under the name "semi-Heusler phases," which reflects solely their structural relationship.

The technically important γ-brass family can be obtained from the simple bcc structure by formation of a superstructure (also with partial occupation of sites by two kinds of atoms) with a unit cell volume which is 27 times that of bcc. Two of the 54 positions are unoccupied and the neighboring atoms are shifted toward the vacancy (Fig. 2.16). Various compositions of the γ-brass phases originate from a mixed occupation of some positions (gray in Cu_9Al_4, Fig. 2.16, bottom) depending on the constituting elements (cf. Hume-Rothery phases, Section 2.3.1).

2.4.4
Crystal Structures Derived from the Packings of the Spheres of Different Sizes

Laves phases form the largest group of intermetallic compounds with the ideal composition AB_2. The crystal structures of the Laves phases can be topologically derived from the closest packings. Eight of the 16 atoms in the unit cell are substituted by four larger atoms corresponding to the composition $16B - 8B + 4A = 4AB_2$. In case of the packing ...$ABAC$... (structure type α-Nd, $(ch)_2$), the replacement takes place in neighboring layers (Fig. 2.17). Pairs of vertex-condensed tetrahedra are replaced by two A atoms, and a pattern of the structure type $MgZn_2$ forms. In a similar way, the structural motif of the Laves phase $MgCu_2$ can be derived from the closest packing ...$ABCABC$... (structure type

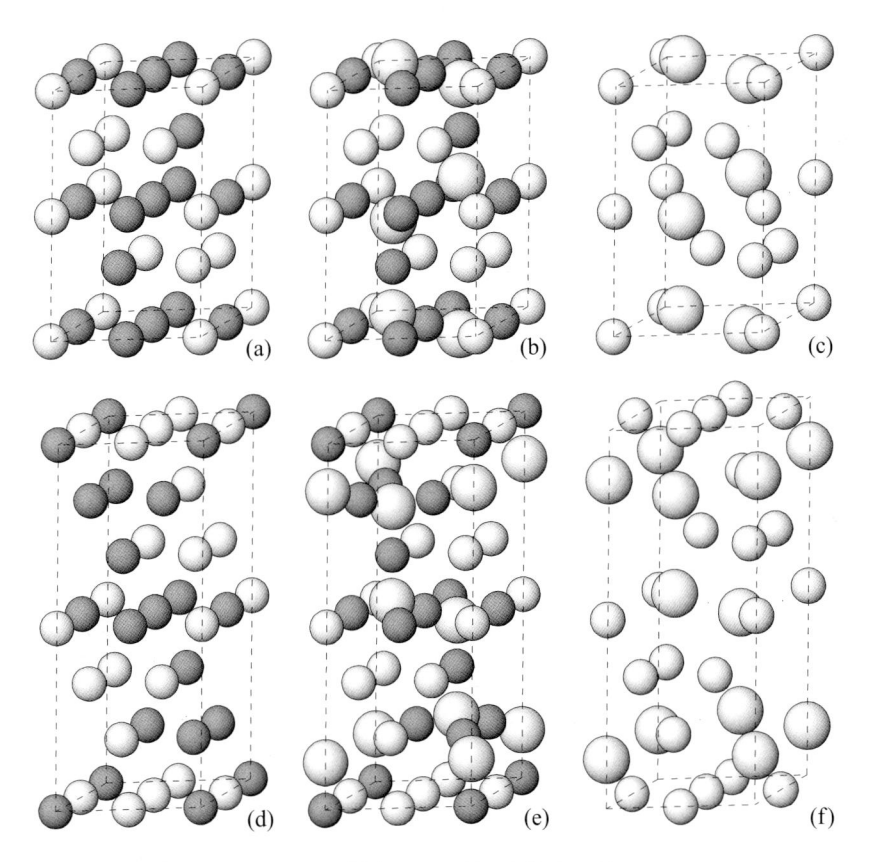

Fig. 2.17 (a) The structure type α-Nd (closest packing $(ch)_2 t$) and its transformation to (c) a crystal structure of the $MgZn_2$ type by (b) substitution of the four dark gray smaller atoms by one larger atom. (d) The structure type of Cu (closest packing $(c_3)_2$) and its transformation to (f) the structure of the $MgCu_2$ type by (e) substitution of the four dark gray smaller atoms by one larger atom.

Cu, $(c_3)_2$). According to this building principle an optimal radius ratio of $1:2^{1/3} = 1.26$ (Nowotny and Mornheim 1939) or $1:\sqrt{1.5}$ (approximately 1:1.225, Laves and Witte 1935; Dehlinger and Schulze 1940) is calculated for the components B and A. The importance of radius ratios for the packing of atoms of different sizes seems to be underlined by the experimental observation of the solid noble gas compound NeHe$_2$ with MgZn$_2$-type crystal structure at high pressures (Loubeyre et al. 1993), but only a few of the known Laves phases have a radius ratio in the interval obtained from these geometrical considerations (Stein et al. 2004; Fig. 2.1). One of the extreme outliers is KAu$_2$, which does exhibit a radius ratio $r(K)/r(Au) = 1.648$.

2.4.5
Selected Crystal Structures with Complex Structural Patterns

The intermetallic compound SmCo$_5$ is the first and best-known representative of a family of magnetic materials with general composition RX_5. The crystal structure of these compounds belongs to the CaCu$_5$ type. This structural pattern can not be derived from the closest- or close-packed structures, but from the structures of the Laves phases. The structure of the MgZn$_2$ and CaCu$_5$ types can be described as frameworks of vertices-condensed tetrahedra X_4 with R atoms embedded in their cavities (Fig. 2.18a,b). The relationship between the two structure types becomes clearly visible when the atomic layers perpendicular to the six-fold axis are compared (Fig. 2.18c,d). Both patterns comprise common Kagome nets (Wells 1968; O'Keeffe and Hyde 1996) with composition X_3 (within a unit cell). The mixed layers of R and X atoms have the same topology in both structures with occupied positions $(0,0)$, $(1/3, 2/3)$ and $(2/3, 1/3)$. However, the compositions are different: R_2X in MgZn$_2$ and RX_2 in CaCu$_5$. They can be easily obtained as $R_2X + X_3 = 2RX_2$ and $RX_2 + X_3 = RX_5$ for MgZn$_2$ and CaCu$_5$, respectively.

Despite the apparent simplicity, the atomic pattern of CaCu$_5$ offers several possibilities for formation of new structural motifs. The presence of the same Kagome net in the structure types of CaCu$_5$ and the Laves phases (e.g., MgCu$_2$) allows a combination of both structural motifs along the hexagonal axis. The Kagome net serves as the common interface. The new structural patterns obtained in this way are called intergrowth structures (Parthé et al. 1985; Grin 1992). The composition of the resulting intergrowth structures is simply the sum of the compositions of the prototypes: $R_2X_4 + RX_5 = 3RX_3$ for the CeNi$_3$ and PuNi$_3$ structure types and $R_2X_4 + 2RX_5 = 2R_2X_7$ for Ce$_2$Ni$_7$ and β-Gd$_2$Co$_7$ structure types (Fig. 2.19).

Another route to the formation of new structural motifs derived from the CaCu$_5$ type is a double replacement of an R atom by an X_2 pair. This leads to the crystal structures of the Th$_2$Ni$_{17}$ and Th$_2$Zn$_{17}$ types ($3RX_5 - R + X_2 = R_2$Ni$_{17}$) or the ThMn$_{12}$ type ($2RX_5 - R + X_2 = RX_{12}$).

The crystal structure of a large group of superconducting materials (e.g., Nb$_3$Ge, which for a long time had the highest known transition temperature of

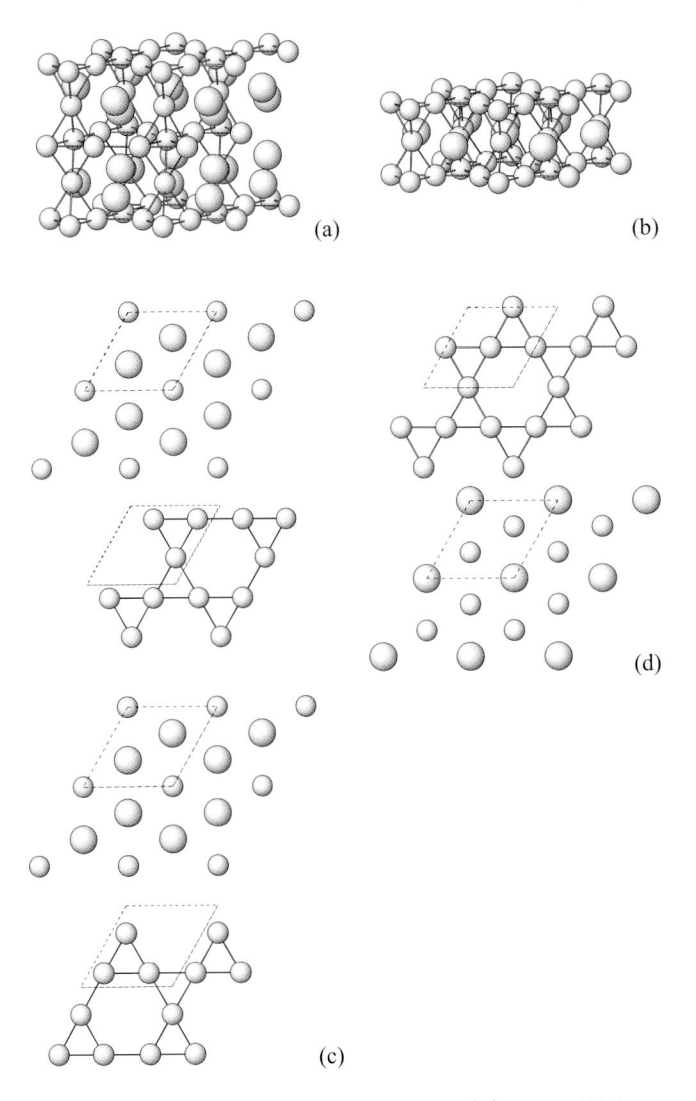

Fig. 2.18 Crystal structures of (a) MgZn$_2$ (RX$_2$) and (b) CaCu$_5$ (RX$_5$) as frameworks of vertices-condensed tetrahedra X$_4$ with embedded R atoms. Atomic layers normal to the hexagonal axis in (c) MgZn$_2$ and (d) CaCu$_5$. Dotted lines show projection of the unit cell.

23 K, in the era before oxide-based superconductors) belongs to the Cr$_3$Si (A15) type (Fig. 2.20). Similarly to the CaCu$_5$ motif, the Cr$_3$Si pattern cannot be derived directly from the closest or close packings. In a first approximation the structure can be described as composed of layers perpendicular to the axes of the cubic lattice (Fig. 2.20b). The mixed Cr/Si layer is a defect closest-packed one. Between these layers, additional Cr atoms are located in front of the vacancies, forming a

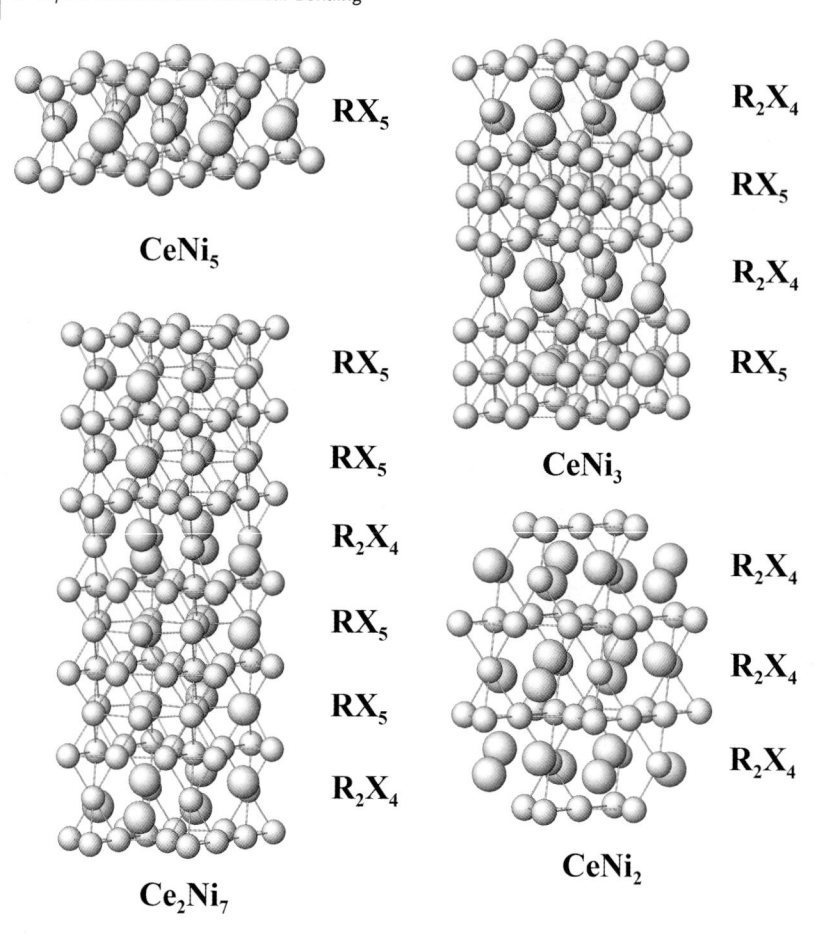

Fig. 2.19 Crystal structures of binary compounds in the Ce–Ni system as intergrowth derivatives of CaCu$_5$ and Laves phases.

very loose network (Fig. 2.20b). On the other hand, the crystal structure of Cr$_3$Si can be understood as Si atoms and Cr$_2$ pairs located in the positions of the fcc pattern (000) and $(\frac{1}{2},\frac{1}{2},0)$, respectively (Fig. 2.20c).

The general complexity of the structural pattern increases further by consideration of the crystal structure of Nd$_2$Fe$_{14}$B (Fig. 2.21). This compound represents the currently most efficient hard magnetic materials. The tetragonal crystal structure can be understood as a stacking of two types of different structural slabs along the four-fold axis. The first one is centered at $z = 0$, the second at $z = \frac{1}{4}$. In both segments we observe atomic arrangements originating from the CaCu$_5$ motif (broken-line parallelogram). Additionally, fragments of the Cr$_3$Si pattern are present (dotted square). Thus, the whole crystal structure can be described as a three-dimensional arrangement of fragments of simpler crystal structures like

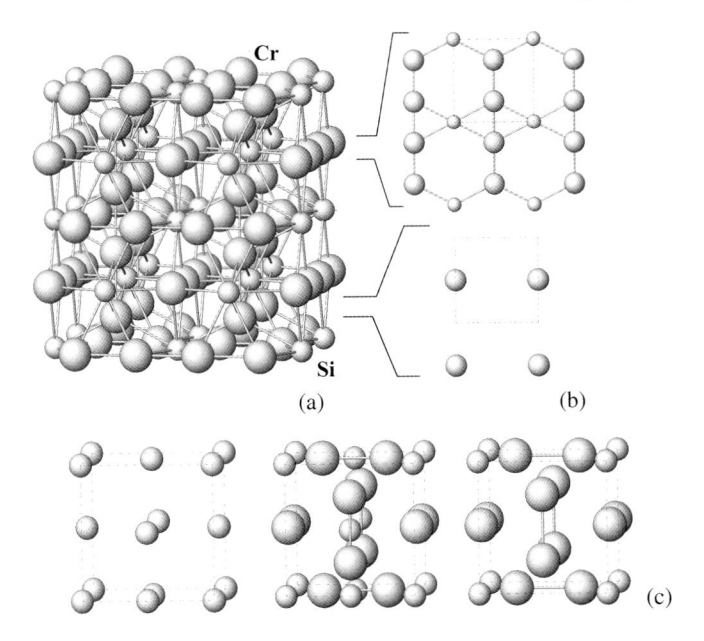

2.4 Structure Types of Intermetallic Compounds | 47

Fig. 2.20 The crystal structure of the Cr$_3$Si type (a) as a stacking of defect closest-packed layers (b) with additional embedded Cr atoms. Derivation of the Cr$_3$Si pattern from fcc by double substitution on the ($\frac{1}{2},\frac{1}{2}$,0) position (c). Dotted lines show the unit cells.

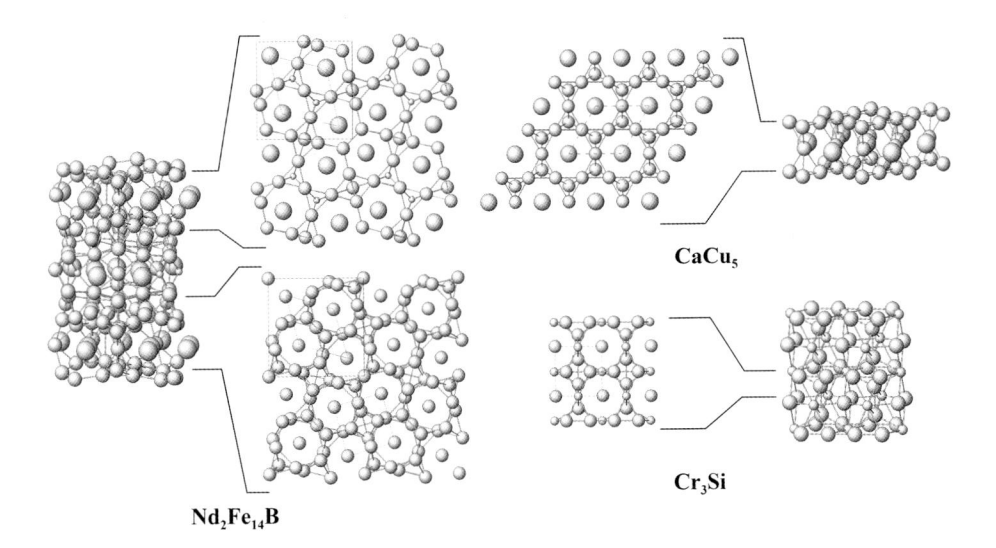

Fig. 2.21 The crystal structure of the hard magnetic material Nd$_2$Fe$_{14}$B as a derivative of the crystal structure motifs of CaCu$_5$ and Cr$_3$Si.

CaCu$_5$ or Cr$_3$Si. A further increase in the structural complexity is observed, proceeding to the quasicrystalline phases and their approximants.

2.5 Quasicrystals

2.5.1 Introduction

All known single-phase solids in thermodynamic equilibrium are crystalline. This means that the symmetry of their crystal structures can be described by an n-dimensional (nD) space group in direct space (cf. Steurer and Haibach 1999) or, fully equivalently, by the corresponding symmetry group in 3D reciprocal space (Mermin 1992). For the standard case of 3D translationally periodic crystal structures, $n = 3$. For the known aperiodic crystals that can all be described geometrically as 3D sections of nD hypercrystals, $4 \leq n \leq 6$ applies. Crystallinity leads to an intrinsic anisotropy of physical properties. Faceted crystals are one of its manifestations (Fig. 2.22).

A universal definition of aperiodic crystals directly related to a diffraction experiment was provided by the IUCr Commission on Aperiodic Crystals (IUCr 1992): "... by *crystal* is meant any solid having an essentially discrete diffraction diagram, and by *aperiodic crystal* is meant any crystal in which three dimensional lattice periodicity can be considered to be absent." "Essentially discrete" means that the diffraction intensities can even be densely distributed as long as they are Bragg reflections. The known aperiodic crystals comprise the classes of quasicrys-

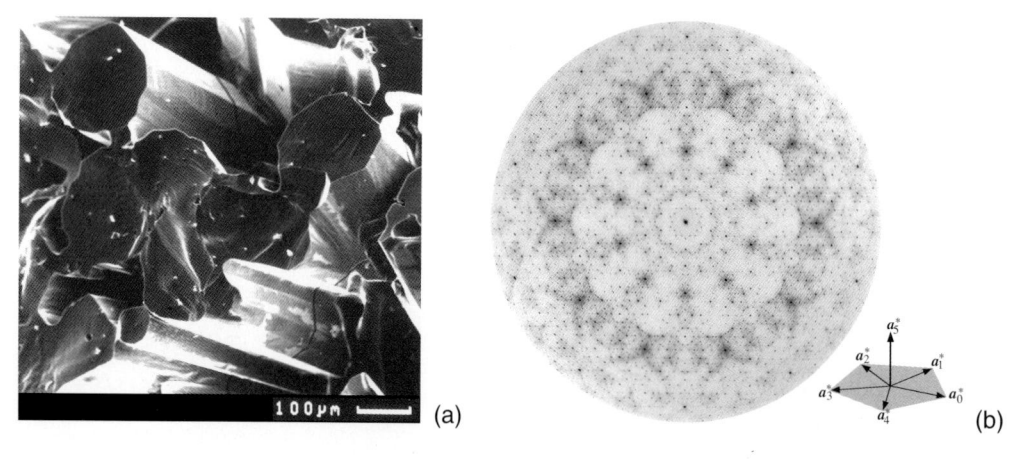

Fig. 2.22 (a) SEM photograph of decagonal Al–Co–Ni with decaprismatic growth morphology; (b) the diffraction pattern and star of reciprocal lattice vectors.

2.5 Quasicrystals

tals (QCs), incommensurately modulated crystals (IMCs), and composite crystals (CCs) (cf. Yamamoto 1996). All three of these classes possess the kind of topological long-range order which leads to Bragg reflections in a diffraction experiment (Fig. 2.22), although their structural building units are arranged aperiodically in 3D space.

3D lattice periodicity is compatible with $n^c =$ one-, two-, three-, four- and six-fold rotational symmetry, consequently called crystallographic symmetry. If one of the lattice points is a point of global n^c-fold symmetry, then all other lattice points are as well. Rotation around any lattice point brings the lattice n^c times into coincidence with itself. Contrariwise, a 3D aperiodic crystal structure cannot contain more than a single axis of global noncrystallographic n^{nc}-fold symmetry. Typical, however, is the existence of an infinite number of axes with only local n^{nc}-fold symmetry (i.e., symmetry of patches of tiles or of atomic clusters) as well as global n^{nc}-fold orientational symmetry (i.e., the edges in a tiling are oriented only along n^{nc} symmetrically equivalent directions).

Noncrystallographic n^{nc}-fold symmetries can become crystallographic in higher-dimensional lattices. For 5-, 8-, 10- and 12-fold rotational symmetry this is the case in 4D lattices and for 7-, 9-, 14- and 18-fold symmetry in 6D lattices (Hermann 1949; Hiller 1985). This property underlies the higher-dimensional approach (Wolff 1974; Janssen 1986; cf. Steurer and Haibach 1999), a powerful and elegant way to describe aperiodic crystals. The information on symmetry and metrics of an aperiodic structure is coded in the parameters of the nD lattice and its orientation to the 3D physical space. The full information on atomic positions, site occupancy, stoichiometry, and all interatomic distances is coded in the $(n-3)$D atomic surfaces and their positions in the nD unit cell (Fig. 2.23, below). The nD embedding space consists of the physical or parallel space (par-space) and the perpendicular space (perp-space) orthogonal to it, $\mathbf{V} = \mathbf{V}^{\parallel} \oplus \mathbf{V}^{\perp}$. The point

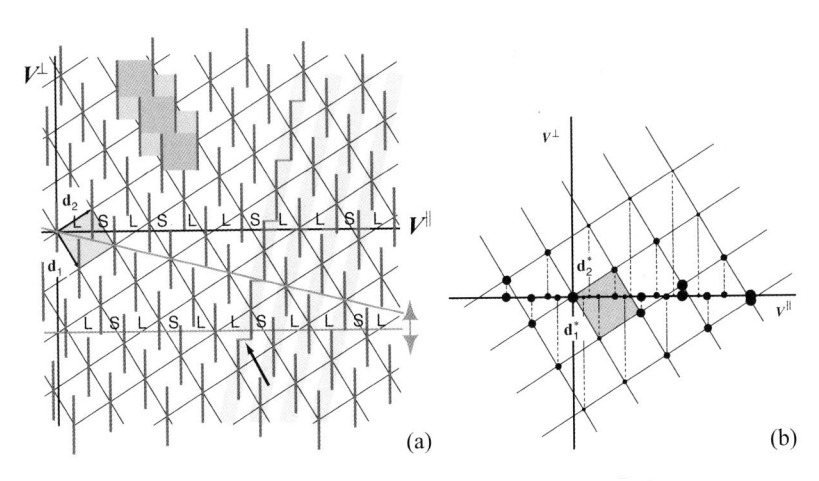

Fig. 2.23 (a) Fibonacci sequence and (b) its reciprocal space in the 2D description.

symmetry group of the nD hypercrystal has to be isomorphous with the point group of the 3D aperiodic crystal.

Special degrees of freedom that are characteristic of aperiodic crystals can be easily visualized by this approach. Moving the hypercrystal along the perp-space generates an infinite number of locally isomorphous (homometric and isoenergetic) quasiperiodic structures. Shearing it along the perp-space generates periodic rational approximants. Periodic or random fluctuations of the perp-space coordinates lead to phason modes or phason fluctuations, respectively (Socolar et al. 1986; Socolar 1988; Bancel 1989; Janssen and Radulescu 2004).

2.5.2
Quasiperiodic Structures in Direct and Reciprocal Space

The simplest example of a QC is the 1D Fibonacci sequence (FS). It can be generated either by substitution rules or as a cut of a 2D hypercrystal (Fig. 2.23). The substitution rule $\sigma : S \mapsto L, L \mapsto LS$ replaces a letter S by L and the letter L by the word LS (Table 2.1) (cf. Luck et al. 1993). An infinite FS remains invariant under this substitution rule. The Fibonacci numbers F_{n+1} and F_n, with $F_n = F_{n-1} + F_{n-2}$, give the frequency of L and S, respectively, in the chain. Their ratio [Eq. (4)] approaches the golden mean for n going to infinity.

$$\lim_{n \to \infty} \frac{F_{n+1}}{F_n} = \tau = 1.618\ldots = (1 + \sqrt{5})/2 = 2\cos(\pi/5) \tag{4}$$

The number τ is the solution of the algebraic equation $\tau^2 - \tau - 1 = 0$ and it is therefore a quadratic irrationality. If we assign to L and S long and short atomic distances with the ratio τ [Eq. (5)], then we get a quasiperiodic structure, which is invariant under scaling with τ^n.

$$\frac{L}{S} = \frac{LS}{L} = \frac{LSL}{LS} = \frac{LSLLS}{LSL} = \cdots = \tau \tag{5}$$

Table 2.1 Generation of the Fibonacci sequence by the substitution rule σ: the frequency of the letters L and S is given by the respective Fibonacci numbers F_{n+1} and F_n.

n	Word	Number of $S(F_n)$	F_{n+1}	F_{n+1}/F_n
0	L	0	1	$0/1 = 0$
1	LS	1	1	$1/1 = 1$
2	LSL	1	2	$2/1 = 2$
3	LSLLS	2	3	$3/2 = 1.5$
4	LSLLSLSL	3	5	$5/3 = 1.666\ldots$
5	LSLLSLSLLSLLS	5	8	$8/5 = 1.6$
	$\ldots$			
∞	$\ldots$			$\tau = 1.618\ldots$

2.5 Quasicrystals

Embedding the FS in 2D space allows its description as a 1D cut of a periodic 2D hypercrystal (cf. Steurer and Haibach 1999) (Fig. 2.23). The lattice is spanned by the basis vectors $\mathbf{d}_1$, $\mathbf{d}_2$. The slope of $\mathbf{d}_1$ is equal to τ. The vertices of the FS are generated where the physical space $\mathbf{V}^\parallel$ cuts the atomic surfaces (line segments). The size of the atomic surface results from the projection of a unit cell upon the perp-space. The frequency of distances L and S is proportional to the height of the shaded areas in the upper left part of Fig. 2.23(a). Cutting the hypercrystal with the physical space with a rational slope (for instance, the inclined gray line through lattice points 00, 11, 22 ...,) gives the n/m-approximant (in our example 1/1). Moving the physical space (horizontal gray line) through the end points of the atomic surface (marked by an arrow) leads to vertex jumps (phason flips) of the type $LS \Leftrightarrow SL$. Projecting the hypercrystal along the light-gray shaded lines gives the periodic average structure (PAS) of the FS.

In 1D reciprocal space, the FS shows Bragg reflections at positions given by a $\mathbb{Z}$-module of rank 2 [Eq. (6)], with $\mathbf{a}_2^* = \tau \mathbf{a}_1^*$.

$$\mathbf{H} = \sum_{i=1}^{2} h_i \mathbf{a}_i^*, \quad h_i \in \mathbb{Z} \tag{6}$$

By embedding the FS in 2D space the periodicity of the reciprocal lattice is restored (Fig. 2.23b). Since direct and reciprocal spaces are mathematically related by Fourier transformation, the cut in direct space, generating the aperiodic crystal structure, corresponds to a projection in reciprocal space yielding the 1D diffraction pattern. Reflections which are close to each other in the 1D diffraction pattern originate from regions of the reciprocal lattice which are far apart with strongly differing perp-space components of their diffraction vectors $\mathbf{H} = (\mathbf{H}^\parallel, \mathbf{H}^\perp)$. They have very different intensities because they decrease rapidly with increasing $|\mathbf{H}^\perp|$. This is the reason why an experimentally observed diffraction pattern of a QC shows clearly separated Bragg reflections although they have basically a dense distribution.

The Penrose tiling (PT; Penrose 1974) is a good example of a quasilattice in two dimensions (Fig. 2.24). It can be constructed from two unit tiles: a fat rhomb (acute angle $\alpha = 2\pi/5$) and a skinny rhomb (acute angle $\alpha = \pi/5$) with equal edge lengths a_r. The ratios of their areas as well as of their frequencies in the infinite tiling are τ:1. The construction of the PT has to obey matching rules of the tiles that can be derived from its scaling properties (cf. Steurer and Haibach 1999). A special decoration of the unit tiles with line segments yields a PT with quasiperiodically spaced lines in five orientations, the so called Ammann lines. These lines can be seen as the equivalent of the lattice lines in 2D or lattice planes (net planes) in 3D lattices. A quasiperiodic structure resulting from an appropriate decoration of the unit tiles with atoms or atomic clusters can reasonably well describe some characteristics of real decagonal QCs (DQCs), which are geometrically stackings of quasiperiodic layers.

A 3D quasilattice, the 3D Penrose tiling or Ammann tiling, is based on two rhombohedral unit tiles (Fig. 2.25) (cf. Steurer and Haibach 1999). The acute an-

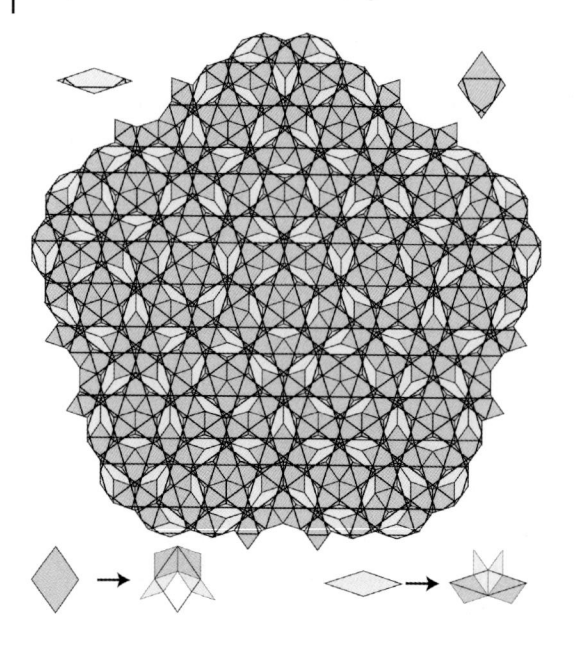

Fig. 2.24 Penrose tiling with Ammann lines. Top: the two unit tiles with line segments giving infinite straight lines in the ideal PT. Bottom: substitution of the unit tiles by tiles which are smaller by a factor τ.

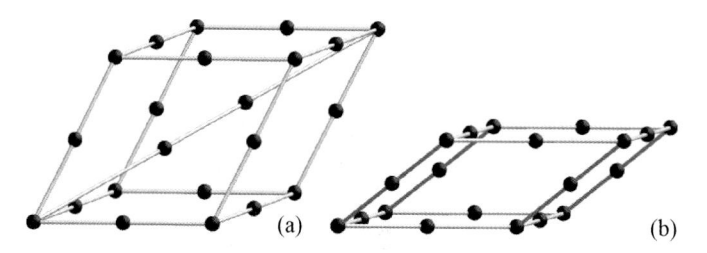

Fig. 2.25 The two unit tiles of the 3D Penrose tiling: (a) a prolate and (b) an oblate rhombohedron with equal edge lengths a_r. Decoration sites by atoms or clusters are marked by spheres.

gles of the rhombs amount to $\alpha_r = \theta = \arctan(2) = 63.44°$. Their volumes and frequencies are in a ratio of τ:1.

2.5.3
Formation and Stability

QCs are binary or ternary intermetallic phases. They are hard and brittle, similarly to almost all known complex intermetallic phases. Due to their special cluster-based structure, the electronic density of states of QCs shows a pseudogap at the Fermi edge. Although their constituents are electronically very good metallic conductors, their electrical and thermal conductivity decreases with temperature. It falls even below the Mott conductivity threshold in some cases. For a detailed introduction into the physics and (still only potential) applications of QCs, see Dubois (2005).

More than 100 stable or metastable representatives are known so far. Most QCs have been discovered by a search strategy based on the concept of electron compounds (Hume-Rothery phases) (Tsai 2003). Most of them contain aluminum as main component, some titanium, magnesium, zinc, or cadmium. Some are line phases such as icosahedral Cd–Yb (i-Cd–Yb), others show broad compositional stability ranges such as decagonal Al–Co–Ni (d-Al–Co–Ni). With the exception of i-Ti–Zr–Ni, they are stable up to the melting temperature. Only a few QCs melt congruently.

2.5.4
Structures of Decagonal Quasicrystals (DQCs)

DQCs, with diffraction symmetry $\overline{10}/mmm$ or $\overline{10}/m$, can be geometrically seen as periodic packings of quasiperiodic atomic layers. More physical, however, is the description as quasiperiodic packing of decagonal columnar clusters that are periodic along the column axis. The needle-like morphology of DQCs indicates faster growth along the ten-fold axis, i.e., the periodic direction (Fig. 2.22).

A typical example of the structure of a DQC is depicted in Fig. 2.26(a) (Cervellino et al. 2002). The projected electron density of decagonal Al–Co–Ni is overlaid with different tilings and coverings to show the cluster structure. A pentagonal

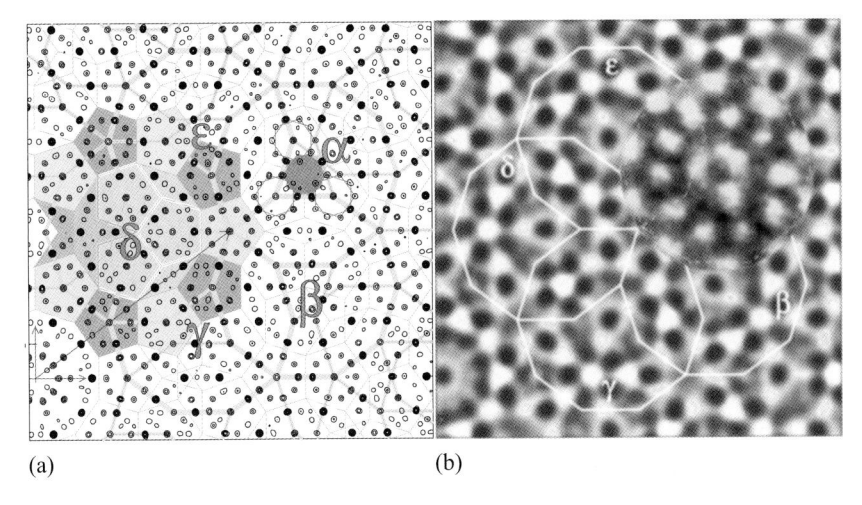

Fig. 2.26 45 Å × 45 Å section of the projected electron density of decagonal Al–Co–Ni with (a) high resolution and (b) a resolution comparable with that of the HAADF image (Abe and Tsai 2001) copied on the position of Gummelt decagon α. In (b) black contrasts correspond to zero density, white to maximum density (from Steurer 2004).

PT with edge length $a_r = 4.625$ Å is shown by thin lines. A group of five decagonal Gummelt clusters (Gummelt 1996) with 20.6 Å diameters is drawn in, labeled α to ε. The centers of the Gummelt decagons are located on a τ^2-inflated version of the pentagon tiling. In Fig. 2.26(b), the projected electron density of Fig. 2.26(a) is now shown at a resolution comparable with that of the HAADF-STEM (High-Angle Annular Dark Field Scanning Transmission Electron Microscopy) image (Abe and Tsai 2001) copied on the position of Gummelt decagon α. There is a good qualitative agreement between the distribution of transition metal atom related contrasts of the calculated and observed images.

Intermetallic phases with structures closely related to those of quasicrystals are called approximants. "Closely related" means that they consist of the same (more or less distorted) structural building units (clusters). In the special case where the structure can be described by a rational section of the hypercrystal structure that also generates the quasicrystal structure, it is called a rational n/m-approximant. An example of an approximant of d-Al–Co–Ni is shown in Fig. 2.27. One layer shows a decorated pentagon tiling, the other a hexagon tiling dual to it (black lines). All the vertices of these tilings are decorated by Co atoms. The hexagons have the same edge lengths as those of Fig. 2.26.

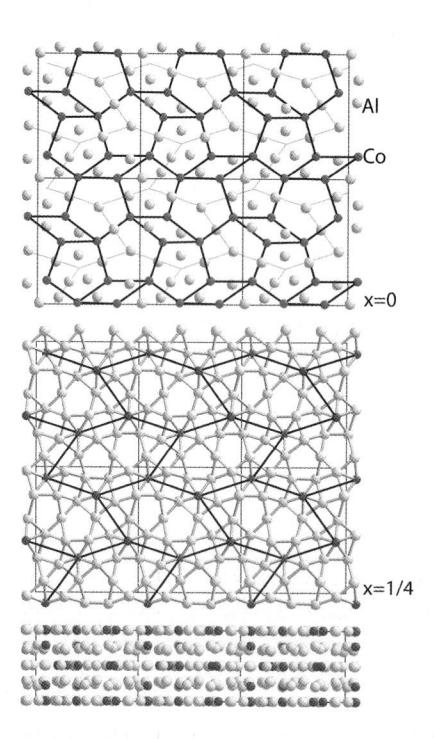

Fig. 2.27 (011) atomic layers of o-$Al_{13}Co_4$ and structure projected along [001] (Grin et al. 1994). Grin, J., Burkhardt, U., Ellner, M., Peters, K. (1994), *J. Alloy. Compd.* **206**, 243.

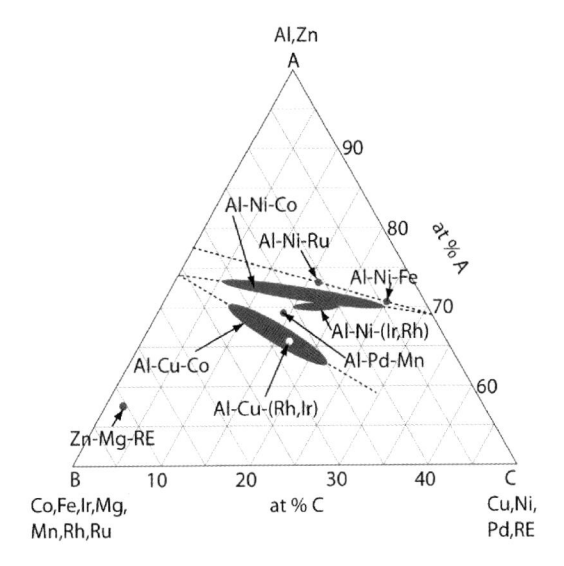

Fig. 2.28 Compositions of stable decagonal phases. Only the upper half diagram ($A_x B_y C_z$, $x + y + z = 100$, $50\% \leq x \leq 100\%$, $0\% \leq y \leq 50\%$, $0\% \leq z \leq 50\%$) is shown.

More than 20 stable decagonal phases with n-layer periodicity along the tenfold axis have been found so far in the following systems (Steurer 2004).
- *Two-layer periodicity (sometimes with two-fold superstructure)*: Al–Cu–Me (Me = Co, Rh, Ir), Al–Ni–Me (Me = Co, Fe, Rh, Ru), Zn–Mg–RE (RE = Y, Dy, Ho, Er, Tm, Lu)
- *Six-layer periodicity*: Al–Mn–Pd, Al–Mn–Fe–Ge, Ga–Co–Cu, Ga–Cu–Fe–Si, Ga–V–Ni–Si
- *Eight-layer periodicity*: Al–Ni–Ru, Al–Pd–Me (Me=Fe, Ru, Os).

The approximate stability regions of the stable DQCs are shown in Fig. 2.28 (based on Grushko and Velikanova 2004).

2.5.5
Structures of Icosahedral Quasicrystals

There are different ways of classifying icosahedral quasicrystals (IQCs). Historically the first approach was to classify them as Frank–Kasper-type QCs (FK type; sp-QCs) and Al-transition-metal type (Al-TM type; Mackay type; spd-QCs) QCs (Henly and Elser 1986). The first case is characterized by a tetrahedral building principle (tetrahedral close packing) and the occurrence of Bergman clusters (Pauling triacontahedra). In the second case TM atoms are commonly coordinated icosahedrally, Al atoms often form octahedra. The larger clusters which occur are of the Mackay type.

(a)   (b)   (c)   (d)   (e)

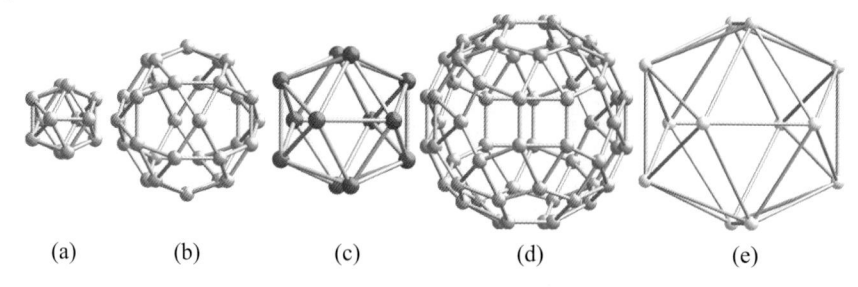

Fig. 2.29 Shells of the double-Mackay cluster building $cP138$-α-Al–Mn–Si (projected along [−1 0 5.7]) (Sugiyama et al. 1998). The shells (a)–(e) are centered at the origin of the unit cell and shells (a)–(c) in the body center. Shells (a)–(c) together form the 54-atom Mackay cluster.

With an increasing number of IQC structures known, the 6D hypercrystal structure type (position, size, shape, and chemical partition of atomic surfaces) could be used for classification. The 6D Bravais-lattice type can be either hypercubic primitive (iP) or hypercubic face-centered (iF). Note that the last case is frequently just a 2^6-fold superstructure of the first, i.e., all 6D lattice parameters are doubled. In the cases where both Bravais types are found for one and the same composition, the iP phase is just the chemically disordered variant of the iF phase.

Sticking to the 3D IQC structure, the fundamental cluster types building the structures of quasicrystals and their rational approximants can be used for a classification. There are three main types, the Mackay cluster (MC), the Bergman cluster (BC), and the Tsai cluster (TC) (Figs. 2.29–2.31):

- *Mackay cluster* (Fig. 2.29): The innermost cluster, an Al/Si icosahedron, is surrounded by an Al icosidodecahedron. Its pentagons are capped by Mn atoms at the corners of an icosahedron. These three shells form the Mackay cluster. Adding two further shells, a complex Al shell and an icosahedral Al/Si shell, gives the double-Mackay cluster (Sugyama et al. 1998). If the icosahedral symmetry is broken by the innermost shell, then one gets a pseudo-Mackay cluster (Boudard et al. 1992).
- *Bergman cluster* (Fig. 2.30): The Bergman cluster in $cI160$-R-Al$_5$CuLi$_3$ is built from an Al/Cu icosahedron surrounded by a Li pentagondodecahedron. The pentagons are capped again by Al/Cu atoms yielding an icosahedron with a similar diameter to the dodecahedron. These two shells together form a triacontahedron. The fourth shell is a truncated icosahedron of Al/Cu atoms, and the fifth shell is a Li triacontahedron with Li atoms capping the hexagonal faces of the truncated icosahedron. The diameters of the large and

(a)   (b)   (c)   (d)   (e)

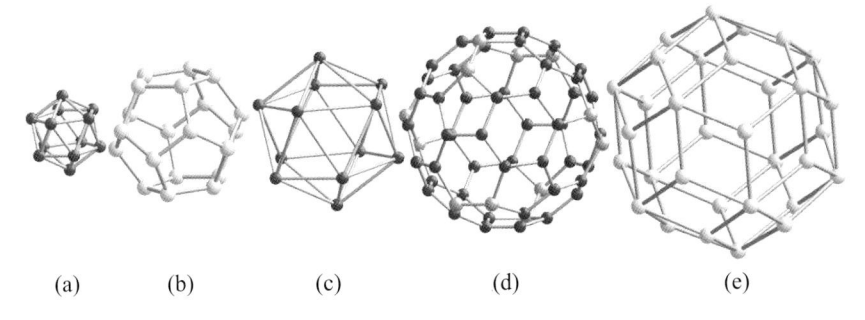

Fig. 2.30 Shells of the 104-atom Samson cluster building cI160-R-Al$_5$CuLi$_3$ projected along [−1 −1 0.7] (Audier et al. 1988). All the shells are centered at the origin. Shells (a)–(c) together form the 44-atom Bergman-cluster.

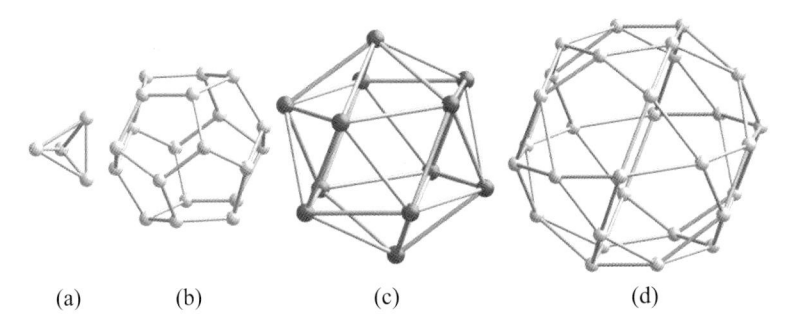

(a)   (b)   (c)   (d)

Fig. 2.31 Shells of the 66-atom Tsai-cluster building cI168-Cd$_6$Yb projected along [−1 −1 0.7] (Palenzona 1971). All the shells are centered at the origin.

the small triacontahedra are in a ratio equal to τ (Audier et al. 1988).

- *Tsai cluster* (Fig. 2.31): The Tsai cluster in cI168-Cd$_6$Yb consists of an orientationally disordered Cd tetrahedron inside a Cd pentagondodecahedron with an average Cd–Cd distance of 2.98 Å. The pentagons are capped by Yb forming an icosahedron with edge length 5.83 Å. The fourth shell, an icosidodecahedron, is formed by edge-centering Cd atoms with Cd–Cd distances of 3.97 Å (Takakura et al. 2001).

One should bear in mind, however, that a clear decision is not always possible on whether a QC is built mainly from one or the other cluster type. A careful analysis of the structure of i-Al–Mn–Pd demonstrated, for instance, that it can be covered up to 77.1% by MCs and up to 72.8% by BCs, respectively (Loreto et al. 2003). A similar statistic was found for i-Al–Cu–Fe (Gratias et al. 2001).

The cluster centers of MCs and BCs are empty or partially filled in the case of the Al-based QCs, and filled in the case of the Ti-based ones (Kim et al. 1998). Empty Al_{12}-cluster shells (e.g., in MCs) have a covalent bonding nature while the centered Al_{13} ones (e.g., in Mg–Al–Zn quasicrystals) are metallic (Kimura et al. 1997). Typical of Al_{12} clusters are very short nearest-neighbor distances below 2.55 Å (in fcc Al, ≈ 2.86 Å), which are indicative of strong bonding. In free Al_6 clusters, the atomic distances can even reach 2.435 Å (Jia et al. 2002).

There are typical ratios, $a_r/\bar{d}$, of the quasicrystal lattice constants (Elser 1985) to the average interatomic distances for QCs built from MCs (1.65–1.75), TCs (≈ 1.75), and BCs (≈ 2.0), respectively (Chen et al. 1987; Guo et al. 2002). Typical values for the optimum electron concentrations amount to 1.7–1.9 for the MC-type, 2.0–2.1 for the TC-type, and 2.0–2.2 for the BC-type QCs (Trambly de Lassadiere et al. 2005).

i-QCs descriptions can be based on cluster-decorated 3D Penrose tilings with edge lengths of the rhombohedra $\tau^3 a_r$. The quasilattice constant and the lattice parameter of the hypercubic lattice are related by $a_r = a\sqrt{2}/2$. The prototype structures and representatives based on 6D structure analyses (Yamamoto and Takakura 2004) and the three cluster types, respectively, are:

- i-Al–Pd–Mn type: $a = 9.14$ Å, $a_r = \tau^3 a/2$, $Fm\bar{3}\bar{5}$
  (2^6-fold superstructure; Al-TM type; 3D PT with edge length a_r decorated by pseudo-icosahedral MCs)

  Al–Pd–Me (Me = Mn, Re), Al–Cu–Me (Me = Fe, Ru, Os), Ti–Zr–Ni.

- i-Zn–Mg–Ho type: $a = 10.28$ Å, $a_r = \tau^3 a/2$, $Fm\bar{3}\bar{5}$
  (2^6-fold superstructure; Frank–Kasper type; 3D PT with edge length a_r decorated by pseudo-BCs)

  Mg–Zn–RE (RE = Y, Nd, Gd, Ho, Dy, La, Pr, Tb, Ce), Zn–Mg–Hf, Zn–Mg–Zr

  and Al–Cu–Li, Mg–Ga–Zn, Mg–Al–M (M = Rh, Pd, Pt) with symmetry $Pm\bar{3}\bar{5}$ (i.e., these are disordered variants of the F-type).

- i-Cd–Yb type: $a = 5.689$ Å, $a_r = \tau^3 a/2$, $Pm\bar{3}\bar{5}$
  (3D PT with edge length a_r decorated by TCs with a disordered first shell)

  Cd–Me (Me=Ca, Yb), Cd–Mg–Ca, Cd–Mg–RE (RE = Y, Nd, Eu, Gd, Tb, Dy, Ho, Er, Tm, Yb, Lu), Zn–Me–Sc (Me = Ag, Au, Co, Cu, Fe, Mg, Mn, Ni, Pd, Pt), Cu–Ga–Mg–Sc, Zn–Mg–Ti, Ag–In–(Mg)–Me (Me = Ca, Yb)

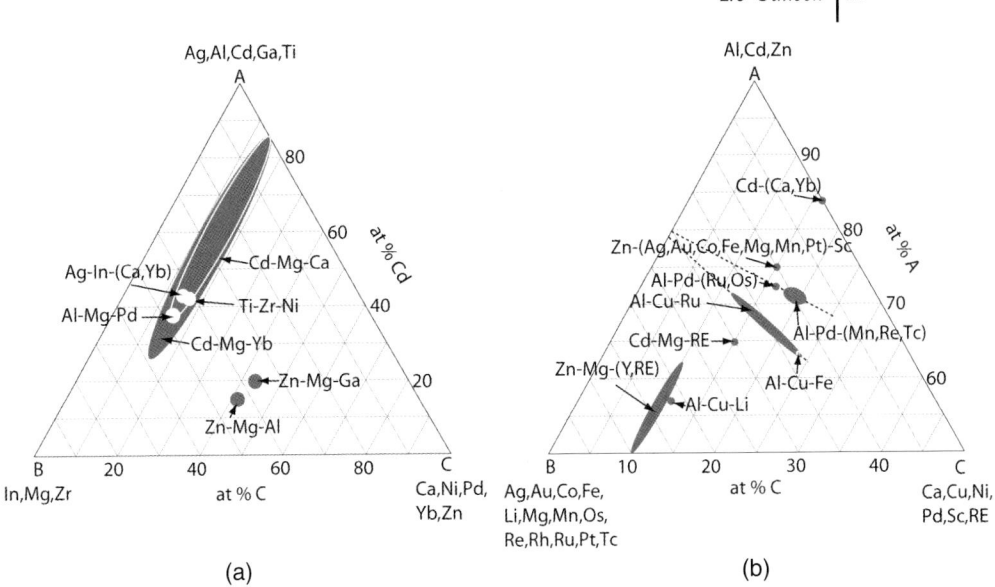

Fig. 2.32 Compositions of stable icosahedral phases. Note that only the upper half ($A_xB_yC_z$, $x + y + z = 100$, $50\% \leq x \leq 100\%$, $0\% \leq y \leq 50\%$, $0\% \leq z \leq 50\%$) is shown in (b).

The approximate stability regions of stable IQCs are shown in Fig. 2.32 (based on Grushko and Velikanova 2004).

2.6 Outlook

One of the most important recent trends in the knowledge of intermetallic compounds is the disappearance of the border between the chemistry and physics of this class of materials. Being a part of inorganic chemistry, intermetallic compounds reveal a very rich spectrum of physical properties which qualify them for prospective applications. In this respect, the structure–property relationships appear to be an issue of ongoing interest. Despite a vast number of investigations, a causal link between atomic organization and physical behavior is still not finally established. Involving the analysis of chemical bonding into these considerations appears as an emerging trend in the research on intermetallic compounds.

As an example, materials such as intermetallic clathrates or filled skutterudites are interpreted as covalently bonded framework anions with embedded cations. Consequently, they combine low charge-carrier density with phonon-glass behavior, so they meet the criteria for thermoelectric applications. Moreover, the analysis of chemical bonding offers the possibility of recognizing low-dimensional structural units which may cause pronounced anisotropy in, e.g., magnetic or mechanical behavior.

Quasicrystals are special representatives of complex intermetallic phases with inherent low-dimensional atomic arrangements. Their intensive study will increasingly stimulate the search for compounds with more pronounced cluster structure, i.e., significantly different chemical bonding and/or composition of cluster and embedding matrix. This would be the metallic counterpart to supramolecular structures or fullerenes. One of the goals may be the development of a modular approach of forming crystalline, quasicrystalline, fractal, or amorphous structures which represent a bulk homogeneous counterpart to materials with nanocrystalline precipitates. Novel physical properties will materialize in cases where the length scales of the giant unit cells or even of the cluster diameters are larger than those relevant to the physical effect. Finally, a more precise understanding of atomic interactions is expected to open new routes fo preparation of intermetallic compounds.

References

Abe, E., Tsai, A. P. (2001), *JEOL News* **36E**, 18.

Abegg, R. (1904), *Z. Anorg. Chem.* **39**, 330.

Anon. (1931–1943), *Strukturbericht*, Supplement issues of *Z. Kristallogr.*

Anon. (1940–1990), *Structure Reports*, published for the Int. Union of Crystallography, Cambridge.

Asahi, R., Sato, H., Takeuchi, T., Mizutani, U. (2005a), *Phys. Rev. B* **71**, 165 103.

Asahi, R., Sato, H., Takeuchi, T., Mizutani, U. (2005b), *Phys. Rev. B* **72**, 125 102.

Asaumi, K., Kondo, Y. (1981), *Solid State Commun.* **40**, 715.

Audier, M., Pannetier, J., Leblanc, M., Janot, C., Lang, J. M., Dubost, B. (1988), *Physica B* **153**, 136.

Bader, R. F. W. (1999), *Atoms in Molecules: A Quantum Theory*, OUP, Oxford.

Bancel, P. A. (1989), *Phys. Rev. Let.* **63**, 2741.

Becke, A. D., Edgecombe, K. E. (1990), *J. Chem. Phys.* **92**, 5397.

Boudard, M., Deboissieu, M., Janot, C., Heger, G., Beeli, C., Nissen, H. U., Vincent, H., Ibberson, R., Audier, M., Dubois, J. M. (1992), *J. Phys.-Condens. Matt.* **4**, 10 149.

Budnyk, S., Weitzer, F., Kubata, Ch., Prots, Yu., Akselrud, L. G., Schnelle, W., Hiebl, K., Nesper, R., Wagner, F. R., Grin, Yu. (2006), *J. Solid State Chem.* **179**, 2329.

Burdett, J. K. (1995), *Chemical Bonding in Solids*, OUP, New York.

Cervellino, A., Haibach, T., Steurer, W. (2002), *Acta Crystallogr. B* **58**, 8.

Chen, H. S., Phillips, J. C., Villars, P., Kortan, A. R., Inoue, A. (1987), *Phys. Rev. B* **35**, 9326.

Dehlinger, U., Schulze, G. E. R. (1940), *Z. Kristallogr.* **102**, 377.

Demchyna, R., Leoni, S., Rosner, H., Schwarz, U. (2006), *Z. Kristallogr.* **221**, 420.

Drude, P. (1904), *Ann. Phys.* **14**, 677.

Dubois, J.-M. (2005), *Useful Quasicrystals*, World Scientific, Singapore.

Duthie, J. C., Pettifor, D. G. (1977), *Phys. Rev. Lett.* **38**, 564.

Elser, V. (1985), *Phys. Rev. B* **32**, 4892.

Emsley, J. (1991), *The Elements*, 2nd edition, Clarendon Press, Oxford.

Freeth, W., Raynor, G. V. (1953/54), *J. Inst. Met.* **82**, 569.

Friedel, J. (1969), in: *The Physics of Metals* (Ed.: J. M. Ziman), Cambridge University Press, New York, chapter 8.

Gatti, C. (2005), *Z. Kristallogr.* **220**, 399.

Goldschmidt, V. M. (1926), *Skr. Norske Vidensk.-Akad. Oslo, I. Mat.-Naturvidensk. Kl.* no. 2.

Goldschmidt, V. M. (1927), *Ber. Dtsch. Chem. Ges.* **60**, 1263.

Gratias, D., Puyraimond, F., Quiquandon, M., Katz, A. (2001), *Phys. Rev. B* **63**, 024 202.

Grimm, H. G., Sommerfeld, A. (1926), *Z. Physik* **36**, 36.

Grin, J., Burkhardt, U., Ellner, M., Peters, K. (1994), *J. Alloy. Compd.* **206**, 243.

Grin, Yu. (1992), *Modern Perspectives in Inorganic Chemistry* (Ed.: E. Parthé), Kluwer, Dordrecht.

Grin, Yu. (2000), *Wiss. Z. TU Dresden* **49**(1), 16.

Grin, Yu., Schnelle, W., Cardoso Gil, R., Sichevich, O., Müllmann, B. D., Mosel, G., Kotzyba, G., Pöttgen, R. (2003), *J. Solid State Chem.* **176**, 567.

Grin, Yu., Wagner, F. R., Armbrüster, M., Kohout, M., Leithe-Jasper, A., Schwarz, U., Wedig, U., von Schnering, H. G. (2006), *J. Solid State Chem.* **179**, 1707.

Grushko, B., Velikanova, T. Y. (2004), *J. Alloy. Compd.* **367**, 58.

Gummelt, P. (1996), *Geometr. Ded.* **62**, 1.

Guo, J. Q., Abe, E., Tsai, A. P. (2002), *Phil. Mag. Lett.* **82**, 27.

Henley, C. L., Elser, V. (1986), *Phil. Mag. B* **53**, L59.

Hermann, C. (1949), *Acta Crystallogr.* **2**, 139.

Hiller, H. (1985), *Acta Crystallogr. A* **41**, 541.

Hoppe, R. (1970), *Angew. Chem.* **82**, 7; *Angew. Chem. Int. Ed.* **9**, 25.

Hoppe, R. (1979), *Z. Kristallogr.* **150**, 23.

Hume-Rothery, W. (1925), *J. Inst. Met.* **35**, 209.

Hume-Rothery, W. (1930), *Phil. Mag.* **9**, 65.

Hume-Rothery, W., Raynor, G. V. (1938), *Phil. Mag.* **26**, 143.

IUCr (1992), Report of the Executive Committee for 1991, *Acta Crystallogr. A* **48**, 922.

Jagodzinski, H. (1954), *Acta Crystallogr.* **7**, 17.

Janssen, T. (1986), *Acta Crystallogr. A* **42**, 261.

Janssen, T., Radulescu, O. (2004), *Ferroelectrics* **305**, 179.

Jia, J., Wang, J. Z., Liu, X., Xue, Q. K., Li, Z. Q., Kawazoe, Y., Zhang, S. B. (2002), *Appl. Phys. Lett.* **80**, 3186.

Jones, H. (1937), *Proc. Phys. Soc. A* **49**, 250.

Karpov, A., Nuss, J., Wedig, U., Jansen, M. (2003), *Angew. Chem.* **115**, 4966; *Angew. Chem. Int. Ed.* **42**, 4818.

Ketelaar, J. A. A. (1958), *Chemical Constitution: An Introduction to the Theory of the Chemical Bond*, 2nd edition, Elsevier, Amsterdam.

Kienast, G., Verma, J., Klemm, W. (1961), *Z. Anorg. Allg. Chem.* **310**, 143.

Kim, W. J., Gibbons, P. C., Kelton, K. F., Yelon, W. B. (1998), *Phys. Rev. B* **58**, 2578.

Kimura, K., Takeda, M., Fujimori, M., Tamura, R., Matsuda, H., Schmechel, R., Werheit, H. (1997), *Solid State Chem.* **133**, 302.

Kohout, M. (2004), *Int. J. Quantum Chem.* **97**, 651.

Kohout, M., Savin, A. (1996), *Int. J. Quant. Chem.* **60**, 875.

Kohout, M., Wagner, F. R., Grin, Yu. (2002), *Theor. Chem. Acc.* **108**, 150.

Kossel, W. (1916), *Ann. Phys.* **49**, 229.

Krypyakevich, P. I. (1977), *Structure Types of Intermetallic Compounds*, Nauka, Moscow (in Russian).

Laves, F., Witte, H. (1935), *Metallwirtschaft* **14**, 645.

Lewis, G. N. (1916), *J. Am. Chem. Soc.* **38**, 762.

Loreto, L., Farinato, R., Catallo, S., Janot, C., Gerbasi, G., DeAngelis, G. (2003), *Physica B* **328**, 193.

Loubeyre, P., Jean-Louis, M., LeToullec, R. (1993), *Phys. Rev. Lett.* **70**, 178.

Luck, J. M., Godreche, C., Janner, A., Janssen, T. (1993), *J. Phys. A* **26**, 1951.

Mermin, N. D. (1992), *Rev. Mod. Phys.* **64**, 3.

Mooser, E., Pearson, W. B. (1959), *Acta Crystallogr.* **12**, 1015.

Mott, N. F., Jones, H. (1936), *The Theory of the Properties of Metals and Alloys*, Clarendon Press, Oxford.

Nesper, R. (1990), *Prog. Solid State Chem.* **20**, 1.

Nesper, R. (1991), *Angew. Chem.* **103**, 805; *Angew. Chem. Int. Ed. Engl.* **30**, 789.

Nesper, R. (2003), in: *Silicon Chemistry* (Ed.: P. Jutzi, U. Schubert), Wiley-VCH, Weinheim, pp. 171–180.

Nowotny, H., Mohrnheim, A. (1939), *Z. Kristallogr.* **100**, 540.

O'Keeffe, M., Hyde, B. G. (1996), in: *Crystal Structures*, Mineralogical Society of America, Washington DC.

Ormeci, A., Rosner, H., Wagner, F. R., Kohout, M., Grin, Yu. (2006), *J. Phys. Chem. A* **110**, 1100.

Palenzona, A. (1971), *J. Less-Common Met.* **25**, 367.

Parthé, E., Chabot, B. A., Censual, K. (1985), *Chimia* **39**, 164.

Parthé, E., Chabot, B., Penzo, M., Censual, K., Gladyshevskii, R. (1993/94), *Typix Standardized Data and Crystal Chemical Characterization of Inorganic Structure Types*, Springer, Heidelberg.

Pauling, L. (1960), *The Nature of the Chemical Bond*, 3rd edition, Cornell University Press, Ithaca.

Paxton, A. T., Methfessel, M., Pettifor, D. G. (1997), *Proc. R. Soc. Lond. A* **453**, 1493.

Pearson, W. B. (1964), *Acta Crystallogr.* **17**, 1.

Penrose, R. (1974), *Bull. Inst. Math. Appl.* **10**, 266.

Pettifor, D. G. (1990), *J. Chem. Soc. Faraday Trans.* **86**, 1209.

Phillips, J. C. (1968), *Phys. Rev. Lett.* **20**, 550.

Phillips, J. C. (1970), *Rev. Mod. Phys.* **42**, 317.

Reichlin, R., Ross, M., Martin, S., Goettel, K. A. (1986), *Phys. Rev. Lett.* **56**, 2858.

Rosner, H., Koudela, D., Schwarz, U., Handstein, A., Hanfland, M., Opahle, I., Koepernik, K., Kuzmin, M. D., Müller, K.-H., Mydosh, J. A., Richter, M. (2006), *Nature Physics* **2**, 469.

Savin, A., Flad, H. J., Flad, J., Preuss, H., von Schnering, H. G. (1992), *Angew. Chem.* **104**, 185; *Angew. Chem. Int. Ed. Engl.* **31**, 185.

Savin, A., Nesper, R., Wengert, S., Fässler, T. F. (1997), *Angew. Chem.* **109**, 1892; *Angew. Chem. Int. Ed. Engl.* **36**, 1808.

Schäfer, H., Eisenmann, B., Müller, W. (1973), *Angew. Chem.* **85**, 742; *Angew. Chem. Int. Ed. Engl.* **12**, 694.

Schmider, H. L., Becke, A. D. (2000), *J. Mol. Struct. (TheoChem)* **527**, 51.

Schubert, K. (1964), *Kristallstrukturen zweikomponentiger Phasen*, Springer, Berlin.

Schwarz, U., Bräuninger, S., Syassen, K., Kniep, R. (1998), *J. Solid State Chem.* **137**, 104.

Simon, A. (1983), *Angew. Chem.* **95**, 94; *Angew. Chem. Int. Ed. Engl.* **22**, 95.

Skriver, H. L. (1985), *Phys. Rev. B* **31**, 1909.

Socolar, J. E. S. (1988), *Bull. Am. Phys. Soc.* **33**, 463.

Socolar, J. E. S., Lubensky, T. C., Steinhardt, P. J. (1986), *Phys. Rev. B* **34**, 3345.

Söderlind, P., Ahuja, R., Eriksson, O., Wills, J. M., Johansson, B. (1994), *Phys. Rev. B* **50**, 5918.

Sommer, A. H. (1943), *Nature* **152**, 215.

Spicer, W. E., Sommer, A. H., White, J. G. (1959), *Phys. Rev.* **115**, 57.

Stein, F., Palm, M., Sauthoff, G. (2004), *Intermetallics* **12**, 713.

Steurer, W. (2004), *Z. Kristallogr.* **219**, 391.

Steurer, W. Haibach, T. (1999), Crystallography of quasicrystals, in: *Physical Properties of Quasicrystals* (Ed.: Z. M. Stadnik), Springer, Heidelberg, pp. 51–89.

St. John, J., Bloch, A. N. (1974), *Phys. Rev. Lett.* **33**, 1095.

Sugiyama, K., Kaji, N., Hiraga, K. (1998), *Acta Crystallogr. C* **54**, 445.

Takakura, H., Guo, J. Q., Tsai, A. P. (2001), *Phil. Mag. Lett.* **81**, 411.

Teatum, E., Gschneidner, K., Waber, J. (1960), Report LA-2345, US Dept. of Commerce, Washington DC.

Trambly de Lassadière, G., Nguyen-Manh, D., Mayou, D. (2005), *Progr. Mater. Sci.* **50**, 679.

Tsai, A. P. (2003), *Acc. Chem. Res.* **36**, 31.

van Arkel, A. E. (1956), *Molecules and Crystals in Inorganic Chemistry*, Interscience, New York.

van Vechten, J. A. (1969a), *Phys. Rev.* **182**, 891.

van Vechten, J. A. (1969b), *Phys. Rev.* **187**, 1007.

Villars, P. (1983), *J. Less Common Met.* **92**, 215.

Villars, P. (1984a), *J. Less Common Met.* **99**, 33.

Villars, P. (1984b), *J. Less Common Met.* **102**, 199.

Villars, P., Calvert, L. D. (1996), *Pearson's Handbook of Crystallographic Data*, 2nd edition, ASM International, Materials Park, Ohio.

von Schnering, H. G., Cardoso-Gil, R., Hönle, W. (1995), *Angew. Chem.* **107**, 81; *Angew. Chem. Int. Ed.* **34**, 103.

Wade, K. (1976), *Adv. Inorg. Chem. Radiochem.* **18**, 1.

Wells, A. F. (1968), *Acta Crystallogr. B* **24**, 50.

Wolff, P. M. de (1974), *Acta Crystallogr. A* **30**, 777.

Wondratschek, H. (1983), *International Tables for Crystallography* (Ed.: T. Hahn), Reidel, Dordrecht, chapter 8.3.2.

Wosylus, A., Prots, Yu., Burkhardt, U., Schnelle, W., Schwarz, U., Grin, Yu. (2006), *Solid State Sciences* **8**, 773.

Yamamoto, A. (1996), *Acta Crystallogr. A* **52**, 509.

Yamamoto, A., Takakura, H. (2004), *Ferroelectrics* **305**, 223.

Zintl, E. (1939), *Angew. Chem.* **52**, 1.

Zintl, E., Kaiser, H. (1933), *Z. Anorg. Allg. Chem.* **211**, 113.

Zunger, A. (1980), *Phys. Rev. B* **22**, 5839.

3
Solidification and Grown-in Defects

Thierry Duffar

This chapter is dedicated to my colleague Marie-Danielle Dupouy, who died in September 2004. She devoted her scientific activity to the study of the effect of liquid convection on metallurgical alloy structures. We were working together, with the help of numerous colleagues and students, on an academic project that ultimately fed the present text.

3.1
Introduction: the Solid–Liquid Interface

Most solid alloys are prepared from the liquid phase, and the liquid-to-solid transformation, called solidification, is the topic of this chapter.

Obviously, the structure, defects, and properties found in the final alloy are the result of what happened in the elaboration process and therefore first in the liquid phase, where transport phenomena, diffusion and convection, have a strong influence on the final structure. The main and special solidification features occur at the transition between the solid and the liquid phases, the so-called solid–liquid interface. The physics of what happens later in the solid phase is covered by other chapters in this book: this one will deal essentially with the solid–liquid interface and its interaction with the liquid.

This introduction will be completed with some fundamental notions relating to this interface. Section 3.2 will present the classical theories of structures in alloy solidification: stability of the solid–liquid interface, growth of dendrites and eutectics and rapid solidification; it will be seen that the diffusion in the liquid state plays a major role in these processes. Section 3.3 will be focused on the defects resulting from solidification processes, and the role of convection in the liquid will be stressed.

3.1.1
Structure of the Solid–Liquid Interface

There are essentially two different structures of a solid–liquid interface. The "faceted" interface is represented by a flat, smooth surface, at the atomic scale.

Alloy Physics: A Comprehensive Reference. Edited by Wolfgang Pfeiler
Copyright © 2007 WILEY-VCH Verlag GmbH & Co. KGaA, Weinheim
ISBN: 978-3-527-31321-1

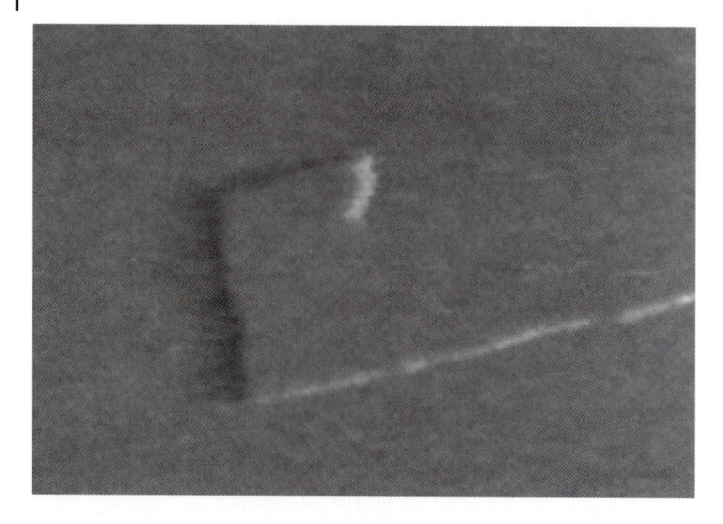

Fig. 3.1 AFM observation of a (101) facet on a growing monoclinic lysozyme crystal. An unsaturable spiral step, likely to be associated with a dislocation, is visible (Chernov et al. 2004).

Therefore there is a sharp, total change in the degree of order when crossing the interface, from the fully crystalline solid side to the liquid (Fig. 3.1).

In "rough" interfaces, the transition between the fully ordered solid and the disordered liquid is less abrupt, extending over several atomic distances (Fig. 3.2). In this layer, atoms, molecules, or building units are experiencing an environment of which the degree of order is fluctuating with time. Results of molecular dynamics simulations have shown that the unit movements do not change markedly when crossing the interface (at the melting temperature). Structural and kinetic models of such diffuse interfaces have been proposed; see for example the pioneering work of J. W. Cahn (1960), and more sophisticated models of

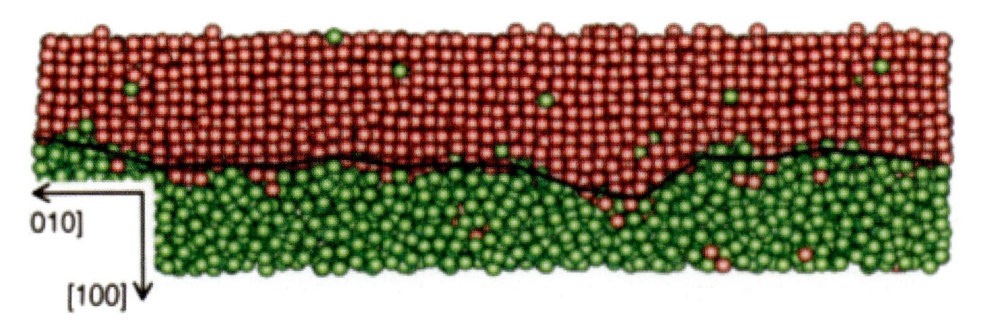

Fig. 3.2 Liquid–solid transition through a diffuse zone with a thickness of a few structural units (Hoyt et al. 2004).

Table 3.1 Observed morphology of the crystal surface during growth of various materials in several processes.

Reduced entropy α	Material	Feeding phase	Morphology
~1	metal	molten liquid	rough
~1	crystalline polymer	molten liquid	rough
2–3	semiconductor	solution	rough to faceted
~10	metal	vapor	faceted

the interface can be found in the review article by Bennema (1993), but for the purpose of the present discussion simplified surface representations will be used.

Theoretical models have been proposed in order to predict the interface roughness. The most classical, from K. Jackson (1958) takes into account the transformation entropy (vaporization, solution, melting) reduced to the ideal gas constant R [Eq. (1)].

$$\alpha = \frac{\Delta S_t}{R} \tag{1}$$

Faceting occurs if $\alpha > 2$.

3.1.2
Kinetics of the Solid–Liquid Interface

Solidification generally occurs under thermal or chemical gradients and cannot be considered as an equilibrium process. However, a local thermodynamic equilibrium on the interface can be considered in the case where, on arrival, a building unit can easily find a satisfactory position in its surroundings. By denoting the surface diffusion coefficient as D_S and the typical size of a structural unit as a, the surface velocity V_u of the unit can be compared with the interface velocity v_i by Eq. (2)

$$\frac{V_u}{v_i} = \frac{D_S}{a v_i} \tag{2}$$

With typical values of D_S and a (5×10^{-10} m² s⁻¹ and 5×10^{-10} m) it can be seen that for interface velocities higher than 1 m s⁻¹ the local thermodynamical equilibrium does not apply. This case will be treated in Section 3.2.3 (Rapid Solidification).

For lower velocities, thermodynamic data, especially the phase diagram in the case of alloys, can be used safely. For example, the temperature T_i of an alloy

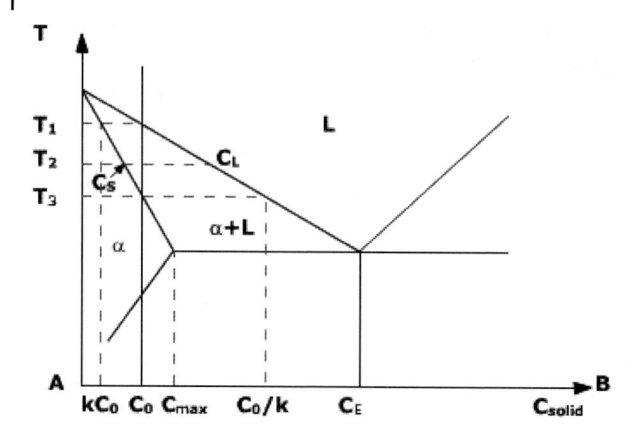

Fig. 3.3 Typical phase diagram used in this chapter.

interface can be computed from the liquidus slope (which is generally negative; see Fig. 3.3), by Eq. (3).

$$T_i = T_m + m_l c_i \tag{3}$$

However, thermodynamic equilibrium is established only in the case of a resting interface. In order to make it move, it is necessary to introduce a slight deviation from equilibrium. In the case of solidification this is kinetic undercooling, measured as the difference between the thermodynamic melting temperature T_m and the actual temperature T_i of the moving interface [Eq. (4)].

$$\Delta T_k = T_m - T_i \tag{4}$$

In the case of the faceted interface, growth can occur only by nucleation of a cluster on the perfectly flat surface. When this cluster is created, the whole surface is quickly covered, so that the factor limiting the velocity is the cluster creation step. This gives an interface velocity varying exponentially with undercooling, and then growth only occurs at high values of undercooling, when the cluster size is small enough to have some chance of appearing spontaneously thanks to the density fluctuations of the liquid phase.

Undercooling values of several 10 K result; experiments show, however, that the undercooling for faceted interfaces is generally of the order of a few degrees only. In practice, facets contain defects such as steps and kinks or unsaturable defects such as screw dislocations (see Fig. 3.1). Following the pioneering work of Burton, Cabrera, and Frank (Burton et al. 1951), numerous authors have derived relationships, such as Eq. (5), between the velocity and the undercooling for an imperfect faceted interface.

$$v_i = K_f \Delta T_k^n \quad 1 < n < 4 \tag{5}$$

K_f is the kinetic coefficient, depending on the number of defects on the surface and also on its crystallographic orientation.

In the case of the rough interface, the interface velocity is calculated from the balance of the structural units coming from the liquid to the interface and leaving the interface for the liquid phase. To a first order, with some simplifying assumptions ([111] surface composed of cubic structural units), Eq. (6) can be shown to hold (Kurz and Fisher 1998).

$$v_i = K_r \Delta T_k \tag{6}$$

K_r is of the order of 1, showing that for classical metallurgical or crystal growth rates (10^{-6}–10^{-4} m s^{-1}) the kinetic undercooling is very low for rough interfaces and can be neglected.

3.1.3
Chemistry of the Solid–Liquid Interface: the Segregation Problem

Figure 3.3 shows a typical, simplified phase diagram of a binary alloy. For a given temperature, T_2, the composition of the solid is related to the composition of the liquid by the segregation coefficient k [Eq. (7)].

$$k = \frac{c_s}{c_l} \tag{7}$$

In most cases, the solid is less concentrated than the liquid ($k < 1$) and this situation will be considered throughout this chapter; the reverse situation ($k > 1$) leads to symmetric results. In case of faceted growth, k is likely to depend on the crystallographic orientation of the interface.

It follows that the advancing interface rejects solute in the liquid and a balance of the incorporated and rejected solute gives the flux toward the liquid [Eq. (8)].

$$-D_l \left(\frac{\partial c}{\partial z}\right)_i = v_i c_i^l (1 - k) \tag{8}$$

This flux generates a boundary layer in the liquid, close to the interface. If diffusive conditions prevail in the liquid, the boundary layer thickness is of the order of δ [Eq. (9)].

$$\delta = \frac{D_l}{v_i} \tag{9}$$

In this case, after an initial transient (whose length is of the order of D_i/kv_i) corresponding to the building of the solute boundary layer, Tiller et al. (1953) have shown that a steady state is obtained, with a growing solid of initial composition

c_0 in equilibrium with a liquid of composition c_0/k. This result is valid as far as the diffusive solute boundary layer is not affected by convection. According to Eq. (9), this is more likely for high interface velocities (even with a growth rate of 10^{-3} m s^{-1}, for a typical diffusion coefficient of 10^{-8} m^2 s^{-1}, the boundary layer is 10 μm thick, and likely to be affected by the convection, which vanishes very close to the interface where solid is growing).

The very general case, then, is that the liquid is mixed by some natural or forced convection and the boundary layer thickness depends on the intensity of convection. If the mixing is strong enough, the solute boundary layer can be neglected and the liquid can be considered as homogeneous. Scheil and Gulliver derived Eq. (10) for the relationship between solid composition c_s and solidified fraction (Scheil, 1942).

$$c_s = kc_0(1 - f_s)^{(k-1)} \tag{10}$$

Figure 3.4 gives the shape of this segregation behavior for different values of k. This equation is not valid right up to the end of the ingot, where the concentration tends toward infinity. In practice different phenomena may occur when the concentration increases, such as precipitation of eutectic material, variation of k with c, or destabilization of the interface (see Section 3.2.1).

For a moderate convection, Burton, Prim and Schlichter (Burton et al. 1953) have shown that Eq. (10), and consequently Fig. 3.4, can be used provided that k is replaced by an "effective" segregation coefficient, k_{eff}, depending on the boundary layer thickness. Its value varies between 1 (the diffusive case seen above) and k (fully mixed liquid, Scheil–Gulliver law); for the interested reader there is a general discussion of these matters, with the calculation of k_{eff} in some configurations, in Garandet et al. (1994).

In the above explanations, cases have been considered where the diffusion of the solute in the solid is negligible, which generally is the case (substitutional alloys). However, this is not valid for solutes of small atoms in lattices of large atoms, as for example in the case of C in Fe (interstitial alloys). If solid diffusion is rapid, on the scale of the solidification time, the solid can be considered homogeneous at any time, especially at the end of solidification, and in equilibrium with the liquid interface composition. In the case of ideal mixing of the liquid, this leads to the well known lever rule, Eq. (11).

$$c_s f_s + c_l f_l = c_s f_s + \frac{c_s}{k}(1 - f_s) = c_0 \tag{11}$$

The chemical segregation problem is unavoidable. Strong mixing of the melt increases the local homogeneity of the solid but increases the macrosegregation all along the ingot. The only way to reduce the segregation is to feed the crucible continuously with fresh solvent in order to keep the liquid composition constant, but this leads to process complexity and this solution is seldom used.

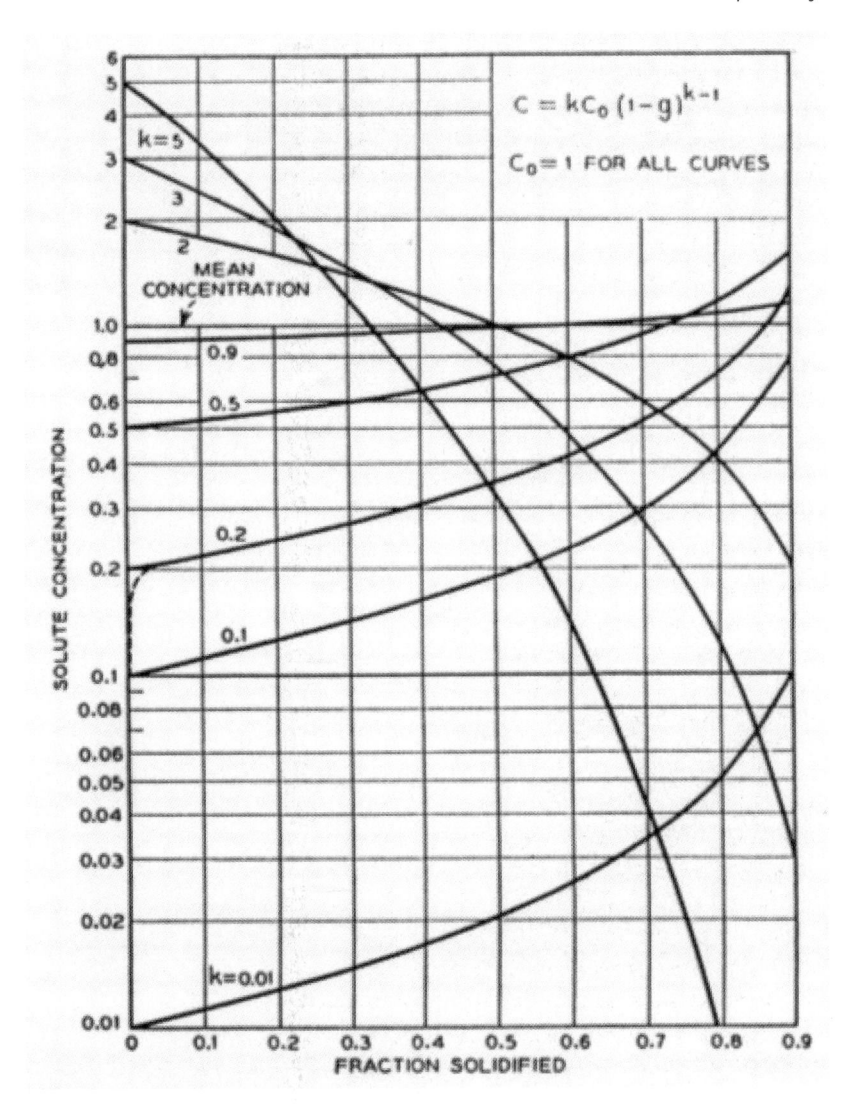

Fig. 3.4 Concentration profile along a solidified ingot for various values of the segregation coefficient k (Garandet et al. 1994).

$$C = kC_0(1-g)^{k-1}$$

$C_0 = 1$ FOR ALL CURVES

3.1.4
Temperature of the Solid–Liquid Interface

The nucleation theory (for details, see Chapter 7) can be applied to solidification and, from Gibbs's classical treatment, it appears that the critical nucleation radius in a liquid showing a certain undercooling can be expressed as Eq. (12), where Γ is the Gibbs coefficient (close to 1×10^{-7} K m for metals).

$$r^* = \frac{2\gamma_{ls}}{\Delta S_m \Delta T} = \frac{2\Gamma}{\Delta T} \qquad (12)$$

It follows that any curved solid–liquid interface (even at rest) in equilibrium with its liquid shows a capillary undercooling depending on the local curvature κ, according to Eq. (13).

$$\Delta T_{cap} = \Gamma \kappa. \qquad (13)$$

This effect is significant only for a radius of curvature of the interface less than 10 μm.

From the above considerations, it follows that the temperature of the solid–liquid interface depends on the chemical composition, on its velocity, and on its curvature. For example, the interface temperature of an alloy growing with a rough interface should be expressed as Eq. (14), from Eqs. (3), (6), and (13), where T_m is the thermodynamic melting point of the pure substance.

$$T_i = T_m + m_l c_i - \frac{v_i}{K_r} - \Gamma \kappa \qquad (14)$$

This equation of state of the solid–liquid interface defines its position and shape. The next Section aims to present some classical studies of the solid alloy structures that result from the interaction of Eq. (14) with the thermal and chemical fields surrounding the interface.

3.2
Solidification Structures

The solid–liquid interface morphology changes with the process parameters alloy composition, thermal gradient, and solidification rate. Figure 3.5 shows how it changes when the solidification rate increases. Similar diagrams can be plotted as a function of the alloy composition or thermal gradient.

Derivations of the most important characteristics of the diagram in Fig. 3.5, i.e., the interface destabilization, which corresponds to the transition between planar (a) and cellular (b) interface, the periodicity of the cellular structure (b), the characteristics of the dendrite field (c), and the transition toward rapid solidification (d), are given below. Another case is included, when an eutectic structure is obtained with two different solid phases growing simultaneously from a single liquid.

In practical applications, three regimes are used in order to grow materials.
- Planar interface growth is used to get high-quality single crystals for electronics, optics, detectors, and so on.

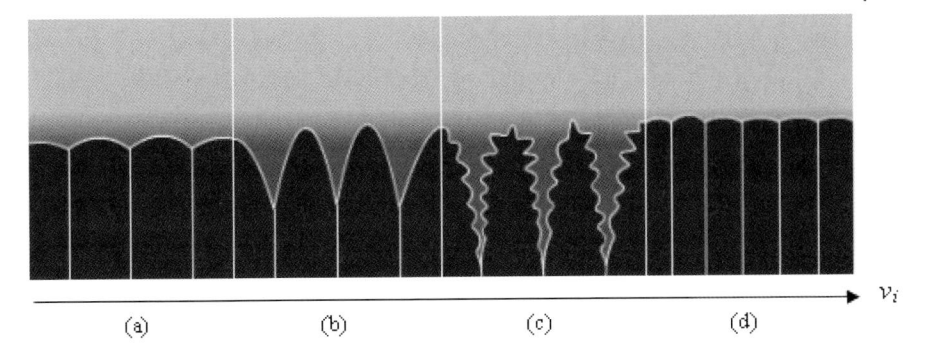

Fig. 3.5 Evolution of the solid–liquid interface morphology with increasing velocity. (a) Planar interface generally observed in single-crystal growth; (b) cellular structures; (c) dendrites observed in classical metallurgical processes; (d) flat interface obtained in rapid solidification.

- Dendritic growth is the most common because it is used universally for all metallurgical processes: steel, cast iron, Al- and Cu-based alloys, and others. The reason is that the dendritic regime leads to tiny microstructures, furthermore accompanied by eutectic areas, and this gives excellent mechanical properties to these alloys.
- Rapid solidification is used in order to obtain very small microstructures and amorphous materials. It is also a way to get metastable phases that cannot be solidified at lower growth rates.

3.2.1
The Interface Stability and Cell Periodicity

Morphological destabilization of the solid–liquid interface is a mechanism aiming to increase its area in order to exchange solute or heat better with the surrounding liquid. This generally leads to important modifications of the structural quality of an alloy and has deserved continuous attention, considering the practical importance of solidification in material processes (Coriell and Mc Fadden 1993).

Destabilization occurs when a perturbation of the interface is likely to expand toward the liquid instead of decreasing and vanishing. To the first order, this can be checked by comparing the actual thermal field in the sample and the local melting temperature, which depends essentially, for a flat interface, on the local chemical composition [Eq. (3)]. Due to the chemical boundary layer at the interface, the chemical composition decreases, and the melting temperature increases, with the distance to the interface. The variation of melting temperature at the interface, toward the liquid, is given by Eq. (15), from Eqs. (3) and (8).

$$\left(\frac{dT_m}{dz}\right)_i = \frac{dT_m}{dc}\left(\frac{dc}{dz}\right)_i = -m_l\frac{v_i c_i^l(1-k)}{D_l} \qquad (15)$$

If the thermal gradient in the liquid, at the interface, is lower than this value, a liquid layer close to the interface is at a temperature lower than its melting temperature and any perturbation of the interface will grow toward the liquid. On the contrary, in a high enough liquid thermal gradient, there is no undercooled liquid layer. It follows that the interface stability is written as Eq. (16).

$$\nabla T_i^l > -m_l \frac{v_i c_i^l(1-k)}{D_l} \qquad (16)$$

As explained in the Introduction, the composition at the interface, c_i, is likely to change all through the solidification process (see for example Eq. (10) in the case of complete mixing). For a solute rejected into the liquid, the concentration increases continuously and destabilization occurs toward the end of the ingot. It is only in the case of pure diffusion in the liquid that the composition at the interface reaches a steady-state value, c_0/k.

These destabilization mechanisms and criteria were initially proposed by Tiller et al. (1953) and are in acceptable agreement with the experimental observation of destabilization for slow growth rates. However they do not take into account the energy cost linked to the increase in the interface area; taking this into account will make it possible to find the typical size of the resulting structure. This analysis has been performed by Mullins and Sekerka (1964) by studying the effect of a sinusoidal perturbation of the interface, of amplitude ε and period λ, on the thermal and chemical field around it and taking into account the effect of the interface curvature on its temperature [Eq. (13)]. This gives the rate of variation of the perturbation amplitude as in Eq. (17), where Λ and ∇T are mean values taken from the liquid and solid regions, and the ξ are positive functions of the growth rate.

$$\frac{1}{\varepsilon}\frac{d\varepsilon}{dt} = m_l \nabla c \xi_c - \Lambda \nabla T \xi_T - \Gamma \frac{4\pi^2}{\lambda^2} \qquad (17)$$

Looking at the right-hand side of Eq. (17), it appears that the first term, related to the solute field, is positive and therefore is the source of the destabilization, whereas the other two terms, related to the thermal field and to the capillarity, are negative and thus stabilize the interface.

From the physical point of view, the interface deformation is an increase in its area in order to improve the rejection of solute toward the liquid. However it is then more difficult to extract the latent heat toward the solid: this explains the stabilizing effect of the thermal field. Furthermore, creating surface has an energetic cost, reflected in the capillary term, which is also stabilizing.

From this expression an improved stability criterion, Eqs. (18) and (19), can be deduced (Sekerka 1965).

$$\left(\frac{\Delta H_m}{2\Lambda_l} + \frac{\nabla T_i^l}{v_i}\right) > -\frac{(\Lambda_s + \Lambda_l)}{2\Lambda_l} m_l \frac{c_i^l(1-k)}{D_l} S(A) \tag{18}$$

with:

$$A = -\frac{\Gamma k v_i}{m_l D_l (1-k) c_i^l} \tag{19}$$

This expression differs from Eq. (16) by taking into account a mean thermal conductivity and the latent heat of transformation but, more importantly, by the function $S(A)$, which takes into account the stabilizing effect of capillarity.

This function is plotted in Fig. 3.6. When $A = 1$, $S(A) = 0$ and the interface is stable. This occurs for high values of the interface velocity and will be discussed

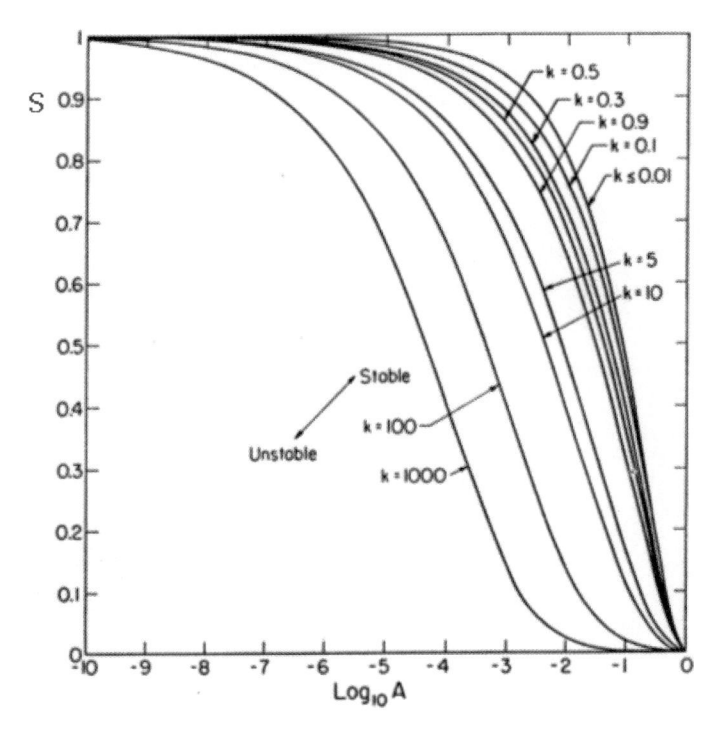

Fig. 3.6 Variation of Sekerka's stability function with the dimensionless parameter A (Sekerka, 1965).

in Section 3.2.3. At low growth rate, the latent heat term becomes negligible, $A \to 0$, $S(A) \to 1$, and the stability criterion takes a form similar to Eq. (16).

Finally, the periodicity of the fluctuation can be obtained by setting Eq. (17) equal to zero, which, for small growth rates, can be written as Eq. (20) (Kurz and Fischer 1998).

$$0 = m_l \nabla c - \nabla T - \Gamma \frac{4\pi^2}{\lambda^2} \tag{20}$$

Equation (21) follows.

$$\lambda = 2\pi \sqrt{\frac{\Gamma}{m_l \nabla c - \nabla T}} \tag{21}$$

It can be noted that λ takes a real value if Eq. (22) is true; this is the condition for instability already given in Eq. (16).

$$m_l \nabla c > \nabla T \tag{22}$$

For typical growth parameters used in metallurgy, the periodicity is of the order of 1 μm.

3.2.2
Dendrites

Destabilization of the interface leads to the formation of cell and dendrite structures. From the practical point of view, cells create defects in semiconductor crystals and crystal growers carefully avoid working under destabilizing conditions. Dendrites have been studied intensively in materials processing because they affect the properties of the alloy. This influence can be either negative (e.g., the disappearance of the transparency of silicate glasses in the case of dendrite generation within them), or positive (e.g., the increase in the elastic limit of eutectoid steels when the microstructure decreases in size).

The interface destabilization studied in Section 3.2.1 corresponds to the transition between Fig. 3.7(a) and (b). If the destabilization is increased (in this case by increasing the growth rate), the interface shows cells, then fully developed dendrites. The tips of the perturbations reject solute very efficiently and the regions between them trap solute, then grow more slowly. The resulting structure has a periodicity of its chemical composition, very often associated with precipitates or eutectics between the dendrites or cells.

It should be borne in mind that the structure obtained during solidification, close to the melting point, is likely to be significantly modified during the natural cooling of the solid, or during subsequent mechanical or thermal treatments. This modification depends on the alloy of interest: in the case of steels, the original structure of which can be totally lost after forging, whereas the solidification

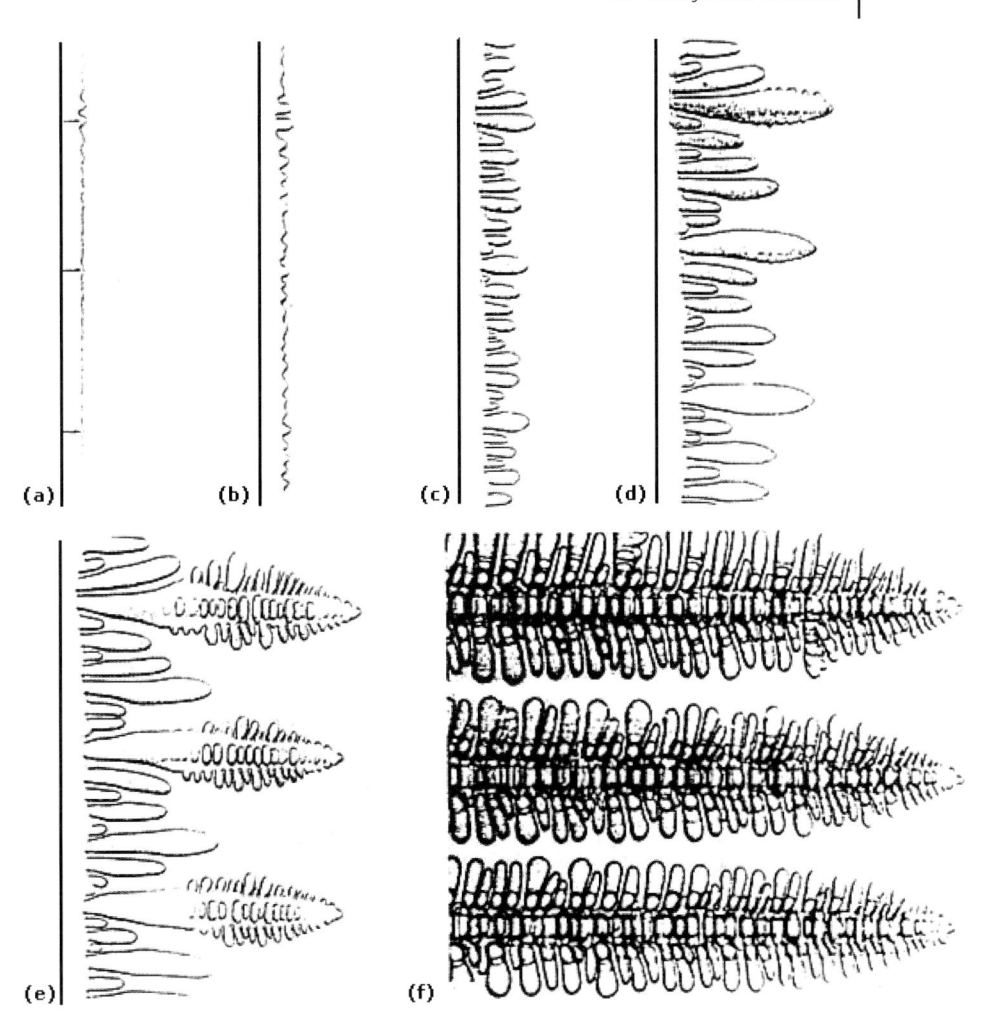

Fig. 3.7 From planar interface to dendrites: evolution of the solid–liquid interface structure when the growth rate is increased. The alloy is a transparent succinonitrile/acetone (4% mixture), solidified between two glass plates under a microscope. The liquid is on the right in all the pictures (Trivedi and Somboonsuk 1984).

structures in aluminum-based alloys do not change significantly during subsequent processes. These solid-state transformations are covered in Chapter 7.

3.2.2.1 Different Types of Dendrites

The dendrites obtained by destabilization of a macroscopic solid–liquid interface are called "columnar dendrites". They appear only in the case of an alloy processed under conditions of interface destabilization and their growth is con-

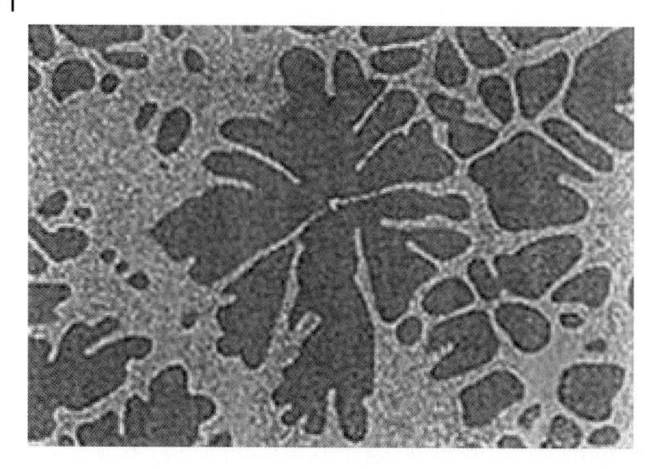

Fig. 3.8 Equiaxed dendrites in an Al–Ni alloy; each dendrite constitutes an individual grain (typical size 200 μm). The eutectic structure between the dendrites cannot be seen at this magnification.

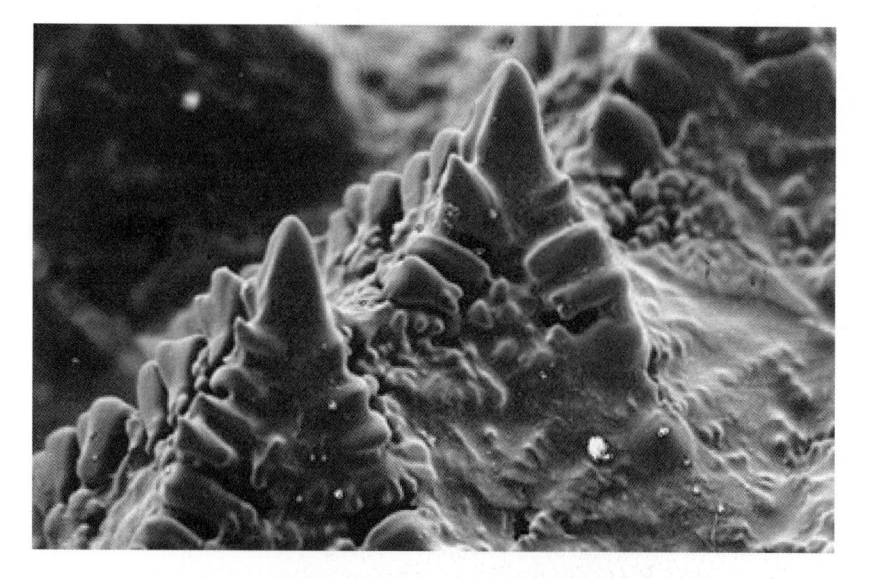

Fig. 3.9 Typical columnar dendrites in an Al–7%Si alloy: SEM observation of the interface after quickly removing the liquid. These dendrites have a typical size of 200 μm. Their faces are all oriented in the same direction, because they belong to the same single crystal grain (Access-Aachen).

strained (oriented) by the thermal gradient, so that the latent heat is evacuated by the solid. A common error is to associate structural grains with the dendrites. Generally speaking, the solid–liquid interface of a single grain is composed of several columnar dendrites: for example, single-crystal turbine blades are grown in the columnar dendritic mode.

Another type is called an "equiaxed dendrite". It is obtained when crystal seeds nucleate inside the liquid phase. This happens when the liquid is undercooled. Each seed grows on its own, at the beginning as a sphere, which may be maintained throughout the solidification (globular structures); more generally, however, dendrite arms appear on the crystal by destabilization. Growth is free (in all directions) and controlled by both solute and thermal diffusion into the liquid in the case of alloys. Equiaxed dendrites occur for pure elements as well, in which case the growth is controlled by thermal diffusion alone. Due to the release of latent heat, the solidified dendrite becomes warmer than the surrounding liquid and the latent heat is evacuated toward the liquid. Each equiaxed dendrite gives birth to an individual grain in the final structure.

Fig. 3.10 Shape of the columnar interface of an Al_2Cu intermetallic compound: SEM observation after quickly removing the liquid (Dupouy, 1986).

Fig. 3.11 Typical equiaxed carbon flakes in gray cast iron: optical metallography, typical size 500 µm.

The shape of the dendrite depends on the material. In the case of solid solutions based on simple crystalline metals with high symmetry (bcc, fcc) numerous growth directions are available and the structure shows an axis and secondary and even ternary arms typical of dendrites (Fig. 3.9). In general, the axis and the arms have simple crystallographic orientations, [001] in the case of cubic crystals, and the growth rate anisotropy is low.

In the case of intermetallic compounds with crystallographic structures of lower symmetry, only a few growth directions are available; the dendrites are faceted and the growth rate anisotropy is high (Fig. 3.10).

Non metallic compounds have high melting entropies and they grow as thin plates, flakes, or needles (Fig. 3.11).

3.2.2.2 Kinetics of Columnar Dendrites

The fundamental hypothesis of all models for the study of dendrite dynamics considers that the growth is controlled by what happens at the tip of the dendrite. This rather surprising assumption (the dendrite tip radius is typically 1 µm and the dendrite length may be several hundreds of micrometers) is based on observations performed by Huang and Glicksman (1981) and Esaka (1986) and has been verified for the columnar dendrite.

For a given velocity, the problem is to find the most appropriate curvature of the dendrite tip to evacuate the amount of solute rejected. It is supposed that the solute field around the dendrite tip is purely diffusive. As shown in Fig. 3.12,

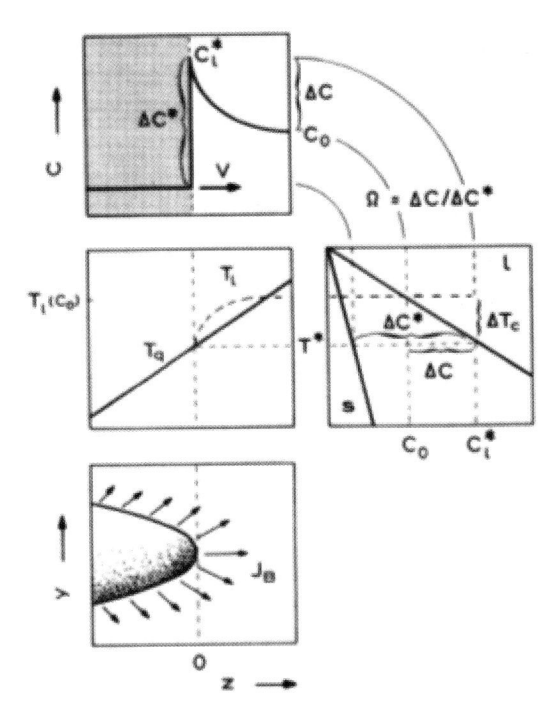

Fig. 3.12 Chemical and thermal fields in front of the tip of the dendrite (Kurz and Fischer 1998).

supersaturation and undercooling are related by Eq. (23) and depend on the chemical composition at the tip.

$$\Omega = \frac{c_i^l - c_0}{c_i^l(1-k)} = \frac{-\Delta T}{m_l k c_l^i} \tag{23}$$

It is therefore necessary to compute the solute field around the tip in order to solve the problem.

A parabolic shape of the dendrite is generally assumed and fits well with the observation of dendrites in transparent media in the diffusive approximation (see Fig. 3.13b).

The diffusion equation in this case has been solved by Ivantsov (1947) and gives Ivantsov's function Q [Eq. (24)].

$$\Omega = Iv(Pe) = e^{Pe} Pe \int_{Pe}^{\infty} \frac{e^{-z}}{z} dz \tag{24}$$

The leading parameter is the Peclet number, Pe [Eq. (25)].

$$Pe = \frac{v_i r}{2 D_l} \tag{25}$$

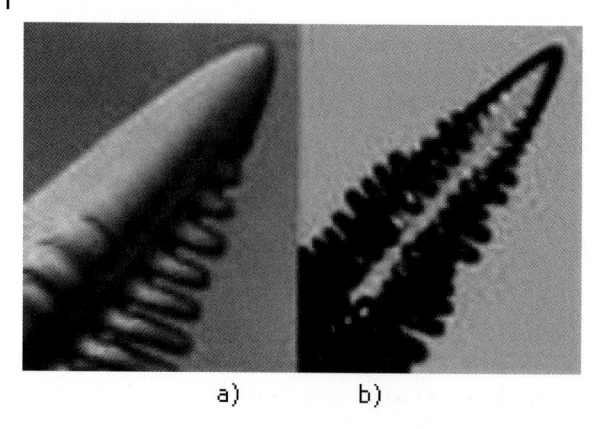

Fig. 3.13 Shape of succinonitrile dendrites: (a) under a liquid flow which is coming from the left and stabilizes the exposed side; (b) under microgravity conditions where no convection occurs (Huang and Glicksman 1981). Dendrite tip radius is a few µm.

If Pe is low (low growth rate), the supersaturation tends toward Pe, which is also the solution obtained by considering that the dendrite is a cylinder ending in a half-sphere, instead of a parabola.

This solution does not take into account the effect of capillarity, which has been added by Huang and Glicksman (1981). For example, in the low-Pe approximation, Eq. (24) is modified to Eq. (26).

$$\Omega = Pe + \frac{2\Gamma}{c_0 m_l (k-1) r} \tag{26}$$

The relationship imposed by Eq. (26) between tip radius and velocity at given undercooling is shown in Fig. 3.14, in which the Ivantsov solution is the straight line on the right-hand side and departure from this law under the effect of capillarity is seen to occur for small radii of curvature.

For a given supersaturation (undercooling in the thermal case) the number of couples (v_i, r) is practically infinite and another criterion is necessary to solve the problem. It was first considered that the tip radius is given by the extremum of the curve in Fig. 3.14, because it corresponds to the maximum velocity and minimum undercooling of the dendrite. However this leads to tip radii which are too small compared with the experimental values.

Langer and Müller-Krumbhaar (1977) have postulated that the tip radius is of the order of the periodicity of the interface destabilization. Their argument is that if the dendrite tip is smaller than this value the dendrite is likely to disappear, and if it is higher the dendrite tip will be destabilized and two dendrites will appear. From Eq. (21) we get Eq. (27), where the unknown variable is the solute gradient at the tip, which depends on the concentration at the interface.

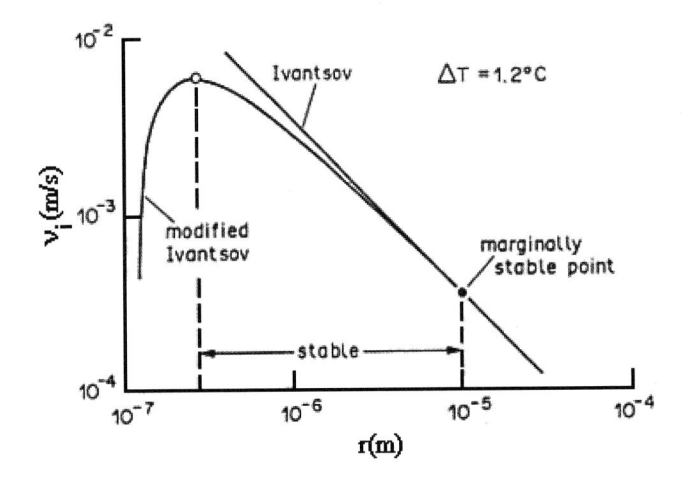

Fig. 3.14 Relationship between the velocity and the tip radius for a given undercooling/supersaturation (Kurz and Trivedi, 1990).

$$r = 2\pi \sqrt{\frac{\Gamma}{m_l \nabla c - \nabla T}} \tag{27}$$

Combination of Eqs. 8, 23, 26, and 27 gives a relationship between the growth rate and the tip radius which, in the case of a low growth rate and a negligible thermal gradient, can be written as Eq. (28) (Kurz and Fischer 1998).

$$r^2 v_i = 4\pi^2 \frac{D_l \Gamma}{c_0 m_l (k-1)} \tag{28}$$

This criterion is in good agreement with the dendrite tips measured by Glicksman.

3.2.2.3 Kinetics of Equiaxed Dendrites

For dendritic structures of pure metal growing in an undercooled melt (thermal equiaxed dendrites), the treatment of the columnar dendrite is applied, but the solute diffusion equation is replaced by the heat transfer equation, and Eq. (24) is replaced by Eq. (29).

$$\Delta T = \left(\frac{\Delta H_m}{C_p}\right) Iv(Pe) \tag{29}$$

The growth is not constrained, the velocity depends on the undercooling of the melt, and, in the simplified case of low Pe, the kinetic equations are (Kurz and Fischer, 1998) (30a) and (30b).

$$v_i = \frac{D_{th}C_p}{2\pi^2 \Delta H_m \Gamma} \Delta T^2 \tag{30a}$$

$$r = 4\pi^2 \frac{\Gamma}{\Delta T} \tag{31a}$$

In the case of an alloy, it is necessary to take into account simultaneously the thermal and the solute fields around the dendrite (Kurz and Fischer, 1998). This gives Eqs. (30b) and (31b).

$$v_i = \frac{D_l}{4C_0 m_l(k-1)\Gamma} \Delta T^2 \tag{30b}$$

$$r = 4\frac{\Gamma}{\Delta T} \tag{31b}$$

The equiaxed dendrite arms are very sensitive to sedimentation and convection, as shown by the comparison of experiments performed on Al–Ni alloys under microgravity (no sedimentation and convection), those under normal gravitational conditions (Dupouy and Camel, 2001), and those on transparent materials (Gerardin et al. 2001). Figure 3.15 shows the phase field simulation of the growth of an equiaxed dendrite in a liquid flow (phase field simulation techniques are

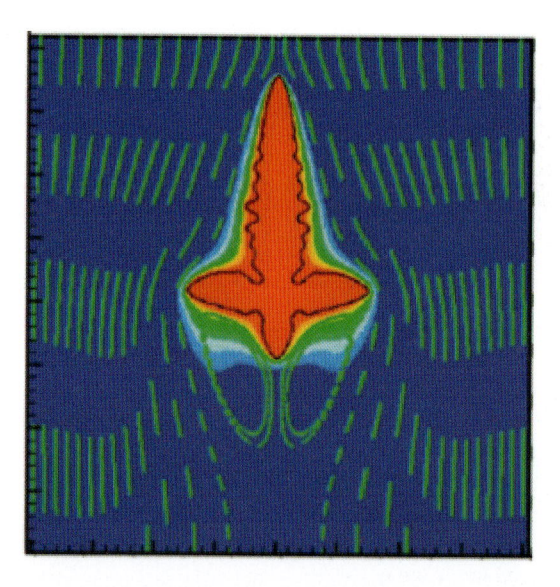

Fig. 3.15 Two-dimensional phase field simulation of free dendritic growth in a fluid flow. The flow enters at the top and leaves through the bottom. Colours indicate temperatures and the light lines are the streamlines (Boettinger et al. 2002).

3.2 Solidification Structures

discussed in Chapter 10). The dendrite arm facing the incoming liquid is accelerated, so that the equiaxed crystal becomes elongated in the direction of flow.

3.2.2.4 Characteristic Dimensions of the Dendrite

Knowing now the geometry of the dendrite tip, it is possible to compute the characteristic dimensions of the dendrite field: height, a, and periodicity, λ_1, under the hypothesis of an elliptical shape and regular hexagonal arrangement of the dendrite field. With a and b as half-major and -minor axis, respectively, the geometrical properties of the ellipse give Eq. (32).

$$r = \frac{b^2}{a} \tag{32}$$

Neglecting the tip undercooling, Eq. (33) follows from Fig. 3.16.

$$\lambda_1 = 1.7b = 1.7\sqrt{ra} = 1.7\sqrt{r\frac{\Delta T_0}{\nabla T}} \tag{33}$$

Substituting r from Eq. (28) gives λ_1 according to Eq. (34).

$$\lambda_1 = \frac{4.3}{\sqrt{\nabla T}} \sqrt[4]{\frac{D_l \Gamma c_0 m_l (k-1)}{k v_i}} \tag{34}$$

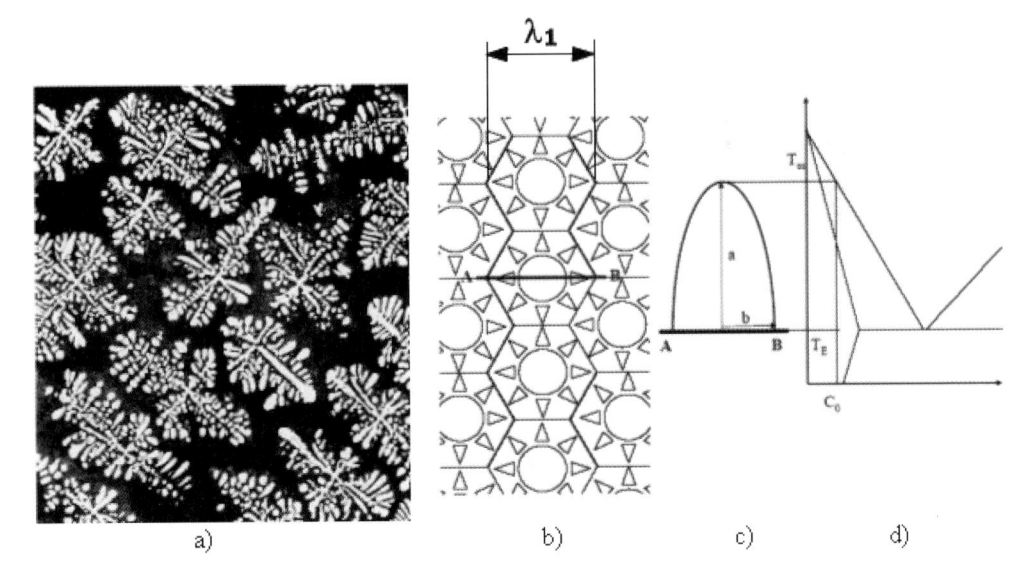

Fig. 3.16 (a) Columnar Al dendrites (0.5 mm diameter) in an Al–26wt%Cu alloy grown under microgravity in order to prevent convective perturbation (Dupouy et al. 1992); (b) their schematic representation as a hexagonal packing of ellipses; c) elevation along the line A–B; d) corresponding temperatures in the phase diagram.

Table 3.2 Experimental growth rate exponents for the primary spacing of dendritic arrays, for various alloys: $\lambda_1 \approx v_i^m \nabla T^n$.

Dendrite	Solute [wt%]	m	n
Al	2.2 to 10.1 Cu	−0.43	−0.44
	5.7 Cu	−0.36	−
	0.15 Mg, 0.33 Si, 0.63 Mg, 1.39 Si (at %)	−0.28	−0.55
	0.1 to 8.4 Si	−0.28	−0.55
	1 to 27 Cu	−0.5	−0.5
	1 to 5 Sn	−0.5	−0.5
	0.1 to 4.8 Ni	−0.5	−0.5
	5 Ag	−0.5	−0.5
Pb	2 to 7 Sb	−0.42	
	5 to 10 Sb	−0.75	−0.45
	10 to 50 Sn	−0.45	−0.33
	8 Au	−0.44	
	10 to 40 Sn	−0.39 to −0.43	−0.3 to −0.41
Fe	0.4 C, 1 Cr, 0.2 Mo	−0.2	−0.4
	8 Ni	−0.19	−
	0.6 to 1.5 C; 1.1 to 1.4 Mn	−0.25	−0.56
	0.035 C, 0.3 Si	−0.26	−

This parameter gives the main periodicity of the alloy structure and also of the defects that are generally trapped between the dendrites, precipitates, minority phases, microsegregations, and so on. Table 3.2 gives experimental values of the exponents of the growth rate and the thermal gradient, which are the principal process control parameters. It follows that Eq. (34) describes well the tendency but somewhat underestimates the effect of the growth rate.

Another important parameter is the secondary spacing, λ_2, which is the spacing between the secondary arms along the dendrite side. As follows from Figs 3.8 and 3.16(a), because of the evolution of the structure during cooling, these arms are degenerated and often are the only measurable structures. The original size of these arms, close to the dendrite tip, is the destabilization periodicity but, as can be seen from Fig. 3.13, it increases rapidly because of diffusion in the liquid between the arms, the solute going from the smaller arms toward the larger ones under the effect of capillarity (Ostwald ripening). This occurs as long as the sides of the dendrites are in contact with liquid. This phenomenon has been studied by Kattamis and Flemings (1965) and Feurer and Wunderlin (1977); they found an expression [Eq. (35)] for the spacing of secondary arms.

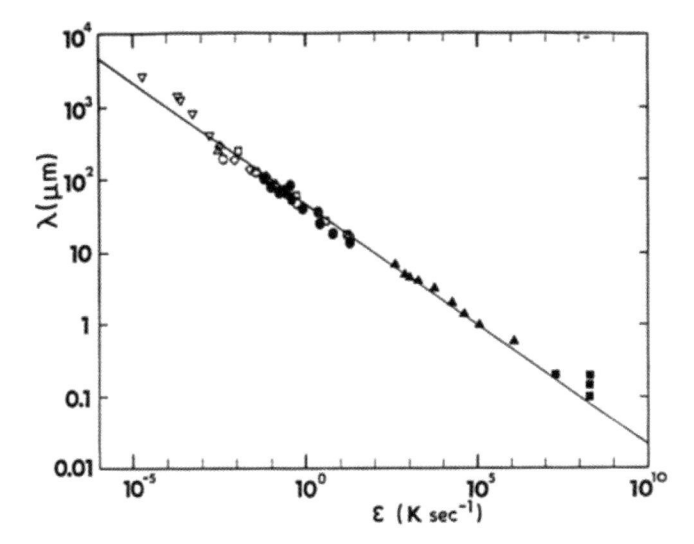

Fig. 3.17 Secondary spacing λ_2 as a function of cooling rate ε for Al–4%Cu and Al–11%Si alloys (Jones 1984).

$$\lambda_2 = 5.5 \left(-\frac{\Gamma D_l \ln\left(\frac{c_b}{c_0}\right)}{m_l(1-k)(c_b-c_0)} \frac{(T_0 - T_b)}{v_i \nabla T} \right)^{1/3} \tag{35}$$

In Eq. (35) the subscript b stands for the bottom of the dendrite, often at the eutectic composition and temperature (as shown in Fig. 3.16d). This expression has been obtained with the help of crude hypothesis and simplifications, yet it gives a good description of the phenomenon and compares well with experimental observation (see Fig. 3.17).

3.2.2.5 Microsegregation

The cellular or dendritic morphology is the source of chemical heterogeneities in the material, as solute is trapped between the cells or dendrites and the composition varies periodically through the microstructure.

Equation (10) can be used to account for this periodic variation of solute, by considering the solidification of the dendrite thickness in the y direction, perpendicular to its z axis [Eq. (36)]. The tip of the dendrite solidifies at a composition close to the melt composition (Fig. 3.12) and solidification proceeds along a distance $\lambda_1/2$ till it reaches the side of the facing dendrite:

$$c(y) = kc_i^l \left(1 - \frac{2y}{\lambda_1}\right)^{(k-1)} \tag{36}$$

When this concentration reaches the solubility limit, other solid phases precipitate, often as eutectic structures.

3.2.3
Rapid Solidification

The study of rapid solidification processes began in the 1960s, as it is a way to obtain materials in nonequilibrium and metastable states: microcrystalline or amorphous phases, such as metallic glasses. They are obtained by a rapid cooling of the sample, at 10^2–10^6 K s^{-1} (10^{11} K s^{-1} in certain laboratories). In this section the three major aspects of rapid solidification are explained: the stabilization of the solid–liquid interface, how it is possible to obtain metastable phases, and in which case an amorphous solid can be obtained. There are no direct relationships between these phenomena: amorphous and stable or metastable crystalline phases can be obtained below, as well as above, the absolute stability velocity.

3.2.3.1 Absolute Stability and Diffusionless Solidification

From Eqs. (18) and (19), it can be seen that, during directional solidification, the interface becomes unconditionally stable when the parameter A equals 1, which gives immediately the expression for the absolute stability velocity, Eq. (37).

$$v_a = -\frac{m_l D_l (1-k) C_0}{\Gamma k} \qquad (37)$$

It is independent of the thermal gradient because its value becomes negligible compared with the solute gradient in front of the interface. With classical material parameters, the absolute velocity is in the range 0.1–1 m s^{-1}.

At very high growth rates, the solute boundary layer in the liquid decreases so much [Eq. (9)] that it becomes of the order of the interface thickness. Coming back to the argument developed in the Introduction [Eq. (2)] it follows that the interface velocity becomes so high that the atoms or molecules do not have enough time to rearrange at the interface and the liquid is solidified without any segregation; the solute segregation coefficient tends toward 1. Of course, the chemical potential of the solid and the liquid are no longer equal, as they are not in equilibrium. Aziz (1982) has proposed a relationship [Eq. (38)] for the velocity-dependent segregation coefficient:

$$k_v = \frac{k + \dfrac{av_i}{D_s}}{1 + \dfrac{av_i}{D_s}} \qquad (38)$$

As k_v approaches 1 with increasing velocities, this favors the absolute stability of the front in the case of directional solidification. In the case of growth from the undercooled melt, dendrites become purely thermal once diffusionless solidification conditions are reached at the tip ($k_v = 1$).

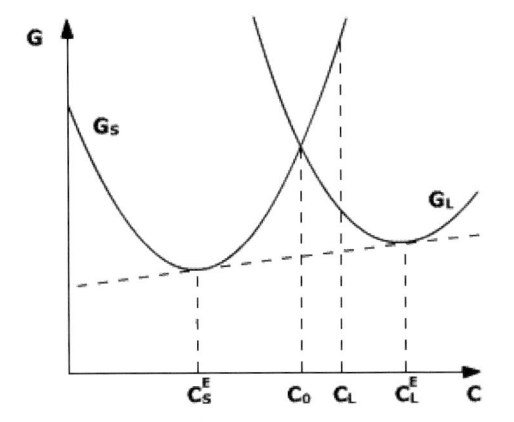

Fig. 3.18 Free energy of the solid and the liquid phases at equilibrium. The dotted line is a tangent to both energy curves corresponding to the solidus and liquidus lines ("double tangent construction").

In any case, the interface velocity cannot be higher than the velocity of sound in the liquid, which is of the order of 10^3 m s^{-1} for metals.

3.2.3.2 Nonequilibrium Phase Diagrams

An interesting fact in rapid solidification is that the undercooling of the interface increases with the growth rate (see Eqs. (5) and (6) for the kinetic effect, or Eqs. (23)–(25) for a columnar dendrite) and liquids can be solidified far from their equilibrium melting temperature, in regions of the phase diagram where metastable phases can be obtained.

At equilibrium, for a given temperature T, the compositions of the solid and liquid phases are defined through the free energy diagram (Fig. 3.18) (see Chapter 7 for calculation of phase diagrams).

When rapid solidification occurs, the chemical compositions of the solid and of the liquid can be equal, as explained above, but they cannot be higher than c_0, otherwise the energy of the solid would be higher than the energy of the liquid. Therefore, for the given temperature T, it is possible to get a solid with compositions ranging from c_e to c_0. The corresponding T–c_0 curve, known as the T_0 line, can be plotted in the phase diagram of the alloy, as shown in Fig. 3.19. It is located between the liquidus and solidus lines.

Provided that the undercooling is large enough, it can be seen from Fig. 3.19 that the composition domain of the equilibrium phases can be increased. But the most important application is that other phases, which cannot be solidified at equilibrium, can be obtained.

3.2.3.3 Structure of the Rapidly Solidified Phase

Crystals show well-ordered piles of atoms. On the contrary, amorphous solids, such as glasses, show atomic arrangements that are closer to the structure of

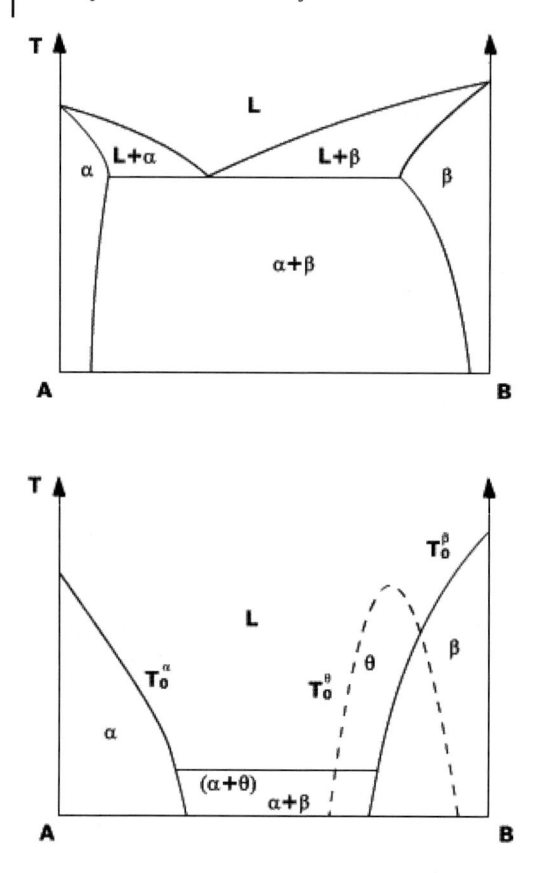

Fig. 3.19 Comparison between the equilibrium and the kinetic phase diagram of a given alloy. The eutectic temperature is substantially lowered, the eutectic composition width decreases, and a new solid phase θ appears.

liquid phases, with absence of topological long-range order. Obtaining one or the other structure depends on the probability of nucleation of crystal seeds in the cooling liquid.

The nucleation rate of seed crystals depends on two factors. In order to nucleate, a seed should be large enough, so that enough solid is created to counterbalance the energy cost of creating the seed surface. Once the seed has nucleated, it should be fed by diffusion through the liquid in order to grow. Both mechanisms are energetically activated and the nucleation rate, i.e., the number of growing seeds nucleated per second, is given by Eq. (39), with ΔG_n depending on the undercooling of the liquid according to Eq. (40) (see Chapter 7 for a detailed discussion of the nucleation mechanisms).

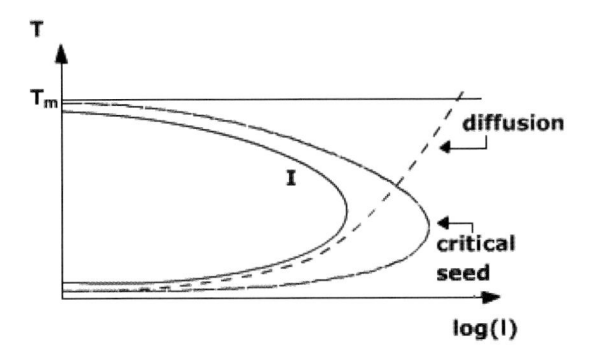

Fig. 3.20 Evolution of nucleation rate below the melting temperature.

$$I = I_0 e^{-\Delta G_n/k_B T} e^{-\Delta G_D/k_B T} \tag{39}$$

$$\Delta G_n = \frac{16\pi}{3} \frac{\gamma_{sl} \Gamma^2}{\Delta T^2} \tag{40}$$

Figure 3.20 shows both contributions and the corresponding nucleation rate.

In general, this curve is presented in a plot of temperature versus time; see Fig. 3.21 (known as a TTT diagram, for Time–Temperature–Transformation), where the curve gives the time after which a given nucleus density or solid fraction is obtained.

Three domains can be distinguished in this diagram. Above the melting temperature, the sample is liquid. Below the melting temperature, the crystalline phase is obtained in the domain on the right of the "nose"-shaped curve. In the case of a small undercooling, the time to obtain the first crystal seed may be very long (path 1). When the undercooling increases, the probability of seed nucleation increases as well and the critical nucleus size decreases. However, after pass-

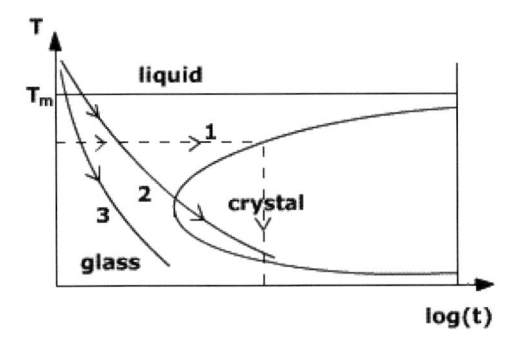

Fig. 3.21 TTT diagram, showing how the cooling rate selects either the crystalline or the amorphous structure of an alloy.

ing a certain critical undercooling, it becomes difficult to get crystals because the nucleation and diffusion processes are both thermally activated.

It follows from this diagram that, whatever the material or alloy, there always exist cooling conditions permitting either the crystalline (path 2) or the amorphous (path 3) phase to be obtained. However, the critical cooling rate to get crystals may be extremely low for materials such as silica glasses. On the contrary, extremely high cooling rates, of the order of 10^6 K s^{-1} and obtainable only on thin materials (ribbons, wires and suchlike), are usually necessary to get amorphous metallic binary alloys. In practice, the difficulty of obtaining a crystal seed increases with the complexity of the crystallographic structure, which in turn increases with the number of different atoms in the structure. It follows that it is easier to get metallic glasses of quaternary or quinary alloys: centimetric bulk metallic glasses of Zr–Ti–Ni–Cu–Be are currently produced with cooling rates of the order of 10 K s^{-1} (Johnson 1999).

3.2.4
Eutectic Structures

Eutectic structures occur when two or more solid phases grow simultaneously from the liquid. This generally leads to tiny structures, of the order of 1 µm. Eutectics are composite materials and often show very good mechanical properties.

3.2.4.1 Size of the Eutectic Structure
The derivation of the period of a eutectic structure has been introduced by Jackson and Hunt (1966). The calculation presented here is a simplified version for a symmetric phase diagram and equivalent physical parameters of both phases, together with linearized fluxes.

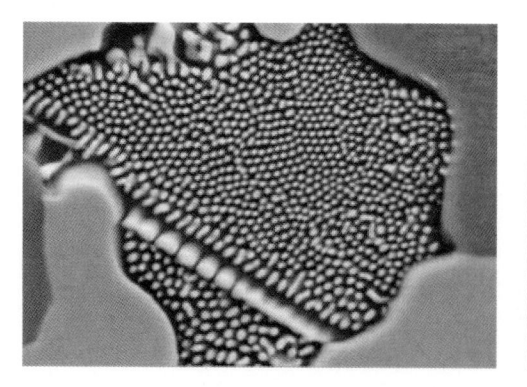

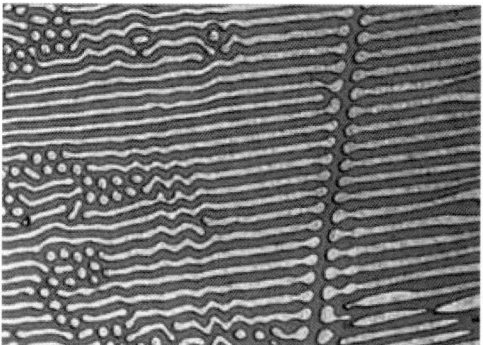

Fig. 3.22 Typical fibrous (Ag$_2$Al phase in a Cu–Zn–Al–Ag alloy) and lamellar (Al–23%Cu alloy) eutectic structures. Their characteristic dimension is 1 µm (Access-Aachen).

3.2 Solidification Structures

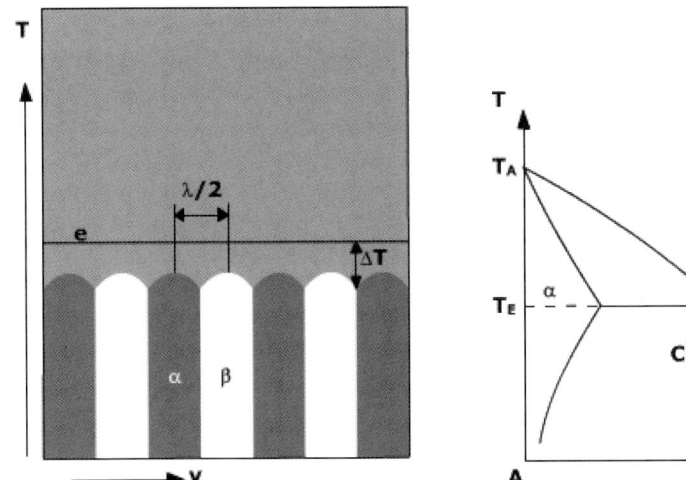

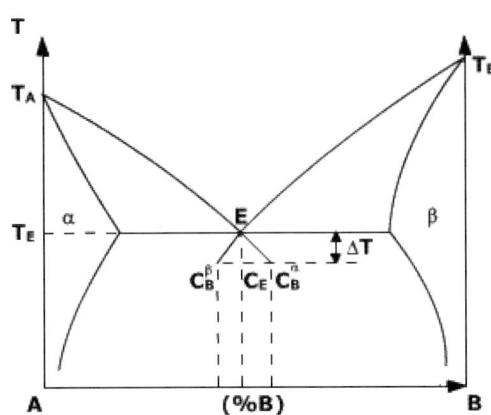

Fig. 3.23 Lamellar eutectic growth and corresponding phase diagram.

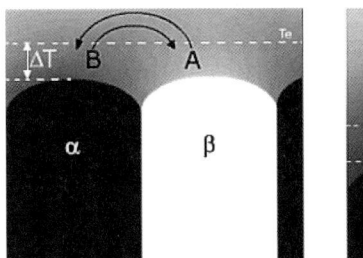

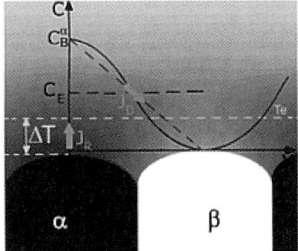

Fig. 3.24 Diffusion of solute rejected by one phase in order to feed the adjacent one.

Each phase rejects the solute corresponding to the growth of the other phase, which then will diffuse in order to feed the adjacent lamella (Fig. 3.24).

In a first approximation, the flux of B atoms from the α to the β phase can be written as Eq. (41).

$$J_B = -D_l \frac{(C_B^\alpha - C_B^\beta)}{\frac{\lambda}{2}} \tag{41}$$

From [Eq. (8)], the flux of solute B rejected by the α phase is given by Eq. (42).

$$J_B = v_i c_B^\alpha (1 - k) \tag{42}$$

Under conditions of stationary growth, both fluxes are equal and the composition is close to the eutectic composition, which gives Eq. (43).

3 Solidification and Grown-in Defects

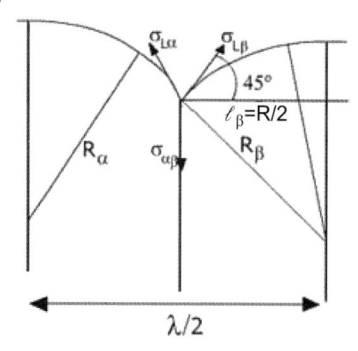

Fig. 3.25 Geometry of a junction of two lamellae with the equilibrium of the surface energies leading to the curvature of the solid–liquid interface at the lamella top.

$$(C_B^\alpha - C_B^\beta) = \frac{\lambda v_i c_E (1-k)}{2D} \tag{43}$$

Equation (44) is the corresponding expression in terms of chemical undercooling (see Fig. 3.23).

$$\Delta T_{ch} = \frac{c_E(1-k)}{2D\left(\dfrac{1}{m_\beta} - \dfrac{1}{m_\alpha}\right)} \lambda v_i \tag{44}$$

The geometry of the junction between two lamellae gives another expression [Eq. (45), derived from Eq. (13)] for the undercooling (Fig. 3.25). It is supposed that the top of the lamella is a portion of a circle and that the geometry is symmetric.

$$\Delta T_{cap} = \frac{2\Gamma}{\lambda} \tag{45}$$

The temperature of the eutectic interface T_i is plotted in Fig. 3.26 with the two undercooling terms ΔT_{ch} and ΔT_{cap} [Eq. (46)].

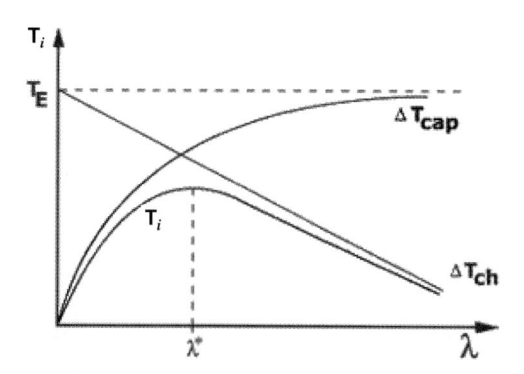

Fig. 3.26 Temperature of the eutectic interface versus eutectic spacing.

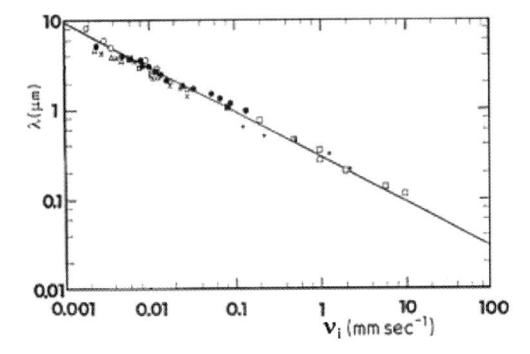

Fig. 3.27 Variation of the eutectic spacing with the growth rate for the Al–Al$_2$Cu eutectic (Jones 1984).

$$T_i = T_E - \Delta T_{ch} - \Delta T_{cap} \tag{46}$$

The system is likely to adopt the position where the energy, and thus the undercooling, are minimal.

The eutectic spacing is then obtained by derivation of Eq. (46) with respect to λ and finally gives Eq. (47).

$$\lambda^2 v_i = 4 \frac{D\Gamma\left(\dfrac{1}{m_\beta} - \dfrac{1}{m_\alpha}\right)}{c_E(1-k)} \tag{47}$$

This simplified relationship captures the basic physical concepts leading to the eutectic spacing: the periodicity of the structure is a compromise between the interdiffusion of the species, which works better when the distance between lamellae is decreased, and the surface energy, which tends to increase the thickness of the lamellae. It should be noted that in a first approximation the thermal gradient and the alloy composition have no effect on the eutectic structure. Equation (47) is in good agreement with results for an Al–Al$_2$Cu eutectic (compare Fig. 3.27).

3.3
Defects in Single and Polycrystals

Section 3.2 has reviewed the main theories developed in order to explain the structures obtained after solidification. However, the reality is generally more complicated and materials and structures are not as ideal and perfect as could be expected from the previous explanations.

This section is therefore devoted to analysis of the defects which are generated during the solidification of alloys. This applies essentially to two classes of alloys and growth regimes: single crystals, which are expected to have a highly perfect atom arrangement, and dendritic structures, because perturbations of the dendritic field are likely to create defects in the material which may cause the mechanical properties of the alloy to deteriorate and restrict considerably its technical application. Section 3.3.2 is devoted to the development of the grain structure of an alloy.

3.3.1
Defects in Single Crystals

A convenient way to classify defects is to use their characteristic dimension.

3.3.1.1 Point Defects

A full treatment of point defects, such as vacancies, interstitials, antisites and so on, is provided in Chapter 5.

Solidification by itself has no direct effect on point defects in pure elements. As the liquid phase is not constrained and electrically neutral, the basic principle is that the interface acts as an ideal source and sink of point defects that are incorporated in the solid at their equilibrium value, which depends essentially on the temperature. Considering Si at its melting temperature, vacancies and interstitials show a level of 10^{15} cm^{-3} in the solid, while holes and electrons are at the 10^{19} cm^{-3} level. However, only a rough interface is an ideal surface, faceted interfaces may decrease the level of defects because the energy of one atom absorbed on or desorbed from the interface is high in this case.

The situation is more complicated in compound phases. For GaAs, vacancies and interstitials are at the level of 10^{19} cm^{-3} (0.05%) for a perfectly stoichiometry and neutral material, but it is difficult to obtain stoichiometry. Figure 3.28 shows an enlarged view of the phase diagram of GaAs. It appears that the congruent point is shifted to the As-rich side, which means that only $Ga_{0.497}As_{0.503}$ can be grown without segregation effects. However, during cooling, this will cause As to precipitate. The situation is complicated by the fact that As has a high vapor pressure above liquid GaAs (2 atm at the melting point). Perfectly stoichiometric GaAs can be grown from the liquid phase, provided that its composition is controlled continuously during growth, which is technically (but not easily) obtained by controlling the As pressure in the gas in equilibrium with the liquid.

In turn, antisite defects (for definition and more details, see Chapter 5) are controlled by the stoichiometry of the compound. An important defect in GaAs, known as EL2, is As$_{Ga}$. The reaction governing this defect is represented by (Lagowski et al. 1982):

$$As_{As}^{0} + h_{Ga}^{-} \rightarrow As_{Ga}^{2+} + h_{As}^{+} + 4e^{-}$$

It is found that the concentration of EL2 is dependent on stoichiometry, but also on the electrical state of the semiconductor [Eq. (48)].

$$[As_{Ga}] = K(T) \frac{[h_{Ga}]}{[h_{As}][e^{-}]^{4}} \tag{48}$$

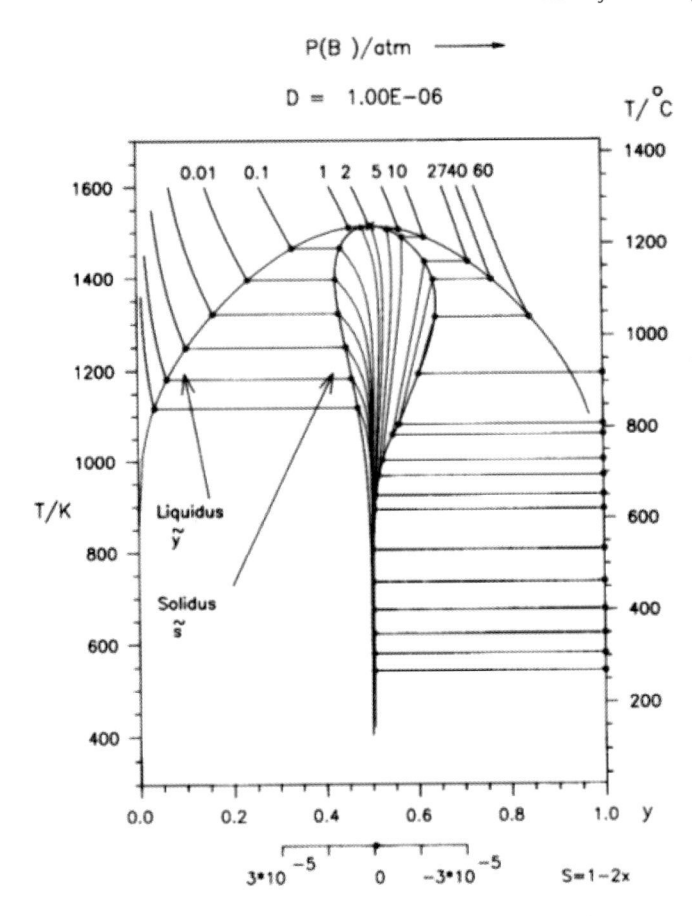

Fig. 3.28 GaAs phase diagram with an enlarged view of the solidus close to the stoichiometric point superposed on it (Wenzl et al. 1991).

1D: Dislocations Dislocations are covered in detail in Chapter 6, and only a few comments on their interaction with solidification will be given here.

The only stress applied on the growing solid–liquid interface is the hydrostatic pressure. For materials which are very soft at the melting temperature, for example, HgI_2, this may have an effect on the dislocations created in the crystal. However, this pressure is negligible in more conventional materials and a no-stress mechanical boundary condition is typically used on the interface. It follows that the solidification process by itself does not generate dislocations.

Stresses occurring after solidification, during the cooling process, or because of possible adhesion on crucible walls easily generate dislocations in the crystal; especially close to the melting point where plasticity is enhanced (Gondet et al. 2003). Since this process is not directly linked to solidification matters, it will not be covered here, but the reader should be aware that it is the main cause of dislocation generation in industrial single crystals (see Völkl, 1994).

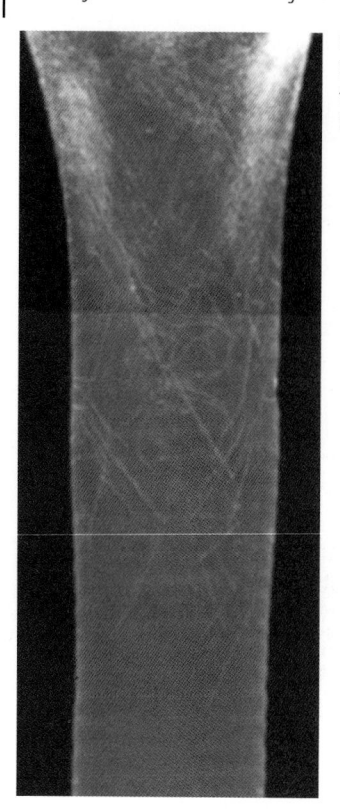

Fig. 3.29 X-ray topography of the seed of a zone-melting Si single crystal. Dislocations appear as white lines and their number decreases as growth proceeds, from top to bottom (Hurle and Rudolph 2004).

Solidification may be used to decrease the number of dislocations in crystals. If dislocation lines are not perfectly perpendicular to the growing interface, they will move, as growth proceeds, toward the edge of the crystal, where they will disappear. This property is used for "cleaning" seeds in the pulling of Si single crystals (Fig. 3.29) (Dash 1959).

Last but not least, the reader should keep in mind, as mentioned in the Introduction, that the existence of dislocations at the growing interface has a strong effect on the undercooling, especially in the case of faceted growth.

2D: Twins and grain boundaries Although point defects and dislocations can, to a certain extent, be accepted in single-crystal production, this is not the case for twins and grain boundaries which separate the sample into two or more parts with different crystallographic orientations, making the material unsuitable for a use as single crystal.

It is well known that twins and grains are easily produced during mechanical deformation of materials (see Chapter 6). The thermal stresses occurring during the cooling phase of crystal growth processes are not large enough to produce twins, but they produce dislocations that may align in sub-grain boundaries and, ultimately, give grains (Boiton et al. 1999). Since this is not directly linked to

3.3.1.2 Twins

Twins are specifically linked to single crystals, because they are due to a defect in the stacking of atom piles, leading to a rotation of the crystal structure (Fig. 3.30). The energy associated with the twin plane is due to the stacking faults and is relatively low compared with a grain boundary, so that twins can grow easily under certain conditions.

The twinning of a crystal during Czochralski pulling will be discussed, but the case can be extended to other crystal growth processes where a crucible wall is in contact with both liquid and solid. A first condition for the twinning to occur is that the solid–liquid interface is faceted at the crystal–liquid–gas triple line (Fig. 3.31). It can be shown that a twinned seed cannot nucleate in the middle of the facet, because its energy, increased by the twin energy boundary, is higher than the energy of a regular seed.

Hurle (1995) has shown that twinning will occur on the facet at the triple line, if a twinned seed can nucleate. Such a seed is a truncated disk and its energy is lower than for a regular, circular, seed. The question is under which conditions a truncated twinned seed has a lower energy than a truncated regular (well-oriented) seed. The key point is that the lateral surface of the seed (facing the gas) should have a given orientation, generally $\langle 111 \rangle$. This happens if the angle α in Fig. 3.31 is 70.5°, which is likely to occur in the conical part of the crystal, when its diameter is increased. Temperature oscillations, leading to diameter fluctuations, are also a cause of reaching this particular value.

In order to make possible the growth of the twinned seed, it should also be small enough; this gives a value for its undercooling according to Eq. (49), where subscripts TB, C and T respectively stand for the Twin Boundary, for a Circular seed, and for the Truncated seed.

$$\Delta T_s = \frac{\gamma_{TB}}{a \Delta S_m} \frac{\Delta G_C}{\Delta G_C - \Delta G_T} \tag{49}$$

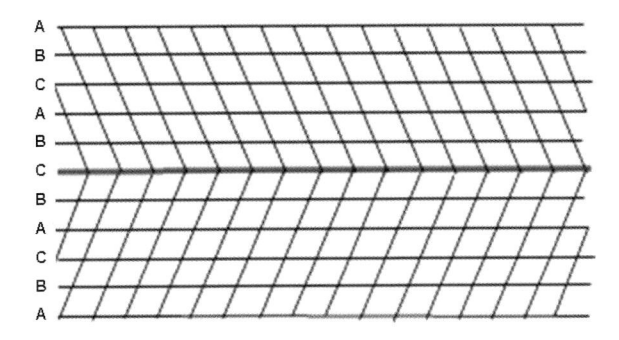

Fig. 3.30 Typical twin boundary, due to an error of piling A, B, and C atomic planes.

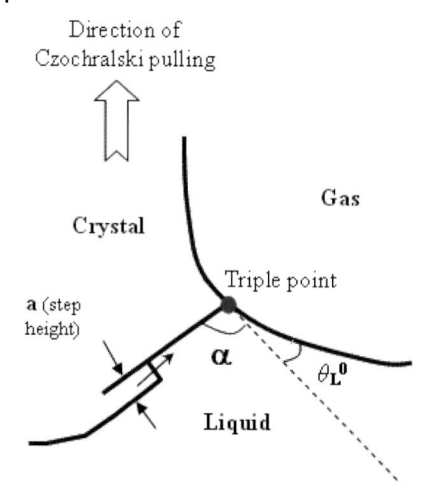

Fig. 3.31 Normal growth in the Czochralski process: the facet at the triple line grows step by step. A twin may occur if a seed of this orientation nucleates at the triple line.

It follows that the facet on which the twinned seed is nucleating should be highly undercooled. As dislocations facilitate the growth of faceted crystals and then decrease their undercooling, twins are more likely to appear in crystals with a very low dislocation density, which is in good agreement with the experimental observation.

3.3.1.3 Grains

Single-crystal growth often fails because the occurrence of grains. Due to the energy associated with the grain boundary, the nucleation of a randomly disoriented seed directly on the growing interface is not possible and the mechanism should be different from that of twin nucleation. The basic principle is still the decrease in energy of the seed of a grain with respect to a seed correctly oriented on the interface. This may occur if foreign surfaces, on which a disoriented seed can grow, are present in the system. Such surfaces are typically precipitates or crucible walls. In addition, the grain seed should nucleate ahead of the interface and then a certain undercooling is necessary.

As shown in Fig. 3.32, two configurations may lead to grain nucleation in the layer of undercooled melt ahead of the interface: nucleation on a crucible wall, and in holes of a rough crucible. Only the case of a seed nucleating on the crucible wall (Duffar et al. 1999) will be discussed here, but the treatment of the other configuration is equivalent. In processes without crucibles, grains nucleate on foreign particles or on precipitates in the melt.

From the nucleation theory, there is a critical radius for nucleation of a 2D nucleus on a faceted interface [see Eq. (12)] and the associated energy is given by Eq. (50).

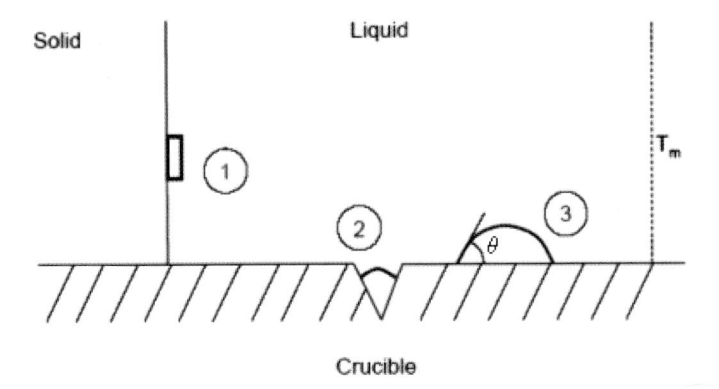

Fig. 3.32 Nucleation of solid seeds: (1) on the interface; (2) in a hole in a rough crucible. (3) Spurious nucleation on the crucible wall.

$$\Delta G_I = \frac{\pi a \gamma_{sl}^2}{\Delta S_m \Delta T} \qquad (50)$$

On the crucible wall, the critical seed has a similar radius, but the energy is decreased by a factor depending on the contact angle of the seed on the wall [Eq. (51)].

$$\Delta G_W = \frac{16 \pi \gamma_{sl}^3}{3 \Delta S_m^2 \Delta T^2} \frac{(2 + \cos \theta)(1 - \cos \theta)^2}{4} \qquad (51)$$

Figure 3.33 shows that if the contact angle is low enough, spurious nucleation may occur, even for low values of undercooling.

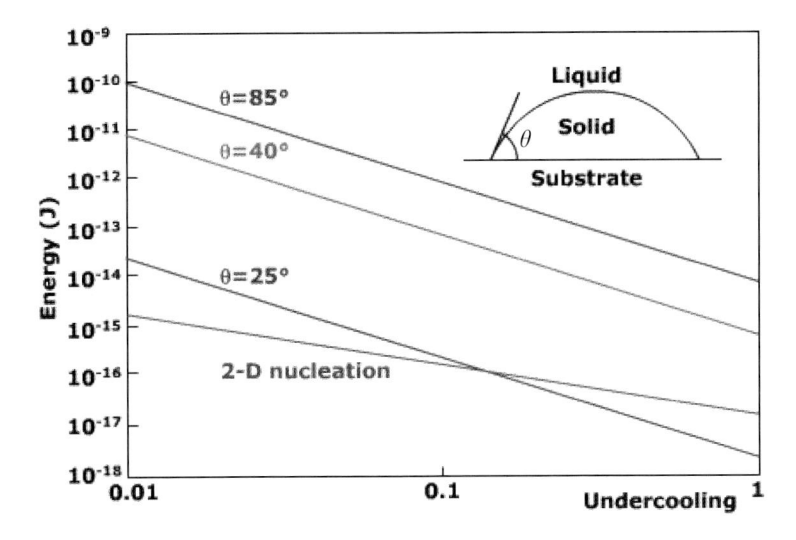

Fig. 3.33 Nucleation energy of a 2D nucleus on the interface and on walls with various contact angles, plotted as a function of the undercooling.

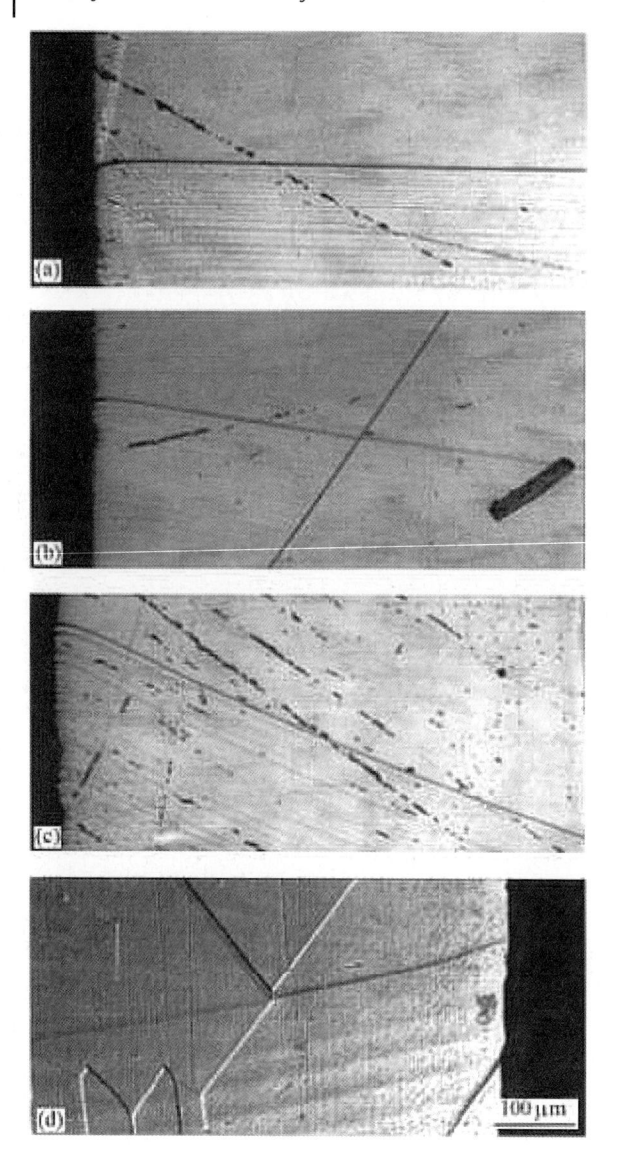

Fig. 3.34 Metallographs of marked GaSb solid–liquid interfaces, in contact with the crucible wall: (a) silica ($\theta = 100°$); (b) vitreous carbon ($\theta = 90°$); (c) BN with carbon coating ($\theta = 90°$); (d) BN ($\theta = 40°$, the crucible wall is on the right).

Figure 3.34 shows the angle made by a solid–liquid interface on a crucible wall. The measured angles, introduced into Eq. (51), are in agreement with the observation that spurious nucleation in GaSb crystal growth never occurs in silica crucibles, but always happens in BN crucibles.

3.3.2
Grain Structure of an Alloy

As already mentioned in Section 3.2.2 with respect to dendrites, the grain structure of an alloy is of primary importance to its mechanical properties. Grain boundaries are defect zones with a high energy (0.5 J m^{-2}, close to the energy of a solid–liquid interface), attracting stresses and impurities, and helping the precipitation of foreign phases. On the one hand, this results in a cleaning effect of the intragranular material but, on the other hand, grain boundaries may be locations of corrosion, mechanical weakness, or crack initiation. Although this is extremely alloy-dependent, it is generally recognized that the quality of an alloy is improved when grains are smaller.

Grains can be columnar, when they are elongated in the direction of the growth, and in this case each grain is constituted by several columnar dendrites (see Section 3.2.2). Grains with an isotropic shape are called equiaxed and they are generally produced from individual equiaxed dendrites. As shown on Fig. 3.35, it is possible to change the growth conditions in order to get one or the other type of grain structure.

Typical defects are associated with the grain structure.
- Chemical segregation may occur between dendrites and grains, and on the full ingot scale. Precipitation of oxides, sulfides, or eutectic phases, or nucleation of dissolved gases (H_2) often follows.
- Shrinkage may also occur between dendrites and grains, and on the ingot scale. It is a major defect that deserves much attention. Its distribution across the solid depends strongly on the solidification conditions.

Fig. 3.35 Transition from columnar (left) to equiaxed (right) grain growth through an increase in the growth rate from 2 to 15 μm s^{-1} (refined Al–3.5%wtNi, $VT = 20$ K cm^{-1}) (Dupouy and Camel 2000). Diameter 8 mm.

Fig. 3.36 Typical grain structure in a cast alloy.

In practice, all metallurgical processes consist in pouring the liquid into a mold. The extraction of heat toward the walls leads to a typical grain structure in the cast ingot which is shown in Fig. 3.36. There are three zones: the outer dendrites, the columnar region, and an equiaxed area in the center.

The outer equiaxed zone is composed of small grains that grew by heterogeneous nucleation in the undercooled liquid at the mold walls (step 1; compare Fig. 3.37). As there is a great heat extraction by the mold, their growth rate is

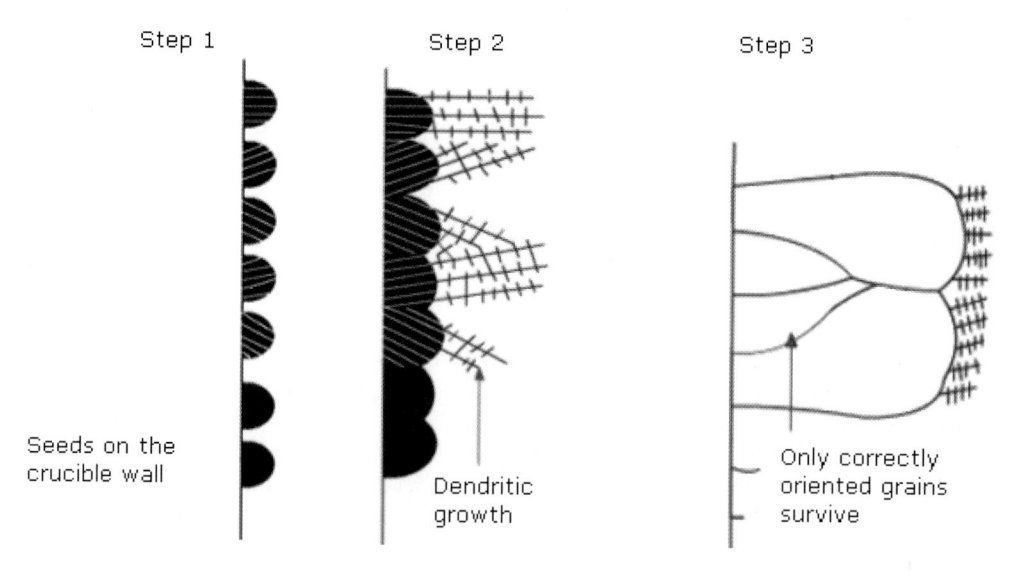

Fig. 3.37 Three steps of the transition from the outer equiaxed zone to the columnar structure.

Fig. 3.38 Grain structure of an alloy as predicted by Gandin's stochastic model (Gandin and Rappaz, 1994). The growth proceeds from the left to the right and the effect of melt convection can be seen. The colors are representations of the various crystallographic orientations of the grains.

high and destabilization occurs, leading to columnar dendritic growth (step 2). Dendrites follow the crystallographic orientation of the initial grain and those oriented opposite to the thermal flux "overgrow" the others, so that grain selection occurs (step 3).

So far, these nucleation, growth, and grain selection mechanisms have been modeled only through stochastic numerical simulation, with realistic and predictive results (Fig. 3.38). The method and some results are given in Rappaz and Gandin (1993) and Gandin and Rappaz (1994).

The central equiaxed region appears when the thermal gradient in the liquid vanishes and seeds transported from the front (detached dendrite arms) are able to grow in the liquid. Depending on the application of the alloy, it is possible to avoid the equiaxed structure (photovoltaic silicon, turbine blades) or on the contrary to favor it, by mixing the liquid or using refining particles.

In Section 3.3.2.1 the equiaxed grain growth regime is studied, in an attempt to find out the typical size of the grain structure. The parameters involved in the columnar to equiaxed transition are explained in Section 3.3.2.2.

3.3.2.1 Equiaxed Growth in Presence of Refining Particles

Grain refining is used to get a homogeneous and tiny equiaxed structure. It consists in enhancing the heterogeneous grain nucleation with the help of foreign solid particles introduced in the melt. The principle is based on the fact that heterogeneous nucleation on foreign particles is easier than homogeneous nucleation, especially if the contact angle of the seed on the particle is low [Eq. (51)].

The first condition is to have nucleating particles (refiners) in the melt. Finding adequate particles is not easy because they should not dissolve in the melt and they should have nucleation properties only for a sufficiently undercooled melt. In practice, two refinement processes are used (Jackson et al. 1966).

- Introduction of the appropriate particles in an undercooled melt. A very typical example is the introduction of Ti (1–10%) and B (0.1–3%) in Al alloys.
- Strong mixing of the melt, for example by electromagnetic forces. The common assumption is that the melt flow is strong enough to detach secondary arms of columnar dendrites (this detachment occurs by chemical dissolution rather than mechanical effect) and these solid particles are seeds for the equiaxed growth if they remain in or pass into an undercooled liquid.

The second condition means that the melt should be, and should remain, undercooled as long as possible. However, because of the equiaxed dendrite growth, latent heat is released in the system and the liquid temperature increases. It is only in the case of efficient solute rejection by the grains that the liquid remains undercooled (chemical undercooling, Eq. (3)) and that the growth of the already existing grains is restricted. This ability is measured by the parameter of growth restriction, q [Eq. (52)].

$$q = m_l(k-1)c_0 = k\Delta T_0 \tag{52}$$

The number of grains in the alloy is a direct consequence of the competition between the heterogeneous nucleation rate, which increases with the undercooling, and the growth rate, which decreases as the undercooling increases.

In the model proposed by Maxwell and Helawell (1975), three different regimes are considered.

In a first step, the hot liquid cools down, passes the melting temperature, then is undercooled. The energy barrier for nucleation of a seed on a foreign particle is given by Eq. (51) and therefore decreases when the undercooling increases and nucleation starts at a given undercooling. The nucleation rate is given by Eq. (39) and is proportional to the number of foreign particles N_p. Considering the low undercooling, the diffusion term is taken to be constant and Eq. (53) results.

$$\frac{dN}{dt} = N_p I_0 e^{-\Delta G_n/k_B T} \tag{53}$$

In the second step, particles are growing. It is supposed that the undercooled liquid is isothermal and that the solute fields around the particles are not interacting, so that the liquid between two grains is at the initial composition. These assumptions are based on the fact that the thermal diffusivity is two to three orders of magnitude higher than the chemical diffusivity. Then the characteristic length for chemical diffusion is smaller than the typical grain size (200 µm) but the characteristic length for the heat transport corresponds to hundreds of grains.

The crystals are spherical and their growth rate is limited by the solute rejection and then varies inversely with q. The time evolution of the radius of a particle growing and rejecting solute has been obtained by Aaron et al. (1970) as Eqs. (54).

$$r = \lambda \sqrt{D_l t}$$

$$\lambda = \sqrt{\frac{S^2}{4\pi} - S - \frac{S}{\sqrt{2\pi}}}$$

$$S = -\frac{2\Delta T_S}{m_l(k-1)(C_0 - \Delta T_S/m_l)}, \quad \Delta T_S = m_l(c_i^l - c_0) \tag{54}$$

The undercooling of the melt is the sum of the solute undercooling [Eq. (54)] and of the capillary undercooling [Eq. (13)], or as stated in Eq. (55).

$$\Delta T_S = \Delta T - \Delta T_{cap} \tag{55}$$

During the growth, the system is cooled at a rate P and the latent heat is released proportionally to the increase of the solid fraction, so the variation of liquid undercooling is given by Eq. (56).

$$\frac{d\Delta T}{dt} = P - \frac{\Delta H_m}{C_p} \frac{df_s}{dt} \tag{56}$$

Equations (53–56) form a system into which the number of particles, their size, and the undercooling enter as unknown variables. This system can be solved numerically if an initial value is given for the radius r_0.

The third step starts when the temperature of the melt increases enough to stop the grain nucleation. No more grains are formed, but the other grains continue growing and it has been shown that it is during this step that they are destabilized and show dendrite arms.

This model is able to reproduce the melt undercooling, the undercooling breakdown, and the recalescence, and, most importantly, gives the tendencies concerning the final number of grains. Some interesting results are shown in Fig. 3.39. The contact angle of the seed on the foreign particle has a huge influence on the number of grains. It is also shown that the cooling rate is likely to help to control the number of grains.

Finally, the model is able to explain the experimental fact that the grain size decreases with the number of refining particles, but only to a certain extent. If too many particles are present, the number of growing grains increases quickly at the beginning and the latent heat released prevents the nucleation of other grains. It is then possible to determine the number of foreign particles leading to the smallest grains. However, the model remains only qualitative.

Greer has proposed an improvement of the previous model, taking into account the statistical distribution of particle size and a free growth of the dendrites

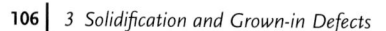

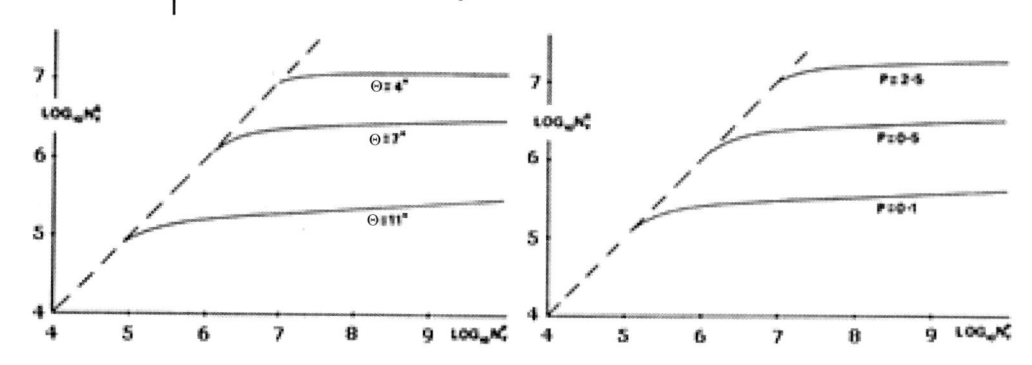

Fig. 3.39 Equiaxed grain density versus foreign particle density. Parameter: contact angle for $P = 0.5$ K s^{-1} (left), and cooling rate for $\theta = 7°$ (right) (Al–Ti, $r_0 = 1$ µm) (Maxwell and Hellawell 1975).

(Greer et al. 2000). This analysis has been performed in the case of Al alloy casting with Al-Ti5-1B as refiner.

Suppose that Al seeds nucleate on the large [0001] faces of TiB$_2$ particles: if the wetting angle is low enough, this is likely to happen at low undercoolings. The Al seed grows laterally and when the face is covered, it keeps on growing, but with an increase in the curvature of the solid–liquid interface (Fig. 3.40). The radius of curvature of the nucleus cannot be lower than the critical radius for nucleation, r^*, which depends on the undercooling [Eq. (12)]. Therefore free growth occurs only when the critical radius becomes lower than the particle radius. It follows that, for a particle population ranging from 0.1 to 10 µm, this kind of free growth occurs for undercoolings from 6–0.06 K, which indeed corresponds to the undercoolings observed in the processing of refined Al alloys.

It follows also that during the liquid cooling seeds appear first on large particles then on smaller and smaller particles when the undercooling increases.

Greer's model then adopts Maxwell–Hellawell's treatment, where the growth of existing seeds causes a decrease in the undercooling and of the subsequent recalescence. This means that all the particles under a certain diameter do not lead to

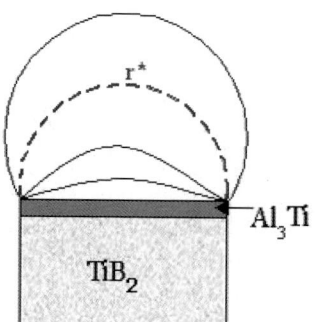

Fig. 3.40 Growth of Al seed on the Al-Ti5-1B refiner particle.

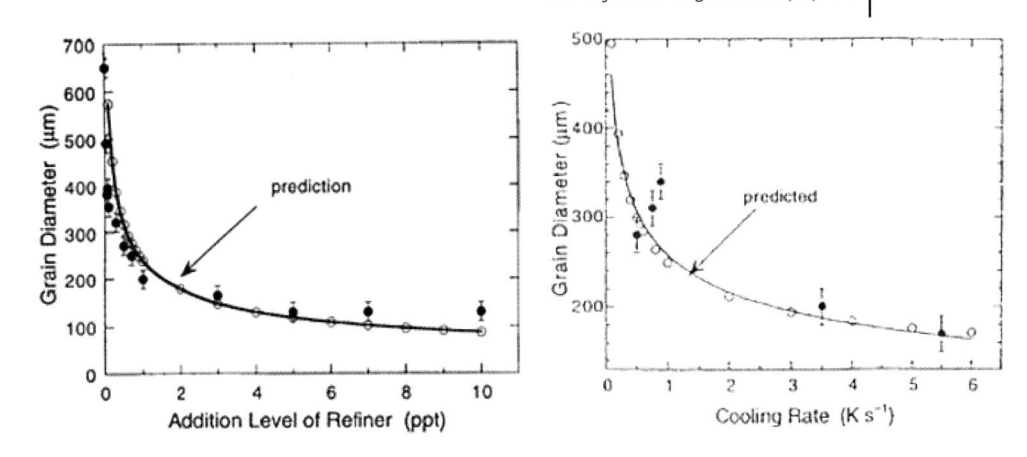

Fig. 3.41 Variation of grain size with refiner quantity and with cooling rate, as predicted by Greer's model and compared to experimental values (Greer et al. 2000).

seed nucleation because the undercooling does not decrease sufficiently. In practice only 5% of the particles are useful for seeding, but, as they are the biggest ones, these represent a significant part of the refiner mass.

The prediction of the grain size versus quantity of refiner and cooling rate agrees quantitatively with the experimental observations (Fig. 3.41) and shows that above a quantity of about 1 ppt (part per thousand) (by weight) of refiner, or a cooling rate of 1 K s^{-1}, a further increase is not useful for decreasing the grain size. It is worth noting that the model does not use any adjustable parameter.

It should be noted that the model is based on the hypothesis that the nucleation on the particle is easy (i.e., the contact angle is low) and therefore cannot be applied if the nucleation barrier is too large. It has been remarked that certain solutes, such as Zr, have a strong effect on the nucleation properties and thus act as killers of the refining effect.

3.3.2.2 Columnar to Equiaxed Transition

As shown in the introduction to Section 3.3.2, columnar and equiaxed transitions are competing at some moment of the ingot solidification. It therefore is important to know when the transition between the two structures will occur and what the parameters influencing this transition are.

The classical model has been proposed by Hunt (1984). A columnar dendritic interface is growing; its growth rate is imposed by the heat extraction through the mold (constrained growth). The dendrite tip undercooling is self-adjusted to the given velocity (see Section 3.2.2.2). A region of undercooled liquid is thus created ahead the columnar front, whose extent and undercooling depend on the solidification rate and temperature gradient. It is supposed that solid seeds are nucleating (in the presence of refining particles) in this undercooled region and grow as equiaxed dendrites. The structure will remain columnar if the volume of

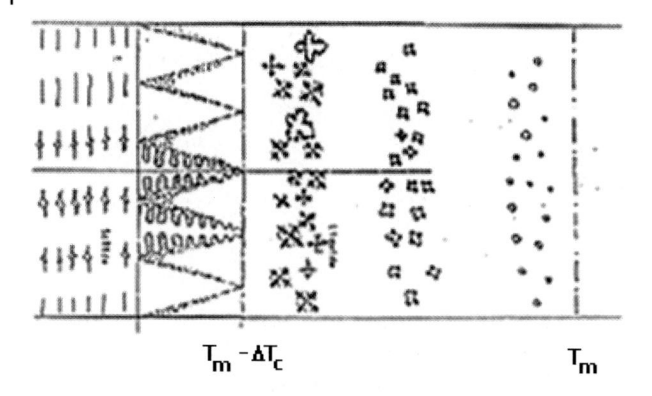

Fig. 3.42 Schematic plot of the competition between columnar and equiaxed growths. The columnar front is undercooled by ΔT_c and seeds nucleate in the liquid below the melting temperature T_m and then grow, as the columnar front is approaching.

equiaxed grains is negligible compared with the advancement of the columnar front. Conversely, if the equiaxed grains grow fast enough to occupy the major part of the volume before being passed by the columnar font, the resulting structure will be equiaxed.

The nucleation rate is expressed, from Eq. (53), taking into account the remaining number of free nucleating particles, as Eq. (57).

$$\frac{dN}{dt} = (N_0 - N) I_0 e^{-\Delta G_n/k_B T} \tag{57}$$

An undercooling is necessary for nucleation [ΔG_n varies as $1/\Delta T_N^2$; compare Eq. (3.51)]. At the columnar dendrite tip, the undercooling is maximal and equal to ΔT_c and it follows that there are two situations:
- if $\Delta T_N > \Delta T_c$, the equiaxed dendrites cannot nucleate and no equiaxed grain will appear
- if $\Delta T_c > \Delta T_N$, seeds will nucleate at the forefront of the columnar tips, and as Eq. (30b) states, the growth rate of these equiaxed dendrites is proportional to the square of the local undercooling [Eq. (58)].

$$\frac{dr}{dt} = \frac{D_l}{4 C_0 m_l (k-1) \Gamma} \Delta T^2 \tag{58}$$

This local undercooling varies between ΔT_N and ΔT_c at a rate depending on the velocity of the columnar front and on the thermal gradient in the liquid [Eq. (59)].

$$\frac{d\Delta T}{dt} = v_i \nabla T \tag{59}$$

3.3 Defects in Single and Polycrystals

When the equiaxed dendrite comes into contact with the columnar front, its radius is given by Eq. (60) and the volume of equiaxed grains formed per unit of time by Eq. (61).

$$r = \frac{D_l}{4C_0 m_l(k-1)\Gamma} \int_0^t \Delta T^2 \, dt = \frac{D_l}{4C_0 m_l(k-1)\Gamma} \int_{\Delta T_N}^{\Delta T_C} \frac{\Delta T^2}{v_i \nabla T} d\Delta T \quad (60)$$

$$V_{eq} = N_0 \frac{4}{3}\pi r^3 \quad (61)$$

Hunt considered that the structure will be equiaxed if this volume is more than 66% of the total volume solidified per unit time (at the velocity v_i) and columnar if it is less than 0.66%. It finally follows that the structure is columnar if Eq. (62a) is satisfied, and equiaxed if Eq. (62b) is true.

$$\nabla T > 0.617(100 N_0)^{1/3}(1 - \Delta T_N^3/\Delta T_c^3)\Delta T_c \quad (62a)$$

$$\nabla T < 0.617 N_0^{1/3}(1 - \Delta T_N^3/\Delta T_C^3)\Delta T_c \quad (62b)$$

Between these two values, the structure is a mixture of columnar and equiaxed grains. ΔT_c is linked to the growth rate v_i through Eqs. (27) and (28).

The effect of the leading parameters is plotted in Fig. 3.43. For a given thermal gradient, the equiaxed growth is possible only at high velocity, because at low velocity the undercooling is too small to allow for seed nucleation. Increasing the thermal gradients helps the columnar growth, because the thickness of the undercooled layer depends on the thermal gradient. This is the reason why equiaxed grains are observed frequently at the end of the solidification, when the liquid becomes almost isothermal.

The limiting velocity below which no equiaxed grain can nucleate decreases when the solute concentration increases and when the nucleation undercooling decreases (Fig. 3.44).

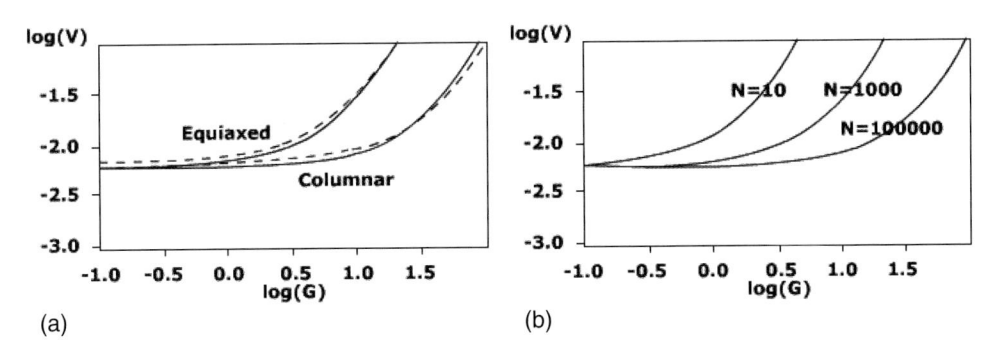

Fig. 3.43 (a) Equiaxed and columnar domains in a $v_i/\nabla T$ diagram, and (b) effect of the number of refining particles on the equiaxed side of the transition. Al–3%Cu (Hunt 1984).

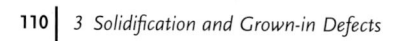

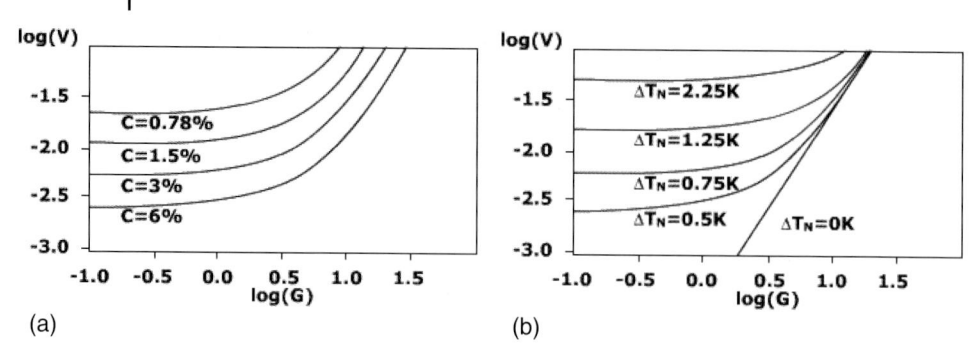

Fig. 3.44 Equiaxed and columnar domains in diagrams showing growth rate versus thermal gradient. (a) Effect of the alloy composition and (b) of the nucleation undercooling on the equiaxed side of the transition. Al–3%Cu (Hunt, 1984).

The model neglects convective effects, which are known to be strong, because in reality there is transport of the smallest equiaxed grains by the convective flow, into regions where they may melt back. Larger grains tend toward sedimentation. In any case, the model is useful as it helps an understanding of the leading parameters acting on the columnar equiaxed transition. Experiments performed under microgravity conditions in order to avoid any convective perturbations have made it possible to discuss the validity of the model, which is limited by its sensitivity to the initial parameters (undercooling and number of seeds) (Dupouy and Camel 2001).

3.3.3
Macro- and Mesosegregation

In this section, perturbations of the columnar dendritic field are discussed. The region between the tip and the bottom of the dendrite, where solid and liquid exist simultaneously, is often called the "mushy zone". Neglecting the undercooling of the tip, its thickness can be calculated through Eq. (63) (see Fig. 3.16).

$$a = \frac{\Delta T_0}{VT} \tag{63}$$

This mushy zone in certain solidification processes may reach several decimeters. The thermal and chemical characteristics of the mushy zone are directly related to the dendrite, as studied in Section 3.2.2. However it also has a strong interaction with the flow field, acting as a porous medium (Darcy flow), with the particularity that the permeability depends on the solidification conditions, which in turn are influenced by the flow field (see Fig. 3.13). By changing the thermal and solutal fields around the dendrite, and inside the dendritic field, convective flows have a strong influence on the solidification conditions. The coupling of

3.3 Defects in Single and Polycrystals

the interdendritic flow with hydrodynamics in the bulk liquid leads to defects in the material and chemical segregations on the ingot scale. The chemical segregation is called "positive" where the local chemical composition is higher than the mean ingot concentration, and "negative" where it is lower.

Independently of forced convection that may be imposed on the system (through mechanical, electromagnetic, or other processes), natural convection always exists because of the thermal field perturbation by the solidification front, but also, and even more so, because of the chemical gradients that induce variations in the liquid density. Both thermal and solutal fields are acting on the liquid density through their expansion coefficients [Eq. (64)].

$$\rho_l = \rho_l^0 (1 + \beta_T \Delta T + \beta_C c_l). \tag{64}$$

Since the chemical and thermal fields are related by the thermodynamic equilibrium hypothesis along the dendrite side [Eq. (3)], Eq. (65) follows.

$$\rho_l = \rho_l^0 \left(1 + \left(\beta_T + \frac{\beta_C}{m_l}\right)\Delta T\right) \tag{65}$$

The thermal expansion coefficient is of the order 10^{-4} K^{-1} and the chemical term in Eq. (64) is of the order of 10^{-2}–10^{-3} K^{-1}. Therefore the thermal convection within the mush can generally be neglected. The resulting Archimedes force is calculated through Eq. (66).

$$F_p = \beta_C \Delta T g / m_L \tag{66}$$

It follows that the convective effects will depend on the orientation of the solute gradients relative to gravity.

From a solute balance at the local scale, Flemings and Nereo (1967) derived the equation of solute distribution [Eq. (67)], taking into account the local convective flow.

$$\frac{\partial f_l}{\partial c_l} = -\frac{(1-\delta_p)}{(1-k)}\left(1 + \frac{v_l.\nabla T}{v_i.\nabla T}\right)\frac{f_l}{c_l}, \tag{67}$$

Here δ_p is the solidification shrinkage, and v_l the liquid velocity, which is given by Darcy's law, with a permeability factor depending on the liquid fraction according to Eq. (68).

$$v_l = \frac{-K}{\mu f_l}(\nabla p - \rho_l g) \quad K = a f_l^2 \tag{68}$$

If there is no convection and no shrinkage, Eq. (67) is the differential form of the classical Scheil's law, Eq. (10). If convection is insignificant or parallel to the thermal gradient, integration of Eq. (67) gives a modified Scheil's law, Eq. (69).

$$c_l = c_0(1 - f_s)^{(k-1)/(1-\delta_p)} \tag{69}$$

Transition toward the convective regime occurs when the fluid velocity becomes of the same order as the growth rate.

In order to show the effect of the orientation of the thermal gradient (or equivalently, of the growth direction) versus gravity, examples of solidification of Al–Cu alloys grown for various configurations are shown below. Considering the complexity of the interactions and coupled mechanisms involved, only qualitative arguments will be given.

Six configurations exist, depending on the direction of solidification (horizontal, vertical upward, and vertical downward) and on the alloy concentration: hypo-eutectic (the rejected Al is lighter than the Cu solvent) or hyper-eutectic (the rejected Cu is heavier than the Al solvent). These can be grouped into three different convective situations: horizontal, vertical with solute stabilization of the liquid (Al-rich downward or Cu-rich upward), and vertical with solute destabilization of the liquid (Al-rich upward or Cu-rich downward).

The hypo-eutectic horizontal case is shown in Fig. 3.45. Dendrites are located at the bottom of the ingot cross-section because the flow, driven by the rising of the Al-enriched liquid, increases the chemical composition at the top, where the eutectic grows. Conversely, in the case of a hyper-eutectic alloy, the eutectic is located at the bottom and the dendrites at the top.

In the vertically destabilizing case (Fig. 3.46) the liquid is unstable and driven away from the interface, which creates convective chimneys in the dendritic field. The solute is carried through the mush to those particular places, where the concentration increases and a purely eutectic structure may eventually form. This causes freckles in the solid; such freckles are extremely detrimental to the mechanical properties of the alloy.

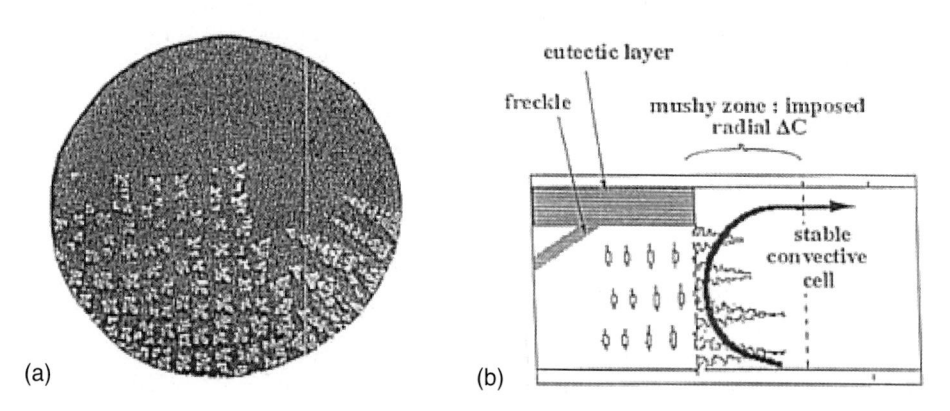

Fig. 3.45 Effect of convection on the structure of a horizontally solidified Al-rich Al–Cu alloy; (a) metallographic picture in cross-section; (b) sketch of the flow and its effect on the microstructure (Dupouy and Camel 2000).

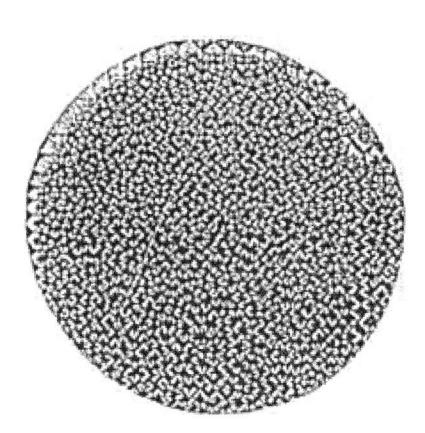

Fig. 3.46 Effect of convection on the structure of an Al–Cu alloy grown under vertically destabilized configuration: downward solidification of a hyper-eutectic alloy. One freckle appears here as a region with smaller dendrites (Dupouy and Camel 2000).

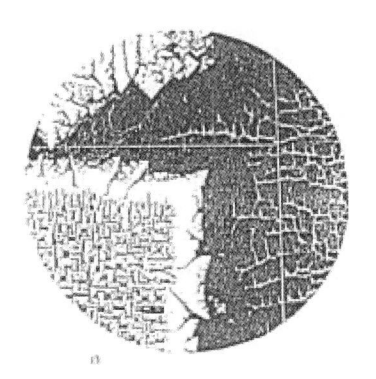

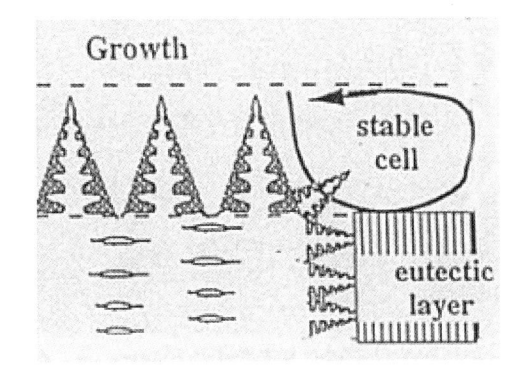

Fig. 3.47 Effect of convection on the structure of an Al–Cu alloy grown under vertically stabilizing configuration: upward solidification of a Cu-rich alloy (Dupouy and Camel, 2000).

The vertically stabilizing case is shown in Fig. 3.47. No interdendritic flow would be expected if the isoconcentrations were strictly planar and horizontal. However, defects inside the dendritic field induce radial solute gradients which force convective loops in the mushy zone, and cause radial segregation, possibly leading to the formation of purely eutectic regions as in the example shown. However, contrarily to the previous case, the average concentration in a cross-section remains equal to the nominal concentration (absence of longitudinal segregation).

3.4
Outlook

The objectives of this chapter have been to introduce the reader to the fundamental aspects of solidification in a simplified way, to show the basic physical phenomena involved in structure formation of alloys, and to explain the generation of defects during solidification.

The most important parameter determining all the solidification-related phenomena is the undercooling of the solid–liquid interface, which depends essentially on the local chemical composition and curvature. The kinetic undercooling is negligible in the case of metallic alloys which present a rough interface, but may reach several Kelvins for materials exhibiting faceted growth, such as intermetallics or semiconductors.

Therefore the structuring of the alloy is the result of the competition between:
- diffusion in the liquid, which governs the chemical field close to the interface: solutal undercooling decreases as microstructure size decreases
- capillary forces: capillary undercooling decreases as microstructure size increases.

This competition is active continuously in all structuring processes: interface destabilization and formation of dendrites, cells, eutectics, and grains.

Simultaneously, the solid–liquid interface interacts with the surroundings. The essential external parameter acting on solidification is the thermal field applied to the alloy through the heat fluxes extracted, or generated, in the growth facility. This heat extraction is responsible for the spatial temperature distribution, acting on most of the structuring processes through the thermal gradient, and for the temperature variation with time, which fixes the growth rate.

Defects are related to foreign elements, which lead to chemical segregation, nucleation of particles or bubbles, or act as nucleating agents. A second effect concerns the interaction of convection with the solidification process through perturbation of the thermal and chemical fields. It should be kept in mind that the convective patterns are often coupled with the solidification process. This is a consequence of the solute rejection in the liquid, which makes the problem extremely difficult to tackle.

Owing to such complexity, the present development in solidification research tends essentially toward numerical simulation. This is the only way to get explanations or predictive conclusions in the industrial processes, which are geometrically complex and time-dependent. Currently, software is being developed in order to fully solve the problem in real configuration, computing the temperatures, velocities, solute field, and structures (dendrite or eutectic sizes, etc.). The models generally take into account the mechanical evolution of the material after solidification and the associated defect generation.

Numerical simulation is also at the forefront of more fundamental research, especially by the use of phase field simulation techniques, which give new insight into the structuring phenomena and their interaction with the surroundings.

Fig. 3.48 Phase field simulation of equiaxed growth. Three equivalent seeds are introduced into an undercooled melt. At the end of the growth, the equiaxed grains are closely connected to each other. The blue color in the last picture represents liquid which is unsolidified due to solute rejection and trapping in the dendrite arms or between the grains.

This numerical technique is particularly well adapted to take into account the complex shape of the solid–liquid interface. An example of simulation of equiaxed growth is given in Fig. 3.48. A comprehensive review of this contribution can be found in Boettinger et al. (2002) and partially in the introduction to phase field methods given in Chapter 10.

Acknowledgments

The author expresses his gratitude to Dr. Denis Camel and Dr. Carmen Stelian for their help in carefully reading and correcting this chapter. The help of the Editor, Prof. Pfeiler, has also been very much appreciated.

Nomenclature

a	lattice parameter, structural unit size
b	Burgers vector
c	chemical composition
f	fraction (of solid or liquid)
g	gravity vector
h	vacancy or hole
k	segregation coefficient
k_B	Bolzmann constant
m	liquidus or solidus slopes
p	pressure
r	radius (of nuclei, of curvature, of dendrite tip ...)
u	fluid velocity
v	solid–liquid interface velocity
x	mole fraction
x, y, z	axes
C_p	heat capacity
D	diffusion coefficient
E	Young's modulus
Gr	Grashof number
G	Gibbs free energy
H	enthalpy
I	nucleation rate (s^{-1})
J, j	flux and flux density
K	kinetic coefficient or permeability coefficient in Darcy's law
N	a number (of nuclei, particles etc.)
Pe	Peclet number
R	ideal gas constant
Re	Reynolds number
S	entropy
T	temperature
V	volume (V_a, atomic volume)
α	Jackson's reduced transformation entropy
α	thermal dilatation coefficient
β	solutal dilatation coefficient
γ	surface energy (solid–liquid, liquid–gas etc.)
δ	boundary layer thickness
δ_p	relative density jump aty the interface $(\rho_s - \rho_l)/\rho_l$
ε	fluctuation, strain, perturbation
κ	curvature ($1/r$)
λ	characteristic solidification structure periodicity or length
μ	viscosity
σ	stress

Γ	Gibbs parameter ($\gamma/\Delta S$, Km)
Λ	thermal conductivity
$\Delta G, \Delta H, \Delta S$	formation/migration/etc. energy values (differences)
ΔT	undercooling
Δc	supersaturation
θ	contact angle
Ω	reduced supersaturation
∇	gradient

Subscripts or superscripts

cap	capillary
s	solid
l	liquid
i	at the solid–liquid interface
m	melting
0	reference
ch	chemical
S	surface
T	thermal

References

Aaron H. B., Fainstein D., Kotler G. R. (1970), *J. Appl. Phys.* **41**, 4404–4410.

Access-Aachen, from the CD-Rom *Materials in Focus*, Access-RWTH, Intzestr., 5, Aachen, Germany.

Aziz M. J. (1982), *J. Appl. Physics* **53**, 1158–1168.

Bennema P. (1993), "Growth and morphology of crystals" in *Handbook of Crystal Growth Vol. 1a*, D. T. J. Hurle Ed., North Holland, pp. 477–581.

Boettinger W. J., Warren J. A., Beckermann C., Karma A. (2002), *Annu. Rev. Mater. Res.* **32**, 163–194.

Boiton P., Giacometti N., Duffar T., Santailler J. L., Dusserre P., Nabot J. P. (1999), *J. Crystal Growth* **206**, 159–165.

Burton W. K., Cabrera N., Frank F. C. (1951), *Phil. Trans. Roy. Soc.* **A243**, 299–358.

Burton J. A., Prim R. C., Schlichter W. P. (1953), *J. Chem. Phys.* **21**, 1987–1991.

Cahn J. W. (1960), *Acta Met.* **8**, 554–562.

Chernov A. A., De Yoreo J. J., Rashkovich L. N., Vekilov P. G. (2004), *MRS Bull.* **29**, 927–934.

Coriell S. R., McFadden G. B. (1993), "Morphological stability" in *Handbook of Crystal Growth Vol. 1b*, D. T. J. Hurle Ed., North Holland, pp. 785–857.

Dash W. C. (1959), *J. Appl. Phys.* **30**, 459–474.

Duffar T., Dusserre P., Giacometti N., Boiton P., Nabot J. P., Eustathopoulos N. (1999), *J. Crystal Growth* **198–199**, 374–378.

Dupouy M. D. (1986), PhD Thesis, INP-Grenoble (March 27).

Dupouy M. D., Camel D. (2000), *Recent Developments in Crystal Growth Research*, Ed., **2**, 179–210, Transworld Research Network.

Dupouy M. D., Camel D. (2001), *J. Phys. IV* **11**, 119–126.

Dupouy M. D., Camel D., Favier J. J. (1992), *Acta Metall. Mater.* **40**, 1791–1801.

Dupouy M. D., Camel D., Mazille J. E. (2003), *Proc. Int. Conf. Modelling of Casting, Welding and Adv. Solidification Process X*, Destin, Florida, USA, May 25–30, D. Stefanescu et al. Ed.,TMS, pp. 685–692.

Esaka H. (1986), "Dendrite growth and spacing in succinonitrile–acetone alloys," PhD Thesis, École Polytechnique Fédérale de Lausanne, Switzerland.

Feurer U., Wunderlin R. (1977), *Fachb. Deuts. Ges. Metallk.*, Oberursel, FRG, cited in Kurz and Fischer (1998).

Flemings M. C. (1974), *Solidification Processing*, Mc Graw-Hill, New York.

Flemings M. C., Nereo G. E. (1967), *Trans. Met. Soc. AIME* **239**, 1449, cited in Flemings (1974).

Gandin C. A., Rappaz M. (1994), *Acta Metall. Mater.* **42**, 2233–2246.

Garandet J. P., Favier J. J., Camel D. (1994), "Segregation phenomena in crystal growth from the melt" in *Handbook of Crystal Growth Vol. 2b*, D. T. J. Hurle Ed., North Holland, pp. 659–707.

Gerardin S., Combeau H., Lesoult G. (2001), *J. de Physique IV* **Pr6**-143.

Gondet S., Louchet F., Théodore F., Van den Bogaert N., Santailler J. L., Duffar T. (2003), *J. Crystal Growth* **252**, 92–101.

Greer A. L., Bunn A. M., Tronche A., Evans P. V., Bristow D. J., *Acta Mater.* (2000) **48**, 2823–2835.

Hoyt J. J., Asta M., Haxhimali T., Karma A., Napolitano R. E., Trivedi R., Laird B. B., Morris J. R. (2004), *MRS Bull.* **29**, 935–939.

Huang S. C., Glicksman M. E. (1981), *Acta Met.* **29**, 701–715 and 717–734.

Hunt J. D. (1984), *Mater. Sci. Eng.* **65**, 76–83.

Hurle D. T. J. (1995), *J. Crystal Growth* **147**, 239–250.

Hurle D. T. J., Rudolph P. (2004), *J. Crystal Growth* **264**, 550–564.

Ivantsov G. P. (1947), *Dokl. Akad. Nauk USSR* **58**, 567–569.

Jackson K. A. (1958) in *Liquid Metals and Solidification*, American Society of Metals.

Jackson K. A., Hunt J. D. (1966), *Trans. Met. Soc. AIME* **236**, 1129–1142.

Jackson K. A., Hunt J. D., Uhlmann D. R., Seward T. P. (1966), *Trans. Met. Soc. AIME* **236**, 149–158.

Johnson W. L. (1999), *Mater. Res. Soc. Symp. Proc.* **554**, 311–339.

Jones H. (1984), *J. Mater. Sci.* **19**, 1043–1076.

Kattamis T. Z., Flemings M. C. (1965), *Trans. Met. Soc. AIME* **233**, 992–999.

Kurz W., Fischer D. J. (1998), *Fundamentals of Solidification*, TransTech Publications.

Kurz W., Trivedi R. (1990), *Acta Metall. Mater.* **38**, 1–17.

Lagowski J., Gatos H. C., Parsey J. M., Wada K., Kaminska M., Walukiewicz W. (1982), *Appl. Phys. Lett.* **40**, 342–344.

Langer J. S., Müller-Krumbhaar H. (1977), *J. Crystal Growth* **42**, 11–14.

Maxwell I., Helawell A. (1975), *Acta Met.* **23**, 229–237.

Mullins W. W., Sekerka R. F. (1964), *J. Appl. Phys.* **35**, 444–451.

Rappaz M., Gandin C. A. (1993), *Acta Metall. Mater.* **41**, 345–360.

Scheil E. (1942), *Zeit. Metall.* **34**, 70–72.

Sekerka R. F. (1965), *J. Appl. Phys.* **36**, 264–268.

Tiller W. A., Jackson K. A., Rutter J. W., Chalmers B. (1953), *Acta Met.* **1**, 428–437.

Trivedi R., Somboonsuk K. (1984), *Mater. Sci. Eng.* **65**, 65–74.

Völkl J. (1994), "Stress in the cooling crystal" in *Handbook of Crystal Growth Vol. 2b*, D. T. J. Hurle Ed., North Holland, pp. 821–874.

Wenzl H., Dahlen A., Fattah A., Petersen S., Mika K., Henkel D. (1991) **109**, 191–204.

4
Lattice Statics and Lattice Dynamics

Véronique Pierron-Bohnes and Tarik Mehaddene

4.1
Introduction: The Binding and Atomic Interaction Energies

The cohesion of solids is mainly due to the electrostatic attraction between the negatively charged electrons and the positively charged nuclei. The magnetic forces and the repulsions between electrons have a weak effect on cohesion but are to be taken into account to give a good description of the total energy of the crystal, and are responsible for the relative stability of the different phases. The cohesive energy is the energy that must be added to the crystal to separate its different components into neutral atoms at rest at an infinite distance (Kittel 1976).

Crystals of inert gases have a cohesion based on polarization effects leading to polar interactions between induced dipole moments (van der Waals forces). Their cohesive energy is some tens of meV per atom. Ionic crystals are composed of positive and negative ions with strong electrostatic interactions (the Madelung energy) and a strong cohesion of tens of eV per atom. Covalent crystals also have strong bonds (some eV per atom). Two atoms share a pair of electrons, one electron coming from each atom. Covalent bonds are very directional and the pair of electrons can be found with high probability near the bond line. This makes possible the stability of very open structures, such as the diamond structure.[1]

In metals, the atoms are not completely ionized but the outer-shell electrons are free to move across the sample, giving rise to a high electrical conductivity.

---

1) An open structure is a non-compact structure. The compactness is defined as the ratio between the volume occupied by the atoms, considered as spheres in contact, and the total volume of the crystal cell. The two most compact structures are the face-centered cubic (fcc) structure and the hexagonal close-packed structure (hcp, with a c/a ratio of 1.633). Both have a compactness of 0.74, whereas the diamond and the body-centered cubic (bcc) structures have a compactness of 0.34 and 0.68, respectively.

Alloy Physics: A Comprehensive Reference. Edited by Wolfgang Pfeiler
Copyright © 2007 WILEY-VCH Verlag GmbH & Co. KGaA, Weinheim
ISBN: 978-3-527-31321-1

In addition to the electrostatic energy between all the charges present, a repulsive interaction between electrons has to be taken into account because of Pauli's exclusion principle: two electrons at the same position cannot have all their quantum numbers equal. When two atoms come close, their electronic distributions overlap and an electron initially belonging to one atom can occupy a free orbital of the other atom. Since free states have comparatively high energy levels this induces a strong repulsive force between atoms at a very close distance.

Moreover, the electronic levels of the atoms, initially all at the same energy, undergo a degeneracy break for the outer electronic shell when the atoms come close (as in a diatomic molecule, an atomic level at E separates in a bonding level at $E - \delta E$ and an anti-bonding level at $E + \delta E$). This induces a lowering of the energy of the valence electrons, because the valence band is not full. In the transition metals and their alloys, the binding energy is due to the outer s and p shells but also contains a contribution of the inner more localized d or f shells. The cohesion energy varies considerably depending on the metal: it is typically of 1 eV in alkali metals and as high as 8–9 eV in refractory metals such as Ta, W, Re, and Os. These aspects will be discussed in more detail in Chapter 11.

The crystal structures of metals are generally compact: fcc or bcc. As the atoms approach to a close distance, the energy decreases due to hybridization and to the Madelung energy. This binding energy is thus responsible for an attractive force whereas Pauli's exclusion principle is responsible for a repulsive force. The equilibrium distance corresponds to the exact compensation of these two forces. Metal bonds have no definite directions relative to the central atom; as in the covalent crystals, there are no forces perpendicular to the lines connecting two neighboring atoms. The relative stability of the structures is determined by the values of the total energy in the different possible structures. This induces a great variability of the metal structures in comparison to ionic, van der Waals, or covalent crystals: the phase diagrams of metallic alloys are thus often richer and more complex than the phase diagrams of systems forming, for example, ionic compounds. For this reason too, metals and alloys are less brittle, more elastic, and more plastic than covalent structures. In alloys, bonding is often mixed and there is some ionic or covalent contribution due to the electropositivity difference between the different metals, for example.

The evaluation of the total energy $\mathscr{U}$ of an intermetallic alloy is quite complex because there is a multitude of electrons interacting with many nuclei. It is necessary to make some approximations when writing the Hamiltonian and when calculating the total energy from this Hamiltonian in the different accessible states.

On the basis of the reasoning of Born and Oppenheimer (1927), many quantum mechanical calculations of the electronic structure of a system of atoms treat the nuclei as classical point masses. The energy of a system of atoms is adiabatically parameterized by the positions of the nuclei (adiabatic approximation). Accurate methods exist for calculating the energy of a few atoms in some configurations; an overall evaluation of the energy hypersurface, however, is impossible and, even if it was carried out, one would not gain physical insight.

Several approximations to the many-body problem have been developed (Carlsson 1990; Moriarty et al. 2002; Pettifor et al. 2002) designed to describe certain aspects of interest of a given system. The different approaches used to approximate the adiabatic energy hypersurface may be classified accordingly in three groups.

- The total energy is decomposed into many-body potentials, i.e., pair, three-body, four-body ..., N-body potentials. Carlsson (1990) denotes these many-body potentials as cluster potentials. However, there has been no unique prescription for constructing a many-body potential which can be used for any configuration of a given system. Moreover, the many-body potentials in general exhibit a rather slow convergence.
- Effective potentials show a faster convergence. Carlsson (1990) refers to this large class of potentials as cluster functionals. The energy is written as a sum of a pair interaction term and a rest term. The rest term is written formally as the sum of contributions of single atoms, the contribution of a single atom depending on the surrounding of the atom as a function of atom pairs, triplets etc. Again no clear prescription of how to construct cluster functionals exists. They are often constructed according to physical intuition and therefore their transferability is often limited. It has been demonstrated recently (Drautz et al. 2004) that by using an adequate definition of perfectly transferable many-body potentials, a systematic and objective comparison of the existing potentials could be performed.
- The cluster expansion method writes the total energy as a sum of contributions over all conceivable clusters in the system. This method has been used mainly in the past with a frozen lattice: the atoms are fixed on certain lattice positions and there are only chemical degrees of freedom. The cluster expansion method was introduced by Sanchez et al. (1984). Extensive work was also done by Zunger and coworkers (Lu et al. 1991; Zunger 1994) in an effort to handle the effect of local relaxations and long-range elastic interactions. For practical calculations the cluster expansion has to be terminated at a maximum cluster; Finel (1987, 1994) has proposed a method to produce a convergent series of approximations. More details can be found in Chapter 11 or in Mueller (2003).

In any case, the simplest interaction Hamiltonian [Eq. (1)] is based on empirical pairwise energies and has been shown to be valid in many metals (Ducastelle 1970; Bieber and Gautier 1984).

$$H = \frac{1}{2} \sum_{I,J,l,m} \varepsilon_{IJ}^{lm} p_I^l p_J^m \qquad (1)$$

Here the summation runs on the atomic species I, J and the sites l, m. p_I^l is the occupation operator of site l by atom I. ε_{IJ}^{lm} is the pair interaction energy between atoms I, J at sites l, m. It depends on the positions of the sites. In disordered alloys, an isotropic approximation has to be made with an average lattice parameter. Often, due to the size mismatch of the constituents of an alloy, the atoms do not take their ideal lattice positions, but relax into a lower energy state where the atomic positions are displaced from their ideal lattice sites. The relaxation energy of a given configuration is defined as the energy difference between the energy of the configuration of atoms on an ideal undistorted lattice and the energy of the configurations with the atomic positions locally relaxed in the whole crystal. For many systems, the calculation of phase diagrams requires local relaxations in the cluster expansion coefficients to be taken into account (Lu et al. 1991; Zunger 1994). The relaxations can be taken into account in molecular dynamics, for example, for pure metals or ordered alloys, by using periodic conditions. In disordered alloys, the size of the system is quite limited and the calculation is performed by averaging on several small samples with random filling of the lattice sites (Grange et al. 2000; Galanakis et al. 2000).

This point will be developed in more detail in Chapter 10 and we will only give here the general form of the total energy as a function of the lattice parameter in Pt and CoPt. The calculation can be made using molecular dynamics in the second moment approximation (Tréglia et al. 1988; Goyhenex et al. 1999). In the tight binding formalism (Friedel 1969), this energy is written as the sum of two terms, an attractive band energy and a repulsive pair interaction. The band term is obtained by integrating the local electronic density of states up to the Fermi level and gives rise to the many-body character of the potentials (Legrand et al. 1990). The realistic density of states is replaced by a schematic rectangular density having the same second-moment [Eq. (2)], where the summation is made on sites m within a fixed radius of neighbors of site l (usually up to the second neighbors), r_0 is the nearest-neighbor distance, β_{IJ} are the hopping integrals for atoms I and J, and q_{IJ} are parameters.

$$E_l^b = -\left\{ \sum_m \beta_{IJ}^2 \exp(-2q_{IJ}(r_{lm}/r_0 - 1)) \right\}^{1/2} \qquad (2)$$

The repulsive term is described by a sum of Born–Mayer ion–ion repulsions (Ducastelle 1970) with a schematically parameterized exponential form [Eq. (3)]. The summation is made on sites m within the same radius of neighbors of site l, A_{IJ}, and p_{IJ} are parameters.

$$E_l^i = \sum_m A_{IJ} \exp(-p_{IJ}(r_{lm}/r_0 - 1)) \qquad (3)$$

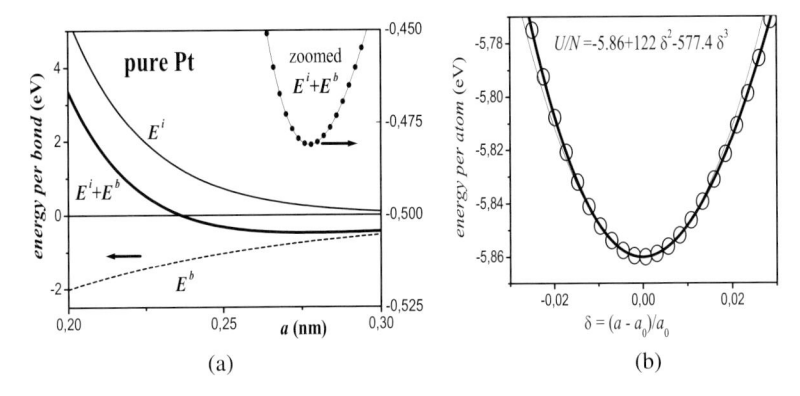

Fig. 4.1 (a) Band E^b and ionic repulsive E^i energy contributions between two atoms in the Pt system as a function of the bond distance. Top right: zoom on the sum showing the minimum at the nearest-neighbor distance 0.277 nm. (b) Calculations using molecular dynamics in pure Pt: variation of the energy per atom with the relative change in lattice parameter (o) and its adjustment to a third-degree polynomial (thick line). Thin line: second-degree part of the polynomial.

The values of the parameters are the same as those used in a recent study on Pt/Co(0001) (Goyhenex et al. 1999). For each pure metal species (Pt, Co), the parameters were determined by fitting the cohesive energy, the lattice parameter, and bulk modulus (Kittel 1976). For mixed pairs (Co, Pt), the parameters are adjusted to fit the order–disorder transition temperature or the dissolution energy of atom I in the pure metal J.

Figure 4.1(a) shows the E^b and E^i contributions per bond in pure Pt and Fig. 4.1(b) the variation of the total energy with the lattice parameter in pure Pt. We see from Fig. 4.1(b) that the total energy is mainly quadratic with a small asymmetry that will be responsible for anharmonicity (see Section 4.4).

In intermetallic compounds and alloys, in ordered phases, the structure may be tetragonal due to either the anisotropy of the Bravais lattice itself or the anisotropic distribution of atoms on the lattice sites. The $L1_0$ structure, for example CoPt shown on the insert in Fig. 4.2(a), consists of a succession of pure Co and pure Pt planes along the [001] direction. This direction is thus no longer equivalent to the two other $\langle 001 \rangle$ directions. The lattice parameter a in the [100] and [010] directions will be a little smaller than c, the lattice parameter in the [001] direction, because of the pure (001) planes of the largest atoms (here Pt).

Figure 4.2 shows the variation of the energy per atom as a function of a and c around its minimum. For each a or c value, the energy per atom value plotted corresponds to the local minimum of $U(c)$ (or $U(a)$) at constant a or c, respectively. On the same graph is plotted the variation of c or a, respectively corresponding to this minimum.

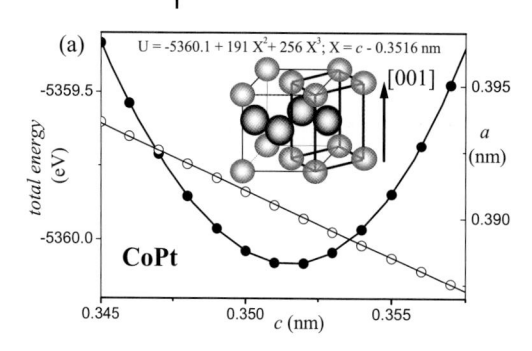

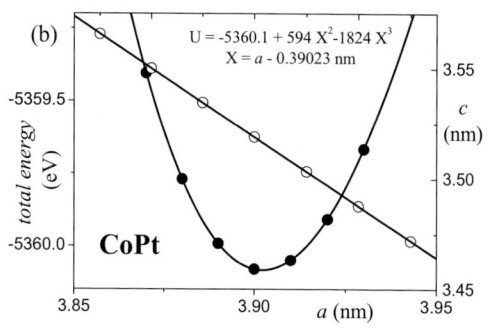

Fig. 4.2 Calculations using molecular dynamics in long-range ordered CoPt. •: Variation of the energy per atom when varying c (a) or $a = b$ (b) and its adjustment to a third-degree polynomial. ○: Variation of one lattice parameter when varying the other one. In (a), the slope $-0.50(1)$ gives $v_a = 0.50(1)$. In (b), the slope $-2v_c/(1-v_c) = -1.56(1)$ gives $v_c = 0.44(1)$. Insert in (a): CoPt L1$_0$ phase: pure planes of platinum atoms (dark) and cobalt atoms (light) alternate along the [001] direction of the initial fcc phase.

4.2
Elasticity of Crystalline Lattices

The elasticity of a metal is directly related to the chemical bonding between atoms and can be considered at both macroscopic and microscopic scales. In this section, we will describe both approaches and make the link between them.

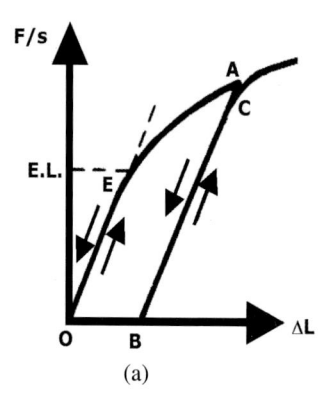

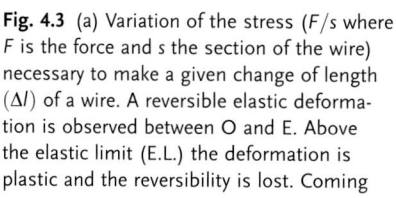

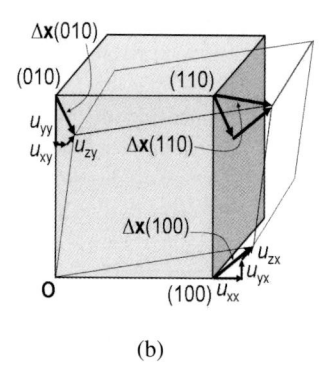

Fig. 4.3 (a) Variation of the stress (F/s where F is the force and s the section of the wire) necessary to make a given change of length (Δl) of a wire. A reversible elastic deformation is observed between O and E. Above the elastic limit (E.L.) the deformation is plastic and the reversibility is lost. Coming back to a zero stress, an elastic range is again observed with the same slope (ABC) but not necessarily with the same range. The slope is a characteristic of the material and corresponds to a deformation of the lattice without any bond change. (b) Example of homogeneous deformation of a solid.

4.2.1
Linear Elasticity

When a stress (force per unit surface) is applied to a solid, its external form is changed. If the intensity of this stress is small enough, this change is reversible (Fig. 4.3). The deformation is then only due to a deformation of the lattice cells without any bond break. But the energetic cost of such a deformation of all lattice cells is very high. Consequently, for higher stresses and deformations, plasticity is observed. It corresponds to a deformation by slipping some lattice cells along high-density atomic planes: this is the motion of a dislocation (see Chapter 6). Very few chemical bonds are broken and reconstructed in the elementary step. The energetic cost of this elementary step is thus not high and the elastic limit is therefore low (generally less than 1%) and depends very much on the dislocation density in the sample, which in turn depends on the mechanical and thermal history of the sample.

4.2.2
Elastic Constants

If the lattice is homogeneously deformed (Fig. 4.3b), we can calculate the position of a point A of the lattice after deformation x' from its initial position x from Eq. (4), where α and β are x, y, or z indices; $u_{\alpha\beta}$ is the strain tensor.

$$x'_\alpha = x_\alpha + \sum_\beta u_{\alpha\beta} x_\beta \qquad (4)$$

Its components are assumed to be small compared with unity. The energy can then be expressed as a Taylor expansion in $u_{\alpha\beta}$. Limiting this expansion to second-order terms and taking into account the symmetry of the problem as well as the different relationships expressing that the energy is minimum, it can be shown (Born and Huang 1968) that the energy density depends only on the symmetrical parameters $u_{\alpha\beta} + u_{\beta\alpha}$.

Voigt proposed to introduce:

$$s_p = u_{pp} \quad \text{for } p = 1 \text{ to } 3 \ (x \text{ to } z)$$
$$s_p = u_{\alpha\beta} + u_{\beta\alpha}, \quad \text{with } p = 9 - \alpha - \beta \text{ for } \alpha \neq \beta \qquad (5)$$

$$\text{(for example: } s_4 = u_{23} + u_{32}\text{)}$$

The components s_p with Voigt's indices are the elastic strains in the classical elasticity theory. They specify the size and shape of a macroscopic specimen as well as of the lattice cell.

In alloys and intermetallic compounds, there may be several atoms within a lattice cell. Equations (4) and (5) describe the deformation of the lattice cell. The components w_α^k of translation of atom k inside the cell obey the translation symmetry but may have a different deformation law than the cell (e.g., the piezoelectric effect). The energy density $\mathscr{U}/V$ contains contributions from both types of deformation:

$$\mathscr{U}/V = \frac{1}{2}\sum_{\alpha,k,\beta,k'} \mathscr{A}_{\alpha\beta}^{kk'} w_\alpha^k w_\beta^{k'} + \frac{1}{2}\sum_{\alpha,k,p} \mathscr{B}_{\alpha p}^k w_\alpha^k s_p + \frac{1}{2}\sum_{p,\sigma} \mathscr{C}_{p\sigma} s_p s_\sigma \qquad (6)$$

k and k' scan the atoms in the lattice cell (the basis of the crystal); $\mathscr{A}_{\alpha\beta}^{kk'}$, $\mathscr{B}_{\alpha p}^k$, $\mathscr{C}_{p\sigma}$ are constants.

The energy density is not affected if an arbitrary vector is added to all the w_α^k. This shows that the energy density depends only on the difference of translation between an origin atom and the other atoms or between the different sublattices. These shifts are microscopic and do not affect the macroscopic dimensions of the sample. They are described as internal strains, as opposed to the external strains s_p. The internal and external strains are coupled through the mixed term in the energy density. Thus, when a body is elastically deformed, some internal strains can be induced. Their values will minimize the total energy. Canceling the derivatives of the energy density as a function of the w_α^k provides linear expressions of the w_α^k as a function of the s_p. Finally, the energy density can be written as Eq. (7), without loss of generality.

$$\mathscr{U}/V = \frac{1}{2}\sum_{p,\sigma} c_{p\sigma} s_p s_\sigma \qquad (7)$$

In the presence of internal strains, the macroscopic elastic constants $c_{p\sigma}$ defined from Eq. (7) contain a contribution due to internal strain. In any case, it is not possible to observe the external strains while keeping the induced internal strains at zero.

This result is the classical expression of elasticity. The stresses are given by the first derivatives of the energy density with respect to the elastic strains [Eq. (8)].

$$S_p = \sum_\sigma c_{p\sigma} s_\sigma \qquad (8)$$

The first three components are the forces per surface unit applied along direction p perpendicularly to the surface normal to the p axis of surface $\mathscr{A}_p$ (Fig. 4.4b): for example, $S_1 = T_{11}/\mathscr{A}_1$. The other three components are the shear stresses: they correspond to forces applied parallel to both opposite surfaces (Fig. 4.4c): for example, $S_6 = T_{12}/\mathscr{A}_1 = T_{21}/\mathscr{A}_2$.

Equation (8) is called Hooke's law: the elastic stresses are linear functions of the elastic strains. The tensor of the elastic constants $c_{p\sigma}$ is symmetric: $c_{p\sigma} = c_{\sigma p}$. There are altogether 21 independent coefficients in the general case. If there are

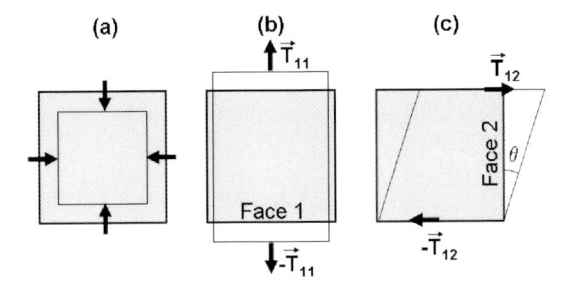

Fig. 4.4 Examples of deformations of a solid: (a) compression, (b) elongation, (c) shear.

no internal strains (this is the ideal defect-free case because the presence of a center of symmetry in the crystal implies the absence of internal strains), the elastic constants verify the Cauchy relationships.

$$c_{23} = c_{44}, \quad c_{13} = c_{55}, \quad c_{12} = c_{66}, \quad c_{14} = c_{56}, \quad c_{25} = c_{64}, \quad c_{36} = c_{45} \tag{9}$$

There are then only 15 independent coefficients in the general case without internal strains (but the Cauchy relationships are generally not fulfiled in metals and fulfiled in some ionic crystals only).

4.2.3
Cases of Cubic and Tetragonal Lattices

In a cubic lattice, Eqs. (10) apply.

$$c_{11} = c_{22} = c_{33}, \quad c_{44} = c_{55} = c_{66}, \quad c_{23} = c_{13} = c_{12},$$
$$c_{56} = c_{46} = c_{45} = c_{24} = c_{34} = c_{35} = c_{15} = c_{16} = c_{26} = 0 \tag{10}$$

There are only three nonzero independent elastic constants.

In the case of Pt, the plot of the energy density, $\mathcal{U}/V$, as a function of c keeping $a = b$ constant (not shown), gives Eq. (11) and induces $c_{11} = 3.468(15)$ in good agreement with the value found in the *CRC Handbook of Chemistry and Physics* (1997): 3.467×10^{11} N m^{-2}.

$$\mathcal{U}/V(10^{11} \text{ N m}^{-2}) = -0.624 + 1.7338(c/c_0 - 1)^2. \tag{11}$$

In a tetragonal lattice, index 3 is no longer equivalent to indices 1 and 2, there are six nonzero independent elastic constants:

$$c_{11} = c_{22} \neq c_{33}, \quad c_{44} = c_{55} \neq c_{66}, \quad c_{23} = c_{13} \neq c_{12} \tag{12}$$

4.2.4
Usual Elastic Moduli

The most famous macroscopic elastic modulus is the bulk modulus K, the inverse of the compressibility κ_T, which is the resistance of the crystal against a volume change under pressure (Fig. 4.4a). This resistance is due to the repulsive term in the interatomic energy: when the atoms are brought nearer, the exclusion principle energy and the exchange energy increase rapidly.

$$K = \frac{1}{\kappa} = -\left[\frac{1}{V}\left(\frac{\partial V}{\partial P}\right)_T\right]^{-1} = V_0\left(\frac{\partial^2 \mathscr{U}}{\partial V^2}\right)_{T, V=V_0} \tag{13}$$

K is directly related to the curvature of the cohesion energy as a function of the volume, or the interatomic distance r, or the lattice parameter a (Fig. 4.1b). If we develop the cohesion energy at the third order in $\delta = (r - r_0)/r_0 = (a - a_0)/a_0$ and write:

$$\mathscr{U} = \mathscr{U}_0 + \xi\delta^2 - \zeta\delta^3 \tag{14}$$

we get $K = 2\xi/9V_0$, where V_0 is the atomic volume.

In the case of Pt in Fig. 4.1(b), the energy per atom,

$$\mathscr{U}/N(eV) = -5.86 + 122\delta^2 - 577\delta^3 \tag{15}$$

and $a_0 = 0.3917$ nm give $K = 2.887 \times 10^{11}$ N m^{-2}, in perfect agreement with the value tabulated by Kittel (1976): 2.88×10^{11} N m^{-2}.

The other elastic constants are also directly related to the interatomic forces, but their calculation is not as easy. Young's modulus E relates the relative change in length of a wire to the tensile force applied to it (Fig. 4.4b); a priori it depends on the direction:

$$s_1 = \frac{T_{11}}{\mathscr{A}_1} = E_1\frac{\Delta x_1}{x_1} = E_1 u_{11} = E_1 s_1 \tag{16}$$

where $\mathscr{A}_1$ is the surface of the section of the wire and T_{11} the force applied to the wire.

Poisson's ratio ν describes the associated change in size in the transverse direction

$$u_{22} = \frac{\Delta x_2}{x_2} = -\nu_{12}\frac{\Delta x_1}{x_1} = -\nu_{12} u_{11} = -\nu_{12} s_1. \tag{17}$$

A priori Poisson's ratios are different if the structure has no cubic symmetry. Writing Eq. (18),

$$s_1 = E_1 s_1 = c_{11} s_1 + c_{12} s_2 + c_{13} s_3 = (c_{11} - \nu_{12} c_{12} - \nu_{13} c_{13}) s_1 \tag{18}$$

leads to Eq. (19)

$$E_1 = c_{11} - \nu_{12}c_{12} - \nu_{13}c_{13} \tag{19}$$

and equivalent relationships for the other directions.

From Eq. (20) we get more relationships, making it possible in simple cases to write Es and νs from the cs.

$$S_2 = c_{21}s_1 + c_{22}s_2 + c_{23}s_3 = (c_{21} - \nu_{12}c_{22} - \nu_{13}c_{23})s_1 = 0 \tag{20}$$

In a cubic structure, we get Eqs. (21), (22), and finally (23).

$$E = c_{11} - 2\nu c_{12} \tag{21}$$

$$\nu = \frac{c_{12}}{c_{11} + c_{12}} \tag{22}$$

$$E = \frac{(c_{11} - c_{12})(c_{11} + 2c_{12})}{c_{11} + c_{12}} \tag{23}$$

Young's modulus and bulk modulus can be related by writing that the change in size in direction ρ is due to the stress S applied on all three faces, with strains additivity.

$$\frac{s_1}{S} = \frac{1}{E_1} - \frac{\nu_{12}}{E_2} - \frac{\nu_{13}}{E_3} \tag{24}$$

and

$$\frac{1}{K} = \frac{-1}{p}\frac{\delta V}{V} = \frac{s_1 + s_2 + s_3}{S} = \sum_\rho \frac{1 - \sum_{\sigma \neq \rho} \nu_{\sigma\rho}}{E_\rho} \tag{25}$$

In the cubic symmetry case, this leads to:

$$K = \frac{E}{3(1 - 2\nu)} \tag{26}$$

In Pt we get $\nu = 0.42$, in good agreement with the values found in tables (Le Neindre): $\nu = 0.396$.

When a shear stress is applied to a cube of the crystal, it is deformed according to Fig. 4.4(c). The shear modulus G quantifies the angle θ describing this deformation: $s_6 = u_{12} + u_{21} = 2\theta$. In first approximation, the deformation occurs at constant volume and we can define the shear modulus G by:

$$s_6 = \frac{T_{12}}{\mathcal{A}_1} = G\frac{\Delta x_2}{x_1} = Gu_{12} = Gs_6 \tag{27}$$

To describe correctly the shear modulus in a simulation, we have to take into account more neighbors than the first two coordination shells.

4.2.5
Link with Sound Propagation

The pressure modulations of sound induce vibrations in solids. These vibrations can propagate, thanks to the elastic properties of materials.

Consider an elemental rod of section $\mathscr{A}$ made of a material with density ρ (Fig. 4.5). The elemental mass $\rho\mathscr{A}.dx_\alpha$, positioned between x_α and $x_\alpha + dx_\alpha$ at rest, is submitted to $-T_{\alpha\alpha}(x_\alpha)$ and $T_{\alpha\alpha}(x_\alpha + dx_\alpha)$ at x_α and $x_\alpha + dx_\alpha$ respectively. We can write the equation of motion for this mass element:

$$T_{\alpha\alpha}(x_\alpha + dx_\alpha) - T_{\alpha\alpha}(x_\alpha) = \rho\mathscr{A}\, dx_\alpha \frac{\partial^2 x'_\alpha}{\partial t^2} \tag{28}$$

and the elasticity law, assuming that the only deformation is along α, is:

$$T_{\alpha\alpha}(x_\alpha) = c_{\alpha\alpha}\mathscr{A}\frac{\partial x'_\alpha}{\partial x_\alpha} \tag{29}$$

Thus we get the propagation equation:

$$\frac{\partial^2 x'_\alpha}{\partial x_\alpha^2} - \frac{1}{v_\alpha^2}\frac{\partial^2 x'_\alpha}{\partial t^2} = 0 \tag{30}$$

where $v_\alpha = (c_{\alpha\alpha}/\rho)^{1/2}$ is the sound propagation velocity in the α direction of the solid. We will see in (Section 4.3) that it is related to the phase velocity of longitudinal phonons in the α direction. In a system without cubic symmetry, the sound

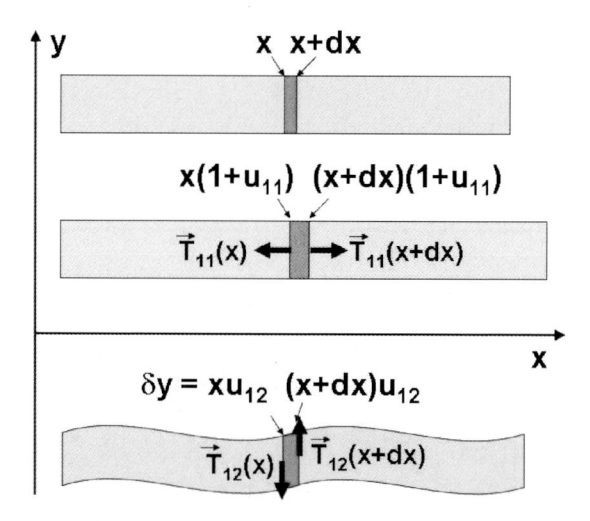

Fig. 4.5 Top: elementary rod at rest; definition of the parameters before the deformation. Middle: changes in the case of a longitudinal wave. Bottom: case of a transversal wave.

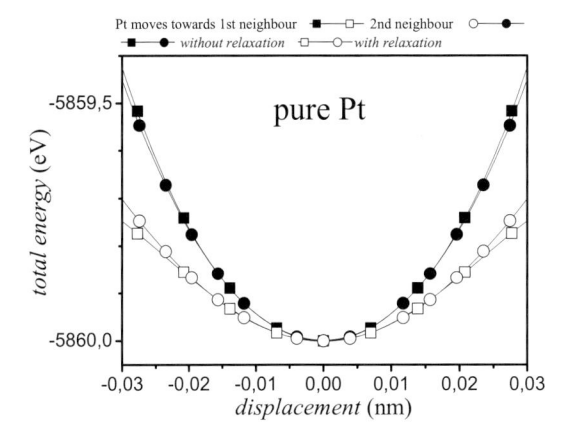

Fig. 4.6 Variation of the total energy when a Pt atom moves in pure Pt. We can see that the relaxation has an important effect on the total energy (black: without relaxation; white: with relaxation) and that the energy well is quite isotropic (circles: toward first neighbor along [1/2 1/2 0] and squares: toward second neighbor along [1 0 0]).

propagates with different speed in different directions, as shown by the different curves of total energy $\mathcal{U}(x)$ when an atom is moved to different positions around its equilibrium position in the lattice.

Whereas in pure Pt the curve is identical in any direction (Fig. 4.6), in an anisotropic alloy such as CoPt the potential varies very differently according to whether it is in the plane or out of the plane (Fig. 4.7). Nevertheless, in the (001)

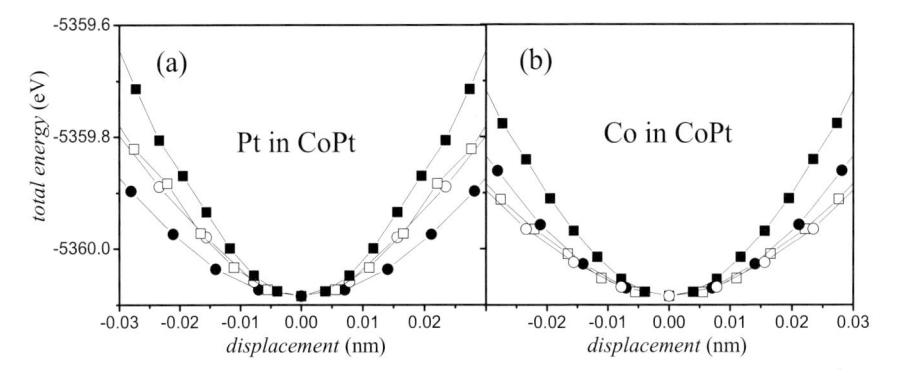

Fig. 4.7 Variation of the total energy in CoPt when a Pt (a) or a Co (b) atom moves in plane (white) or out of plane (black), toward a first (squares) or second (circles) neighbor. (a) For Pt atoms, the energy well is quite isotropic in the plane, but not out of the plane. The flattest is toward second neighbors along [0 0 1]: toward the window of four small cobalt atoms. (b) For Co atoms, the energy well is quite isotropic, except toward first neighbors out of the plane. The elastic energy is the largest when cobalt atoms move toward neighboring platinum atoms.

planes, the [110] (toward first neighbors) and [100] (toward second neighbors) directions are equivalent.

Whereas gases and liquids can propagate only longitudinal waves, solids can also propagate transversal waves, thanks to their shear properties (Fig. 4.5, bottom). The calculations are very similar to the case of a compression wave. We get a propagation velocity $(c_{\alpha\beta}/\rho)^{1/2}$ for transversal wave propagation along α with a vibration along β.

4.3
Lattice Dynamics and Thermal Properties of Alloys

In this section, we present an overview of the theory of lattice vibrations in crystals. First, we describe the semiclassical approach to the lattice dynamics for a three-dimensional crystal in the harmonic approximation. The lattice dynamics theory being relatively heavy with a multitude of indices (Cartesian coordinates, atom numbers, unit cell numbers, etc.), the theory is also applied in the case of a simple diatomic linear chain to point out the important features. The concept of the normal modes of vibration or phonons is then deduced from the quantum description of the lattice vibrations. Calculations of the related thermodynamic quantities such as specific heat, thermal conductivity, and thermal expansion are also presented.

At first glance, a theory of lattice vibrations would appear daunting for a system containing typically 10^{23} ions interacting with each other and with an even huger number of electrons. However, there is a natural expansion parameter for this problem, the ratio of the electron mass to the ionic mass ($m_e/M_i \ll 1$), which allows us to make reliable approximations and to derive a good theory. If we imagine that, at least for small displacements, the forces binding the electrons and the ions can be modeled as harmonic oscillators, then

$$\text{Force} \propto e^2/a^2 \propto m_e \omega_e^2 a \simeq M_i \omega_I^2 a \tag{31}$$

where e is the elementary charge; ω_e and ω_I are the electron and the ion frequencies respectively; a is the lattice parameter. The ratio of the ion frequency to the electron frequency would then be

$$\frac{\omega_I^2}{\omega_e^2} \simeq \left(\frac{m_e}{M_i}\right) \simeq 10^{-3} \text{ to } 10^{-4} \tag{32}$$

This means that ions are essentially stationary during the period of electronic motion. One therefore can assume that the ions are stationary at fixed locations in the lattice and that the electrons will be in their ground state for this particular instantaneous ionic configuration. This approximation, formulated by Born and Oppenheimer (1927), is known as the adiabatic approximation. It is based on the fact that the typical electronic velocities are much greater than the ion velocities,

4.3.1
Normal Modes of Vibration in the Harmonic Approximation

4.3.1.1 Classical Theory

We consider a general three dimensional lattice containing N unit cells and p atoms per unit cell. A unit cell, indexed by l, is situated at position $\mathbf{r}_l = l_1\mathbf{a}_1 + l_2\mathbf{a}_2 + l_3\mathbf{a}_3$, where $\mathbf{a}_1$, $\mathbf{a}_2$, $\mathbf{a}_3$ are the primitive translation vectors. The equilibrium position of the kth atom in the lth unit cell is given by: $\mathbf{r}_{l,k} = \mathbf{r}_l + \mathbf{r}_k$, where $\mathbf{r}_k$ gives the equilibrium position of the k-th atom in the considered unit cell. If now the atom is vibrating around its equilibrium position, its coordinate will be defined by $\mathbf{R}_{l,k} = \mathbf{r}_{l,k} + \mathbf{u}_{l,k}$ where $\mathbf{u}_{l,k}$ is the relative displacement of the kth atom in the lth unit cell around its equilibrium position.

Let us now derive the equation of motion of the atom considered. If we assume that the total potential energy $\mathscr{U}$ of the crystal is a function of the instantaneous positions of all the atoms, we can expand the potential energy in a Taylor series in powers of small displacements around its minimum $\mathscr{U}_0$:

$$\mathscr{U} = \mathscr{U}_0 + \sum_{lk\alpha} \left.\frac{\partial \mathscr{U}}{\partial u_\alpha(lk)}\right|_0 u_\alpha(lk) + \frac{1}{2}\sum_{lk,l'k'}\sum_{\alpha\beta} \Phi_{\alpha\beta}(lk,l'k')u_\alpha(lk)u_\beta(l'k') + \cdots$$

$$= \mathscr{U}_0 + \mathscr{U}_1 + \mathscr{U}_2 + \cdots \tag{33}$$

where the indices α and β stand for the Cartesian coordinates x, y, and z. The equilibrium value, $\mathscr{U}_0$, is a constant and is unimportant for the dynamical problems. The second term of the expansion vanishes in the equilibrium configuration and

$$\Phi_{\alpha\beta}(lk,l'k') = \left.\frac{\partial^2 \mathscr{U}}{\partial u_\alpha(lk)\partial u_\beta(l'k')}\right|_0 \tag{34}$$

If the atomic displacements are now assumed to be small, so that the forces may be regarded as essentially linear functions of the atomic displacements (Hooke's law), the higher orders in Eq. (33) can be neglected. The consideration of only the quadratic term in the above expansion is known as the harmonic approximation. In this scenario, the lattice is treated as a collection of coupled simple harmonic oscillators and the potential energy of the crystal reduces to

$$\mathscr{U} = \frac{1}{2}\sum_{lk,l'k'}\sum_{\alpha\beta} \Phi_{\alpha\beta}(lk,l'k')u_\alpha(lk)u_\beta(l'k'). \tag{35}$$

The force constant matrix elements $\Phi_{\alpha\beta}(lk,l'k')$ have to obey a number of symmetry requirements. First of all, the infinitesimal translational invariance of the crystal (i.e., when all the atoms are equally displaced, there is no force on any atom) leads to

$$\sum_{lk} \Phi_{\alpha\beta}(lk, l'k') = 0 \tag{36}$$

The invariance of the potential energy $\mathscr{U}$ and its derivatives against symmetry operations, such as translations, ensures $\Phi_{\alpha\beta}(lk, l'k') = \Phi_{\alpha\beta}(0k, l - l'k')$. The symmetry operations do not leave only the potential and its derivatives invariant but also the undistorted lattice. Under these symmetry operations, therefore, not only the numerical value of the potential energy but also the form of the expression for it, Eq. (35), is invariant. This invariance implies that the force constant of the ideal lattice does not depend on the the absolute positions of the primitive cells but only on their distance vector. The symmetry of the force constants is technically very important because it reduces significantly the calculation efforts.

If we take the derivative of $\mathscr{U}$ with respect to the relative atomic displacement $u_\alpha(lk)$, we get the force $F_{lk\alpha}$ acting on the atom (lk) in the direction α. In the absence of any external applied field (stress, electric or magnetic field) the equation of motion reads

$$F_{lk\alpha} = m_k \ddot{u}_\alpha(lk) = -\frac{\partial \mathscr{U}}{\partial u_\alpha(lk)} = -\sum_{l'k'\beta} \Phi_{\alpha\beta}(lk, l'k') u_\beta(l'k') \tag{37}$$

where m_k is the mass of the kth atom and Φ is the interatomic force constant matrix, also called the matrix of coupling parameters, defined in Eq. (34). In fact, $\Phi_{\alpha\beta}(lk, l'k')$ represents the force, in the direction α, acting on the kth atom of the lth unit cell due to a unit displacement of the k'th atom of the l'th unit cell in the direction β with all other atoms in their equilibrium positions.

The equation of motion (37) can be written in a symmetrized form as

$$\ddot{w}_\alpha(lk) = -\sum_{l'k'\beta} \mathscr{D}_{lk\alpha}^{l'k'\beta} w_\beta(k'l') \tag{38}$$

where $w_\alpha(lk) = \sqrt{m_k} u_\alpha(lk)$ and the matrix

$$\mathscr{D}_{lk\alpha}^{l'k'\beta} = \frac{1}{\sqrt{m_k m_{k'}}} \frac{\partial^2 \mathscr{U}}{\partial u_\alpha(kl) \partial u_\beta(k'l')} \tag{39}$$

is called the dynamical matrix. The matrix $\mathscr{D}$ is real and symmetric. For the same symmetry reasons as for the force constant matrix, the matrix elements $D_{lk\alpha}^{l'k'\beta}$ do not depend on the absolute positions of the primitive cells but on their relative distance. Thus $\mathscr{D}_{lk\alpha}^{l'k'\beta} = \mathscr{D}_{k\alpha}^{k'\beta}(\mathbf{r}_{l'} - \mathbf{r}_l) = \mathscr{D}_{k\alpha}^{k'\beta}(l' - l)$.

To solve the system of Eq. (38), we try a solution in the form of plane waves:

$$w_\alpha(lk) = e_{k\alpha}(\mathbf{q}) \exp[i(\mathbf{q}.\mathbf{r}_l - \omega t)] \tag{40}$$

4.3 Lattice Dynamics and Thermal Properties of Alloys

Inserting Eq. (40) in Eq. (38), we get

$$\omega^2 e_{k\alpha}(\mathbf{q}) = \sum_{l'k'\beta} \mathscr{D}_{k\alpha}^{k'\beta}(l'-l) \exp[i\mathbf{q}(\mathbf{r}_{l'}-\mathbf{r}_l)]. \tag{41}$$

As we can see, the plane waves make it possible to simplify the problem from a coupled system of $3pN$ differential equations to N systems of $3p$ linear equations. For a crystal with p atoms per unit cell and for each value of $\mathbf{q}$, there are $3p$ solutions with eigenvalues $\omega_j^2(\mathbf{q})$ and eigenvectors $e_k(j,\mathbf{q})$ to this problem. A complete solution of the eigenproblem of Eq. (41) is sought in terms of $\omega = \omega_j(\mathbf{q})$, called the phonon dispersion relation for eigenvalues, and $\mathbf{e} = \mathbf{e}_j(\mathbf{q})$ (also denoted $\mathbf{e} = \mathbf{e}(j,\mathbf{q})$), called the dispersion relation for eigenvectors. An eigenmode with frequency ω_j and eigenvector $\mathbf{e}_k(j,\mathbf{q})$ is called the jth normal mode, or phonon, of the system. The graphical representation of the $3p$ functions $\omega_j(\mathbf{q})$ as a function of $\mathbf{q}$ is called the vibrational dispersion spectrum of the system. The index $j = 1, 2, 3, \ldots, 3p$ stands for the phonon branch. Since the dynamical matrix is real and symmetric, the eigenvalues are real. The harmonic system is stable if all eigenvalues are positive. It is sometimes convenient to use, instead of real eigenvectors, complex ones. For example, in a translationally invariant lattice, the normal modes are plane waves which are mostly characterized by eigenvectors of the form $\exp(i\mathbf{q}.\mathbf{r})$. The fact that the atomic displacement is real requires the condition: $\mathbf{e}_k^*(j,\mathbf{q}) = \mathbf{e}_k(-j,\mathbf{q})$; $\omega_j(\mathbf{q}) = \omega_j(-\mathbf{q})$. Orthogonal eigenvectors can be chosen:

$$\sum_{k\alpha} e_{k\alpha}^*(j,\mathbf{q}).e_{k\alpha}(j',\mathbf{q}) = \delta_{jj'} \tag{42}$$

and

$$\sum_j e_{k\alpha}^*(j,\mathbf{q}).e_{k'\alpha'}(j,\mathbf{q}) = \delta_{kk'}\delta_{\alpha\alpha'} \tag{43}$$

Out of the $3p$ phonon branches, there will be three *acoustic* branches such that $\omega(\mathbf{q}) \to 0$ when $\mathbf{q} \to 0$ and $3p - 3$ *optical* branches such that $\omega(\mathbf{q}) \to$ constant $\neq 0$ when $\mathbf{q} \to 0$. Atomic vibrations corresponding to any of the branches, acoustic or optical, can either be longitudinal such that $\mathbf{e} \parallel \mathbf{q}$ or transverse such that $\mathbf{e} \perp \mathbf{q}$, or a mixture of longitudinal and transverse. The elastic continuum theory shows that in an isotropic crystal, it is always possible to construct three mutually independent polarization modes for a given $\mathbf{q}$: $\mathbf{e}_L \parallel \mathbf{q}$ and $\mathbf{e}_{T_1} \perp \mathbf{e}_{T_2} \perp \mathbf{q}$. In an anisotropic crystal, a clear relationship between $\mathbf{e}$ and $\mathbf{q}$ does not exist, except when $\mathbf{q}$ is along a high-symmetry direction. For example, in cubic crystals the concept of pure longitudinal and transverse polarization modes is defined only when $\mathbf{q}$ is along the symmetry directions [100], [110], and [111]. In tetragonal crystals even the [111] branches become mixed.

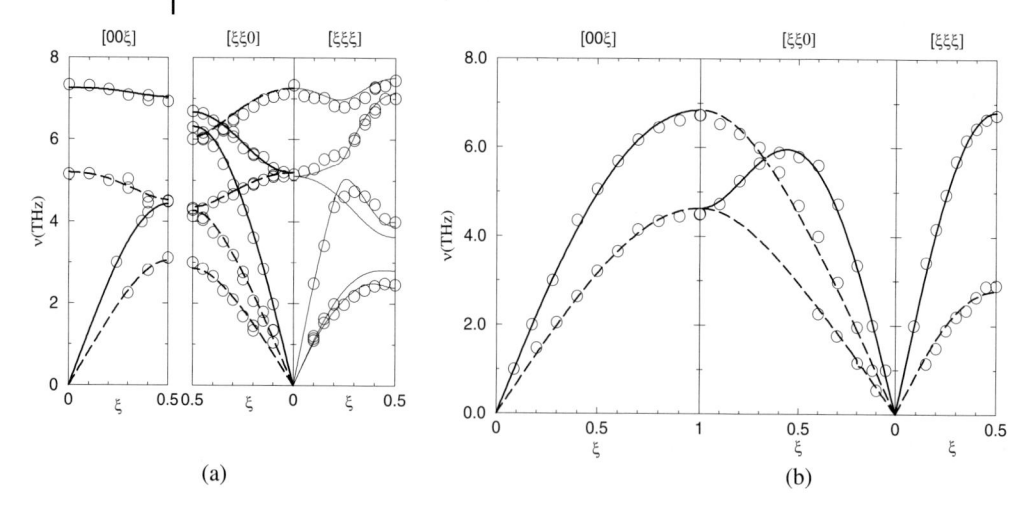

Fig. 4.8 Phonon dispersion for [001], [110], and [111] directions in L1$_0$ ordered FePd at 300 K (a) and in disordered FePd at 1020 K (b). Solid and broken curves indicate dispersion curves of longitudinal and transverse branches. The thin solid curves indicate branches of mixed polarization.

Figure 4.8 shows a vibrational spectrum of long-range ordered L1$_0$ FePd (see insert of Fig. 4.2a) measured at room temperature by inelastic neutron scattering along the [001], [110], and [111] directions (Mehaddene et al. 2004). The underlying lattice of the L1$_0$ structure is tetragonal with two atoms per unit cell; thus, we get six phonon branches ($3 \times 2 = 6$), three acoustic and three optical, for each value of **q**. Longitudinal and transverse branches are displayed as thick solid and dashed curves, respectively. In crystals of tetragonal structure, however, some of the branches are mixed in character, and those branches are plotted as thin solid curves. In some high-symmetry directions such as [001] and [110], all branches are purely either longitudinal or transverse. In the [001] direction, the transverse branches are doubly degenerated, similarly to the case of cubic crystals. The vibrational spectrum is much more simple in the same FePd sample measured at 1020 K (Fig. 4.8), where the structure is disordered. In the disordered state, all atoms are equivalent; the underlying lattice is thus an fcc Bravais lattice with a single atom per unit cell, giving rise to only three normal modes for each **q**, two transverse (degenerate along [100] and [110]) and one longitudinal.

4.3.1.2 Diatomic Linear Chain

The essential features of the lattice dynamics can be understood most simply by studying the dimensionally reduced systems such as planes or atomic chains. In the latter case, this would be equivalent to looking in a given direction of space of the three-dimensional system. In the case of FePd, for instance, the [011] direction can be viewed as a linear chain with a basis of two atoms with masses m_1

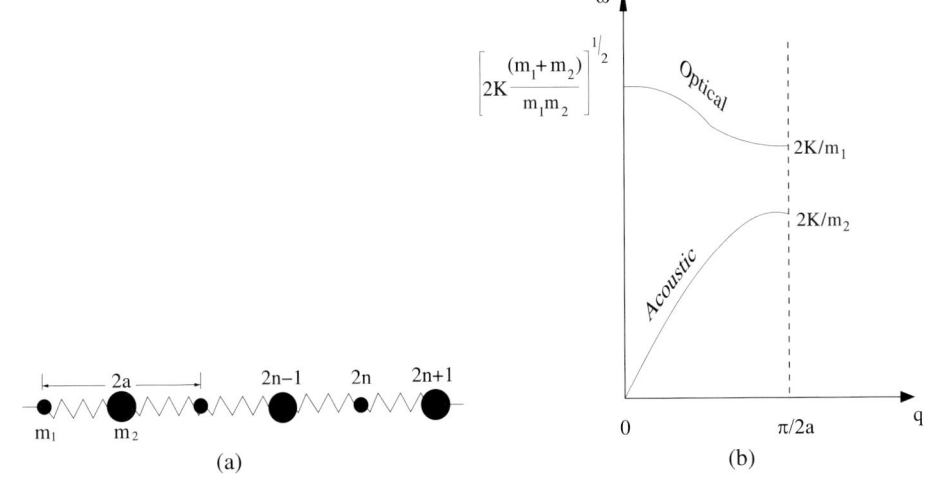

Fig. 4.9 (a) Diatomic linear chain with alternating m_1 and m_2 masses. The unit cell has length $2a$. (b) Phonon dispersion spectrum of a diatomic chain showing the acoustic and the optical branches. The gap in between is due to the different atomic masses ($m_1 < m_2$) in the unit cell.

and m_2. Considering that each plane interacts only with its nearest-neighbor planes and that the force constants are identical between all pairs of nearest-neighbor planes, the system can be modeled by an infinite chain of $2N$ atoms forming N unit cells, each of length $2a$ (Fig. 4.9a), where a is related to the lattice parameters of the L1$_0$ structure by: $a = \sqrt{a_{L1_0}^2 + c_{L1_0}^2}/2$.

Suppose that, at a particular time, the neighboring $2n$th and $2n+1$th atoms in the chain have displacements u_{2n} and w_{2n+1} from their equilibrium position respectively. The equation of motion then reads, in the harmonic approximation:

$$m_1 \ddot{u}_{2n} = k[w_{2n+1} + w_{2n-1} - 2u_{2n}] \tag{44}$$

$$m_2 \ddot{w}_{2n+1} = k[u_{2n+2} + u_{2n} - 2w_{2n+1}] \tag{45}$$

where k is the force constant between the atoms m_1 and m_2. Let us look for solutions of the form

$$u_{2n} = e_1 \exp[i(2nqa - \omega t)] \tag{46}$$

$$w_{2n+1} = e_2 \exp[i(2n+1)qa - \omega t] \tag{47}$$

Combining Equations (44–47), we get the coupled eigenvalue equations

$$-\omega^2 e_1 m_1 = k[e_2 e^{iqa} + e_2 e^{-iqa} - 2e_1] \tag{48}$$

$$-\omega^2 e_2 m_2 = k[e_1 e^{-iqa} + e_1 e^{iqa} - 2e_2] \tag{49}$$

Nontrivial solutions to this system are found by solving the secular equation $|\mathscr{D}_{ij} - \omega^2 \delta_{ij}| = 0$ where $\mathscr{D}_{ij}$ is the (2×2) dynamical matrix given by

$$\mathscr{D}_{ij} = \begin{pmatrix} 2k/m_1 & -(2k/m_1) \cos qa \\ -(2k/m_2) \cos qa & 2k/m_2 \end{pmatrix} \tag{50}$$

The solutions are

$$\omega^2 = k \frac{m_1 + m_2}{m_1 m_2} \pm k \sqrt{\left(\frac{m_1 + m_2}{m_1 m_2}\right)^2 - \frac{4}{m_1 m_2} \sin^2 qa} \tag{51}$$

The two signs in Eq. (51) correspond to two branches of the phonon dispersion curve. From Eqs. (46) and (47), the ratio of the vibration amplitudes e_1 and e_2 is

$$\frac{e_1}{e_2} = \frac{2k \cos qa}{2k - m_1 \omega^2} = \frac{2k - m_2 \omega^2}{2k \cos qa} \tag{52}$$

For small q values (long wavelength), we have $\cos qa \approx 1 - \frac{1}{2} q^2 a^2 + \cdots$ The two roots are

$$\omega_O^2 = 2k \left(\frac{m_1 + m_2}{m_1 m_2}\right) \quad \text{and} \quad \omega_A^2 = \frac{1}{2} \frac{k}{m_1 + m_2} q^2 a^2 \tag{53}$$

which correspond to the upper and lower branches in Fig. 4.9(b) respectively. The upper branch in Fig. 4.9(b) corresponds to a negative ratio e_1/e_2, meaning that the two atoms are moving in opposite directions. Such a mode can be excited by an electric field of the appropriate frequency in the case of ionic crystals; this branch is thus called an optical branch. The lower branch has a positive ratio e_1/e_2, equal to 1 at small q, meaning that the two atoms in the unit cell are moving in phase with each other. This is a characteristic of a sound wave. Hence the lower branch is called an acoustic branch. We can deduce from the above results the shape of the dispersion curve in the case of monoatomic chains. When $m_2 \to m_1$, the gap between the optical and the acoustic branch disappears at the zone boundary and the upper branch can be unfolded to cover the region $\pi/2a$ to π/a and $-\pi/2a$ to $-\pi/a$, which is indeed the Brillouin zone of the monoatomic lattice.

4.3.1.3 Quantum Theory

So far we have treated the atomic vibrations within classical mechanics. For many applications, however, quantum properties are needed. In this case, the energy function [Eq. (35)] is the potential part of the Hamiltonian of the system, $\mathscr{H}$. The vibrational wavefunctions can then be derived by the standard methods to

4.3 Lattice Dynamics and Thermal Properties of Alloys

treat the harmonic oscillator. The Hamiltonian of the system in the harmonic approximation reads:

$$\mathcal{H} = \sum_{lk} \frac{\mathbf{p}^2(lk)}{2m_k} + \frac{1}{2} \sum_{lk,l'k'} \sum_{\alpha\beta} \Phi_{\alpha\beta}(lk, l'k') u_\alpha(lk) u_\beta(l'k') \tag{54}$$

where $\mathbf{p}(lk)$ is the momentum operator of the kth atom in the lth unit cell. The form (54) of the Hamiltonian is quite complicated; it can be simplified by introducing normal coordinate operators $\mathbf{X}(\mathbf{q})$ and $\mathbf{P}(\mathbf{q})$, Fourier transforms of the coordinate $\mathbf{u}$ and $\mathbf{p}$:

$$\mathbf{u}(lk) = \frac{1}{\sqrt{V}} \sum_{\mathbf{q}} \mathbf{X}(\mathbf{q}, k) \exp(i\mathbf{q}\cdot\mathbf{r}_l) \tag{55}$$

$$\mathbf{p}(lk) = \frac{1}{\sqrt{V}} \sum_{\mathbf{q}} \mathbf{P}(\mathbf{q}, k) \exp(-i\mathbf{q}\cdot\mathbf{r}_l) \tag{56}$$

where V is the volume of the crystal. Since $\mathbf{u}(lk)$ and $\mathbf{p}(lk)$ are Hermitian, we have

$$\mathbf{X}^\dagger(\mathbf{q}, k) = \mathbf{X}(-\mathbf{q}, k) = \frac{1}{\sqrt{V}} \sum_l \mathbf{u}(lk) \exp(i\mathbf{q}\cdot\mathbf{r}_l) \tag{57}$$

$$\mathbf{P}^\dagger(\mathbf{q}, k) = \mathbf{P}(-\mathbf{q}, k) = \frac{1}{\sqrt{V}} \sum_l \mathbf{p}(lk) \exp(-i\mathbf{q}\cdot\mathbf{r}_l) \tag{58}$$

i.e., the new coordinate operators are non-Hermitian. These operators satisfy the commutation relations:

$$[X(\mathbf{q}, k), P(\mathbf{q}', k')] = i\hbar \delta_{\mathbf{q}\mathbf{q}'} \delta_{kk'} \mathbf{I} \tag{59}$$

where $\mathbf{I}$ is the identity operator and $\hbar = h/2\pi$, the reduced Planck constant. The substitution of Eq. (55) into Eq. (54) gives

$$\mathcal{H} = \frac{1}{V} \sum_{lk\mathbf{q}\mathbf{q}'} \frac{\mathbf{P}(\mathbf{q}, k)\cdot\mathbf{P}(\mathbf{q}', k)}{2m_k} \exp[-i(\mathbf{q}+\mathbf{q}')\mathbf{r}_l]$$

$$+ \frac{1}{2} \frac{1}{V} \sum_{lk,l'k'\mathbf{q}\mathbf{q}'} \sum_{\alpha\beta} \Phi_{\alpha\beta}(lk, l'k') X_\alpha(\mathbf{q}, k) X_\beta(\mathbf{q}', k') \exp[i(\mathbf{q}\cdot\mathbf{l} + \mathbf{q}'\cdot\mathbf{l}')] \tag{60}$$

The Hamiltonian (60) can be simplified by performing the summation over l in the first term, introducing $h = l - l'$ and defining in the second term $\Phi_{\alpha\beta}(kk', \mathbf{q})$:

$$\Phi_{\alpha\beta}(kk',\mathbf{q}) = \sqrt{m_k m_{k'}}\mathscr{D}_{\alpha\beta}(kk',-\mathbf{q})$$

$$= \sum_h \Phi_{\alpha\beta}(0k,hk')\exp(-i\mathbf{q}\cdot\mathbf{r}_h) \tag{61}$$

The Hamiltonian can be then written

$$\mathscr{H} = \sum_{qk}\frac{\mathbf{P}(\mathbf{q},k)\cdot\mathbf{P}^\dagger(\mathbf{q},k)}{2m_k} + \frac{1}{2}\sum_{\mathbf{q},k,k'\alpha\beta}\Phi_{\alpha\beta}(kk',\mathbf{q})X_\alpha(\mathbf{q},k)X_\beta^\dagger(\mathbf{q},k') \tag{62}$$

At this stage, the crystal Hamiltonian is viewed in terms of the coordinates $\mathbf{X}(k\mathbf{q})$ and the momenta $\mathbf{P}(k\mathbf{q})$ of pN atoms, coupled by a set of harmonic force constants $\Phi(kk',\mathbf{q})$. Thus for each value of $\mathbf{q}$ the problem of finding the normal modes of the system is equivalent to finding the eigenstates of the Hamiltonian. For this purpose, following Section 4.3.1 we introduce the polarization vector $\mathbf{e}_k(j,\mathbf{q})$ to represent the magnitude and direction of vibration mode $(j,\mathbf{q})$, where j denotes the polarization branch. The $\mathbf{e}_k(j,\mathbf{q})$ obey the orthogonality relation (42). With the introduction of $\mathbf{e}_k(j,\mathbf{q})$, we do another set of normal coordinate transformations:

$$X(j,\mathbf{q}) = \sum_k \sqrt{m_k}\mathbf{e}^*_k(j,\mathbf{q})\cdot\mathbf{X}(\mathbf{q},k) \quad \text{and} \quad P(j,\mathbf{q}) = \sum_k \frac{1}{\sqrt{m_k}}\mathbf{e}_k(j,\mathbf{q})\cdot\mathbf{P}(\mathbf{q},k) \tag{63}$$

We introduce the phonon annihilation and creation operators defined respectively by Eqs. (64) and (65):

$$b_{j,\mathbf{q}} = \frac{1}{\sqrt{2\hbar\omega_j(\mathbf{q})}}P(j,\mathbf{q}) - i\sqrt{\frac{\omega_j(\mathbf{q})}{2\hbar}}X^\dagger(j,\mathbf{q}) \tag{64}$$

$$b^\dagger_{j,\mathbf{q}} = \frac{1}{\sqrt{2\hbar\omega_j(\mathbf{q})}}P^\dagger(j,\mathbf{q}) + i\sqrt{\frac{\omega_j(\mathbf{q})}{2\hbar}}X(j,\mathbf{q}) \tag{65}$$

It can be verified that these operators obey the commutation relations $[b_{j,\mathbf{q}},b^\dagger_{j',\mathbf{q}'}] = \delta_{\mathbf{q}\mathbf{q}'}\delta_{j,j'}$. Equation (64) can be inverted. Using $\omega_j(\mathbf{q}) = \omega_j(-\mathbf{q})$ and Eqs. (58) and (63), we have:

$$X(k\mathbf{q}) = -i\sum_j \sqrt{\frac{\hbar}{2m_k\omega_j(\mathbf{q})}}\mathbf{e}_j(\mathbf{q},l)(b^\dagger_{j,\mathbf{q}} - b_{j,-\mathbf{q}}) \tag{66}$$

$$P(k\mathbf{q}) = \sum_j \sqrt{\frac{m_k\hbar\omega_j(\mathbf{q})}{2}}\mathbf{e}^*_j(\mathbf{q},l)(b^\dagger_{j,\mathbf{q}} - b_{j,-\mathbf{q}}) \tag{67}$$

We now have expressions of the coordinate and the momenta vector $\mathbf{X}(k\mathbf{q})$ and $\mathbf{P}(k\mathbf{q})$ in terms of the phonon creation and annihilation operators and the polarization vectors. If we substitute Eq. (66) in Eq. (62), we get a simple expression of the Hamiltonian:

$$\mathcal{H} = \sum_{qj} \hbar\omega_j(\mathbf{q}) \left(b^\dagger_{j,\mathbf{q}} b_{j,\mathbf{q}} + \frac{1}{2} \right) \tag{68}$$

At this stage we expressed the Hamiltonian of the system as a sum of $3pN$ Hamiltonians of independent harmonic oscillators. The eigenvalues of $\mathcal{H}$ are well known from the quantum treatment of the harmonic oscillator which can be found in many text books (Cohen-Tanoudji et al. 1977). Let us denote by $|n_{j\mathbf{q}}\rangle$ an eigenstate which has n phonons of wave vector $\mathbf{q}$ and polarization j. The effect of the operators $b_{j,\mathbf{q}}$, $b^\dagger_{j,\mathbf{q}}$ and $b^\dagger_{j,\mathbf{q}} b_{j,\mathbf{q}}$ on the state $|n_{qj}\rangle$ are given as follows:

$$\begin{aligned} b^\dagger_{j,\mathbf{q}} |n_{j\mathbf{q}}\rangle &= \sqrt{n_{j\mathbf{q}} + 1} |n_{j\mathbf{q}} + 1\rangle \\ b_{j,\mathbf{q}} |n_{j\mathbf{q}}\rangle &= \sqrt{n_{j\mathbf{q}}} |n_{j\mathbf{q}} - 1\rangle \\ b^\dagger_{j,\mathbf{q}} b_{j,\mathbf{q}} |n_{j\mathbf{q}}\rangle &= n_{j\mathbf{q}} |n_{j\mathbf{q}}\rangle \end{aligned} \tag{69}$$

$b^\dagger_{j,\mathbf{q}} b_{j,\mathbf{q}}$ is a phonon number operator which gives the number of phonons in the $(j, \mathbf{q})$ state. Using Eqs. (68) and (69), we get

$$\mathcal{H} |n_{qj}\rangle = \sum_{j\mathbf{q}} \hbar\omega_j(\mathbf{q}) \left(n_{j\mathbf{q}} + \frac{1}{2} \right) |n_{j\mathbf{q}}\rangle = \sum_{j\mathbf{q}} \bar{\epsilon}_{j\mathbf{q}} |n_{j\mathbf{q}}\rangle \tag{70}$$

and the average energy of the phonons in the mode $(j, \mathbf{q})$: $\bar{\epsilon}_{j\mathbf{q}} = \hbar\omega_j(\mathbf{q}) \bar{n}_{j\mathbf{q}}$ where the thermal average $\bar{n}_{j\mathbf{q}}$ is the Bose–Einstein distribution function. The term $\frac{1}{2}$ in Eq. (70) is due to the zero-point energy.

4.3.1.4 Phonon Density of States

In the following sections, we will perform sums or integrals of functions of the dispersion over the crystal momentum state $\mathbf{q}$ within the reciprocal lattice. However, the translational symmetry of the crystal often greatly reduces the set of points for the summation. Let us once more consider a three-dimensional crystal with $N = L^3$ unit cells, where L is the linear size of the crystal along x, y, and z. If we apply the periodic boundary conditions along the three directions, we seek phonon states of the type $\psi(\mathbf{r}) = A \exp[i(\mathbf{q}\cdot\mathbf{r} - \omega t)]$ obeying $\psi(\mathbf{r}) = \psi(\mathbf{r} + \mathbf{r}_L)$ where $\mathbf{r}_L = L\mathbf{a}_1$. Thus

$$q_x, q_y, q_z = 0; \pm \frac{2\pi}{L}; \pm \frac{4\pi}{L}; \ldots; \pm \frac{N\pi}{L} \tag{71}$$

i.e., the allowed $\mathbf{q}$ values form a cubic mesh in $\mathbf{q}$ space, with one $\mathbf{q}$ confined to a volume $(2\pi/L)^3$. Therefore, for each phonon polarization index, a unit volume in the $\mathbf{q}$ space contains $(L/2\pi)^3$ values of $\mathbf{q}$. $(L/2\pi)^3$ is called the "phonon density of states," DOS, in the **q** space and gives the number of vibrational states or normal modes between wave vectors **q** and **q** + d**q**:

$$g(\mathbf{q})\, d\mathbf{q} = \frac{V}{(2\pi)^3} d\mathbf{q} \tag{72}$$

where $V = L^3$ is the volume of the crystal. Equivalently, we can define $g(\omega)$, the number of normal modes which have a frequency between ω and $\omega + d\omega$. The partial phonon density of states of the branch j is

$$g(\omega_j) = \frac{V}{(2\pi)^3} \int \frac{dS_\omega}{|\nabla_\mathbf{q} \omega_j|}$$

To get the total density of states, we sum over all branches:

$$g(\omega) = \sum_j g(\omega_j) = \sum_j \frac{V}{(2\pi)^3} \int \frac{dS_\omega}{|\nabla_\mathbf{q} \omega_j|} \tag{73}$$

$v_g = \nabla_\mathbf{q} \omega_j$ is the group velocity. For some **q** values, v_g approaches zero and leads to $g(\omega) \to \infty$. Such points in the **q** space are called critical points and the singularities in the density of states are known as van Hove singularities.

The phonon density of states is one of the most important quantities in lattice dynamical studies. Its determination, however, requires the calculation of the integral which appears in Eq. (73) over the whole Brillouin zone. In general, such an integration requires knowledge of phonon frequencies of wave vectors **q** in the entire Brillouin zone. However, in most experimental determinations of the normal mode of vibration such a task is not feasible with regard to the time and to the experimental effort required. In practice we proceed by an interpolation of the normal modes of vibration using some phenomenological models. Furthermore, the point symmetry of the system is taken into account to reduce the phonon frequency calculation to wave vectors within the irreducible part of the Brillouin zone.

Figure 4.10 shows an example of phonon density of state in the L1$_0$ ordered FePd at 300 K and in the fcc disordered state at 1020 K (Mehaddene et al. 2004). The occurrence of van Hove singularities is due, as explained below, to the maxima in the dispersion curves, which are numerous in the ordered phase because of the greater number of branches (Fig. 4.8a). Actually, in the fcc disordered state it is easy to link the van Hove singularities to the responsible phonon modes in the dispersion curve, which are mainly due to the zone boundary modes. The cut-off frequency decreases from 7.5 THz at room temperature to 6.8 THz at 1020 K, reflecting the increasing anharmonicity and the softening of the lattice at high temperature. The shape of the DOS at 1020 K is common to all fcc crystals.

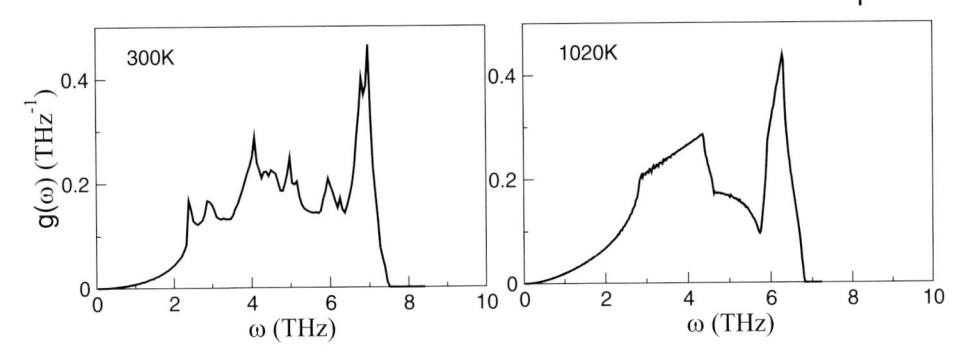

Fig. 4.10 Phonon DOS of FePd in the L1$_0$ ordered state at 300 K and in the disordered state at 1020 K.

4.3.1.5 Lattice Specific Heat

The heat capacity at constant volume is an important quantity when dealing with the thermal properties of alloys. It defines the change in the internal energy $\mathcal{U}$ of the system when the temperature T is increased by an infinitesimal amount:

$$C_V = \left(\frac{\partial \mathcal{U}}{\partial T}\right)_V \tag{74}$$

At this stage, we should keep in mind that not only the phonons contribute to the internal energy $\mathcal{U}$. The electrons, as well as the magnetic moments, have a contribution to $\mathcal{U}$ but their dependence on T is different. In the following, we derive an expression of the phonon contribution to C_V.

We consider first the high-temperature limit. According to the equipartition theorem, which holds at high temperature, the thermal energy of a given system is simply given by $k_B T$ per quadratic term in the Hamiltonian ($\frac{1}{2}k_B T$ for both kinetic and potential energies). In the case of a crystal, with pN atoms, each of them is free to move in the three directions of space, we get $\mathcal{U} = 3pNk_B T$ and $C_V = 3Npk_B$. The molar specific heat of any solid at high temperature is thus $C_V^m = 3\mathcal{N}_A k_B = 3R$, which is constant and independent on T. $\mathcal{N}_A$ is Avogadro's number and R is the ideal gas constant. This is known as the law of Dulong and Petit.

To derive an expression for C_V at low temperatures, where the classical theory does not hold any more, we consider the phonons as a set a undiscernible particles that a priori can occupy any possible $(j, \mathbf{q})$ state with energy $\hbar\omega_j(\mathbf{q})$. The partition function of such a system is $\mathcal{Z} = \sum_s \exp(-\beta \mathcal{U}_s)$ where $\beta = (k_B T)^{-1}$ and the index s stands for the different states. The energy $\mathcal{U}_s$ of the phonon gas is given by Eq. (68); we have

$$\mathcal{Z} = \prod_{j\mathbf{q}} \sum_{n_{j\mathbf{q}}} \exp\left[-\beta\left(n_{j\mathbf{q}} + \frac{1}{2}\right)\hbar\omega_j(\mathbf{q})\right] = \prod_{j\mathbf{q}} \frac{\exp(-\beta\hbar\omega_j(\mathbf{q})/2)}{1 - \exp(-\beta\hbar\omega_j(\mathbf{q}))} \tag{75}$$

From mechanical statistics (Gallavotti, 1999) we know that the internal energy can be deduced from the partition function as

$$\mathcal{U} = -\frac{\partial}{\partial \beta} \ln \mathcal{Z} = \sum_{jq} \frac{\hbar \omega_j(\mathbf{q})}{2} + \sum_{jq} \bar{n}_{jq} \hbar \omega_j(\mathbf{q}) \tag{76}$$

where the thermal average $\bar{n}_{jq}$ is the Bose–Einstein distribution function and represents the average number of phonons in the state $(j, \omega_j(\mathbf{q}))$:

$$\bar{n}_{qj} = \frac{1}{\exp(\beta \hbar \omega_j(\mathbf{q})) - 1} \tag{77}$$

From Eq. (74), we can deduce C_V:

$$C_V = \left(\frac{\partial \mathcal{U}}{\partial T}\right)_V = \sum_{jq} \frac{(\hbar \omega_j(\mathbf{q}))^2}{k_B T^2} \frac{\exp(\beta \hbar \omega_j(\mathbf{q}))}{(\exp(\beta \hbar \omega_j(\mathbf{q})) - 1)^2} \tag{78}$$

In the previous derivations, we have assumed a finite model crystal. For an infinite periodic lattice, the allowed $\mathbf{q}$ values become dense and $\mathbf{q}$ a continuous variable. The summations over $\mathbf{q}$ are replaced by integrals:

$$\sum_{\mathbf{q}} \rightarrow \int g(\mathbf{q}) \, d^3 \mathbf{q} \rightarrow \int g(\omega) \, d\omega \tag{79}$$

The final expression for C_V is

$$C_V = \int_0^{\omega_{max}} g(\omega) \frac{(\hbar \omega)^2}{k_B T^2} \frac{\exp(\beta \hbar \omega)}{(\exp(\beta \hbar \omega) - 1)^2} \, d\omega \tag{80}$$

If we now consider the high-temperature limit, we can derive the Dulong and Petit law obtained using the classical theory:

$$C_V = k_B \int_0^{\omega_{max}} g(\omega) \, d\omega = 3pNk_B \tag{81}$$

The integral in Eq. (81) gives the total number of normal modes of vibration, which is equal to the total number of degrees of freedom of the crystal.

4.3.1.6 Debye's Model

The Dulong and Petit law has been corroborated by many experimental observations performed on metals and insulators as well as on semiconductors at high temperature. In the low-temperature limit, the experimental observations show that C_V varies with T^3 in insulators and with T in metals. The behavior in metals is dominated by the electronic contribution. A satisfactory explanation of the experimental results, for both high- and low-temperature variations of C_V in insulators, emerged from the model of Debye for the phonon density of states.

In 1912, Debye used the isotropic continuum approximation of the phonon dispersion and considered the contributions from acoustic phonons in all polarization branches with a frequency lower than a cut-off frequency ω_D, which is called the Debye frequency (Debye, 1912). In this approximation, the velocity of sound is taken as constant and equal for all polarization types, as it would be for a classical elastic continuum. Thus the dispersion relation reduces to

$$\omega = vq \tag{82}$$

where v is the constant velocity of sound. In the Debye model, considering a uniform density in the reciprocal space and taking Eq. (82) as valid, the phonon density of states reads

$$g_D(\omega) = \begin{cases} f\omega^2 : \omega \leq \omega_D \\ 0 : \omega > \omega_D \end{cases} \tag{83}$$

where f is a constant. The singularities in the real density of states are smeared out. The value of the constant f is chosen so that the integration in ω gives the total number of normal modes: we get $f = 9pN/\omega_D^3$.

If we replace the expression for the Debye density of states (83) in Eq. (80), we get the specific heat in the Debye model:

$$C_V = 9pNk_B \left(\frac{T}{\theta_D}\right)^3 \int_0^{\theta_D/T} \frac{x^4 e^x}{(e^x - 1)^2} dx = 9pNk_B \left(\frac{T}{\theta_D}\right)^3 \zeta(\theta_D/T) \tag{84}$$

where $x = \hbar\omega/k_B T$; $\theta_D = \hbar\omega_D/k_B$ is called the Debye temperature and ζ is the Riemann "zeta" function. At high temperature, $x \ll 1$, we get $C_V = 3pNk_B$, in agreement with the Dulong and Petit law. At low temperature, $x \gg 1$, $\theta_D/T \approx \infty$ and $\zeta(\infty) = 4\pi^4/15$ give

$$C_V = \frac{36\pi^4}{15} pNk_B \left(\frac{T}{\theta_D}\right)^3$$

The Debye model successfully explains both low- and high-temperature dependence of the lattice specific heat of insulating crystals. It should be emphasized that real crystals cannot be treated adequately as an elastic continuum and that their atomic nature must be considered. This requires an accurate evaluation of C_V from $g(\omega)$ using a realistic lattice dynamical model. The Debye formula of C_V can nevertheless be used; θ_D is then treated as an empirical parameter in Eq.

(84). A usual way to choose the Debye temperature as a temperature-dependent parameter $\theta_D(T)$ is to adjust the measured heat capacity with Eq. (84).

4.3.1.7 Elastic Waves in Cubic Crystals

In the long-wavelength limit, $\mathbf{q} \to 0$, the displacement of the atoms will vary slowly in space. In this limit, the acoustic phonons, for which the atoms move in phase as shown in the case of the diatomic chain, correspond to sound waves, which are usually described by differential equations on an elastic continuum instead of motion Eq. (44) of lattice dynamics. In this limit, it can be shown that, for cubic crystals, the equation of motion can be written (Kittel 1976) as:

$$\rho \ddot{u}_x = c_{11} \frac{\partial^2 u_x}{\partial x^2} + c_{44}\left(\frac{\partial^2 u_x}{\partial y^2} + \frac{\partial^2 u_x}{\partial z^2}\right) + (c_{12} + c_{44})\left(\frac{\partial^2 u_y}{\partial x \partial y} + \frac{\partial^2 u_z}{\partial x \partial z}\right) \tag{85}$$

where ρ is the mass density and u_x is the x component of the displacement $\mathbf{u}$. The corresponding equations of motion along y and z can be found by cyclic permutation.

We can look for the solution corresponding to a longitudinal wave. For a wave along [100], i.e., $\mathbf{q} \parallel [100]$, we can write: $u_x = u_0 \exp[i(q_x x - \omega t)]$; $u_y = 0$; $u_z = 0$. Substituting in Eq. (85), we get $\omega^2 \rho = c_{11} q^2$. The velocity of a longitudinal wave in the [100] direction is thus: $v_L[100] = \omega/q = (c_{11}/\rho)^{1/2}$.

To obtain the speed of a transverse wave, we can try either $u_y = u_0 \exp[i(q_x x - \omega t)]$; $u_x = 0$; $u_z = 0$ or $u_z = u_0 \exp[i(q_x x - \omega t)]$; $u_x = 0$; $u_y = 0$. On substitution in Eq. (85) we get the same velocity of a transverse wave in the [100] direction, $v_T[100] = (c_{44}/\rho)^{1/2}$.

Thus, for a wave propagating along the x axis, we have a longitudinal wave with particle displacement along the x axis, and two identical but independent transverse waves with particle displacements along y and z.

For a wave along [110], $q_x = q_y = q/\sqrt{2}$, a little bit of algebra is required to determine the different velocities. The longitudinal wave with atomic displacement along [110] has the speed $v_L[110] = [(c_{11} + c_{12} + 2c_{44})/2\rho]^{1/2}$ and there are two non-degenerate transverse waves with the speeds $v_{T_1}[110] = (c_{44}/\rho)^{1/2}$; $v_{T_2}[110] = [(c_{11} - c_{12})/2\rho]^{1/2}$, which correspond to atomic displacements along [001] and [1$\bar{1}$0] respectively. The wave propagation in the [110] direction is of special interest because the three cubic elastic constants, c_{11}, c_{12}, and c_{44} can be deduced from the propagation velocities in this direction: c_{44}, for example, can be obtained from the slope at the origin of the Brillouin zone of the phonon dispersion curves in the [110] direction or it can be measured by ultrasonic resonance spectrometry.

For a wave along [111], $q_x = q_y = q_z = q/\sqrt{3}$, the longitudinal wave has the speed $v_L[111] = \omega/q = [(c_{11} + 2c_{12} + 4c_{44})/3\rho]^{1/2}$. The two transverse waves have the identical speed $v_{T_1}[111] = v_{T_2}[111] = [(c_{11} - c_{12} + c_{44})/3\rho]^{1/2}$ and correspond to displacements along [1$\bar{1}$0] and [$\bar{1}\bar{1}$2].

In directions other than [100], [110], and [111], the waves in a cubic crystal cannot be interpreted as purely longitudinal or purely transverse unless the material is isotropic.

4.3.1.8 Vibrational Entropy

From a structural point of view, lattice vibrations or phonons can be viewed as defects inducing a disorder in a perfect crystal. This disorder can be quantified through the vibrational entropy, which plays an important role in phase stability. To understand the effect of lattice vibrations on phase stability, it is useful to write the free energy of a phase $F = \mathscr{U} - TS$ as the sum of its configurational and vibrational parts:

$$\mathscr{F} = \mathscr{U}_{conf} + \mathscr{U}_{vib} - T(S_{conf} + S_{vib}) \tag{86}$$

Doing this, we neglect any other contribution, such as magnetic and electronic entropies. From statistical mechanics, we know that the vibrational entropy S_{vib} is the negative temperature derivative of the vibrational free energy (Dalvit et al. 1999) $S_{vib} = -\partial \mathscr{F}_{vib}/\partial T$ where $\mathscr{F}_{vib} = -k_B T \ln \mathscr{Z}$ is the free energy of the system and $\mathscr{Z}$ the partition function, given in the case of a phonon gas by Eq. (75). The vibrational entropy per atom in the harmonic approximation reads

$$S_{vib} = -3k_B \int g(\omega)[\bar{n}_{jq}(\omega) \ln(\bar{n}_{jq}(\omega)) - (1 + \bar{n}_{jq}(\omega)) \ln(1 + \bar{n}_{jq}(\omega))]\, d\omega \tag{87}$$

where $\bar{n}_{jq}(\omega) = [\exp(\beta \hbar \omega_j(\mathbf{q})) - 1]^{-1}$. Basically, the vibrational contribution $F_{vib} = \mathscr{U}_{vib} - TS_{vib}$ has two distinct effects: change in the phase stability of an alloy or ordered compound in a given configuration with respect to the pure solid constituents (formation energy), and change in the stability of ordered compound with respect to the random alloy at the same composition (ordering energy). The first effect will affect the free energy of formation and may lead to changes in the structure. Previous studies have shown that in phase-separating alloys such as GaP–InP (Silverman et al. 1995), Cu–Ag (Sanchez et al. 1991; Asta and Foiles 1996), or Ni–Cr (Craivich and Sanchez 1997), one often finds a vibrational stabilization of compounds, leading to increased solubilities and lowered miscibility gap temperatures. The second effect will affect the order–disorder transition temperature. One can obtain an expression that relates the transition temperature with configurational entropy only ($T_{conf}^{p_1 \to p_2}$) to the transition temperature with configurational and vibrational entropy ($T_{conf,vib}^{p_1 \to p_1}$) by writing the free energy for phases p_1 and p_2, with and without vibrations, and solving the two sets of equations. This final expression, which does not include thermal expansion effects, is (Ozolins et al. 1998):

$$T_{conf,vib}^{p_1 \to p_2} = T_{conf}^{p_1 \to p_2}\left(1 + \frac{\Delta S_{vib}^{p_1 \to p_2}}{\Delta S_{conf}^{p_1 \to p_2}}\right)^{-1} \tag{88}$$

As expected, $\Delta S_{vib}^{p_1 \to p_2}/\Delta S_{conf}^{p_1 \to p_2}$ determines the effect of lattice vibrations on calculated phase diagrams. The configurational entropy per atom for a binary solid solution depends on the state of short-range order. Its maximum value for an ideal (random) solution is

$$S_{conf} = k_B[c \ln(c) + (1-c) \ln(1-c)] \tag{89}$$

The maximum value of ΔS_{conf} is $0.693 k_B/$atom: this value occurs when a fully ordered state ($S_{conf} = 0$) at $c = 0.5$ transforms into a fully random solid solution. In most cases, $\Delta S_{conf}^{p_1 \to p_2}$ will be smaller, owing to short-range order in the disordered state and also to some disorder in the low-temperature phase. Thus, even a relatively small $\Delta S_{vib}^{p_1 \to p_2}$ can lower $T_{conf,vib}^{p_1 \to p_2}$ by a significant amount. Indeed, experimental values of $\Delta S_{vib}^{p_1 \to p_2} = 0.14 k_B/$atom for Cu$_3$Au have led to expectation of a large lowering of the order–disorder temperature transition. However, if there is a large contribution of anharmonicity involved in $\Delta S_{vib}^{p_1 \to p_2}$, then the vibrational entropy increases rapidly with temperature and the effect of vibration will be smaller. Equation (88) is then incorrect.

One typical example of the major role played by the vibrational entropy is the stability of the β-phase of many metals and intermetallic compounds at high temperature. Because metallic bonding is essentially nondirectional, the greatest stability is expected for those crystallographic arrangements that maximize the coordination number of each atom (Zangwill and Bruinsma 1987). It may be concluded therefore that the bcc phase is energetically unstable in comparison to close-packed structures. However, the large entropy of the bcc phase results in a lower free energy, which is responsible for the stability of this phase. In the late 1940s, Zener (1947) predicted that the large entropy of the β-phase has a mostly vibrational origin. Forty years later, phonon dispersions of group III and IV metals have been investigated by Petry et al. by growing single crystals in situ at high temperature in the neutron scattering furnace (Flottmann et al. 1987). The dispersion curves are characterized by low-energy phonons in the [112] direction (Petry et al. 1991a,b; Trampenau et al. 1991; Güthoff et al. 1993). This makes them different from group V and VI metals, where low-energy phonons are absent and where the bcc phase stability has an electronic origin. In terms of free energy, it is mainly the increased entropy due to these low-energy phonon modes which stabilizes the bcc structure of group IV metals. The anomalous decrease in entropy in the bcc phase is caused by the increase in frequency of the low-energy phonons with increasing temperature. We note that this behavior is anomalous because with increasing temperature metals usually become softer and their phonon frequencies decrease. Phonon softening with *decreasing* temperature is a common behavior of many bcc materials and reflects their ability to undergo a phonon-triggered martensitic transformation from the high-temperature bcc phase to a close-packed structure at low temperature.

The link between the phonon softening and the martensitic transition will be discussed in Section 4.4.3. The martensitic transition takes place when the entropic contribution to the free energy is compensated by the energy difference between open and close-packed structures. For instance, the gains in vibrational entropy ΔS_{vib} at the transition for Ti and Zr are found to be equal to 0.29 and $0.26 k_B/$atom, respectively. For comparison, the total change in entropy is $\Delta S = 0.42$ and $0.40 k_B/$atom, i.e., the electronic contribution can be estimated to be $\Delta S_{el} = 0.13$ and $0.14 k_B/$atom, respectively. This means that for Ti and Zr the lattice vibrational entropy accounts for 65–70% of the excess entropy at the martensitic transition.

The same tendency has been observed in the Cu-based martensitic alloys. Assuming that the phonon modes in the close-packed phase can be described within the Debye theory, Romero and Pelegrina (1994) have calculated the $\Delta S_{vib} = 0.15 k_B$/atom for the Cu-based martensitic alloys. Owing to the small value of the electronic contribution (L. Mañosa et al. 1993a), to a very good approximation, this excess of entropy of the β-phase, responsible of its high-temperature stability, has a vibrational origin that arises from the low-energy $TA_2[110]$ phonons and is practically independent of composition. Therefore, the martensitic transition in Cu-based alloys can be considered as a transition driven purely by vibrational entropy.

4.4
Beyond the Harmonic Approximation

Up to now, we have treated the lattice vibrations in the harmonic approximation, which gives a satisfactory description of vibration properties in solids at low temperature when the vibrational amplitudes are small in comparison to the nearest-neighbor atomic distance. The harmonic approximation gives the simple picture of non-interacting phonons but has, however, a number of consequences. Being eigenstates of the harmonic Hamiltonian, the phonons have infinite lifetime, their energy linewidths are zero, and their mean free paths are infinite. In reality, lattice forces are anharmonic for all crystals at finite temperature, a fact which corresponds to the experimental observations of thermal expansion and finite values of the thermal conductivity due to multiphonon scattering. To correct for these shortcomings, anharmonicity has to be taken into account. However, the well-resolved experimental techniques used for lattice vibration investigations, such as neutron scattering, indicate that one-phonon scattering processes are clearly dominant. This suggests that anharmonicity, expressed by the third and higher-order terms in the expansion (33), can be viewed as a perturbation of the non-interacting phonon states of a crystal.

The third term $\mathscr{U}_3$ in the expansion (33) can be transformed using Eq. (55) into Eq. (90),

$$\mathscr{U}_3 = \frac{1}{3!} \frac{1}{V^{3/2}} \sum_{qq'q''} \delta_{G,q+q'+q''} \sum_{\alpha\beta\gamma} \phi_{\alpha\beta\gamma}(\mathbf{q},\mathbf{q}',\mathbf{q}'') X_\alpha(\mathbf{q}) X_\beta(\mathbf{q}') X_\gamma(\mathbf{q}'') \tag{90}$$

where

$$\phi_{\alpha\beta\gamma}(\mathbf{q},\mathbf{q}',\mathbf{q}'') = \sum_{h',h''} \phi_{\alpha\beta\gamma}(0,h',h'') \exp(i\mathbf{q}'.\mathbf{r}_{h'}) \exp(i\mathbf{q}''.\mathbf{r}_{h''}) \tag{91}$$

and **G** is a reciprocal lattice vector. The anharmonic term $\mathscr{U}_3$ involves three phonon wave vectors, which, due to the translational invariance, have to satisfy the conservation of momentum:

$$\mathbf{G} = \mathbf{q} + \mathbf{q}' + \mathbf{q}'' \tag{92}$$

The first anharmonic term introduces a coupling process between three phonons. Processes with $\mathbf{G} = 0$ are called "normal" processes and with $\mathbf{G} \neq 0$ "umklapp" processes. Similarly, the fourth term, $\mathscr{U}_4$, causes in first-order perturbation four-phonon interactions, and so on. For many purposes, the temperature-dependent properties of anharmonic crystals can be studied using the quasi-harmonic approximation, in which the equilibrium positions are taken as parameters which are determined by the minimum of the temperature-dependent free energy. The expansion is then, as in the harmonic approximation, limited to the quadratic terms in the displacement from these minimum positions. The free energy takes the same form as in the harmonic approximation, i.e., the sum of potential energy and vibrational free energy.

4.4.1
Thermal Expansion

In the previous section, we discussed the effect of anharmonicity in terms of phonon–phonon coupling. In this section we show that anharmonicity has another effect too: it gives rise to the thermal expansion of the crystal.

The thermal expansion $\alpha_P = (1/V_0)(\partial V/\partial T)_P$ is due to the anharmonicity of the total energy: the curve of Fig. 4.1(b) is slightly flatter toward the high values of a than at the small values of a. For a given energy increase $k_B T$, the average lattice parameter will therefore be larger than r_0. In the simple model of Eq. (14),

$$\langle \delta \rangle = \frac{\int_{-1}^{+\infty} \delta \exp(-\mathscr{U}(\delta)/k_B T)\, d\delta}{\int_{-1}^{+\infty} \exp(-\mathscr{U}(\delta)/k_B T)\, d\delta}$$

is calculated by expanding $\exp(\zeta\delta^3/k_B T) \approx 1 + \zeta\delta^3/k_B T$ for small δ; thus we get the thermal expansion coefficient $\alpha_P = (9\zeta/4\xi^2)k_B$ (Kittel 1976; Philibert et al. 1998). In pure Pt, we can compare this with the tabulated value: $\alpha_P = 8.90 \times 10^6$ K^{-1}. From the adjustment of the curve $\mathscr{U}(a)$ in Fig. 4.1(b), we get 7.50×10^6 K^{-1}.

The thermal expansion can also be related to the specific heat via the bulk modulus. The pressure of the system, expressed in the frame of the canonical description in statistical mechanics, is $P = -(\partial \mathscr{F}/\partial V)|_T$ where $\mathscr{F} = -k_B T \ln \mathscr{Z}$ is the free energy of the system and $\mathscr{Z}$ is the partition function. Thus the vibrational free energy is

$$\mathscr{F} = k_B T \sum_{j,\mathbf{q}} \ln[1 - \exp(-\beta\hbar\omega_j(\mathbf{q}))] + \sum_{j,\mathbf{q}} \hbar\omega_j(\mathbf{q})/2 \tag{93}$$

The derivative of $\mathscr{F}$ with respect to V gives

$$P = -\left(\frac{\partial \mathscr{F}}{\partial V}\right)\bigg|_T = -\left(\sum_{j,\mathbf{q}} \frac{\partial \mathscr{F}}{\partial \omega_j(\mathbf{q})} \frac{\partial \omega_j(\mathbf{q})}{\partial V}\right)\bigg|_T \qquad (94)$$

From Eq. (94), we see that the pressure and the volume of the system, depend on T only if the term $(\partial \omega_j(\mathbf{q})/\partial V)$ is not equal to zero. Thus the harmonic approximation, in which the frequencies $\omega_j(\mathbf{q})$ are volume-independent, cannot account for the thermal expansion.

Inserting Eq. (93) in Eq. (94) we get

$$P = \frac{1}{V} \sum_{j\mathbf{q}} \gamma_j(\mathbf{q}) \left(\hbar\omega_j(\mathbf{q}) \left(\bar{n}_{j\mathbf{q}} + \frac{1}{2}\right)\right)\bigg|_T \quad \text{with } \gamma_j(\mathbf{q}) = -\frac{\partial \ln[\omega_j(\mathbf{q})]}{\partial[\ln V]} \qquad (95)$$

where $\bar{n}_{j\mathbf{q}}$ is the equilibrium phonon distribution function. The Grüneisen parameter $\gamma_j(\mathbf{q})$ for the mode $(j, \mathbf{q})$ characterizes the variation of the normal mode frequencies $\omega_j(\mathbf{q})$ with the volume V of the crystal. To obtain the Grüneisen parameters of the mode, measurements or calculations of the phonon dispersion at different temperatures are needed. Usually neither is available. Therefore, averaged Grüneisen parameters are often used:

$$\gamma = -\frac{V}{\omega_D}\frac{\partial \omega_D}{\partial V} = -\frac{V}{\theta_D}\frac{\partial \theta_D}{\partial V} = -\frac{\partial(\ln \theta_D)}{\partial(\ln V)} \qquad (96)$$

The experimental values of γ typically vary between 1 and 2 for most solids.

The thermal expansion coefficient, defined by $\alpha = (1/V)(\partial V/\partial T)_P$, is proportional to the specific heat and is given by (Ashcroft and Mermin 1976) $\alpha = \gamma C_V/K$, where K is the bulk modulus defined by $K = -V(\partial P/\partial V)_T$. Grüneisen considered γ as independent of temperature; the thermal expansion parameter thus vanishes as T^3 when $T \to 0$. This normal behavior is violated in a number of substances where, at increasing temperature, the lattice constant decreases at low temperatures and increases at high temperatures. Such a behavior can occur if some groups of phonons have opposite temperature dependences. This has been observed even in simple structures, such in the bcc phase of group IV metals (β-Zr, β-Ti). In these systems, the majority of the phonons soften with increasing temperature, as expected for a normal metal. The phonons in a certain region of the Brillouin zone, however, show the opposite behavior: they harden considerably. This inverse temperature behavior is related to the martensitic transition which, in this case, is a phonon-driven first-order displacive transition and will be discussed in detail in Section 4.4.3.

4.4.2
Thermal Conductivity

If a temperature gradient is imposed across a solid, the elementary excitations such as free electrons and phonons will acquire more energy in the hottest part than in the coldest part. This will induce a spontaneous flow of energy (heat) from the hottest to the coldest part of the specimen. In metals both electrons and phonons contribute to the thermal conductivity. In insulators, most of the heat is conducted by the phonons. If ∇T is the applied temperature gradient, in a steady state the rate of heat flow, per unit area normal to the gradient, is given by

$$\mathcal{Q} = -\mathcal{K}\nabla T \tag{97}$$

where $\mathcal{K}$ is the thermal conductivity. The form of Eq. (97) implies that the thermal energy transfer is a random process. The energy, carried by the phonons, does not proceed directly in a straight path from the hottest spot to the coldest spot, but diffuses through the specimen. If the energy were propagating directly, the flux would not depend on the gradient but on the difference in temperature between the hottest and the coldest spots of the specimen. We have seen in the previous section that anharmonicity induces a phonon–phonon coupling process which reduces the phonon lifetime. Besides, real crystals are of finite size and contain defects and impurities. This leads to phonon scattering mechanisms that can be taken into account by introducing a mean free path into the expression of the thermal conductivity. If we assume that the gradient is applied along the x axis only, we get $\mathcal{Q}_x = -\mathcal{K}(dT/dx)$.

We assume a phonon gas is confined in a box in which the particles are free to move. If v_x is the x component of the particle velocity, the flux of particles in the x direction is $\frac{1}{2}n\langle|v_x|\rangle$, where n is the particle density and the brackets $\langle \ \rangle$ stand for the average value. At equilibrium, without a T gradient, there is a flux of equal magnitude in the opposite direction. We suppose now that heat is provided at one end of the box and we consider two isothermal cross-sections at T and $T + \Delta T$. Because of the gradient in temperature now established, the particles will move from the region at temperature $T + \Delta T$ toward the region at T. If we assume that the particle is scattered in the two regions $T + \Delta T$ and T, and is free between them, then

$$\Delta T = \frac{dT}{dx}l_x = \frac{dT}{dx}v_x\tau \tag{98}$$

where τ is the average time between the two scattering processes. While moving, the particle will give up the energy $c\Delta T$ where c is the particle heat capacity. Thus the flux balance between the two particle fluxes is

$$\mathcal{Q}_x = -n\langle v_x^2\rangle c\tau\frac{dT}{dx} = -\frac{1}{3}\langle v^2\rangle nc\tau\frac{dT}{dx} = -\frac{1}{3}Cvl\frac{dT}{dx} \tag{99}$$

with $l = v\tau$ being the mean free path and $C = nc$. Thus $\mathcal{K} = \frac{1}{3}Cvl$. The specific heat and the collision rate τ^{-1}, related to the phonon mean free path by $v = l\tau^{-1}$, are the main quantites which determine the temperature dependence of the thermal conductivity. Let us discuss qualitatively the temperature dependence of $\mathcal{K}$. At high temperature, the Bose–Einstein distribution reduces to an expression proportional to the temperature; thus, the total number of phonons present in the system is proportional to T. Since a given phonon is more likely to be scattered if there are other phonons present to reduce the mean free path by scattering processes, we expect that the collision rate τ^{-1} increases when increasing temperature. Furthermore, since at high temperature the phonon specific heat obeys the Dulong and Petit law and therefore is temperature-independent, we expect the thermal conductivity to decrease with increasing temperature. This statement is confirmed by many experimental results. Generally, the decrease of $\mathscr{K}$ at high temperature obeys a power law $\mathscr{K} \propto T^{-r}$ where r is between 1 and 2.

At low temperature, the average number of phonons with energy ω, according to the Bose–Einstein distribution, is given by $\exp(-\beta\hbar\omega)$ [see Eq. (77)]. In this temperature regime, the fraction of phonons with energies comparable to or smaller than $k_B T$ will be dominant. In particular, when $T \ll \theta_D$, the phonons will have $\omega_j(\mathbf{q}) \ll \omega_D$ and $k \ll k_D$, where k_D is the Debye wave vector defined by $k_D = \omega_D/v$. In the case of three-phonon scattering, since the total energy is conserved during the scattering process, the energy of the emerging phonon will also be small compared to $\hbar\omega_D$. This requires that the wave vector of each phonon, and hence the sum of the wave vectors, are small compared to k_D. Thus the additive reciprocal lattice vector $\mathbf{G}$ appearing in the crystal conservation of momentum law [Eq. (92)] is zero. At very low temperature, the only collisions occurring with an appreciable probability are those conserving the total momentum transfer exactly, the so-called "normal" processes. The freezing out of the *umklapp* processes is of crucial significance for the low-temperature thermal conductivity. Normal processes alone cannot bring the system to full thermodynamic equilibrium because they leave the heat energy flow $\mathscr{Q}$ unchanged. If we start with a distribution of hot phonons down a rod with $\mathscr{Q} \neq 0$, the distribution will propagate down the rod with $\mathscr{Q}$ unchanged. Therefore there is no thermal resistance and the thermal conductivity will be infinite. Crystals have finite thermal conductivity at low temperature because there will still be some small probability of crystal-momentum-destroying *umklapp* processes that degrade the thermal conductivity. This requires the presence of phonons with wave vectors comparable to k_D. The mean number of such phonons is given by the factor $\exp(-\theta_D/T)$ where θ_D is the Debye temperature. Thus, as the temperature drops, the number of phonons that can participate in *umklapp* processes drops exponentially, and the collision rate is expected to vary as $\exp(-T_0/T)$ where T_0 is a temperature of the order of θ_D. At very low temperature, however, we expect the thermal conductivity to be limited by temperature-independent scattering processes determined by the geometry and purity of the sample.

4.4.3
Soft Phonon Modes and Structural Phase Transition

Many metallic and intermetallic compounds undergo a martensitic transition, which is a first-order, diffusionless solid–solid transformation from a symmetric high-temperature phase called austenite to the low-symmetry low-temperature phase called martensite. Some of them have the ability to recover large stress-induced deformations by heating the alloy through the martensitic transition temperature. They are known as shape-memory materials (Funakubo 1987). This property makes them potential candidates for use as actuators, clamps, and sensors. Recent studies even suggest that the characteristic distortions of such materials can be exploited to create micrometric machines (Bhattacharya and James 2005). It is now well known that these displacive transitions are phonon-triggered and can be explained by the soft-phonon modes. The suggestion that certain kinds of solid–solid phase transitions might be triggered by phonon instabilities was first made by Anderson (1960) and Cochran (1964); since then many experimental studies on martensitic materials have been made.

Usually, the frequencies of normal modes of vibration are expected to decrease for increasing temperature due to anharmonic interactions at high temperature, but in martensitic materials, such as in group III and IV metals, some phonon modes show the opposite behavior (Petry et al. 1991a,b; Trampenau et al. 1991). Phonon dispersions of the bcc phase of group III and group IV metals have been investigated using one high-temperature furnace for both growth of single crystals and neutron scattering measurement (Flottmann et al. 1987). The dispersions are dominated by (a) a valley of low-energy phonons with transverse polarization and propagation along [112] and [110] directions and (b) a strong dip in the longitudinal branch in the [111] direction for a value of two-thirds of the reduced wave vector. The atomic motions corresponding to this mode can be viewed as bringing together two neighboring (111) planes, while the third stays at rest. This atomic motion gives rise, for a specific wave vector, to the $\beta \to \omega$ phase transition (Fig. 4.11). Furthermore, the phonons along these branches are strongly damped and show at large q a typical lifetime as short as one vibrational period. This feature seems to be common to other bcc-based materials and is due to their natural instability toward the formation of close-packed structures.

In the case of intermetallic compounds, Ni-based (Predel 1991; Saburi 1998) and Cu-based (Mañosa et al. 1993a,b, 1999; Nicolaus, 2000) materials are among the most investigated ones. Shape-memory alloys have an open bcc structure, denoted as the β phase, only stable at high temperatures. For kinetic reasons, the β phase can be retained as a metastable phase below its stability region by means of fast enough cooling. In this process, the bcc structure is configurationally ordered in the B2, $L2_1$, or DO_3 superstructures. On further cooling, the martensitic transition takes place at a temperature that depends strongly on composition. Because this transition is diffusionless, martensite inherits the ordered arrangement of the high-temperature β phase. Martensite is a close-packed phase that can be described by application of a combination of a shear mechanism and a shuffle to the high-temperature bcc phase (Burgers 1934; Kajiwara 1986) together with a superimposed static modulation (or shuffle) which corresponds to a phonon mode on the $TA_2[110]$ with a specific wave vector. The low-energy $TA_2[110]$ phonon branch shows a hollow in these alloys. For example, lattice dynamics of the Heusler alloy Ni_2MnGa has been investigated by inelastic neutron

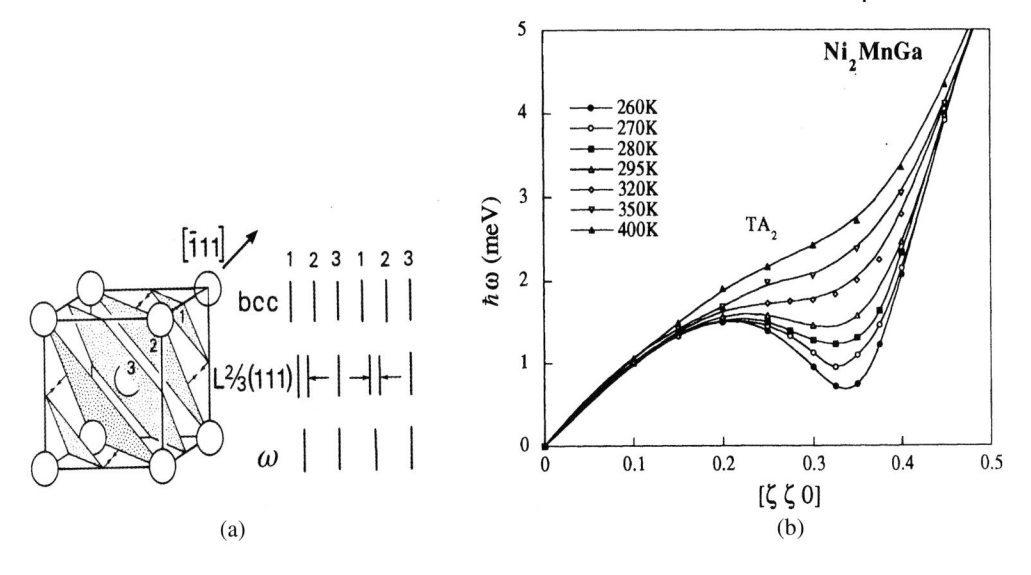

Fig. 4.11 (a) $\beta \to \omega$ transformation scheme. The ω phase has a hexagonal structure with three atoms per unit cell with $a_\omega = a_\beta \sqrt{2}$ and $c_\omega = a_\beta \sqrt{3}/2$, where a_β is the lattice parameter of the bcc Bravais lattice. The c and a axes of the ω-phase point along the [111] and [110] directions of the bcc lattice respectively. (b) Phonon dispersion in the [110] direction of L2$_1$ Ni$_2$MnGa at different temperatures showing the wiggle around $\xi = 0.33$ (Zheludev et al. 1995).

scattering (Zheludev et al. 1995). A wiggle in the TA$_2[\xi\xi 0]$ branch deepens with decreasing temperature and results in a distinct minimum at $\xi_0 = 0.33$ below 300 K (Fig. 4.11b).

Furthermore, a diffuse elastic scattering develops at $\xi_0 = 0.33$ when the martensitic transition temperature is approached, and is accompanied by an elastic peak at the same position in the **q** space. Such a behavior, which has been found in other Ni-based alloys such as NiAl (Shapiro et al. 1991) and NiTi (Mercier et al. 1980), is known as a precursor effect of the martensitic transition. From first-principle calculations (Zhao and Harmon 1992), the elastic peak has been attributed to a strong electron–phonon coupling and specific nesting properties of the Fermi surface, whereas Halperin and Varma (1976) considered it as an extrinsic property attributed to crystal imperfections. Although premonitory effects in the form of anomalously soft phonons are frequently observed, the soft-phonon frequencies may remain finite at the temperature where the austenitic phase becomes unstable (incomplete phonon softening). The crystal surface can then play an important role. Landmesser et al. (2003) have studied such phase transitions between two and three dimensions by X-ray scattering under grazing angles in Ni$_2$MnGa. They have shown that a free surface should be considered as a defect for nucleation if the soft bulk phonon couples to the surface. The resulting surface phonon promotes the phase transformation by lowering the nucleation barrier at the surface.

4.5
Experimental Investigation of the Normal Modes of Vibration

4.5.1
Raman Spectroscopy

Raman scattering is a powerful light scattering technique used to diagnose the internal structure of molecules and crystals. In a light scattering experiment, light of a known frequency and polarization is scattered by a sample and analyzed in frequency and polarization. Raman scattered light is frequency-shifted with respect to the excitation frequency due to atomic or molecular vibrations, but the magnitude of the shift is independent of the excitation frequency. This "Raman shift" is an intrinsic property of the sample. Additional information, related to the spatial form of the excitation, derives from the polarization dependence of the Raman scattered light. The shape of an excitation in a material, for example a vibration pattern of the atoms in a crystal, and the polarization dependence of the scattering are determined by the equilibrium structure of the material through the rules of group theory. By this route one gets valuable and unambiguous structural information from the Raman polarization dependence.

The conservation laws for energy and crystal momentum transfers during such processes require

$$\hbar\omega_f = \hbar\omega_i \pm \hbar\omega_j(\mathbf{q})$$
$$\hbar n \mathbf{k}_f = \hbar n \mathbf{k}_i \pm \hbar \mathbf{q} + \hbar \mathbf{G} \tag{100}$$

where $\mathbf{G}$ is a reciprocal lattice vector. $\mathbf{k}_i$, $\mathbf{k}_f$, and ω_i, ω_f are the wave vectors and frequencies of the incident and the scattered photons respectively. Note that the photon wave vector inside the crystal will differ from its free space value due to the index of refraction n of the crystal. The upper (plus) sign in Eqs. (100) refers to the scattering process in which a phonon is created (the Stokes component of the scattered light) and the lower (minus) sign refers to the process in which a phonon is annihilated (known as the anti-Stokes component). If photons of visible light are used, their wave vectors are small compared to the Brillouin zone dimensions, and the excited phonons lie only near the immediate neighborhood of the center of the Brillouin zone ($\mathbf{q} = 0$). In this case, the crystal momentum conservation law (100) is obeyed only if the reciprocal lattice vector $\mathbf{G}$ is zero. To a good approximation, Raman scattering occurs from zero wave vector phonons. However, to the extent that the phonon wave vector differs from zero, phonon selection rules will deviate from the zero wave vector rules and will depend on the angle between the directions of propagation of the incident and scattered light. If optical phonons, which have no dispersion at the zone center, are excited, the direction dependence of the Raman shift is quite small. On the other hand, for the acoustic phonons, which have a linear dispersion near the zone center, the angular dependence of the Raman shift is more pronounced. Because the phonon wave vector is small, the acoustic phonon will have a small energy. Raman scattering from low-energy acoustic phonons is known as Brillouin scattering. The essential difference between Raman and Brillouin scattering is the sensitivity of the "Brillouin shift" to the relative angle of scattering.

The frequency spectrum of the Raman scattered light maps out only a part of the excitation spectrum. Other spectroscopic techniques, such as infrared absorption, are used to map out the non-"Raman-active" excitations. As will be seen in the next section, Raman spectroscopy is a less powerful probe of the phonon dispersion spectra than neutron scattering. The great virtue of neutrons is that the dispersion is accessible over the whole Brillouin zone and once the scattered energies have been resolved, the highly informative one-phonon scattering processes are clearly identifiable.

4.5.2
Inelastic Neutron Scattering

The energy of thermal neutrons is in the same order of magnitude as that of the dynamical excitations in solids. So when the neutron is inelastically scattered by the creation or the annihilation of a phonon, the change in the energy of the neutron is a large part of its initial energy. Measurement of neutron energy thus provides accurate information on energies of phonons and hence on interatomic forces. Thermal inelastic neutron scattering is one of the major tools to measure the phonon dispersion relationship, i.e., the frequency ω_j as a function of $\mathbf{q}$ and polarization index j.

Consider a neutron, of momentum $\mathbf{k}_i$ and energy $E_i = \hbar^2 k_i^2/2m$, that is incident upon a crystal. After the interaction, the neutron emerges with momentum $\mathbf{k}_f$ and energy $E_f = \hbar^2 k_f^2/2m$. During the neutron–crystal interaction, a dynamical excitation of wave vector $\mathbf{q}$ and energy $\hbar\omega_j(\mathbf{q})$ is created or annihilated. The conservation laws of energy and momentum imply that the scattering process, in the case of phonon annihilation, obeys the conditions $E_f - E_i = \hbar\omega_j(\mathbf{q})$ and

$$\mathbf{k}_f - \mathbf{k}_i = \mathbf{q} + \mathbf{G} \tag{101}$$

where $\mathbf{G}$ is a reciprocal lattice vector. These conditions are so restrictive that for given scattering angles only a few phonons of particular $\mathbf{q}$ and $\omega_j(\mathbf{q})$ can be involved in the scattering process. We can make use of this to determine the phonon dispersion $\omega_j(\mathbf{q})$. Generally, a monochromatic beam arrives on the crystal and the energy of the neutrons scattered through a given angle is measured by using a time-of-flight apparatus or another crystal as analyzer. When a phonon is detected, the scattered energy gives access to $\omega_j(\mathbf{q})$ and Eq. (101) to the change in wave vector. As $\mathbf{q}$ has to lie on the first Brillouin zone, both $\mathbf{q}$ and $\mathbf{G}$ are obtained. One point on the phonon dispersion curve is determined.

In an inelastic neutron scattering experiment, the measured quantity is the partial differential cross-section which gives the fraction of neutrons of incident energy E scattered into an element of solid angle $d\Omega$ with an energy between E and $E + dE$:

$$\left(\frac{d^2\sigma}{d\Omega\, dE}\right)(\mathbf{q},\omega) = \frac{(2\pi)^3}{2v_0} \frac{|\mathbf{k}_f|}{|\mathbf{k}_i|} \sum_{j,\mathbf{G}} \frac{|F_j(\mathbf{Q},\mathbf{q})|^2}{\omega_j(\mathbf{q})} (\Delta_- + \Delta_+) \tag{102}$$

where (Squires 1978; Lovesey 1984)

$$\Delta_- = (\bar{n}_{j\mathbf{q}} + 1)\delta(\omega - \omega_j(\mathbf{q}))\delta(\mathbf{Q} - \mathbf{q} - \mathbf{G}), \tag{103}$$

$$\Delta_+ = \bar{n}_{j\mathbf{q}}\delta(\omega + \omega_j(\mathbf{q}))\delta(\mathbf{Q} + \mathbf{q} - \mathbf{G}) \tag{104}$$

and v_0 is the volume of the unit cell, $\mathbf{Q} = \mathbf{k}_i - \mathbf{k}_f$ the scattering vector, and $\bar{n}_{j\mathbf{q}}$ is the Bose–Einstein distribution. $F_j(\mathbf{Q},\mathbf{q})$ is the dynamical structure factor, given by:

$$F_j(\mathbf{Q},\mathbf{q}) = \sum_{k=1}^{p} \frac{b_k}{\sqrt{m_k}} \exp(-W_k \mathbf{Q}^2) \exp(i\mathbf{Q}\cdot\mathbf{r}_k) \times (\mathbf{Q}\cdot\mathbf{e}_k(j,\mathbf{q})) \tag{105}$$

b_k is the coherent scattering length of the kth atom in the unit cell. $W_k = \langle u(k)^2 \rangle/2$ is the mean square displacement of atom k. The delta functions in Eqs. (103) and (104) express the conservation laws. Equation (104) describes a scattering process in which a phonon is created. The neutron energy decreases by an amount equal to the energy of a phonon of type j with a wave vector $\mathbf{q}$. Similarly Eq. (104) describes a phonon annihilation process where the energy of the neutron is increased.

The intensity of the phonon observation by neutron scattering is determined by different factors. The term $1/\omega_j(\mathbf{q})$ in Eq. (102) indicates, independently of all other terms, that the intensity is inversely proportional to the frequency of the mode. This attenuation of the intensity results from the quantum mechanics of the harmonic oscillator. High-energy modes are thus always more difficult to observe than the low-energy ones. Since neutron experiments suffer from the low flux of the existing sources, this effect frequently prohibits the study of the high-frequency part of the phonon dispersion curves. The term $\bar{n}_{j\mathbf{q}}$ results from the Bose–Einstein statistics of a given mode with frequency $\omega_j(\mathbf{q})$; the occupation number is given by Eq. (77) and tends toward zero for $T \to 0$. For $T \to \infty$, $\bar{n}_{j\mathbf{q}}$ approaches $\hbar\omega_j(\mathbf{q})/k_B T$, i.e., the classical relation. Equation (103) indicates that the Bose–Einstein function is increased by $+1$ in the case of a phonon creation. At low temperature, where $\bar{n}_{j\mathbf{q}}$ is close to zero, the phonon can be observed only in the creation mode, the cross-section for the annihilation process becomes vanishing since the phonon states are no longer occupied. At finite temperature, the Bose–Einstein distribution further simplifies the observation of phonons with low frequencies. One more factor in expression (105) is the scalar product $\mathbf{Q}.\mathbf{e}_k(j,\mathbf{q})$, which means that by measuring the intensity of the scattered signal, it is possible, in principle, to deduce the polarization vectors $\mathbf{e}_k(j,\mathbf{q})$. In general the polarization vectors corresponding to a given wave vector $\mathbf{q}$ are not related in a simple way to the direction of $\mathbf{q}$. But in certain cases there is a simple relationship. For example, if $\mathbf{q}$ lies in the (001) plane of a cubic crystal, one of the $\mathbf{e}_k(j,\mathbf{q})$ is along the [001] axis. If the scattering vector $\mathbf{Q}$ is arranged so that $\mathbf{Q}$ is in the (001) plane, $\mathbf{Q}.\mathbf{e}_k(j,\mathbf{q})$ is zero for this mode. This can be used to extinguish one polarization branch. Inelastic neutron scattering experiments are usually performed using either a time-of-flight setup or a three-axis spectrometer. In the time-of-flight setup, the energy transfer of the inelastically scattered neutrons is deduced from the time the neutrons spend in flight over the chopper–sample–detector distance and compared to that of the elastically scattered neutrons. The shift in time-of-flight can then be converted into a gain or a loss in energy. The time-of-flight method is a useful technique for determining the phonon density of states by measuring the incoherent one-phonon scattering as a function of energy. A basic difficulty of the method is that incoherent scattering from multiphonon processes also occurs. Both one-phonon and multiphonon processes give incoherent scattering, and it is not even easy to estimate the contribution of the latter even if the use of an appropriate sample environment, such as hollow cylinders, may reduce it. Furthermore, the time-of-flight method to measure phonon frequencies suffers from the disadvantage that the $\mathbf{q}$ value of the phonon cannot be preselected. The use of three-axis spectrometers, developed originally by Brockhouse (1960), overcomes this disadvantage. On a three-axis spectrometer, the energy and wave vector of the incident neutron are selected from the white beam by a first Bragg reflection on a monochromator (first axis, angle θ_M; see Fig. 4.12). They will then

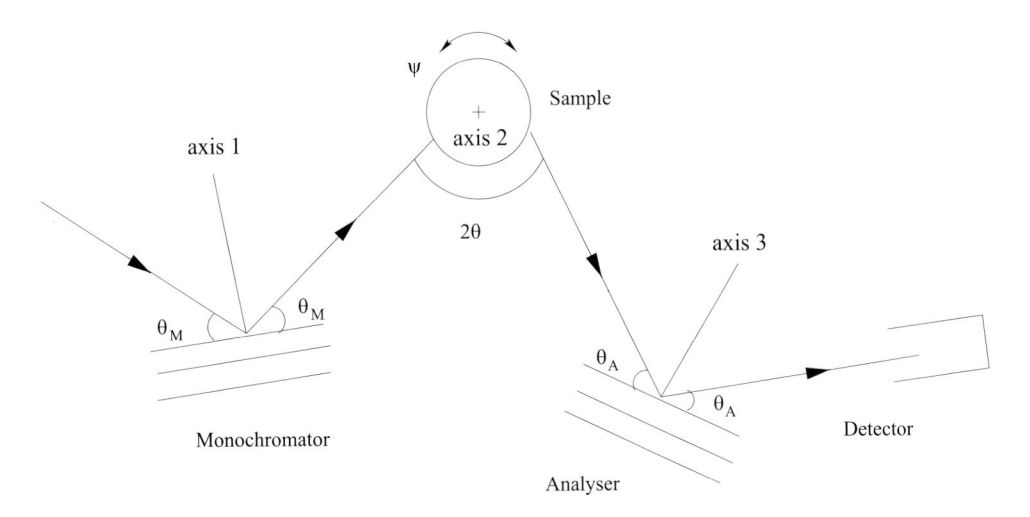

Fig. 4.12 Three-axis spectrometer scheme with the different angles.

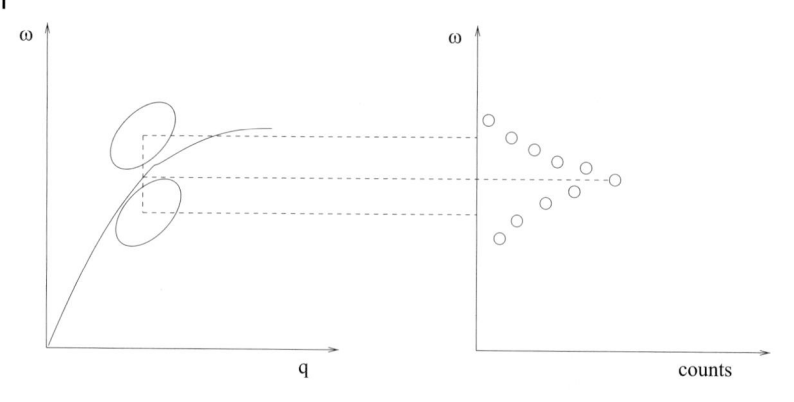

Fig. 4.13 Measurement of a phonon $(\mathbf{q}, \omega)$ in a **q-constant** mode. The ellipse shows the resolution function of the spectrometer.

interact with the sample and will be scattered in a direction (second axis, angle ψ, defining the wave vector direction of the scattered neutrons) along which a second crystal, called the analyzer, is placed to select the energy by another Bragg reflection (third axis, angle θ_A). Finally the neutrons are detected by a detector. The momentum transfer is chosen by rotating the sample and the analyzer to get the good exit wave vector direction and length. One point in the $(\mathbf{q}, \omega)$ space corresponds to one configuration of the spectrometer, i.e., the angles θ_M, θ_A, and ψ of the three axes. The idea is to scan the $(\mathbf{q}, \omega)$ space using different configurations of the spectrometer; when a given point $(\mathbf{q}, \omega_j)$ verifies the dispersion relationship $\omega = \omega_j(\mathbf{q})$ of the crystal, the number of counted neutrons will substantially increase.

Either q-constant or ω-constant scans are made depending on the slope of the dispersion curve at the measured point (Fig. 4.13). Strictly speaking, because of the mosaicity spreads of the sample itself, of the monochromator, and of the analyzer, the scattering will not occur at a point but rather on a volume centered on $(\mathbf{q}, \omega_j)$. If a neutron is detected, it has a probability $R(d\mathbf{q}, d\omega)$ to have in reality a transfer $(\mathbf{q} + d\mathbf{q}, \omega_j + d\omega)$. $R(d\mathbf{q}, d\omega)$ is called the resolution function of the spectrometer. The measured signal is the convolution of the inelastic scattering cross-section by the resolution function of the spectrometer.

4.6
Phonon Spectra and Migration Energy

Of the two parameters that govern diffusion and ordering kinetics, i.e., E_F the vacancy formation energy and E_M the vacancy migration energy, E_M is generally less known and more difficult to measure for ordered intermetallic compounds. In pure metals and random alloys, it can be deduced, for example, from stage III of resistivity recovery during annealing after low-temperature irradiation

(Schultz 1991), or from a thorough analysis of residual resistometry along isothermal and isochronal annealing series (Balanzat and Hillairet 1981; Schulze and Lücke 1972). Those two methods are, however, very sensitive to the microstructure of the samples and to any impurities or defects.

An alternative method of determining E_M is an evaluation from lattice-dynamical properties, i.e., elasticity and phonon dispersion. Flynn (1968) proposed a model for estimating the migration energy in metals of cubic structures from elastic constants. Recently, a model relating the migration energy for nearest-neighbor jumps in cubic metals to the phonon density of states (DOS) has been proposed by Schober et al. (1992).

Both models assume that the total energy is well approximated by a sinusoidal form so that the knowledge of the phonon properties (the bottom of the well) gives valuable information on the saddle point (the top of the ridge) of atomic jump processes. The validity of this approximation has been verified in various fcc metals (Schober et al. 1992) and L1$_2$ compounds (Kentzinger and Schober 2000) by molecular dynamics. In the different phases, there is a geometrical factor that makes it possible to take into account the departure from a sinusoidal form. This factor is determined by making the calculation in molecular dynamics in different systems within the phase concerned.

Figure 4.14 shows the result of such a calculation in the ordered phase CoPt (Montsouka et al. 2006). The well corresponds to a second-order polynomial with equation: $\mathcal{U} = \mathcal{U}_0 + \xi' x^2$, with ξ' values of 410, 276, 345, and 281 eV nm^{-2} in a–d, respectively. We clearly see that the departure from the sinusoidal form is present. All parameters of the sinusoidal function are fixed by the nearest-neighbor distance d and the saddle point energy E_S: $\mathcal{U} = \mathcal{U}_0 + E_S/2(1 - \cos(\pi x/d))$. In the case of a jump of the atoms out of the (001) plane, we have to add a slope due to the formation energy of the anti-sites (0.235 eV in both cases here). A comparison between different L1$_0$ systems will allow generalization to all L1$_0$ phases. The comparison of the different figures clearly shows that there is a strong anisotropy of the vacancy migration in the L1$_0$ phase. Surprisingly, the migration within the (001) planes is more difficult for stereographic reasons.

Flynn's model has been developed for monoelemental cubic lattices and assumes that the migration energy is dominated by the contribution of acoustical modes at the center of the Brillouin zone. Its obvious drawback is that the other low-energy phonons are not directly taken into account. The migration energy derived by Flynn (1968) in the continuum limit reads:

$$E_M^{Fly} = c V_0 \delta^2$$

Here,

$$\frac{15}{2c} = \frac{3}{c_{11}} + \frac{2}{c_{11} - c_{12}} + \frac{1}{c_{44}}$$

where c_{11}, c_{12}, and c_{44} are the three independent elastic constants of the cubic crystal and V_0 is the atomic volume. δ is a structure-dependent constant, which

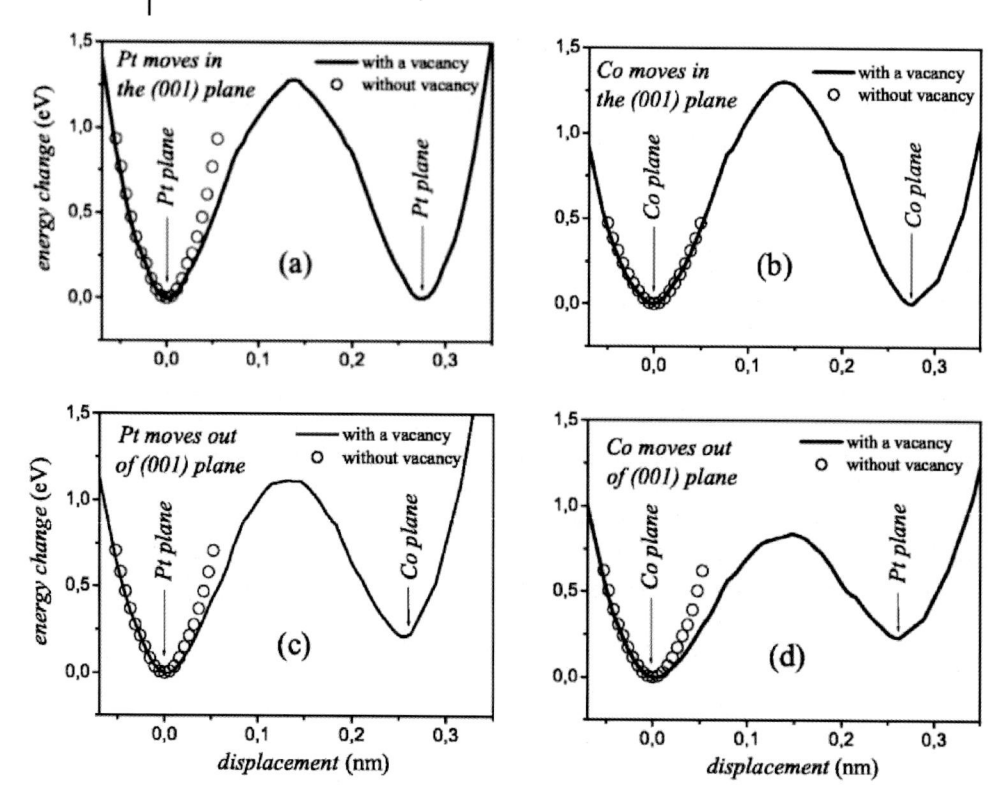

Fig. 4.14 Line: variation of the total energy in CoPt when a Pt (a,c) or a Co (b,d) atom moves toward a nearest-neighbor vacancy within the same (001) plane (a,b) or not (c,d). ○: Variation of the total energy in CoPt when a Pt (a,c) or a Co (b,d) moves around its equilibrium position toward a nearest-neighbor atom within the same (001) plane (a,b) or not (c,d) (Montsouka et al. 2006).

is to be determined by minimizing the deviations between the experimental values and E_M^{Fly} in different materials of the structure concerned. Schober et al. (1992) have obtained $\delta^2 = 0.081$ for the fcc structure. Using the elastic constants deduced from the phonon dispersion slope at the origin of the Brillouin zone in the [110] direction of the fcc disordered FePd, the migration energy in Flynn's model has been found equal to 0.78 eV, a value which compares quite well with that obtained by the ultrasonic resonance spectrometry (Ichitsubo 2000) (0.67 eV).

Based on earlier work of Flynn, Schober et al. (1992) have developed a model relating the migration energy to the whole phonon spectrum through the phonon DOS. In this model, the migration energy is derived from Eq. (106):

$$E_M = a^2 \alpha \mathcal{G}_0^{-1} \quad \text{where} \quad \mathcal{G}_0 = \int (g(\omega)/m\omega^2)\, d\omega. \tag{106}$$

In Eq. (106), E_M separates into a structural factor $(a^2\alpha)$, common to all fcc, and an electronic factor $(\mathcal{G}_0)$, which is the static Green's function and reflects the electronic particularities. a is the lattice parameter and α a geometrical constant; α has been determined by computer simulation to be equal to 0.0135 in fcc metals (Schober et al. 1992).

The underlying assumption of the model is that diffusion takes place via nearest-neighbor jumps and that the potential has a sinusoidal form along the atom trajectory. In fcc metals, it is commonly accepted that self-diffusion is dominated by the simplest vacancy mechanism, namely $\frac{1}{2}(1,1,0)$ atomic jumps into a nearest-neighbor vacancy (Peterson 1978; Vogl et al. 1989; Petry et al. 1991b). In pure fcc metals, excellent agreement between calculated and measured values of E_M was found, whereas in bcc metals, where the experimental values are less well known, predictions were obtained that show a pronounced systematic behavior with chemical group. Strictly speaking, this method to determine E_M is only valid for elemental crystals but as the model was not yet developed for the superstructures, this method has been used in $Fe_{1-x}Si_x$ (Randl 1994; Randl et al. 1995) and $Fe_{1-x}Al_x$ alloys (Kentzinger et al. 1996) to estimate the migration enthalpies assuming an average mass of the component. Recently, the same model has been used to deduce the migration energy in the L1$_0$-FePd (Mehaddene et al. 2004). Since the model was not yet extended to L1$_0$ phases, as a first approximation, the FePd L1$_0$ structure was considered as an fcc lattice with an atomic mass equal to the averaged mass of the two atoms. The migration energy at room temperature in FePd (0.91 eV) is close to the migration barriers obtained by Kentzinger and Schober (2000) in FePd$_3$ (0.97 eV for a Pd atom migrating within its own lattice and 1.14 eV when it jumps on the Fe lattice; 0.90 eV for a Fe atom leaving its own lattice). With an average atomic mass and the total DOS, a 0.94 eV value was obtained, very comparable to the value obtained within the same approximation in FePd. Recently, 0.7 eV was obtained as the average migration energy in the ordered paramagnetic phase of FePd, using residual resistivity measurements (Partyka et al. 2000; Kulovits et al. 2005), in good agreement with the values deduced from phonon measurements using neutron inelastic scattering in the same chemical and magnetic state of order.

In a further step, Schober's model was extended to A$_3$B compounds with L1$_2$ structure using the static lattice Green's functions of the two constituent species (Kentzinger and Schober 2000). The Green's function matrix elements $\mathcal{G}_{\alpha\beta}^{kk}$ ($\alpha, \beta = x, y$ or z) of a given atom k can be expressed in terms of the eigenvalues $\omega_j(\mathbf{q})$ and eigenvectors (i.e., polarization vectors) $\mathbf{e}_k(j,\mathbf{q})$ ($j = 1,\ldots,3p$ where p is the number of atoms in the unit cell) of the dynamical matrix at wave vector $\mathbf{q}$ which are known from Born–von Karman (BVK) fits of the measured phonon dispersion curves: $\mathcal{G}_{\alpha\beta}^{kk} = \int (g_{\alpha\beta}^{kk}(\omega)/m_k\omega^2)\,d\omega$, where m_k is the mass of atom k and $g_{\alpha\beta}^{kk}(\omega)$ is the partial density of states:

$$g_{\alpha\beta}^{kk}(\omega) = \frac{1}{3p}\frac{v_0}{(2\pi)^3}\sum_j \int e_k^\alpha(j,\mathbf{q}) e_k^\beta(j,\mathbf{q}) \delta(\omega - \omega_j(\mathbf{q}))\,d\mathbf{q} \qquad (107)$$

with v_0 being the volume of the unit cell; the integration extends over the first Brillouin zone.

In the case of elemental fcc or bcc metals, the static lattice Green's function matrix of an atom $k(\mathscr{G}^{kk})$ is diagonal and reduces to a single number ($\mathscr{G}_0$). In A_3B compounds with $L1_2$ structure, the B atoms have cubic point symmetry and the Green's function matrix $\mathscr{G}^{BB}$ also reduces to a single number, $\mathscr{G}^{BB}_{xx}$. The point symmetry of the A atoms is tetragonal. For an A atom at position $(\frac{1}{2},\frac{1}{2},0)$ in the unit cell, $\mathscr{G}^{AA}$ is diagonal with $\mathscr{G}^{AA}_{xx} = \mathscr{G}^{AA}_{yy} \neq \mathscr{G}^{AA}_{zz}$. In this case, starting from the fully ordered lattice with one vacancy, three types of nearest-neighbor jumps can be considered: the jump of an A atom into an A vacancy (A → V_A), the jump of an A atom into a B vacancy (A → V_B), and the jump of a B atom into an A vacancy (B → V_A).

The Green's matrix elements of many $L1_2$ compounds which have been calculated from the phonon spectra, via the densities of states using the force constants obtained in the BVK fits, are given and discussed in Kentzinger and Schober (2000). Here we report on recent calculation of the Green's function elements in $CoPt_3$ and $FePd_3$. The Green's functions and the deduced migration energies for each kind of atomic jump and at the different temperatures are compiled in Table 4.1. In order to compare the different systems, the $L1_2$ phases have also been treated within Schober's model for fcc pure metals using an average atomic mass; the corresponding migration energies are denoted E_M^{av} in Table 4.1. The average migration energies calculated for these systems show a tendency to increase with the average atomic mass of the system (Table 4.1). The migration energies in the $L1_2$ ordered alloys also show, as expected, an increase with the size of the atom involved in the jump. Moreover, in each alloy, A → V_B is associated with the highest E_M and B → V_A with the lowest (Table 4.1). The migration energy is expected to decrease with increasing temperature as a consequence of increasing anharmonicity. In fact, Schober's model assumes that the diffusion jump – the most anharmonic conceivable event – follows a trajectory parallel to the direction of low-harmonic-restoring forces. In other words, in the directions where the harmonic part of the potential is low, we expect the anharmonic part, i.e., the migration barrier, to be low as well. Using a Monte Carlo model, based on

Table 4.1 Static lattice Green functions (in 10^{-2} m/N) of $CoPt_3$ and $FePd_3$ in the $L1_2$ phase and migration enthalpies (in eV) deduced from the model developed by Kentzinger and Schober (2000).*

Alloy	T [K]	$\mathscr{G}^{AA}_{xx}$	$\mathscr{G}^{AA}_{zz}$	$\mathscr{G}^{BB}_{xx}$	$E_M^{A \to V_A}$	$E_M^{A \to V_B}$	$E_M^{B \to V_A}$	E_M^{av}
$CoPt_3$	300	0.968	0.741	1.154	1.28	1.47	1.20	1.36
$CoPt_3$	930	1.048	0.734	1.101	1.35	1.43	1.19	1.33
$FePd_3$	80	1.264	0.929	1.553	0.97	1.14	0.90	0.94

*A refers to the majority atoms (Pt or Pd) and B to the minority ones (Co or Fe).

a nearest-neighbor vacancy jump mechanism and on an Ising Hamiltonian with effective pair interaction energies for first- and second-neighbors in L1$_2$ and L1$_0$ (Kerrache et al. 2000) compounds, the contribution of the ordering energy to the mean migration energy has been found equal to $\Delta E_M^{av}/T_C = 0.28$ meV K^{-1} and 0.19 meV K^{-1}, respectively. ΔE_M^{av} is the difference between the migration energy in the ordered and in the disordered phases and T_C is the order–disorder transition temperature. The migration energy is larger in the ordered state because, at least in some directions, the migration of the vacancies is accompanied by a change of chemical long-range order that costs some additional energy. It is always the case when the vacancy changes sublattice, contrary to the case where the vacancy remains within one sublattice (the sublattice of majority atoms in the L1$_2$ and both planar sublattices in L1$_0$) (Oramus et al. 2001; Kozubski et al. 2001). These values are close to $\Delta E_M^{av}/T_C = 0.23$ meV K^{-1} and 0.18 meV K^{-1}, respectively, obtained experimentally from the phonon measurements via the Schober model considering the total phonon DOS and an average atom of the L1$_0$-FePd and the L1$_2$-CoPt$_3$.

4.7
Outlook

Elastic and dynamic properties are still studied in many "modern" materials. The electron–phonon couplings have appeared as of prime importance in superconducting materials, for instance.

Interesting examples are also the magnetic martensitic materials which exhibit, beyond the conventional thermal shape-memory effect, a shape recovery under applied magnetic field and large magnetic-field-induced strains. They are known as magnetic shape-memory materials. In these magnetic materials, the motion of the twin boundaries of the different martensite variants costs less energy than the reorientation of magnetic moments within a magnetic domain, inducing a change in the shape depending on the direction of the applied magnetic field (Suorsa et al. 2004). This suggests the existence of a strong magnetoelastic coupling in these materials. This coupling has been found to be responsible of the appearance of a premartensitic phase in Ni$_2$MnGa (Planes et al. 1997). A better understanding of this coupling along with the investigation of other candidate systems is one of the challenging tasks in this field for the coming years. Lattice dynamics investigations under applied magnetic field and/or mechanical stresses on these systems is a powerful technique for investigating the interatomic potential and the interplay between the magnetic order and the normal modes of vibration; it is a necessary step for any extensive applied research on these systems.

Along with these developments, a new technique, based on time-resolved phonon measurements, has been developed recently and used successfully for the study of the kinetics of unmixing in silver alkali halogenides (Eckold et al. 2004). It can be extended to investigate the kinetics of the phonon-driven structural phase transitions, offering a unique tool for examining the phonon–time dependence across the transition.

The study of elastic and dynamic properties of materials has moreover encountered renewed interest with the development of thin films and self-organized systems. As a matter of fact, the relaxation of strains is one of the most important phenomena which has to be taken into account to understand the different epitaxial growth modes and the self-organization.

When the lattice mismatch between a substrate and an epitaxial film cannot be neglected, the elastic energy accumulated in both the strained film and the substrate is one of the driving forces for the film relaxation, in addition to the surface energies and the cohesive energies. If the substrate is flat and perfect enough (with a single crystal buffer layer, for example), up to a certain thickness dislocations are not generated as long as the accumulated elastic energy is not greater than the nucleation energy of a dislocation (Pan et al. 1995). This activation energy is large because a supplementary atomic plane has to be introduced, starting from the surface that is the only possible source. In the case of intermetallic ordered films, the antiphase boundary energy is large and prevents the propagation of perfect dislocations in ordered alloys. In this case, other processes appear to relax the system: stacking faults and microtwins form preferentially to relax a strained layer (Halley et al. 2002, 2004), or exchanges between substrate and layer atoms can occur (Goyhenex et al. 1999).

For the semiconductor optoelectronic industry, the optical properties of the systems are strongly deteriorated by any defect. Compliant substrates have therefore been developed to grow zero defect – and zero strain – layers. These substrates are made of two atomically flat GaAs substrates that are twist-bonded so that a bicrystal is formed (Patriarche and Le Bourhis 2000; Le Bourhis and Patriarche 2005). One of the substrates is then thinned down to a nanometer-sized thickness. When the angle between the respective lattices of the two crystals is equal to 45°, the bonding between the two crystals is very soft. An epitaxial layer on the substrate will be pseudomorphic: epitaxial dislocations will be located at the grain boundary of the bicrystal only. Until now, these types of substrate are developed for optoelectronics, but with the increasing miniaturization of the magnetic dots necessary for high-density magnetic storage it may be also used in intermetallic systems in the future as it will then be necessary to avoid defects.

A very surprising result of these studies is that the macroscopic laws of elasticity (as described in this chapter) are valid down to a few atomic distances (typically less than 1 nm) of the interfaces or surfaces. For example, a combination of TEM and X-ray diffraction allowed Kegel et al. (2001) to determine shape, strain fields, and interdiffusion in semiconductor quantum dots grown in the Stranski–Krastanov mode. In metallic systems, Muller et al. (2001) have shown that the oscillations of in-plane lattice spacing observed during two-dimensional (2D) homo- and heteroepitaxial growth can be interpreted well by using the classical elasticity theory and the bulk elastic constants to formulate the problem of the elastic relaxation of a coherent 2D epitaxial deposit. They have predicted a dependence of the amplitude of in-plane lattice spacing oscillation on the nucleation density in good agreement with experimental observations in several intermetallic systems: V/Fe(0 0 1), Mn/Fe(0 0 1), Ni/Fe(0 0 1), Co/Cu(0 0 1), and V/V(0 0 1).

Similarly, deformations around a dislocation core (several tens of nanometers) are well described by the macroscopic laws of elasticity down to about 3 nm from the dislocation line. Until now, these studies have also been made mainly on semiconductors, because defects have a direct influence on the band structure properties of these systems. Moreover, the high-resolution images necessary to compare calculations with experiments are easier in these systems (Snoeck et al. 1998). Ab initio calculations give a good description of the dislocation core structure (Pizzagalli et al. 2003) whereas the method of finite elements yields the long-range strains (Kret et al. 1999). New static theories of gradient elasticity, nonlocal elasticity, gradient micropolar elasticity, and nonlocal micropolar elasticity have been developed to describe the dislocation altogether (Lazar et al. 2005).

Elasticity measurements have been performed by Faurie et al. (2004) on W and Au thin films. They found the Young's modulus to be slightly smaller in W and larger in Au compared to bulk materials, probably due to the many defects and particular texture (isotropic in W and columnar texture in Au), whereas the Poisson ratios were in both cases very similar to the bulk values. Another important elastic constant in films is the thermodilatation constant, because a large difference in the constant of the substrate will induce large strains and defects during temperature changes in operation.

Recently, elastic strains have been used extensively to grow nanostructured surfaces in a self-organized manner. For example, the surface of new compliant substrates, with an angle between the respective lattices of the two crystals lower than 15°, contains a dense network of pure screw dislocations and presents a modulation of strains on its surface (Patriarche and Le Bourhis 2002). The growth of a film on it will favor either the compressive or the extended regions, depending on the relative lattice constants of the substrate and the film. A regular mesolattice (a lattice of dots with a periodicity of several nanometers) will thus be obtained.

A similar phenomenon can appear in the case of a reconstructed surface of the substrate. For example, the gold (111) surface relaxes by forming a herringbone structure with alternating fcc and hcp stacking within the last atomic layers (Padovani et al. 1999; Fruchart et al. 1999; Bulou et al. 2004). At the kinks of this fishbone pattern appear compressed and dilated ranges where Co or Fe grows preferentially. Fruchart et al. have grown some Co columns using the preferential growth of Au on unstrained Au and preferential growth of Co on strained Au. The strain propagates through only a few atomic layers, but columns several nanometers long can be obtained if Au and Co are deposited alternatively in the right proportions. Equivalent phenomena are at the origin of vertical arrangement of the quantum dot formation in InAs/GaAs systems (Xie et al. 1995). When some InAs dots are grown and covered by a film of GaAs as thick as 40 monolayers, the strain propagates up to the surface and favors the nucleation of the next dot on top of the previous one. Samples containing such coupled InGaAs/GaAs columns of dots are suitable for normal incidence quantum dot infrared photodetectors (Adawi et al. 2003). In other systems such as GaN/InGaN, blue high-brightness light-emitting diodes can be developed (Lee et al. 2004).

The strains and bond changes around the steps on the surfaces are also used to self-organize subatomic films. On the surface of a Au single crystal cut along a (7 8 8) plane, a new periodicity appears along the [2 1 1] direction: the terrace width is 3.9 nm due to the miscut angle. The surface is thus spontaneously patterned in two dimensions at a nanometer scale with a macroscopic coherence length. All the sites around a terrace edge are not equivalent for adatoms. A self-organized growth can thus be obtained. In this way, using a Au(7 8 8) vicinal surface, long-range-ordered cobalt nanodots could be grown on a substrate cooled down to 130 K (Repain et al. 2002).

On the other hand, phonons are studied extensively at present: to understand in detail the behavior of superconducting alloys the coupling of phonons with electrons and plasmons must be investigated (see Chapter 14b). The surface reconstruction can also be sensitive to phonons: Meunier et al. (2002) have shown that in the Ag/Cu (111) system there are two competing superstructures: the Moiré and the triangular structures, with a strong effect of temperature. Calculating the local phonon density of states by means of the recursion method gave evidence for a "strong correlation between local pressure and local vibrational entropy." The phase transition from a triangular (10 × 10) structure toward a Moiré (9 × 9) structure when the temperature increases is due to "a subtle balance between vibrational entropy and anharmonic effects in the internal energy."

Phonon excitations within crystals and at their surfaces have also been investigated by X-ray, neutron, and light diffraction methods, because of possible applications of surface acoustic waves as tunable monochromators for X-rays or neutrons (Sauer et al. 1999).

It can be expected that due to the increasing precision of electronic structure calculations, in the future lattice vibrations will be taken into account more and more. The development of nanostructures will continue and implies an increasing importance of the elastic and phonon contributions to the total energy of these nanometer systems.

References

Adawi, A.M., Zibik, E.A., Wilson, L.R., Lematre, A., Cockburn, J.W., Skolnick, M.S., Hopkinson, M., Hill, G., Liew, S.L., Cullis, A.G. (2003), *Appl. Phys. Lett.* **82**, 3415.

Anderson, P.W. (1960), *Fizika Dielektrikov*, ed. G.I. Skanavi, Akademia Nauk, Moscow, p. 290.

Ashcroft, N.W., Mermin, N.D. (1976), *Solid State Physics, International Edition*, p. 826.

Asta, M., Foiles, S.M. (1996), *Phys. Rev. B* **53**, 2389.

Balanzat, E., Hillairet, J. (1981), *J. Phys. F: Met. Phys.* **11**, 1977.

Bhattacharya, K., James, R.D. (2005), *Nature* **307**, 53.

Bieber, A., Gautier, F. (1984), *J. Phys. Jpn.* **53**, 2061.

Born, M., Huang, K. (1968), *Dynamical Theory of Crystal Lattices*, International Series of Monographs on Physics, Oxford University Press, London.

Born, M., Oppenheimer, J.R. (1927), *Ann. Phys., Lpz.* **84**, 457.

Brockhouse, B.N. (1960), *Bull. Am. Phys. Soc.* **5**, 462.

Bulou, H., Scheurer, F., Ohresser, P., Barbier, A., Stanescu, S., Quirós, C. (2004), *Phys. Rev. B* **69**, 155 413.

Burgers, W.G. (1934), *Physica* **1**, 561.
Carlsson, A.E. (1990), *Solid State Physics* **43**, ed. H. Ehrenreich, D. Turnbull, Academic Press, Boston, MA, p 1.
Cochran, W. (1964), *Phys. Rev.* **133**, 412.
Cohen-Tanoudji, C., Diu, B., Laloe, F. (1977), *Quantum Mechanics*, Wiley-Interscience, New York, p. 898.
Craivich, J., Sanchez, J.M. (1997), *Comput. Mater. Sci.* **8**, 92.
CRC Handbook of Chemistry and Physics (1997), 78th edition, ed. D.R. Lide.
Dalvit, D., Frasati, J., Lawrie, I.D. (1999), *Problems on Statistical Mechanics*, Graduate Student Series in Physics, IOP Publishing, London, p. 3.
Debye, P. (1912), *Ann. Phys., Lpz.* **39**, 789.
Drautz, R., Faehnle, M., Sanchez, J.M. (2004), *J. Phys.: Condens. Matter* **16**, 3843.
Ducastelle, F. (1970), *J. Phys. (Paris)* **31**, 1055.
Eckold, G., Caspary, D., Elter, P., Güthoff, F., Hoser, A., Schmidt, W. (2004), *Physica B* **350**, 83.
Faurie, D., Renault, P.O., Le Bourhis, E., Villain, P., Goudeau, P., Badawi, F. (2004), *Thin Solid Films* **469–470**, 201.
Finel, A. (1987), Ph.D. Thesis, Université Paris VI.
Finel, A. (1994), *Statistics and Dynamics of Alloy Phase Transformation*, ed. P.E.A. Turchi, A. Gonis, Vol. 319 of NATO Advanced Study Institute, Series B: Physics, Plenum Press, New York, p. 495.
Flottmann, T., Petry, W., Serve, R., Vogl, G. (1987), *Nuclear Instruments and Methods in Physics Research A* **260**, 165.
Flynn, C.P. (1968), *Phys. Rev.* **171**, 682.
Friedel, J. (1969), *The Physics of Metals*, ed. J.M. Ziman, Cambridge University Press, Cambridge, p. 340.
Fruchart, O., Klaua, M., Barthel, J., Kirschner, J. (1999), *Phys. Rev. Lett.* **83**, 2769.
Funakubo, H. (1987), *Shape Memory Alloys*, Gordon and Breach Science Publishers, New York, p. 30.
Galanakis, I., Ostanin, S., Alouani, M., Dreyssé, H., Wills, J.M. (2000), *Phys. Rev. B* **61**, 599; *Phys. Rev. B* **61**, 4093.
Galavotti, G. (1999), *Statistical Mechanics*, Springer-Verlag, New York, p. 62.
Goyhenex, C., Bulou, H., Deville, J.P., Tréglia, G. (1999), *Phys. Rev. B* **60**, 2781.
Grange, W., Galanakis, I., Alouani, M., Maret, M., Kappler, J.-P., Rogalev, A. (2000), *Phys. Rev. B* **62**, 1157.
Guethoff, F., Petry, W., Stassis, C., Heiming, A., Hennion, B., Herzig, C., Trampenau, J. (1993), *Phys. Rev. B* **47**, 2563.
Halley, D., Samson, Y., Marty, A., Bayle-Guillemaud, P., Beigné, C., Gilles, B., Mazille, J.E. (2002), *Phys. Rev. B* **65**, 205 408.
Halley, D., Marty, A., Bayle-Guillemaud, P., Gilles, B., Attane, J.P., Samson, Y. (2004), *Phys. Rev. B* **70**, 174 438.
Halperin, B.I., Varma, C.M. (1976), *Phys. Rev. B* **14**, 4030.
Ichitsubo, T. (2000), PhD Thesis, Kyoto University.
Kajiwara, S. (1986), *Mater. Trans. JIM* **17**, 435.
Kegel, I., Metzger, T. H., Lorke, A., Peisl, J., Stangl, J., Bauer, G., Nordlund, K., Schoenfeld, W. V., Petroff, P. M. (2001), *Phys. Rev. B* **63**, 035 318.
Kentzinger, E., Schober, H.R. (2000), *J. Phys.: Condens. Matter* **12**, 8145.
Kentzinger, E., Cadeville, M.C., Pierron-Bohnes, V., Petry, W., Hennion, B. (1996), *J. Phys.: Condens. Matter* **8**, 5535.
Kerrache, A., Parasote, V., Kentzinger, E., Pierron Bohnes, V., Cadeville, M.C., Kerrache, A., Zemirli, M., Bouzar, H., Hennion, B. (2001), *Defect and Diffusion Forum* **194–199**: Diffusion in Materials DIMAT2000 p. 403.
Kittel, C. (1976), *Introduction to Solid State Physics*, 7th edition, Wiley, New York, p. 87.
Kozubski, R., Oramus, P., Pfeiler, W., Cadeville, M.C., Pierron-Bohnes, V., Massobrio, C. (2001), *Archiv. Metall.* **46**, 145.
Kret, S., Benabbas, T., Delamarre, C., Androussi, Y., Dubon, A., Laval, J.Y., Lefebvre, A. (1999), *J. Appl. Phys.* **86**, 1988.
Kulovits, A., Soffa, W.A., Pueschl, W., Pfeiler, W. (2005), *Intermetallics* **13**, 510.
Landmesser, G., Finayson, T.R., Johnson, R.L., Aspelmeyer, M., Plech, A., Peisl, J., Klemradt, U. (2003), *J. Phys. IV* **112**, 123.
Lazar, M., Maugin, G.A., Aifantis, E.C. (2005), *Phys. Stat. Sol. (B)* **242**, 2365.
Le Bourhis, E., Patriarche, G. (2005), *Acta Materialia* **53**, 1907.

Lee, D.S., Florescu, D.I., Dong, L., Ramier, J.C., Merai, V., Parekh, A., Begarney, M.J., Armour, E.A. (2004), *Phys. Stat. Solidi A: Appl. Res.* **201**, 2644.

Legrand, B., Guillopé, M., Luo, J.S., Tréglia, G. (1990), *Vacuum*, **41**, 311.

Le Neindre, B. (1991), in *Techniques de l'Ingénieur, Traité Constantes Physico-chimiques Constantes Mécaniques Form. K 486 Coefficients d'Élasticité*.

Lovesey, S.W. (1984), *Theory of Neutron Scattering from Condensed Matter*, Oxford University Press. London: Clarendon Press.

Lu, Z.W., Wei, S.H., Zunger, A., Frota-Pessoa, S., Ferreira, L.G. (1991), *Phys. Rev. B* **44**, 512.

Mañosa, L., Planes, A., Ortin, J., Martinez, B. (1993a), *Phys. Rev. B* **48**, 3611.

Mañosa, L., Zarestky, J., Lograsso, T., Delaney, D.W., Stassis, C. (1993b), *Phys. Rev. B* **48**, 15 708.

Mañosa, L., Zarestky, J., Bullock, M., Stassis, C. (1999), *Phys. Rev. B* **59**, 9239.

Mehaddene, T., Kentzinger, E., Hennion, B., Tanaka, K., Numakura, H., Marty, A., Parasote, V., Cadeville, M.C., Zemirli, M., Pierron-Bohnes, V. (2004), *Phys. Rev. B* **69**, 024 304.

Mercier, O., Brüesch, P., Bührer, W. (1980), *Helv. Phys. Acta* **53**, 243.

Meunier, I., Tréglia, G., Tétot, R., Creuze, J., Berthier, F., Legrand, B. (2002), *Phys. Rev. B* **66**, 125 409.

Montsouka, R.V.P., Goyhenex, C., Schmerber, G., Ulhaq-Bouillet, C., Derory, A., Faerber, J., Arabski, J., Pierron-Bohnes, V. (2006), submitted to *Phys. Rev. B* **74**, 144409 (2006).

Moriarty, J.A., Belak, J.F., Rudd, R.E., Söderlind, P., Streitz, F.H., Yang, L.H. (2002), *J. Phys.: Condens. Matter* **14**, 2825.

Mueller, S. (2003), *J. Phys.: Condens. Matter* **15**, R1429.

Muller, P., Turban, P., Lapena, L., Andrieu, S. (2001), *Surf. Sci.* **488**, 52.

Nicolaus, K. (2000), PhD Thesis, Technical University of Munich, Germany.

Oramus, P., Kozubski, R., Pierron-Bohnes, V., Cadeville, M.C., Pfeiler, W. (2001), *Phys. Rev. B* **63**, 174 109.

Ozolins, V., Wolverton, C., Zunger, A. (1998), *Phys. Rev. B* **58**, 5897.

Padovani, S., Scheurer, F., Bucher, J.P. (1999), *Europhys. Lett.* **45**, 327.

Pan, G.Z., Michel, A., Pierron-Bohnes, V., Vennegues, P., Cadeville, M.C. (1995), *J. Mater. Res.* **10**, 1539.

Partyka, E., Kmiec, D., Czekaj, S., Kozubski, R. (2000), in: *Proceedings of the VIII Seminar Diffusion and Thermodynamics of Materials*. Brno: Masaryk University, Brno, 2002. ISBN 80-210-2934-X. (the university is the publisher) p. 83.

Patriarche, G., Le Bourhis, E. (2000), *Philos Mag A* **80**, 2899.

Patriarche, G., Le Bourhis, E. (2002), *J. Phys.: Condens. Matter* **14**, 12 967.

Peterson, N.L. (1978), *J. Nucl. Mater.* **69–70**, 3.

Petry, W., Heiming, A., Trampenau, J., Alba, M., Herzing, C., Schober, H.R., Vogl, G. (1991), *Phys. Rev. B* **43**, 10 933.

Petry, W., Heiming, A., Herzig, C., Trampenau, J. (1991), *Defect and Diffusion Forum* **75**, 211.

Pettifor, D.G., Oleinik, I.I., Nguyen-Manh, D., Vitek, V. (2002), *Comput. Mater. Sci.* **23**, 33.

Philibert, J., Vignes, A., Bréchet, Y., Combrade, P. (1998), *Métallurgie: du Minerai au Matériau*, ed. Paris: Masson (1998).

Pizzagalli, L., Cicero, G., Catellani, A. (2003), *Phys. Rev. B* **68**, 195 302.

Planes, A., Obradó, E., Gonzàles-Comas, A., Mañosa, L. (1997), *Phys. Rev. Lett.* **79**, 3926.

Predel, B. (1991), *Landolt-Börnstein*, New Series IV/Sa, ed. O. Madelung, Springer-Verlag, Berlin, p. 212.

Randl, O.G. (1994), PhD Thesis, University of Vienna.

Randl, O.G., Vogl, G., Petry, W., Hennion, B., Sepiol, B., Nembach, K. (1995), *J. Phys.: Condens. Matter* **7**, 5983.

Repain, V., Baudot, G., Ellmer, H., Rousset, S. (2002), *Europhys. Lett.* **58**, 730.

Romero, R., Pelegrina, J.L. (1994), *Phys. Rev. B* **50**, 9046.

Saburi, T. (1998), in *Shape Memory Materials*, ed. K. Otsuka, C.M. Waymann, Cambridge University Press, Cambridge, Cambridge, p. 97.

Sanchez, J.M., Ducastelle, F., Gratias, D. (1984), *Physica A* **128**, 334.

Sanchez, J.M., Stark, J.P., Moruzzi, V.L. (1991), *Phys. Rev. B* **44**, 5411.

Sauer, W., Streibl, M., Metzger, T.H., Haubrich, A.G.C., Manus, S., Wixforth, A.,

Peisl, J., Mazuelas, A., Härtwig, J., Baruchel, J. (1999), *Appl. Phys. Lett.* **75**, 1709.

Schober, H.R., Petry, W., Trampenau, J. (1992), *J. Phys.: Condens. Matter* **4**, 9321 and corrigendum: (1992), *J. Phys. Condens. Matter* **5**, 993.

Schultz, H. (1991), *Atomic Defects in Metals*, Landolt-Börnstein New Series Group III, Vol. 25, ed. H. Ullmaier, Berlin.

Schulze, A., Lücke, K. (1972), *Acta Met.* **20**, 529.

Shapiro, S.M., Yang, B.X., Noda, Y., Tanner, L.E., Schryvers, D. (1991), *Phys. Rev. B* **44**, 930.

Silverman, A., Zunger, A., Kalish, R., Adler, J. (1995), *J. Condens. Matter.* **17**, 1167; (1995), *Phys. Rev. B* **51**, 107 95.

Snoeck, E., Warot, B., Ardhuin, H., Rocher, A., Casanove, M.J., Kilaas, R., Hytch, M.J. (1998), *Thin Solid Films* **319**, 157.

Squires, G.L. (1978), *Introduction to the Theory of Thermal Neutron Scattering*, Cambridge University Press, Cambridge.

Suorsa, I., Pagounis, E. (2004), *J. Appl. Phys.* **95**, 4958.

Trampenau, J., Heiming, A., Petry, W., Alba, M., Herzig, C., Mikeley, W., Schober, H.R. (1991), *Phys. Rev. B* **43**, 10 963.

Tréglia, G., Legrand, B., Ducastelle, F. (1988), *Europhys. Lett.* **7**, 575.

Vogl, G., Petry, W., Flottman, T., Heiming, A. (1989), *Phys. Rev. B* **39**, 5025.

Zangwill, A., Bruinsma, R. (1987), *Comments, Condens. Mater. Phys.* **13**, 1.

Zener, C. (1947), *Phys. Rev.* **71**, 846.

Zhao, G.L., Harmon, B.N. (1992), *Phys. Rev. B* **45**, 2818.

Zheludev, A., Shapiro, S.M., Wochner, P., Schwartz, A., Wall, M., Tanner, L.E. (1995), *Phys. Rev. B* **51**, 11 310.

Zunger, A. (1994), *NATO Advanced Study Institute on Statics and Dynamics of Alloy Phase Transformations*, ed. P. Turchi, A. Gonis, Plenum, New York, p. 361.

Xie, Q., Madhukar, A., Chen, P., Kobayashi, N. (1995), *Phys. Rev. Lett.* **75**, 2542.

5
Point Defects, Atom Jumps, and Diffusion

Wolfgang Püschl, Hiroshi Numakura, and Wolfgang Pfeiler

5.1
Point Defects

All crystalline solids contain structural imperfections (cf. Chapter 3), and those of atomic scale are called point defects or atomic defects. Defects in solids play essential roles in many properties of materials, through their characteristic production of local distortion and mobility. The effects are sometimes undesirable but very often beneficial, contrary to what the name "defect" suggests: for example, the atomic transport (diffusion and ionic conduction) in solids occurs by movements of point defects, and mechanical strength (hardness, flow stress, and creep resistance) is strongly influenced by their presence. In this section, after outlining the fundamentals, we discuss the equilibrium properties of point defects in metallic crystals: pure metals, dilute alloys, and ordered alloys. We aim to form the basis for a quantitative understanding of diffusion processes from the microscopic viewpoint. Since atomic diffusion in crystalline solids is governed by the concentrations and the mobilities of mediating defects, detailed knowledge on the properties of point defects is essential. The elementary processes and phenomenological aspects of diffusion are described in Sections 5.2–5.4.

5.1.1
A Brief Overview

5.1.1.1 Types of Point Defects
Figure 5.1 shows atomic defects commonly found in elemental crystals. Atomic vacancies (vacant lattice sites) and self-interstitial atoms are intrinsic defects, and a pair of them is called a Frenkel defect. Foreign atoms, i.e., substitutional and interstitial solute atoms, constitute extrinsic defects. Complex defects may be formed by association of these defects, such as pairs of vacancies (divacancies), pairs of a vacancy and a solute atom, triplets of vacancies or larger clusters, and so on.

Intrinsic atomic defects in an ordered AB alloy are illustrated in Fig. 5.2. Atoms at regular lattice sites are indicated as A_α and B_β, i.e., A atoms on α sublattice sites

Alloy Physics: A Comprehensive Reference. Edited by Wolfgang Pfeiler
Copyright © 2007 WILEY-VCH Verlag GmbH & Co. KGaA, Weinheim
ISBN: 978-3-527-31321-1

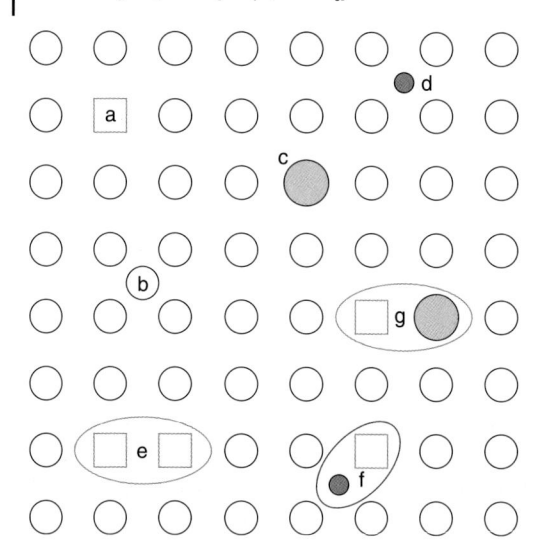

Fig. 5.1 Atomic defects in an elemental crystal. a: atomic vacancy (customarily indicated by a square), b: self-interstitial atom, c: substitutional solute atom, d: interstitial solute atom, e: vacancy–vacancy pair (divacancy), f: vacancy–interstitial solute atom pair, g: vacancy–substitutional solute atom pair. If 'a' and 'b' are viewed as a pair, it is called a Frenkel defect.

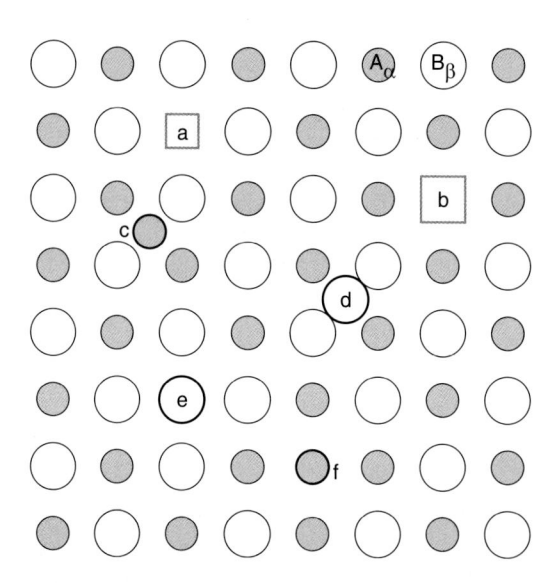

Fig. 5.2 Atomic defects in an ordered AB alloy, in which A atoms (shaded circles) and B atoms (open circles) are arranged on two sublattices α and β. A_α and B_β: atoms at regular positions. a: vacancy V_α (vacant α site), b: vacancy V_β (vacant β site), c: self-interstitial atom I_A, d: self-interstitial atom I_B, e: antisite atom B_α, f: antisite atom A_β.

and B atoms on β sublattice sites, respectively. Vacancies in each sublattice and self-interstitial atoms of each atom species may occur, with the relative ease of formation depending on the material. The intrinsic defect species characteristic of ordered alloys are antisite atoms (also called antistructure atoms or substitutional defects), which are atoms that occupy sublattice sites for the other atom species. Increase in their numbers brings about disordering of the atomic arrangement, leading eventually to disruption of the ordered structure, i.e., an order–disorder transition.

These defects in ordered alloys cannot be present in arbitrary numbers, since the numbers of the sublattice sites must be maintained at a ratio specific to the structure, such as 1:1 or 3:1. Because of this constraint, vacancies often tend to be formed at a fixed proportion, e.g., in pairs in AB alloys, or in a set of four vacancies in A_3B alloys, with three A vacancies, V_α, and one B vacancy, V_β. Such a set of vacancies is referred to as a Schottky defect. Complexes of these defects are also possible. Note, however, that a Schottky defect does not necessarily indicate that the vacancies are closely associated in space, and neither does a Frenkel defect; they are concepts in statistical thermodynamics.

5.1.1.2 Formation of Equilibrium and Nonequilibrium Defects

Point defects exist in thermal equilibrium owing to the significant increase in the configurational entropy associated with their presence. This makes a strong contrast to such extended defects as dislocations and grain boundaries, which are essentially unstable but remain in the crystal because of the lack of sufficient mobility. Here we shall briefly outline the formation of point defects and introduce fundamental parameters characterizing their theromodynamic properties.

Figure 5.3 illustrates schematically the formation of a vacancy in an elemental crystal: an atom is removed from the interior and then attached to the external surface. In terms of the simple pair-interaction model of cohesion, on removal of the atom the internal energy E of the crystal is raised by the bond energy times the coordination number, and roughly one half of this energy is recovered if the atom is then attached at a favorable position, e.g., at a ledge of a surface step. At the same time, the volume of the crystal is increased by about one atomic volume, increasing slightly the enthalpy of the crystal. On the other hand, the con-

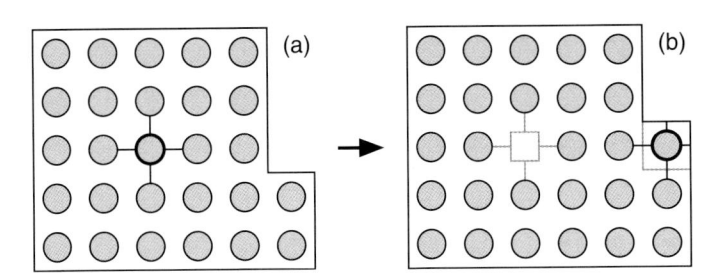

Fig. 5.3 Formation of a vacancy in an elemental crystal. An atom in the interior (a) is removed and attached to the external surface (b).

figurational entropy of the vacancies lowers the Gibbs free energy $G = E + PV - TS$ at any temperature $T > 0$. The configurational entropy accounts for the number of possible ways of placing the vacancies in the crystal, and for a crystal consisting of N atoms and n vacancies it is given by

$$S_c = k_B \ln \frac{(N+n)!}{N!n!} \tag{1}$$

where k_B is the Boltzmann constant. Assuming that the vacancy concentration is low enough for their interactions to be neglected, the change in the Gibbs energy due to the formation of vacancies is expressed as

$$\Delta G(N,n) = n\Delta g_f - TS_c \tag{2}$$

where Δg_f is the change in the free energy per vacancy, excluding the configurational entropy term. The equilibrium concentration can be derived readily by substituting Eq. (1) into Eq. (2) and by minimizing it with respect to n:

$$c_v = \frac{n}{N+n} = \exp\left(-\frac{\Delta g_f}{k_B T}\right) \tag{3}$$

Here, the Stirling approximation $\ln x! \approx x \ln x - x$ ($x \gg 1$) is used for both N and n. The principle is the same for interstitial atoms and other atomic defects, and thus their equilibrium concentrations are given by similar formulae.

The free energy of formation Δg_f can be expressed by the parameters that describe the characteristics of the defect in more detail:

$$\begin{aligned}\Delta g_f &= \Delta h_f - T\Delta s_f \\ &= \Delta e_f + P\Delta v_f - T\Delta s_f\end{aligned} \tag{4}$$

Here, Δh_f, Δs_f, Δe_f, and Δv_f are the enthalpy, entropy, energy, and volume of formation, respectively, and P is pressure. As noted earlier, Δs_f is the contribution (per defect) to the change in the entropy of the crystal excluding the configurational entropy, and is referred to as the nonconfigurational entropy of formation. Since it is due largely to the influence on the lattice vibration, it is often called the vibrational entropy of the defect. The formation volume of a point defect is of the order of the atomic volume, Ω, which is typically 10^{-29} m^3. Under the atmospheric pressure, the $P\Delta v_f$ term amounts to only 10^{-24} J, or 10^{-5} eV, and contributes little to the formation enthalpy. Hence the formation enthalpy Δh_f and the formation energy Δe_f are virtually equal except at very high pressures.

Complexes of point defects may form by association of individual defects, and the associated defects are often energetically more favored. The balance between the associated and separated defects may be described by a chemical reaction between the defect α and defect β:

$$\alpha + \beta \rightleftharpoons \alpha\beta : \Delta g_{a,\alpha\beta} \tag{5}$$

The free energy of association $\Delta g_{a,\alpha\beta}$ is defined as the difference in the formation energy between the isolated and associated states:

$$\Delta g_{a,\alpha\beta} = \Delta g_{f,\alpha\beta} - (\Delta g_{f,\alpha} + \Delta g_{f,\beta}) \tag{6}$$

Association is favored when $\Delta g_{a,\alpha\beta}$ is negative, and vice versa. The equilibrium concentration of the complexes is expressed in the form

$$c_{\alpha\beta} \approx A c_{\alpha} c_{\beta} \exp\left(-\frac{\Delta g_{a,\alpha\beta}}{k_B T}\right) \tag{7}$$

where A is a number of the order of 10 depending on the geometrical configuration of the associated defect. The free energy of binding, $\Delta g_{b,\alpha\beta} \equiv -\Delta g_{a,\alpha\beta}$, is often used to characterize the preference for the complex. The cases of divacancies and vacancy–solute atom pairs are discussed in Section 5.1.2.1 in more detail.

As the reader might be aware, the process of creating a vacancy illustrated in Fig. 5.3 is unrealistic; indeed, it is only to explain the change in the internal energy on introducing a vacancy in a crystal. In reality, intrinsic point defects – vacancies, self-interstitial atoms, and their clusters – are produced at dislocations, grain boundaries, and surfaces, which serve as sources and at the same time as sinks, and the defects move around in the crystal to establish a desired spatial distribution. Intrinsic defects are created and annihilated incessantly at various places; the equilibrium concentrations are not quiescent but are maintained dynamically.

There are artificial means of introducing intrinsic point defects into the crystal, other than the thermal, or spontaneous, formation described above. Plastic deformation and particle irradiation are the most common among the others. Jogs in edge dislocations act as sources and sinks of vacancies and self-interstitial atoms (Hirth and Lothe 1982; see also Chapter 6) and thus severe plastic deformation produces these point defects. Irradiation with energetic particles, such as electrons, neutrons, and various ions, causes "knock-on" of atoms, i.e., bombarded atoms are displaced out of the regular positions, generating vacancies and self-interstitial atoms. In metallic crystals, knock-on events occur when the transferred kinetic energy exceeds a certain value, ca. 20–30 eV, which is referred to as the displacement threshold energy (Gibson et al. 1960; Lucasson and Walker 1962; Erginsoy et al. 1964; Vajda 1977). Accelerated electrons having an energy of 100 keV to 1 MeV deliver roughly this amount of kinetic energy to the atoms of the target material and introduce pairs of vacancies and self-interstitial atoms. This is the technique that has widely been used in experimental studies of Frenkel defects in metals (Meechan and Brinkman 1954; Corbett et al. 1957; Lucasson and Walker 1962). In contrast, heavier particles with similar kinetic energies produce much more significant disorder in the crystal, because the energy transferred to the primary target atoms is much higher than in the case of electrons; the atoms displaced in the first event (the primary knock-on atoms) bring about multiple knock-on events (Marx et al. 1952; Blewitt et al. 1957; Thomas et al. 1969). This leads to serious deterioration of the material, and the effect, known as radiation damage, has been one of the key issues in nuclear reactor technology (Thompson 1969; Odette and Lucas 2001).

5.1.1.3 Mobility

Point defects migrate in the crystal by successive translational movements between equivalent positions. The elementary process is a jump to a neighboring site, passing through an intermediate high-energy state, which is accompanied by local distortion. Figure 5.4 illustrates the atomic configurations in the course of a jump of an interstitial solute atom and the variations of the free energy. As will be discussed in detail in Section 5.2.1, the jump over the energy barrier usually occurs by thermal activation, and the rate, or frequency, of the jump obeys the Arrhenius law,

$$\omega = \nu \exp\left(-\frac{\Delta g_m}{k_B T}\right) \tag{8}$$

Here, ν is the attempt frequency, which is of the order of the Debye frequency (10^{13}–10^{14} s^{-1}). The free energy of migration, Δg_m, can be expressed using the enthalpy, entropy, energy, and volume of migration, similarly to the case of the free energy of formation, Eq. (4).

The jump frequency ω and the free energy of migration Δg_m of a particular defect species can be determined by measuring its diffusion coefficient D as a func-

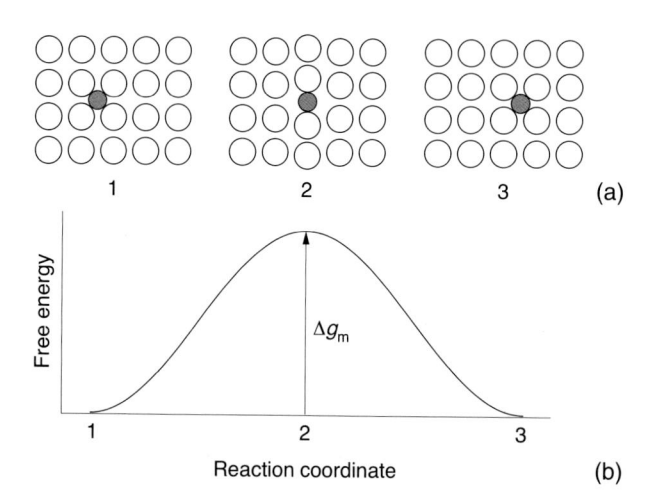

Fig. 5.4 (a) Motion of an interstitial solute atom to a neighboring site. (b) The associated variation of the free energy of the crystal.

tion of temperature, as D is proportional to ω. One of the common techniques is to analyze the kinetics of annihilation or precipitation of the defects. The decay of excess defects during annealing may be monitored by, e.g., measuring the electrical resistivity, and from the temperature dependence of the kinetics the activation energy of the rate-controlling process can be evaluated (Bauerle and Koehler 1957; Koehler et al. 1957; Corbett et al. 1959a,b). The next section describes the method in more detail.

5.1.1.4 Experimental Techniques

Differential dilatometry The absolute concentration of vacancies can be determined by measuring simultaneously macroscopic and microscopic thermal expansion (Feder and Nowick 1958). The macroscopic volume of an elemental crystal consisting of N atoms and n vacancies at temperature T is expressed as

$$V(T) = N\Omega^\circ(T) + n\Delta v_{f,v}(T) \tag{9}$$

where $\Omega^\circ(T)$ is the atomic volume in a hypothetical defect-free crystal, and $\Delta v_{f,v}$ is the formation volume. The actual atomic volume, $\Omega(T)$, which can be determined by measuring the lattice parameter, reflects the relaxation around each vacancy and is given by the fractional average (Eshelby 1956):

$$\Omega(T) = \frac{N}{N+n}\Omega^\circ(T) + \frac{n}{N+n}\Delta v_{f,v}(T) \tag{10}$$

Owing to this formula, the volume of the crystal can be written simply as

$$V(T) = (N+n)\Omega(T) \tag{11}$$

Let us consider the difference in volume between temperatures T_0 and T. Denoting $V(T_0)$, $\Omega(T_0)$, and $n(T_0)$ as V_0, Ω_0, and n_0, respectively, the relative difference is expressed as

$$\frac{V(T) - V_0}{V_0} = \frac{(N+n)\Omega(T) - (N+n_0)\Omega_0}{(N+n_0)\Omega_0}$$

$$\approx \frac{n}{N} - \frac{n_0}{N} + \frac{\Omega(T) - \Omega_0}{\Omega_0} \tag{12}$$

Here, all three terms on the right-hand side are assumed to be small, and second-order terms are omitted. If, moreover, n_0/N can be neglected, we arrive at

$$\frac{\Delta V}{V_0} - \frac{\Delta \Omega}{\Omega_0} \approx \frac{n}{N} = \frac{c_v}{1-c_v} \approx c_v \tag{13}$$

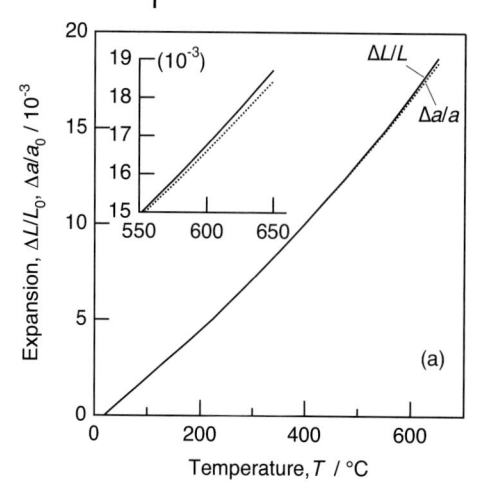

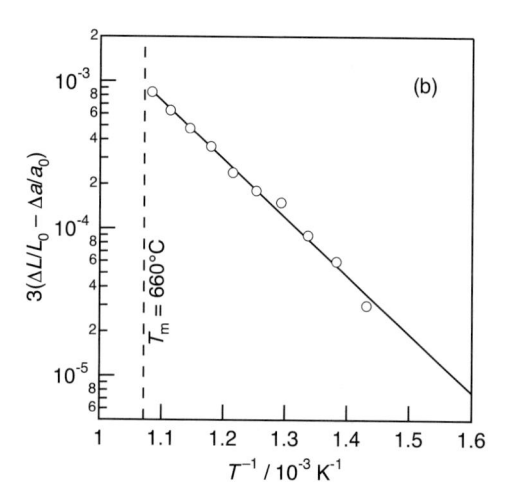

Fig. 5.5 Macroscopic and microscopic thermal expansion of pure aluminium (a), and an Arrhenius plot of the vacancy concentration, $c_v = 3(\Delta L/L_0 - \Delta a/a_0)$ (b). After Simmons and Balluffi (1960).

where notations $\Delta V \equiv V(T) - V_0$ and $\Delta \Omega \equiv \Omega(T) - \Omega_0$ are introduced. For crystals of cubic structure, the relative increase in the macroscopic volume $\Delta V/V_0$ and in the atomic volume $\Delta \Omega/\Omega_0$ can be determined by measuring simultaneously the linear dimension of the sample, L, and the lattice parameter, a, respectively. From the two sets of data, the absolute concentration of vacancies is obtained from Eq. (14).

$$c_v \approx 3\left(\frac{\Delta L}{L_0} - \frac{\Delta a}{a_0}\right) \tag{14}$$

Note that the magnitude of the formation volume $\Delta v_{f,v}$ need not be known. Figure 5.5 shows the experimental results for pure aluminum by Simmons and Balluffi (1960). The deviation of $\Delta L/L_0$ from $\Delta a/a_0$ at the temperatures close to the melting temperature (660 °C) demonstrates the production of an excess volume due to vacancies. Figure 5.5(b) is an Arrhenius plot of the vacancy concentration calculated by Eq. (14), from which the formation enthalpy $\Delta h_{f,v}$ and the formation entropy $\Delta s_{f,v}$ were evaluated (0.76 eV and 2.4 k_B, respectively).

When self-interstitial atoms are considered together with vacancies, Eq. (14) is modified to Eq. (15),

$$c_v - c_i \approx 3\left(\frac{\Delta V}{V_0} - \frac{\Delta \Omega}{\Omega_0}\right) \tag{15}$$

where c_i is the concentration of self-interstitial atoms. If self-interstitial atoms dominated over vacancies, the macroscopic expansion would negatively deviate from

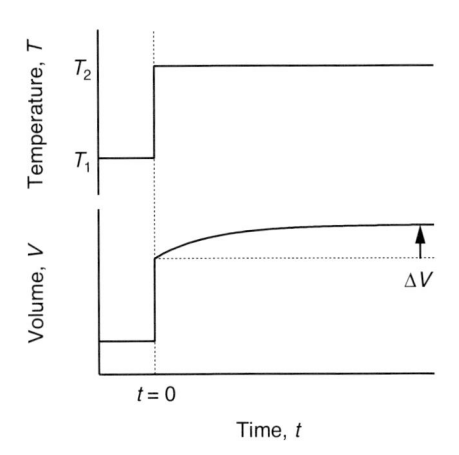

Fig. 5.6 Transient increase in volume due to vacancy production (lower curve) at temperature T_2, after an abrupt rise in temperature from T_1 to T_2 (upper curve).

the microscopic expansion. Experimental results on pure metals indicate that this is not the case; the principal defect species is always vacancies (Siegel 1978). As mentioned in Section 5.1.2, the formation enthalpy of a self-interstitial atom is much higher than that of a vacancy; the equilibrium concentration of self-interstitial atoms in metals is never high enough to be detectable.

The fact that point defects increase or decrease the volume of the material in proportion to their number allows us to determine the migration energy by measuring the transient volume change after a rapid change in temperature. As illustrated in Fig. 5.6, if a sample is first equilibrated at temperature T_1 and is then heated abruptly to T_2 and kept constant, the volume of the sample changes first without significant delay by thermal expansion of the lattice, and then it gradually increases with time. This evolution originates from equilibration of the concentration(s) of intrinsic point defects at the new temperature. On the basis of Eq. (9), the variation of the volume with the holding time t is expressed as

$$\Delta V(T_1, T_2; t) = [n(T_2, t) - n(T_1)]\Delta v_f(T_2) \tag{16}$$

where n is the number of the defect species in question and Δv_f is its formation volume. The number of the defects at $t = 0$ is $n(T_1)$, which is supposed to be the equilibrium value at T_1. While being kept at temperature T_2, point defects are either generated from sources or annihilated to sinks and eventually attain an equilibrium spatial distribution through diffusion. The approach to the equilibrium is controlled by the migration of the defects, and thus the kinetics of the volume change, i.e., the rate constant, is characterized by the migration enthalpy Δh_m. This method was first attempted for vacancies in CoGa by van Ommen and de Miranda (1981), who measured the length change by mechanical probing. Schaefer and coworkers improved the method by utilizing a laser interferometer and evaluated the migration enthalpy of vacancies in FeAl and NiAl (Schaefer et al. 1999). The volume change due to the difference in the defect concentration over a limited temperature range is generally small and is difficult to detect. Neverthelss, these experiments turned out successfully owing to the unusually high vacancy concentrations in the materials studied. (See Section 5.1.3.4.)

Electrical resistometry The presence of point defects can be detected sensitively by measuring electrical resistivity at low temperatures, as they contribute to the residual resistivity in proportion to their concentration (Lucasson and Walker 1962). Relative changes in the concentration can be monitored accurately, whereas the absolute value is difficult to determine unless the magnitude of the specific contribution is known by other means. The measurements are to be made at low temperatures to eliminate the effect of thermal vibration, for the changes in resistivity due to the defects are usually small.

The formation enthalpy of vacancies in various metals has been determined by measuring the electrical resistivity of samples quenched from high temperatures (Bauerle and Koehler 1957). A sample equilibrated at temperature T_q is rapidly cooled to a temperature that is low enough to prevent any motion of vacancies, and the resistivity ρ is measured at a cryogenic temperature T_r. The procedure is repeated for several quenching temperatures, and the values of the excess resistivity $\Delta\rho(T_q; T_r) = \rho(T_q; T_r) - \rho_0(T_r)$ due to retained vacancies are plotted against the reciprocal of T_q. Here, $\rho_0(T_r)$ is the resistivity of a reference sample that is free from excess vacancies. The plot would fall on a straight line, similarly to Fig. 5.5(b), whose slope gives the enthalpy of formation, $\Delta h_{f,v}$. Many of the experimental data for $\Delta h_{f,v}$ in metals are due to this method (Siegel 1978; Ullmaier 1991; Wollenberger 1996).

The migration enthalpy of point defects can also be determined by resistivity measurements (Bauerle and Koehler 1957; Koehler et al. 1957; Balluffi 1978; Young 1978). After freezing-in the defects formed at a high temperature by quenching, the resistivity of the sample is measured as a function of time during the course of isothermal or isochronal annealing at an intermediate temperature where the defects move at a measurable rate. The supersaturated defects tend to annihilate to sinks or precipitate to form clusters, and the decay of the resistivity due to the reduction in the number of the defects is monitored through resistivity. Since the kinetics of the decay is controlled by the migration of the defects, the enthalpy of the migration can be determined from the values of the rate constant at different temperatures. For example, if the defects annihilate at unsaturable sinks and the rate is controlled by diffusion of individual defects, the kinetics of annihilation is described by Eq. (17),

$$\frac{d\Delta c}{dt} = -K\Delta c \tag{17}$$

where Δc is the excess concentration of the defects. The rate constant K is proportional to the diffusion coefficient of the defect and the density of the sinks. The decay of the excess concentration thus follows a simple exponential time law [Eq. (18)],

$$\Delta c(t) \sim \exp(-Kt) \tag{18}$$

which can be traced by measuring the resistivity. As K is proportional to the diffusion coefficient and therefore to the jump frequency, one can determine the migration enthalpy from the temperature dependence of the decay rate (Bauerle and Koehler 1957).

As mentioned earlier, excess point defects can be introduced not only by quenching but also by plastic deformation and by particle irradiation. While the formation and migration of vacancies in metals can be studied most conveniently by quenching experiments, the properties of self-interstitial atoms can be investigated almost solely by electron irradiation, in which pairs of a vacancy and a self-interstitial atom (Frenkel pairs) can be introduced in a controllable manner, by adjusting the energy and the fluence of the incident electrons. The mobility of the self-interstitial atoms is then studied by measuring the resistivity (Lucasson and Walker 1962; Balluffi 1978; Young 1978). In annealing experiments, characteristic "stages" appear in the resistivity versus annealing time or temperature curve, which arise from various processes such as annihilation of close vacancy–interstitial pairs, clustering of vacancies and of interstitial atoms, annihilation to sinks, etc. With proper interpretations and analyses, information on the mobility of the relevant defects can be acquired (Koehler et al. 1957; Damask and Dienes 1963).

Positron annihilation Positrons injected into a solid are annihilated with electrons and emit two characteristic γ rays of 511 keV. The lifetime of positrons, which is of the order of 100 ps in solids, can readily be measured as the time interval between the injection[1] and the detection of the γ rays. When the solid contains defects at which electron density is reduced, the lifetime is much extended. Positrons that encounter such defects are localized there, i.e., trapped, and have statistically long lives: for the case of monovacancies in metals, the lifetime is increased from a typical value of 100 ps to 200 ps. By measuring the extended lifetime, variations of the density of vacancy-type defects as low as 1 ppm can be monitored. Alternatively one may analyze the energy of the emitted γ rays to obtain information on how, or where, the positron terminated its life. Since the momentum of the positron–electron pair is conserved before and after the annihilation, the energy of the emitted γ ray is shifted by the Doppler effect, which reflects the energy of the electron. Positrons trapped at a vacancy-type defect suffer less from the Doppler effect because they are less likely to encounter the high-

---

[1] The most convenient positron source ^{22}Na emits a γ ray of 1.28 MeV when it undergoes the β^+ decay, signaling the injection.

energy, inner-shell electrons of an atom. This leads to characteristic sharpening of the overall photon-energy spectrum, from which the defect concentration may be evaluated. Measurements of lifetime and of Doppler shift are basic techniques of positron annihilation spectroscopy (PAS), and have been applied successfully to investigations of vacancy-type defects in a variety of solid materials: pure metals (Siegel 1978, 1982; West 1979; Schaefer 1982, 1987), intermetallic compounds (Schaefer et al. 1999; Sprengel et al. 2002), semiconductors and ceramics (Rempel et al. 2002). Since experiments under high pressure are not too difficult, PAS has been conveniently applied to determination of the formation volume of vacancies (Dickman et al. 1977, 1978; Wolff et al. 1997; Müller et al. 2001). Migration of vacancies can also be studied by PAS, through the kinetics of equilibration on abrupt changes in temperature similar to that in Fig. 5.6 (Schaefer and Schmid 1989; Würschum et al. 1995).

The lifetime changes according to how many of the injected positrons are annihilated at defects. If we consider only one defect species, e.g., monovacancies in an elemental crystal, a simple two-state trapping model shows that the trapping rate at the defects, κ_1, is given by Eq. (19) (see, e.g., Siegel 1978):

$$\kappa_1 = \sigma_1 c_1 = I_1 \left(\frac{1}{\tau_0} - \frac{1}{\tau_1} \right) \tag{19}$$

where σ_1 is the specific trapping rate and c_1 is the defect concentration. On the right-hand side, I_1 is the relative intensity of the extended-lifetime component in the spectrum, and τ_0 and τ_1 are lifetimes of free and trapped positrons, respectively, all of which can be evaluated from an observed lifetime spectrum. The specific trapping rate σ_1 is generally unknown, but if it is only assumed to be independent of temperature one can determine the formation enthalpy of the defect from an Arrhenius plot of κ_1. For materials for which data of absolute defect concentrations are available from other experiments (most probably by differential dilatometry), one may compare them with the measured values of κ_1 to evaluate σ_1, which may then be adopted for other materials of the same class, i.e., of similar electronic structures, to estimate the concentration of the particular defect species from PAS experiments.

Anelastic relaxation It is easy to envisage from the illustration of Fig. 5.1 that vacancies and substitutional solute atoms induce isotropic displacements of neighboring atoms[2] and thus a volume change (shrinkage or dilatation) of the material. Other defects such as pairs of point defects, some interstitial solute atoms (Figs. 5.13c,d below) and also self-interstitial atoms (Figs. 5.11b,c below) produce, in addition, anisotropic distortion because of their low-symmetry structure. Such

---

2) The displacements around a vacancy or a substitutional solute atom are not always isotropic in ordered structures, such as L1$_2$ (Numakura et al. 1999).

low-symmetry defects can be in several orientations in the crystal which are crystallographically equivalent but orientationally distinguishable: for example, a self-interstitial atom in the "split" configuration shown in Fig. 5.12(b) is of tetragonal symmetry and thus can be in any one of the three orientations, [100], [010], and [001]. Usually the defects are equally distributed over all the possible orientations, but when external stress is applied some of the orientations may become more favored over others because the external stresses interact differently with defects in different orientations. This results in an increase in the population of the favored orientation(s), which in turn produces a macroscopic, anisotropic deformation of the sample material. This is the phenomenon called "anelastic relaxation," which manifests itself as a creep deformation under constant stress, a stress relaxation under constant strain, delay of the strain from the stress under dynamic loading, and so on. All of these pertain to dissipation of the elastic energy due to stress-induced redistribution of the defects, and provide a useful means of studying the defect concentration and mobility through the magnitude of the energy dissipation and the rate of the redistribution, respectively (Nowick and Berry 1972; Weller 2001).

The anisotropic distortion of the host crystal is determined by the characteristic strain, called the "λ tensor", defined as the strain produced by a unit concentration of the defect in orientation p [Eq. (20)].

$$\lambda_{ij}^{(p)} \equiv \frac{d\epsilon_{ij}}{dc_p} \tag{20}$$

While the volume change due to defect is given by the trace of the λ tensor, the magnitude of the relaxation is determined by the dispersion with respect to the stress axis. For example, the λ tensor of a defect of tetragonal symmetry in a cubic crystal is expressed as Eq. (21),

$$(\lambda_{ij}) = \begin{pmatrix} \lambda_1 & 0 & 0 \\ 0 & \lambda_2 & 0 \\ 0 & 0 & \lambda_2 \end{pmatrix} \tag{21}$$

and the magnitude of the relaxation in the elastic compliance is shown to be

$$\delta S' = \frac{2}{3} \frac{c\Omega}{k_B T} |\lambda_1 - \lambda_2|^2 \tag{22}$$

where c is the total defect concentration ($c = \sum_p c_p$) and Ω is the atomic volume of the host crystal. If either the concentration c or the so-called the "shape factor" $|\lambda_1 - \lambda_2|^2$ is known from another experiment, the other can be evaluated by measuring the relaxation magnitude.

A useful feature is that the intensity of the relaxation depends on the orientation of the external stress. For example, one can readily see from Fig. 5.12(b) (below) that a tetragonal defect in a cubic crystal would respond to a tensile stress if the stress is applied in either of the $\langle 100 \rangle$ directions, but no reorientation, and thus no redistribution, would occur if the stress axis is parallel to the $\langle 111 \rangle$ directions, for the stress interacts identically with the defects of the three orientations. In general, how a point defect responds to external stress is determined by its symmetry, and is an important property governing the relaxation behavior. By nature, defects of tetragonal symmetry in cubic crystals respond only to $\{110\}$ $\langle 1\bar{1}0 \rangle$ shear stress, while those of trigonal symmetry (having a principal axis in $\langle 111 \rangle$) only to $\{001\}$ $\langle \bar{1}00 \rangle$ shear stress, producing relaxation in the compliance S' $(= 2(S_{11} - S_{12}))$ and S $(= S_{44})$, respectively. Such rules, called "selection rules," have been derived from group theory and summarized by Nowick and Heller (1965). For cubic crystals, the relaxation of the compliance under tensile stress of an arbitrary orientation is given as Eq. (23).

$$\delta J = \left(\frac{1}{3} - \Gamma\right)\delta S' + \Gamma \delta S \tag{23}$$

Here, Γ is defined as $\Gamma \equiv \gamma_2^2\gamma_3^2 + \gamma_3^2\gamma_1^2 + \gamma_1^2\gamma_2^2$ with γ_i being the directional cosines between the stress axis and the crystal axes i, and ranging from 0 for the stress axis in $\langle 100 \rangle$ to $1/3$ for $\langle 111 \rangle$. The dependence of the relaxation intensity on the stress direction for tetragonal defects is therefore the opposite of that for trigonal defects, as illustrated in Fig. 5.7. Examination of the relaxation strength in single crystals thus allows identification of the defect symmetry, which can be an important clue to determining the atomic configuration of the defect.

Measurement of the relaxation rate is also useful, as it gives us information on the mobility of the defect. If an appropriate stress is applied abruptly and is kept constant just as in the temperature change experiment illustrated in Fig. 5.6, an extra strain (the anelastic strain) appears gradually with time, similarly to the volume change shown in Fig. 5.6. The evolution of the anelastic strain in such a

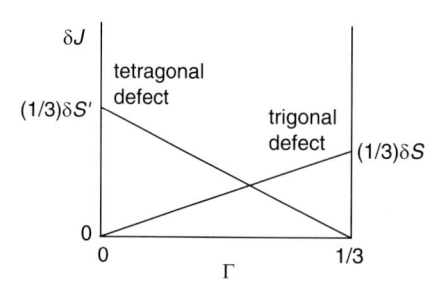

Fig. 5.7 Dependencies of the magnitude of anelastic relaxation in compliance J on the orientation of the tensile stress for the case of tetragonal and trigonal defects in cubic crystals. Γ indicates the orientation of the stress (see text); $\Gamma = 0$ and $1/3$ corresponds to a stress parallel to $\langle 100 \rangle$ and to $\langle 111 \rangle$, respectively.

"quasi-static" experiment is most often described by the simple exponential time law, $1 - \exp(-t/\tau)$, with the amplitude given by $\sigma_0 \delta J$, i.e., proportional to the stress amplitude σ_0. On the other hand, if a sinusoidal stress is applied the strain lags behind the stress, and the phase lag ϕ, or its tangent, which is a measure of the loss of mechanical energy, is expressed by the Debye equation (24)

$$\tan \phi = \frac{\delta J}{J} \frac{\omega \tau}{1 + (\omega \tau)^2} \tag{24}$$

with ω being the angular frequency of the applied stress. As this function shows a maximum at $\omega \tau = 1$, one can readily determine the relaxation time by measuring the phase lag as a function of vibration frequency. From the relaxation time τ obtained by any such experiments, the frequency of the reorientation jump of the defect can be evaluated. Since the jump is most often the elementary process of diffusion, the relaxation rate τ^{-1} is, in such cases, directly proportional to the diffusion coefficient D. If the mechanism of the reorientation and that of the diffusion are both known, an exact relationship of the two quantities may be found. There, the measurement of the relaxation rate provides a unique tool for determining the diffusion coefficient out of the ranges usually accessible by conventional techniques such as interdiffusion and radiotracer experiments. For more details of anelastic relaxation and its practice, "mechanical spectroscopy," see Nowick and Berry (1972), Weller (2001), and Numakura (2003, 2006).

5.1.2
Point Defects in Pure Metals and Dilute Alloys

5.1.2.1 Vacancies

Single vacancies The atomic structure of a vacancy in pure metal crystals is a simple vacant lattice site, with the neighboring atoms being displaced, or "relaxed," to some extent toward the vacant site (Johnson and Brown 1962; Johnson 1964). This simple structure has been found to be stable by a number of computer simulation studies (see, e.g., Finnis and Sinclair 1984; Daw et al. 1993), and is supported by the agreement between theory and experiment of positron annihilation (Schaefer 1987). The formation energy $\Delta e_{f,v}$ is the net increase in the internal energy on introducing a vacancy. If one assumes the nearest-neighbor pair interaction model, it equals to $z/2$ times the bond energy e_1, as seen in Fig. 5.3: ze_1 is lost around a vacancy, and $(z/2)e_1$ is recovered around the atom at the surface site. Since the internal energy of the perfect crystal consisting of N atoms is given by $-(Nz/2)e_1$ in this model, the vacancy formation energy is expected to be equal to the binding energy per atom, i.e., the cohesive energy, E_c, neglecting possible effects of the atomic relaxation around a vacancy. In real metals, E_c and $\Delta h_{f,v}$ ($\approx \Delta e_{f,v}$) are both known to be roughly proportional to the melting temperature T_m, but the latter is only about one third of the former. Fig-

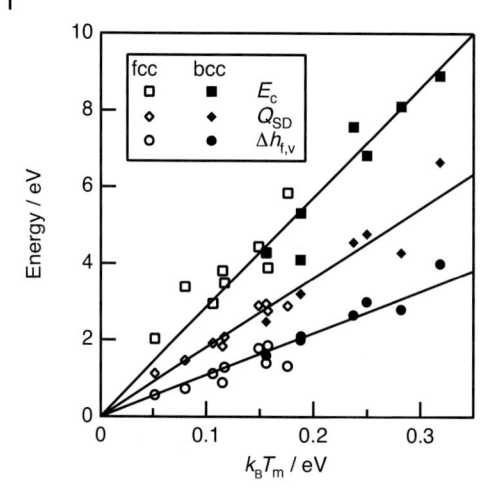

Fig. 5.8 Correlations of the cohesive energy E_c, activation energy for self-diffusion Q_{SD}, and vacancy formation enthalpy $\Delta h_{f,v}$ with the melting temperature T_m for pure metals. k is the Boltzmann constant.

ure 5.8 displays the cohesive energy and the vacancy formation enthalpy of fcc and bcc metals with $k_B T_m$ on the abscissa. These relationships are expressed respectively as Eqs. (25) and (26).

$$E_c \approx (29 \pm 1) k_B T_m \tag{25}$$

$$\Delta h_{f,v} \approx (11 \pm 1) k_B T_m \tag{26}$$

The disparity of the prediction of the pair interaction from experiment is due partly to the neglect of the atomic relaxation but to a greater extent to the simplistic model. The experimental data of the vacancy formation enthalpy are listed in Table 5.1, together with the migration enthalpy $\Delta h_{m,v}$ and those of a self-interstitial atom.

The formation volume of a vacancy, $\Delta v_{f,v}$, must be smaller to some extent than one atomic volume, Ω, because the atoms around the vacancy relax their positions inward on the average. There are not many experimental data for the formation volume, however. The limited data for fcc metals fall in the range between 0.5 Ω to 0.95 Ω (Ullmaier 1991; Wollenberger 1996), but the uncertainties are not small and no systematic trend has been recognized. On the other hand, theoretical values have been accumulating recently, by computer simulation with fairly reliable many-body interatomic potentials and first-principle calculations (Finnis and Sinclair 1984; Daw et al. 1993; Mishin et al. 2001a,b). They reveal that the formation volume is correlated with Poisson's ratio of the host crystal (Kurita and Numakura 2004). Figure 5.9 shows the linear relation between the relaxation volume Δv_r and Poisson's ratio v. According to this relationship, the

Table 5.1 The melting temperature T_m, cohesive energy E_c (Kittel 1996), formation enthalpy Δh_f and migration enthalpy Δh_m of a vacancy (v) and a self-interstitial atom (i) in pure metals (Ullmaier 1991; Wollenberger 1996).

Material	T_m [K]	E_c [eV]	Vacancy		Self-interstitial	
			$\Delta h_{f,v}$ [eV]	$\Delta h_{m,v}$ [eV]	$\Delta h_{f,i}$ [eV]	$\Delta h_{m,i}$ [eV]
fcc						
Al	934	3.39	0.67	0.61	3.0–3.6	0.112–0.115
γ-Fe	1811	4.28	1.40	1.26		
Ni	1728	4.44	1.78	1.04		0.15
Cu	1358	3.49	1.28	0.70	1.6–4.2	0.117
Pd	1828	3.89	1.85	1.03		
Ag	1235	2.95	1.11	0.66		0.088
Pt	2041	5.84	1.35	1.43	1.1–1.5	0.063
Au	1337	3.81	0.93	0.71		
Pb	601	2.03	0.58	0.43		0.01
bcc						
V	2183	5.31	2.2	0.5–0.7		
Cr	2180	4.1	2.0	0.95		
α-Fe	1811	4.28	1.59–1.73[a] 1.79–1.85[b]	1.11	4.7–5.0	0.3
Nb	2750	7.57	3.07	0.55		
Mo	2896	6.82	3.0	1.35		0.083
Ta	3269	8.1	3.1	0.7		
W	3695	8.9	3.6	1.70		0.054

[a] Ferromagnetic state. [b] Paramagnetic state.

relaxation is large in crystals whose Poisson's ratio is high and, therefore, the formation volume, given by $\Omega + \Delta v_r$, is significantly smaller than Ω. Since knowledge of the formation volume is important in studies of diffusion under high pressure, this relationship must be useful as a rough estimate when no experimental value is available.

The nonconfigurational entropy associated with a vacancy, $\Delta s_{f,v}$, is expected to be positive, since the introduction of a vacancy would bring about low-frequency modes in the lattice vibration. It has been evaluated from experimental data for the absolute concentration for some metals; for example (in k_B), it is 0.7 for aluminum, 2.8 for copper, 1.5 for silver, 0.72 for gold, 1.6 for molybdenum, and 3.2 for tungsten (Wollenberger 1996).

With the formation parameters discussed above, one can estimate the equilibrium vacancy concentration in typical metals. At the melting temperature T_m, for example,

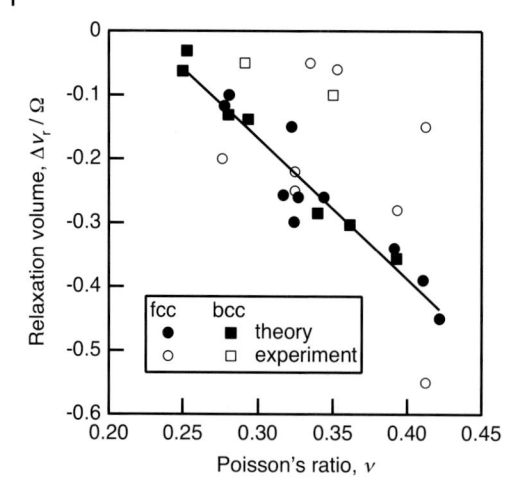

Fig. 5.9 Relaxation volume of a vacancy Δv_r in pure metals evaluated by computer simulation and first-principles calculations as a function of Poisson's ratio v of the host crystal.

$$c_v(T_m) = \exp\left(\frac{\Delta s_{f,v}}{k_B}\right) \exp\left(-\frac{\Delta h_{f,v}}{k_B T_m}\right)$$

$$\approx \exp(2 \pm 2) \exp(-11 \pm 1)$$

$$\approx 10^{-4 \pm 1} \tag{27}$$

It decreases with decreasing temperature, and is reduced to $10^{-8\pm1}$ at $T_m/2$. Vacancy concentrations in solid-solution alloys may also be estimated in the same manner by taking the solidus temperature as T_m.

The translational motion of a vacancy to a neighboring site is in fact a jump of the neighboring atom to the vacant site, and is associated with the energy barrier at the intermediate state (Johnson and Brown 1962; Johnson 1964), similarly to the case of the interstitial atom depicted in Fig. 5.4. The migration enthalpy of a vacancy can be determined from the decay of quenched-in excess vacancies as explained in Section 5.1.1.3. It can also be estimated from the activation energy of tracer self-diffusion. The self-diffusion in metals occurs by the vacancy mechanism, and the tracer self-diffusion coefficient, D^*, is given by the product of the jump frequency of a vacancy ω_v, its concentration c_v and the tracer correlation factor f, which is a constant (cf. Section 5.2). Therefore, if both the activation energy of the tracer self-diffusion Q_{SD} and the vacancy formation enthalpy $\Delta h_{f,v}$ are known, their difference must give a good estimate of the migration enthalpy $\Delta h_{m,v}$. The activation energy of self-diffusion in metals is also known to be well correlated with the melting temperature (Brown and Ashby 1980), expressed as $Q_{SD} \approx (18 \pm 1)k_B T_m$; this correlation is also shown in Fig. 5.8. Together with Eq. (26), therefore, we have $\Delta h_{m,v} \approx (7 \pm 2)k_B T_m$. In fact, Q_{SD} and $\Delta h_{f,v}$ are directly correlated with each other and are expressed as $\Delta h_{f,v} = (0.60 \pm 0.01)Q_{SD}$. The migration enthalpy is thus related to Q_{SD} as $\Delta h_{m,v} = (0.40 \pm 0.01)Q_{SD}$. Here we have yet another empirical formula, $\Delta h_{m,v}/\Delta h_{f,v} \approx 2/3$, for pure metals. A similar relationship is expected for disordered solid-solution alloys, while the behavior can be different in cases where, for example, strong interaction between vacancies and solute atoms exists, or solute atoms do not migrate by the ordinary vacancy mechanism (Hood 1993).

Divacancies At elevated temperatures where the equilibrium concentration of vacancies is fairly high, close pairs of vacancies at neighboring sites, i.e., divacancies, may occur. By considering the reduction in the number of missing bonds and also in lattice strain energy (see Section 5.1.2.3), association of two vacancies is expected to reduce the enthalpy of the crystal. The free energy of association is defined from the free energies of formation as

$$\Delta g_{a,vv} = \Delta g_{f,vv} - 2\Delta g_{f,v} \tag{28}$$

and the concentration of divacancies is given by Eq. (29).

$$c_{vv} \approx \frac{z}{2} \exp\left(-\frac{2\Delta g_{f,v}}{k_B T}\right) \exp\left(-\frac{\Delta g_{a,vv}}{k_B T}\right) = \frac{z}{2} \exp\left(-\frac{g_{f,vv}}{k_B T}\right) \tag{29}$$

The total concentration of vacant lattice sites is then

$$c_{v,\text{total}} = c_v + 2c_{vv}$$

$$\approx \exp\left(-\frac{\Delta g_{f,v}}{k_B T}\right)\left[1 + z \exp\left(-\frac{\Delta g_{f,v} + \Delta g_{a,vv}}{k_B T}\right)\right] \tag{30}$$

The second term in the bracket can be of appreciable magnitude if there is a strong binding between two vacancies, i.e., the association energy $\Delta g_{a,vv}$ is negative and its magnitude is not negligible in comparison to $\Delta g_{f,v}$.

Divacancies are known to become noticeable in fcc metals at high temperatures and play a role in diffusion (Mehrer 1978). Calculations for the enthalpy of association in the noble metals have reported a range of values from -0.1 to -0.3 eV (Balluffi 1978). Computer simulation studies suggest that divacancies in fcc metals are more mobile than single vacancies (Johnson 1965, and references therein). The Arrhenius plot of the tracer self-diffusion coefficient is often found to be curved upward at high temperatures, and this trend is attributed to the contribution of divacancies to the diffusion (Mehrer 1978; Peterson 1978).

Much less is known about divacancies in bcc metals. The stable structure was suggested to be a pair of vacancies at the second-neighbor positions, and the migration energy was similar to the case of the single vacancy (Johnson 1964). The Arrhenius plots of the tracer self-diffusion coefficient in some bcc metals were known to be appreciably curved upward at high temperatures, and were assumed to be due to the contribution of divacancies (Peterson 1978). However, those curved Arrhenius plots are now considered to be due largely to the dynamical properties of the crystal (Gilder and Lazarus 1975; Herzig 1993). In this context, the role of divacancies in the self-diffusion in fcc metals has been questioned and critically re-analyzed (Mundy 1987).

Vacancy–Solute Complexes In dilute solid-solution alloys, complexes of vacancies and solute atoms may be formed, owing to electronic and/or elastic interaction between the two defect species. Formation of pairs of a vacancy and a solute atom in metals has been studied theoretically and experimentally (Le Claire 1978; March 1978), and the diffusion behavior of the solute atoms, as well as the solvent atoms, has been analyzed (Le Claire 1978, 1993). As the concentration of the solute atoms can be (and usually is) much higher than that of vacancies and may have significant effects on the properties of the material, we discuss the formation of vacancy–solute atom pairs in slightly more detail.

Let us consider a dilute binary A–B alloy consisting of N_A and N_B of A and B atoms, respectively ($N_A \gg N_B$). The number of lattice sites next to a B atom is zN_B and the number of other sites is $N_A - zN_B$. The probabilities of finding a vacancy at these sites are given respectively by $\exp(-\Delta g_{f,v}/k_B T) \times \exp(-\Delta g_{a,vs}/k_B T)$ and $\exp(-\Delta g_{f,v}/k_B T)$. The numbers of isolated vacancies and those associated with a solute atom are thus written as

$$n_v = (N_A - zN_B) \exp\left(-\frac{\Delta g_{f,v}}{k_B T}\right) \tag{31}$$

$$n_{vs} = zN_B \exp\left(-\frac{\Delta g_{f,v} + \Delta g_{a,vs}}{k_B T}\right) \tag{32}$$

Their concentrations are obtained by dividing by the total number of lattice sites, $N_A + N_B + n_v + n_{vs}$.

Figure 5.10 shows Arrhenius plots of calculated concentrations of isolated vacancies, c_v (dotted line), vacancy–solute atom pairs, c_{vs} (broken line), and their sum, $c_{v,\text{total}}$ (solid line), for an exemplary case of $z = 12$, $x_B \equiv N_B/(N_A + N_B) = 1 \times 10^{-2}$ and $\Delta g_{a,vs} = -0.15 \Delta g_{f,v}$. Since the solute concentration is low in this example, c_v is virtually equal to the vacancy concentration in an unalloyed crystal. The attractive interaction between a vacancy and the solute atom gives rise to vacancies bound to the solute atoms at low temperatures. With decreasing temperature the number of vacancy–solute atom pairs is progressively increased and becomes significantly larger than the number of isolated vacancies, leading to a higher total concentration of vacancies than without the solute atoms. If we take aluminum, for which $\Delta h_{f,v} \approx 0.7$ eV and $\Delta s_{f,v}/k_B \approx 0.7$, the melting temperature (934 K) and room temperature correspond respectively to 8 and 26 on the abscissa, $\Delta g_{f,v}/(k_B T)$. The association energy $\Delta g_{a,vs} = -0.15 \Delta g_{f,v}$ (≈ -0.1 eV) applies to such solute atoms as magnesium. In this case, below 520 K, which corresponds to $\Delta g_{f,v}/(k_B T) = 15$ (where c_{vs} equals to c_v), the majority of vacancies are bound to the solute atoms.

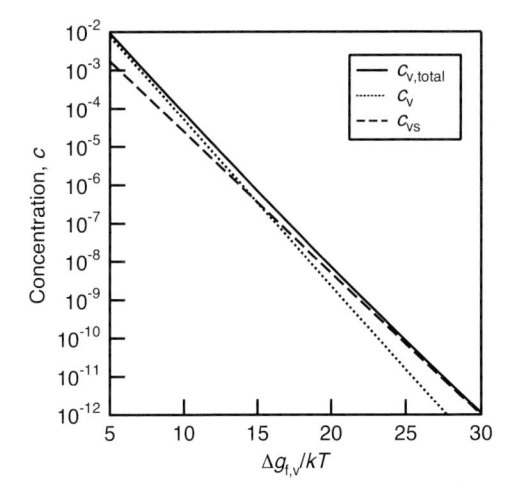

Fig. 5.10 Arrhenius plots of calculated concentrations of isolated vacancies c_1, vacancy–solute atom pairs c_2, and the total concentration c_t in a dilute A–B solid-solution alloy. The co-ordination number $z = 12$, the solute concentration (mole fraction) $x_B = 1 \times 10^{-2}$, and $\Delta g_{a,vs}/\Delta g_{f,v} = -0.15$.

An unusually large number of vacancies, amounting to 10 at.%, has been discovered in metals that contain hydrogen as interstitial solutes (Fukai and Okuma 1993; Fukai 2003). These vacancies, called "super-abundant vacancies," are a somewhat excessive example of the enhancement of the vacancy concentration by the attractive solute–vacancy interaction described above. There is ample evidence that interstitial hydrogen atoms attract vacancies to form clusters, with a binding energy of 0.3 to 0.4 eV, and considerably reduce the effective formation energy of vacancies (Fukai 2003). The immediate impact of a high concentration of vacancies on diffusion is particularly interesting (Hayashi et al. 1998; Iida et al. 2005).

5.1.2.2 Self-Interstitial Atoms

The atomic configuration of a self-interstitial atom in fcc metals was first envisaged as an extra atom located at the body-centered position of the unit cell, (Fig. 5.11a) (Huntington and Seitz 1942), which is referred to as the octahedral configuration. The configuration shown in Fig. 5.11(b), called a $\langle 100 \rangle$ split, or "dumbbell", interstitial, was originally believed to be an intermediate state in the translational motion of the octahedral interstitial. However, this configuration was suggested to be stable by later theoretical studies (Huntington 1953; Johnson and Brown 1962; Seeger et al. 1962), and was proven to exist in irradiated metals by anelastic relaxation experiments (Nowick 1978; Robrock 1990). Self-interstitial atoms in bcc metals are expected to be of a similar form: the $\langle 110 \rangle$ split configuration of Fig. 5.11(c) (Johnson 1964, 1973).

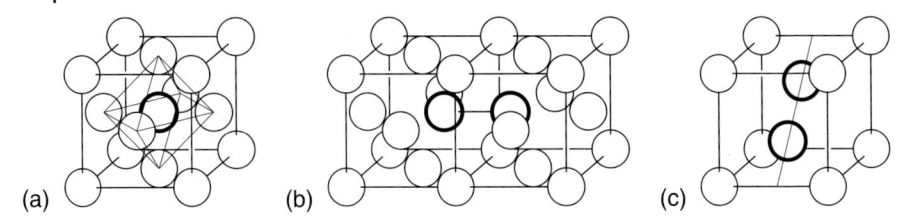

Fig. 5.11 Atomic structure of a self-interstitial atom in metals. (a) The octahedral configuration in fcc crystals. (b) The $\langle 100 \rangle$ split (or "dumbbell") configuration in fcc crystals. (c) The $\langle 110 \rangle$ split configuration in bcc crystals.

The formation enthalpy of a self-interstitial atom, $\Delta h_{f,i}$, in metals has been evaluated as the difference between the formation enthalpy of a Frenkel pair, $\Delta h_{f,F}$, and that of a vacancy, $\Delta h_{f,v}$ (Schilling 1978; Ullmaier 1991; Wollenberger 1996). The former can be determined by calorimetric measurements, from the release of the stored energy of a sample irradiated by electrons. The results for aluminum, copper, and iron show that the formation enthalpy of the self-interstitial atom is considerably higher than that of the vacancy, by a factor of 2 to 4.

The formation volume $\Delta v_{f,i}$ is expected to be $-\Omega$ if no relaxation occurs, but positive values ranging from 0.1 Ω to 1.1 Ω have been reported by experiment and simulation (Schlling 1978; Wollenberger 1996). A positive formation volume means that, when an extra atom is inserted into a crystal, the crystal dilates more than the specific volume of the inserted atom, even though the atom is somehow accommodated within. This occurs because the distortion around the self-interstitial atom is very large; it is in fact the origin of the high energy of formation.

For the entropy of formation, $\Delta s_{f,i}$, virtually no experimental evaluations have been made, whereas data are available for vacancies in some metals which were derived from measurements of the absolute concentration (Wollenberger 1996). Theoretical calculations of the formation entropy are not straightforward (de Koning et al. 2002) and numerical evaluation is demanding. Nevertheless, a novel scheme for computing the formation entropy of point defects has been developed recently (Mishin et al. 2001a), which is expected to promote detailed studies and produce reliable estimations.

The equilibrium concentration of self-interstitial atoms in metals is negligibly low because of the high formation enthalpy. Assuming $\Delta h_{f,i} \approx 33 k_B T_m$ ($= 3\Delta h_{f,v}$) and tentatively neglecting the effect of the formation entropy, the concentration at the melting temperature is estimated to be below 10^{-14}, which is hardly detectable by any experimental method. At the temperature $T_m/2$, the concentration would be further reduced to below 10^{-28}, which means that not a single self-interstitial atom would be found in a macroscopic piece of metal.

The mobility of a self-interstitial atom is known to be much higher than that of a vacancy (Johnson 1973; Young 1978). Figure 5.12 illustrates an elementary step

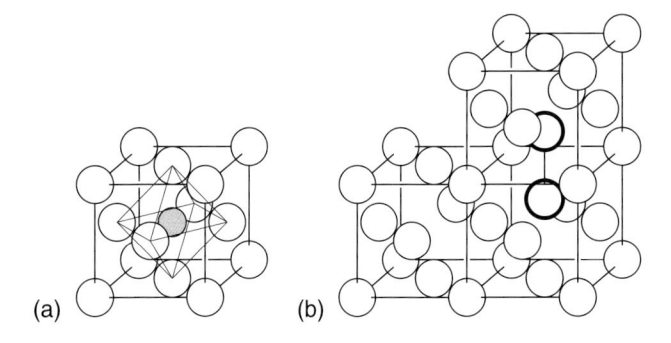

Fig. 5.12 Motion of a $\langle 100 \rangle$ split self-interstital in an fcc crystal.

of the motion of a $\langle 100 \rangle$ split interstitial in an fcc crystal. As it consists of two identical atoms only partly displaced from the regular position, the motion of the interstitial is achieved by the slight movement of one of the members to the next displaced position (displacing a regular atom ahead) and the accompanying movement of the other member to the regular atomic site. This process occurs with a low activation enthalpy in the range between 0.05 and 0.2 eV in aluminum, copper, nickel, and iron (Ullmaier 1991; Wollenberger 1996).

5.1.2.3 Solute Atoms

Solute or impurity atoms constitute atomic defects by themselves. Moreover, they may influence the properties of intrinsic point defects by forming close pairs as discussed in Section 5.1.2.1, or globally changing the electronic properties of the host material.

Light elements with small atoms dissolve in metals as interstitial solute atoms, producing significant lattice distortion. For instance, hydrogen and carbon in nickel and other fcc metals occupy the octahedral or tetrahedral interstitial sites shown in Figs. 5.13(a,b), and give rise to isotropic dilatation. Substitutional solute atoms also produce isotropic distortion, but either dilatation or contraction, de-

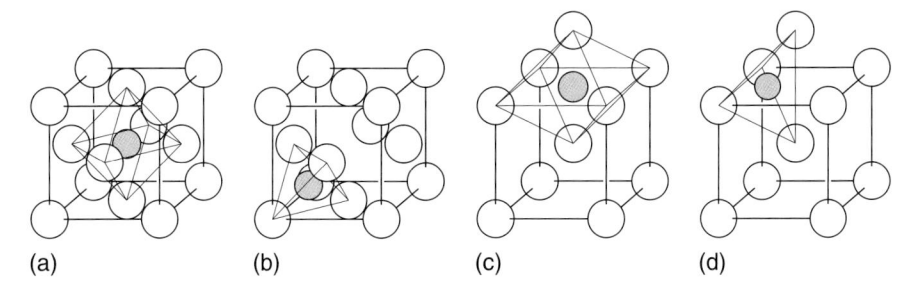

Fig. 5.13 An interstitial solute atom (shaded circle) occupying octahedral (a) and tetrahedral (b) interstitial sites in an fcc crystal, and octahedral (c) and tetrahedral (d) interstitial sites in a bcc crystal.

pending on the relative atomic size in comparison to the host atoms. The volume change is given by the trace of the λ tensor and can be evaluated, for cubic crystals, by the relative change in the lattice parameter caused by the addition of the solute [Eq. (33)].

$$\frac{\Delta V}{V} = \text{tr}(\lambda_{ij}) = \frac{3}{a}\frac{da}{dc} \tag{33}$$

According to anisotropic elasticity theory, two centers of isotropic distortion interact with each other through the elastic anisotropy of the host crystal (Eshelby 1955). The interaction energy in cubic crystals is expressed as

$$E_{int} = \frac{15}{4\pi} \frac{C' - C_{44}}{[3(1-v)/(1+v)]^2} \frac{\Delta V_1 \Delta V_2}{r^3} \Lambda \tag{34}$$

where C' ($= (C_{11} - C_{12})/2$) and C_{44} are elastic stiffness constants corresponding respectively to $\{110\} \langle 1\bar{1}0 \rangle$ and $\{001\} \langle 100 \rangle$ shear, v is Poisson's ratio, ΔV_1 and ΔV_2 are the changes in volume associated with defect 1 and defect 2, respectively, and r is the distance between them. The last factor Λ is defined by the orientation of the pair as

$$\Lambda \equiv \alpha_1^4 + \alpha_2^4 + \alpha_3^4 - \frac{3}{5} \tag{35}$$

with the α_i being the directional cosines between the pair axis and the three principal crystal axes. According to this formula, in crystals whose elastic anisotropy $A = C_{44}/C'$ is greater than 1, for example, two point defects producing volume changes of the same sense attract each other if aligned in $\langle 100 \rangle$ but repel in $\langle 111 \rangle$. A defect pair thus formed may produce, in turn, uniaxial distortion along the axis of the pair, and cause an anelastic relaxation effect, which is called Zener relaxation for the case of substitutional atom pairs (Nowick and Berry 1972; Weller 2001). Similar relaxation effect is anticipated for divacancies and interstitial solute atom pairs. The latter is in fact observed for carbon in fcc metals (Numakura et al. 2000).

Interstitial solute atoms in bcc metals bring about anisotropic distortion in addition to the dilatation, since the octahedral and tetrahedral sites that they occupy, shown in Figs. 5.13(c,d), are of tetragonal symmetry, which is lower than the symmetry of the host lattice. The uniaxial distortion is again the origin of the anelastic relaxation effect, known as the Snoek relaxation, which is caused by reorientation of the anisotropic defect under external stress. By measuring the magnitude and the rate of the relaxation effect, the concentration and the mobility of the interstitial solute atoms can be determined (Nowick and Berry 1972; Numakura 2003).

5.1.3
Point Defects in Ordered Alloys

5.1.3.1 Point Defects and Properties of the Material

There has been growing interest in ordered alloys and intermetallic compounds as novel structural and functional materials. In particular, transition-metal aluminides, gallides, and silicides are attractive as high-strength lightweight alloys for high-temperature use. The interesting and sometimes puzzling properties of these materials often originate from point defects in the ordered crystal structure (de Novion 1995). Ordered alloys and intermetallic compounds exhibit a certain range of stability in composition, in which the deviation from the stoichiometry must be realized by intrinsic point defects, i.e., antisite atoms and vacancies. Quite often these defects strongly influence material properties. For instance, in the aluminides of iron and nickel of the B2 structure, the hardness and the chemical diffusion coefficient increase strongly with deviation of the composition from the stoichiometry (Pike et al. 1997; Kim and Chang 2000), which originate from increasing point-defect density.

The chemical diffusion in stoichiometric NiAl is slower than in off-stoichiometric alloys, which is natural in view of restrictions on atom movements in the ordered structure (Bakker 1984; Koiwa 1992; see also Section 5.2.3). However, it is not as slow as it might be; the tracer diffusion coefficient of Ni in NiAl, shown in Fig. 5.14, is similar in magnitude to the tracer self-diffusion coefficient in pure nickel, when comparison is made with temperature scaled to melting temperatures T_m. The diffusion of Ni in Ni_3Al of the fcc-based $L1_2$ structure is slower than in B2 NiAl even though the Ni atom sites in the $L1_2$ structure are mutually connected, unlike the case of the B2 structure. In strong contrast, the diffusivity of Ni in Ni_3Sb of the bcc-based DO_3 structure is higher by several orders of magnitude; Ni atoms diffuse in this compound as fast as interstitial carbon atoms in nickel. This variety of diffusion behavior can only be understood through quantitative knowledge of diffusion-mediating defects. In fact, Ni_3Sb is known to have a large number of vacancies in the Ni sublattice (Heumann and Stüer 1966). Experiments have shown that the defect and diffusion properties are "normal" in Ni_3Al (Numakura et al. 2001); experimental data (Wang et al. 1984; Frank et al. 1995; Badura-Gergen and Schaefer 1997) show that the empirical laws for the vacancy formation enthalpy ($\Delta h_{f,v} \approx 11 k_B T_m$), the diffusion activation energy ($Q \approx 18 k_B T_m$), and the proportions of the formation and migration enthalpies of vacancies roughly hold. In contrast, compounds based on the bcc structure tend to exhibit anomalous behavior. In particular, the formation energy of vacancies in transition-metal aluminides and gallides is markedly low and, on the other hand, the migration energy is somewhat high (Sprengel et al. 2002).

In order–disorder alloys, the degree of order is gradually lowered with increasing temperature and finally falls to zero at the transition temperature. This lowering of the long-range order occurs by progressive enhancement of antistructural disorder, i.e., increase in the concentrations of antisite atoms. Deviations

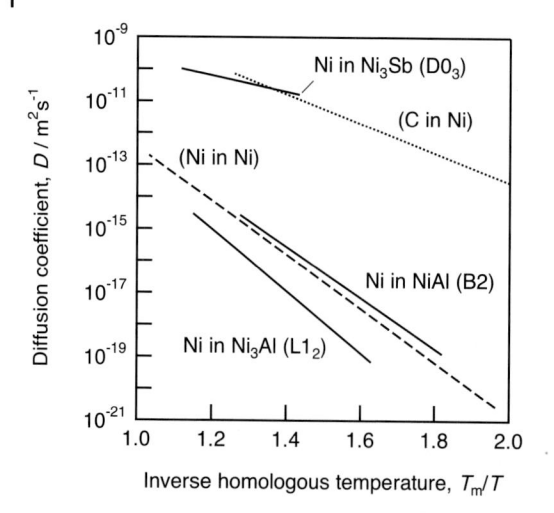

Fig. 5.14 Tracer diffusion coefficient of Ni in B2 ordered NiAl (Divinski et al. 2001), L1$_2$ ordered Ni$_3$Al (Frank et al. 1995), D0$_3$ ordered Ni$_3$Sb (Heumann and Stüer 1966) as a function of inverse temperature normalized to the melting temperature of each material. The broken line is the tracer self-diffusion coefficient in nickel (Bakker 1968; Maier et al. 1976), and the dotted line is the tracer diffusion coefficient of carbon in nickel (Smith 1966; Čermák and Mehrer 1994).

from stoichiometry within the range of stable compositions are usually materialized by introduction of antisite atoms, similarly to the case of thermal disordering. In some intermetallic compounds, on the other hand, this is not always the case. Among others, FeAl, NiAl, CoAl, NiGa, and CoGa are known to bear an unusually high concentration of vacancies, particularly at Al-rich or Ga-rich compositions (Chang and Neumann 1983). Figure 5.15 shows the variations of the lattice parameter of Ni$_3$Al and NiAl with composition (Taylor and Doyle 1972; Aoki and Izumi 1975). In Ni$_3$Al, the lattice parameter increases steadily with Al concentration as it passes through the stoichiometric composition, x_{Al} (mole fraction) $= 0.25$, reflecting the larger specific volume of Al than Ni. The predominant defect species was found to be antisite Ni atoms at $x_{Al} < 0.25$ and antisite Al atoms at $x_{Al} > 0.25$ (Aoki and Izumi 1975). In NiAl, the lattice parameter also increases linearly with Al content in the Ni-rich composition range, but at the stoichiometric composition it turns sharply to decrease. The negative slope at the Al-rich compositions indicates that vacancies are introduced in the Ni sublattice instead of antisite Al atoms. This behavior suggests that the formation enthalpy of vacancies is extremely low in this compound, and thus their concentration could be notably high. The not-too-slow self-diffusion in NiAl in spite of the ordered structure can be explained by the high concentration of vacancies, which are certainly the defects mediating diffusion.

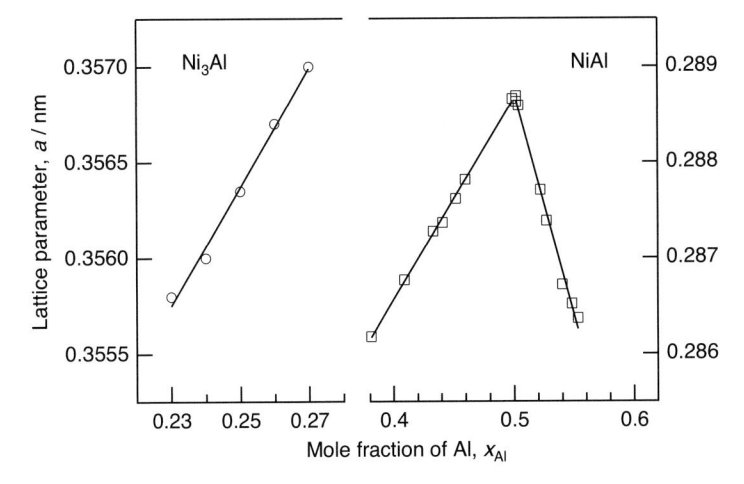

Fig. 5.15 Variation of the lattice parameter of Ni$_3$Al (Aoki and Izumi 1975) and of NiAl (Taylor and Doyle 1972) with composition.

As exemplified above, detailed knowledge of point defects is important for a good understanding of a great variety of diffusion behavior and other properties of ordered alloys and intermetallic compounds. For this purpose, in what follows we first describe the statistical thermodynamics of point defects in binary ordered alloys and then present some interesting examples of equilibrium concentrations of vacancies and antisite atoms. We restrict our discussion to thermodynamic principles, focusing on relationships between interatomic interactions and trends in the defect formation. For experimental facts, we refer the reader to recent reviews on the subject by de Novion (1995) and Sprengel et al. (2002).

5.1.3.2 Statistical Thermodynamics

Theoretical analyses of point defects in ordered alloys date back to the pioneering work of Wagner and Schottky (1931), who established the statistical thermodynamic treatment of vacancies, self-interstitial atoms, and antistructure defects in binary compounds. In their model, each defect is assumed to be in a defect-free local environment and is characterized by a specific energy (the defect energy). The equilibrium concentrations are then calculated in the grand canonical formalism with the approximation that the defect concentrations are so low that each defect finds no other defects around it. The theory of Bragg and Williams (1934, 1935) allowed the presence of other defects in the vicinity with probabilities proportional to their concentrations, which is known as the mean-field approximation. The theory was able to describe the co-operative nature of the order–disorder transition reasonably well. Bragg and Williams considered only antistructural defects, but it is straightforward to include vacancies in their theory; the studies by Cheng et al. (1967) and Schapink (1969) are examples of

Table 5.2 Constituent species in a binary A–B alloy with Schottky defects (vacancies) and substitutional defects (antisite atoms).

Label	Symbol	Species
1	V_α	vacancies in α sublattice
2	V_β	vacancies in β sublattice
3	B_α	B atoms in α sublattice
4	A_β	A atoms in β sublattice
5	A_α	A atoms in α sublattice
6	B_β	B atoms in β sublattice

such an extension. Although a variety of more sophisticated statistical methods are available today (to name a few, the cluster variation method, Calphad approach, and Monte Carlo simulation), the Bragg–Williams (B–W) theory is simple and thus useful to see how the structure and the atomic interaction affect the defect formation.[3]

Definition of the problem We consider formation of vacancies and antisite atoms in a binary ordered alloy $A_m B_n$. The sublattices for the species A and B are referred to as α and β, respectively. The system consists of six species (four point defects and two regular atoms), as listed in Table 5.2. We will refer to these species by the labels 1 to 6 as given in the table, and denote their numbers by N_i ($i = 1, 2, \ldots, 6$). Let us consider a system consisting of N unit cells. Here, a unit cell is not necessarily a crystallographic unit cell but is defined as a structural unit containing m α sites and n β sites. Letting the number of A and B atoms be N_A and N_B, respectively, the composition (i.e., the mole fractions) is defined as

$$x_A = \frac{N_A}{N_A + N_B} = \frac{c_A}{c_A + c_B} \tag{36}$$

$$x_B = \frac{N_B}{N_A + N_B} = \frac{c_B}{c_A + c_B} \tag{37}$$

where $c_A \equiv N_A/N$ and $c_B \equiv N_B/N$.

---

[3] The B–W model is known to predict a much too high order–disorder transition temperature, but it is not too bad at describing the thermodynamics of ordered phases as long as the degree of order is fairly high. The overestimation of the transition temperature comes primarily from inaccuracy in the free energy of the disordered phase; short-range order must be important in the disorderd phase but is not taken into account in the theory.

The problem to be solved is to determine the equilibrium numbers of the four defect species at a given composition and temperature. The number of components in the system is two, but is formally three with vacancies as the tertiary component. The problem is one in the statistical thermodynamics of a ternary system, where we are to find equilibrium distributions of the three components in a crystal comprising two sublattices. It should be noted here that not all of the N_i are independent. First, the numbers of atoms are conserved:

$$N_4 + N_5 = N_A \tag{38}$$

$$N_3 + N_6 = N_B \tag{39}$$

On the other hand, the numbers of vacancies are not conserved, i.e., the size of the system, N, is not constant. Nevertheless, for the ordered structure to be maintained, the ratio of the numbers of the sublattice sites must satisfy Eqs. (40) and (41).

$$N_1 + N_3 + N_5 = mN \tag{40}$$

$$N_2 + N_4 + N_6 = nN \tag{41}$$

There are formally seven variables, N_i and N, but only three are independent because of these four constraints. If the last two conditions, Eqs. (40) and (41), are combined to a single equation,

$$\frac{N_1 + N_3 + N_5}{m} = \frac{N_2 + N_4 + N_6}{n} \tag{42}$$

we have six formal variables (N_i) with three constraints, again leaving three independent variables.

Governing equations There are a few different ways of determining the defect equilibria, all of which indeed give identical results (Schott and Fähnle 1997; Hagen and Finnis 1998; Mishin and Herzig 2000). The methods of the canonical ensemble and the grand canonical ensemble are most commonly employed, in which the numbers of atoms and the chemical potentials, respectively, are given as inputs. Here we outline the first formalism to derive the governing equations, along the same lines as Allnatt and Lidiard (1993) and Hagen and Finnis (1998).

First we write the free energy of a solid that consists of N unit cells and contains vacancies and antisite atoms at temperature T:

$$G(N, N_i; T) = G_0(N; T) + \sum_{i=1}^{4} N_i g_i^\infty - TS_c \tag{43}$$

The first term on the right-hand side is the free energy of a defect-free crystal, the second term is the change in the free energy due to the defects, and the last term

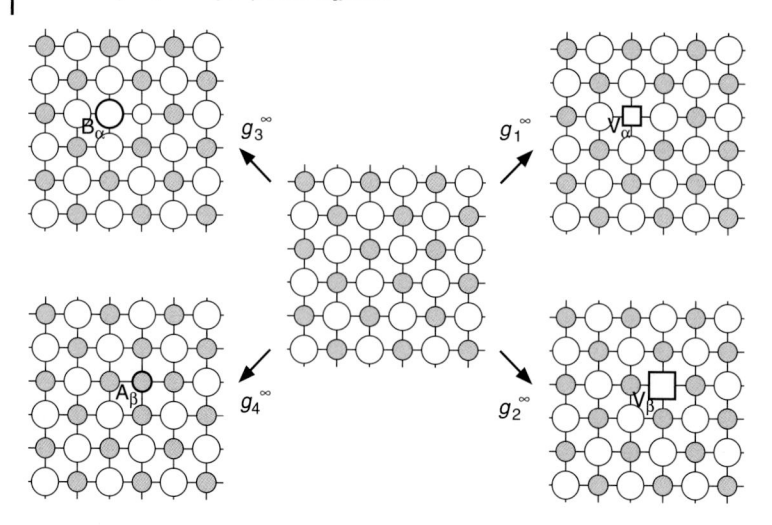

Fig. 5.16 Changes in atomic configuration for which the free energy of formation g_i^∞ of a point defect species i is defined in the grand canonical ensemble formalism.

is the contribution of the configurational entropy. The free energy of formation of a point defect g_i^∞ is defined, as illustrated in Fig. 5.16, as the difference in the free energy between a defect-free crystal and a crystal consisting of the *same* number of cells with a single defect i. On the assumption that the defects are distributed randomly in each of the sublattices, the configurational entropy S_c is given by Eq. (44).

$$S_c = k_B \ln \frac{(mN)!}{N_1!N_3!N_5!} \frac{(nN)!}{N_2!N_4!N_6!} \tag{44}$$

After substituting this expression into the free energy with the Stirling approximation, we incorporate the three constraints, Eqs. (42), (38) and (39), with Lagrange multipliers λ_1, λ_2, λ_3:

$$G = G_0 + \sum_{i=1}^{4} N_i g_i^\infty - k_B T \left(mN \ln mN + nN \ln nN - \sum_{i=1}^{6} N_i \ln N_i \right)$$

$$+ \lambda_1 \left(\frac{N_1 + N_3 + N_5}{m} - \frac{N_2 + N_4 + N_6}{n} \right)$$

$$+ \lambda_2(N_4 + N_5 - N_A) + \lambda_3(N_3 + N_6 - N_B) \tag{45}$$

The condition for thermal equilibrium, $(\partial G/\partial N_i)_{N_{j \neq i}} = 0$, gives a set of six simultaneous equations for the six numbers N_i. By eliminating λ_1, λ_2, and λ_3 from those equations, one obtains, after some mathematical manipulations,

$$\left(\frac{N_1}{mN}\right)^m \left(\frac{N_2}{nN}\right)^n = \exp\left(-\frac{g_S}{k_B T}\right) \tag{46}$$

$$\frac{N_2 N_3}{N_1 N_6} = \exp\left(-\frac{g_{XB}}{k_B T}\right) \tag{47}$$

$$\frac{N_1 N_4}{N_2 N_5} = \exp\left(-\frac{g_{XA}}{k_B T}\right) \tag{48}$$

where

$$g_S \equiv \mu_0 + m g_1^\infty + n g_2^\infty \tag{49}$$

$$g_{XB} \equiv g_2^\infty + g_3^\infty - g_1^\infty \tag{50}$$

$$g_{XA} \equiv g_1^\infty + g_4^\infty - g_2^\infty \tag{51}$$

Equations (46), (47), and (48) are the basic formulae determining the equilibrium defect concentrations, called respectively the Schottky product equation, B-antisite disorder equation, and A-antisite disorder equation. They describe the rates of the reactions

$$(\text{null}) \rightarrow mV_\alpha + nV_\beta \tag{52}^{4)}$$

$$V_\alpha + B_\beta \rightarrow B_\alpha + V_\beta \tag{53}$$

$$A_\alpha + V_\beta \rightarrow V_\alpha + A_\beta \tag{54}$$

We may note that combining Eqs. (47) and (48) yields

$$\frac{N_3 N_4}{N_5 N_6} = \exp\left(-\frac{g_{XB} + g_{XA}}{k_B T}\right) \tag{55}$$

which pertains to production of a pair of antisite atoms,

$$A_\alpha + B_\beta \rightarrow B_\alpha + A_\beta \tag{56}$$

This equation is apparently the same as the problem discussed in the original work of Bragg and Williams (1934, 1935) on order–disorder transition. However, we are to consider vacancies in calculating g_{XB} and g_{XA}. The energies g_S, g_{XB}, and g_{XA} govern the defect equilibria and are functions of interatomic interaction energies. They are at the same time functions of the defect concentrations, since the energy to create a point defect at a particular site depends on the atomic configuration around it.

---

4) Note that a Schottky defect in a compound $A_m B_n$ consists of a set of m of V_α and n of V_β.

5 Point Defects, Atom Jumps, and Diffusion

The Bragg–Williams approximation Now we shall derive expressions for the governing equations and the energy parameters g_S, g_{XB}, and g_{XA} using the B–W approximation.[5] For simplicity we take into account the atomic interactions between nearest neighbors only. Let us denote the interaction energy between a neighboring pair of constituent species p and q by e_{pq}, where p and q are either atom A, atom B, or a vacancy. An analytic expression for the free energy can readily be derived by summing the pair interaction energies e_{pq} for the energy term and adopting Eq. (44) for the configurational entropy. For example, for an AB alloy of the L1$_0$ structure, it is found to be

$$G = -\frac{2}{N}\{(N_A N_B + N_3 N_4 + N_5 N_6)v_{AB} + [N_A(N_1 + N_2) + N_1 N_4 + N_2 N_5]v_{AV}$$

$$+ [N_B(N_1 + N_2) + N_2 N_3 + N_1 N_6]v_{BV}\}$$

$$+ 6[N_A u_{AA} + N_B u_{BB} + (N_1 + N_2)e_{VV}]$$

$$- k_B T \left(2N \ln N - \sum_{i=1}^{6} N_i \ln N_i \right) \qquad (57)$$

Here, we have introduced the effective pair interaction (EPI) energies

$$v_{AB} \equiv e_{AA} + e_{BB} - 2e_{AB} \qquad (58)$$

$$v_{AV} \equiv e_{AA} + e_{VV} - 2e_{AV} \qquad (59)$$

$$v_{BV} \equiv e_{BB} + e_{VV} - 2e_{BV} \qquad (60)$$

Each of them is a parameter that dictates the thermodynamic properties of the p–q binary system. Figure 5.17(a) explains schematically the energy v_{AB} in an ordered alloy AB (cf. Chapter 7). When a pair of antisite atoms B$_\alpha$ and A$_\beta$ is introduced into an ideally ordered crystal, the internal energy changes by zv_{AB}. Since v_{AB} is positive for a system of ordering tendency, this change in energy is positive. In Fig. 5.17(b) is shown a corresponding process for a hypothetical "binary compound" AV, in which the atoms B in Fig. 5.17(a) have been replaced by vacancies; the energy change is accordingly zv_{AV}. The EPI energy between an atom and a vacancy is assumed to be negative in most cases, but this is not guaranteed.

To determine the equilibrium defect concentrations, we apply the standard condition that partial derivatives of G with respect to pertinent variables must be simultaneously zero. As three independent variables, it is convenient to choose the number of unit cells, N, the number of B$_\alpha$, N_3, and the number of A$_\beta$, N_4. The conditions $(\partial G/\partial N)_{N_3, N_4} = 0$, $(\partial G/\partial N_3)_{N, N_4} = 0$ and $(\partial G/\partial N_4)_{N, N_3} = 0$ give, in

---

[5] The mathematical treatment given here is formally identical to the "smeared displacement model" of Cheng et al. (1967).

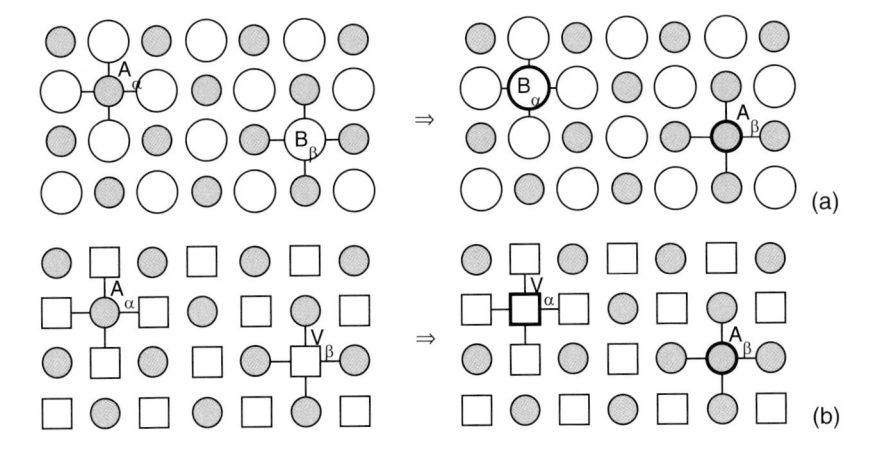

Fig. 5.17 Schematic illustation of the effective pair interaction (EPI) energy v_{pq}. (a) In a compound AB of co-ordination number z, creating a pair of antisite atoms increases the energy of the crystal by $zu_{AA} + zu_{BB} - 2zu_{AB} = zv_{AB}$. (b) The change in energy by an exchange of an atom A and a vacancy V in a hypothetical binary compound AV is similarly given by zv_{AV}.

fact, equations corresponding to (46), (47), and (48), respectively. The expressions for the energy parameters g_S, g_{XB}, and g_{XA} are found from these three conditions. For AB alloys of the L1$_0$ structure, they are given as

$$g_S = 2\{(c_A c_B + c_3 c_4 + c_5 c_6)v_{AB} + [c_A(c_1 + c_2 - 3) + c_1 c_4 + c_2 c_5]v_{AV}$$
$$+ [c_B(c_1 + c_2 - 3) + c_2 c_3 + c_1 c_6]v_{BV}\} + 12e_{VV} \tag{61}$$

$$g_{XB} = 2[(c_5 - c_4)(v_{AB} - v_{AV}) + (c_6 - c_3 + c_1 - c_2)v_{BV}] \tag{62}$$

$$g_{XA} = 2[(c_6 - c_3)(v_{AB} - v_{BV}) + (c_5 - c_4 - c_1 + c_2)v_{AV}] \tag{63}$$

where c_i are site fractions defined as $c_i \equiv N_i/N$. In Eq. (61), there is a term containing e_{VV}, but e_{VV} can be taken to be zero, which is equivalent to defining the vacuum as the reference energy level. We now see that the equilibrium concentrations of the defects are determined completely by v_{AB}, v_{AV}, and v_{BV}. For a given set of the three EPI energies, composition, and temperature, numerical solutions of c_i that satisfy the governing equations can be found by iteration or by trial and error. If desired, one may empirically take into account possible effects of vibrational entropy, by adding to g_S a term $-T(m+n)\hat{s}_{f,v}$, where $\hat{s}_{f,v}$ is the average contribution per vacancy.

From Eqs. (62) and (63), we find that the energy controlling the production of antisite defects, reaction (56), is given as

$$g_{XB} + g_{XA} = 2\{[c_5 + c_6 - (c_3 + c_4)]v_{AB} + (c_1 - c_2)(v_{BV} - v_{AV})\} \tag{64}$$

In the B–W theory of order–disorder transition, the long-range order (LRO) parameter is used to describe the antisite disorder. For a binary alloy containing vacancies, the LRO parameter may be defined as (Cheng et al., 1967)

$$\eta \equiv \frac{1}{2} \frac{m+n}{c_A + c_B} \left(\frac{c_5 - c_3}{m} + \frac{c_6 - c_4}{n} \right) \tag{65}$$

which describes the regularity in the distribution of atoms. With this definition, the site fractions of the antisite atoms are written as Eqs. (66) and (67).

$$c_3 = \frac{mn}{m+n} \left[\frac{c_B}{n} - \frac{c_A + c_B}{m+n} \eta - \frac{1}{2}\left(\frac{c_1}{m} - \frac{c_2}{n} \right) \right] \tag{66}$$

$$c_4 = \frac{mn}{m+n} \left[\frac{c_A}{m} - \frac{c_A + c_B}{m+n} \eta + \frac{1}{2}\left(\frac{c_1}{m} - \frac{c_2}{n} \right) \right] \tag{67}$$

Similar expressions for c_5 and c_6 can readily be derived. Using η, Eq. (64) is rewritten as

$$g_{XB} + g_{XA} = 4\left(1 - \frac{c_1 + c_2}{2}\right)\eta v_{AB} - 2(c_1 - c_2)(v_{AV} - v_{BV}) \tag{68}$$

If c_1 and c_2 are neglected, it indeed reduces to the familiar B–W expression for the energy governing the $L1_0$–$A1$ transition. The formation energy for the Schottky defect can also be expressed using η, as

$$\begin{aligned}
g_S = &\left[3c_A c_B + \left(1 - \frac{c_1 + c_2}{2}\right)^2 \eta^2 - \left(\frac{c_1 - c_2}{2}\right)^2 \right] v_{AB} \\
&- 2\left[\left(3c_A + \frac{c_1 - c_2}{2}\eta\right)\left(1 - \frac{c_1 + c_2}{2}\right) - \left(\frac{c_1 - c_2}{2}\right)^2\right] v_{AV} \\
&- 2\left[\left(3c_B - \frac{c_1 - c_2}{2}\eta\right)\left(1 - \frac{c_1 + c_2}{2}\right) - \left(\frac{c_1 - c_2}{2}\right)^2\right] v_{BV} + 12 e_{VV}
\end{aligned} \tag{69}$$

As mentioned earlier, v_{AB} is positive in ordering systems, and v_{AV} and v_{BV} are usually negative and of similar magnitude to v_{AB}. Hence we see from these expressions that, at low defect concentrations ($c_1, c_2 \to 0, \eta \to 1$), the formation energy of a Schottky defect ($\approx 4v_{AB} - 6v_{AV} - 6v_{BV}$) is several times as high as that of a pair of antisite atoms ($\approx 4v_{AB}$) in $L1_0$ ordered alloys.

The exact expressions for the energies of the governing equations depends on the crystal structure, namely the chemical formula, m, and n, and the co-ordination numbers. Figure 5.18 shows some common ordered structures, and Table 5.3 summarizes the crystal-geometry parameters m, n and nearest-neighbor co-ordination numbers z_{pq}. In the $D0_3$ structure, the β sublattice sites in the B2

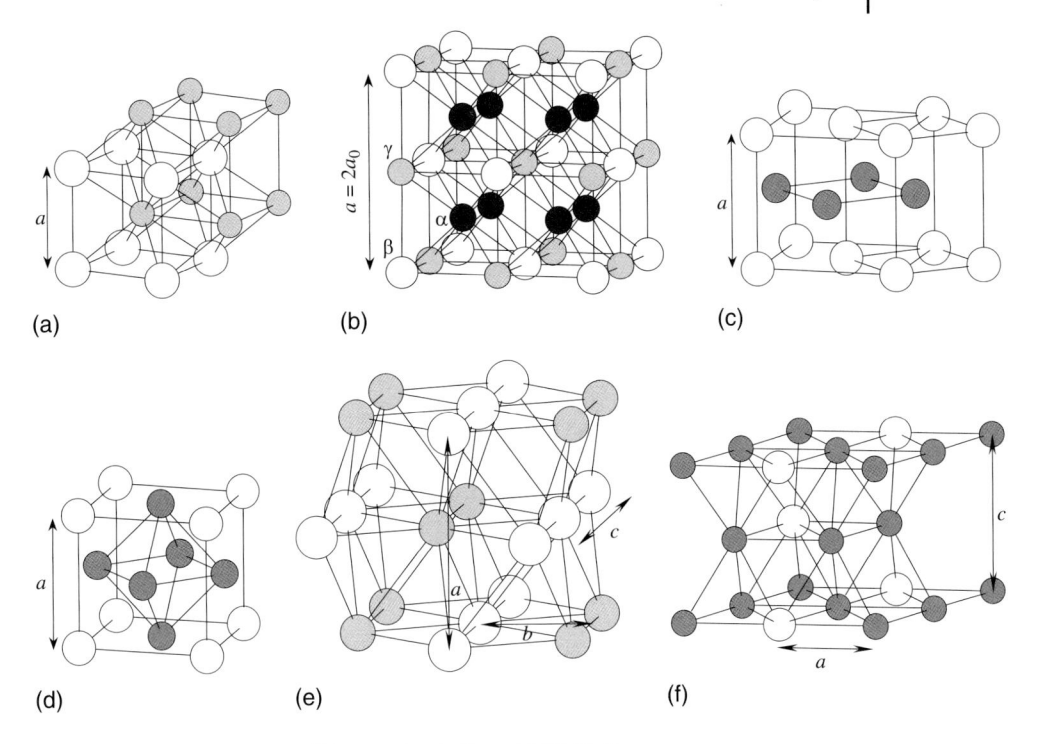

Fig. 5.18 Common structures of binary ordered alloys: (a) B2 (CsCl-type, cubic), (b) D0$_3$ (Fe$_3$Al-type, cubic), (c) L1$_0$ (CuAu-type, tetragonal), (d) L1$_2$ (Cu$_3$Au-type, cubic), (e) B19 (MgCd-type, orthorhombic), (f) D0$_{19}$ (Ni$_3$Sn-type, hexagonal). a, b, and c are the lattice parameters.

Table 5.3 Crystal-geometry parameters of common ordered structures.[a]

Structure	m	n	z_{AA}	z_{AB}	z_{BA}	z_{BB}
L1$_0$ (CuAu)	1	1	4	8	8	4
B19 (AuCd)	1	1	4	8	8	4
B2 (CsCl)	1	1	0	8	8	0
D0$_3$ (Fe$_3$Al)	3	1	4 (α), 8 (γ)	4 (α), 0 (γ)	8	0
D0$_{19}$ (Ni$_3$Sn)	3	1	8	4	12	0
D0$_{22}$ (Al$_3$Ti)	3	1	8	4	12	0
L1$_2$ (Cu$_3$Au)	3	1	8	4	12	0

[a] z_{pq} are the numbers of atoms q around an atom p.

Table 5.4 Effective pair interaction energies used for example calculations of the defect concentrations.

Model	v_{AV}/v_{AB}	v_{BV}/v_{AB}	Note
I	−1	−0.5	Model for TiAl, Ni$_3$Al
II	−1	−0.05	NiAl, NiGa, CoAl, CoGa

structure are subdivided into β sites for B atoms and γ sites for A atoms, and z_{AA} and z_{AB} are different for A atoms in the α sites and in the γ sites. In quite a few structures we find $z_{BB} = 0$, which, as we will see below, gives rise to unusual trends in point defect formation. It is also seen in the table that these geometrical parameters are identical for L1$_0$ and B19 structures, and this is also the case for L1$_2$, D0$_{19}$, and D0$_{22}$ structures. Therefore, the equations determining the statistical distribution of point defects are the same in each of these groups of structures in the nearest-neighbor pair interaction model.[6]

5.1.3.3 Equilibrium Concentrations – Examples

In this section we present the equilibrium defect concentrations in L1$_0$ and B2 ordered AB alloys calculated by the extended B–W theory described above. The L1$_0$ and B2 structures are chosen as typical examples of the distinctive behavior of defect formation in ordered alloys. The purpose is to see the effects of atomic arrangements (i.e., coordinations) and the EPI energies on defect formation. As an example, we choose two sets of the EPI energies listed in Table 5.4. Model I would represent transition-metal aluminides of close-packed structures such as TiAl and Ni$_3$Al. The atom–atom interaction energy v_{AB} can be evaluated from the thermodynamic properties of the ordered alloy, such as the formation enthalpy and thermodynamic activity (Chang and Neumann 1983; Ikeda et al. 1998). The vacancy–atom interaction energy, on the other hand, may be conveniently estimated from the vacancy formation energy of the pure solid (Kim 1991). By these procedures the parameters for Ni$_3$Al were evaluated as $v_{AB} = 0.232$ eV, $v_{AV} = -0.293$ eV and $v_{BV} = -0.122$ eV, which give $v_{AV}/v_{AB} = -1.264$ and $v_{BV}/v_{AB} = -0.524$ (Numakura et al. 1998). The choice of the much smaller v_{BV}/v_{AB} for Model II, i.e, −0.05, is somewhat arbitrary; it is intended to repro-

---

6) The D0$_3$ structure involves atomic ordering at the second-neighbor positions and thus requires introduction of interactions beyond the nearest-neighbor distance in the free energy. Such analyses of the D0$_3$ structure and the closely related L2$_1$ structure (A$_2$BC Heusler structure) are found in the literature (Schapink 1968; Bakker and van Winkel 1980; Hasaka et al. 1973; Ipser et al. 2002; Murakami and Kachi 1982). Also, the L1$_2$ and D0$_{22}$ structures are degenerate in the nearest-neighbor interaction model; it is necessary to consider at least the second-neighbor interaction to distinguish these two (Richards and Cahn 1971).

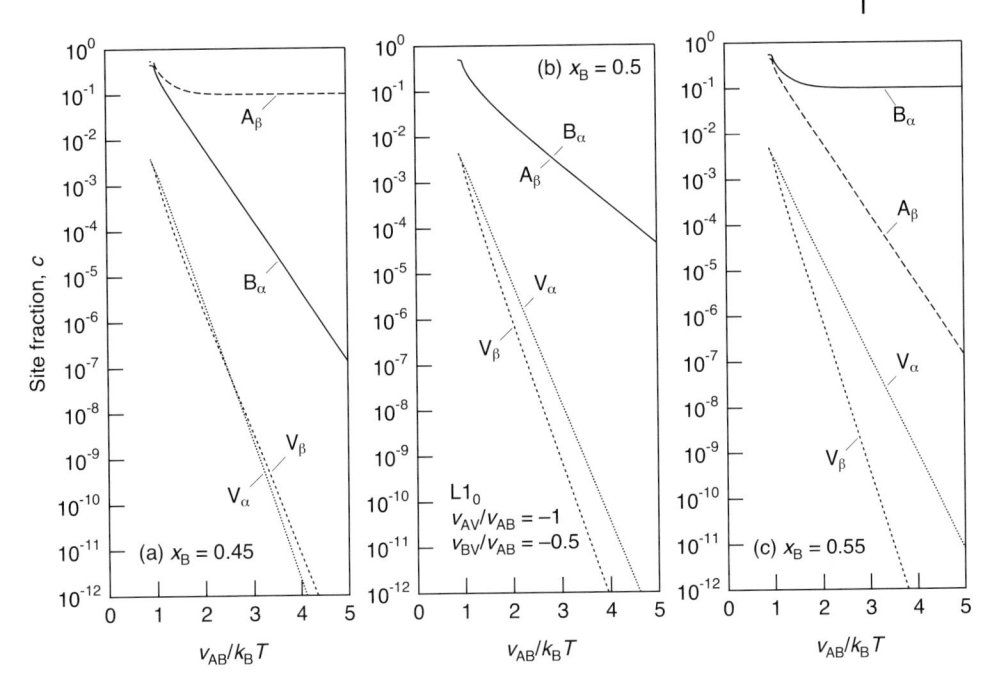

Fig. 5.19 Site fractions of vacancies and antisite atoms in L1$_0$ ordered AB alloys, with EPI energies of Model I, at composition x_B of (a) 0.45, (b) 0.5, (c) 0.55.

duce the anomalously high vacancy concentration of vacancies in B-rich B2 ordered alloys.

L1$_0$ ordered alloys Figures 5.19 and 5.20 present the concentrations of vacancies and antisite atoms in L1$_0$ ordered AB alloys as calculated with the EPI energies of Model I. The three diagrams of Fig. 5.19 display the site fractions c_i at compositions x_B of 0.45 (a), 0.5 (b), and 0.55 (c) as a function of reciprocal temperature $k_B T/v_{AB}$.[7] The slope of each curve in Fig. 5.19 gives the effective formation energy of the defect. It is seen that the slope depends sensitively on composition, particularly for the antisite defects. The defect species with very low effective formation energy may be recognized as "constitutional" defects. In this example, obviously it is A$_\beta$ at A-rich compositions and B$_\alpha$ at B-rich compositions.

In turn, Fig. 5.20 shows the site fractions at temperatures $k_B T/v_{AB}$ of 0.3 (a), 0.5 (b), and 0.7 (c) as a function of composition. We clearly see that the concen-

---

7) In the B–W model with the nearest-neighbor interaction, the order–disorder transition from L1$_0$ to A1 (fcc solid solution) occurs at $k_B T/v_{AB} = 1$ for the stoichiometric composition, but other advanced methods show that the transition must take place at a temperature as low as $k_B T/v_{AB} \approx 0.5$. It is not very useful to discuss the results of the B–W theory concerning temperatures above this "correct" transition temperature.

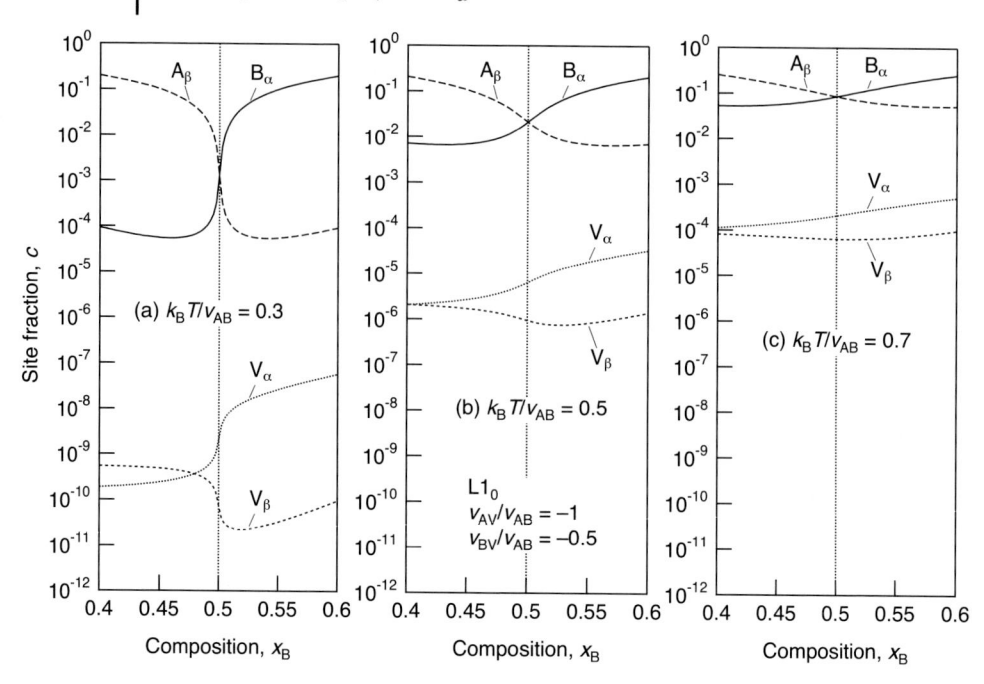

Fig. 5.20 Site fractions of vacancies and antisite atoms in L1$_0$ ordered AB alloys, with EPI energies of Model I, at temperature $k_B T/v_{AB}$ of (a) 0.3, (b) 0.5, (c) 0.7.

trations of V$_\alpha$ and V$_\beta$ vary with composition asymmetrically, while the curves of the concentrations of A$_\beta$ and B$_\alpha$ are almost symmetrical. As already presented in Fig. 5.19(b), vacancies are more populated in the α sublattice at the stoichiometric composition. In terms of the B–W theory, the ratio of the concentrations of the two vacancy species is described by Eq. (70),

$$\left(\frac{c_1}{c_2}\right)^2 \frac{c_4 c_6}{c_3 c_5} = \exp\left(\frac{g_{XB} - g_{XA}}{k_B T}\right) \tag{70}$$

which is obtained from Eqs. (47) and (48). At low defect concentrations the energy $g_{XB} - g_{XA}$ approximately equals $4(v_{BV} - v_{AV})$ for the L1$_0$ structure. This quantity is positive in the present case and thus biases vacancies to occupy α sublattice sites.

B2 ordered alloys Figures 5.21 and 5.22 show the site fractions of the four defect species in B2 ordered AB alloys calculated using the same set of EPI parameters (Model I) as in the previous example of L1$_0$ ordered alloys. A remarkable feature is that the concentrations of vacancies are of the same order of magnitude as those of antisite atoms. (The B2–A2 transition occurs at $k_B T/v_{AB} = 2$ in the nearest-neighbor B–W theory without vacancies, but here the transition tempera-

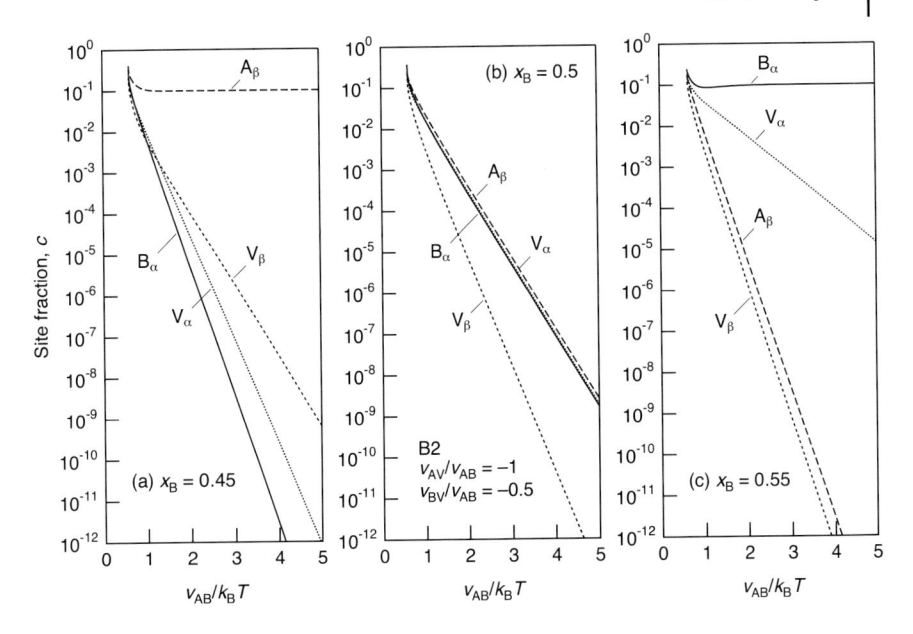

Fig. 5.21 Site fractions of vacancies and antisite atoms in B2 ordered AB alloys, with EPI energies of Model I, at composition x_B of (a) 0.45, (b) 0.5, (c) 0.55.

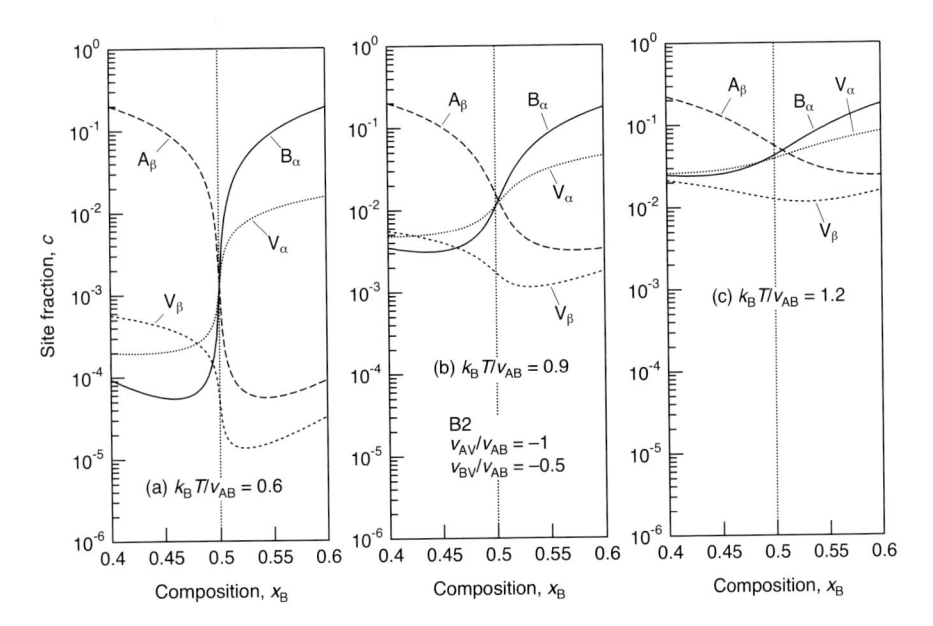

Fig. 5.22 Site fractions of vacancies and antisite atoms in B2-ordered AB alloys, with EPI energies of Model I, at temperature $k_B T / v_{AB}$ of (a) 0.6, (b) 0.9, (c) 1.2.

ture is lowered to about 1.6 by the high concentration of vacancies.) The concentrations of vacancies that exist in thermal equilibrium are much higher than in the L1$_0$ structure, even though the EPI energies are identical. In the present framework, this striking difference is due solely to the differences in the coordination numbers (Table 5.3). The energies controlling the formation of a pair of antisite atoms and a Schottky defect in the B2 ordered alloy are given as

$$g_{XB} + g_{XA} = 8\left(1 - \frac{c_1 + c_2}{2}\right)\eta v_{AB} - 4(c_1 - c_2)(v_{AV} - v_{BV}) \tag{71}$$

and

$$g_S = 2\left[c_A c_B + \left(1 - \frac{c_1 + c_2}{2}\right)^2 \eta^2 - \left(\frac{c_1 - c_2}{2}\right)^2\right] v_{AB}$$
$$- 4\left[\left(c_A + \frac{c_1 - c_2}{2}\eta\right)\left(1 - \frac{c_1 + c_2}{2}\right) - \left(\frac{c_1 - c_2}{2}\right)^2\right] v_{AV}$$
$$- 4\left[\left(c_B - \frac{c_1 - c_2}{2}\eta\right)\left(1 - \frac{c_1 + c_2}{2}\right) - \left(\frac{c_1 - c_2}{2}\right)^2\right] v_{BV} + 8e_{VV} \tag{72}$$

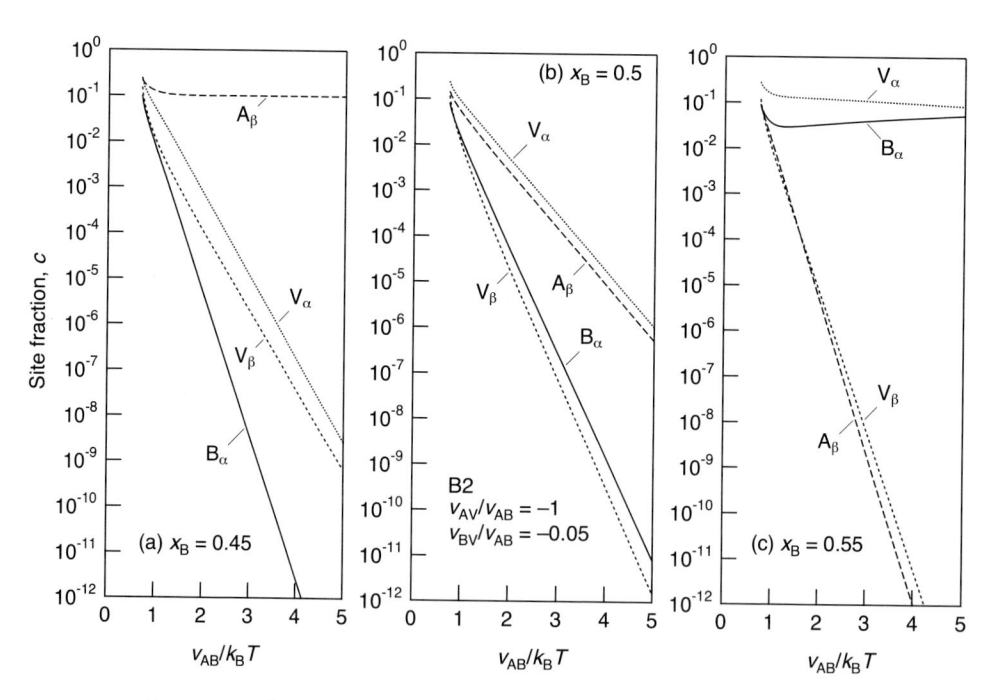

Fig. 5.23 Site fractions of vacancies and antisite atoms in B2 ordered AB alloys, with EPI energies of Model II, at composition x_B of (a) 0.45, (b) 0.5, (c) 0.55.

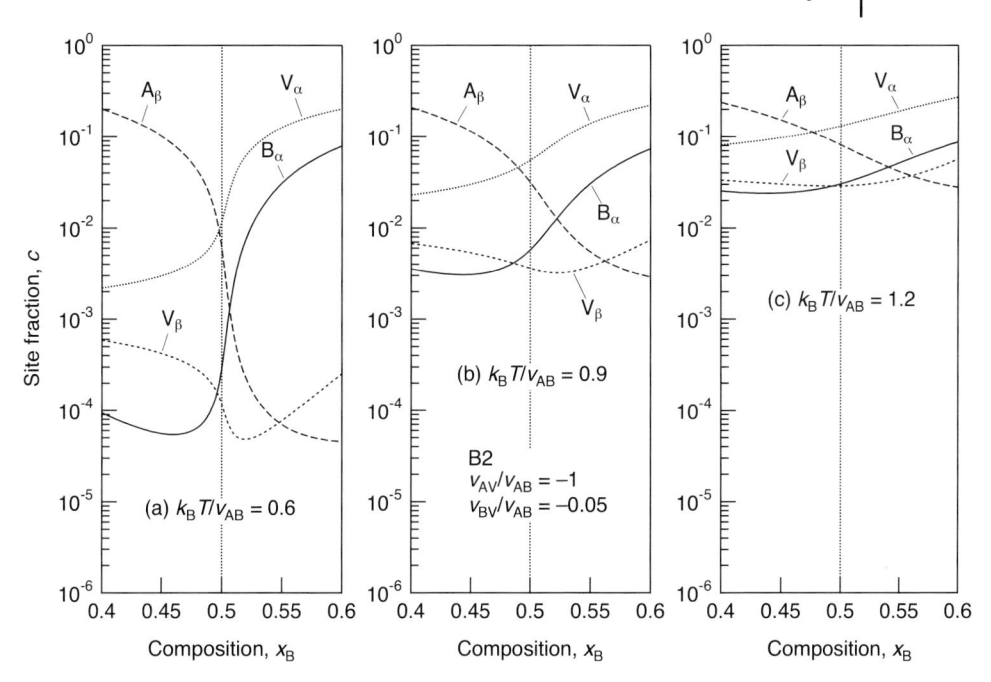

Fig. 5.24 Site fractions of vacancies and antisite atoms in B2-ordered AB alloys, with EPI energies of Model II, at temperature $k_B T/v_{AB}$ of (a) 0.6, (b) 0.9, (c) 1.2.

They are approximated to $8v_{AB}$ and $4v_{AB} - 4v_{AV} - 4v_{BV}$, respectively, at low defect concentrations. With the present set of parameters the latter equals $10v_{AB}$, so that the formation energy of a Schottky defect is only 20% higher than that of an antisite-atom pair. It is also to be noted in Fig. 5.22 that the concentrations of vacancies are not very low at A-rich compositions. In particular, the concentration of V_β, which is often neglected in simplified models, is no lower than those of V_α and B_α.

Figures 5.23 and 5.24 show the equilibrium defect concentrations for the EPI energies of Model II, where v_{BV}/v_{AB} is still negative but is one tenth of that for Model I. The "constitutional" defect species at B-rich compositions is now V_α, and it is the predominant defect species even at the stoichiometric composition. Physical origins of such anomalously high vacancy concentrations are discussed in the next section.

5.1.3.4 Abundant Vacancies in some Intermetallic Compounds

Preference of defect species Intermetallic compounds of the B2 structure used to be classified into two types according to the trends in point defect formation: of substitutional disorder; and of "triple defects" (Chang and Neumann 1983). In the former class of materials, antisite atoms are always the predominant defect

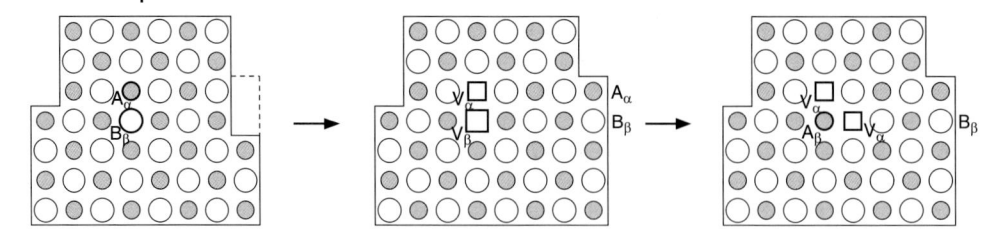

Fig. 5.25 Schematic illustration of the formation of a triple defect in an AB compound. A new unit cell is created at the surface, leaving a V_α and a V_β in the interior, but the latter is then filled by an A atom to convert it to A_β, leaving another V_α.

species. In the latter, on the other hand, while the major defect species is A_β at A-rich compositions, it is V_α at B-rich compositions. These two trends are related to the relative ease of the reaction

$$A_\alpha + B_\beta \rightarrow B_\alpha + A_\beta \tag{73}$$

and an alternative,

$$A_\alpha + B_\beta \rightarrow 2V_\alpha + A_\beta + B_\beta \tag{74}$$

In the latter, three defects, i.e., "triple defects", are created, avoiding V_β and B_α, as illustrated in Fig. 5.25. By comparing the right-hand sides of the two reactions, we see that the trends in defect formation are essentially determined by the preference between B_α and V_α, pertaining to the balance of the reaction $B_\alpha \leftrightarrow 2V_\alpha + B_\beta$. This balance is formally represented by the following reaction, which is derived by combining reaction (52) and the inverse reaction of (53):

$$nB_\alpha + nV_\beta \leftrightarrow (m+n)V_\alpha + nV_\beta + nB_\beta \tag{75}$$

The energy governing this reaction is given by

$$\Delta E_\alpha = g_S - ng_{XB} \tag{76}$$

In the nearest-neighbor pair interaction model, the right-hand side is approximately given by $-zv_{BV}$, or $-8v_{BV}$ for the B2 structure, when defect concentrations are low. Antisite atoms B_α are dominant as long as ΔE_α is negative, but when it is small in magnitude, vacancies V_α are formed as much as B_α. If ΔE_α turns positive, vacancies V_α become the major defect species.

NiAl, NiGa, CoAl, and CoGa, which are said to be typical triple-defect compounds, are chemically stable intermetallic phases which remain ordered right up to the melting temperature. These materials have large formation enthalpy, and thus large positive EPI energy v_{AB}. The atom–vacancy EPI energy v_{AV} is ex-

Table 5.5 The energy governing the preference of point defect species in α and β sublattices in an ordered alloy $A_m B_n$. Antisite atoms or vacancies predominate if ΔE is positive or negative, respectively.

Structure	$\Delta E_\alpha = g_S - n g_{XB}$	$\Delta E_\beta = g_S - m g_{XA}$
B2 ($m = n = 1$)	$-8v_{BV}$	$-8v_{AV}$
L1$_0$ ($m = n = 1$)	$2(v_{AB} - 2v_{AV} - 4v_{BV})$	$2(v_{AB} - 4v_{AV} - 2v_{BV})$
L1$_2$ ($m = 3, n = 1$)	$4(v_{AB} - 4v_{AV} - 2v_{BV})$	$-24v_{AV}$

pected, from their thermodynamic properties (Chang and Neumann 1983), to be negative and its magnitude is as large as v_{AB}. On the other hand, v_{BV} could be much smaller in magnitude, or even slightly positive in these materials (Chang and Neumann 1983; Kim 1992); the bonding between B atoms at the nearest-neighbor positions is suggested to be very weak, or the interaction could possibly be repulsive, because a β site (surrounded by B atoms) might be too small to accommodate a B atom (Cottrell 1995; Meyer and Fähnle 1999; Korzhavyi et al. 2000). This leads to preference for V_α over B_α according to the criterion $\Delta E_\alpha = -8v_{BV}$, and explains the high concentration of α vacancies observed in these materials.

Preference of defect species in the β sublattice can be discussed similarly on the basis of a reaction analogous to (75). The energy that controls the preference between A_β and V_β is given as

$$\Delta E_\beta = g_S - m g_{XA} \tag{77}$$

Approximate expressions for ΔE_β at low defect concentrations are listed in Table 5.5 for the B2, L1$_0$, and L1$_2$ structures. The expression for ΔE_β for the L1$_2$ structure is similar to the case of the B2 structure discussed above: it is determined solely by v_{AV}. This predicts that, in an L1$_2$ ordered A_3B alloy composed of elements A and B of low and high vacancy formation energies such as Al$_3$Ti,[8] high concentrations of vacancies on β sublattice sites may occur. From these two examples, it is seen that abundant vacancies arise from the characteristic atomic configurations: when a particular site is surrounded by a single species of atoms, an antisite atom of the same species might be unfavored, and at high temperatures, where gaining configurational entropy is important, the site could rather be made vacant. By referring to Table 5.5, such a configuration is found also in the D0$_3$ structure, in which $z_{BB} = 0$ as in the B2 and L1$_2$ structures. By experiment, it is

---

[8] The structure of Al$_3$Ti is D0$_{22}$, which is described by stacking two L1$_2$ unit cells with an anti-phase boundary, but the nearest-neighbor coordination numbers are identical with L1$_2$.

known that Ni$_3$Sb exists at high temperatures over a certain range of Sb-rich compositions, in which the defects responsible for the deviation from stoichiometry are almost entirely vacancies on Ni sites (Heumann and Stüer 1966). This is the reason for the anomalously fast diffusion in this compound shown in Fig. 5.14.

"Structural vacancies" – a remark on terminology The abundant vacancies observed in the "triple-defect" B2 compounds, NiAl, NiGa, CoAl, and CoGa, are essential in realizing the wide range of compositional stability at high temperatures, and are often referred to as "structural," "constitutional," or "compositional" vacancies, contrasting with "thermal" vacancies (de Novion 1995; Chang and Neumann 1983). When a compound phase exists over a certain composition range, the defects responsible for the deviation from the stoichiometry are generally called "constitutional defects" or "structural defects." In the transition-metal (TM) aluminides and gallides mentioned above, A_β is always the primary defect species (i.e., the constitutional defects) at TM-rich compositions. On the other hand, at TM-deficient compositions, vacancies V_α are often found to exceed antisite atoms B_α. There have been discussions on which of these defects is more favored in a particular material, or whether those constitutional defects are stable at all, in the framework of statistical thermodynamics (Kim 1988; Ren and Otsuka 2000; Kim 1992; Breuer et al. 2002). However, it should be noted that statistically distributed point defects cannot exist at absolute zero temperature according to the third law of thermodynamics; statistical thermodynamics is not very useful for discussing "stability" of point defects.

The composition of a real material is never exactly stoichiometric. A single-phase compound would eventually separate into two phases on lowering the temperature, most probably to the original compound phase with a more ordered structure and a neighboring phase in the phase diagram. The latter can be another compound phase, or a terminal elementary phase of one of the two components. In the Ni–Al system, there exists a compound phase Ni$_2$Al$_3$ next to the B2 ordered NiAl. The structure of Ni$_2$Al$_3$ is described as a slightly distorted B2 structure in which every third plane of nickel atoms perpendicular to the $\langle 111 \rangle$ direction is systematically absent (Taylor and Doyle 1972). Similar vacancy-ordered phases are known to occur in the Ni–Ga system: Ni$_3$Ga$_4$ (Ellner et al. 1969) and Ni$_2$Ga$_3$ (Hellner 1950). The Ni vacancies in these phases deserve the name of "structural" vacancies, as they are elements of specific crystal structures.

The presence of these vacancy-ordered phases, in which vacancies continue to exist, is consistent with the predominance of vacancy defects over antisite atoms in these B2 compounds at high temperatures, but those vacancies in the B2 phase are in principle not allowed to remain at low temperatures. In practice, however, the compound may remain single-phase with supersaturated point defects, since the defects present at high temperatures may not be annihilated completely because of the loss of mobilities. Often those defects are regarded as "structural" or "constitutional" defects, but they are actually nothing more than "frozen-in" defects. After all, the defects present in thermal equilibrium at high temperatures are formed to realize a thermodynamic state whose free energy attains the minimum for the given composition and temperature. Since it is impossible to determine if a particular point defect has been formed thermally or constitutionally, it does not seem adequate to employ such a distinction as "thermal defects" and "constitutional defects".

5.2 Defect Migration: Microscopic Diffusion

5.2.1 The Single Atom Jump

In the previous section the presence of point defects in a crystal at finite temperature was seen to be a consequence of fundamental thermodynamic principles. In their turn, point defects are the prerequisite for diffusion and all changes in atom configuration, for whenever an atom leaves its position a defect of some kind is created, displaced, or annihilated. By far the most important mechanism, natural for substitutional solid solutions, is the interchange of an atom with a neighboring vacancy, and it is the vacancy mechanism which we will mainly consider in the following. Diffusion may take place also by hopping of atoms from one interstitial lattice site to the next, a diffusion mode important for small impurity atoms such as hydrogen or carbon. In a third mechanism of some relevance, the interstitialcy mechanism, an atom on an interstitial site displaces a neighboring atom from a regular lattice which it afterwards occupies (cf. Section 5.2.2.2 on correlation.) All these mechanisms lend themselves to the same treatment in the framework of statistical mechanics which will be expounded below.

5.2.1.1 Transition State Theory

Transport of atoms by single atom jumps is a thermally activated process, which means that some energy barrier has to be overcome by thermal energy fluctuation. Such an event is, of course, the elementary step of a multitude of kinetic rate processes considered in physics and chemistry. Time-honored though it is, the transition state theory remains in many cases a viable description of the single activation event. Originally it was formulated for chemical reaction rates (Glasstone et al. 1941) and refined on the basis of statistical mechanics by Wert (1950), Zener (1952), and Vineyard (1957). We give a short account of this method below.

Let us consider a solid with N atoms. The Born–Oppenheimer approximation is invoked as usual (cf. Chapters 4, 10, and 11), the electrons as the component with much lower mass adapting almost instantaneously to the position of the lattice ions. The ions (henceforth also called atoms, for simplicity) can then be regarded as subject to forces created by the interaction with the electrons and dependent only on ion coordinates. Then we have to deal with $3N$ degrees of freedom with ion positions and momenta as coordinates. As in a classical system positions and momenta are distributed independently, we can restrict our

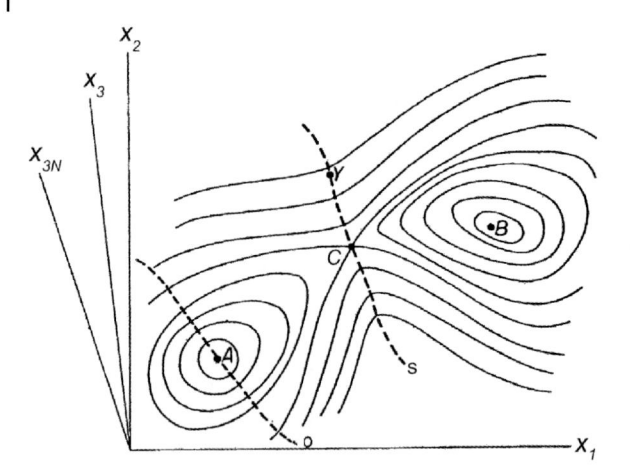

Fig. 5.26 Potential Φ as a function of $3N$ atom coordinates. A is the original equilibrium state of the system, B the new equilibrium state after the atom has jumped, C saddle point, s dividing hypersurface. Adapted from Vineyard (1957).

treatment to the $3N$ dimensional configuration space subtended by the position coordinates. The total internal energy of the system can be represented as a potential function Φ of the position coordinates. Let us now regard an elementary event of diffusion, say the jump of an atom into a neighboring vacancy.

The original equilibrium position A (Fig. 5.26) of the atom corresponds to a potential minimum, as does the new equilibrium position B after the jump. In between, there has to be at least one transition state, corresponding to a saddle point C in the potential surface where the curvature is negative along a certain direction which the transition path follows. The system, excited by thermal fluctuations, will move about in the phase space region around A and eventually, aided by a favorable fluctuation, pass over C into the region around B. Instead of following its path in time an ensemble of systems distributed in the vicinity of A is considered, and we calculate ensemble averages instead of time averages. A hypersurface of dimension $3N - 1$ which we call s, dividing the A and B basins, passes through C (the broken line in Fig. 5.26) and is perpendicular to the contours of constant potential Φ everywhere else. The simplifying assumption is made that the ensemble of representative system points is in thermal equilibrium right up to the saddle point, as well as along the relevant part of s, corresponding to a canonical ensemble under isothermal conditions. The second important assumption of transition state theory is that any system which has reached s and has a positive velocity component normal to s will definitely cross over into the B basin. This means that the system point does not follow a random, reversible motion at the saddle point, and that any friction forces are sufficiently small. The transition rate will be given by the average number of system points crossing over s in unit time, divided by the total number of systems in the A basin. An orthogonal rotation of the Cartesian coordinates can be performed everywhere along the dividing hypersurface s so that a given coordinate axis, which we may call x_1, is perpendicular to it. Only the positive part of the corresponding velocity u_1 is then contributing to the flux across s, and from the Maxwellian velocity distribution Eq. (78) follows for the average velocity $\bar{u}$ across s,

$$\bar{u} = \frac{\int_{u_1=0}^{\infty} u_1 \exp\left(-\frac{mu_1^2}{2k_BT}\right) du_1}{\int_{u_1=-\infty}^{\infty} \exp\left(-\frac{mu_1^2}{2k_BT}\right) du_1} = \sqrt{\frac{k_BT}{2\pi m_1}} \tag{78}$$

m_1 being the mass associated with degree of freedom number 1. Now we introduce a transition region by adding a small strip of width δ on both sides of the hypersurface s. If we write $N_{s,\delta}$ for the number of systems of the ensemble in this region, and N_A for the number of systems in the A basin, the transition rate can be written as (Schoeck 1980):[9]

$$\omega = \frac{\bar{u} N_{s,\delta}}{\delta N_A} = \frac{\sqrt{\frac{k_BT}{2\pi m_1}}}{\delta} \frac{\int_{s,\delta} \exp\left(-\frac{\Phi}{k_BT}\right) d^{3N}x}{\int_A \exp\left(-\frac{\Phi}{k_BT}\right) d^{3N}x} \tag{79}$$

It remains to calculate the partition functions $N_{s,\delta}$ and N_A. Provided the potential barrier $\Delta\Phi = \Phi(C) - \Phi(A) \gg k_BT$ is high enough and the curvature of the saddle surface normal to the transition path is strong enough, then most of the representative system points will be found near A, and the contribution to the flux across s will be largest at C. We may then expand the potential around A and C into Taylor series, and after suitable orthogonal transformations write Eqs. (80) in terms of the normal frequencies v_i at A and v_i' at the saddle point C:

$$\Phi_A = \Phi(A) + \frac{1}{2}\sum_{i=1}^{3N} m_i(2\pi v_i)^2 q_i^2, \quad \Phi_C = \Phi(C) + \frac{1}{2}\sum_{i=2}^{3N} m_i(2\pi v_i')^2 q_i'^2 \tag{80}$$

where at the saddle point the dependence on the transition coordinate, associated with a negative curvature, may safely be omitted as it is constrained to an infinitesimal strip of width δ, a quantity which finally cancels out. With Eq. (80) the integration in Eq. (79) can be performed, easily yielding the simple result

$$\omega = \left(\frac{\prod_{i=1}^{3N} v_i}{\prod_{i=2}^{3N} v_i'}\right) \exp\left(-\frac{\Phi(C) - \Phi(A)}{k_BT}\right) = v_{\text{eff}} \exp\left(-\frac{\Phi(C) - \Phi(A)}{k_BT}\right) \tag{81}$$

---

9) In contrast to some previous studies, we use ω here instead of Γ, because Γ will be reserved, in the context of this chapter, for an effective atom jump frequency that contains a factor (e.g., vacancy concentration) accounting for the availability of the defect mediating diffusion. The transition rate considered here has the role of an exchange frequency with a defect already lying adjacent to the jumping atom.

The effective frequency v_{eff}, in rate theory usually referred to as an "attempt frequency" and often identified liberally with the Debye frequency, here turns out as the product of all $3N$ normal frequencies in the equilibrium state, divided by the $(3N-1)$ normal frequencies in the constrained saddle point state. Note that, at least in principle, all the degrees of freedom of the system enter in this expression. Although v_{eff} is very difficult to estimate in the first place, some significant progress has recently been made on the basis of quantum mechanical first-principles calculations where phonon spectra can now be obtained in the activated, constrained state (Wimmer et al. 2006; cf. also Chapter 11).

For a system kept at constant temperature and pressure P, work has to be done against the external pressure when changing the configuration of the system. The potential Φ then contains an additional PV term and Eq. (81) can be rewritten in a more compact form familiar from Arrhenius analysis as

$$\omega = v_{eff} \exp\left(-\frac{\Delta H_m}{k_B T}\right) \tag{82}$$

Equation (82) can be cast into yet another useful form. Following Wert and Zener (1949) we introduce at the original equilibrium state a hypersurface similar to the one at the saddle point. This hypersurface, called o (Fig. 5.26), is normal at A to the line of force leading to C, perpendicular to the contours of constant Φ everywhere else, and should also be fitted with a strip of width δ on both sides. Then Eq. (79) can be modified according to Eq. (83):

$$\omega = \frac{\bar{u} N_{s,\delta}}{\delta N_A} = \frac{\bar{u}}{\delta} \frac{N_{o,\delta}}{N_A} \frac{N_{s,\delta}}{N_{o,\delta}} = \tilde{v} \frac{N_{s,\delta}}{N_{o,\delta}} = \tilde{v} \exp\left(-\frac{\Delta G_m}{k_B T}\right) \tag{83}$$

for a system at constant temperature and pressure, where now

$$\tilde{v} = v_{eff} \exp\left(-\frac{\Delta S_m}{k_B}\right) \tag{84}$$

and

$$\Delta S_m = k_B \ln\left(\frac{\prod_{i=2}^{3N} v_j^0}{\prod_{i=2}^{3N} v_j'}\right) \tag{85}$$

The v_j^0 and the v_j' denote the constrained frequencies in the equilibrium state A and the saddle point state C, respectively. Equation (84) follows from comparison of Eq. (83) with Eq. (82). As $\exp[-\Delta G_m/(k_B T)]$ is the quotient of the two constrained partition functions at C and A, we see that ΔG_m is identified as the Gibbs free energy needed to carry a system constrained to the (moving) hypersurface reversibly from A to C. Specifically, ΔS_m reflects the change in the vibration spectrum connected with this transfer.

5.2.1.2 Alternative Methods

Transition state theory is associated with two critical assumptions which contradict each other to a certain extent. On the one hand, the activated systems should interact with the heat bath strongly enough to acquire thermal equilibrium; on the other hand, the interaction must be weak enough not to reverse a positive velocity across the saddle hypersurface.

Dynamical Theories Several authors have argued that it is altogether questionable whether the atom remains at the saddle point long enough to acquire thermal equilibrium (especially if the driving force for diffusion is very large or there is strong co-operation between successive or simultaneous atom jumps). Dropping the assumption of thermal equilibrium, it was consequently attempted to derive the atom jump frequency from a different argument involving lattice dynamics. In these "dynamical" theories of thermal activation (Rice 1958; Slater 1959; Manley 1960) the motion of a jumping atom was considered as a superposition of n harmonic normal modes of the crystal. A transition event is supposed to take place once the amplitude exceeds a certain limit, and at the same time atoms surrounding the transition path "move out of the way" enabling the jumping atom to pass. A statistical treatment of the atom displacements in the framework of classical thermodynamics yields a transition frequency where the energy barrier for the jumping atom and the energies needed to displace the surrounding atoms figure in Boltzmann factors. The attempt frequency which appears as a prefactor has in the numerator the normal frequencies of the system, and in the denominator the normal frequencies with the coordinate q_1 of the jumping atom constrained to the saddle point, making this expression very similar to Eq. (81) in transition state theory.

"Diffusion" Theory of Thermal Activation Sometimes in thermal activation the saddle point of the energy surface in phase space corresponding to the critical state is not a sharp maximum as a function of the transition variable. Then it is to be expected that the representative system points, by their stochastic interaction with the heat bath, perform a random motion in a region of phase space near the saddle point with an energy difference not exceeding $k_B T$. In this case the second assumption of transition state theory – that a positive velocity at the saddle point is not reversed – cannot be upheld any more and we have to resort to "diffusion" theories of thermal activation (Kramers 1940; Chandrasekhar 1943; Brinkman 1956). In analogy with classical nucleation theory, a steady-state flow of representative system points over the saddle point can then be derived. As a result, that strong damping together with a "soft" potential (low curvature at the saddle point) decreases the transition rate (jump frequency) in comparison with the value predicted by transition state theory (Kramers 1940; Schoeck 1980).

Molecular Dynamics Most classical treatments of thermal activation such as transition state theory and dynamical theory rest on the harmonic approximation of the crystal and therefore neglect nonlinear behavior near the saddle point. By employing modern, computer-based methods this limitation and any restrictive assumptions about the activated state of the system can be overcome. The most widely used method is molecular dynamics (MD; cf. Chapter 10), which amounts to numerical solution of the equations of motion of the jumping atom and its surroundings. A valid description of interaction potentials is of course necessary, the EAM (embedded-atom method) potentials (Daw and Baskes 1984) being the most popular form besides the Finnis–Sinclair type (Finnis and Sinclair 1984). Molecular dynamics is the preferred method of numerical simulation for diffusion processes in materials which have no regular crystal structure like, of course, liquid, amorphous or organic materials, or in cases where the diffusion pathways are dimensionally restricted as on the crystal surface, in grain boundaries, interfaces, and dislocation cores (cf. Section 5.4.3).

Nudged Elastic Band Method Activation energies are inferred from MD simulations only indirectly from an Arrhenius analysis of the temperature dependence of the transition rates. More direct approaches to numerical determination of the migration barriers consist in static energy calculations where the jumping atom is displaced from its original equilibrium position toward a new equilibrium state by small steps, relaxing at each step toward the configuration of minimum energy. The interactions are calculated by fitting various potentials or employing ab-initio density functional methods (cf. Chapter 11). Besides the fact that relaxed configurations may represent a poor picture of the actual, dynamical diffusion process, the main disadvantage is that in most of these calculations the jump path is unnecessarily restricted. Allowance is made for all degrees of freedom in the nudged elastic band (NEB) method (Mills et al. 1995), where between the initial equilibrium state of the system (A in Fig. 5.26) and the final equilibrium state (B in Fig. 5.26) a set of replicas of the system is interposed which reflect the changing state of the system during the transition. The replicas are linked by spring forces in an additional dimension of phase space. All coordinates of all replicas are numerically relaxed until a minimum energy configuration is reached. The set of connected replicas has then settled like an elastic band across a saddle surface to a "most comfortable fit" where it is least stretched. This method can even be applied when calculating the energy on an ab-initio basis if the number of degrees of freedom is not too large.

5.2.2
Solid Solutions

5.2.2.1 Random Walk
Atoms migrate in solids by random jumps of defects between sites of a generally three-dimensional lattice.[10] The symmetry properties and topological connectivity

---

10) In some situations, the jumps may be restricted to two dimensions or one dimension (cf. Section 5.4.3).

of the crystal lattice can be almost arbitrarily complex. In many cases (e.g., long-range ordered alloys) it will be necessary to classify the lattice sites as belonging to sublattices, and the jump probability generally depends on in which sublattice the jump starts or ends. Nevertheless, we will talk of a random walk proper[11] as long as the jump probability does not depend on local composition or composition gradients and is independent of any previous steps (i.e., uncorrelated walk). In this section we will for the sake of simplicity assume that the concept of a random walk applies to a migrating atom as well as to a migrating defect.

The random walk is a well-known mathematical problem (Rudnick and Gaspari 2004) with many applications in statistical physics. It is employed in the study of organic systems, and even more abstract fields such as sociology. Some of its elementary properties can already be gleaned from the simplest case, a one-dimensional random walk, where the atom makes jumps of equal length along a straight line. As discussed in Section 5.2.1, the jump probabilities are determined by the properties of the atom potential. In Fig. 5.27 (Manning 1968) different possibilities are displayed schematically. In Fig. 5.27(a) jumps to the left or right are equally probable; in Fig. 5.27(b) there is a constant driving force such as on ions in an electric field, so the probability is biased toward one side; in Fig. 5.27(c) the properties of the potential and thus the diffusion coefficient are changing with position. In this latter case we will not talk of a random walk but of a diffusion walk.

Macroscopic behavior may now be deduced from the random jump events by following two possible approaches. Either (1) we consider the net atom flux between parallel lattice planes of the crystal or (2) we trace the random walk of atoms along some distance and learn about the transport of macroscopic quantities of matter by calculating averages.

Following approach (1) it is an easy exercise to connect microscopic jump properties with the macroscopic diffusion laws. Consider for simplicity a simple cubic lattice with jump distance l and diffusion normal to the $\{100\}$ planes, which reduces diffusion to a one-dimensional problem. If the number of, say, A atoms in plane 1 per unit area is n_1 and in the adjacent plane n_2 and the respective jump frequencies according to the scenario of Fig. 5.27(b) are $\Gamma_{12} = \Gamma_0(1+\varepsilon)$ and $\Gamma_{21} = \Gamma_0(1-\varepsilon)$, then the net flux per unit area (i.e., flux density) will be[12]

$$j = n_1\Gamma_{12} - n_2\Gamma_{21} = \Gamma_0(n_1 - n_2) + \Gamma_0\varepsilon(n_1 + n_2)$$

$$= -\Gamma_0 l^2 \frac{\partial C}{\partial x} + 2\Gamma_0\varepsilon Cl \qquad (86)$$

---

11) Terminology in the literature may draw the line of distinction in different places. Some authors refer to a random walk only when all jump probabilities are equal, others consider stochastic motion by any set of probabilities a random walk. We take an intermediate viewpoint.

12) Here, as everywhere else in Chapter 5, upper-case C will be used for concentrations as atoms per unit volume whereas lower-case c will be reserved for site fractions. Note that $c = \Omega C$, where Ω is the atomic volume.

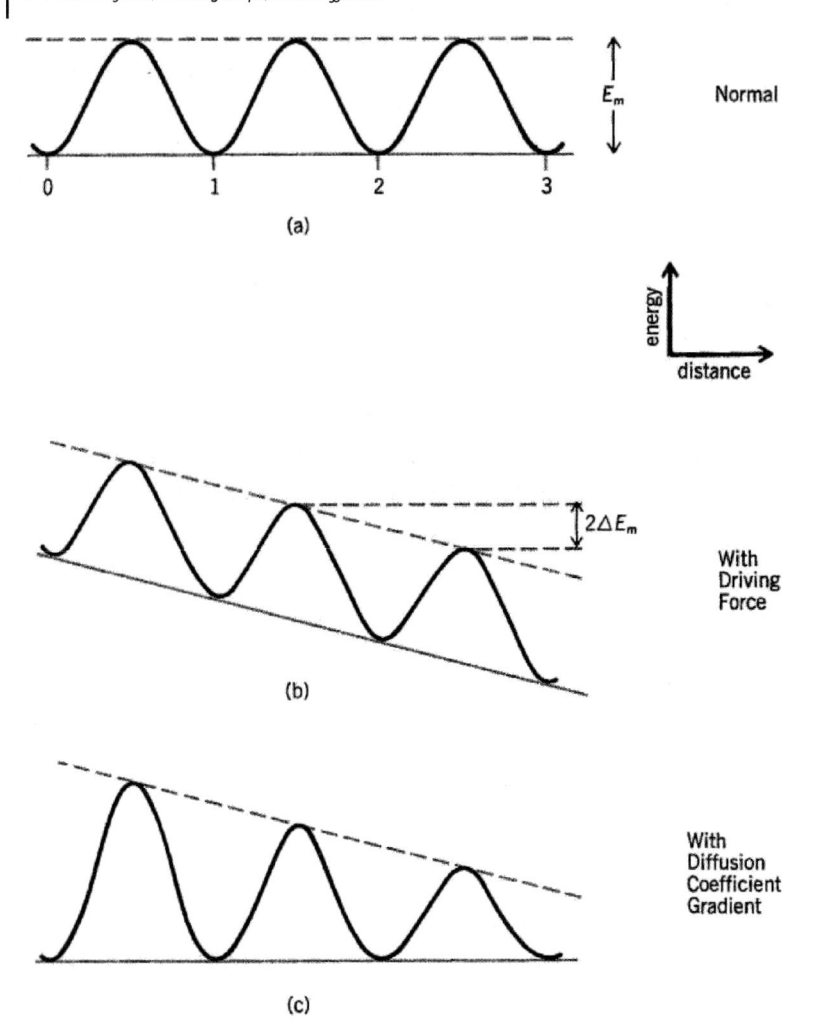

Fig. 5.27 Three possible potential landscapes for random walks (Manning 1968). (a) Unbiased random walk; (b) random walk with driving force; (c) diffusion walk in nonhomogeneous environment.

where we have used the relationship $Cl = (n_1 + n_2)/2$ for the volume concentration C of A atoms.

A driving force F will lead to an energy difference Fl within the distance between adjacent lattice planes, and $\varepsilon = Fl/(2k_B T)$. Defining the diffusion coefficient as $D = \Gamma_0 l^2$ eventually puts the flux in the form

$$j = -D\frac{\partial C}{\partial x} + C\frac{DF}{k_B T} \tag{87}$$

in which we recognize Fick's first law of diffusion with a drift term. If we consider jumps that do not coincide with the diffusion direction but subtend angles α_i with it, one has to sum up the projections. Allowing only nearest-neighbor jumps to happen, we get Eq. (88).

$$D = \frac{1}{2}\Gamma_0 l^2 \sum_{i=1}^{z} \cos^2 \alpha_i \qquad (88)$$

For cubic crystal structures, this can be shown to result in Eq. (89),

$$D = \frac{1}{6}l^2 \Gamma_0 z = \frac{1}{6}l^2 \Gamma \qquad (89)$$

with the total jump frequency $\Gamma = \Gamma_0 z$, independent of the diffusion direction. z is the number of nearest neighbors, the coordination number. An expression for the hcp structure can be found in Manning (1968), p. 48.

Applying the continuity equation for the one-dimensional flux

$$\frac{\partial C}{\partial t} = -\frac{\partial j}{\partial x} \qquad (90)$$

to Eq. (87) (or directly considering the change of atom numbers in the middle one of three adjacent planes) we get Eq. (91),

$$\frac{\partial C}{\partial t} = D\frac{\partial^2 C}{\partial x^2} - \frac{DF}{k_B T}\frac{\partial C}{\partial x} \qquad (91)$$

having thus derived Fick's second law for one-dimensional diffusion.

Let us now have a closer look at the second approach, in which we follow an atom along its path. Steps will be made in random sequence to the left or to the right with probabilities p and $q = 1 - p$. Assuming that the right and left probabilities differ ever so slightly, we set $p = (1+\varepsilon)/2 + O(\varepsilon^2)$ and $q = (1-\varepsilon)/2 + O(\varepsilon^2)$. Then after N steps the probability of having advanced a total distance of m to the right can be expressed by the binomial distribution, which can be shown to converge in the limit of a large number of steps to a Gaussian probability distribution. We suppose now that the particle on average suffers $2\Gamma_0$ displacements per unit time. Introducing the continuous variable $x = ml$ when the step width is l, we can write for the differential probability $W(x,t)\Delta x$ that the particle has arrived in the interval $[x, x + \Delta x]$ at time t,

$$W(x,t) = \frac{1}{2\sqrt{\pi Dt}} \exp\left[-\frac{(x - \bar{u}t)^2}{4Dt}\right] \qquad (92)$$

where we have written $D = \Gamma_0 l^2$ (diffusion coefficient) and $\bar{u} = 2\Gamma_0 l\varepsilon$ (drift velocity). With Eq. (92) we have at the same time found Green's function for solving the initial-value problem of diffusion (Glicksman 2000) under the restrictive assumptions made here: no dependence of diffusion coefficient or driving force on x. As Fick's second law Eq. (91) is a linear differential equation, solutions can be constructed by superposition. Equation (92) gives the solution of Eq. (91) for an initial distribution of one A atom exactly at $x=0$, which in continuum language means a Dirac δ function centered at $x=0$. The general solution of Eq. (92) for any initial distribution $C_0(x) = C(x,0)$ is therefore found as the convolution

$$C(x,t) = \int_{\xi=-\infty}^{\xi=+\infty} C_0(x-\xi)W(\xi,t)\,d\xi \tag{93}$$

where W is defined by Eq. (92). Since the random behavior of the atom as considered in this section does not depend on previous history, we may write quite generally

$$C(x',t+\tau) = \int_{x=-\infty}^{x=+\infty} C(x'-x,t)W(x,\tau)\,dx \tag{94}$$

which means that the state at time $t+\tau$ develops by random walks of all atoms with duration τ from the state at a previous time t. Expanding the left-hand side of Eq. (94) in t and the integrand in x we obtain Eq. (95).

$$C(x',t) + \tau\frac{\partial C}{\partial t} + \cdots$$
$$= \int_{x=-\infty}^{x=+\infty} \left(C(x',t) - x\frac{\partial C}{\partial x} + \frac{1}{2}x^2\frac{\partial^2 C}{\partial x^2} + \cdots\right) W(x,\tau)\,dx \tag{95}$$

Defining the nth moment of the probability distribution W as

$$\langle x^n \rangle = \int_{-\infty}^{+\infty} x^n W(x,\tau)\,dx \tag{96}$$

we can write Eq. (95) for short times τ as

$$\frac{\partial C}{\partial t} = \frac{\langle x^2 \rangle}{2\tau}\frac{\partial^2 C}{\partial x^2} - \frac{\langle x \rangle}{\tau}\frac{\partial C}{\partial x} \tag{97}$$

Comparing this form with Fick's second law, Eq. (91), we can give the constants appearing therein a statistical meaning (Einstein 1905):

$$D = \lim_{\tau \to 0}\frac{\langle x^2 \rangle}{2\tau}, \quad \frac{DF}{k_B T} = \bar{u} = \frac{\langle x \rangle}{\tau} \tag{98}$$

Equations (98) are general formulations, in the sense that they apply even when the probability distribution W does not represent true random walk behavior.

We note that when directly calculating the first and second moments of the probability distribution, Eq. (92), for finite time t, then $\langle x \rangle = [FD/(k_B T)]t = \bar{u}t$, but $\langle x^2 \rangle = 2Dt$ only if there is no driving force or in the limit for short times.

Definition (98) in the case of a random walk refers to the tracer diffusion coefficient, which is usually denoted as D^* when it is necessary to distinguish it from other types of diffusion coefficient (cf. Section 5.4). To get an idea of mean random walk distances we note that in substitutional fcc alloys near the melting point a typical value is $D \approx 10^{-12}$ m^2 s^{-1}. The mean diffusion distance for one day results as $\sqrt{\langle x^2 \rangle} \approx 0.4$ mm. For small interstitial atoms a typical value of D near the melting point is 10^{-9} m^2 s^{-1}. This would yield a mean diffusion distance after one day of $\sqrt{\langle x^2 \rangle} \approx 13$ mm.

The expansion according to Eq. (95) can also be performed in the general three-dimensional case of diffusion leading to a general three-dimensional formulation of Fick's second law (small τ is again implied so as to make higher-order terms negligible):

$$\frac{\partial C}{\partial t} = -\sum_i \frac{\langle x_i \rangle}{\tau} \frac{\partial C}{\partial x_i} + \sum_i \sum_j D_{ij} \frac{\partial^2 C}{\partial x_i \partial x_j} \qquad (99)$$

where the diffusion coefficient D is defined as the second-rank tensor

$$D_{ij} = \lim_{\tau \to 0} \frac{\langle x_i x_j \rangle}{2\tau} \qquad (100)$$

The off-diagonal elements vanish if an appropriate set of principal axes is chosen, the diagonal elements then corresponding to the diffusion coefficients along those axes. When choosing these axes we can make use of the fact that the symmetry properties of the diffusion tensor D_{ij} reflect the symmetry of the crystal. Cubic crystals exhibit a very high degree of symmetry, with the consequence that any three orthogonal axes are principal axes and diffusion is isotropic: $\langle x_1^2 \rangle = \langle x_2^2 \rangle = \langle x_3^2 \rangle$. Then we have for the three-dimensional displacement $\mathbf{R} = (x_1, x_2, x_3)$

$$D = \lim_{\tau \text{ small}} \frac{\langle \mathbf{R}^2 \rangle}{6\tau} \qquad (101)$$

so that for isotropic diffusion we may write three-dimensional forms of Fick's first law:

$$\mathbf{j} = -D\nabla C + \frac{FDC}{k_B T} \qquad (102)$$

and Fick's second law:

$$\frac{\partial C}{\partial t} = D\left[\nabla^2 C - \nabla\left(\frac{FC}{k_B T}\right)\right] \qquad (103)$$

The three-dimensional generalization of Eq. (92) follows if we consider that the random walk can be decomposed into three independent walks in the three principal diffusion directions, so the probability density is just the product of the probability densities in the three directions:

$$W(x_1, x_2, x_3) = W(x_1, t) W(x_2, t) W(x_3, t) \tag{104}$$

In the absence of driving forces,

$$W(x_1, x_2, x_3) = \frac{1}{8\sqrt{\pi^3 D_1 D_2 D_3 t^3}} \exp\left[-\left(\frac{x_1^2}{4D_1 t} + \frac{x_2^2}{4D_2 t} + \frac{x_1^2}{4D_3 t}\right)\right] \tag{105}$$

We see that the surfaces of constant concentration (or probability) are ellipsoids whose axes are determined by the principal diffusion coefficients D_1, D_2, and D_3. In the case of isotropic diffusion (as in cubic crystals) the ellipsoids turn into spheres.

Heterogeneity of Phase Transformations by Limited Vacancy Mobility Equation (105) tells us that the random walk of atoms or defects such as vacancies defines ellipsoids or spheres of probability which grow with time. Let us consider a system where diffusion takes place via a vacancy mechanism, which is not in equilibrium for $t = 0$. Any change in atom configuration will begin in restricted areas within the ellipsoids defined above. If the mobility of the vacancies is low (low temperature, high enthalpy of migration ΔH_m), and their density not too high, these areas will remain separate during reasonable observation times. This means that structural changes such as phase transformations or changes of order parameter, which should by thermodynamic arguments proceed homogeneously in all volume elements of the sample, may by reasons of vacancy mobility lead to a heterogeneous configuration, at least in the initial stages of the transformation. This effect, usually called "local ordering," has been recognized by various authors (Beeler 1965; Allen and Cahn 1976; Athènes et al. 1996; Belashchenko and Vaks 1998; Soffa et al. 2003). The initially "disjunct" state of kinetics cannot end before the probability ellipsoids [Eq. (105)] impinge on one another. This criterion is however not sufficient because the migrating vacancy does not fill out the probability ellipsoid densely enough (Reith et al. 2005).

5.2.2.2 Correlated Walk – the Interaction of Defect and Atom

We consider a walk in the absence of driving forces, using Eq. (101) for the definition of the tracer diffusion coefficient. The displacement $\mathbf{R}$ results from a succession of n jumps $\mathbf{r}_1, \mathbf{r}_2, \ldots \mathbf{r}_n$ which are not necessarily of equal length or direction. The averaged squared displacements can be written as Eq. (106).

$$\langle \mathbf{R}^2 \rangle = \left\langle \left(\sum_{i=1}^n \mathbf{r}_i\right)^2 \right\rangle = \sum_{i=1}^n \langle \mathbf{r}_i^2 \rangle + 2\sum_{i=1}^{n-1}\sum_{j=i+1}^n \langle \mathbf{r}_i \cdot \mathbf{r}_j \rangle \tag{106}$$

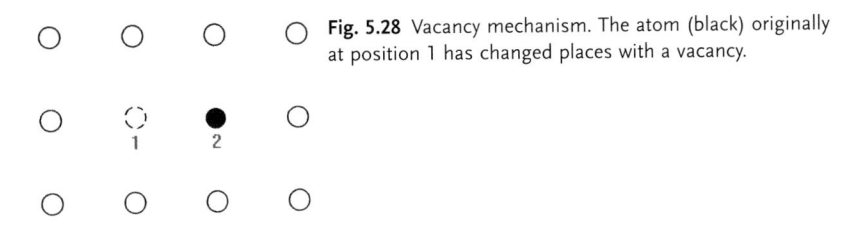

Fig. 5.28 Vacancy mechanism. The atom (black) originally at position 1 has changed places with a vacancy.

For a completely uncorrelated random walk the cross-terms cancel out on the average since steps in any given direction are equally probable to those in the reverse direction. If they do not, it is an indication that there is some "memory" of the previous path of the diffusing atom. How this comes about is easily demonstrated, taking the vacancy mechanism as an example.

A tracer atom (black) in Fig. 5.28 has just changed places with a vacancy, advancing from position 1 to 2. There is a more than equal probability that the next jump of the atom will be an exchange with the same vacancy, putting it back to position 1. Note that the motion of the atom is correlated even if there is no preference of the vacancy to lie in the vicinity of any atom, which means that the movement of the vacancy itself is completely uncorrelated. Another example is the correlation arising from the interplay of interstitial and regular site atoms in an interstitialcy mechanism (Fig. 5.29). The tracer atom (black), initially in an interstitial position, moves on to a regular lattice site, and the atom originally sitting there now becomes an interstitial. There is an enhanced probability that the next jump of the black atom will be just the reverse, because the necessary defect (an interstitial atom) is still at hand.

The usual method of quantitatively describing the correlation effects in atom diffusion is to compare $\langle R^2 \rangle$ of Eq. (106) with $\langle R^2 \rangle_{random}$ of a totally random walk. The quotient of both quantities is called the correlation factor:

$$f = \frac{\langle R^2 \rangle}{\langle R^2 \rangle_{random}} = \lim_{n \to \infty} \left(1 + \frac{2 \sum_{i=1}^{n-1} \sum_{j=i+1}^{n} \langle r_i \cdot r_j \rangle}{\sum_{i=1}^{n} \langle r_i^2 \rangle} \right) \tag{107}$$

Much effort has been put over the years into calculation and measurement of correlation factors in systems with various symmetries and diffusion mechanisms and several jump types. Elaborate systematic techniques based on probability con-

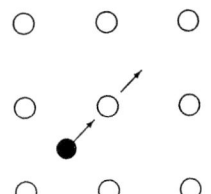

Fig. 5.29 Interstitialcy mechanism. An interstitial atom (black) moves to a regular lattice site (white). The atom which is sitting there now becomes an interstitial atom.

cepts (Bardeen and Herring 1952), matrix methods (Mullen 1961; LeClaire and Lidiard 1956), and analogies to electric circuits (Compaan and Haven 1956) and different numerical methods have been reviewed by Manning (1968). In less symmetric lattices, correlation factors in the principal diffusion directions may be different and must be defined by equations analogous to Eq. (107) with displacements in the respective directions. In cubic lattices, the high symmetry reduces correlation factors to rather simple expressions involving the average cosine of the angle between two successive jumps, yielding for the vacancy mechanism

$$f = \frac{1 + \langle \cos \theta \rangle}{1 - \langle \cos \theta \rangle} \tag{108}$$

and for the interstitialcy mechanism

$$f = 1 + \langle \cos \theta' \rangle \tag{109}$$

θ' being the angle between the interstitial-to-lattice jump and the subsequent lattice-to-interstitial jump of the tracer atom.

Table 5.6 (Murch 2001) lists various correlation factors which result for different diffusion mechanisms, different lattices, and different dimensionalities. The correlation factor for vacancy diffusion is a characteristic number always smaller than unity depending on the specific lattice and its connectivity. Interstitial diffusion of an infinitely dilute solute is uncorrelated, as diffusion does not depend on an ancillary defect. If, however, the interstices are filled with solute atoms so that only a few free places remain where the atoms can go, we revert to the situation of vacancy-assisted diffusion with the same correlation factors as found there.

Table 5.6 Correlation factors for various lattice structures and diffusion mechanisms.

Lattice	Mechanism	Correlation factor, f
Honeycomb	vacancy	$\frac{1}{3}$
Square	vacancy	$1/(\pi - 1)$
Triangular	vacancy	0.56006
Diamond cubic	vacancy	$\frac{1}{2}$
Body-centered cubic	vacancy	0.72714
Simple cubic	vacancy	0.65311
Face-centered cubic	vacancy	0.78146
Face-centered cubic	divacancy	0.4579 ± 0.0005
All lattices	interstitial	1
NaCl structure	collinear interstitialcy	$\frac{2}{3}$
CaF$_2$ structure (F$^-$ ion)	noncollinear interstitialcy	0.9855
CaF$_2$ structure (Ca^{2+} ion)	collinear interstitialcy	$\frac{4}{5}$
CaF$_2$ structure (Ca^{2+} ion)	noncollinear interstitialcy	1

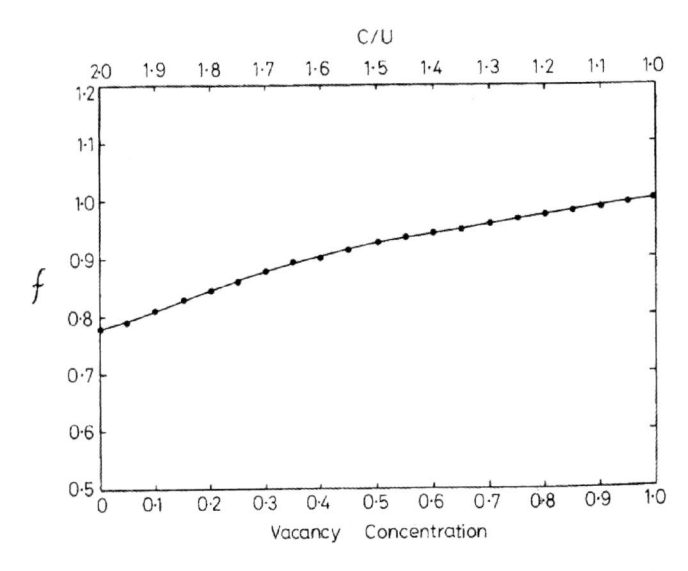

Fig. 5.30 Correlation factor for vacancy diffusion in the fcc lattice (Murch 1975).

Conversely, if in substitutional diffusion by a vacancy mechanism more and more vacancies are introduced, correlation with specific vacancies is lost as the atom becomes free to move almost anywhere, because a vacancy is always around. Both cases are characteristic of a situation where a large number of objects fill an available space until they behave collectively like the one vacant lot they have left free.

Figure 5.30 (Murch 1975) illustrates this change of regimes in terms of the correlation factor for vacancy diffusion in the fcc lattice. The correlation factor goes from the value for single vacancies (near 0.8) up to 1 as the vacancy concentration rises from 0 to 1. The same diagram can in principle be read from right to left for interstitial diffusion as the solute atom concentration increases from 0 at the right-hand side of the diagram to 1 at the left-hand side. In the latter case f decreases as the number of available sites is depleted.

In cubic crystals all jumps are of the same length, and we have for the correlation factor

$$f = \lim_{n \to \infty} \frac{\langle \mathbf{R}^2 \rangle}{nl^2} \tag{110}$$

an especially useful form if one wants to obtain f by direct computer simulation. With the help of Eq. (101) and the total jump frequency $\Gamma = n/t$ the diffusion coefficient can therefore be written as

$$D^* = \frac{\Gamma l^2 f}{6} \tag{111}$$

which is the usual representation of the diffusivity as an uncorrelated part and a factor taking account of the correlation effects. We note that in alloys with a high concentration of defects, for instance nonstoichiometric intermetallic compounds or fast ion conductors, the apparent activation energy of D^* may contain an important contribution from the temperature dependence of f (via the T-dependence of the defect concentrations).

There are alloys where any diffusion jump leads to a different sublattice of the long-range ordered structure (cf. Section 5.2.3). An example is B2 (FeAl, NiAl,...) with all nearest neighbors of a given atom belonging to a different sublattice. If we assume nearly perfect order, and allow only nearest-neighbor jumps to occur (which in many cases is the most probable hypothesis) then a diffusion jump of a "regular" atom is bound to create an antisite defect. As the new configuration in general is at a higher level of energy, there will be an enhanced probability of an immediate back-jump, leading sometimes to very low correlation factors. Although we note a jump reversal effect as in the usual case of correlation due to the cooperation with a vacancy, the cause is now different. For the special situation of diffusion and the correlation effects in highly ordered alloys, refer to Section 5.2.3.

Correlation effects in impurity diffusion: the five-frequency model Whereas in tracer diffusion f can be given as a simple number depending on lattice geometry, the correlation factor of impurity atoms depends on the relative value of their jump frequency as compared to the jump frequency of the solvent. The obvious reason for the particular temperature dependence of impurity diffusion is that both ω values are a function of temperature. In Fig. 5.31 (Murch and Thorn 1978) f is plotted as a function of ω_2/ω_0, the ratio of the vacancy exchange frequency ω_2 of the impurity and ω_0 of the host atom. The curves have been calculated for the walk of a sufficiently dilute impurity atom in the fcc lattice.

If $\omega_0 \gg \omega_2$, the vacancy exchanges much more rapidly with a host atom than with the impurity atom. It will therefore move away quickly from the impurity atom, whose next jump will probably not involve the same vacancy. The consequence is weak correlation, viz., $f \approx 1$. Vice versa, if $\omega_0 \ll \omega_2$ there is little inducement for the vacancy to exchange with one of the sluggish host atoms, and therefore the next exchange of the vacancy will be with the impurity atom again. The result is strong correlation and $f \ll 1$.

In order to describe the impurity diffusion and to derive the correlation factors plotted in Fig. 5.31, the phenomenological so-called five-frequency model (Murch and Thorn 1978) was used. The influence of different atomic neighborhoods is accounted for by introducing five distinct exchange frequencies, as explained in Fig. 5.32. Here ω_1 is the frequency for the exchange of the host atom with a vacancy if both are nearest neighbors to the impurity, and ω_2 is the exchange frequency of the impurity with the vacancy, as above; ω_3 is the frequency of a dissociative jump of the host atom, which takes the host atom away from the impurity atom, and finally ω_4 refers to the opposite movement which brings the host atom

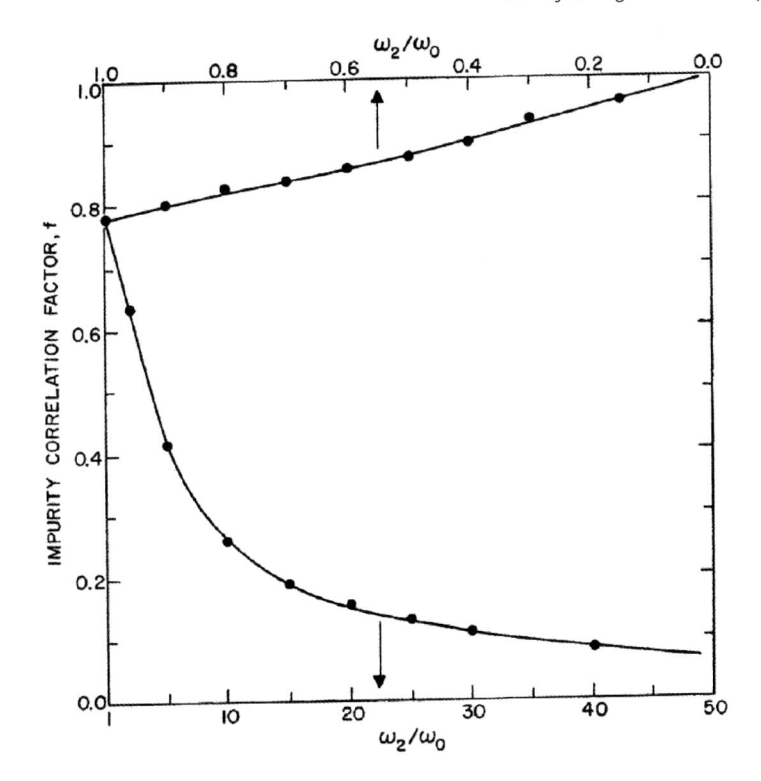

Fig. 5.31 Correlation factor for impurity diffusion. The upper and lower curves refer to the upper and lower x axis labels and represent the two regimes $\omega_2/\omega_0 < 0$ and $\omega_2/\omega_0 > 0$. After Murch and Thorn (1978).

into a nearest-neighbor position to the impurity. The fifth frequency is ω_0 for the exchange of a host atom with a vacancy in the absence of impurities. We note that

$$\frac{\omega_4}{\omega_3} = \exp\left(-\frac{E_B}{k_B T}\right) \tag{112}$$

E_B being the impurity–vacancy binding energy (negative if the vacancy is attracted to the impurity). Manning (1959), within the framework of this model, has made a rigorous derivation of the impurity correlation factor:

$$f_2 = \frac{2\omega_1 + 7F\omega_3}{2\omega_1 + 2\omega_2 + 7F\omega_3} \tag{113}$$

where the dependence on ω_0 enters through the fraction F of the vacancies permanently lost from a site. F is given by the very good approximate expression (114).

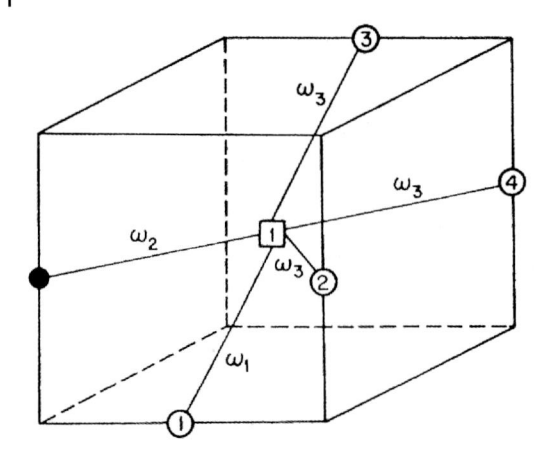

Fig. 5.32 Five-frequency model for impurity diffusion (Murch and Thorn 1978). Open square: vacancy; open circle: host atom; filled circle: impurity. ω_3 is the frequency of a dissociative jump of the host atom away from the impurity. The fourth frequency ω_4 is the corresponding associative jump, the fifth frequency, ω_0, refers to the exchange of a host atom far from an impurity, with the vacancy.

$$7F = 7 - \frac{10\alpha^4 + 180.5\alpha^3 + 927\alpha^2 + 1341\alpha}{2\alpha^4 + 40.2\alpha^3 + 254\alpha^2 + 597\alpha + 436}, \quad \alpha \equiv \frac{\omega_4}{\omega_0} \tag{114}$$

Analogous expressions have also been derived for other lattice structures, e.g., for bcc and the diamond structure (Manning 1964), the hcp structure (Huntington and Ghate 1962; Ghate 1964) and for impurity diffusion by interstitialcy jumps (Manning 1959).

5.2.2.3 Diffusion Walk with Chemical Driving Force

When deriving the diffusion flux in Eq. (87) we assumed that the enthalpy of the crystal was unchanged after an atom jump had taken place, a premise that can be upheld in a homogeneous alloy or in the case of an ideal solution. In Fick's first law the concentration gradient formally plays the role of a driving force. The net flux arises, however, only because there are more atoms on plane 1 than on the adjacent plane 2. We might also say that it is the gradient of the entropy of mixing which drives diffusion in an ideal solution.[13] In the general case of a non-ideal solution we have to expect, however, an enthalpy difference if the environment of the jumping atom has changed as a result of the jump, for instance if the number of neighboring A atoms has increased. The jump probability according to the usual ansatz of transition state theory, Eq. (83), depends on the Gibbs

---

[13] It is readily verified that this is equivalent to a concentration gradient, if the gradient is not too large.

free energy difference between the initial state and the saddle point state.[14] We should remember that the Gibbs free energy difference used here contains a contribution from lattice vibration entropy, but does not contain mixing entropy.[15]

Any kinetic model must be formulated consistently so that the system approaches a true thermodynamic equilibrium state at long times. It is usual to invoke dynamical equilibrium for every single pair of an atom jump and its reverse, a well-known requirement known as the principle of detailed balance. Regarding an ensemble of systems and calling n_i^∞ the number of systems in state i in equilibrium, we have for the fluxes of the ensembles between both states

$$n_1^\infty \omega_{12} = n_2^\infty \omega_{21} \tag{115}$$

or

$$n_1^\infty \tilde{\nu} \exp\left(-\frac{G_S - G_1}{k_B T}\right) = n_2^\infty \tilde{\nu} \exp\left(-\frac{G_S - G_2}{k_B T}\right) \tag{116}$$

from which

$$\frac{n_2^\infty}{n_1^\infty} = \exp\left(-\frac{G_2 - G_1}{k_B T}\right) \tag{117}$$

Thus it is seen that the occupation numbers of systems in equilibrium in states 1 and 2 correspond to the respective ensemble densities, as demanded by statistical mechanics.

If equilibrium has not yet been reached, the net diffusion flux depends on driving forces toward equilibrium. An obvious driving force is the Gibbs free energy difference incurred by the moving atoms. This difference is conveniently described by a change in local chemical potential, which is defined as the Gibbs free energy change when an atom is added at a given lattice site.[16] For the chemical potential of atom species i we write

$$\mu_i = \frac{\partial G}{\partial N_i} = \mu_0 + \mu_i^{\text{ideal}} + \mu_i^{\text{exc}} \tag{118}$$

---

14) Whether we choose Eq. (82) or (83) for the jump probability depends on if we want to make allowance for the change in lattice vibration spectrum on displacing the atom from the equilibrium position to the saddle point (and on to the new equilibrium position). To remain in accordance with the main body of literature, we decide to use Eq. (83). There still remains the delicate question of whether the attempt frequencies $\tilde{\nu}$ of Eq. (83) are the same for a jump and its reverse. Keeping in mind the approximations made in transition state theory, instead of overstretching this model we invoke the principle of detailed balance, requiring the attempt frequencies to be the same if the proper Gibbs free energy differences have been used.

15) The Gibbs free energy level of the new equilibrium state after the jump enters only insofar as it determines the probability for the back jump.

16) Sometimes, especially in chemical physics, chemical potential is defined per mol.

where μ_0 is the chemical potential of a reference state and the second contribution is due to mixing entropy and is the only part remaining in an ideal solution. If we regard the jump of one particular atom, we deliberately take the change of mixing entropy out of our consideration because it just comes about by the statistics of many atoms jumping. The free energy difference which enters the jump probability explicitly is therefore only the third contribution in Eq. (118), an excess chemical potential which arises only for non-ideal solutions. It yields the following Gibbs free energy change, equivalent to the effect of a driving force F acting along a jump of length l:

$$\Delta G_{12} = G_2 - G_1 = \frac{\partial \mu_i^{\text{exc}}}{\partial x} l = -Fl \tag{119}$$

According to the conventions of chemical thermodynamics μ_i is often written in terms of the activity a_i and the activity coefficient γ_i describing the deviation from the behavior of an ideal solution:

$$\mu_i - \mu_0 = k_B T \ln a_i = k_B T \ln a_i^{\text{ideal}} + k_B T \ln \gamma_i \tag{120}$$

Thus by comparison with Eq. (118),

$$\mu_i^{\text{exc}} = k_B T \ln \gamma_i \tag{121}$$

the activity coefficient being nothing else than the excess chemical potential appearing in a different guise. The bias in jump frequencies due to a driving force F is given by $\varepsilon = Fl/(2k_B T)$. In the case of a chemical driving force we therefore get Eq. (122).

$$\varepsilon = -\frac{l}{2} \frac{\partial \ln \gamma_i}{\partial x} \tag{122}$$

Following the same procedure which has led us to Eq. (87), but now using the chemical driving force F in Eq. (119) together with Eq. (121), we obtain Eq. (123) directly from Eq. (87).

$$j_i = -D_i^* \frac{\partial C_i}{\partial x} - D_i^* C_i \frac{\partial \ln \gamma_i}{\partial x} = -D_i^* \left(1 + C_i \frac{\partial \ln \gamma_i}{\partial C_i}\right) \frac{\partial C_i}{\partial x}$$

$$= -D_i^* \left(1 + c_i \frac{\partial \ln \gamma_i}{\partial c_i}\right) \frac{\partial C_i}{\partial x} \tag{123}$$

Here we have introduced for later convenience the site fraction $c_i = \Omega C_i$, Ω being the atomic volume. We can therefore describe diffusion in a chemical driving force by the introduction of a thermodynamic factor φ [Eq. (124)] to the diffusion coefficient.

$$\varphi = 1 + c_i \frac{\partial \ln \gamma_i}{\partial c_i} = 1 + \frac{\partial \ln \gamma_i}{\partial \ln c_i} \tag{124}$$

We have indicated by the asterisk and the index that the diffusion coefficient appearing in Eq. (123) is a tracer diffusion coefficient valid for species i. For the diffusion coefficient, e.g., in cubic crystal structures, we have to take the expression of Eq. (89),

$$D_i^* = \frac{z\Gamma_0 l^2}{6} \tag{125}$$

where the effective jump frequency Γ_0 follows from the atom–defect exchange frequency by incorporating the appropriate factor for defect concentration and another factor for correlation so that

$$\Gamma_0 = c_V f \omega = c_V f \tilde{\nu} \exp\left(-\frac{G_S - G_1}{k_B T}\right) \tag{126}$$

which is to be compared with Eq. (116).

5.2.2.4 Diffusion Walk in an Inhomogeneous Crystal

It goes without saying that not only the thermodynamic factor by its dependence on composition, but also the other quantities involved, make diffusivity inhomogeneous in the general case. This is especially true if crystal defects or boundaries constitute enhanced diffusion pathways (see Section 5.4.3) or if for energetic reasons the diffusion-mediating defect is confined in some region of the crystal such as vacancies in the antiphase boundaries of a long-range ordered structure, an effect which becomes strikingly evident when studying domain growth by kinetic Monte Carlo simulations. Mass transport in a general diffusion field then depends on the local properties of the random walk, and for a time interval τ we can write Eq. (127) in terms of the flux in one dimension (Manning 1968; Le Claire 1958):

$$j = \frac{C\langle x \rangle}{\tau} - \frac{\partial}{\partial x}\left(\frac{C\langle x^2 \rangle}{2\tau}\right) \tag{127}$$

The composition change it causes consequently results as (Alnatt and Rice 1960):

$$\frac{\partial C}{\partial t} = -\frac{\partial}{\partial x}\left[\frac{C\langle x \rangle}{\tau} - \frac{\partial}{\partial x}\left(\frac{C\langle x^2 \rangle}{2\tau}\right)\right] \tag{128}$$

The concentration dependence of the diffusion coefficient implied in Eq. (128) can lead to unexpected diffusion behavior in alloys, such as transient sharpening of an interface concentration profile, as was recently discovered (Erdélyi et al. 2004).

Up to now we have treated the diffusion of the atom species considered with respect to a fixed reference frame, and especially the behavior of the diffusion-mediating defect (most often a vacancy) as independent of the diffusion of other species. If, however, several atom species diffuse with different rates, a net flow of vacancies is to be expected, and the vacancy presents itself with unequal probability on different sides of the moving atom. The effects of this vacancy flow, questions of the reference frame against which the fluxes have to be measured, and the necessity to distinguish between different types of diffusion coefficients will be discussed in Section 5.4.1.

5.2.3
Atom Migration in Ordered Alloys

5.2.3.1 Experimental Approach to Atom Kinetics in Ordered Alloys

Ordered intermetallic compounds (intermetallics) are favorite subjects of study in alloy physics because of their interesting strength and electromagnetic properties, which are linked to the long-range ordered structure. For thermodynamic reasons the degree of long-range order (LRO) depends on temperature. Out of equilibrium, the stability of the ordered state is limited by the kinetic processes altering it. It is therefore of great practical relevance to study and characterize atom kinetics in ordered intermetallics. How can details of the atomic movement be observed in these alloys? The usual diffusion experiments, e.g., by measuring tracer diffusivity (Frank et al. 1997; Divinski et al. 1998) or recently by secondary ion mass spectroscopy (SIMS) analysis of the diffusion of a corresponding stable isotope (Frank et al. 1995, 2001), give information on atom jumps during long-range diffusion at a constant average degree of LRO ("steady-state diffusion"). The same holds for the observation of the atomic jump vector by various scattering methods under equilibrium conditions (Sepiol et al. 1998a; Thiess et al. 2001).

A completely different approach concerns so-called "order–order" relaxation experiments (Kozubski 1997; Pfeiler 2000; Pfeiler and Sprusil 2002). In this case, the system under observation is pushed slightly out of its thermodynamic equilibrium state by the application of a small temperature shift, and the subsequent relaxation into a new equilibrium is observed. During such order relaxation experiments just those atom jumps are observed which entail a change in the degree of LRO.

Interestingly, there is up to now only one experimental method with the necessary high precision so that the very small changes of the degree of order corresponding to slight changes of temperature can be resolved: residual resistometry (REST), which is known for its ultrahigh resolution of structural changes (Pfeiler 1997). Here the sample is cooled to low temperature for each measuring point so as to virtually eliminate the phonon contribution to electrical resistivity, thereby improving the signal-to-noise ratio. REST measurements yield an integral signal composed of several effects, and considerable experience is necessary to separate them correctly. We are of course interested to interpret the macroscopic kinetics in the light of what is happening at the atomic scale. What is needed is firstly information which is as detailed as possible on the single atom jump in different environments, and secondly a method to sum up the individual atom jumps statistically and therefore to arrive at the mesoscopic and/or macroscopic kinetic behavior. The first requirement can be met by employing ab-initio jump profile calculations based on modern density functional methods (Pfeiler et al. 2004). A natural statistical method of evaluating the cumulative effect of all single atom jumps is Monte Carlo simulation as it is practically free from additional assumptions (Pfeiler et al. 2004).

5.2.3.2 Jumps Within and Between Sublattices

Some special considerations apply for atom kinetics in long-range ordered alloys. Any approximation based on the solution being dilute is no longer valid since the stoichiometry usually corresponds to a simple numerical proportion (or at least approximately so), putting these alloys in the intermediate composition range. The elementary cell of ordered alloys contains atoms sitting on different sublattices. Jumps may carry an atom from one sublattice to another or to a neighboring position on the same sublattice. Therefore in the general case a certain number of nonequivalent jump types arises, which we have to account for when we write down the diffusion coefficient in terms of atomistic kinetic properties. Taking Eq. (101) as a definition of D we again regard, as in Eq. (106), the atom displacement $\mathbf{R}$ as composed of possibly different jump vectors $\mathbf{r}_i$ ($i = 1, \ldots, n$) which can be classified into M different jump types with respective jump length r_μ and jump number n_μ. For cubic crystal structures, Eq. (129) can then be shown (Manning 1968) to apply,

$$D = \frac{f}{6t} \sum_{\mu=1}^{M} n_\mu r_\mu^2 \tag{129}$$

with the correlation factor f measuring the effectiveness of this ensemble of jumps for long-range atom transport. This formulation was used by Bakker and Westerveld (1988) (see also Bakker 1987) to obtain expressions for the diffusion coefficient in ordered intermetallics in terms of the jump frequencies involved, assuming nearest-neighbor (NN) exchange with a single vacancy. For instance, in the B2 structure, Fig. 5.18(a), we can write for the total rate of A atom jumps

$$\frac{n_A}{t} = \frac{N_A^\alpha}{N_A} \Gamma_A^{\alpha\beta} + \frac{N_A^\beta}{N_A} \Gamma_A^{\beta\alpha} \tag{130}$$

Here N_A^α means the number of A atoms sitting on the α sublattice, N_A the total number of A atoms, and $\Gamma_A^{\alpha\beta}$ the jump rate of an A atom from the α to the β sublattice. The jump rate will depend on the availability of a vacancy according to Eq. (131),

$$\Gamma_A^{\alpha\beta} = 8 p_{AV}^{\alpha\beta} \omega_A^{\alpha\beta} \tag{131}$$

$p_{AV}^{\alpha\beta}$ being the probability of finding a vacancy on β next to an A atom on α and $\omega_A^{\alpha\beta}$ the atom–vacancy exchange rate under this premise. Invoking the principle of detailed balance for jumps in equilibrium,

$$N_A^\alpha \Gamma_A^{\alpha\beta} = N_A^\beta \Gamma_A^{\beta\alpha} \tag{132}$$

the total rate of A atom jumps can be written as Eq. (133).

$$\frac{n_A}{t} = 2 \frac{N_A^\alpha}{N_A} \Gamma_A^{\alpha\beta} \tag{133}$$

Given these ingredients it is easy to write down the diffusion coefficient for species A:

$$D_A = f_A \frac{c_A^\alpha}{c_A} p_{AV}^{\alpha\beta} \omega_A^{\alpha\beta} a^2 \tag{134}$$

with f_A the relevant correlation factor, c_A^α and c_A the fraction of A atoms on the α sublattice and in total, respectively, and a the lattice constant. It is to be noted that the quantities appearing here, especially the correlation factor and the concentration of defects, are very sensitive to the state of long-range order and to deviations from ideal stoichiometry. Figure 5.33 illustrates the strong dependence of the cor-

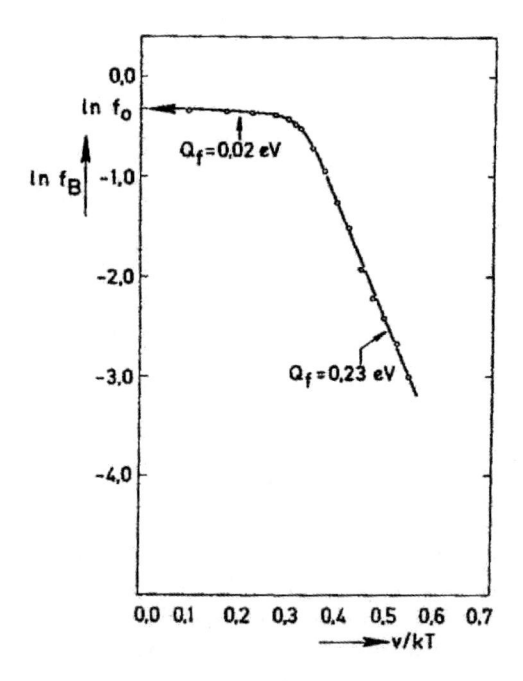

Fig. 5.33 Logarithm of correlation factor for diffusion of one atom species in the B2 structure versus reduced reciprocal temperature (Bakker et al. 1976). $v/k_B T = 0.3$ corresponds to the order–disorder transition temperature T_t.

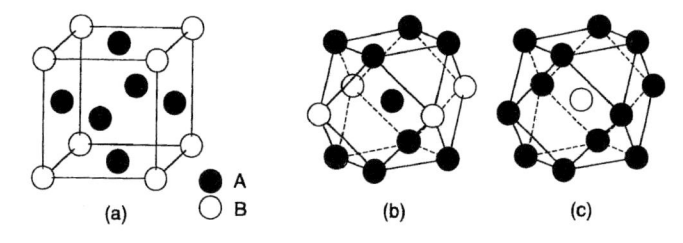

Fig. 5.34 (a) L1$_2$ ordered structure; (b) nearest neighbors for a majority A atom; (c) nearest neighbors of a minority B atom (After Numakura et al. 1998).

relation factor on the degree of order. It is the result of computer simulation of diffusion in the B2 structure (Bakker et al. 1976). The system considered here has an order–disorder transition temperature T_t given by $v/(k_B T_t) = 0.3$ where v is the mixed interaction energy, or "ordering energy" cf. p. 244. Above this temperature the correlation factor f is almost constant and close to 1, indicating a weakly correlated diffusion mechanism. Down from T_t, the correlation factor decreases sharply with temperature, because strong correlation effects hamper diffusion to an ever-increasing extent. The reason is that any mass transport in B2 has to start with a local disruption of LRO. In an almost perfectly ordered state at low temperature, the Gibbs free energy cost of disordering is high as the entropy increase associated with generating an antisite defect will then no longer compensate for the enthalpy expenditure.

Regarding the L1$_2$ structure (stoichiometry A$_3$B, Fig. 5.34(a)) as another example of ordered intermetallics, we note an important qualitative difference in comparison to B2. In the B2 structure, all nearest neighbors (NN) of an atom belong to the other sublattice. In L1$_2$, the minority B atoms are likewise surrounded by 12 A atoms, but majority A atoms are surrounded by eight A atoms and only four B atoms as nearest neighbors. This means that any NN jump in B2 ends on the other sublattice, but in L1$_2$ majority atoms can diffuse within their own sublattice by NN jumps. The different possibilities for atom jumps are represented schematically in Fig. 5.35. We have to decide between atom jumps on the majority α sublattice (intra-sublattice) which are neutral with respect to order ("0" in Fig. 5.35, arrows in the vertical direction) and jumps between the sublattices (inter-sublattice; Fig. 5.35(a): horizontal arrows, "+" increases the number of antisites by one, "−" decreases the number of antisites by one). The restricted possibilities for the B2 lattice are displayed in Fig. 5.35(b) as a contrast.

It is clear that any experimental method such as REST, which is sensitive to changes in the degree of order, will see only the effect of order–disorder inter-sublattice atom jumps ("+−"), and classical diffusion measurements will see mainly the effect of long-range atom transport by intra-sublattice jumps along the majority sublattice. Adequate caution must therefore be exercised when comparing activation enthalpies obtained from diverse methods of measurement. For example, Arrhenius analysis of order–order relaxation times observed in residual resistivity measurements on Ni$_3$Al yields an activation energy of 4.6 eV (Kozubski

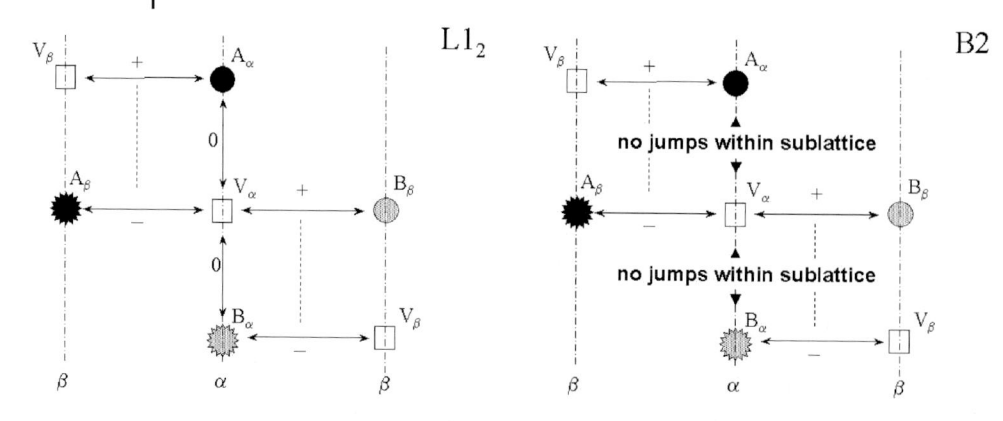

Fig. 5.35 (a) Atom jump types in L1$_2$; (b) atom jump types in B2. Schematic representation of sublattices as vertical lines.

and Pfeiler 1996), whereas in standard Ni tracer diffusion experiments in Ni$_3$Al an activation energy of about 3 eV is found (Hancock 1971; Hoshino et al. 1988). (See Section 5.2.3.3 for a discussion of a possible co-operative jump mechanism.)

Distinguishing between $\alpha \leftrightarrow \alpha$ intra-sublattice jumps and $\alpha \leftrightarrow \beta$ inter-sublattice jumps, the expressions analogous to Eqs. (130)–(134) read for the L1$_2$ lattice

$$\frac{n_A}{t} = \frac{N_A^\alpha}{N_A}\Gamma_A^{\alpha\alpha} + \frac{N_A^\alpha}{N_A}\Gamma_A^{\alpha\beta} + \frac{N_A^\beta}{N_A}\Gamma_A^{\beta\alpha} \tag{135}$$

with

$$\Gamma_A^{\alpha\alpha} = 8p_{AV}^{\alpha\alpha}\omega_A^{\alpha\alpha}, \quad \Gamma_A^{\alpha\beta} = 4p_{AV}^{\alpha\beta}\omega_A^{\alpha\beta} \tag{136}$$

Applying the principle of detailed balance again leads to

$$D_A = \frac{1}{2}f_A \frac{c_A^\alpha}{c_A}(p_{AV}^{\alpha\alpha}\omega_A^{\alpha\alpha} + p_{AV}^{\alpha\beta}\omega_A^{\alpha\beta})a^2 \tag{137}$$

D_B follows by replacing A by B,

$$D_B = \frac{1}{2}f_B \frac{c_B^\alpha}{c_B}(p_{BV}^{\alpha\alpha}\omega_B^{\alpha\alpha} + p_{BV}^{\alpha\beta}\omega_B^{\alpha\beta})a^2 \tag{138}$$

Equations (137) and (138) correspond to a six-frequency model of diffusion in the ordered L1$_2$ structure, where two of the frequencies do not appear explicitly as they have been eliminated by the principle of detailed balance. Analogous formulae for the D0$_3$ (Fe$_3$Si), the A15 (Cr$_3$Si), and the C1 (CaF$_2$) structures and the Laves phase C15 (MgCu$_2$) can be found in the original work of Bakker and Westerveld (1988).

Of course expressions like Eqs. (134), (137), and (138) must remain of academic value only if the quantities appearing therein cannot be determined by theoretical or experimental methods. The probability of finding a vacancy next to a given atom follows from the vacancy concentration on the respective sublattice and the binding energy of the vacancy to the atom species considered. Both can be obtained from solution models on various levels of approximation (details in Section 5.1), or from more direct, modern ab-initio methods based on density functional theory (cf. Chapter 11 and Section 5.2.3.3), followed by a treatment according to the principles of statistical thermodynamics in order to obtain the temperature dependence (cf. Chapter 10). The atom–vacancy interchange frequency ω can be assessed by a calculation of the energy profile of atom migration, using approximate methods based on potentials such as the EAM or again by an ab-initio treatment. The activation process of the atom–vacancy interchange may be analyzed either in a static formulation, if possible allowing for the relaxation of a fairly large number of coordinates (nudged elastic band method; cf. Chapter 11 and Section 5.2.1.2), or in a molecular dynamics version (see Chapter 12 and Section 5.2.1.2). A more difficult problem is the determination of the tracer correlation factors f_A and f_B. One possible strategy is kinetic Monte Carlo simulation (Bakker et al. 1976; Murch 1984), others are random-walk approaches (Manning 1971; Bakker 1979; Stolwijk 1981) or the path probability method (Sato et al. 1985).

In an extended study comprising a series of papers, Kikuchi and Sato (1969, 1970, 1972) have presented a thorough analysis of diffusion in B2 ordered binary alloys by means of the path probability method, revealing the role of the various factors in the tracer diffusion coefficient of a component, and how the two sublattices contribute to them. A direct calculation of the correlation factor f was achieved. It proved convenient to consider two different correlation factors f_{AB} and f_{BA}, dependent on the tracer atom (always belonging to the same species, e.g., a B atom) jumping from a site on either the α or the β sublattice. Equation (108) is applied separately to each of the two cases. The overall correlation factor is then the harmonic mean

$$f^{-1} = \frac{1}{2}(f_{AB}^{-1} + f_{BA}^{-1}) \tag{139}$$

and shows, especially for a stoichiometric alloy, the expected sharp decrease (tantamount to strong correlation) as perfect LRO is established. The tracer diffusion coefficient of one of the components, say the B component, can be factorized according to Eq. (140),

$$D_B^* = 4w_3 V_B W_B f \tag{140}$$

w_3 being an attempt frequency, V_B a vacancy availability factor, and W_B an effective jump frequency factor, which is in effect a Boltzmann factor containing in the exponent the energy of the bonds broken by the atom when jumping.

While Eq. (140) is formally equivalent to what appears in Eq. (126) together with Eq. (125), the interesting information is, however, contained in the behavior of the individual factors. Both V_B and W_B turn out to be quite sensitive to stoichiometry and state of order, in fact even more so than the correlation factor f itself. Like the correlation factor, they can be decomposed into contributions from the two sublattices, as a weighted arithmetic mean in the case of V_B and another harmonic mean in the case of W_B. The behavior of the three contributing factors is impressively illustrated in Fig. 5.36 (Kikuchi and Sato 1972). Here reduced versions of the quantities V_B and W_B are plotted together with the correlation factor f versus the reciprocal temperature normalized to the order–disorder transition temperature. The parameter $\alpha = c_A - 0.5$ is a measure of deviation from the stoichiometric composition, and the parameter U is a measure of energy asymmetry and is defined as $U = (e_{AA} - e_{BB})/v$ with $v = e_{AA} + e_{BB} - 2e_{AB}$, e being the respective pair interaction energy between atom species. The normalizations are defined according to $\tilde{V}_B = V_B/c_V$, c_V being the overall vacancy concentration and $\tilde{W}_B = W_B \exp(-7e_{AB}\beta)$, $\beta = 1/(k_B T)$.

In L1$_2$ ordered intermetallic compounds such as Ni$_3$Al, Ni$_3$Ga and Ni$_3$Ge, diffusion takes place mainly on the majority sublattice, where even the minority atoms are able to diffuse rapidly as an antistructural atom population (Numakura et al. 1998). This explains also why the dependence on stoichiometry of the minority atom diffusion coefficient is practically identical to that of the fraction of B atoms in antisite position (Belova and Murch 1999). A possible role of six-jump cycles (see Section 5.2.3.2) could be ruled out because this would require B atoms to diffuse faster than A atoms, a fact which is not supported by the experimental findings.

Another ordered structure which has gained much attention in recent years is L1$_0$, Fig. 5.18(c), also derived from the fcc structure, but with stoichiometry 1:1. The structure consists of alternate layers of A and B atoms. The lattice cell is tetragonally distorted, making the tetragonal c axis (normal to the layers) an easy axis for magnetization in the case of ferromagnetic alloys such as FePd, FePt, CoPt. These materials are under investigation as future high-density magnetic or magneto-optic recording media, envisioned in the form of thin films or nanostructures. As in L1$_2$, intra-sublattice jumps provide an easy diffusion path. Diffusion in the direction normal to the A/B layers requires, however, inter-sublattice jumps, as the sublattices are not connected in this direction. Likewise, inter-sublattice jumps are necessary to produce any change of LRO. The anisotropy of diffusion and the role of different jumps for both constituent elements in L1$_0$ ordered FePt and TiAl have recently been discussed by Nakajima et al. (2005), generally confirming the expected behavior (see also Kushida et al. 2003; Nosé et al. 2005). Experiments comprising residual resistivity measurements, X-ray diffraction, magnetization, and Mössbauer absorption measurements have been done on FePd, FePt, and CoPt thin films and foils by Issro et al. (2006) with the aim of shedding light on order kinetics. Thin films of FePd deposited by MBE show originally an orientation of the L1$_0$ c axis normal to the film plane. After a temperature treatment, an increasing fraction of domain variants with the

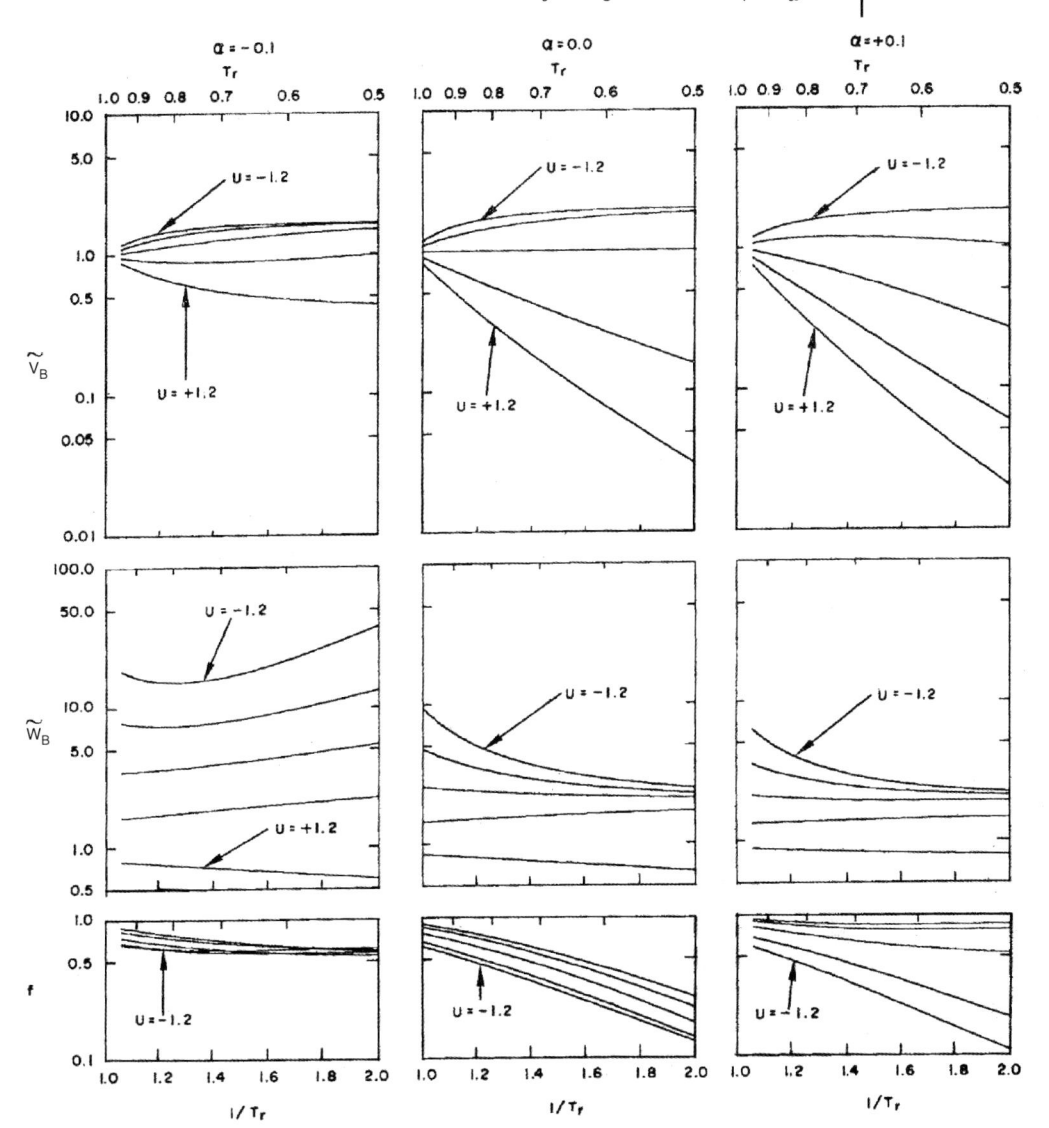

Fig. 5.36 Reduced vacancy availability factor $\tilde{V}_B$, reduced effective frequency factor $\tilde{W}_B$ and correlation factor f (all in logarithmic scales) as functions of the reciprocal reduced temperature. α stoichiometry parameter, U energy asymmetry parameter (see text). After Kikuchi and Sato (1972).

c axis in-plane was found. Monte Carlo computer simulations of atom kinetics in FePt (Kozlowski et al. 2005) show the same trend, which can be understood as relaxation toward an equilibrium position of lower chemical and magnetic field energy.

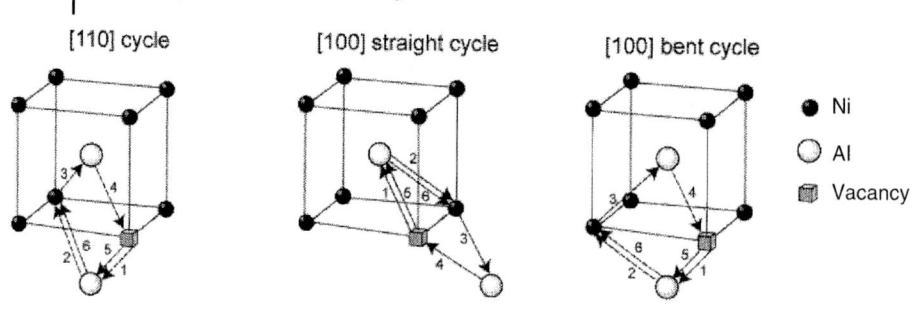

Fig. 5.37 Six-jump cycles (6JC) proposed for the B2 lattice, here NiAl.

5.2.3.3 Jump Cycles and Cooperative Atom Jumps

Long-range atom transport in highly ordered alloys can be imagined by several possible mechanisms. A primary requirement, arising from the high price in terms of Gibbs free energy, is that any disturbance of order be local and transitory only. If a sublattice is continuous by nearest-neighbor connections it offers an easy "highway" of diffusion. Such is the case, for instance, in the $L1_2$ structure (majority sublattice), and for the $L1_0$ structure, although in only two of the three directions in space. In B2, however, NN jumps necessarily affect order. (See below for a discussion of next-nearest-neighbor (NNN) jumps.)

Six-jump cycles Co-operative jump cycles have therefore been proposed, most notably the six-jump cycle (6JC), originally introduced by Elcock and McCombie (1958), after a suggestion by H. B. Huntington, which on completion leave the pristine ordered state and an atom interchanged with a vacancy. In Fig. 5.37 we show different kinds of possible six-jump cycles proposed for the B2 structure. The arrows in Fig. 5.37 indicate the path the vacancy takes around a quadrangle of lattice sites, completing one and a half turns. Having finished the six-jump cycle the vacancy has changed places with a Ni atom distant by a face diagonal ([110] cycle) or by a cube edge ([100] cycle). The state of order is destroyed locally during the first part of the cycle and restored by the second, symmetric part. The six successive steps need not really be separate atom jumps. Thus Mishin et al. (2003) found by computer calculations with EAM potentials in NiAl that after the first NN jump of the Ni vacancy an unstable state follows (which is to be expected considering the extreme scarcity of Al vacancies in this alloy; see below) so that the first and the second jump are performed simultaneously and collectively.[17] For the [110] cycle, even the middle configuration corresponds to a local maximum of energy instead of a minimum, making this 6JC effectively a three-jump cycle by merging together of pairs of successive jumps in a collective motion. Because of an altogether lower energy barrier, the [110] cycle seems to be

---

17) For reasons of symmetry, this is also true for the fifth and sixth jumps.

the predominant jump cycle in NiAl. Essentially the same result was obtained in the computer simulations of Farkas and Soulé des Bas (2001), who discovered also a minor contribution of more complex correlated cycles with 10 and 14 jumps.

As the mean square displacement of an A atom in a series of six-jump cycles of an α vacancy is twice that of a B atom and vice versa, it can be reasoned (Elcock and McCombie 1958) that the ratio of the tracer diffusion coefficients of the two species must for low enough temperatures obey the inequality in Eq. (141).

$$\frac{1}{2} < \frac{D_A^*}{D_B^*} < 2 \qquad (141)$$

The validity of Eq. (141) is confirmed by experiment for a series of B2 alloys, such as AgMg (Domian and Aaronson 1964), AuCd (Huntington et al. 1961; Gupta et al. 1967), AuZn (Gupta and Lieberman 1971), CoGa (Bose et al. 1979), making the 6JC eligible as a mechanism for self-diffusion in these materials. Results conflicting with this requirement were, however, reported for AuZn (Hilgedieck et al. 1983), CoGa (Stolwijk et al. 1977), NiGa (Donaldson and Rawlings 1976) and NiAl (Hancock and McDonnell 1971; Lutze-Birk and Jacobi 1975), inducing some workers to propose alternative diffusion mechanisms like the antistructure bridge mechanism or the triple defect mechanism (see below).

Generally, it has to be kept in mind that the importance of clearly defined correlated jump cycles rapidly diminishes when the number of defects increases and the state of almost perfect order or the region of ideal stoichiometry is left. This is perfectly plausible as a jump cycle is a very low-entropy path allowing few alternative variants. Even so, according to the statistical behavior of atom jumps which can be observed in kinetic Monte Carlo (KMC) simulations, correlated jump cycles should not be imagined as a strict step-by-step sequence of events. Rather, they result from many futile attempts and reversals of incomplete cycles. KMC investigations with improved residence-time algorithms (Athènes et al. 1997) have revealed in the B2 structure the importance of alternate, asymmetrical cycles branching off from conventional 6JC, together with new antisite-assisted cycles combining properties of the 6JC and the antistructure bridge mechanism (see below). The relative contribution of these mechanisms is strongly dependent on asymmetry of the like-atom interaction energies, deviation from ideal stoichiometry, and temperature. As was shown by Athènes et al. (1997), their interplay can produce a negative curvature of the Arrhenius plot of the diffusion coefficient, without having recourse to a divacancy (triple-defect) mechanism.

Six-jump cycles have also been proposed for the L1$_2$ structure by Hancock (1971). Figure 5.38 shows two variants of such cycles for Ni$_3$Al. Debiaggi et al. (1996) showed by simulation based on EAM potentials that, at least in Ni$_3$Al, long-range diffusion of both atom species on the majority sublattice is by energy reasons preferred over inter-sublattice diffusion via 6JC. The same conclusion was drawn also by other authors (Numakura et al. 1998; Belova and Murch 1999) on the basis of experimental data (dependence of diffusivity on stoichiome-

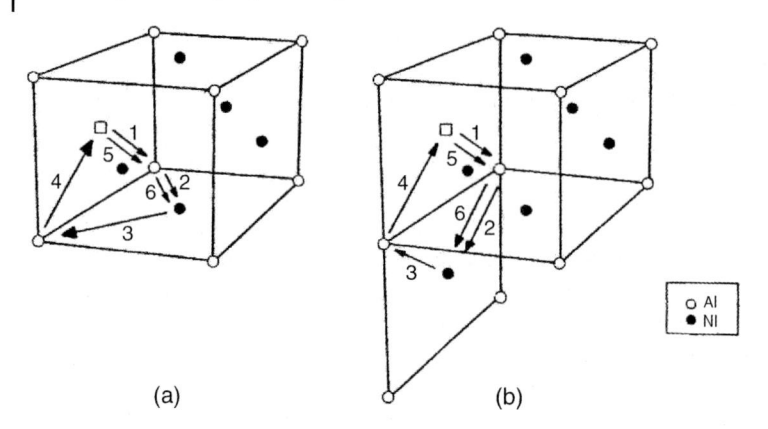

Fig. 5.38 Six-jump cycle in the L1$_2$ structure: (a) bent; (b) straight (From Debiaggi et al. 1996).

try). The condition imposed on the tracer diffusion coefficients (Eq. (141) for B2) reads, for the L1$_2$ lattice (Maeda et al. 1993):

$$0.1064 < \frac{D_A^*}{D_B^*} < 0.8524 \qquad (142)$$

It is on this basis that 6JC diffusion can be excluded for Ni$_3$Al.

Cooperative Generation of Anti-site Pairs Not only atom jumps leading to long-range atom transport but also those responsible for a change of order (cf. Section 5.2.3.1) may involve co-operative atom movements. In Ni$_3$Al it is known from KMC simulations (Oramus et al. 2000) that the faster of two exponential processes of disordering consists mainly of the generation of antisite pairs. A cooperative mechanism of two atom jumps was therefore proposed (Schweiger et al. 2001), Fig. 5.39.

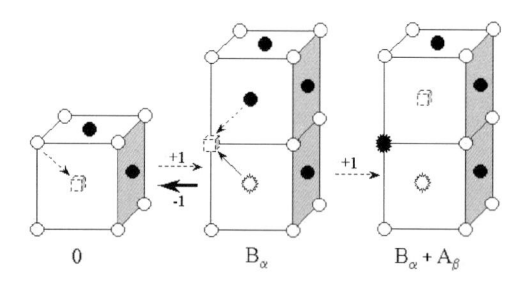

Fig. 5.39 Cooperative jump sequence leading to an antisite pair (Schweiger et al. 2001). From left to right: An Al atom (open circle) jumps into a Ni vacancy. A Ni atom (filled circle) occupies the Al vacancy and prevents the Al atom from jumping back to its original position.

First, an Al atom jumps into a vacancy on the majority (Ni) lattice. For energy reasons, there is a high probability of immediate reversal of this jump. If, however, a Ni atom almost simultaneously jumps into the Al vacancy, the back jump is blocked, and an antisite pair results. Ab-initio calculation of the energy surface for simultaneous jumps of the Al and the Ni atom shows that the activation energy of such cooperative jumps following representative paths is significantly higher than in the case of single atom jumps (Schweiger et al. 2001), and closely approaches the value measured in the resistivity measurements when the energy of formation of the initial Ni vacancy is superimposed. It is well known that the density of Al (β) vacancies for near-stoichiometric Ni_3Al is by several orders of magnitude smaller than the density of Ni (α) vacancies. By taking advantage of an Al vacancy in statu nascendi a Ni atom can create an antisite, but at an energy cost which the system is willing to pay, as it were, because there is no alternative mechanism available.

Antistructure Bridge (ASB) Mechanism   The conventional six-jump cycle (Elcock and McCombie 1958), although able to explain diffusion behavior in B2 ordered compounds in the vicinity of the ideal stoichiometric composition at low enough

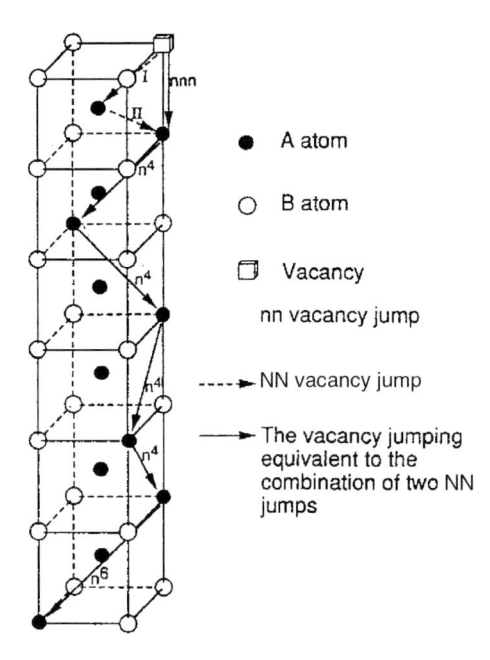

Fig. 5.40 Antistructure bridge (ASB) mechanism of diffusion in an ordered B2 lattice (Kao and Chang 1993). Uppermost unit cell: arrows I and II designate two NN jumps leading to an effective NNN jump. In the cells below other possibilities of effective jumps to more distant neighborhood positions are shown. Each effective jump consists of two NN jumps of the vacancy, the second involving a pre-existing antistructure (antisite) atom.

temperatures, quickly loses importance at large deviations from stoichiometry. As an alternative in this region the antistructure bridge (ASB) mechanism was introduced as a model by Kao and Chang (1993), Fig. 5.40. It depends on a sufficient concentration of antistructure (antisite) atoms, which serve as "bridges" by which a vacancy can migrate through the lattice without changing the ordered state. In the topmost unit cell of Fig. 5.40, two NN jumps of the vacancy result in an effective jump to an NNN position. In the elementary cells below, other possible effective jumps are shown, which lead to lattice positions in higher-order coordination shells of the initial position of the vacancy. All of these jumps, however, are equivalent to two NN vacancy jumps, which together leave the state of LRO unchanged.

Triple-Defect (TD) Mechanism This mechanism was proposed by Stolwijk et al. (1980) to explain the tracer diffusion data in B2 ordered CoGa, where the requirement of Eq. (141) for the ratio of the tracer diffusion coefficients of the two species was not fulfilled. Frank et al. (2001) reverted to the TD mechanism to explain the diffusion behavior of NiAl, especially on the Ni-rich side. In this alloy, triple defects (cf. Section 5.1) play a very prominent role, Ni vacancies occurring more frequently than Al vacancies by several orders of magnitude. Figure 5.41 shows the steps of the TD mechanism (Frank et al. 2001). By a series of vacancy ex-

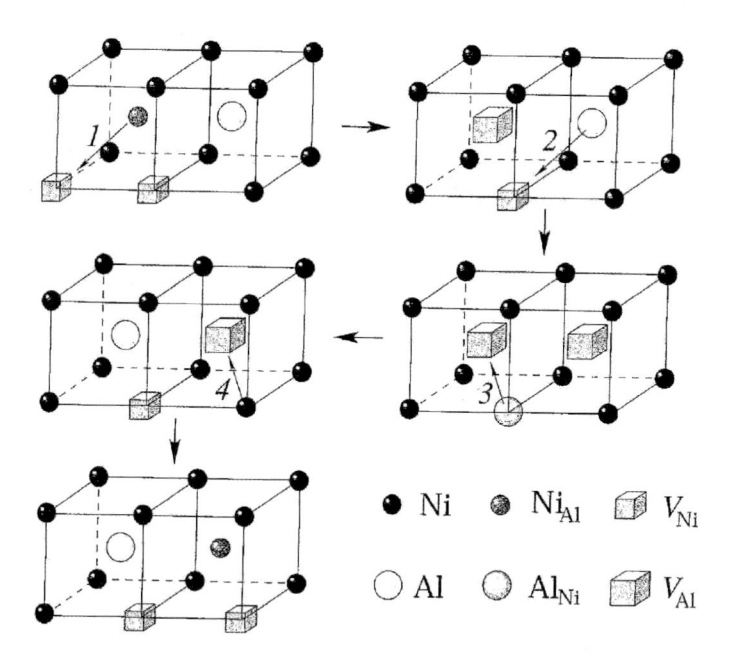

Fig. 5.41 Triple-defect (TD) mechanism in NiAl (Frank et al. 2001). An atom in an antisite position changes places with an NNN atom in a regular position with the help of a vacancy pair. The antisite atom together with the two vacancies constitutes a triple defect.

changes, a Ni atom in a triple defect position is exchanged with an adjacent Al atom.

Next-nearest-neighbor (NNN) jumps We have hitherto made the tacit assumption that diffusion jumps in intermetallics always end up in an NN vacancy position. The possibility of next-nearest-neighbor (NNN) jumps in the B2 structure cannot be excluded a priori, however, in view of the great advantage of keeping the ordered state intact, and bearing in mind that the NNN jump distance is only 15% larger than the NN distance. This mechanism was first proposed by Donaldson and Rawlings (1976) for NiGa. Exchange of an atom with an NNN vacancy represents a viable alternative to correlated jump cycles if the energy of migration is smaller than the energy needed to create disorder. In Table 5.7 (Neumann 1980) we have arranged some common B2 alloys according to their enthalpies of formation which are correlated with the energy needed to crate an antisite defect. The higher this value, the more frequent NNN jumps should become as they are relatively more favored.

In B2 ordered FeAl, the results of quasielastic Mössbauer spectroscopy (Vogl and Sepiol 1994) and nuclear resonant scattering (Sepiol et al. 1998b) indicated that diffusion of Fe takes places via NN jumps, which was further supported by Monte Carlo simulations (Weinkamer et al. 1999). Measurements of quasielastic neutron scattering in alloys with intermediate formation energies, namely CoGa (Kaisermayr et al. 2001) and NiGa (Kaisermayr et al. 2000), could not be reconciled with the assumption of NNN jumps of Ni, neither could the diffusion mechanism be attributed to any specific correlated cycle. Rather, a great variety of jump modes involving several defects seems to be operative. (For a more detailed discussion concerning the application of nuclear scattering methods to investigate elementary steps of diffusion, see Chapter 13.1.)

In NiAl, NNN jumps were suggested as a plausible diffusion mechanism especially on the Al-rich side, where there is an ample supply of constitutional Ni vacancies (Mishin and Farkas 1997, 1998). An activation energy of about 2 eV found in previous experiments by Hancock and McDonnell (1971) agreed well with NNN migration energies computed with EAM potentials (Mishin and Farkas 1997, 1998). This conclusion was, however, contested by Frank et al. (2001) on the basis of their own experimental results showing an almost constant diffusivity

Table 5.7 Formation enthalpies for some B2 intermetallics (Neumann 1980).

Material	Formation enthalpy, ΔH_f [eV]
FeAl	0.34
CoGa	0.37
NiGa	0.47
NiAl	0.72

at Al-rich compositions, at odds with an NNN jump mechanism, where diffusivity should increase markedly as more constitutional Ni vacancies become available. Frank et al. (2001) instead proposed the TD mechanism as the principal mode of diffusion at the stoichiometric composition and on the Al-rich side. An increase of diffusivity on the Ni-rich side was explained by an increased contribution of the ASB mechanism. As a part of the ongoing controversy on diffusion mechanisms in NiAl, NNN jumps were recently defended with better EAM and ab-initio calculations (Mishin et al. 2003).

In conclusion, it might be said that at the present state of knowledge details of diffusion in intermetallics, especially in the B2 ordered structure, are still very much a matter of active research and animated discussion.

5.3
Statistical Methods: from Single Jump to Configuration Changes

Modern materials science is pursuing the ambitious but ever more realistic aim of describing and understanding the development of microstructure on a coherent progression of scales, reaching from the atomistic details to the macroscopic world (multiscale approach). The problem is related to the classical question of statistical physics: How does a multitude of elementary events (e.g., collisions of molecules) determine the behavior of the very few observables (e.g., pressure, temperature) we are really interested in? The difference is, however, that in alloy physics we have to deal with subsystems like nuclei, precipitate particles, domains of order, diffusion zones, defects of various dimensionality and extent, etc. Therefore, statistical behavior is very often the result of cooperation of not quite so many atoms or molecules, sometimes only a few hundred, in most cases definitely less than Avogadro's number. It may well be doubted that it is legitimate to transfer concepts liberally from macroscopic thermodynamics, such as entropy and free energy, to very small subsystems, as has often been done in the past. There are some indications of cases where the difference might count. For instance, Miyazaki et al. (2005) in various alloys observed a precipitate nucleation rate deviating from the predictions of classical theory, leading him to adopt a new entropy concept by Tsallis (1988, 2003), which takes account of long-range interactions and therefore is nonextensive. The peculiar properties of very small systems have become a natural topic in the treatment of nanostructures (see also Chapter 9) for which surface and interface interactions cannot be disregarded in comparison to volume interactions, and certain approximations such as using a simplified version of Stirling's approximation formula must be applied with caution.[18]

---

18) From the simplified version of Stirling's formula $\ln(N!) \approx N \ln N - N$ we get for 100 atoms 360.5, whereas the exact value is $\ln(100!) \approx 363.7$. The approximate value thus deviates by about 1%. For 30 atoms the respective values are 72.0 and 74.7, corresponding to an error of about 4%.

5.3.1
Master Equation Method

Quite generally a master equation gives the rate of change in the occupation of a certain state of a system as a balance between transitions leading to it from other possible states, and of transitions away from it to other states, so that we can write

$$\frac{dP_m}{dt} = \sum_n (W_{nm}P_n - W_{mn}P_m) = \hat{\mathbf{S}}\mathbf{P} \tag{143}$$

where the occupation number P_m of state m is changed by transitions with frequency W_{nm} leading toward it and W_{mn} leading away from it. This form of kinetics is valid whenever a system changes its states according to a Markov process (see Section 12.2.1.2). The transition can be written in a more compact form as a transfer matrix applied to the state vector $\mathbf{P}$.

In an advanced formulation of the master equation method (Vaks 1996; Belashchenko et al. 1999) the occupation of each lattice site is left free to vary individually. A high-dimensional configuration space is subtended by all these degrees of freedom of occupation, and the time evolution of an ensemble density on this configuration space is considered. The transfer matrix $\hat{\mathbf{S}}$ of Eq. (143) can be expressed in terms of single atom jumps taking place according to transition state theory (cf. Section 5.2.1.1). Vaks and coworkers adopted a direct-exchange model after showing its equivalence to the vacancy-exchange model in advanced stages of phase transformation (Belashchenko and Vaks 1998). The underlying difficulty of this method is, of course, how to express the transfer matrix $\hat{\mathbf{S}}$ as a function of the current state of the systems, that is, by means of local concentrations and correlations. This cannot be done without some kind of averaging if one wants to avoid outright simulation. Vaks and coworkers used various kinds of cluster mean-field approximations. Ordering kinetics and the development of long-range order domains in two dimensions could be described very well by this technique. Taking averages over statistical distributions deprives the method, however, of the possibility of describing phenomena that rest upon kinetic paths of nonmaximum probability, such as nucleation (cf. Chapter 7 and Section 5.4.2.2).

5.3.2
Continuum Approaches to Microscopic Diffusion and their Interrelationship with Atom Jump Statistics

Another possible way to obtain microscopic diffusion equations is to start from continuum arguments and to set changes of field variables proportional to thermodynamic driving forces. A field variable can be any local variable describing the state of the system, such as an order parameter (nonconserved) or a local concentration (conserved). This approach leads to the time-dependent Ginzburg–Landau equations (TDGL; see Chapter 10, Eq. (124)). If the field variable is concentration, then the diffusion fluxes can be shown to obey Eq. (144) (Chapter 10, Eq. (128)):

$$\mathbf{j} = -M\nabla\left(\frac{\delta F}{\delta c}\right) \tag{144}$$

i.e., the flux is proportional, by a mobility M, to the gradient of the local chemical potential. Together with the continuity equation and a suitable form of the free energy functional comprising a gradient-energy term (see Eq. (118) in Chapter 10), this expression for the fluxes results in the well-known Cahn–Hilliard diffusion equation. This formalism is also the point of departure for a host of phase-field methods which in recent times have become popular in the investigation of complex phase transformations (see Chapter 12).

Martin (1990) has given a revealing analysis of the interconnection between a stochastic model based on individual atom jumps in a master equation formulation and the TDGL approach, Eq. (144) (see Chapter 8). In a one-dimensional discrete diffusion model, an analogue of the local chemical potential could be constructed, which must be uniform for an equilibrium state. The dynamic steady state of diffusion jumps was identified with thermodynamic equilibrium and shown to have a probability density equal to that derived from the canonical ensemble. Special care was devoted to the question of under what conditions it might be justified to set the mobility M constant, as is routinely done in the Cahn–Hilliard formulation. It was recognized that M contained a second-order term in the deviation from uniform concentration, which should become important whenever the gradient-energy contribution is nonzero.

An atomistic application of the TDGL equations was used by Khachaturyan and coworkers (Chen et al. 1994), where the change in occupation of a lattice site at position $\mathbf{r}$ was made proportional to the variation of free energy by atom jumps to neighboring lattice sites:

$$\frac{dn(\mathbf{r},t)}{dt} = \sum_{\mathbf{r}'} L(\mathbf{r}-\mathbf{r}')\frac{\delta F}{\delta n(\mathbf{r}',t)} \tag{145}$$

The kinetic coefficients L should be proportional to the probability of elementary diffusion jumps. The free energy was formulated so as to account for elastic strain energy, and the kinetic equation (145) was put in Fourier form. The model was able to explain both ordering and segregation within the same formalism and reproduce a wealth of precipitate morphologies including such interesting effects as the splitting due to elastic interaction of overgrown coherent particles in Ni–Al alloys (Miyazaki et al. 1982).

Both in this variant of TDGL kinetics and in the more continuum-oriented methods of Cahn–Hilliard and the phase-field family, the same caveats have to be kept in mind: (1) these approaches are essentially based on linearization valid in the vicinity of equilibrium and must be viewed with caution in highly nonequilibrium situations; and (2) they incorporate mean-field concepts that have to be revised in complex alloys with non-negligible higher-order correlations.

5.3.3
Path Probability Method

Kikuchi (1951) devised the cluster variation method (CVM) in order to include multibody correlations up to arbitrary order in the configuration entropy expression for an alloy. The most probable (i.e., equilibrium) state of the system was then determined by the interplay, via free energy, between internal energy and CVM entropy. A further logical step was to extend this concept by carrying it into the time dimension, thus creating the path probability method (PPM) described in Section 10.2.6; see also Kikuchi (1966): the most probable kinetic path between two states for a given energy difference is the one with the highest number of microscopic transition possibilities. Transitions between a hierarchy of cluster configurations are expressed by path variables such as $Y_{ij,kl}$ giving the probability of changing the atom pair ij to kl by an elementary kinetic step (modeled as a spin flip in an Ising system). The total number of transition possibilities at each step is measured by a path probability function

$$P = P_1 P_2 P_3 \tag{146}$$

where the first factor is connected with the unbiased spin flip probability per unit time, the second is a Boltzmann factor of the energy difference before and after the transition, and the third a combinatorial expression, akin to CVM entropy, dependent on the path variables and taking account of the number of microscopic transition paths. P is now maximized at each step with respect to the path variables, which in their turn describe the change of the configuration.

The advantage of the PPM is that it can be used even far from thermodynamic equilibrium since it does not explicitly use the concept of free energy. However, the complexity of the formalism limits it practically to homogeneous systems and precludes the treatment of a vacancy or direct-exchange diffusion mechanism.

5.3.4
Monte Carlo Simulation Method

This method, which is described in detail in Chapter 12, can be viewed as a way of doing statistics by computer experiment. No averaging is carried out before or on the level of basic calculation, which consists of performing random atom jumps on a lattice. Statistical evaluation takes place only after the jump sequence, in at least two ways: first by accumulating the results of several runs with the same starting configuration and the same macroscopic parameters, but each time a different sequence of random numbers, and second by calculating integral quantities by summing over the calculation cell.

At the microscopic level, however, there is no restriction on the freedom of atom jumps. In principle all correlations and all interactions are described as realistically as the jump probabilities and the underlying Hamiltonian can be formulated. (For a discussion see Chapter 12 and Section 5.2.1.1.) Specifically, low-probability paths do get chosen by the system with their respective statistical weights, and thermally activated processes are observed regularly if the activation enthalpy is in a viable range for the given temperature. Therefore, a nucleation process need not be introduced into the kinetics artificially. Likewise, the establishment and coarsening of ordered domains happen quite naturally together with the ordering process itself. The power to describe rich microstructural evolution out of a single jump process without any intervening additional assumptions is certainly one of the reasons for the immense popularity of this method.

5.4
Macroscopic Diffusion

5.4.1
Formal Description

5.4.1.1 Fick's Laws

When studying the random walk of atoms in a crystal (Section 5.2.2.1) we have already seen that for a concentration gradient in an ideal solid solution Fick's first diffusion law follows directly

$$\mathbf{j} = -D\nabla C \tag{147}$$

Applying the continuity equation (mass conservation),

$$\frac{\partial C}{\partial t} = -\nabla \cdot \mathbf{j} \tag{148}$$

Fick's second diffusion law results,

$$\frac{\partial C}{\partial t} = \nabla \cdot (D\nabla C) \tag{149}$$

which describes changes of a concentration field due to the diffusion process.

For a constant diffusion coefficient D, Eq. (149) simplifies to the linear diffusion equation

$$\frac{\partial C}{\partial t} = D\nabla^2 C \tag{150}$$

For the case of mixing in isotropic solids the diffusion coefficient D is a scalar. In general, however, atom flux is determined by a symmetric second-rank diffusion tensor D_{ij}.

Solving the linear diffusion equation under specific boundary conditions is a standard problem in technical applications for which we refer to the corresponding literature (see, for instance, Carslaw and Jaeger 1984; Crank 1994). An example of special relevance for the measurement of diffusion properties in alloys is the behavior of diffusion couples. When a thin film of solute is attached to an alloy under investigation, for example to determine tracer diffusion coefficients, the "thin-film solution" of Fick's second law for an initial condition of $C(x, t=0) = \delta(x)$ is

$$C(x,t) = \frac{M}{\sqrt{\pi D t}} \exp\left(-\frac{x^2}{4Dt}\right) \tag{151}$$

M being the mass of the thin-film diffusion source (mass per area). Note that Eq. (151) corresponds to Eq. (92) of Section 5.2 for the case of a half-space. In practical situations, chemical diffusion coefficients often turn out as composition-dependent. For specific boundary conditions Fick's second law can be integrated in this case. A classical method is to solve the inverse diffusion problem: if the concentration field $C(x,t)$ is known from experiment, the diffusion coefficient can be determined as a function of composition. Matano (1933) has demonstrated a method to solve the corresponding transformed diffusion equation. Let us consider the concentration profile of a diffusion couple which was step-like at $t=0$ (Fig. 5.42). The solution for the diffusion coefficient can be written as Eq. (152).

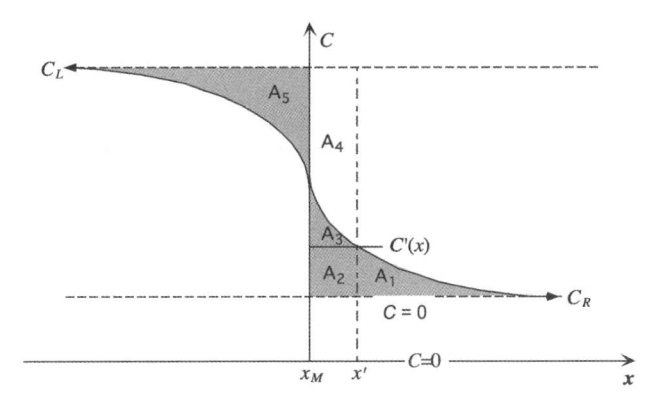

Fig. 5.42 Boltzmann–Matano geometry for a diffusion couple (after Glicksman (2000)).

$$D(C') = -\frac{1}{2t}\frac{dx}{dC}\bigg|_{C'}\int_{C_R}^{C'}(x-x_M)\,dC \tag{152}$$

Here x_M is the coordinate of the Matano plane, for which the shaded areas in Fig. 5.42 balance: $A_5 = A_1 + A_2 + A_3$. Equation (151) can be solved graphically if x_M is taken as the origin of the x axis.

Very recently, in a simple but very effective way, equilibrium intermetallic compounds and solid-solution phases of various compositions were created by long-term annealing of junctions of three or more metals or alloy phases (Zhao 2005; Zhao et al. 2005; Zhao 2006). This "diffusion-multiple" approach enables the generation of a large number of alloy phases and compositions for efficiently mapping phase diagrams, phase properties, and kinetics. Many critical alloy data are obtained in this way with a high efficiency: phase diagrams, diffusion coefficients, hardness, elastic modulus, and thermal conductivity, but also precipitation kinetics and solution and precipitation strengthening effects. It is planned to extend the diffusion-multiple approach in order to determine optical, magnetic, mechanical, and other properties, by the development of corresponding microscale screening tools. It seems that thereby an extremely useful method for alloy design has been generated, as shown in Fig. 5.43.

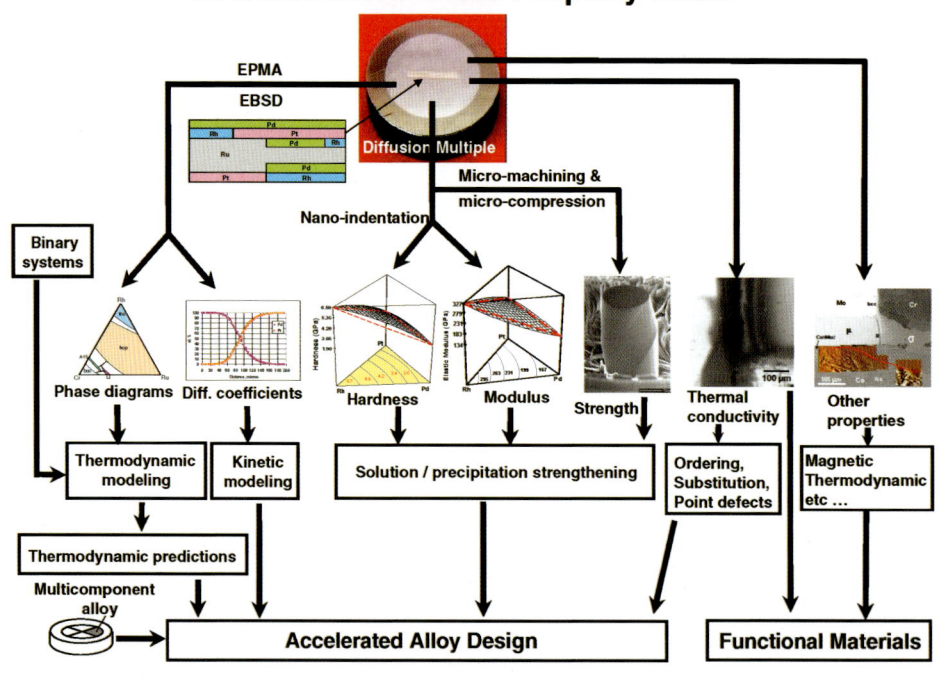

Fig. 5.43 Schematic representation showing how the "diffusion-multiple" approach can be used for alloy design and discovering new functional materials (Zhao et al. 2005). EPMA: Electron Probe Micro-Analysis; EBSD: Electron Back Scatter Diffraction.

5.4.1.2 Nonreciprocal Diffusion, the Kirkendall Effect

Since diffusion coefficients of constituent species in alloys are usually of the same order of magnitude, any mixing of atoms due to concentration gradients is brought about by jumps of all atoms. It is well known that the interdiffusion (or chemical diffusion) of atoms in a binary alloy can still be characterized by a single diffusion coefficient (the interdiffusion coefficient or chemical diffusion coefficient $\tilde{D}$). This is straightforward if both atoms, A and B, diffuse at the same rate (reciprocal diffusion), where Fick's first law reads

$$\mathbf{j}_i = -D_i \nabla C_i \quad (i = A, B) \tag{153}$$

with $D_A = D_B = \tilde{D}$. If, on the other hand, the two species in a diffusion couple do not move at the same rate (nonreciprocal diffusion), atomic planes are shifted (lattice flow) to compensate for a net flux of vacancies (vacancy wind). This phenomenon is known as the Kirkendall effect (Smigelskas and Kirkendall 1947).

If a net flux of atoms occurs in a diffusion couple due to nonequal diffusion coefficients, two frames of reference can be used. (a) In the laboratory reference frame, the ends of the sample remain unchanged, but lattice positions within the diffusion zone identified by surface markers due to the unbalanced diffusion move with a velocity v_M (marker velocity). Here, as before, a single diffusion coefficient, the interdiffusion coefficient $\tilde{D}$, describes the macroscopic interdiffusion of both species. (b) In the lattice reference frame, which is fixed to the atomic planes in the diffusion zone, the ends of the couple move and two intrinsic diffusion coefficients, D_A and D_B, describe the microscopic diffusion process.

At any instant of time t, the fluxes of A and B atoms within the lattice frame are given by Fick's first law:

$$j_A = -D_A \frac{\partial C_A}{\partial x} \quad \text{and} \quad j_B = -D_B \frac{\partial C_B}{\partial x} \tag{154}$$

If $D_A \neq D_B$, a net atomic flux results as $j_{net} = j_A + j_B$. The rate at which mass is accumulated in the diffusion zone can be related to the velocity of a marker in the diffusion couple,

$$v_M = -j_{net} \Omega \tag{155}$$

where Ω is the atomic volume of the alloy, which is assumed constant. Noting that with this assumption $C_A + C_B$ is constant, it can be shown that the velocity can be expressed in terms of the concentration gradients:

$$v_M = (D_A - D_B)\Omega \frac{\partial C_A}{\partial x} = (D_B - D_A)\Omega \frac{\partial C_B}{\partial x} \tag{156}$$

The occurrence of a marker velocity, verified in the famous experiment by Smigelskas and Kirkendall (1947) using a diffusion couple of pure copper and 70–30 brass, gives evidence for the motion of lattice planes (advective motion). Assuming a conservation of lattice sites in the crystal, a net flux of vacancies has to compensate for the unbalanced atom fluxes:

$$j_V + j_A + j_B = 0 \tag{157}$$

Inserting the atom fluxes according to Fick's first law, Eq. (154), and considering that $C_A + C_B$ can still be regarded constant for the usual low concentrations of vacancies, the vacancy flux in the lattice frame is obtained:

$$j_V = (D_A - D_B)\frac{\partial C_A}{\partial x} = (D_B - D_A)\frac{\partial C_B}{\partial x} \tag{158}$$

It is seen that the net flux of vacancies is proportional to the difference in the intrinsic diffusion coefficients. The flux of vacancies compensating for the flow of crystal lattice is called the vacancy wind, which "blows" through the lattice against the faster-diffusing species. This means that in the diffusion zone the vacancy concentration changes locally to values below or above the local equilibrium value. As a consequence, lattice planes are gradually created by generation of vacancies and in turn are destroyed by vacancy annihilation at sinks (e.g., jogs, dislocation loops and stacking faults). The motion of lattice planes is therefore high where the concentration of vacancies is far from equilibrium which is a consequence of the different atoms diffusing at different rates.

The Kirkendall effect has of course a consequence for Fick's second law, which has to account for the advective motion of the lattice planes. It can be shown that Fick's second law can be modified to give the diffusion advection equation:

$$\frac{\partial C}{\partial t} = \nabla \cdot (\tilde{D}\nabla C) + \mathbf{v}_M \cdot \nabla C \tag{159}$$

For the time-independent case and for constant $\tilde{D}$, Eq. (159) simplifies to the steady-state diffusion advection equation:

$$\nabla^2 C + \frac{\mathbf{v}_M \cdot \nabla C}{\tilde{D}} = 0 \tag{160}$$

Several applications, especially in the field of solidification and crystal growth, for a steadily moving interface can be treated using Eq. (160) when transforming to the lattice reference frame (cf. Chapter 3).

Shortly after the discovery by Smigelskas and Kirkendall, Darken (1948) analyzed the effect in a purely phenomenological treatment. He obtained the first Darken equation,

$$v_M = D_A \frac{\partial c_A}{\partial x} + D_B \frac{\partial c_B}{\partial x} \tag{161}$$

corresponding with Eq. (156).[19] In addition, he obtained an essential relationship between the "overall" interdiffusion coefficient $\tilde{D}$ and the intrinsic diffusion coefficients (the second Darken equation):

$$\tilde{D} = c_B D_A + c_A D_B \tag{162}$$

This means that, by determining the interdiffusion coefficient $\tilde{D}$, e.g., by the Boltzmann–Matano analysis, and simultaneous measurement of the marker velocity v_M, the intrinsic diffusion coefficients D_A and D_B involved can be obtained.

5.4.1.3 Nonideal Solutions

In an alloy, each atom species and vacancy diffuses with its own diffusivity. We can define a generalized diffusion driving force, which causes a certain drift velocity of a species i:

$$\mathbf{u}_i \equiv \mathbf{B}_i \mathbf{F}_i^{\text{diff}} \tag{163}$$

The mobility **B** of the specific atom is in general a tensor, which relates the drift velocity and the driving force. The generalized force driving the diffusion of a species i can be written as a gradient of the corresponding chemical potential,

$$\mathbf{F}_i^{\text{diff}} = -\nabla \mu_i \tag{164}$$

linked to the concentration gradient. The diffusion flux, being the product of local concentration and drift velocity, can be written as

$$\mathbf{j}_i = C_i \mathbf{u}_i = C_i \mathbf{B}_i \mathbf{F}_i^{\text{diff}} = -C_i \mathbf{B}_i \nabla \mu_i \tag{165}$$

The chemical potential μ_i is not directly accessible to experiment but can be written in terms of the activity a_i, a measurable thermodynamic quantity, conveniently expressed by the activity coefficient γ_i ($\mu_i = \mu_0 + k_B T \ln a_i, a_i \equiv c_i \gamma_i$) describing the deviation from the behavior of an ideal solution ($\gamma = 1$). Then we can write Eq. (166) for the driving force using the site fraction c_i ($= \Omega C_i$) instead of the concentration C_i:

$$\nabla \mu_i = k_B T \nabla \ln a_i = k_B T (\nabla \ln c_i + \nabla \ln \gamma_i)$$

$$= k_B T \left(\frac{\partial \ln c_i}{\partial c_i} + \frac{\partial \ln \gamma_i}{\partial c_i} \right) \nabla c_i$$

$$= \frac{k_B T}{c_i} \left(1 + \frac{\partial \ln \gamma_i}{\partial \ln c_i} \right) \nabla c_i \tag{166}$$

---

19) The site fractions c_i are identical to the mole fractions in disordered alloys of low vacancy concentrations and are related to the concentrations as $c_i = \Omega C_i$.

This yields Eq. (167) for the atom flux:

$$j_i = -k_B T B_i \left(1 + \frac{\partial \ln \gamma_i}{\partial \ln c_i}\right) \nabla C_i \qquad (167)$$

If this is compared with Fick's first law, Eq. (147), the corresponding intrinsic diffusion coefficient D_i can be identified:

$$\tilde{D}_i = k_B T B_i \left(1 + \frac{\partial \ln \gamma_i}{\partial \ln c_i}\right) \qquad (168)$$

The term within the parentheses is called the thermodynamic factor φ and takes into account that for diffusion in nonideal solutions the driving concentration gradient has to be modified. Note that this result was obtained independently in Section 5.2 [Eq. (123)] using the bias in individual atom jump probabilities.

The thermodynamic factor entering the diffusion coefficient for nonideal, concentrated solid solutions has a remarkable consequence for unmixing systems. We show below that the thermodynamic factor is proportional to the second derivative of a local Gibbs free energy per atom with respect to concentration, $g''(c)$, if an adequate local concentration can be defined. First we write the local Gibbs energy for an A–B alloy ($c_A + c_B = 1$) as $g = (1 - c_B)\mu_A + c_B \mu_B$. The first derivative, $g' \equiv \partial g/\partial c_B$, is given as

$$g' = -\mu_A + (1 - c_B)\frac{\partial \mu_A}{\partial c_B} + \mu_B + c_B \frac{\partial \mu_B}{\partial c_B}$$

$$= \mu_B - \mu_A \qquad (169)$$

Note that in the first expression the sum of the second and the fourth term is zero due to the Gibbs–Duhem relationship $c_A \, d\mu_A + c_B \, d\mu_B = 0$. Expressing the chemical potentials by activities, i.e.,

$$g' = \mu_B^0 + k_B T \ln a_B - \mu_A^0 - k_B T \ln a_A \qquad (170)$$

the second derivative, $g'' \equiv \partial^2 g/\partial c_B^2$, is found to be

$$g'' = k_B T \frac{\partial}{\partial c_B}(\ln a_B - \ln a_A)$$

$$= k_B T \left(\frac{1}{c_A} + \frac{1}{c_B} + \frac{\partial \ln \gamma_B}{\partial c_B} - \frac{\partial \ln \gamma_A}{\partial c_B}\right)$$

$$= k_B T \left(\frac{1}{c_A} + \frac{1}{c_B} + \frac{1}{c_B}\frac{\partial \ln \gamma_B}{\partial \ln c_B} + \frac{1}{c_A}\frac{\partial \ln \gamma_A}{\partial \ln c_A}\right)$$

$$= k_B T \frac{1}{c_A c_B}\left(1 + \frac{\partial \ln \gamma_B}{\partial \ln c_B}\right) = \frac{k_B T}{c_A c_B} \varphi \qquad (171)$$

The interdiffusion coefficient can therefore be written as:

$$\tilde{D} = [c_B D_A + (1-c_B) D_B] \frac{c_A c_B}{k_B T} g'' = M_D g'' \tag{172}$$

The thermodynamic factor φ and in consequence the diffusion coefficient $\tilde{D}$ is proportional to g'', which can be either positive or *negative*. For a decomposing alloy system the two curves of $g_A(c)$ and $g_B(c)$ (where c is either c_A or c_B) may be connected, having a maximum between the two minima. The two inflection points ($g'' = 0$) of the combined $g(c)$ curve lie on the spinodal connecting the inflection points for all temperatures. For all concentrations within the spinodal ($g'' < 0$) the diffusion flux turns from downhill (down the concentration gradient) to uphill diffusion (up the concentration gradient); i.e., the diffusion coefficient changes its sign to $D < 0$. This can be interpreted as an instability of alloys in this composition range with respect to infinitesimally small concentration fluctuations: any concentration fluctuation starts the decomposition process, however small. This phenomenon of spontaneous decomposition process is called spinodal decomposition and is in contrast to the decomposition mode by nucleation and growth when $D > 0$. A detailed treatment of this distinction is given in Chapter 7 on phase transformations.

5.4.2
Phase Transformations as Diffusion Phenomena

5.4.2.1 Spinodal Decomposition
As we discussed above, if $g''(c) < 0$ we expect phase separation to happen spontaneously by "uphill" diffusion. Cahn and Hilliard (1958) write the free energy of a crystal as a functional of concentration and concentration gradient[20] (cf. Sections 7.5 and 10.2.1):

$$\Delta G = \int_V [g(c) - g(c_0) + \kappa (\nabla c)^2] \, dV \tag{173}$$

In the classical treatments of spinodal decomposition the diffusion flux density is set proportional (by a mobility M) to the gradient of the chemical potential.[21] Generalizing the chemical potential as a variational derivative of the integrand in Eq. (173), the flux density results as

$$\mathbf{j}(\mathbf{x}, t) = -M \nabla \left(\frac{\partial g}{\partial c} - 2\kappa \nabla^2 c \right) \tag{174}$$

---

[20] Here we have omitted the term taking account of elastic strain energy.
[21] That M is taken to be independent of composition c can be justified in the vicinity of the critical point but is certainly not true in the general case.

Applying the continuity equation to the atom flux we finally arrive at a diffusion equation, which is a generalized version of Fick's second law:

$$\frac{\partial c(\mathbf{x}, t)}{\partial t} = \nabla \cdot \left[M \nabla \left(\frac{\partial g}{\partial c} - 2\kappa \nabla^2 c \right) \right] \tag{175}$$

For small concentration amplitudes this equation becomes linear and is solved by sinusoidal concentration modulations growing or decaying exponentially in time. For a detailed discussion, see Chapter 7.

Spinodal ordering The long-range ordered state can be seen as a concentration fluctuation with a very short wavelength equal to the periodicity of the ordered lattice, a so-called ordering wave. Instability of a homogeneous, disordered alloy with respect to such a fluctuation with infinitesimal amplitude leads to a behavior analogous to spinodal decomposition, appropriately called spinodal ordering (de Fontaine 1975; Khachaturyan 1978). Of course, the diffusion jumps connected with the growth of a fluctuation with such a short wavelength cannot be adequately rendered by concepts like continuum diffusion fluxes. In a discrete model of spinodal decomposition (Hillert 1961; Cook et al. 1969) the possibility of such short-wavelength concentration instabilities was already recognized and it was remarked that in phase-separating systems the instability was driven by negative g'', originating from the first term in Eq. (175), whereas in spinodal ordering g'' is positive, and the instability is driven by negative κ, the gradient-energy coefficient. Statistical concepts like the path probability method (PPM) have also been profitably applied to the kinetics of spinodal ordering: In a computer experiment (Mohri 1994) an AB alloy with stoichiometric composition was quenched to a temperature below the ordering temperature and an initial ordering composition fluctuation imposed. It was demonstrated how above the spinodal ordering temperature only an initial fluctuation with sufficient amplitude could grow (nucleation regime) whereas below the spinodal ordering temperature even a very small amplitude of fluctuation spontaneously led to a long-range ordered state (spinodal regime).

5.4.2.2 Nucleation, Growth, Coarsening

If the composition of an alloy is in the two-phase field but outside the spinodal region, i.e., $g''(c) > 0$, small-amplitude spontaneous fluctuations decay rather than grow, and the kinetics of the phase transformation can no longer be described by macroscopic diffusion flows. Instead we have to consider the jumps individually and allow for kinetic paths of nonmaximum probability. This is done in classical nucleation theory, which regards a sequence of subcritical clusters (embryos) where a cluster of size $i + 1$ is created by the addition of one atom to a cluster of size i. Since the Gibbs free energy of the system increases as a subcritical cluster grows, the net local diffusion flux is always directed outward from the cluster. Nevertheless, after a certain incubation time a population of nuclei of each size is established so that there is a practically equal flux of growing and shrinking nuclei $N(i)\omega_i^+ = N(i+1)\omega_{i+1}^-$ where ω^+ and ω^- are the probabilities per unit time of growth and shrinkage. The important point is that although $\omega_i^+ < \omega_{i+1}^-$ the net flux between cluster size classes is not directed from $i+1$ toward i but because $N(i) > N(i+1)$ the smaller nuclei restore the balance by sheer numerical superiority. The kinetics of the size distribution function $N(i,t)$ can be cast in the form of a Fokker–Planck-like generalized diffusion equation in phase space,

$$\frac{\partial N(i,t)}{\partial t} = \frac{\partial}{\partial i}\left[\omega_i^+ N_e(i) \frac{\partial}{\partial i}\left(\frac{N(i,t)}{N_e(i)}\right)\right] \qquad (176)$$

where $N_e(i)$ means an ideal equilibrium distribution of nuclei. For a more detailed discussion including the computation of the nucleation rate (at which the nuclei grow beyond critical size) as a steady-state solution of Eq. (176), see Chapter 7.

Once supercritical nuclei have been created, their growth takes place by the usual macroscopic diffusion. In a steady-state regime a parabolic growth law is followed as long as the depletion zones around the growing particles do not yet impinge on one another. For spherical particles the radius then grows as $R = \lambda\sqrt{Dt}$.

When the equilibrium volume fraction of the precipitate phase has been reached, large particles grow at the expense of smaller ones, exchanging atoms via the matrix phase. The driving force of this coarsening process is interfacial energy. In the classical treatments of Lifshitz and Slyozov (1961) and Wagner (1961) (LSW theory), the particle size distribution function $f(r)$ (r being the particle radius) is shown to reach under certain conditions a normalized standard form. The average particle size then grows as $r \propto t^{1/3}$. The $t^{1/3}$ law changes to other power law dependencies if coagulation and splitting of precipitate particles are enabled (Binder and Heermann 1985). For a detailed discussion, again see Chapter 7.

Summing up this section, we recognize that, in the phase transformation scenarios mentioned, (a) local diffusion fluxes may very well be directed up the concentration gradient; (b) for nucleation, atom jumps have to be considered as to their effect upon the cluster size distribution function instead of taking local averages; and (c) cluster and precipitate size distribution functions evolve in phase space according to generalized diffusion equations.

5.4.3
Enhanced Diffusion Paths

Whereas a homogeneous and infinitely extended single crystal containing only point defects is a welcome simplification for the theoretical treatment of diffusion, real materials rarely meet such a condition. What is more, the extended defects that they are endowed with not only produce correction terms to the ideal behavior but very often they constitute fast diffusion paths (short-circuit paths) which exceed bulk diffusion by several orders of magnitude and thus influence in a qualitative and quantitative manner the diffusion properties.

Mechanisms of diffusion along extended defects are less thoroughly understood than volume diffusion, and there is often no obvious connection between the structure of a defect and the rate of diffusion within it. In any case, it is plausible that in regions of the crystal where the lattice structure is loosened or broken up atom transport may happen more easily. In experiments, defect diffusion is always intertwined with volume diffusion taking place simultaneously, and is therefore difficult to separate from it. An early treatment of this interconnection by Hart (1957) relied on random-walk arguments and resulted in the hardly surprising formula for an effective diffusion coefficient

$$D_{\text{eff}} = (1-f)D_V + fD_D \qquad (177)$$

with a volume fraction f of the defect(s), a diffusivity D_D in the defected volume, and a diffusivity D_V in the matrix volume.

Solute tends to segregate toward defects, especially high-angle grain boundaries, so that in equilibrium the concentrations in the defect and the matrix volume are related by a segregation coefficient $k = C_d/C_V$, entering the equations whenever we consider solute diffusion by defects (as distinguished from self-diffusion or tracer diffusion).

An important distinction between moving and static defects should not be overlooked. Often defects such as dislocations transport solute atoms into a matrix where they have to be distributed subsequently by volume diffusion. If the defects are at rest, a region of higher solute concentration builds up in the matrix surrounding the defect, thus hindering effectively the flow of the solute along the defect. A moving defect, on the other hand, is able to shed solute copiously into ever-fresh crystal volume, or to effect fast phase transformation where pure volume diffusion would be too slow. Driving forces exerted by concentration fields lead to chemically induced dislocation or grain boundary motion, which is also part of the mechanism of discontinuous precipitation (Purdy 1990).

5.4.3.1 Dislocation-Core Diffusion

Computer simulation studies Diffusion along dislocations (pipe diffusion) can be imagined as taking place along a cylinder with a radius of a few ångstroms where the lattice is widened, especially in edge dislocations. Experimental access to pure defect diffusion is notoriously difficult. For an explanation and a quantitative understanding of the diffusion mechanism along dislocations, atomistic calculations, and especially Monte Carlo and MD simulations, have turned out to be indispensable. MD simulations of diffusion along dissociated edge dislocations in copper were made by Huang et al. (1989, 1990), a kinetic Monte Carlo study for the same material using an EAM potential was done by Hoagland et al. (1998), and a hyper-MD simulation by Fang and Wang (2000). These works were criti-

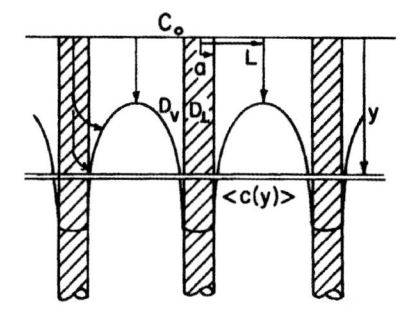

Fig. 5.44 Dislocation pipe diffusion model by Le Claire and Rabinovitch: solute is transported by volume diffusion with coefficient D_V and pipe diffusion with $D_\perp$. Dislocations are modeled as cylinders of radius a set at mutual distance $2L$ normal to the surface in a hexagonal array through which the figure shows a cut. Surface concentration is C_0, average solute concentration in a thin slice at depth y is $\langle c(y) \rangle$ (Purdy 1990).

cally discussed recently by Vegge and Jacobsen (2002), who performed a nudged elastic-band (NEB) study on the same system.

Experimental determination A well-established method of determining diffusion coefficients experimentally is to apply a thin coating of the diffusing species (often a radioactive tracer atom) to the surface of the specimen. After diffusion annealing, the specimen is cut into thin slices. Measuring the amount of diffusing species that has penetrated to depth y then allows evaluation of the diffusivity according to the appropriate solution [Section 5.4.1.1, Eq. (151)] of Fick's second law. Striving to adapt this method to the case where dislocations modify diffusion behavior, Le Claire and Rabinovitch (1981, 1982, 1983, 1984) in a series of papers developed a classical model of this situation, which is depicted in Fig. 5.44. Under the condition that at the boundaries between dislocations and matrix both the concentrations and the diffusion fluxes are continuous,[22] analytical solutions could be obtained for the two standard cases: (a) of a constant surface concentration and (b) of a thin surface film containing a specified initial amount of the diffusing species per unit area. In plots of $\langle c(y) \rangle$ vs. normalized penetration depth $\eta = y/\sqrt{Dt}$ the dislocation contribution shows up as characteristic linear "tails" as in Fig. 5.45, the slope of which is a function of the ratio of diffusion coefficients D_V (volume) and $D_\perp$ (dislocation core, i.e., pipe). Knowing D_V, the tail can therefore be used to estimate dislocation pipe diffusivity, which in the case

---

22) If we regard solute diffusion, due to the segregation factor k the continuity condition at the boundary reads $C_d = kC_V$ instead of $C_d = C_V$.

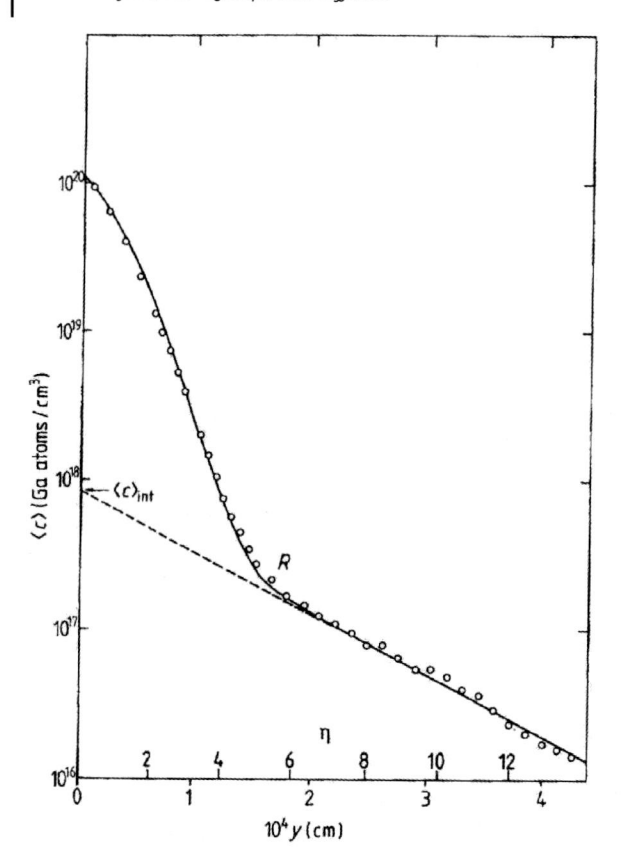

Fig. 5.45 Average concentration vs. penetration depth plot of an annealing experiment with boundary condition (b), showing diffusion of Ga atoms in a single crystal of germanium (Ahlborn 1979). The linear tail shows the contribution of dislocation pipe diffusion.

of solute diffusion contains the segregation coefficient k (Le Claire and Rabinovitch 1984).

5.4.3.2 Grain-Boundary Diffusion

Grain boundaries can be distinguished qualitatively by the amount of misorientation between the grains.[23] Symmetric tilt boundaries with up to about 15° misorientation (low-angle grain boundaries) can be represented by equivalent arrays of edge dislocations (Fig. 5.46), where the spacing λ of dislocations with Burgers vector b is connected to the tilt angle θ by $\lambda = b/[2 \sin(\theta/2)]$. The diffusion properties of these grain boundaries can be broken down into those of the corresponding dislocation pipes with the boundary diffusion coefficient $D_b = D_\perp/\lambda$ and the

---

23) For a review on grain boundaries see, for instance, Gleiter (1982).

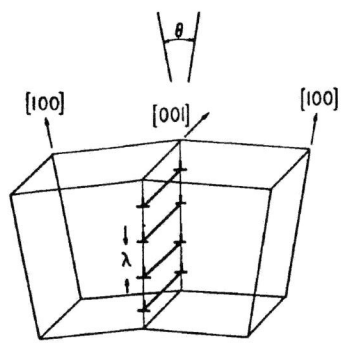

Fig. 5.46 Small-angle tilt boundary consisting of an array of edge dislocations with spacing λ. After Bocquet et al. (1983).

obvious consequence that diffusion within the boundary along the dislocation pipes (in the direction of the tilt axis) is faster than in a direction normal to them. As can be seen from these simple relationships, diffusivity in low-angle grain boundaries is generally proportional to the misorientation angle.

For larger misorientation, the cores within the grain boundaries begin to overlap, and the grain boundary structure, although periodic, can no longer be described in terms of lattice dislocations. The detailed structure has been unraveled by atomistic simulation (Vitek 1984), in some instances supported by high-resolution electron microscopy. Diffusivities in high-angle grain boundaries of a given system tend to be in the same range, especially if the orientations do not correspond to high coincidence, and as a rule of thumb the activation enthalpy may be assumed to be roughly one half of the value in the bulk phase (Peterson 1983).

As regards experimental measurement, a standard geometry of combined grain boundary/volume diffusion has been treated by Whipple (1954) after an approximate solution by Fisher (1951) according to Fig. 5.47: diffusion takes place from a surface where a constant concentration C_0 is being maintained, into a bicrystal with a grain boundary of width δ normal to the surface. Taking again the average concentration $\langle c(y) \rangle$ of a slice at distance y from the surface, the experimental results are most conveniently plotted in the form of $\langle c \rangle$ versus $y^{6/5}$ curves, in which the contribution of grain boundary diffusion is visible as linear tails, from which the grain boundary diffusion coefficient can be evaluated (Whipple 1954).

In conclusion, it can be stated that according to a distinction by Harrison (1961) short-circuit networks of dislocations or grain boundaries can contribute to diffusion in one of three regimes. (a) When the bulk diffusion penetration depth $\sqrt{Dt}$ is larger than a characteristic length L of the network (grain size or link length of a dislocation network) then the diffusion zones of neighboring short circuits overlap and Hart's formula, Eq. (177), is valid. (b) When there is bulk diffusion but the penetration depth is smaller than L the "tails" in a $\langle c \rangle$ versus y plot (cf. Fig. 5.45) can be used to deduce defect-diffusion properties. (c) When bulk diffusion is completely negligible but the penetration depth along the network is larger than

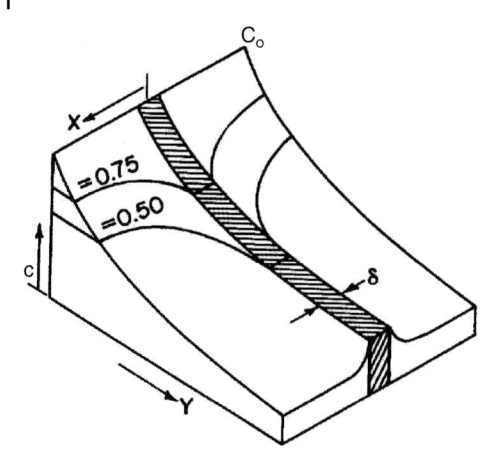

Fig. 5.47 Iso-concentration plots for diffusion into a bicrystal containing a grain boundary of width δ (Whipple 1954).

L, the concentration profile is similar to bulk diffusion with an effective diffusivity $D' \propto D_b/L^2$ instead of D_V.

5.4.3.3 Diffusion along Interfaces and Surfaces

Inner surfaces such as those along a crack or a void can offer short-circuit diffusion paths and may be treated formally along the lines of grain-boundary diffusion (cf. Fig. 5.47) where one of the grains is missing, if a reasonable value for the thickness of the surface layer can be chosen.

The diffusion properties at free surfaces play an pre-eminent role in vacuum physics and for epitaxial growth processes. At the same time, surface structures are ideally accessible by modern analytical techniques such as scanning tunneling microscopy (STM), atomic force microscopy (AFM) and updated versions of field-ion microscopy (FIM). Surface diffusion is largely determined by the geometry of the surface, which in turn depends on its orientation. Figure 5.48 shows the relevant structures of a surface near a low-index orientation.

The surface consists of terraces (T) along close-packed planes. The deviation of the surface from a close-packed plane is accommodated by ledges (L) of atomic height which themselves contain kinks (K). Single ad-atoms can be created by detaching them from a kink or a ledge or extracting them from a terrace, leaving behind an ad-vacancy. The first of these processes is by far the most likely to become thermally activated. Formation energies of both defects are comparable for most orientations, so both are contributing to mass transport at the surface to a significant extent. Migration energies for fcc crystals increase approximately with surface roughness and are therefore higher for less close-packed surfaces. It must be remembered, however, that migration is quite anisotropic. It is markedly

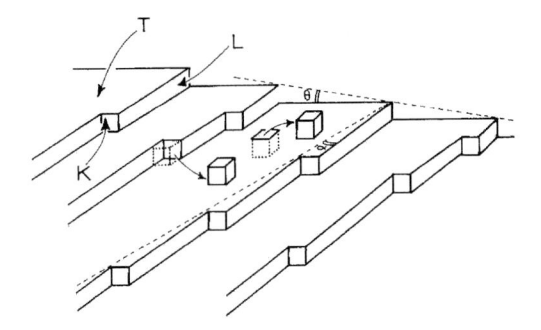

Fig. 5.48 Terrace (T)–ledge (L)–kink (K) model of a crystal surface. The processes of creating an ad-atom out of a kink or out of a terrace with the simultaneous creation of an ad-vacancy are shown. After Bocquet et al. (1983).

easier along dense rows or channels than in a direction crossing them. Above a certain temperature $T_R \approx 0.5\varepsilon/k_B$, where ε is the energy of the nearest-neighbor bond, ledges can be created freely by thermal activation and the surface becomes delocalized (roughness transition (Leamy and Gilmer 1974)). Motion of atoms at the surface takes place by single-atom jumps if the temperature is sufficiently low ($T \leq 0.15 T_m$). At higher temperatures new mechanisms come into play, such as correlated and collective jumps of multidefects, and finally evaporation–condensation-like scenarios, where the moving atom is flying over the surface.

An overview of recent developments in surface diffusion research can be found in a monograph (Tringides 1996). Experimental approaches to measuring the surface diffusion coefficient fall into either of two large classes. In one of them, the random walk of single ad-atoms is followed by direct imaging methods such as FIM and STM, and the tracer diffusion coefficient is derived from the mean square distance covered by them during time t, $D_S = \langle R^2 \rangle/(4t)$, in direct analogy to Eq. (101) for the three-dimensional case. The second large class of methods deduces surface diffusivity from the collective behavior of many atoms. Due to the interactions between ad-atoms a "collective" or interdiffusion coefficient is thus obtained [cf. Eq. (168)]. A multitude of methods falls into this class; see the table in the introductory chapter of Tringides (1996). It must be remembered in any case that surface diffusion is extremely sensitive to impurities, which must therefore be kept at a low level.

Mass transport along interfaces and surfaces is only one of the many processes by which atoms interact with surfaces and thus has many cross-references with neighboring topics like gas adsorption and desorption, growth of solids from liquid or gas phases, mixing and segregation of thin layers with substrates or of multiple layers, establishment of ordered structures, and development of nanostructures in various environments and conditions. There is an immense and steadily growing literature on all these questions to which we cannot make reference in this concise chapter on diffusion in alloys, and which should therefore be sought under the headings of surface science, nanostructures, and phase transformations.

5.5
Outlook

After about a century of research on diffusion, we can claim a fairly good understanding of diffusion mechanisms in solids provided the conditions are not too exotic. This means that established methods generally rely on aggregates of atoms whose number is large enough to apply classical statistical mechanics, and field gradients small enough to allow linearization and expansion. In the treatment of atomistic diffusion processes, some kind of adiabatic approximation such as the Born–Oppenheimer behavior of electrons with respect to the ions or the steady maintenance of equilibrium in large parts of the system, even in a transition state, is usually invoked.

Modern developments of the theory will therefore go together with the successive removal of these restrictions and the introduction of new paradigms and principles. One recent example of this evolution is the introduction of the nudged elastic band method of path optimization instead of calculating the migration profile along a fixed path. Another example is ab-initio molecular dynamics in those cases where a static profile calculation will not suffice, for instance in highly nonequilibrium situations, for weakly bound atoms at surfaces, and in catalysis.

Soaring computer power has put kinetic simulation of comparatively large, complex structures comprising millions of atoms within easy reach, and progress in ab-initio calculations delivers the necessary potentials. The path has been opened to treat diffusion in special environments such as grain boundaries, dislocations, and complex nanostructures. More economical simulation techniques for KMC further increase the ability to handle large aggregates, e.g., by exploiting statistics to find shortcuts, and more realistic jump frequencies based on a more detailed knowledge about jump paths are on the verge of being introduced (Pfeiler et al. 2005). In the mesoscale range, advanced phase field methods enable the treatment of intricate structure evolutions in systems with rich phase transformation behavior.

The application of simulation methods is already ubiquitous in present-day research, and it is to be expected that these tools will go on solving very specific diffusion problems in the emerging materials of interest, among which nanostructures and thin-layer structures seem to be most important in the near future. It is very clear that with nanoscale microstructures where the atomistic nature is beginning to be felt, continuum-based calculations must give way to customized models based on individual atom jumps, making simulation methods such as MC and MD a natural choice. Ever more realistic simulation of complex problems in realistic materials will undoubtedly dominate diffusion theory in the foreseeable future, although some surprising effects in seemingly well-understood "classical" diffusion scenarios may still lie in wait, like the unexpected sharpening of concentration profiles due to concentration dependence of the diffusion coefficient which was discovered recently (Erdélyi et al. 2004).

Whereas in past years diffusion research centered on the determination of fundamental parameters such as diffusion coefficients, activation energies, and correlation factors, attention during the last 15 years or so has shifted toward the treatment of complex processes. The topics involve multicomponent systems, diffusion flows between different phases in various shapes, and the interaction with other physical properties, e.g., diffusion under mechanical stress, under the influence of magnetic fields, and concomitant with chemical reaction. A recent conference volume (Danielewski et al. 2005) reflects this shift of research interest to a remarkable diversity of applied problems.

On the experimental side, modern development is characterized by a steady increase in new methods appearing as new acronyms in the scientific language, which give insight on diffusion processes on ever smaller spatial and temporal scales. Many of them are scattering methods with nuclear particles or radiation, such as neutron spin echo (NSE) spectroscopy, quasi-elastic neutron scattering (QNS), and nuclear resonant scattering (NRS), or are based on nuclear reaction, such as quasielastic Mössbauer scattering (QMS), positron annihilation, and perturbed angular correlation (PAC). The burgeoning field of surface science is contributing a host of modern analysis techniques which can be chosen according to the requirements of the alloy system and the boundary conditions, such as field ion microscopy (FIM), atomic force microscopy (AFM), scanning tunneling microscopy (STM), magnetic force microscopy (MFM), photoelectron energy microscopy (PEEM), scanning Auger microscopy (SAM), secondary-ion mass spectrometry (SIMS), scanning electron microscopy (SEM), and low-energy electron diffraction (LEED). Modern microscopic tools such as high-resolution transmission electron microscopy (HRTEM) offer glimpses on configuration changes and phase transformations in an unprecedented resolution and will continue to be improved. Without doubt, powerful radiation sources such as high-flux reactors or synchrotrons contribute to the expansion of experimental ability. They also open a temporal window on the very initial stages of phase transformations or ordering when the scattering amplitude is still very weak (time-resolved X-ray scattering).

Owing to computer and software progress and a cross-fertilization from the biomedical sciences, tomography by X-rays or neutrons has gained a firm standing in materials science also, offering a more integral view on the evolving microstructure in three dimensions.

As a conclusion, we venture to state the view that the basic problems of diffusion in alloys may be considered as solved by now. It is in complex, not so familiar situations found in real materials, where many parameters interact, and in submicroscopic detail that exciting discoveries are to be expected in the future.

Further Reading

Point Defects

Damask, A. C., Dienes, G. J. (1963), *Point Defects in Metals*, Gordon and Breach, New York.

Peterson, N. L., Siegel, R. W. (Eds.) (1978), *Properties of Atomic Defects in Metals*, North-Holland, Amsterdam (*J. Nucl. Mater.* **69 & 70**).

Seeger, A., Schumacher, D., Schilling, W., Diehl, J. (Eds.) (1970), *Vacancies and Interstitials in Metals*, North-Holland, Amsterdam.

Ullmaier, H. (Ed.) (1991), *Atomic Defects in Metals*, Landolt-Börnstein New Series. Group III: Crystal and Solid State Physics, Vol. 25, Springer, Berlin.

Wollenberger, H. (1996), in: *Physical Metallurgy*, Cahn, R. W., Haasen, P. (Eds.), Elsevier Science BV, Oxford, Chapter 18.

Atomic Diffusion

Allnatt, A. R., Lidiard, A. B. (1993), *Atomic Transport in Solids*, Cambridge University Press, Cambridge.

Balluffi, R. W., Allen, S. M., Carter, W. C. (2005), *Kinetics of Materials*, John Wiley & Sons, Hoboken, NJ.

Glicksman, M. E. (2000), *Diffusion in Solids: Field Theory, Solid-State Principles, and Applications*, John Wiley & Sons, New York.

Mehrer, H. (Ed.) (1990), *Diffusion in Solid Metals and Alloys*, Landolt-Börnstein New Series. Group III: Crystal and Solid State Physics, Vol. 26, Springer, Berlin.

Philibert, J. (1991), *Atom Movements – Diffusion and Mass Transport in Solids*, Les Editions de Physique, Les Ulis.

Shewmon, P. (1989), *Diffusion in Solids*, 2nd ed., TMS, Warrendale.

References

Ahlborn, K. (1979), *J. Phys.* (Paris) **40** (Coll. C6), 185.

Alnatt, A. R., Rice, S. A. (1960), *J. Chem. Phys.* **33**, 573.

Allen, S. M., Cahn, J. W. (1976), *Acta Metall.* **24**, 425.

Allnatt, A. R., Lidiard, A. B. (1993), *Atomic Transport in Solids*, Cambridge University Press, Cambridge, Chapter 3.

Aoki, K., Izumi, O. (1975), *Phys. Stat. Sol. (a)* **32**, 657.

Athènes, M., Bellon, P., Martin, G., Haider, F. (1996), *Acta Mater.* **44**, 4739.

Athènes, M., Bellon, P., Martin, G. (1997), *Phil. Mag. A* **76**, 565.

Badura-Gergen, K., Schaefer, H.-E. (1997), *Phys. Rev. B* **56**, 3032.

Bakker, H. (1968), *Phys. Stat. Sol.* **28**, 569.

Bakker, H. (1979), *Phil. Mag. A* **40**, 525.

Bakker, H. (1984), in: *Diffusion in Crystalline Solids*, Murch, G. E., Nowick, A. S. (Eds.), Academic Press, New York, p. 189.

Bakker, H. (1987), *Mater. Sci. Forum* **15–18**, 1155.

Bakker, H., van Winkel, A. (1980), *Phys. Stat. Sol. (a)* **61**, 543.

Bakker, H., Westerveld, J. P. A. (1988), *Phys. Stat. Sol. (b)* **145**, 409.

Bakker, H., Stolwijk, N. A., van der Meij, L., Zuurendonk, T. J. (1976), *Nucl. Metall.* **20**, 96.

Balluffi, R. W. (1978), *J. Nucl. Mater.* **69–70**, 240.

Bardeen, J., Herring, C. (1952), in: *Imperfections in Nearly Perfect Crystals*, Shockley, W. (Ed.), John Wiley & Sons, New York, p. 261.

Bauerle, J. E., Koehler, J. S. (1957), *Phys. Rev.*, **107**, 1493.

Beeler, J. R. (1965), *Phys. Rev. A* **138**, 1259.

Belashchenko, K. D., Vaks ,V. G. (1998), *J. Phys.: Cond. Matter* **10**, 1965.

Belashchenko, K. D., Dobretsov, V. Yu., Pankratov, I. R., Samolyuk, G. D., Vaks,

V. G. (1999), *J. Phys.: Cond. Matter* **11**, 10 593.
Belova, I. V., Murch, G. E. (1999), *Defect Diffusion Forum* **177–178**, 59–68.
Binder, K., Heermann, D. W. (1985), in: *Scaling Phenomena in Disordered Systems*, Pinn, R., Skjeltorp, T. (Eds.), Plenum Press, New York, pp. 207–230.
Blewitt, T. H., Coltman, R. R., Klabunde, C. E., Noggle, T. S. (1957), *J. Appl. Phys.*, **28**, 639.
Bocquet, J. L., Brébec, G., Limoge, Y. (1983), Diffusion in metals and alloys, in: *Physical Metallurgy*, Cahn, R. W., Haasen, P. (Eds.), North-Holland, Amsterdam, p. 385.
Bose, A., Frohberg, G., Wever, H. (1979), *Phys. Stat. Sol.* **52**, 509.
Bragg, W. L., Williams, E. J. (1934), *Proc. Roy. Soc.* (London) **A145**, 699.
Bragg, W. L., Williams, E. J. (1935), *Proc. Roy. Soc.* (London) **A151**, 540.
Breuer, J., Sommer, F., Mittemeijer, E. J. (2002), *Phil. Mag. A* **82**, 479.
Brinkman, H. C. (1956), *Physica* **22**, 149.
Brown, A. M., Ashby, M. F. (1980), *Acta Metall.* **28**, 1085.
Cahn, J. W., Hilliard, J. E. (1958), *J. Chem. Phys.* **28**, 259.
Carslaw, H. S., Jaeger, H. C. (1984), *Conduction of Heat in Solids*, 2nd ed., Oxford University Press, Oxford.
Čermák, J., Mehrer, H. (1994), *Acta Metall. Mater.* **42**, 1345.
Chandrasekhar, S. (1943), *Rev. Mod. Phys.* **15**, 1.
Chang, Y. A., Neumann, J. P. (1983), *Prog. Solid State Chem.* **14**, 221.
Chen, L. Q., Wang, Y. Z., Khachaturyan, A. G. (1994), in: *Statics and Dynamics of Alloy Phase Transformations*, Turchi, P. E. A., Gonis, A. (Eds.), Plenum Press, New York, p. 587.
Cheng, C. Y., Wynblatt, P. P., Dorn, J. E. (1967), *Acta Metall.* **15**, 1045.
Compaan, K., Haven, Y. (1956), *Trans. Faraday Soc.* **52**, 786; and (1958), *Trans. Faraday Soc.* **54**, 1498.
Cook, H. E., de Fontaine, D., Hilliard, J. E. (1969), *Acta Metall.* **17**, 765.
Corbett, J. W., Denney, J., Fiske, M. D., Walker, R. M. (1957), *Phys. Rev.* **108**, 954.
Corbett, J. W., Smith, R. B., Walker, R. M. (1959), *Phys. Rev.*, **114**, 1452.

Cottrell, A. H. (1995), *Intermetallics* **3**, 341.
Crank, J. (1994), *The Mathematics of Diffusion*, 2nd ed., Clarendon Press, Oxford.
Damask, A. C., Dienes, G. J. (1963), *Point Defects in Metals*, Gordon and Breach, New York.
Danielewski, M., Filipek, R., Kozubski, R., Kucza, W., Zieba, P., Zurek, Z. (Eds.) (2005), *Diffusion in Materials DIMAT 2004* (Proceedings of the 6th International Conference on Diffusion in Materials, Cracow, Poland), Trans Tech Publ., Uetikon-Zürich; *Defect and Diffusion Forum* **237–240**.
Daw, M. S., Baskes, M. I. (1984), *Phys. Rev. B* **29**, 6443.
Daw, M. S., Foiles, S. M., Baskes, M. I. (1993), *Mater. Sci. Rep.* **9**, 251.
Debiaggi, S. B., Decorte, P. M., Monti, A. M. (1996), *Phys. Stat. Sol. (b)* **195**, 37.
de Fontaine, D. (1975), *Acta Metall.* **23**, 553.
de Koning, M., Miranda, C. R., Antonelli, A. (2002), *Phys. Rev. B* **66**, 104110.
de Novion, C. (1995), in: *Intermetallic Compounds: Principles and Practice*, Westbrook, J. H., Fleischer, R. L. (Eds.), Vol. 1, *Principles*, John Wiley & Sons, Chichester, p. 559.
Dickman, J. E., Jeffery, R. N., Gustavson, D. R. (1977), *Phys. Rev. B* **16**, 3334.
Dickman, J. E., Jeffery, R. N., Gustavson, D. R. (1978), *J. Nucl. Mater.* **69–70**, 604.
Divinski, S. V., Frank, St., Södervall, U., Herzig, Chr. (1998), *Acta Mater.* **46**, 4369.
Divinski, S. V., Frank, S., Södervall, U., Herzig, C. (2001), *Defect Diffusion Forum* **194–199**, 487.
Domian, H. A., Aaronson H. I. (1964), *Trans. AIME* **230**, 44.
Donaldson, A. T., Rawlings, R. D. (1976), *Acta Metall.* **24**, 285.
Einstein, A. (1905), *Ann. Phys.* **17**, 549.
Elcock, E. W., McCombie, C. W. (1958), *Phys. Rev.* **109**, 605.
Ellner, M., Best, K. J., Jacobi, H., Schubert, K. (1969), *J. Less-Common Metals* **19**, 294.
Erdélyi, Z., Sladecek, M., Stadler, L. M., Zizak, I., Langer, G. A., Kis-Varga, M., Beke, D. L., Sepiol, B. (2004), *Science* **306**, 1913.
Erginsoy, C., Vineyard, G. H., Englert, A. (1964), *Phys. Rev.* **133**, A595.
Eshelby, J. D. (1955), *Acta Metall.* **3**, 487.
Eshelby, J. D. (1956), *Sol. State Phys.* **3**, 79.

Fang, Q. F., Wang, R. (2000), *Phys. Rev. B* **62**, 9317.

Farkas, D., Soulé des Bas, B. (2001), *Mater. Res. Soc. Symp. Proc.* **646**, N6.7.1.

Feder, R., Nowick, A. S. (1958), *Phys. Rev.* **109**, 1959.

Finnis, M. W., Sinclair, J. E. (1984), *Phil. Mag. A* **50**, 45.

Fisher, J. C. (1951), *J. Appl. Phys.* **22**, 74.

Frank, St., Södervall, U., Herzig, Chr. (1995), *Phys. Stat. Sol. (b)* **191**, 45.

Frank, St., Södervall, U., Herzig, Chr. (1997), *Intermetallics* **5**, 221.

Frank, St., Divinski, S. V., Södervall, U., Herzig, Chr. (2001), *Acta Mater.* **49**, 1399.

Fukai, Y. (2003), *J. Alloys Comp.* **356–357**, 263.

Fukai, Y., Okuma, N. (1993), *Jpn. J. Appl. Phys.* **32**, L1256.

Ghate, P. B. (1964), *Phys. Rev. A* **113**, 1167.

Gibson, J. B., Goland, A. N., Milgram, M., Vineyard, G. H. (1960), *Phys. Rev.* **120**, 1229.

Gilder, H. M., Lazarus, D. (1975), *Phys. Rev. B* **11**, 4916.

Glasstone, S., Laidler, K. J., Eyring, H. (1941), *The Theory of Rate Processes*, McGraw-Hill, New York.

Gleiter, H. (1982), *Mater. Sci. Eng.* **52**, 91.

Glicksman, M. E. (2000), *Diffusion in Solids*, John Wiley & Sons, New York.

Gupta, D., Lieberman, D. S. (1971), *Phys. Rev. B* **4**, 1070.

Gupta, D., Lazarus, D., Lieberman, D. S. (1967), *Phys. Rev.* **153**, 863.

Hagen, M., Finnis, M. W. (1998), *Phil. Mag. A* **77**, 447.

Hancock, G. F. (1971), *Phys. Stat. Sol. (a)* **7**, 535.

Hancock, G. F., McDonnell, B. R. (1971), *Phys. Stat. Sol. (a)* **4**, 143.

Harrison, L. G. (1961), *Trans. Farad. Soc.* **57**, 1191.

Hart, E. (1957), *Acta Metall.* **5**, 597.

Hasaka, M., Oki, K., Eguchi, T. (1973), *J. Japan Inst. Metals* **37**, 1101.

Hayashi, E., Kurokawa, Y., Fukai, Y. (1998), *Phys. Rev. Lett.* **80**, 5588.

Hellner, E. (1950), *Z. Metallk.* **41**, 480.

Herzig, C. (1993), *Defect Diffusion Forum* **95–98**, 203.

Heumann, T., Stüer, H. (1966), *Phys. Stat. Sol.* **15**, 95.

Hilgedieck, R., Herzig, Chr. (1983), *Z. Metallk.* **74**, 38.

Hillert, M. (1961), *Acta Metall.* **9**, 525.

Hirth, J. P., Lothe, J. (1982), *Theory of Dislocations*, 2nd ed., John Wiley & Sons, New York, Section 12–3.

Hoagland, R., Voter, A., Foiles, S. (1998), *Scr. Mater.* **39**, 589.

Hood, G. M. (1993), *Defect Diffusion Forum* **95–98**, 755.

Hoshino, K., Rothman, S. J., Averback, R. S. (1988), *Acta Metall.* **36**, 1271.

Huang, J., Meyer, M., Pontikis, V. (1989), *Phys. Rev. Lett.* **63**, 628.

Huang, J., Meyer, M., Pontikis, V. (1990), *Phys. Rev. B* **42**, 5495.

Huntington, H. B. (1953), *Phys. Rev.* **91**, 1092.

Huntington, H. B., Ghate, O. B. (1962), *Phys. Rev. Lett.* **8**, 421.

Huntington, H. B., Seitz, F. (1942), *Phys. Rev.* **61**, 315.

Huntington, H. B., Miller, N. C., Nerses, V. (1961), *Acta Metall.* **9**, 749.

Iida, T., Yamazaki, Y., Kobayashi, T., Iijima, Y., Fukai, Y. (2005), *Acta Mater.* **53**, 3083.

Ikeda, T., Numakura, H., Koiwa, M. (1998), *Acta Mater.* **46**, 6605; erratum (1999), *Acta Mater.* **47**, 1993.

Ipser, H., Semenova, O., Krachler, R. (2002), *J. Alloys Comp.* **338**, 20.

Issro, Ch., Abes, M., Püschl, W., Sepiol, B., Pfeiler, W., Rogl, P. F., Schmerber, G., Soffa, W. A., Kozubski, R., and Pierron-Bohnes, V. (2006), in: *Proceedings of the 4th International Alloy Conference*, Met. Mat. Trans. A. **37A**, 3415.

Johnson, R. A. (1964), *Phys. Rev.* **134**, A1329.

Johnson, R. A. (1965), *J. Phys. Chem. Solids* **26**, 75.

Johnson, R. A. (1973), *J. Phys. F* **3**, 295.

Johnson, R. A., Brown, E. (1962), *Phys. Rev.* **127**, 446.

Kaisermayr, M., Combet, J., Ipser, H., Schicketanz, H., Sepiol, B., Vogl, G. (2000), *Phys. Rev. B* **61**, 12038.

Kaisermayr, M., Combet, J., Ipser, H., Schicketanz, H., Sepiol, B., Vogl, G. (2001), *Phys. Rev. B* **63**, 054303.

Kao, C. R., Chang, Y. A. (1993), *Intermetallics* **1**, 237.

Khachaturyan, A. G. (1978), *Prog. Mater. Sci.* **22**, 1.

Kikuchi, R. (1951), *Phys. Rev.* **81**, 988.

Kikuchi, R. (1966), *Prog. Theor. Phys. Suppl.* **35**, 1.

Kikuchi, R., Sato, H. (1969), *J. Chem. Phys.* **51**, 161.
Kikuchi, R., Sato, H. (1970), *J. Chem. Phys.* **53**, 2702.
Kikuchi, R., Sato, H. (1972), *J. Chem. Phys.* **57**, 4962.
Kim, S. M. (1988), *J. Phys. Chem. Solids* **49**, 65.
Kim, S. M. (1991), *J. Mater. Res.* **6**, 1455.
Kim, S. M. (1992), *Acta Metall. Mater.* **40**, 2793.
Kim, S., Chang, Y. A. (2000), *Metall. Mater. Trans. A* **31**, 1519.
Kittel, C. (1996), *Introduction to Solid State Physics*, 7th ed., John Wiley & Sons, New York, p. 57.
Koehler, J. S., Seitz, F., Bauerle, J. E. (1957), *Phys. Rev.*, **107**, 1499.
Koiwa, M. (1992), in: *Ordered Intermetallics – Physical Metallurgy and Mechanical Behavior*, Liu, C. T., Cahn, R. W., Sauthoff, G. (Eds.), Kluwer Academic, Dordrecht, p. 449.
Korzhavyi, P. A., Ruban, A. V., Lozovoi, A. Y., Vekilov, Y. K., Abrikosov, I. A., Johansson, B. (2000), *Phys. Rev. B* **61**, 6003.
Kozlowski, M., Kozubski, R., Pierron-Bohnes, V., Pfeiler, W. (2005), *Comput. Mater. Sci.* **33**, 287–295.
Kozubski, R. (1997), *Prog. Mater. Sci.* **41**, 1.
Kozubski, R., Pfeiler, W. (1996), *Acta Mater.* **44**, 1573.
Krachler, R., Ipser, H. (1999), *Intermetallics* **7**, 141.
Kramers, H. A. (1940), *Physica* **7**, 264.
Kurita, N., Numakura, H. (2004), *Z. Metallk.* **95**, 876.
Kushida, A., Tanaka, K., Numakura, H. (2003), *Mater. Trans.* **44**, 59.
Le Claire, A. D. (1958), *Phil. Mag.* **3**, 921.
Le Claire, A. D. (1978), *J. Nucl. Mater.* **69–70**, 82.
Le Claire, A. D. (1993), *Defect Diffusion Forum* **95–98**, 19.
Le Claire, A. D., Lidiard, A. B. (1956), *Phil. Mag.* **1**, 518.
Le Claire, A. D., Rabinovitch, A. (1981), *J. Phys. C: Solid State Phys.* **14**, 3863.
Le Claire, A. D., Rabinovitch, A. (1982), *J. Phys. C: Solid State Phys.* **15**, 3455.
Le Claire, A. D., Rabinovitch, A. (1983), *J. Phys. C: Solid State Phys.* **16**, 2087.
Le Claire, A. D., Rabinovitch, A. (1984), *J. Phys. C: Solid State Phys.* **17**, 991.
Leamy, H. J., Gilmer, G. H. (1974), *J. Cryst. Growth* **24/25**, 499.

Lifshitz, I. M., Slyozov, V. V. (1961), *J. Phys. Chem. Solids* **19**, 35.
Lucasson, A. D., Walker, R. M. (1962), *Phys. Rev.* **127**, 485.
Lutze-Birk, A., Jacobi, H. (1975), *Scr. Metall.* **9**, 761.
Maeda, S., Tanaka, K., Koiwa, M. (1993), *Defect Diffusion Forum* **95–98**, 855.
Maier, K., Mehrer, H., Lessmann, E., Schüle, W. (1976), *Phys. Stat. Sol. (b)* **78**, 689.
Manley, O. P. (1960), *J. Phys. Chem. Solids* **13**, 244.
Manning, J. R. (1959), *Phys. Rev.* **113**, 1445.
Manning, J. R. (1964), *Phys. Rev.* **136**, 1758.
Manning, J. R. (1968), *Diffusion Kinetics for Atoms in Crystals*, Van Nostrand, Princeton, NJ.
Manning, J. R. (1971), *Phys. Rev. B* **4**, 1111.
March, N. H. (1978), *J. Nucl. Mater.* **69 & 70**, 490.
Martin, G. (1990), *Phys. Rev. B* **41**, 2279.
Marx, J. W., Cooper, H. G., Henderson, J. W. (1952), *Phys. Rev.*, **88**, 106.
Matano, C. (1933), *Japan. J. Phys.* **8**, 109.
Mayer, J., Fähnle, M. (1997), *Acta Mater.* **45**, 2207.
Meechan, C. J., Brinkman, J. A. (1954), *Phys. Rev.* **103**, 1193.
Mehrer, H. (1978), *J. Nucl. Mater.* **69 & 70**, 38.
Meyer, B., Fähnle, M. (1999), *Phys. Rev. B* **59**, 6072; erratum (1999), *Phys. Rev. B* **60**, 717.
Mills, G., Jónsson, H., Schenter, G. (1995), *Surf. Sci.* **324**, 305.
Mishin, Y., Farkas, D. (1997), *Phil. Mag. A* **75**, 187.
Mishin, Y., Farkas, D. (1998), *Scr. Mater.* **39**, 625.
Mishin, Y., Herzig, Chr. (2000), *Acta Mater.* **48**, 589.
Mishin, Y., Sørensen, M. R., Voter, A. F. (2001a), *Phil. Mag. A* **81**, 2591.
Mishin, Y., Mehl, M. J., Papaconstantopoulos, D. A., Voter, A. F., Kress, J. D. (2001b), *Phys. Rev. B* **63**, 224106.
Mishin, Y., Lozovoi, A. Y., Alavi, A. (2003), *Phys. Rev. B* **67**, 014 201.
Miyazaki, T., Kozakai, T., Schoen, C. G. (2005), in: *Phase Transformations in Inorganic Solids (PTM 2005)*, The Minerals, Metals and Materials Society, Warrendale, PA, p. 271.
Miyazaki, T., Imamura, H., Kozakai, T. (1982), *Mater. Sci. Engng.* **54**, 9.

Mohri, T. (1994), in: *Statistics and Dynamics of Alloy Phase Transformations*, Turchi, E. A., Gonis, A. (Eds.), NATO ASI Series, Plenum Press, New York, p. 665.

Mullen, J. G. (1961), *Phys. Rev.* **124**, 1723.

Müller, M. A., Sprengel, W., Major, T., Schaefer, H.-E. (2001), *Mater. Sci. Forum* **363–365**, 85.

Mundy, J. N. (1987), *Phys. Stat. Solidi (b)* **144**, 233.

Murakami, Y., Kachi, Y. (1982), *J. Japan Inst. Metals* **46**, 8.

Murch, G. E. (1975), *J. Nucl. Mater.* **57**, 239.

Murch, G. E. (1984), in: *Diffusion in Crystalline Solids*, Murch, G. E., Nowick, A. S. (Eds.), Academic Press, New York, p. 379.

Murch, G. E. (2001), Diffusion kinetics in solids, in *Phase Transformations in Materials*, Kostorz, G. (Ed)., Wiley-VCH, Weinheim, p. 195.

Murch, G. E., Thorn, R. J. (1978), *Phil. Mag. A* **38**, 125.

Nakajima, H, Nosé. Y., Terashita, N., Numakura, H. (2005), *Defect and Diffusion Forum* **237–240**, 7.

Neumann, J. P. (1980), *Acta Metall.* **28**, 1165.

Nosé, Y., Ikeda, T., Nakajima, H., Numakura, H. (2005), *Defect and Diffusion Forum* **237–240**, 450.

Nowick, A. S. (1978), *J. Nucl. Mater.* **69–70**, 215.

Nowick, A. S, Berry, B. S. (1972), *Anelastic Relaxation in Crystalline Solids*, Academic Press, New York.

Nowick, A. S., Heller, W. R. (1965), *Adv. Phys.* **14**, 101.

Numakura, H. (2003), *Solid State Phenom.* **89**, 93.

Numakura, H. (2006), in: *Diffusion Study in Japan 2006*, Iijima, Y. (Ed.), Research Signpost, Kerala, Chapter 13.

Numakura, H., Ikeda, T., Koiwa, M., Almazouzi, A. (1998), *Phil. Mag. A* **77**, 887.

Numakura, H., Kurita, N., Koiwa, M., Gadaud, P. (1999), *Phil. Mag. A* **79**, 943.

Numakura, H., Kashiwazaki, K., Yokoyama, H., Koiwa, M. (2000), *J. Alloys Comp.*, **310**, 344.

Numakura, H., Ikeda, T., Nakajima, H., Koiwa, M. (2001), *Mater. Sci. Engng. A* **312**, 109.

Odette, G. R., Lucas, G. E. (2001), *JOM* **53**(7), 18.

Oramus, P., Kozubski, R., Cadeville, M. C., Pierron-Bohnes, V., Pfeiler, W. (2000), *Solid State Phenom.* **72**, 209.

Peterson, N. (1978), *J. Nucl. Mater.* **69–70**, 3.

Peterson, N. L. (1983), *Int. Met. Rev.* **28**, 65.

Pfeiler, W. (1997), in: *Properties of Complex Inorganic Solids*, Gonis, A., Meike, A., Turchi, E. A. (Eds), Plenum Press, New York, p. 219.

Pfeiler, W. (2000), *JOM* **52**, 14.

Pfeiler, W., Sprusil, B. (2002), *Mater. Sci. Eng. A* **324**, 34.

Pfeiler, W., Püschl, W., Podloucky, R. (2004), *J. Mater. Sci.* **39**, 3877–3887.

Pfeiler, W., Vogtenhuber, D., Houserova, J., Wolf, W., Pike, L. M., Chang, Y. A., Liu, C. T. (1997), *Acta Metall.* **45**, 3709.

Podloucky, R., Püschl, W. (2005), *Mater. Res. Soc. Symp. Proc.* **842**, S5.28.1–6.

Purdy, G. R. (1990), in: *Diffusion in Materials*, NATO ASI Series, Laskar, A. L., Bocquet, J. L., Brebec, G., Monty, C. (Eds.), Kluwer Academic Publishers, Dordrecht, p. 309.

Reith, D., Püschl, W., Pfeiler, W., Haider, F., Soffa, W. A. (2005), in: *Phase Transformations in Inorganic Solids* (PTM 2005), The Minerals, Metals and Materials Society, Warrendale, PA, p. 765.

Rempel, A. A., Sprengel, W., Blaurock, K., Reichle, K. J., Major, J., Schaefer, H.-E. (2002), *Phys. Rev. Lett.* **89**, 185501.

Ren, X., Otsuka, K. (2000), *Phil. Mag. A* **80**, 467.

Rice, S. A. (1958), *Phys. Rev.* **112**, 804.

Richards, M. J., Cahn, J. W. (1971), *Acta Metall.* **19**, 1263.

Robrock, K.-H. (1990), *Mechanical Relaxation of Interstitials in Irradiated Metals*, Springer, Berlin.

Rudnick, J., Gaspari, G. (2004), *Elements of the Random Walk*, Cambridge University Press, Cambridge.

Sato, H., Akbar, S. A., Murch, G. E. (1985), in: *Diffusion in Solids: Recent Developments*, Dayananda, M. A., Murch, G. E. (Eds.), The Metallurgical Society, Warrendale, p. 67.

Schaefer, H.-E. (1982), in: *Positron Annihilation*, Coleman, P. G., Sharma, S. C., Diana, L. M. (Eds.), North-Holland, Amsterdam, p. 369.

Schaefer, H.-E. (1987), *Phys. Stat. Sol. (a)* **102**, 47.

Schaefer, H.-E., Schmid, G. (1989), *J. Phys. Cond. Matter* **1**, SA 49.

Schaefer, H.-E., Frenner, K., Würschum, R. (1999), *Phys. Rev. Lett.* **82**, 948.
Schapink, F. W. (1968), *Scripta Metall.* **2**, 635.
Schapink, F. W. (1969), *Statistical Thermodynamics of Vacancies in Binary Alloys*, PhD thesis, Delft University of Technology, Delft.
Schilling, W. (1978), *J. Nucl. Mater.* **69–70**, 465.
Schoeck, G. (1980), Thermodynamics and thermal activation of dislocations, in: *Dislocations in Solids*, Nabarro, F. R. N. (Ed.), Vol. 3, North-Holland, Amsterdam, p. 65.
Schott, V, Fähnle, M. (1997), *Phys. Stat. Sol. (b)* **204**, 617.
Schweiger, H., Podloucky, R., Püschl, W., Pfeiler, W. (2001), *Mater. Res. Soc. Symp. Proc.* **646**, N5.11.1–6.
Seeger, A., Mann, E., v. Jan, R. (1962), *J. Phys. Chem. Solids* **23**, 639.
Sepiol, B., Meyer, A., Vogl, G., Franz, H., Rüffer, R. (1998a), *Phys. Rev. B* **57**, 10 433.
Sepiol, B., Czihak, C., Meyer, A., Vogl, G., Metghe, J., Rüffer, R. (1998b), *Hyperfine Interact.* **113**, 449.
Siegel, R. W. (1978), *J. Nucl. Mater.* **69–70**, 117.
Siegel, R. W. (1982), in: *Positron Annihilation*, Coleman, P. G., Sharma, S. C., Diana, L. M. (Eds.), North-Holland, Amsterdam, p. 351.
Simmons, R. O., Balluffi, R. W. (1960), *Phys. Rev.* **117**, 52.
Slater, N. B. (1959), *Theory of Unimolecular Reactions*, Cornell University Press, Ithaca, NY.
Smigelskas, A. D., Kirkendall, E. O. (1947), *Trans. AIME* **171**, 130.
Smith, R. P. (1966), *Trans. Amer. Inst. Min. Engrs.* **236**, 1224.
Soffa, W. A., Püschl, W., Pfeiler, W. (2003), *Intermetallics* **11**, 161.
Sprengel, W., Müller, M. A., Schaefer, H.-E. (2002), in: *Intermetallic Compounds: Principles and Practice*, Westbrook, J. H., Fleischer, R. L. (Eds.), Vol. 3, *Progress*, John Wiley & Sons, Chichester, Chapter 15.
Stolwijk, N. A. (1981), *Phys. Stat. Sol. (b)* **105**, 223.
Stolwijk, N. A., Spruijt, T., Hoetjes-Eijkel, M. A., Bakker, H. (1977), *Phys. Stat. Sol.* **42**, 537.
Stolwijk, N. A., van Gend, M., Bakker, H. (1980), *Phil. Mag. A* **42**, 783.

Taylor, A., Doyle, N. J. (1972), *J. Appl. Crystallogr.* **5**, 201.
Thiess, H., Kaisermayr, M., Sepiol, B., Sladecek, M., Rüffer, R., Vogl, G. (2001), *Phys. Rev. B* **64**, 104305.
Thomas, L. E., Schober, T., Balluffi, R. W. (1969), *Radiat. Eff.* **1**, 257.
Thompson, M. W. (Ed.) (1969), *Defects and Radiation Damage in Metals*, Cambridge University Press, London.
Tringides, C. (Ed.) (1996), *Surface Diffusion: Atomistic and Collective Processes*, NATO ASI Series, Plenum, New York.
Tsallis, C. (1988), *J. Statist. Phys.* **52**, 479.
Ullmaier, H. (Ed.), *Atomic Defects in Metals*, Landolt-Börnstein New Series, *Group III: Crystal and Solid State Physics*, Vol. 25, Springer, Berlin 1991, Chapter 2.
Tsallis, C. (2003), in: *Nonextensive Statistical Mechanics and its Application*, Abe, S., Okamoto, Y. (Eds.), Springer Series Lecture Notes in Physics, Springer, Berlin, p. 3.
Vajda, P. (1977), *Rev. Mod. Phys.* **49**, 481.
Vaks, V. G. (1996), *JETP Lett.* **63**, 477.
van Ommen, A. H., de Miranda, J. (1981), *Phil. Mag. A* **43**, 387.
Vegge, T., Jacobsen, K. W. (2002), *J. Phys.: Cond. Matter* **14**, 2929.
Vineyard, G. H. (1957), *J. Phys. Chem. Solids* **3**, 121.
Vitek, V. (1984), in: *Core Structure and Physical Properties of Dislocations*, Veyssière, P., Kubin, L., Castaing, L. (Eds.), Editions du CNRS, Paris, p. 435.
Vogl, G., Sepiol, B. (1994), *Acta Metall. Mater.* **42**, 3175.
Wagner, C. (1961), *Z. Elektrochem.* **65**, 581.
Wagner, C., Schottky, W. (1931), *Z. Phys. Chem. B* **11**, 163.
Wang, T. M., Shimotomai, M., Doyama, M. (1984), *J. Phys. F: Metal Phys.* **14**, 37.
Weinkamer, R., Fratzl, P., Sepiol, B., Vogl, G. (1999), *Phys. Rev. B* **59**, 13, 8622.
Weller, M. (2001), in: *Mechanical Spectroscopy Q^{-1} 2001*, Schaller, R., Fantozzi, G., Gremaud, G. (Eds.), Trans Tech Publ., Uetikon–Zürich, Chapter 2.
Wert, C. A. (1950), *Phys. Rev.* **79**, 601.
Wert, C. A., Zener, C. (1949), *Phys. Rev.* **76**, 1169.
West, R. N. (1979), in: *Positrons in Solids*, Hautojärvi, P. (Ed.), Springer, Berlin, Chapter 3.

Whipple, R. (1954), *Phil. Mag.* **45**, 1225.
Wimmer, E., Wolf, W., Sticht, J., Saxe, P., Geller, C. B., Najafabadi, R., Young, G. A. (2006), unpublished work.
Wolff, J., Broska, A., Franz, M., Köhler, B., Hehenkamp, Th. (1997), *Mater. Sci. Forum* **255–257**, 593.
Wollenberger, H. (1996), in: *Physical Metallurgy*, Cahn, R. W., Haasen, P. (Eds.), Elsevier Science BV, Oxford, Chapter 18.
Würschum, R., Grupp, C., Schaefer, H.-E. (1995), *Phys. Rev. Lett.* **75**, 97.

Young, F. W., Jr. (1978), *J. Nucl. Mater.* **69–70**, 310.
Zener, C. (1952), in: *Imperfections in Nearly Perfect Crystals*, Shockley, W. (Ed.), Wiley, New York, p. 289.
Zhao, J.-C. (2005), *Ann. Rev. Mater. Res.* **35**, 51.
Zhao, J.-C., Zheng, X., Cahill, D. G. (2005), *Materials Today* **8**(10), 28.
Zhao, J.-C. (2006), *Progr. Mater. Sci.* **51**, 557.

6
Dislocations and Mechanical Properties

Daniel Caillard

6.1
Introduction

It has been known for a long time that metals and alloys – and more generally all crystalline materials – deform by shear along dense lattice planes and dense lattice directions. This property was easily interpreted by assuming that the shear displacement is a multiple of one of the smallest translation lattice vectors. It was also demonstrated that a global shear would require a very high stress, of the order of $\mu/2\pi$, where μ is the shear modulus, i.e., several tens of gigapascals (GPa). Under such conditions, realistic deformation stresses could be explained only by introducing dislocations.

Plastic deformation can be described in terms of dislocation motion, multiplication, annihilation, and storage. However, the multiplicity of the microscopic mechanisms involved, and the rather long-range elastic interactions between neighboring dislocations, make a complete calculation of collective dislocation behavior almost impossible. Fortunately, the mechanical properties are usually dominated by a few mechanisms, in such a way that reasonable approximations can be made yielding surprisingly nice results.

Dislocations can be introduced in any material by the Volterra process described in Fig. 6.1. The solid is cut along the surface S ending along the curved line d. The two lips are then displaced by a uniform translation b, and pasted together again. Note that the latter process generally requires the addition (or the removal) of atomic layers to maintain the continuity of the structure across the cut. All the elastic distortions are then concentrated around the line d, which is called a dislocation with Burgers vector b. In crystals, the cut surface is perfectly healed, and the corresponding dislocation is perfect, if b is a translation vector of the lattice. If b is not a translation vector a fault is created along the cut surface and the dislocation is imperfect. Figure 6.1 also shows that the motion of the dislocation increases the size of the cut surface, which increases the total amount of plastic deformation, ε. Planar dislocation motion involving only shear displacements (b parallel to the cut plane, which is also the plane of motion) is called

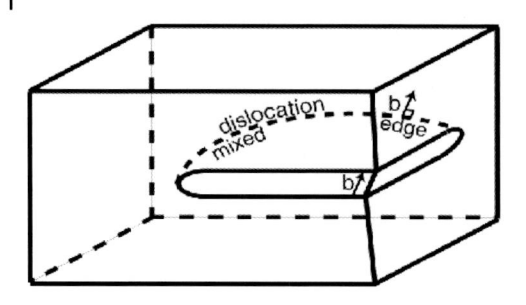

Fig. 6.1 Creation of a dislocation d by the Volterra process.

"glide," whereas motion involving addition or removal of matter is called "climb" (b perpendicular to the plane of motion) or "mixed climb" (intermediate situations).

The plastic strain rate, projected in the plane of dislocation motion and along the Burgers vector direction, $\dot{\varepsilon}$, can then be expressed by the Orowan law [Eq. (1)], where ρ is the dislocation density (in m^{-2}), b the Burgers vector, and v the average velocity of the moving dislocations.

$$\dot{\varepsilon} = \rho b v \qquad (1)$$

The determination of dislocation velocities is the key point of the study of plasticity mechanisms. Dislocation velocity is controlled by two types of obstacles.

- Long-range elastic interactions with extended obstacles, such as elastic stress fields of other dislocations, precipitates etc.: generally these are alternately attractive and repulsive, in the form of an oscillating stress $\tau_i^{(loc)}$ of amplitude τ_i (Fig. 6.2a). τ_i is called the "internal stress."

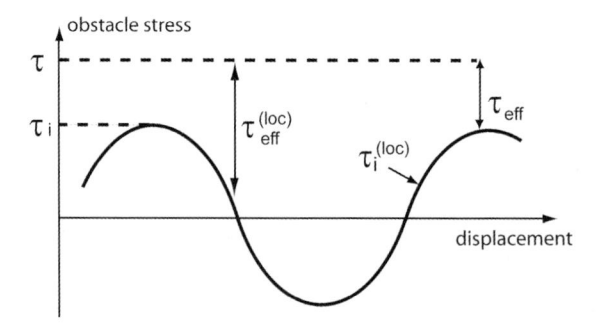

Fig. 6.2 Schematic description of the local stress as a function of displacement.

- Short-range interactions with local energy barriers: if τ is the applied stress,[1] the effective stress available to overcome these barriers is $\tau_{\mathit{eff}}^{(loc)} = \tau - \tau_i^{(loc)}$. Where the resistance against dislocation movement exerted by local stress is a maximum, a minimum velocity results. Since the average dislocation velocity is only a few times larger than its minimum value,[2] we can assume that Eq. (2) holds.

$$v = v(\tau_{\mathit{eff}}) \quad \text{with} \quad \tau_{\mathit{eff}} = \tau - \tau_i \tag{2}$$

Short-range interactions, which are largely responsible for the dependence of the deformation stress on temperature and strain-rate, are the subject of Section 6.2. Long-range interactions, which are the result of the history of the sample (thermal treatments and deformation-induced dislocation structure) are described in Section 6.3. The latter are especially important in fatigue, discussed in Section 6.4.2. Both types of interactions generally tend to decrease at increasing temperature, which accounts for the usual loss of mechanical strength at increasing temperature. However, the reverse behavior, called "stress anomaly," is sometimes observed in ordered intermetallic alloys. This will be discussed in Section 6.4.1. Other properties, such as small-size effects (Section 6.4.3) are also directly related to long-range and short-range interactions.

6.2
Thermally Activated Mechanisms

6.2.1
Introduction to Thermal Activation

Short-range interaction of dislocations with energy barriers takes place in such a small volume – a few hundreds of atoms only – that it is strongly influenced by thermal vibrations. Such a "thermal activation" helps dislocations to overcome barriers of height ranging between a fraction of an electronvolt and a few electronvolts. This results in a reduction of the effective stress τ_{eff} necessary to reach a given dislocation velocity $v(\tau_{\mathit{eff}})$ and average strain rate $\dot{\varepsilon} = \rho b v$, as the temperature increases. These short-range thermally activated processes govern almost all the temperature-dependent mechanical properties of materials. The reader can refer to Schöck (1980) for a complete description of the theory of thermal activa-

---

[1] All stresses are expressed in terms of local or average shear stresses. The applied shear stress τ can be deduced from the normal applied stress by multiplication with the Schmid, factor which is usually between 0.2 and 0.5.

[2] The proportionality coefficient is the total displacement distance divided by the distance at which the local stress is close to its minimum value.

tion, and to Caillard and Martin (2003) for a description of the thermally activated dislocation mechanisms in crystals.

Although plastic deformation is an irreversible process, because a dislocated sample is not in equilibrium, the microscopic processes of dislocation motion can be analyzed in terms of Eyring thermodynamics of viscous flow (Eyring 1936). If v is the vibration frequency of a dislocation segment held up by the stress against an energy barrier, the probability per unit time of jumping over the barrier can be expressed using a Boltzmann factor as Eq. (3), where k_B is the Boltzmann constant, T is the absolute temperature, ΔG is the variation of Gibbs energy during the process (commonly called the activation energy), and v is the attempt frequency.

$$P = v \exp\left(-\frac{\Delta G}{k_B T}\right) \tag{3}$$

Taking $v = v_D \dfrac{b}{l}$, where v_D is the Debye frequency ($v_D \approx 10^{13}$ s^{-1}) and l is the wavelength of the vibrating dislocation, and considering that the distance between obstacles in the direction of motion is y, the average dislocation velocity can be written as Eq. (4).

$$v = v_D \frac{by}{l} \exp\left(-\frac{\Delta G}{k_B T}\right) \tag{4}$$

The increase in velocity with increasing stress lies in the decrease of ΔG with increasing effective stress τ_{eff}, defined in Section 6.1 [Eq. (2)].

Defining the activation volume V as in Eq. (5), the activation energy can indeed be written as Eq. (6), where ΔG_0 is the energy barrier at zero stress, and $\tau_{\mathit{eff}} V$ is the work done by the effective stress during the crossing event. V represents b times the area swept by the dislocation during this event.

$$V = -\left(\frac{\partial \Delta G}{\partial \tau_{\mathit{eff}}}\right)_T \tag{5}$$

$$\Delta G = \Delta G_0 - \tau_{\mathit{eff}} V \tag{6}$$

The activation energy is deduced from the slope of Arrhenius plots, $\ln v = f\left(\dfrac{1}{T}\right)$. The slope is equal to the change in activation enthalpy, ΔH, which is related to ΔG by Eq. (7), where $\Delta S = -\left(\dfrac{\partial \Delta G}{\partial T}\right)_{\tau_{\mathit{eff}}}$ is the activation entropy.

$$\Delta H = \Delta G + T \Delta S, \tag{7}$$

The estimation of ΔS is a key step toward the determination of ΔG from ΔH. It is generally considered that the main contribution to ΔS is the variation of the elastic shear modulus μ with temperature. Under such conditions, Schöck (1980) has derived the following relation to deduce ΔG from ΔH:

$$\Delta G = \left[\Delta H + \frac{T}{\mu}\frac{d\mu}{dT}\tau_{eff}V\right]\left[1 - \frac{T}{\mu}\frac{d\mu}{dT}\right] \quad (8)$$

Dislocation mechanisms can in principle be identified by activation energy measurements. However, writing ΔG in the inverse form [Eq. (9)] deduced from Eqs. (1) and (4), and considering that the logarithm term ranges between 20 and 30 in the usual experiments, one can see that ΔG adjusts itself to a value depending essentially on T [Eq. (9)], whatever the controlling mechanisms.

$$\Delta G = k_B T \ln \frac{v_D b^2 \gamma \rho}{l\dot{\varepsilon}}, \quad (9)$$

It is thus usually impossible to identify dislocation mechanisms through the determination of ΔH and ΔG only. A more suitable parameter is the activation volume V, which can vary by several orders of magnitude for different controlling mechanisms.

The most important thermally activated mechanisms are now described in more detail. In this section (6.2), which is devoted to thermally activated mechanisms, the effective stress τ_{eff} is often considered equal to the total stress τ, which is justified at small deformation where the internal stress τ_i is small.

6.2.2
Interactions with Solute Atoms

6.2.2.1 General Aspects

Four domains can generally be identified in the stress–temperature curves of dilute alloys, as shown in Fig. 6.3. Domain 1, which involves inertial effects due to very high dislocation velocities, has been described by Granato (1971). Domain 2 corresponds to dislocation interaction with fixed solute atoms and will be described in Section 6.2.2.2. Domain 3 is characterized by stress instabilities (Section 6.2.2.3) and domain 4 is controlled by the dragging of solute atmospheres surrounding dislocations (Section 6.2.2.4).

The main contribution to dislocation–solute interaction is the difference in atomic volume between solvent and solute, also called the size effect. In polar coordinates, the radial elastic interaction force between a pure edge dislocation and a solute atom can be written as Eq. (10) (Haasen, 1979), where $\Delta\Omega$ is the local change of atomic volume due to a solute atom, $\theta = 0$ in the direction of the Burgers vector, $\theta = \pi/2$ in the direction perpendicular to the slip plane, and r is the distance between the solute atom and the dislocation.

$$F_{int} = \frac{\mu b}{\pi}\Delta\Omega\frac{\sin\theta}{r^2}, \quad (10)$$

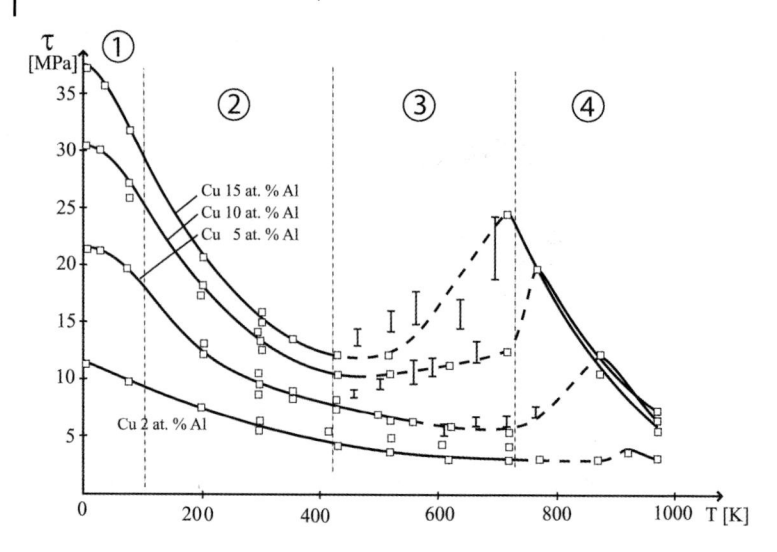

Fig. 6.3 Deformation shear stress of CuAl alloys as a function of temperature (from a review by Neuhauser and Schwink, 1993).

The maximum value of this interaction force at one interatomic distance is given by Eq. (11), where Ω is the atomic volume.

$$F_{\text{int}}^{(\max)} = \frac{\mu b^3}{\pi} \frac{\Delta \Omega}{\Omega} \tag{11}$$

6.2.2.2 Low Temperatures (Domain 2, Interaction with Fixed Solute Atoms)

When a dislocation is pinned, the force exerted on the pinning point in the isotropic approximation (Fig. 6.4) is given by Eq. (12), where T is the dislocation-line tension, equal to the increase in dislocation energy per increase of length.

$$F = 2\hat{T} \sin \alpha \tag{12}$$

The line tension is not exactly equal to the dislocation energy per unit length, for several reasons discussed by Friedel (1964) and Kocks et al. (1975). Its order of magnitude is $\hat{T} = \frac{1}{2}\mu b^2$.

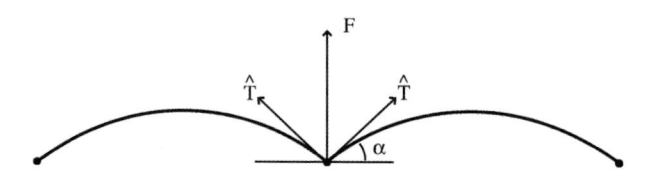

Fig. 6.4 Force exerted by a dislocation on a pinning point.

In the case of a low density of weak obstacles (small α), this force can also be written as Eq. (13), where L_F is the distance between obstacles along the dislocation line.

$$F = \tau b L_F \tag{13}$$

In this approximation by Fleischer (1961), the net force acting on L_F is transferred to the obstacle point. Combining Eqs. (12) and (13) yields Eq. (14).

$$\tau = \frac{\mu b \sin \alpha}{L_F}. \tag{14}$$

L_F is stress-dependent because more strongly curved dislocations interact with a higher density of obstacles. It has been estimated by Friedel (1964) as given by Eq. (15), where d is an in-plane average distance between obstacles and c_{ob} is their atomic concentration (number of obstacles per atom).

$$L_F = d\left(\frac{\mu b}{\tau d}\right)^{1/3} = b\left(\frac{\mu}{\tau c_{ob}}\right)^{1/3} \tag{15}$$

Note that c_{ob} can be smaller than the average concentration of solute atom, if obstacles are clusters. Combining Eqs. (12), (14), and (15) yields Eq. (16).

$$\tau = \mu \frac{b}{d}\left(\frac{F}{\mu b^2}\right)^{3/2}. \tag{16}$$

Another expression [Eq. (17)] has been proposed by Mott and Labusch (Labusch 1970; Neuhauser and Schwink 1993) for obstacles of width $w \gg b$.

$$\tau = \mu \left(\frac{b}{d}\right)^{4/3}\left(\frac{F}{\mu b^2}\right)^{4/3}\left(\frac{w}{2b}\right)^{1/3} \tag{17}$$

Several phenomenological expressions can be taken for the energy necessary to cross the small-size obstacles as a function of the force F (Kocks et al. 1975; Caillard and Martin 2003). They can be expressed as Eq. (18), where ΔG_{max} is the maximum energy at $F = 0$, and F_{max} is the force the obstacle opposes to the dislocation motion.

$$\Delta G(F) = \Delta G_{max}\left[1 - \left(\frac{F}{F_{max}}\right)^a\right]^b, \tag{18}$$

F_{max} is given approximately by Eq. (11). The exponents a and b are of the order of unity, e.g., $a = 1$ and $b = 1$ for a rectangular force–distance profile; $a = 1$ and $b = 3/2$ for a parabolic force–distance profile; and $a = 0.69$ and $b = 3/2$ for the

Cottrell–Bilby potential which correctly describes dislocation–solute atom interactions (Cottrell and Bilby 1949, Wille et al. 1987).

Combining Eqs. (16) and (18) yields Eq. (19), which is the expression proposed by Kocks et al. (1975) with $0 < 2a/3 < 1$ and $1 < b < 2$.

$$\Delta G(\tau) = \Delta G_{max} \left[1 - \left(\frac{\tau}{\tau_{max}} \right)^{2a/3} \right]^{b} \tag{19}$$

Taking Eq. (20) from Eq. (8), with Eq. (21) where T_0 is the temperature above which the obstacles are crossed under a negligible stress, (i.e., the "athermal" temperature above which the deformation stress no longer decreases), yields Eq. (22), which becomes Eq. (23) for the Cottrell–Bilby potential.

$$\Delta G(\tau) = k_B T \ln \frac{\dot{\varepsilon}_0}{\dot{\varepsilon}}, \tag{20}$$

$$\Delta G_{max} = k_B T_0 \ln \frac{\dot{\varepsilon}_0}{\dot{\varepsilon}}, \tag{21}$$

$$\frac{\tau}{\tau_{max}} = \left[1 - \left(\frac{T}{T_0} \right)^{1/b} \right]^{3/2a}. \tag{22}$$

$$\frac{\tau}{\tau_{max}} = \left[1 - \left(\frac{T}{T_0} \right)^{2/3} \right]^{2.17}. \tag{23}$$

This law has been verified experimentally in CuMn alloys by Wille et al. (1987). The Cu–Mn system has been selected because the tendency to form short-range order is thought to be negligible below 5 at.% Mn (see Flor et al. 2003), the stacking-fault energy is almost independent of the Mn concentration, and the size effect at the origin of the dislocation–solute interaction is very large. Figure 6.5 shows the critical resolved shear stresses (CRSS), in (a) as a function of temperature, and in (b) in a suitable form to check the validity of Eq. (23). The effective stress τ in (b) has been deduced from the total stress in (a) by subtracting the internal stress τ_i, assumed to be the value of the CRSS at the athermal temperature (T_0 ranges between 530 and 610 K, depending on the solute concentration; see Fig. 6.5). τ_{max} is the value of τ at 0 K. The results are in excellent agreement with Eq. (23). ΔG_{max} is estimated using Eq. (21), F_{max} is deduced from ΔG_{max} using the approximation $F_{max} = \frac{\Delta G_{max}}{w}$ where $w \approx 2.5b$ (w is the width of the obstacle), and $c_{ob} \approx \left(\frac{b}{d} \right)^2$ is deduced from Eq. (16) using the experimental values of τ_{max} and ΔG_{max}. The results indicate that the concentration of obstacles c_{ob} is 20 times smaller than the average solute concentration. In addition, the experimental value $\Delta G_{max} = 1.3$ eV is at least three times that expected from interactions with individual solute atoms. Therefore the authors conclude that the obstacles consist of doublets or triplets of solute atoms.

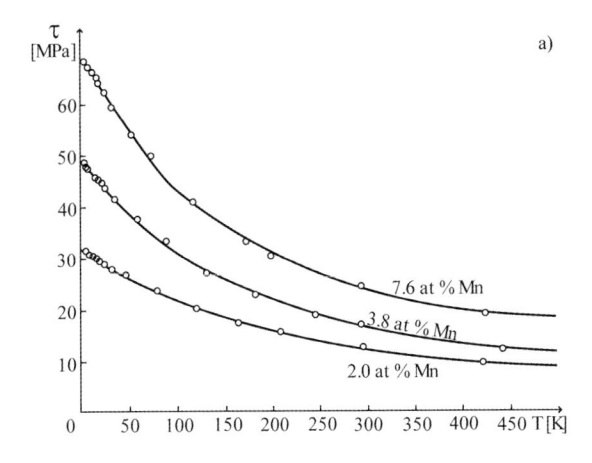

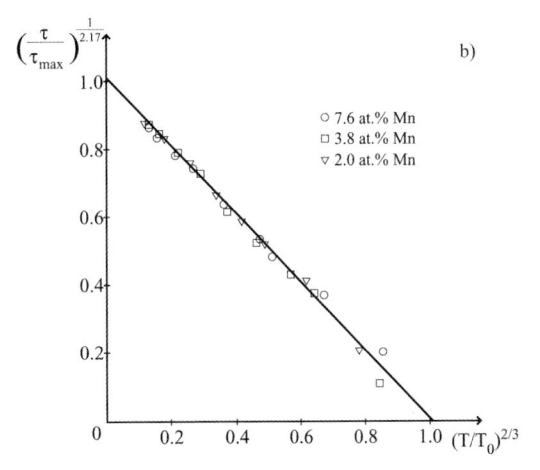

Fig. 6.5 Deformation shear stress in CuMn alloys plotted (a) as a function of temperature, and (b) to check the validity of Eq. (23). From Wille et al. (1987).

6.2.2.3 Intermediate Temperatures (Domain 3, Stress Instabilities)

As the temperature rises, solute atoms become sufficiently mobile to diffuse toward dislocations. Immobile dislocations are then surrounded by an "atmosphere" of solute atoms with local concentration c given by Eq. (24), where U_{int} is given by the integration of Eq. (10), i.e., $U_{int} = \frac{1}{\pi} \mu b \Delta \Omega \frac{\sin \theta}{r}$, and c_0 is the average solute concentration.

$$c = c_0 \exp\left(-\frac{U_{int}}{k_B T}\right) \tag{24}$$

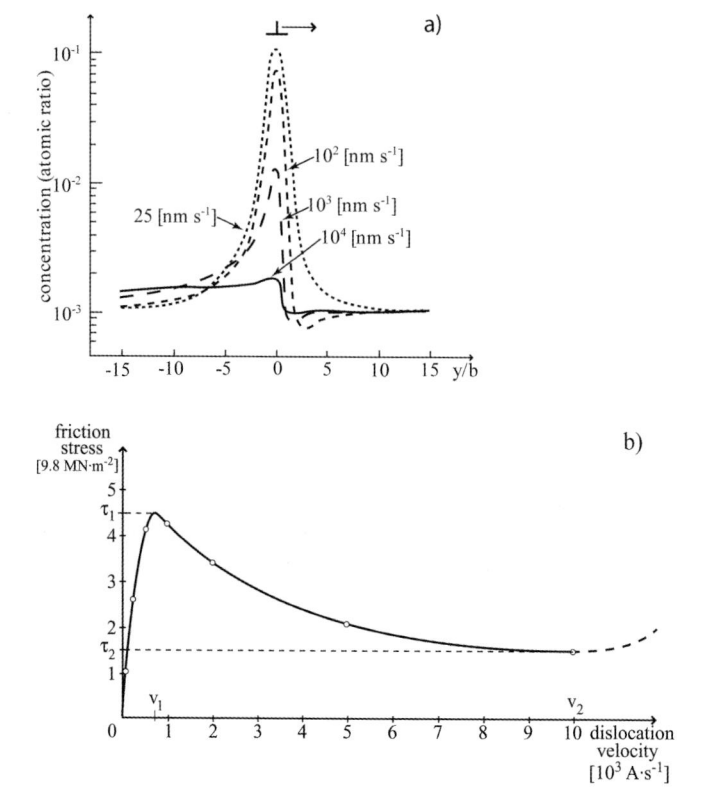

Fig. 6.6 Interaction between a moving edge dislocation and solute atoms. (a) Computed solute atom density in a plane located one interatomic distance from the slip plane (the dislocation moves to the right). (b) Corresponding friction stress. From Yoshinaga and Morozumi (1971).

The concentration c can be either larger or smaller than c_0 according to the sign of $\sin\theta$.

Moving dislocations tend to drag their atmosphere, and a dynamic equilibrium is established which depends on temperature and dislocation velocity. This equilibrium has been computed and the results are shown in Fig. 6.6(a). Figure 6.6(b) shows that the corresponding friction stress increases with dislocation velocity as long as the concentration profile remains unchanged. It decreases, however, when the density of the atmosphere decreases with increasing dislocation velocity. It increases again when the dislocation velocity is so high that solute diffusion becomes negligible (which is the case in Section 6.2.2.2).

This behavior is at the origin of stress instabilities. Figure 6.7 reproduces the variation of the velocity dependence of the friction stress plotted in Fig. 6.6(b). Let us consider a sample containing a density of mobile dislocations ρ, deformed

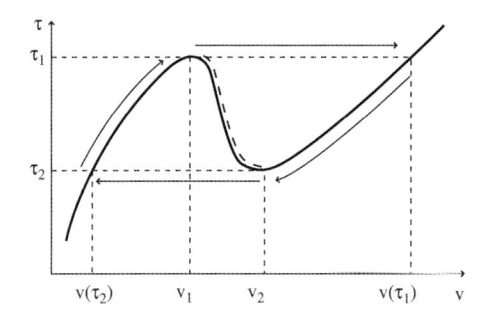

Fig. 6.7 Schematic description of the origin of stress instabilities at intermediate temperatures (domain 3).

at an imposed strain-rate $\dot{\varepsilon}$ ranging between $\rho b v_1$ and $\rho b v_2$. Figure 6.7 shows that when the stress increases to the critical value τ_1, dislocations suddenly accelerate from v_1 to $v(\tau_1) > v_2$. Since $\rho b v(\tau_1) > \dot{\varepsilon}$, the tensile machine relaxes and the applied stress decreases to the second critical value τ_2. The dislocation velocity then instantaneously decreases to $v(\tau_2) < v_1$. Since $\rho b v(\tau_2) < \dot{\varepsilon}$, the applied stress increases again and another cycle starts. This behavior leads to stress instabilities, also called the Portevin–le Châtellier (PLC) effect. Increasing the applied strain-rate increases the total time spent by the dislocations in the high-velocity regime but the average flow stress keeps oscillating between τ_1 and τ_2. This implies that the stress dependence of the strain rate must be zero.

An anomalous increase in average deformation stress with increasing temperature can even be obtained when $v_1 \approx 0$. As a matter of fact, the unlocking stress τ_1 increases with locking time (because of a more intense diffusion process), i.e., with decreasing strain-rate. This effect is called dynamic strain aging. More details on stress instabilities can be found in several reviews (see, e.g., MacCormick 1972; Van den Beukel 1975; Mulford and Kocks 1979; Estrin and Kubin 1989; Kubin and Estrin 1990).

6.2.2.4 High Temperatures (Domain 4, Diffusion-Controlled Glide)

In this temperature range, dislocation glide is controlled by the diffusion of solute atoms present in their core.

Two different approximations can be made corresponding to $|U_{\text{int}}| \gg k_B T$, and $|U_{\text{int}}| < k_B T$.

For $|U_{\text{int}}| \gg k_B T$, Eq. (24) shows that the solute cloud around the dislocation is highly asymmetrical, with a high concentration of solute atoms in the region corresponding to $U_{\text{int}} < 0$, and a weak depletion in the opposite one ($U_{\text{int}} > 0$). According to Friedel (1964, p. 410), the solute atoms are then all very close to the dislocation core, forming a row of pinning points with average distance λ (Fig. 6.8). The work done by the applied stress during the diffusion of one solute atom over one interatomic distance is $\tau b^2 \lambda$.

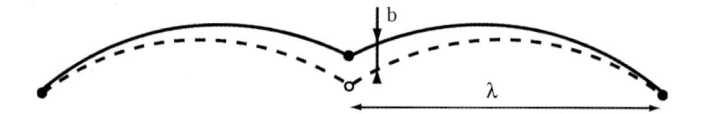

Fig. 6.8 Elementary dislocation motion controlled by the diffusion of a solute atom over the distance b.

The frequency of this event is $\frac{1}{2}\nu_D \frac{b}{\lambda} \exp{-\frac{U_d - \tau b^2 \lambda}{k_B T}}$ in the forward direction and $\frac{1}{2}\nu_D \frac{b}{\lambda} \exp{-\frac{U_d + \tau b^2 \lambda}{k_B T}}$ in the backward one, where U_d is the activation energy of solute diffusion. The dislocation velocity, accordingly, is given by

$$v = \frac{1}{2}\nu_D \frac{b^2}{\lambda} \left[\exp\left(-\frac{U_d - \tau b^2 \lambda}{k_B T}\right) - \exp\left(-\frac{U_d + \tau b^2 \lambda}{k_B T}\right)\right]$$

or, assuming that $\tau b^2 \lambda \ll k_B T$, by Eq. (25a).

$$v = \nu_D \frac{\tau b^4}{k_B T} \exp{-\frac{U_d}{k_B T}} \tag{25a}$$

This yields Eq. (25b), where $D = b^2 \nu_D \exp{-\frac{U_d}{k_B T}}$ is the diffusion coefficient of solute atoms.

$$\tau = \frac{k_B T}{b^2 D} v \tag{25b}$$

The friction force appears to be independent of the solute concentration in the dislocation core.

For $|U_{\text{int}}| < k_B T$, the calculations are more complex. As shown by Caillard and Martin (2003), the resulting friction force is however close to the value given by Eq. (25).

Equation (1), where the velocity is given by Eq. (25a), and where the mobile dislocation density is taken proportional to τ^2 (Taylor law; cf. Section 6.3.2), is considered to account for the creep properties of "class I" alloys. Class I alloys are indeed characterized by a creep rate proportional to the third power of stress, an activation energy equal to that of solute diffusion, and viscous dislocation motion (see, e.g., Takeuchi and Argon 1976). In many cases, however, the creep properties are considered too complex to be explained so easily (see, e.g., Poirier, 1976).

6.2.3
Forest Mechanism

The interaction between dislocations and other dislocations intersecting their slip plane is called the "forest" mechanism, because each intersecting dislocation behaves like a "tree". The intersection of two dislocations with Burgers vectors at right angles, which only requires the formation of a pair of jogs, is similar to the interaction between dislocations and fixed solute atoms, treated in Section 6.2.2.2, and the same equations hold. The activation energy is then given by Eq. (18), where ΔG_{max} is twice the jog energy U_j, and the activation volume is given by Eq. (5). Taking $a = b = 1$ (rectangular force–distance profile), the activation volume is $V = L_F bw$, where $w = \dfrac{\Delta G_{max}}{F_{max}}$ is the obstacle width, of the order of the dislocation dissociation width, and the activation energy is $\Delta G(\tau) = \Delta G_0 - \tau_{eff} L_F bw$. At a given strain-rate and temperature, ΔG and thus the product $\tau_{eff} L_F$ must remain constant along deformation curves. According to Eq. (15) τ_{eff} is then inversely proportional to the average dislocation distance, d. In other words, τ_{eff} can be written in the form of the Taylor law, i.e., $\dfrac{\tau_{eff}}{\mu} = \alpha_{eff} \dfrac{1}{\sqrt{\rho}}$ (cf Section 6.3.3.2), where α_{eff} depends on temperature and strain rate, but is constant along a deformation curve. The ratio $\Delta G / T$ must also remain constant, according to Eq. (9), which implies that α_{eff} at constant strain rate must decrease linearly with increasing temperature.

Experimental evidence for the forest mechanism results from the "Cottrell–Stokes" law (Cottrell and Stokes 1955) which states that effective and internal stresses both verify the Taylor law, and as a result they remain proportional to each other for various amounts of strain. These experiments and the corresponding interpretations will be described in Section 6.3.3.

6.2.4
Peierls-Type Friction Forces

6.2.4.1 The Kink-Pair Mechanism

Peierls (1940) and Nabarro (1947) were the first to remark that the dislocation energy per unit length, E, has necessarily the same periodicity as the crystal lattice. This variation can be described by a periodic "Peierls" potential with hills and valleys of amplitude ΔE and wavelength h. The dislocation core structure is determined by local atomic displacements which depend on the anisotropy of the lattice and corresponding atomic bonds. It tends to extend along the directions of easier shear displacements.

When the core is spread in the slip plane, the core smoothes the lattice periodicity and the friction stress is low (small ΔE). This is the case for the $1/2[110]$ dislocations in the $\{111\}$ planes of face-centred cubic (fcc) metals and alloys, and for the $1/6[11\bar{2}0]$ dislocations in the basal plane of some hexagonal closea-packed (hcp) metals and alloys. When the core is extended out of the slip plane, dislocation motion requires either very energetic collective atomic displacements, or periodic transformations into a more compact and more glissile core, which results in a high Peierls hill and high friction stress (large ΔE).

Dislocations which are locked in the bottom of a Peierls valley ($y = 0$) can move to the neighboring one ($y = h$) by the nucleation and propagation of a kink-pair (Fig. 6.9).

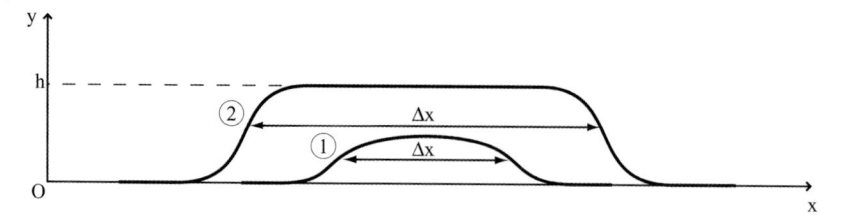

Fig. 6.9 Two steps of the nucleation of a kink-pair on a dislocation.

According to Eqs. (3) and (4), the nucleation rate of kink-pairs per unit dislocation length can be written as Eq. (26) (see Guyot and Dorn 1967), where $\Delta G_{kp}(\tau)$ is the energy of the threshold position, $\nu_D \frac{b}{l}$ is the dislocation vibration frequency of wavelength l, and $1/l$ is the corresponding density of antinodes of vibration (i.e., the density of possible kink-pair nucleation sites).

$$P_{kp} = \nu_D \frac{b}{l^2} \exp\left(-\frac{\Delta G_{kp}(\tau)}{k_B T}\right), \qquad (26)$$

The wavelength l is usually taken equal to the critical value of the kink separation, Δx, at the threshold position. The threshold energy (or activation energy) $\Delta G_{kp}(\tau)$ decreases with increasing shear stress τ. The corresponding dislocation velocity v_{kp} is given by Eq. (27), where L is the dislocation length. It is proportional to L because the longer the dislocation, the larger the total number of possible nucleation sites.

$$v_{kp} = \nu_D \frac{bL}{\Delta x^2} h \exp\left(-\frac{\Delta G_{kp}(\tau)}{k_B T}\right) \qquad (27)$$

The difficult point is to estimate the activation energy $\Delta G_{kp}(\tau)$. Two different approximations can be made.

(a) When the two kinks are well separated at their threshold position (Fig. 6.9, case 2), the kink-pair energy $\Delta G_{kp}(\tau)$ is the energy of two isolated kinks, G_k, minus their elastic interaction energy, and minus the work done by the stress over the area swept by the kinks [Eq. (28)].

$$\Delta G_{kp}(\tau) = 2G_k - \frac{\mu h^2 b^2}{8\pi \Delta x} - hb\tau \Delta x \qquad (28)$$

Elastic calculations yield Eq. (29), where E_0 is the average dislocation-line energy per unit length.

$$G_k \approx h(E_0 \Delta E)^{1/2} \qquad (29)$$

This energy goes through a maximum value (the activation energy) [Eqs. (30, 31)].

$$\Delta G_{kp}(\tau) = 2G_k - (hb)^{3/2}\left(\frac{\mu\tau}{2\pi}\right)^{1/2} \tag{30}$$

for

$$\Delta x = \left(\frac{hb}{8\pi}\frac{\mu}{\tau}\right)^{1/2} \tag{31}$$

The decrease in $\Delta G_{kp}(\tau)$ with increasing stress τ [Eq. (30)] is at the origin of the stress dependence of the dislocation velocity [Eq. (27)].

This approximation is valid till Δx decreases below the width of a single kink, i.e., for $\tau < \dfrac{\Delta E}{2\pi hb}$.

(b) When the critical kink-pair separation is small (Fig. 6.9, case 1), which happens at large stresses according to Eq. (31), the two kinks are not well separated and the threshold configuration is like a bowed-out dislocation arc. Several approximations have been proposed, in the framework of elastic calculations by Seeger (1956) and others (see Caillard and Martin 2003). Atomistic calculations have also been carried out by Duesberry (1983), which confirm that $\Delta G_{kp}(\tau)$ decreases with increasing τ.

6.2.4.2 Locking–Unlocking Mechanism

Vitek (1966) was the first to remark that if the kink-pair formation involves the thermally activated transformation of a sessile dislocation (in the bottom of a Peierls valley) into a more glissile one (close to the hill of the Peierls valley), the reverse transformation may also be hindered by an energy barrier, and be thermally activated. In contrast to the situation described in Fig. 6.9, it cannot be considered that dislocations extracted from a Peierls valley fall spontaneously into the neighboring one. In other words, they can keep a glissile configuration for a substantial length of time, sufficient to move over a large distance. The reverse transformation takes place subsequently and locks the dislocation (Fig. 6.10).

In this case, the motion can be treated as a series of locking and unlocking events with respective probabilities P_l and P_{ul}, per unit time and dislocation length. If the dislocation velocity in the glissile configuration is v_g, the average velocity is then given by Eq. (32).

$$v = v_g \frac{P_{ul}}{P_l} \tag{32}$$

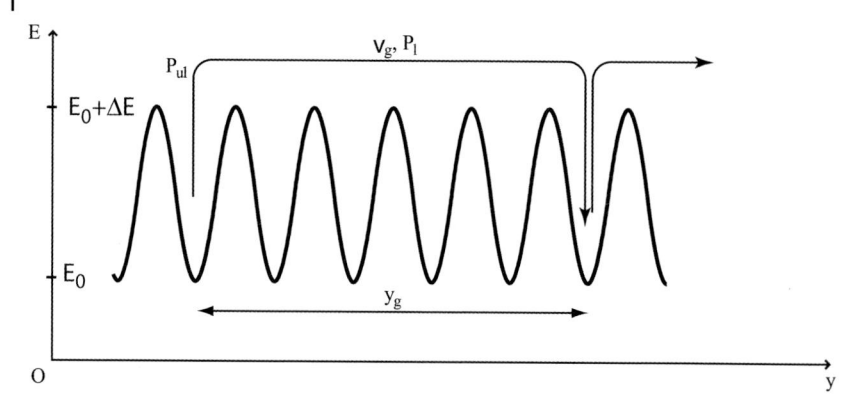

Fig. 6.10 Schematic description of the locking–unlocking mechanism.

Both probabilities can be written in the same way as Eq. (26), which yields Eqs. (33a) and (33b).

$$P_l = v_D \frac{b}{l^{(l)^2}} \exp\left(-\frac{\Delta G_l}{k_B T}\right) \tag{33a}$$

$$P_{ul} = v_D \frac{b}{l^{(ul)^2}} \exp\left(-\frac{\Delta G_{ul}}{k_B T}\right) \tag{33b}$$

The corresponding dislocation velocity is v [Eq. (34)].

$$v = v_g \left(\frac{l^{(l)}}{l^{(ul)}}\right)^2 \exp\left(-\frac{\Delta G_{ul} - \Delta G_l}{k_B T}\right) \tag{34}$$

Since ΔG_l is expected to be fairly small, of the order of the energy of a constriction on the glissile configuration, the dislocation velocity has the positive activation energy $\Delta G_{ul} - \Delta G_l \approx \Delta G_{ul}$. Note that the pre-exponential term does not have the usual form of Eqs. (4), (25a), or (27), because it does not contain the Debye vibration frequency, v_D. This results from the fact that thermal vibrations have two canceling effects, one contributing to increase the unlocking probability, P_{ul}, one contributing to increase the locking probability, P_l [Eqs. (32) and (33)]. The activation energy of unlocking, ΔG_{ul}, is estimated below.

The critical configuration for unlocking has the shape of an arc of a circle of extension y_c, which can be larger than the distance h between Peierls valleys. If y_c is large enough, elastic calculations can be used. The bulge energy is then, according to Fig. 6.11:

$$G_b = 2R \sin \theta \Delta E + (E_0 + \Delta E)(2R\theta - 2R \sin \theta) - \tau b R^2 (\theta - \sin \theta \cos \theta)$$

where R is the radius of curvature of the bowing dislocation.

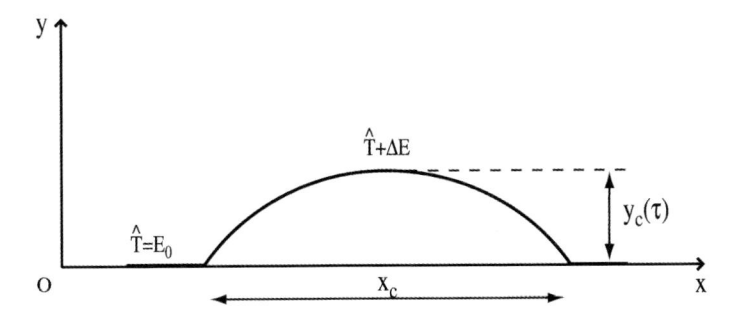

Fig. 6.11 Dislocation bowing-out of a Peierls valley.

After a Taylor expansion of $\sin\theta$ and $\cos\theta$, and taking $\tau = \hat{T}/R$, where $\hat{T}$ is the dislocation-line tension [see Eq. (12)], G_b goes through a maximum for the critical angle

$$\theta_c = \left(\frac{2\Delta E}{2\hat{T} - E_0}\right)^{1/2} \approx \left(\frac{2\Delta E}{E_0}\right)^{1/2}$$

This maximum energy of G_b (i.e., the activation energy of unlocking) is given by Eq. (35a), and the corresponding activation volume by Eq. (35b).

$$\Delta G_{ul} = \frac{2^{5/2}}{3} \frac{E_0^{1/2}\Delta E^{3/2}}{\tau b}, \tag{35a}$$

$$V_{ul} = \frac{2^{5/2}}{3} \frac{E_0^{1/2}\Delta E^{3/2}}{\tau^2 b}. \tag{35b}$$

6.2.4.3 Transition between Kink-Pair and Locking–Unlocking Mechanisms

The mean dislocation jump length of the locking–unlocking mechanism [Eq. (36)] decreases with increasing temperature, according to a more efficient locking [larger P_l, in agreement with Eq. (33a)].

$$y_g = \frac{v_g}{P_l L} \propto \exp\left(\frac{\Delta G_l}{k_B T}\right). \tag{36}$$

Therefore the locking–unlocking mechanism can take place only at low enough temperatures to have $y_g > h$. The decrease of y_g tends to slow down dislocations, but is largely compensated by the increase of the unlocking probability P_{ul} [Eq. (33b)]. As a result, the dislocation velocity increases in accordance with $\Delta G_{ul} - \Delta G_l > 0$. When the temperature is sufficiently high for y_g decreasing to h, dislocations move by the kink-pair mechanism. The latter thus appears as the high-temperature limit of the locking–unlocking mechanism. Details on the transition mechanism have been published by Farenc et al. (1995) and Caillard and

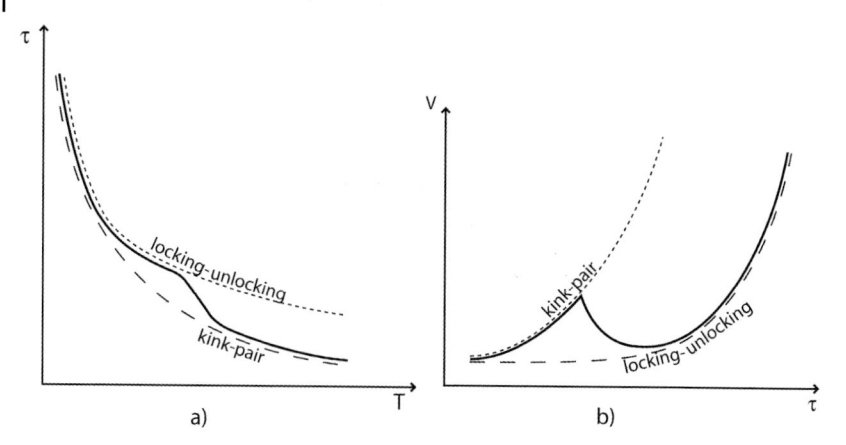

Fig. 6.12 Transition between kink-pair and locking–unlocking mechanisms: (a) CRSS as a function of temperature; (b) activation volume as a function of stress.

Martin (2003). The expected variations of yield stress and activation volume as a function of temperature are shown schematically in Fig. 6.12.

6.2.4.4 Observations of Peierls-Type Mechanisms

The deepest Peierls valleys are expected to lie along the densest atomic rows of the structure. They are also expected to be especially deep along screw dislocations that are surrounded by several directions of easier shear displacement, on account of their high degree of symmetry. Dislocations are expected to remain straight along Peierls valley directions, and to move either smoothly (kink-pair mechanism) or by series of small jumps (locking–unlocking mechanism).

Peierls-type friction stresses control the motion of dislocations in many metals and alloys with various structures: bcc ones at low temperatures, hcp ones also at low temperature (in prismatic slip), intermetallic alloys with various structures ($L1_2$, $L1_0$, B2, $D0_{19}$ etc.). In fact, easy glide in close-packed planes of fcc metals appears to be more an exception rather than the rule, although fcc alloys such as steel are the most familiar to us.

(a) Body-centered cubic metals, such as Fe, Nb, Mo, and their alloys, were the first ones in which the kink-pair mechanism was clearly identified. The CRSS of pure Fe with two orientations, C-doped Fe, and Mo with three orientations are shown in Fig. 6.13. The strong decrease in stress at increasing temperature and the small corresponding activation volumes (of the order of a few tens of b^3) are typical of this mechanism. Alloys with the same structure exhibit similar properties.

Discontinuities in the behavior of stress and activation volume with temperature, however, are still unexplained.

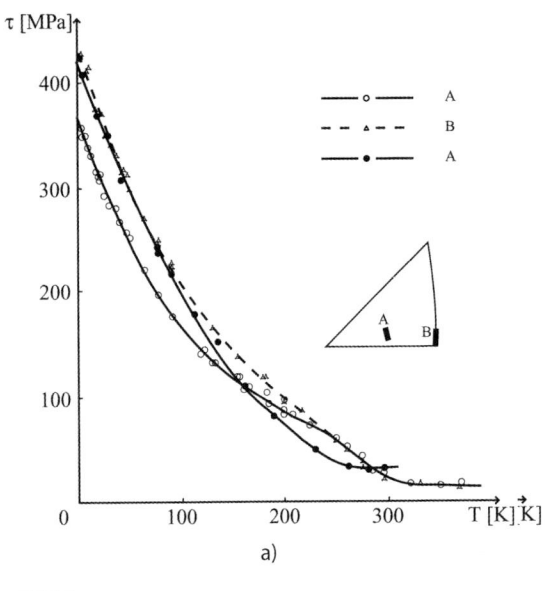

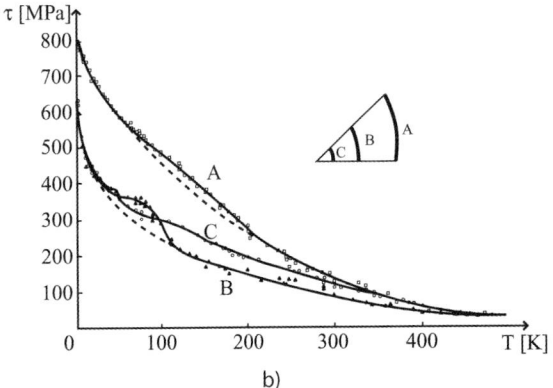

Fig. 6.13 Deformation shear stress versus temperature in bcc metals: (a) pure iron with two orientations (open symbols) and carbon-doped iron (full symbols), from Kuramoto et al. (1979); (b) molybdenum with various orientations (from Aono et al. 1983).

Strong orientation effects are observed in Fig. 6.13, in agreement with a complex stress orientation dependence of the core structure (Duesberry 1989; Duesberry and Vitek 1998). In-situ experiments show the corresponding steady and viscous motion of straight screw dislocations in iron alloys (Furubayashi 1969) and in Nb (Ikeno and Furubayashi 1972; Louchet and Kubin 1979). The nonplanar core structure of screw dislocations at the origin of Peierls-type

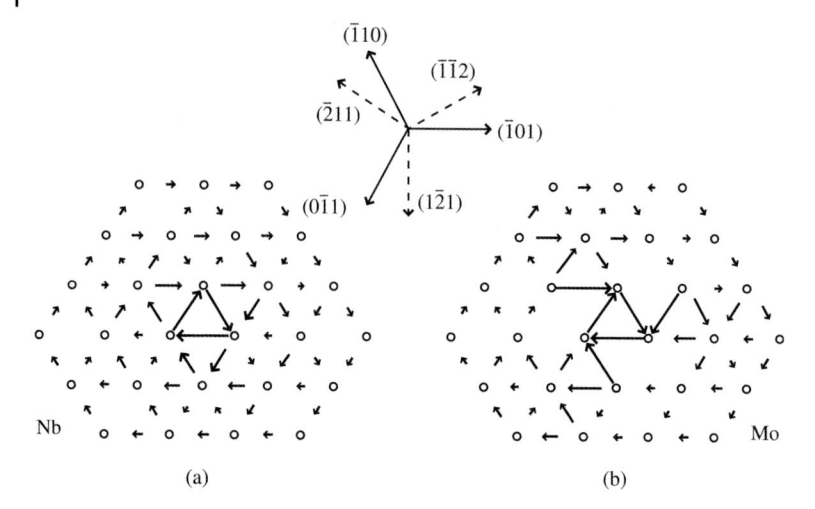

Fig. 6.14 Computed core structure of screw dislocations: (a) non-degenerate core in niobium; (b) degenerate core in molybdenum. The relative atomic displacements between adjacent atoms are perpendicular to the figure, and proportional to the length of the arrows connecting them. From Duesberry and Vitek (1998).

friction forces has been computed by several authors, e.g., Duesberry and Vitek (1998) (Fig. 6.14).

(b) Hexagonal close-packed metals and alloys deformed in prismatic slip also exhibit important Peierls friction stresses up to room temperature. In Mg, in-situ experiments reveal straight screw dislocations moving steadily and viscously at room temperature in agreement with a kink-pair mechanism. The dislocation velocity is proportional to length in agreement with Eq. (27) (Fig. 6.15). The motion becomes jerkier at lower temperatures, in agreement with a transition toward the locking–unlocking mechanism (Couret and Caillard 1989). The latter has been extensively studied in Be, for which histograms of waiting times and jump lengths are fully consistent with Eqs. (33a,b) (Couret and Caillard 1989). The out-of-plane core structure at the origin of the Peierls-type friction stress in the prismatic planes of Mg and Be is an extension (or dissociation into Shockley partials) in the intersecting basal plane. It is more complex in Ti (Legrand 1985). The transition between kink-pair and locking–unlocking mechanisms has been studied by means of in-situ experiments in Ti by Farenc et al. (1995). Figure 6.16 shows that the temperature variation of the activation volumes varies as expected from Fig. 6.12(b), and that the

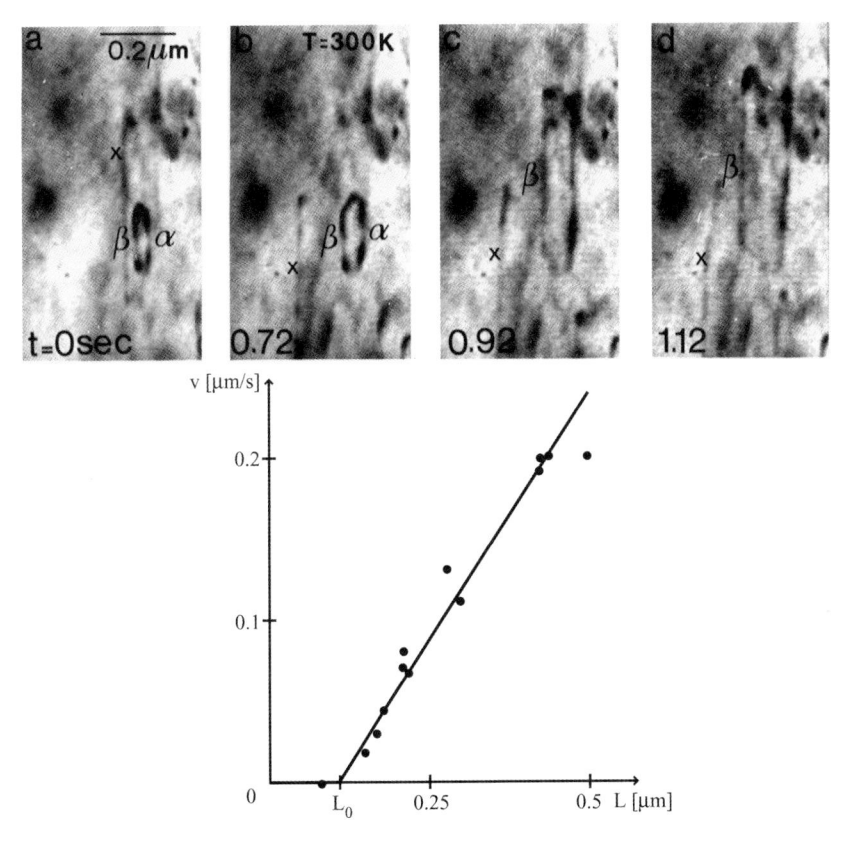

Fig. 6.15 Motion of a screw dislocation (denoted by ×) and expansion of a screw dipole (denoted by $\beta\alpha$), and corresponding screw dislocation velocity as a function of length. In-situ experiment in magnesium, 300 K. From Couret and Caillard (1985).

discontinuity corresponds exactly to the transition between locking–unlocking (large γ_g) and kink-pair (very small γ_g) mechanisms. One can see from Fig. 6.17 that activation volumes at low temperature vary as expected from Eq. (35b). Similar results have been obtained in Ti–5 at.% Al alloys, Zr, and zircalloy polycrystals (Caillard and Martin 2003).

(c) Intermetallic alloys exhibit extensive jerky dislocation motion controlled by the locking–unlocking mechanism, e.g., ordinary dislocations in TiAl (Couret 1999), super-dislocations in cube planes of Ni_3Al (Molénat and Caillard 1992), prismatic and basal planes of Ti_3Al (Legros et al. 1996), {110} planes of Fe_3Al (Molénat et al. 1998). Atomistic calculations in TiAl reveal two possible core configurations for ordinary dislocations, one sessile and one glissile, which

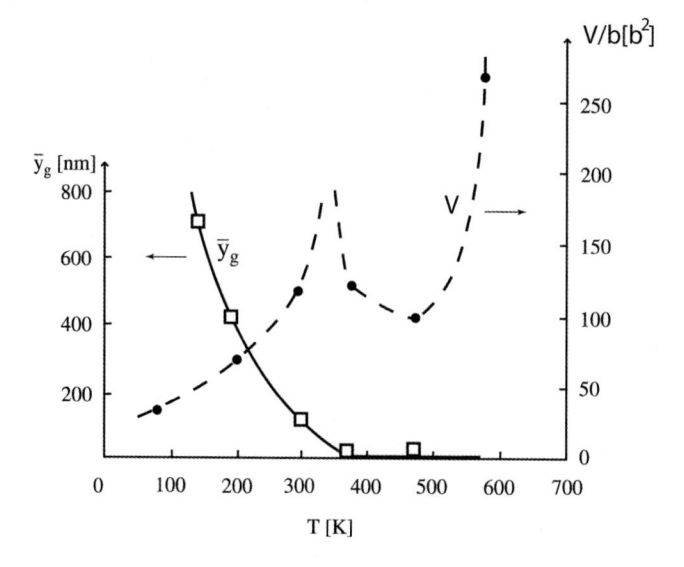

Fig. 6.16 Average jump distance $\bar{y}_g$ as a function of temperature (in-situ experiments in titanium deformed in prismatic slip, from Farenc et al. 1985), and comparison with the corresponding activation volume V, measured in a conventional deformation test by Naka et al. (1988).

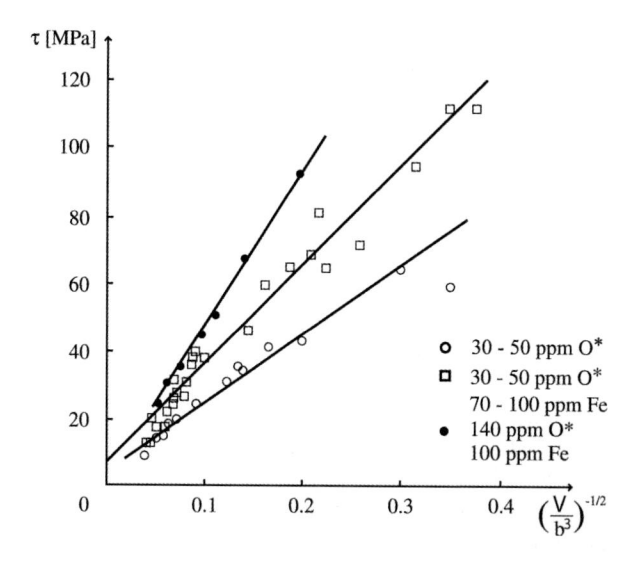

Fig. 6.17 Stress dependence of the activation area of prismatic slip in titanium at $T < 300$ K. Open symbols are from Biget and Saada (1989); full symbols are from Levine (1966).

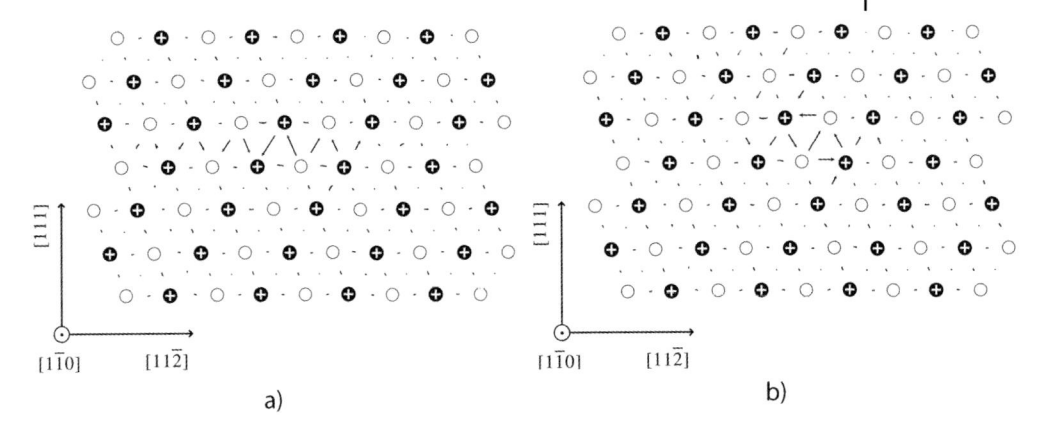

Fig. 6.18 Two possible core structures for a screw ordinary dislocation in TiAl: (a) planar and glissile in the {111} plane; (b) nonplanar and subjected to a high friction force. Arrows have the same meaning as in Fig. 6.14. From Simmons et al. (1997).

may account for the locking–unlocking mechanism observed experimentally (Simmons et al. 1997) (Fig. 6.18).

(d) In fcc metals, dislocations glide mostly in close-packed {111} planes where their cores are extended (or dissociated in the case of a low stacking fault energy), and where the Peierls-type friction stress is accordingly very weak. However, the same dislocations can also glide in the non-close-packed planes, where the Peierls-type frictional forces are much higher. Close-packed and non-close-packed glide can thus be compared, respectively, with basal and prismatic glide in hcp Mg or Be. Extensive glide in non-close-packed {110} and {100} planes has been observed in various fcc pure metals, for specific orientations of the straining axis (Fig. 6.19). It is characterized by coarse and wavy slip lines at variance with the usual fine slip lines in {111} planes. Figure 6.19(a) shows that the reduced temperature above which non-close-packed slip becomes important varies linearly with the inverse of the dissociation width, $\mu b/\gamma$ (where γ is the stacking fault energy). The critical stress for {100} slip measured by Carrard and Martin (1988), shown in Fig. 6.19(b), has the same general properties as those shown in Fig. 6.12(a). The replacement by non-close-packed slip of the – a priori, easier – {111} slip is probably due to an easier cross-slip, which enhances dislocation multiplication.

Glide in non-close-packed planes is generally ignored in plasticity models of fcc metals and alloys. Recent results,

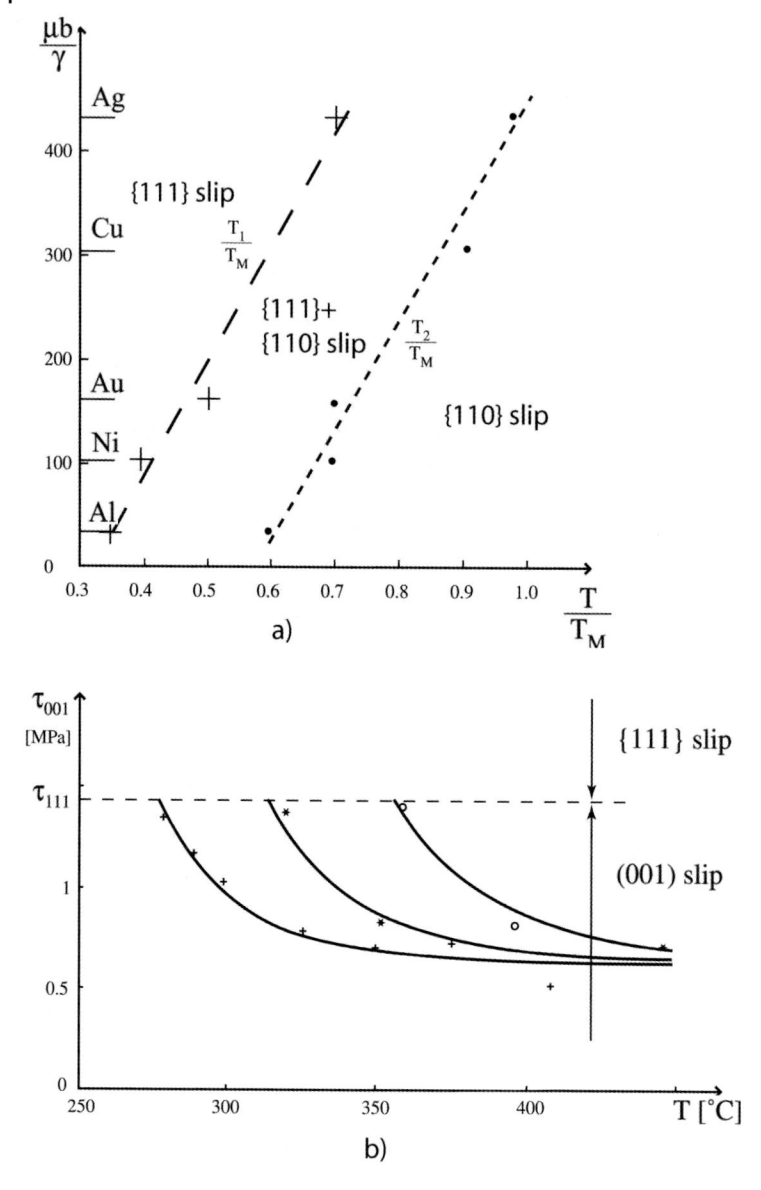

Fig. 6.19 Properties of glide in non-close packed planes of fcc metals: (a) onset of {110} glide as a function of temperature and stacking fault energy (from Le Hazif et al. 1973); (b) CRSS of {001} glide in aluminum as a function of temperature, for different strain rates (from Carrard and Martin 1988).

however, indicate that the mechanical properties of precipitation-hardened aluminum alloys are related to {100} non-close-packed slip (Majimel et al. 2004).

6.2.5
Cross-Slip in fcc Metals and Alloys

Cross-slip is a thermally activated change of dislocation slip plane, which plays an important role in by-passing of obstacles, source operation, and recovery. When dislocation motion is controlled by Peierls-type friction forces as in bcc alloys in the low-temperature regime, cross-slip occurs via a change of plane in which bulges or kink-pairs are nucleated, which does not require any supplementary activation energy. Cross-slip is thus especially important in fcc metals and alloys where dislocations glide freely in {111} planes, and where any change of plane is thermally activated. Under such conditions, it may determine the onset of deformation stage III, where hardening becomes dependent on temperature and strain-rate (Section 6.3.3.2). A review of cross-slip has been published recently by Puschl (2002).

6.2.5.1 Elastic Calculations

Cross-slip is generally considered to take place by the Friedel–Escaig mechanism described in Fig. 6.20. Elastic calculations have been proposed to estimate the constriction energy (Stroh 1954), and the activation energy and activation volume of cross-slip (Escaig 1968). The results predict that the activation volume at low stress should be of the order of V_{cs} in Eq. (37), where d is the dissociation width, τ is the applied shear stress, τ_d and τ'_d are respectively the components of the applied stress which tend to constrict the dislocation in the primary plane and to widen it in the cross-slip plane, and α ranges between 0.5 and 1.

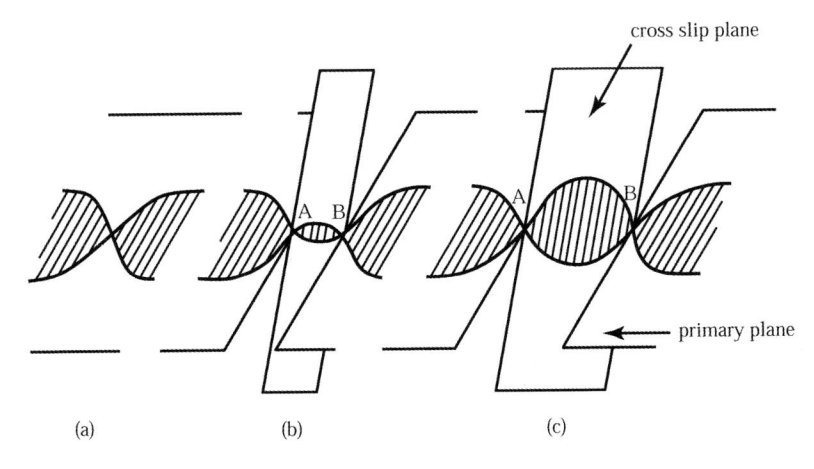

Fig. 6.20 Schematic description of the Friedel–Escaig cross-slip mechanism.

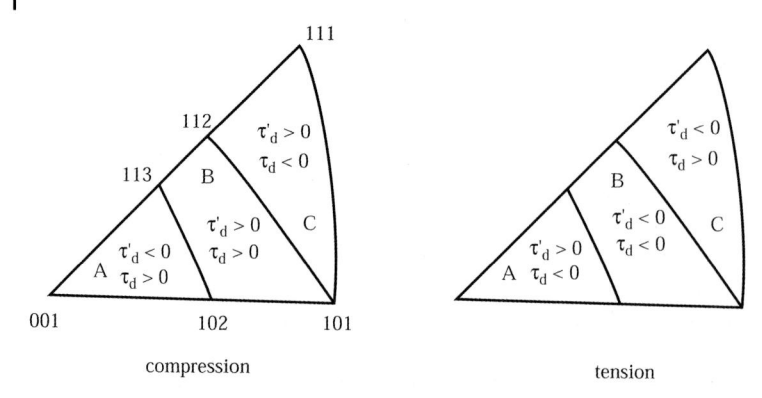

Fig. 6.21 Tension–compression asymmetry of the different stresses involved in cross-slip. The dissociation width in the primary and cross-slip planes changes according to the direction of the applied stress (three zones in the stereographic triangle), in tension and compression.

$$V_{cs} = 3d^2 b \frac{\alpha \tau_d + \tau'_d}{\tau} \tag{37}$$

Equation (37) shows that the stress dependence of the cross-slip probability is complex, and Fig. 6.21 shows that important tension–compression asymmetries are expected, in domain B. Taking $\tau_d \approx \tau'_d \approx \tau$ and $\alpha \approx 1$ yields Eq. (38).

$$V_{cs} \approx 6d^2 b \tag{38}$$

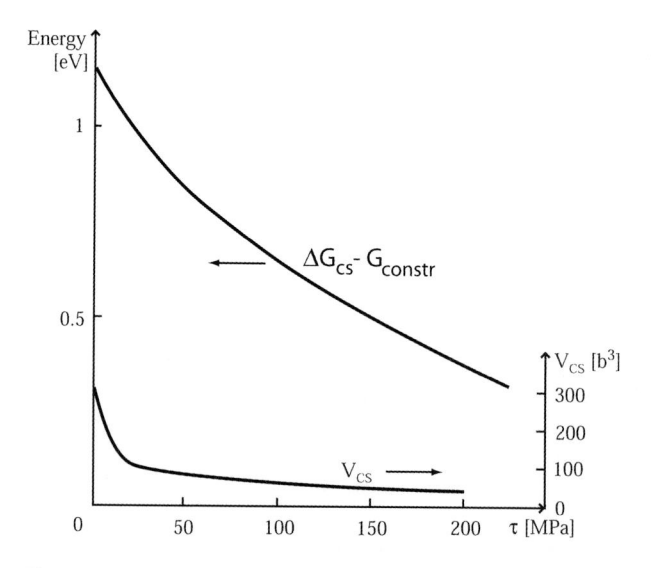

Fig. 6.22 Activation parameters of cross-slip in copper (from Escaig 1968).

6.2 Thermally Activated Mechanisms

Figure 6.22 shows the variations of the cross-slip activation energy, ΔG_{cs}, and activation volume, V_{cs}, over a large range of stress, still assuming $\tau_d \approx \tau'_d \approx \tau$. G_{constr} is the constriction energy calculated by Stroh (1954).

6.2.5.2 Atomistic Calculations

Rao et al. (1999) simulated the core structure of cross-slipping dislocations in the framework of the embedded-atom method. The results show that the constriction is extended in the primary and cross-slip planes, as opposed to the point constriction considered so far (Fig. 6.23). Vegge and Jacobsen (2002) used molecular dynamics calculations and the "nudged elastic band" method. The resulting cross-slip process is of the Escaig type as shown in Fig. 6.24. Activation volumes and energies are of the same order of magnitude as those estimated in the elastic calculations above.

6.2.5.3 Experimental Results

The tension–compression asymmetry predicted in Fig. 6.21 has been verified experimentally by Bonneville and Escaig (1979) (Fig. 6.25). By a suitable pre-deformation and change of straining axis of their samples, the authors were able to induce a large amount of cross-slip at yield, identified by slip line observations

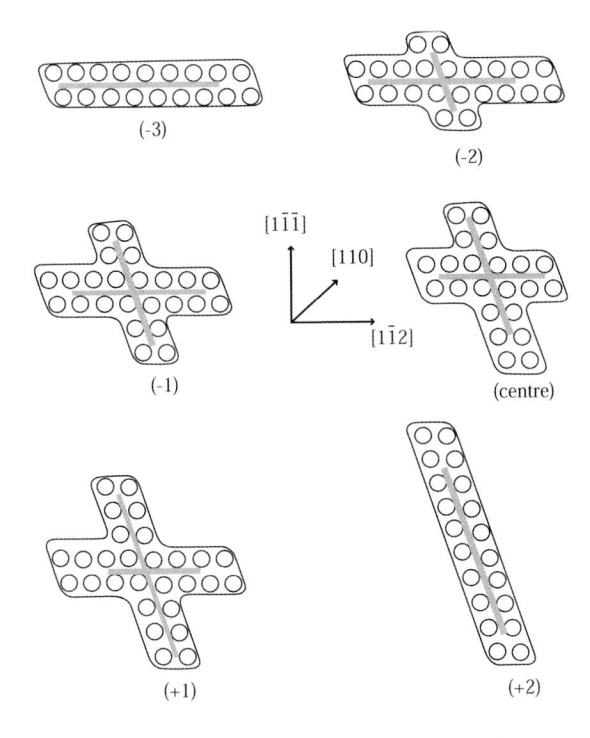

Fig. 6.23 Series of cross-sections through a screw dislocation constriction. Atomistic calculations by Rao et al. (1999).

(a)   (b)   (c)   (d)   (e)   (f)

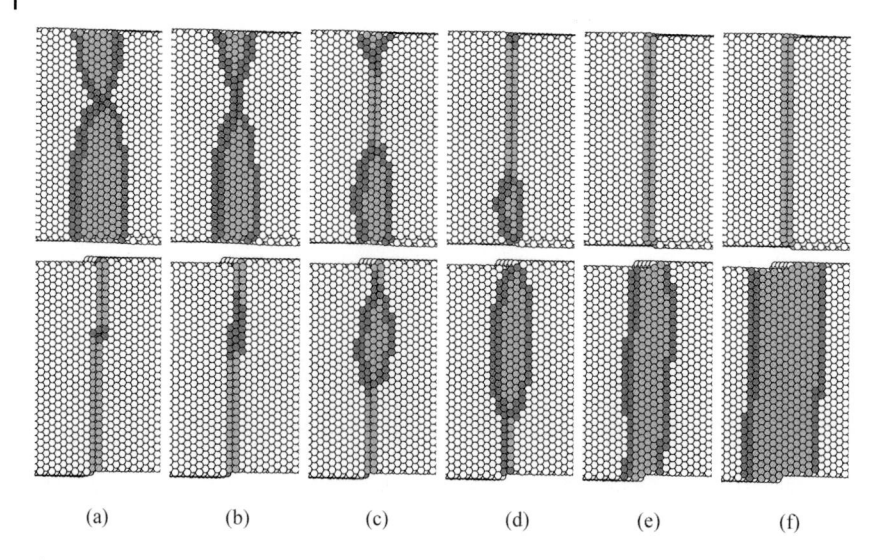

Fig. 6.24 Successive core configurations during cross-slip. The primary dissociation plane is in the plane of the figure, whereas only the trace of the cross-slip plane can be seen. From Vegge and Jacobsen (2002).

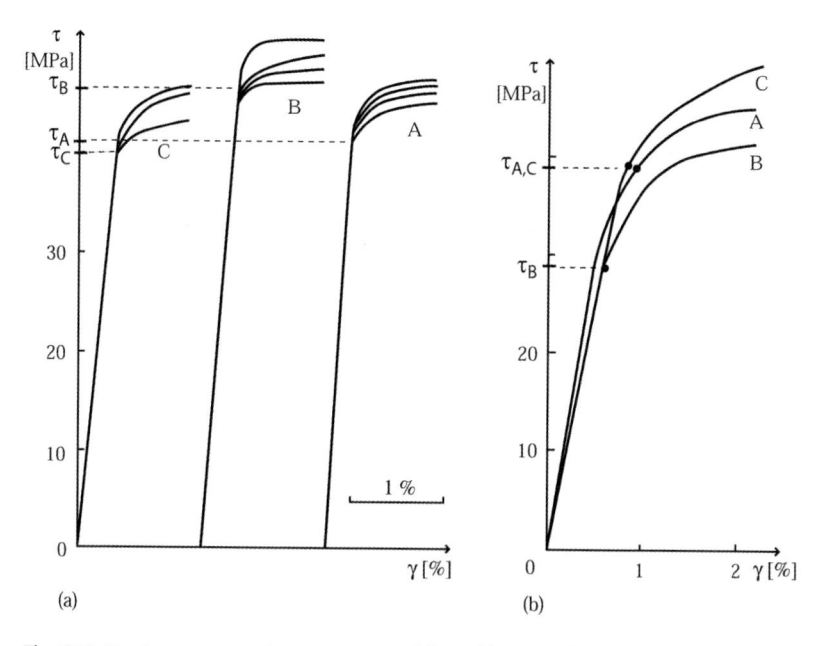

Fig. 6.25 Tension–compression asymmetry of the yield stress of pre-deformed copper, at 300 K: (a) tension; (b) compression. Curves A–C refer to the sample orientations of Fig. 6.21. From Bonneville and Escaig (1979).

and activation volume measurements. The corresponding stresses are $\tau_B > \tau_{A,C}$, in tension, and $\tau_B < \tau_{A,C}$, in compression, as expected from Fig. 6.21. Bonneville et al. (1988) measured $V_{cs} = 280b^2 \pm 65b^2$ in copper, a value close to the prediction of Eqs. (37) and (38) ($V_{cs} = 294b^2$ with $d/b = 7$). They also deduced from their experiments an activation enthalpy $\Delta H = 0.47eV \pm 0.16eV$ which yields the activation energy $\Delta G_0 = 1.15eV \pm 0.37eV$ when extrapolated to 0 K. This value corresponds to $\Delta G_{cs} - G_{constr}$ at 0 K given by Fig. 6.22 (case of a pre-existing constriction). Similar results have been obtained in aluminum (Bonneville and Vanderschaeve 1985).

6.2.6
Dislocation Climb

Dislocations are climbing when their motion involves the diffusion of atoms over long distances, i.e., when their Burgers vector lies out of their plane of motion (see Section 6.1). Climb plays a fundamental role in the mechanical properties of metals and alloys at high temperature. It is generally coupled with glide but sometimes can account for the whole plastic deformation, as discussed in Section 6.2.6.6. Since it is the only process able to annihilate edge dislocations and to build regular sub-boundary networks, it also plays a fundamental role in high-temperature recovery and creep. In spite of its importance, climb is poorly known, however, mainly because of a lack of clear experimental data.

6.2.6.1 Emission of Vacancies at Jogs
Climb takes place by emission or absorption of vacancies or interstitial atoms at jogs. Only vacancy-mediated climb is treated here, because of its lower activation energy and higher velocity. The results can be transposed easily to the case of interstitials, however (Fig. 6.26).

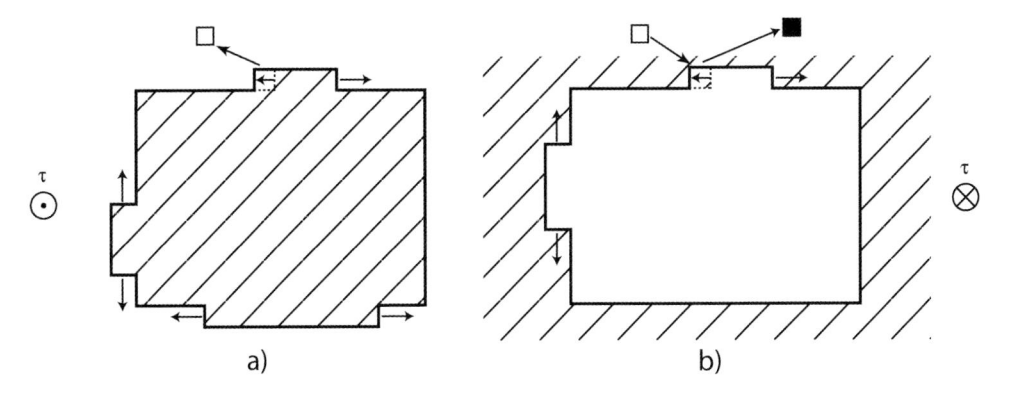

Fig. 6.26 Growth of (a) interstitial and (b) vacancy loops by exchange of vacancies (open symbol) or interstitials (full symbol).

According to Eqs. (3–6) (where y and h are taken equal to one interatomic distance, a), the frequency of jog motion by vacancy emission in the direction favored by the stress is given by Eq. (39a), where n is the number of first nearest neighbors in the lattice where the vacancy can go ($n = 12$ in fcc metals and alloys), $\Delta G_f^{(v)}$ and $\Delta G_d^{(v)}$ are respectively the formation and diffusion energies of vacancies, $D_{sd}^{(v)} = a^2 \nu_D \exp -\dfrac{\Delta G_{sd}^{(v)}}{k_B T}$ is the self-diffusion coefficient of vacancies, $\Delta G_{sd}^{(v)} = \Delta G_f^{(v)} + \Delta G_d^{(v)}$ is the activation energy of self-diffusion, Ω is the atomic volume ($\Omega \approx a^3$), and $\tau\Omega$ is the work done by the stress during the jog motion.

$$P_j^+ = n\nu_D \exp -\left(\dfrac{\Delta G_f^{(v)} + \Delta G_d^{(v)} - \tau\Omega}{k_B T}\right) = \dfrac{nD_{sd}^{(v)}}{a^2} \exp \dfrac{\tau\Omega}{k_B T} \quad (39\text{a})$$

The probability per unit time of the reverse motion by vacancy absorption is given by Eq. (39b), where $c_v^{(0)} = \exp -\dfrac{\Delta G_f^{(v)}}{k_B T}$ is the equilibrium vacancy concentration, and $c_v^{(j)}$ is the local vacancy concentration near the jog.

$$P_j^- = n\nu_D c_v^{(j)} \exp -\left(\dfrac{\Delta G_d^{(v)}}{k_B T}\right) = \dfrac{nD_{sd}^{(v)}}{a^2} \dfrac{c_v^{(j)}}{c_v^{(0)}} \quad (39\text{b})$$

The dislocation velocity can then be expressed as Eq. (40), where x is the average distance between jogs; $c_v^{(j)}$ and x are estimated in Sections 6.2.6.2 and 6.2.6.3 below.

$$v = \dfrac{\Omega}{bx}(P_j^+ - P_j^-) = \dfrac{n\Omega}{ba^2 x} D_v^{(sd)} \left[\exp \dfrac{\tau\Omega}{k_B T} - \dfrac{c_v^{(j)}}{c_v^{(0)}}\right] \quad (40)$$

6.2.6.2 Diffusion of Vacancies from Jogs

The vacancies which have a local concentration of $c_v^{(j)}$ around the jogs tend to diffuse away at the average velocity given by Eq. (41) (compare Chapter 5) where D_v is the diffusion coefficient of vacancies, defined in Eq. (42).

$$v_v = -D_v \dfrac{\text{grad } c_v}{c_v} \quad (41)$$

$$D_v = a^2 \nu_D \exp -\dfrac{\Delta G_d^{(v)}}{k_B T} \quad (42)$$

Considering a low jog density and a spherical diffusion around each jog (Fig. 6.27), the diffusion flux away from the jog is $\Phi_d = \dfrac{c_v(r)}{\Omega} 4\pi r^2 v_v$. Using Eq. (41), the steady-state condition $\dfrac{\partial \Phi_d}{\partial r} = 0$ yields the vacancy concentration at the

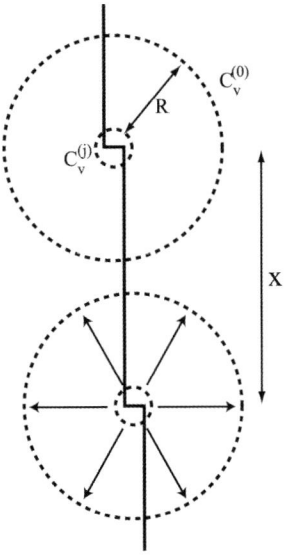

Fig. 6.27 Diffusion of vacancies from climbing jogs.

distance r from the jog, $c_v(r) = c_v^{(0)} + \dfrac{a(c_v^{(j)} - c_v^{(0)})}{r}$, and the total diffusion flux away from the jog, $\Phi_d = 4\pi \dfrac{aD_v^{(sd)}}{\Omega}\left(\dfrac{c_v^{(j)}}{c_v^{(0)}} - 1\right)$.

This flux must be equal to $P_j^+ - P_j^-$ given by Eqs. (39a,b), which allows one to eliminate $c_v^{(j)}$. The corresponding dislocation velocity is then given by Eq. (43a) or, assuming $n \approx 4\pi$, $\Omega \approx a^3$, $b \approx a$, and $\Omega\tau \ll k_BT$, by Eq. (43b).

$$v = \frac{4\pi n}{n + 4\pi} \frac{a}{bx} D_v^{(sd)} \left[\exp\left(\frac{\tau\Omega}{k_BT}\right) - 1\right], \tag{43a}$$

$$v = 2\pi \frac{D_v^{(sd)}}{x} \frac{\tau\Omega}{k_BT} \tag{43b}$$

This method, proposed by Edelin (1971), yields a dislocation climb velocity substantially different from that often found in the literature [Eq. (43c)].

$$v = 2\pi \frac{D_v^{(sd)}}{a} \frac{\tau\Omega}{k_BT} \tag{43c}$$

The latter expression is justified only when $x \approx a$, which is difficult to satisfy in practice. The average jog spacing x is estimated in Section 6.2.6.3.

6.2.6.3 Jog Density and Jog-Pair Mechanism

When the climb velocity is very low, the jog concentration a/x remains close to thermal equilibrium, according to Eq. (44) where G_j is the energy of an isolated jog.

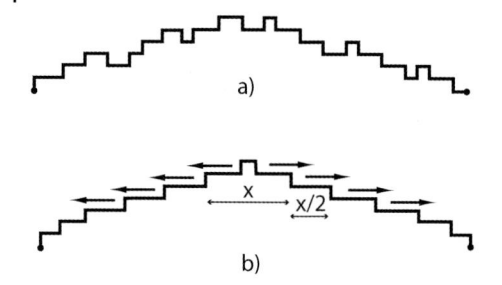

Fig. 6.28 Dislocations containing jogs: (a) thermal jogs on a dislocation at rest; (b) jog-pair nucleation and jog motion under stress.

$$\frac{a}{x} = \exp - \frac{G_j}{k_B T} \tag{44}$$

This is generally not the case, however, because jogs are rapidly swept by the stress along dislocation lines, and jog-pair nucleation must be considered (Fig. 6.28b). Jog-pair nucleation is different from the kink-pair nucleation discussed in Section 6.2.3.1, because jog motion is controlled by diffusion, whereas kinks can glide freely. Jog-pair nucleation can be better compared with the kink-pair nucleation in semiconductors, where the motion of kinks is slowed down by the cutting of covalent bonds. In fact, as pointed out by Hirth and Lothe (1982), the same equations can be used. This yields the mean free path of jogs issued from a pair between two jog-pair nucleations (Fig. 6.28b), Eq. (45) where $\Delta G_{jp}(\tau)$ is given by Eq. (30) containing G_j instead of G_k.

$$x = a \exp \frac{\Delta G_{jp}(\tau)}{2k_B T} \tag{45}$$

This value can be inserted in Eq. (43), which yields (for a dislocation length larger than x, as in Fig. 6.27) the climb velocity from Eq. (46a) or, at sufficiently low stresses, Eq. (46b).

$$v = 2\pi \frac{D_v^{(sd)}}{a} \left[\exp\left(\frac{\tau\Omega}{k_B T}\right) - 1 \right] \exp \frac{\Delta G_{jp}(\tau)}{2k_B T}, \tag{46a}$$

$$v = 2\pi \frac{D_v^{(sd)}}{a} \frac{\tau\Omega}{k_B T} \exp \frac{\Delta G_{jp}(\tau)}{2k_B T}. \tag{46b}$$

The corresponding activation energy is $\Delta G_v^{(sd)} + \Delta G_{jp}^{(c)}(\tau)$, which is thus larger than the self-diffusion energy, $\Delta G_v^{(sd)}$.

Fast diffusion in the core of climbing dislocations may increase the dislocation velocity at low temperature (see, e.g., Hirth and Lothe 1982).

6.2.6.4 Effect of Over- (Under-) Saturations of Vacancies: Chemical Force

The diffusion of vacancies discussed in Section 6.2.6.2 is modified when for some reason the average vacancy concentration is not the thermal equilibrium one. Then, $c_v^{(0)}$ must be replaced by a higher or lower value, $c_v^{(s)}$, which results in a modified dislocation climb velocity [Eq. (47a)]; x is given by Eq. (44) or (45).

$$v = 2\pi \frac{D_v^{(sd)}}{x} \left[\exp\left(\frac{\tau\Omega}{k_B T}\right) - \frac{c_v^{(s)}}{c_v^{(0)}} \right], \tag{47a}$$

Defining the chemical force as $F_c = \frac{bk_B T}{\Omega} \ln \frac{c_v^{(s)}}{c_v^{(0)}}$, Eq. (47a) can be written in the more symmetrical form of Eq. (47b) or, at sufficiently low stresses, as Eq. (47c).

$$v = 2\pi \frac{D_v^{(sd)}}{x} \left[\exp\left(\frac{\tau\Omega}{k_B T}\right) - \exp\left(\frac{F_c \Omega}{bk_B T}\right) \right], \tag{47b}$$

$$v = 2\pi \frac{D_v^{(sd)}}{x} \frac{\Omega}{k_B T} \left[\tau - \frac{F_c}{b} \right]. \tag{47c}$$

Equation (47) reduces to Eq. (43) when $F_c = 0$.

Chemical forces can originate from quenching or from the climb mechanism itself. As a matter of fact, pure climb deformation by emission of vacancies from system 1 (Fig. 6.29) strongly increases the average vacancy concentration. This induces a high chemical force, opposing the first system, but activating the second one, which is not activated by the applied stress. The second system thus contributes to absorbing the vacancies emitted by the first one. A steady state is rapidly established, for which the number of emitted vacancies, which is proportional to $\dot{\varepsilon} = \rho_1 b_1 v_1$, is equal to the number of absorbed ones, which is proportional to $\rho_2 b_2 v_2$. If $\rho_1 \approx \rho_2$, and $b_1 = b_2$, the two velocities are equal, which according to Eq. (47c) implies that $\tau - \frac{F_c}{b} = \frac{F_c}{b}$, i.e., $F_c = \frac{\tau b}{2}$.

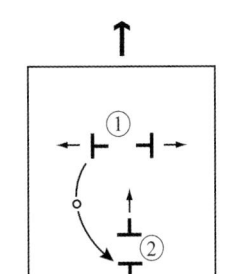

Fig. 6.29 Pure climb tensile deformation, by exchange of vacancies between system 1 activated by the applied stress, and system 2 activated by a chemical stress.

6.2.6.5 Stress Dependence of the Dislocation Climb Velocity

The dislocation climb velocity is usually considered to be proportional to stress, in agreement with Eq. (43b) or (43c), where x is assumed to be constant. Using Eq. (46b), the stress dependence of the dislocation velocity, $n = \dfrac{\Delta \ln v}{\Delta \ln \tau}$, can more precisely be expressed as in Eq. (48), and is higher than unity.

$$n = \frac{\Delta \ln v}{\Delta \ln \tau} = 1 + \frac{\tau}{2k_B T} \frac{\partial \Delta G_{jp}(\tau)}{\partial \tau} \tag{48}$$

6.2.6.6 Experimental Results

Quantitative experimental studies of dislocation climb are rather scarce, because climb is difficult to isolate. Available results concern the growth and shrinkage of prismatic loops during annealing, and pure climb of c dislocations in hcp metals.

Experiments have been carried out on quenched metals (essentially Al) where thermal vacancies coalesce to form vacancy-type perfect and faulted prismatic loops. The main results have been reviewed by Washburn (1972) and Smallman and Westmacott (1972).

In thin foils, where vacancies can easily be created or eliminated at the surfaces, the chemical stress remains close to zero. The shrinking velocity is then given by Eq. (43b) where $x \approx b$ (very high jog density) and where the stress originates from the line tension and the surface energy of the inner fault as expressed in Eq. (49), where R is the loop radius and γ is the stacking fault energy.

$$\frac{dR}{dt} = -\frac{D_v^{(sd)}}{b^2} \frac{\Omega}{k_B T} \left(\frac{\hat{T}}{R} + \gamma \right) \tag{49}$$

For perfect loops ($\gamma = 0$), this expression can be integrated to yield Eq. (50), where t_0 is a constant.

$$R = R_0 \left(1 - \frac{t}{t_0} \right)^{1/2} \tag{50}$$

This variation has been observed in experiments, as shown in Fig. 6.30, which demonstrates that the climb velocity is actually proportional to stress when the jog density is high.

The most comprehensive study of climb in deformed materials has been carried out in hcp magnesium by Edelin and Poirier (1973). When compressed at high temperature along the c axis, deformation takes place by pure climb of dislocations with Burgers vector parallel to the c axis. The density of dislocations has been measured as a function of strain, and the corresponding velocity has been

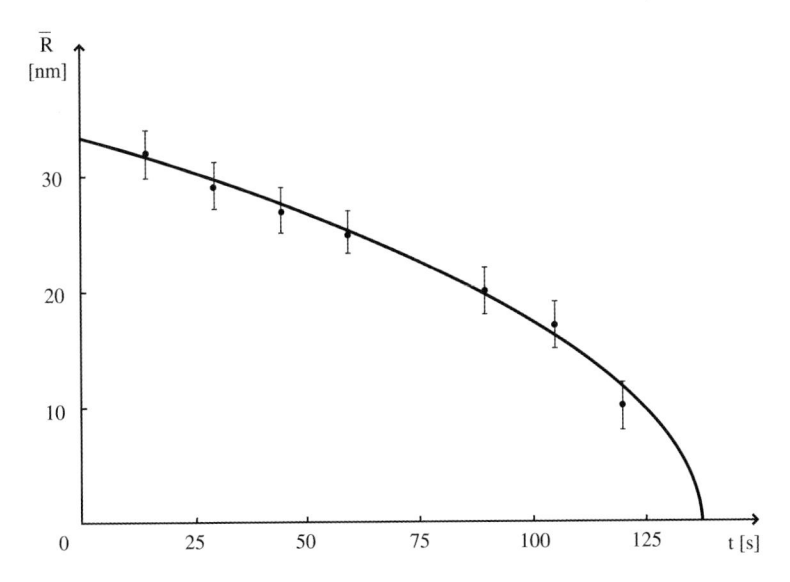

Fig. 6.30 Kinetics of shrinking of perfect prismatic loops in aluminum. The curve corresponds to Eq. (50). Annealing experiment in TEM by Silcox and Whelan (1960).

deduced from the applied strain-rate using the Orowan law [Eq. (1)]. The results shown in Fig. 6.31 yield an activation energy larger than that of self-diffusion (1.80 eV instead of 1.43 eV) and $n = \dfrac{\Delta \ln v}{\Delta \ln \tau} = 2.8$. These parameters are not consistent with Eq. (43). They may however be explained by Eq. (46b), provided $\frac{1}{2}\Delta G_{jp}$ is of the order of 0.37 eV, and by Eq. (48), provided $\dfrac{\partial \Delta G_{jp}(\tau)}{\partial \tau}$ is large

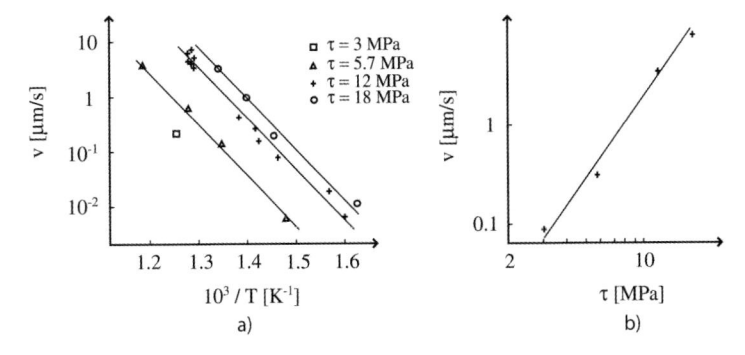

Fig. 6.31 Dislocation climb velocity in magnesium: (a) as a function of temperature; (b) as a function of stress. From Edelin and Poirier (1973).

enough. The latter value can be estimated from Eq. (30). Using Eqs. (30) and (48), and taking $\mu = 1.75 \times 10^4$ MPa, $\tau = 10$ MPa, $b = 0.55$ nm, and $h \approx b$, we find $n = 1.8$, which is definitely greater than 1 although smaller than the experimental value ($n = 2.8$).

Even if the theoretical estimates do not compare perfectly with the experimental results, these experiments confirm that the activation energy of climb can be larger than the self-diffusion one, and that the stress dependence of the corresponding dislocation velocity, n, can be greater than unity, at least at high stresses.

Dislocation motion by pure climb also takes place in intermetallic alloys, such as NiAl (Fraser et al. 1973a,b; Srinivasan et al. 1997), and in the γ' phase of superalloys (Louchet and Ignat 1986; Eggeler and Dlouhy 1997; Epishin and Link 2004). However, these results have never been incorporated in any model of plastic deformation or creep at high temperature.

Other examples of pure climb dislocation motion with similar properties are described in Section 6.4.5, on quasicrystals.

6.2.7
Conclusions on Thermally Activated Mechanisms

This review of the principal thermally activated dislocation mechanisms in metals and alloys shows that the temperature and strain-rate dependence of mechanical properties can result from various origins. Peierls-type stresses and interactions with fixed solute atoms are generally important at low temperature, dynamic interactions with mobile solute atoms and cross-slip become rate-controlling at intermediate temperatures, and diffusion-controlled glide and climb are restricted to high temperatures. Other thermally activated mechanisms such as grain boundary sliding, dislocation emission at free surfaces and crack tips, etc. are less well documented and will not be described here. The identification of these mechanisms requires a fine analysis of the experimental activation parameters (especially the activation volume) as well as detailed TEM analyses of the dislocation substructure and in-situ strain experiments.

6.3
Hardening and Recovery

The mechanisms described in this section are related to the long-range elastic interactions between dislocations, which lead to the formation of complex dislocation structures. Internal stresses appear progressively as a result of the accumulation of back-products of dislocation motion: stored dislocations, debris, tangled zones, etc. It is thus important to make the distinction between mobile dislocations which account for the strain rate, and stored ones which are obstacles. Several aspects of hardening and recovery processes are summarized below.

6.3.1
Dislocation Multiplication versus Exhaustion

As already discussed in Section 6.1, plastic deformation can be expressed by the Orowan law:

$$\varepsilon = \rho_m b x$$

Here ε is the plastic strain, ρ_m is the density of mobile dislocations (here assumed to be only a fraction of the total), b is their Burgers vector, and x is their traveling distance. The plastic strain-rate can then be expressed as Eq. (51a), where v is the dislocation velocity.

$$\dot{\varepsilon} = \rho_m b v + \dot{\rho}_m b x \qquad (51a)$$

Equation (51a) yields Eq. (1) when the second term of the right-hand side is neglected. The exact meaning of this second term can be understood better in the more explicit form of Eq. (51b), where $\Delta\rho_m = \dfrac{\partial \rho_m}{\partial x} x$ is the variation of the original density ρ_m after the displacement x.

$$\dot{\varepsilon} = \rho_m b v + \frac{\partial \rho_m}{\partial x}\frac{\partial x}{\partial t} b x = (\rho_m + \Delta\rho_m) b v \qquad (51b)$$

The value of $\Delta\rho_m$ can be either positive, as a result of dislocation multiplication, or negative, as a result of dislocation exhaustion. At constant strain rate $\dot{\varepsilon}$, any in-

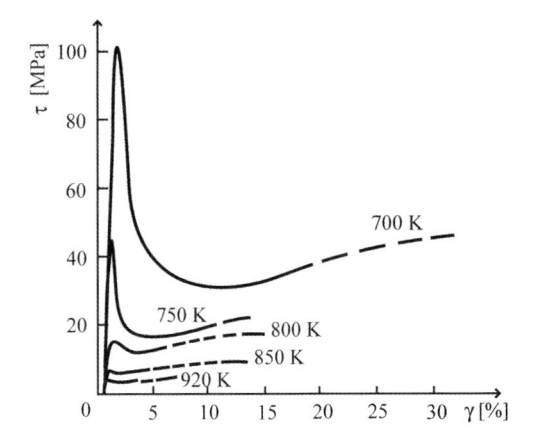

Fig. 6.32 Decrease in stress after yield ("yield drop") as a result of intensive dislocation multiplication in germanium. (From Dupas et al. 2002.)

crease in p_m results in a decrease in the dislocation velocity v, and a decrease in the corresponding effective applied stress. This accounts for the high yield drop observed in materials such as semiconductors (Fig. 6.32). On the contrary, any decrease in p_m must be compensated by the activation of new sources, which requires an increase in stress, called exhaustion hardening. The difference between exhaustion hardening and classical internal stress hardening will be discussed in Section 6.3.2.

6.3.1.1 Dislocation Sources

Various types of dislocation sources have been observed by means of TEM of in-situ strain experiments.

- Spiral sources consisting of one dislocation revolving around a single anchoring point have been observed in most cases (Fig. 6.33): these occur more frequently than the classical Frank–Read source, which requires the presence of two strong anchoring points at a suitable distance.
- Open and closed loops forming on mobile screw dislocations: the principles of this mechanism are as follows. A screw dislocation subjected to extensive cross-slip meets a weak obstacle, such as a small cluster or a solute atom. A jog is formed, provided cross-slip takes place independently on both sides of the obstacle (Fig. 6.34); then, the two arms can

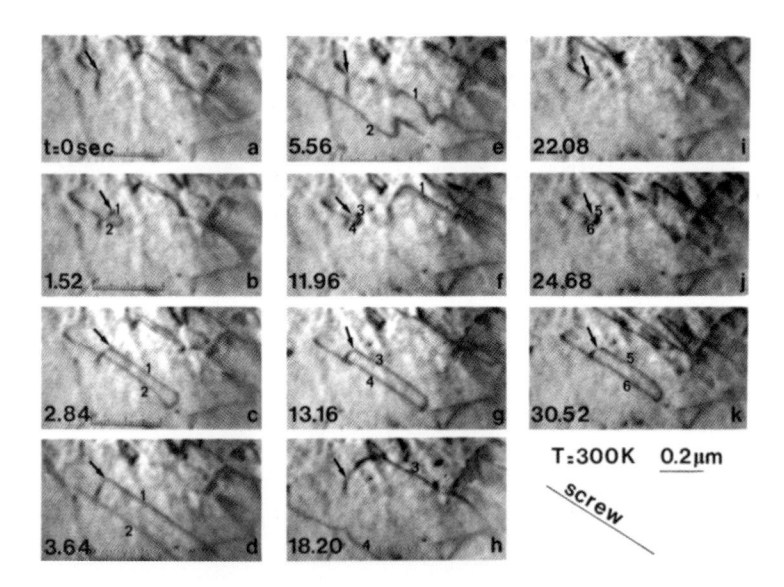

Fig. 6.33 Dislocation source rotating around a single anchoring point (arrowed). Note the elongation of the emitted loops in the screw orientation. In-situ experiment in magnesium at 300 K by Couret and Caillard (1985).

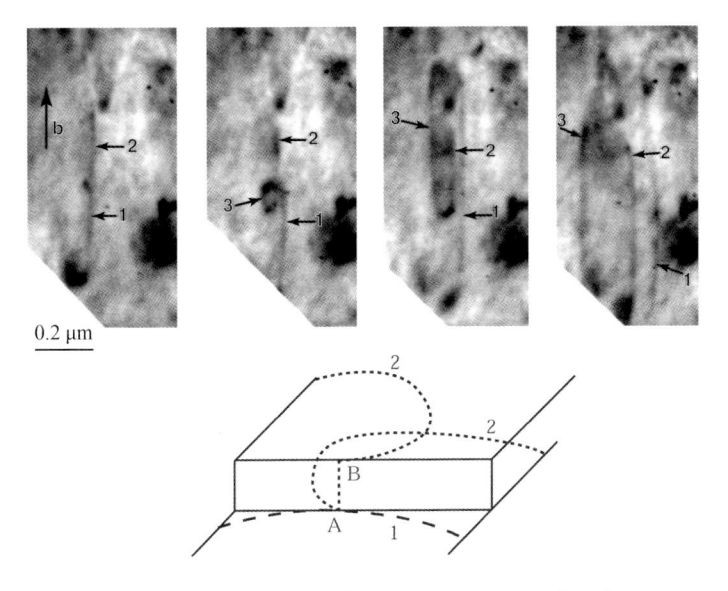

Fig. 6.34 Dislocation multiplication, by formation and expansion of an open loop. Magnesium deformed in situ in prismatic slip at 300 K. From Couret and Caillard (1985).

rotate around the jog in two parallel planes. The lifetime of such sources is generally very short because the two arms can cross-slip back in their original slip plane after one revolution. The same process can lead to the formation of a dipole (or closed loop) which subsequently expands as shown in Fig. 6.15(a)–(d) (where the closed loop $\beta\alpha$ has been nucleated by the dislocation $\times$). Multiplication at closed and open loops is restricted to materials where straight screw dislocations subjected to Peierls-type forces exhibit frequent cross-slip (bcc and hcp metals and alloys at low temperatures).

Several multiplication laws can be proposed:
- Increment of dislocation density $\delta\rho$ proportional to the area swept by the dislocations: this corresponds to the open- and closed-loop mechanism described above. This yields $\delta\rho \propto \rho v\, dt$, or, using Eq. (1) (the Orowan law), yields Eq. (52).

$$\delta\rho \propto \dot{\varepsilon}\, dt \qquad (52)$$

- Natural loop expansion: the length increment of an individual dislocation loop is proportional to its traveling distance. This yields $\delta\rho \propto Nv\, dt \propto \dfrac{\dot{\varepsilon}}{\bar{R}} dt$, where $N \propto \rho/\bar{R}$ is the number of loops per unit volume and $\bar{R}$ is their average radius. If the geometry of glide is such that $\bar{R}$ remains constant but inversely proportional to stress, the expression proposed by Alexander and Haasen (1968) for semiconductors is obtained, i.e., Eq. (53).

$$\delta\rho \propto \dot{\varepsilon}\tau\, dt \tag{53}$$

- Multiplication at Frank–Read sources: if dislocations move at a constant velocity, and if N is the number of dislocations emitted by the source, with respective radii of curvature $R_1 \propto 1/\tau$, $R_2 = 2R_1$, $R_3 = 3R_1$, etc., the dislocation density can be expressed as $\rho \propto R_1 + R_2 + \cdots \propto R_1 \dfrac{N^2}{2}$, whence $N = \left(\dfrac{2\rho}{R_1}\right)^{1/2}$. Then, the same calculation as above yields the expression already proposed by Moulin et al. (1999), i.e.:

$\delta\rho \propto Nv\, dt \propto N \dfrac{\dot{\varepsilon}}{\rho} dt$, and Eq. (54).

$$\delta\rho \propto \dot{\varepsilon}\left(\dfrac{\tau}{\rho}\right)^{1/2} dt \tag{54}$$

Many more expressions could be proposed as well, depending on the actual situation observed in the electron microscope.

6.3.1.2 Dislocation Exhaustion and Annihilation

Dislocation storage results in a decrease of mobile dislocation density, ρ_m, also called exhaustion. Exhaustion tends to increase the effective stress $\tau_{e\!f\!f}$ (i.e., the thermally activated component of the total stress defined in Section 6.1), because any decrease of ρ_m must be compensated by an increase of dislocation velocity v, in order to keep the applied strain rate $\dot{\varepsilon}$ constant [Eq. (1)]. At low strain and low dislocation density, exhaustion can also be compensated by the activation of new sources, which requires an additional increase in applied stress (see the early part of Section 6.3.1). This effect, and the variation of internal stress τ_i (which will be studied in Section 6.3.2), contribute to strain hardening.

Dislocation storage can result from interactions with extrinsic obstacles such as other dislocations or cell walls, from cross-slip (in the case of Ni_3Al; cf. Section 6.4.1.2), or from solute diffusion (Section 6.2.2.3). As discussed in Section 6.4.1, the two latter processes, which are thermally activated, can be at the origin of stress anomalies.

Several empirical exhaustion laws can be used, for example:

$$\delta\rho \propto -\rho\, dt$$

(constant probability of locking per unit time), or:

$$\delta\rho \propto -\rho x \propto -\dot{\varepsilon}\, dt$$

(probability of dislocation locking proportional to the traveling distance).

The first expression can be used to describe thermally activated dislocation storage by cross-slip in Ni_3Al (Section 6.4.1.2), and the second corresponds to dislocation locking on fixed obstacles, as in stage II work-hardening in fcc metals and alloys (Section 6.3.3).

On the other hand, dislocation annihilation can take place by cross-slip or climb, at a rate given by:

$$\delta\rho \propto -\dot{\varepsilon}\sqrt{\rho}\, dt$$

(probability of dislocation annihilation proportional to the traveling distance, with additional effect of the annihilation stress between neighboring dislocations of opposite signs).

Here again, it is seen that different laws can be used, depending on the exact mechanisms involved.

6.3.2
Dislocation–Dislocation Interaction and Internal Stress: the Taylor Law

At high strain, dislocation multiplication and storage increase the internal stress τ_i according to the Taylor law, Eq. (55a), where ρ is the total dislocation density ($1/\sqrt{\rho} = d$, average dislocation length or spacing), and α_i ranges between 0.2 and 0.5.

$$\frac{\tau_i}{\mu} = \alpha_i b \sqrt{\rho} \tag{55a}$$

This law originates from the long-range interactions between moving dislocations and all other (mobile or stored) ones. These interactions are of two kinds:

- When two parallel dislocations are close to each other, they are subjected to an elastic interaction stress of the order of $b/2\pi d$ (Eq. (55b); see Fig. 6.35a).

$$\frac{\tau_i}{\mu} \approx \frac{b}{2\pi d} \tag{55b}$$

- When gliding dislocations intersect "forest" dislocations, they are often pinned by the formation of attractive junctions (Fig. 6.35b). They can bow out and escape when their radius of curvature, R, is small enough to reach the breaking angle, θ necessary to recombine the junction. For $\theta = 0$, the

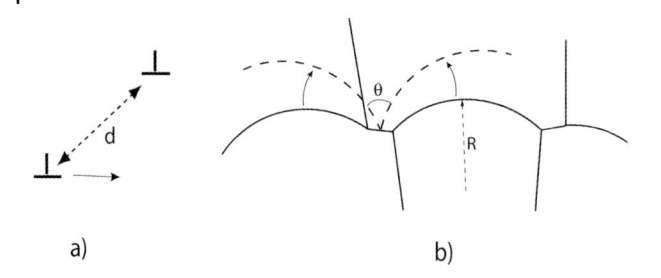

Fig. 6.35 Schematic description of the origin of internal stresses: (a) long-range elastic interaction; (b) forest mechanism.

corresponding internal stress is $\tau_i \approx \dfrac{\hat{T}}{bR}$, with $R \approx d/2$ and $\hat{T}$ (line tension) $\approx \tfrac{1}{2}\mu b^2$, whence Eq. (55c) results. Smaller values of τ_i are obtained for more realistic values of θ.

$$\tau_i \approx \frac{\mu b}{d} \tag{55c}$$

These various interactions have been calculated in detail by Saada (1960) and Schöck and Frydman (1972) using elasticity theory. More recently, all the possible situations have been simulated at the mesoscopic scale by Wickham et al. (1999), Shenoy et al. (2000), and Madek et al. (2000). They confirm the validity of the Taylor law and show that the stability of junctions (which determines the internal stress) is independent of the core structure of dislocations. Madek et al.'s calculations also confirm that τ_i depends essentially on the average dislocation density, not on its spatial arrangement. As a matter of fact, the proportionality coefficient α is found to be only 10% lower when dislocations are rearranged from a uniform distribution to the inhomogeneous structure of cells and walls observed in experiments. The origin of this surprising property will be discussed in Section 6.3.3.1.

At low strain and low average dislocation density, Eq. (55c) gives the stress necessary to activate a source as a function of its length, d. However, since the lengths of potential sources are generally not equal to the average value $1/\sqrt{\rho}$, the proportionality coefficient of the Taylor law adapted to this situation can be different. This describes "exhaustion hardening", as opposed to "internal stress hardening" operating at higher strain, and considered above. An example of exhaustion hardening is the pre-yielding of bcc metals, where shorter and shorter edge segments trailing screw dipoles must be activated, at the expense of an increasing stress. Another example is the origin of yield stress anomalies, treated in Section 6.4.1. Exhaustion hardening and internal stress hardening are not fundamentally different processes, however, and a transition occurs between them when the dislocation structure is sufficiently homogeneous to have a critical source of size $d \approx 1/\sqrt{\rho}$.

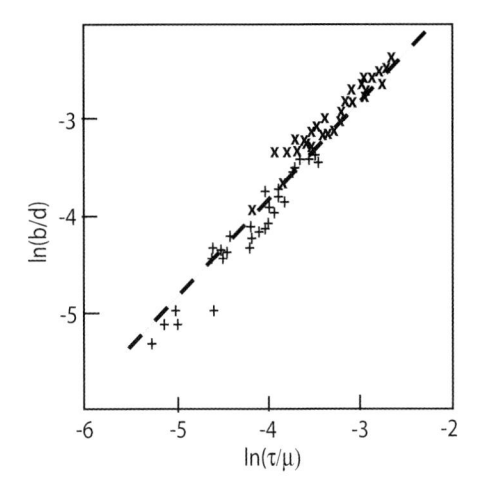

Fig. 6.36 Experimental evidence for the Taylor law in copper at room temperature. Data collected by Mecking and Kocks (1981).

Sometimes, the total stress $\tau = \tau_{eff} + \tau_i$ can be written in the same form, i.e., as Eq. (55d), where $\alpha = \alpha_{eff} + \alpha_i$ (Fig. 6.36).

$$\frac{\tau}{\mu} = \alpha b \sqrt{\rho} \tag{55d}$$

This form is justified in case of forest mechanisms where $\frac{\tau_{eff}}{\mu} = \alpha_{eff} b \sqrt{\rho}$ (cf. Section 6.2.3).

6.3.3
Hardening Stages in fcc Metals and Alloys

Most studies of hardening have been carried out in fcc metals and alloys. However, the results should be transposable to bcc metals, which exhibit the same stages as fcc ones (see, e.g., Kuhlmann-Wilsdorf 2002). The dislocation density and corresponding deformation stress [(given by Eq. (55a)] increase with increasing deformation, as a result of the dislocation multiplication and storage described in Sections 6.3.1.1 and 6.3.1.2, respectively. Hardening is defined by the coefficient $\theta = \left.\frac{\partial \tau}{\partial \varepsilon}\right|_{T,\dot{\varepsilon}}$, often plotted as a function of stress.

Reviews of strain hardening in fcc metals and alloys have been published by Basinski and Basinski (1979) and by Nabarro (1989), and more recently by several authors in Volume 11 of *Dislocations in Solids* devoted to work-hardening, and by Kocks and Mecking (2003). Some important conclusions on this complex topic are discussed below.

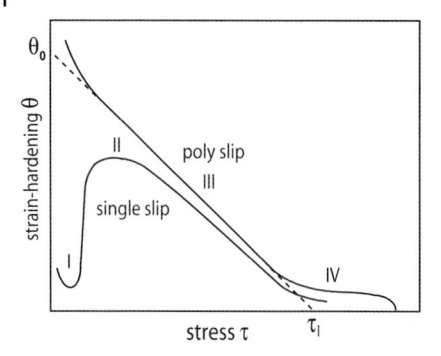

Fig. 6.37 The different strain-hardening stages in fcc metals.

Hardening is usually divided into several stages, observed at increasing strain, and described schematically in Fig. 6.37. Stage I of "easy glide" is observed at the beginning of the deformation of single crystals in single slip. The corresponding value of hardening coefficient θ is fairly low, on account of the weak elastic interaction stress given by Eq. (55b). TEM observations reveal a homogeneous distribution of dipoles in this stage. Stage II is observed as soon as multiple slip forms a three-dimensional dislocation structure with junction reactions, building tangles that subsequently evolve to cell walls surrounding cells (see Fig. 6.40, below). The corresponding deformation stress is then described by Eq. (55c). It is characterized by a constant value of θ, i.e., a linear increase of stress with strain ("linear hardening"). Stage III starts when θ decreases with increasing strain and stress (Fig. 6.37). The onset of stage II is strongly dependent on temperature and strain rate, in such a way that stage II can disappear at high temperatures. Lastly, stage IV corresponds to a stabilization of θ at a fairly low value. Stages II to IV are now described in some more detail.

6.3.3.1 Stage II (Linear Hardening)

The strain-hardening coefficient in stage II has a more or less constant value, $\theta_{II} \approx \mu/200$, which is almost independent of material, temperature, and strain rate (within a factor of 2). Figure 6.38 shows the relative variation of the Taylor factor α, defined by Eq. (55d), as a function of temperature. The rapid decrease shows that the total flow stress involves a thermally activated component, τ_{eff}, proportional to α_{eff}, which depends on stacking fault energy. Since the Cottrell-Stokes law is verified by experiments, the ratio $\dfrac{\tau_{eff}}{\tau_i} = \dfrac{\alpha_{eff}}{\alpha_i}$ is constant along deformation curves. Then, the Taylor law which is valid for τ is also valid for τ_{eff}, which demonstrates that the forest mechanism is active. This conclusion is corroborated by Fig. 6.39 showing the "Haasen plot" of the inverse of the activation volume (deduced from strain-rate jumps) as a function of the applied stress corresponding to various amounts of deformation. It is linear at small stresses, which shows that

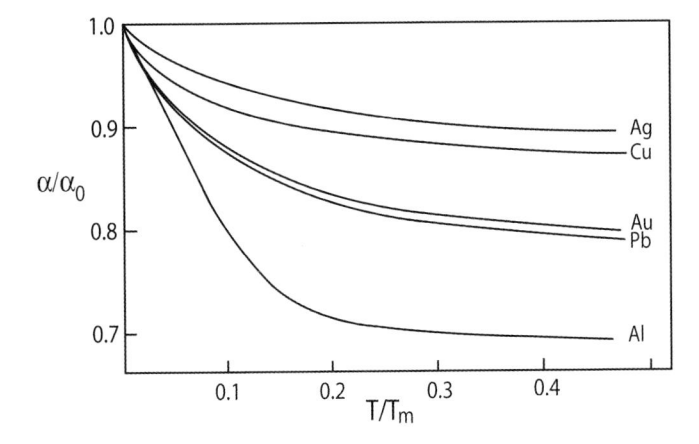

Fig. 6.38 Effect of thermal activation on the proportionality coefficient of the Taylor law. From Kocks and Mecking (2003).

the activation volume varies as expected from the forest mechanism (Section 6.2.3).

Stage II is observed as long as the Cottrell–Stokes law is verified, i.e., as long as $\frac{\tau_{eff}}{\tau_i}$ remains constant. The flow stress in stage II is thus thermally activated (and controlled by the forest mechanism), although the corresponding strain-hardening coefficient is independent of temperature and strain rate.

The formation and the evolution of the cell structure shown in Fig. 6.40 result from a very complex mixture of multiplications, reactions, and annihilations of dislocations belonging to several slip systems. It could be the result of statistical

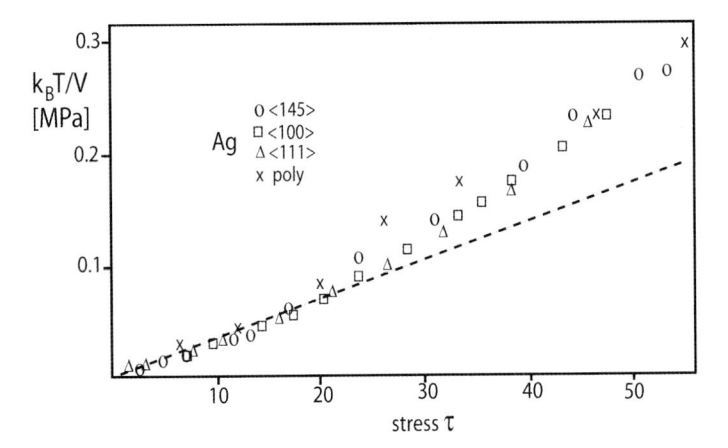

Fig. 6.39 Haasen plot of the inverse of the activation volume V as a function of stress, in silver. From Mecking and Kocks (1981).

fluctuations in an out-of-equilibrium system, according to Aifantis (1986) and Hahner (1996). The heterogeneous structure could also be driven by the decrease of the total elastic energy aone, according to Hansen and Kuhlmann-Wilsdorf (1986) and Kuhlmann-Wilsdorf (2002). As a matter of fact, the total elastic energy stored in the deformed crystal must decrease upon dislocation tangling, because the dislocation-line energy is proportional to the logarithm of the dislocation distance.

The simplest explanation for dislocation patterning is based on the composite model of Mughrabi (see, e.g., Mughrabi and Ungar 2002). The deformed crystal is treated as a two-phase material with hard zones (tangles or walls) and soft zones (cell interiors) which are subjected to different local stresses. Since these zones must deform plastically at the same rate, a heterogeneous internal stress must develop, positive in the hard zones (forward stress), and negative in the soft ones (backward stress). Assuming now that local stresses are related to local dislocation densities according to the Taylor law [Eq. (55a)], the applied stress can be written as Eq. (56), where $\Delta\rho$ is the fraction of the initial dislocation density which is transferred from the soft zones (volume fraction $1 - K$) to the hard ones (volume fraction K).

$$\tau = \alpha\mu b \left[K\left(\rho + \frac{\Delta\rho}{K}\right)^{1/2} + (1 - K)\left(\rho - \frac{\Delta\rho}{1 - K}\right)^{1/2} \right] \tag{56}$$

The dislocation density in cell walls, $\rho + \frac{\Delta\rho}{K}$, is hereafter denoted ρ_w, and the corresponding dislocation-line length is $d_w = 1/\sqrt{\rho_w}$.

Assuming that the cell interior contains a negligible dislocation density,[*] i.e., that $\Delta\rho = (1 - K)\rho$, the expression of the flow stress reduces to Eq. (57) and the dislocation density in cell-walls reduces to $\rho_w = \rho/K$, whence Eq. (58) follows.

$$\tau = \alpha\mu b K^{1/2}/d \tag{57}$$

$$\frac{d_w}{d} = \sqrt{K} \tag{58}$$

The crossing of the wall-network, of mesh size d_w, requires a local stress $\tau_w = \alpha\frac{\mu b}{d_w}$, which, using Eqs. (57) and (58), requires a stress concentration factor $\frac{\tau_w}{\tau} = \frac{1}{K}$. This stress concentration can be provided by pile-ups of dislocations emitted by sources (Fig. 6.40).

In the case of cubic cells of size D and thickness w, the volume fraction of hard zones is given by Eq. (59).

$$K \approx 3w/D \tag{59}$$

[*] Note that under such conditions, the cell interior can deform under a zero stress.

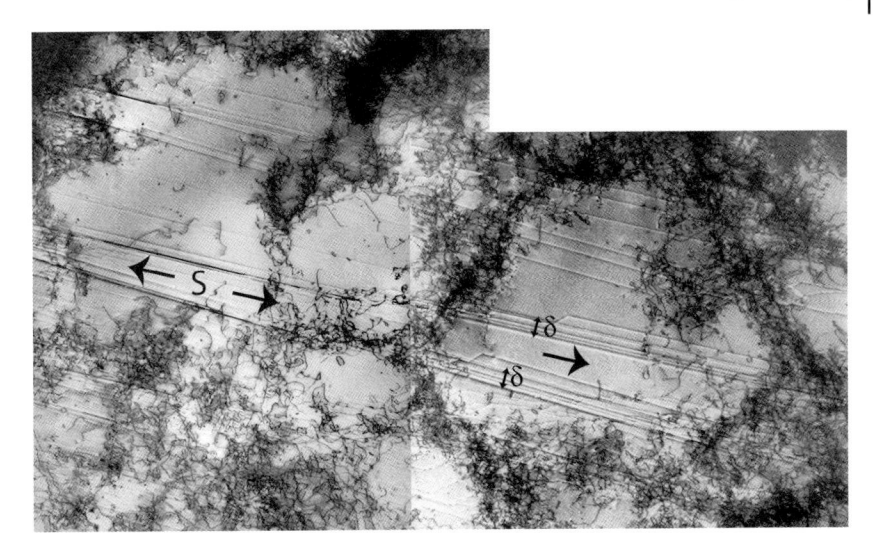

Fig. 6.40 Dislocation emitted by a source S, and crossing through cell walls. Note the spreading δ of the slip planes after wall crossing. In-situ experiment in aluminum, at 80 °C.

According to Kocks and Mecking (2003), this coefficient is not much lower than unity, because the proportionality factor of the Taylor law for a uniform dislocation distribution in stage I ($\tau = \alpha\mu b/\sqrt{\rho}$) is only 10% higher than for a heterogeneous dislocation distribution in stages II and III ($\tau = \alpha K^{1/2}\mu b/\sqrt{\rho}$).

Using Eq. (1), the strain-hardening coefficient in stage II can be expressed as Eq. (60), where the increase in the average dislocation density with strain is $\frac{d\rho}{d\varepsilon} = \frac{2}{b\Lambda}$, and $\Lambda/2$ is the dislocation mean free path.

$$\theta = \frac{d\tau}{d\varepsilon} = \alpha\mu b K^{1/2}\frac{d(\sqrt{\rho})}{d\varepsilon} = \frac{\alpha}{2}K^{1/2}\frac{\mu b}{\sqrt{\rho}}\frac{d\rho}{d\varepsilon} \tag{60}$$

Then the resulting strain-hardening coefficient takes the very simple form of Eq. (61a).

$$\frac{\theta}{\mu} = \alpha K^{1/2}\frac{d}{\Lambda} \tag{61a}$$

However, since it is difficult to estimate the ratio d/Λ directly, it is interesting to express the strain-hardening coefficient as Eq. (61b), where D is the cell size.

$$\frac{\theta}{\mu} = \alpha K^{1/2}\frac{d}{D}\frac{D}{\Lambda} \tag{61b}$$

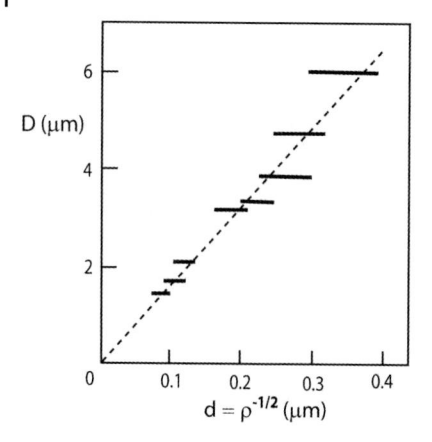

Fig. 6.41 Dislocation cell size D versus average dislocation spacing in copper polycrystals strained to 10% at various temperatures. From Kocks and Mecking (2003).

As a matter of fact, experiments show that d/D is more or less constant and of the order of 1/15 in fcc pure metals (Fig. 6.41), and the slip-line length Λ is a few times larger than the cell size, D (cf. Fig. 6.42, Mughrabi and Ungar, 2002). This yields $\dfrac{\theta}{\mu} \approx \dfrac{K^{1/2}}{120}$, which, considering that $K^{1/2}$ is slightly lower than unity, is in reasonably good agreement with the average experimental value (1/200).

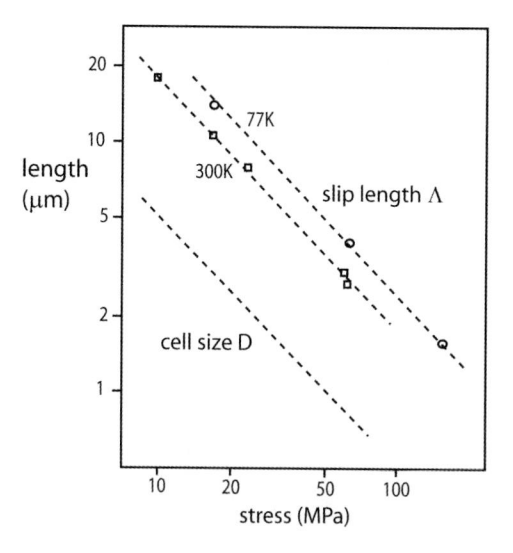

Fig. 6.42 Average slip length Λ and cell-size D as a function of stress, in a single crystal of copper. From Ambrosi and Schwink (1978).

A constant value of $D/d \approx 15$ has been explained tentatively by Nabarro (2000), Kuhlmann-Wilsdorf (2002), and Kocks and Mecking (2003), on the basis of geometrical and energy considerations.

The experimental finding that Λ is in the region of a few D is more surprising and difficult to justify, because only two alternative situations should be met a priori: either wall crossing is impossible, and Λ/D should be equal to unity, or wall crossing is possible, and Λ/D could take any large value. Figure 6.40, which shows that the relationship "$\Lambda \sim$ a few D" is reproduced in an in-situ experiment in Al, yields an answer to this problem. The dislocations emitted by the source S can indeed cross through neighboring walls, as long as the required stress concentration (by a factor of K) can be provided by pile-up effects. This is easily realized close to the source when dislocations move in the same glide plane. However, dislocation reaction and cross-slip at walls result in their spreading in parallel slip planes, which progressively inhibits pile-up formation and subsequent wall crossing.

6.3.3.2 Stage III

In stage III, θ versus τ curves exhibit a more or less linear decrease which can be described by the Voice law (Fig. 6.37) stated in Eq. (62), where θ_0 is of the order of the initial constant strain-hardening coefficient θ_{II}, and where the limit stress τ_l strongly depends on strain rate and temperature.

$$\theta = \theta_0 \left(1 - \frac{\tau}{\tau_l}\right) \tag{62}$$

Equations (60) and (61) show that the decrease of the strain-hardening coefficient θ results from the decrease of α, the proportionality coefficient of the Taylor law, and from an increase of the dislocation mean free path, Λ. The transition between stages II and III is described by the generic term "dynamic recovery". It can be observed on the Haasen plot of Fig. 6.40 that the linear relationship breaks down, which shows that τ_{eff} no longer increases at the same rate as τ_i. In other words, a new thermally activated mechanism, which can no longer be described by the classical equations of the forest mechanism, allows an easier forest crossing (Mecking and Kocks 1981).

The transition between stages II and III is traditionally attributed to the onset of recovery by cross-slip. The main arguments in favor of this interpretation are based on slip-line observations, and on the strong dependence of the onset of stage III on stacking fault energy. However, Basinski (1968) reported experimental observations of cross-slip operating long before the onset of stage III, from halfway through stage II to halfway through stage III. In addition, the similarity of stage III in fcc and bcc metals is difficult to reconcile with the cross-slip described in Section 6.2.5 which is restricted to fcc metals (Nabarro 1986). Several cross-slip and climb mechanisms have been compared by Nabarro (1989).

Most authors consider now that cross-slip is necessary but not sufficient to account for the whole of stage III. Under such conditions, dynamic recovery could be helped by the glide of dislocations in non-close-packed planes (Section 6.2.4.4), which has been observed abundantly at the end of stage III in copper (Anongba et al. 1993 – see also Fig. 6.47, below), and which offers interesting possibilities of decreasing the strength of attractive junctions. Indeed, it can be shown that the only way to move three reacting dislocations in three glide planes intersecting along a common direction is to push one of them in a {100} plane. For instance, many reaction segments in the form of Lomer–Cottrell locks, which are formed in stage II (Basinski and Basinski 1979), have been shown to move in {100} planes by Karnthaler (1978). A similar mechanism has also been observed by Caillard and Martin (1982) in aluminum during creep.

6.3.3.3 Stage IV

Stage IV corresponds to the transformation of thick cell walls into thin regular dislocation networks called sub-boundaries. The deformation stress is stabilized at a value slightly higher than τ_I, which more or less corresponds to the so-called stage II (or steady-state stage) of creep. Dislocation climb is expected to play an important role, at least in the formation and evolution of sub-boundaries. Intensive glide in non-close-packed planes (Section 6.2.4.4) has also been observed at the onset of stage IV in copper (Anongba et al. 1993).

6.3.3.4 Strain-Hardening in Intermetallic Alloys

Several intermetallic alloys such as Ni_3Al and $TiAl$ exhibit the so-called yield stress anomaly, which will be discussed in Section 6.4.1 below. In these materials, strain hardening can be much higher than in fcc metals, and values as high as $\mu/10$ to $\mu/2$ have been reported. Viguier (2003) has shown that Eqs. (55a) (the Taylor law) remains valid, which means that high values of strain hardening θ are related to short dislocation mean free paths Λ. At low strain, the proportionality coefficient α of the Taylor law can be higher than expected. This corresponds to the transition between exhaustion and internal stress hardening discussed in Section 6.3.2.

6.4
Complex Behavior

Many materials exhibit a complex behavior which can be related to mixtures of thermally activated glide, dislocation multiplication, hardening, and recovery processes. In that sense, stage III of hardening could be considered as the first example of a complex behavior. We describe in this section a few other examples, the list of which is of course nonexhaustive.

6.4.1
Yield Stress Anomalies

Many alloys and intermetallic alloys, in their intermediate temperature range, exhibit an anomalous increase of yield stress with increasing temperature, called yield stress anomaly (YSA) (Caillard 2001). This effect accounts for good mechanical properties at high temperatures because it compensates for the natural decrease in yield stress at increasing temperature. YSAs originate from several thermally activated mechanisms of dislocation locking. In contrast to what occurs in the locking–unlocking mechanisms described in Section 6.2.4.2, unlocking is assumed to be impossible or at least very difficult. This results in a small dislocation mean free path Λ decreasing with increasing temperature. According to Eq. (61), this accounts for a high strain-hardening coefficient θ increasing with increasing temperature. Note, however, that the proportionality coefficient α of Eq. (61) may be different from that expected from the Taylor law, because hardening at low strain is of the "exhaustion type" (see Sections 6.3.2 and 6.3.3.4). Lastly, high cumulated strain hardening results in an increase in flow stress at given strain (0.2% for the yield stress). YSA can be classified in two groups, with two different origins for the thermally activated locking mechanism, which are described below.

6.4.1.1 Dynamic Strain Aging

Several alloys exhibit a small YSA associated with stress instabilities and negative values of the stress–strain rate sensitivity. According to the mechanism described in Section 6.2.2.3, there is a range of dislocation velocities where dislocations are subjected to an increasing friction force at increasing temperature, because more

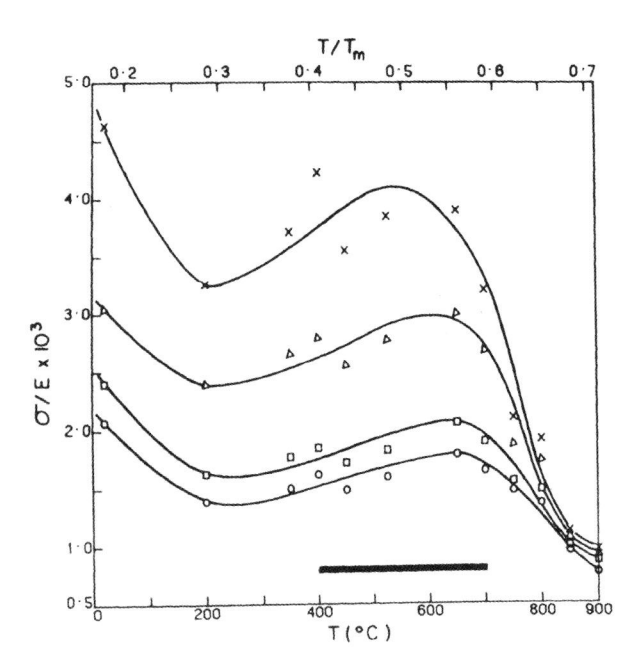

Fig. 6.43 Yield stress anomaly in stainless steel. The bar corresponds to the observation of stress instabilities (or jerky flow). From Kashyap et al. (1988).

solute atoms can diffuse to them and hinder their motion (Fig. 6.6). This effect, which accounts for stress instabilities, can also induce YSAs. An example of this behavior in stainless steel is shown in Fig. 6.43. Other examples are CuAl alloys (see Fig. 6.3), L1$_2$-stabilized titanium trialuminide alloys (see Caillard 2001; Caillard and Couret 1996), and Fe$_3$Al alloys (Molénat et al. 1998).

6.4.1.2 Cross-Slip Locking

The most remarkable example of YSA as a result of cross-slip locking is provided by Ni$_3$Al. In this intermetallic alloy with the L1$_2$ structure, superdislocations are dissociated in two superpartials according to $[110] = 1/2[110] + APB + 1/2[110]$, where APB denotes a stacking fault ribbon of an antiphase boundary. These dislocations glide freely in {111} planes. However, they rapidly lock themselves by thermally activated cross-slip and change of APB plane from {111} to {001} where the APB energy is lower. The locked configuration is called complete Kear–Wilsdorf lock when the APB is completely in the {001} plane, and incomplete Kear–Wilsdork lock when the amount of APB in the {001} plane is very small.

The resulting high strain-hardening coefficients, which are at the origin of the YSA, have been measured by several authors. They can be as high as $\dfrac{\theta}{\mu} \approx \dfrac{1}{4}$ at small strain, according to Neveu (1991). Figure 6.44 shows the yield stress, and the strain hardening, at a larger strain (3%), as a function of temperature. It can be noted that θ increases up to the temperature $T(\theta_{\max})$ and decreases above it. This shows that dislocation locking becomes less efficient for temperatures

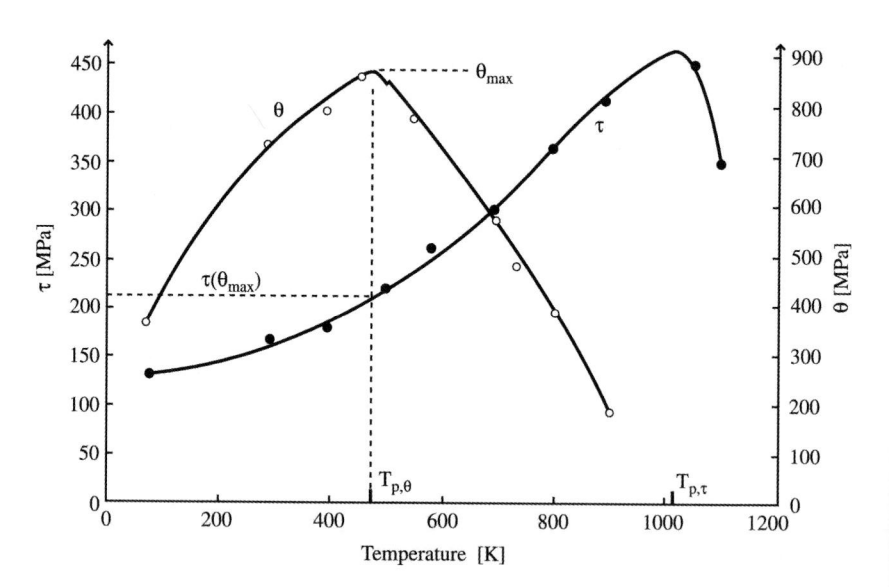

Fig. 6.44 Temperature dependence of the deformation stress (τ) and work-hardening coefficient (θ) in Ni$_3$(Al, 1.5% Hf). From Ezz and Hirsch (1994).

higher than $T(\theta_{max})$, and stresses higher than $\tau(\theta_{max})$. In other words, stresses higher than $\tau(\theta_{max})$ can release a substantial amount of locked screw dislocations. In that sense, the unlocking stress $\tau(\theta_{max})$ is similar to the yield stress in bcc metals for which there is a transition between microyielding (locked screws) and macroyielding (mobile screws). The unlocking stress is characteristic of the obstacle involved. Measurements by Conforto et al. (2005) showed that they are incomplete Kear-Wilsdorf locks, for which the width of the APB in the {001} plane is of the order of one interatomic distance.

The formation of incomplete Kear–Wilsdorf locks, and their breaking above $T(\theta_{max})$ (which leads to the so-called "APB jumps"), have been observed in deformed specimens and in experiments in situ (see Caillard 2001; Conforto et al. 2005).

6.4.2
Fatigue

A fatigue test involves a series of small amounts of deformation by tension and compression over a very large number of cycles. Although the average final deformation is zero at the end of the test, a very large cumulative strain is at the origin of irreversible damage of the sample. A steady state with stress amplitude τ_s (saturation stress) is reached after a certain number of cycles.

Figure 6.45 is the schematic representation of the variation of the saturation stress as a function of the strain amplitude. A homogeneous dislocation structure is built in region A, whereas deformation concentrates in the so-called "persistent slip bands" (PSBs), in region B. The density of PSBs increases with increasing strain amplitude in region B, until deformation is again homogeneous in region C. PSB play a very important role in the lifetime of fatigued samples because they

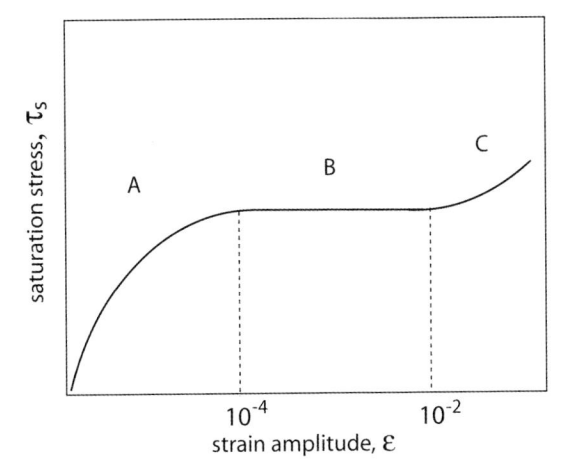

Fig. 6.45 The three stages of fatigue.

are privileged sites for crack initiation. The main aspects of fatigue are developed below. Many more details can be found in the review paper of Basinski and Basinski (1992).

6.4.2.1 Microstructure of Fatigued Metals and Alloys

Most observations have been carried out in the PSBs developing in region B of Fig. 6.45. PSBs are slices of material parallel to the primary glide plane. They are divided into channels bounded by walls in which screw dislocations move back and forth by glide and cross-slip (Fig. 6.46).

One can find a very good correspondence in several materials between the saturation stress, τ_s, and the corresponding channel width, D. It can be expressed by $\tau_s = 1.5$ to $2\tau_{bend}$, where $\tau_{bend} = \dfrac{2\hat{T}}{bD}$ is the stress necessary to bend a screw dislocation in a channel [the same equation as (55c)] (Basinski and Basinski 1992). Since $\tau_s > \tau_{bend}$, screw dislocations have the possibility of overcoming obstacles between the walls, in agreement with TEM observations of Mughrabi (1983) showing that they do interact with many dipoles and debris.

Walls are made of dipoles of height h and average distance d. Although they were too small to be observed by conventional TEM, the dipoles were first assumed to be vacancy-type. Their density was estimated from resistivity measurements (Basinski and Basinski 1992), and the stress $2\hat{T}/bd$ necessary to cross through the walls was accordingly considered to be a few times τ_s. More recent measurements by weak-beam TEM have shown that (a) vacancy and interstitial dipoles are both present in the walls, and (b) dipole sizes and densities vary only

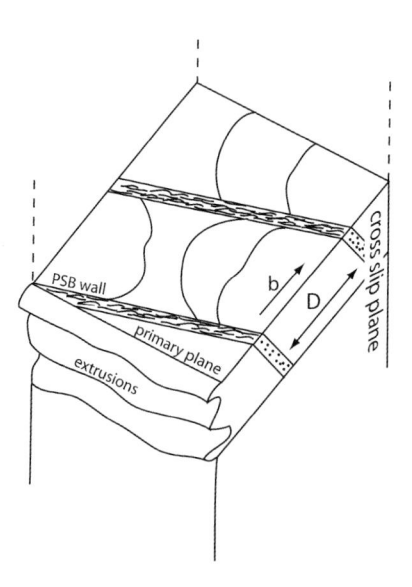

Fig. 6.46 The microstructure of fatigue.

weakly with temperature and stress (Tippelt et al. 1997). The minimum dipole height seems to increase slightly with increasing temperature, which gives some indication in favor of a climb-controlled recovery mechanism. However, all other quantities, average and maximum dipole heights, and average dipole distances ($d \approx 8$–$10h$), have too a small variation to be related to the saturation stress τ_s. Under such conditions, the stress necessary to cross through the walls is considered to be larger than the saturation stress, but not proportional to it.

6.4.2.2 Comparison with Stages II and III of Monotonic Strain Hardening

It has been noted that the saturation stress is very close to the stress τ_{III} at which stage III begins in tension (Section 6.3.3.2) (Basinski and Basinski 1992). This fairly good correspondence is illustrated in Fig. 6.47.

The activation energy varies between 0.2 eV at 77 K and 0.7 eV at 300 K in copper, which a priori excludes climb as a rate controlling process (Basinski and Basinski 1992). Activation volumes have been measured by means of strain-rate jumps by Kaschner and Gibeling (2002). When inserted in a Haasen plot (Fig. 6.48), they follow the same variation as in stage III of hardening, which corroborates the link between the two mechanisms.

6.4.2.3 Intrusions, Extrusions and Fracture

Many extrusions and intrusions are visible along the emerging PSBs (see the schematic representation in Fig. 6.46). They are usually interpreted as the result of the partly irreversible motion of screw segments subjected to cross-slip. The role of the vacancies generated by the annihilation of the dipoles inside the walls

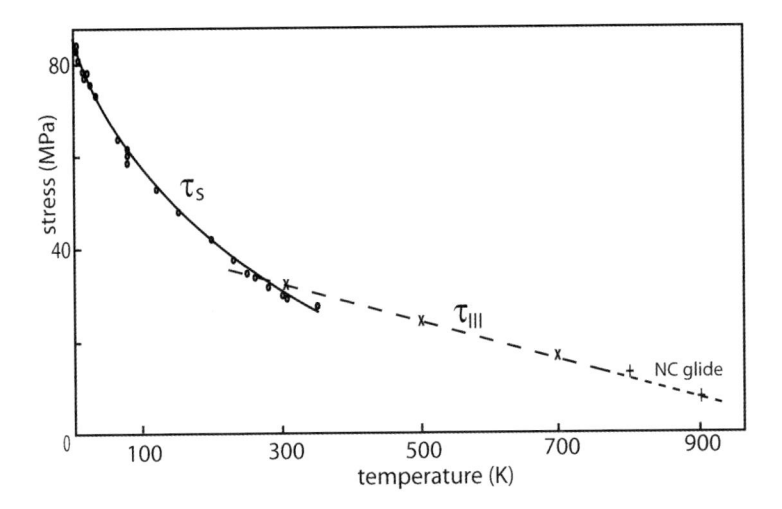

Fig. 6.47 Comparison between the saturation stress in fatigue, τ_s, and the onset of stage III, in copper. Non-close-packed (NC) glide is observed in the highest temperature range. From Basinski and Basinski (1989) and Anongba et al. (1993).

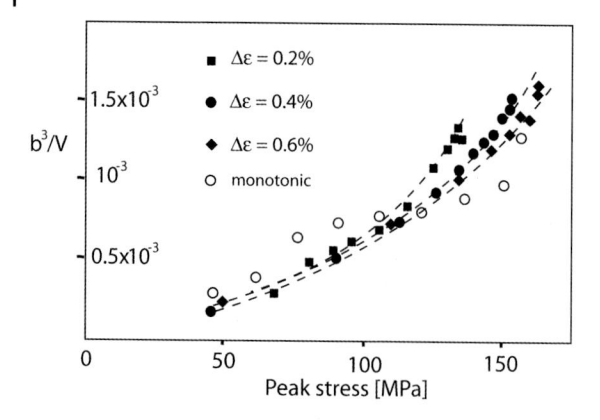

Fig. 6.48 Haasen plot of the activation volume of fatigue in copper, and comparison with data from monotonic tests. From Kaschner and Gibeling (2002).

is unclear. Long-range diffusion of vacancies may account for extrusions being more pronounced than intrusions. However, Basinski and Basinski (1992) concluded that there is no direct evidence for the existence of single vacancies formed as the result of fatigue. They also pointed out that intrusions are more difficult to observe than extrusions, and that vacancies are not sufficiently mobile to account for the formation of intrusion/extrusion at very low temperatures.

In any case, intrusions are responsible for crack initiation leading to failure. On the contrary, the homogeneous deformation taking place in region A of Fig. 6.46 inhibits sample fracture. More details can be found in the "Proceedings of Fundamentals of Fracture" published in *Phil. Mag. A* **82**(17–18) (2002).

6.4.2.4 Conclusions

Fatigue appears to be controlled by the motion of screw dislocations in the channels rather than by the interaction of the edge segments with the walls. The role of dipole recovery by climb and point defect diffusion is unclear, especially at low temperature. The strong connection between the saturation stress in fatigue and the onset of stage III in monotonic tension indicates that the cross-slip and/or forest mechanism should be rate-controlling.

6.4.3
Strength of Nanocrystalline Alloys and Thin Layers

Nanocrystalline metals and alloys have been studied extensively in recent years because of their ultra-high deformation stress. The Hall–Petch law, which accounts for the strength of polycrystals with grain sizes down to 20 nm, is introduced in Section 6.4.3.1. Then, the possible origins for the Hall–Petch law breakdown which occurs at very small grain size is discussed in Section 6.4.3.2. Results are transposable to thin layers although they have a reduced size in one direction.

6.4.3.1 The Hall–Petch Law (Grain Size $D \geq 20$ nm)

It has been known for a long time that the yield stress of polycrystal, τ, varies as a function of grain size D according to the Hall–Petch law [Eq. (63), where k and τ_0 are constants for a given material].

$$\tau(D) = \tau_0 + k/\sqrt{D} \tag{63}$$

Different elaboration processes (sintering, large deformation, etc.) generate different types of grain boundaries (low angle sub-boundaries, various types of high-angle boundaries, twins, etc.) which oppose different resistances to dislocation movements and hence give rise to different values of k (see, e.g., Hansen 2004).

It has been shown that the plasticity results mostly from dislocation movements (see, e.g., Kumar et al. 2003: in Ni with D as low as 30 nm). Under such conditions, the Hall–Petch law was first explained in terms of dislocation pile-ups emitted by sources and blocked against grain boundaries. However, since this model cannot describe large-grain materials where long dislocation pile-ups have never been observed, a more general explanation has been proposed on the basis of the strain-hardening laws discussed in Section 6.3.3 (Saada 2005; Lefebvre et al. 2005). The results are summarized below.

The increase in dislocation density with increasing strain is given by $\dfrac{d\rho}{d\varepsilon} = \dfrac{1}{b\Lambda}$

[Eqs. (60–61)], where Λ is the mean free path of mobile dislocations. Considering that Λ at yield is determined by the grain size D, we obtain, after integration:

$$\rho = \frac{\varepsilon}{bD} + cte$$

and, using the Taylor law [Eq. (55d)],

$$\tau = \alpha\mu b\sqrt{\rho} + cte = \alpha\mu\sqrt{\varepsilon b}/\sqrt{D} + cte$$

This expression is similar to Eq. (63) with $k = \alpha\mu\sqrt{\varepsilon b}$, which, taking $\varepsilon = 0.2\%$, has the right order of magnitude. It should be noted that Λ can be slightly larger than the grain size D in nanomaterials processed by heavy deformation, where grains are often limited by walls and low-angle sub-boundaries (cf. Section 6.3.3.1). This probably explains why the corresponding values of k can be smaller than expected.

6.4.3.2 Hall–Petch Law Breakdown (Grain Size $D \leq 20$ nm)

Although the experimental results are not always fully convincing, the Hall–Petch law seems to break down below a critical grain size of the order of 20 nm (Takeuchi 2001; Zhao et al. 2003; Nieh and Wang 2005). A transition toward an inverse behavior is observed instead. Obviously, this results from the increasing fraction of atoms that are contained in grain boundaries. Considering that the grain boundary thickness is of the order of 0.5 nm, Saada (2005) indeed estimates this fraction as 10% for $D = 30$ nm. Assuming that grain boundaries behave like amorphous materials, Nieh and Wang (2005) interpret the properties of nanocrystalline BeB alloys (in particular their lack of ductility) by a progressive transition toward a 100% amorphous material.

Although some dislocation activity is still sometimes reported, it is clear that deformation arises mostly from grain boundary sliding (Markmann et al. 2003; Kumar et al. 2003). However, the term "sliding" should not be interpreted in the sense of a pure shear mechanism, because it is controlled by grain boundary self-diffusion. Considering that the grain boundary self-diffusion activation energy is smaller than the bulk one, substantial amounts of plasticity can be achieved by grain boundary sliding even at yield and at room temperature. Simulations taking all these effects into account have been carried out by Derlet et al. (2003).

6.4.4
Fracture

The fracture behavior of alloys changes from brittle to ductile as the temperature rises. This is a key issue in many structural applications. Fracture is ductile as soon as dislocations can be emitted at the crack front, thus blunting the crack tip, releasing the stress concentration, and slowing down the crack propagation. Rice and Thompson (1974) performed the first attempts at modeling dislocation activity at the crack tip. They considered the different forces acting on a dislocation emitted at the crack surface: the crack stress field, the surface tension force corresponding to the creation of a step, and the image force repelling the dislocation. The authors concluded that materials are the most ductile when they have a small $\mu b/\gamma$ (where γ is the surface energy) and wide dislocation core. It is clear, however, that this process is helped by thermal activation via an easier dislocation nucleation at free surface and easier propagation away from the crack tip.

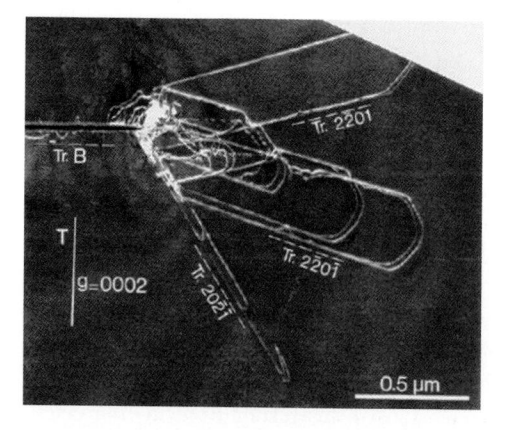

Fig. 6.49 Dislocation emission from a crack tip, in Ti_3Al. In-situ experiment at 200 °C. From Legros et al. (2006).

Dislocation emission is also enhanced at crack tip heterogeneities (Hirsch et al. 1989). Dislocation emissions in complex but realistic situations have been calculated by Schöck (1996) and Xu and Argon (1997). Fracture is easily initiated at surface heterogeneities such as extrusions created in fatigue (Section 6.4.2.3), voids appearing in highly deformed polycrystals and multiphased materials, etc. Experiments in situ can yield information on the emission of dislocations at crack tips (Caillard et al. 1999; Legros et al. 2006). One example is shown in Fig. 6.49. This experiment shows that both dislocation emission from the crack tip and dislocation motion in the tangled area close to the crack tip are difficult processes.

6.4.5
Quasicrystals

The mechanical properties of quasicrystals are different from those of ordinary alloys in several aspects. First, they are highly brittle up to three-quarters of their melting temperature and become very ductile above. Second, the stress–strain curves exhibit a pronounced yield drop followed by a stage of very low (sometimes negative) strain hardening. Third, although it is well known since the work of Wollgarten et al. (1993) that deformation proceeds by dislocation motion, specific mechanisms are expected on account of the lack of translational periodicity.

The structure of quasicrystals has been described in Chapter 2 (Section 2.6). Let us just recall what is necessary to understand the structure of dislocations. Figure 6.50 shows a two-dimensional aperiodic tiling defined by the cut of a pile-up of cubes along an irrational plane close to {111}, where each tile is the projection of one face of an emerging cube. Important to note is the presence of wavy bands of constant thickness (one is highlighted on Fig. 6.50b) which correspond to the emergence of the {100} planes of the three-dimensional cubic structure. These bands are equivalent to the parallel (but not equidistant) dense planes of the quasicrystalline structure. The real icosahedral quasicrystalline structure is defined in the same way, by cut and projection of a six-dimensional hypercubic lattice. It contains corrugated "dense planes" which are equivalent to the wavy bands of Fig. 6.50(b). The space which is lost during the projection (i.e., the direction perpendicular to the cut plane in Fig. 6.50) is called "perpendicular space".

Dislocations can be introduced by the Volterra process described in Fig. 6.1, via two distinct procedures which are described schematically in Fig. 6.50(c)–(f).
- If the Volterra process is performed in the aperiodic two-dimensional lattice (i.e., the cut plane in Fig. 6.50), for instance by removing half a wavy band and pasting the two lips together again, a partial dislocation is obtained (Fig. 6.50c). As a matter of fact, a fault is created on account of the lack of translational periodicity, which is seen in Fig. 6.50(d) as a step in the truncated three-dimensional pile-up of cubes.

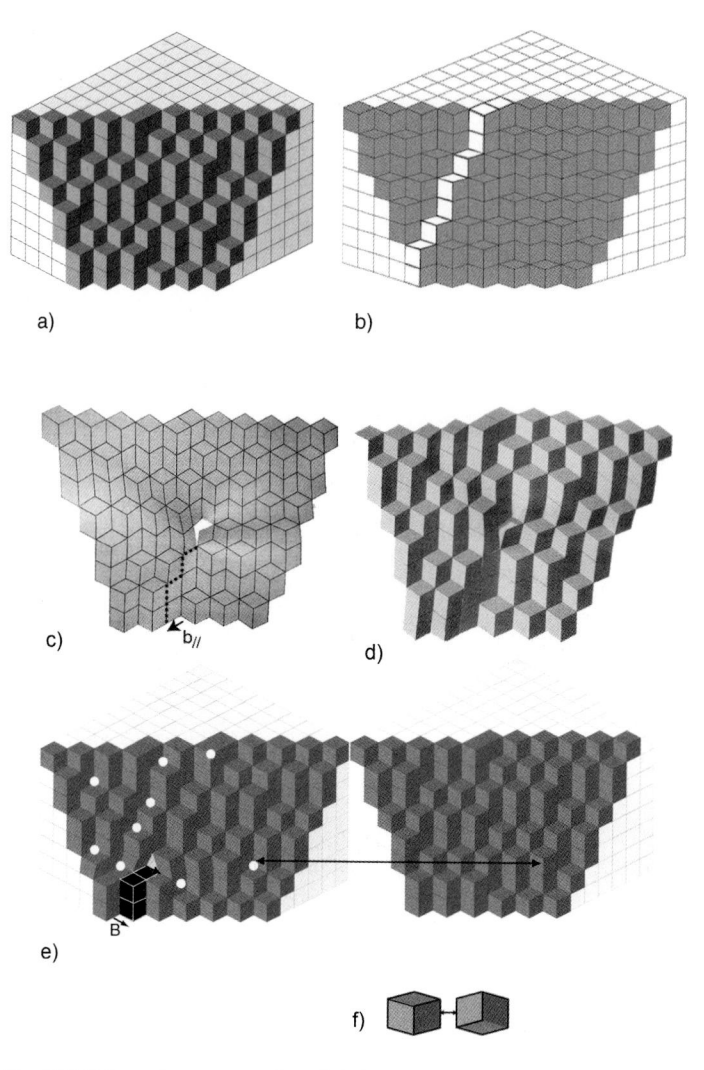

Fig. 6.50 Schematic representation of dislocations in quasicrystals: (a,b) definition of a two-dimensional aperiodic tiling from a three-dimensional periodic network; (c,d) creation of an imperfect dislocation by the Volterra process in two-dimensional tiling; (e) creation of a perfect dislocation by the Volterra process in the three-dimensional network; (f) phason point defect created at dots in (e). From Mompiou and Caillard (2005).

- If the Volterra process is performed in the periodic three-dimensional lattice, by insertion of a supplementary half {100} plane (dark in Fig. 6.50e), and if the dislocated structure is subsequently cut and projected in the two-dimensional plane, a perfect dislocation is obtained. In this case, the dislocation is no longer connected to a fault. However, since the supplementary {100} half-plane has distorted the periodic cubic lattice along every direction, in particular the direction perpendicular to the figure, the cut plane does not intersect the very same cubes, which results in local tiling changes all around the dislocation. These point defects, which are typical of quasicrystals, are called phasons.

In summary, quasicrystals can contain partial dislocations connected to "phason walls" (which can be compared to partial dislocations connected to stacking faults in crystals). However, these partial dislocations can transform into perfect ones if their phason walls evaporate and evolve to a cloud of individual phasons surrounding the dislocations. These two kinds of dislocations have been observed abundantly in TEM (see, e.g., Mompiou and Caillard 2005).

The mechanical properties of quasicrystals have been reviewed by several authors (Feuerbacher et al. 1999, Takeuchi 1999). They are characterized by low activation volumes and high activation energies which do not yield any clear indication on the rate-controlling process.

Microscopic observations indicate that the back-stress due to the formation of phason walls in the wake of moving dislocations is a substantial but not preponderant fraction of the flow stress. The remaining part is due to internal stresses and to the dislocation–lattice interaction. The latter has been explained in a first step by obstacles opposing dislocation glide. More recent investigations, however, showed that dislocation motion takes place by climb (Fig. 6.51; see Caillard et al. 2000; Mompiou et al. 2003). Conversely, glide appears almost impossible, probably on account of the very high corrugation of dense planes, which is an intrinsic property of quasicrystals, as shown in Fig. 6.50b.

Under such conditions, the low-temperature brittleness can easily be interpreted by the absence of glide, which is the usual mode of deformation of conventional alloys at low temperature. However, interpretations of the high-temperature mechanical properties are facing the same difficulties as in Mg (Section 6.2.6.6), because the activation parameters are not those expected a priori from the simplest theories of climb. As a matter of fact, the dislocation velocity varies with stress as $v \propto \tau^n$, with n ranging between 4 and 8 (Mompiou 2004),

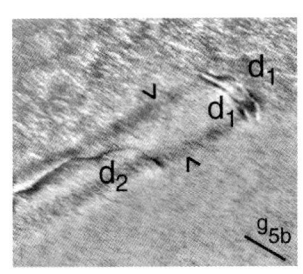

Fig. 6.51 Dislocations d_1 and d_2 moving by climb in an AlPdMn quasicrystal. The contrast of the phason wall is arrowed. From Mompiou et al. (2003).

whereas Eq. (48) accounts for smaller values not exceeding 3. Further work is thus necessary to reconcile the microscopic and macroscopic aspects of climb, in crystals and in quasicrystals.

6.5
Outlook

Although several aspects of dislocation plasticity may appear complex, major progress has been made toward rationalizing mechanical properties in terms of a few elementary processes. Indeed, the number of different thermally activated mechanisms controlling the plasticity of metals and alloys – even the most complex ones – appears fairly small, and the main tendencies of dislocation patterning in monotonic deformation and in fatigue are now well understood.

Since several elementary mechanisms are contributing, in parallel and/or in series, to the mobility of each moving dislocation, the determination of the most important one – which will determine the macroscopic mechanical properties – is a delicate step. It can be done unambiguously only when there is a fairly good agreement between results of several complementary techniques, e.g., conventional "post mortem" TEM, in-situ TEM, and transient deformation experiments.

Research into new materials is still very active in the direction of lower weight, higher strength (especially at high temperature), and higher ductility. In this respect, it is expected that small-grain and nanograin alloys and intermetallic alloys will be developed extensively in the near future. However, and without any doubt, further progress in this field requires a better understanding of almost all the fundamental properties described in this chapter, in particular diffusion-controlled climb processes, grain boundary sliding, and nucleation of dislocations at surfaces and interfaces.

References

Aifantis, E.C., 1986, *Mater. Sci. Eng.* **81**, 563.

Alexander, H., Haasen, P., 1968, *Solid State Phys.* **22**, 28.

Ambrosi, P., Schwink, Ch., 1978, *Scripta Met.* **12**, 303.

Anongba, P.N.B., Bonneville, J., Martin, J.L., 1993, *Acta Met. Mater.* **41**, 2897 and 2907.

Aono, Y., Kuramoto, E., Kitajima, K., 1983, in *Proc. 6th Int. Conf. on the Strength of Metals and Alloys*, ed. R.C. Gifkins, Pergamon, New York, p. 9.

Basinski, Z., 1968, in *Dislocation Dynamics*, ed. A.R. Rosenfeld, G.T. Hahn, A.L. Bement, R.I. Jaffee, Mc Graw-Hill, New York, p. 674.

Basinski, S.J., Basinski, Z.S., 1979, in *Dislocations in Solids*, Vol. 4, Ch. 16, ed. F.R.N. Nabarro, North-Holland, Amsterdam, p. 261.

Basinski, Z.S., Basinski, S.J., 1989, *Acta Met.* **37**, 3255.

Basinski, Z.S., Basinski, S.J., 1992, *Progr. Mater. Sci.* **36**, 89.

Biget, M.P., Saada, G., 1989, *Phil. Mag.* **59**, 747.

Bonneville, J., Escaig, B., 1979, *Acta Met.* **27**, 1477.

Bonneville, J., Vanderschaeve, G., 1985, *Strength of Metals and Alloys*, Vol. 1, ed. H.J. McQueen, J.P. Bailar, J.I. Dickson, J.J. Jonas, M.G. Akben, p. 9.

Bonneville, J., Escaig, B., Martin, J.L., 1988, *Acta Met.* **36**, 1989.

Caillard, D., 2001, *Mater. Sci. Eng. A* **319–321**, 74.

Caillard, D., Couret, A., 1996, in *Dislocations in Solids*, Vol. 10, Ch. 50, ed. F.R.N. Nabarro, M.S. Duesberry, Elsevier, Amsterdam, p. 69.

Caillard, D., Martin, J.L., 1982, *Acta Met.* **30**, 791.

Caillard, D., Martin, J.L., 2003, *Thermally Activated Mechanisms in Crystal Plasticity*, Pergamon Materials Series, ed. R.W. Cahn, Pergamon, Amsterdam.

Caillard, D., Vailhé, C., Farkas, D., 1999, *Phil. Mag. A* **79**, 723.

Caillard, D., Vanderschaeve, G., Bresson, L., Gratias, D., 2000, *Phil. Mag. A* **80**, 237.

Carrard, M., Martin, J.L., 1988, *Phil. Mag. A* **58**, 491.

Conforto, E., Molénat, G., Caillard, D., 2005, *Phil. Mag.* **85**, 117.

Cottrell, A.H., Bilby, B.A., 1949, *Proc. Phys. Soc. London A* **62**, 49.

Cottrell, A.H., Stockes, R.J., 1955, *Proc. Roy. Soc. A* **233**, 17.

Couret, A., 1999, *Phil. Mag. A* **79**, 1977.

Couret, A., Caillard, D., 1985, *Acta Met.* **33**, 1447 and 1455.

Couret, A., Caillard, D., 1989, *Phil. Mag. A* **59**, 783 and 801.

Derlet, P.M., Hasnaoui, A., Swygenhoven, H., 2003, *Scripta Mat.* **49**, 629.

Duesberry, M.S., 1983, *Acta Met.* **31**, 1759.

Duesberry, M.S., 1989, in *Dislocations in Solids*, Vol. 8, Ch. 39, ed. F.R.N. Nabarro, North-Holland, Amsterdam, p. 67.

Duesberry, M.S., Vitek, V., 1998, *Acta Mater.* **46**, 1481.

Dupas, C., Zuodar, N., Coddet, O., Kruml, T., Martin, J.L., 2002, *J. Phys: Cond. Matter.* **14**, 12 989.

Edelin, G., 1971, *Phil. Mag.* **23**, 1547.

Edelin, G., Poirier, J.P., 1973, *Phil. Mag.* **28**, 1203 and 1211.

Eggeler, G., Dlouhy, A., 1997, *Acta Met.* **45**, 4251.

Epishin, A., and Link, T., 2004, *Phil. Mag.* **84**, 1979.

Escaig, B., 1968, *J. Physique* **29**, 225.

Estrin, Y., Kubin, L.P., 1989, *J. Mech. Behav. Mater.* **2**, 255.

Eyring, J., 1936, *J. Chem. Phys.* **4**, 283.

Ezz, S.S., Hirsch, P.B., 1994, *Phil. Mag. A* **69**, 105.

Farenc, S., Caillard, D., Couret, A., 1995, *Acta Met. Mater.* **43**, 3669.

Feuerbacher, M., Klein, H., Schall, P., Bartsch, M., Messerschmidt, U., Urban, K., 1999, *MRS Symp. Proc.* **553**, 307.

Fleischer, R.L., 1961, *Acta Met.* **9**, 996.

Flor, H., Nortmann, A., Dierke, H., Neuhauser, H., 2003, *Z. Metallkde* **94**, 572.

Fraser, H.L., Smallman, R.E., Loretto, M.H., 1973a, *Phil. Mag.* **28**, 651.

Fraser, H.L., Loretto, M.H., Smallman, R.E., 1973b, *Phil. Mag.* **28**, 667.

Friedel, J., 1964, *Dislocations*, Pergamon, Oxford.

Furubayashi, E., 1969, *J. Phys. Soc. Japan* **27**, 130.

Granato, A.V., 1971, *Phys. Rev. B* **4**, 2196; *Phys. Rev. Lett.* **27**, 660.

Guyot, P., Dorn, J.E., 1967, *Can. J. Phys.* **45**, 983.

Haasen, P., 1979, in *Dislocations in Solids*, Vol. 4, Ch. 15, ed. F.R.N. Nabarro, North-Holland, Amsterdam, p. 155.

Hahner, P., 1996, *Acta Mat.* **44**, 2345.

Hansen, N., 2004, *Scripta Mat.* **51**, 801.

Hansen, N., Kuhlman-Wilsdorf, D., 1986, *Mater. Sci. Eng.* **81**, 141.

Hirsch, P.B., Roberts, S.G., Samuels, J., Warner, P.D., 1989, in *Advances in Fracture Research*, Vol. 1, ed. K. Salama, K. Ravi-Chandar, D.M.R. Taplin, P. Rama Rao, Pergamon, Oxford, p. 139.

Hirth, J.P., Lothe, J., 1982, *Theory of Dislocations*, Krieger, Malabar, FL.

Ikeno, S., Furubayashi, E., 1972, *Phys. Stat. Sol. (a)* **12**, 611.

Karnthaler, H.P., 1978, *Phil. Mag. A* **38**, 141.

Kaschner, G.C., Gibeling, J.C., 2002, *Acta Mat.* **50**, 653.

Kashyap, B.P., Taggart, K., Tangri, K., 1988, *Phil. Mag. A* **57**, 97.

Kocks, U.F., Mecking, H., 2003, *Progr. Mater. Sci.* **48**, 171.

Kocks, U.F., Argon, A.S., Ashby, M.F., 1975, *Thermodynamics and Kinetics of Slip*, Pergamon, Oxford.

Kubin, L.P., Estrin, Y., 1990, *Acta Met. Mater.* **38**, 697.

Kuhlmann-Wilsdorf, D., 2002, in *Dislocations in Solids*, Vol. 11, Ch. 59, ed. F.R.N. Nabarro, M.S. Duesberry, Elsevier, Amsterdam, p. 211.

Kumar, K.S., Suresh, S., Chisholm, M.F., Horton, J.A., Wang, P., 2003, *Acta Mat.* **51**, 387.
Kuramoto, E., Aono, Y., Kitajima, K., 1979, *Scripta Met.* **13**, 1039.
Labusch, R., 1970, *Phys. Stat. Sol.* **41**, 659.
Lefebvre, S., Devincre, B., Hoc. T., 2005, *Mater. Sci. Eng. A* **400–401**, 150.
Legrand, B., 1985, *Phil. Mag. A* **52**, 83.
Legros, M., Couret, A., Caillard, D., 1996, *Phil. Mag. A* **73**, 61 and 81.
Legros, M., Couret, A., Caillard, D., 2006, *J. Mater. Sci.* **41**, 2647.
Le Hazif, R., Dorizzi, P., Poirier, J.P., 1973, *Acta Metall.* **21**, 903.
Levine, E.D., 1966, *Trans. Met. Soc. AIME*, **236**, 1558.
Louchet, F., Ignat, M., 1986, *Acta Met.* **34**, 1681.
Louchet, F., Kubin, L.P., 1979, *Phys. Stat. Sol. (a)* **56**, 169.
MacCormick, P.G., 1972, *Acta Met.* **20**, 351.
Madek, R., Devincre, B., Kubin, L.P., 2002, *Phys. Rev. Lett.* **89**, 255 508.
Majimel, J., Casanove, M.J., Molénat, G., 2004, *Mater. Sci. Eng. A* **380**, 110.
Markmann, J., Bunzel, P., Rösner, H., Liu, K.W., Padmanabhan, K.A., Birringer, R., Gleiter, H., Weissmüller, J., 2003, *Scripta Mat.* **49**, 637.
Mecking, H., Kocks, U.F., 1981, *Acta Met.* **29**, 1865.
Molénat, G., Caillard, D., 1992, *Phil. Mag. A* **65**, 1327.
Molénat, G., Rösner, H., Caillard, D., 1998, *Mater. Sci. Eng. A* **258**, 196.
Mompiou, F., 2004, Ph.D Thesis, Université Paul Sabatier, Toulouse.
Mompiou, F., Caillard, D., 2005, *Mater. Sci. Eng. A* **400–401**, 283.
Mompiou, F., Bresson, L., Cordier, P., Caillard, D., 2003, *Phil. Mag.* **83**, 3133.
Moulin, A., Condat, M., Kubin, L.P., 1999, *Acta Mat.* **47**, 2879.
Mughrabi, H., 1983, *Acta Met.* **31**, 1367.
Mughrabi, H., Hungar, T., 2002, in *Dislocations in Solids*, Vol. 11, Ch. 60, ed. F.R.N. Nabarro, M.S. Duesberry, p. 343.
Mulford, R.A., Kocks, U.F., 1979, *Acta Met.* **27**, 1125.
Nabarro, F.R.N., 1947, *Proc. Phys. Soc.* **59**, 256.
Nabarro, F.R.N., 1986, in *Strength of Metals and Alloys*, Vol. 3, ed. H.J. Mc Queen, J.P. Bailar, J.I. Dickson, J.J. Jonas, M.G. Akben, Pergamon, New York, p. 1661.
Nabarro, F.R.N., 1989, *Acta Met.* **37**, 1521.
Nabarro, F.R.N., 2000, *Phil. Mag. A* **80**, 759.
Naka, S., Lasalmonie, A., Costa, P., Kubin, L.P., 1988, *Phil. Mag. A* **57**, 717.
Neuhauser, H., Schwink, C., 1993, in *Materials Science and Technology*, Vol. 6, ed. R.W. Cahn, P. Haasen, E.J. Kramer, VCH Verlag, Weinheim, p. 191.
Neveu, C., 1991, PhD Thesis no. 1623, Orsay University, France.
Nieh, T.G., Wang, J.G., 2005, *Intermetallics* **13**, 377.
Peierls, R.E., 1940, *Proc. Phys. Soc.* **52**, 34.
Poirier, J.P., 1976, *Plasticité à Haute Température des Solides Cristallins*, Eyrolles, Paris.
Puschl, W., 2002, *Progr. Mater. Sci.* **47**, 415.
Rao, S., Parthasarathy, T.A., Woodward, C., 1999, *Phil. Mag.* **79**, 1167.
Rice, J.R., Thompson, R., 1974, *Phil. Mag.* **29**, 73.
Saada, G., 1960, *Acta Met.* **8**, 841.
Saada, G., 2005, *Mater. Sci. Eng. A* **400–401**, 146.
Schöck, G., 1996, *Phil. Mag.* **74**, 419.
Schöck, G., 1980, in *Dislocations in Solids*, Vol. 3, Ch. 10, ed. F.R.N. Nabarro, Elsevier, p. 63.
Schöck, G., Frydman, R., 1972, *Phys. Stat. Sol. (b)* **53**, 661.
Seeger, A., 1956, *Phil. Mag.* **1**, 651.
Shenoy, V.B., Kukta, R.V., Phillips, R., 2000, *Phys. Rev. Lett.* **84**, 1491.
Silcox, J., Whelan, M.J., 1960, *Phil. Mag.* **5**, 1.
Simmons, J.P., Rao, S.I., Dimiduk, D.M., 1997, *Phil. Mag.* **A75**, 1299.
Smallman, R.E., Westmacott, K.H., 1972, *Mater. Sci. Eng.* **9**, 249.
Srinivasan, R., Savage, M.F., Daw, M.S., Noebe, R.D., Mills, M.J., 1997, *MRS Symp. Proc.* **460**, 505.
Stroh, A.N., 1954, *Proc. Phys. Soc. B* **67**, 427.
Takeuchi, S., 1999, *MRS Symp. Proc.* **553**, 283.
Takeuchi, S., 2001, *Scripta Mat.* **44**, 1483.
Takeuchi, S., Argon, A.S., 1976, *Acta Met.* **24**, 883.
Tippelt, B., Bretschneider, J., Hähner, P., 1997, *Phys. Stat. Sol. (a)* **163**, 11.
Van der Beukel, A., 1975, *Phys. Stat. Sol. (a)* **30**, 197.

Vegge, T., Jacobsen, K.W., 2002, *J. Phys: Condens. Matter* **14**, 2929.

Viguier, B., 2003, *Mater. Sci. Eng. A* **239**, 132.

Vitek, V., 1966, *Phys. Stat. Sol.* **18**, 687.

Washburn, J., 1972, in *Radiation Induced Voids in Metals*, ed. J.W. Corbett, L.C. Ianello, Nat. Techn. Inf. Service, US Dpt. of Commerce, Springfield (VI), p. 647.

Wickham, L.K., Schwarz, K.W., Stölken, J.S., 1999, *Phys. Rev. Lett.* **83**, 4574.

Wille, T.H., Gieseke, W., Schwink, C.H., 1987, *Acta Met.* **35**, 2679.

Wollgarten, M., Beyss, M., Urban, K., Liebertz, H., Koster, U., 1993, *Phys. Rev. Lett.* **71**, 549.

Xu, G., Argon, A.S., 1997, *Phil. Mag.* **75**, 341.

Yoshinaga, H., Morozumi, S., 1971, *Phil. Mag.* **23**, 1367.

Zhao, M., Li, J.C., Jiang, Q., 2003, *J. Alloys Compounds* **361**, 160.

7
Phase Equilibria and Phase Transformations

Brent Fultz and Jeffrey J. Hoyt

Metallic alloys are composed of more than one chemical species of atom, and these atoms have preferred spatial arrangements at fixed temperatures and pressures. This chapter explains the thermodynamic origin of these spatial preferences, using simple models and some advanced concepts.[1] It comprises five main sections. Section 7.1 is devoted to thermodynamic equilibrium, with emphasis on how phase diagrams can be understood with generic interatomic interactions. Section 7.2 covers departures from equilibrium owing to kinetic constraints, and also how interatomic interactions bias the atom movements that lead to equilibrium structures. The last three sections (7.3–7.5) cover the major categories of phase transformations in alloys, with emphasis on microstructural evolution from a continuum viewpoint.

7.1
Alloy Phase Diagrams

This section presents the alloy physics of phase diagrams. It begins with the development of phase diagrams of binary alloys as functions of temperature and composition, using concise arguments that should be understood by all students of materials science. The section continues with an overview of more advanced phenomena that are needed for a quantitative understanding of alloy phase diagrams. Some of these topics are covered in greater depth in Chapters 4, 10, and 11.

7.1.1
Solid Solutions

Consider a binary alloy with atom species A and B arranged randomly on the sites of a crystal. The concentration of B atoms is c, and the concentration c (in

---

1) Some phenomena are described in more depth in Chapters 3, 4, 5, 8, 10, 11, and 12.

Alloy Physics: A Comprehensive Reference. Edited by Wolfgang Pfeiler
Copyright © 2007 WILEY-VCH Verlag GmbH & Co. KGaA, Weinheim
ISBN: 978-3-527-31321-1

mole fractions) varies from 0 to 1. When $c = 0$ the alloy is pure element A, when $c = 1$ the alloy is pure element B. The concentration of A atoms is $1 - c$. The crystal has N sites, and therefore it has cN B atoms and $(1 - c)N$ A atoms. Each site is surrounded by z sites as first-nearest-neighbors (1NN). For a random solid solution, on average each A atom is surrounded by zc B atoms, and its other $(1 - c)z$ neighbors are A atoms. We start with a basic thermodynamic analysis of a solid solution using two key assumptions:

- All the energy of atom configurations depends on first-nearest-neighbor pairs of atoms.[2] Each A–A pair has energy e_{AA}, each A–B pair has energy e_{AB}, and each B–B pair has energy e_{BB}. To obtain the total energy of an alloy configuration, we count the number of pairs with each energy.
- The alloy is truly random, so the species of atom on any lattice site does not depend on the species of atoms occupying its neighboring sites.

Consider an A atom on a lattice site. Its energy depends on the number of its A atom neighbors, $z(1 - c)$, and the number of its B atom neighbors, zc. The energy for this average A atom, and likewise for the average B atom, is

$$e_A = \frac{z}{2}(1 - c)e_{AA} + \frac{z}{2}ce_{AB} \tag{1}$$

$$e_B = \frac{z}{2}(1 - c)e_{AB} + \frac{z}{2}ce_{BB} \tag{2}$$

The division by 2 corrects for the double counting that occurs when finding the average by summing over all the atoms in the crystal (because the bonds from the neighbor atoms back to our central atom are counted two times, in both directions).

We can now write the partition function for a single site, $Z_{1\text{site}}$:

$$Z_{1\text{site}} = e^{-\beta e_A} + e^{-\beta e_B} \tag{3}$$

where $\beta \equiv 1/(k_B T)$. Using the second assumption above, all site occupancies are independent, so the total partition function of the alloy with N sites is

$$Z_N = Z_{1\text{site}}^N \tag{4}$$

We can evaluate Z_N with the binomial expansion,

---

[2] One way to justify this assumption is to argue that all the chemical bonding energy is associated with neighboring atoms only.

7.1 Alloy Phase Diagrams

$$Z_N = \sum_{n=0}^{N} \left(\frac{N!}{(N-n)!n!}\right)(e^{-\beta e_A})^{N-n}(e^{-\beta e_B})^n \tag{5}$$

When N is very large, Z_N becomes a sharply peaked function with a maximum at $n = cN$. It is a standard practice in statistical mechanics to replace Z_N with its maximum value (Kittel 1969),

$$Z_N(c) \simeq \left(\frac{N!}{(N-n)!n!}\right)(e^{-\beta e_A})^{N-n}(e^{-\beta e_B})^n \tag{6}$$

Obtaining the Helmholtz free energy from the expression $F = -k_B T \ln Z$, and using the Stirling approximation $\ln(x!) \simeq x \ln x - x$, we obtain Eqs. (7–10)

$$F(c) = [(N-n)e_A + n e_B](z/2)$$
$$\quad - k_B T[N \ln N - n \ln n - (N-n)\ln(N-n)] \tag{7}$$

$$F(c) = [e_{AA} + 2c(e_{AB} - e_{AA}) + c^2(4V)](zN/2)$$
$$\quad - k_B T[\ln N - c \ln n - (1-c)\ln(N-n)]N \tag{8}$$

$$E_{\text{conf}}(c) = [e_{AA} + 2c(e_{AB} - e_{AA}) + c^2(4V)](zN/2) \tag{9}$$

$$S_{\text{conf}}(c) = k_B[-c \ln c - (1-c)\ln(1-c)]N \tag{10}$$

Here $E_{\text{conf}}(c)$ and $S_{\text{conf}}(c)$ are the configurational energy and configurational entropy of the alloy. The word "configurational" refers to arrangements of atoms on the crystal sites, and here $E_{\text{conf}}(c)$ and $S_{\text{conf}}(c)$ were derived for a random arrangement.

In Eq. (8) we defined an important combination of pair potentials:

$$V \equiv (e_{AA} + e_{BB} - 2e_{AB})/4 \tag{11}$$

This parameter V is sometimes called the "interchange energy." Its physical meaning is made clear when we consider the change in configuration of an A and B atom pair as shown in Fig. 7.1. For the initial configuration of the central

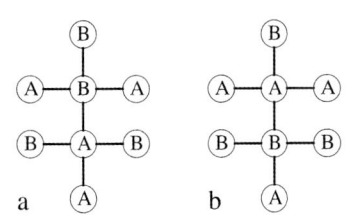

Fig. 7.1 (a) Initial configuration around central A and B atoms. (b) New configuration after exchange of the central A–B pair.

A and B atoms in Fig. 7.1(a), the figure shows a total of five A–B pairs, one A–A pair, and one B–B pair. After interchange of only the central A and B atoms,[3] we have three A–B pairs, two A–A pairs, and two B–B pairs (Fig. 7.1(b)). After interchange, the change in pair energy is precisely $4V = e_{AA} + e_{BB} - 2e_{AB}$ (or an integer multiple of $4V$ for different configurations).

Consider first the case where $V < 0$, meaning that the average of A–A and B–B pairs is more favorable than A–B pairs. This gives the alloy a tendency to unmix chemically. The opposite sign of $V > 0$ means that at low temperatures the alloy will tend to maximize the number of its A–B pairs. It does so by developing chemical order, defined by order parameters presented in Section 7.1.7.

The energy and entropy terms of Eqs. (9) and (10) are plotted in Fig. 7.2(a). When $E_{config}(c)$ is added to $-TS_{config}(c)$ to obtain the free energy curves in Fig. 7.2(b), the logarithmic singularity of the entropy always forces a downward slope away from $c = 0$ and $c = 1$. The energy term, positive in this case with $V < 0$, can dominate at intermediate concentrations, especially at low temperatures. For small magnitudes of $zV/k_B T$ (high temperatures), the minimum in free energy is obtained for a random solid solution, which has a simple shape in Fig. 7.2(b) when the ratio is -0.667. The free energy $F_{config}(c)$ has an interesting curvature in c when the ratio is more negative than -1, or equivalently for temperatures

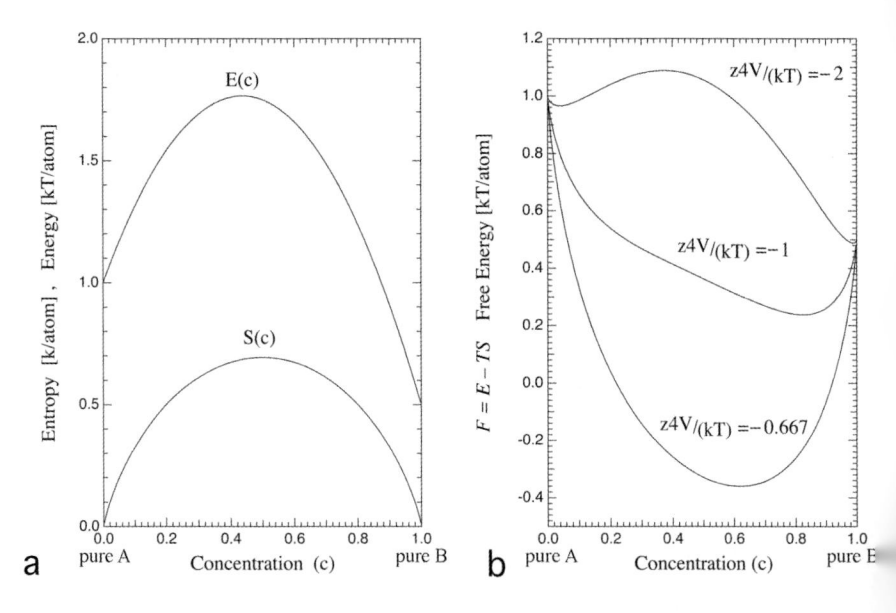

Fig. 7.2 (a) Configurational energy and entropy for $V < 0$ [Eqs. (9) and (10)]. (b) Free energy for different values of $z4V/(kT)$ as labeled.

---

3) Note that this interchange conserves the total number of atoms of each type. Also note that an interchange of like pairs, A–A or B–B, has no effect on the total number of like or unlike pairs.

below $T_c = -4.0zV/k_B$. This T_c proves to be the "critical temperature" for "spinodal decomposition" at a concentration of $c = 0.5$. Further interpretation of these curves, the subject of the next sections, leads to the unmixing phase diagram of Fig. 7.6(a), below.

7.1.2
Free Energy and the Lever Rule

When working with alloy thermodynamics, the analysis of free energy versus composition curves is an essential skill because it is the basis for understanding equilibrium phase diagrams of temperature and composition. In what follows we deduce the basic forms of alloy T–c phase diagrams. Furthermore, modifications of this analysis can sometimes be used when equilibrium is constrained by kinetic processes. Some considerations for the Helmholtz free energy, $F(c) = E(c) - TS(c)$, for each phase are listed below:

- The phase of maximum entropy will dominate at the highest temperatures. For metallic systems this is a gas of isolated atoms, but at lower temperatures most metals form a liquid phase with continuous solubility of A and B atoms. There are cases of chemical unmixing in the liquid phase, however, for which the thermodynamics of Eq. (8) is again relevant.
- At the lowest temperatures, the equilibrium case for a general composition is a combination of crystalline phases differing in their chemical compositions.
- At low to intermediate temperatures, the equilibrium phases and their chemical compositions depend in detail on the free energy versus composition curves $F_\xi(c)$ for each phase ξ.

A simple but important phase diagram is for a binary alloy with just two phases: a liquid phase of continuous solubility, and one crystalline solid phase of continuous solid solubility. The free energy versus composition curves, $F_S(c)$ and $F_L(c)$, are similar to the lowest curve of 7.2(b), but they differ in detail. Figure 7.3 shows a typical case. At the highest temperature, T_4, the liquid curve has the lowest free energy for any composition. The alloy is a liquid at all compositions, as expected, since the liquid has the higher entropy. The $-TS$ part of the free energy for the liquid, $F_L(c)$, changes quickly with temperature, however, because S_L is large. With decreasing temperature, the free energy curve for the liquid rises relative to the solid curve (and may flatten somewhat). At the lowest temperature, T_1, the crystalline solid curve has the lowest free energy, F, for any composition, consistent with its lower energy, E. At the temperature T_1, the alloy is a solid at all compositions. The intermediate temperatures T_2 and T_3 require further analysis. It is not simply a matter of picking the lowest of the free energy curves at a particular composition because the alloy can unmix chemically. The free energy is minimized by selecting an optimal combination of liquid and solid phases of different chemical compositions.

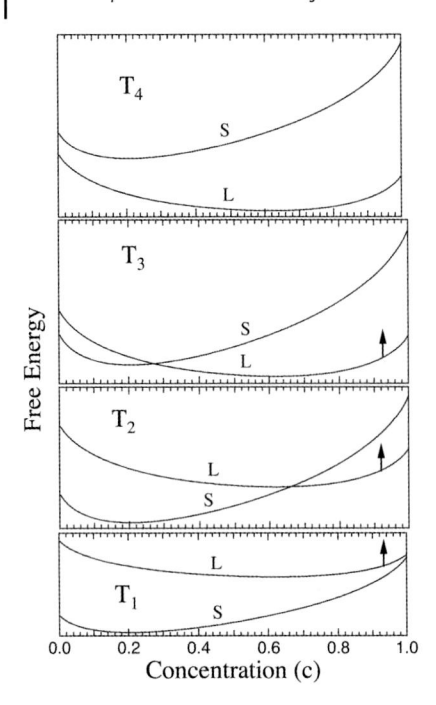

Fig. 7.3 Free energy versus composition curves for solid and liquid phases at four temperatures $T_4 > T_3 > T_2 > T_1$.

The conservation of solute provides a constraint between the fractional amounts of phases and the chemical compositions of these phases. Consider the material depicted in Fig. 7.4, which has an overall composition of $c_0 = 0.3$. If it decomposes into a mixture of phases having different compositions, conservation of solute requires that one of the phases has more solute than c_0, and the other less. The composition deviations from c_0, indicated by the gray zones in Fig. 7.4, must average to zero. The condition is stated in Eqs. (12) and (13).

$$(c_\alpha - c_0)f_\alpha + (c_\beta - c_0)f_\beta = 0 \tag{12}$$

$$\frac{c_0 - c_\alpha}{f_\beta} = \frac{c_\beta - c_0}{f_\alpha} \tag{13}$$

The symbols α and β are used here as general names for phases, and f_α and f_β are the mole fractions of the α and β phases. Because $f_\alpha + f_\beta = 1$, Eqs. (14) and (15) follow:

$$f_\alpha = \frac{c_\beta - c_0}{c_\beta - c_\alpha} \tag{14}$$

$$f_\beta = \frac{c_0 - c_\alpha}{c_\beta - c_\alpha} \tag{15}$$

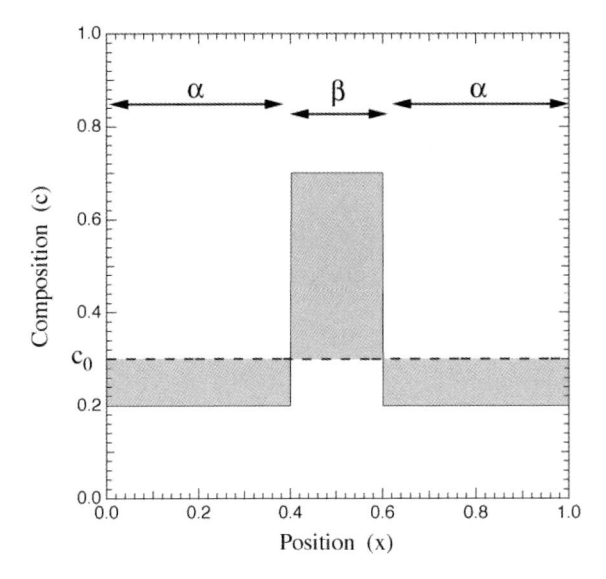

Fig. 7.4 Composition profile of a two-phase alloy with overall composition $c_0 = 0.3$. The area of the upper gray band is $(0.7 - 0.3) \times 0.2 = 0.8$, whereas the areas of the two lower gray bands are each $(0.2 - 0.3) \times 0.4 = -0.4$.

Recall the balancing of a lever, where the heavier mass (larger f) placed on the shorter arm of the lever (smaller Δc) balances a smaller mass on the longer lever arm. Equation (13) is therefore known as the "lever rule." The lever rule is a consequence only of solute conservation, and is useful when working either with curves of free energy versus composition, or with T–c phase diagrams.

7.1.3
Common Tangent Construction

One more tool is needed to relate curves of free energy versus composition to phase diagrams – the "common tangent construction." It is based on the reasonable assumption that the total free energy, F_T, is the sum of contributions from the fractions of the α and β phases. In Eq. (18) below we transform this to a dependence on composition [through Eq. (15)] to make it more applicable to curves of free energy versus composition [Eqs. (16–19)]

$$F_T(f_\alpha, c_\alpha, f_\beta, c_\beta) = f_\alpha F_\alpha(c_\alpha) + f_\beta F_\beta(c_\beta) \tag{16}$$

$$F_T(c_\alpha, c_\beta, f_\beta) = (1 - f_\beta) F_\alpha(c_\alpha) + f_\beta F_\beta(c_\beta) \tag{17}$$

$$F_T(c_\alpha, c_\beta, c_0) = \left(1 - \frac{c_0 - c_\alpha}{c_\beta - c_\alpha}\right) F_\alpha(c_\alpha) + \left(\frac{c_0 - c_\alpha}{c_\beta - c_\alpha}\right) F_\beta(c_\beta) \tag{18}$$

$$F_T(c_\alpha, c_\beta, c_0) = F_\alpha(c_\alpha) + \left(\frac{c_0 - c_\alpha}{c_\beta - c_\alpha}\right) (F_\beta(c_\beta) - F_\alpha(c_\alpha)) \tag{19}$$

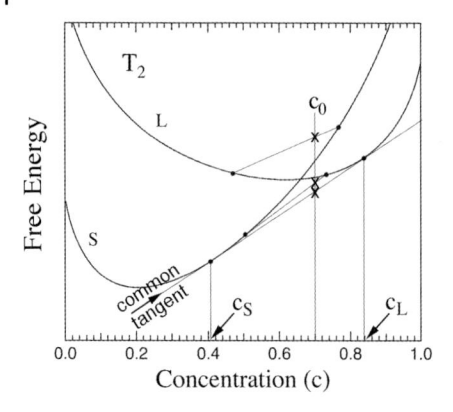

Fig. 7.5 Free energy versus composition curves for solid (S) and liquid (L) phases at the temperature T_2. For a selected c_0, the total free energy is obtained on the straight line between selected points on the two free energy curves. The minimum free energy is found on the tangent common to the two curves.

One way to interpret Eq. (19) is as the equation of a straight line of free energy versus the composition c_0 in the range $c_\alpha < c_0 < c_\beta$. The values of F_T at the endpoints of the line are $F_\alpha(c_\alpha)$ and $F_\beta(c_\beta)$, and F_T remains on a straight line of F versus c_0 at any intermediate composition. This linear relationship of Eq. (19) can be used to optimize the chemical compositions of a two-phase mixture. One of the curves of Fig. 7.3 (for temperature T_2) is redrawn in Fig. 7.5. This figure includes three sloping straight lines between pairs of endpoints on the curves F_L and F_S. A chemical composition of $c_0 = 0.7$ is picked for illustration.

For thermodynamic equilibrium, we minimize the total free energy F_T by moving the endpoints of the sloping lines in Fig. 7.5, keeping them on the curves of the two allowable phases. From the linear relationship of Eq. 7.19, we know that F_T will be at the three intersections marked by the crosses, "×". From Fig. 7.5 we see that the lowest possible position for the "×" is found on the "common tangent" between the two curves of free energy versus composition. The common tangent minimizes F_T. The equilibrium compositions of the solid and liquid phases are therefore at the compositions of the common tangent, and are labeled "c_L" and "c_S" in Fig. 7.5.

7.1.4
Unmixing and Continuous Solid Solubility Phase Diagrams

We now construct phase diagrams for alloys with the free energy versus composition curves of Figs. 7.2 and 7.3. First note from Fig. 7.5 that the compositions c_L and c_S from the common tangent construction serve to minimize the free energy of a mixture of solid and liquid phases for any intermediate composition (here from $0.4 < c_0 < 0.84$). With $F_\xi(c)$ curves at various temperatures for each phase

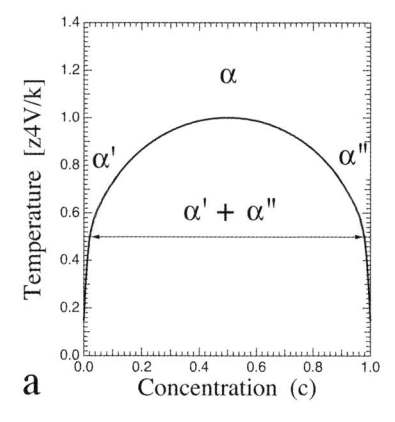

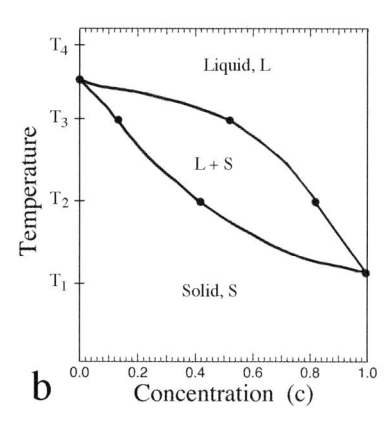

Fig. 7.6 (a) Unmixing phase diagram derived from curves of Fig. 7.2. Arrow marks composition limits obtained by common tangent construction from top curve of Fig. 7.2(b). (b) Continuous solid solubility phase diagram derived from curves of Fig. 7.3. Dots mark compositions obtained by common tangent constructions.

ξ, it is a straightforward exercise to construct common tangents and identify the compositions of the phase boundaries. Results are presented in Fig. 7.6.

The unmixing phase diagram in Fig. 7.6(a) shows a solid solution at high temperatures, denoted the "α-phase." At low temperatures this solution becomes unstable to chemical unmixing, and forms an A-rich "α'-phase" and a B-rich "α''-phase." Beneath the arching curve in Fig. 7.6(a), the chemical compositions of the two phases can be read directly from the curve itself for any temperature.[4] The horizontal arrow points to these two compositions for a normalized temperature of 0.5 (which is consistent with the common tangent applied to the top free energy curve of Fig. 7.2(b)). In this unmixing phase diagram of Fig. 7.6(a), the composition segregation is greater at lower temperatures, and is infinitesimally small at $c = 0.5$ when at the critical temperature $T_c = 4.0z|V|/k_B$.

The phase diagram shown in Fig. 7.6(b) is for the case of continuous solid solubility. The dots on this diagram were obtained directly from the common tangents of the free energy versus composition curves of Fig. 7.3 and 7.5. Some discretion was exercised in interpolating between the points. Between the temperatures T_3 and T_4, perhaps halfway between them, Fig. 7.3 shows that the last of the A-rich solid phase has melted. The pure element A has a single melting point, of course, and this is also true for the element B. The phase diagram shows that an A–B alloy does not have a single melting temperature, however. In equilibrium at temperatures above the "solidus" (the lower curve in this phase diagram), the alloy tends to form a little liquid that is enriched in the element B. Just below

---

[4] The fractional amounts of the phases are set by the lever rule or Eqs. (14) and (15).

the "liquidus" (the upper curve in this phase diagram), the alloy is mostly liquid with a small amount of solid phase enriched in the element A.

7.1.5
Eutectic and Peritectic Phase Diagrams

It is possible to have equilibrium between three phases in a binary A–B alloy at a fixed temperature and pressure. The canonical examples involve a liquid phase in equilibrium with two solid phases. For example, Pb dissolves in Sb in the liquid phase, but at low temperatures the Pb-rich phase is fcc, whereas the Sb-rich phase is hexagonal. A separate free energy versus composition curve is needed for each crystalline phase.

There are two possible forms of phase diagrams involving one high-temperature phase and two low-temperature phases – "eutectic" and "peritectic" phase diagrams. The form depends on the ordering in composition of the minima of the three $F_\xi(c)$ curves. The eutectic case occurs when the $F_L(c)$ curve of the liquid phase lies between the curves for the two solid phases as shown in Fig. 7.7(a). This figure shows four liquid free energy curves for four temperatures. For simplicity, the solid phase curves are assumed constant with temperature, since they have the lower entropy. With increasing temperature, the liquid free energy curve falls relative to the solid curves owing to the large $-TS_L$ term

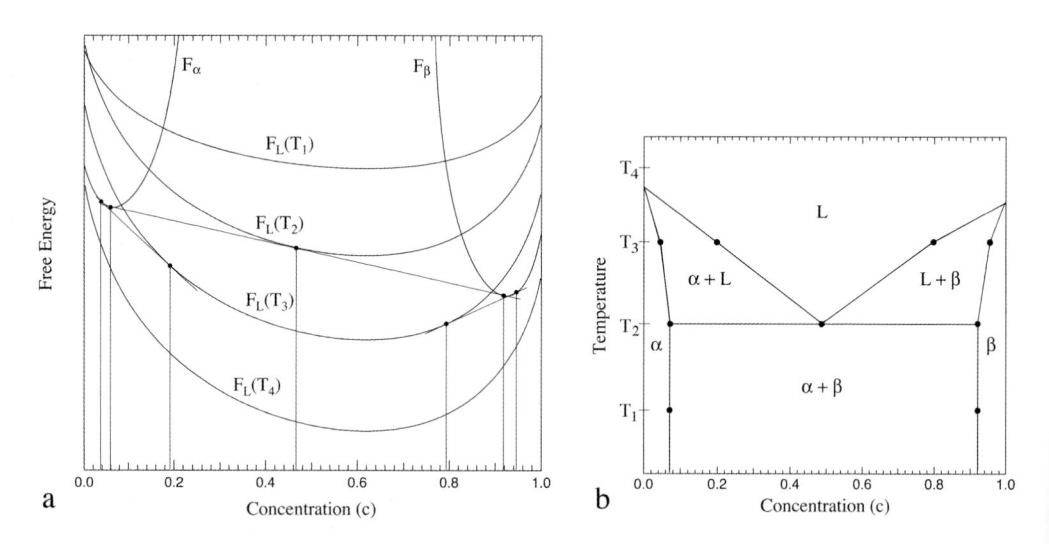

Fig. 7.7 (a) Free energy curves for two solid phases, α and β, and a liquid phase drawn for four temperatures $T_1 < T_2 < T_3 < T_4$. Three common tangent constructions are shown, from which seven compositions are found. (b) Eutectic phase diagram derived for curves of part (a). Dots mark compositions from common tangent constructions.

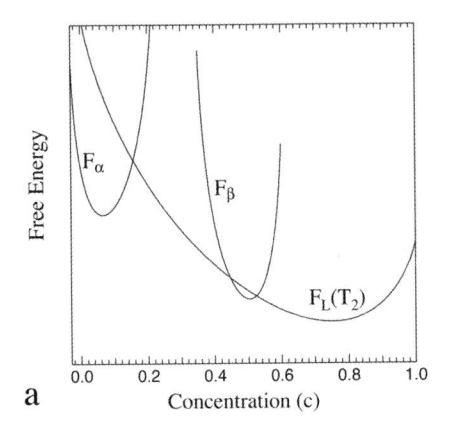

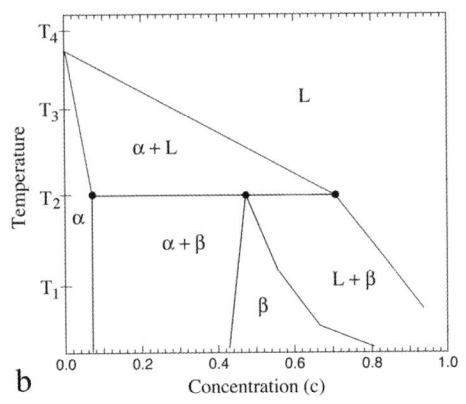

Fig. 7.8 (a) Free energy curves for two solid phases, α and β, and a liquid phase drawn for one temperature. (b) Peritectic phase diagram estimated from curves of part (a).

in the liquid free energy. Three common tangent constructions are shown in the figure. The most interesting one is at the temperature T_2, where the common tangent touches all three phases, with the liquid in the middle. This temperature T_2 is the "eutectic temperature." These points of contact are shown in Fig. 7.7(b), which is a complete eutectic phase diagram.

The other possible ordering in composition for three free energy curves occurs when the liquid curve has its minimum on one side of the minima of the two curves for the solids. This case is depicted schematically in Fig. 7.8(a), with the corresponding "peritectic" phase diagram in Fig. 7.8(b). This phase diagram can be constructed by the same common tangent exercise as was used for the eutectic case in Fig. 7.7 – lowering with temperature the $F_L(c)$ curve relative to $F_\alpha(c)$ and $F_\beta(c)$. Peritectic phase diagrams are common when one crystalline phase, here the β-phase, has a much lower melting temperature than the α-phase. Note also that if the liquid free energy curve were to lie to the left of the two solid curves, the common tangent construction would produce a peritectic phase diagram with the higher melting temperatures to the right.

7.1.6
More Complex Phase Diagrams

Eutectic, peritectic, and continuous solid solubility phase diagrams can be assembled into more complex diagrams with multiple phases and two-phase regions (see Okamoto 2000). In various combinations, however, the forms of Figs. 7.6, 7.7(b), and 7.8(b), plus the ordering diagrams discussed below, account for almost all features found in real binary alloy phase diagrams. For example, two eutectic phase diagrams over composition ranges $0 < c < 0.5$ and $0.5 < c < 1.0$ could be attached together at the composition $c = 0.5$, where a third solid phase,

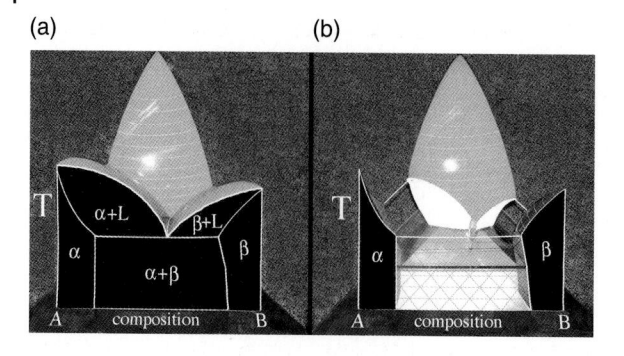

Fig. 7.9 (a) Ternary phase diagram of elements A, B, and C (to back). (b) Sections removed to show that the ternary eutectic composition has a lower melting temperature than any of the three binaries. The triangular grid at the bottom depicts the alloy composition.

perhaps an ordered compound, is stable.[5] Alternatively, a peritectic phase diagram could be attached to one side of a eutectic diagram. In another variation, the liquid phase in a eutectic diagram could be replaced by a random solid solution, perhaps the disordered solid solution in the phase diagram of Fig. 7.6(b). (When a solid solution replaces the liquid region in Figs. 7.7(b) or 7.8(b), the diagram is called a "eutectoid" or "peritectoid" diagram, respectively.)

The same principles apply to "ternary" alloys containing three elements. Instead of one independent composition variable as for a binary alloy, a ternary alloy has two, so a two-dimensional T–c diagram is no longer sufficient. Figure 7.9 is an example of a compound eutectic diagram for a ternary alloy. Here the temperature is plotted vertically, and the concentration of the third element, C, increases with distance from the front to the back of the structure. This ternary diagram is a special case where each pair of elements, A–B, B–C, and A–C, has a eutectic phase diagram. The regions of the A–B eutectic diagram are labeled in Fig. 7.9(a). The Gibbs phase rule (Gibbs 1876) states that adding a component to the system (the element C in the ternary alloy) gives an additional degree of freedom in the intensive variables of temperature, pressure or mole fractions. By the Gibbs phase rule, the two-phase areas of the binary T–c eutectic diagram at constant pressure are now volumes in the ternary diagram. Two of these volumes are removed in Fig. 7.9(b), showing more clearly the solidus surfaces and the low-temperature ternary eutectic point of equilibrium between the liquid and three solid phases.

Handbooks of ternary phase diagrams usually do not show three-dimensional structures as in Fig. 7.9, but instead present various cuts through the structure. A cut with a vertical plane includes all temperatures, but only a constrained set

---

[5] This double eutectic diagram could be generated with the free energy versus composition curves of Fig. 7.17.

of compositions. The intersections of the phase boundaries on such vertical cuts are called "pseudo-binary" diagrams. More commonly, horizontal cuts are used to depict isothermal sections of the ternary diagram. These are shown with a triangular composition boundary, with the pure elements at each corner. A stack of these diagrams is required to depict the ternary diagram at multiple temperatures. A subtlety of ternary isothermal sections concerns the compositions in equilibrium across two-phase regions. The lever rule is still valid, but for an alloy composition in a two-phase region it is necessary to know the "tie-lines" that identify the compositions of the two equilibrium phases. Ternary alloys allow for phases and microstructures beyond those possible for binary alloys, but today our knowledge of ternary phase diagrams is much less complete. Many compilations are missing phases, temperatures, or tie-lines, and it remains a challenge to obtain this important information by experiment and theory.

7.1.7
Atomic Ordering

We return [Eq. (20)] to the issue of atom interaction preferences expressed in Eq. (11):

$$V \equiv \frac{e_{AA} + e_{BB} - 2e_{AB}}{4} \qquad (20)$$

An alloy with unmixing tendencies has $V < 0$, so A–A and B–B pairs are preferable to A–B pairs. An alloy with the opposite tendency, $V > 0$, tends to maximize the number of A–B pairs at low temperatures. At high temperatures both types of alloys are expected to become random solid solutions if the configurational entropy is dominant, but their phases at low temperatures are fundamentally different because of their different signs of V.

For illustration of an ordering transformation we choose the square lattice shown in Fig. 7.10, and an equiatomic A–B alloy. For $V > 0$, we expect the struc-

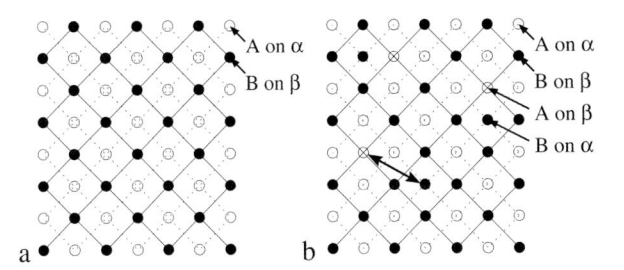

Fig. 7.10 (a) Ordered square lattice with two species $\{A, B\}$ and two interpenetrating square sublattices $\{\alpha, \beta\}$. (b) Misplaced atoms A and B (antisites) on generally ordered sublattices. The double arrow indicates how a duo of A and B atoms can be exchanged to eliminate antisites.

ture of Fig. 7.10(a) at $T = 0$. Figure 7.10 identifies sublattices for the two species of atoms; for a crystal with N sites, each of these sublattices contains $N/2$ sites.

Finite temperatures favor some disorder in the structure – putting a few atoms on the wrong sublattice gives a big increase in the configurational entropy,[6] driving some mixing of A atoms onto the β-sublattice. By conservation of atoms and conservation of sublattice sites, an equal number of B atoms must move onto the α-sublattice. The number of these "wrong" atoms (or "antisites") on each sublattice is W, and the number of "right" atoms is R. The sublattice concentrations are used to define a "long-range order" (LRO) parameter, L, for the alloy (Warren 1990), given by Eq. (21).

$$L \equiv \frac{R - W}{N/2} \tag{21}$$

This L can range from -1 to $+1$. The case of $L = 1$ corresponds to Fig. 7.10(a), where $R = N/2$ and $W = 0$. The disordered alloy has as many right as wrong atoms on each sublattice, i.e., $R = W$, so $L = 0$ for a disordered alloy.

We now calculate the temperature dependence of the order parameter, $L(T)$, again by counting the number of A–A, A–B, and B–B pairs of atoms as in Section 7.1.1. As we move atoms between sublattices, we must be sure that we conserve the number of each species. We ensure this by considering two atoms at a time – we transform between a "right duo" which comprises an A atom and a B atom on their proper sublattices, and a "wrong duo" which comprises an A atom and a B atom on the wrong sublattices (antisites). This corresponds to swapping the two atoms at the ends of the double arrow in Fig. 7.10. To calculate the equilibrium value of the LRO parameter, we seek the ratio of right and wrong duos as a function of temperature. The energy of the two atoms in the wrong duo is given by Eq. (22).

$$e_W = \frac{zR}{N/2} e_{AA} + \frac{zR}{N/2} e_{BB} + 2\frac{zW}{N/2} e_{AB} \tag{22}$$

site: wrong A   wrong B   wrong A (B)

neighbor: right A   right B   wrong B (A)

To understand Eq. (22), remember that the neighbors of an atom are on the other sublattice, so the first term is from the wrong A which has right A atoms as neighbors, the second term is from the wrong B which has right B atoms as neighbors, and the third term is from the other neighbors of the wrong A which are wrong B atoms (plus the wrong A neighbors of the wrong B). Exchanging the

---

6) In fact the entropy has a logarithmic singularity in sublattice concentrations analogous to the concentration dependence of Eq. 7.10 near $c = 0$.

positions of the wrong A and B atoms (as indicated by the double arrow in Fig. 7.10) puts the two atoms on their proper sublattices. The analogous expression for the energy of the "right" duo is Eq. (23), where the A neighbors of the right A are wrong, the B neighbors of the right B are wrong, but the B (A) neighbors of the right A (B) are right.

$$e_R = \frac{zW}{N/2} e_{AA} + \frac{zW}{N/2} e_{BB} + 2\frac{zR}{N/2} e_{AB} \tag{23}$$

On the average, the energy of a single atom will be half that of Eq. (22) or (23). We calculate the ratio of right to wrong duos (equal to R/W) as the ratio of their Boltzmann factors, Eqs. (24) and (25).

$$\frac{R}{W} = \frac{e^{-\beta e_R/2}}{e^{-\beta e_W/2}} \tag{24}$$

$$\frac{R}{W} = \exp\left[-\beta \frac{1}{2}(R-W)(2e_{AB} - e_{AA} - e_{BB})\frac{z}{N/2}\right] \tag{25}$$

With Eq. (21) and noting that $R + W = N/2$, we find that $R = (1+L)N$ and $L = (1-L)N$, so Eq. (25) becomes Eq. (26), where again $V \equiv (e_{AA} + e_{BB} - 2e_{AB})/4$.

$$\frac{1+L}{1-L} = \exp\left[\frac{L2Vz}{k_B T}\right] \tag{26}$$

Equation (26) cannot be reduced further into a simple analytical expression for $L(T)$. It can be solved numerically, and the result for $L(T)$ is shown in Fig. 7.11. The critical temperature is found analytically by taking L as infinitesimally small, for which Eq. (26) becomes Eq. (29) after rearrangement via Eqs. (27) and (28).

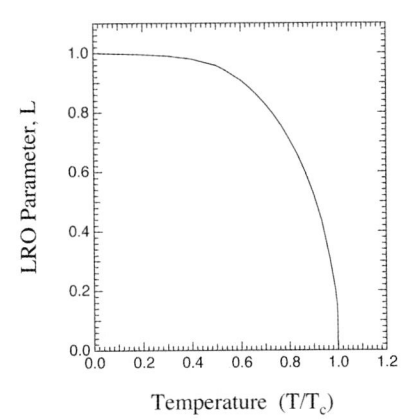

Fig. 7.11 Long-range order versus temperature in the Bragg–Williams approximation.

$$\ln(1+2L) = \frac{L2Vz}{k_B T_c} \tag{27}$$

$$2L = \frac{L2Vz}{k_B T_c} \tag{28}$$

$$T_c = \frac{Vz}{k_B} \tag{29}$$

The theory presented in this section is sometimes called the "Bragg–Williams" or "mean field" theory of ordering (Bragg and Williams 1934, 1935). Equation (22) makes a simplifying assumption about counting pairs – if we pick an A-atom, its neighbors on the adjacent sublattice sites will be determined only by the overall concentration on the sublattice. There is no provision for any preference for the neighbors of a wrong A-atom to be B-atoms, for example. To describe the atom configurations, the theory uses only a simple sublattice concentration at each site (or "point") in the crystal. The Bragg–Williams approach is a theory in the "point approximation." Because it keeps track only of the average sublattice concentrations, there is no spatial scale. The sublattices are infinite in size. It is therefore appropriate to call L a long-range order parameter. A better thermodynamic theory is possible by defining "pair variables" to account for the short-range preference of neighboring atoms (Bethe 1935). This is discussed further in Sections 7.1.9 and 7.2.6, and Chapter 10.

7.1.8
Beyond Simple Models

The models presented above in Section 7.1 give a valuable conceptual understanding, but any quantitative understanding requires additional physical concepts and complexity. This section presents an overview of "advanced" concepts and methods which are essential for understanding phase diagrams of real alloys. The most reliable concept is the oldest – minimizing the free energy of multi-phase mixtures is the way to obtain equilibrium alloy phase diagrams. The more advanced concepts are needed for obtaining accurate free energy functions for these minimizations.

- *Enthalpy*: At low temperatures the enthalpy originates with static structures of the nuclei and electrons, which are the essential components of matter. Both the bonding energy and the effects of pressure can be obtained from the electronic structure of the alloy. Most quantitative electronic structure calculations of today are based on the local density approximation for the electron exchange and correlation energy. The reliability of these methods is now known, and is often satisfactory for many predictions. Improvements are expected over the next decade, especially for calculations on magnetic systems and rare earth alloys.

- *Entropy*: The configurational entropy can be calculated with more accuracy than in Eq. (10). The problem with Eq. (10) is that at intermediate temperatures it is not correct to assume that the occupancy of a crystal site is independent of the occupancy of neighboring sites. There are short-range correlations between atom positions, and these reduce the configurational entropy. (Likewise, these correlations alter the configurational enthalpy.)

    Configurational entropy is not the only source of entropy, and in some systems it is only the minor contribution. For crystals of pure elements there is no configurational entropy, and for shear transformations there is no change in the configurational entropy. Dynamical sources of entropy from atom vibrations must always be considered, and motions of magnetic spins and electronic excitations can be important, too. Metals have a modest electronic entropy that is calculated from the density of electron states near the Fermi level.

- *High temperatures and quasiparticle interactions*: The configurational entropy has no intrinsic temperature dependence – it depends only on atom positions. On the other hand, the dynamical sources of entropy have several sources of temperature dependence. From the quasiparticle viewpoint, there are more interactions between phonon and electron excitations when more of them are present at elevated temperatures. These anharmonic sources of entropy can be large, and are currently an active area of research.

The topic of electronic structure is covered in Chapter 11, and will not be discussed further in the present chapter. Further discussion of the statistical mechanics of atom configurations is presented in Chapter 10, so only a brief overview is presented here. The subject of lattice dynamics is covered in Chapter 4, but here we present the thermodynamic consequences of phonons. Analogous arguments pertain to magnons.

7.1.9
Entropy of Configurations

Section 7.1.7 presented a thermodynamic analysis in the point approximation, which assumed that all atoms on a sublattice were distributed randomly. This assumption is best in situations when (a) the temperature is very high, so the atoms are indeed randomly distributed on the sublattice; (b) the temperature is very low, and only a few antisite atoms are present; or (c) in a hypothetical case when the coordination number of the lattice goes to infinity. For more interesting temperatures around the critical temperature, for example, it is possible to improve upon this analysis by systematically allowing for short-range correlations between the positions of atoms. In the pair approximation, the state of order is parameterized in more detail by using independent variables to count the numbers of pairs of atoms of the different types. In a solid solution these can express the local tendency for preferences of like or unlike neighbor pairs, which is "short-range order" (SRO). In an ordered alloy the SRO can change with some independence of the long-range order (LRO). In fact, considerable SRO exists in alloys above the critical temperature where the LRO is zero.

A pair variable p_{ij} is the probability that an arbitrarily chosen 1NN (first-nearest-neighbor) pair of sites contains an i and j atom. In the pair approximation, the free energy of a disordered solid solution is formulated with an energy of pairs according to Eq. (30), and the configurational entropy in the pair approximation for an alloy without the sublattices of LRO is given by Eq. (31). (Sato 1970; Kikuchi 1974):

$$E_{conf2} = N \frac{z}{2} \sum_{i,j} e_{ij} p_{ij} \tag{30}$$

$$S_{conf2} = k_B N \left[(z-1) \sum_i [p_i \ln p_i - p_i] \right.$$
$$\left. - \frac{z}{2} \sum_{i,j} [p_{ij} \ln p_{ij} - p_{ij}] + \left(\frac{z}{2} - 1\right) \right] \tag{31}$$

The double sum in the entropy expression is expected from the combinatorics of arranging pairs of atoms on pairs of lattice sites, assuming they are placed randomly without consideration of the neighboring pairs. The double sum comes from a random mixing of pairs, not individual atoms (points) as in Eq. (6).[7] The p_{ij} are the independent variables in the pair approximation. This leads to a difficulty with the point variables, p_i, which are derived as sums over pair variables as: $p_i = \sum_j p_{ij}$. Each site in a square lattice belongs to four pairs, however. There is an overcounting problem that needs to be addressed to ensure that the numbers of atoms are conserved as order evolves.[8] This is done with the "Fowler–Guggenheim" correction factor to the combinatorics (Guggenheim 1935; Fowler and Guggenheim 1940), and with the Stirling approximation it completes the expression of Eq. (31). (Note the opposite sign of the first term with point variables in Eq. (31)). With this correction, Eq. (31) for a truly random solid solution without local atomic preferences is the same as the entropy of mixing of Eq. (10).

For further accuracy, it is natural to extend this hierarchy of approximations from point to pair to larger clusters such as squares on the square lattice, for ex-

---

7) The double sum in Eq. (31) is obtained analogously to the entropy expression of Eq. (10), which involved the logarithm of the factorials in Eq. (6), and the Stirling approximation for $\ln(x!)$.

8) Counting pairs is not equivalent to counting points. Consider the replacement of one B-atom with an A-atom in an equiatomic alloy. In a disordered alloy this will create $z/2$ new A–A pairs, but in an ordered alloy z A–A pairs will be created.

ample. A systematic "cluster variation method" to do this was developed by Kikuchi (1951), and has been in widespread use since the late 1970s (Van Baal 1973; de Fontaine 1979). Variables for the populations of larger clusters, such as first-nearest-neighbor square variables, p_{ijkl}, are placed randomly on the square lattice, for example, consistently with their concentrations. In this case the pair variables are overcounted, and the handling of this problem is the key to the method. In essence, it is necessary to account for the overcounting of all subclusters within the largest cluster, using weights that depend on how many subclusters are present in the large, independent cluster. The systematics to do this are elegant and straightforward, and are the subject of Chapter 10 of this book (see also (Kikuchi 1951; Sanchez and de Fontaine 1980)). One measure of the success of these higher-order approximations is their convergence on the actual critical temperature for ordering, as determined, for example, by Monte Carlo methods. In the pair approximation the exchange potential, V, sets the critical temperature at $T_c = 0.869zV/k_B$, which is lower than the critical temperature of the point approximation of Eq. (29), and is more accurate. Discrepancies of 1% or so are possible with modest cluster sizes, but the analysis of accuracy is itself a specialty.

An important conceptual point is addressed by cluster approximations for the problem of chemical unmixing, or spinodal decomposition. The phase diagram for unmixing, Fig. 7.6, is based on the free energy versus composition curves of Fig. 7.2. At intermediate compositions, these curves obtained in the point approximation show a characteristic "hump" at temperatures below the critical temperature for unmixing. In the pair approximation, the magnitude of this hump is decreased considerably, and it continues to decrease as the cluster approximation extends to higher order (Cenedese and Kikuchi 1994). In the limit of very large clusters, the hump in the free energy versus composition curve vanishes, and a straight line connects two solid solutions of different compositions. What happens is that the populations of higher-order clusters are themselves biased toward unmixed structures, where all the A-atoms are on one side of the cluster and the B-atoms on the other side, for example. With these internal degrees of freedom accounting for much of the unmixing enthalpy, the composition variations have less effect on the free energy. Composition is therefore insufficient for characterizing the state of the alloy, and the free energy versus composition curve in the point approximation is naïve. This viewpoint helps to overcome a conceptual problem with curves such as in Fig. 7.2: "How reliable is an unstable free energy function for calculating equilibrium phase diagrams?" Such difficult questions are avoided in part by treating spinodal decomposition as a kinetic process with a modified diffusion equation, as in Section 7.4.

7.1.10
Principles of Phonon Entropy

The partition function for a single harmonic oscillator of frequency $\omega_i = \varepsilon_i/\hbar$ is given by Eq. (32).

$$Z_i = \sum_{n}^{\infty} e^{-\beta(n+1/2)\varepsilon_i} \tag{32}$$

$$Z_i = \frac{e^{-\beta\varepsilon_i/2}}{1 - e^{-\beta\varepsilon_i}} \tag{33}$$

Equation (33) was obtained by identifying Eq. (32) as a geometric series times the constant factor $\exp(-\beta\varepsilon_i/2)$. The partition function for a harmonic solid with N atoms and $3N$ independent oscillators is the product of these individual oscillator partition functions, Eq. (34), from which we can calculate the phonon free energy as $F = -k_B T \ln Z$ [Eq. (35)], and the phonon entropy by differentiating with respect to T [Eq. (36)].

$$Z_N = \prod_{i}^{3N} \frac{e^{-\beta\varepsilon_i/2}}{1 - e^{-\beta\varepsilon_i}} \tag{34}$$

$$F_{ph} = \frac{1}{2}\sum_{i}^{3N} \varepsilon_i + k_B T \sum_{i}^{3N} \ln(1 - e^{-\beta\varepsilon_i}) \tag{35}$$

$$S_{ph} = k_B \sum_{i}^{3N} \left[-\ln(1 - e^{-\beta\varepsilon_i}) + \frac{\beta\varepsilon_i}{e^{\beta\varepsilon_i} - 1} \right] \tag{36}$$

It is often useful to work with a phonon density of states (DOS), $g(\varepsilon)$, where $3N g(\varepsilon)\,d\varepsilon$ phonon modes are in an energy interval $d\varepsilon$. For a DOS acquired as digital data in m intervals of width $\Delta\varepsilon$ (so $\varepsilon_j = j\Delta\varepsilon$), the partition function can be computed numerically [Eq. (37)].

$$Z_N = \prod_{j=1}^{m} \left(\frac{e^{-\beta\varepsilon_j/2}}{1 - e^{-\beta\varepsilon_j}} \right)^{3Ng(\varepsilon_j)\Delta\varepsilon} \tag{37}$$

With no derivation we present Eq. (38), a useful expression for the harmonic phonon entropy of a material at any temperature, where $g(\varepsilon)$ is normalized to 1 and $n(\varepsilon)$ is the Planck distribution for phonon occupancy at the temperature of interest.

$$S_{ph} = 3k_B \int_0^{\infty} g(\varepsilon)[(n(\varepsilon) + 1)\ln(n(\varepsilon) + 1) - n(\varepsilon)\ln(n(\varepsilon))]\,d\varepsilon \tag{38}$$

A handy expression for the high-temperature limit of the difference in phonon entropy between two harmonic phases, α and β, Eq. (39), can be obtained readily from Eq. (36).

$$S_{ph}^{\beta-\alpha} = 3k_B \int_0^{\infty} (g^{\alpha}(\varepsilon) - g^{\beta}(\varepsilon))\ln(\varepsilon)\,d\varepsilon \tag{39}$$

The important point about Eq. (37) is that the only material parameter relevant to the thermodynamic partition function is the phonon DOS, $g(\varepsilon)$. Measuring or calculating the phonon DOS is the key to understanding the phonon contributions to alloy thermodynamics. This must be done with high accuracy. For example, if we apply Eq. (39) to a case where the phonon DOS curves of the α and β phases have the same shape, but differ in energy scaling by 10%, we obtain a change in phonon entropy of $\Delta S_{ph} = 3k_B \ln(1.1) \simeq 0.3 k_B$/atom. This change in phonon entropy is almost half of the maximum possible change in configurational entropy of the order–disorder transformation of a binary alloy, which is $k_B \ln 2$/atom. When comparing the phonon entropies of different alloy phases, a 1% accuracy in the difference in logarithmic-averaged phonon energy is often required. Although the role of phonon entropy in phase transformations has been discussed for many years (Slater 1939; Booth and Rowlinson 1955), only in recent times have measurements and calculations become adequate for assessing trends (Anthony et al. 1993; Wallace 2003).

At low temperatures it is possible to use electronic structure calculations with the local density approximation to calculate phonon frequencies with reasonable accuracy, but a 1% accuracy remains a challenge. Inelastic neutron scattering (see Chapter 4) faces its challenges too (Squires 1978; Lovesey 1984). A triple axis spectrometer at a reactor neutron source is highly accurate in its energy and momentum measurements, and when a full set of phonon dispersions are measured on single crystals of ordered compounds, the phonon DOS derived from a lattice dynamics model is highly reliable. This technique is not so reliable for disordered solid solutions, unfortunately, because the dispersions are inherently broadened, and this broadening need not be symmetrical in energy. Direct measurements of the full phonon DOS of disordered alloys are possible with chopper spectrometers at pulsed neutron sources, but other difficulties arise. In general, the neutron scattering from phonon excitations does not receive an equal contribution from the different elements in an alloy, and the displacement amplitudes of the different elements may be different in different phonons. This causes a "neutron weight" that is difficult to correct, although differential measurements are often possible.

7.1.11
Trends of Phonon Entropy

Our knowledge about the phonon entropy of alloy phases is still being organized, but some trends are clear, and some correlations between phonon entropy and alloy properties are known. The Hume–Rothery rules of alloy thermodynamics are based on the atomic properties of: (a) electronegativity, (b) metallic radius, and (c) electron-to-atom ratio, with the first factors being the most important (see Chapter 2). For phonon entropy, we should perhaps add (d) atom mass to this list. It is found that across the periodic table, where atomic mass varies from 1 to 238 and beyond, for matrices having atoms of mass M_0, and solutes having mass M_S, the phonon entropy upon alloying tends to scale with $\ln(M_0/M_S)$ (Bogdanoff and

Fultz 1999). This is a general consequence of the relationship for an oscillator of $\omega = \sqrt{k/M}$, where the frequency, ω, scales as the inverse root of the mass, M, assuming that the spring constant, k, is in fact a constant. There is a large scatter in this correlation, however, and it is not a reliable one for most alloys. The spring constants, or more specifically the interatomic force constants Φ_{ij} (see Chapter 4), are in fact not constant in alloys. In particular, a larger solute atom will cause a local compression. Since interatomic forces generally become stiffer under compression, alloying with an atom having both a large size and a large mass will produce an uncertain result for the vibrational entropy.

Progress has been made, however, by van de Walle and Ceder (2002), who proposed a model based upon a bond stiffness versus bond length argument. In their model, an atom pair in different local atomic configurations will have different bond stiffnesses, with greater stiffness for shorter interatomic distances. These characteristics seem to be transferable when atom pairs are in different crystal structures. The model is useful for semiquantitative arguments, and when calibrated for specific elements by ab-initio calculations, for example, this model can be used for comparing phonon entropies of different alloy phases.

In alloying there are correlations between phonon entropy and electronegativity. In a systematic study of transition metal solutes in vanadium, it was found that the phonon entropy had a robust correlation with the difference in electronegativity between the solute atom and the vanadium atom. Results shown in Fig. 7.12 indicate that this correlation works well for solutes from the 3d, 4d, and 5d rows of the periodic table, in spite of their large differences in mass and atomic size. Electronic structure calculations have shown that the large differences in electronegativity cause large charge transfers between the solute and its first-nearest-neighbor atoms that alter the charge screening sufficiently to vary the interatomic forces (Delaire and Fultz 2006).

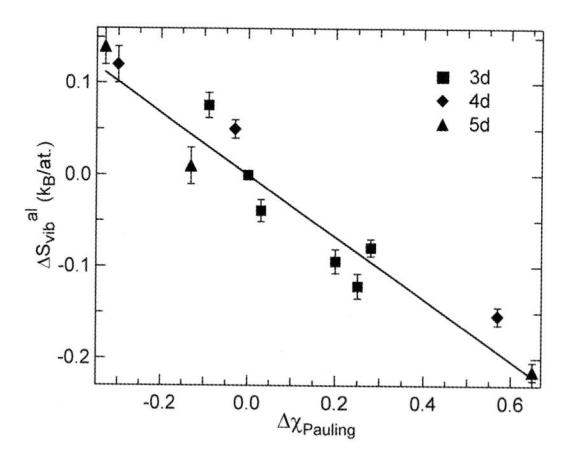

Fig. 7.12 Vibrational entropy of alloying of 6 at% solutes in bcc vanadium, plotted versus Pauling's electronegativity of the solute element.

7.1.12
Phonon Entropy at Elevated Temperatures

A harmonic oscillator has generalized forces that are linear with the displacement of a generalized coordinate. This is consistent with Hooke's law. The word "anharmonic" describes any oscillator with forces that deviate from this linearity. In thermodynamics, the word "anharmonic" is used more restrictively, and the word "quasiharmonic" describes some effects from classical thermodynamics that are not strictly harmonic, but can be interpreted as altered harmonic behavior. Here we use the following words to describe how phonon frequencies change with the intensive variables of temperature and pressure.

- Harmonic phonons undergo no change in frequency with T or P.
- Quasiharmonic phonons have frequencies that depend on volume only. At a fixed volume, however, they behave as harmonic oscillators. Their frequencies can change with temperature, but only because thermal expansion alters the volume of the solid.
- Anharmonic phonons change their frequencies for reasons other than volume. For example, the interatomic potential may change if temperature drives electronic excitations across the Fermi surface. Alternatively, quartic terms in the interatomic potential change the phonon frequencies with phonon occupancy, but need not cause thermal expansion. A pure temperature dependence of phonon frequencies, independent of volume effects, is typical of anharmonic effects.

With increasing temperature, energy goes into phonon creation, but energy also goes into the expansion of a crystal against its bulk modulus. It is necessary to minimize a free energy function that includes phonon energy, phonon entropy, and the elastic energy of thermal expansion.

We can test for anharmonicity by the following method, which compares the thermal shifts of phonons to predictions from classical thermodynamics. It is possible to calculate a free energy $F(V, T)$, with an elastic energy that depends on thermal expansion (a quadratic function of $\beta\Delta T$), and a quasiharmonic phonon entropy that increases by $\gamma\beta\Delta T$ (where γ is the Grüneisen parameter and β is the coefficient of linear thermal expansion) as the phonon frequencies soften.[9] With the elastic energy and phonon entropy, optimizing the amount of volume expansion at a fixed temperature gives an equality from the energy term, $B\beta v$, and the phonon entropy term, γC_V, where v is the specific volume. B is the bulk

---

9) It is also possible to add a contribution from the phonon energy that softens with volume, but it is found that this effect is small compared to the change in elastic energy.

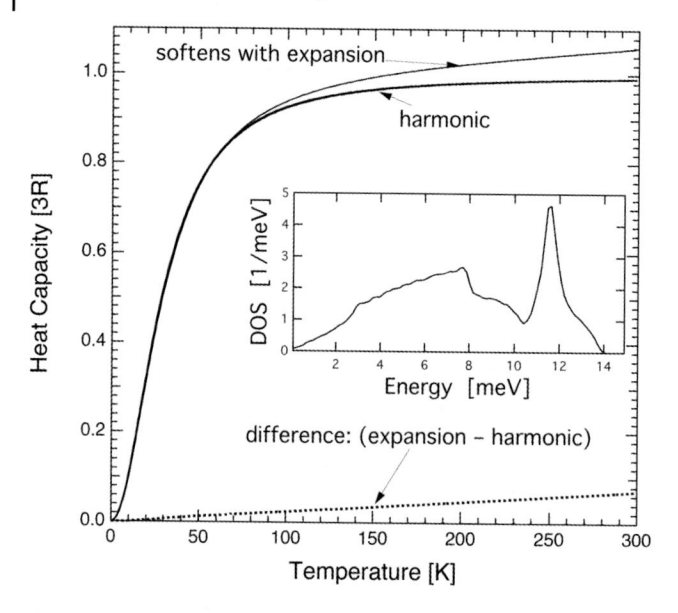

Fig. 7.13 Heat capacity versus temperature for two assumed physical models. Inset is the phonon DOS at 0 K – all modes were assumed to soften by 6.5% at 300 K. A simple harmonic heat capacity with a phonon DOS unchanged with temperature is also shown. 3R is the classical limit. The full heat capacity also includes an electronic contribution that may be comparable to the difference curve.

modulus, and C_V is the specific heat at constant volume. The thermal expansion is therefore given by Eq. (40).

$$\beta = \frac{\gamma C_V}{Bv} \tag{40}$$

(With electronic entropy, an additional term, $\gamma_{el} C_{el} V/Bv$, would be added to the thermal expansion coefficient.) Equation (40) is useful for identifying quasiharmonic behavior. Suppose we know the phonon softening from the parameter γ, or better yet we know the softening for all phonon modes in the DOS. With the phonon DOS we can calculate C_V, and conventional measurements can provide B, v, and β, accounting for all unknowns in Eq. (40). If Eq. (40) proves to be true, we say that the solid is "quasiharmonic." Typical quasiharmonic effects on the heat capacity are shown in Fig. 7.13.

Anharmonic behavior is phonon softening or stiffening inconsistent with Eq. (40). Anharmonicity involves phenomena beyond those of independent phonons. Electron–phonon interactions, changes in electronic structure with temperature, or higher-order phonon–phonon interactions can cause anharmonic behavior. Today, unfortunately, it is difficult to differentiate between these anharmonic effects (Wallace 1998). Nevertheless, substantial anharmonic effects are known for Cr

7.2
Kinetics and the Approach to Equilibrium

A phase diagram is a construction for thermodynamic equilibrium, and therefore contains no information about how much time is needed before the phases appear with their equilibrium fractions and compositions. It is usually true that the phases found after practical times of minutes or hours will be consistent with the phase diagram, since most phase diagrams were deduced from experimental measurements on such time scales.[10] For rapid heating or cooling, however, the kinetic processes of atom rearrangements often cause deviations from equilibrium, and these are described below. The last part of this section addresses atomistic theories of kinetics by kinetic extension of the statistical mechanics of Sections 7.1.7 and 7.1.9.

7.2.1
Suppressed Diffusion in the Solid (Nonequilibrium Compositions)

We start with equilibrium predictions for phase fractions and compositions for an alloy of composition c_0, cooled from the liquid and allowed to freeze (Fig. 7.14). The relevant graphs to examine are on the right of Fig. 7.14, specifically the thick solid curves (the thin dotted lines indicate the correspondence between the cool-

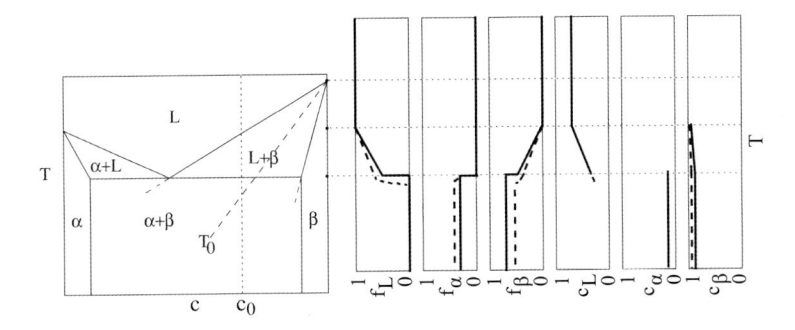

Fig. 7.14 Analysis of eutectic phase diagram for phase fractions, f, and compositions, c, upon cooling under equilibrium conditions (thick curves on the right), suppressed diffusion in the solid phases (thick broken curves), and very rapid cooling with suppressed diffusion in the liquid (thin curves). See text for discussion.

---

10) On the other hand, this does not necessarily mean that all phases on phase diagrams are in fact equilibrium phases. Exceptions are found, especially at temperatures below about half the liquidus temperature.

ing curves and the eutectic phase diagram). The six graphs are the fractions and compositions for each phase, L, α, and β, over a range of temperatures. At high temperatures the alloy is entirely liquid, with composition c_0, but below the liquidus temperature the fraction of liquid decreases as crystals of the β-phase grow in the liquid. Using the lever rule [Eq. (13)], the fractions of β-phase and liquid phase are approximately equal at the eutectic temperature of Fig. 7.14. Their chemical compositions are simply read from lines drawn in the phase diagram. Note that the α-phase does not exist above the eutectic temperature for this alloy of composition c_0, but the lever rule shows that immediately below the eutectic temperature the α-phase should exist with a phase fraction, f_α, of approximately 0.3.

Kinetic deviations from this equilibrium analysis of Fig. 7.14 can be predicted approximately. First consider the effects of kinetic processes with the longest time scales, since these will be the first to alter phase formation as cooling or heating becomes more rapid. Atom diffusion in the solid phases is a kinetic process that is typically slow. Solid-state diffusion will modify the equilibrium cooling curves of Fig. 7.14 as follows. At the highest temperature where the β-phase first forms during cooling (the liquidus temperature for the composition c_0 in Fig. 7.14), the β-phase is especially rich in B atoms. At lower temperatures the phase diagram shows that this initial β-phase is too rich in B atoms, and some B atoms in the solid should exchange with A atoms in the liquid. This often does not occur because it requires atoms in the interior of the crystals of the β-phase to migrate to the outside of the crystal to reach the liquid. Diffusion in the solid is often too slow to ensure equilibrium during cooling. In essence, the extra B atoms in the β-phase are unable to participate in the equilibrium process, and the effective alloy composition can be considered to decrease below c_0 (see Chapter 3). The effects are shown in the thick broken curves of Fig. 7.14. Although the average composition of the β-phase is too rich in B atoms, at the surface of β-phase crystals the new layers of solid are forming with their equilibrium composition, and the adjacent liquid composition is given by the equilibrium liquidus line. Conservation of solute requires that the total amount of β-phase be somewhat suppressed in comparison to the amount of liquid (consistent with the effective overall composition decreasing below c_0). The same argument holds for temperatures below the eutectic temperature – the B-rich cores of the β-phase cause less β-phase to be present with the α-phase at low temperatures.

A common phenomenon that occurs during freezing of eutectic systems is the formation of "dendritic" or tree-like microstructures of intertwined α- and β-phases. There are two reasons why the interface between the liquid and solid may not be stable to asperities in the solid, which grow forward rapidly into the liquid (Chalmers 1959; Mullins and Sekerka 1964). First, there is latent heat release during freezing, and this makes the interface warmer than the solid and liquid immediately around the interface. The interface, however, is at the melting temperature, and therefore the nearby liquid must be below the melting temperature and especially unstable to solid formation. A protuberance that juts into this undercooled liquid can grow quickly. Second, the new solid phase at the interface rejects A atoms into the liquid, making a liquid that freezes at a lower temperature. This chemical contribution to the liquid instability to protruberances is called "constitutional supercooling." Solid dendritic structures grow rapidly into the liquid, often with side branches. Between these dendrites is a liquid of a composition which solidifies at lower temperatures. Dendritic freezing produces coarse compositional inhomogeneities in the solidified material (see Chapter 3), and processings such as forging may be required to eliminate them.

7.2.2
Nucleation Kinetics

Nucleation of a solid phase frequently slows the kinetics of formation of a new phase during cooling. Nucleation is a broad topic, which is discussed in Section 7.3, but suffice it to say that some undercooling is necessary before the new solid phase can form. Undercooling is especially important for solid → solid phase transformations, although some undercooling is also typical of liquid → solid phase transformations.[11] There is an energy penalty associated with the solid–liquid interface, and undercooling is needed to boost the thermodynamic driving force for nucleating a new volume of solid. The undercooling depends on many factors, including contact between the liquid and the walls of its container, so it is not easy to predict. Nevertheless, at exactly the critical temperature of a phase transformation, the free energies of the two phases are equal, so it takes an infinite amount of time to nucleate the new phase at the critical temperature.

The number of nuclei of the new phase that are able to grow increases rapidly with the undercooling below the critical temperature of the phase transformation (Section 7.3). On the other hand, the diffusional processes needed for growth are thermally activated, and are slower at lower temperatures. The general kinetics of a nucleation-and-growth phase transformation is indicated on the left in Fig. 7.15. The rate of the phase transformation is zero at the critical temperature and above (owing to nucleation), and sluggish at low temperatures (owing to diffusion). On the right-hand side of Fig. 7.15 is a kinetics map called a "time–temperature–transformation," or TTT, diagram. It pertains to a particular phase transformation at a particular chemical composition. The assumption in using a TTT diagram is that the high-temperature phase is cooled instantaneously to the temperature on the ordinate of the plot, and held at this temperature for various lengths of time. The degree of completion of the phase transformation (often given as 10% and 90%) varies with temperature and time, as shown in the TTT diagram. A typical nucleation-and-growth transformation has a "nose" at a temperature where the transformation occurs in the shortest time.

Finally, at the top of Fig. 7.15 are drawn microstructures for two nucleation-and-growth transformations that came to the same degree of completion in the

---

11) On the other hand, superheating tends to be much less of an effect – an alloy will usually begin to melt close to the temperatures predicted from the phase diagram, even for moderately fast heating rates.

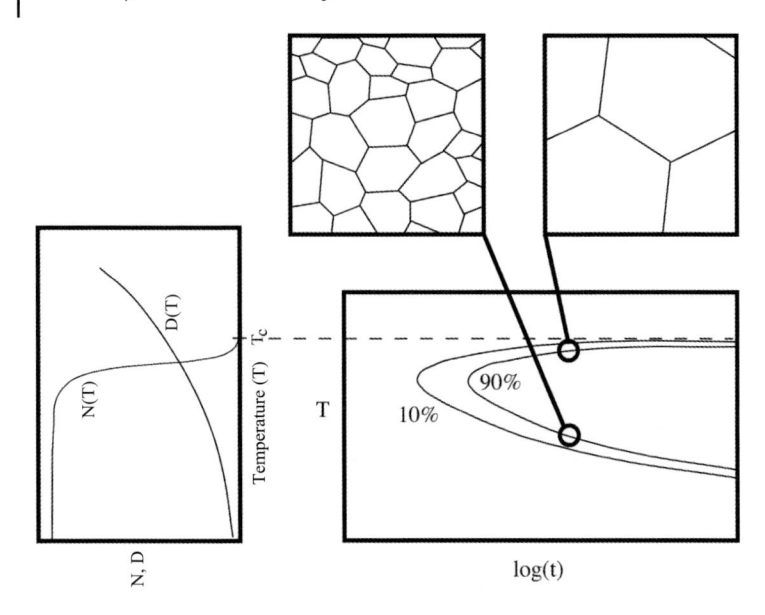

Fig. 7.15 Left: the number of nuclei, N, and diffusivity, D, at temperatures near the critical temperature, T_c, of a nucleation-and-growth type of phase transformation. Right: the TTT diagram, with typical microstructures above.

same time, consistent with the TTT diagram. For the microstructure on the left, which formed at the lower temperature, the number of nuclei was high, but their growth rates were low. The opposite case is true for the microstructure on the right. In general, a phase transformation that occurs at a lower temperature is finer-grained than one occurring at a higher temperature.

7.2.3
Suppressed Diffusion in the Liquid (Glasses)

Liquid diffusion is much faster than diffusion in a solid, but liquid-phase diffusion can become an issue at much higher cooling rates. The partitioning of B atoms between the solid and liquid can be suppressed at high cooling rates, and in this case the β-phase forms with a lower concentration of B atoms than predicted with the phase diagram of Fig. 7.14. An effective extension of A-atom solubility in the β-phase is expected owing to this kinetic constraint. An extreme case, amenable to convenient analysis, can occur for cooling rates in excess of order 10^6 K s^{-1}. Here we assume that solute partitioning between the solid and liquid is impossible, and all regions of the alloy maintain a composition of c_0 at all temperatures. A set of free energy versus composition curves are shown in Fig. 7.16 (consistent with Fig. 7.7). The common tangent construction for thermodynamic equilibrium predicts the coexistence of two solid phases (α and β) at this temperature, but without solute partitioning the liquid phase is preferable over a wide

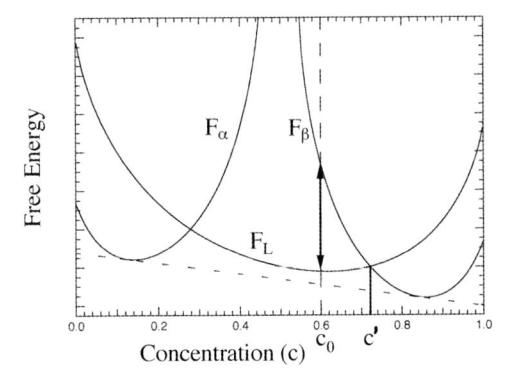

Fig. 7.16 Free energy versus composition curves, similar to those for the phase diagram of Fig. 7.7 at a temperature below the eutectic temperature. If solute partitioning between phases is suppressed by kinetics, at this temperature we may expect liquid phase to exist for compositions between $0.28 < c < 0.72$.

range of compositions. Without solute partitioning between the α- and β-phases, the formation of a crystal of β-phase of composition c_0 would cause an increase in free energy, as indicated by the line with the arrowheads. Such crystallization does not occur. The liquid phase is especially stable for compositions in the middle of Fig. 7.16. Near the eutectic composition, deep undercoolings of the liquid are possible at high quench rates. At low temperatures, however, the liquid becomes increasingly viscous, and as the viscosity increases past 10^{12} Pa s, the alloy will not crystallize in measurable times. This frozen liquid is a "metallic glass" (Klement et al. 1960). Many alloys having phase diagrams with low eutectic temperatures can be quenched into a metallic glass by rapid cooling.

Figure 7.14 shows a broken line labeled "T_0" in the two-phase region of liquid and β-phase coexistence. This line denotes the locus of intersections of the solid and liquid free energy curves. One such point for one temperature is indicated in Fig. 7.16 as the composition labeled c'. To the right of the composition c' in Fig. 7.16 the free energy curve of the solid β-phase is lower than the curve for the liquid at the same composition, but the situation reverses to the left of c'. The composition of the crossing of the two curves is roughly halfway between the liquidus and β solidus lines on the phase diagram. The T_0 line denotes the limit of solubility of A atoms in the β-phase, or equivalently the limiting composition where metallic glass could form upon rapid cooling (see Chapter 3).

7.2.4
Suppressed Diffusion in a Solid Phase (Solid-State Amorphization)

Another example of a kinetic constraint is demonstrated with Fig. 7.17. Free energy versus composition curves are shown for three solid phases. The new solid phase, the γ-phase, is typically an ordered compound of A and B atoms. Many

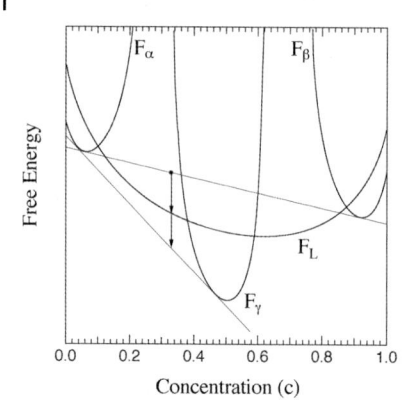

Fig. 7.17 Free energy versus composition curves, similar to those of Fig. 7.16, but with a curve for an additional γ-phase. Arrows indicate reductions in free energy that are possible from a starting mixture of α- and β-phases for overall composition $c = 0.33$.

such compounds have large unit cells, requiring substantial diffusion for their nucleation. Their formation can be suppressed by rapid cooling, and if γ-phase formation is ignored, the analysis of glass formation upon cooling is the same as that used above for 7.16. More interesting, however, is the possibility that intermetallic phases can be suppressed upon heating, because this can cause crystalline phases to become amorphous. Suppose that solid phases of A-rich α-phase and B-rich β-phase are brought into contact at low temperatures, and atoms of A and B are allowed to interdiffuse across the interface. An important consideration is that there are often large differences in the mobilities of A and B atoms. As an extreme case, suppose that only the A atoms are diffusively mobile in the solid phases. The formation of the γ-phase, with its large unit cell, usually requires the coordinated movements of B atoms. If these movements are suppressed kinetically, the γ-phase formation is suppressed. Without the possibility of nucleating the γ-phase, the free energy may reduce to the next best phase, the liquid phase, as indicated by the first arrow in Fig. 7.17. This process of "solid-state amorphization" produces an amorphous phase at the interface between the α- and β-phases which is stable at low temperatures. Solid-state amorphization was first reported in 1983 (Schwarz and Johnson 1983), but many examples have since been found (Johnson 1986). In the time required for the layer of metallic glass to grow to macroscopic dimensions, however, the γ-phase usually nucleates and then grows quickly, consuming the amorphous layer.

7.2.5
Combined Reactions

Microstructures of engineering materials are often complex, having compositional inhomogeneities, different morphologies for different phases, and different defect structures within the different phases. Many such microstructural features can be altered independently. For example, a phase with fast solid-state diffusion kinetics may undergo recovery processes at a lower temperature than a phase with slower atom mobilities. The consequent difference in defect concentrations can lead to differences in free energies of the two phases, and this can affect the relative stabilities of the two phases.

When multiple microstructural features relax toward equilibrium, the kinetics are usually altered in comparison to independent relaxation processes. In the simplest case, two transformations such as defect recovery and chemical unmixing may occur sequentially. For sequential reactions it may be possible to consider the steps independently, but the earlier transformations are altered because they occur in material that is not relaxed in its other microstructural parameters.

More complex "combined reactions" may have simultaneous relaxations of microstructural parameters. Usually such transformations are first-order or discontinuous, because second-order transformations must satisfy symmetry relationships between the new phase and the parent phase (Landau and Lifshitz 1969). These symmetry constraints are difficult to satisfy when multiple parameters are free to change. A common combined reaction occurs at a reaction front where the parent microstructure is consumed, and a new phase (or phases) grows into it. Behind the moving interface is a microstructure that is closer to equilibrium in its chemical composition and defect concentrations owing to atom movements that bring equilibrium to the interface. For such simultaneous combined reactions, the decrease in free energy is larger as the interface moves further. This reduction in free energy has contributions from both chemical rearrangement and defect annihilation. The driving force for interface motion has similar factors, and this kinetic process involving combined, simultaneous reactions can be quite complicated when the relaxations of one microstructural feature, for example vacancy annihilation, alters the kinetics of the other relaxation process, for example diffusional unmixing. Reactions in steels are often characterized by such complexity, as described in a review article (Hornbogen 1979). A phase field approach to coupled evolution of composition and dislocation fields was reported recently (Haataja and Leonard 2004).

7.2.6
Statistical Kinetics of Phase Transformations

Section 7.1.9 explained in general how pair variables account for the atomic structure and the thermodynamics of alloy phase transformations in a more detailed way, and hinted at how higher-order cluster variables could be defined and used for thermodynamics. This section describes an atomistic kinetic theory that uses such a level of structural detail. Any full kinetics treatment requires a mechanism for atom movement. Here the kinetic mechanism is the exchange of an atom with a neighboring vacancy. The moving atom breaks its initial pairs and forms new ones after it moves to its new site. Rate equations based on activated-state rate theory account for how the chemical neighborhood of a moving atom affects the rate of its diffusive jump. Specific developments of rate equations and their solution for chemical ordering on different lattices are given by Kikuchi and Sato (1969, 1972), Sato and Kikuchi (1976), Gschwend et al. (1978), and Fultz (1990). With all these theories, a steady-state solution is found by setting the rates of change equal to zero, and the state variables become identical to those of thermodynamic equilibrium for the same level of approximation. The path to equilibrium, however, depends on the individual atom mobilities and chemical preferences, and proves rich in its own phenomena.

7.2.7
Kinetic Pair Approximation

We define a pair variable as p_{ij}, where, for example, the two sites on an arbitrary nearest-neighbor pair have the probability p_{AB} of being occupied by an A atom and a B atom. With long-range order (LRO) and sublattice formation as in Fig. 7.10, it is necessary to define sublattice preferences for A atoms on the α-sublattice, for example, so $p_{AB}^{\alpha\beta} \neq p_{BA}^{\alpha\beta}$ when LRO exists. Conditional pair probabilities, such as the probability of having an A on α, given a B on β, $p(A\alpha|B\beta)$, are also useful for describing short-range order (SRO). For conciseness, in this section the exponents of Boltzmann factors will be written without the required factors of $(k_B T)^{-1}$.

Figure 7.18 depicts the pairs undergoing change as an A atom changes sites with a vacancy. This example pertains to the square lattice of Fig. 7.10, or to the B1 or B2 ordered structures, for example. We make two assumptions.

- There are no vacancy–atom interactions. These can be handled easily (Fultz 1992), but we neglect them here for clarity.
- In this activated-state rate theory, the moving A atom surmounts an activation barrier. If it surmounts the barrier, it always moves to the new configuration on the right of Fig. 7.18, irrespective of the types of neighbors in the new environment. The jump process is not influenced by the new pairs formed by the jump, but the initial pairs do influence the jump. For example, if the A atom initially has an energetically unfavorable chemical neighborhood, it is more likely to jump.

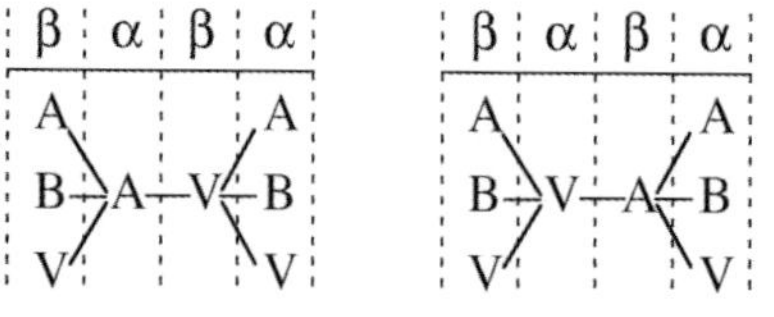

Fig. 7.18 Left: an A–V pair before the jump, showing the initial bonds to neighboring pairs on the sublattices of a square lattice. Right: a V–A pair after the jump, showing the pairs that were formed by the jump.

7.2 Kinetics and the Approach to Equilibrium

The kinetic master equation developed in Chapter 5 is applied here to show how pair variables evolve with time. Rate equations for the different pair variables must be developed independently, and a set of coupled ordinary differential equations must be solved.[12] For example, the exchange of a B atom with a vacancy, V, has no effect on the number of A–A pairs. The kinetic master equation allows us to write an expression [Eq. (41)] for how the jump process of Fig. 7.18 alters the number of A–A pairs.

$$\frac{d}{dt} p^{\alpha\beta}_{AA;\alpha A\beta} = p^{\alpha\beta}_{AV} \Gamma(\{p(i\beta|A\alpha)\}, T)(z-1)$$

$$\times \left(p(A\alpha|V\beta) - \frac{p(A\beta|A\alpha)e^{e_{AA}}}{p(V\beta|A\alpha) + p(A\beta|A\alpha)e^{e_{AA}} + p(B\beta|A\alpha)e^{e_{AB}}} \right) \quad (41)$$

The second subscript on the left-hand side of Eq. (41) identifies the kinetic step of Fig. 7.18, which is $\alpha \to A \to \beta$ (a reverse motion of a vacancy is implied). The right-hand side has two main factors. The first is a frequency factor, which depends on the number of A–V pairs that are able to exchange, and a jump rate, Γ, that is set by the initial configuration of atoms around the A atom (described by $\{p(i\beta|A\alpha)\}$, where i may be A or B).

The second main factor in the large parentheses ( ) is the statement of the kinetic master equation, that the change in the number of A–A pairs has a positive contribution from the new A–A pairs that are formed by the jump, and a negative contribution from the broken A–A pairs that existed before the jump. Since the neighborhood of the A atom after the jump is assumed to have no influence on the jump rate, the positive term is a simple probability that does not depend on the temperature or energetics of the final configuration, i.e., $(z-1)p(A\alpha|V\beta)$.[13] The jump frequency does depend, however, on the initial chemical environment around the A atom on the α-sublattice. The more complicated fraction in Eq. (41) is the rate at which the moving A atom leaves its surrounding neighborhood owing to the influence of its A neighbors. This contains a probability of the different numbers of A neighbors, and a Boltzmann factor for how these A neighbors alter the initial energy of the moving A atom. (The denominator is a normalization for the rate from the full environment of the A atom.) This factor shows the level of detail provided by the pair approximation. It is obtained from a weighted average of all types of A–A pairs. With a binomial distribution of probabilities for independent pairs, the weight is simply $p(A\beta|A\alpha)$. The presence of one A–A pair does not correlate with the presence of other pairs. A similar type of averaging is performed in Eq. (42) to obtain the rate $\Gamma(\{p(i\beta|A\alpha)\}, T)$ of Eq. (41).

$$\Gamma(\{p(i\beta|A\alpha)\}, T) = e^{e^*_{AA}}[p(V\beta|A\alpha) + p(A\beta|A\alpha)e^{e_{AA}} + p(B\beta|A\alpha)e^{e_{AB}}]^{z-1} \quad (42)$$

---

12) For low vacancy concentrations the set is "stiff," and can be numerically challenging to solve efficiently.

13) Note that the final neighbors of the A atom are the initial neighbors of the vacancy on the β-sublattice. The coordination number is reduced by 1 because one of the pairs is the moving A atom.

One way to understand Eq. (42) is by expanding the trinomial. Each term in this expansion corresponds to a possible configuration of A, B, and V in the neighborhood about the A atom, with its average Boltzmann factor for the exchange. The trinomial expansion is consistent with the assumption of randomness in the placements of pairs with respect to other pairs. The factor $e^{e^*_{AA}}$ in Eq. (42) is the activation barrier height without chemical interactions. Substituting Eq. (42) into (41) gives a rate equation for the kinetic process $\alpha \to A \to \beta$. This is but one of 32 such rate equations (Fultz 1990) for the change in the nine pair variables $\{p_{ij}^{\alpha\beta}\}$ through changes of the 18 conditional pair probabilities $\{p(i\beta|j\alpha), p(i\alpha|j\beta)\}$, where i and j denote $\{A, B, V\}$.

7.2.8
Equilibrium State of Order

Equilibrium at a given temperature is obtained by setting the rates of change equal to zero. For example, from Eq. (41) we obtain Eq. (43).

$$0 = \left(p(A\alpha|V\beta) - \frac{p(A\beta|A\alpha)e^{e_{AA}}}{p(V\beta|A\alpha) + p(A\beta|A\alpha)e^{e_{AA}} + p(B\beta|A\alpha)e^{e_{AB}}} \right) \quad (43)$$

To recover the thermodynamics of a binary A–B alloy, we neglect terms like $p(V\beta|A\alpha)$ because the vacancy concentration is small. The result from Eq. (43) is Eq. (44), and likewise for expressions involving B-atom motions it is Eq. (45), where the conditional probability $p(i|j)$ is the probability of finding an i atom at a specific 1NN site near a known j atom.

$$\frac{e^{e_{AB}}}{e^{e_{AA}}} = \frac{p(A|A)}{p(B|A)} \quad (44)$$

$$\frac{e^{e_{AB}}}{e^{e_{BB}}} = \frac{p(B|B)}{p(A|B)} \quad (45)$$

As mentioned in Section 7.1.9, the pair approximation accounts for the chemical SRO that exists even above the critical temperature for LRO. Equations (44) and (45) for the SRO are monotonic through the critical temperature of the phase transition, but they will be altered when sublattices are formed.

7.2.9
Kinetic Paths

The formation of a sublattice, which involves long-range atom coordinations, is a relatively slow process. The statistical kinetics approach as developed above predicts three regimes of time scale (Gschwend et al. 1978). The first is the equilibration of vacancies, which respond quickly to any change in atom configurations. The second, longer time scale is for SRO evolution, which is faster in proportion

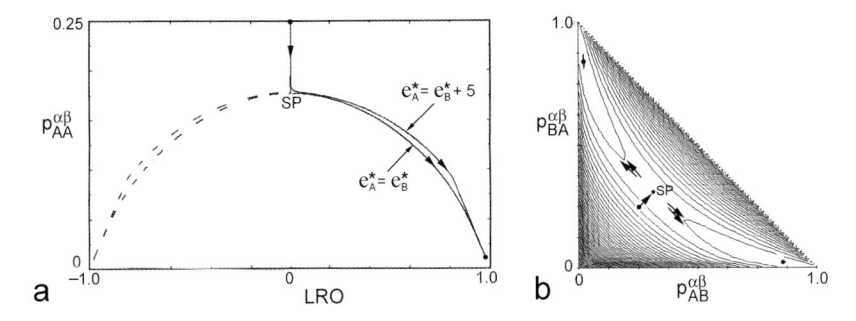

Fig. 7.19 (a) Kinetic paths for $V = 0.3$, $U = 1$, $e_{AA} = 0.3$ for the two cases $e_A^* = e_B^* + 5$ and $e_A^* = e_B^*$. (b) Free energy surface showing intitial and final states of panel (a). Here LRO $= (p_{AB}^{\alpha\beta} - p_{BA}^{\alpha\beta})/4$.

to the number of vacancies in the alloy. The slowest time scale is that for LRO, but its rate also scales with the number of vacancies.

With multiple state variables, it is often useful to make parametric plots of one variable versus another, termed "kinetic paths" (Fultz 1989). Plots of kinetic paths will be independent of vacancy concentration, so long as the vacancy concentration is low. Examples are presented in Fig. 7.19(a). The solid curves were calculated for identical chemical interactions, but with different activation barrier energies. The end states of the kinetic paths are the same, having the SRO and LRO parameters of thermodynamic equilibrium for the selected value of $V = (e_{AA} + e_{BB} - 2e_{AB})/4$. For the case with $e_A^* \neq e_B^*$, the A and B atoms have different mobilities, and this imbalance helps drive the sublattice imbalance and the more rapid growth of LRO for the same initial conditions. Similarly, the individual exchanges of A and B atoms with vacancies require more information than the thermodynamic parameter. Kikuchi and Sato (Kikuchi and Sato 1969, 1972; Sato and Kikuchi 1976) define the new parameter for chemical preferences of individual atom species, $U \equiv (e_{BB} - e_{AB})/(e_{AA} - e_{AB})$. This U alters the kinetic paths, but does not alter the final equilibrium state of the alloy.

If the initial conditions for LRO have no bias of A atom occupancy for one of the sublattices, the kinetic master equation will have a detailed balance of rates of A-atom flux on and off both the α- and β-sublattices, and LRO cannot evolve. With some initial bias, an incubation time for the evolution of LRO is observed. To some extent this is a characteristic of a theory that has only pair variables and LRO parameters, and no state variables of intermediate length scales. Larger cluster variables do not solve this problem, however, because the formation of LRO still requires a breaking of symmetry as sublattice formation occurs, and such incubations are also found in kinetic Monte Carlo simulations.

An advantage of statistical kinetic theories is that they allow mapping of the kinetic path onto a corresponding free energy surface. This makes it possible to see, for example, that the change in activation energy can readily alter the time scale and kinetic path when LRO is evolving because this part of the kinetic path lies in a broad minimum of the free energy surface of Fig. 7.19(b). It is also clear that there is rapid variation of the free energy with SRO, whereas large changes in LRO have weaker effects on reducing the free energy.

In theories of kinetic evolution containing multiple state variables, a free energy surface may show considerable structure beyond a simple thermodynamic minimum. The existence of local minima is widely known; they are termed "metastable states." Saddle points become increasingly common in functions with more independent variables. A saddle point in a free energy surface is marked "SP" in both Figs. 7.19(a) and (b). As mentioned previously, the initial conditions of the calculations need to be biased so that $p_{AB}^{\alpha\beta} \neq p_{BA}^{\alpha\beta}$. Even with this bias, the breaking of symmetry caused by atom preferences for α- or β-sublattices is a slow process because it occurs near a saddle point of the free energy surface, where there are no gradients. A kinetic arrest often occurs at such points, and this is confirmed with Monte Carlo simulations (Anthony and Fultz 1994). Such states, which have been termed "pseudostable" (Fultz 1993), are an example of many kinetic processes where the initial breaking of symmetry is driven only weakly by the free energy.

7.3
Nucleation and Growth Transformations

Section 7.1 described states of thermodynamic equilibrium and how they are summarized in phase diagrams, and Section 7.2 explained how kinetic processes alter the equilibrium phase fractions and compositions of alloys. The arguments of these previous sections tended to use generic interatomic interactions, since these are the basis for alloy thermodynamics. Ordering transformations, which occur at the spatial scale of atomic separations, are treated well with the arguments of Sections 7.1 and 7.2. Nevertheless, the formation of ordered compounds often occurs by nucleation and growth, which requires longer-range atom redistributions that can be treated classically. The present section 7.3 tends toward classical continuum theories for understanding microstructural evolution as a nucleus of a new phase grows within a parent phase.

7.3.1
Definitions

Consider a binary A–B alloy with the temperature–composition equilibrium diagram of Fig. 7.20. Shown schematically on the diagram are two quenching experiments denoted by the downward arrows labeled "I" and "II." In both cases a single phase is brought quickly to a final temperature, the tips of the arrows, where nucleation of another phase is imminent. In quench I, pure A is cooled from the liquid state to a temperature below its melting point where the resulting nuclei will be crystals of pure A. In quench II, the single phase α is brought into the two-phase "$\alpha + \beta$ region," where now the β nuclei will, in general, possess a

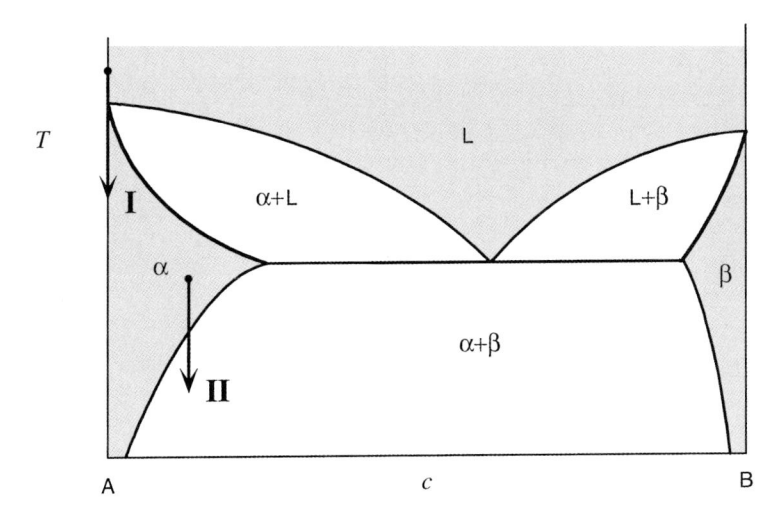

Fig. 7.20 Binary phase diagram depicting two quench paths, labeled I and II, that lead to nucleation of a new phase. L liquid phase, α, β solid phases, $\alpha + \beta$ miscibility gap. Shading denotes regions of a single stable phase.

different crystal structure from the parent α. Moreover, the emerging β-phase will be of a composition much richer in component B than that of the α-phase.

Although cases I and II can be treated within the same general theoretical framework, it will sometimes be necessary to distinguish between them. The driving force for nucleation in case I depends on the free energy difference between the pure solid and liquid at the final temperature, whereas the nucleation driving force for β particles involves free energy changes that also depend on concentration. Path II better illustrates a "supersaturated phase." Immediately after this quench, the single phase α finds itself possessing a concentration of B greater than the solubility limit for the final temperature. The α-phase will relieve its supersaturation by precipitating B-rich particles.

"Heterogeneous" nucleation refers to the formation of nuclei at specific sites in a material. In solid → solid transformations, heterogeneous nucleation occurs on grain boundaries, dislocation lines, stacking faults or any other crystalline defect. For liquids, the wall of the container is a common heterogeneous site. "Homogeneous" nucleation refers to nuclei that form randomly throughout the bulk material, i.e., without preference for location. In solids, heterogeneous nucleation is usually the favored mechanism.

Solid phases precipitating in solids are classified as "coherent" and "incoherent" nucleation. Figure 7.21(a) illustrates, in two dimensions, an incoherent nucleus. The precipitating β-phase has a crystal structure different from the parent α-phase. In contrast the coherent nucleus in Fig. 7.21(b) exhibits lattice planes that are continuous with the α matrix. The β particle in this case has a different composition and different lattice parameter from the α, but there is no structural

| (a) incoherent | (b) coherent | (c) semicoherent |

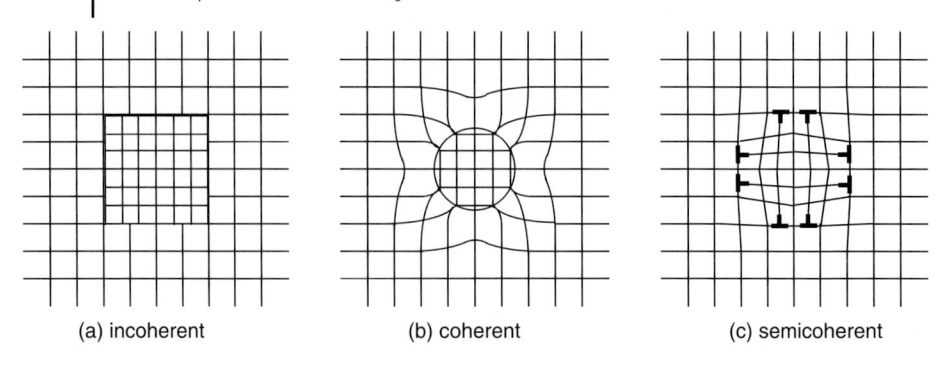

Fig. 7.21 Schematic representation of (a) an incoherent precipitate, (b) a coherent precipitate, and (c) a semicoherent interface. For the coherent case crystal lattice planes are continuous throughout the matrix and particle, whereas for a semicoherent boundary a periodic array of dislocations relieves some of the elastic strain.

discontinuity between the two phases. Heterogeneous, incoherent nucleation is the most common form of precipitation reaction in solids. Homogeneous, coherent nucleation is observed in some systems, however, including well-documented cases of Co-rich precipitates in the Cu–Co system, metastable δ' (Al_3Li) nuclei in Al–Li alloys, and the γ' phase in Ni-based superalloys. To date, there have been no reported instances of homogeneous, incoherent nucleation in any crystalline solid.

The key difference between coherent and incoherent nucleation is that, in the incoherent case, a definite surface of discontinuity between the nucleating and parent phases can be identified and a surface energy can be assigned to it. In the coherent case, a concentration gradient and elastic distortion exist between the two phases. The elastic field usually has a significant effect on the thermodynamics of nucleation.

There is a type of interface intermediate between coherent and incoherent. To relieve the elastic strain of a lattice mismatched interface, an array of edge dislocations may form on the boundary as shown in Fig. 7.21(c). An interface that is coherent except for a periodic sequence of dislocations is termed "semicoherent." With increasing mismatch of lattice parameters, the spacing of dislocations decreases and the interfacial energy increases. The same can be said of a low-angle grain boundary, which is in fact a special case of a coherent boundary.

7.3.2
Fluctuations and the Critical Nucleus

The probability of occurrence of a nucleus, $\mathscr{P}$, is given by Eq. (46).

$$\mathscr{P}(T) \propto \exp\left(-\frac{W}{k_B T}\right) \tag{46}$$

Since the volume of the nucleus is very small, the work of formation W must include an extra term $A\sigma$, the area of interface between the two phases times the surface energy. The introduction of surface costs energy, and σ, the reversible work required to form a unit area of surface, is a positive quantity. From the change in Gibbs' free energy, the work of formation is calculated via Eq. (47), where ΔG_V is the Gibbs free energy per unit volume of forming the new phase from the parent phase, and V is the volume of the new phase.

$$W = V\Delta G_V + A\sigma \qquad (47)$$

Assume that the surface energy of the second phase is isotropic, i.e., σ is independent of crystallographic directions and planes.[14] For all β particles of a specified volume, the preferred shape will be a sphere of radius R, which minimizes the surface area-to-volume ratio, and Eq. (48) is obtained.

$$W = \frac{4}{3}\pi R^3 \Delta G_V + 4\pi R^2 \sigma \qquad (48)$$

In most cases low-index planes such as $\{100\}$, $\{111\}$, etc., will possess a lower surface energy than higher-index planes, e.g. $\{221\}$. In these instances the shape that minimizes the total energy will have a larger surface area than a sphere, but will expose a greater area of low-index planes.

When $\Delta G_V > 0$, the work of formation rises very quickly with crystal size, and nucleation does not occur. When $\Delta G_V < 0$, nucleation of the β-phase is possible for large R where the contribution from the surface energy is relatively small. As shown in Fig. 7.22 for $\Delta G_V < 0$, the negative term varying as R^3 is added to a positive surface contribution varying as R^2. The result of this competition is a maximum in the work of formation given by W^*, located at the size R^*. For small $R < R^*$, W increases with R. Because Boltzmann statistics govern the fluctuation process, it is more probable that a nucleus smaller than R^* will dissociate, rather than grow by atom attachment. For sizes greater than R^*, growth is more likely. The quantity R^* locates the maximum in the $W(R)$ function, and R^* is called the "critical radius." Similarly, W^* is the critical work of formation, and represents an energy or activation barrier that must be overcome for the stable phase to nucleate and grow. The values of the critical quantities can be determined by a straightforward differentiation [Eqs. (49) and (50)].

$$\left.\frac{dW}{dR}\right|_{R=R^*} = 0 \Rightarrow R^* = -\frac{2\sigma}{\Delta G_V} \qquad (49)$$

$$W^* \equiv W(R^*) \Rightarrow W^* = \frac{16\pi}{3}\frac{\sigma^3}{(\Delta G_V)^2} \qquad (50)$$

---

14) This is not expected to be true in crystalline solids, especially when low-index crystallographic interfaces favor lower surface energies. Determination of the energy-minimizing shape for anisotropic cases requires the "Wulff construction," which can be found in advanced texts on the thermodynamics of solids.

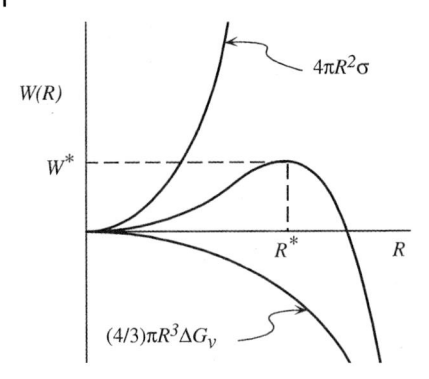

Fig. 7.22 Work of formation vs. nucleus size for for $T < T_m$. The positive surface energy contribution and the negative volume energy term lead to a critical nucleus radius, R^*, and critical work of formation, W^*.

The critical radius and the critical work of formation are key quantities of nucleation theory. Using more precise definitions, a fluctuation whose size is less than the critical radius is called a "cluster" (some texts prefer the word "embryo"), and a fluctuation greater than critical size is called a "nucleus."

For a system in equilibrium, the variation of its energy with respect to any thermodynamic variable must be zero. This leads to an alternative and often very useful definition of the critical nucleus; it is that nucleus which is in thermodynamic equilibrium with the parent phase. This equilibrium at R^* is unstable, of course.

For the case of a binary system quenched from a single phase to a two-phase region of the equilibrium diagram, quench II, the expression for the work of formation is exactly the same as Eq. (47), but we must reconsider the meaning of ΔG_V. The average concentration of the system is denoted by c_0 and, for nucleation to occur, c_0 must lie between the values c_α and c_β. Consider the effect of the internal pressure in the nucleus of the new phase, which originates from the positive surface tension. If we know that the pressure inside a critical nucleus is P, larger than the ambient P_0, we can draw the broken curve of Fig. 7.23 for $G^\beta(P)$ above the $G^\beta(P_0)$ curve. This $G^\beta(P)$ is the free energy of the β-phase at the pressure of the critical nucleus, P.

Let us assume that the volume of the critical nuclei of β-phase is so small that the loss of B atoms from the α-phase causes a negligible change in the composition of the parent α-phase. Then, since the critical nucleus is in thermodynamic equilibrium with the parent phase, the concentration of the nucleus can be found immediately by a second common tangent construction (Section 7.1.3) between the G^α and the broken $G^\beta(P)$ curves of Fig. 7.23. The quantity ΔG_V is found by an additional equilibrium condition, Eq. (51).

$$\frac{\partial W}{\partial c} = 0 \tag{51}$$

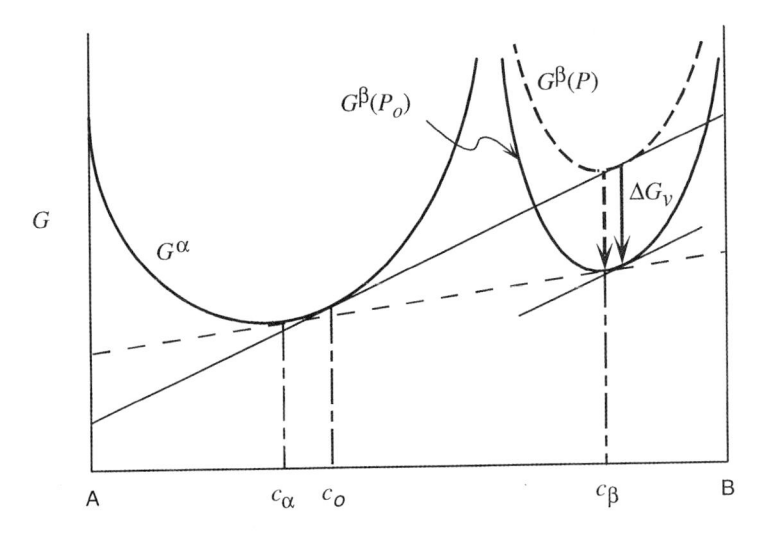

Fig. 7.23 Graphical illustration of the quantity ΔG_V for a binary A–B solution. The solid arrow is the distance between a common tangent line constructed from the average composition c_0 and a parallel line tangential to the $G^\beta(P_0)$ curve. The approximation given by the broken arrow is found by the distance from the $G^\beta(P)$ and the $G^\beta(P_0)$ curves evaluated at the equilibrium concentration c^β.

Invoking this constraint leads to a graphical determination of ΔG_V as shown in Fig. 7.23. It is the vertical distance between the common tangent line described above and a second line parallel to the first, but tangential to the $G^\beta(P_0)$ curve. The vertical distance is depicted in Fig. 7.23 as the downward pointing, solid arrow. Notice that the concentration of the critical nucleus is very close to that of the β-phase at equilibrium pressure. Furthermore, the value of ΔG_V is very nearly equal to the vertical distance between the solid and broken common tangent curves evaluated at the concentration c_β; i.e., the broken and solid arrows in Fig. 7.23 are practically the same length.

7.3.3
The Nucleation Rate

The thermodynamic concept of the critical nucleus of the preceding section, due to Gibbs (Gibbs 1948), provides key insights into the nucleation process, but it does not address a quantity of primary importance: the nucleation rate. Throughout the 1920s to 1940s, the Gibbs droplet picture of nucleation was extended to include kinetic considerations. Of primary importance in the subsequent discussion is the work of Volmer and Weber (1926), Farkas (1927), Becker and Döring (1935), and Zeldovich (1943).

In what follows, clusters and nuclei are not described by their radii, but instead by the number of molecules (or atoms) in them, termed the "number of monomers," g, in the cluster. Equation (48) then gives rise to Eq. (52).

$$W(g) = \Delta G_V \Omega g + \sigma s_1 g^{2/3} \tag{52}$$

Here Ω is the volume and s_1 is the surface area of the monomer.

To establish a nucleation rate it will be important to define two cluster size distributions. The cluster size distribution given by $n(g)$ refers to the number of clusters of size g that exist at equilibrium. From the preceding discussion of Boltzmann statistics $n(g)$ is obtained from Eq. (53) where n_1 is the *total* number of monomers in the system.

$$n(g) = n_1 \exp\left(-\frac{W}{k_B T}\right) \tag{53}$$

The number of clusters of critical size can be found from Eq. (53) by replacing W with W^*, where the critical work of formation in terms of the new spatial variable g is found in a procedure analogous to that indicated in Eqs. (49) and (50). Consider the different ways in which a cluster can change its size. A cluster containing five monomers can, by fluctuation processes, combine with a monomer to produce a cluster of size six, a $g = 2$ cluster can attach to a $g = 7$ cluster creating a size $g = 9$, etc. Here we consider only the most probable reactions – those occurring by single monomer attachment. Define $\beta(g)$ as the rate of the reaction $g \to g + 1$. Noting that the nucleation rate is simply the rate at which critical clusters are promoted to a larger size class, it seems reasonable that the nucleation rate is the number of critical clusters times $\beta(g^*)$ [Eq. (54)]

$$J_{ss} = \beta(g^*) n_1 \exp\left(-\frac{W^*}{k_B T}\right) \tag{54}$$

Equation (54) is the nucleation rate first proposed by Volmer and Weber. The subscript "ss" refers to steady state.

There is one serious problem with the Volmer and Weber prediction of the nucleation rate. The formation of clusters is through fluctuations. Therefore, it is possible for a cluster of critical size to lose monomers and thus be demoted to a lower size class, i.e., $g^* \to g^* - 1$. The arguments leading to Eq. (54) do not include such reverse reactions.

The cluster size distribution evolves in time, and we let $f(g, t)$ represent the actual cluster size distribution at a given time t. An equation for the evolution of f can be established by first examining the flux of clusters from a given size g to size $g + 1$. Using β for the rate of monomer attachment and γ for the rate of monomer detachment, the net flux of clusters of size g is given by Eq. (55).

$$J(g) = \beta(g) f(g, t) - \gamma(g) f(g + 1, t) \tag{55}$$

Equation (55) can be transformed to a partial differential equation of the Fokker–Planck form, Eq. (56).

$$\frac{\partial f}{\partial t} = \frac{\partial}{\partial g} \beta n \frac{\partial}{\partial g}\left(\frac{f}{n}\right) \tag{56}$$

It is the flux evaluated at the critical size, g^*, that sets the nucleation rate. The solution of this Fokker–Planck equation with the time derivative set equal to zero will yield the steady-state nucleation rate, but first boundary conditions must be specified. Let a factor Z be given by Eq. (57).

$$Z = \sqrt{-\frac{1}{2\pi k_B T} \frac{\partial^2 W}{\partial g^2}\bigg|_{g=g^*}} \tag{57}$$

With a Taylor expansion of $W(g)$ about the critical size and the definition of the complementary error function, the approximate result of Eq. (58) for the steady-state form of the cluster size distribution is obtained, a result found by Zeldovich.

$$f(g) = \frac{1}{2} n(g)\, \mathrm{erfc}[\sqrt{\pi} Z(g - g^*)] \tag{58}$$

The quantity Z of Eq. (57) is known as the "Zeldovich factor." It accounts for the possibility that a critical nucleus can either grow or shrink with equal probability, and larger nuclei have some probability of shrinking. The behavior of the function f is shown by the curve labeled "Z" in Fig. 7.24. The actual distribution of clusters lies below the equilibrium value, but approaches n at small g. At g^*,

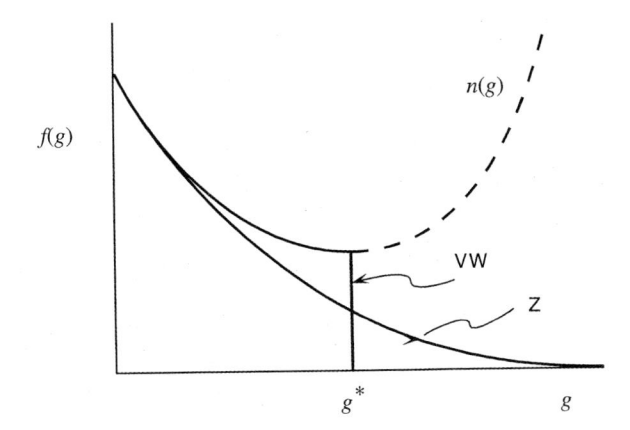

Fig. 7.24 Steady-state cluster size distributions from Volmer–Weber (VW) and Zeldovich (Z) theories. Equilibrium distribution is the broken line, and g^* is the critical cluster size.

$f(g) = 1/2n(g)$. The Zeldovich curve in Fig. 7.24 is nonzero past the critical size, a result consistent with the fact that the reverse reactions of the type $g \to g-1$ have been included in the formalism.

The steady-state nucleation rate of Eq. (59) can now be found by combining Eq. (58) with flux expressions at $g = g^*$.

$$J_{ss} = Z\beta(g^*)n_1 \exp\left(-\frac{W^*}{k_B T}\right) \tag{59}$$

This differs from the nucleation rate of Volmer and Weber by the factor Z.

To complete the picture of the nucleation rate we need to evaluate the only remaining unknown, $\beta(g^*)$. Here the functional form of the rate constant will depend on the type of system, liquid from vapor, solid–solid, etc. In the case of a liquid droplet nucleating from a supersaturated vapor it has been shown, using the kinetic theory of gases (Frenkel 1955), that β has the form of Eq. (60), where m is the mass of monomer. Note that the rate of attachment of monomer to the critical nucleus is proportional to the total surface area through the $s_1(g^*)^{2/3}$ term.

$$\beta(g^*) = n_1 s_1(g^*)^{2/3} \sqrt{\frac{k_B T}{2\pi m}} \tag{60}$$

For nucleation of a crystalline solid from an amorphous or liquid phase, Turnbull and Fisher (1949) obtained the attachment rate from Eq. (61), where h is Planck's constant, ΔG_A represents the activation energy required for an atom to jump across the interface between the parent phase and the nucleus, and λ is the jump distance.

$$\beta(g^*) = n_1 s_1(g^*)^{2/3} \lambda \frac{k_B T}{h} \exp\left(\frac{\Delta G_A}{k_B T}\right) \tag{61}$$

The above two expressions for the β parameter are actually of the same general form. In both cases the rate at which atoms are hopping across the interface of the nucleus is multiplied by the number of atoms that are poised to make the jump. For example, in Eq. (61), $k_B T/h$ represents a characteristic attempt frequency for the monomer and the exponential term represents the fraction of attempts that are successful. The term $s_1(g^*)^{2/3}\lambda$ is the volume of a thin spherical shell immediately adjacent to the nucleus. Within this shell there are n_1 monomers per unit volume poised to make the jump onto the nucleus.

The nucleation rate of Eq. (59) is a very strong function of temperature; to illustrate the sensitivity, consider again quench I of Fig. 7.20. The quantity ΔG_V is approximately linear with temperature in the vicinity of the transition temperature. If the undercooling is defined as $\Delta T = T_m - T$, where T_m is the melting temperature, Eq. (62) can be shown to hold, where L is the latent heat of fusion and the approximation is valid for small ΔT.

$$\Delta G_V \simeq \frac{L \Delta T}{T_m} \tag{62}$$

The critical work of formation can now be found with Eq. (50) and the result when substituted into Eq. (59) yields the nucleation rate [Eq. (63)]

$$J_{ss} = J_0 \exp\left(-\frac{16\pi\sigma^3 T_m^2}{3L^2 k_B T} \frac{1}{(\Delta T)^2}\right) \tag{63}$$

Here all the pre-exponential factors have been lumped into the quantity J_0.

Consider a typical system in which $T_m = 1000$ K, $L = 1.0 \times 10^9$ J m^{-3}, and $\sigma = 1$ J m^{-2}. Furthermore, an order of magnitude estimate for J_0 is 10^{40} m^{-3} s^{-1}, and for the purposes of illustration we will neglect the temperature dependence of J_0. At an undercooling of 126 K below the melting point the system will, after 1 s, contain 100 nuclei per cubic meter, a number so low that the β-phase is essentially undetectable. Just four degrees lower in temperature, $\Delta T = 130$ K, the nucleation rate jumps by a factor of 100. The extreme sensitivity to temperature suggests that for practical purposes a specific temperature can be identified as the onset of nucleation. The onset is referred to as the critical undercooling.

7.3.4
Time-Dependent Nucleation

Consider again path II of Fig. 7.20. Immediately after the quench a metastable α-phase matrix exists with B atoms arranged more or less randomly on the lattice; there are few B-rich clusters of sizes greater than $g = 1$. For a nucleation event to occur at some time after $t = 0$, a critical nucleus must form by atom transport through the α crystal. So before the steady-state nucleation rate can be observed, one has to wait for a time before a critical-size cluster can form by fluctuations. The "incubation time" or "nucleation time lag" is the characteristic time before which the nucleation rate is very low, and after which the nucleation rate is approximately the steady-state value.

To calculate the incubation time, and to map out the full time-dependent behavior of J, requires the solution to the Fokker–Planck equation [Eq. (56)]. Although the partial differential equation is linear, obtaining its analytic solution is a formidable task. Two approximate solutions exist, but the mathematical techniques employed are beyond the scope of this chapter and only the final results will be presented. For the initial condition, assume only clusters of size $g = 1$ exist at $t = 0$. Trinkaus and Yoo (1987) used a Green's function approach to arrive at the time-dependent nucleation rate according to Eq. (64), where E is defined in Eq. (65), and τ represents a natural time scale for the kinetic problem [Eq. (66)].

$$J(t) = J_{ss} \frac{1}{\sqrt{1-E^2}} \exp\left(-\frac{\pi Z^2 E^2 (1-g^*)^2}{1-E^2}\right) \tag{64}$$

$$E = \exp\left(-\frac{t}{\tau}\right) \tag{65}$$

$$\tau = \frac{1}{2\pi Z^2 \beta(g^*)} \tag{66}$$

Shi et al. (1990) obtained an analytic solution to Eq. (56) by employing the technique of singular perturbation. The solution is Eq. (67), where λ is given by Eq. (68).

$$J(t) = J_{ss} \exp\left\{-\left[\exp\left(-2\frac{t-\lambda\tau}{\tau}\right)\right]\right\} \tag{67}$$

$$\lambda = (g^*)^{-1/3} - 1 + \ln\{3\sqrt{\pi}Zg^*[1-(g^*)^{-1/3}]\} \tag{68}$$

Although Eqs. (64) and (67) have different mathematical forms, the two approximate solutions are in fact quite close (see Fig. 7.25). The two predictions of the time-dependent nucleation rate are also consistent with the qualitative picture described above – the rate is quite low for a time period on the order of τ, and rises rapidly to the steady-state value at later times. From the definition of τ it is clear that the incubation time is a function of both the critical size and the attachment rate term $\beta(g^*)$.

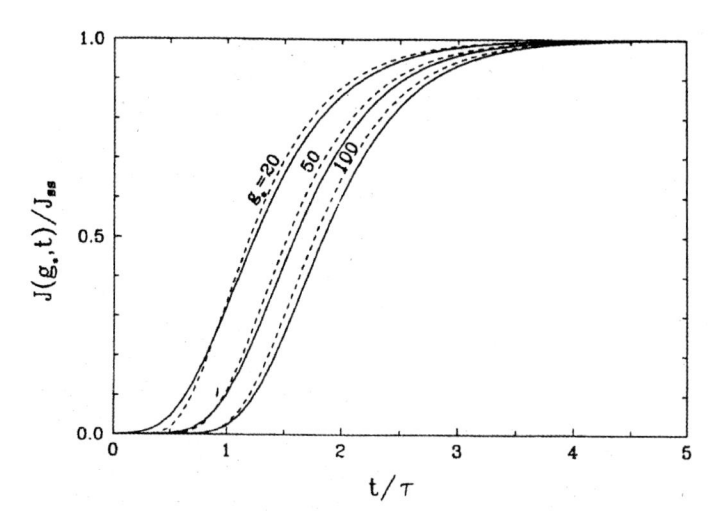

Fig. 7.25 Time-dependent nucleation rates normalized by the steady-state value, labeled by the cluster size g. The Trinkaus and Yoo (1987) solutions are the broken lines, and the Shi et al. (1990) solutions are solid curves, with labels for three critical sizes.

7.3.5
Effect of Elastic Strain

When a droplet of liquid condenses from gaseous vapor, a considerable volume change takes place, but this is easily accommodated by the flow of vapor. When a solid nucleates inside another solid, as illustrated in Fig. 7.21(b), the volume change may be smaller, but the nucleus is constrained by a stiff matrix. Accompanying the formation of a β nucleus will be a buildup of stresses and strains in both the matrix and precipitate. The elastic strain energy of the system must be added to the work of formation of the nucleus, altering the kinetic picture presented in the previous section.

The nucleation of a cubic, elastically isotropic β-phase from an infinite cubic isotropic α matrix can consist of four steps:

1. A sphere of α is cut out of the matrix material and allowed to transform unconstrained to β. The free energy change in this first step of the transformation is again ΔG_V.
2. Surface forces (tractions) are applied to the sphere to bring it back to the volume of α originally removed. The change in volume required of the sphere may be either positive or negative.
3. The β sphere is inserted into the α hole and the two phases are "welded" together.
4. Finally, the system is allowed to relax to mechanical equilibrium. In addition to the volume free energy decrease and the surface area increase, the total elastic strain energy of the nucleus–matrix system represents an extra term that must be included in the work of formation [Eq. (48)].

The displacements, stresses, strains, and energy of the misfitting sphere for the elastic continuum problem described above were first solved in a classic paper by Eshelby (1954) (see also Eshelby 1956).

We now modify the work of formation by adding to it the total strain energy associated with the nucleus–matrix system. The strains and hydrostatic pressure of the β sphere generate a strain energy density of the form of Eq. (69), where K is the bulk modulus, δ is a misfit strain arising from the difference in lattice parameters between the α and β phases, $A \equiv 3K_\beta/(3K_\beta + 4\mu_\alpha)$, and μ is the shear modulus. The total strain energy for the nucleus E_β is given by Eq. (70):

$$e_\beta = \frac{1}{2} P \frac{\Delta V}{V} = \frac{9}{2} K_\beta (A - 1)^2 \delta^2 \tag{69}$$

$$E_\beta = \int_0^R 4\pi r^2 e_\beta \, dr = \frac{4\pi}{3} R^3 \frac{9}{2} K_\beta (A - 1)^2 \delta^2 \tag{70}$$

A similar procedure applied to the α-phase results in Eq. (71).

$$E_\alpha = \frac{4\pi}{3} R^3 6\mu_\alpha A^2 \delta^2 \tag{71}$$

The total strain energy for the α matrix containing a spherical β nucleus is the sum of Eqs. (70) and (71).

$$E = E_\alpha + E_\beta = \frac{4\pi}{3} R^3 6\mu_\alpha A \delta^2 \tag{72}$$

Since the total strain energy is proportional to the volume of the nucleus, Eq. (48) can be modified in the following way to Eq. (73).

$$W = \frac{4\pi}{3} R^3 (\Delta G_V + 6\mu_\alpha A \delta^2) + 4\pi R^2 \sigma \tag{73}$$

All of the parameters appearing in Eq. (73) are positive quantities. If the cost of elastic energy overrides the decrease due to the ΔG_V term, i.e., $\Delta G_V + e > 0$, homogeneous nucleation of the β-phase cannot occur. Because the strain energy contribution is positive, the nucleation rate will be much slower than predicted by Eq. (59). The bottom two curves of Fig. 7.26 show the effect on the volume contribution to $W(R)$ when strain energy is included. An increase in the coefficient of the R^3 term leads to an increase in both the critical radius and the critical work of formation. The shift to larger W^* and R^* is shown by the middle two functions in Fig. 7.26. By reference to Eq. (59) it can be seen that even small increases in the work of formation cause drastic reductions in the steady-state nucleation rate.

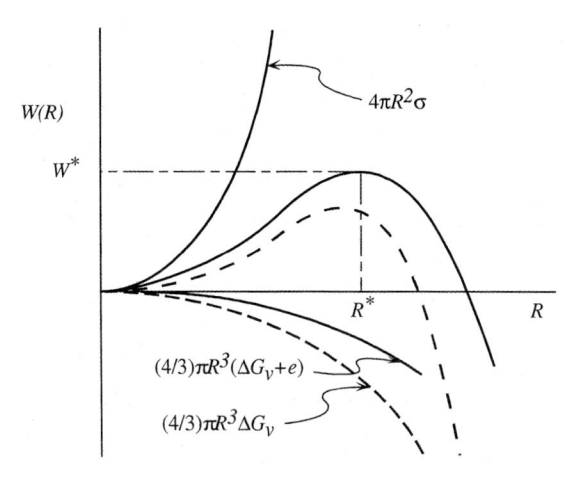

Fig. 7.26 Work of formation vs. cluster radius for the case where elastic strain energy is included. An increase in the volume energy term (lower two curves), due to positive elastic energy, results in an increase in W^* and R^*.

7.3.6
Heterogeneous Nucleation

Homogeneous nucleation in solid systems is rarely observed. The usual type of nucleation is heterogeneous, where the second phase nucleates at dislocations or at grain boundaries. To extend classical nucleation theory to heterogeneous nucleation, we first discuss how the size, shape, and work of formation of the critical nucleus change in the presence of a defect.

In homogeneous nucleation, a spherical nucleus was considered because it is the shape that minimizes the surface area for a given volume. In the case of heterogeneous nucleation on a grain boundary, the elimination of grain boundary area causes the shape of minimum surface area to become a double spherical cap. The surface area of a double spherical cap having a contact angle θ (see Fig. 7.27) is $4\pi(1 - \cos\theta)R^2$, and its volume is $2\pi(2 - 3\cos\theta + \cos^3\theta)R^3/3$, where R is the (constant) radius of curvature. With these expressions, the work of formation as a function of R can be formulated in Eq. (74) along the same lines as in the derivation of Eq. (48).

$$W = \frac{2\pi}{3}(2 - 3\cos\theta + \cos^3\theta)R^3 \Delta G_V$$
$$+ 4\pi(1 - \cos\theta)R^2\sigma - \pi R^2 \sin^2\theta \sigma_{gb} \qquad (74)$$

The new last term on the right-hand side contains the grain boundary surface energy, σ_{gb}, and has a negative sign. When a grain boundary nucleus is formed, an area of pre-existing α–α interface is eliminated, as shown by the broken line in Fig. 7.27. The energy saving of $\pi R^2 \sin^2\theta \sigma_{gb}$ decreases the work of formation. Since this decrease is not realized for a nucleus forming in the bulk, heterogeneous nucleation is preferred energetically. Critical quantities can again be found by taking the derivative of W with respect to the radius of curvature and setting the result equal to zero. Interestingly the value of R^* is identical to that derived in the homogeneous case [Eq. (75)], whereas the critical work of formation W^* is now given by Eq. (76).

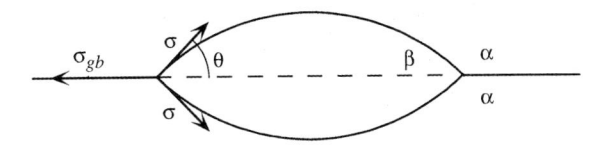

Fig. 7.27 Schematic representation of a grain boundary β-phase nucleus as a double spherical cap. The contact angle θ must satisfy a balance of surface energy forces depicted by the three vectors of lengths σ and σ_{gb}. The broken line shows the region of pre-existing grain boundary that is eliminated by the nucleus.

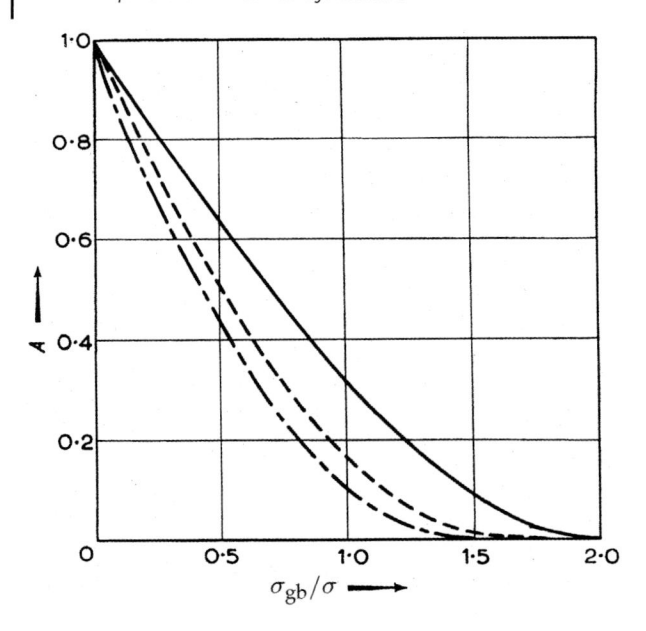

Fig. 7.28 Critical work of formation for grain boundary nucleation (top curve), grain edge nucleation (middle curve) and grain corner nucleation (bottom curve) vs. the quantity σ_{gb}/σ. Work of formation is normalized by the value obtained for homogeneous nucleation. From Cahn (1956).

$$R^* = -\frac{2\sigma}{\Delta G_V} \tag{75}$$

$$W^* = \frac{16\pi\sigma^3}{3(\Delta G_V)^2} \frac{2 - 3\cos\theta + \cos^3\theta}{2}$$

$$= W^*_{hom} \frac{2 - 3\cos\theta + \cos^3\theta}{2} \tag{76}$$

Here W^*_{hom} is the critical work of formation for homogeneous nucleation. The term $(2 - 3\cos\theta + \cos^3\theta)/2$ can range from zero to one, so the work of formation of a heterogeneous grain boundary nucleus is less than that of a homogeneous nucleus. The behavior of W^* normalized by the homogeneous value is shown as the top (solid) curve in Fig. 7.28.

Figure 7.27 was drawn assuming an isotropic surface energy for the α–β interface. Lee and Aaronson (1975) have investigated the grain boundary nucleation problem and concluded that the double spherical cap morphology is very unlikely. A faceted nucleus is generally expected. Figure 7.29 shows a faceted β-phase nucleus that is expected when the crystallographic planes of the β- and α-phases have low-energy orientation relationships. The nucleus in Fig. 7.29 is called a "grain boundary allotriomorph."

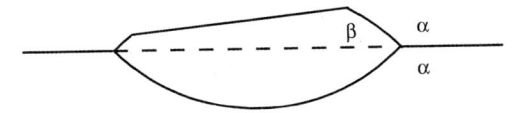

Fig. 7.29 A faceted grain boundary nucleus can lower its work of formation by exposing a large area of a crystallographic plane with low surface energy. The shape is referred to as an allotriomorph.

7.3.7
The Kolmogorov–Johnson–Mehl–Avrami Growth Equation

For long aging times the steady-state nucleation rate of Eq. (59) ceases to be valid. For quench II of Fig. 7.20, as more and more β nuclei form and grow, the matrix phase α becomes depleted in solute. Since the work of formation of a critical nucleus, and hence the nucleation rate, depend strongly on supersaturation, J should decrease dramatically with time in the later stages of the phase transformation. A description of the competition between solute depletion, nucleation, and particle growth is a difficult mathematical undertaking, but significant progress was made by Langer and Schwartz (1980) (see also Binder and Stauffer 1976).

The full time dependence of the transformation after quench I of Fig. 7.20 is also a challenging problem. Fortunately the time dependence of a quantity of central interest, the volume fraction of β-phase, can be found readily by an ingenious construction developed by Kolmogorov (1937), Johnson and Mehl (1939), and Avrami (1939, 1940, 1941). Figure 7.30 depicts the microstructure of the $\alpha + \beta$ mixture at three representative times, where $t_1 < t_2 < t_3$. At t_1 a number of nuclei appear as governed by the nucleation rate J, and each β particle is approximately the critical size at this early time. At t_2 nucleation leads to an increased number of β-phase particles, and all the nuclei that had formed at earlier times have grown in size. If v denotes the rate of change of the particle radius, then the total volume of β-phase at early times can be written as Eq. (77), where V is the total volume of the system. The dummy variable t' accounts for the growth of all particles that nucleated at earlier times.

$$V^\beta(t) = \frac{4\pi}{3} V \int_0^t J v^3 (t - t')\, dt' \qquad (77)$$

Unfortunately, the kinetic equation (77) cannot possibly be valid for all times. As shown by the microstructure at t_3, eventually the β particles will impinge on one another, halting the transformation. A single growth rate v is not possible after impingement, so the trick now is to account for the impingement effects.

Imagine that the β-phase can grow unimpeded by the presence of other particles; in other words the particles are free to overlap as shown in the bottom microstructure of Fig. 7.30. In addition let us allow nucleation of β to take place in

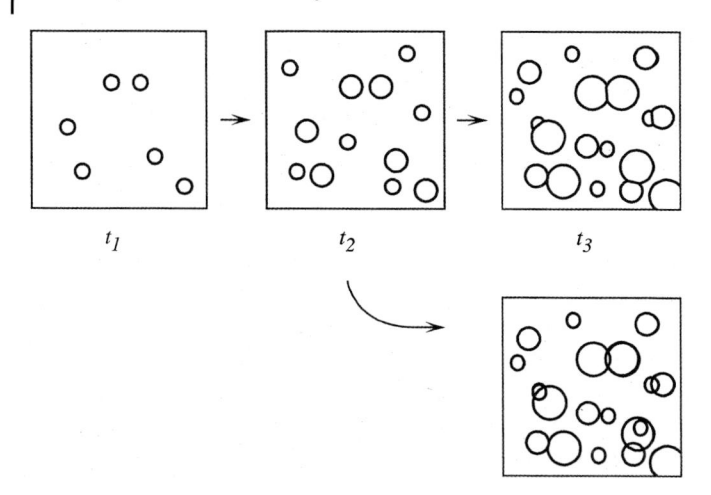

Fig. 7.30 Evolution of the microstructure of a single-phase α transforming to the single-phase β at three times, where $t_1 < t_2 < t_3$. At time t_3, β-particles begin to impinge on one another; the extended volume concept allows the β-phase to grow and nucleate in previously transformed material as shown in the bottom diagram.

regions that have already transformed! The volume of β under this clearly artificial construction is called the extended volume and is denoted by V_e^β. The actual volume of β must approach the total volume as $t \to \infty$, but the extended volume increases without bound as time progresses. The quantity V_e^β is given precisely by the right-hand side of Eq. (77) for all time t.

After a volume V^β has transformed, the volume fraction that remains untransformed is $(1 - V^\beta/V)$. In the next time interval dt a fraction of the increase in extended volume dV_e^β will occur in regions of untransformed material and will thus contribute to the increase of the volume dV^β. The remaining fraction of dV_e^β will necessarily lie in already transformed regions and thus will add nothing to the volume of the β-phase. Now we make the key assumption that the β-phase nucleates randomly in the parent α matrix. On average, the fraction of increase in V_e^β that lies in untransformed material is simply that volume fraction that has not yet been transformed, i.e. dV^β from Eq. (78), or by a straightforward integration, V_e^β from Eq. (79).

$$dV^\beta = \left(1 - \frac{V^\beta}{V}\right) dV_e^\beta \tag{78}$$

$$V_e^\beta = -V \ln\left(1 - \frac{V^\beta}{V}\right) \tag{79}$$

Since the extended volume is given by Eq. (77), the time dependence of the true volume fraction of β-phase is obtained in Eq. (80), the Kolmologorov–Johnson–

Mehl–Avrami (KJMA) equation, which predicts that the volume fraction at $t = 0$ is equal to zero, but increases in a sigmoidal shape, and asymptotically approaches 1 as $t \to \infty$.

$$\frac{V^\beta}{V} = 1 - \exp\left[-\frac{4\pi}{3}\int_0^t Jv^3(t-t')\,dt'\right] \qquad (80)$$

The exact form of V^β/V depends, of course, on the precise description of the time-dependent nucleation rate term and the growth rate. Two limiting cases are typically analyzed, and both assume a constant v. First, consider a system in which all nucleation takes place instantaneously at heterogeneous sites distributed randomly throughout the material. If all nucleation sites are assumed saturated at $t = 0$, Eq. (80) can be integrated to give Eq. (81), where N is the number of sites per unit volume which become instantaneously filled with nuclei.

$$\frac{V^\beta}{V} = 1 - \exp\left[-\frac{4\pi}{3}v^3 Nt^3\right] \qquad (81)$$

The second limiting case is one in which the nucleation rate, here J_{ss}, is assumed to be a constant with time. Integration of the KJMA equation (80) yields Eq. (82).

$$\frac{V^\beta}{V} = 1 - \exp\left[-\frac{\pi}{3}v^3 J_{ss} t^4\right] \qquad (82)$$

Notice the change in exponent from t^3 to t^4 between Eqs. (81) and (82). Since the above two examples are limiting cases of nucleation behavior, Avrami suggested that a plot of $\ln[\ln(1 - V^\beta/V)]$ versus t will result in a straight line with a slope between 3 and 4, and the exponent of t will indicate the dominance of homogeneous or heterogeneous nucleation. In deriving Eq. (81) Avrami assumed that heterogeneous sites were distributed randomly throughout the bulk material. This is a dubious assumption for polycrystalline solids, however. Second-phase nuclei tend to cluster on specific grain boundaries or along dislocation lines. Equation (81) can be modified to accommodate these processes (Cahn 1956), but this is beyond the scope of the present treatment.

7.4
Spinodal Decomposition

In his classic work on equilibrium in heterogeneous substances, Gibbs (1948) distinguished between two types of concentration fluctuations that give rise to the formation of a new stable phase: (a) those that are small in spatial size but large in concentration change, and (2) those that are large in spatial size but small in concentration change. The first pertains to nucleation (the subject of

Section 7.3), the second to "spinodal decomposition" (the subject of this section). The process of spinodal decomposition by the growth of infinitesimally small-amplitude concentration fluctuations was elucidated by Cahn (1961).

7.4.1
Concentration Fluctuations and the Free Energy of Solution

Consider the idealized, one-dimensional concentration fluctuation shown in Fig. 7.31 for a sample of average concentration c_0. Over the spatial range, x, from 0 to $+\zeta$, the composition of the sample is slightly greater than the average, i.e., $c(x) = c_0 + \delta c$ for $0 < x < +\zeta$, where δc is small compared to c_0. Similarly, in the range from $-\zeta$ to 0 the concentration is less than average. By conservation of solute, we must have $c(x) = c_0 - \delta c$ for $-\zeta < x < 0$. Take the cross-sectional area of the sample to be unity, and the total length to be L. Then the total free energy of homogeneous material before introducing the composition fluctuation is simply $Lf(c_0)$, where f is the Helmholtz free energy per unit volume. Of interest here is the total free energy change, ΔF, from an initial state of uniform composition to the state depicted in Fig. 7.31. Since the amplitude of fluctuation is assumed small we can find the free energy of the material by Taylor-expanding about c_0. Therefore, ΔF for the region 0 to $+\zeta$ becomes ΔF^+ [Eq. (83)], where, for the time being, the gradient energy contribution has been neglected.

$$\Delta F^+ = \zeta f(c_0) + \zeta \left.\frac{\partial f}{\partial c}\right|_{c=c_0} \delta c + \zeta \frac{1}{2}\left.\frac{\partial^2 f}{\partial c^2}\right|_{c=c_0} (\delta c)^2 + \cdots - \zeta f(c_0) \qquad (83)$$

Similarly, the free energy change over the region where $x < 0$ is given by Eq. (84).

$$\Delta F^- = \zeta f(c_0) - \zeta \left.\frac{\partial f}{\partial c}\right|_{c=c_0} \delta c + \zeta \frac{1}{2}\left.\frac{\partial^2 f}{\partial c^2}\right|_{c=c_0} (\delta c)^2 + \cdots - \zeta f(c_0) \qquad (84)$$

The total free energy change upon introduction of the fluctuation is, of course, the sum of Eqs. (83) and (84), minus $f(c_0)$ for the homogeneous alloy, according

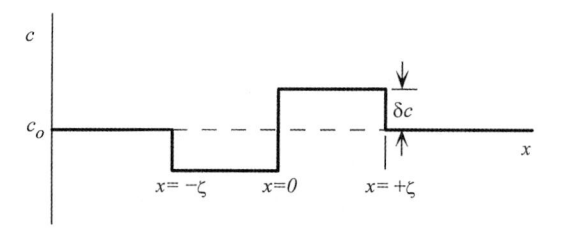

Fig. 7.31 Idealized, one-dimensional concentration fluctuation in a material of average composition c_0. The positive and negative parts of the fluctuation account for the same amount of solute.

7.4 Spinodal Decomposition

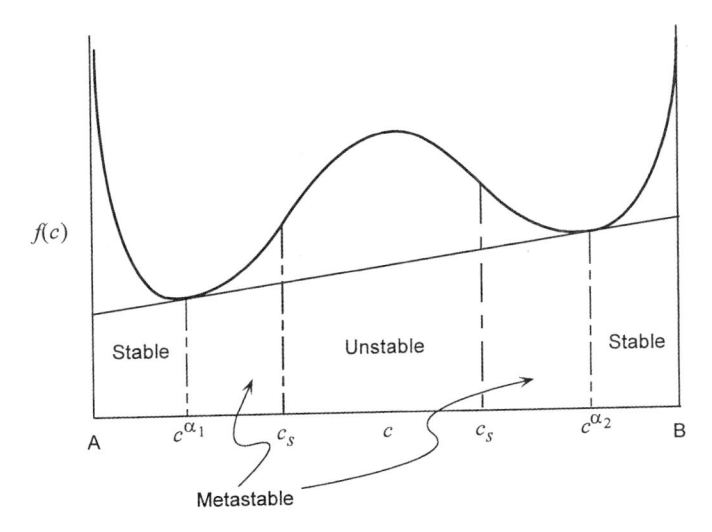

Fig. 7.32 Free energy per unit volume as a function of composition for a system exhibiting a miscibility gap. Also shown are the ranges of concentration for which the solution is stable, metastable, and unstable.

to Eq. (85), where the notation f'' denotes the second derivative of the free energy with respect to composition, evaluated at c_0.

$$\Delta F = \zeta f''(\delta c)^2 \qquad (85)$$

In Eq. (85) both the terms ζ and δc^2 are positive, and therefore the sign of the total free energy change is governed by f''. Figure 7.32 shows the behavior of f versus c for a binary system exhibiting a miscibility gap. For the ranges of average composition labeled "stable" and "metastable," the second derivative of f is positive. In these ranges, small fluctuations in concentration will increase the free energy of solution, so the unfavorable fluctuation will decay, and the system will return to a homogeneous state. On the other hand, for a range of composition in the unstable region, f'' is in fact negative. In this unstable range, concentration disturbances, no matter how small in amplitude, cause the free energy to decrease. Furthermore, Eq. (85) suggests that if the fluctuation continues to grow, the free energy can decrease more. The formation of a stable two-phase mixture from a solution for which $f'' < 0$ is known as spinodal decomposition.

The meaning of the labels of Fig. 7.32 is now clear. For the metastable region, small concentration fluctuations (for which the Taylor expansion of Eq. (85) remains valid) will decay, but large composition deviations (i.e., those that create a critical nucleus resembling the stable α_2-phase) will grow. In the unstable range where $f'' < 0$, even infinitesimally small fluctuations will increase spontaneously in amplitude. Finally, in the concentration regions labeled "stable," the equilibrium state is a single-phase solid solution. Here any departure from a homogeneous solution of concentration c_0, no matter how large, will eventually return to the single-phase state. The concentration marked c_s in Fig. 7.32, corresponding to the point at which $f'' = 0$, represents the limit of stability of the solution and is called the "spinodal composition." In the continuum theory of nucleation, the quantity c_s coincides with the vanishing of the critical work of formation. In light of Eq. (85) and the curve shown in Fig. 7.32, the fact that $W^* = 0$ at c_s is no surprise. At this composition there is no longer an initial increase in energy of the system, or activation barrier on forming a "nucleus," but rather a decrease in total energy.

7.4.2
The Diffusion Equation

The gradient in chemical potential drives diffusion, not the gradient in concentration *per se*. Fick's law is valid for ideal solutions, but the free energy in the unstable region of Fig. 7.32 is certainly not ideal. We seek a diffusion equation appropriate for the case of spinodal decomposition. When lattice sites are conserved, the gradients of chemical potential can be replaced by a single gradient in the difference of chemical potentials between the two species A and B in solution. One can then show that the flux of solute becomes $\vec{J}$ [Eq. (86)], where M is the mobility of solute atoms, a positive quantity, and N_v is the number of atoms per unit volume. Here F is a functional of c. The term in parentheses in Eq. (86) is the variation of F with respect to the function $c(\vec{r})$.

$$\vec{J} = -\frac{M}{N_v} \vec{\nabla} \left(\frac{\delta F}{\delta c} \right) \tag{86}$$

With an approach used by Cahn and Hilliard (1958, 1959), the variation is then given by Eq. (87), where κ is the gradient energy coefficient, here assumed independent of concentration.

$$\frac{\delta F}{\delta c} = \int_{\text{vol}} \left\{ \frac{\partial f}{\partial c} - 2\kappa \nabla^2 c \right\} d^3\vec{r} \tag{87}$$

For a small element of volume, the variation of F with respect to c is given by an Euler–Lagrange equation. Using Eq. (87), Eq. (88) is obtained for the flux.

$$\vec{J} = -\frac{M}{N_v} \vec{\nabla} \left\{ \frac{\partial f}{\partial c} - 2\kappa \nabla^2 c \right\} \tag{88}$$

When combined with the continuity equation, Eq. (88) yields the general diffusion equation describing any diffusional transformation in which the gradient energy contribution is important. The result is Eq. (89), where in anticipation of a stability analysis, we used the function $u(r, t) \equiv c(r, t) - c_0$, describing the deviation of the concentration from its average value.

$$\frac{\partial u}{\partial t} = \vec{\nabla} \cdot \left[\frac{M}{N_v} \vec{\nabla} \left\{ \frac{\partial f}{\partial u} - 2\kappa \nabla^2 u \right\} \right] \qquad (89)$$

To perform a stability analysis of the concentration, we must specify the spatial dependence of the small perturbation, and examine how this specific form of the fluctuation evolves with time. Consider a one-dimensional concentration disturbance of the form in Eq. (90), where $a(t)$ is the amplitude of the concentration wave.

$$u(\vec{r}, t) = a(t) \cos(kx) \qquad (90)$$

The time dependence of a determines the stability. Here k is the wavenumber, which can also be written as $2\pi/\lambda$, λ being the wavelength of the concentration profile. Substitution of Eq. (90) into the diffusion equation (89) leads to Eq. (91).

$$\begin{aligned}
\frac{\partial}{\partial t}[a(t) \cos(kx)] = & \frac{1}{N_v} \frac{\partial}{\partial x} [M_0 + M' a(t) \cos(kx) + M'' a^2(t) \cos^2(kx) + \cdots] \\
& + \frac{\partial}{\partial x} \left[f' + f'' a(t) \cos(kx) + f''' a^2(t) \cos^2(kx) + \cdots \right. \\
& \left. - \frac{\partial \kappa}{\partial u} a^2(t) k^2 \sin^2(kx) + 2\kappa a(t) k^2 \cos(kx) + \cdots \right]
\end{aligned} \qquad (91)$$

Two functions, M and $\partial f / \partial u$, which are generally nonlinear functions of u, have been replaced by their Taylor expansions, and primes indicate differentiation evaluated at $u = 0$. In keeping with the linear stability analysis, only small-amplitude perturbations are considered. Therefore, any term in Eq. (91) that is multiplied by a factor $a^2(t)$, $a^3(t)$, etc., is neglected. The small-amplitude assumption results in the simplification to Eq. (92), to which Eq. (93) has the solution.

$$\frac{\partial a(t)}{\partial t} = -\frac{M_0}{N_v}[f'' k^2 - 2\kappa k^4] a(t) \qquad (92)$$

$$a(k, t) = a(k, t = 0) \exp\left[-\frac{M_0}{N_v} k^2 (f'' + 2\kappa k^2) t \right] \qquad (93)$$

The result, Eq. (93), shows that if $f'' > 0$, all small-amplitude concentration fluctuations will decay exponentially with time, and the system is stable. On the other hand, if $f'' < 0$, then some perturbations are stable, i.e., those wavenumbers for which $|f''| > 2\kappa k^2$, but all others are unstable.

Figure 7.33 shows the behavior of the amplification factor, $\alpha(k)$, versus k where α is the term multiplying the time in the exponential function of Eq. (93), i.e., $\alpha(k) = -(M_0/N_v) k^2 (f'' + 2\kappa k^2)$. For wavenumbers less than k_c, $\alpha(k)$ is positive and the amplitude $\alpha(t)$ grows exponentially with time, whereas for $k > k_c$ the

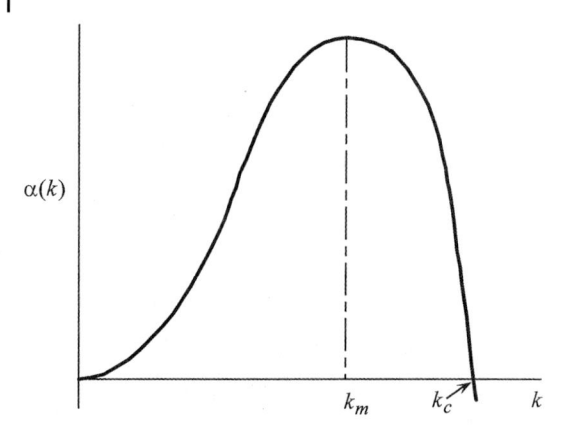

Fig. 7.33 Amplification factor $\alpha(k)$ vs. k. The value k_c corresponds to the point $\alpha = 0$ and k_m refers to the maximum in α.

amplitude diminishes. The value of k_c [Eq. (94)] is known as the "critical wavenumber."

$$k_c = \sqrt{-\frac{f''}{2\kappa}} \tag{94}$$

In addition, let the point k_m correspond to the maximum in the curve of Fig. 7.33. The exponential dependence of the solution, Eq. (93), and the peak in $\alpha(k)$ at $k = k_m$, suggest that for any initial concentration fluctuation, those Fourier components in the vicinity of k_m will be amplified most rapidly. This selective amplification implies that, shortly after a quench to the unstable region of the phase diagram, the Fourier spectrum will be dominated by a single component. It suggests a real-space concentration profile characterized by a nearly periodic, three-dimensional interconnected network of regions with high and low solute concentrations.

7.4.3
Effects of Elastic Strain Energy

In general, the molar volumes of the two species in a binary solution are different. As a result, a concentration fluctuation creates an elastic strain field in the material. We have seen in the case of nucleation that the positive elastic strain energy from a misfitting spherical nucleus must be added to the work of formation, thereby dramatically changing the predicted kinetics. The strain energy generated by growing concentration fluctuations also plays an important role in the spinodal decomposition of crystalline solids.

The elastic strain energy for an arbitrary composition fluctuation can be found in much the same way as was done in solving the diffusion equation by Fourier transform methods. By Fourier-transforming the composition, computing the elastic energy of each Fourier component, and utilizing the fact that each of the components does not interact, one obtains a general elastic energy contribution [Eq. (95)], where E is Young's modulus, v is Poisson's ratio and η is the fractional change of lattice parameter with composition.

$$e = \frac{E}{1-v}\eta^2 \int_{vol} u^2 \, d^3\vec{r} \qquad (95)$$

To include the elastic strain energy in either the continuum theory of nucleation or that of spinodal decomposition, the Cahn–Hilliard free energy expression must be modified as Eq. (96).

$$F = \int_{vol} \left[f(c) + \frac{E}{1-v}\eta^2(c-c_0)^2 + \kappa(\nabla c)^2 \right] d^3\vec{r} \qquad (96)$$

It is a straightforward exercise to show that the limit of stability of a solid solution is no longer $f'' = 0$ but is given by Eq. (97).

$$f'' + \frac{2\eta^2 E}{1-v} = 0 \qquad (97)$$

The concept of a spinodal can now be illustrated with Fig. 7.34. The heavy solid line shows the equilibrium miscibility gap (cf. Fig. 7.6). Below the equilibrium phase boundary is a broken curve corresponding to the locus of points such that $f'' = 0$. This curve is called the "chemical spinodal." The dash-dot curve, the "coherent spinodal," is the limit of stability with elastic contributions included. It is

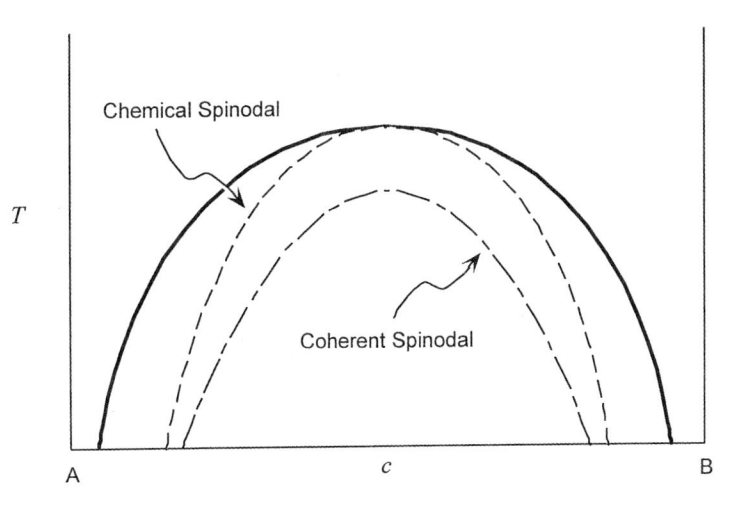

Fig. 7.34 Phase diagram of a binary system with a miscibility gap (solid curve). Also shown are the chemical spinodal where $f'' = 0$ and the coherent spinodal for which $f'' + 2E\eta^2/(1-v) = 0$.

found by displacing the chemical spinodal downward by an amount equal to the additional term in Eq. (97). Cahn showed that the $2\eta^2 E/(1-v)$ term is large in many solid solutions.

Not only does the additional elastic term change the criterion for the limit of stability, it alters the kinetics of decomposition too. With Eq. (96), the linearized form of the diffusion equation (89) has a solution in Fourier space given by Eq. (98).

$$a(k,t) = a(k,t=0) \exp\left[-\frac{M_0}{N_v} k^2 \left(f'' + 2\eta^2 \frac{E}{1-v} + 2\kappa k^2\right) t\right] \qquad (98)$$

The presence of the term involving elastic constants implies a decrease in the rate at which concentration waves are amplified. As for the nucleation of a misfitting sphere, the positive elastic energy contribution slows the reaction. The assumption of an elastically isotropic solid is often inaccurate. Spinodal decomposition in anisotropic systems is characterized by preferential rapid growth of concentration waves along elastically soft directions (Cahn 1962).

7.5
Martensitic Transformations

The previous two sections described phase transformations involving the diffusion of atoms over moderate distances, for which continuum diffusion equations give key insights. This section describes martensitic transformations, in which all atoms in a crystal distort cooperatively into a new shape. Martensitic transformations are often called "diffusionless," and depend on features of crystal geometry.

A plain carbon steel at a temperature of 950 °C exists in a face-centered cubic (fcc) phase known as "austenite." In austenite the Fe atoms occupy the sites of an fcc lattice, and the C atoms are squeezed into octahedral interstices. After the steel has been quenched rapidly to room temperature, the material transforms into a body-centered tetragonal (bct) phase known as "martensite." Because of the technological importance of steel, the martensite transformation has been well studied, but that is not to say it is completely understood. The martensite transformation also occurs in Fe–Ni, Au–Cd, Ti–Nb, In–Tl, Cu–Zn, Cu–Al, and many other alloys; in superconductors, e.g., V_3Si and Nb_3Sn, and in ceramic systems such as ZrO_2 and $BaTiO_3$. The formation of martensite is different from the types of phase transformation discussed previously: eight characteristic features are enumerated in Section 7.5.1.

7.5.1
Characteristics of Martensite

1. Martensitic transformations are "diffusionless." Atoms move by less than one interatomic distance and the product phase

is formed by a cooperative motion of many atoms. Since no long-range, solid-state diffusion occurs, the product martensite phase must have the same composition as the parent (austenite) phase. The diffusionless nature of the transition was proved by Kurdjumov and coworkers (Kaminsky and Kurdjumov 1936; Kurdjumov et al. 1939), who demonstrated that the ordered alloys Cu–Zn and Cu–Al remain ordered after the martensitic transformation. The transformation therefore causes a negligible change in the configurational entropy of the alloy. The entropy of the transformation originates from changes in vibrational modes, or in some Fe alloys, from changes in magnetic order.

2. Martensite usually appears in the form of thin plates in the austenite phase, but other morphologies such as needles and laths are also observed. For a given composition, the plates lie on distinct crystallographic planes in the parent phase. The preferred plane is known as the "habit plane." Sometimes the habit plane is a low-index or close-packed plane, but sometimes the Miller indices of the habit plane are irrational numbers. Martensite plates grow at velocities approaching the speed of sound, and the motion is terminated when the plate encounters a grain boundary, the specimen surface, or another martensite plate. The formation of a martensite platelet can trigger the formation of other plates. This autocatalytic process is often accompanied by an audible "click."

Figure 7.35 is a schematic diagram of the plate formation process; notice how successively smaller and smaller plates appear as more of the specimen is transformed to martensite. Figure 7.36 shows an almost fully transformed Fe–Ni alloy. Martensite initially forms as very thin plates that subsequently thicken during the rapid growth stage. The thickening process is inferred from optical and electron microscopy, which reveal the presence of midribs running

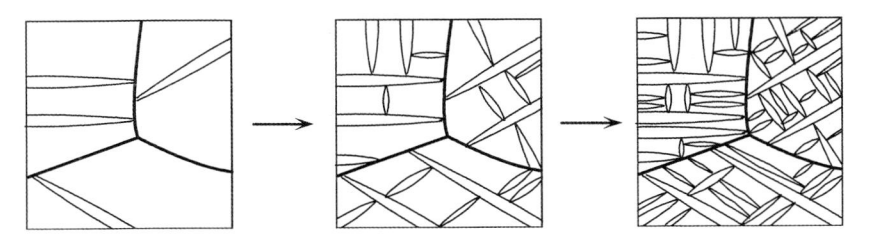

Fig. 7.35 Sequence of martensite plate formation. Growth of plates is terminated at grain boundaries and at other martensite plates.

Fig. 7.36 Martensite plates in an almost completely transformed Fe–Ni specimen (magnification ×500). From Shewmon (1969).

along the center line of the plate. Midribs, as seen in Fig. 7.37, are thought to be the original nuclei of the martensite phase.

3. The volume fraction of martensite is a function of temperature. The first martensite plates form at the "martensite start" temperature denoted by M_s, whereas the specimen is completely transformed for temperatures below M_f, the "martensite finish" temperature. The volume fraction may or may not change with time. In the "isothermal" case

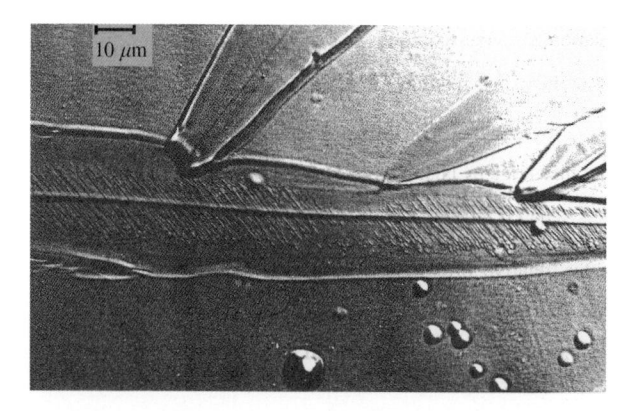

Fig. 7.37 Intersecting martensite plates in an Fe–32%Ni specimen. The straight line running through the center of the plates is known as a midrib. The fine lines running across the midrib are twins. From Shewmon (1969).

the volume fraction of martensite increases with time, whereas for "athermal" martensites the volume fraction of the product phase changes almost instantaneously at any temperature in the range $M_s > T > M_f$, and holding at a temperature for any length of time does not cause the formation of more martensite.

4. There exists a definite crystallographic relationship between the martensite and austenite phases. In plain carbon steels, Kurdjumov and Sachs (1930) determined that the (111) plane in the austenite phase lies parallel to the (011) plane in the martensite.[15] In addition, the $[\bar{1}01]$ direction in austenite, which lies in the (111) plane, is oriented parallel to the $[\bar{1}\bar{1}1]$ direction in the martensite plates. If we employ the standard notation of "γ" representing the fcc austenite and "α'" denoting the bct product phase, the Kurdjumov–Sachs relationships are expressed by Eqs. (99).

$$(111)_\gamma \| (011)_{\alpha'}, \quad [\bar{1}01]_\gamma \| [\bar{1}\bar{1}1]_{\alpha'} \tag{99}$$

The two planes that lie parallel to one another are the close-packed planes in each structure. In Fe–Ni alloys, with the Ni content greater than 28 wt.% Ni, the orientation relationships, first measured by Nishiyama (1934), are as in Eqs. (100).

$$(111)_\gamma \| (011)_{\alpha'}, \quad [\bar{1}\bar{1}2]_\gamma \| [0\bar{1}1]_{\alpha'} \tag{100}$$

Often the relative orientations of parent and martensite phases cannot be expressed in simple forms such as those of Eqs. (99) and (100). In the Au–47.5 at.%Cd alloy, the high-temperature phase is the ordered B2 (CsCl) structure, denoted β, and the martensite phase is an ordered orthorhombic crystal structure (β') (Lieberman et al. 1955). The low-index planes are misoriented by a couple of degrees.

5. The macroscopic distortion caused by the formation of martensite is a homogeneous shear. Figure 7.38(a) shows a series of straight parallel lines that were inscribed on the surface of the parent phase. An illustration of the surface relief phenomenon is depicted in Fig. 7.38(b), which represents a cross-section of the specimen shown in Fig. 7.38(a). After a martensite plate forms and intersects the

---

15) Orientation relationships are measured by taking two diffraction patterns, X-ray or electron: one from a single martensite plate and another from an adjacent austenite grain. By indexing both patterns one can formulate the relative spatial orientations of the two crystals.

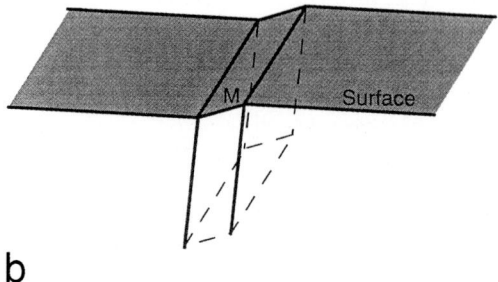

Fig. 7.38 (a) View normal to the surface of a specimen exhibiting a martensite plate. Parallel lines inscribed on the surface remain straight, but a macroscopic shear due to the transformation is observed.
(b) Surface upheaval due to the plate intersecting the surface of the sample.

sample surface, the lines remain straight in the martensite (labeled M), but make a distinct angle with respect to the lines of the austenite. The habit plane, i.e., the plane separating the parent and martensite phases, is one of zero macroscopic distortion. The emphasis on the word "macroscopic" is important. As will be seen in Section 7.5.4, the presence of a plane of zero distortion in the martensite phase necessarily implies that the plate is microscopically sheared or twinned.

6. Martensite transformations are often reversible in the sense that the product phase will transform back to the austenite phase upon heating. The temperature at which the reverse reaction first begins is denoted by A_s, and the last remnant of martensite disappears at the temperature A_f. In general, however, there is thermal hysteresis such that $A_s > M_f$ and $A_f > M_s$. The magnitude of the hysteresis can be quite large, hundreds of degrees Celsius in Fe-based alloys, but it can be small, as for Cu–Zn–Al. In a study of the Cu–Al–Ni system, Kurdjumov and Khandros (1949; see also Tong and Wayman

1974) demonstrated the microscopic reversibility of martensite reactions. The first plate to form at M_s during the cooling cycle is the last one remaining upon heating. Also, the last plate to appear, at M_f, is the first plate to retransform at the temperature A_s.

7. Temperature is not the only external field that can trigger the martensite reaction. An applied stress or high magnetic field can also initiate the transformation. For any temperature in the range $M_s > T > M_f$, the amount of martensite will usually be increased by plastic strain. A plastic strain can also cause the formation of martensite plates even at temperatures above M_s. The highest temperature at which the product phase can form under stress is denoted M_d. In contrast, cold working of the parent phase at temperatures above M_d often inhibits the martensite transformation. The M_s temperature is depressed and the amount of martensite is reduced at any temperature within the transformation range.

8. Consider the following cooling sequence. A material is rapidly quenched to a temperature below M_s but above M_f and then held at this intermediate temperature for some period of time. After the hold time the temperature of the specimen is then lowered again to some final temperature, say T_f. It is found that the martensite does not form immediately during the second cooling step; instead, the partially transformed sample must be cooled to a new start temperature M_s' before any new martensite plate formation is observed. Furthermore, the amount transformed at T_f is less than that obtained if the material had been directly quenched to this final temperature. This process is called "stabilization" (Nishiyama 1978).

7.5.2
Massive and Displacive Transformations

Of the characteristics listed above, the diffusionless nature of the transformation (item 1) taken together with the macroscopic shape change (item 5) distinguishes martensite from similar transformations. For example, a Cu–Zn alloy with Zn concentration in the range 36.8–38.3 at.% exists as a single-phase β (disordered bcc) at high temperatures. A rapid quench to lower temperatures brings the material into another single-phase field, disordered fcc α, such that the $\beta \to \alpha$ transformation involves no change in composition and hence no long-range diffusion. The morphology of the new α-phase contains blocky, or "massive," precipitates that form at grain boundaries. The growth of the α-phase involves the rapid advance of relatively flat interphase boundaries. Despite its diffusionless nature, the $\beta \to \alpha$ reaction in Cu–Zn is not martensitic because it does not occur with a macroscopic shape change based on a homogeneous shear. This type of phase transformation is called a "massive transformation."

Finally, martensite reactions are a subset of a general category known as "displacive transformations." The product phase of a displacive transformation is formed by the cooperative motion of atoms in the initial phase, where the extent of the individual displacements is less than one interatomic distance. In contradistinction to displacive reactions are replacive-type transformations that involve the diffusion of atoms. An example of a transformation that is displacive but not martensitic is the bcc to "ω-phase" in Ti, Zr, and Hf alloys. The formation of the hexagonal ω-phase is accomplished, not by a shear mechanism, but by a periodic collapse of adjacent $\{111\}$ planes in the parent lattice.

7.5.3
Bain Strain Mid-Lattice Invariant Shear

In 1924 Bain (1924) described the distortions required to form a bct martensite from an fcc austenite crystal. As demonstrated in Fig. 7.39(a), a bct unit cell can be constructed from an fcc lattice by taking as the top and bottom of the cell the squares formed by connecting two face and two corner atoms. In this representation the [110] direction of the austenite becomes the [010] direction in the tetragonal phase and, if a_0 is the original austenite lattice parameter, the short dimension of the bct unit cell in Fig. 7.39(a) is $a_0\sqrt{2}$. The atom located at $a[\frac{1}{2}\ 0\ \frac{1}{2}]$ in the fcc phase becomes the body-center atom under the new construction. Carbon atoms, randomly dispersed on interstitial sites in the austenite phase, occupy the base center, $a[\frac{1}{2}\ \frac{1}{2}\ 0]$ positions and the $a[0\ 0\ \frac{1}{2}]$ edge positions of the new bct cell. To obtain the correct cell size, a contraction of about 20% in the vertical [001] direction must occur, along with an expansion of 12% in both the [100]

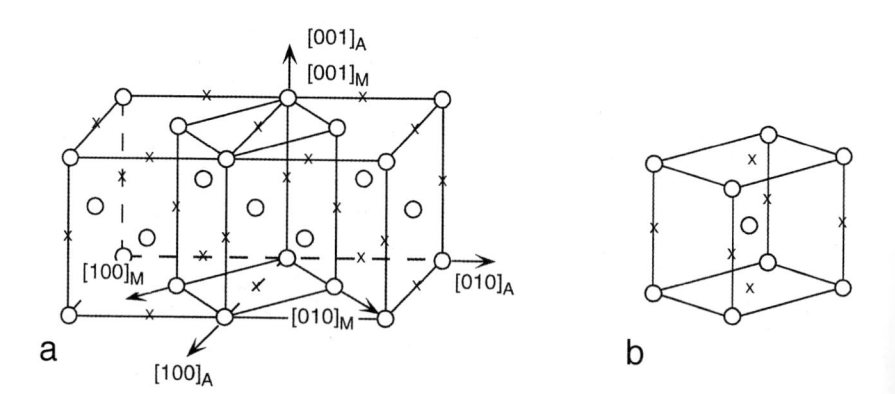

Fig. 7.39 (a) Illustration of how a bct unit cell of martensite can be constructed from an fcc austenite cell. (b) Final unit cell of martensite obtained after application of the Bain strain.

and [010] directions.[16] If a and c denote the lattice parameters of the martensite, then the so-called "Bain strain" or "Bain distortion" is given by Eqs. (101).

$$\eta_1 = \frac{\sqrt{2}a}{a_0} \quad \eta_2 = \frac{c}{a_0} \tag{101}$$

The unit cell produced by the Bain distortion is depicted in Fig. 7.39(b). One can envision other distortions that will transform the fcc parent phase into the martensite structure, but Bain argued that the correspondence shown in Fig. 7.39 and its strains of Eq. (101) represent the lowest elastic energy.

It is tempting to describe the formation of martensite simply as the Bain distortion of plates of material. However, such a straightforward description of the process is inconsistent with the experimental facts. A key feature of the martensite phase transformation is the "habit plane" (item 2), a plane that remains undistorted and unrotated during the reaction. It turns out that there is no way the Bain strain can be applied to a volume of material so that an undistorted plane is preserved.

7.5.4
Martensite Crystallography

Wechsler, Lieberman, and Read (Wechsler et al. 1953), and independently Bowles and Mackenzie (Bowles and Mackenzie 1954; Mackenzie and Bowles 1954, 1957) proved that both the habit plane and the crystallographic relationships can be computed from a knowledge of only the lattice parameters and structure of the austenite and martensite phases. Although the two approaches differ somewhat, Christian (1956) showed that their results are identical. The crystallographic theory of martensite can also predict the magnitude of the macroscopic shear and can provide some details of the microscopic lattice invariant shear. These two theoretical contributions from the mid-1950s tied together many seemingly unrelated experimental facts concerning the martensite reaction and are crucial to our present day understanding of the transformation.

Imagine a martensite plate to consist of twins as shown in Fig. 7.40. One region, with thickness $1 - x$, and a second region, with thickness x, represent two crystallographic variants of the martensite phase. Each of the two twinned volumes has associated with it a pure Bain distortion matrix written as Eqs. (102).

$$\underline{T}_1 = \begin{bmatrix} \eta_1 & 0 & 0 \\ 0 & \eta_2 & 0 \\ 0 & 0 & \eta_1 \end{bmatrix} \quad \underline{T}_2 = \begin{bmatrix} \eta_2 & 0 & 0 \\ 0 & \eta_1 & 0 \\ 0 & 0 & \eta_1 \end{bmatrix}. \tag{102}$$

16) The necessary strains are only approximate because the lattice parameters of both austenite and martensite in steels are functions of carbon content.

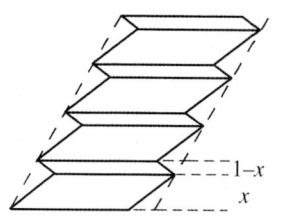

Fig. 7.40 A microscopically twinned martensite plate gives a macroscopic shear.

If the total distortion is described by Eq. (103), then the matrix $\underline{E}$ is given by Eq. (104).

$$\vec{r}\,' = \underline{E}\vec{r} \tag{103}$$

$$\underline{E} = (1-x)\underline{\Phi}_1\underline{T}_1 + x\underline{\Phi}_2\underline{T}_2 \tag{104}$$

Here two rotation matrices, $\underline{\Phi}_1$ and $\underline{\Phi}_2$, have been introduced. These two matrices are, at this point, unknowns in the problem. However, the relative rotation between regions 1 and 2 can be established. A vector $\vec{r} = [110]$ transforms into the vector $r'_1 = [\eta_1, \eta_2, 0]$ in region 1 by the Bain distortion $\underline{T}_1$. The resulting vector is shown in Fig. 7.41. On the other hand, the same vector in region 2 becomes $r'_2 = [\eta_2, \eta_1, 0]$ and, as demonstrated in Fig. 7.41, application of the Bain distortions to the two regions creates a gap. To "fill in" the gap, notice that the magnitude of r'_1 is equal to that of r'_2, meaning that both vectors must lie in the coherent twin plane.

With the relative rotations between the twins established, the habit plane normal can be found from an algebraic analysis of the transformation strain distor-

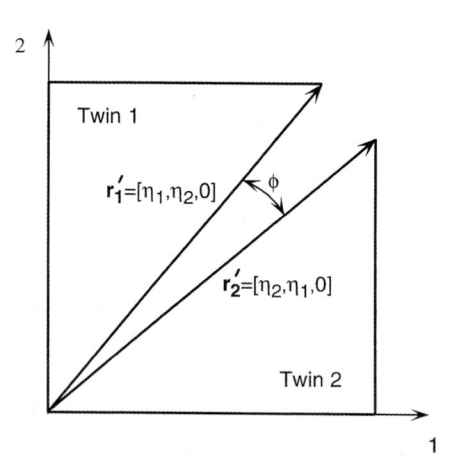

Fig. 7.41 Two twin variants within a martensite plate. To maintain a coherent twin plane, region 2 must be rotated by an amount ϕ relative to region 1.

tions and the requirement that an invariant plane exists. The derivation of the habit plane crystallography is quite lengthy and we shall only quote the final result here. The components h, k, and l of the habit plane normal, $\vec{n}$, are found to be Eqs. (105), where $\eta_3 = \eta_1$.

$$h = \frac{1}{2\eta_1}\left(\sqrt{\frac{\eta_1^2+\eta_2^2-2\eta_1^2\eta_2^2}{1-\eta_2^2}} + \sqrt{\frac{2-\eta_1^2-\eta_2^2}{1-\eta_2^2}}\right)$$

$$k = \frac{1}{2\eta_1}\left(\sqrt{\frac{\eta_1^2+\eta_2^2-2\eta_1^2\eta_2^2}{1-\eta_2^2}} - \sqrt{\frac{2-\eta_1^2-\eta_2^2}{1-\eta_2^2}}\right)$$

$$l = \frac{1}{\eta_1}\sqrt{\frac{\eta_1^2-1}{1-\eta_2^2}} \tag{105}$$

Notice that the habit plane will, in general, have irrational indices. The only input required in the Wechsler–Lieberman–Read/Bowles–MacKenzie crystallographic theory is the lattice parameters of the austenite and martensite phases, both of which can be easily and accurately measured. An fcc to bcc martensite reaction takes place in Fe–30.9 wt.%Ni such that the lattice parameters are $a_0 = 3.591$ Å, and $a = c = 2.875$ Å. Equation (101), combined with the central result, Eq. (105), predicts a habit plane of $h = 0.1848$, $k = 0.7823$, and $l = 0.5948$. In an X-ray diffraction experiment, Breedis and Wayman (1962) found the habit plane to be close to (3 14 10); a result that is a scant 1.5° different from the theoretical prediction. Excellent agreement between theory and experiment has also been reported in the Fe–Pt system (again a fcc → bcc transition) (Nishiyama 1978) and the cubic to orthorhombic transformation in ordered Au–47.5 at.%Cd (Lieberman et al. 1955). Another success of the theory is for In–20.72 at.%Tl. The habit plane in this alloy is the low index (011) whereas the predicted normal is (0.013, 0.993, 1.00), a discrepancy of only 0.43°.

7.5.5
Nucleation and Dislocation Models of Martensite

The Wechsler–Lieberman–Read/Bowles–Mackenzie theory does not address the questions of why a habit plane forms in the first place, how an individual martensite plate forms initially, or how the atoms move cooperatively as the plate shoots through the material. To extend the discussion beyond the phenomenological stage, a mechanism for the transformation is needed, and perhaps the logical starting point is homogeneous nucleation theory. Unfortunately, experiments have confirmed that martensite reactions are not a result of homogeneous nucleation. Small-particle experiments (Cech and Turnbull 1956; Huizing and Klostermann 1966; Easterling and Swann 1971), pioneered by Cech and Turnbull (1956), studied the transformation behavior in spheres of Fe–Ni alloys whose diameters varied from below a micron to a fraction of a millimeter. The experiments demonstrated that, even after cooling to 4 K, an undercooling of 600–700 °C, not all particles transformed to martensite, a finding that rules out both homogeneous nucleation and nucleation at the surface. Experiments also showed that the number of nuclei is independent of grain size, meaning grain boundaries are not a preferred site for plate formation. If the martensite phase initiates at some defect site in the material that is neither grain boundaries nor surfaces, then the most logical heterogeneous site is dislocations.

Perhaps the easiest example with which to illustrate the role of dislocations is the fcc to hexagonal close-packed transformation that occurs in pure cobalt. An fcc and an hcp structure both contain close-packed layers of atoms, but the two lattices differ in their stacking sequence; fcc has an ABCABC stacking, whereas hcp has an ABABAB pattern. A partial dislocation of the type $a/6\langle 112\rangle$ that glides along a (111) plane will leave behind a stacking fault, i.e., a disruption in the stacking sequence such that an A plane is shifted to the B position (and B → C, C → A). Passage of a partial dislocation across every other (111) fcc plane creates the hcp structure (see Chapter 6). An additional very small dilatation perpendicular to (111) is required to contract the ideal c/a ratio (1.633) produced by the dislocation mechanism to the value of 1.623 observed in Co. From this process one can establish the orientation relationships in Co [Eqs. (106)], which are known as the Shoji–Nishiyama relations (Nishiyama 1978).

$$(111)_{\text{fcc}} \| (0001)_{\text{hcp}} \quad [11\bar{2}]_{\text{fcc}} \| [1\bar{1}00]_{\text{hcp}} \tag{106}$$

Also, a shear angle of $\tan^{-1}(\sqrt{2}/4) = 19.5°$ is readily found. Unlike the fcc → hcp transformation, however, it is impossible to construct a bcc martensite product phase using only the dislocations commonly encountered in fcc materials.

In the mid 1970s, Olson and Cohen published a series of papers describing a dislocation-assisted theory of martensite plate formation that has gained some acceptance in the subsequent years (Olson and Cohen 1976a,b,c). The model extends the dislocation concepts of shear transformations to obtain a critical nucleation event. The Olson–Cohen model uses energy concepts of partial dislocations. Recall the situation in an fcc structure where a perfect dislocation of Burgers vector dissociates spontaneously into two partials, for example as in Eq. (107), where the total strain energy of the dislocation on the left-hand side of the reaction (proportional to b^2) is greater than the sum of the strain energies for the two partials ($a^2/2 > a^2/3$).

$$\frac{a}{2}[\bar{1}10] \rightarrow \frac{a}{6}[\bar{2}11] + \frac{a}{6}[\bar{1}2\bar{1}] \tag{107}$$

The two partials in Eq. (107) are oriented to repel each other, but between them is a stacking fault that is unfavorable energetically. An equilibrium spacing is achieved when the stacking fault energy compensates the elastic energy effect.

In the Olson–Cohen model, the stacking fault energy is a function of temperature. At the temperature where the stacking fault energy goes to zero, the separation between the partial dislocations can grow without limit. The character of the partial dislocations in the Olson–Cohen model is unusual, however, with one set having a very small Burgers vector, giving a local kinked region spread over a few crystal planes. A three-step process is required to obtain the transformation product.

7.5.6
Soft Mode Transitions, the Clapp Lattice Instability Model

In Section 7.4 it was established that some materials can become unstable with respect to infinitesimally small fluctuations in concentration. Using a completely analogous approach, one can establish conditions for the stability of a crystal with respect to a set of strain components. In terms of the elastic constants, the stability criteria for a cubic crystal are stated in Eqs. (108) (compare Chapter 4).

$$C_{11} > 0$$
$$C_{44} > 0$$
$$C_{11}^2 > C_{12}^2$$
$$\frac{1}{2}(C_{11} - C_{12}) > 0 \tag{108}$$

Each of the individual stability criteria listed in Eqs. (108) corresponds to a different mode of deformation. For example, the last inequality, which is of primary interest in the study of martensite, defines the stability with respect to small-amplitude shears on (110)-type planes in ⟨110⟩ directions. If the difference $(C_{11} - C_{12})$ vanishes at the M_s temperature, the material will shear spontaneously. This mode of deformation is known as a "soft mode," and the reaction is termed a "soft mode instability." The very first explanation of the martensite phenomenon (Scheil 1932; Zener 1948) was based on a soft mode instability. In 1973 Clapp extended the soft mode concept one step further. The free energy per unit volume of a crystalline solid can be written in the form of Eq. (109) (Clapp 1973), where F_0 is the free energy of the strain-free material. Here the convenient Voigt notation for the strains η has been employed.[17]

$$F = F_0 + \frac{1}{2}\sum_{i=1}^{6}\sum_{j=1}^{6} C_{ij}\eta_i\eta_j + \frac{1}{3!}\sum_{i=1}^{6}\sum_{j=1}^{6}\sum_{k=1}^{6} C_{ijk}\eta_i\eta_j\eta_k + \cdots \tag{109}$$

Finally the last sum in Eq. (109) represents contributions to the free energy due to higher-order, anharmonic strains and the C_{ijk} variables refer to higher-order elastic stiffness constants. We have seen in the theory of spinodal decomposition that the stability of a homogenous solution is determined by the sign of the second derivative of the free energy with respect to composition. Stability with re-

---

[17] Here $\eta_1 = \epsilon_{11}, = \epsilon_{22}, \eta_3 = \epsilon_{33}, \eta_4 = \epsilon_{12}, \eta_5 = \epsilon_{13}$, and $\eta_6 = \epsilon_{23}$ (compare Chapter 4).

spect to strain is governed by the sign of each of the six eigenvalues of the 6×6 matrix of the second derivatives [Eq. (110)].

$$\frac{\partial^2 F}{\partial \eta_i \partial \eta_j} = F_{ij} \quad (i, j = 1, 6) \tag{110}$$

If a particular eigenvalue vanishes at a given temperature, the material is unstable with respect to the deformation mode, described by the associated eigenvector. The stability analysis applied to a free energy written in terms of only the first summation in Eq. (109) will reproduce the inequalities presented in Eq. (108). However, Clapp argues that the anharmonic contribution may be important in certain cases and should not be neglected.

Dislocation models of the martensite transformation describe the reaction in terms of a collective dissociation and splitting of specific arrays of defects, whereas the lattice instability model views the transition as the onset of an elastic instability in the crystal. Which interpretation is correct? In agreement with the aforementioned small-particle experiments, both models identify dislocations as the heterogeneous site associated with the initiation of martensite plates. The lattice instability approach works well for a number of systems, but fails in the important case of steels and other ferrous alloys. The Olson–Cohen description appears to capture the main features of the fcc to hcp transformation and correctly predicts the Kurdjumov–Sachs relationship found in certain steels, but it cannot account for the variety of habit planes observed in different Fe alloys. It is safe to say that we are many years away from a complete theory of the martensite transformation.

7.6
Outlook

Studies of phase transformations and alloy phase equilibria have considerable history and well-established scientific principles behind them. Alloy phase stability and kinetic processes remain central to modern metals physics, however, because they are essential to understanding the microstructures that control alloy properties. Fortunately, the outlook for studies of alloy phases over the next decade shows a dynamic field that is poised for progress in several important areas:

- *Thermodynamics.* Over the past decades, electronic structure calculations have become accessible to more scientists, practical on more systems, and better understood in their reliability. Improvements in computing hardware and software will further these trends, but fundamental advances are also expected. Specific progress is expected in predictive capabilities for magnetic systems of transition metals and rare earths. It will be interesting to see the directions taken to understand the thermodynamics of actinide systems.

At high temperatures anharmonicity becomes increasingly important, and understanding anharmonicity has remained a challenge for many years. Progress on electron–phonon and phonon–phonon interactions will further our understanding of high-temperature thermodynamics.

The study of alloy thermodynamics under extreme conditions is also motivated by research in geophysics and high-pressure physics. Today new experimental tools are available for pressures beyond a megabar, and these are expected to stimulate work on the thermodynamics of alloy phases under pressure, and work on materials at high pressures and temperatures.

- *Kinetics.* A fundamental understanding of kinetics will always be at least as difficult as that of thermodynamics. A kinetic process must accommodate the thermodynamic end state, and the mechanism of atom movements adds richness to the process. Developments in analytical theory are likely to continue, but there is no doubt that the opportunities for computational modeling will expand significantly. Monte Carlo methods and phase-field modeling will continue to address kinetic phenomena of increasing complexity, and will allow more rigorous results on the systems studied today. The more detailed results from this computational work will facilitate experimental tests of key predictions.
- *Nanostructured materials.* Much of what is device engineering at the micrometer scale becomes atom engineering at the nanometer spatial scale. The effects of interfaces and quantum confinement alter the thermodynamics of alloy phases in nanostructures, and these phenomena also offer opportunities for new device functionality. The widespread interest in applications of thin films, nanowires, and atom clusters offers the real possibility of the discovery of new phenomena in alloy phase equilibria and phase transformations.

The field of phase equilibria and phase transformations began with Gibbs, and classical concepts from his work (Gibbs 1876) remain vital some 130 years later. Over time, our increased understanding of electrons, bonding, and atomic structure of alloys has enabled a detailed understanding of individual components of the enthalpy and entropy that underlie the thermodynamics. Free energy functions can now be constructed from these fundamental parts. This level of detail is migrating toward our understanding of kinetic processes and more complex phase transformations. Progress in phase equilibria and phase transformations continues to produce practical results, and good rewards for intellectual effort. We see innovation and discovery for decades to come.

Acknowledgments

B.F. acknowledges the support of the US National Science Foundation under grant DMR 0520547.

Further Reading

Christian, J.W. (1975) *The Theory of Transformations in Metals and Alloys*, Pergamon Press, Oxford.
Ducastelle, F. (1991) *Order and Phase Stability in Alloys*, North-Holland, Amsterdam.
Frenkel, J. (1955) *Kinetic Theory of Liquids*, Dover, Mineola, NY.
Hollomon, J.H., Turnbull, D. (1953) *Prog. Metal Phys.* **4**, 333.
Kelly, A., Groves, G.W. (1970) *Crystallography and Crystal Defects*, Addison-Wesley, Reading, MA.
Khachaturyan, A.G. (1983) *Theory of Structural Transformations in Solids*, John Wiley & Sons, New York.
Lifschitz, E.M., Pitaevskii, L.P. (1980) Statistical physics, Part 1, 3rd edition. In Landau, L.D., Lifschitz, E.M., editors, *Course of Theoretical Physics*, Vol. 5, Pergamon Press, Oxford.
Nishiyama, Z. (1978) *Martensitic Transformation*, Academic Press, New York.
Porter, D.A., Easterling, K. (1992) *Phase Transformations in Metals and Alloys*, Chapman & Hall, New York.
Russell, K.C. (1980) *Adv. Colloid Interface Sci.* **13**, 205.
Sato, H. (1970) *Order–Disorder Transformations*. In Jost, W., editor, *Physical Chemistry: An Advanced Treatise*, Vol. X, Academic Press, New York, Chapter 10, p. 579.
Shewmon, P.G. (1969) *Transformations in Metals*, McGraw-Hill, New York.

References

Anthony, L., Fultz, B. (1994) *J. Mater. Res.* **9**, 348.
Anthony, L., Okamoto, J.K., Fultz, B. (1993) *Phys. Rev. Lett.* **70**, 1128.
Avrami, M. (1939) *J. Chem. Phys.* **7**, 1103.
Avrami, M. (1940) *J. Chem. Phys.* **8**, 212.
Avrami, M. (1941) *J. Chem. Phys.* **9**, 177.
Bain, E.C. (1924) *Trans. AIME*, **70**, 25.
Becker, R., Döring, W. (1935) *Ann. Phys.* **24**, 1.
Bethe, H.A. (1935) *Proc. Roy. Soc. London A* **150**, 552.
Binder, K., Stauffer, D. (1976) *Adv. Phys.* **25**, 343.
Bogdanoff, P.D., Fultz, B. (1999) *Philos. Mag.* **79**, 753.
Bogdanoff, P.D., Fultz, B., Robertson, J.L., Crow, L. (2002) *Phys. Rev. B* **65**, 014303.
Booth, C., Rowlinson, J.S. (1955) *Trans. Faraday Soc.* **51**, 463.
Bowles, J.S., Mackenzie, J.K. (1954) *Acta Metall.* **2**, 129, 224.
Bragg, W.L., Williams, E.J. (1934) *Proc. Roy. Soc. London A* **145**, 699.
Bragg, W.L., Williams, E.J. (1935) *Proc. Roy. Soc. London A* **151**, 540. Ibid. **152**, 231.
Breedis, J.F., Wayman, C.M. (1962) *Trans. AIME*, **224**, 1128.
Cahn, J.W. (1956) *Acta Metall.* **4**, 449.
Cahn, J.W. (1961) *Acta Metall.* **9**, 795.
Cahn, J.W. (1962) *Acta Metall.* **10**, 179.
Cahn, J.W., Hilliard, J.E. (1958) *J. Chem. Phys.* **28**, 258.
Cahn, J.W., Hilliard, J.E. (1959) *J. Chem. Phys.* **31**, 688.
Cech, R.E., Turnbull, D. (1956) *Trans. AIME*, **206**, 124.
Cenedese, P., Kikuchi, R. (1994) *Physica A* **205**, 747.
Chalmers, B. (1959) *Physical Metallurgy*, John Wiley, New York.
Christian, J.W. (1956) *Inst. Met.* **84**, 385.
Clapp, P.C. (1973) *Phys. Stat. Sol. b* **57**, 561.
de Fontaine, D. (1979) *Solid State Phys.* **34**, 73.
Delaire, O., Fultz, B. (2006) *Phys. Rev. Lett.* **97**, 245701.

Easterling, K.E., Swann, P.R. (1971) *Acta Metall.* **19**, 117.
Eshelby, J.D. (1954) *J. Appl. Phys.* **25**, 255.
Eshelby, J.D. (1956) *Solid State Phys.* **3**, 79.
Farkas, Z. (1927) *Z. Phys. Chem. A* **125**, 236.
Fowler, R.H., Guggenheim, E.A. (1940) *Proc. Roy. Soc. London A* **174**, 189.
Frenkel, J. (1955) *Kinetic Theory of Liquids*, Dover, Mineola, NY.
Fultz, B. (1989) *Acta Metall.* **37**, 823.
Fultz, B. (1990) *J. Mater. Res.* **5**, 1419.
Fultz, B. (1992) *J. Mater. Res.* **7**, 946.
Fultz, B. (1993) *Philos. Mag. B* **67**, 253.
Gibbs, J.W. (1876) *Trans. Connecticut Acad.* **3**, 108.
Gibbs, J.W. (1948) *Collected Works*, Yale University Press, New Haven, CT.
Gschwend, K., Sato, H., Kikuchi, R. (1978) *J. Chem. Phys.* **69**, 5006.
Guggenheim, E.A. (1935) *Proc. Roy. Soc. London A* **148**, 304.
Haataja, M., Leonard, F. (2004) *Phys. Rev. B* **69**, 081 201.
Hornbogen, E. (1979) *Metall. Trans A* **10**, 947.
Huizing, R., Klostermann, J.A. (1966) *Acta Metall.* **14**, 1963.
Johnson, W.A., Mehl, P.A. (1939) *Trans. AIME*, **135**, 416.
Johnson, W.L. (1986) *Prog. Mater. Sci.* **30**, 81.
Kaminsky, E.Z., Kurdjumov, G.V. (1936) *Zh. Tekh. Fiz.* **6**, 984.
Kikuchi, R. (1951) *Phys. Rev.* **81**, 988.
Kikuchi, R. (1974) *J. Chem. Phys.* **60**, 1071.
Kikuchi, R., Sato, H. (1969) *J. Chem Phys.* **51**, 161.
Kikuchi, R., Sato, H. (1972) *J. Chem Phys.* **57**, 4962.
Kittel, C. (1969) *Thermal Physics*, John Wiley, New York, Chapter 2.
Klement, W., Willens, R.H., Duwez, P. (1960) *Nature* **187**, 869.
Kolmogorov, A.N. (1937) *Bull. Acad. Sci. USSR, Phys. Ser. 1* 355.
Kurdjumov, G.V., Sachs, G. (1930) *Z. Phys.* **64**, 325.
Kurdjumov, G.V., Khandros, G. (1949) *Dokl. Nauk. SSSR*, **66**, 211.
Kurdjumov, G.V., Miretzskii, V.I., Stelletskaya, T.I. (1939) *Zh. Tekh. Fiz.* **2**, 1956.
Landau, L.D., Lifshitz, E.M. (1969) *Statistical Physics*, Addison-Wesley, Reading, MA.
Langer, J.S., Schwartz, A.J. (1980) *Phys. Rev. A* **21**, 948.
Lee, J.K., Aaronson, H.I. (1975) *Acta Metall.* **23**, 799.
Lieberman, D.S., Weschler, M.S., Read, T.A. (1955) *J. Appl. Phys.* **26**, 473.
Lovesey, S.W. (1984) *Theory of Neutron Scattering from Condensed Matter*, Vol. 1, Clarendon Press, Oxford.
Mackenzie, J.K., Bowles, J.S. (1954) *Acta Metall.* **2**, 138.
Mackenzie, J.K., Bowles, J.S. (1957) *Acta Metall.* **5**, 137.
Manley, M.E., McQueeney, R.J., Fultz, B., Osborn, R., Kwei, G.H., Bogdanoff, P.D. (2002) *Phys. Rev. B* **65**, 144 111.
Mullins, W.W., Sekerka, R.F. (1964) *J. Appl. Phys.* **35**, 444.
Nishiyama, Z. (1934) *Sci. Rep. Tohoyu Univ.* **23**, 637.
Nishiyama, Z. (1978) *Martensitic Transformation*, Academic Press, New York.
Okamoto, H. (2000) *Desk Handbook Phase Diagrams for Binary Alloys*, ASM International, Materials Park, OH.
Olson, G.B., Cohen, M. (1976a) *Metall. Trans.* **7A**, 1897.
Olson, G.B., Cohen, M. (1976b) *Metall. Trans.* **7A**, 1905.
Olson, G.B., Cohen, M. (1976c) *Metall. Trans.* **7A**, 1915.
Sanchez, J.M., de Fontaine, D. (1980) *Phys. Rev. B* **21**, 216.
Sato, H. (1970) *Order–Disorder Transformations*. In Jost, W., editor, *Physical Chemistry: An Advanced Treatise*, Vol. X, Academic Press, New York, Chapter 10, p. 579.
Sato, H., Kikuchi, R. (1976) *Acta Metall.* **24**, 797.
Scheil, E.S. (1932) *Anorg. Allg. Chem.* **207**, 21.
Shewmon, P.G. (1969) *Transformations in Metals*, McGraw-Hill, New York.
Schwarz, R.B., Johnson, W.L. (1983) *Phys. Rev. Lett.* **51**, 415.
Shi, G., Seinfeld, J.H., Okuyama, K. (1990) *Phys Rev. A* **41**, 2101.
Slater, J.C. (1939) *Introduction to Chemical Physics*, McGraw-Hill, New York, Chapter 13.
Squires, G.L. (1978) *Introduction to the Theory of Thermal Neutron Scattering*, Dover, Mineola, NY.
Tong, H.C., Wayman, C.M. (1974) *Acta Metall.* **22**, 887.
Trampenau, J., Petry, W., Herzig, C. (1993) *Phys. Rev. B* **47**, 3132.

Trinkaus, H., Yoo, M.H. (1987) *Philos. Mag. A* **55**, 269.
Turnbull, D., Fisher, J.C. (1949) *J. Chem. Phys.* **17**, 71.
Van Baal, C.M. (1973) *Physica* **64**, 571.
van de Walle, A., Ceder, G. (2002) *Rev. Mod. Phys.* **74**, 11.
Volmer, M., Weber, A. (1926) *Z. Phys. Chem.* **119**, 277.
Wallace, D.C. (1998) *Thermodynamics of Crystals*, Dover, Mineola, NY.
Wallace, D.C. (2003) *Statistical Physics of Crystals and Liquids: A Guide to Highly Accurate Equations of State*, World Scientific, Singapore.
Warren, B.E. (1990) *X-Ray Diffraction*, Dover, Mineola, NY.
Wechsler, M.S., Lieberman, D.S., Read, T.A. (1953) *Trans. AIME*, **197**, 1503.
Zeldovich, J.B. (1943) *Acta Physicochim.* **18**, 1.
Zener, C. (1948) *Elasticity and Anelaticity of Metals*, University of Chicago Press, Chicago.

8
Kinetics in Nonequilibrium Alloys

Pascal Bellon and Georges Martin

The alloys (more broadly, materials as well as devices) which we live with are mostly used in nonequilibrium states: they have been processed in such a way as to achieve useful properties. The built-in microstructure is retained either because of sluggish kinetics back to equilibrium, or because it is clamped into the nonequilibrium state by some difficult nucleation process. An example of the former is the slow coarsening rate of a finely dispersed precipitate population in light alloys for the aircraft industry; examples of the latter are the nucleation of graphite in iron–carbon alloys, and of graphite in diamond.

Whatever the case, we know – or at least, we know how we could find – what the alloy configuration should be at equilibrium: the latter minimizes the Gibbs free energy (if we deal with isothermal, isobaric, isoconcentration processes). The problem we are left with is to find the "kinetic pathway" back to equilibrium: a kinetic pathway is a sequence of configurations parameterized with time.

A quite distinct situation is that of materials subjected to some form of energy input, such as an imposed gradient of chemical potential, or a flow of energetic particles in an irradiation environment, or dynamical mechanical straining. For sure, the alloy configuration adapts to the environment, but we lack a universal principle to tell us whether such a driven material will achieve some stationary state or not, whether that state is unique or not, and if it is obtained, what functional, if any, that state optimizes.[1]

The above considerations have dramatic consequences on alloy design strategy, e.g., for nuclear materials which operate in an irradiation environment. Unlike the classical case of relaxation toward equilibrium, we ignore the configuration to which the driven alloy should evolve. The only way we have to anticipate the future evolution is to model it. Imagine the situation if, in order to know whether H_2O is a gas or a liquid, we had to model the nucleation of the liquid out of a

---

1) In the early 1970s, Prigogine and his school tried to promote such general principles, based on the excess of entropy production rate [P. Glansdorff, I. Prigogine, *Structure Stabilité et Fluctuations*, Paris, Masson et Cie, 1971]; soon afterward, they acknowledged the limitations of such an approach [G. Nicolis, I. Prigogine: *Self Organization in Non Equilibrium Systems*, Wiley, New York, 1977].

Alloy Physics: A Comprehensive Reference. Edited by Wolfgang Pfeiler
Copyright © 2007 WILEY-VCH Verlag GmbH & Co. KGaA, Weinheim
ISBN: 978-3-527-31321-1

vapor, in the absence of thermodynamic concepts! One consequence of this is that a reliable theory and modeling of driven alloys must rest on a robust description of the kinetics. The description must be such that, in the absence of external forcing, we recover the classical thermodynamic behavior of the material. In some particular cases (to be discussed in Section 8.2) one may define a functional as a minimum under stationary conditions, and as giving information on the respective stability of competing stationary states. The counterpart of phase diagrams could therefore be built, which we named "dynamical equilibrium phase diagrams" in order to make it clear that we deal with *stationary dynamical* states, rather than equilibrium states in the thermodynamic sense.

The purpose of this chapter is to summarize the state of affairs in that field. Because of the importance of diffusion-controlled kinetics in the evolution of configurations, we begin with a complementary account of the classical modeling techniques for nucleation, growth, and coarsening: the latter have been addressed in Chapter 7. The complements we give are useful for the second part, which deals with driven alloys.[2]

8.1
Relaxation of Nonequilibrium Alloys

The aim here is to emphasize the complementarities of the various descriptions of the kinetic pathway for phase separation: one is based on the time evolution of concentration fields, the other on the time evolution of the precipitate distribution. The latter splits into two subgroups. In the dilute case, one may describe the time evolution of atomic clusters, as seen by three-dimensional atom probe, or as simulated by kinetic Monte Carlo (KMC) methods: this is the so-called cluster dynamics method. In the concentrated case, isolated atomic clusters can no longer be safely identified: one speaks of precipitates in a solid solution and the appropriate modeling tool is the classical nucleation theory.

Such tools and the relevant concepts have been introduced in Chapter 7, but here we discuss several points which are of key importance for understanding the modeling of driven alloys: these include the definition of the diffusion coefficients which enter the models, and the way fluctuations are built in the models. The modeling of driven alloys, which, as already stressed, is based on kinetics, must be such that in the absence of forcing, classical thermodynamics is recovered. In the following, we give a fully kinetic interpretation of the classical nucleation theory so as to prepare the link with driven alloys. We restrict ourselves to diffusion-controlled *coherent* precipitation with or without concomitant ordering.

---

2) Historically, the reverse did occur: some difficulties we met when modeling driven alloys stimulated us to reconsider the classical theories.

8.1.1
Coherent Precipitation: Nothing but Solid-State Diffusion

Coherent precipitation is indeed nothing but a diffusion process: atoms exchange lattice sites (e.g., because of vacancy jumps) at a frequency which depends on their local environment. In a solid solution with a clustering tendency, bonds between like atoms are energetically favored, so that bonds between pairs of like atoms undo less easily than they form. Qualitatively, one may say that, in an undersaturated solid solution, because solute atoms are rare, it takes a long time for one solute to reach a region where solute has clustered; as a consequence, a solute pair can dissociate before it is reached by one more solute atom. In a supersaturated solid solution, the reverse is true: the time lag for one solute atom to reach one cluster is short compared with the lifetime of a given solute atom at a cluster. Without going into more detail for the time being, it is seen that the very same atomic jumps are at the origin of the fluctuations in the undersaturated solid solution and control the kinetic pathway for phase separation.[3] This is well exemplified by KMC simulations, where the input parameters define only the frequency of atomic jumps as a function of the local environment; because of microreversibility, the set of jump frequencies defines the thermodynamics for the alloy.[4] Indeed, if $P_e(i)$ is the equilibrium probability of configuration (i), and $W(i \rightarrow j)$ is the transition probability per unit time from configuration (i) to (j), a detailed balance leads to Eq. (1a)

$$P_e(i) W(i \rightarrow j) = P_e(j) W(j \rightarrow i) \tag{1a}$$

From Eq. (1a) we get:

$$\frac{P_e(i)}{P_e(j)} = \exp -\beta(E(i) - E(j)). \tag{1b}$$

$E(i)$ is the internal energy of configuration (i) and $\beta = 1/k_B T$. To go from Eq. (1a) to (1b), we have used the fact that the atomic jump is a thermally activated process, so Eq. (1c) holds

$$W(i \rightarrow j) = \nu \exp[-\beta(E_{ij}^{sp} - E(i))], \tag{1c}$$

where E_{ij}^{sp} ($= E_{ji}^{sp}$) is the energy of the system at the saddle point between configurations (i) and (j) [ν is an attempt frequency, which we take as a constant for the sake of simplicity.[5]

---

3) This is an example of the fluctuation dissipation theorem.
4) The reverse is not true: for a given thermodynamics, several kinetic pathways are possible.
5) According to the transition state theory (see Chapter 5), $\nu_{ij} = \frac{\prod_\alpha \omega_i^\alpha}{\prod_{\alpha'} \omega_{ij}^\alpha}$ where the ω's are the stable eigenmode vibration frequencies in configuration (i) and at the saddle point between (i) and (j). At the saddle point position, one mode is unstable, so that the product in the denominator contains one factor fewer than in the numerator. The denominator cancels out of the ratio in Eq. (1b), and the numerator can be incorporated into the E's in the form of a vibrational entropy (cf. C.P. Flynn, *Point Defects and Diffusion*, Clarendon Press, Oxford, 1972, p. 319).

Once it is realized that the set of all possible jump frequencies (i.e., transition probabilities between all the possible configurations) defines both the alloy thermodynamics and the kinetic pathway for the relaxation toward equilibrium, we are left with the following question: to what extent do the classical theories (the Cahn–Hilliard equation for spinodal decomposition, cluster dynamics, nucleation theory) relate to the latter set of jump frequencies? This question is a key one for modeling driven alloys, since, as we will see in Section 8.2, the forcing imposed on the configuration will be modeled by a set of extra transition probabilities superimposed on that we have just discussed and which we call "intrinsic" in the following.

8.1.2
Cluster Dynamics, Nucleation Theory, Diffusion Equations: Three Tools for Describing Kinetic Pathways

Experimentalists have access to a variety of tools for observing and describing a kinetic pathway. Scattering techniques (neutron or X-rays; see Chapter 13.1) operate in reciprocal space and give the Fourier transform of the composition field; they allow for in-situ determination of time evolution. Other techniques operate in real space, at coarse spatial resolution (classical metallography), or at a finer resolution by transmission electron microscopy (TEM) [1], or at atomic resolution by three-dimensional tomographic atom probe (3DTAP) [2] (see Chapter 13.2). These latter techniques give a post mortem image at a given time. Depending on the spatial resolution, they exhibit either clusters of solute atoms (3DTAP, with a yield of about 60%), or precipitates. From such data, various types of statistical information can be extracted: size distribution, shape distribution, correlations in position and size, composition, solute distribution in the vicinity and within the precipitate, etc.

From the theoretical and modeling side, three distinct tools are available: cluster dynamics, the output of which is a time-dependent size distribution of clusters; the classical theory of nucleation, which yields the incubation time, the nucleation rate of precipitates out of a supersaturated solution; and diffusion equations, to model the time evolution of the concentration (and order) fields of the various components of the alloy.

Cluster dynamics (CD) This relies on a representation of the alloy as a lattice gas of solute clusters: a cluster is characterized by the number of solute atoms from which it is made. For instance, if the phase to precipitate is a disordered solid solution, a cluster with size n (an "n-mer") is a set of n solute atoms, each of which has at least one nearest neighbor belonging to the cluster. An "n-mer" achieves several distinct configurations, ranging from the most diffuse one, n atoms in a dense row, to the most compact; each cluster therefore has a free energy of its own. Hence, at equilibrium, the gas of clusters exhibits a free energy with two distinct contributions: the sum of the free energies of the n-mers and the mixing entropy of the n-mers on the lattice. Cluster dynamics therefore implies a thermodynamic model for the alloy, viewed as a lattice gas of clusters. For such a representation to be meaningful, the alloy must be sufficiently dilute to be far below the percolation limit.

Classical nucleation theory (CNT) Despite its elegance, cluster dynamics fails to account for the relaxation of concentrated solid solutions, where clusters cannot be defined unambiguously, and for diffusion mechanisms more complex than single-atom impurity. Also, since it relies on thermodynamics of its own, CD cannot be cast into the classical thermodynamic models, e.g., in order to define a driving force for nucleation. The classical nucleation theory, on the contrary, has none of these drawbacks. However, at some stage it requires a link with CD, which is stated in all textbooks, but not established precisely. In the following we propose a technique to establish the latter link formally.

Diffusion equations At some scale (to be discussed), the alloy can be characterized by a set of concentration fields for the solvent and for the various solutes. Precipitates appear as a local change in concentration, delineated by a more or less diffuse interface. Such a picture can be obtained using low spatial resolution characterization techniques (e.g., conventional microscopy), or from a proper averaging, over several precipitates, of atomic resolution pictures (see below). Modeling a kinetic pathway implies solving the appropriate diffusion equation. To what extent the latter equation relies on the set of atomic jump frequencies introduced above will be discussed in the corresponding section.

8.1.3
Cluster Dynamics

As stated above, we describe the alloy as a set of clusters dispersed on the lattice. This picture, even in its crudest form, gives a very efficient way to describe both the fluctuations in the alloy at equilibrium and the relaxation toward equilibrium.

8.1.3.1 Dilute Alloy at Equilibrium

Under equilibrium conditions, the number density of n-mers is stationary and minimizes the free energy of the gas of clusters. We first notice that an n-mer may exhibit several configurations $\{i, n\}$, each with an energy $E^i{}_n$. An n-mer therefore exhibits a free energy given by Eq. (2), with g_i the degeneracy of configuration i.

$$F_n = -\beta^{-1} Ln \left(\sum_{\{i\}} g_i \exp -\beta E_n^i \right) \qquad (2)$$

The free energy of the whole gas of clusters is the sum of the free energies of the individual clusters, minus the contribution of the configurational entropy of the gas of clusters [Eq. (3a), where C_n is the number of n-mers per lattice site and W

is the number of ways to arrange N_n clusters of size n (n ranging form 1 to infinity) on N_s lattice sites (in the limit of infinite N_s).]

$$F_{gas} = \sum_n C_n F_n - k_B T \mathrm{Ln}(W) \tag{3a}$$

Assuming the gas of clusters to be very dilute, i.e., neglecting the constraint that clusters should not overlap, W is given by Eq. (3b).

$$W \approx \frac{N_{sites}!}{\prod_{n=1}^{\infty} N_n!} \tag{3b}$$

Minimizing F_{gas} [Eq. (3a)] with the constraint that the total number of solute is a constant [Eq. (4a)] yields the equilibrium concentration of n-mers, Eq. (4b).

$$\frac{\partial (F_n - \mu \sum n C_n)}{\partial n} = 0 \tag{4a}$$

$$\overline{C_n} = \exp -\frac{F_n - n\mu}{k_B T}. \tag{4b}$$

Here μ is the free energy per solute atom in the gas at equilibrium. Provided F_n is known for each value of n, μ is obtained by solving Eq. (4c), where c_{tot} is the total solute concentration in the gas.

$$\sum_n n\overline{C_n}(\mu, T) = c_{tot} \tag{4c}$$

Equation (4c) gives $\mu(T, c_{tot})$, the chemical potential as a function of temperature and of the solute content of the alloy.

In summary, the concept of a gas of clusters, together with the approximations used so far, constitute a thermodynamic model of the alloy.

As shown by Eq. (3a), the key quantity in the cluster description is the set of F_n, the free energies of the n-mers. The latter can be evaluated numerically from solute–solute and solute–solvent interactions, by thermodynamic integration using Monte Carlo techniques to compute $\partial F_n / \partial n$. According to Perini et al. [3], in the Ising model, F_n is given by Eq. (4d).

$$F_n = an + bn^{2/3} + cn^{1/3} + d + \tau \mathrm{Ln}(n). \tag{4d}$$

In Eq. (4d) the parameters (a, b, c, d, τ) have been computed for the simple cubic lattice. For sufficiently large n, we recover the classical model with a bulk and a surface contribution. For small clusters, the next two terms in the expansion can be understood as an edge and a vertex contribution. The last term accounts for the fact that, since the cluster is of finite size, the cluster/solvent interface cannot exhibit the full spectrum of undulations, which contribute to the macroscopic interfacial free energy embedded in b [4]. The term $cn^{1/3}$ can be viewed as a curvature correction to the interfacial free energy; it was found to be negative.

8.1.3.2 Fluctuations in the Gas of Clusters at Equilibrium

In the cluster gas at equilibrium, solute atoms permanently evaporate from one cluster, diffuse through the solvent, and stick to another cluster. But on average, the cluster distribution [Eq. (4b)] is stationary. If we assume that only single atoms evaporate and migrate, the probability per unit time of an n-mer becoming an $(n+1)$-mer $(\{n\} + \{1\} \Rightarrow \{n+1\})$ is given by Eq. (5a)

$$\overline{\beta_{n,n+1}} = g_n D_1 \overline{C_1} \tag{5a}$$

Indeed, the probability of a cluster catching a solute atom per unit time is higher, the higher the concentration of monomers, C_1, and the more frequently the latter jump. In this equation, D_1 is the diffusion coefficient of single solute atoms in the pure solvent (called the impurity diffusion coefficient in the diffusion literature), and g_n is a geometrical factor. One usually chooses $g_n \approx 4\pi R_n / \Omega$ where R_n is the radius of a sphere with n solute atoms and Ω is the atomic volume. The latter expression for g_n is obtained by computing the stationary flux of solute from infinity to a sphere of radius R_n, at the surface of which the concentration is kept at zero. More advanced formulations are under development [5]. In Eq. (5a) and hereafter, an overbar on a quantity implies that the latter is given its equilibrium value.

The probability per unit time of the reverse process, i.e., where an $(n+1)$-mer becomes an n-mer by the evaporation of a single solute $(\{n+1\} \Rightarrow \{n\} + \{1\})$, is such that the equilibrium distribution of clusters is stationary [Eq. (5b)].

$$\overline{\alpha_{n+1,n}}\,\overline{C_{n+1}} = \overline{\beta_{n,n+1}}\,\overline{C_n} \Rightarrow \overline{\alpha_{n+1,n}} = \frac{\overline{C_n}}{\overline{C_{n+1}}} g_n D_1 \overline{C_1} \tag{5b}$$

In summary, describing the alloy as a dilute gas of clusters yields a thermodynamic model [Eqs. (4)], with fluctuations built into it; indeed, the probability of an n-mer changing size per unit time is $(\alpha_{n,n-1} + \beta_{n,n+1})$. Also, note that at infinite temperature the model correctly predicts a cluster distribution, which does not reduce to isolated single solutes [6]. The gas of clusters therefore yields a thermodynamic model with some correlations built into it, at variance with, e.g., the Bragg–Williams approximation.

8.1.3.3 Relaxation of a Nonequilibrium Cluster Gas

If the cluster gas is not at equilibrium, i.e., if the concentration of n-mers in the gas is distinct from that given by Eq. (4b), the clustering process will evolve, i.e., monomers will evaporate preferentially from those n-mers of which the number density is too large, and condense preferentially on those n-mers which are too rare for the gas to be at equilibrium. The net balance of such processes is described by the master equation, Eq. (6).

$$\frac{\partial C_n}{\partial t} = J_{n-1,n} - J_{n,n+1}; \quad J_{n,n+1} = \beta_{n,n+1} C_n - \alpha_{n+1,n} C_{n+1} \tag{6}$$

In Eq. (6), $J_{n,n+1}$ is the net number density of n-mers which catch one solute atom per unit time. Note that the quantities in Eq. (6) do not have bars, since they have nonequilibrium values. The value to be given to the impingement rates β, is given by Eq. (5a), where the concentration of monomers is given its actual value at time t, $C_1(t)$. As for the evaporation rates, α, one assumes that the evaporation rate is a characteristic of the cluster itself, independently of the cluster gas which surrounds it. In other words, α is given the value defined in Eq. (5b). The physical picture behind this assumption is that, on average, n-mers explore their many configurations at a rate much higher than the impingement or evaporation rate. This argument is consistent with the expression we used for g_n. More advanced formulations are under development [5].

As can be seen, once D_1 and the set of F_n are known, Eq. (6) can be integrated for any initial distribution of clusters $C_n(0)$ up to any time t. In the case where the initial cluster distribution corresponds to a supersaturated solid solution, the cluster size distribution will evolve toward a bimodal distribution: the first part peaks at C_1 and decreases to some critical size n^*; beyond n^*, $C_n(t)$ increases, goes through a maximum, decreases again, and vanishes for large values of n. The cluster size distribution evolves continuously in time, and depicts the nucleation stage at the beginning (with an increase in the number of clusters, keeping the average size approximately constant), the growth stage (with a constant number of clusters with increasing average size), and later the coalescence stage (with a decrease in the number of clusters as the average size increases).

Note that in the framework of cluster dynamics, no strict distinction can be made between the solid solution and the precipitates: whatever its degree of decomposition, the alloy is nothing but clusters dispersed on a lattice.

Provided the above definition of cluster dynamics is strictly followed [including the first three terms of the cluster free energy in Eq. (4d)], it is possible to reproduce, by CD, the results of lattice KMC (LKMC) simulations, using the very same set of input parameters, at least in alloys with low solubility limits. This latter point has been demonstrated carefully by Clouet et al., for Al(Sc) alloys [7]. Moreover, without further adjustment, the master equation can be integrated to much longer times; As illustrated by Fig. 8.1, it is found that CD results do reproduce the experimentally measured mean radii of precipitates, despite the fact the experiments are done at times which are three to four orders of magnitude greater than the longest affordable LKMC simulation [7].

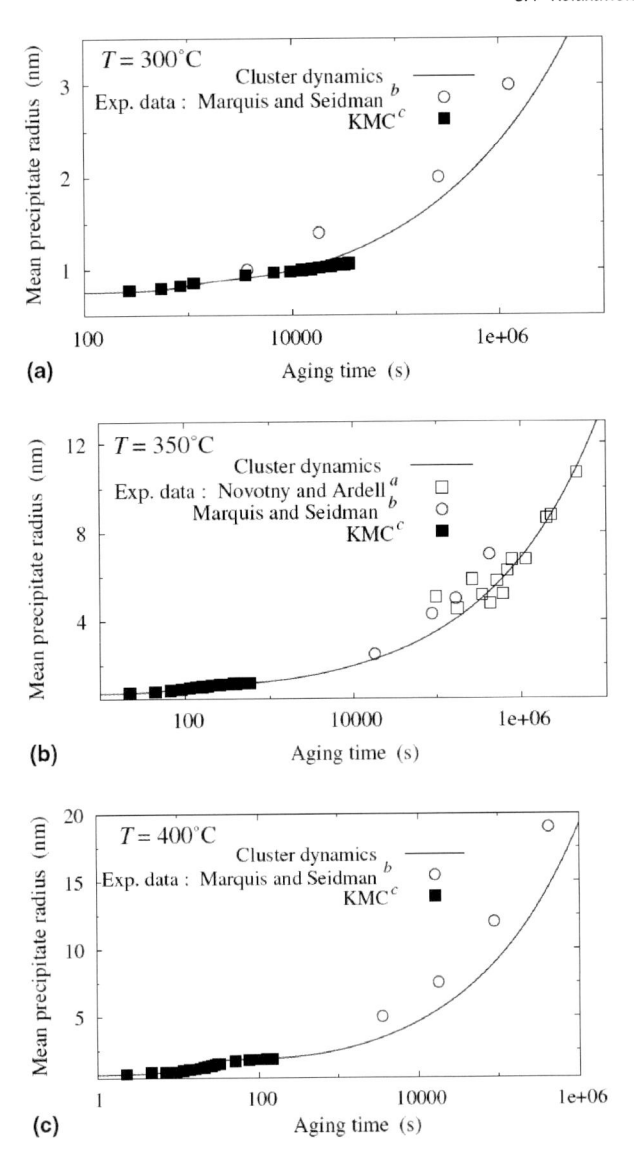

Fig. 8.1 Mean precipitate radius as a function of the aging time for an Al–Sc solid solution of composition $x^0{}_{Sc} = 0.18$ at.% at temperatures $T = 300, 350$, and $400\,°C$; the point symbols are LKMC results [^{c}: E. Clouet, M. Nastar, C. Sigli, Nucleation of Al$_3$Zr and Al$_3$Sc in aluminum alloys: from kinetic Monte Carlo simulations to classical theory, *Phys. Rev. B* **69** 064 109 (2004)]; the curve is obtained from CD based on the same set of parameters as LKMC. Open symbols: experimental results obtained at much longer times [^{a}: G.M. Novotny, A.J. Ardell, Precipitation of Al$_3$Sc in binary Al–Sc alloys, *Mater. Sci. Eng. A* **318** 144–154 (2001); ^{b}: E.A. Marquis, D.N. Seidman, Nanoscale structural evolution of Al$_3$Sc in a dilute Al–Sc Alloy, *Acta Mater.* **49** 1909–1919 (2001)]. The cutoff radius used for CD and KMC is $r^*{}_X \approx 0.75$ nm ($n^*{}_X = 27$) (From ref. [7]).

The CD formalism, as described above, is of current use for studying materials under irradiation. It is called the "rate theory" of defect accumulation, and has been enriched by many extensions: solute diffusion proceeds both by vacancy and interstitialcy mechanisms; vacancies and self-interstitials (Frenkel pairs) are produced by irradiation and may be annihilated by recombination or on lattice discontinuities (dislocations, grain boundaries, internal surfaces, etc.); they may also agglomerate into point defect clusters, etc. [8]. As will be discussed in Section 8.2, cluster dynamics has been used successfully by Cauvin et al. [9] to account for the decrease in solubility observed in certain solid solutions under irradiation (so-called "irradiation-induced precipitation").

8.1.4
Classical Nucleation Theory

As discussed in Section 8.1.3, cluster dynamics is based on an assumption of high dilution; it implies that the alloy has a thermodynamic description of its own; and it assumes that solute diffusion proceeds by jumps of single atoms. Such limitations are not met by the classical nucleation theory, which we summarize now. Full details are to be found in Chapter 7.

8.1.4.1 Summary of CNT

The picture behind CNT is that, unlike in CD, we can unambiguously define precipitates embedded in a solid solution (the latter being super- or undersaturated). It is then claimed that first-order transitions proceed by the formation (nucleation) and the growth of well-defined domains of the second phase (precipitates) in the bulk of the mother phase (supersaturated solid solution). According to classical thermodynamics, heterophase fluctuations (precursors of the precipitates) have an equilibrium number density given by Eq. (7), where ΔF is the reversible work to form the fluctuation in the mother phase.

$$P(n) \propto \exp(-\beta \Delta F) \tag{7}$$

The lower ΔF is, the more probable the fluctuation. As a consequence, the most frequent fluctuations exhibit:
- a concentration close to that of the precipitate in equilibrium with the solid solution
- the equilibrium shape (which minimizes the total interfacial energy, e.g., a sphere in the case of isotropic interfacial energy)
- a size dependence of the form (spherical fluctuation) given by Eq. (9).

$$\Delta F(R) = \frac{4}{3} \frac{\pi R^3}{\Omega} (f_{\alpha'} - f_\alpha) + 4\pi R^2 \sigma \tag{8}$$

In Eq. (8), R is the radius of the spherical fluctuation, σ the interfacial energy, Ω the atomic volume (assumed to be the same in both phases), and $f_{\alpha'}$ and f_α respectively the free energy per atom in the α' precipitate and in the α solid solution.

In the case where $f_{\alpha'} > f_\alpha$, the free energy per atom is greater in the second phase than in the mother phase: the latter is undersaturated; as shown by Eq. (8), reversibly decreasing the radius R releases work, so that the dissolution of the fluctuations into the mother phase is a natural process. In the opposite case ($f_{\alpha'} < f_\alpha$), one finds a critical radius R^* below which the reversible work is negative and above which it is positive: subcritical fluctuations should dissolve spontaneously in the mother phase, while supercritical ones would grow. The critical fluctuation with size R^* is in an unstable equilibrium with the solid solution.

Strictly speaking, the above statements imply that the equilibrium distribution, as given by Eqs. (7) and (8), is *not* stationary! The reason for this apparent contradiction is that the above presentation rests on macroscopic concepts and therefore fails to describe the *source* which generates the fluctuations, the equilibrium distribution of which is given correctly by Eqs. (7) and (8). In order to circumvent this difficulty, classical presentations of CNT state that the fluctuations can be modeled by a polymerization chain of the type: $\{n\} + \{1\} \Leftrightarrow \{n+1\}$, i.e., one similar to that used in cluster dynamics; however, unlike in CD, the clusters as well as the rate constants in the master equation are, here, defined intuitively. Below, we derive an expression for the source of fluctuations which is fully consistent with CNT and which makes explicit the missing link between CNT and CD.

8.1.4.2 Source of Fluctuations Consistent with CNT

We give here the principle of the demonstration. We start from the macroscopic formalism used in CNT. The growth rate of a spherical fluctuation of radius R is written classically as Eq. (9a), with $J(R)$ the flux of solute at the interface (for the sake of simplicity, we assume, without loss of generality, that the fluctuation consists of pure solute).

$$\partial R/\partial t = -J(R)\Omega \tag{9a}$$

The flux is proportional to the gradient of diffusion potential ($\mu_{diff} = \mu_B - \mu_A$) and is computed in the stationary regime [Eq. (9b)] in the spherical cell with the inner radius equal to R and the outer one equal to R_{ext}, and with the boundary conditions given by Eqs. (9b) and (9c).

$$J(r) = -L\Omega^{-1}\nabla\mu_{diff}; \quad \partial c/\partial t = -\nabla J = 0 \tag{9b}$$

$$r = R: \mu_{diff}(R) = \bar{\mu} + 2\sigma\Omega/R; \quad r = R_{ext}: \mu_B - \mu_A = \mu_{diff}(t) \tag{9c}$$

Note that the intrinsic diffusion coefficient is $D = L\delta\mu/\delta c$. The solution of the problem, in the limit $R/R_{ext} \ll 1$, is Eq. (9d), with n the number of solutes in a precipitate of radius R_n, where the value of L is computed in the equilibrium solid solution.

$$\partial n/\partial t = \frac{4\pi R_n}{\Omega} L(\bar{c}_\alpha)[\mu_{diff}(R_{ext}) - \mu_{diff}(R_n)] \qquad (9d)$$

Following a technique introduced by Landauer [10], we rewrite Eq. (9d) as Eq. (10).

$$\partial n/\partial t = \partial n/\partial t|^+ - \partial n/\partial t|^-$$

$$\partial n/\partial t|^{+/-} = \frac{4\pi R_n}{\Omega} L(\bar{c}_\alpha) k_B T \exp\left[\frac{\mu_{diff}(R_{+/-}) - \bar{\mu}_{diff}}{k_B T}\right]; \qquad (10)$$

$$R_+ = R_{ext}; \quad R_- = R_n$$

What we have done is to arbitrarily interpret the difference, which appears on the right-hand side (RHS) of Eq. (9d), as a difference between a growth and a decay rate [Eq. (10)] and express the latter as a first-order expansion of exponentials close to equilibrium $\left(\frac{\mu_{diff}(R_{+/-}) - \bar{\mu}_{diff}}{k_B T} \ll 1\right)$. In agreement with the above interpretation, we expect the equilibrium distribution of fluctuations to obey Eq. (11).

$$\overline{C_n} \partial n/\partial t|^+_{R_n} = \overline{C_{n+1}} \partial n/\partial t|^-_{R_{n+1}}$$

$$\mathrm{Ln}\left(\frac{\overline{C_{n+1}}}{\overline{C_n}}\right)_{kin} = -\frac{1}{3(n+1)} - (k_B T)^{-1}\left[\bar{\mu}_{diff} - \mu_\alpha + \frac{2\sigma\Omega}{R_n(1+1/3n)}\right] \qquad (11)$$

The subscript *kin* reminds us that the equilibrium distribution is that expected from Eq. (10). From Eqs. (7) and (8), the equilibrium distribution of fluctuations used in CNT is given by Eq. (12)

$$\mathrm{Ln}\left(\frac{\overline{C_{n+1}}}{\overline{C_n}}\right)_{CNT} = -(k_B T)^{-1}\left[\bar{\mu}_{diff} - \mu_\alpha + \frac{2\sigma\Omega}{R_n}\right] \qquad (12)$$

Equations (11) and (12) are identical in the limit of large n ($3n \gg 1$). We conclude that the artificial decoupling of the net growth rate [Eq. (9d)] into a difference between a growth- and a decay-rate [Eq. (10)], together with the particular way we write the latter, provides us with the equilibrium distribution of fluctuations used in CNT. Based on Eq. (10), we can now write the master equation, which governs the distribution of fluctuations in the CNT: it is identical to Eq. (6) for n large enough, but with quite distinct expressions [Eqs. (13a) instead of (13b)] for the rate constants.

$$\beta_{n,n+1}^{CNT} = g_n L(\bar{c}_\alpha) k_B T \, \exp\left[\frac{\mu_\alpha - \bar{\mu}_{diff}}{k_B T}\right] \tag{13a}$$

$$\alpha_{n,n-1}^{CNT} = g_n L(\bar{c}_\alpha) k_B T \, \exp\left[\frac{2\sigma\Omega}{R_n k_B T}\right]$$

$$\beta_{n,n+1}^{CD} = g_n D_1 C_1 \tag{13b}$$

$$\alpha_{n+1,n}^{CD} = g_n D_1 \overline{C_1} \, \overline{C_n/C_{n+1}}$$

Note also that, unlike in CD, the master equation (6) cannot be integrated from $n = 1$ since, for small values of n, our scheme yields a wrong equilibrium distribution. But the link between CNT and CD is now established formally.

We now explore Eqs. (13a) for a specific diffusion model, i.e., a specific expression for $L(c)$.

8.1.4.3 A First Application

Elaborating an atomistic theory of Onsager's transport coefficients L_{ij} is a long-standing issue, which is making steady progress [11–14]. Ideally, for a nucleation model to be fully self-consistent, the expression of L should be consistent with the thermodynamic model used to compute the diffusion potentials μ_{diff} in Eqs. (13a). The simplest case where the above requirement is met is the regular solution model, in the point approximation (Bragg–Williams) with a direct exchange diffusion mechanism [15]. In this model, L is written as in Eq. (14), where v is the attempt frequency, z is the plane-to-plane coordination, c is the solute concentration in the solid solution, μ is the diffusion potential in the solid solution (a bar means the two-phase equilibrium value), E_{sp} is the binding energy of the exchanging pair to the saddle point and the ε_{ij}s are the usual pair interactions.

$$L = vz(1-\bar{c})^{1+\alpha}(\bar{c})^{1-\alpha} \, \exp\left[-\frac{E_{sp} - \bar{\mu}}{k_B T}\right] \tag{14}$$

$$\alpha = u/2\omega; \quad \omega = \varepsilon_{AB-} - \frac{\varepsilon_{AA} + \varepsilon_{BB}}{2}; \quad u = \varepsilon_{AA} - \varepsilon_{BB}$$

This expression for L, when introduced in Eqs. (13a), yields the kinetic parameters of the master equation and can be used, for example, to estimate the Zeldovich constant, the incubation time, etc.

Moreover, one can easily check that in the limit of very low solubility, the rate constants we developed for CNT [Eqs. (13a)], together with Eq. (14) for L, do converge to the classical expression used in CD [Eqs. (13b)].

To summarize this section, the expressions we propose for the rates of solute impingement and evaporation [Eqs. (13)] provide the correct equation for the growth rate of precipitates and the correct equilibrium distribution of heterophase fluctuations, and do converge, in the dilute limit, for the direct exchange mechanism in the regular solution model, to the classical expressions used in cluster dynamics. They are not restricted, however, to that simple case; work remains to be done to develop the above formalism in detail for the vacancy diffusion mechanism in a multicomponent alloy. The point we wanted to establish is that the link of the CNT to the details of atomic jumps is now established, as long as the link between the L_{ij}s [Eq. (9b)] and atomic jump frequencies is sorted out.

8.1.5
Kinetics of Concentration Fields

At some scale (see below), the solid solution, together with the coherent precipitates, can be represented by a three-dimensional concentration field; the sharp precipitate/matrix interface is replaced by a boundary layer with a steep but smooth variation of the concentrations [16]. Nucleation, growth, and coarsening are described by the time evolution of the concentration field. To be more specific, a multicomponent alloy with S chemical species and vacancy-mediated diffusion is specified by $(S+1)$ concentrations: that of vacancies C_V and those for the alloy components C_s ($s = 1$ to S). The scale where nucleation and growth proceed is much finer than the mean distance between vacancy sources and sinks; vacancies are thus conserved. Because of the conservation of lattice sites the $(S+1)$ concentrations sum to unity: we are left with S independent concentration fields. For the same reason, we define S diffusion potentials, $(\mu_s - \mu_V)$, where μ_s is the chemical potential of the component s and μ_V is the chemical potential of the vacancy. The chemical fluxes are proportional to the diffusion potentials via the $(S \times S)$ Onsager matrix, which is symmetric, definite, and positive; the fluxes are therefore specified by $S(S+1)/2$ Onsager coefficients, L_{ij}, $(i, j = 1$ to $S)$. The flux of matter in the lattice frame of reference $\tilde{J}$ is given by $\tilde{J} = -\bar{\bar{L}}\nabla\tilde{\mu}(\Omega kT)^{-1} = -\bar{\bar{D}}\nabla\tilde{C}\Omega^{-1}$, where $\tilde{J}$ is the column vector with elements J_s, ($s = 1$ to S), similarly to the diffusion potential $\tilde{\mu}$, and the concentration of atomic species, $\tilde{C}$; Ω is an atomic volume. The diffusion matrix is given by the product of the Onsager matrix and the susceptibility matrix:

$$\bar{\bar{D}} = \frac{\bar{\bar{L}}}{kT}\overline{\left(\frac{\partial(\mu_i - \mu_V)}{\partial C_j}\right)} \quad (i, j = 1 \text{ to } S)$$

The time evolution of the concentration field, $\tilde{C}$, is given by the conservation equation, Eq. (15).

$$\partial_t \tilde{C} = -\nabla \cdot \tilde{J} \tag{15}$$

As discussed in Chapters 7 and 10, for Eq. (15) to give the equilibrium concentration field at infinite time, the chemical potentials must include a contribution of the inhomogeneity of the concentration field: to first order in inhomogeneity, the latter reduces to a term proportional to the square of the concentration gradient, first introduced in this field of physics by Cahn; this term can be viewed as the second moment of the interatomic potential [17].

As will be seen below, because the effect of external forcing is best described at the atomic scale, it is important to clarify at which scale the above concentration field is defined.

The classical phase field approach defines the concentration by a coarse graining procedure: the volume is divided into cells within which the details of the atomic configuration are ignored. $\tilde{C}$ is defined at the coarse grain scale. Equation (15) can be made dimensionless, the length scale being given by the cell size. However, the size of the cell plays a key role whenever fluctuations are to be taken into account [18–20], which is the case for driven alloys.

Another approach is sometimes called the "lattice phase field" approach as was first proposed, to our knowledge, by Vaks and coworkers [21, 22]. Here the concentration at each time is defined on each single site as the first moment of the (time-dependent) probability of occupation of that site by the various species. This approach allows the direct link to be made between Eq. (15) and the atomic jump frequencies, whatever their origin (thermal fluctuations or/and external forcing).

As an illustration, Fig. 8.2 shows the concentration profiles in a two-phase model superalloy, as observed by 3DTAP and as modeled by LKMC. The concentration here is defined by averaging on many precipitates, and by a weighted average on several neighboring sites. With that procedure, a direct link can be established between observations (with 3DTAP), simulations (using LKMC technique) at the atomic scale, and the theory of diffusion [23].

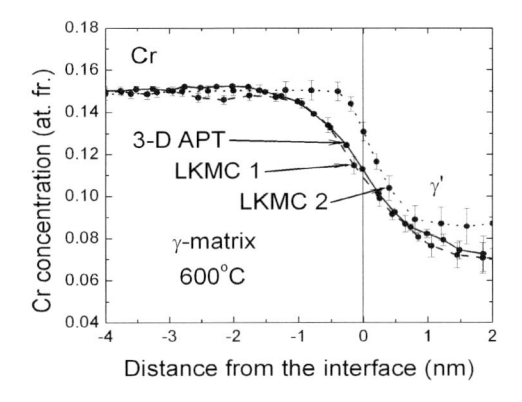

Fig. 8.2 Decomposition of Ni (5.2 Al at.%–14.2 Cr at.%) solid solutions at 873 K: concentration of chromium as a function of distance from the γ/γ' iso-ordering interface, when the mean precipitate radius $\langle R \rangle$ reaches 1.25 nm, as observed from 3DAPT and as simulated by LKMC, using two sets of parameters; LKMC1 reproduces all the features of the experimentally observed kinetic pathway well, while LKMC2, which generates the same alloy thermodynamics but distinct diffusion correlation effects in the ternary alloy, yields much sharper interfaces; this is shown to have noticeable effects on the early-stage decomposition microstructure [from ref. 23].

8.1.6
Conclusion

In the absence of any macroscopic formalism for taking into account external forcing, one is left with modeling-driven alloys from the atomic scale where the mechanism of forcing is understood. Any reliable modeling must be such that, in the absence of forcing, a nonequilibrium state will decay to the proper equilibrium state via a realistic kinetic pathway. Section 8.1 was aimed at presenting (in the simplest case of coherent transformations) the limitations of the classical approaches to nucleation, growth, and coarsening, when we want to make the link between the phenomenological kinetic coefficients they rely on and the atomic jump frequencies one would like to start from. Important improvements were achieved in the last decade; much work remains to be done.

8.2
Driven Alloys

In the first part of this chapter, we have considered materials that, once brought into some nonequilibrium state, are relaxing toward their equilibrium state through diffusion-controlled processes. For the sake of simplicity here we will refer to these processes as intrinsic. In many practical situations, however, materials are driven and maintained away from equilibrium by some external forcing, as already indicated in the Introduction. The external forcing may not only affect the rate of the intrinsic dynamics, for instance by creating an excess of point defects, it often leads to the introduction of additional dynamical processes, which we will refer to as extrinsic. We refer to materials under external forcing as "driven system materials", or, more concisely, as "driven alloys", since, in most situations of interest, materials contain more than one chemical species, and the spatial distribution of these species is modified by the sustained external forcing. Under appropriate conditions, extrinsic dynamical processes compete with intrinsic ones, and, as a result of this competition, the material may undergo nonequilibrium phase transitions and microstructural evolutions, as well as self-organization reactions. These evolutions are not driven by thermodynamic forces, and, therefore, they cannot be predicted or rationalized by equilibrium thermodynamics and near-equilibrium relaxation kinetics. In 1997, we published a comprehensive account of a general framework specifically designed to study driven alloys [24]. The two main types of forcing discussed in detail in that review were the sustained irradiation of alloys with energetic particles, and the processing of powders by high-energy ball milling. We showed that, despite the diversity of the external forcing situations, the response of these materials could be studied, understood, and to some extent predicted by relying on one common approach. The starting point of this framework is to establish a consistent kinetic description that encompasses all the relevant dynamical processes, from the intrinsic to the extrinsic ones. The spatio-temporal evolution of these materials, with several dynamics in parallel, can then be analyzed by applying and extending concepts and tools developed in the physics community for the study of dynamical dissipative systems [10, 25–32].

In the past decade, our knowledge of driven alloys has grown significantly in terms of the range of situations where the above framework has been applied, as well as in terms of our fundamental understanding of the evolution of these materials. This growth is partly reflected in the contributions that appeared in a Viewpoint Set on "Materials under external forcing" which appeared in 2003 in *Scripta Materialia* [33]. In the past few years, additional important results have been published, and thus we provide here an up-to-date review of the salient features of driven alloys, with an emphasis on the recent experimental, theoretical, and computer simulation results. We start by providing examples of driven alloys in Section 8.2.1. We then discuss the concepts of control parameters and steady-state dynamical equilibrium phase diagrams in Section 8.2.2. Theoretical approaches and simulation techniques are reviewed and illustrated in Section 8.2.3, and in Section 8.2.4 we discuss the role of external length scales in self-organization reactions. Finally in Section 8.2.5 we present some practical applications of the driven alloy framework, and we indicate some open questions that should deserve attention in the future.

8.2.1
Examples of Driven Alloys

The initial questions that led to the development of the driven alloy framework [34, 35] were motivated by puzzling experimental observations of alloys subjected to sustained irradiation by energetic projectiles, such as electrons, ions, and neutrons, and in alloys subjected to sustained plastic deformation during fatigue. The intriguing point is that one material subjected to a given type of forcing can undergo one type of evolution or its very opposite, depending upon the exact forcing conditions. Typical examples, which are reviewed in ref. [24], include precipitation–dissolution [36], order–disorder [37], and crystal-to-amorphous [38] reactions. Later, similar paradoxical observations were reported for powders processed by high-energy ball milling and for thin films grown from a vapor phase. In the case of alloys under irradiation, Adda et al. [34] proposed rationalization of these effects by introducing a generalized phase diagram, which, in addition to thermodynamic variables, includes the irradiation flux, a specific example of forcing intensity. More recently, it has become clear that, in addition to the forcing rate(s), it is also important to identify and include the characteristic lengths of the elementary intrinsic and extrinsic processes. From that perspective, we review below important examples of driven alloys.

8.2.1.1 Alloys Subjected to Sustained Irradiation
We first focus on the case of metallic alloys under irradiation. This choice is motivated by three factors, not mentioning the practical interest of this question. First, this situation is the one for which the driven alloy theory is the most ad-

vanced, since the atomic-scale physical processes involved in particle–matter interactions are well understood. Second, our ability to perform controlled experiments in this field makes it possible to compare experimental results critically with predictions from analytical models and computer simulations. Third, this situation is also interesting from a pedagogical viewpoint since it lends itself to a progressive introduction of the concepts and tools useful for the study of driven alloys.

The study of the elementary effects induced by the slowing down of energetic projectiles in solids is a mature field (see ref. [39] for a recent review). An energetic projectile, or a target atom that has been set in motion, may lose energy by interacting with the nuclei of the target or its electronic system (for charged particles), or through resonant nuclear reactions. Since we are mostly concerned with metallic alloys here, we restrict our discussion to nuclear interactions, though electronic excitations can be relevant as well at very high projectile energies, typically >1 MeV amu^{-1}. Nuclear energy loss, which results from elastic collisions, and its resulting primary recoil spectrum can be calculated using universal expressions for the nuclear stopping power, which rely on a binary collision approximation. In crystalline targets, when the kinetic energy transferred to an atom exceeds a threshold energy E_d, typically 25 eV for metals, this atom is displaced from its initial lattice site, thus creating a vacancy. At low recoil energy, single Frenkel pairs are produced: one interstitial atom is created near the vacancy, so as to keep the total number of atoms constant. For higher recoil energies, the primary knocked atom may displace several other target atoms. In particular, for recoil energies above ≈ 1 keV, a cascade of displacement reactions takes place, resulting in the production of many vacancies and interstitials. The modified Kinchin–Pease formula, also known as the NRT formula, was introduced to determine the number of Frenkel pairs n_{FP} produced in such cascades as a function of the damage energy E_D, which is defined as the fraction of the recoil energy involved in nuclear collisions [Eq. (16)].

$$n_{FP}^{NRT} = 0.8 \frac{E_D}{2E_d} \qquad (16)$$

However, it is now well recognized that the binary collision approximation fails to account properly for the interactions and recombination of defects during the few picoseconds of the cascade lifetime [39–41]. As a result of these many-body interactions, the actual number of Frenkel pairs produced is lower than that predicted by the NRT formula (see Fig. 8.3). Bacon and coworkers [42, 43] obtained a general expression [Eq. (17)] by fitting molecular simulations results.

$$n_{FP} = A(E_D)^m \qquad (17)$$

E_D is expressed in keV, and $A \approx 5$ and $m \approx 0.75$ are weakly material-dependent constants. In addition, a significant fraction of these defects may agglomerate and form interstitial or vacancy clusters, so that the number of defects free to migrate in the material at the end of the displacement cascade is reduced even further. We note also that when the recoil energy exceeds a certain value, typically ≈ 50 keV for metals, a cascade splits into subcascades; provided correlation effects between subcascades may be ignored, it is sufficient to evaluate the damage produced by recoil energies less than that subcascade formation threshold.

In addition to the production of point defects, nuclear collisions also result in the forced relocation of atoms from their initial site to some final location. At low recoil energy, these atomic replacements occur in replacement collision sequences (RCS) along dense crystallographic directions, while, at higher energies, they result from the dynamically correlated transport of atoms within the displacement cascade. The number of replacements per Frenkel pair produced ranges from a few (for low recoil energy) to more than hundred at high energies. These replacements play a critical role in driving alloys away from equilibrium: they may result in forced atomic mixing, and chemical disordering if the initial microstructure is ordered. Since the kinetic energy per atom in a cascade during most of its lifetime exceeds ≈ 1 eV, it is much larger than typical enthalpy changes involved in phase transformations, ≈ 0 to 0.1 eV, and, as a first approximation, one may assume that these replacements are ballistic, i.e., unaffected by the thermodynamics of the irradiated alloy [44]. This simplification may break down in alloy systems with a very large enthalpy of mixing, ΔH_m. For large and negative ΔH_m, the core of the cascade may retain some chemical short-range order (e.g., Ni_3Al, [45, 46]). For large and positive ΔH_m, little or no mixing at all is observed in molecular dynamics (MD) simulations. Within the liquid-like model introduced by Averback and coworkers for displacement cascades [39], little or no forced mixing is expected to take place in alloy systems that are immiscible in the liquid state, such as Ni–Ag, Fe–Ag, Cu–Mo, and Cu–W. This rule is in good agreement with experimental findings, and with MD simulations [47, 48]. Finally, in certain alloy systems the disorder induced by nuclear collisions leads to the formation of amorphous regions [39].

The atomic mixing and disordering forced by displacement cascades is characterized by a rate, the rate of forced replacements, but also by two length scales. Firstly, a cascade extends over a finite region, with an average diameter L ranging from about 1 to 10 nm. Cascade sizes can be measured indirectly by dark-field transmission electron microscopy, using alloys that are long-range ordered, e.g., Cu_3Au, and assuming that a cascade transforms fully into a chemically disordered zone [49]. These sizes are in good agreement with those directly determined from MD simulations [39]. We will indicate in Section 8.2.5 that L plays a determining role in the possible self-organization of the chemical order in irradiated alloys. Secondly, within a cascade, atomic relocations are characterized by a distance R, or more precisely by a distribution of distances. MD simulations indicate that most atoms are relocated to first-nearest-neighbor sites. Some atoms, however, are relocated further away [39]. This distribution was investigated in detail for a $Cu_{50}Ag_{50}$ alloy subjected to ion irradiation, with ion masses ranging

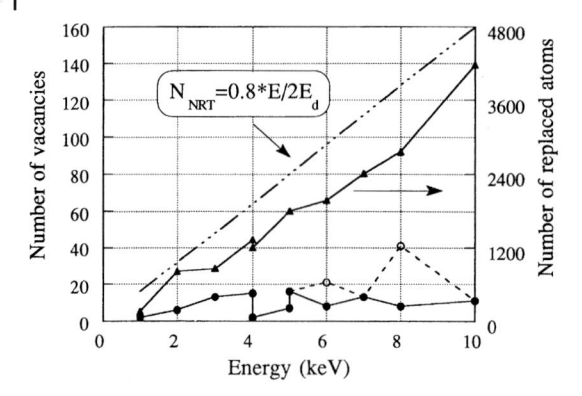

Fig. 8.3 Total number of vacancies (empty circles) and number of isolated vacancies (solid circles) as a function of the primary recoil energy from MD simulations using a generic Ni interatomic potential. The mixed line indicates the defect production predicted by the NRT formula [see Eq. (16)]. The solid triangles give the total number of replaced atoms, to be read on the right-hand ordinate (after ref. [41]).

from low to high, and the distribution of relocation distances was found to follow an exponential decay [50]. Moreover, the decay length increased from half a nearest-neighbor distance for light-ion irradiation to about one nearest-neighbor distance for heavy-ion irradiation. While this variation is rather modest, it will be shown in Section 8.2.4 that it may nevertheless be sufficient to trigger compositional patterning. L and R characterize the spatial correlations of the forced mixing, and, in the context of dissipative systems, these length scales relate to the structure of the "noise," or fluctuations, introduced by the external forcing.

As a result of the elementary effects described above, an alloy under sustained irradiation is under continuous siege by many processes. Point defects and point defect clusters are produced, migrate through thermally activated processes, disappear by interstitial-vacancy recombination and by elimination in sinks such as grain or phase boundaries, surfaces, or dislocations; chemical species undergo forced relocation and thermally activated migration. The lifetime of displacement cascades is typically a few picoseconds, a time scale much shorter than those involved in thermally activated diffusion, even at high irradiation temperatures. This decoupling in time scale makes it possible to simplify the kinetics of the system and assume that the dynamics resulting from forced atomic mixing and defect production takes place in parallel with the internal dynamics of the alloy, by thermally activated jumps of point defects. Nearly all kinetic models for alloys under irradiation rely on such a superposition principle, where the total rate of evolution of any particular variable describing the alloy microstructure is written as the sum of the rates due to each contributing process. Whereas this approach appears to be well founded for atomistic models, one should realize that problems can arise in continuum descriptions, since they rely on quantities averaged in time over many events. A first example is found in assessing the flux of point defects which are eliminated into a sink. In this case, the rate of point defect elimination in sinks is a function of the recombination rate between defects [51, 52], and more generally is influenced directly by the presence of other sinks, in particular biased ones [53]. Another example is the possible modification of diffusion coefficients by the forced ballistic jumps [54]. When any given atom undergoes alternately thermally activated and ballistic jumps, at similar rates, standard expressions for diffusion coefficients miss certain coupling effects between the two types of jumps.

A first effect of irradiation is to increase intrinsic atomic diffusion proportionally to the ratio of nonequilibrium vacancy concentration to its thermal equilibrium value. It is important to emphasize that, in a first approximation, this dynamics is controlled by the thermodynamics of the alloy, i.e., the gradient of chemical potential. As noted by Adda et al. [34], the presence of large point defect supersaturations may, however, modify the chemical potential of other species. Second, the presence of interstitials adds another process which contributes to atomic diffusion. Self-interstitials have quite small migration energies in most metallic systems, from 0.1 to 0.3 eV, and MD simulations indicate that small interstitial clusters may even migrate almost athermally [55]. In some alloy systems, however, large interstitial migration energies have been reported, sometimes even approaching the vacancy migration energies [56, 57]. The thermodynamic potentials controlling the evolution of vacancy and interstitial fluxes are different, as well as their coupling to chemical fluxes. Interstitials can thus lead to the formation of nonequilibrium phases distinct from those stabilized by vacancy fluxes. Finally, a third dynamics is the forced atomic mixing produced by nuclear collisions. Within the approximation of ballistic jumps, the atoms involved as well as the direction of the jumps are randomly sampled. Such a forced dynamics will tend to erase composition gradients, and it is thus equivalent to an "infinite-temperature" dynamics. Systems with infinite-temperature dynamics have sometimes been considered in the nonequilibrium statistical physics community, in particular for recovering hydrodynamic equations [58], or as a generic case of nonequilibrium dynamical systems [59].

Depending upon the alloy and the irradiation conditions, these various dynamics may work synergistically, or they may compete with one another. Under sustained irradiation conditions, it is commonly observed that the composition and chemical order fields reach a stationary state for a sufficiently large irradiation dose. More accurately, these states are often only quasi-steady states, since other elements of the microstructure, such as the density and the distribution of dislocations, are still slowly evolving. An important question is whether, for a given set of irradiation conditions, the same steady state is reached regardless of the initial conditions. While this property holds for alloy systems at thermal equilibrium – even though the system may be trapped in a metastable state for a very long but finite time, there is no principle that guarantees that the property holds for nonequilibrium dissipative systems [60, 61]. In nearly all the experimental results available to date, however, it appears that alloys under irradiation do reach steady states which are independent of their initial state [24, 62, 63].

One exception reported during the electron irradiation of Ni_4Mo ordered alloy may be due to the very slow coarsening of the $D1_a$ long-range ordered phase at rather low temperatures [64]. A large part of the theoretical effort presented in this chapter focuses on the prediction of these steady states. The practical application of this knowledge cannot be underestimated since one crucial question in the nuclear industry is the assessment of the long-term stability of engineering materials in irradiation environments, over periods of time ranging from 50 years (nuclear reactor) to thousands of years (nuclear waste storage). The first such steady-state diagrams for alloys under irradiation were reported by Barbu and Martin for the Ni–Si system [65], and by Cauvin and Martin for the Al–Zn system [9].

Before techniques that can be used to determine steady states stabilized by irradiation or other external forcing are presented in Sections 8.2.3 and 8.2.4, it is useful to introduce some simple qualitative arguments on the evolution of materials under the three dynamics identified above. If one ignores for the time being the interstitial-mediated dynamics, it is useful to distinguish three regimes [33, 34].

At high temperatures, the vacancy-mediated dynamics is dominant, and the alloy is expected to evolve toward a steady state which is close to its equilibrium state. Irradiation can in fact be used to accelerate the rate of relaxation toward equilibrium, for instance, to achieve a high degree of chemical order [36, 66], one that would require prohibitively long thermal annealing in the absence of irradiation.

At low enough temperatures, on the other hand, the evolution of the alloy microstructure is dominated by the forced jumps, and if they are ballistic, they will drive any alloy toward a chemically disordered and homogeneous state. The dynamical stabilization of a solid solution to phase separation is observed, for instance, in Ag–Cu multilayers undergoing ion-beam mixing at low temperatures, typically below room temperature [63]. As discussed above, however, no mixing, or only a little, is observed in alloy systems with large positive heats of mixing, such as Ag–Co, Ag–Ni, Ag–Fe, or Cu–Mo [39, 67, 68]. In alloy systems which undergo both decomposition and ordering at equilibrium, the dissolution and disordering of ordered precipitates have also been observed often during low temperature irradiations [37]. Recent work which combined atom probe field ion microscopy (APFIM) with TEM has revealed that, in the Ni–Al system, disordering may or may not precede dissolution, depending upon the irradiation conditions [69, 70]. This result is in agreement with an earlier prediction [71] based on a mean-field kinetic model [24]. Matsumura et al. [72] have also reached a similar conclusion using phase field modeling.

At intermediate temperature, the two dynamics are of similar magnitude and their competition may drive the alloy into steady states which have no counterpart in the equilibrium phase diagram. In Section 8.2.4, for instance, we discuss patterning reactions which develop in this intermediate temperature regime in the Ag–Cu and Cu–Co systems. It should be clear also that the boundaries between these three regimes depend upon the irradiation conditions, and in particular upon the displacement and replacement rates, since the competition between ballistic and intrinsic dynamics is determined by the jump frequency of forced jumps relative to thermal ones.

Let us now turn to situations where the contribution of ballistic jumps is negligible, which usually requires an elevated irradiation temperature. Due to the sustained production of point defects, which takes place homogeneously in the solid, and their subsequent elimination in localized sinks, irradiation leads to the build-up of permanent nonzero net fluxes of point defects. These fluxes are responsible, in particular, for the development of the defect microstructure, including dislocation loops and voids. Since the present chapter is focused on alloys, we will not discuss these microstructural evolutions per se. Defect fluxes, however, may couple to chemical fluxes, and thus result in the preferential transport of certain chemical species to sinks, the so-called inverse Kirkendall effect. One well-known and technologically relevant outcome of these couplings is the phenomenon of radiation-induced segregation (RIS) [73–81]. A remarkable fact is that this coupling may lead to local chemical enrichments at sinks [82, 83], or homogeneously [9], and these enrichments can be large enough to lead to the precipitation of second phases, even when the nominal composition of the alloy is such that the material is an undersaturated solid solution at equilibrium. The production and the migration of interstitials also introduce a new pathway for alloy evolution. For instance, a solid solution at thermal equilibrium may develop precipitates through the preferential transport by interstitials of a chemical species with a negative size effect. These evolutions can be captured by the rate equations discussed below in Section 8.2.3.

We now return to the more general case where the three dynamics operate together. A wealth of microstructural evolutions is possible, and is indeed observed. While it is often possible to rationalize experimental results by invoking effects produced by the three types of dynamics, it is often challenging to predict these evolutions a priori, or to extrapolate known results to other irradiation environments, because these evolutions are quite sensitive to the details of each dynamics and to their interplays. We will see in particular in Section 8.2.3 that, in the control variable space, there exist boundaries that delimit very different kinetics and steady states. As a result, a small variation of irradiation temperature or of the alloy composition may lead to dramatic changes in the microstructure which will evolve under irradiation. It is proposed, however, that kinetic modeling based on the driven alloy theory, coupled with selected experiments, has the potential to make safe predictions and extrapolations.

While the theory of alloys driven by irradiation was initially developed for metallic alloys, many results on nonmetallic systems subjected to irradiation can be rationalized within the same theoretical framework. In silicon, in semiconductor compounds, and in oxides, irradiation can lead to the amorphization of initially crystalline phases [84, 85]. Even though amorphization may occur directly in displacement cascades or by accumulation of disorder [38, 86], we note that the existence of critical temperatures above which amorphization no longer takes place has been reported for several systems [24]. The presence of these critical temperatures suggests that, near these temperatures, the steady-state microstructure results from a dynamical competition between irradiation damage and thermal recovery [87–91]. Irradiation effects in ceramic materials have received renewed attention since these materials may be used for nuclear waste immobilization and in Generation-IV fission reactors. Recent experimental results on irradiation-induced amorphization in $A_2B_2O_7$ pyrochlore compounds demonstrated the existence of such a critical temperature for amorphization [92, 93]. This critical temperature varies with the composition of the compound, in particular the ratio of the atomic radii of the A and B cations, and with the irradiation conditions. Recent MD simulations of damage production in $La_2Zr_2O_7$ pyrochlore [94] supported the idea that amorphization results from the accumulation of cation antisites. This ensemble of results therefore suggests that one should be able to predict the outcome of an irradiation in such compounds by modeling the dynamical competition between antisite production by nuclear collisions and antisite elimination by thermally activated processes. A similar approach has also been proposed for the case of irradiated $SrTiO_3$ perovskites [95].

Unexpected irradiation-induced phase transformations have also been reported for carbon nanostructures such as buckyballs and nanotubes (see ref. [96] for a review). In particular, it has been reported that electron irradiation in transmission electron microscopes operating at 400 and 1250 keV can alter the respective stability of graphite and diamond phases [97, 98]. As shown in Fig. 8.4, in the parameter space spanned by the electron dose rate and the irradiation temperature, irradiation at intermediate temperature induces the growth and the stabilization of the diamond phase over the graphite one. Key to this inversion of stability is the fact that the average displacement threshold energy is lower in the graphite

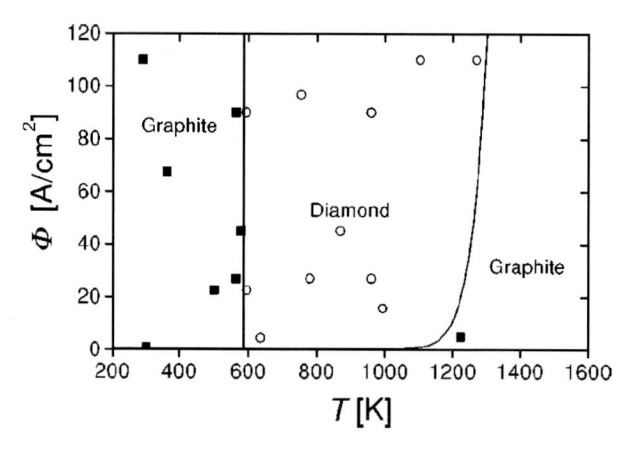

Fig. 8.4 Experimentally determined nonequilibrium phase diagram (irradiation intensity versus temperature) for carbon under irradiation with 1250 keV electrons. Open circles: diamond growth; black squares: graphite growth; full line: theoretical curve (from ref. [98]).

phase, largely because it is easy to displace atoms along the c direction due to the sp^2 bonding. Zaiser and Banhart [97, 98] have demonstrated that the above effects cannot be simply rationalized by a higher defect concentration in the graphite phase under irradiation. Indeed such an argument, if it were to account for the stabilization of diamond at intermediate temperature, fails to predict a restabilization of the graphite phase at even lower temperatures. Banhart and Zaiser have proposed a kinetic model where interstitials produced near a graphite/diamond interface feed one phase or the other, depending upon the irradiation conditions. Following an approach based on a master equation (see Section 8.2.3 and Chapter 5), they calculated an effective free energy for the graphite–diamond system and showed (see Fig. 8.4) that this effective free energy can be used quantitatively to calculate the stability boundaries of these phases under irradiation at steady state, in excellent agreement with experimental results.

Surfaces of materials subjected to energetic ion beams constitute another related example of materials driven by irradiation. Clearly, the projectile energy does not need to be as large as for bulk irradiation, and typical projectile energies range from a few electronvolts to a few kiloelectronvolts. While the low-energy range is used in beam-assisted deposition to provide additional mobility to adatoms and suppress the roughness of growing films, the higher energies can lead to morphological instabilities of initially planar surfaces. In particular, one-dimensional or two-dimensional ripples have been reported. It has been proposed to take advantage of these instability and self-organization reactions during the synthesis of functional surfaces. For the sake of conciseness, we will not discuss further surfaces under irradiation, for which several recent papers and reviews are available (see, e.g., ref. [99]). It is also interesting to note that ripple formation and other surface instability reactions may take place unintentionally, for instance during the preparation by ion milling of thin foils suitable for TEM.

8.2.1.2 Alloys Subjected to Sustained Plastic Deformation

Plastic deformation provides another way to drive materials away from equilibrium, and it is commonly found in many materials processing techniques, including rolling, extrusion, drawing, friction-stir welding, high-energy ball milling, and laser shocking (also called laser peening). In addition, materials in service may also be subjected to plastic deformation, for instance in fatigue and during frictional wear. Our focus, as in Section 8.2.1.1 above, is on the steady states reached in alloys subjected to sustained plastic deformation. As for irradiated materials, there is a wealth of experimental observations demonstrating that these systems can be driven into nonequilibrium states (see ref. [24] for some examples). Under sustained plastic deformation, precipitates can be disordered and dissolved, crystalline phases may become amorphous, and the composition and the degree of order may organize spontaneously into patterns. These reactions are in addition to other well-documented nonequilibrium evolutions which take place even in pure systems, such as dislocation patterning [100–104] and dynamic recrystallization [105] (see also ref. [106] for additional references on patterning in driven alloys).

In the late 1980s, Martin and Gaffet pointed out that powder materials processed by high-energy ball milling should be regarded as driven alloys, due to the repeated fracture and deformation of the powders during processing, all events which force changes of the local environment of atoms, in a way reminiscent of the ballistic jumps. Several examples of phase transformations induced by this sustained plastic deformation were discussed in detail in our previous review [24]. In particular, detailed analysis of the extensive results available for the crystal-to-amorphous transitions in Ni–Zr compounds and order–disorder transitions in B2 Fe–Al compounds induced by high-energy ball milling revealed the following important points. First, after extended milling times, these alloy systems always reach a stationary state, e.g., as defined by the degree of long-range order, which is furthermore independent of the initial state. Secondly, the transitions do not correspond to a minimization of the equilibrium free energy of the materials, even by imposing some constraint on the free energy. Thirdly, there exist boundaries in the dynamical equilibrium phase diagrams separating these steady states; thus dynamical phase transitions, i.e., transitions from one steady state to another, can be induced by varying the temperature and milling parameters, in particular the milling intensity (see Section 8.2.2). We showed in our previous review that these steady states and dynamical transitions, as well as their kinetics, could be fully rationalized by extending the framework initially introduced for alloys under irradiation. Additional results published since then have been rationalized successfully as well [107–111].

Another important parallel was brought to the materials science community's attention by Rigney and coworkers [112–114]. They showed that the microstructures found in powders processed by high-energy ball milling were very similar to those observed in the mechanically mixed layers which form at the contact between bodies undergoing frictional wear, also referred to as transfer layers or third bodies. Phase transformations induced by wear are in fact well known, and one classical example is the destabilization of pearlite to the benefit of carbon-supersaturated bcc solid solution at the surface of rail tracks [115]. This phase transformation is linked to the formation of the Bielby layer, a white layer that is revealed by etching. More recently, Le Bouar et al. have shown that, by extending a model developed for transitions induced by ball milling, they could predict quantitatively the wear rate of fast train wheels [116], as discussed in Section 8.2.5.

More recently, the theoretical framework for alloys driven by plastic deformation found another and unexpected area of application in the processing of organic compounds by ball milling for pharmaceutical applications, as briefly discussed in Section 8.2.5.

From a fundamental viewpoint, it is helpful to compare the processes involved in alloys subjected to sustained plastic deformation with those discussed above for alloys under irradiation. Plastic deformation in crystalline systems, whether by dislocation slip or by twinning, induces a forced mixing of atomic species, as each dislocation glide shifts atoms in glide planes with respect to their initial neighbors. Furthermore, plastic deformation creates point defects [117], and thus the rate of processes controlled by the internal dynamics is accelerated in proportion to the point defect supersaturation [118], as for alloys under irradiation. However, this forced atomic mixing presents three significant differences from the forced mixing induced by irradiation. First, the local shearing rate may depend significantly on the local microstructure of the alloys. Indeed, dislocation activity can be reduced by the presence of obstacles such as other dislocations, precipitates, grain boundaries, and interphase interfaces. One extreme situation is when, in a two-phase material, the imposed macroscopic deformation is accomplished by the plastic deformation of one phase only, for instance when the second phase consists of hard, nonshearable precipitates. In that case, as in other cases of localized plasticity, the deformation will not force atomic mixing between the two phases [119]. Recent MD simulation results [120] suggest that this straightforward effect is the principle reason for the lack of mixing by high-energy ball milling which is observed at low temperatures in alloy systems such as Ag–Ni [121], Ag–Fe [122, 123], and Cu–W [124] (see ref. [119] for an extensive review). Now restricting our discussion to cases where plastic deformation produces fairly homogeneously atomic mixing, the second difference from the irradiation case is that, based on currently available data, the nonconservative glide and crossing of dislocations produce vacancies but no interstitials [117, 125]. Recent molecular dynamics simulations in bcc metals [126] indicated that interstitial clusters can be produced, but, due to the high binding energy between interstitials and an interstitial cluster, no isolated interstitials can be re-emitted by these clusters. Finally, the third main difference resides in the intrinsic characteristics of the mixing produced by the glide of a dislocation (see Chapter 6). One glide event of a dislocation produces a perturbation, the sheared area, which is characterized by three lengths: the amount of shear, i.e., the modulus of the Burgers vector; the length of the moving segment of the dislocation; and the distance traveled by this segment, i.e., the sheared length L_{sh}. While the particular values of the latter two lengths are a function of the alloy and its microstructure, they may be very large. In some cases one expects the shear length to be of the order of the grain size. As discussed in detail in Section 8.2.4, this results in a forced atomic mixing whose efficiency is scale-dependent, and this property appears to play a key role in the spontaneous formation of compositional patterns.

8.2.1.3 Alloys Subjected to Sustained Electrochemical Exchanges

A third broad range of driven materials is the case of materials undergoing corrosion due to chemical exchanges with their environment, e.g., a gas phase, or electrochemical exchanges with an electrolyte. While corrosion is generally and rightly perceived as a problem, subjecting alloys to forced electrochemical exchanges with their environment can also be used to synthesize new structures, for instance by dealloying [127, 128]. This type of driven alloy systems will not be discussed further here, although we believe that the concepts and tools discussed in the following sections could be transferred to materials undergoing electrochemical exchanges.

8.2.2
Identification of the Relevant Control Parameters: Toward a Dynamical Equilibrium Phase Diagram

The first important step in the study of a driven alloy is to identify the relevant control parameters. Consider, for instance, the case of an alloy under irradiation discussed in the previous section. The parameters set for an experiment include the mass, the energy, and the flux of the projectiles, the irradiation time, the temperature and the composition of the target, and the environment of the chamber – typically a high vacuum. These experimental parameters, however, are not necessarily the parameters relevant to the driven alloy framework. The mass, the energy, and the flux of the projectiles for a given target determine the nature and the rate of the damage produced by nuclear collisions, i.e., the rate of production of freely migrating defects and of clustered defects, the rate of chemical mixing, the size of the displacement cascades, and the characteristic rate for atomic relocation in the cascades. It is therefore this second set of parameters which directly determines the spatial and temporal characteristics of the several dynamics active in the driven alloy. In the modeling of driven alloys simplifications are often made, thus leading to a smaller set of control parameters. For instance, if one focuses on order–disorder transformations, it may be acceptable to ignore clustered defects. Furthermore, if one is interested in steady-state properties, kinetic equations can be simplified further by the use of reduced, often dimensionless, quantities. In the previous example of an ordered alloy under irradiation, if the dominant effects can be reduced to the disordering produced by ballistic mixing and the reordering promoted by thermal diffusion, a reduced forcing intensity can then be defined as the ratio of the ballistic jump frequency to the thermally activated jump frequency. Once control parameters have been correctly identified, the outcomes of different irradiation experiments can be directly compared.

As discussed in Section 8.2.1.1, alloys under irradiation often reach a steady or a quasisteady state (see Fig. 8.5 for an example). By analogy with equilibrium phase diagrams, it is of fundamental and practical interest to construct the map of these steady states as a function of the control parameters. We refer to these maps as dynamical equilibrium phase diagrams, to emphasize the fact that these steady states are dynamical states stabilized by the external forcing. By construction, for a given alloy system, a dynamical phase diagram contains the equilibrium phase diagram, since when the forcing intensity goes to zero the stationary state reached is the equilibrium state. One of the main goals of the study of driven alloys is to develop tools, analytical ones if possible, that make it possible to construct dynamical equilibrium phase diagrams. We will illustrate in Section 8.2.3 how this goal can be achieved in simple cases, for instance for microstructures fully described by their composition field. In the most general case, however, there are fundamental obstacles, such as the lack of detailed balance, and one is often reduced to using computer simulations to construct dynamical equilibrium phase diagrams.

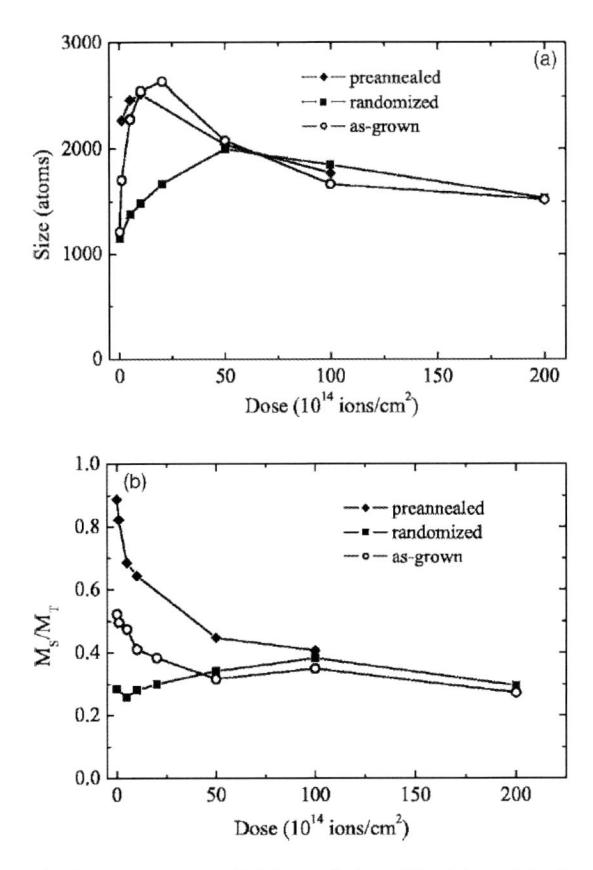

Fig. 8.5 Comparison of (a) the evolution of Co-rich particle sizes and (b) normalized magnetizations in $Cu_{90}Co_{10}$ irradiated at 200 °C with 1 MeV Kr ions as a function of the dose for three sets of thin films: as-grown by co-sputtering, randomized by pre-irradiation at room temperature, and preannealed at elevated temperature. Magnetic measurements (SQUID) were performed at T = 100 K (from ref. [197]).

The characterization of the physical processes introduced by the external forcing is less advanced in the case of alloys subjected to sustained plastic deformation than it is for alloys under irradiation. This results partly from the fact that the local rate of deformation can couple with the microstructure, whereas the rate of defect production by irradiation is largely independent of that microstructure. Nevertheless, one natural variable to characterize the intensity of the forcing is the plastic strain rate. It is often difficult, however, to determine this rate, in particular for powders processed by ball milling [129, 130]. Analyses of the mechanical system consisting of the powders, the milling balls, and the milling container were performed using instrumented mills, direct visualization, and modeling [131, 132]. They have provided useful estimates of the frequency and velocity

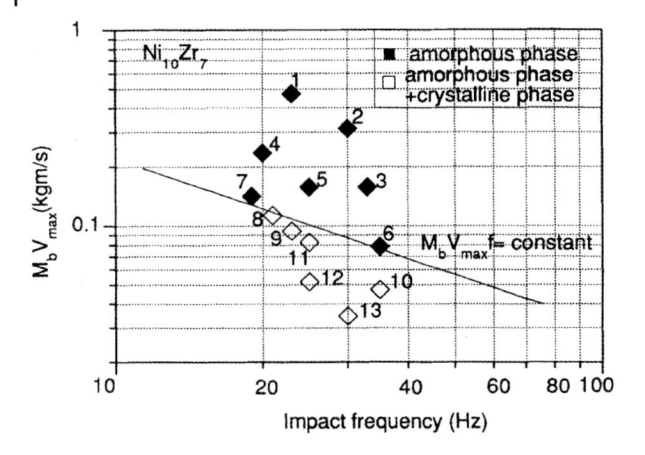

Fig. 8.6 Steady-state structure of $Ni_{10}Zr_7$ as a function of milling conditions as defined by the transfer of momentum at impact and the impact frequency (see the text for definition). Filled and open squares correspond to fully amorphous and partly amorphous steady states, respectively. The boundary separating the two types of steady-state microstructures fits well a line of constant value for $M_b V_{max} f$ (from ref. [133]).

of impacts between milling tools and powder particles, as a function of milling parameters such as number, size, and mass of balls, ball-to-powder mass ratio, and ratio of velocity of vial to velocity of main disc in planetary mills. These parameters, however, are not independent as far the degree of amorphization or the degree of chemical order is concerned. An alternative approach developed by Martin and coworkers has been to perform parametric studies to assess the effect of the experimental parameters on a phase transformation, e.g., crystal-to-amorphous [133] or order–disorder [134], and to define a milling intensity. One can then propose and assess reduced variables which can rationalize the observed phase transformations consistently. A point of debate in the community has been to identify whether the milling intensity is related to the kinetic energy or to the momentum transferred per impact per unit mass of powder. Chen et al. [133] for amorphization reactions in the Ni–Zr system (see Fig. 8.6), and Pochet et al. [134] for order–disorder reactions in the Fe–Al system, used an instrumented vibratory ball mill and concluded that the correct specific milling intensity is in fact related to the momentum transferred and is defined by Eq. (18), where M_b and M_p are the ball and powder masses respectively, V_{max} the maximum velocity of the frame, and f the impact frequency.

$$I = \frac{M_b V_{max} f}{M_p} \tag{18}$$

For the $Ni_{10}Zr_7$ compound, for instance, full amorphization is achieved when the specific milling intensity exceeds a threshold value of ≈ 510 m s^{-2}, regardless of the particular combination of values for M_b, M_p, V_{max}, and f. The practical value of the experimental dynamical equilibrium phase diagrams is that they can be used to anticipate the outcome of other experiments. A practical limitation is that it can be quite time-consuming to construct these phase diagrams relying solely on experiments, in particular if one varies the composition of the alloy. An additional limitation is that results obtained with different mills, e.g., shaker, vibratory, and planetary mills, may differ somewhat due to variations in the mechanical processes involved. It is nevertheless remarkable that the milling intensity as defined by Eq. (18) has been successfully extended to predict quantitatively the wear rate of fast train wheels [116].

It is worth making a connection between dynamical equilibrium phase diagrams and the so-called Ashby maps. Ashby and coworkers have introduced very useful mechanism-based maps for materials subjected to creep, sintering, and dry frictional wear [135–137]. These maps indicate, for a given set of reduced variables, the dominant mechanism in the kinetic evolution of the variable used to monitor the evolution of the materials, i.e., the creep rate, the sintering rate, and the wear rate in the above examples. Ashby's maps, however, do not locate transitions of these evolution rates. The purpose of dynamical equilibrium phase diagrams is to identify these transitions precisely. The consequences of the existence of dynamical phase transitions should not be underestimated. When one performs accelerated tests to assess the resistance of a material to a certain external forcing, for instance, a pressure vessel steel in a nuclear reactor or metallic waste container subjected to corrosion, the extrapolation of results obtained from short-time experiments at high forcing rates to an environment where the material would be subjected to a much lower forcing rate, but for a considerably longer period of time, may be valid as long as one does not cross a dynamical phase transition on going from the high to the low forcing rate. Indeed, if such a dynamical transition takes place, the system at low forcing rate would be evolving toward a different steady state, and the accelerated experiments would be of no predictive value.

It will be recalled in Section 8.2.3 that, under fairly general conditions for time-independent external forcing, if the microstate probability distribution of a driven alloy reaches a steady-state distribution, this steady-state distribution is unique. There is however no guarantee that the microstate probability distribution will reach a steady state, or that the macroscopic variables used to characterize a driven alloy will reach steady state. In fact, it has been known for several decades that dissipative systems may reach time-dependent states such as oscillatory and chaotic ones [30, 138]. Oscillatory states had not been reported for the driven alloys considered in this chapter until recently. El-Eskandarany and coworkers, however, have observed oscillatory regimes in the amorphization–crystallization reactions in $Co_{75}Ti_{25}$ [139], $Co_{50}Ti_{50}$ [140], $Al_{50}Zr_{50}$ [141], and $Cu_{33}Zr_{67}$ [142]. Based on the experimental procedure used in these experiments, and on a careful analysis of the milled powders using XRD, DSC, SEM and TEM, these authors have excluded the possibility that the oscillatory behavior would be due to contamination or temperature variations. A possible explanation for the cyclic response is that the damage created by one collision is a function of the state of the material. The corresponding situation in alloys under irradiation would be a system where the displacement cross-sections vary with the state of the material, so that, for a constant ion flux, various phases would in fact be irradiated at various displacement rates. As discussed in Section 8.2.1, this is the case for the irradiation of carbon, due to a threshold displacement energy for graphite lower than that for diamond. Very recently, Johnson and coworkers introduced a kinetic model which aims at identifying the general conditions required for producing cyclic responses during ball milling [143].

Until now, the forcing intensity has been assumed to be a constant. There are practical situations, however, where this intensity varies in time. This is clearly the case for materials subjected to stress or thermal fatigue. The same is true for nuclear reactor components whenever the power is adjusted to the demand on the electric grids, or whenever a reactor is stopped and restarted. New reactor designs, for instance the pebble-bed nuclear reactor, offer the possibility to do so on rather a short time scale, of the order of one minute. Similarly, fission reactors will operate on a pulsed mode, at short time scales. From a fundamental perspective, one cannot assume that it is sufficient to use the average forcing intensity to describe these situations. The kinetic equations describing the evolution of the material variables, for instance its composition field, are no longer autonomous differential equations. Very little basic work has been devoted to this question in the context of alloys under irradiation. Rauh and coworkers [144] calculated sink strengths under pulsed irradiation conditions and reported that, in particular for free surfaces, the instantaneous sink strength can differ significantly from its average value. More generally, the rate theory of defect accumulation (see below) should be used with caution in order to assess the effect of pulsed irradiation; indeed, the former rests on a time- and space-averaging procedure which is not precisely defined; the applicability of the rate theory depends on the respective values of the duration of pulses, of the time lag between pulses, and on the lifetime of the cascade-produced defects [145].

8.2.3
Theoretical Approaches and Simulation Techniques

The evolution of an alloy subjected to a sustained external forcing is not driven by thermodynamic forces. During this evolution, the free energy of the alloy may in fact decrease or increase, as demonstrated experimentally in ball milling experiments (see Section 8.2.1). In the past, some authors proposed adding a constraint on the free energy to account for the contributions of the defects and the structural or chemical disorder present in a nonequilibrium system. However, this approach fails to rationalize even simple irradiation-induced phase transformations [146]. Due to the continuous production and annealing of disorder and defects in a driven alloy, only a kinetic model can provide a sound description. That kinetic model, furthermore, should include the relevant kinetic processes and their coupling. We review below the various methods which are available to model and simulate these processes, from the atomistic to the continuum level.

8.2.3.1 Molecular Dynamics Simulations

Molecular dynamics (MD), which is discussed in detail in Chapter 12, is a powerful method to identify the short-term response of an ensemble of atoms to a given external forcing. MD simulations have provided unique information on the elementary processes involved during forcing events, in particular for the damage produced by irradiation in solids [39] (see, for instance, Fig. 8.3), and more recently for the plastic deformation in materials subjected to large stresses. Thanks to the development of realistic semiempirical interatomic potentials (see Chapter 11), and to the explosion in computing power, systems containing up to

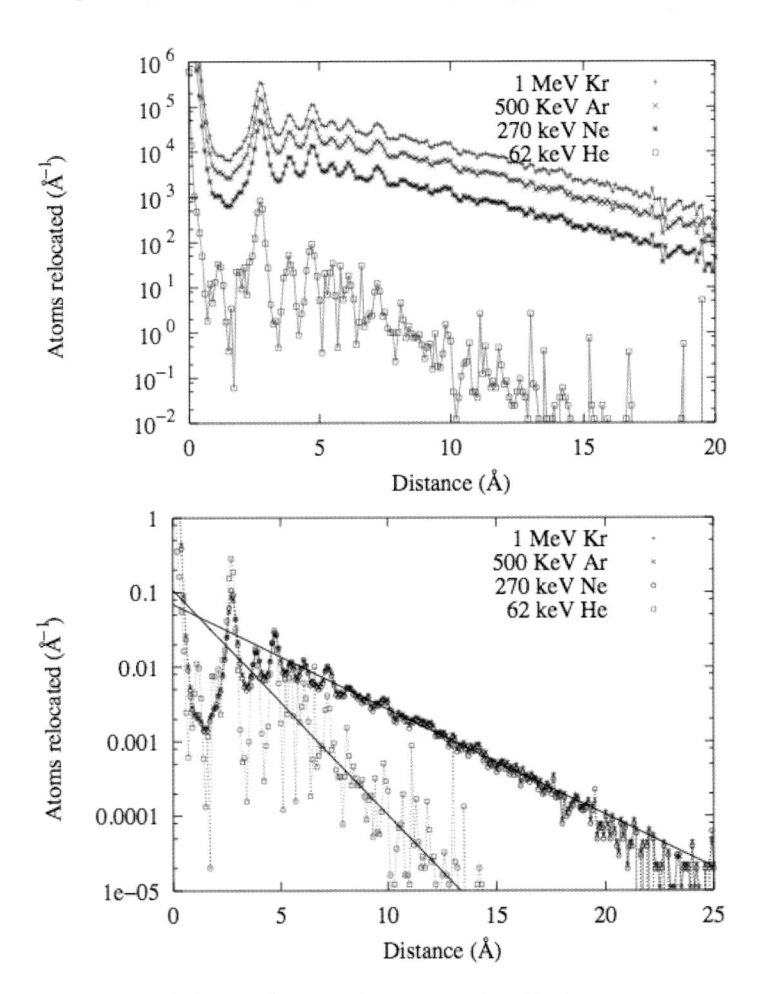

Fig. 8.7 MD calculations of atomic relocations produced by the collision cascades initiated by the indicated ions on a $Ag_{50}Cu_{50}$ solid solution. Top: absolute histograms of relocation distances; bottom: probability densities of atomic relocations. Also shown are fits to the data based on exponential decays: $0.07\,\text{Å}^{-1}\,\exp[-r/(3.08\,\text{Å})]$ for the case of Ne, Ar, and Kr, and $0.08\,\text{Å}^{-1}\,\exp[-r/(1.44\,\text{Å})]$ for the case of He (from ref. [50]).

several millions of atoms can be simulated for several nanoseconds of physical time. These simulated times, in general, do not make it possible to reach steady states which are partly determined by thermally activated diffusion. These limitations can be alleviated by simulating small systems, and by using high temperature or high forcing rate, as long as the extrapolation of these results to more realistic conditions is valid. In the case of alloys under irradiation, in addition to assessing the number of Frenkel pairs and atomic replacements produced by a cascade of a given energy [see Fig. 8.3 and Eq. (8.17)], MD simulations have been employed to reveal the distribution of relocation distances as a function of the mass and energy of the projectiles. Figure 8.7 shows the number density of relocation as a function of the relocation distance for 1 MeV Kr, 500 keV Ar, 270 keV Ne, and 62 keV He ions in a $Ag_{50}Cu_{50}$ solid solution [50]. The energies of the ions are chosen so that they produce the same projected range for the projectiles, ≈ 260 nm. As expected, the heavier the ion, the more relocations are produced. For short relocation distances, the number density displays oscillations which correspond to the neighbor distances allowed by the crystal structure. At longer distances, however, the distribution follows an exponential decay, with a decay length $R \approx 3.08$ Å for the three heavy ions and $R \approx 1.44$ Å for He. The fact that an exponential decay, rather than a Gaussian one, reproduces well the distribution for relocation distances greater than 5 Å suggests that these relocations result from ballistic events, as opposed to a diffusion-like process.

8.2.3.2 Microscopic Master Equation

Kinetic Monte Carlo (KMC) simulations provide a powerful way to assess the effect of the various parameters identified by MD on the long-term evolution of a driven alloy. KMC methods take advantage of transition rate theory for thermally activated processes to calculate directly the frequency of events which have been identified as being relevant for the alloy kinetics. In order to appreciate better the foundations of KMC methods, it is useful first to review a general microscopic description of an alloy where atoms can migrate because of the presence of several dynamical processes. For simplicity, consider a binary alloy system where $\alpha = A, B$ species are distributed on lattice sites $i = 1, N$. We denote the 2^N microconfigurations by the set of N occupation numbers, $\{n_i^\alpha\}$, where $n_i^\alpha = 1$ if the site i is occupied by an α atom, $n_i^\alpha = 0$ otherwise – point defects are not included in this presentation. Consider now an ensemble of equivalent systems, and define the probability of finding one system in a given microconfiguration at time t, $P(\{n_i^\alpha\}, t)$, as the fraction of systems found in that state at time t. Restricting ourselves to Markovian processes, i.e., processes that do not depend on the past history, the temporal evolution of the microconfiguration probability is given by a master equation (19).

$$\dot{P}(\{n_i^\alpha\}, t) = \sum_{\{n_i^\alpha\}'} P(\{n_i^\alpha\}', t) W(\{n_i^\alpha\}' \to \{n_i^\alpha\})$$

$$- P(\{n_i^\alpha\}, t) W(\{n_i^\alpha\} \to \{n_i^\alpha\}') \qquad (19)$$

The summation is performed over all the states $\{n_i^\alpha\}'$ which can be reached from $\{n_i^\alpha\}$ by one event. $W(\{n_i^\alpha\} \to \{n_i^\alpha\}')$ is the transition rate from state $\{n_i^\alpha\}$ to $\{n_i^\alpha\}'$. This transition rate includes all the processes that can induce such a transition. If we assume that forced transitions are taking place independently of the thermally activated ones, we can write Eq. (20), where the "*th*" and "*b*" subscripts correspond to thermal and ballistic events, respectively.

$$W(\{n_i^\alpha\} \to \{n_i^\alpha\}') = W_{th}(\{n_i^\alpha\} \to \{n_i^\alpha\}') + W_b(\{n_i^\alpha\} \to \{n_i^\alpha\}') \tag{20}$$

Following transition rate theory, the thermal rate is expressed as Eq. (21), where $E(\{n_i^\alpha\} \to \{n_i^\alpha\}')$ is the activation energy for that transition.

$$W_{th}(\{n_i^\alpha\} \to \{n_i^\alpha\}') = \nu \exp[-\beta E(\{n_i^\alpha\} \to \{n_i^\alpha\}')], \tag{21}$$

In contrast to the activation energy, the pre-exponential factor ν displays only a weak dependence on the configuration (see Chapter 5), and for simplicity it is assumed here to be constant. In the simple case where all forced transitions correspond to ballistic exchanges of first-nearest-neighbor atoms, the forced transition rate in Eq. (20) reduces to Γ_b, the ballistic jump frequency. More complex expressions may be required to take into account the specifics of a given forcing situation, for instance the relocation distance, as discussed in detail in Section 8.2.4.

In order to identify the steady states reached by a driven alloy, one solves for $dP(\{n_i^\alpha\}, t)/dt = 0$ in Eq. (19). In the absence of forcing, microscopic reversibility implies that detailed balance [Eq. (22)] is satisfied at equilibrium for all $\{n_i^\alpha\}$ and $\{n_i^\alpha\}'$ microconfigurations, where the subscript "*e*" refers to the equilibrium probability distribution.

$$P_e(\{n_i^\alpha\})W(\{n_i^\alpha\} \to \{n_i^\alpha\}') = P_e(\{n_i^\alpha\}')W(\{n_i^\alpha\}' \to \{n_i^\alpha\}) \tag{22}$$

This property makes it straightforward to calculate the probability of any state relative to that of a reference state, and thus it yields the explicit solution [Eq. (23)] for the equilibrium probability distribution, where $E(\{n_i^\alpha\})$ is the internal energy of the state $\{n_i^\alpha\}$, and the normalization factor Z is the partition function.

$$P_e(\{n_i^\alpha\}) = Z^{-1} \exp[-\beta E(\{n_i^\alpha\})] \tag{23}$$

In the presence of external forcing, however, microscopic reversibility is lost, and detailed balance does not have to be satisfied. There is in fact no guarantee that, for a dissipative system, a steady-state probability distribution exists. Lebowitz et al. [60], however, demonstrated under quite general conditions that if one such solution exists, it is unique. One approach which has sometimes been proposed is to assume that this solution exists, and to write it as Eq. (24), where H_{eff} is an effective Hamiltonian [147, 148].

$$P_{ss}(\{n_i^\alpha\}) = Z_{eff}^{-1} \exp[-\beta H_{eff}(\{n_i^\alpha\})] \tag{24}$$

The problem is then shifted to the determination of these effective interactions, which, unlike Ising-type interactions, are a function of temperature and forcing parameters. While there is no general method to solve exactly this problem, effective interactions can be determined by an inverse KMC method [149], and approximate solutions can be obtained using a mean-field approximation (see below).

8.2.3.3 Kinetic Monte Carlo Simulations

The microscopic master equation formalism provides a clear foundation for KMC simulations. As discussed in Chapter 12, one can formally integrate the master equation, and, starting from an ensemble of systems all in the same microscopic state, the probability for a given system to have remained in that state until time t is given by Eq. (25),

$$P_r(\Delta t) = \exp(-\Delta t \sigma(\{n_i^\alpha\})) \tag{25}$$

where $\sigma(\{n_i^\alpha\})$ is defined as the sum of the rate of all the possible transitions from state $\{n_i^\alpha\}$ to any other state, as a result of thermally activated and forced processes [Eq. (26)].

$$\sigma(\{n_i^\alpha\}) = \sum_{\{n_i^\alpha\}'} W(\{n_i^\alpha\} \to \{n_i^\alpha\}'). \tag{26}$$

As seen from Eq. (25), the average residence time of the system in its initial state is given by $\bar{\tau}(\{n_i^\alpha\}) = 1/\sigma(\{n_i^\alpha\})$. The probability of escaping along a given transition is given by the transition rate of that particular transition normalized by $\sigma(\{n_i^\alpha\})$. The so-called residence time algorithm takes full advantage of these relationships (see Chapter 12 for specific details).

As discussed in Chapter 12, this residence time algorithm is equivalent to the so-called Bortz–Khalos–Lebowitz algorithm, and to various algorithms introduced in different contexts. The main advantages of these KMC algorithms is firstly that they are computationally effective, in particular at low temperature and when the system contains only few defects (vacancies) which are responsible for thermal diffusion. Secondly, they rely naturally on the physical time for building the evolution of the system. As a result, these algorithms are easily extended to the case of driven systems, where several dynamical processes are acting in parallel, as in Eq. (20). The price that one pays in using KMC rather than MD is that only the atomistic processes identified a priori will take place during a simulation. Voter and coworkers [150–152], and Henkelman and Jónsson [153] have recently developed hybrid methods where MD is used to uncover all possible transitions, and this information is then fed into a KMC algorithm. Another limitation of nearly all current KMC simulations is that atoms are constrained to reside on the sites of a rigid lattice. This excludes the modeling of amorphous phases. Even for crystalline materials it can be a severe limitation, for instance when stress, dislocations, or grain boundaries play a significant role in the evolution of the driven alloy. Attempts have been made to include lattice relaxations in KMC algorithms [154, 155], but they limit significantly the physical times that can be reached in the simulations. Indeed, in a concentrated binary alloy, there already exists an impossibly large number of distinct atomic configurations around a vacancy, and the situation is of course worse for multicomponent alloys. Furthermore, the direct calculation of activation energies for relaxed configurations is typically 10^3 to 10^6 times longer than for the case of an alloy on a rigid lattice. Various interpolation schemes have been proposed where one first builds a small database of activation energies calculated exactly, on relaxed configurations, and uses this data set to predict the activation energies of all other configurations. Le Bouar and Soisson [156] employed effective pair interaction energies successfully to predict activation energies for vacancy migration in a dilute Fe–Cu alloy described by embedded atom model (EAM) interatomic potentials. For concentrated alloys, Sastry et al. [157] showed that, for two-dimensional systems, genetic algorithms can be used successfully to predict activation energies by symbolic regression of a small subset of exactly calculated activation energies. In some favorable situations, for instance in $Al_{(1-x)}Li_x$ alloys [158], it appears that the migration energies are mostly dependent on the type of diffusing atoms, and nearly independent of the chemical environment at the saddle point. It is then possible to calculate these few barriers, for instance using ab-initio methods, and use them as input for the KMC algorithm. Nevertheless, significant work remains to be done before general large-scale and long-time KMC simulations can be performed on relaxed lattices.

Kinetic Monte Carlo simulations are also employed to evolve the microstructure of alloys under irradiation based on a continuum-space description. The evolution can be based on "events," such as the elimination of a point defect on a defect cluster, as in the JERK code [159–161], or on the diffusion and reaction of objects, such as point defects and defect clusters, as in the BIGMAC code [162, 163]. These codes are quite powerful for elemental systems, but their extension to dilute and concentrated alloys remains a challenge because, in general, one may expect complex interactions between the chemistry-related variables, composition and chemical order, and the defect-related variables, such as point defects, defect clusters, dislocations, and grain boundaries.

Continuum methods are also widely employed. For spatially homogeneous systems, deterministic methods such as rate equations have been used in the past, for instance to calculate the evolution of point defects under irradiation [164, 165], or to model the evolution of the long-range order parameter of chemically ordered alloys under irradiation [64]. A limitation of these techniques is that they lack fluctuations. These can be recovered by using stochastic variables, for instance solving the master equation [Eq. (19)] [166, 167] or its Kubo Ansatz [168], or using the path probability method (PPM): see Chapter 10. For inhomogeneous systems, one can still rely on rate equations, or use phase field models (PFMs; see Chapters 10 and 12). A key difficulty in these continuum methods is in the definition of the continuum variables, which are obtained by some time and space averaging of the atomic-scale, discrete variables. In the absence of a clear physical picture for this averaging process, PFM simulations lack absolute length scales and time scales. Recent developments have been proposed that make it possible to define an absolute length scale (see Chapter 10) and time scale [18]. Finally we note that while the new phase field crystal model [169], where the field variable is defined as a time-averaged atomic density, retains a clear definition of time scale and length scale, its computational efficiency over KMC simulations remains to be explored.

From a broader perspective, while it appears clearly that the need to simulate large-scale systems over long physical time scales will make it necessary to rely on continuum methods, an important challenge is to retain in these models the key physical effects which can influence the long-term evolution of driven systems. In particular, special attention needs to be paid to the thermodynamic and kinetic coupling which may exist between point defects, chemical species, and elements of the microstructure. In addition, one cannot ignore fluctuations in driven systems. In particular, the extrapolation of results obtained for one alloy irradiated with one type of ion and energy to other particles or other energies requires kinetic models that include the effect of the recoil spectrum on damage production.

8.2.3.4 Kinetics of Concentration Fields under Irradiation [24]

We may provide a continuum description for a solid solution under irradiation (the simplest example of a driven alloy) by completing the formalism summarized in Section 8.1.5 above. Indeed Eq. (15) should be complemented by the conservation equations for vacancies and interstitials. In a solid solution with uniform concentration, the latter are expressed as Eq. (27), where K_d ($d = i$ for interstitials, v for vacancies) is the defect production rate (with a linear dependence on the irradiation flux), K_{iv} is the rate constant for the bimolecular recombination reaction between a vacancy and an interstitial, and K_{fd} ($d = i, v$) is the rate constant for the monomolecular elimination reaction of a vacancy or an interstitial on a fixed sink f. The latter elimination has been homogenized (smeared out) over the full volume. All such rate constants are dependent on solute concentration a priori.

$$\partial_t C_d = K_d - K_{iv} C_i C_v - K_{df} C_d; \quad d = i, v \tag{27}$$

We restrict our treatment to a binary solid solution,[6] together with interstitials and vacancies and, assuming that the overall number of lattice sites remains constant (e.g., there is no "swelling" resulting from a net accumulation of vacancies) we are left with three independent concentration fields; we choose the interstitial, vacancy, and solute (hereafter labeled s) concentrations as independent variables.

---

6) Extending the formalism to multicomponent alloys is still an open challenge.

8.2 Driven Alloys

In a non-uniform solid solution, the conservation equations are Eqs. (28).

$$\partial_t C_i = K_i - K_{iv} C_i C_v - K_{if} C_i - \nabla \cdot J_i$$
$$\partial_t C_v = K_v - K_{iv} C_i C_v - K_{vf} C_v - \nabla \cdot J_v \tag{28}$$
$$\partial_t C_s = -\nabla \cdot J_s$$

$$J_\alpha = -\sum_\beta D_{\alpha\beta} \cdot \nabla C_\beta; \quad \alpha, \beta = i, v, s$$

The diffusion coefficients can be expressed, as a first approximation (see below), as Eqs. (29), where D^{th} is the diffusion coefficient which results of the thermally activated jumps of point defects (as in Section 8.1), while D^{bal} is the "ballistic diffusion coefficient," which results from the collisions of atoms with the irradiating particles. The latter tends to erase concentration gradients, while the former erases the gradients of chemical potentials; as discussed in Section 8.1, D^{th} is the product of Onsager's matrix and the susceptibility matrix, $f_{\alpha\beta}$ [Eq. (30)].

$$D_{\alpha\beta} = D^{th}_{\alpha\beta} + D^{bal}_{\alpha\beta} \tag{29}$$

$$D^{th}_{\alpha\beta} = \sum_\gamma L'_{\alpha\gamma} \cdot f_{\gamma\beta}$$

$$f_{\alpha\beta} = \frac{\partial(\mu_\alpha - \mu_{Solvent})}{\partial C_\beta}; \quad \alpha, \beta = i, v, s \tag{30}$$

Note that the Onsager coefficients are L' rather than L, as before. Indeed, the solute diffusion parameters are proportional to the defect concentrations, which are higher under irradiation compared to thermal equilibrium; also, solute now diffuses via both vacancy and dumbbell mechanisms. In the latter case, two atoms share one lattice site, forming a dumbbell. While a dumbbell may rotate around its center, its migration requires a rotation–translation mechanism, which leaves one atom of the dumbbell on its initial lattice site and transports the second atom to a neighboring site where it forms a new dumbbell with the atom found at that site.

As shown by Eqs. (28) and (29), the kinetics of concentration fields under irradiation differs from that discussed in Section 8.1.5 by two distinct contributions:
- ballistic diffusion promotes random mixing
- defect fluxes, which are sustained by irradiation, drive a solute flux (the so-called "inverse Kirkendall effect"), which appears in Eq. (28) via the solute–defect cross-diffusion coefficients.

From Eqs. (28) and (29), we can identify three regimes, depending on whether ballistic diffusion dominates, the inverse Kirkendall effect dominates, or both effects are of the same order of magnitude.

Ballistic mixing dominant We can neglect defect fluxes and we are left with a single conservation equation, Eq. (31), where $\dfrac{\delta F}{\delta C_s}$ stands for the functional derivative of the free energy functional with respect to the concentration profile C_s.

$$\partial_t C_s = -\nabla \cdot J_s$$
$$J_s = -L' \nabla \frac{\delta F}{\delta C_s} - D^{bal} \nabla C_s = -L' \left(\frac{\delta^2 F}{\delta C_s^2} + \frac{D^{bal}}{L'} \right) \nabla C_s \tag{31}$$

As is well known, in the absence of irradiation the equilibrium concentration profile (flat for a uniform solid solution, S-shaped for a two-phase state) corresponds to an extremum of the free energy functional. The latter equation can be written formally as Eq. (32).

$$J_s = -L' \nabla \frac{\delta F_{eff}}{\delta C_s}$$
$$\frac{\delta^2 F_{eff}}{\delta C_s^2} = \frac{\delta^2 F}{\delta C_s^2} + \frac{D^{bal}}{L'} \tag{32}$$

F_{eff} has some of the properties of a free energy functional: the stationary concentration profiles correspond to extrema of F_{eff}. The local stability of a stationary solution is given by the curvature of the functional. However, the relative stability of competing stationary states cannot be assessed from F_{eff}. Such a functional is indeed a Lyapunov function of a deterministic problem; to assess the full stability of a dynamic regime, we need a correct description of the noise it is subjected to (see below).

In the simple case of the regular solution model, with some simplifying assumptions, it is a simple matter to show that Eq. (33) applies.

$$F_{eff} = \int \{ f_{eff}[C_s(x)] + \kappa |\nabla C_s|^2 \} \, dx$$
$$f_{eff}(C_s) = U(C_s) - T_{eff} S(C_s) \tag{33}$$

U and S are the contributions of a slab of matter between x and $x + dx$ (with concentration C_s), to the internal energy and to the entropy respectively; κ is the gradient energy coefficient (see chapter 10), In the regular solution model, the latter is linked to the second moment of the ordering potential.

The effective temperature T_{eff}, is given by Eq. (34), where D'_{ch} is the enhanced chemical diffusion coefficient.

$$T_{eff} = T \left(1 + \frac{D^{bal}}{D'_{ch}} \right) \tag{34}$$

The details of the demonstration are given in ref. [44].

As a first guess, one may say that the equilibrium state of a solid solution under irradiation is what it would achieve in the absence of irradiation at a temperature, $T_{\textit{eff}}$, higher than the physical temperature T, by a factor of $(1 + D^{bal}/D'_{ch})$. As a consequence, the effective temperature is very high in the case of low-temperature irradiations (D'_{ch} is very low): the latter will promote second phase re-solution. On the other hand, for high-temperature low-flux irradiation, ballistic mixing will not affect the respective stability of the precipitates with respect to the solid solution (D'_{ch} is large and D^{bal} is small). Such a simple criterion proves very useful as a first guess, in many circumstances.

Inverse Kirkendall effect dominant Ballistic mixing, and therefore ballistic fluxes, can be neglected, but we must consider in detail all the couplings between solute and defect fluxes. We find an entirely new class of phenomena, well identified experimentally: these imply irradiation-induced segregation and irradiation-induced precipitation in *under*saturated solid solutions, etc. Indeed, when the phenomenological coefficients in Eq. (28) are dependent on solute concentration, permanent point defect fluxes may be triggered by irradiation due to solute concentration heterogeneity in an undersaturated solid solution. There are many possible mechanisms for that. As an example, point defects may be produced uniformly (K_i, K_v do not depend on C_s), but they recombine preferentially in solute-enriched (or -depleted) regions, due to enhanced defect mobility (which enters K_{iv}) in the latter regions: a uniform defect production together with a localized enhanced recombination sustain point defect fluxes to those regions where recombination is enhanced. Even if K_{iv}, K_i, and K_v do not depend on C_s, a local solute overconcentration may trap point defects: the recombination rate, $K_{iv}C_iC_v$, will be locally enhanced. Hence, it is no surprise that fluxes of point defects may be triggered under irradiation, by solute concentration heterogeneity. Provided the latter defect fluxes couple to the solute flux with the proper sign, they will trigger an uphill diffusion of the solute, which may overcome the thermodynamic force, which tends to erase solute concentration heterogeneities in the undersaturated solid solution. Such an effect is therefore the opposite of ballistic effects: it may promote solute segregation, even precipitation, in an undersaturated solid solution.

The above effects can be modeled using a linear stability analysis in the following way. Consider a solid solution with a uniform solute concentration, $\bar{C}_s$; under irradiation, according to Eq. (28) the system accepts a uniform defect concentration, $\bar{C}_i$, $\bar{C}_v$, with zero flux, since all concentrations are uniform in space ($\nabla \bar{C}_\alpha = 0, \alpha = i, v, s$). Now we assume that the uniform solute concentration is slightly perturbed by a concentration oscillation with amplitude c_s^0 and wave vector q, giving Eq. (35).

$$C_s(x,t) = \bar{C}_s + c_s; \quad c_s = c_s^0(t) \cos qx \tag{35}$$

The concentration gradient will initiate defect and solute fluxes and changes in concentrations at a rate given by Eq. (28). Thus we write Eq. (36).

$$C_d = \bar{C}_d + c_d; \quad d = i, v \tag{36}$$

We note that C_i, C_v vary rapidly compared to C_s, since the solute mobility, unlike the defect mobility, scales with the point defect concentrations, which are much less than unity. We can therefore assume that the fast-relaxing defect concentration fields, C_i, C_v, adjust instantaneously to the slowly varying solute concentration field, C_s, a technique called adiabatic elimination of fast variables. As a consequence, we split the concentration space (i, v, s) into two subspaces (i, v) and s, each governed by its own diffusion equations. The defect fluxes are given by Eq. (37), where the bold letters stand for column vectors (lower case, e.g., $\mathbf{c}_d$) and rank two tensors (upper case, e.g., $\mathbf{D}_d$).

$$\mathbf{j}_d = -\mathbf{D}_d \nabla \mathbf{c}_d - \mathbf{d}_{ds} \nabla c_s$$

$$\mathbf{j}_d = \begin{pmatrix} J_i \\ J_v \end{pmatrix}; \quad \mathbf{D}_d = \begin{pmatrix} D_{ii} & D_{iv} \\ D_{vi} & D_{vv} \end{pmatrix}; \quad \mathbf{c}_d = \begin{pmatrix} c_i \\ c_v \end{pmatrix}; \quad \mathbf{d}_{ds} = \begin{pmatrix} D_{is} \\ D_{vs} \end{pmatrix} \quad (37)$$

Equation (38) gives the solute flux.

$$J_s = -\hat{\mathbf{d}}_{sd} \nabla \mathbf{c}_d - D_{ss} \nabla c_s$$

$$\hat{\mathbf{d}}_{sd} = (D_{si} \quad D_{sv}) \quad (38)$$

The elimination of fast variables consists in computing the stationary solution of Eq. (37) with c_s assumed to be independent of time. We restrict it to a one-dimensional problem, so that $\nabla \cdot \equiv \partial_x \cdot$; assuming $\partial_t \mathbf{c}_d = 0$, Eq. (28) gives Eq. (39).

$$\partial_x J_d = K_d - K_{iv} C_i C_v - K_{df} C_d; \quad d = i, v \quad (39)$$

Since c_s is a small quantity, a first-order expansion around the uniform stationary state, yields Eq. (40).

$$\partial_{x^2}^2 \mathbf{j}_d = \delta(\partial_x c_s) - \mathbf{K}^2(\partial_x \mathbf{c}_d)$$

$$\delta = \begin{pmatrix} \delta_i \\ \delta_v \end{pmatrix}; \quad \delta_d = \frac{\partial}{\partial C_s}(K_d - K_{iv} \bar{C}_i \bar{C}_v - K_{df} \bar{C}_d); \quad d = i, v \quad (40)$$

$$\mathbf{K}^2 = \begin{pmatrix} K_{iv} \bar{C}_v + K_{if} & K_{iv} \bar{C}_i \\ K_{iv} \bar{C}_v & K_{iv} \bar{C}_i + K_{vf} \end{pmatrix}$$

After some manipulation, we obtain Eq. (41), where $\mathbf{I}$ is the identity matrix, and all terms in σ, δ, $\mathbf{K}^2$ are evaluated for the stationary uniform state, i.e., with all the concentrations given their uniform stationary value, $\bar{C}_\alpha$, $\alpha = i, v, s$.

$$J_s = -\nabla c_s \{(D_{ss} - \hat{\mathbf{d}}_{sd} \cdot \sigma) + \hat{\mathbf{d}}_{sd} \cdot (\mathbf{D}_d^{-1}[\mathbf{K}^2 \mathbf{D}_d^{-1} + q^2 \mathbf{I}]^{-1})(\delta + \mathbf{K}^2 \sigma)\}$$

$$\sigma = \mathbf{D}_d^{-1} \mathbf{d}_{ds} = \begin{pmatrix} D_{is} D_{vv} - D_{vs} D_{iv} \\ D_{vv} D_{ii} - D_{iv} D_{vi} \\ D_{vs} D_{ii} - D_{is} D_{vi} \\ D_{vv} D_{ii} - D_{iv} D_{vi} \end{pmatrix} \quad (41)$$

The *first* term in the bracket { } in Eq. (41) represents the interdiffusion coefficient in the absence of irradiation, assuming point defects are conservative. To make the connection with the classical treatment of diffusion, and with Eq. (31), we write the latter interdiffusion flux as Eq. (42).

$$J_s^{th} = -\nabla c_s (D_{ss} - \hat{\mathbf{d}}_{sd} \cdot \sigma) = -L' f_{ss} \nabla c_s; \quad f_{ss} = \frac{\partial^2 F}{\partial C_s^2} \tag{42}$$

Equation (41) can be cast in the same form as Eq. (42), with f_{ss} substituted by f_{ss}^{eff} [Eq. (43)].

$$f_{ss}^{eff} = f_{ss} + \hat{\mathbf{d}}_{sd} \cdot (\mathbf{D}_d^{-1} [\mathbf{K}^2 \mathbf{D}_d^{-1} + q^2 \mathbf{I}]^{-1})(\delta + \mathbf{K}^2 \sigma). \tag{43}$$

Double integration of Eq. (43) with respect to $\bar{C}_s$, yields an *effective* free energy, F, functional for the solid solution under irradiation. F is a function of the solute concentration, of the temperature, and also of the irradiation conditions (which determine the defect concentrations entering the $D_{\alpha\beta}$s), of the microstructure (sink density which enters K_{df} in $\mathbf{K}^2$) and of the wavenumber, q, of the composition modulation. The fact that the wavenumber enters the irradiation-induced correction to the free energy implies that the interfacial effective free energy is distinct from the thermodynamic one.

Equation (43) allows for a simple discussion of the effect of irradiation on the stability of the solid solution with respect to unmixing. Setting f_{ss}^{eff} to zero defines the spinodal surface in the control parameter space: temperature, composition, and irradiation flux (which scales $\bar{C}_d$, $d = i, v$, via the defect production rates, K_d, which are proportional to the irradiation flux). Depending on the signs of the cross-diffusion coefficients, which themselves depend on the signs of the solute–defect interactions [via $\mathbf{d}_{ds}$ in Eq. (42)], and of the solute concentration dependence of defect mobilities and defect production rate (the δ matrix), the spinodal line may be shifted up or down by irradiation: an upward (or downward, respectively) shift implies irradiation-induced precipitation (or re-solution, respectively). A quantitative evaluation of the foregoing effects is intricate but has been done with a simple diffusion model [170].

As a summary, the irradiation-induced inverse Kirkendall effect [the second term on the RHS in Eq. (41)] may either stabilize or destabilize a solid solution with respect to unmixing. The possibility for irradiation-induced formation of large fluctuations in amplitude concentration in undersaturated solid solutions is clearly demonstrated by this model. Such fluctuations have been reported, after irradiation, in austenitic steels close to the Invar composition [171] (see Chapter 14b).

Ballistic mixing and inverse Kirkendall effect in competition Provided we only retain the solute diffusion coefficient in the ballistic diffusion matrix [D_{ss}^{bal} in Eq. (29)], it is a simple matter to obtain Eq. (44).

$$f_{ss}^{eff} = f_{ss} + \frac{D_{ss}^{bal}}{L'} + \hat{\mathbf{d}}_{sd} \cdot (\mathbf{D}_d^{-1} [\mathbf{K}^2 \mathbf{D}_d^{-1} + q^2 \mathbf{I}]^{-1})(\delta + \mathbf{K}^2 \sigma) \tag{44}$$

The physical meaning of the three terms on the RHS of Eq. (44) is as follows.
- The first term defines the instability limit of the solid solution in the absence of forcing: $f_{ss} = 0$ is the equation for the spinodal line.
- The second term (always positive) is the contribution of random ballistic mixing to the stabilization of the solid solution: indeed, adding a positive contribution to f_{ss} shifts the inflexion point of F (i.e., the spinodal limit) to higher concentrations, or lower temperatures; according to Eq. (34), Eq. (45) applies.

$$T'_{spi} \approx \frac{T_{spi}}{1 + \dfrac{D^{bal}}{D'_{ch}}} \qquad (45)$$

- The third term results from the point defect fluxes sustained by the external forcing; this term does not reduce to the contribution of the point defects to the internal energy and entropy of the alloy (an effect taken into account by f_{ss}). On top of the point defect supersaturation, the external forcing sustains defect fluxes, because defects may be eliminated away from their site of generation. Such is the case whenever the defect properties (formation energy, migration energy, production yield) vary with the solute concentration (as reflected in Eq. (44) by a nonvanishing $\mathbf{K}^2$ matrix). The forcing-induced defect fluxes ($\mathbf{D}_d$ matrix) may drive solute up- or downhill diffusion ($\hat{\mathbf{d}}_{sd}$), which destabilizes or stabilizes, respectively, the uniform solid solution.

The above formalism was first proposed in the mid-1990s, and many improvements have been made, e.g., in the formulation of ballistic mixing, or in the development of modeling techniques. But this general formalism deserves more work, in particular to address multicomponent alloys, precipitation of ordered compounds, fluctuations in solid solutions under irradiation, etc.

8.2.3.5 Nucleation Theory under Irradiation [9, 172]

Section 8.2.3.4 dealt with the effects of irradiation on the spinodal line, but did not address the question of the solvus line under irradiation. A large body of literature is available on the nucleation of defect clusters under irradiation; since the defect concentration under irradiation is very small (but the defect supersaturation is large), such phenomena are best described by cluster dynamics [173, 174]. This topic is beyond the scope of this chapter.

Here we focus on solid solutions and their solvus line under irradiation. Forgetting about defect agglomeration, we wonder whether the nonconservative nature of irradiation-induced point defects may alter the solubility limit, much in the same way that it alters the instability limit (spinodal line) as discussed in Section 8.2.3.4.[7] We restrict our discussion, as before, to irradiation-induced *coherent* unmixing, as was observed, e.g., in dilute solid solutions of zinc in aluminum under high-energy electron irradiation (Cauvin et al. [9])[8]. Since we restrict our consideration to dilute solutions, cluster dynamics and nucleation theory stricto sensu as defined in Sections 8.1.3 and 8.1.4 above can be used equivalently. We represent the solid solution as a gas of solute clusters, embedded in an extremely dilute gas of point defects (vacancies and self-interstitials). Point defects may be trapped at the cluster, because of the solute–defect binding energy. For simplicity, we assume that a vacancy and a self-interstitial recombine instantaneously when trapped on the same solute cluster: as a consequence, a cluster made of n solute atoms contains a small number of defects ($\pm d$, $+$ for interstitials, $-$ for vacancies); the reaction paths by which an $(n, \pm d)$ cluster may evolve are sketched in Fig. 8.8: on top of reversible paths, which would prevail if the point defect were a conservative species (the broken double arrows in Fig. 8.8), irreversible reactions (the solid single arrows in Fig. 8.8) drive the cluster to higher solute contents. As an example, whenever a solute vacancy pair reaches a solute cluster where a dumbbell is trapped, the vacancy–interstitial recombination deposits one solute at the cluster, and at the same time destroys the defects, which would allow the extra solute to escape from the cluster.

Such reaction paths can be included in the master equation for the cluster size distribution (Eq. (6), Section 8.1.3). Without going into detail, two new parameters play a key role: the respective trapping probability (p_+/p_-) of defects at the solute atoms, and their respective arrival rates at solute clusters (β_+/β_-), where the cluster size dependence of the latter factorizes out. With some simplifying assumptions, it is found that the solubility under irradiation is reduced by a factor B [Eq. (46)].

$$C_s^{irr} = \frac{C_s^{therm}}{B}; \quad B = \begin{cases} (1 + \beta_-/\beta_+) \\ (1 + \beta_+/\beta_-) \\ 1/2 \end{cases} \text{ for } \begin{cases} p_+ > p_- \\ p_+ < p_- \\ p_+ = p_- \end{cases} \tag{46}$$

Note that $\beta_+/\beta_- = D_s^i/D_s^v$, the ratio of the solute diffusion coefficients via the interstitial and the vacancy mechanism respectively. As shown by Eq. (46), the reduction of solubility is more pronounced if solute atoms diffuse to the cluster predominantly via the defect which is less trapped at the cluster. Under such a con-

---

[7] Several attempts have been made to assess the contribution of the point defect supersaturation to the free energy function of solid solutions. At least for metals, the shift of the solvus curve resulting for such a contribution is negligible compared to what is observed experimentally.

[8] Incoherent precipitation under irradiation has been addressed by Russell et al. [172]: assuming that the precipitate is oversized, vacancy elimination at the interface relaxes the strain field built up by the growth of the precipitate. A "linked flux" analysis of the growth or decay rate of a precipitate made of n solute atoms and a net number of vacancies, v, is performed and leads to an irradiation-dependent critical nucleus size.

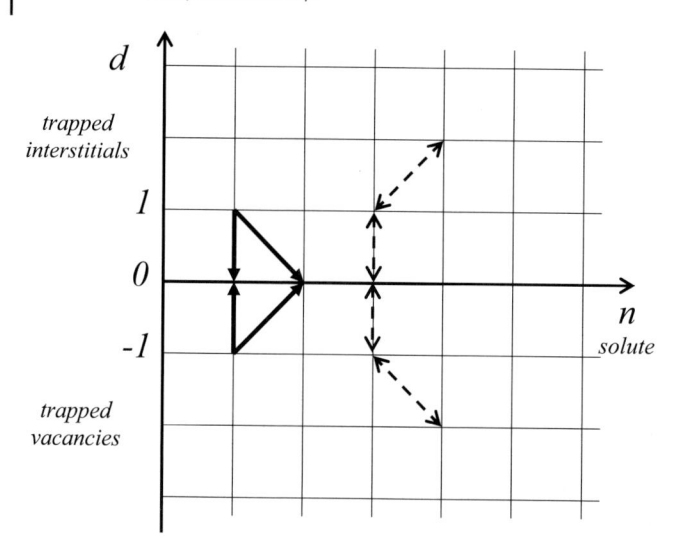

Fig. 8.8 A priori possible evolutions of a cluster in the (solute × defect) space: the broken double arrows show (some of) the reversible condensation–evaporation processes which operate in the absence of irradiation in a solid solution where solute diffusion proceeds either by vacancy (lower part of the chart) or by interstitial (upper part). The solid single arrows show (some of the) new processes which are active under irradiation, resulting from the irreversible vacancy–interstitial recombination. Among the latter, some drive the cluster to a larger solute content, n, and a smaller number of trapped defects, d, thus delaying solute re-emission to the surrounding (see the text).

dition, most solute arrival at the cluster is accompanied by the destruction of the defect which promoted solute migration: the solute atom is temporarily trapped at the cluster. Such a simple model explains, at least qualitatively, the known cases of irradiation-induced coherent precipitation, as well as the irradiation flux dependence of the solvus, in Al(Zn) solid solutions under high-energy electron irradiation [9].

Rate theory of solute defect clustering under irradiation The master equation for the time evolution of the size distribution of solute clusters [Eq. (6) of Section 8.1.3] can be extended to clusters made of solute and vacancies, or of solute and interstitials, in the limit of very dilute systems. A full numerical treatment of such an equation is not yet available, but work is in progress [175].

8.2.4
Self-Organization in Driven Alloys: Role of Length Scales of the External Forcing

As discussed in Section 8.2.1, the external forcing may introduce new length scales. Displacement cascades produced by large recoil energies in an irradiated alloy can be characterized by two lengths: the spatial extension of the cascade, L, and the average atomic relocation distance within a cascade, R. In solids subjected to plastic deformation, individual shearing events are characterized by the area swept by moving Burgers dislocations in crystals, or by Somigliana dislocations in amorphous solids [176, 177]. We show in Sections 8.2.4.1–8.2.4.3 that, whenever there is a competition between thermal and forced dynamics, self-organization of the microstructure may take place provided that the different dynamics operate at different length scales.

8.2.4.1 Compositional Patterning under Irradiation

Materials under irradiation are dissipative systems and, as such, they are capable of undergoing self-organization. In 1972 Nelson, Hudson, and Mazey reported on the evolution of precipitate size under irradiation [178]. A Ni–Al alloy containing $L1_2$-type Ni_3Al precipitates embedded in a Ni-rich matrix was irradiated with Ni ions at a flux producing $\approx 10^{-2}$ dpa s^{-1}, at various temperatures. Room-temperature irradiation led to the disordering and dissolution of the precipitates, owing to the predominance of the ballistic mixing over thermally activated diffusion at low homologous temperature. Irradiation at 550 °C, however, led to the stabilization of an unexpected microstructure composed of nanoscale Al-rich ordered precipitates, apparently randomly distributed in the disordered Al-lean matrix. This self-organized microstructure, with nanoscale precipitates, was also stabilized by irradiation under the same conditions but starting from a quenched solid solution instead of a two-phase microstructure. Very similar observations were recently reported by Schmitz and coworkers [179], who irradiated a two-phase Ni–12 at.% Al alloy with 300 keV Ni ions, producing a displacement rate of $\approx 10^{-3}$ dpa s^{-1}, at 550 °C. These authors, taking advantage of improvements in electron microscopy, measured the evolution of the average ordered precipitate diameter as a function of the irradiation dose. For the above conditions, they reported that this diameter, 5 nm initially, decreased and reached a steady-state value of 2 nm. These recent observations seem to indicate that the nanoscale microstructure is not simply the result of irradiation-induced segregation and precipitation on microstructural defects such as dislocations. An important practical and fundamental problem is thus to determine the origin and the characteristics of the self-organization of precipitates in irradiated alloys.

In order to rationalize their observations, Nelson et al. proposed a model in which irradiation-induced dissolution competes with the thermally activated growth of the precipitates, assuming that the precipitate number density remains constant [178]. These authors made the ad hoc assumption that the irradiation-induced forced chemical mixing contributes only to the transport of solute atoms from the precipitates to the matrix, neglecting the contribution of the forced mixing to the transport of solute atoms from the matrix to the precipitates. Under such an assumption, the dissolution rate of a precipitate is proportional to its interfacial area, i.e., its radius squared, instead of being proportional to its radius when both forward and backward forced solute fluxes are included and treated as a diffusion process. Neglecting the backward forced solute flux, an assumption that has no particular justification, thus leads to a pathological model. Indeed, the dissolution and thermal growth components in Nelson's model have different precipitate–radius dependence, and the model predicts that precipitates will always reach a stationary size under irradiation. Furthermore, there is no limit to this steady-state size. We will show below that the approximation made by Nelson and coworkers is in fact equivalent to assuming that the relocation distance of ballistic mixing is arbitrarily large. This is in contradiction to all the MD simulation results, which demonstrate that most of the forced mixing takes place between nearest-neighbor (NN) atoms [39]. If we assume for the moment that the forced mixing is restricted to first-NN atoms, it is then equivalent to a forced diffusion. The dissolution rate and the growth rate would now have the same linear dependence on the precipitate radius. Such a model would predict that precipitates would either grow to a macroscopic size at elevated temperature or low forced mixing rate, or shrink and dissolve at low temperature or high forced mixing rate. This model, which is based on a physical picture very similar to the one used by Martin in his derivation of the effective temperature criterion [44], does not predict that irradiation can lead to the dynamical stabilization of nanoscale precipitates.

A key element in the resolution of this paradox is recognition that, while most of the forced mixing takes place between nearest-neighbor atoms, some forced mixing extends beyond that distance. MD simulations carried out in a Ag–Cu solid solution indicate that the histogram of relocation distances can be well approximated by an exponential decay. The decay length, R, increases from ≈ 1.4 Å for light-ion irradiation (He) to ≈ 3.1 Å for heavy ion irradiation (Ne, Ar, Kr) (see Fig. 8.7). As a result of this finite-range mixing, at scales of the order or smaller than the mixing distance the process cannot be reduced to a forced diffusion. The challenge is then to study the evolution of the composition field of an alloy under irradiation when the forced mixing takes place over a finite characteristic distance R. We present below several continuum models, as well as KMC simulations, demonstrating that R plays a critical role in triggering the self-organization of the composition field, as well as in limiting the maximum length scale of this self-organization. While the results discussed below draw largely from Enrique's PhD Thesis [180], early simulation results of Haider [181] and modeling results of Vaks and Kamyshenko [148] and Pavlovitch [182] emphasized the importance of the finite-range character of the forced mixing on the possible self-organization of the composition field.

We consider a binary alloy $A_{1-c}B_c$ with a miscibility gap at low temperature, and the temperature and composition are such that, at equilibrium, the alloy would be decomposed into two phases, an A-rich and a B-rich solid solution. KMC simulations reveal that, for short relocation distances (typically, when the forced mixing is restricted to near neighbors), the alloy reaches a steady state that is a solid solution when the relative rate of ballistic to thermal jumps is large; at lower relative forcing rates, the steady state is a macroscopically decomposed structure, with some solubility enhancement due to the ballistic mixing. For sufficiently large relocation distances, however, a third steady state is found at intermediate relative forcing intensity: it corresponds to a decomposed state, albeit at a

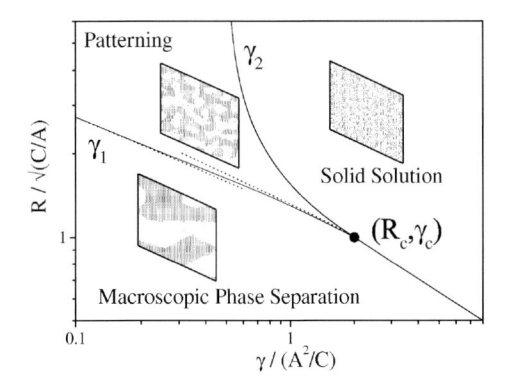

Fig. 8.9 Dynamical equilibrium phase diagram for an $A_{50}B_{50}$ alloy that undergoes phase separation at thermodynamic equilibrium. $\gamma = \Gamma_b/M$ is a dimensionless forcing intensity, and R is the average atomic relocation distance. These variables are expressed in terms of the coefficients A, B, C, entering the free energy functional (see Eq. (48), with $f = -A\psi^2 + B\psi^4$). The transition lines are calculated from the effective free energy given in Eq. (54). Insets: typical microstructures simulated by KMC simulations, observed at steady state here in a (111) plane (from ref. [184]).

finite, mesoscopic, scale (see Fig. 8.9). The characteristic scale of these compositional patterns increases as the relative forcing intensity decreases. However, a maximum size for these compositional patterns exists, and a further reduction of the forcing intensity then leads to the stabilization of the macroscopically decomposed steady state.

In order to better understand the origin of this compositional patterning reaction, and the conditions required for it, a continuum kinetic model was proposed [149, 183, 184]. The temporal evolution of the local deviation from the nominal composition $\psi(x) = c(x) - \bar{c}$ is composed of two terms [Eq. (47)].

$$\frac{\partial \psi}{\partial t} = M \nabla^2 \left(\frac{\delta \Omega F}{\delta \psi} \right) - \Gamma_b (\psi - \langle \psi \rangle_R) \tag{47}$$

The first term corresponds to the thermally activated dynamics for the conserved order parameter ψ. M is the thermal atomic mobility and is related to the chemical diffusion coefficient through $M = c(1-c)\tilde{D}/k_B T$, Ω stands for the atomic volume, and F is the free energy of the alloy, which we express using the Cahn–Hilliard form, Eq. (48), where f is the free energy density of a homogeneous alloy and C the gradient energy term.

$$F = \frac{1}{\Omega} \int (f(\psi) + C|\nabla \psi|^2) \, dV \tag{48}$$

The second term in Eq. (47) expresses the forced mixing induced by ballistic jumps, which occur at a frequency Γ_b, with relocation distances distributed according to the normalized weight function w_R. Due to conservation of the chemical species, the ballistic rate of change of the order parameter is proportional to the difference between the local composition and the nonlocal average denoted by the brackets and defined by Eq. (49).

$$\langle \psi \rangle_R = \int w_R(x - x')\psi(x')\,dx' \tag{49}$$

When the average relocation distance is of the order of the length used in the coarse graining to define the continuum variable $c(x)$, the contribution of the ballistic mixing reduces to Eq. (50).

$$\left(\frac{\partial \psi}{\partial t}\right)_b = \Gamma_b \nabla^2 (\psi) \tag{50}$$

In this specific case, one recovers the model introduced by Martin in 1984 [44] to discuss the effect of ballistic jumps on phase stability in alloys under irradiation.

Linear stability analysis It is often interesting from a practical and fundamental viewpoint to start with the control parameters which stabilize a homogeneous steady state, and then to carry out a linear stability analysis to determine the domain of local stability of this homogeneous solution as the control parameters are varied [185]. In ref. [183], such a linear stability analysis is performed for a one-dimensional equiatomic system, with a specific choice for the distribution of ballistic jumps [Eq. (51)].

$$w_R(x) = \frac{1}{2R} \exp\left(-\frac{|x|}{R}\right) \tag{51}$$

Distributions of ballistic jumps for Ag–Cu alloys, measured by MD simulations, are indeed fitted well using such a decay function. Furthermore, such a function lends itself very well to analytical calculations. For small perturbations of the form $e^{\omega t + ikx}$ around a homogeneous composition profile $c(x) = \bar{c} = 0.5$, i.e., $\psi(x) = 0$ everywhere, we can linearize Eq. (47). Transforming to Fourier space, denoted by the carets, we obtain Eq. (52), where the amplification factor is related to the wave vector of the perturbation through the dispersion equation Eq. (53); f'' is the second derivative, with respect to composition, of the free energy density in the homogeneous alloy.

$$\left(\frac{d\hat{\psi}}{dt}\right)_{tot} = \omega(k)\hat{\psi} \tag{52}$$

$$\frac{\omega(k)}{M} = -f''(\bar{c})k^2 - 2Ck^4 - \gamma \frac{k^2 R^2}{1 + k^2 R^2} \tag{53}$$

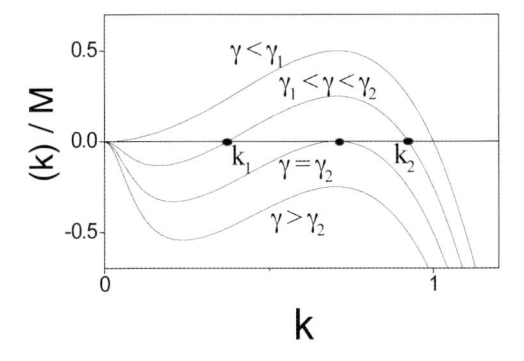

Fig. 8.10 Linear stability analysis of the kinetic model given by Eqs. (47) and (48). The solid solution is unstable with respect to fluctuations of wavenumber **k** when the amplification factor $\omega(k)$, normalized by the thermal mobility M, is positive (from ref. [184]).

Within a regular solution model (see Chapter 10), $f'' = u'' + k_B T/(\bar{c}(1-\bar{c}))$, where u is the internal energy contribution to the free energy density f. A dimensionless forcing parameter $\gamma = \Gamma_b/M$ has been introduced in Eq. (53).

Figure 8.10 displays a set of dispersion curves for various γ values, assuming that the irradiation temperature is such that $f''(\bar{c}) < 0$, i.e., inside the equilibrium miscibility gap. Figure 8.10 also illustrates the main results of this linear stability analysis:

1. For large forcing intensity ($\gamma > \gamma_2$), the homogeneous steady state is always locally stable since $\omega < 0$ for all **k** wavenumbers.
2. For forcing intensity satisfying $\gamma_1 < \gamma < \gamma_2$, the homogeneous steady state is unstable for wave vectors such that $k_1 < k < k_2$.
3. For low forcing ($\gamma < \gamma_1$), the homogeneous steady state is unstable in a well-defined range of wave vectors ($0 < k < k_2$).

Cases 1 and 3 resemble what would be found for an alloy above and below its spinodal instability in the absence of irradiation. The remarkable and new regime is found in case 2. The fact that the amplification factor remains negative in the limit of long wavelength ($k \to 0$) suggests that irradiation may stabilize a microstructure which has decomposed, albeit not at the macroscopic scale. It is found that there is a critical value for the relocation distance R, R_c, below which this second case ceases to exist, and the alloy goes directly from case 1 to case 3.

Global stability analysis neglecting fluctuations In order to determine the global stability of the various steady states encountered so far, one needs to go beyond a linear stability analysis. If one first neglects fluctuations, one can define an effective free energy functional, $E\{\psi(x)\}$ for the steady states under irradiation by imposing Eq. (54).

$$\frac{\partial \psi}{\partial t} = M\nabla^2 \left(\frac{\delta \Omega F}{\delta \psi}\right) - \Gamma_b(\psi - \langle \psi \rangle_R) = M\nabla^2 \left(\frac{\delta \Omega E}{\delta \psi}\right) \quad (54)$$

For the above choice of distribution of relocation distances, w_R, the effective free energy E can be calculated analytically, and it can then be used to build a dynamical equilibrium phase diagram [183] if one assumes that the most stable steady state corresponds to the deepest minimum in E. The resulting diagram is shown in Fig. 8.9. In this particular case, the homogeneous free energy density was written using a Landau form, $f = -A\psi^2 + B\psi^4$. The relocation distance R and the forcing intensity γ are then naturally expressed in $\sqrt{C/A}$ and A^2/C units, respectively. This analysis confirms the existence of the three steady states suggested by the linear stability analysis, and it also confirms the existence of a threshold relocation distance R_c for compositional patterning to be induced by ballistic jumps. These predictions are in excellent agreement with the results of KMC simulations (insets in Fig. 8.9).

A clear weakness of the above analysis, however, is that we have neglected fluctuations, whereas it is well known that the respective stability of competing steady states is controlled by fluctuations [61, 27, 166]. As a result, the effective free energy defined above displays several deficiencies [180, 186]. In the regime $R \to 0$, in particular, one does not recover the free energy of the alloy at the effective temperature introduced by Martin (see Section 8.34). We now extend the analysis so as to include fluctuations, thus transforming Eq. (54) into a Langevin equation, closely following the procedure proposed in ref. [186].

Langevin equation with thermal dynamics only We briefly recall here the procedure introduced by Cook [187] to introduce thermal fluctuations. In the absence of irradiation, the deterministic Eq. (54) is transformed into a Langevin equation by adding a random noise term $\xi_{th}(x, t)$, with suitably defined statistical properties (to be specified below), in Eq. (55).

$$\left(\frac{\partial \psi}{\partial t}\right)_{th} = M\nabla^2 \left(\frac{\delta \Omega F}{\delta \psi}\right) + \nabla^2 \xi_{th} \quad (55)$$

For small fluctuations around $c(x) = \bar{c}$, we can again linearize Eq. (55) and transform it to Fourier space [Eq. (56)].

$$\left(\frac{d\hat{\psi}}{dt}\right)_{th} = -\alpha_{th}(k)\hat{\psi} - k^2 \hat{\xi}_{th} \quad (56)$$

The amplification factor, now defined with a negative sign for convenience, is given by $\alpha_{th}(k) = Mk^2(f''(\bar{c}) + 2Ck^2)$. The noise term is assumed to be uncorrelated in time, and has a spatial structure given by the quantity $Q_{th}(k)$ [Eq. (57)].

$$\langle \hat{\xi}_{th}(k, t)\hat{\xi}_{th}^*(k, t')\rangle = Q_{th}(k)\delta(t - t') \quad (57)$$

8.2 Driven Alloys

The statistical quantity $Q_{th}(k)$ can be obtained by calculating the equilibrium structure factor $S_{eq}(k) = \langle|\hat{\psi}(k, t \to \infty)|^2\rangle$. Following a standard procedure to solve Eq. (56), Cook obtains Eq. (58).

$$S_{eq}(k) = \frac{k^4 Q_{th}(k)}{2\alpha_{th}(k)} \tag{58}$$

This structure factor can also be evaluated through a fluctuation–dissipation relationship, Eq. (59) [187].

$$S_{eq}(k) = \frac{k_B T}{f''(\bar{c}) + 2Ck^2} \tag{59}$$

Therefore the functional dependence of $Q_{th}(k)$ which reproduces the structure factor properly is given by Eq. (60).

$$Q_{th}(k) = \frac{2k_B TM}{k^2} \tag{60}$$

Langevin equation with ballistic dynamics only Following a similar strategy, we can write an equation including fluctuations for the irradiation-induced mixing dynamics [Eq. (61)].

$$\left(\frac{\partial \psi}{\partial t}\right)_b = -\Gamma_b\left(\psi - \int w_R(x - x')\psi(x')\,dx'\right) + \nabla^2 \xi_b. \tag{61}$$

This equation is easily transformed into an ordinary differential equation in Fourier space, i.e., Eq. (62) where the ballistic amplification factor is given by Eq. (63)

$$\left(\frac{d\hat{\psi}}{dt}\right)_b = -\alpha_b(k)\hat{\psi} - k^2\hat{\xi}_b \tag{62}$$

$$\alpha_b(k) = \Gamma_b(1 - \hat{w}_R(k)) \tag{63}$$

Now the steady-state structure factor $S_b(k)$ which we expect from the ballistic dynamics alone is that of a random solid solution, i.e., $S_b(k) = \bar{c}(1 - \bar{c})$. Applying Cook's solution to this case, we find that the spatial correlations of the ballistic fluctuations must be given by Eq. (64).

$$Q_b(k) = \frac{2\bar{c}(1 - \bar{c})\alpha_b}{k^4} \tag{64}$$

Langevin equation with two dynamics in parallel For the model with competing dynamics with fluctuations, we add the terms describing the deterministic evolution of the composition, as in Eq. (54), as well as the noise terms. We assumed earlier that the two deterministic dynamics act in parallel, i.e., independently. Within this assumption, the corresponding fluctuations are uncorrelated and therefore additive [Eq. (65)].

$$\frac{d\hat{\psi}}{dt} = -(\alpha_{th}(k) + \alpha_b(k))\hat{\psi} - k^2(\hat{\xi}_{th} + \hat{\xi}_b) \tag{65}$$

Equation (65) yields Eq. (66) for the steady-state structure factor.

$$S(k) = k^4 \frac{Q_{th}(k) + Q_b(k)}{2(\alpha_{th}(k) + \alpha_b(k))} \tag{66}$$

After simple algebraic manipulations we obtain Eq. (67), which is the central result of this model.

$$S(k) = \frac{k_B T \left(1 + \frac{\alpha_b}{\tilde{D}k^2}\right)}{2Ck^2 + u''(\bar{c}) + \frac{k_B T}{\bar{c}(1-\bar{c})}\left(1 + \frac{\alpha_b}{\tilde{D}k^2}\right)} \tag{67}$$

We will now consider specifically two cases of ballistic dynamics.

Case 1: Short-range ballistic jumps In this case, as discussed above, the dynamics can be approximated by a diffusional process with a diffusion coefficient D_b, and thus $\alpha_b(k) = D_b k^2$. Plugging this equation into Eq. (67) yields an equation that admits a straightforward interpretation in terms of Martin's effective temperature criterion: the effective temperature $T_{\text{eff}} = T(1 + D_b/\tilde{D})$ describes *both* the effective driving force *and* the effective fluctuations of the alloy under irradiation at steady state. We should also note that in Martin's derivation of the effective temperature criterion, the ratio $D_b/\tilde{D}$ is assumed to be independent of composition. It thus implies that the fluctuations also are independent of composition. Therefore, for short-range ballistic jumps, Martin's nonequilibrium potential has the desired properties of a nonequilibrium potential, and the common tangent rule is a valid construction to determine the globally stable steady states.

Case 2: Finite-range ballistic jumps Following MD results [50], we choose w_R to be an exponential decay with decay length R. Its Fourier transform is then given by $\hat{w}_R(k) = 1/(1 + k^2 R^2)$, and thus $\alpha_b(k) = \Gamma_b k^2 R^2/(1 + k^2 R^2)$. Plugging this equation into Eq. (67) demonstrates that the same rescaling applies to the temperature in the numerator, which corresponds to the amplitude of the fluctuations, and in the denominator, which corresponds to the second derivative of the nonequilibrium potential. The concept of effective temperature, however, is no longer appropriate since the effective temperature would be a function of the wavenumber k. A better approach is to divide both the numerator and the denominator in

Eq. (67) by the rescaling factor of the temperature. It can then be shown [186] that one can re-interpret Eq. (67) for that particular case as the structure factor, evaluated at the physical temperature T, of an alloy system with effective pairwise interactions defined by Eq. (68).

$$\hat{v}_2^{\text{eff}}(k) = \frac{\hat{v}_2(k)}{1 + \frac{R^2 \Gamma_b}{\tilde{D}} \frac{1}{1 + k^2 R^2}} \qquad (68)$$

A similar equation was originally derived by Vaks and Kamyshenko [148], albeit from a discrete description. An analysis of Eq. (68) shows that finite-range ballistic mixing is equivalent to the introduction of effective medium-range interactions which are repulsive between like atoms. Patterning can then be rationalized as the result of the competition between the short-range attractive physical interactions and the medium-range repulsive effective interactions. It is worth noting that competition between short-range attractive and medium-range repulsive interactions in equilibrium systems, such as block copolymers with dipolar moments [188–191], also leads to patterning and to the stabilization of mesoscopic phases. In the patterning regime, the steady-state structure factor $S(k)$ peaks at a finite k value, k_P. The microstructure has thus developed compositional patterns at a preferred length scale, which is proportional to $1/k_P$.

Confrontation between model predictions and experiments The above models offer a clear rationalization of the origin of compositional patterning under irradiation. They also set a limit to the scale of decomposition. The largest decomposition scale, reached on the γ_1 boundary in Fig. 8.9, is equal to $4\pi R$. In the limit where R is arbitrarily large, as assumed implicitly by Nelson and coworkers, the steady-state scale of decomposition, and the size of the precipitates, can thus be arbitrarily large. In the limit $R \to \infty$, it is in fact possible to show directly that the forced ballistic exchanges can be seen as producing electrostatic-like repulsive effective interactions between like atoms [182]. It is well established that such repulsive long-range interactions, when competing with short-range attractive interactions, lead to patterning [149].

Heinig and Strobel studied the possible compositional patterning of precipitates during ion-beam implantation and ion-beam mixing [192, 193]. They introduced a kinetic model for the evolution of the precipitate size which takes into account both the forward and backward solute fluxes forced by irradiation. They used an effective surface energy to rationalize the size reached by precipitates at steady state. While this effective surface energy is positive at low ballistic jump frequency or at high temperature, it reaches a negative value for a range of precipitate size at greater ballistic jump frequency or at low temperature. The flux or the temperature at which the effective surface energy goes to zero is interpreted as the onset of inverse ripening, or, in other terms, compositional patterning. While Heinig and Strobel's model bears some similarity to the models presented in this section, it relies on several approximations and consequently the results are not valid when $R \to 0$ or when the irradiation flux is large. In particular, the model predicts that compositional patterning can take place when R is arbitrarily small, in contradiction to KMC simulations [184].

Returning to metallic systems, the observations reported by Nelson and coworkers and by Schmitz and coworkers appear to be qualitatively consistent with the predictions from the above models and summarized in the dynamical equilibrium phase diagram, Fig. 8.9. The decomposition scale observed experimentally, several nanometers, seems however to be at the higher end of the spectrum expected from the analytical model. It is unclear whether the presence of chemical order in the precipitates could have influenced the patterning reaction. In Section 8.2.4.2, we will briefly review recent work indicating that the other characteristic length introduced by displacement cascades, their size L, can in fact trigger a patterning of the chemical order in alloys. In off-stoichiometric alloys, this patterning of order produces a compositional patterning also, and additional experimental work is needed to establish whether the patterning reported by Nelson and coworkers and by Schmitz and coworkers is triggered by the relocation distance R, by the cascade size L, or by both.

Averback and coworkers have recently performed experiments designed specifically to test the main predictions of Enrique's model. They have studied alloy systems which are immiscible in the solid state only, e.g., Ag–Cu [194, 195] and Co–Cu [196], or in the solid and in the liquid states such as Ag–Co [197]. The results obtained for the former systems are consistent with Enrique's model predictions. In the Co–Cu case in particular, 1 MeV ion irradiation can induce a stabilization of the size of Co-rich precipitates to a value that increases with the irradiation temperature, up to a threshold temperature of 350 °C, above which precipitates appear to coarsen continuously. In the case of alloys that are immiscible in both the solid and the liquid state, however, no saturation of the size of the precipitates is observed. Averback and coworkers rationalized this result by noting that displacement cascades in highly immiscible alloys produce little or no chemical mixing [39]. Quantitative analysis of the coarsening rate in this alloy under irradiation led the authors to propose that, in fact, the mechanism of precipitate coarsening did not involve atomic diffusion, but rather particle coalescence.

8.2.4.2 Patterning of Chemical Order under Irradiation

Turning now to the case of alloy systems which undergo chemical ordering, as recalled in the introduction (Section 8.2.1.1) it is well documented that irradiation at low temperature or high flux can disorder partly or fully an ordered alloy, while irradiation at elevated temperature or low flux can induce chemical ordering. Recent KMC simulations and analytical modeling, however, have predicted that irradiation can also drive a chemically ordered alloy into a steady state whose microstructure consists of well-ordered domains of finite size, typically in the nanometer range [198–200]. The stabilization of this new steady state, which we referred to as patterning of order, requires three conditions. Firstly, the irradiation temperature should be below the equilibrium order–disorder transition temperature, T_c. Secondly, the irradiation flux should be such that the rate of disordering imposed by irradiation is similar to the rate of reordering promoted by the thermally activated migration of point defects. A third condition was uncovered by atomistic simulations: the size of the disordered zones resulting from the dense displacement cascades produced by the energetic ions has to exceed a threshold value, L_c. This value is such that, during the thermal annealing of a disordered zone, several new ordered domains form and grow in that volume. The continuous disordering of existing ordered domains and the continuous creation of small new domains lead to the destabilization of pre-existing macroscopic ordered domains, and to the dynamical stabilization of finite-size ordered domains at steady state. In the case of smaller cascade sizes ($L < L_c$), as well as for large but dilute cascades, reordering proceeds instead by migration of the boundary between the ordered matrix and the disordered zone. It does not produce new domains, therefore, and no patterning of order is observed. It has been proposed that the state of patterning of order can be defined by analyzing quantitatively the intensity and the nature of the fluctuations of order using a scaling approach [200]. In the absence of patterning, fluctuations of order obey a scaling that is consistent with the effective temperature criterion. This scaling, however, breaks down as the cascade size is increased, and the fluctuations of order then obey a different scaling. This change in scaling can be used to define rigorously the steady state of patterning of order, and to build dynamical equilibrium phase diagrams. One example is given in Fig. 8.11 for an $A_{50}B_{50}$ alloy developing an $L1_0$ ordered structure.

It is important to note that, according to the above simulations and models, the largest average size of the ordered domains at steady state is close to that of the

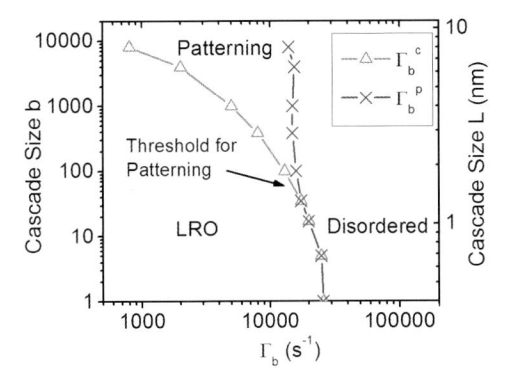

Fig. 8.11 Dynamical equilibrium phase diagram for an $A_{50}B_{50}$ alloy that develops $L1_0$ long-range order (LRO) at thermodynamic equilibrium, as a function of the ballistic exchange frequency between nearest-neighbor atoms, Γ_b, and the number of pairs exchanged at once in each cascade, b (or the cascade size L, calculated using the lattice parameter of FePt). The KMC simulations are performed at a constant temperature $T = 0.09$ eV $\approx 0.6 T_c$. The threshold for patterning of order is around $b = 100$ (from ref. [201]).

disordered zones. Since these zones can extend up to 10 nm, according to dark-field TEM results [49] and MD simulations [39, 45, 46], it appears that the patterning of order could reach larger dimensions than the compositional patterns induced by a finite relocation distance R. From an application viewpoint, it has been proposed that one could take advantage of the dynamical stabilization of nanoscale ordered domains under irradiation to synthesize nanostructures optimized for mechanical or magnetic applications which required the presence of ordered domains at the nanoscale with a high degree of chemical order [201, 202]. Such microstructures are difficult to achieve by conventional thermal processing since ordering is then accompanied by domain growth [203]. Systematic experiments need to be carried out to test the model predictions on patterning of order, and, if this patterning reaction is indeed observed, to evaluate its usefulness for the synthesis of functional materials.

8.2.4.3 Compositional Patterning under Plastic Deformation

Alloys subjected to sustained plastic deformation can undergo nonequilibrium phase transformations and microstructural evolutions. As pointed out in the early 1980s [35], these evolutions shared many similarities with those observed in alloys subjected to irradiation. In particular, plastic deformation at low temperature can dissolve pre-existing precipitates, while at elevated temperature it may induce the formation of equilibrium or even nonequilibrium precipitates. In the mid 1990s, generic atomistic computer simulations indicated furthermore that sustained plastic deformation can lead to the dynamical stabilization of compositional patterns [204], and of patterns of order [205]. The scale of these steady-state patterns was predicted to increase continuously as the temperature or the shearing frequency is increased. While these predictions have received experimental support from systematic studies employing high-energy ball milling, it is only recently that a theoretical approach has been introduced to offer a clear insight into the origin of plasticity-induced compositional patterning [120, 206].

This analysis, which parallels the standard approach to study of the mixing of passive markers transported by convective and possibly turbulent flow in fluids, focuses on the mean-squared relative displacement (MSRD) of pairs of atoms. In contrast to the case where the two atoms of a pair undergo uncorrelated migration, for instance due to thermally activated diffusion, the MSRD produced by plastic deformation is a function of the separation distance, R, of the atom pair considered. By analogy with the works of Taylor and Richardson for turbulent flows (see ref. [207] for a recent review), one can define an effective diffusion coefficient, $D_{eff}(R)$ by Eq. (69), where L is the glide length of dislocations, and is thus bounded by the grain size, b the magnitude of the Burgers vector, and Γ_{sh} the atomic jump frequency due to shearing events (a factor 2 appears because the two atoms comprising a pair may jump).

$$D_{eff}(R) \propto \frac{b^2}{6} \frac{RL^2}{V} \Gamma_{sh} \quad \text{for } R \leq L \tag{69}$$

The R-dependence of D_{eff} comes from the fact that the probability of shearing a pair of atoms separated by a distance R increases linearly with that separation distance. Equation (69) is derived by assuming that plastic deformation is homogeneous and isotropic, and that the mobile dislocation segments and the glide distance of these segments are larger than R. Equation (69) thus applies only to separation distances $R \leq L$. For separation distances $R \geq L$, the length R becomes irrelevant, and the effective diffusion coefficient becomes independent of R [Eq. (70)].

$$D_{eff}(R) \propto \frac{b^2 L^3}{6 V} \Gamma_{sh} \quad \text{for } R \gg L \tag{70}$$

The R dependence of D_{eff} for $R \leq L$ leads to a superdiffusive mixing by plastic deformation. It also provides a simple and powerful rationalization for the stabilization of compositional patterns. The outcome of the competition between the mixing forced by deformation and the decomposition promoted by thermodynamic forces is now scale-dependent. At small length scales, thermal decomposition dominates provided that temperature is high enough, and the microstructure is thus decomposed; at larger length scales, the forced mixing dominates, and thus the microstructure, at these scales, is homogeneous. Such a microstructure corresponds in fact to the mesoscale patterns observed in KMC simulations [204]. Recent 3DTAP characterization of ball-milled Ag–Cu alloys [206] offers support for this framework; in particular, in agreement with an earlier prediction, it is observed that the length scale of the patterns increases continuously as the temperature of the materials during plastic deformation is increased. It is furthermore proposed that the crossover length scale, which is obtained by equating the forced mixing diffusion coefficient to the thermal one, should correspond to the length scale of the compositional patterns. More work is necessary to evaluate systematically these predictions, which rely on a very simplified description of the alloy system undergoing plastic deformation. In particular, this description ignores stress effects, interactions between dislocations, and any possible feedback between the microstructure and the plastic deformation. These limitations can be overcome by using MD, although one then needs to extrapolate the results from the high strain rates required in MD simulations to the range of experimental strain rates, and from the high temperatures required to allow for thermally activated diffusion to take place over the simulation times to the experimental temperature range.

8.2.5
Practical Applications and Extensions

8.2.5.1 Tribochemical Reactions
It is a well reported fact that high-carbon steels, when subjected to repeated loading, develop a so-called "white phase", a supersaturated solution of carbon in

iron, resulting from the dissolution of the cementite, which is brittle. Such is the case in ball bearings, in railway tracks, and in the wheels of express trains. As discussed in Section 8.2.1, alloys under repeated mechanical loading are one example of driven alloys, much in the same way as alloys under ball milling. How should the forcing intensity, e.g., for a train wheel, be defined?

For ball milling, on the basis of experimental measurements the forcing intensity I_{bm} was identified as the momentum transferred from the ball to unit mass of powder per unit time [see Eq. (18)]. This latter intensity can be interpreted, also, as the force imposed on the material multiplied by the fraction of time for which the material is being loaded: indeed, V = acceleration × duration of impact, so that $M_b V$ = force × duration of impact and f/M_p = frequency at which a given grain of powder is being strained. For a train wheel, we can estimate from the theory of elastic contact between a cylinder (the wheel) and a plane (the track), the size of the zone affected by the load [208]. The duration of loading is that size divided by the circumferential velocity of the wheel, and the frequency of loading is the frequency of rotation of the wheel. As a consequence, the forcing intensity of the material at the periphery of the wheel is written as Eq. (71), where F is the load on the contact, l is the length of the contact between the track and the wheel, E^* is an effective Young's modulus, and R is the wheel radius [116].

$$I_w = \left[\frac{4F^3}{\pi^3 l E^* R}\right]^{1/2} \tag{71}$$

As can be seen, for a perfectly circular wheel the forcing intensity does not depend on the velocity. Assuming that the rate of wear of a wheel is proportional to the intensity, we conclude that a perfectly circular wheel remains circular over the course of time.

This is no longer the case if the wheel has local changes of curvature, as most real wheels do: a wheel may, for instance, present two local maxima of curvature (in an elliptical shape) or three or more. In such a situation, because of inertial effects, the load on the contact depends both on the location on the wheel and on the velocity. For a given symmetry of the wheel (two-, three-, fourfold), the higher the velocity, the higher the intensity.

An interesting property is revealed by this analysis: for a given symmetry of the wheel, the location of the maxima of forcing intensity depends on the velocity. At lower velocities, the forcing intensity is a maximum at the points of maximum curvature; above a crossover velocity, the reverse is true: the intensity is a maximum in the flatter regions. Assuming the rate of wear increases with the forcing intensity, we identify a critical velocity below which self-healing of the defects in curvature will dominate; above that velocity, the defects will amplify: with typical values for French TGV trains, the predicted velocity is 55 km h^{-1} (or 165 km h^{-1}) for threefold (or two-fold) symmetry.

If we assume that the rate of wear is proportional to the forcing intensity, it is easily found that the amplitude of the oscillation of the wheel radius increases exponentially with the mileage; this is observed to be true on actual trains.

The practical implications of this approach are numerous: developing an appropriate miniaturized wear test, developing cost-effective control techniques for the circularity of the wheels, optimizing the policy for remachining of the wheels, etc.

8.2.5.2 Pharmaceutical Compounds Synthesized by Mechanical Activation

Grinding and micronization are techniques commonly employed to reduce the size of particles of powder materials in the pharmaceutical, agrochemical, and dye industries. More recently, however, it has been realized that grinding can induce phase transformations in such compounds. Descamps and coworkers, in particular, have studied systematically the phase transformations induced by controlled high-energy ball milling in several organic compounds such as trehalose, lactose, D-sorbitol, and indomethacin (IMC) [209–212]. Using a planetary ball mill, these authors report that, regardless of the initial state, one and only one steady state is reached for a given milling temperature and rotation disc speed, Ω. Furthermore, they observe that lowering the milling temperature at constant Ω, or increasing Ω at constant temperature, favors the stabilization of an amorphous phase at steady state. Such effects can thus be rationalized, at least qualitatively, by invoking the effective temperature criterion (see Section 8.2.4). In the case of indomethacin, however, the equilibrium crystalline phase γ is first transformed to an amorphous phase, before crystallizing into a phase that is only metastable at equilibrium.

This sequence of transformations is reminiscent of results reported for some metallic [134] and ceramic materials. Within the effective temperature approach, the rationalization of this sequence and of the stabilization of a phase that has no stability domain in the equilibrium phase diagram would imply, at least, that one should define different effective temperatures for different structures, and that the effective temperature of the metastable crystalline phase would be much lower than that of the γ phase. On a phenomenological basis, since the same conditions of mechanical processing can produce different types or amounts of damage in different phases for the same overall chemical composition, and since the annealing kinetics of that damage can be quite different, it may not be too surprising that the dynamical stability of different compounds should be given by different effective temperatures. This approach, however, raises practical questions since it would imply that one can no longer predict phase stability during milling by simple inspection of the equilibrium phase diagram. More fundamentally, it would require that the various phases in competition share a parent phase which can serve as a common reference, here the amorphous phase, for comparison of the values of the effective temperatures to be meaningful. We note that the more recent analysis of alloys driven by plastic deformation indicates that the concept of effective interactions provides a sounder approach than that of the effective temperature. The difficulties noted above are no longer problematic if one uses this concept of effective interactions. From a practical viewpoint, however, this concept is somewhat difficult to apply since there is no simple formula for rescaling of the physical interactions which would describe the steady states stabilized by mechanical activation.

8.3
Outlook

Constructing a phenomenology of driven materials, i.e., an extended database and a theoretical framework to rationalize the latter, is of both academic and practical interest. The diversity and the complexity of the responses of materials to external forcing remain to be revealed; identifying the control parameters, finding the proper way to define the forcing intensity, and identifying the new characteristic lengths introduced by the external forcing, are fascinating challenges with direct practical implications (process optimization, prediction of time evolution of materials properties, alloy design, etc.). On the theoretical side, the development of models for driven materials often challenges the existing descriptions of the kinetics of the intrinsic thermal relaxation (i.e., that which occurs in the absence of forcing) of nonequilibrium states. It also forces one to understand better the characteristics of the external dynamics, and, in particular, the effect of space and time correlations in the latter. Since the physics of forcing is better understood at the atomic scale, we are facing the need to make better links between the macroscopic theories and the atomistic processes.

Much work remains to be done, which requires altogether, *systematic* accumulation of experimental data over a broad range of forcing conditions and materials, computer experiments on model systems, and the development of appropriate theoretical approaches.

Acknowledgments

P.B. acknowledges the support of the US National Science Foundation under grants DMR 03-04942 and DMR 04-07958, and the support of the DOE Basic Energy Sciences Office under grant DE-FG02-05ER46217.

References

1 C. K. Sudbrack, K. Y. Yoon, Z. Mao, R. D. Noebe, D. Isheim, D. N. Seidman, in *Electron Microscopy: Its Role in Materials Research – The Mike Meshii Symposium*, Ed. J. R. Weertman, M. Fine, K. Faber, D. Quesnel, W. King, P. Liaw, TMS (The Minerals, Metals & Materials Society), Warrendale, PA, 2003, pp. 43–50.

2 C. Pareige, F. Soisson, G. Martin, D. Blavette, *Acta Mater.* **47**, 1889 (1999).

3 A. Perini, G. Jaccuci, G. Martin, *Surf. Sci.* **144**, 53 (1984).

4 A. Perini, G. Jaccuci, G. Martin, *Phys. Rev. B* **29**, 2689 (1984).

5 J. Lepinoux, *Philos. Mag.* **86**, 5053 (2006).

6 E. Clouet, PhD Thesis (Report CEA-R-6062, ISSN 0429-3460, CEA–Saclay, 2004).

7 E. Clouet, A. Barbu, L. Laé, G. Martin, *Acta Mater.* **53**, 2313 (2005).

8 F. Christien, A. Barbu, *J. Nucl. Mater.* **324**, 90 (2004).

9 R. Cauvin, G. Martin, *Phys. Rev. B* **23**, 3322 and 3333 (1981); **25**, 3385 (1982).

10 R. Landauer, *J. Appl. Phys.* **33**, 2209 (1962).
11 A. R. Allnatt, A. B. Lidiard, *Atomic Transport in Solids*, Cambridge University Press, 1993, pp. 380–537.
12 M. Nastar, E. Clouet, *Phys. Chem. Chem. Phys.* **6**, 3611 (2004).
13 V. Barbe, M. Nastar, *Philos. Mag.* **86**, 1513 (2006).
14 V. Barbe, M. Nastar, *Philos. Mag.* **86**, 3503 (2006).
15 G. Martin, *Phys. Rev. B* **41**, 2279 (1990).
16 J. W. Cahn, in *Interfacial Segregation*, Eds. W. C. Johnson, J. M. Blakely, ASM International Materials Park, OH, 1979, p. 3.
17 J. W. Cahn, *J. Chem Phys.* **28**, 258 (1958).
18 A. Finel, private communication.
19 See T. Mohri, this book, Chap. 10.
20 V. Yu. Dobretsov, I. R. Pankratov, V. G. Vaks, *JETP Lett.* **80**, 602 (2005).
21 V. Y. Dobretsov, F. Soisson, G. Martin, V. G. Vaks, *Europhys. Lett.* **31**, 417 (1995).
22 V. Y. Dobretsov, V. G. Vaks, G. Martin, *Phys Rev. B* **54**, 3227 (1996).
23 Zugang Mao, C. K. Sudbrack, K. E. Yoon, G. Martin, D. N. Seidman, *Nature Mat.*, (2007) in press.
24 G. Martin, P. Bellon, *Solid State Physics* **50**, 189 (1997).
25 G. Nicolis, I. Progogine, *Self-organization in Nonequilibrium Systems: from Dissipative Structures to Order through Fluctuations*, Wiley, New York, 1977.
26 H. Haken, *Advanced Synergetics*, Springer, Berlin, 1983.
27 N. G. Van Kampen, *Stochastic Processes in Physics and Chemistry*, North-Holland, Amsterdam, 1981.
28 F. Schlögl, *Physics Reports*, **62**, 267 (1980).
29 M. C. Cross, P. C. Hohenberg, *Reviews of Modern Physics* **65**, 851 (1993).
30 P. Bergé, Y. Pomeau, C. Vidal, *Order within Chaos: Towards a Deterministic Approach to Turbulence* New York, Wiley, 1984.
31 R. Kubo, K. Matsuo, K. Kitahara, *J. Stat. Phys.* **9**, 51 (1973).
32 M. Suzuki, *Prog. Theoret. Phys.* **53**, 1657 (1975); *ibid.* **55**, 383 (1976); *ibid.* **55**, 1064 (1976).
33 See preface in P. Bellon, R. S. Averback, *Scripta Mater.* **49**, 921 (2003).
34 Y. Adda, M. Beyeler, G. Brebec, *Thin Solid Films* **25**, 107 (1975).
35 G. Martin, *Ann. Chim. Fr.* **6**, 46 (1981).
36 K. C. Russel, *Prog. Mater. Sci.* **18**, 229 (1984).
37 E. M. Schulson, *J. Nucl. Mater.* **83**, 239 (1979).
38 P. R. Okamoto, N. Q. Lam, L. E. Rehn, *Solid State Physics* **52**, 1 (1999).
39 R. S. Averback, T. D. de la Rubia, *Solid State Physics* **51**, 281 (1998).
40 N. V. Doan, H. Tietze, *Nucl. Instrum. Meth. B* **102**, 58 (1995).
41 R. S. Averback, R. Benedek, K. L. Merkle, *Phys. Rev. B* **18**, 4156 (1978).
42 D. J. Bacon, A. F. Calder, F. Gao, V. G. Kapinos, S. J. Wooding, *Nucl. Instrum. Meth. B* **102**, 37 (1995).
43 D. J. Bacon, Yu. N. Osetsky, R. Stoller, R. E. Voskoboinikov, *J. Nucl. Mater.* **323**, 152 (2003).
44 G. Martin, *Phys. Rev. B* **30**, 1424 (1984).
45 T. Diaz de la Rubia, A. Caro, M. Spaczer, *Phys. Rev. B* **47**, 11 483 (1993).
46 M. Spaczer, A. Caro, M. Victoria, T. Diaz de la Rubia, *Phys. Rev. B* **50**, 13 204 (1994).
47 K. Nordlund, R. S. Averback, *Phys. Rev. B* **59**, 20 (1999).
48 T. J. Colla, H. M. Urbassek, K. Nordlund, R. S. Averback, *Phys. Rev. B* **63**, 104206 (2000).
49 M. L. Jenkins, C. A. English, *J. Nucl. Mater.* **108.109**, 46 (1982).
50 R. Enrique, K. Nordlund, R. S. Averback, P. Bellon, *J. Appl. Phys.* **93**, 2917 (2003).
51 A.D. Brailsford, R. Bullough, *Philos. Trans. R. Soc. London, Ser. A* **302**, 87 (1981).
52 H. Rauh, M. H. Wood, R. Bullough, *Philos. Mag. A* **44**, 1255 (1981).
53 N. V. Doan, G. Martin, *Phys. Rev. B* **67**, 134 107 (2003).
54 J.-M. Roussel, P. Bellon, *Phys. Rev. B* **65**, 144 107 (2002).

55 Yu. N. Osetsky, D. J. Bacon, A. Serra, B. N. Singh, S. I. Golubov, *Philos. Mag.* **83**, 61 (2003).
56 M. Halbwachs, J. Hillairet, *Phys. Rev. B* **18**, 4927 (1978).
57 A. Benkaddour, C. Dimitrov, O. Dimitrov, *J. Nucl. Mater.* **217**, 118 (1994).
58 G. L. Eyink, J. L. Lebowitz, H. Spohn, *J. Stat. Phys.* **83**, 385 (1996).
59 B. Schmittmann, R. K. P. Zia, *Phase Transitions and Critical Phenomena*, Vol. 17, Academic Press, London, 1995.
60 Lebowitz, P. G. Bergmann, *Ann. Phys.* **1**, 1 (1957).
61 andauer, *Physics Today*, November, 23 (1978).
62 S. Nelson, J. A. Hudson, D. J. Mazey, *J. Nucl. Mater.* **44**, 318 (1972).
63 C. Wei, R. S. Averback, *J. Appl. Phys.* **81**, 613 (1997).
64 Banerjee, K. Urban, M. Wilkens, *Acta Metall.* **32**, 299 (1984).
65 Barbu, G. Martin, *Script. Metall.* **11**, 771 (1977).
66 H. Bernas, J.-Ph. Attane, K.-H. Heinig, D. Halley, D. Ravelosona, A. Marty, P. Auric, C. Chappert, Y. Samson, *Phys. Rev Lett.* **91**, 077 203 (2003).
67 K. Nordlund, M. Ghaly, R. S. Averback, *J. Appl. Phys.* **83**, 1238 (1998).
68 S. J. Kim, M. A. Nicolet, R. S. Averback, D. Peak, *Phys. Rev. B* **37**, 38 (1988).
69 F. Bourdeau, E. Camus, Ch. Abromeit, H. Wollenberger, *Phys. Rev. B* **50**, 16205 (1994).
70 E. Camus, Ch. Abromeit, F. Bourdeau, N. Wanderka, H. Wollenberger, *Phys. Rev. B* **54**, 3142 (1996).
71 G. Martin, F. Soisson, P. Bellon, *J. Nucl. Mater.* **205**, 301 (1993).
72 S. Matsumura, S. Mueller, Ch. Abromeit, *Phys. Rev. B* **54**, 6184 (1996).
73 R. A. Johnson, N. Q. Lam, *Phys. Rev. B* **13**, 4364 (1976).
74 P. R. Okamoto, H. Wiedersich, *J. Nucl. Mater.* **53**, 336 (1974).
75 W. Wagner, L. E. Rehn, H. Wiedersich, V. Naundorf, *Phys. Rev. B* **28**, 6780 (1983).
76 G. Martin, R. Cauvin, A. Barbu, in *Phase Transformations During Irradiation*, Ed. F. V. Nolfi, Applied Science, London, 1983), p. 47.
77 Y. Grandjean, P. Bellon, G. Martin, *Phys. Rev. B* **50**, 4228 (1994).
78 T. R. Allen, G. S. Was, *Acta Mater.* **46**, 3679 (1998).
79 T. R. Allen, E. A. Kenik, G. S. Was, *J. Nucl. Mater.* **278**, 149 (2000).
80 M. Nastar, P. Bellon, G. Martin, J. Ruste, *Mater. Res. Soc. Symp. Proc.* **481**, 383 (1997).
81 M. Nastar, *Philos. Mag.* **85**, 641 (2005).
82 A. Barbu, A. J. Ardell, *Scripta Metall.* **9**, 1233 (1975).
83 A. Barbu, G. Martin, *Scripta Metall.* **11**, 771 (1977).
84 L. W. Hobbs, F. W. Clinard, Jr., S. J. Zinkle, R. C. Ewing, *J. Nucl. Mater.* **216**, 291 (1994).
85 A. Motta, *J. Nucl. Mater.* **244**, 227 (1997).
86 C. Massobrio, V. Pontikis, G. Martin, *Phys. Rev. Lett.* **62**, 1142 (1989).
87 K. A. Jackson, *J. Mater. Res.* **3**, 1218 (1988).
88 J. S. Im, H. A. Atwater, *Nucl. Instrum. Methods Phys. Res. B* **59**, 760 (1991).
89 H. A. Atwater, J. S. Im, W. L. Brown, *Nucl. Instrum. Methods Phys. Res. B* **59**, 386 (1991).
90 N. Hecking, K. F. Heidemann, E. T. E. Kaat, *Nucl. Instrum. Methods Phys. Res. B* **51**, 760 (1986).
91 F. Gao, W. J. Weber, *Phys. Rev. B* **63**, 054 101 (2000).
92 J. Lian, X. T. Zu, K. V. G. Kutty, J. Chen, L. M. Wang, R. C. Ewing, *Phys. Rev. B* **66**, 054108 (2002).
93 J. Lian, J. Chen, L. M. Wang, R. C. Ewing, J. M. Farmer, L. A. Boatner, K. B. Helean, *Phys. Rev. B* **68**, 134107 (2003).
94 A. Chartier, C. Meis, J.-P. Crocombette, W. J. Weber, L. R. Corrales, *Phys. Rev. Lett.* **94**, 025505 (2005).
95 Y. Zhang, J. Lian, C. M. Wang, W. Jiang, R. C. Ewing, W. J. Weber, *Phys. Rev. B* **72**, 094112 (2005).
96 F. Banhart, *Phil. Trans. R. Soc. Lond. A* **362**, 2205 (2004).
97 M. Zaiser, F. Banhart, *Phys. Rev. Lett.* **79**, 3680 (1997).

98 M. Zaiser, Y. Lyutovich, F. Banhart, *Phys. Rev. B* **62**, 3058 (2000).
99 E. Chason, M. J. Aziz, *Scripta Mater.* **49**, 953 (2003).
100 D. Walgraef, E. C. Aifantis, *Res Mechanica* **23**, 161 (1988).
101 L. P. Kubin, in *Plastic Deformation and Fracture of Materials*, Ed. H. Mugrhabi, Volume 6 of *Materials Science and Technology: A Comprehensive Treatment*, Eds. R. W. Cahn, P. Haasen, E. J. Kramer, VCH Weinheim, 1993, pp. 137–190.
102 P. Hähner, *Scripta Mater.* **47**, 415 (2002).
103 M. Zaiser, *Mater. Sci. Eng. A* **309–310**, 304 (2001).
104 A. El-Azab, *Phys. Rev. B* **61**, 11 956 (2001).
105 F. J. Humphreys, M. Hatherly, *Recrystallization and Related Annealing Phenomena*, Elsevier, Oxford, 1995.
106 *Non-linear Phenomena in Materials Science*, Ed. L. Kubin, G. Martin, Trans Tech Pub., Aedermannsdorf, 1988; *Non-linear Phenomena in Materials Science II*, Ed. G. Martin, L. Kubin, Trans Tech Pub., Aedermannsdorf, 1992; *Non-linear Phenomena in Materials Science III*, Ed. G. Ananthakrishna, G. Martin, L. Kubin, Trans Tech Pub., Aedermannsdorf, 1995.
107 Fang Wu, P. Bellon, T. Lusby, A. Melmed, *Acta Mater.* **49**, 453 (2001).
108 S. Zghal, M. J. Hÿtch, J.-P. Chevalier, R. Twesten, F. Wu, P. Bellon, *Acta Mater.* **50**, 4695 (2002).
109 S. Zghal, R. Twesten, F. Wu, P. Bellon, *Acta Mater.* **50**, 4711 (2002).
110 F. Wu, D. Isheim, P. Bellon, D. N. Seidman, *Acta Mater.* **54**, 2605 (2006).
111 J. D. Hahn, Fang Wu, P. Bellon, *Metall. Mater. Trans.* **35A**, 1105–1111 (2004).
112 D. A. Rigney, L. H. Chen, M. G. S. Naylor, A. R. Rosenfield, *Wear* **100**, 195 (1984).
113 D. A. Rigney, *Wear* **245**, 1 (2000).
114 D. A. Rigney, X. Y. Fu, J. E. Hammerberg, B. L. Holian, M. L. Falk, *Scripta Mater.* **49**, 977 (2003).
115 S. B. Newcomb, M. W. Stobbs, *Mater. Sci. Eng.* **66**, 195 (1984).
116 Y. Le Bouar, L. Chaffron, G. Saint-Ayes, G. Martin, *Scripta Mater.* **49**, 985 (2003).
117 J. Friedel, *Dislocations*, Pergamon, New York, 1964.
118 H. Meckin, Y. Estrin, *Scripta Metall.* **14**, 815 (1980).
119 C. Suryanarayana, *Prog. Mater. Sci.* **46**, 1 (2001).
120 S. Odunuga, Y. Li, P. Kraschnotchekov, P. Bellon, R. S. Averback, *Phys. Rev. Lett.* **95**, 045901 (2005).
121 J. Xu, U. Herr, T. Klassen, R. S. Averback, *J. Appl. Phys.* **79**, 3935 (1996).
122 M. Angiolini, M. Krasnowski, G. Mazzone, A. Montone, M. Urchulutegui, M. Vittori-Antisari, *Mater. Sci. Forum* **195**, 13 (1995).
123 E. Ma, J.-H. He, P. J. Schilling, *Phys. Rev. B* **55**, 5542 (1997).
124 E. Gaffet, C. Louison, M. Harmelin, F. Faudot, *Mater. Sci. Eng.* **A134**, 1380 (1991).
125 F. R. N. Nabarro, *Theory of Crystal Dislocations*, Dover Publications, New York, 1987.
126 J. Marian, W. Cai, Vasily Bulatov, *Nature Mater.* **3**, 158 (2004).
127 J. Erlebacher, M. J. Aziz, A. Karma, N. Dimitrov, K. Sieradzki, *Nature* **410**, 450 (2001).
128 J. Erlebacher, K. Sieradzki, *Scripta Mater.* **49**, 991 (2003).
129 C. C. Koch, *Materials Science and Technology Vol. 15: Mechanical Milling and Alloying*, Ed. R. W. Cahn, P. Haasen, E. J. Kramer, VCH, Weinheim, 1991, p. 193.
130 G. Martin, E. Gaffet, *J. Phys. (Paris)* C-4 **15**, 71 (1990).
131 G. Cocco, F. Delogu, L. Schiffini, *J. Mater. Synth. Proc.* **8**, 167 (2000).
132 F. Delogu, G. Cocco, *J. Mater. Synth. Proc.* **8**, 271 (2000).
133 Y. Chen, M. Bibole, R. Le Hazif, G. Martin, *Phys. Rev. B* **48**, 14 (1993).
134 P. Pochet, E. Tominez, L. Chaffron, G. Martin, *Phys. Rev. B* **52**, 4006 (1995).
135 H. J. Frost, M. F. Ashby, *Deformation-Mechanism Maps*, Pergamon Press, Oxford, 1982.

136 S. C. Lim, M. F. Ashby, *Acta Metall.* **35**, 1 (1987).
137 M. F. Ashby, S. C. Lim, *Scripta Metall.* **24**, 805 (1990).
138 E. Ott, *Chaos in Dynamical Systems*, Cambridge University Press, Cambridge, UK, 1993.
139 M. Sherif El-Eskandarany, K. Aoki, K. Sumiyama, K. Suzuki, *Appl. Phys. Lett.* **70**, 1679 (1997).
140 M. Sherif El-Eskandarany, K. Aoki, K. Sumiyama, K. Suzuki, *Scripta Mater.* **36**, 1001 (1997).
141 M. Sherif El-Eskandarany, K. Aoki, K. Sumiyama, K. Suzuki, *Metall. Mater. Trans. A* **30**, 1877 (1999).
142 M. Sherif El-Eskandarany, K. Suzuki, *Metall. Mater. Trans. A* **33**, 135 (2002).
143 William C. Johnson, Jong K. Lee, G. J. Shiflet, *Acta Mater.* **54**, 5123 (2006).
144 H. Rauh, R. Bullough, M. H. Wood, *J. Nucl. Mater.* **114**, 334 (1983).
145 G. Martin et al. Note technique SRMP-00-12 (2000), p. 74. CEA, Saclay, France.
146 J.-L. Bocquet, G. Martin, *J. Nucl. Mater.* **83**, 186 (1979).
147 P. L. Garrido, J. Marro, *Phys. Rev. Lett.* **62**, 1929 (1989).
148 V. G. Vaks, V. V. Kamyshenko, *Phys. Lett. A* **177**, 269 (1993).
149 R. Enrique, P. Bellon, *Phys. Rev. B*, **60** 14649 (1999).
150 A. Voter, *Phys. Rev. Lett.* **78**, 3908 (1997).
151 M. D. Sorensen, A. F. Voter, *J. Chem. Phys.* **112**, 9599 (2000).
152 A. F. Voter, F. Montalenti, T. C. Germann, *Annu. Rev. Mater. Res.* **32**, 321 (2002).
153 G. Henkelman, H. Jónsson, *Phys. Rev. Lett.* **90**, 11 6101 (2003).
154 J. L. Bocquet, *Defect and Diffusion Forum* **203–205**, 81 (2002).
155 D. R. Mason, R. E. Rudd, A. P. Sutton, *J. Phys. Condens. Matter* **16**, S2679 (2004).
156 Y. Le Bouar, F. Soisson, *Phys. Rev. B* **65**, 094 103 (2002).
157 K. Sastry, D. D. Johnson, D. E. Goldberg, P. Bellon, *Phys. Rev. B* **72**, 085438 (2005).
158 A. Van der Ven, G. Ceder, *Phys. Rev. Lett.* **94**, 045 901 (2005).
159 J.-M. Lanore, *Rad. Effects* **22**, 153 (1974).
160 J. Dalla Torre, J.-L. Bocquet, N. V. Doan, E. Adam, A. Barbu, *Philos. Mag.* **85**, 549 (2005).
161 C.-C. Fu, J. Dalla Torre, F. Willaime, J.-L. Bocquet, A. Barbu, *Nature Mater.* **4**, 68 (2005).
162 M. J. Caturla, N. Soneda, E. Alonso, B. D. Wirth, T. Diaz de la Rubia, J. M. Perlado, *J. Nucl. Mater.* **276**, 13 (2000).
163 M. J. Caturla, N. Soneda, T. D. de la Rubia, M. Fluss, *J. Nucl. Mater.* **351**, 78 (2006).
164 R. Sizman, *J. Nucl. Mater.* **69&70**, 386 (1978).
165 H. Wiedersich, *J. Nucl. Mater.* **205**, 40 (1993).
166 P. Bellon, G. Martin, *Phys. Rev. B* **38**, 2570 (1988).
167 P. Bellon, G. Martin, *Phys. Rev. B* **39**, 2403 (1989).
168 R. Kubo, K. Matsuo, K. Kitahara, *J. Stat. Phys.* **9**, 51 (1973).
169 Kuo-An Wu, A. Karma, J. J. Hoyt, M. Asta, *Phys. Rev. B* **73**, 094101 (2006).
170 W. G. Wolfer, *J. Nucl. Mater.* **114**, 292 (1983).
171 H. R. Brager, F. Garner, in *Effects of Radiation on Materials: Twelfth International Symposium STP 870*; ASTM, Philadelphia, 1985, p. 139.
172 K. C. Russel, *Prog. Mater. Sci.* **18**, 229 (1984).
173 L. K. Mansur, *J. Nucl. Mater.* **216**, 97 (1994).
174 A. Barbu, G. Martin, *Solid State Phenomena* **30–31**, 179 (1993).
175 A. Barbu, private communication.
176 J. C. M. Li, *Metallic Glasses*, Am. Soc. Met., Metals Park, OH, 1978, p. 224.
177 B. Escaig, *Polym. Eng. Sci.* **24**, 737 (1984).
178 R. S. Nelson, J. A. Hudson, D. J. Mazey, *J. Nucl. Mater.* **44**, 318 (1972).
179 G. Schmitz, J. C. Ewert, F. Harbsmeier, M. Uhrmacher, F. Haider, *Phys. Rev. B* **63**, 224 113 (2001).
180 R. A. Enrique, PhD Thesis, University of Illinois at Urbana-Champaign, 2001.
181 F. Haider, Habilitation Thesis, University of Göttingen, 1995.

182 A. Pavlovitch, V. Yu. Dobretsov, Commissariat à l'Energie Atomique Report No. CEA-SRMP-97-85, 1997, p. 50.
183 R. A. Enrique, P. Bellon, *Phys. Rev. Lett.* **84**, 2885 (2000).
184 R. A. Enrique, P. Bellon, *Phys. Rev. B* **63**, 134 111 (2001).
185 G. Martin, in *Diffusion in Materials*, Ed. A. L. Laskar, Kluwer Academic Press, Dordrecht, 1990, p. 129.
186 R. A. Enrique, P. Bellon, *Phys. Rev. B* **70**, 224 106 (2004).
187 H. E. Cook, *Acta Metall.* **18**, 297 (1970).
188 H. M. McConnell, *Annu. Rev. Phys. Chem.* **42**, 171 (1991).
189 S. L. Keller, H. M. McConnell, *Phys. Rev. Lett.* **82**, 1602 (1999).
190 A. Stoycheva, S. J. Singer, *Phys. Rev. Lett.* **84**, 4657 (2000).
191 G. Malescio, G. Pellicane, *Nature Mater.* **2**, 97 (2003).
192 G. C. Rizza, M. Srtobel, K. H. Heinig, H. Bernas, *Nucl Instr. Meth. B* **178**, 78 (2001).
193 K. H. Heinig, T. Müller, B. Schmidt, M. Strobel, W. Möller, *Appl. Phys. A* **77**, 17 (2003).
194 L. C. Wei, R. S. Averback, *J. Appl. Phys.* **81**, 613 (1997).
195 R. Enrique, P. Bellon, *Appl. Phys. Lett.* **78**, 4178 (2001).
196 P. Krasnochtchekov, R. S. Averback, P. Bellon, *Phys. Rev. B* **72**, 174102 (2005).
197 See Wee Chee, R. S. Averback, submitted.
198 Jia Ye, P. Bellon, *Phys. Rev. B* **70**, 094 104 (2004).
199 Jia Ye, P. Bellon, *Phys. Rev. B* **70**, 094 105 (2004).
200 Jia Ye, P. Bellon, *Phys. Rev. B* **73**, 224 121 (2006).
201 E. F. Kneller, R. Hawig, *IEEE Transactions on Magnetics* **27**, 3588 (1991).
202 R. Skomski, J. M. D. Coey, *Phys. Rev. B* **48**, 15 812 (1993).
203 J. P. Liu, C. P. Luo, Y. Liu, D. J. Sellmyer, *Appl. Phys. Lett.* **72**, 483 (1998).
204 P. Bellon, R. S. Averback, *Phys. Rev. Lett.* **74**, 1819 (1995).
205 P. Pochet, PhD Thesis, University of Lille 1, France, 1997.
206 F. Wu, D. Isheim, P. Bellon, D. N. Seidman, *Acta Mater.* **54**, 2605 (2006).
207 B. I. Shraiman, E. D. Siggia, *Nature* **405**, 639 (2000).
208 K. L. Johnson, *Contact Mechanics*, Cambridge University Press, Cambridge, UK, 1985.
209 M. Descamps, J.-F. Willart, S. Desprez, in *Solid-to-Solid Phase Transformations in Inorganic Materials*, Ed. J. M. Howe, D. E. Laughlin, J. K. Lee, U. Dahmen, W. A. Soffa, TMS (The Minerals, Metals, and Materials Society), Warrendale, PA, 2005, p. 835.
210 J. F. Willart, A. De Gusseme, S. Hemon, G. Odou, F. Danède, M. Descamps, *Solid State Commun.* **119**, 501 (2001).
211 J. F. Willart, V. Caron, R. Lefort, F. Danède, D. Prévost, M. Descamps, *Solid State Commun.* **132**, 693 (2004).
212 J.-F. Willart, J. Lefebvre, F. Danède, S. Comini, P. Looten, M. Descamps, *Solid State Commun.* **135**, 519 (2005).

9
Change of Alloy Properties under Dimensional Restrictions

Hirotaro Mori and Jung-Goo Lee

9.1
Introduction

Small particles in the size range from a few to several nanometers have attracted considerable attention from both scientific and technological viewpoints in recent years. The reason is that these nanometer-sized particles (hereafter NPs) often exhibit structures and properties which are significantly different from those of the corresponding bulk materials and therefore have potential for use as advanced materials with new electronic, magnetic, optic, and thermal properties [1, 2]. These NPs consist of 10^2 to 10^4 atoms and the surface-to-volume ratio becomes remarkably large with decreasing size of the NP (Table 9.1). The surface atoms have some broken bonds and are in a high-energy state compared with the inner atoms.

The surface atoms therefore make significant contributions to the unique properties of NPs. This is called the "surface effect." There is another effect in relation to the finite size of particles. When the particle size becomes as small as the wavelength of electrons, the energy state of the electrons can be changed; this results in unique, measurable physical properties in NPs, and is known as the "quantum size effect" [3]. To get a fundamental understanding of these effects on the properties of NPs, intensive research has been carried out. It is well known, for example, that the stable structure of indium undergoes a change from body-centered tetragonal (bct) for bulk to face-centered cubic (fcc) when the size of the particles is reduced to several nanometers [4]. It is also well established that such phase transition temperatures as the melting temperature (T_m) are significantly reduced with decreasing size of particles [5–7]. All these examples are concerned with phase transitions in nanometer-sized *pure* substances, and in related experiments the temperature (T) and size (D) of particles were employed as experimental parameters. On the other hand, studies on alloy NPs examined as a function of T, D, and composition (X), are quite limited [8–11], although in recent years much attention has been paid to alloy NPs as candidates for new functional materials [12, 13]. This limitation of studies on alloy NPs is mainly due to the fact

Alloy Physics: A Comprehensive Reference. Edited by Wolfgang Pfeiler
Copyright © 2007 WILEY-VCH Verlag GmbH & Co. KGaA, Weinheim
ISBN: 978-3-527-31321-1

Table 9.1 The number of total atoms, the number of surface atoms, and the ratio of surface atoms in nanometer-sized particles.

The diameter of particle	The number of total atoms	The number of surface atoms	The ratio of surface atoms (%)
10 nm	30000	6000	20
5 nm	4000	1600	40
2 nm	250	200	80
1 nm	30	~29	~99

that it was rather difficult to control and measure the three parameters, T, D, and X, at the same time in a particle.

However, recent remarkable progress in transmission electron microscopy (TEM) enables us to study not only the structure, but also the chemical composition of an isolated nanometer-sized target material at a fixed temperature. With the use of this technique it now becomes possible to examine such alloy properties as the alloy phase formation in isolated NPs as a function of T, D, and X of the system. This situation opens a wide, unexplored research field on the structural stability of nanometer-sized condensed matter in two- (or multi-)component systems.

In this chapter, an overview on recent work on alloy phase formation in NPs will be presented, with emphasis placed on studies carried out by in-situ TEM in combination with thermodynamic calculations. Factors affecting the phase equilibrium of alloy NPs are described.

9.2
Instrumentation for in-situ Observation of Phase Transformation of Nanometer-Sized Alloy Particles

It is well recognized that TEM is one of the most powerful experimental tools for the characterization of NPs, especially for microstructural (or phase) analysis, and much work concerning NPs has been carried out using TEM. However, special attention has to be paid to keeping the surface of NPs as clean as possible during the experiments, because most properties of NPs are very sensitive to surface cleanliness.

For this reason, such in-situ experiments with TEM, in which both production and observation of NPs are possible in the same vacuum chamber without exposing particles to any undesired atmosphere, have been developed [14, 15]. Two types of evaporation systems have been designed. In one type, a double-source evaporator is installed directly inside the TEM, as shown in Fig. 9.1. It consists of two tungsten filaments (three or four are possible if required).

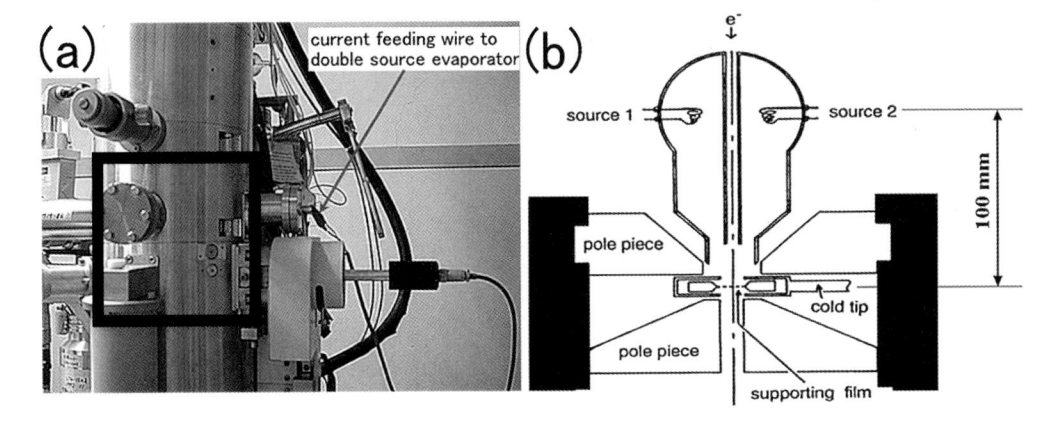

Fig. 9.1 Hitachi H-800 type 200 kV electron microscope as modified for analyzing nanoparticles [14]: (a) column; (b) schematic cross-sectional illustration of a double-source evaporator installed in the specimen chamber (in the square in (a)).

The distance between the filaments and a TEM sample holder is approximately 100 mm. An amorphous carbon film was mainly used as a supporting film, mounted on a TEM grid. Using this evaporator, one of the source materials was first evaporated from one filament onto the supporting film kept at a certain temperature, and NPs of the material were produced on the film. Next, the other source material was evaporated from the second filament onto the same film kept at another fixed temperature. The filaments were heated by the Joule effect. Alloy phase formation in the NPs associated with the deposition of the second source material was studied by taking bright-field images (BFIs) and dark-field images (DFIs) and selected area electron diffraction (SAED) patterns. For quantitative analyses of the intensity of Debye–Scherrer rings, imaging plates (IPs) were employed instead of conventional films. A heating or a cooling holder was used to control the temperature of the amorphous carbon film. The TEM was equipped with a turbo-molecular pumping system to achieve a base pressure of around 3×10^5 Pa in the specimen chamber.

In the other type, a side-entry TEM sample holder equipped with a miniature double-source evaporator has been developed; actually, it is an improved version of the hot holder designed by Kamino and Saka [16], and is shown schematically in Fig. 9.2: it consists essentially of three spiral-shaped tungsten filaments. The middle filament is attached with a flake of graphite used as the supporting substrate, and the outer two filaments with source materials. Before experiments, the graphite flake was baked at 1070 K for 60 s by feeding an appropriate amount of current into the middle filament. After being baked, the graphite substrate was cooled to a certain temperature. Subsequently, one source material was evaporated from the second filament onto the graphite substrate, where pure NPs were produced. The other source material was then evaporated from the third fil-

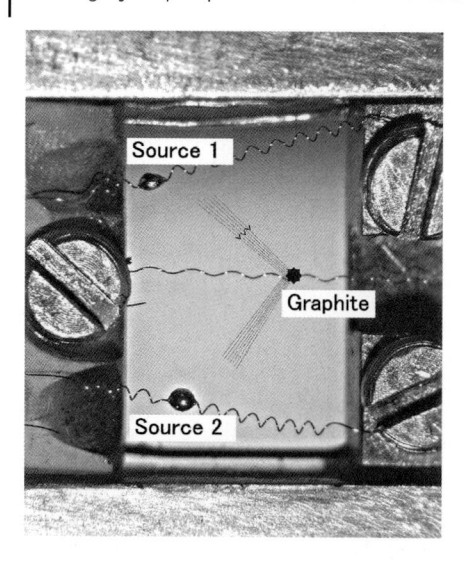

Fig. 9.2 The tip of a side-entry TEM holder equipped with a miniature double-source evaporator (see the text for details).

ament onto the pure NPs. Microstructural changes in an isolated particle associated with alloying of solute atoms were studied in situ in the microscope, using a supersensitive television camera (Gatan 622SC) with a time resolution of 30 frames s^{-1}. Images were recorded on videotape through a video recorder system connected to the television camera. Some photographs were reproduced from the videotapes through an image processor (Image Sigma) operating in an accumulation mode (three frames were accumulated for one photograph). The base pressure in the microscope was around 5×10^7 Pa. The chemical composition of alloy NPs was examined with energy-dispersive X-ray spectroscopy (EDS), if necessary.

9.3
Depression of the Eutectic Temperature and its Relevant Phenomena

9.3.1
Atomic Diffusivity in Nanometer-Sized Particles

The atomic diffusivity in NPs is a subject of fundamental importance, since it plays a key role in determining the kinetics of all diffusional phase transformations in NPs. Experimental studies on self-diffusion and on interdiffusion in NPs in the literature will be described briefly.

Studies on the self-diffusion in NPs of pure substances are rather limited. Dick et al. investigated in an indirect manner the size-dependent self-diffusion coefficient of gold in gold NPs [17]. First, the melting point of gold NPs was determined as a function of particle size by a combination of differential thermal analysis and thermal gravimetric analysis techniques. Then, based upon the size-dependent melting point data, the size-dependent self-diffusion coefficient was calculated under the following assumptions: (a) the diffusion coefficient exhibits a Boltzmann–Arrhenius dependence on temperature; (b) at T_m, the diffusion coefficient is the same for all gold particles, independently of the size; and (3) the activation enthalpy of diffusion is proportional to the average number of bonds that need to be broken in the melting process. It has been shown that the self-diffusion coefficient obtained increases significantly with decreasing size of particles [17]. For example, the coefficient was reported to be 10^{-28} m^2 s^{-1} for a 2 nm-sized particle at room temperature (RT) [17], which is far higher than the bulk diffusion coefficient at RT, i.e., around 3.6×10^{-36} m^2 s^{-1} [18]. Discussions on the assumptions employed will be interesting. Unfortunately, direct experimental evidence for an enhanced self-diffusion in NPs of pure substances is lacking until now.

On the other hand, studies on the interdiffusion (or alloying) in NPs are not rare. For example, Mori et al. have examined the size-dependent alloying of copper into gold NPs by electron microscopy [19, 20], and the following observations were made:

- when vapor-deposited copper atoms came in contact with gold NPs of diameter approximately 4 nm, supported on an amorphous carbon film, they quickly (in less than 20 s) dissolved into gold NPs to form a Au–Cu solid solution even at RT
- in gold NPs of diameter approximately 10 nm, alloying of copper took place only in a shell-shaped surface region of individual NPs – no alloying occurred at the central core region of NPs at RT
- in gold particles of approximately 30 nm diameter such alloying of copper atoms was not detected at RT.

It has been confirmed that the rapid alloying in approximately 4 nm-sized particles took place not via a melting process, but via a solid-state process [20]. It has also been confirmed that such rapid alloying is not an artifact originating from a temperature rise due to electron beam heating, heat of condensation, heat of mixing, or impingement of flying solute atoms with a kinetic energy of the order of kT, but is an intrinsic property of NPs [21]. These observations clearly indicate that the interdiffusion (or alloying) becomes markedly enhanced with decreasing size of particles.

A simple relationship $x = (Dt)^{1/2}$, can be used as a rough estimate for the diffusion coefficient D, where t is the time needed to achieve appreciable diffusion of solute atoms over a distance x [22]. With $x = 2.0$ nm (half the diameter of 4 nmsized gold NPs) and $t = 20$ s, a value of 8×10^{-19} m^2 s^{-1} is obtained for D. This value gives an estimate for the lower limit of the diffusivity of copper in 4 nm-sized gold NPs (or of the copper atom mixing rate with gold), since the alloying took place in less than 20 s. On the other hand, the diffusion coefficient of copper in bulk gold, D_{Cu-Au}, can be written (in m² s⁻¹) as $D_{Cu-Au} = (1.2 \times 10^{-7}) \exp(-14\,300/T)$, where T is the temperature [22]. Extrapolation of this relationship to room temperature gives a value of 2.4×10^{-28} m² s⁻¹ for D_{Cu-Au} at 300 K. These calculations suggest that the observed copper diffusivity in 4 nm-sized gold NPs is many orders of magnitude faster than that in bulk gold. It has been confirmed experimentally that the enhanced alloying takes place not only in the solid-solution-forming alloy systems (e.g., Au–Cu) but also in the compound-forming alloy systems [21].

The rapid alloying (or interdiffusion) mentioned above is concerned with the atom mixing in NPs supported on a solid substrate, and is investigated by electron microscopy. It is of interest that enhanced atom mixing has also been observed in core–shell-type Au–Ag NPs in a completely different experiment. Shibata et al. have studied size-dependent alloy formation of silver-coated gold NPs in aqueous solution at ambient temperature using X-ray absorption fine structure spectroscopy (XAFS) [23]. Various Au core sizes and Ag shell thicknesses were prepared using radiolytic wet techniques. The authors observed remarkable size dependence in the room-temperature interdiffusion of the two metals: for the very small particles (≤4.6 nm initial Au-core size), evidence for the interdiffusion between the Au core and the Ag shell was obtained, whereas for the larger particles the core/shell boundary was maintained to within one monolayer, indicating the absence of interdiffusion. It was proposed that defects at the boundary may play an important role in the enhancement of the atom migration (or atom mixing) in small NPs [23].

Since similar enhanced atom mixing has been obtained in completely different kinds of experiments, it seems safe to consider that size-dependent alloying of two metals does take place rather generally in NPs and a marked enhancement of intermixing (for example, room-temperature alloying of Cu into Au) has been achieved in NPs. The mechanism of the enhanced atom intermixing remains a subject for future studies.

9.3.2
Eutectic Temperature in Nanometer-Sized Alloy Particles

It is well established that the T_m of pure substances is reduced with decreasing size of particles. Similarly, the eutectic temperature (T_{eu}) decreases with decreasing size of alloy particles. It should be noted that in the eutectic temperature suppression, not only the surface energy but also the interfacial energy plays an essential role. In the following, the extent of the eutectic temperature suppression will be described, together with the phase equilibrium induced by such deep suppression of T_{eu}, which is characteristic to alloy NPs in the eutectic system.

Figure 9.3 shows a typical sequence of the alloying process of indium atoms into an approximately 5.2 nm-sized tin particle at RT [15]. These pictures were taken from a video recording. The three numbers inserted in each micrograph indicate the time in units of minutes, seconds, and one-sixtieth seconds.

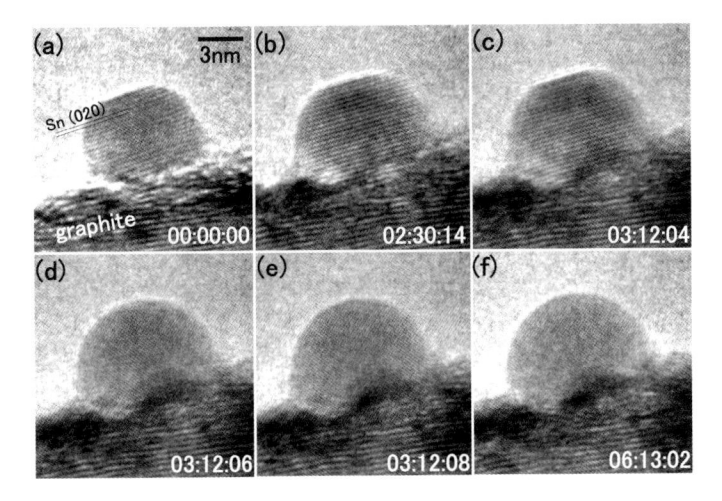

Fig. 9.3 In-situ observation of the alloying process of indium into an approximately 5.2 nm-sized tin particle. A crystalline-to-liquid (C–L) transition took place during deposition of indium onto the tin particle kept at room temperature (compare (c) with (e)). The time is indicated on each micrograph in units of minutes, seconds, and sixtieths of seconds.

Figure 9.3(a) shows an as-produced, pure tin particle on a graphite substrate (bottom). The 0.29_1 nm-spaced fringes seen in this particle are the (020) lattice fringes of β-Sn. Figures 9.3(b)–(e) show the same particle during indium-atom deposition. In spite of the indium deposition, the particle remains a single crystal, as shown in Figs. 9.3(b) and (c), which indicates the formation of a solid solution in the particle. The diameter of the particle has increased from approximately 5.2 nm (Fig. 9.3a) to 6.7 nm (Fig. 9.3c) during indium deposition. From this size increment, the indium concentration in the solid solution can be estimated at approximately 53 at.% In. With continued deposition of indium, the particle underwent a crystalline-to-liquid (C–L) phase transition, as shown in Figs. 9.3(c) and (e): that is, all the lattice fringes in the particle disappeared abruptly and only a uniform contrast typical of the liquid state appeared (Fig. 9.3e). The time interval between Figs. 9.3(c) and (e) was $\frac{1}{15}$ s, indicating that the C–L phase transition took place very quickly. Figure 9.3(f) shows the same particle (liquid droplet) after the power of the tungsten filament attached to the indium had been turned off. It is evident from Fig. 9.3(f) that the C–L phase transition in Fig. 9.3(c) to (e) did not result from the temperature increase induced by radiation from the neighboring filament attached with indium during heating for indium deposition onto the tin NP. In view of the fact that the T_{eu} of bulk alloys is 393 K, it seems safe to conclude that the melting temperature of In–Sn alloy NPs with compositions close to the eutectic one is below RT. In other words, a depression of T_{eu} by as much as 100 K took place in association with the particle-size reduction.

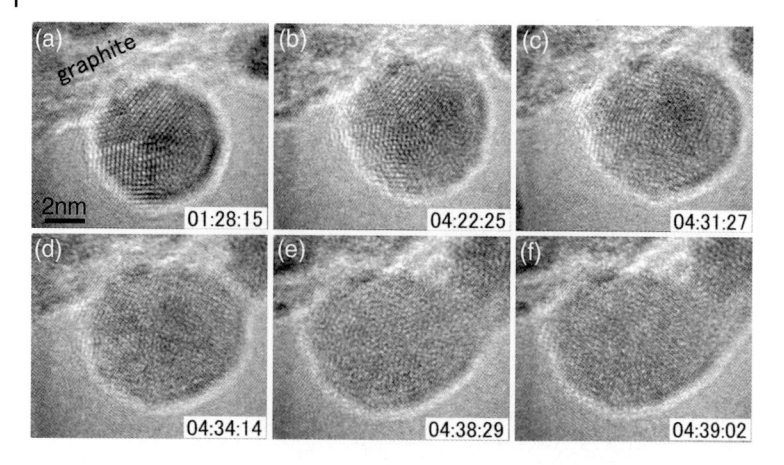

Fig. 9.4 A sequence from a video recording of the alloying process of tin atoms into a nanometer-sized gold particle. A crystalline-to-amorphous (C–A) transition took place during deposition of tin onto the gold particle. The time is indicated as in Fig. 9.3.

Figure 9.4 shows a typical sequence of an alloying process of tin atoms into an approximately 6 nm-sized gold particle at RT. In Fig. 9.4(a) the as-produced pure gold particle lies on a graphite substrate (top) [24]. Figures 9.4(b)–(f) show the same particle under tin-atom deposition. In Figs. 9.4(b)–(d), the (111) lattice fringes of the gold particle become more and more distorted and segmented with an increasing amount of tin, and eventually the salt-and-pepper contrast appears as a characteristic reminiscent of an amorphous structure (Figs. 9.4e,f).

The observations shown in Fig. 9.4 indicate that tin atoms vapor-deposited onto a gold NP rapidly dissolve into the gold particle, which is consistent with the results on In–Sn in Fig. 9.3. A difference is that the alloyed gold particle transforms into an amorphous phase.

In an attempt to study the thermal stability of the amorphous phase such as that shown in Figs. 9.4(e) and (f), additional experiments on annealing of Au–36 at.% Sn alloy particles were carried out. Figure 9.5 shows a sequence of structural changes in an Au–36 at.% Sn alloy particle during heating and subsequent cooling [25]. In Fig. 9.5(a) the as-produced amorphous alloy particle is sitting on a graphite substrate (bottom). (As shown in Fig. 9.4, Au–Sn alloy NPs formed at RT are in an amorphous state, when the composition is in a range of 30–40 at.% Sn [24, 25].) With increasing temperature, the salt–pepper contrast, which is characteristic of the amorphous phase, starts to fluctuate; that is, at about 473 K individual black and bright spots in the salt–pepper contrast begin to move to and fro slowly in the particle, suggesting that atoms in the particle have started a long-range migration in a noncrystalline state (Figs. 9.5b,c). With continued heating, the particle melted, and in the molten state only a uniform contrast typical of the liquid state appeared (Fig. 9.5d). It should be noted here

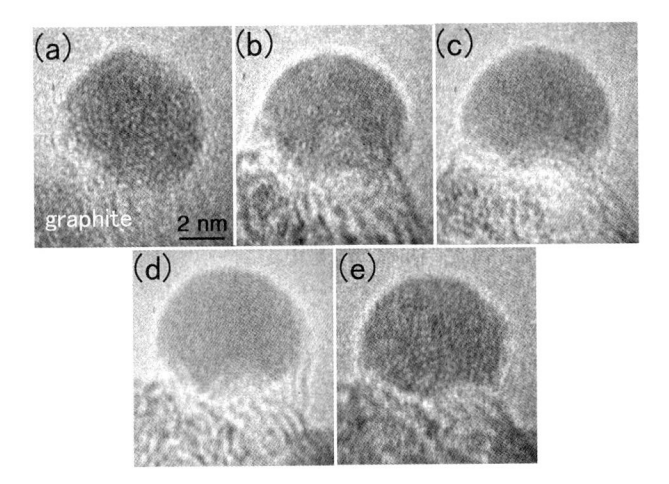

Fig. 9.5 A typical annealing process sequence in an approximately 5 nm-sized Au–36 at.% Sn alloy particle.

that before melting no traces of crystallization were observed. Upon cooling to RT, the liquid particle solidified again into the amorphous phase (Fig. 9.5e). From the results in Fig. 9.5, it is evident that the amorphous phase which appears over a composition range near the eutectic composition possesses an extremely high thermal stability so that upon heating it melts without crystallization and the melt solidifies directly into an amorphous solid upon cooling.

As shown above, when the particle diameter was approximately 6 nm, a liquid phase and a thermodynamically stable solid amorphous phase were formed at RT in particles in the In–Sn and the Au–Sn system, respectively. We speak here of a thermodynamically stable amorphous phase, because the above result in the Au–Sn system suggests that the Gibbs free energy of the amorphous phase is lower than that of a crystalline counterpart, at least at temperatures near and above RT. This situation is a consequence of the large depression of T_{eu} across the glass transition temperature (T_g) induced by the size reduction, which to the best of our knowledge never occurs in bulk materials.

It is known that there is an empirical rule that T_g is located at a temperature near two-thirds of the melting temperature of a bulk material [26, 27]. According to this rule, T_g is expected to be located at a temperature higher than RT in the Au–Sn system, and lower than RT in the In–Sn system, respectively. T_g is considered to be directly related to the cohesive energy of the material [28], and is thought to be less sensitive to the size of particles in the size region around 10 nm. In the case when the three temperatures, T_{eu}, T_g, and RT (where the observation is carried out), lie in such an order as $T_g > RT > T_{eu}$ in NPs in the Au–Sn system, it is postulated that a crystalline-to-amorphous (C–A) phase change is induced by simply adding solute atoms (i.e., Sn atoms) to crystalline NPs of pure gold. This postulate is in agreement with what is shown above in Fig. 9.4. In the case, however, when the two temperatures T_g and RT lie in such an order as RT > T_g in NPs in the In–Sn system, it is postulated that a C–L phase transition would be induced instead by simply adding solute atoms (i.e., In atoms) to crystalline NPs of pure tin. Again, this postulate is in agreement with what is shown above in Fig. 9.3. It should be emphasized here that neither the amorphous phase in the former system nor the liquid phase in the latter system is the equilibrium phase at RT in bulk materials. These phases can be present at RT as the equilibrium phase only when the size of the system is in the nanometer range. One of the reasons for the enhanced suppression of T_{eu} with size reduction may come from the fact that the relative contribution of interfacial energy associated with a distinct hetero-interface to the total Gibbs free energy of a crystalline two-phase particle increases with decreasing size of the particle, and therefore such an interface destabilizes a crystalline, two-phase structure relative to a noncrystalline, single-phase structure. In this manner, a noncrystalline phase becomes more stable than a mixture of crystalline phases.

9.3.3
Structural Instability

It is established that NPs often exhibit a structural instability under observation in a TEM; three different explanations have been proposed. The first was suggested by Iijima and Ichihashi [29, 30], who proposed that a particle and a local area of the insulating substrate might deviate temporarily from electrical neutrality under electron-beam irradiation and the coulombic forces would then induce structural fluctuations [30]. In accordance with this suggestion, it has been reported that the charging effect becomes weak when an electrically conductive substrate such as amorphous carbon is used [29]. In the second explanation, Ajayan and Marks [31, 32] proposed that the free energy barrier between different atomic configurations in a NP might be so small that the thermal activation is large enough even at RT to allow random fluctuations to occur among different configurations; this is called the quasi-molten state [32]. It has been confirmed experimentally that once initiated, the particle remained in a quasi-molten state even after the electron-beam intensity is reduced to almost zero [32]. The quasi-melting model will be particularly suited to explaining such an observation. However, it is emphasized that for a stable particle on the substrate to transform into the quasi-molten state, decoupling of the particle from the substrate and reduction of the contact area between the particle and the substrate are required [32]. The third explanation was suggested by Mitome et al. [33]. Structural fluctuations in approximately 3 nm-sized gold particles occurred only at temperatures above about 773 K and the temperature became lower the smaller the particle was [33]. They also reported that under electron irradiation at an intensity of 10–15 A cm^{-2} the particle temperature increased by about 20 K at most [33]. Based upon these observations, they suggested that the fluctuations are caused by a thermal effect [33].

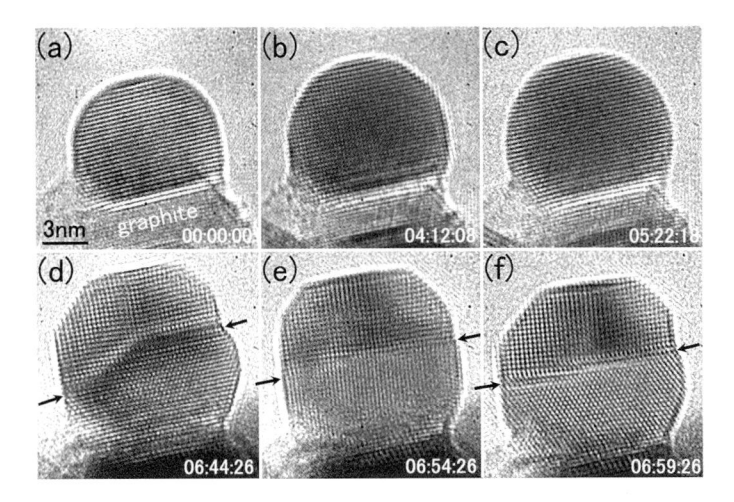

Fig. 9.6 A typical alloying process sequence of bismuth atoms into a nanometer-sized tin particle, taken from a video recording: (a) as-produced pure tin particle on a graphite substrate; (b), (c) the same particle after bismuth deposition; (d) the same particle after additional bismuth deposition, with a two-phase microstructure; (e), (f) the same particle after turning off the power of the tungsten filament attached with the bismuth. The arrows in (d)–(f) indicate the position of a Bi/Sn hetero-interface, and the time is indicated as in Fig. 9.3.

Figure 9.6 presents a typical sequence of alloying process of bismuth atoms into a tin NP [34]. Figure 9.6(a) shows an as-produced pure tin particle on a graphite substrate (bottom). The pure tin particle was a single crystal with a diameter of approximately 8 nm, in which structural fluctuations were completely absent. Figures 9.6(b) and (c) show the same particle after bismuth deposition. The particle size increased from ca. 8 to ca. 9.5 nm by bismuth deposition but the particle remained a single crystal, which indicates that a solid solution was formed in it. The bismuth concentration in the solid solution, as estimated from the size increment, was approximately 40 at.% Bi. In this solid solution particle, fluctuations in the particle shape took place at a rate of approximately one per 1–3 s. The change in shape was always accompanied with the change in orientation. Therefore, the fluctuations observed here were orientational fluctuations of the crystallite, as seen from a comparison of Figs. 9.6(b) and (c). It should be noted that the fluctuations became more frequent with increasing bismuth concentration in the solid solution. With continued bismuth deposition, the structure of the particle changed from a single crystalline solid solution to a mixture of two solid phases, as shown in Fig. 9.6(d). The particle in Fig. 9.6(d) was approximately 12 nm in diameter. The bismuth concentration of this particle evaluated from the size increment was approximately 70 at.% Bi. It was also confirmed by an analysis of the Fourier transform pattern taken from each area in Fig. 9.6(d) that the particle was composed of a bismuth grain and a tin grain [34]. (Arrows in Figs. 9.6(d)–(f) indicate the position of a Bi/Sn hetero-interface.) This is the same phase relationship as in a bulk alloy with the corresponding composition at RT [35]. It can therefore be concluded that the phase relationship in an alloy particle of approximate diameter 12 nm is essentially the same as that in a bulk alloy in the Bi–Sn system. Figures 9.6(e) and (f) show the particle as in Fig. 9.6(d) after the power of the tungsten filament attached with bismuth had been truned off. For the 12 nm-sized alloy particle with a two-phase microstructure, the shape and position of the Bi–Sn hetero-interface changed continually with high frequency (i.e., a few times per second), in contrast to the solid solution particle shown in Figs. 9.6(a)–(c). It should be noted here that the evaporation of bismuth from the source filament had been stopped right before the micrograph in Fig. 9.6(d) was taken. The structural fluctuations observed here are therefore not induced by the arrival of bismuth atoms on the particle.

It is evident from the observations in Fig. 9.6 that even at the fixed temperature where the experiments were carried out (i.e., RT), the fluctuations became more frequent with increasing bismuth concentration in the alloy particle. It is of interest here that the difference in Gibbs free energy between a 12 nm-sized solid particle and the corresponding liquid particle of a tin-rich Bi–Sn alloy at RT decreases continuously from a high value of 1620 J mol^{-1} for pure tin to almost zero for an alloy with a near-eutectic composition [36]. The reason why the Gibbs free energy of a 12 nm-sized solid particle with a two-phase microstructure becomes almost equal to that of the corresponding liquid particle, even at RT, can be found in the large contribution of the Bi–Sn interfacial energy to the total Gibbs free energy of the solid particle [36]. As a result, T_{eu} of a 12 nm-sized alloy particle can be lowered to a temperature slightly above RT [36]. In fact, it has been confirmed that T_{eu} is below RT when the size of a particle is further reduced to approximately 10 nm or below [37]. All these results suggest the view that a solid NP, no matter whether it consists of a single phase or of multiple phases, undergoes structural fluctuations, when the Gibbs free energy difference between the solid particle and the corresponding liquid particle is reduced to a value close to zero.

According to an estimate based upon Thermo-Calc [38], in a 12 nm-sized particle with a near-eutectic composition in the Bi–Sn system, the Gibbs free energy difference between a liquid particle and a solid particle (with a two-phase microstructure) changes with temperature at a rate of approximately 100 J mol^{-1} per 5 K. This implies that if the particle is warmed up, for example, by 5 K under electron-beam irradiation, then the relative stability of the particle in the solid state in comparison with the liquid state deteriorates by 100 J mol^{-1} when measured in terms of the free energy. This value of 100 J mol^{-1} is quite small compared to the free energy difference in a 12 nm-sized particle of pure tin, i.e., 1620 J mol^{-1} as mentioned before. It is postulated that even if a temperature rise of 5 K is induced by electron irradiation, it would not affect the phase stability in a 12 nm-sized pure tin particle and therefore it would not bring about any structural fluctuations in the solid particle. On the other hand, the value of 100 J mol^{-1} is not negligibly small compared to the free energy differences between a solid and a liquid particle with a near–eutectic composition [36]. It is therefore postulated that if a temperature rise of, for example, 5 K is induced by electron irradiation, it would produce an appreciable alteration in the relative phase stability and would result in the occurrence of structural fluctuations in the solid particle. This postulate is in agreement with the observations shown in Fig. 9.6. These results provide evidence for the view that a solid NP undergoes structural fluctuations when the Gibbs free energy difference between the solid NP and the corresponding liquid NP becomes almost zero. This view is consistent with the suggestion by Mitome et al., who observed the onset of structural fluctuations in an approximately 3 nm-sized gold particle only when the substrate temperature was raised to about 773 K [33]. According to the experimental data of Buffat and Borel [6], the melting temperature of such a gold particle is around 773 K. This fact therefore indicates that the Gibbs free energy difference between the solid and the liquid particle of that dimension does take a value close to zero when the substrate temperature is increased to about 773 K, where the structural fluctuations set in. On the other hand, in the experiment shown in Fig. 9.6, a flake of graphite, a good electric conductor, was used as a substrate. In fact, in no case was the detachment of a particle from the substrate encountered during observation, a typical event which is expected for the charging effect; the contact area between the particle and the substrate remained almost the same, as shown in Fig. 9.6. It seems safe, then, to consider that the structural fluctuations shown in Fig. 9.6 were not caused by a charging effect. Therefore, the structural fluctuations shown in Fig. 9.6 are not compatible with the quasi-melting model.

In conclusion, it is reasonable to assume that the structural fluctuations in an NP take place when the phase stability of a solid particle under observation becomes almost equal to that of the particle in the liquid state.

9.3.4
Thermodynamic Discussion

9.3.4.1 Gibbs Free Energy in Nanometer-Sized Alloy Systems
In order to specify the phase equilibrium of NPs, not only the Gibbs free energy in the bulk (i.e., the volume free energy), but also the surface free energy has to be taken into account due to the extremely large surface-to-volume ratio. Therefore the total Gibbs free energy of a nanometer-sized alloy system can be described by Eq. (1) [39].

$$G^{total} = G^{bulk} + G^{surface} \qquad (1)$$

Calculations of G^{total} for a simple eutectic system (which consists of a liquid phase and two pure solid phases) can be done in the following way.

The Gibbs free energy of a bulk A–B binary alloy, G^{bulk} in Eq. (1), which corresponds to G^{total} in bulk materials, is expressed by Eqs. (2a) and (2b), where N_A and N_B are the mole fractions of components A and B.

Table 9.2 Temperature dependence of the surface tension of pure liquids [41].

Element	Surface tension of pure liquid [N m^{-1}]	$\frac{\partial \gamma_X^L}{\partial T}$ [N m^{-1} T^{-1}] clarify
Bi	0.378–0.00007 $(T - 1336.15)$	−0.00007
Sn	0.560–0.00009 $(T - 505)$	−0.00009

$$G^{bulk, L} = N_A G_A^{0, L} + N_B G_B^{0, L} + RT(N_A \ln N_A + N_B \ln N_B)$$
$$+ G^{excess} \quad \text{(for liquid)} \tag{2a}$$

$$G^{bulk, S} = N_A G_A^{0, S} + N_B G_B^{0, S} \quad \text{(for solid)} \tag{2b}$$

$G_A^{0, S}$ and $G_B^{0, S}$ (or $G_A^{0, L}$ and $G_A^{0, L}$) are the Gibbs free energies of pure A and B relative to their reference phases (i.e., solid phases) at 298.15 K. G^{excess} is the excess Gibbs free energy of mixing, which represents the interaction between the two components. G^{excess} can be written as Eq. (3), with the interaction parameters L_i $(i = 0, 1, 2, \ldots)$ [38].

$$G^{excess} = N_A N_B \{L_0 + L_1(N_A - N_B) + L_2(N_A - N_B)^2 \cdots\} \tag{3}$$

It is known that the surface energy depends upon composition [40]. Conventionally, a linear relationship is assumed for a liquid phase, for simplicity. Then, the second term of Eq. (1) can be described by Eqs. (4) and (5), where γ_X^S and γ_X^L $(X = A$ or $B)$ are the surface energy of pure solid and liquid particles, r is the radius of a particle, and V_A^S and V_A^S are the molar volumes of pure solid A and B. As an example, the temperature dependence of γ^L for Bi and Sn is listed in Table 9.2 [41].

$$G^{surface, S} = \frac{2}{r}(N_A \gamma_A^S V_A^S + N_B \gamma_B^S V_B^S) \tag{4}$$

$$G^{surface, L} = \frac{2V^L}{r}(N_A \gamma_A^L + N_B \gamma_B^L) \tag{5}$$

The value of γ_X^S of pure solid at the melting temperature is known to be approximately 25% greater than the surface tension of the pure liquid [39]. This way the surface energy γ_X^S of a pure solid X as a function of temperature T can be expressed as Eq. (6), where $\gamma_{X, mp}^L$ is the surface energy of the pure liquid X at its melting temperature $T_{X, mp}$.

$$\gamma_X^S = 1.25 \gamma_{X, mp}^L + (T - T_{X, mp}) \frac{\partial \gamma_X^L}{\partial T} \tag{6}$$

The temperature dependence of γ_X^S in Eq. (6) is assumed to be the same as that of γ_X^L. Finally, it is also assumed that a surface with a free energy γ (in J m^{-2}) exerts a surface tension of γ (in N m^{-1}) in both solid and liquid.

9.3.4.2 Result of Calculations

With the use of Eqs. (1–6), it is possible to construct a Gibbs free energy–temperature–composition (G–T–X) diagram for an alloy NP. An example of such a G–T–X diagram is depicted in Fig. 9.7, which is constructed based upon the result of calculations for alloys in the Bi–Sn system. The solid and dotted lines indicate the Gibbs free energies of the bulk material and a 10 nm-sized particle, respectively. The relatively large increase in the Gibbs free energy in a solid NP compared to that in a liquid NP leads to the melting temperature depression ΔT of 51 K from 544 K to 493 K in pure bismuth and of 112 K from 505 K to 393 K in pure tin particles, respectively.

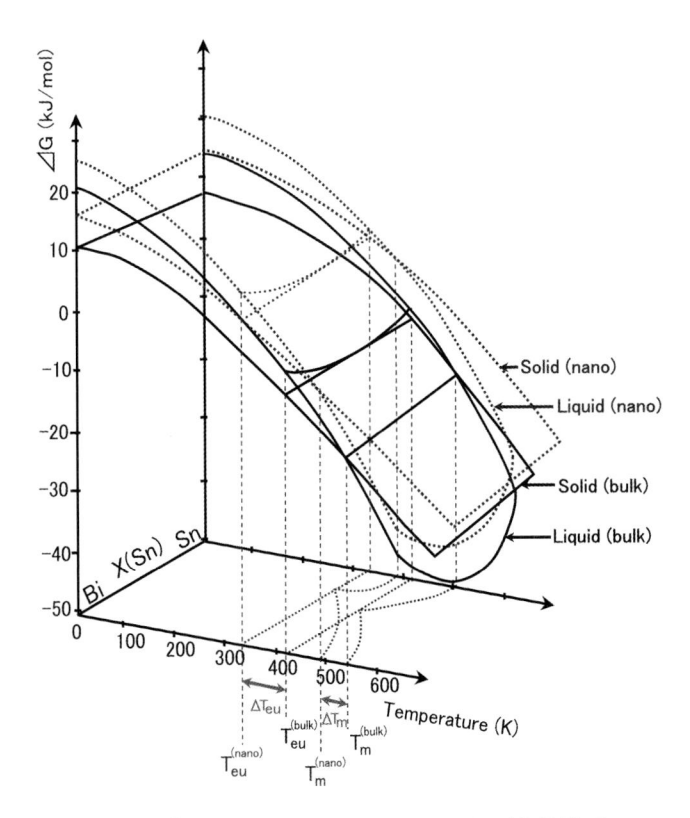

Fig. 9.7 Gibbs free energy–temperature–composition (G–T–X) diagram constructed from the result of calculations. The solid and dotted lines indicate the Gibbs free energy of bulk material and 10 nm-sized particles, respectively.

Figure 9.8 shows the Gibbs free energy–composition (G–X) diagrams of a 10 nm-sized Bi–Sn particle at several different temperatures; clearly, according to the calculation, T_m of 10 nm-sized particles of pure bismuth and of pure tin are 493 K (Fig. 9.8b) and 393 K (Fig. 9.8c), respectively. The melting temperature depressions ΔT obtained here, i.e., $\Delta T = 51$ K for bismuth, and $\Delta T = 112$ K for tin, are somewhat smaller than those confirmed by experiments [7]. A possible reason for this discrepancy may be that such effects as the presence of a hetero-interface and the intrinsic lattice distortion in a solid NP are not included in the present calculation. An improved calculation, taking these effects into account, would predict larger values for the melting temperature suppression, ΔT.

As mentioned above, the effect of the interface between the different phases is not included in the calculation. In the following, the further enhancement of the suppression of the eutectic temperature, ΔT_{eu}, in alloy NPs by the influence of the interfacial energy associated with a hetero-interface in a solid particle is discussed. The relative contribution of interfacial energy to the total Gibbs free energy of a system becomes increasingly large with decreasing size of the system. In the case where the composition of an alloy particle falls in the two-phase region and the particle becomes decreasingly small, the relative contribution of interfacial energy therefore becomes so high that the total Gibbs free energy of a two-phase mixture cannot be expressed as a straight line connecting points A_0 and B_0 in Fig. 9.9, as can be done for bulk materials, but the total Gibbs free energy of the mixture should be expressed as a convex curve, such as those shown by dotted lines in Fig. 9.9 [42]. The upward deviation of these convex curves (denoted as $\Delta G^{interface}$ in Fig. 9.9) from the straight line $A_0 B_0$ increases with decreasing size of the system.

In general, the interfacial energies between two solid phases range from 0.2 J m^{-2} (in a coherent interface) to 1.0 J m^{-2} (in an incoherent interface) [43]. In the literature it is reported that the interfacial energy depends on the size of a system, but that the dependence is rather weak over the size range discussed here [44–46]. In the present calculation the size dependence was therefore ignored, for simplicity. Based upon these premises, the contribution of the interfacial energy to the total Gibbs free energy has been evaluated as a function of the size of particles in the Bi–Sn system, and examples of the results are listed in Table 9.3. The values in this table have been calculated assuming that the interfacial energy is 0.2 J m^{-2}. It is seen from Table 9.3 that the contribution of the interfacial energy, $\Delta G^{interface}$, becomes higher than the Gibbs free energy difference between the single-phase liquid and the two-phase mixture in bulk at RT when the particles become smaller than about 10 nm (this Gibbs free energy difference corresponds to the gap between curve L and the straight line $A_0 B_0$ in Fig. 9.9). In other words, T_{eu} becomes lower than RT in 10 nm-sized alloy particles. This means that the total Gibbs free energy of an isolated Bi–Sn alloy particle 10 nm in diameter and of eutectic composition, even at RT, becomes lower when the particle is in the liquid single-phase state than in the two-phase (i.e., Bi and Sn) state, when a hetero-interface is considered. This is consistent with the experimental observations [37].

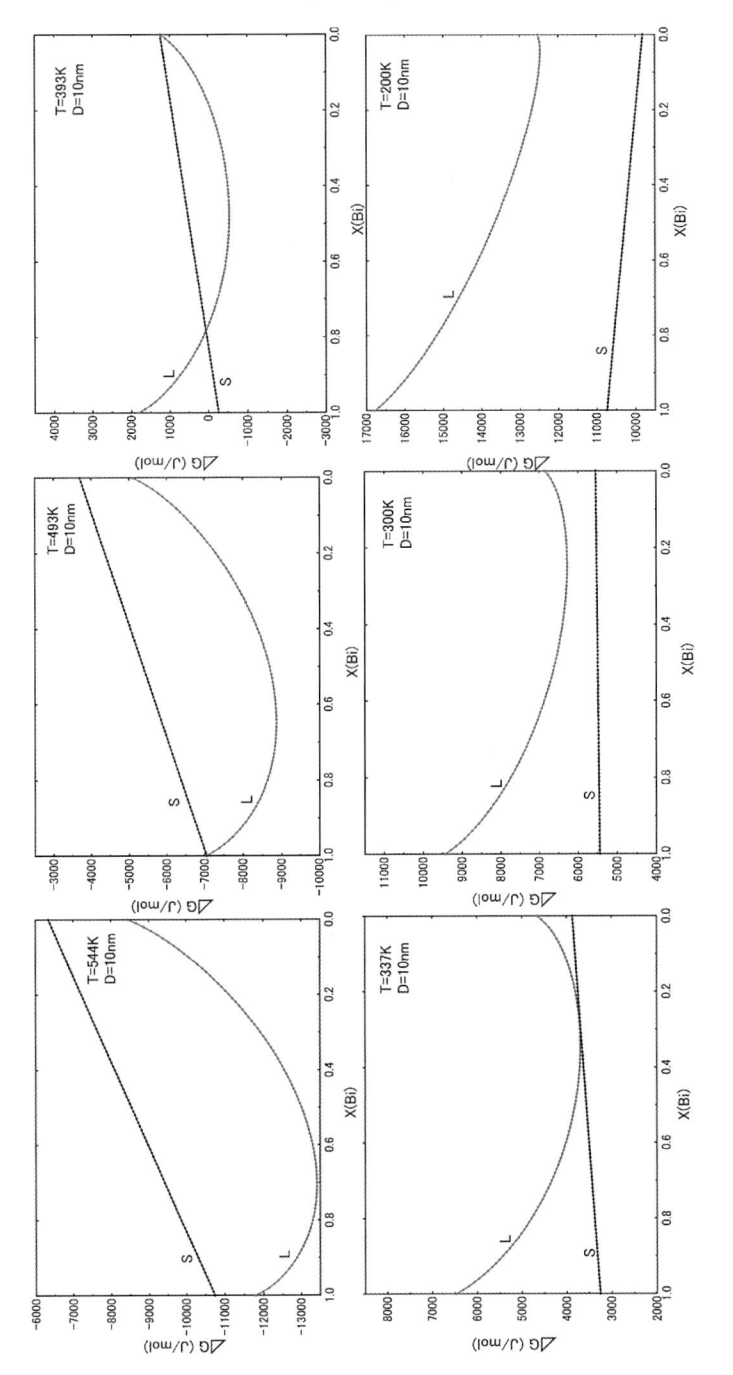

Fig. 9.8 Gibbs free energy–composition (G–X) diagrams at various temperatures constructed from the result of calculations.

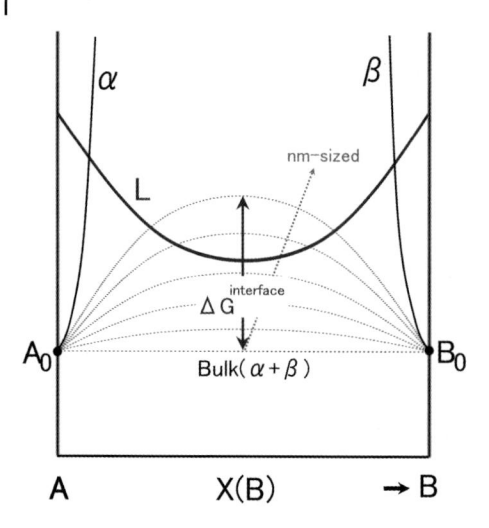

Fig. 9.9 Schematic diagram showing the increase in Gibbs free energy due to the contribution of the interfacial energy in a nanometer-sized alloy particle with a solid/solid two-phase microstructure.

Table 9.3 Size dependence of the contribution of the interfacial energy to the total Gibbs free energy of an alloy particle with a two-phase microstructure.

No. of atoms	Alloy particle diameter [nm]	Max. area of interface [nm^2]	$\Delta G^{interface}$ [J mol^{-1}][a]	Gibbs free energy difference between the liquid phase and the two-phase mixture in bulk at RT [J mol^{-1}]
2170	5	20	4357	
17 000	10	79	2224	2252
30 000	12	113	1815	
139 000	20	314	1088	

[a] Contribution of interfacial energy to total Gibbs free energy.

9.4
Solid/Liquid Two-Phase Microstructure

9.4.1
Solid–Liquid Phase Transition

Recently, it became possible to observe directly an alloy NP in the solid/liquid two-phase state and to analyze the phase relation in a direct manner.

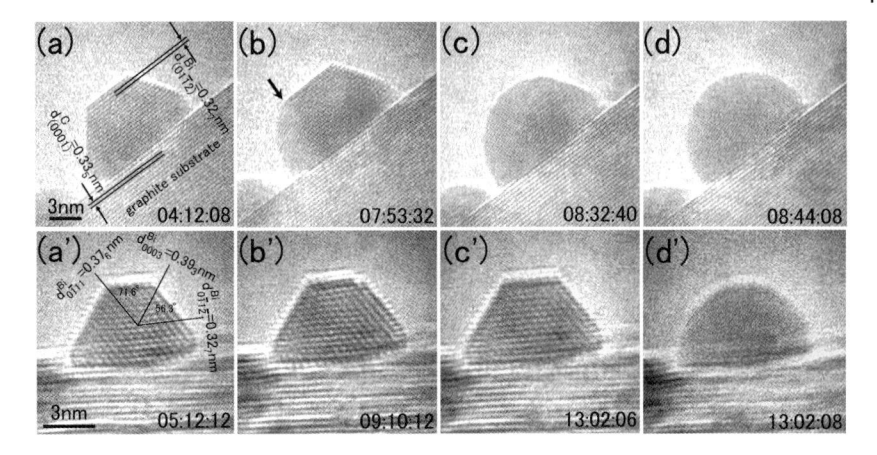

Fig. 9.10 A sequence from a video recording of the alloying process of tin atoms into two differently sized (8 nm and 5 nm) bismuth particles: (a) and (a′) as-produced bismuth particle on the graphite substrate; (b)–(d), (b′)–(d′) the same particle during tin deposition. In the 8 nm-sized bismuth particle during tin deposition, (b) a crystalline/liquid interface was first formed in the interior which then moved right and upward at the expense of the crystalline phase (cf. (b) and (c)) until the whole particle became liquid (d). However, for the 5 nm-sized bismuth particle during tin deposition, first a bismuth-rich solid solution was formed (cf. (b′) and (c′)), which with continued deposition abruptly changed directly into the liquid phase (d′). The arrow in (b) indicates the location of a crystal/liquid interface, and the time is indicated as in Fig. 9.3.

Figure 9.10 shows typical sequences of the alloying process of tin atoms in two bismuth NPs of different sizes at around 350 K [42, 47]. The three numbers inserted in each micrograph indicate the time in units of minutes, seconds, and one-sixtieth seconds. Figure 9.10(a) shows an as-produced pure bismuth particle. The particle was approximately 8 nm in diameter, and the 0.32_7 nm-spaced fringes seen in the particle correspond to the $(01\bar{1}2)$ lattice fringes of pure bismuth. The 0.33_5 nm-spaced fringes correspond to the (0001) lattice fringes of the graphite substrate. Figure 9.10(b) shows the same particle after tin deposition, which changed the structure of the particle from a single crystal to a mixture of a crystalline phase and a liquid phase. The liquid phase corresponds to the portion in uniform contrast in the lower left part of the particle (Fig. 9.10b). A faceted surface of the crystalline bismuth (shown in Fig. 9.10a) was replaced by a round, curved surface of the liquid phase (the arrow in Fig. 9.10(b) indicates the location of a crystal/liquid interface). With continued deposition of tin atoms, the crystal/liquid interface moved right and upward at the expense of the crystalline phase (Fig. 9.10c), and eventually the whole particle became liquid (Fig. 9.10d). An EDS spectrum taken from the central portion of the particle depicted in Fig. 9.10(d) showed that the composition of the particle was Bi–50 at.% Sn (spectrum not shown here). It did take about 4.5 min to form the liquid alloy particle (Fig. 9.10d) by alloying of tin atoms into the pure bismuth particle (Fig. 9.10a). This period can be considered long enough for the alloy particle to reach its equilibrium state. The change in the microstructure mentioned above (i.e., crystalline → (crystalline + liquid) → liquid) is qualitatively consistent with that predicted from the phase diagram for bulk materials between the eutectic temperature and the melting temperature of bismuth [35].

Figure 9.10(a′) shows another as-produced pure bismuth particle, which is approximately 5 nm in diameter, i.e., smaller than the particle in Fig. 9.10(a). The 0.32_7 nm-spaced fringes seen in the particle can be assigned consistently as the $(0\bar{1}12)$ lattice fringes of bismuth, and the 0.39_3 and 0.37_6 nm-spaced fringes as the (0003) and $(0\bar{1}11)$ lattice fringes, respectively. Figure 9.10(b′) shows the same particle after the onset of tin deposition. In spite of the deposition and concomitant alloying of tin atoms, the structure of the particle remained single-crystalline, as shown in Fig. 9.10(b′), suggesting the formation of a terminal bismuth solid solution. With continued tin deposition, the structure of the particle still remained unchanged for a while (Fig. 9.10c′), but then the particle underwent a C–L phase change abruptly, as shown in Fig. 9.10(d′). The C–L phase change occurred within $\frac{1}{30}$ s, the time resolution of the present video recording system, and formation and movement of a crystal/liquid interface were not detected. This is in sharp contrast to the manner of phase change observed in the larger (8 nm) particle shown in Figs. 9.10(a)–(d). From these results, it seems reasonable to consider that the phase change shown in Figs. 9.10(a′)–(d′) took place not by a heterogeneous mechanism (i.e., by the movement of a crystal/liquid interface) but by a homogeneous mechanism. It took approximately 8 min to form the liquid alloy particle (Fig. 9.10d′) by alloying of tin atoms into the pure, crystalline bismuth particle (Fig. 9.10a′). This period of about 8 min is considered again to be long enough to achieve an equilibrium state. In short, when the bismuth particle was 5 nm or below, a direct C–L phase change took place with increasing concentration of tin, but when the bismuth particle was 8 nm or larger, the C–L transition proceeded via the migration of a solid/liquid interface.

In an attempt to compare the phase change in bismuth NPs induced by tin alloying (i.e., Figs. 9.10a–d) with that in tin NPs induced by bismuth alloying, a C–L phase change in a tin NP associated with alloying of bismuth is depicted in Fig. 9.11, which shows a typical sequence of the alloying process of bismuth atoms into a tin NP at around 350 K [47].

Figure 9.11(a) shows an as-produced pure tin particle, approximately 8 nm in diameter; the 0.29_1 nm-spaced fringes seen in the particle correspond to the (020) lattice fringes of β-Sn. Figure 9.11(b) shows the same particle after bismuth deposition. In spite of the deposition and alloying of bismuth atoms, the structure of the particle remained single-crystalline. With continued bismuth deposition, the structure of the particle was still unchanged for a while (Figs. 9.10c, d), but it then underwent an abrupt C–L phase change (Fig. 9.10e). The C–L phase change occurred again within $\frac{1}{30}$ s, the time resolution of the present video recording system, and formation and movement of a crystal/liquid interface were not observed. An EDS spectrum of the central portion of the particle depicted in Fig. 9.11(e) showed that the composition of the particle was Sn–6 at.% Bi (Fig. 9.11f).

9.4 Solid/Liquid Two-Phase Microstructure

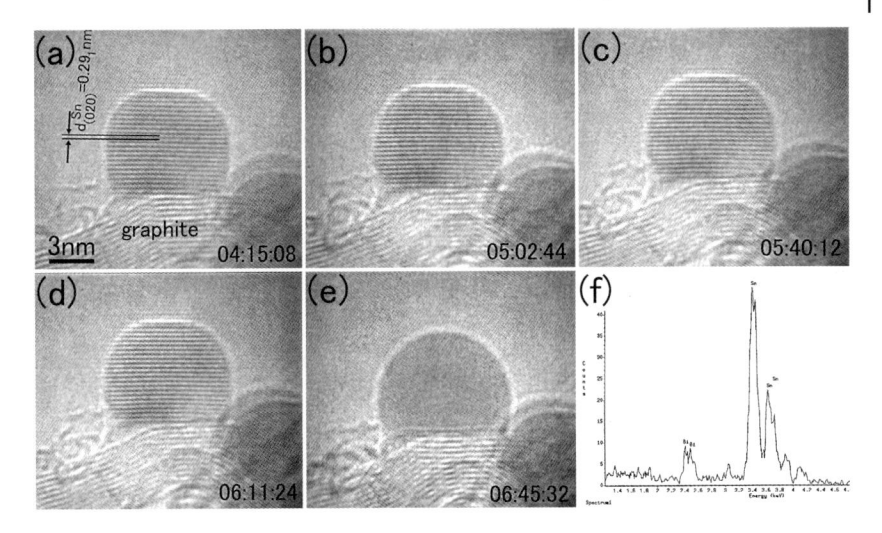

Fig. 9.11 An alloying process sequence of bismuth atoms into a nanometer-sized tin particle: (a) as-produced tin particle on a graphite substrate; (b)–(e) the same particle during bismuth deposition. First (b)–(d) a tin-rich solid solution was formed, which with continued bismuth deposition changed abruptly into the liquid phase (e). (f) EDS spectrum of the central portion of the particle depicted in (e).

It has been shown above that the change in the microstructure associated with alloying on the bismuth side (see Figs. 9.10a–d) differs from that on the tin side (Fig. 9.11) in the 8 nm-sized particle: in the former a sequence of phase changes of crystalline → (crystalline + liquid) → liquid has taken place, whereas in the latter a direct change from crystalline to liquid has occurred without passing through a two-phase crystalline/liquid mixture state. It is also shown that even on the bismuth side, the microstructure change induced by alloying of tin atoms has taken place in different manners depending upon the size of particle (compare Figs. 9.10(a)–(d) with Figs. 9.10(a′)–(d′)). It is of interest to analyze these results from the thermodynamic viewpoint.

Figure 9.12 shows the Gibbs free energy curves for the liquid phase, the Bi-based rhombohedral phase, and the Sn-based bct phase in a 10 nm-sized particle at 350 K over a composition range from pure tin to pure bismuth. The Gibbs free energy curves were calculated in such a manner that curves for bulk materials available in the Thermo-Calc database system [38] were modified by taking the contribution from the surface energy into account [47], which was already mentioned in Section 3.4.1. It is evident from the figure that in a 10 nm-sized alloy particle at 350 K, the conjugate phase of the Bi-based rhombohedral phase is not the Sn-based bct phase but the liquid phase. This fact suggests that T_{eu} is below 350 K in this reduced-size system, which is in agreement with the present observations shown in Figs. 9.10 and 9.11. In the case of alloy NPs, in addition to the contribution of the surface energy mentioned above, the contribution of the solid/

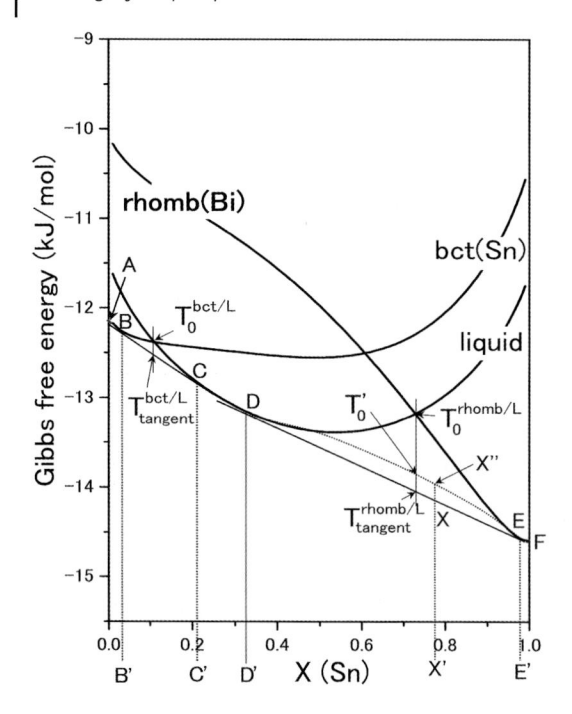

Fig. 9.12 Calculated Gibbs free energy curves for the liquid phase (liquid), the Sn-based bct phase (bct) and the Bi-based rhombohedral phase (rhomb) in a 10 nm-sized alloy particle at 350 K. See the text for the other labels and lines.

liquid interfacial energy, γ^{sl}, to the total Gibbs free energy has also to be taken into account. This contribution from γ^{sl} is not included in the curves in Fig. 9.12. Therefore, for example, the total Gibbs free energy of a 10 nm-sized alloy particle of composition X′ between compositions D′ and E′, consisting of the bismuth-based rhombohedral phase and the liquid phase in Fig. 9.12, cannot be given by point X on the common tangent DE but by point X″ located somewhat above point X. The difference between points X and X″ corresponds to the contribution of γ^{sl}. When it is assumed that there is only one, rather flat, solid/liquid interface in a 10 nm-sized particle as observed in the above experiment, then point X″ will be on a convex curve joining points D and E. If the shape of the particle is postulated to be hemispherical, the curve will be a parabolic one with the maximum at a composition near the midpoint between compositions D′ and E′. Such a parabolic curve is shown in Fig. 9.12 (dotted line). $T_0^{rhomb/L}$ is the point where the total Gibbs free energies for the bismuth solid solution and for the liquid phase become equal. $T_{tangent}^{rhomb/L}$ and T_0' are the points where the vertical line through $T_0^{rhomb/L}$ intersects with the common tangent DE and the parabolic curve (dotted line), respectively. A preliminary specific estimate shows that in a 10 nm-sized alloy particle, the Gibbs free energy difference between points $T_{tangent}^{rhomb/L}$ and

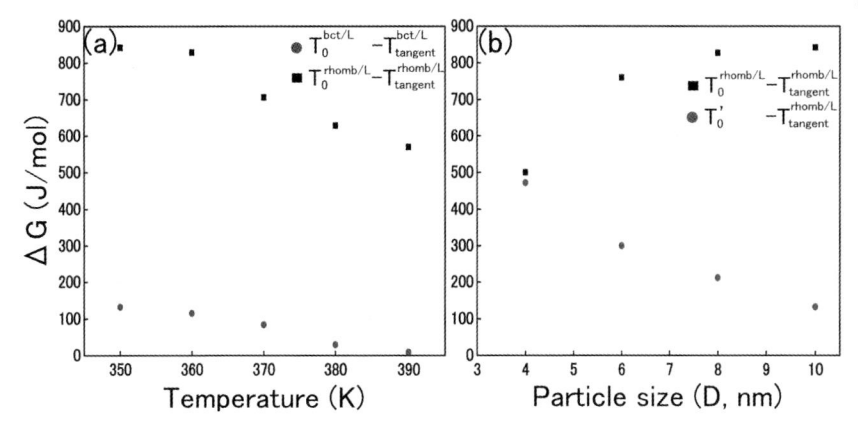

Fig. 9.13 Dependence of Gibbs free energy difference in a 10 nm-sized alloy particle at 350 K on (a) temperature and (b) particle size. See Fig. 9.12 for the meanings of the labels.

T_0' is small compared to the Gibbs free energy difference between points $T_{tangent}^{rhomb/L}$ and $T_0^{rhomb/L}$ if a value of 0.061 J m^{-2} is used for γ^{sl} [48]. In fact, the parabolic curve (dotted line) in this figure. was calculated using this value. The height of the maximum of this curve relative to the common tangent DE is approximately 165 J mol^{-1}. This means that in a 10 nm-sized alloy particle, the interfacial energy γ^{sl} of 0.061 J m^{-2} would make a maximum contribution of about 165 J mol^{-1} to the total Gibbs free energy at a composition near the midpoint between compositions D' and E'.

The Gibbs free energy difference between points $T_{tangent}^{rhomb/L}$ and $T_0^{rhomb/L}$ and that between points $T_{tangent}^{bct/L}$ and $T_0^{bct/L}$ over a temperature range from 350 to 390 K are shown in Fig. 9.13(a). It should be noted that the Gibbs free energy difference between points $T_{tangent}^{rhomb/L}$ and $T_0^{rhomb/L}$ is much larger than that between $T_{tangent}^{bct/L}$ and $T_0^{bct/L}$ and is also larger than that between points $T_{tangent}^{rhomb/L}$ and T_0', which is approximately 165 J mol^{-2}, over the 350–390 K range. In this case, it is predicted that a phase change essentially similar to that for bulk material will take place upon alloying tin atoms into a bismuth particle. That is, a pure bismuth particle would first change to a particle with the crystal/liquid two-phase microstructure and then eventually to a particle of the liquid phase. This prediction is in agreement with the observation depicted in Figs. 9.10(a)–(d). On the other hand, as seen from Figs. 9.12 and 9.13(a), the Gibbs free energy difference between points $T_0^{bct/L}$ and $T_{tangent}^{bct/L}$ on the tin-rich side, where $T_0^{bct/L}$ is the point at which the total Gibbs free energies for the tin solid solution and for the liquid phase become equal, and $T_{tangent}^{bct/L}$ is the point where the vertical line through $T_0^{bct/L}$ intersects with the common tangent BC, is smaller than that between points $T_{tangent}^{bct/L}$ and T_0' (not shown here); the latter is approximately 155 J mol^{-1} if a value of 0.055 J m^{-2} is used for γ^{sl} [48], over a 350–390 K range. In this case it is therefore predicted that with increasing concentration of bismuth, a particle of the terminal tin solid solution would change directly into a particle of the liquid phase, without having a stage of solid/liquid coexistence. This prediction is consistent with the observation that the phase change from the terminal tin solid solution to the liquid phase took place very quickly and in the process no movement of a crystal/liquid interface was confirmed, as shown in Fig. 9.11. The contribution of the solid/liquid interfacial energy, γ^{sl}, toward the total Gibbs free energy is expected to increase with decreasing size of the system.

The Gibbs free energy difference between points $T_{\text{tangent}}^{\text{rhomb/L}}$ and $T_0^{\text{rhomb/L}}$ and that between points $T_{\text{tangent}}^{\text{rhomb/L}}$ and T_0' at 350 K over a size range from 4 to 10 nm are shown in Fig. 9.13(b). It should be noted that the Gibbs free energy difference between points $T_{\text{tangent}}^{\text{rhomb/L}}$ and $T_0^{\text{rhomb/L}}$ slowly decreases, but that the contribution of the solid/liquid interfacial energy, γ^{sl}, toward the total Gibbs free energy increases with decreasing size of the system, and these two values become almost equal at an approximately 4 nm-sized alloy particle as shown in Fig. 9.13(b). This means that for the system smaller than approximately 4 nm in size, an alloy with a bismuth solid-solution/liquid two-phase microstructure is not necessarily thermodynamically favorable compared to the alloy with a single-phase microstructure (i.e., either a liquid phase or a terminal bismuth solid solution). Therefore, in such a case, it is predicted that with increasing concentration of tin, a particle of the terminal bismuth solid solution would directly change into a particle of the liquid phase, without having a stage of solid/liquid coexistence. This prediction is qualitatively consistent with the observation that in an approximately 5 nm-sized alloy particle the phase change from the terminal bismuth solid solution to the liquid phase took place very quickly and in the process no movement of a crystal/liquid interface was confirmed, as shown in Figs. 9.10(a′)–(d′). From these discussions, it seems reasonable to consider that the different manners in which the phase change occurs as observed in Figs. 9.10 and 9.11 can be explained consistently in terms of the magnitude of the contribution of γ^{sl} relative to the Gibbs free energy difference between points $T_{\text{tangent}}^{\text{solid/L}}$ and $T_0^{\text{solid/L}}$.

9.4.2
Two-Phase Microstructure

The morphology of the two-phase microstructure in alloy NPs is a function of the surface energy and the interfacial energy. Details will be discussed in this section using examples from Au–Sn and Bi–Sn systems.

Figure 9.14 shows a typical sequence of microstructural changes in an isolated gold NP during alloying of tin at 430 K. Here 430 K is located below T_{eu} of the bulk alloy materials (i.e., 551 K for the Au–Sn system [35]), but above T_{eu} of 10 nm-sized alloy particles [11]. Figure 9.14(a) shows an as-produced pure gold particle on the surface of graphite at 430 K. The particle is approximately 8 nm in diameter, and the 0.23_5 nm-spaced fringes seen in it correspond to the (111) lattice fringes of pure gold. The 0.33_5 nm-spaced fringes correspond to the (0001) lattice fringes of the graphite substrate. Figures 9.14(b)–(h) show the same particle undergoing continuous deposition of tin atoms. With tin deposition, a thin liquid layer appeared in a form of spherical shell around the crystalline gold solid

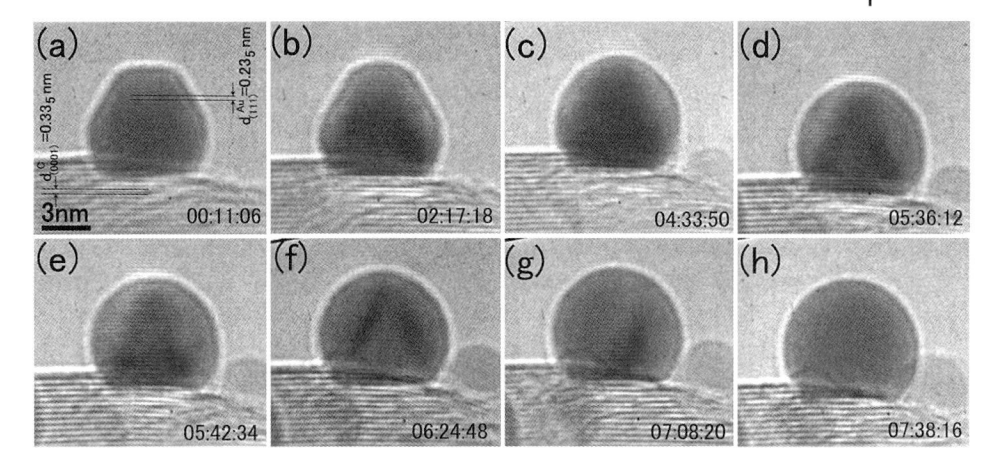

Fig. 9.14 Sequence from a video recording of the alloying process of tin atoms into a nanometer-sized gold particle: (a) as-produced gold particle on a graphite substrate; (b)–(h) the same particle undergoing continuous tin deposition. (c)–(e) First a thin liquid shell was formed over the gold–tin solid solution; (f)–(g) the liquid layer grew inward at the expense of crystalline solid solution until (h) the whole particle became liquid. The time is indicated as in Fig. 9.3.

solution, as shown in Figs. 9.14(c)–(g). The formation of a thin liquid shell can be seen from the fact that the shape of the particle changed from a faceted polyhedron (Fig. 9.14a) to a truncated sphere (Fig. 9.14c) and from the fact the (111) lattice fringes of the gold–tin solid solution disappeared in the outermost peripheral region of the particle. With continued deposition of tin atoms, the interface between the crystalline gold solid solution and the liquid shell moved inward at the expense of the crystalline core (i.e., the central region appearing dark) as shown in Figs. 9.14(d)–(g), and eventually the whole particle became liquid as shown in Fig. 9.14(h). In view of a previous observation that in isolated NPs alloying reaction takes place quite rapidly compared to that in bulk alloys [25], alloy phase formation could occur even at RT in a particular Au–Sn system [24, 25]. It seems reasonable to consider that the core (crystal)–shell (liquid) two-phase microstructure seen in Figs. 9.14(c)–(g) is in an equilibrium state at this temperature of around 430 K [49].

On the other hand, it is known that in alloy NPs in the Bi–Sn system, a quite different crystal–liquid two-phase microstructure developed in an equilibrium state as shown in Figs. 9.10(b) and (c). That is, in the Bi–Sn system, a two-phase structure appeared in which a single, rather flat, planar hetero-interface connected a crystalline phase (the bismuth solid solution) and a liquid phase (see Fig. 9.10b), with deposition of tin atoms onto a particle of bismuth solid solution. With continued tin deposition, the interface moved right and upward at the expense of the crystalline phase until the whole particle became liquid (Figs. 9.10c,d). As mentioned above, both a gold NP and a bismuth NP undergo a

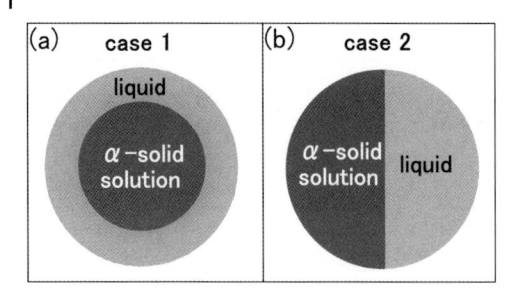

Fig. 9.15 Schematic illustrations of possible two-phase microstructures in an isolated nanometer-sized alloy particle: (a) case 1; (b) case 2. See the text for details.

sequence of phase changes such as crystal (solid solution) → crystal + liquid → liquid, with increasing tin concentration. This sequence itself is consistent with the sequence which an alloy would exhibit when the solute concentration increases across the cystal + liquid two-phase region bounded by the solidus and liquidus curves. However, the microstructure developed in the crystal + liquid two-phase region in particles in the Au–Sn system differed from that in the Bi–Sn system: a core (crystal)–shell (liquid) structure (designated as case 1 hereafter) appeared in the former system, whereas a simple two-phase structure with a planar interface separating the crystalline and the liquid phases (designated as case 2 hereafter) in the latter system. These two microstructures are illustrated schematically in Fig. 9.15.

It has been reported by Hagege and Dahmen that several different microstructures within small Pb–Cd inclusions embedded in an aluminum matrix can be discussed qualitatively in terms of the relative magnitudes of the A–B, A–M, and B–M interfacial energies (here A, B, and M stand for Pb, Cd, and Al, respectively) [50]. A similar analysis would be possible in isolated alloy particles if the crystal–liquid interfacial energy, γ^i, the surface energy of crystal, γ^s, and the surface energy of liquid, γ^l, are employed instead of A–B, A–M, and B–M interfacial energies in Hagege and Dahmen's treatment, respectively. Based upon this premise, the experimental observations shown in Figs. 9.10 and 9.14 are discussed in terms of γ^i, γ^s, and γ^l. Firstly, the difference in Gibbs free energy (per mol), ΔG, between a bulk alloy and an alloy NP is calculated for cases 1 and 2. Of course, in the case of NPs, the surface and interfacial energies make an appreciable contribution to the total Gibbs free energy. We assume here that the volume fraction of liquid phase and crystalline solid solution phase is the same (i.e., 50%) in a particle. Then, the difference in molar Gibbs free energy for case 1, ΔG_1, can be described By Eq. (7), where r, γ^l, and γ^i are the radius of particle, the surface energy of liquid, and the interfacial energy between crystalline solid solution and liquid, respectively.

$$\Delta G_1 = 4\pi r^2 (\gamma^l + 0.63\gamma^i) \tag{7}$$

The difference for case 2, ΔG_2, can be described by Eq. (8), where γ^s is the surface energy of crystalline solid solution. (In Eqs. (7) and (8), a numerical factor of N_0/N is omitted for simplicity, where N_0 and N are the Avogadro number and the number of atoms comprising a particle, respectively.)

$$\Delta G_2 = \pi r^2 (2\gamma^s + 2\gamma^l + \gamma^i) \tag{8}$$

Next, it should be noted that the relative stability between microstructures in case 1 and 2 becomes equal when $\Delta G_1 = \Delta G_2$. For this condition Eq. (9) is satisfied.

$$2\gamma^s - 2\gamma^l - 1.52\gamma^i = 0 \tag{9}$$

If we set $\gamma^s = Y\gamma^l$ and $\gamma^i = X\gamma^l$, then we get a simple but useful equation [Eq. (10)] which enables us to predict the relative stability using two parameters X and Y.

$$Y = 0.76X + 1 \tag{10}$$

This is shown as a straight line in Fig. 9.16. In the region below the line (i.e., $\Delta G_1 > \Delta G_2$) a microstructure as in case 2 is preferred, whereas in the region above the line (i.e., $\Delta G_1 < \Delta G_2$) a microstructure as in case 1 is more favorable.

The values of γ^s, γ^l, and γ^i for actual alloys can be estimated by calculations under the assumption that the surface and interfacial energies are in linear dependence on composition. Values of γ^s, γ^l, and γ^i for pure tin, bismuth, and gold are collected from the literature [41, 48, 51]. It has been confirmed that the tin concentrations in the bismuth solid solution and in the liquid within a particle with the two-phase structure of melt + α-solid solution are approximately 5 and 50

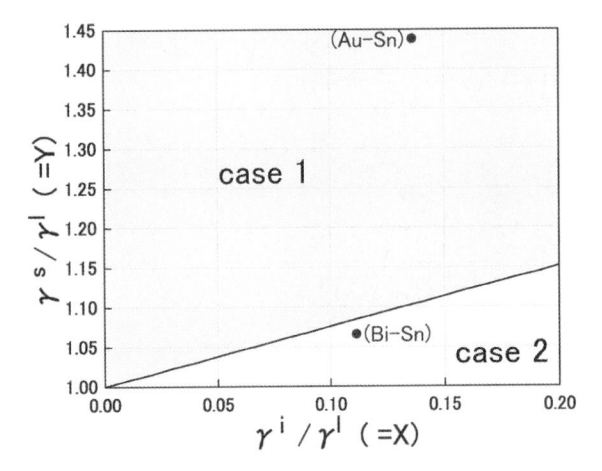

Fig. 9.16 Diagram showing which two-phase microstructure (Fig. 9.15) is more stable as a function of γ^s/γ^l ($= Y$) and γ^i/γ^l ($= X$). See the text for details.

at.%, respectively [47]. Based upon this result, the values of γ^s and γ^l in the Bi–Sn system can be taken as 0.5 and 0.47 J m^{-2}, respectively. The value of γ^i is estimated to be 0.054 J m^{-2}. Consequently, for particles in the Bi–Sn system, γ^s/γ^l (= Y) and γ^i/γ^l (= X) are given as 1.064 and 0.115, respectively, and the point with the coordinates is indicated by a filled circle labeled (Bi–Sn) in Fig. 9.16. Since this point is below the line mentioned beforehand, it is predicted from Fig. 9.16 that in particles in the Bi–Sn system, such a microstructure as in case 2 is thermodynamically more favorable than in case 1. This prediction is consistent with the experimental result shown in Figs. 9.10(b) and (c). On the other hand, in the Au–Sn system, the values of γ^s and γ^l can be taken as 1.43 and 0.99 J m^{-2}, respectively. The value of γ^i is taken as 0.132 J m^{-2}. Then, for particles in the Au–Sn system, γ^s/γ^l (= Y) and γ^i/γ^l (= X) are given as 1.446 and 0.134, respectively. The point with these coordinates is indicated by a filled circle labeled (Au–Sn) in Fig. 9.16. Since this point is above the line, it is predicted from Fig. 9.16 that in particles in the Au–Sn system, such a microstructure as in case 1 is thermodynamically more favorable than in case 2. This prediction is again consistent with the present experimental result shown in Fig. 9.14.

From the discussion, it seems reasonable to consider that the experimental results shown in Figs. 9.10 and 9.14 can be explained consistently in terms of the relative contributions of γ^s, γ^l, and γ^i to the total Gibbs free energy of a particle with the two-phase structure.

9.5
Solid Solubility in Nanometer-Sized Alloy Particles

This section describes the effect of size on the solid solubility in alloy NPs.

Figure 9.17 show a typical sequence of the process of alloying tin atoms into lead NPs at around 383 K [52]. Figures 9.17(a) and (a') show a BFI of as-produced

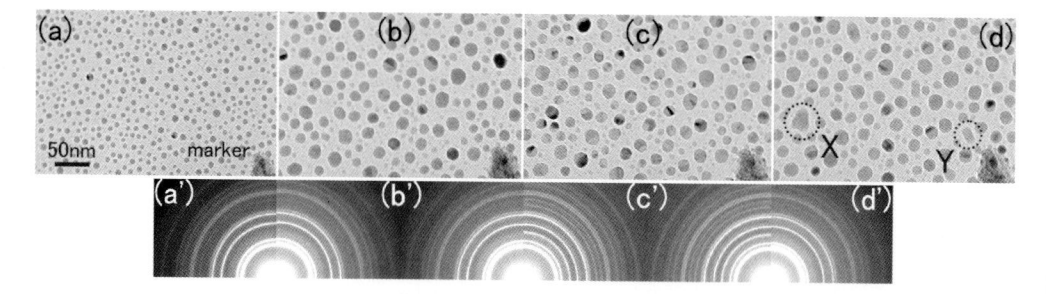

Fig. 9.17 A series of electron micrographs showing the alloy phase formation in nanometer-sized particles in the Pb–Sn system: (a) as-produced lead particles on an amorphous carbon film kept at 383 K; (b) the same area as (a) after tin deposition; (c) the same area as (b) after cooling to RT; (d) the same area as (c) after reheating to 383 K; (a')–(d') the SAEDs corresponding to (a)–(d), respectively. A mordel of dirt at the bottom right-hand corner of each BFI serves to mark a fixed position.

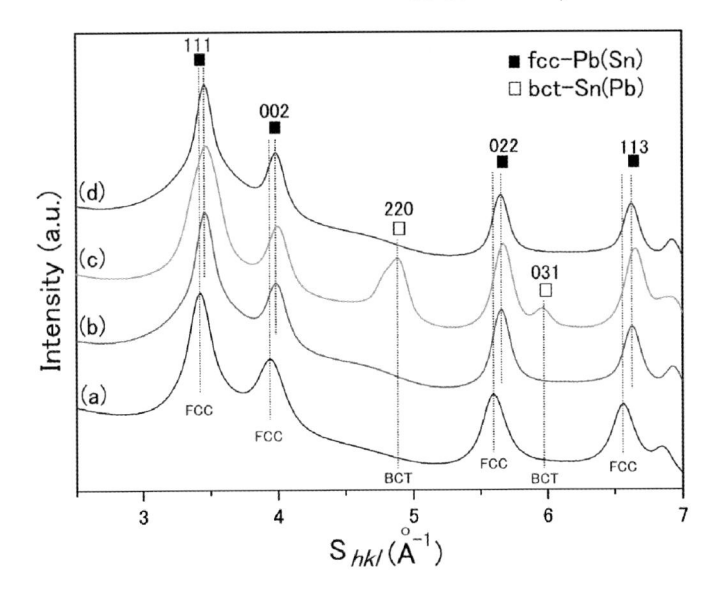

Fig. 9.18 Intensity profiles of the Debye–Scherrer rings shown in Fig. 9.17 (a′)–(d′). See the text for details.

lead particles on an amorphous carbon film at 383 K and the corresponding SAED, respectively. The mean diameter of the lead particles is approximately 6 nm. The corresponding Debye–Scherrer rings can be indexed consistently as those of a crystal with the fcc structure, which is the same as that of pure bulk lead. An intensity profile of the Debye–Scherrer rings is depicted as curve (a) in Fig. 9.18. The profile was obtained by digitalizing the data in IPs. Figures 9.17(b) and (b′) and curve (b) in Fig. 9.18 show a BFI of particles after tin deposition, the corresponding SAED, and an intensity profile of the rings in the SAED, respectively. The particle size increased from 6 to 17 nm on average by tin deposition. The size increment came partially from the coalescence among particles induced by tin deposition. It seems that the tin deposition causes a migration of the particles on the supporting film. All the Debye–Scherrer rings in Fig. 9.17(b′) can be indexed consistently as those of a crystal with the fcc structure. However, as may be seen from a comparison of profile (b) with (a) in Fig. 9.18, the corresponding lattice parameter decreased with respect to that of pure lead particles in Fig. 9.17(a), suggesting the formation of a lead solid solution with undersized tin atoms in NPs. These observations indicate that when tin atoms are vapor-deposited and come in contact with lead particles, they dissolve quickly into the lead particles to form a lead–tin solid solution. It was revealed by EDS that the particles shown in Fig. 9.17(b) contained 56 at.% Sn, on average, which is almost five times higher than the solubility limit of tin in bulk lead–tin solid solution at the same temperature (i.e., 10 at.%) [35].

High resolution electron microscopy (HREM) observations revealed that a fraction of the lead–tin solid solution particles possessed the multiply twinned struc-

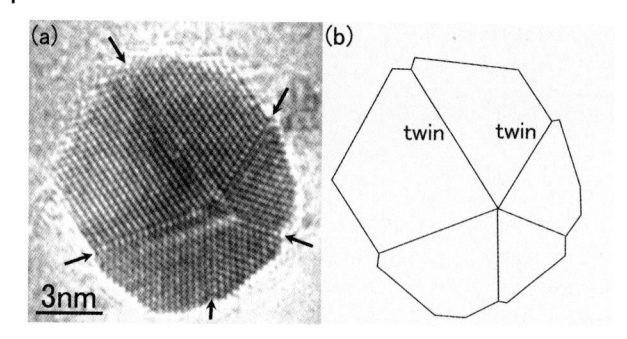

Fig. 9.19 A high-resolution image of a multiply twinned nanometer-sized particle of a lead–tin solid solution. Arrows indicate the locations of twin boundaries.

ture, which has been frequently observed in NPs of pure fcc metals [53, 54]. An example of the multiply twinned particles of the lead–tin solid solution is shown in Fig. 9.19.

Figures 9.17(c), (c′) and curve (c) in Fig. 9.18 show a BFI of the same area as is shown in Fig. 9.17(b) but after cooling to RT, together with the corresponding SAED and an intensity profile of rings in the SAED, respectively. The Debye–Scherrer rings in the SAED can be indexed consistently as those of a crystal with the fcc structure (i.e., lead solid solution) superimposed with those of another crystal with the tetragonal structure (i.e., tin solid solution). These facts indicate that at RT the tin concentration (i.e., 56 at.% Sn) in alloy particles approximately 17 nm in size well exceeded the solubility limit of tin in lead and therefore excess tin was precipitated as a tin solid solution. In other words, a phase change from a single-phase lead solid solution to a two-phase structure of lead solid solution and tin solid solution took place upon cooling from 383 K to RT. This observation is of particular interest since it clearly indicates that even at RT atomic mobility is high enough to induce a phase change in approximately 17 nm-sized particles. The high mobility of atoms in NPs is consistent with the results in the previous sections. It should be noted here that the lattice parameters of the fcc lead solid solution in Fig. 9.17(c) is almost the same as that in Fig. 9.17(b), as seen from a comparison of profiles (c) and (b) in Fig. 9.18. A possible qualitative explanation for this fact is as follows. The lattice parameter of the fcc lead solid solution shown in Fig. 9.17(b) tends to increase due to the loss of dissolved undersized tin atoms (associated with the formation of the tin solid solution within individual NPs). On the other hand, the lattice parameter tends to decrease due to the thermal contraction associated with cooling from 383 K to RT. These two mutually opposite trends apparently cancel each other and bring about an almost constant lattice parameter of the fcc lead solid solution between the two states shown in Figs. 9.17(b) and (c). Quantitative analyses of this point will be an interesting subject of a future study.

Figures 9.17(d) and (d′), and curve (d) in Fig. 9.18, show a BFI of the same area as is shown in Fig. 9.17(c), after heating to 383 K, the corresponding SAED, and an intensity profile of rings in the SAED, respectively. By comparing particles in Fig. 9.17(d) with the corresponding particles in Fig. 9.17(c), it can be seen that in some cases coalescence took place among particles upon reheating to 383 K. Examples are particles X and Y in Fig. 9.17(d). It is considered that migration of particles would be induced due to possible nonisotropic mass flows on the supporting film at an elevated temperature (i.e., 383 K) and that the migration might result in the coalescence of particles. All the Debye–Scherrer rings in the SAED can be indexed consistently as those of a crystal with the fcc structure (i.e., lead solid solution). Furthermore, as seen from a comparison of curves (d) and (b) in Fig. 9.18, the lattice parameter of the fcc lead solid solution in Fig. 9.17(d) is identical with that of the solid solution in Fig. 9.17(b) at the same temperature, 383 K. These facts suggest that a phase change from the two-phase state (a lead solid solution and a tin solid solution) to a single phase of lead solid solution occurred upon heating to 383 K. This is just the reverse of the phase change that took place upon cooling from 383 K (Fig. 9.17b) to RT (Fig. 9.17c). Therefore it can be said that the above-mentioned change from a single- to a two phase state could take place reversibly when the temperature of the system is changed cyclically between 383 K and RT. This observation provides further evidence for the fact that in particles approximately 17 nm in size the solubility limit of tin in lead is higher than 56 at.%Sn at 383 K but lower than 56 at.%Sn at RT.

Recently it was reported that the elastic modulus would decrease with decreasing size of particles. This decrease in the elastic modulus can lower the strain energy in the solid solution. That is, the contribution of strain energy to the heat of mixing (i.e., to the formation enthalpy) will become small when the elastic modulus becomes low. This reduction in the formation enthalpy of a solid solution would inevitably bring about the enhancement of the solid solubility; that is, the solubility limit would be increased when the size of the particles is reduced. This explanation is in accordance with the observation that an enhanced solid solubility of tin ino lead has been achieved in NPs (see Figs. 9.17 and 9.18, for example). However, further study is needed to elucidate quantitatively the mechanism behind the enhanced solid solubility.

9.6
Summary and Future Perspectives

In this chapter properties of alloys under dimensional restrictions have been described, with emphasis on the phase equilibrium in alloy NPs. In particular, the effects of system size on the eutectic temperature, on the crystalline-to-liquid phase change, and on the solid solubility have been presented and discussed in some detail. Furthermore, the formation of a thermodynamically stable amorphous solid, which is a feature characteristic of alloy NPs, has been described.

In order to obtain a comprehensive understanding of alloy properties under dimensional restrictions, studies on the following issues are still needed.

- *Substrate effect on phase stability of nanometer-sized alloy particles*: The crystalline-to-liquid (C–L) transition of NPs has been investigated primarily on solid substrates so far. However, the particle/substrate interfacial tension, as well as the surface tension of the substrate, might also affect such phase transitions as the C–L transition in NPs, which has not been taken into account explicitly in previous studies.
- *Size dependence of the surface tension in NPs*: The surface tension of small particles may change due to the curvature. It may decrease with decreasing size of particle, as suggested by Tolman [55]. Experimental verification for this point is necessary to assess quantitatively the phase diagram of NPs less than a several nanometers in diameter.
- *Evaluation of the solid/liquid interfacial energy*: It is mentioned in this chapter that the interfacial energy for an interface between liquid (or solid) and solid phases can affect the phase equilibrium when the particle is in the two-phase region (liquid–solid or solid–solid). An accurate measurement of the interfacial energy is required to establish the phase equilibrium of alloy NPs.
- *Glass transition*: It is suggested in this chapter that a thermodynamically stable amorphous phase could be formed in a eutectic system when the particles are smaller than approximately 10 nm in diameter. This unique amorphous phase gives us an opportunity to obtain an in-depth understanding of the liquid-to-glass (L-G) transition, which is one of unsolved issues in modern solid-state physics.

References

1 Halperin, W. P. (1986), *Rev. Modern Phys.* **58**, 533.
2 Andres, R. P., Averback, R. S., Brown, W. L., Bras, L. E., Goddard III, W. A., Kaldor, A., Louie, S. G., Moscovits, M., Peercy, P. S., Riley, S. J., Siegel, R. W., Spaepen, F., Wang, Y. (1989), *J. Mater. Res.* **4**, 704.
3 Kubo, R. (1962), *J. Phys. Soc. Japan* **17**, 975.
4 Yokozeki, A., Stein, G. (1978), *J. Appl. Phys.* **49**, 2224.
5 Sambles, J. R. (1971), *Proc. R. Soc. Lond. A* **324**, 339.
6 Buffat, Ph., Borel, J.-P. (1976), *Phys. Rev. A* **13**, 2287.
7 Allen, G. L., Bayles, R. A., Gile, W. W., Jesser, W. A. (1986), *Thin Solid Films* **144**, 297.
8 Palatnik, L. S., Boiko, B. T. (1961), *Phys. Met. Metallogr.* **11**, 119.
9 Allen, G. W., Jesser, W. A. (1984), *J. Cryst. Growth* **70**, 546.
10 Jesser, W. A., Shiflet, G. J., Allen, G. L., Crawford, J. L. (1999), *Mater. Res. Innovations* **2**, 211.
11 Yasuda, H., Mitsuishi, K., Mori, H. (2001), *Phys. Rev. B* **64**, 094 101.

12. Hyeon, T., Lee, S. S., Park, J., Chung, Y., Na, H. B. (2001), *J. Am. Chem. Soc.* **123**, 12 798.
13. Redl, F. X., Cho, K.-S., Murray, C. B., O'Brien, S. (2003), *Nature* **423**, 968.
14. Mori, H., Yasuda, H., Fujii, K. (1996), *Surf. Rev. Lett.* **3**, 1177.
15. Lee, J.-G., Mori, H. (2003), *J. Vac. Sci. Technol. A* **21**, 31.
16. Kamino, T., Saka, H. (1993), *Microsc. Microanal. Microstruct.* **4**, 127.
17. Dick, K., Dhanasekaran, T., Zhang, Z., Meisel, D. (2002), *J. Am. Chem. Soc.* **124**, 2312.
18. Makin, S. M., Rowe, A. H., LeClaire, A. D. (1957), *Proc. Phys. Soc. B* **70**, 545.
19. Mori, H., Yasuda, H. (1994), *Z. Phys. D* **31**, 131.
20. Mori, H., Yasuda, H., Kamino, T. (1994), *Phil. Mag. Lett.* **69**, 279.
21. Yasuda, H., Mori, H. (1997), *Ann. Phys.* **22**, 127.
22. Kubaschewski, O. (1950), *Trans. Faraday Soc.* **46**, 713.
23. Shibata, T., Bunker, B., Zhang, Z., Meisel, D., Vardeman II, C., Gezelter, J. (2002), *J. Am. Chem. Soc.* **124**, 11 989.
24. Mori, H., Yasuda, H. (2001), *Scripta Mater.* **44**, 1987.
25. Mori, H., Yasuda, H. (2001), *Mater. Sci. Eng. A* **312**, 99.
26. Kauzman, W. (1948), *Chem. Rev.* **43**, 219.
27. Jiang, Q., Aya, N., Shi, F. G. (1997), *Appl. Phys. A* **64**, 627.
28. Turnbull, D. (1969), *Contemp. Phys.* **10**, 473.
29. Iijima, S., Ichihashi, T. (1986), *Phys. Rev. Lett.* **56**, 616.
30. Sugano, S., Nishina, Y., Ohnishi, S., *Microclusters*, Springer-Verlag, Berlin, 1987.
31. Ajayan, P. M., Marks, L. D. (1988), *Phys. Rev. Lett.* **60**, 585.
32. Ajayan, P. M., Marks, L. D. (1989), *Phys. Rev. Lett.* **63**, 279.
33. Mitome, M., Tanishiro, Y., Takayanagi, K. (1989), *Z. Phys. D* **12**, 45.
34. Lee, J.-G., Mori, H. (2003), *J. Electron Microsc.* **52**, 57.
35. Massalski, T. B., *Binary Alloy Phase Diagrams*, American Society for Metals, Metals Park, OH, 1986.
36. Lee, J.-G., Diploma Thesis, Osaka University, 2003.
37. Lee, J.-G., Mori, H., Yasuda, H. (2002), *Phys. Rev. B* **66**, 012 105.
38. Thermo-Calc thermodynamic database (software), version L, Stockholm Royal Institute, Sweden, 1998.
39. Tanaka, T., Hara, S. (2001), *Z. Metallkd.* **92**, 467.
40. Tanaka, T., Hack, K., Hara, S. (1999), *MRS Bulletin* **24**, 45.
41. Iida, T., Guthrie, R. I. L., *The Physical Properties of Liquid Metals*, Clarendon Press, Oxford, 1988.
42. Mukhopadhyay, S., Seal, S., Dahotre, N., Agarwal, A., Smugeresky, J., Moody, N. *Surfaces, Interfaces in Nanostructured Materials, Trends in LIGA, Miniaturization, Nanoscale Materials*, TMS, PA, USA, 2004.
43. Porter, D. A., Easterling, K. E., *Phase Transformations in Metals, Alloys*, Chapman and Hall, London, UK, 1992.
44. Broughton, J. Q., Glimer, G. H. (1986), *J. Chem. Phys.* **84**, 5759.
45. Wang, J., Wolf, D., Philpot, S. R., Gleiter, H. (1996), *Philos. Mag. A* **73**, 517.
46. Jiang, Q., Zhao, D. S., Zhao, M. (2001), *Acta Mater.* **49**, 3143.
47. Lee, J.-G., Mori, H. (2004), *Phil. Mag.* **84**, 2675.
48. Murr, L. E., *Interfacial Phenomena in Metals, Alloys*, Addison Wesley, MA, USA, 1975.
49. Lee, J.-G., Mori, H. (2004), *Phys. Rev. B* **70**, 144 105.
50. Hagege, S., Dahmen, U. (1996), *Philos. Mag. Lett.* **74**, 259.
51. Tyson, W. R., Miller, W. A. (1977), *Surf. Sci.* **62**, 267.
52. Lee, J.-G., Mori, H. (2004), *Phys. Rev. Lett.* **93**, 235 501.
53. Ino, S. (1966), *J. Phys. Soc. Japan* **21**, 346.
54. Yagi, K., Takayanagi, K., Kobayashi, K., Honjo, G. (1975), *J. Cryst. Growth* **28**, 117.
55. Tolman, R. (1946), *J. Chem. Phys.* **17**, 333.

10
Statistical Thermodynamics and Model Calculations

Tetsuo Mohri

10.1
Introduction

Electronic structure calculations determine various ground-state properties of an alloy, while at the finite temperature an alloy is brought into excited states, and one needs to consider the entropy and free energy to determine the most stable state. This is basically the main task of statistical thermodynamics in alloy theory. Among various approaches to calculate entropy, the Bragg–Williams approximation (Bragg and Williams 1934) has been the most widely employed in the materials science community. This is mostly due to its physical transparency and mathematical simplicity. In fact, most commercial software for phase equilibria calculations are based on the Bragg–Williams approximation.

   Yet it has been realized that the Bragg–Williams approximation does not provide an accurate result in many thermodynamic calculations including a phase diagram. One of the most serious deficiencies involved in the Bragg–Williams approximation is the neglect of atomic correlations. In general, spatial extension of the atomic correlation exceeds the atomic interaction range. It is well known that atomic interaction in a typical alloy is not confined in a nearest-neighbor pair but extends well over a few distant pairs. Also, many-body interactions cannot be ignored in a less symmetric crystal structure. Hence, it is essential to include wide range of atomic correlations in the entropy formula for an accurate calculation of phase equilibria. The single-site approximate nature of the Bragg–Williams approximation, however, is by no means able to take this into account. The modern theory of alloys, therefore, demands more reliable statistical thermodynamic theory to incorporate atomic correlations.

   The cluster variation method (Kikuchi 1951; hereafter abbreviated to CVM) has been regarded as one of the most reliable theoretical tools in dealing with atomic correlations, and the main focus of the present chapter is placed on the introduction and applications of the CVM. Two advantageous features are particularly emphasized in Section 10.2. The first is that the internal energy calculations are combined coherently with entropy calculations by the CVM through common

Alloy Physics: A Comprehensive Reference. Edited by Wolfgang Pfeiler
Copyright © 2007 WILEY-VCH Verlag GmbH & Co. KGaA, Weinheim
ISBN: 978-3-527-31321-1

configurational variables termed correlation functions. The reader will see how the atomic correlation on a discrete lattice is dealt with by the CVM and how the energy and entropy calculations are combined to construct a free energy in an efficient and accurate manner. The second advantage is that the CVM can be extended *naturally* to a time domain to study kinetics. The kinetics version of the CVM is termed the path probability method (Kikuchi 1966) and a general framework of the theory is introduced. Thereby, the statistical thermodynamics and kinetics on a discrete lattice are discussed consistently in Section 10.2.

Many of traditional text books of statistical thermodynamics do not refer much to the microstructure. However, in view of the fact that most of the macroscopic properties of an alloy, including its mechanical and electric properties, are controlled by microstructure, the discussion of statistical thermodynamics without microstructure is not very useful in alloy physics. Recent advancement of the phase field theory (L.Q. Chen 2002) in describing the microstructure and its time evolution process is quite remarkable. The versatile feature of the phase field method (PFM) has been amply demonstrated in various calculations including crystal growth, twin boundaries, evolution/devolution of magnetic domains, etc. Hence, in Section 10.3, a general framework of the PFM is introduced, which will be applied and extended to the study of order–disorder transition of Fe-based alloys in Section 10.4. The basis of the PFM can be traced back to the celebrated Cahn–Hilliard equation and the time-dependent Ginzburg–Landau equation. After deriving these equations, two kinds of interface properties, the width and the velocity, are calculated. It is shown that, within the PFM, the interface is not a special entity to be described separately but it is merely a spatial localization of field variables.

In order to materialize the theoretical frameworks introduced in Sections 10.2 and 10.3 in real alloy systems, two kinds of calculations recently performed by the author's group on Fe-based binary alloy systems are introduced in Section 10.4. The first is the calculation of phase diagrams. By combination of electronic structure calculations with the CVM, the reader will see that the phase diagram is obtained with surprisingly high accuracy from first principles. In fact, such a fruitful outcome proves the rigorous framework of statistical thermodynamics introduced in this section.

The second is an ongoing subject. As will be seen, the behaviors of electrons and atoms on a discrete lattice are the subjects in Sections 10.2 and 10.3, respectively, while that of microstructure in a continuum medium is described efficiently. This is due to the differences of length scale and time constant involved in each category. How to unify these two kinds of behavior into a single mathematical framework is quite a difficult subject, which is often addressed in terms of multiscale calculation. Although coarse graining operation in statistical physics is known as the clue to settling the issue, there is no fixed recipe for applying the rigorous principle in order to elucidate varied phenomena in alloys. This is one of the most challenging subjects in statistical thermodynamics, and it is realized that a natural consequence expected from the challenge is the performance of the first-principles microstructural evolution calculation. At the end of this section, one potential approach based on the CVM and PFM attempted by the author's group is discussed.

The body of the present chapter is based on the author's own work, but the studies referred to do not cover all the important contributions in this field. Various excellent books and monographs (Khachaturyan 1983; Ducastelle 1991), and review articles (de Fontaine 1979, 1994) exist on the subject matter discussed in this chapter.

10.2
Statistical Thermodynamics on a Discrete Lattice

10.2.1
Description of Atomic Configuration

A microscopic approach to alloy phase equilibria on a discrete lattice has been termed configurational thermodynamics (de Fontaine 1979) and a variable which defines the atomic configuration on a lattice is called a configurational variable. It is essential to define configurational variables properly and efficiently for a systematic study of configurational thermodynamics. For simplicity, the discussion in this article is limited to a binary system in which two atomic species A and B are distributed with a specified composition.

Among configurational variables, the most primitive one is the point probability $x_i(p)$, which specifies the probability of finding a species i (either A or B) on a given lattice point p. It should be noted that this is nothing but a concentration of species i if all the lattice points are indistinguishable. If, on the other hand, lattice points are distinguishable under a symmetry operation, the entire lattice is divided into sublattices. In this case, $x_i(p)$ is the concentration of species i on the sublattice to which p belongs.

The probability, $y_{ij}(p, q)$, of finding species i and j in a pair of lattice points p and q can be defined in a similar way and is termed the pair probability. A special case is the one in which the lattice point q is the nearest neighbor to p, and $y_{ij}(p, p')$ is termed the nearest-neighbor pair probability, where p' is a nearest-neighbor point of p. The distant pair probabilities are defined in a similar way, and it should be noted that a pair probability is nothing but the concentration of pair configurations in a given lattice. When the lattice points are not all equivalent, the pair probability and pair concentration are defined on (between) a relevant sublattice(s). The probability for a multibody configuration such as a triangle, tetrahedron, square, etc. can be defined in a similar way and these are termed cluster probabilities or cluster concentrations.

One should remember that all cluster probabilities are interrelated by conditions. The first is the normalization condition, which is written for point and pair probabilities, respectively, as Eqs. (1) and (2).

$$\sum_i x_i = 1 \tag{1}$$

$$\sum_i \sum_j y_{ij} = 1 \tag{2}$$

The second, the geometrical condition, is given as Eq. (3).

$$x_i = \sum_j y_{ij} \tag{3}$$

The geometrical condition is a constraint imposed by the geometry of a given lattice and is extended to further larger clusters. For an fcc lattice, for instance, x_i and y_{ij} can be related to the regular triangle probability, z_{ijk}, and the nearest-neighbor tetrahedron probability, w_{ijkl}, as in Eq. (4).

$$x_i = \sum_j y_{ij} = \sum_j \sum_k z_{ijk} = \sum_j \sum_k \sum_l w_{ijkl} \tag{4}$$

Because of these two kinds of conditions, the cluster probabilities do not form a set of independent variables in the free energy functional.

An alternative way to describe atomic configuration is to employ correlation functions (Sanchez and de Fontaine 1978; Sanchez et al. 1984; Mohri et al. 1985a; Kikuchi and Mohri 1997; Mohri 2002) which are derived in the following manner. First, in order to specify an atomic species on a lattice point p, we define a spin operator $\sigma(p)$ by Eq. (5).

$$\sigma(p) = \begin{cases} +1 \\ -1 \end{cases} \text{ for } \begin{matrix} A \\ B \end{matrix} \text{ at a lattice point } p \tag{5}$$

Then, by using $\sigma(p)$, an occupation operator $\Gamma_i(p)$ is defined as in Eq. (6), where A and B atoms are specified by $i = +1$ and -1, respectively.

$$\Gamma_i(p) = \frac{1}{2}(1 + i \cdot \sigma(p)) \tag{6}$$

The occupation operator in Eq. (6) suggests that if species i specified in the subscript exists at a lattice point p the operator returns the value 1, while 0 results for the other species at p. For instance, if atom A is specified in the subscript i, Eq. (6) is explicitly written as Eq. (7).

$$\Gamma_1(p) = \frac{1}{2}(1 + 1 \cdot \sigma(p)) \tag{7}$$

When A is actually located at p, the spin operator $\sigma(p)$ is 1 according to the definition in Eq. (5), and $\Gamma_1(p)$ becomes 1. If, on the other hand, B is located at p, $\sigma(p)$ is -1 which yields $\Gamma_1(p) = 0$.

We assume a lattice in which A and B atoms are distributed in numbers $N_1(N_A)$ and $N_{\bar{1}}(N_B)$, respectively. Please note that -1 and $\bar{1}$ are used equivalently in the rest of this chapter. Then, the total number of lattice points, N, is given by the sum of N_1 and $N_{\bar{1}}$ [Eq. (8)] and the concentration of A atoms is given by Eq. (9).

$$N = N_1 + N_{\bar{1}} \tag{8}$$

$$x_1 = \frac{N_1}{N} \tag{9}$$

By operating the occupation operator on the entire lattice points, the number of A atoms, N_1, is written as Eq. (10).

$$N_1 = \sum_{p=1}^{N} \Gamma_1(p) \tag{10}$$

By substituting Eq. (10) into Eq. (9), one obtains Eq. (11).

$$x_1 = \frac{1}{N} \sum_{p=1}^{N} \Gamma_1(p) \tag{11}$$

A further substitution of Eq. (7) into Eq. (11) yields Eq. (12).

$$x_1 = \frac{1}{N} \sum_{p=1}^{N} \Gamma_1(p) = \frac{1}{N} \sum_{p} \frac{1}{2}(1 + \sigma(p)) = \frac{1}{2}\left(1 + \frac{\sum_{p} \sigma(p)}{N}\right) \tag{12}$$

The second term in the parentheses in Eq. (12) is rewritten as Eq. (13) and is called the point correlation function.

$$\xi_1 = \frac{\sum_{p=1}^{N} \sigma(p)}{N} \tag{13}$$

Then, the concentration of A atoms given by Eq. (12) is rewritten as Eq. (14).

$$x_1 = \frac{1}{2}(1 + \xi_1) \tag{14}$$

Likewise, the concentration of B atom is given as Eq. (15).

$$x_{\bar{1}} = \frac{1}{2}(1 - \xi_1) \tag{15}$$

In a more concise form, the concentration of species i is written as Eq. (16).

$$x_i = \frac{1}{2}(1 + i \cdot \xi_1) \tag{16}$$

Next, we consider the pair probability y_{ij} and a particular focus is placed on y_{11}, the concentration of A–A pairs, which is given as Eq. (17), where N_{11} is the number of A–A pairs and $N(2)$ is the total number of nearest-neighbor pairs in the lattice.

$$y_{11} = \frac{N_{11}}{N(2)} \tag{17}$$

One notices that, by employing the occupation operator, N_{11} is given by Eq. (18), where p' is a nearest-neighbor lattice point of p.

$$N_{11} = \sum_{p=1}^{N} \sum_{p'} \Gamma_1(p) \cdot \Gamma_1(p') \tag{18}$$

The substitution of Eq. (18) into Eq. (17) via Eq. (7) leads to Eq. (19).

$$\begin{aligned} y_{11} &= \frac{1}{N(2)} \cdot \sum_{p=1}^{N} \sum_{p'} \frac{1}{2^2} \cdot (1 + \sigma(p)) \cdot (1 + \sigma(p')) \\ &= \frac{1}{2^2} \cdot \left(1 + \frac{\sum_p \sigma(p)}{N} + \frac{\sum_{p'} \sigma(p')}{N} + \frac{\sum_p \sum_{p'} \sigma(p) \cdot \sigma(p')}{N(2)} \right) \end{aligned} \tag{19}$$

It should be noted that the denominators of the second and third terms in the parentheses are N, not $N(2)$, and, therefore, these terms are the point correlation functions. The fourth term is termed the pair correlation function and is formally defined as in Eq. (20).

$$\xi_2 = \frac{\sum_p \sum_{p'} \sigma(p) \cdot \sigma(p')}{N(2)} \tag{20}$$

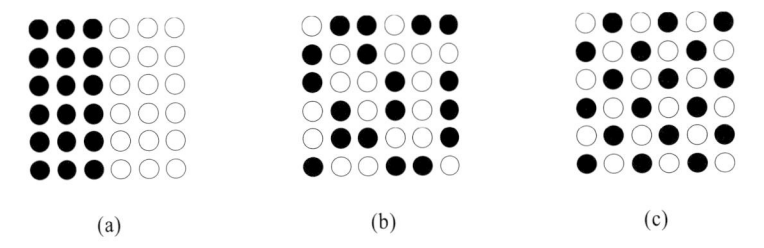

10.2 Statistical Thermodynamics on a Discrete Lattice

Fig. 10.1 Schematic illustration of (a) phase separation, (b) random solid solution, and (c) an ordered phase, on a square lattice (Mohri 2002).

Then, y_{11} is finally given as in Eq. (21).

$$y_{11} = \frac{1}{2^2} \cdot (1 + 2\xi_1 + \xi_2) \tag{21}$$

It is straightforward to derive the general expression for the pair probability in the form of Eq. (22).

$$y_{ij} = \frac{1}{2^2}(1 + (i+j) \cdot \xi_1 + i \cdot j \cdot \xi_2) \tag{22}$$

Before further generalization, it is helpful to consider the examples in Fig. 10.1, in which phase separation (a), random solid solution (b), and ordered arrangement (c) are indicated. The concentration of $A(B)$ atoms is 50 at.% for all three cases; hence Eq. (14) suggests that the point correlation function is 0, while the pair correlation functions for (a), (b), and (c) are, respectively, $+1$, 0, and -1, which are readily derived from Eq. (20). Note that the atomic arrangements on an interface and a surface in Fig. 10.1(a) are negligible for a large system.

The correlation function can be defined for a further larger cluster in a similar way. In fact, it can be easily shown that the cluster concentration of a n-point cluster is written as Eq. (23), where i_j specifies the atomic species at a lattice point j.

$$\begin{aligned} x_{i_1 i_2 \cdot i_j \cdots i_n} = \frac{1}{2^n} \{ &1 + (i_1 + i_2 + \cdots + i_j + \cdots + i_n) \cdot \xi_1 \\ &+ (i_1 \cdot i_2 + i_1 \cdot i_3 + \cdots + i_{n-1} \cdot i_n) \cdot \xi_2 \\ &+ (i_1 \cdot i_2 \cdot i_3 + \cdots + i_{n-2} \cdot i_{n-1} \cdot i_n) \cdot \xi_3 \\ &+ \cdots + (i_1 \cdot i_2 \cdots i_{n-1} \cdot i_n) \cdot \xi_n \} \end{aligned} \tag{23}$$

It is a straightforward extension of Eqs. (13) and (20) to define the n-point correlation function ξ_n as in Eq. (24), where $N(n)$ is the total number of n-point clusters in the given lattice.

$$\xi_n = \frac{\sum_{p_1}\sum_{p_2}\cdots\sum_{p_n} \sigma(p_1)\cdot\sigma(p_2)\cdots\sigma(p_n)}{N(n)} \quad (24)$$

Then, Eq. (23) is conveniently generalized as Eq. (25), where the atomic configuration $\{i_1 i_2 \cdots i_j \cdots i_n\}$ is abbreviated to $\{J\}$, and l indicates the cluster formed by n lattice points, while l' is a subcluster contained in a cluster l. $V_{\{J\}}(l, l')$ is called the V-matrix (Sanchez and de Fontaine 1978; Mohri et al. 1985a) and involves the sum of products of $i_1, i_2, \ldots i_j \ldots, i_n$, as one can see in Eq. (23).

$$X_{\{J\}}(l) = \frac{1}{2^n}\left\{1 + \sum_{l'} V_{\{J\}}(l, l') \cdot \xi_{l'}\right\} \quad (25)$$

As an example, we consider atomic configurations on a nearest-neighbor tetrahedron cluster in the fcc lattice. The total number of atomic arrangements is $2^4 = 16$ and we describe the tetrahedron cluster probability for each configuration by w_{ijkl}. According to the geometrical condition indicated by Eq. (4), the cluster probabilities of subclusters contained in the tetrahedron cluster can be written in terms of w_{ijkl}. Some examples are given in Eqs. (26–28).

$$x_1 = w_{1111} + w_{111\bar{1}} + w_{11\bar{1}1} + w_{11\bar{1}\bar{1}} + w_{1\bar{1}11} + w_{1\bar{1}1\bar{1}} + w_{1\bar{1}\bar{1}1} + w_{1\bar{1}\bar{1}\bar{1}} \quad (26)$$

$$y_{1\bar{1}} = w_{1\bar{1}11} + w_{1\bar{1}1\bar{1}} + w_{1\bar{1}\bar{1}1} + w_{1\bar{1}\bar{1}\bar{1}} \quad (27)$$

$$z_{1\bar{1}\bar{1}} = w_{1\bar{1}\bar{1}1} + w_{1\bar{1}\bar{1}\bar{1}} \quad (28)$$

When all the lattice points are equivalent (nondistinguishable), Eqs. (29–31) hold.

$$w_{111\bar{1}} = w_{11\bar{1}1} = w_{1\bar{1}11} = w_{\bar{1}111} \quad (29)$$

$$w_{11\bar{1}\bar{1}} = w_{1\bar{1}1\bar{1}} = w_{1\bar{1}\bar{1}1} = w_{\bar{1}11\bar{1}} = w_{\bar{1}1\bar{1}1} = w_{\bar{1}\bar{1}11} \quad (30)$$

$$w_{1\bar{1}\bar{1}\bar{1}} = w_{\bar{1}1\bar{1}\bar{1}} = w_{\bar{1}\bar{1}1\bar{1}} = w_{\bar{1}\bar{1}\bar{1}1} \quad (31)$$

Then, only five cluster probabilities, w_{1111}, $w_{111\bar{1}}$, $w_{11\bar{1}\bar{1}}$, $w_{1\bar{1}\bar{1}\bar{1}}$, and $w_{\bar{1}\bar{1}\bar{1}\bar{1}}$, are distinguishable. These five variables, however, should satisfy the normalization condition stated in Eq. (32).

$$w_{1111} + 4w_{111\bar{1}} + 6w_{11\bar{1}\bar{1}} + 4w_{1\bar{1}\bar{1}\bar{1}} + w_{\bar{1}\bar{1}\bar{1}\bar{1}} = 1 \quad (32)$$

Hence, one can conclude that only four $(5 - 1 = 4)$ are independent variables.

By employing Eq. (23), on the other hand, the tetrahedron cluster probability, w_{ijkl}, can be written as Eq. (33), which clearly indicates that all the relevant cluster probabilities can be written with only four correlation functions $\xi_1, \xi_2, \xi_3,$ and ξ_4.

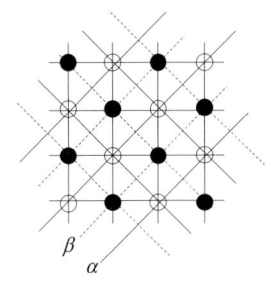

Fig. 10.2 Division of original lattice into two sublattices upon ordering.

$$w_{ijkl} = \frac{1}{2^4}\{1 + (i+j+k+l)\cdot\xi_1 + (i\cdot j + i\cdot k + i\cdot l + j\cdot k + j\cdot l + k\cdot l)$$

$$\cdot\xi_2 + (i\cdot j\cdot k + i\cdot j\cdot l + i\cdot k\cdot l + j\cdot k\cdot l)\cdot\xi_3 + i\cdot j\cdot k\cdot l\cdot\xi_4\} \quad (33)$$

This suggests that the correlation functions form a set of independent variables. Sanchez et al. (1984) further suggested that the correlation functions form an orthonormal basis in the thermodynamic configuration space, which is the basis of the cluster expansion technique described in the next section.

The discussions so far have been limited to a disordered phase in which all the lattice points are indistinguishable. For an ordered phase, on the other hand, a symmetry operation divides the entire lattice into sublattices, as is illustrated schematically in Fig. 10.2 for a two-dimensional square lattice. For such an ordered phase, normalization conditions and geometrical conditions are described by explicitly specifying the sublattices. The normalization condition of Eqs. (1) and (2) can be easily extended, respectively, as Eqs. (34) and (35) and the geometrical condition (3) is likewise extended as Eqs. (36) and (37).

$$\sum_i x_i^\alpha = 1, \quad \sum_i x_i^\beta = 1 \quad (34)$$

$$\sum_{i,j} Y_{ij}^{\alpha\alpha} = 1, \quad \sum_{i,j} y_{ij}^{\alpha\beta} = 1 \quad (35)$$

$$x_i^\alpha = \sum_j Y_{ij}^{\alpha\alpha} = \sum_j y_{ij}^{\alpha\beta} \quad (36)$$

$$x_i^\beta = \sum_j y_{ji}^{\alpha\beta} = \sum_j Y_{ji}^{\beta\beta} \quad (37)$$

It should be noted that in this particular two-dimensional square lattice, $y_{ij}^{\alpha\beta}$ (i on α and j on β) is the nearest-neighbor pair probability in the original lattice, while $Y_{ij}^{\alpha\alpha}$ ($Y_{ij}^{\beta\beta}$) (i, j on α or β) corresponds to the second-nearest neighbor pair probability.

Such distinctions of sublattices are also reflected in the correlation functions, and point and pair cluster probabilities are expressed with the correlation functions defined by explicitly taking the distinction of sublattices into account. One can readily show that Eqs. (38–41) are the extensions of Eqs. (16) and (22), where ξ_1^α and ξ_1^β are point correlation functions on α and β sublattices, respectively, and $\xi_{2,2}^{\alpha\alpha}$ and $\xi_{2,1}^{\alpha\beta}$ are pair correlation functions.

$$x_i^\alpha = \frac{1}{2}(1 + i \cdot \xi_1^\alpha) \tag{38}$$

$$x_i^\beta = \frac{1}{2}(1 + i \cdot \xi_1^\beta) \tag{39}$$

$$Y_{ij}^{\alpha\alpha} = \frac{1}{2^2}(1 + (i+j) \cdot \xi_1^\alpha + i \cdot j \cdot \xi_{2,2}^{\alpha\alpha}) \tag{40}$$

$$y_{ij}^{\alpha\beta} = \frac{1}{2^2}(1 + i \cdot \xi_1^\alpha + j \cdot \xi_1^\beta + i \cdot j \cdot \xi_{2,1}^{\alpha\beta}) \tag{41}$$

It should be noted that the subscript of the pair correlation functions distinguishes the first $(2, 1)$ and second $(2, 2)$ nearest neighbors.

10.2.2
Internal Energy

One of the most primitive descriptions of the internal energy of an alloy is based on the pair interaction model. The internal energy, E, per atom is then given by Eq. (42), where Z is the coordination number, which depends on the crystal structure, and e_{ij} is the atomic pair interaction energy between species i and j.

$$E = \frac{1}{2} \cdot Z \cdot \sum_{i,j} e_{ij} \cdot y_{ij} \tag{42}$$

The functional form of e_{ij} is not unique and often the dependence on the atomic distance is incorporated, for instance, by a Lennard-Jones type potential (Oh et al. 1994). But, for simplicity, we assume in this section that e_{ij} is a constant which depends only on the species i and j.

As was seen in Section 10.2.1, the pair probabilities $\{y_{ij}\}$ do not form a set of independent configurational variables. Hence it is desirable to describe the internal energy, E, in terms of correlation functions which are the independent configurational variables. In fact, according to the cluster expansion method (CEM) (Connolly and Williams 1983) the internal energy of an alloy, E, can be expanded in the form of Eq. (43), where ξ_k indicates a correlation function for a k-type cluster and v_k is termed the effective cluster interaction energy.

$$E(\{\xi\}) = \sum_k v_k \cdot \xi_k \qquad (43)$$

It is essential to correlate effective cluster interaction energies $\{v_k\}$ with atomic interaction energies $\{e_{ij}\}$, and this is attempted in the following way.

By substituting Eq. (22) into Eq. (42), one obtains Eq. (44).

$$\begin{aligned} E &= \frac{1}{2} \cdot Z \cdot \sum_{i,j} e_{ij} \cdot y_{ij} \\ &= \frac{1}{2} \cdot Z \cdot \frac{1}{2^2} \{(e_{AA} + 2e_{AB} + e_{BB}) + 2 \cdot (e_{AA} - e_{BB}) \cdot \xi_1 \\ &\quad + (e_{AA} + e_{BB} - 2e_{AB}) \cdot \xi_2\} \end{aligned} \qquad (44)$$

On the other hand, within the pair interaction model, k takes values 0, 1, and 2, and Eq. (43) is written explicitly as Eq. (45), where the significance of v_0 and ξ_0 will be discussed later.

$$E = \sum_{k=0}^{2} v_k \cdot \xi_k = v_0 \cdot \xi_0 + v_1 \cdot \xi_1 + v_2 \cdot \xi_2 \qquad (45)$$

The comparison of Eq. (44) with Eq. (45) immediately yields Eqs. (46–48)

$$v_0 = \frac{Z}{2^3} \cdot (e_{AA} + 2e_{AB} + e_{BB}) \qquad (46)$$

$$v_1 = \frac{Z}{2^2} \cdot (e_{AA} - e_{BB}) \qquad (47)$$

$$v_2 = \frac{Z}{2^3} \cdot (e_{AA} + e_{BB} - 2e_{AB}) \qquad (48)$$

One can see that v_1 is directly proportional to the difference between the cohesive energies of constituent elements A and B, while v_2 is the interaction parameter (also called the effective interaction energy) often employed in thermochemistry.

It is noticed that in a completely random solid solution, the correlations of atomic arrangement vanish, and the k-point cluster correlation function ξ_k is decomposed into the kth power product of the point correlation function [Eq. (49)].

$$\xi_k = \xi_1^k \qquad (49)$$

By noting that ξ_1 vanishes at 50 at.% as is deduced from Eq. (16), the internal energy of a random solid solution is given as Eq. (50).

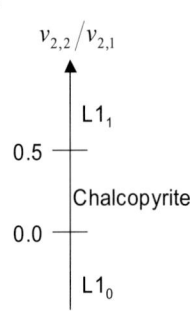

Fig. 10.3 Ground-state ordered structures at 1:1 stoichiometric composition for an fcc-based lattice with first and second nearest-neighbor interactions.

Diagram: vertical axis $v_{2,2}/v_{2,1}$ showing regions $L1_1$ (above 0.5), Chalcopyrite (between 0.0 and 0.5), and $L1_0$ (below 0.0).

$$E^{rand}(50 \text{ at.\%}) = \sum_{k=0}^{2} v_k \cdot \xi_1^k = v_0 + v_1 \cdot \xi_1 + v_2 \cdot \xi_1^2 = v_0 \qquad (50)$$

This suggests that the physical meaning of the effective interaction energy for a null cluster, v_0, is equivalent to the internal energy of a random solid solution at 50 at.%.

Although the physical meaning of v_0 is a general consequence of CEM, it should be noted that the relationships in Eqs. (46–48) are applicable only for the nearest-neighbor pair interaction model. In fact, for the description of the internal energy within the tetrahedron atomic interaction energy, ε_{ijkl}, E can be written as Eq. (51), where ω' is the number of tetrahedron clusters per lattice point, and the comparison with Eq. (43) with $k = 0, 1, 2, 3$, and 4 yields different expressions.

$$E = \omega' \cdot \sum_{i,j,k,l} \varepsilon_{ijkl} \cdot w_{ijkl} \qquad (51)$$

It is left for a reader to derive the relationships between ε_{ijkl} and $\{v_k\}$.

The actual operation of the CEM will be discussed in Section 10.4. It is worth pointing out that a stable ordered phase can be predicted once the effective interaction energies are assigned. Two schemes should be addressed. One is based on linear programming (Richards and Cahn 1971; Sanchez and de Fontaine 1981) and the other is called the method of inequality (Kanamori and Kakehashi 1977). Both yield identical ground-state structures within the pair interaction model which is our main concern. The stable ordered structures predicted at 1:1 stoichiometric composition is demonstrated in Fig. 10.3 for an fcc-based system as a function of $v_{2,2}/v_{2,1}$, which is utilized in the discussions in Section 10.4. The ground-state analysis for bcc-based systems is reported by Allen and Cahn (1972).

10.2.3
Entropy and Cluster Variation Method

The discussions so far have been limited to 0 K, at which interaction energies play a dominant role in determining an equilibrium atomic configuration, while at finite temperatures one needs to consider the entropy contributions. In this section, we focus only on the configurational entropy.

The configurational entropy for a binary system can be formally given as in Eq. (52), where k_B is the Boltzmann constant and W is the number of possible ways of distributing A and B atoms on a given lattice.

$$S = k_B \cdot \ln W \tag{52}$$

In the most primitive approach, W is given as in Eq. (53) and the substitution of Eq. (53) into Eq. (52) yields Eq. (54).

$$W = {}_N C_{N_A} \cdot {}_{N-N_A} C_{N_B} = \frac{N!}{N_A! \cdot N_B!} = \frac{N!}{\prod_i (N_i!)} \tag{53}$$

$$S = k_B \cdot \ln W = k_B \cdot \ln \frac{N!}{\prod_i (N_i!)} \tag{54}$$

By employing Stirling's formula, $\ln x! \cong x \ln x - x$, Eq. (54) is rewritten as Eq. (55).

$$S = -N \cdot k_B \left(\frac{N_A}{N} \cdot \ln \frac{N_A}{N} + \frac{N_B}{N} \cdot \ln \frac{N_B}{N} \right) \tag{55}$$

Finally, division by N yields the entropy per lattice point, given by Eq. (56).

$$\frac{S}{N} = -k_B (x_A \cdot \ln x_A + x_B \cdot \ln x_B) = -k_B \cdot \sum_i x_i \cdot \ln x_i. \tag{56}$$

This celebrated entropy formula has been known as the Bragg–Williams approximation (hereafter abbreviated to BW) (Bragg and Williams 1934).

However, we soon realize that Eq. (53) assumes that each atomic species behaves independently like a rigid sphere. In reality, atoms interact with each other, and therefore their configurations are correlated with each other. Therefore, we focus our attention on the arrangement of atomic pairs instead of individual atoms. W in Eq. (52) for this case can be formulated easily as Eq. (57), where N_2 is equivalent to $N(2)$ defined in Eq. (17).

$$W_{pair} = {}_{N_2} C_{N_{AA}} \cdot {}_{(N_2-N_{AA})} C_{N_{AB}} \cdot {}_{(N_2-N_{AA}-N_{AB})} C_{N_{BA}} \cdot {}_{(N_2-N_{AA}-N_{AB}-N_{BA})} C_{N_{BB}}$$

$$= \frac{N_2!}{N_{AA}! \cdot N_{AB}! \cdot N_{BA}! \cdot N_{BB}!}, \tag{57}$$

Note that $_mC_n$ represents $m!/\{n! \cdot (m-n)!\}$. This equation, however, does not guarantee that the total number of each species is maintained as N_i. In other words, the geometrical condition introduced in the Section 10.2.2 does not necessarily hold. When one attempts to derive W by including atomic arrangements in a bigger cluster to take the wider correlations into account, the problem becomes very complicated. The most efficient and systematic way to circumvent such an inconvenience is to employ the cluster variation method (CVM) devised by Kikuchi (1951). In the following, we reproduce the essential points of the CVM (Kikuchi and Mohri 1997; Mohri 2002) by referring to the smallest cluster, i.e., a pair cluster.

We consider an ensemble which consists of the two-atom systems in Fig. 10.4 (Mohri 2002), and the number of the systems included in the ensemble is specified by L. Then, W_{pair} under a given pair probability y_{ij} is given by Eq. (58); $\prod_{i,j}(L \cdot y_{ij})!$ in this equation appears frequently below, and is abbreviated to $\{Pair\}_L$.

$$W_{pair} = \frac{L!}{\prod_{i,j}(L \cdot y_{ij})!} = \frac{L!}{\{Pair\}_L} \tag{58}$$

Next, we consider W in the left- and right-hand columns of the ensemble independently. One can readily derive Eqs. (59) and (60), where a similar abbreviated notation is defined as $\{Pt\}_L$ (Pt stands for point) for $\prod_i (Lx_i)!$.

$$W_{Left\ Pt} = \frac{L!}{\prod_i (Lx_i)!} = \frac{L!}{\{Left\ Pt\}_L} \tag{59}$$

$$W_{Right\ Pt} = \frac{L!}{\prod_j (Lx_j)!} = \frac{L!}{\{Right\ Pt\}_L} \tag{60}$$

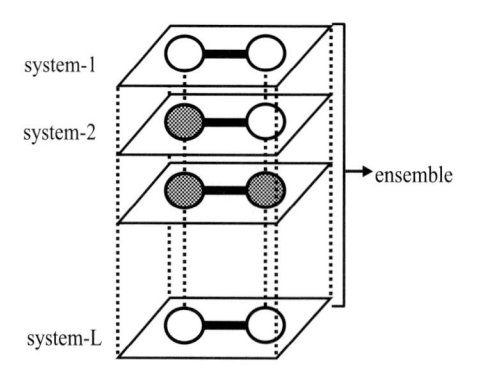

Fig. 10.4 Ensemble of two-atom systems (Mohri 2002).

Furthermore, in Eqs. (59) and (60), the left- and right-hand columns are distinguished by $\{Left\ Pt\}_L$ and $\{Right\ Pt\}_L$, respectively. The product of $W_{Left\ Pt}$ and $W_{Right\ Pt}$ is a special case of W_{Pair} in which a pair of atoms in a system behave independently. Hence, it is necessary to introduce a correction factor to obtain a general expression, Eq. (61), where G_{pair} is the correlation correction factor.

$$W_{pair} = W_{Left\ Pt} \cdot W_{Right\ Pt} \cdot G_{pair} \tag{61}$$

From eqns. (58–61), the correlation correction factor is obtained as in Eq. (62).

$$G_{pair} = \frac{(\{Pt\}_L)^2}{L! \cdot \{Pair\}_L} \tag{62}$$

Note that $\{Left\ Pt\}_L$ and $\{Right\ Pt\}_L$ are not distinguishable in the disordered phase.

The discussion so far has been limited to a two-atom system. We will now extend it to an N-atom system and consider an ensemble as shown in Fig. 10.5 (Mohri 2002). When atoms in each system behave independently, W can be given as in Eq. (63).

$$W = (W_{pt})^N = \left(\frac{L!}{\{Pt\}_L}\right)^N. \tag{63}$$

Noting Eq. (64) holds for large values of N and L, Eq. (63) is substituted into Eq. (52) to obtain Eq. (65). This is the entropy for L systems.

$$\left(\frac{L!}{\{Pt\}_L}\right)^N = \left(\frac{N!}{\{Pt\}_N}\right)^L \tag{64}$$

$$S = k_B \cdot \ln W = k_B \cdot \ln\left(\frac{N!}{\{Pt\}_N}\right)^L = k_B \cdot \ln\left(\frac{N!}{\prod_i (Nx_i)!}\right)^L \tag{65}$$

By dividing Eq. (65) by L, one can obtain the entropy per system from Eq. (66).

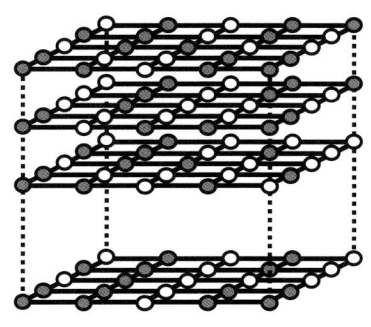

Fig. 10.5 Ensemble of N-atom systems (Mohri 2002).

$$S_{pt} = \frac{S}{L} = \frac{1}{L} \cdot k_B \cdot \ln W = \frac{1}{L} \cdot k_B \cdot \ln \left(\frac{N!}{\{Pt\}_N} \right)^L$$

$$= k_B \cdot \ln \frac{N!}{\prod_i (Nx_i!)} = k_B \cdot \ln \frac{N!}{\prod_i (N_i!)}. \tag{66}$$

This is nothing but an entropy expression given in Eq. (54) which leads to BW approximation in Eq. (56).

Now we extend the above procedures developed for a two-atom system to N-atom systems in Fig. 10.5.

The total number of atomic pairs in the system is given by $N \cdot \omega$, where ω is one-half of the coordination number Z. When atoms in each pair are correlated and cannot behave independently, one needs to make the correction given in Eq. (62) for each pair. W is therefore given by Eq. (67).

$$W = (W_{pt})^N \cdot (G_{pair})^{N\omega} \tag{67}$$

By substituting Eq. (67) into Eq. (52) via Eqs. (62) and (63), one can obtain Eq. (68).

$$S = k_B \cdot \ln \left(\frac{L!}{\{Pt\}_L} \right)^N \left(\frac{(\{Pt\}_L)^2}{L!\{Pair\}_L} \right)^{N\omega} = k_B \cdot \ln \left(\frac{(\{Pt\}_L)^{2\omega-1}}{(L!)^{\omega-1} \cdot (\{Pair\}_L)^{\omega}} \right)^N \tag{68}$$

By noting that Eq. (64) holds also for $\{Pair\}$, the entropy per system is derived as in Eq. (69).

$$S_{pair} = \frac{1}{L} \cdot k_B \cdot \ln \left[\frac{(\{Pt\}_N)^{2\omega-1}}{(N!)^{\omega-1} \cdot (\{Pair\}_N)^{\omega}} \right]^L$$

$$= k_B \cdot \ln \frac{\left(\prod_i (N \cdot x_i)! \right)^{2\omega-1}}{(N!)^{\omega-1} \cdot \left(\prod_{i,j} (N \cdot y_{ij})! \right)^{\omega}} \tag{69}$$

This is a general expression of the entropy formula within the pair approximation. By employing Stirling's approximation, one can readily show that the entropy per atom is given by Eq. (70).

$$S = k_B \cdot \left\{ (2\omega - 1) \sum_i (x_i \cdot \ln x_i - x_i) - \omega \sum_{i,j} (y_{ij} \ln y_{ij} - y_{ij}) + (\omega - 1) \right\} \tag{70}$$

As compared with Eq. (56) of the Bragg–Williams formula, one realizes that the pair probability, y_{ij}, is incorporated in Eq. (70) in addition to the point probability, x_i.

10.2 Statistical Thermodynamics on a Discrete Lattice

The largest cluster contained in the entropy formula is termed the basic cluster, which provides a measure of a level of the approximation. Within the BW approximation, a lattice point is the basic cluster, while for the pair approximation a nearest neighbor pair is the basic cluster. Based on the basic cluster, the BW approximation has often been called a *point approximation*.

The discussions above can be extended to a further larger basic cluster. In fact, the most conventional approximation employed in the fcc-based system is the tetrahedron approximation (Kikuchi 1974), in which the nearest-neighbor tetrahedron cluster is taken as the basic cluster. One can show that the entropy formula in this case can be written as Eq. (71).

$$S_{tetra} = k_B \cdot \ln \frac{\left(\prod_{i,j}(N \cdot y_{ij})!\right)^6 \cdot N!}{\left(\prod_i (N \cdot x_i)!\right)^5 \cdot \left(\prod_{i,j,k,l}(N \cdot w_{ijkl})!\right)^2} \tag{71}$$

It is proved that a cluster concentration of a subcluster, which is not shared by two basic clusters, does not appear in the entropy formula. Indeed, one notices that the nearest-neighbor triangle cluster probability z_{ijk} is not contained in Eq. (71) while point and pair cluster probabilities are both involved in the entropy formula. This is because the triangle is not shared by two tetrahedron clusters in the fcc lattice.

For an fcc lattice, a further larger basic cluster is often employed. This is the cluster which is made up of the combination of a tetrahedron and an octahedron cluster as shown in Fig. 10.6 (Mohri 2002).

The resultant entropy formula is called the tetrahedron–octahedron (TO) approximation (Sanchez and de Fontaine 1978; Mohri et al. 1985a) and is given by Eq. (72), where v_{ijklmn} is the cluster probability of the octahedron cluster.

$$S_{TO} = k_B \cdot \ln \frac{\left(\prod_{i,j,k}(N \cdot z_{ijk})!\right)^8 \cdot \left(\prod_i (N \cdot x_i)!\right)}{\left(\prod_{i,j,k,l,m,n}(N \cdot v_{ijklmn})!\right) \cdot \left(\prod_{i,j,k,l}(N \cdot w_{ijkl})!\right)^2 \cdot \left(\prod_{i,j}(N \cdot y_{ij})!\right)^6} \tag{72}$$

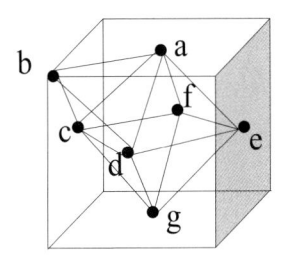

Fig. 10.6 Tetrahedron–octahedron cluster (Mohri 2002).

10.2.4
Free Energy

By combining internal energy and entropy, we derive the free energy expression. The internal energy of a system with N atoms is obtained by multiplying Eq. (42) by N, giving Eq. (73).

$$E = \frac{1}{2} \cdot N \cdot Z \cdot \sum_{i,j} e_{ij} \cdot y_{ij} \tag{73}$$

The combination with the entropy formula within the tetrahedron approximation in Eq. (71) yields the Helmholtz free energy as [Eq. (74)].

$$F = E - T \cdot S_{tetra}$$

$$= \frac{1}{2} \cdot N \cdot Z \cdot \sum_{i,j} e_{ij} \cdot y_{ij} - k_B \cdot T \cdot \ln \frac{\left(\prod_{i,j}(N \cdot y_{ij})!\right)^6 \cdot N!}{\left(\prod_{i}(N \cdot x_i)!\right)^5 \cdot \left(\prod_{i,j,k,l}(N \cdot w_{ijkl})!\right)^2} \tag{74}$$

This is easily extended to an ordered phase; the free energy of the L1$_0$ ordered phase, for instance, is given by Eq. (75)

$$F^{L1_0} = \frac{1}{2} N \cdot Z \sum_{\gamma,\delta} \sum_{i,j} e_{ij} \cdot y_{ij}^{\gamma\delta}$$

$$- T \cdot k_B \cdot \ln \left[\frac{\left\{\prod_{i,j}(Ny_{ij}^{\alpha\alpha})!\right\}\left\{\prod_{i,j}(Ny_{ij}^{\alpha\beta})!\right\}^4\left\{\prod_{i,j}(Ny_{ij}^{\beta\beta})!\right\} N!}{\left\{\prod_{i}(Nx_i^{\alpha})!\right\}^{5/2}\left\{\prod_{i}(Nx_i^{\beta})!\right\}^{5/2}\left\{\prod_{i,j,k,l}(Nw_{ijkl}^{\alpha\alpha\beta\beta})!\right\}} \right] \tag{75}$$

The minimum of Eq. (74) [and Eq. (75)] suggests an equilibrium state. Two major minimization procedures are available for given temperature and atomic pair interaction energies, e_{AA}, e_{AB} and e_{BB}. One is the natural iteration method (NIM) (Kikuchi 1974), which minimizes F in Eq. (74) [and Eq. (75)] under a constraint originating from a normalization condition which is added through the Lagrange multiplier. The other is the Newton–Raphson procedure, for which Eq. (74) [and Eq. (75)] is rewritten in terms of independent variables. The substitution of Eqs. (16), (22), and (33) into Eq. (74) formally provides the free energy F as Eq. (76), and the minimization of F is expressed as Eq. (77).

$$F = F(T, \{e_{ij}\}; \xi_1, \xi_2, \xi_3, \xi_4) = E(\{e_{ij}\}; \xi_1, \xi_2) - T \cdot S(\xi_1, \xi_2, \xi_3, \xi_4) \tag{76}$$

$$\left. \frac{\partial F}{\partial \xi_m} \right|_{T, \{e_{ij}\}, \{\xi_k\}_{k \neq m}} = 0 \tag{77}$$

When the second nearest-neighbor pair interaction energy is considered in the internal energy, E can be given as in Eq. (78), where Z_k, $e_{ij}^{(k)}$, $y_{ij}^{(k)}$, and $\xi_{2,k}$ are the coordination number, atomic pair interaction energy, cluster probability, and pair correlation function for a kth nearest-neighbor pair, respectively.

$$\begin{aligned} E &= \frac{1}{2} \cdot N \cdot \left(Z_1 \cdot \sum_{i,j} e_{ij}^{(1)} \cdot y_{ij}^{(1)} + Z_2 \cdot \sum_{i,j} e_{ij}^{(2)} \cdot y_{ij}^{(2)} \right) \\ &= E(\{e_{ij}^{(1)}\}, \{e_{ij}^{(2)}\}; \xi_1, \xi_{2,1}, \xi_{2,2}) \end{aligned} \tag{78}$$

It is noticed, however, that one cannot add entropy formulae for point [Eq. (56)], pair [Eq. (70)], and tetrahedron approximation [Eq. (71)] to Eq. (78) to construct a free energy. This is because the atomic correlation generally exceeds the range of atomic interactions, and therefore the largest cluster considered in the internal energy term should be involved in the basic cluster of the entropy term. Hence, for the internal energy term given by Eq. (78), the minimum meaningful entropy formula is the TO approximation given by Eq. (72). A combination of Eqs. (72) and (78) provides the free energy as Eq. (79).

$$\begin{aligned} F = &\frac{1}{2} \cdot N \cdot \left(Z_1 \cdot \sum_{i,j} e_{ij}^{(1)} \cdot y_{ij}^{(1)} + Z_2 \cdot \sum_{i,j} e_{ij}^{(2)} \cdot y_{ij}^{(2)} \right) - T \cdot k_B \\ &\cdot \ln \frac{\left(\prod_{i,j,k} (N \cdot z_{ijk})! \right)^8 \cdot \left(\prod_i (N \cdot x_i)! \right)}{\left(\prod_{i,j,k,l,m,n} (N \cdot v_{ijklmn})! \right) \cdot \left(\prod_{i,j,k,l} (N \cdot w_{ijkl;})! \right)^2 \cdot \left(\prod_{i,j} (N \cdot y_{ij})! \right)^6} \end{aligned} \tag{79}$$

This equation can be rewritten formally in terms of correlation functions as Eq. (80), where the subscript m of the correlation function, $\xi_{m,s}$, indicates a pair ($m = 2$), three-body ($m = 3$), four-body ($m = 4$), five-body ($m = 5$) and octahedron ($m = 6$) cluster, and the subscript s distinguishes the different types (shapes) of the cluster.

$$\begin{aligned} F(T, &\{e_{ij}^{(1)}\}, \{e_{ij}^{(2)}\}; \xi_1, \xi_{2,1}, \xi_{2,2}, \xi_{3,1}, \xi_{3,2}, \xi_{4,1}, \xi_{4,2}, \xi_{4,3}, \xi_5, \xi_6) \\ &= E(\{e_{ij}^{(1)}\}, \{e_{ij}^{(2)}\}; \xi_1, \xi_{2,1}, \xi_{2,2}) \\ &\quad - S(\xi_1, \xi_{2,1}, \xi_{2,2}, \xi_{3,1}, \xi_{3,2}, \xi_{4,1}, \xi_{4,2}, \xi_{4,3}, \xi_5, \xi_6) \end{aligned} \tag{80}$$

Table 10.1 Subclusters contained in the tetrahedron–octahedron cluster: see Fig. 10.6 (Mohri 2002).

	ξ_1	$\xi_{2,1}$	$\xi_{3,1}$	$\xi_{4,1}$	$\xi_{2,2}$	$\xi_{3,2}$	$\xi_{4,2}$	$\xi_{4,3}$	$\xi_{5,1}$	ξ_6
Lattice points	a	ab	acd	abcd	ag	ace	acde	cdef	acdeg	acdefg

The type of each correlation function appearing in Eq. (80) can be identified in Fig. 10.6 and Table 10.1 (Mohri 2002).

In summarizing the discussions above, it is necessary that the relationship of Eq. (81) should be obeyed for the free energy formula given by Eq. (82).

$$E_{max} \subset S_{max} \tag{81}$$

$$F = E(\xi_1, \xi_{2,1}, \xi_{2,2}, \ldots \xi_{E_{max}}) - T \cdot S(\xi_1, \xi_{2,1}, \xi_{2,2}, \ldots \xi_{S_{max}}) \tag{82}$$

Recalling that the second nearest-neighbor pair interaction is necessary to distinguish the stability of various ordered phases as shown in Fig. 10.3, it is realized that the TO approximation is necessary for the discussion of the phase equilibria involving these phases at a finite temperature.

The essential task to be achieved by CVM is to determine the exponent of the cluster probability such as $\{Pt\}, \{Pair\}, \ldots \{subcluster\}$ in the entropy formula. We note that there are alternative ways, and Barker's formula (Barker 1953) has been recognized as one of the most powerful formulae with which to determine the exponents in a systematic way.

It is worth pointing out that besides the rigorous CVM-based free energy expressions discussed above, a number of free energy models have been proposed. Among them, the CALPHAD type of model has been recognized as one of the most powerful in dealing with practical multicomponent alloy systems. The essential points are summarized in the Appendix.

10.2.5
Relative Stability and Intrinsic Stability

In the discussion of the stability of a system, one needs to distinguish the *relative* stability and the *intrinsic* stability of the system. The relative stability can be measured by comparing the magnitude of the energy of a phase of interest against those of other competing phases, while the intrinsic stability is defined as the stability of a system against fluctuations. We start the discussion with the relative stability.

Figure 10.7 is a schematic diagram of the heats of formation of various ordered phases at four stoichiometric compositions, A_3B, A_2B, AB, and AB_3. The energy reference state, namely the zero energy level, is the segregation limit, which is

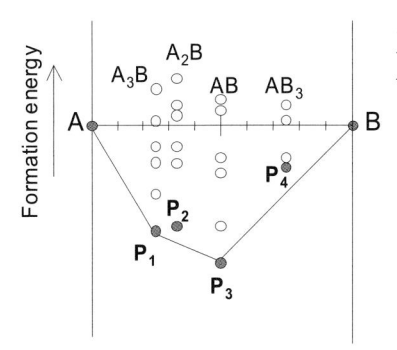

Fig. 10.7 Schematic diagram of the heats of formation of ordered phases. P_1 and P_3 are the table ordered phases.

defined as the concentration average of the total energies of pure A and pure B. At each stoichiometric composition, heats of formation of several ordered phases are plotted and the phase at the lowest energy is the most stable phase at that stoichiometry. For instance, P_1 is the most stable phase among six competing phases at the stoichiometric composition A_3B. The stability of other phases is discussed based on the hierarchy of energies. For instance, the phase just above P_1 in the figure is less stable than P_1 but more stable than all the other phases at the same stoichiometry. Likewise, at each stoichiometry, the phase with the lowest energy is the most stable phase; it is marked by a solid circle and are denoted as P_i, where $i = 1, 2, 3$, and 4 correspond to A_3B, A_2B, AB, and AB_3 stoichiometries, respectively.

Since the formation energies of these phases, P_1–P_4, the lowest at each stoichiometry, are negative with reference to the segregation limit, these phases are stable against the phase separation into A and B. However, the minimum energy state is more carefully examined by connecting each point. One realizes that P_2 (P_4) is located above the connecting line of P_1 (P_3) and P_3 (B), indicating that P_2 (P_4) is unstable against the separation into P_1 (P_3) and P_3 (B). This way, the energy apices being concave downward is the necessary condition of the most stable state which is realized in the system. The difference in the formation energies is a measure of the relative stability. If this is small, the hierarchy may be reversed by some external or internal causes.

In order to discuss intrinsic stability, we start with a thermodynamic potential given as Eq. (83), where $\{\varphi_i\}$ forms an independent set of variables.

$$f = f(\{\varphi_i\}) \tag{83}$$

Among various thermodynamic potentials, we suppose f is an internal energy, and entropy S, volume V, and the number of species (equivalently the number of lattice points for a pure metal) N are assigned to the first three variables, φ_1, φ_2, and φ_3, respectively. Note that these are all extensive variables. Then, the explicit form of the first-order derivative provides Eqs. (84–86), where T, p, and μ are the temperature, pressure, and chemical potential, which are the conjugate variables of the entropy, volume, and number of species, respectively.

$$\frac{\partial f}{\partial \varphi_1} = \frac{\partial f}{\partial S} = T \tag{84}$$

$$\frac{\partial f}{\partial \varphi_2} = \frac{\partial f}{\partial V} = -p \tag{85}$$

$$\frac{\partial f}{\partial \varphi_3} = \frac{\partial f}{\partial N} = \mu \tag{86}$$

When the second-order derivative, $\partial^2 f / \partial \varphi_i^2$, is taken, the result provides a measure of the stability of the system. In fact, a system generally undergoes incessant fluctuations of the variables $\{\varphi_i\}$, and the positive second-order derivative assures the restoring force toward the equilibrium state. It may be easier to grasp the physical significance by examining S, V, and N_i explicitly. Then, one obtains Eqs. (87–89) as the stability criteria.

$$\frac{\partial^2 f}{\partial \varphi_1^2} = \frac{\partial^2 f}{\partial S^2} = \frac{\partial T}{\partial S} = \frac{\Delta T}{\Delta S} > 0, \tag{87}$$

$$\frac{\partial^2 f}{\partial \varphi_2^2} = \frac{\partial^2 f}{\partial V^2} = -\frac{\partial p}{\partial V} = -\frac{\Delta p}{\Delta V} > 0 \tag{88}$$

$$\frac{\partial^2 f}{\partial \varphi_3^2} = \frac{\partial^2 f}{\partial N_i^2} = \frac{\partial \mu_i}{\partial N_i} = \left(\frac{\partial^2 f}{\partial c_i^2}\right) > 0 \tag{89}$$

Eq. (87) claims that if the heat is absorbed ($\Delta S > 0$) in the system, the temperature should increase ($\Delta T > 0$); this is the thermal stability criterion. Likewise, if the volume of the system is decreased ($\Delta V < 0$), the pressure should increase ($\Delta p > 0$); this is the mechanical stability criterion given by Eq. (88). These coincide with our general intuition. Finally, the third equation (89) indicates the chemical stability criterion which, if negative, is the condition of spinodal decomposition.

The arguments can be extended to other variables. As was discussed in the previous section, the configurational free energy is a multivalue function of correlation functions and within the tetrahedron approximation of the CVM, the free energy of a disordered phase given by Eq. (74) is rewritten symbolically in Eq. (90) by converting atomic interactions and cluster probabilities to effective interactions and correlation functions, respectively.

$$f^{dis}[\{\xi_i\}] = f(T, v_2; \xi_1, \xi_2, \xi_3, \xi_4) \tag{90}$$

Likewise, the free energy of the L1$_0$ ordered phase given by Eq. (75) is expressed as Eq. (91).

$$f^{L1_0}[\{\xi_i\}] = f(T, v_2; \xi_1^\alpha, \xi_1^\beta, \xi_2^{\alpha\alpha}, \xi_2^{\alpha\beta}, \xi_2^{\beta\beta}, \xi_3^{\alpha\alpha\beta}, \xi_3^{\alpha\beta\beta}, \xi_4^{\alpha\alpha\beta\beta}) \tag{91}$$

Note that the variables before the semicolon are given a priori, while those after it are unknown, i.e., they are the variational parameters.

When Eqs. (90) and (91) are minimized with respect to each correlation function [Eq. (92)], one can determine the phase equilibria by a common tangent construction.

$$\left(\frac{\partial f^{dis}}{\partial \xi_i}\right)_{T,\xi_j \neq \xi_i} = 0 \quad \text{and} \quad \left(\frac{\partial f^{L1_0}}{\partial \xi_i^{\gamma}}\right)_{T,\xi_j \neq \xi_i} = 0 \tag{92}$$

The $L1_0$-disorder phase diagram calculated in this way is demonstrated in Fig. 10.8 (Mohri 1992, 1994a) in the vicinity of 50 at.%. The temperature axis is normalized with respect to the nearest-neighbor effective pair interaction energy v_2. The transition temperature is determined as shown, and is 1.893. Since the pair interaction energy is independent of concentration, the phase boundary is symmetrical around 50 at.%.

For simplicity, the following discussion is limited to the $L1_0$ ordered phase at a 1:1 stoichiometric composition. Then, due to the symmetry of the $L1_0$ ordered phase, some of the correlation functions in Eq. (91) are mutually related, as $\xi_1^{\beta} = -\xi_1^{\alpha}$, $\xi_2^{\beta\beta} = \xi_2^{\alpha\alpha}$, and $\xi_3^{\alpha\beta\beta} = -\xi_3^{\alpha\alpha\beta}$; hence the resultant free energy is rewritten as Eq. (93).

$$f^{L1_0}[\{\xi_i\}] = f(T, v_2; \xi_1^{\alpha}, \xi_2^{\alpha\alpha}, \xi_2^{\alpha\beta}, \xi_3^{\alpha\alpha\beta}, \xi_4^{\alpha\alpha\beta\beta}). \tag{93}$$

It should be noted that ξ_1^{α} serves as a long-range order (LRO) parameter while all the others are short-range order (SRO) parameters. By following a convention,

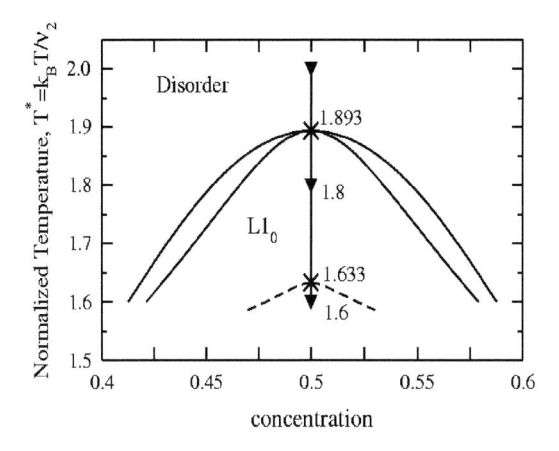

Fig. 10.8 $L1_0$ phase diagram of $L1_0$-disorder system in the vicinity of 50 at.% as calculated by tetrahedron approximation of the CVM. The temperature axis is normalized by the nearest-neighbor pair interaction energy. The broken line indicates the locus of the spinodal ordering temperature (Mohri 1992, 1994a).

the LRO parameter is expressed by η in the following discussion. When the free energy is minimized for a given temperature and LRO parameter, η, with respect to the SRO parameters only [Eq. (94)], one can obtain the free energy, f, as a function of the LRO parameter, η, at each temperature. This is constrained minimization.

$$\left(\frac{\partial f^{L1_0}}{\partial \{\xi_i\}}\right)_{T,\eta} = 0 \tag{94}$$

In Fig. 10.9 (Mohri 2000a,b) $f(\eta)$ is shown at six representative temperatures at 1:1 stoichiometry. A η value of unity corresponds to a perfectly ordered phase, which is realized only in the ground state, while a null value indicates a complete random solid solution, and a finite value suggests some degree of order. At the highest temperature, $T = 2.30$ (Fig. 10.9a), which is in the disordered phase region (see Fig. 10.8), the lowest free energy is found at $\eta = 0$, and at the lowest temperature, $T = 1.60$ (see Fig. 10.9f), which is in the $L1_0$ ordered phase region, the minimum in free energy appears very close to unity. The shifting of

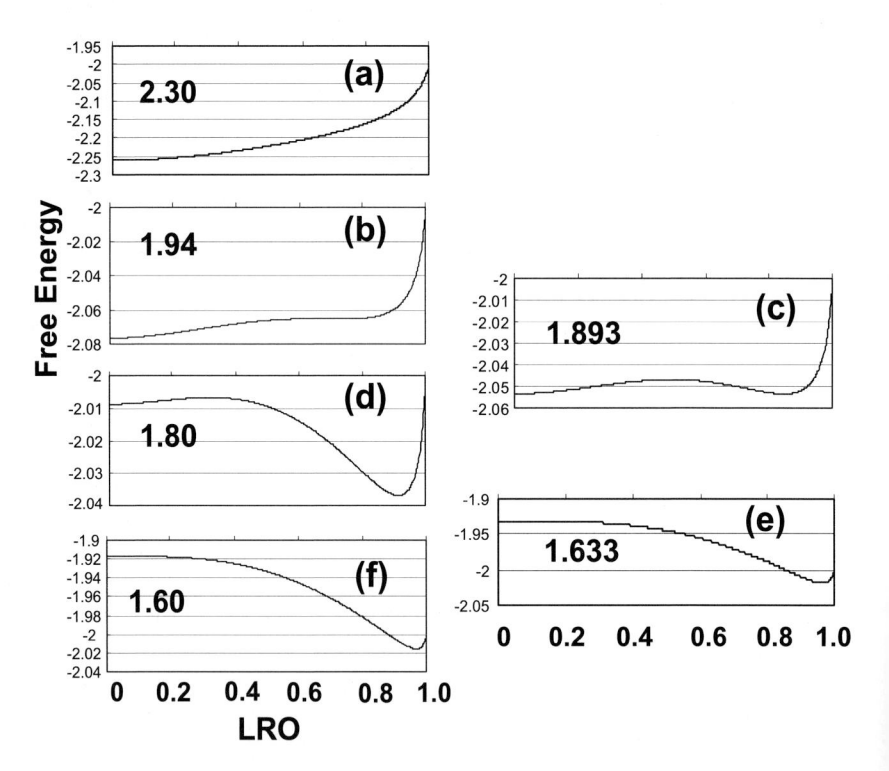

Fig. 10.9 Free energy versus LRO parameter at six temperatures. The number on each graph is the temperature normalized with respect to the nearest-neighbor or pair interaction energy (Mohri 2000a,b).

energy extrema between these two temperatures is demonstrated well in the figures.

Two characteristic temperatures should be addressed. One is the temperature $T = 1.893$ (Fig. 10.9c), at which two free energy minima can be connected by a common tangent, which is the indication of the transition temperature, T_t. The other is $T = 1.633$ (Fig. 10.9e), below which the free energy hump between the disordered and ordered phases disappears and the free energy curve decreases monotonically from the disordered state to the ordered state. This temperature is called the spinodal ordering temperature (de Fontaine 1975; Mohri et al. 1985b), T_{so}, of which the trace is indicated by a broken line in Fig. 10.8. A spinodal disordering temperature, $T_{sd} = 1.94$, corresponding to T_{so} (Mohri 2000a,b) appears in the disordered phase region although it is not shown in the phase diagram in Fig. 10.8.

Let us suppose that a disordered phase is quenched from $T = 2.3$ down to 1.80. The final equilibrium state is the ordered phase characterized by η close to 0.90 (see Fig. 10.9d); however, due to the free energy hump, the system remains in a metastable state at $\eta = 0$. In order for the system to transit to the final equilibrium state, fluctuations are necessary to overcome the free energy barrier. This is an indication of the first-order transition, while, if the system is quenched down to 1.60 (Fig. 10.9f), the system transits to the final ordered phase ($\eta = 0.969$) spontaneously.

It should be clearly noted that the order of the transition is determined at T_t. If T_t coincides with T_{so}, the transition is of the second order. The larger the separation between these two temperatures, the more noticeable is the first-order nature of the transition. Note that for a second-order transition, not only T_{so} but also T_{sd} merges into T_t.

In this connection, metastability of a single system is quite clear. The disordered state at $T = 1.80$ and the ordered state at $T = 1.94$ are metastable states, which eventually transit to an ordered state and disordered state, respectively, with the aid of configurational fluctuations brought about by thermal energy. In fact, fluctuations are detected by diffuse scattering experiments and are predicted theoretically by Fourier transformation of the second-order derivative of the free energy with respect to the correlation functions (de Fontaine 1975; Mohri et al. 1985b). By using the Hermitean properties of the second-order derivative matrix of the free energy, one can diagonalize the matrix, which provides eigenvalues. The temperature at which one of the eigenvalues goes through zero is defined as the spinodal ordering temperature. For this, k-space formulation is essential and interested readers should consult the original articles (de Fontaine 1975; Mohri et al. 1985b).

10.2.6
Atomistic Kinetics by the Path Probability Method

When a system is in a nonequilibrium state, it transits toward an equilibrium state. How the system changes its state to the most stable one is a matter of the kinetics. Kikuchi devised the path probability method (PPM) (Kikuchi 1966) as a natural extension of the cluster variation method to the time domain, and various calculations, including transition kinetics across a phase boundary, relaxation kinetics within an ordered domain, steady-state kinetics under cyclic temperature variations, and fluctuation kinetics in the disordered states, were performed by the present author (Mohri 1990, 1992, 1993, 1994a,b, 1996, 1997, 1999, 2000a,b; Kikuchi et al. 1992; Mohri and Ikegami 1993a,b; Mohri et al. 1995, 1996a,b, 1997; Mohri and Miyagishima 1998).

As has been mentioned, the PPM is an extension of the CVM to the time domain, and there are correspondences between the kinetic variables of the PPM and the thermodynamic functions of the CVM. Corresponding to the cluster probabilities of the CVM are path variables in the PPM, which correlate the cluster probabilities at times t and $t+\Delta t$. Equations (95–98) are examples of the point and pair path variables in a disordered state, where $X_{i,j}(t, t+\Delta t)$ ($Y_{ij,kl}(t, t+\Delta t)$) are point (pair) path variables that describe the configurational transition on a lattice point (nearest-neighbor lattice points) from i (ij) at time t to j (kl) at $t+\Delta t$.

$$x_1(t) = X_{1,1}(t, t+\Delta t) + X_{1,\bar{1}}(t, t+\Delta t) \tag{95}$$

$$x_1(t+\Delta t) = X_{1,1}(t, t+\Delta t) + X_{\bar{1},1}(t, t+\Delta t) \tag{96}$$

$$y_{11}(t) = Y_{11,11}(t, t+\Delta t) + Y_{11,\bar{1}1}(t, t+\Delta t) + Y_{11,1\bar{1}}(t, t+\Delta t) \tag{97}$$

$$y_{11}(t+\Delta t) = Y_{11,11}(t, t+\Delta t) + Y_{1\bar{1},11}(t, t+\Delta t) + Y_{\bar{1}1,11}(t, t+\Delta t) \tag{98}$$

For an ordered phase, the distinction of the sublattice is necessary.

A counterpart of the grand potential (free energy) of the CVM is the path probability function (PPF), P, in the PPM which is given as the product of three terms, P_1, P_2, and P_3. Each term for the disordered phase is given in Eqs. (99–101), where N is the number of atoms, θ is the spin flipping probability per unit time, which corresponds to the diffusivity in an alloy system, and ΔE is the change of the internal energy during Δt, given by Eq. (102).

$$P_1 = (\theta \cdot \Delta t)^{N \cdot X_{1,\bar{1}}} (\theta \cdot \Delta t)^{N \cdot X_{\bar{1},1}} (1 - \theta \cdot \Delta t)^{N \cdot X_{1,1}} (1 - \theta \cdot \Delta t)^{N \cdot X_{\bar{1},\bar{1}}} \tag{99}$$

$$P_2 = \exp\left(-\frac{\Delta E}{2k_B T}\right) \tag{100}$$

$$P_3 = k_B \cdot \ln \frac{\left\{\prod_{ij,kl}(NY_{ij,kl})!\right\}^6 \cdot N!}{\left\{\prod_{i,j}(NX_{i,j})!\right\}^5 \cdot \left\{\prod_{ijkl,mnop}(NW_{ijkl,mnop})!\right\}^2} \tag{101}$$

$$\Delta E = \frac{1}{2} \cdot N \cdot Z \cdot \sum_{i,j} e_{ij} \cdot \{y_{ij}(t+\Delta t) - y_{ij}(t)\} \tag{102}$$

P_1 describes a noncorrelated spin flipping process over the entire lattice, which is the Markov process. P_2 is the conventional thermal activation process and describes the probability of gaining thermal assistance from a heat reservoir. Finally, P_3 most characterizes the PPM, and one may see the similarity with the entropy formula of the CVM as given in Eqs. (69), (71), and (72). In fact, P_3 provides the freedom of the microscopic path, while the CVM entropy describes the configurational freedom. Like the CVM, the PPM forms a hierarchic structure and Eqs. (99–101) in the tetrahedron approximation may be compared with the entropy formula of the CVM given by Eq. (71). It is worth pointing out that the free energy and/or its derivative are not explicitly considered in the path probability function, in marked contrast with other kinetic theories.

The PPM is based on the variational principle and the most probable path of the time evolution is determined so that the PPF, P, is maximized at each time step through Eq. (103), where P is the path probability function given by $P = P_1 \cdot P_2 \cdot P_3$, and $\Xi_{ijk\ldots,lmn\ldots}$ denotes a path variable which describes the transition from $ijk\ldots$ at time t to $lmn\ldots$ at time $t + \Delta t$.

$$\frac{\partial P}{\partial \Xi_{ijk\ldots,lmn\ldots}} = 0 \qquad (103)$$

Figure 10.10 is a schematic illustration of the physical principles involved in the PPM. We suppose there is a transition from some initial state to a final state via various intermediate states. The energy of the initial and final states is indicated as E_{in} and E_f, respectively, while those in the intermediate states are designated E_1, E_2, and E_3. From the initial to final states, there exist various paths which are drawn as tubes in the figure. It should be noted that the energy differences are the same, no matter which paths the system takes. Hence, the P_2 term does not depend on the path. The P_3 term indicates the microscopic freedom of the transition, which is implied by the width of the path or tube. It is natural for the system to take a series of widest paths if the energy difference is the same. The variational principle claimed in the PPM is therefore interpreted as the selection of the quickest passage among the transition paths.

Once the PPF is maximized, the most probable path is determined by optimizing the path variable $\{\Xi_{ijk\ldots,lmn\ldots}\}$. With these path variables, $\{\Xi_{ijk\ldots,lmn\ldots}\}$, the clus-

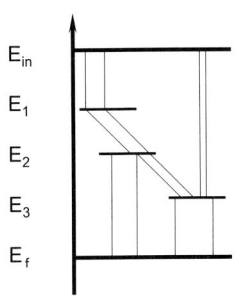

Fig. 10.10 Schematic illustration of the transition paths from the initial state E_{in} to the final state E_f.

ter probability at time $t + \Delta t$ can be uniquely determined. Such a relationship is exemplified in the point and pair probabilities in the disordered phase according to Eqs. (104) and (105), which are readily obtained from Eqs. (95–98).

$$x_1(t + \Delta t) = x_1(t) + X_{\bar{1},1}(t, t + \Delta t) - X_{1,\bar{1}}(t, t + \Delta t) \tag{104}$$

$$y_{11}(t + \Delta t) = x_1(t) + X_{\bar{1},1}(t, t + \Delta t) - X_{1,\bar{1}}(t, t + \Delta t), \tag{105}$$

Once the initial equilibrium state is assigned by the CVM, the nonequilibrium time evolution processes of cluster probabilities are determined by the PPM. Moreover, it has been proved that the final steady state obtained in the PPM in the limit of $t \to \infty$ agrees with the equilibrium state calculated independently by the CVM.

Finally, it should be noted that the formulation of the PPF depends entirely on the type of kinetics assumed in the study. The vacancy-mediated or exchange kinetics (Kawasaki dynamics; Kawasaki 1966) requires a large number of path variables, which makes numerical operations quite intractable. The spin flipping kinetics (Glauber dynamics; Glauber 1953), which is the type used in the above study, generally does not conserve the species with time; however, conservation is assured at 1:1 stoichiometric composition without imposition of any additional constraints. In this regard, the spin system approximates to an alloy system at 1:1 stoichiometry.

PPM is a powerful tool for studying nonequilibrium transition kinetics. In particular, not only near-equilibrium but also far-from-equilibrium transitions can be studied, since the PPM does not explicitly deal with the free energy. However, the number of variables is quite huge and the applicability of the PPM is mostly limited to a homogeneous system.

10.3
Statistical Thermodynamics on Continuous Media

10.3.1
Ginzburg–Landau Free Energy

In the previous section, the free energy was described by taking into account the discrete nature of the crystal lattice. In this section, we ignore the atomistic picture and focus on a continuous medium. Hence, the heterogeneity in the atomistic scale is leveled out, and in this section an inhomogeneity is discussed on a larger scale in terms of field variables, $\{\eta_i\}$, including concentration and order parameters. As was originally derived by Cahn et al. (Cahn and Hilliard 1958; Cahn 1961), the free energy is assumed to be written as Eq. (106), where we focus only on a single field variable, η.

$$f = f(\eta, \nabla\eta, \nabla^2\eta, \ldots) \tag{106}$$

10.3 Statistical Thermodynamics on Continuous Media

Then we attempt a Taylor expansion around the homogeneous system, f_0, characterized by Eq. (107), which yields Eq. (108), in which we ignore the terms higher than the second-order derivative and parabolic terms.

$$f_0(\eta) = f(\eta, 0, 0, 0, \ldots) \tag{107}$$

$$\begin{aligned} f &= f(\eta, \nabla\eta, \nabla^2\eta, \ldots) \\ &\approx f_0(\eta) + \sum_i \left.\frac{\partial f}{\partial\left(\frac{\partial\eta}{\partial x_i}\right)}\right|_0 \cdot \left(\frac{\partial\eta}{\partial x_i}\right) + \sum_{i,j} \left.\frac{\partial f}{\partial\left(\frac{\partial^2\eta}{\partial x_i \partial x_j}\right)}\right|_0 \cdot \left(\frac{\partial^2\eta}{\partial x_i \partial x_j}\right) \\ &\quad + \frac{1}{2}\sum_{i,j} \left.\frac{\partial^2 f}{\partial\left(\frac{\partial\eta}{\partial x_i}\right) \cdot \partial\left(\frac{\partial\eta}{\partial x_j}\right)}\right|_0 \cdot \left(\frac{\partial\eta}{\partial x_i}\right) \cdot \left(\frac{\partial\eta}{\partial x_j}\right) \end{aligned} \tag{108}$$

For simplicity, we confine ourselves to an isotropic system, the symmetry requirement of which eliminates the second term and cross-terms contained in the third and fourth terms on the right-hand side. Equation (108) can then be simplified to Eqs. (109–110), where κ_1 and κ_2 are given as in Eqs. (111) and (112), respectively.

$$\begin{aligned} f &= f(\eta, \nabla\eta, \nabla^2\eta, \ldots) \\ &\approx f_0(\eta) + \sum_i \left.\frac{\partial f}{\partial\left(\frac{\partial^2\eta}{\partial x_i^2}\right)}\right|_0 \cdot \left(\frac{\partial^2\eta}{\partial x_i^2}\right) + \frac{1}{2}\sum_i \left.\frac{\partial^2 f}{\partial\left(\frac{\partial\eta}{\partial x_i}\right)^2}\right|_0 \left(\frac{\partial\eta}{\partial x_i}\right)^2 \end{aligned} \tag{109}$$

$$= f_0(\eta) + \kappa_1 \nabla^2\eta + \kappa_2 (\nabla\eta)^2 \tag{110}$$

$$\kappa_1 = \left.\frac{\partial f}{\partial\left(\frac{\partial^2\eta}{\partial x_i^2}\right)}\right|_0 \tag{111}$$

$$\kappa_2 = \frac{1}{2} \cdot \left.\frac{\partial^2 f}{\partial\left(\frac{\partial\eta}{\partial x_i}\right)^2}\right|_{0_2} \tag{112}$$

Hence, the free energy of the entire system is written as Eq. (113).

$$F = \int_V f(\eta, \nabla\eta, \nabla^2\eta, \ldots) \cdot dV \approx \int_V \{f_0(\eta) + \kappa_1 \nabla^2\eta + \kappa_2 (\nabla\eta)^2\} \cdot dV \tag{113}$$

According to Gauss's divergence theorem, Eq. (114) holds for a vector quantity **F**, where **n** is a unit normal vector on a surface element dS.

$$\int_V \text{div}\, \mathbf{F} \cdot dV = \int_S \mathbf{F} \cdot \mathbf{n} \cdot dS \tag{114}$$

By noting that Eq. (115) applies, with the conventional assumption of vanishing flux at a surface [Eq. (116)], one realizes that Eq. (115) vanishes, which yields Eq. (117).

$$\int_V div(\kappa_1 \nabla \eta) \cdot dV = \int_V \left(\frac{d\kappa_1}{dc} \cdot (\nabla \eta)^2 + \kappa_1 \cdot \nabla^2 \eta \right) \cdot dV = \int_S (\kappa_1 \cdot \nabla \eta) \cdot \mathbf{n} \cdot dS \tag{115}$$

$$\nabla \eta|_S = 0 \tag{116}$$

$$\int_V \kappa_1 (\nabla^2 \eta) \cdot dV = -\int_V \frac{d\kappa_1}{d\eta} (\nabla \eta)^2 \cdot dV \tag{117}$$

By substituting this into Eq. (113), the final form of the free energy of the entire system is derived as Eq. (118), where κ_η is the gradient energy coefficient defined as Eq. (119).

$$F = \int_V \{f_0(\eta) + \kappa_\eta (\nabla \eta)^2\} \cdot dV \tag{118}$$

$$\kappa_\eta = -\frac{d\kappa_1}{d\eta} + \kappa_2 \tag{119}$$

For a multivalued free energy function $f(\eta_1, \eta_2, \ldots) = f(\{\eta_i\})$, one can show that the above free energy functional, Eq. (118), is further generalized as Eq. (120).

$$F = \int_V \left\{ f_0(\{\eta_i\}) + \sum_i \kappa_i \cdot (\nabla \eta_i)^2 \right\} \cdot dV \tag{120}$$

This particular expression of the free energy functional is known as the Ginzburg–Landau type free energy.

10.3.2
Diffusion Equation and Time-Dependent Ginzburg–Landau Equation

Except for a highly nonequilibrium process for which a free energy itself is not suitably defined, a relaxation process can be characterized as the continuous decrease of the free energy with time [Eq. (121)].

$$\frac{\partial F}{\partial t} \leq 0 \tag{121}$$

By applying Eq. (121) to Eq. (118), one obtains Eq. (122).

$$\frac{\partial F}{\partial t} = \int_V \frac{\delta F}{\delta \eta} \cdot \frac{\partial \eta}{\partial t} \cdot dV \leq 0 \tag{122}$$

At this stage, one should distinguish between conservative and nonconservative field variables. Concentration, c, and LRO/SRO parameters, respectively, are typical examples of each category. In the following discussion, we explicitly employ c for the conservative variables, while $\eta(\{\eta_i\})$ is limited to nonconservative variables.

It is soon realized that the necessary condition for Eq. (122) to hold is given in Eq. (123), where L is a positive constant.

$$\frac{\partial \eta}{\partial t} = -L \cdot \frac{\delta F}{\delta \eta} \tag{123}$$

This is generalized to a free energy given as a multivalued function of nonconservative variables as Eq. (124).

$$\frac{\partial \eta_i}{\partial t} = -\sum_j L_{ij} \cdot \frac{\delta F}{\delta \eta_j} \tag{124}$$

Corresponding to Eq. (122) when considering the concentration c as a conservative variable, the derivative of the free energy is explicitly given by Eq. (125).

$$\frac{\partial F}{\partial t} = \int_V \frac{\delta F}{\delta c} \cdot \frac{\partial c}{\partial t} \cdot dV \leq 0 \tag{125}$$

Additionally, the conservative quantity should obey the continuity equation (126).

$$\frac{\partial c}{\partial t} = -\nabla \cdot \mathbf{J} \tag{126}$$

By substituting Eq. (126) into Eq. (125), one obtains Eq. (127), for which Gauss's divergence theorem is applied with the boundary condition of the vanishing flux at the surface, as we did in Eqs. (114–116).

$$\frac{\partial F}{\partial t} = \int_V \frac{\delta F}{\delta c} \cdot (-\nabla \cdot \mathbf{J}) \cdot dV$$

$$= \int_S \frac{\delta F}{\delta c} (-\mathbf{J}) \cdot \mathbf{n} \cdot dS - \int_V \left(\nabla \frac{\delta F}{\delta c}\right)(-\mathbf{J}) \cdot dV = \int_V \left(\nabla \frac{\delta F}{\delta c}\right) \cdot \mathbf{J} \cdot dV \leq 0 \tag{127}$$

Again, one realizes that the necessary condition for Eq. (127) to hold is that the kernel is negative, which claims that Eq. (128) holds, where M is a positive quantity.

$$\mathbf{J} = -M\nabla \left(\frac{\delta F}{\delta c}\right) \tag{128}$$

10 Statistical Thermodynamics and Model Calculations

Substitution of Eq. (128) into Eq. (126) yields Eq. (129).

$$\frac{\partial c}{\partial t} = \nabla \cdot \left(M \cdot \nabla \frac{\delta F}{\delta c} \right) \qquad (129)$$

According to the Euler–Lagrange equation, it is shown that Eq. (130) holds, when R is given by Eq. (131).

$$\delta R = \int \left(\frac{\partial I}{\partial y} - \frac{\partial}{\partial x} \left(\frac{\partial I}{\partial y'} \right) \right) \cdot \delta y \cdot dx \qquad (130)$$

$$R = \int I(x, y, y') \cdot dx \qquad (131)$$

By comparing Eqs. (118) and (131), one can easily identify the correspondences, $y \Leftrightarrow c \,(= \eta)$ and $y' \Leftrightarrow \nabla c \,(= \nabla \mu)$, and derive Eq. (132), where I is given by Eq. (133).

$$\delta F = \int_V \left(\frac{\partial I}{\partial c} - \frac{\partial}{\partial \mathbf{r}} \left(\frac{\partial I}{\partial (\nabla c)} \right) \right) \cdot \delta c \cdot dV = \int_V \left(\frac{\partial f_0}{\partial c} - 2\kappa_c \nabla^2 c \right) \cdot \delta c \cdot dV \qquad (132)$$

$$I(\mathbf{r}, c, \nabla c) = f_0(c) + \kappa_c (\nabla c)^2 \qquad (133)$$

By substituting Eq. (132) into Eq. (129), the kinetic equation for composition, Eq. (134), is obtained. This is the celebrated Cahn–Hilliard equation (Cahn and Hilliard 1958; Cahn 1961).

$$\frac{\partial c}{\partial t} = M \nabla^2 \cdot \left(\frac{\partial f_0}{\partial c} - 2\kappa_c \nabla^2 c \right) \qquad (134)$$

By noting that Eqs. (132) and (133) are also held for nonconservative variables by exchanging c with η, substitution of Eq. (132) into Eq. (124) yields Eq. (135), which has been known as the time-dependent Ginzburg–Landau (TDGL) equation (Fan and Chen 1997) or the Allen–Cahn equation (Allen and Cahn 1979).

$$\frac{\partial \eta_i}{\partial t} = -\sum_j L_{ij} \cdot \left(\frac{\partial f_0}{\partial \eta_j} - 2\kappa_j \nabla^2 \eta_j \right) \qquad (135)$$

We note that these two equations, (134) and (135), couple through the free energy, f_0, of the homogeneous system, and these simultaneous equations have been often termed phase field equations to describe versatile temporal microstructural evolution processes.

10.3.3
Width of an Interface

An inhomogeneous system may be characterized by the existence of an interface; an antiphase boundary (APB) is considered in this chapter as an example. It is shown below that the width of the APB is derived as one of the static properties of an interface. It should be noted that, with a slight modification, the energetics involved in the following arguments can be generalized to other kinds of interface.

The APB is formed upon collision of two ordered domains (the antiphase domain, abbreviated to APD) with opposite phase of atomic arrangements, characterized by the opposite sign of the LRO parameters, $\pm \eta$. An APB is regarded as the region at which the phase is reversed from $+\eta$ to $-\eta$ through $\eta = 0$, which may be realized at the center of the APB. A typical variation of the LRO parameter in a certain direction characterized by g is indicated in Fig. 10.11 (Ohno 2004), where the center of the APB is located at $g = 0$. It should be noted that with the variation of the LRO parameter, the free energy f changes, and a minimum value $\pm \eta_e$ is attained at equilibrium LRO, while the maximum value is reached at $\eta = 0$, which may correspond to a disordered state.

We consider a straight APB normal to the direction g which extends in the x direction. Let $\Delta f_0(\eta)$ be the free energy with reference to the minimum free energy attained at $\eta = \pm \eta_e$. Then, the excess free energy, σ_{APB}, due to the existence of the APB, which separates two ordered domains, is written as Eq. (136).

$$\sigma_{APB} = \int_{-\infty}^{+\infty} \left(\Delta f_0(\eta) + \kappa \left(\frac{d\eta}{dg} \right)^2 \right) \cdot dg \tag{136}$$

The equilibrium shape of the boundary is determined so that σ_{APB} is minimized and the Euler–Lagrange equation applied to Eq. (136) yields Eq. (137), which leads to Eq. (138).

$$\frac{\delta \sigma_{APB}}{\delta \eta} = \frac{\partial \Delta f_0(\eta)}{\partial \eta} - 2\kappa \frac{d^2 \eta}{dg^2} = 0 \tag{137}$$

$$\Delta f_0(\eta) = \kappa \left(\frac{d\eta}{dg} \right)^2. \tag{138}$$

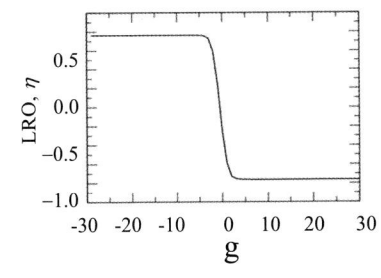

Fig. 10.11 Profile of the LPO parameter normal to the antiphase boundary (Ohno 2004).

By substituting Eq. (138) into Eq. (136), one obtains Eq. (139).

$$\sigma_{APB} = \int_{-\infty}^{+\infty} 2\kappa \left(\frac{d\eta}{dg}\right)^2 \cdot dg \tag{139}$$

Since the equilibrium values of the LRO parameter, $+\eta_e$ and $-\eta_e$, are attained at infinite distance ($\pm\infty$) from the interface, the integrand and the range of integration can be converted as in Eq. (140).

$$\sigma_{APB} = \int_{-\infty}^{+\infty} 2\kappa \left(\frac{d\eta}{dg}\right)^2 \cdot dg = 2\int_{-\eta_e}^{+\eta_e} \kappa \left(\frac{d\eta}{dg}\right) \cdot d\eta \tag{140}$$

The substitution of Eq. (138) yields Eq. (141).

$$\sigma_{APB} = 2\int_{-\eta_e}^{+\eta_e} \sqrt{\kappa \cdot \Delta f_0(\eta)} \cdot d\eta \tag{141}$$

To determine the equililbrium width of the APB, one needs a detailed expression of $\Delta f_0(\eta)$. As an example (Ohno 2004), we consider a double well potential given by Eq. (142).

$$\Delta f_0(\eta) = \Delta f_{max} \left(1 - \frac{\eta}{\eta_e}\right)^2 \left(1 + \frac{\eta}{\eta_e}\right)^2 \tag{142}$$

This potential is symmetric around $\eta = 0$; it takes a minimum, 0, at $+\eta_e$ and $-\eta_e$, and a maximum value, Δf_{max}, is attained at $\eta = 0$. By substituting this into Eq. (138), one can readily obtain a profile of the LRO parameter as a function of g in the form of Eq. (143).

$$\eta(g) = \eta_e \cdot \tanh\left(\frac{1}{\eta_e}\sqrt{\frac{\Delta f_{max}}{\kappa}} \cdot g\right) \tag{143}$$

This is linearized in the vicinity of $g = 0$ as Eq. (144).

$$\eta(g) \approx \sqrt{\frac{\Delta f_{max}}{\kappa}} \cdot g \tag{144}$$

Hence, the width of the interface is obtained as Eq. (145).

$$l_{APB} = 2\eta_e \cdot \sqrt{\frac{\kappa}{\Delta f_{max}}} \tag{145}$$

It is understood that the width of the APB is determined by the gradient energy coefficient κ and the free energy difference Δf_{max} between disordered and ordered phases. The system tends to confine the APB in a smaller region for a larger Δf_{max}, while the bigger the κ is, the more widely is the APB region extended, so that the gradient energy per unit length becomes smaller.

10.3.4
Interface Velocity

In Section 10.3.3, we studied the width of an interface as one of the static properties. In this section, the interface motion driven by a curvature is investigated (Ohno 2004) as a dynamic property. First, we consider a noncurved APB at the steady state at which TDGL equation is given by Eq. (146).

$$\frac{\partial \eta}{\partial t} = -L\left(\frac{\partial f_0}{\partial \eta} - 2\kappa \frac{\partial^2 \eta}{\partial g^2}\right) = 0 \tag{146}$$

Next, we consider a curved APB shown in Fig. 10.12. With the assumption that the LRO parameter η changes only in the direction of g and not in the tangential direction t, one obtains Eq. (147), where **n** is a unit normal vector to the APB.

$$\nabla \eta = \frac{\partial \eta}{\partial g}\mathbf{n} \tag{147}$$

Then, the Laplacian of η, $\nabla^2 \eta$, is transformed as in Eq. (148), and the TDGL equation is rewritten as Eq. (149), where H is the mean curvature, which is equivalent to $\nabla \cdot \mathbf{n}$ in the $-g$ direction.

$$\nabla^2 \eta = \nabla \cdot \nabla \eta = \frac{\partial^2 \eta}{\partial g^2} + \frac{\partial \eta}{\partial g}\nabla \cdot \mathbf{n} \tag{148}$$

$$\frac{\partial \eta}{\partial t} = -L\left(\frac{\partial f_0}{\partial \eta} - 2\kappa \nabla^2 \eta\right) = -L\left(\frac{\partial f_0}{\partial \eta} - 2\kappa \frac{\partial^2 \eta}{\partial g^2} - 2\kappa H \frac{\partial \eta}{\partial g}\right) \tag{149}$$

By noting Eq. (146), Eq. (149) is simplified to Eq. (150).

$$\frac{\partial \eta}{\partial t} = 2\kappa L H \frac{\partial \eta}{\partial g} \tag{150}$$

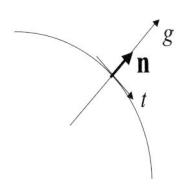

APB **Fig. 10.12** A curved antiphase boundary.

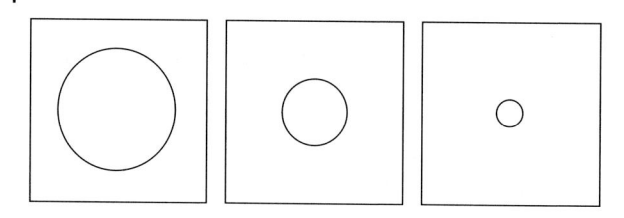

Fig. 10.13 Time evolution of a circular antiphase boundary (Ohno 2004).

For a contour curve of the LRO parameter η, one obtains Eq. (151), and therefore the velocity, v_{APB}, of the APB is given by Eq. (152)

$$d\eta = \left(\frac{\partial \eta}{\partial g}\right)_t dg + \left(\frac{\partial \eta}{\partial t}\right)_g dt = 0 \tag{151}$$

$$\left(\frac{\partial g}{\partial t}\right)_\eta = -\left(\frac{\partial \eta}{\partial t}\right)_g \Big/ \left(\frac{\partial \eta}{\partial g}\right)_t \tag{152}$$

Taking Eq. (152) together with Eq. (150), the final expression of the velocity of APB is obtained as Eq. (153).

$$v_{APB} = -2\kappa L \cdot H \tag{153}$$

This is the typical case of an interface velocity which is driven by a curvature.

As a simple example, one can consider a circular interface with the radius R. Since the mean curvature of a circle is $2/R$, the velocity is given by Eq. (154), integration of which yields Eq. (155), where R_0 is the initial radius of the circle.

$$v_{APB} = \frac{dR}{dt} = -4\kappa L \frac{1}{R} \tag{154}$$

$$R(t)^2 - R_0^2 = -8\kappa L \cdot t \tag{155}$$

It is indicated that the circular interface shrinks with time as shown in Fig. 10.13 (Ohno and Mohri 2002; Ohno 2004).

10.4
Model Calculations

One of the major trends in computational materials science is toward performing multiscale calculations from atomistic to macroscopic scales in a consistent manner. In Section 10.2, atomistic approaches to configurational thermodynamics and kinetics were demonstrated, while in Section 10.3 basic formulae of microstructural statics and dynamics were derived. It is a natural development to combine these two studies to attempt multiscale calculations from atomistic to microstructural scales. In this section, as a summary of the present chapter, we introduce the author's recent work (Ohno and Mohri 2001a,b, 2002, 2003, 2007; Chen et al. 2002a,b; Mohri and Chen 2002, 2004a,b,c; Mohri et al. 2005a,b, 2006; Mohri 2005) on multiscale calculation covering phase stability, phase equilibria, and phase transition by exemplifying three kinds of Fe-based binary alloy systems. Numerous contributions to the subjects dealt with in Sections 10.4.1 and 10.4.2 have been achieved by Zunger and his collaborators (Mbaye et al. 1987; Wei et al. 1987; Zunger et al. 1988; Ferreira et al. 1989, 1999; Bernard and Zunger 1990; Wei et al. 1990, 1992, 1993; Lu et al. 1991, 1995; Laks et al. 1992; Wolverton et al. 1994; Wolverton and Zunger 1997; Ozolins et al. 1998) and Vaks and his collaborators (Vaks and Samolyuk 1999; Bellashchenko et al. 1999a,b; Pankratov and Vaks 2001; Dobretsov et al. 2004, 2005; Khromov et al. 2005), respectively.

10.4.1
Calculation of a Phase Diagram

10.4.1.1 Ground-State Analysis

An example of the ground-state analysis within the first and second nearest-neighbor pair interaction energies was introduced in Section 10.2.2. It was mentioned that the most stable ordered phases at 1:1 stoichiometry were determined in the region specified by the magnitude and sign of the pair interaction energies. Such a ground-state analysis provides sufficient conditions for an ordered phase to be stabilized with specified pair interaction energies. However, for an actual system, there is no guarantee that multibody interaction energies can be neglected or distant pair interaction energies fade away. Moreover, effective interaction energies are not constant but depend on concentration, atomic distance, and atomic configurations. Therefore, examination of the ground-state total energies of possible ordered phases is indispensable. Examples are demonstrated below for three kinds of Fe-based alloy systems: Fe–Ni, Fe–Pd, and Fe–Pt.

Ni, Pd, and Pt belong to the column VIII transition elements in the periodic table. When each element is alloyed with Fe, a quite systematic behavior of phase equilibria is observed (Massalski 1986). For the Fe–Ni system, only one ordered phase appears around 1:3 stoichiometry (FeNi$_3$; L1$_2$), while for the Fe–Pd system an additional ordered phase is found in the vicinity of 1:1 stoichiometric composition (FePd; L1$_0$). Finally, for the Fe–Pt system, a third ordered phase appears near 3:1 stoichiometry (Fe$_3$Pt; L1$_2$).

Based on the FLAPW (full-potential linear augmented plane wave) method (Jansen and Freeman 1984) within the generalized gradient approximation (GGA; Perdew et al. 1992), first-principles total electronic structure calculations were performed on a series of ordered compounds, Fe$_{4-n}$X$_n$ at 1:3 ($n=3$), 1:1 ($n=2$), and 3:1 ($n=1$) for each Fe–X system (X = Ni, Pd, and Pt) as a function of atomic distance r (Chen et al. 2002a,b; Mohri and Chen 2002, 2004a,b,c). In view of the fact that the most stable ground state of Fe is not fcc but is bcc, the

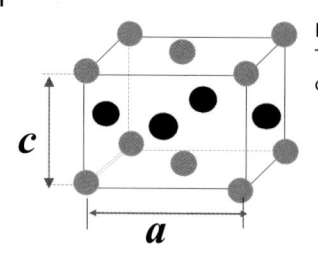

Fig. 10.14 Atomic arrangement of an $L1_0$ ordered phase, The c/a ratio is generally not unity, due to tetragonal distortion.

total energy of bcc Fe has also been calculated. Then, the heat of formation for each compound was obtained according to Eq. (156), where r_{bcc} and r_X are the equilibrium atomic distances of ferromagnetic bcc-based Fe and fcc X, which are the ground states of each species.

$$\Delta E^{(n)}(r) = E_{Fe_{4-n}X_n}(r) - \frac{4-n}{4} E_{Fe_4}^{bcc}(r_{bcc}) - \frac{n}{4} E_{X_4}(r_X) \tag{156}$$

It is understood that the energy reference state assumed in Eq. (156) is the segregation limit. In view of the experimental results in the phase diagram described above, it is natural to adopt the $L1_2$ structure for 1:3 and 3:1 compounds and the $L1_0$ structure for 1:1 compound for the total energy calculations. Also, in order to examine the effects of magnetism, both spin-polarized and nonpolarized calculations were performed.

The $L1_0$ ordered phase is generally characterized by a layered structure in one of the [100] directions, which induces a tetragonal distortion as shown in Fig. 10.14. Shown in Fig. 10.15 are the heats of formation of $L1_0$ ordered compounds for Fe–Pd and Fe–Pt, respectively, as a function of the c/a ratio. These are ob-

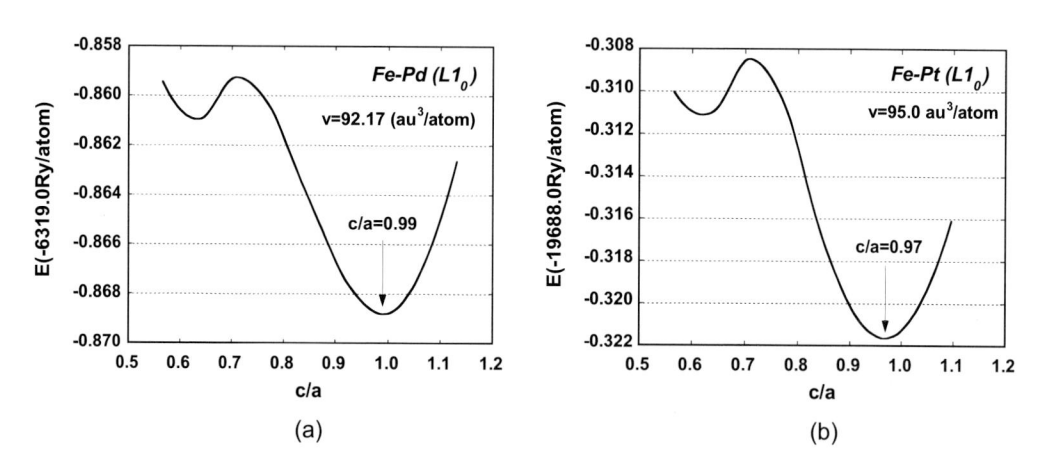

Fig. 10.15 Total energy of the $L1_0$ ordered phase of (a) Fe–Pd (Chen et al. 2002a; Mohri and Chen 2004b) and (b) Fe–Pt (Mohri and Chen 2004b) as a function of c/a.

tained by the spin-polarized calculation. In both systems the energy minimum appears not at $c/a = 1$, but at 0.99 for Fe–Pd and 0.97 for Fe–Pt, indicating that the most stable state of L1$_0$ ordered phases is not a cubic but a tetragonal structure. In fact, the experimental observation suggests that the equilibrium values of c/a are 0.97 and 0.96, respectively, for Fe–Pd and Fe–Pt, with which these ground-state energy calculations agree quite well. However, in view of the fact that the c/a ratios are quite close to unity for both systems, we ignore the tetragonality in the following calculations and assume a fully cubic structure.

Figure 10.16 shows the heats of formation $\Delta E^{(n)}$ of the aforementioned five phases (Mohri 2005): Fe with fcc structure ($n = 0$), Fe$_3$X with L1$_2$ ($n = 1$), FeX

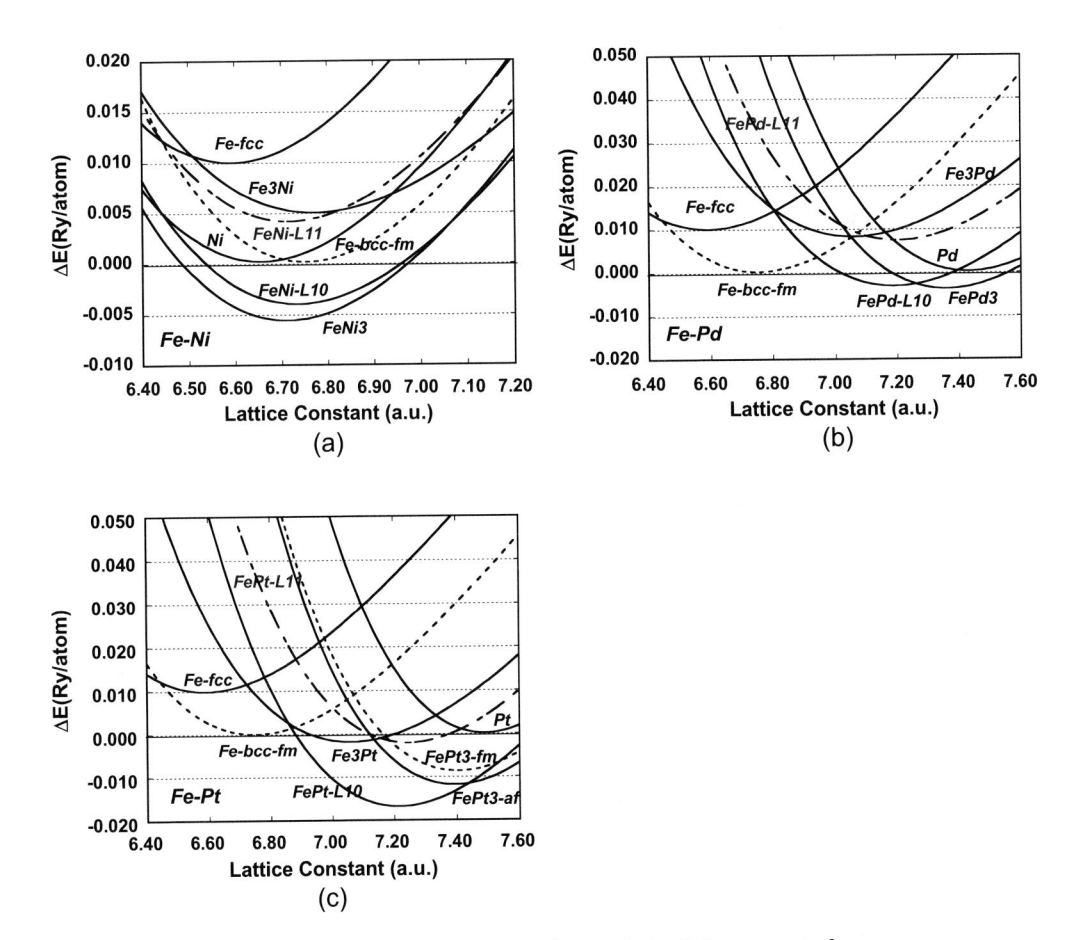

Fig. 10.16 Heats of formation of fcc Fe, bcc Fe (ferro), Fe$_3$X L1$_2$, FeX L1$_0$, FeX L1$_1$, FeX$_3$ L1$_2$, and fcc X, where X is Ni, Pd, or Pt. The energy reference state is the segregation limit, defined as the concentration average of total energy of bcc Fe and fcc X. For FePt$_3$, two results are obtained. One accounts for the ferromagnetic state and the other for the antiferromagnetic state. (a) Fe–Ni (Mohri and Chen 2004b); (b) Fe–Pd (Chen et al. 2002a; Mohri and Chen 2004b); (c) Fe–Pt (Mohri and Chen 2002, 2004b).

with $L1_0$ ($n = 2$), FeX_3 with $L1_2$ ($n = 3$), and X with fcc structure ($n = 4$) obtained from spin-polarized calculations for the Fe–Ni, –Pd, and –Pt systems, respectively. Since the ground state of Fe is bcc with a ferromagnetic state, the heat of formation of bcc Fe is calculated and drawn with a broken line. The lattice constant of bcc Fe (5.36 a.u.) is quite small as compared with other fcc-based phases of interest. In order to facilitate comparison, we converted it to an fcc-equivalent lattice constant, keeping the atomic volume.

For the Fe–Pd system, the heats of formation for FePd ($L1_0$) and $FePd_3$ ($L1_2$) are negative while that of Fe_3Pd($L1_2$) is positive, which implies the possible appearance of the former two phases in a phase diagram as a stable phase and missing the latter phase. This is realized exactly in the phase diagram. For the Fe–Pt system, the heats of formation of all three ordered phases are negative, which is again in good agreement with the phase diagram. Finally for the Fe–Ni system, both Fe_3Ni ($L1_2$) and FeNi ($L1_0$) have negative values. In the phase diagram, however, the latter phase is missing. This contradiction is discussed later.

Two additional features should be noted. The first is that a dot–broken line in each figure indicates the heats of formation of the $L1_1$ ordered phase at a 1:1 stoichiometric composition. As compared with the $L1_0$ ordered phase, one can confirm that the heats of formation are larger and the $L1_0$ ordered phase is more stable. This should be emphasized particularly for the Fe–Pd and Fe–Pt systems. These additional calculations performed for the $L1_1$ ordered phase are utilized in order to extract the second nearest-neighbor effective pair interaction energies through the cluster expansion method (Connolly and Williams 1983) as will be shown soon below.

The second feature is that for $FePt_3$ ($L1_2$) two kinds of magnetic states are considered, ferro- and antiferromagnetic states, which are designated by *fm* and *af* in the figure, respectively. One can see that the magnetic ground state is not ferro- but antiferromagnetic. This is in agreement with the experimental result (Bacon and Crangle 1963), and one will see that this tiny difference in magnetic energy, about 3 mRy atom^{-1}, affects the resultant transition temperature of the $L1_0$ ordered phase significantly.

10.4.1.2 Effective Cluster Interaction Energy

The total energy provides a measure of the relative stability of a phase of interest, while for the calculation of phase equilibria atomic interaction energies are required. According to the cluster expansion method (Connolly and Williams 1983) introduced in Eq. (43), a set of effective cluster interaction energies can be obtained from total energies of ordered phases by matrix inversion. In this section we demonstrate how the effective cluster interaction energies for the Fe-based ordered phases can be extracted from the total energies obtained in Section 10.4.1.1. It should be noted that two terms, total energy and heat of formation, are generally employed interchangeably. But for the application of the CEM in Eq. (43) to obtain effective cluster interaction energies, one needs to make a clear distinction, and the total energy E should be replaced by the heat of formation, ΔE, in Eq. (156). Furthermore, since heats of formation, ΔE, in Section 10.4.1.1

are obtained as a function of atomic distance r, the present CEM is rewritten as Eq. (157).

$$\Delta E(r) = \sum_{m=0}^{M} v_m(r) \cdot \xi_m \tag{157}$$

It should be noted that the correlation function depends on both the type of cluster, m, and the phase, n, while the effective cluster interaction energy is determined once the type of cluster is specified. Hence, Eq. (157) is modified to Eq. (158).

$$\Delta E^{(n)}(r) = \sum_{m=0}^{M} v_m(r) \cdot \xi_m^n \tag{158}$$

For each Fe–X system, the number of fcc-based phases for which the total energies are calculated in the present study is six, including both L1$_0$ and L1$_1$ ordered phases at 1:1 stoichiometry. However, we first include only the L1$_0$ ordered phase at 1:1 and L1$_2$ ordered phases at 1:3 and 3:1 stoichiometries in the CEM. Hence, both $\{\Delta E^{(n)}\}$ and $\{v_m\}$ constitute a 5 × 1 vector and $\{\xi_m^n\}$ represents a 5 × 5 matrix ($M = 4$). The explicit form of Eq. (158) is Eq. (159), and the effective cluster interaction energies are obtained as Eq. (160), or (161), or (162–163).

$$\begin{pmatrix} \Delta E^{Fe} \\ \Delta E^{Fe_3X} \\ \Delta E^{FeX} \\ \Delta E^{FeX_3} \\ \Delta E^{X} \end{pmatrix} = \begin{pmatrix} 1 & 1 & 1 & 1 & 1 \\ 1 & 1/2 & 0 & -1/2 & -1 \\ 1 & 0 & -1/3 & 0 & 1 \\ 1 & -1/2 & 0 & 1/2 & -1 \\ 1 & -1 & 1 & -1 & 1 \end{pmatrix} \cdot \begin{pmatrix} v_0 \\ v_1 \\ v_{2,1} \\ v_3 \\ v_4 \end{pmatrix} \tag{159}$$

$$v_m(r) = \sum_n \{\xi_m^n\}^{-1} \cdot \Delta E^{(n)}(r) \tag{160}$$

$$\begin{pmatrix} v_0 \\ v_1 \\ v_{2,1} \\ v_3 \\ v_4 \end{pmatrix} = \begin{pmatrix} 1 & 1 & 1 & 1 & 1 \\ 1 & 1/2 & 0 & -1/2 & -1 \\ 1 & 0 & -1/3 & 0 & 1 \\ 1 & -1/2 & 0 & 1/2 & -1 \\ 1 & -1 & 1 & -1 & 1 \end{pmatrix}^{-1} \cdot \begin{pmatrix} \Delta E^{Fe} \\ \Delta E^{Fe_3X} \\ \Delta E^{FeX} \\ \Delta E^{FeX_3} \\ \Delta E^{X} \end{pmatrix} \tag{161}$$

$$= \begin{pmatrix} 0.063 & 0.25 & 0.375 & 0.25 & 0.063 \\ 0.25 & 0.50 & 0.0 & -0.5 & -0.25 \\ 0.375 & 0.0 & -0.75 & 0.0 & 0.375 \\ 0.25 & -0.5 & 0.0 & 0.5 & -0.25 \\ 0.063 & -0.25 & 0.375 & -0.25 & 0.063 \end{pmatrix} \cdot \begin{pmatrix} \Delta E^{Fe} \\ \Delta E^{Fe_3X} \\ \Delta E^{FeX} \\ \Delta E^{FeX_3} \\ \Delta E^{X} \end{pmatrix} . \tag{162}$$

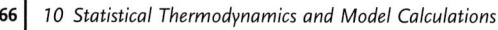

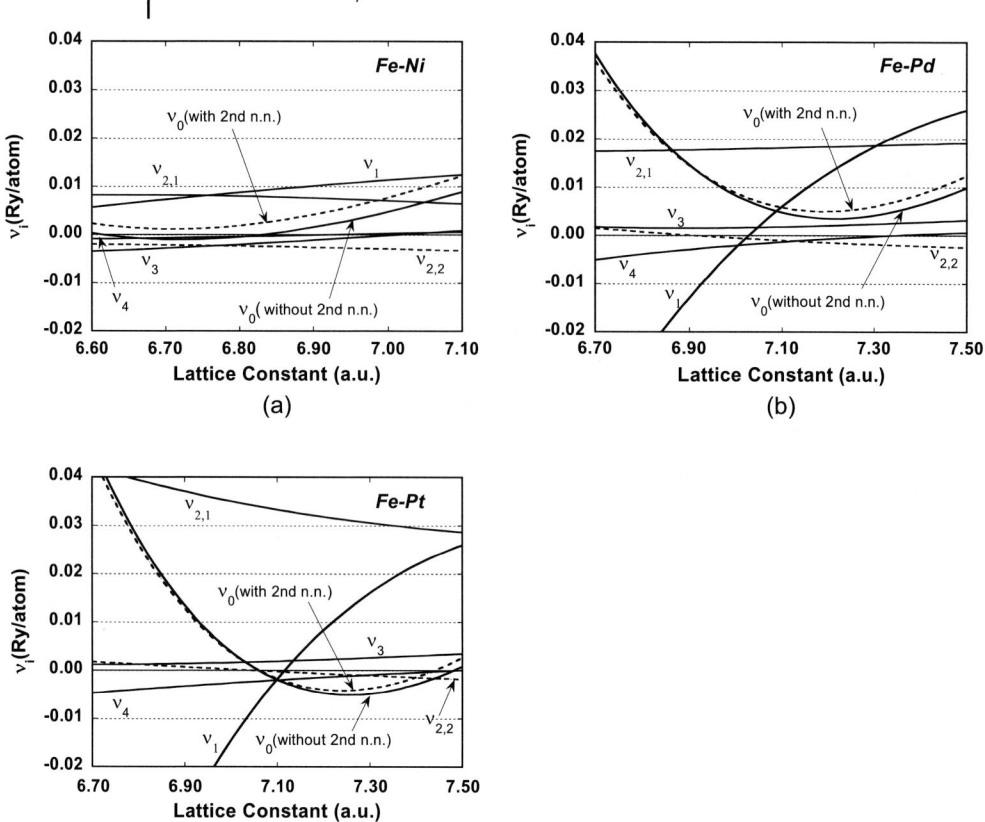

Fig. 10.17 Effective cluster interaction energies obtained for (a) Fe–Ni (Mohri and Chen 2004b); (b) Fe–Pd (Mohri and Chen 2004b); (c) Fe–Pt (Chen, Iwata and Mohri 2002; Mohri and Chen 2004b), by the cluster expansion method. Note that by including the heats of formation of the $L1_1$ ordered phase in Eq. (164), v_0 is altered while all the other interaction energies remain unchanged.

Recall that the subscript of $v_{2,1}$ emphasizes the nearest-neighbor pair interaction energy in order to distinguish it from the second nearest-neighbor pair interaction energy $v_{2,2}$, which will be discussed later.

The extracted effective interaction energies are shown in Fig. 10.17 for Fe–Ni, Fe–Pd, and Fe–Pt systems, respectively (Mohri 2005). The pair interaction energy is dominant for an entire range of lattice constants which virtually covers the concentration range from 0 to 100%. This is a characteristic feature of a metallic alloy system, for which cohesive properties are controlled mostly by central atomic forces.

In order to obtain further interaction energies, more ordered phases are added in vector $\{\Delta E^{(n)}\}$ and, accordingly, the correlation matrix $\{\xi_m^n\}$ is expanded. However, one realizes that when another row and column are added to the $n \times n$ matrix, Φ, to create a $(n+1) \times (n+1)$ matrix, Φ', the element φ'_{kl} of the inverse matrix Φ'^{-1} is generally different from the element φ_{kl} of the inverse matrix of the original Φ ($1 \leq k, l \leq n$). Hence, a simple extension of the energy vector and correlation matrix may alter the resultant elements of effective cluster interaction energies $\{v_m\}$, and one cannot extract a net contribution. As can be seen below, such a mathematical difficulty is circumvented tentatively for the 6×6 case by choosing the $L1_1$ ordered phase as a sixth phase.

By adding the total energy of FeX with the $L1_1$ structure ($n = 6$), Eqs. (159) and (161) are modified to Eqs. (163) and (164–165), respectively.

$$\begin{pmatrix} \Delta E^{Fe} \\ \Delta E^{Fe_3X} \\ \Delta E^{FeX} \\ \Delta E^{FeX_3} \\ \Delta E^{X} \\ \Delta E^{FeX; L11} \end{pmatrix} = \begin{pmatrix} 1 & 1 & 1 & 1 & 1 & 1 \\ 1 & 1/2 & 0 & -1/2 & -1 & 1 \\ 1 & 0 & -1/3 & 0 & 1 & 1 \\ 1 & -1/2 & 0 & 1/2 & -1 & 1 \\ 1 & -1 & 1 & -1 & 1 & 1 \\ 1 & 0 & 0 & 0 & -1 & -1 \end{pmatrix} \cdot \begin{pmatrix} v_0 \\ v_1 \\ v_{2,1} \\ v_3 \\ v_4 \\ v_{2,2} \end{pmatrix} \quad (163)$$

$$\begin{pmatrix} v_0 \\ v_1 \\ v_{2,1} \\ v_3 \\ v_4 \\ v_{2,2} \end{pmatrix} = \begin{pmatrix} 1 & 1 & 1 & 1 & 1 & 1 \\ 1 & 1/2 & 0 & -1/2 & -1 & 1 \\ 1 & 0 & -1/3 & 0 & 1 & 1 \\ 1 & -1/2 & 0 & 1/2 & -1 & 1 \\ 1 & -1 & 1 & -1 & 1 & 1 \\ 1 & 0 & 0 & 0 & -1 & -1 \end{pmatrix}^{-1} \cdot \begin{pmatrix} \Delta E^{Fe} \\ \Delta E^{Fe_3X} \\ \Delta E^{FeX} \\ \Delta E^{FeX_3} \\ \Delta E^{X} \\ \Delta E^{FeX; L11} \end{pmatrix} \quad (164)$$

$$= \begin{pmatrix} 0.063 & -2.75 & 0.375 & 0.25 & 0.063 & 3.0 \\ 0.25 & 0.5 & 0.0 & -0.5 & -0.25 & 0.0 \\ 0.375 & 0.0 & -0.75 & 0.0 & 0.375 & 0.0 \\ 0.25 & -0.5 & 0.0 & 0.5 & -0.25 & 0.0 \\ 0.063 & -0.25 & 0.375 & -0.25 & 0.063 & 0.0 \\ 0.0 & 3.0 & 0.0 & 0.0 & 0.0 & -3.0 \end{pmatrix} \cdot \begin{pmatrix} \Delta E^{Fe} \\ \Delta E^{Fe_3X} \\ \Delta E^{FeX} \\ \Delta E^{FeX_3} \\ \Delta E^{X} \\ \Delta E^{FeX; L11} \end{pmatrix}$$

$$(165)$$

By comparing the elements of the inverse matrices of Eqs. (162) and (165), one recognizes that the second nearest-neighbor effective pair interaction energy $v_{2,2}$ is extracted without altering other interaction energies except v_0, which can be confirmed in Fig. 10.17. Hence, we can examine the stability of the $L1_0$ ordered phase in terms of the sole effects of v_0 and $v_{2,2}$. It should be noted that in the present sign convention, a positive (negative) effective pair interaction enhances the stability of unlike (like) pairs. The negative $v_{2,2}$ values shown in Fig. 10.17 indicate clearly the stabilization of the $L1_0$ ordered phase, since the second nearest-neighbor pairs of the $L1_0$ ordered phase are all like pairs. Therefore, one can expect the transition temperature to be raised by including $v_{2,2}$, which will be discussed in Section 10.4.1.3.

When the elements of the 5 × 5 and the 6 × 6 matrices given in Eqs. (159) and (163), respectively, are compared, one realizes that first and sixth columns have the same elements down to the fifth component. This is a key ingredient to extraction of a new cluster interaction energy by minimizing the effects on predetermined energies. The mathematical foundation and more general conditions are found elsewhere (Mohri and Chen 2004c). Without a clear convergence criterion for the CEM, however, the best strategy is to include a large set of ordered phases. In fact, it is a recent trend that the CEM is operated on more than 100 phases (Sluiter et al. 1995; Santai et al. 2003; Zarkevich and Johnson 2004).

10.4.1.3 Phase Diagram

Within the nearest-neighbor pair interaction model combined with the tetrahedron approximation of the CVM, Helmholtz free energies for disordered and $L1_0$ ordered phases were given in Eqs. (74) and (75), respectively. Since the effective interaction energies were obtained up to the tetrahedron cluster in the Section 10.4.1.2, Helmholtz free energies of formation of the disordered and $L1_0$ ordered phases within the tetrahedron approximation are modified to Eqs. (166) and (167), respectively.

$$\Delta F^{dis} = \sum_{m=0}^{4} v_m(r) \cdot \xi_m - T \cdot k_B \cdot \ln \frac{\left\{\prod_{i,j}(N y_{ij})!\right\}^6 \cdot (N!)}{\left\{\prod_{i}(N x_i)!\right\}^5 \cdot \left\{\prod_{i,j,k,l}(N w_{ijkl})!\right\}^2} \quad (166)$$

$$\Delta F^{L10} = \sum_{m=0}^{4} v_m(r) \cdot \xi_m^{(L10)}$$

$$- T \cdot k_B \cdot \ln \left[\frac{\left\{\prod_{i,j}(N y_{ij}^{\alpha\alpha})!\right\} \left\{\prod_{i,j}(N y_{ij}^{\alpha\beta})!\right\}^4 \left\{\prod_{i,j}(N y_{ij}^{\beta\beta})!\right\} N!}{\left\{\prod_{i}(N x_i^{\alpha})!\right\}^{5/2} \left\{\prod_{i}(N x_i^{\beta})!\right\}^{5/2} \left\{\prod_{i,j,k,l}(N w_{ijkl}^{\alpha\alpha\beta\beta})!\right\}} \right], \quad (167)$$

We note that the E_{max} and S_{max} both refer to the tetrahedron cluster and the condition given by Eq. (81) is satisfied. In view of the fact that the cluster probabilities are transformed to a set of correlation functions, these free energy formulas can be converted to the one written in terms of correlation functions only as in Eqs. (168) and (169).

$$\Delta F^{dis} = f^{dis}(T, \{v_i(r)\}; \xi_1, \xi_2, \xi_3, \xi_4) \quad (168)$$

$$\Delta F^{L10} = f^{L10}(T, \{v_i(r)\}; \xi_1^{\alpha}, \xi_1^{\beta}, \xi_2^{\alpha\alpha}, \xi_2^{\alpha\beta}, \xi_2^{\beta\beta}, \xi_3^{\alpha\alpha\beta}, \xi_3^{\alpha\beta\beta}, \xi_4^{\alpha\alpha\beta\beta}) \quad (169)$$

These are simple extensions of Eqs. (90) and (91), respectively, but it is emphasized that the effective interaction energies are no longer constant, but depend on the atomic distance r. Hence, the minimization conditions given by Eq. (92) are also modified to Eqs. (170) and (171), where Eq. (170) indicates the minimization with respect to the configurational variables, while Eq. (171) provides the optimum volume under an ambient pressure.

$$\frac{\partial \Delta F^{dis}}{\partial \{\xi_m\}}\bigg|_{T,\{\xi_{i\neq m}\},r} = 0 \qquad \frac{\partial \Delta F^{L1_0}}{\partial \{\xi_m^{(\delta)}\}}\bigg|_{T,\{\xi_{i\neq m}\},r} = 0 \qquad (170)$$

$$\frac{\partial \Delta F^{dis}}{\partial r}\bigg|_{T,\{\xi_m\}} = 0 \qquad \frac{\partial \Delta F^{L1_0}}{\partial r}\bigg|_{T,\{\xi_m^{(\delta)}\}} = 0 \qquad (171)$$

It should be noted that Eq. (171) only optimizes the volume by keeping the global lattice symmetry; hence the system is still in an excited state without allowing local lattice distortions. Therefore, in a more rigorous calculation, local atomic displacement should be incorporated into the free energy. This is the main task of the continuous displacement cluster variation method (CDCVM) which has been developed recently (Kikuchi and Beldjenna 1992; Kikuchi 1998; Kikuchi and Masuda-Jindo 1999; Uzawa and Mohri 2001a,b; Uzawa and Mohri 2002). Unfortunately, however, application of the CDCVM is still limited to a primitive case and it is far from being applicable to the present Fe-based alloy systems.

This inconvenience is partly relieved by incorporating lattice vibration effects. According to the Debye–Gruneisen model, which is the quasiharmonic approximation, the vibrational free energy can be given as Eq. (172), where the right-hand side is the sum of the electronic total energy $E_{el}^{(n)}(r)$, the vibrational internal energy $E_{vib}^{(n)}(r,T)$, and vibrational entropy $S_{vib}^{(n)}(r,T)$.

$$F^{(n)}(r,T) = E_{el}^{(n)}(r) + E_{vib}^{(n)}(r,T) - T \cdot S_{vib}^{(n)}(r,T) \qquad (172)$$

Among these three terms, $E_{el}^{(n)}(r)$ is equivalent to $\Delta E^{(n)}(r)$ in Eq. (156), and the vibrational internal energy and entropy, respectively, are written as Eqs. (173) and (174), where $\Theta_D^{(n)}$ is the Debye temperature and $D(x)$ is the Debye function, and the first term of the vibrational energy represents the zero-point energy.

$$E_{vib}^{(n)}(r,T) = \frac{9}{8}k_B \cdot \Theta_D^{(n)} + 3k_B \cdot T \cdot D\left(\frac{\Theta_D^{(n)}}{T}\right) \qquad (173)$$

$$S_{vib}^{(n)}(r,T) = 3k_B \cdot \left[\frac{4}{3}D\left(\frac{\Theta_D^{(n)}}{T}\right) - \ln\left\{1 - \exp\left(-\frac{\Theta_D^{(n)}}{T}\right)\right\}\right] \qquad (174)$$

The essential ingredients proposed by Morruzi et al. (1988) is that thermal properties such as the bulk modulus, the Gruneisen constant, and the Debye temperature are obtained by fitting to a four-parameter Morse potential. They reproduced the coefficient of thermal expansion of various metals. Then, Becker et al. (1990) extended the procedure to ordered compounds. The present author also employed the method extensively to phase equilibria calculations (Mohri et al. 1991, 1993; Nakamura and Mohri 1993, 1994c; Chen et al. 2005). The reader interested in the procedure is referred to the original articles.

For the calculations of phase diagrams, the CEM operates with $\Delta F^{(n)}(r, T)$ which is derived through Eq. (175).

$$\Delta F^{(n)}(r, T) = F_{Fe_{4-n}X_n}(r, T) - \frac{4-n}{4} F_{Fe_4}^{bcc}(r_{bcc}, T) - \frac{n}{4} F_{X_4}(r_X, T) \qquad (175)$$

It should be noted that by the introduction of lattice vibrational effects, the energy reference state is no longer independent of temperature as assumed in Eq. (156), which accounts for the temperature dependences of the free energy of formation in Eq. (175). Then the resultant effective cluster interaction energies are obtained as a function not only of lattice parameter but also of temperature [Eq. (176)].

$$v_m(r, T) = \sum_n \{\xi_m^n\}^{-1} \cdot \Delta F^{(n)}(r, T) \qquad (176)$$

The calculated $L1_0$-disorder phase boundaries with and without lattice vibration effects are displayed in Fig. 10.18 for Fe–Ni, Fe–Pd, and Fe–Pt systems, respectively (Mohri 2005). In these calculations, the second nearest-neighbor pair interaction energy is not included. One can see that the incorporation of the lattice vibration effects reduces the transition temperature considerably. This is interpreted as relative stabilization of the disordered phase. As mentioned above, the local distortion effects are not included in the model. Under such a condition, the extra energy is higher for a disordered phase since atoms of different sizes are more often encountered in the disordered phase than in the ordered phase. The lattice softening introduced by lattice vibration effects accommodates atoms of different sizes, and, therefore, the extra energy is more effectively released in the disordered phase.

For the Fe–Pd system (Fig. 10.18b), the transition temperature is reduced to 1030 K, which is in close agreement with the experimental value of 1023 K. For the Fe–Pt system (Fig. 10.18c), two calculations are performed without vibration effects. In one, an antiferromagnetic state is assumed for ΔE^{FePt_3} in Eqs. (158–162); the other is calculated with ΔE^{FePt_3} in the ferromagnetic state. By introducing the correct antiferromagnetic state, the transition temperature is decreased significantly. The lattice vibration effects further reduce the transition temperature to 1610 K, which is in excellent agreement with the experimental value of 1600 K. Both lattice vibrations and magnetic structure influence the resultant transition temperature quite markedly.

For Fe–Ni, the existence of an $L1_0$ ordered phase has not yet been reported in a conventional phase diagram. The calculation shown in Fig. 10.18(a), however, indicates that this ordered phase can be stabilized below approximately 500 K. In

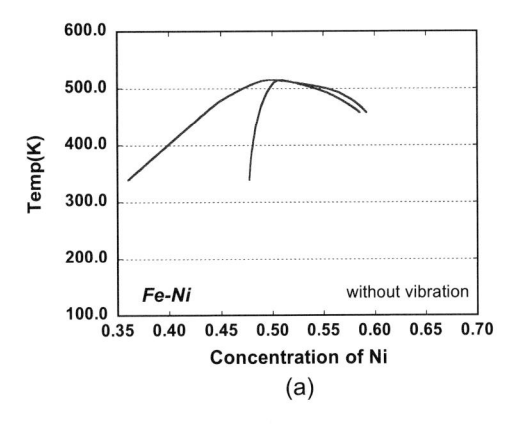

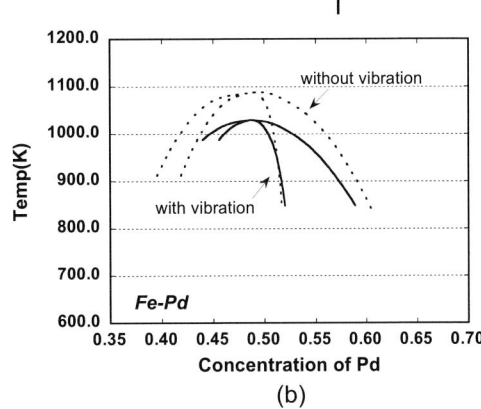

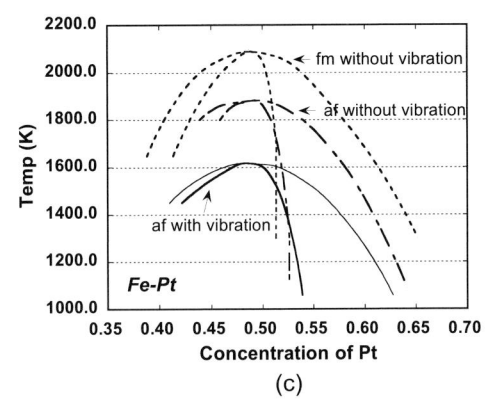

Fig. 10.18 L1$_0$-disorder phase boundaries for the (a) Fe–Ni (Mohri and Chen 2004b); (b) Fe–Pd (Mohri and Chen 2004a,b); (c) Fe–Pt systems (Mohri and Chen 2002, 2004b). For the Fe–Pd and Fe–Pt systems, comparisons are made among phase boundaries with and without lattice vibration effects.

fact, the electronic structure ground-state energy calculations in Fig. 10.16(a) clearly indicated the stability of this phase. Experimental examination of phase equilibria at low temperatures is not an easy task, but the recent experimental confirmation (Ohnuma and Ishida 2004) of the existence of an Fe–Ni L1$_0$ phase is quite encouraging.

Finally, two points should be addressed. One is that the congruent composition of the L1$_0$ ordered phase in Fe–Pd is not at 50 at.% and a considerable deviation is reported. This is not reproduced by the above calculations. This inconvenience may be attributed to the fact that tetragonal distortion of the L1$_0$ ordered phase and the magnetic entropy at finite temperatures were neglected. This should be clarified in the future study.

Secondly, it is interesting to calculate the transition temperature, T_t, by incorporating the second nearest-neighbor effective pair interaction energy, which is extracted by extending the CEM as was discussed in Section 10.4.1.2. By noting the criterion suggested in Eq. (81), the tetrahedron approximation of the CVM is not enough for the inclusion of the second nearest-neighbor pair interaction energy. It is indispensable to introduce the tetrahedron–octahedron approximation described in Eq. (72) for the entropy term. Hence, the free energy of a disordered phase is written as Eq. (177).

$$\Delta F^{dis} = \sum_{m=0}^{5} v_m(r) \cdot \xi_m - T \cdot k_B \cdot \ln \frac{\left(\prod_{i,j,k}(N \cdot z_{ijk})!\right)^8 \cdot \left(\prod_{i}(N \cdot x_i)!\right)}{\left(\prod_{i,j,k,l,m,n}(N \cdot v_{ijklmn})!\right) \cdot \left(\prod_{i,j,k,l}(N \cdot w_{ijkl;})!\right)^2 \cdot \left(\prod_{i,j}(N \cdot y_{ij})!\right)^6}$$

(177)

In a similar manner, the free energy of the L1$_0$ ordered phase can be derived, although this is not demonstrated explicitly here.

The transition temperatures calculated earlier are significantly higher for both Fe–Pd (Chen et al. 2002b) and Fe–Pt systems (Mohri and Chen 2004a,b,c). The increase is attributed to the negative $v_{2,2}$ which stabilizes the structure of the L1$_0$ ordered phase; however, the effect of v_0 should not be overlooked. As was suggested by Eq. (50), v_0 is equivalent to the heat of formation of a completely random solid solution at an equiatomic composition. Then, the increase in v_0 by incorporation of the L1$_1$ ordered phase as shown in Fig. 10.17 results in the destabilization of the disordered phase. It is therefore more reasonable to attribute the increase in the transition temperature to the combined effects of $v_{2,2}$, which stabilizes the L1$_0$ ordered structure, and v_0, which destabilizes the disordered phase.

10.4.2
Microstructural Evolution Calculated by the Phase Field Method

10.4.2.1 Hybrid Model
Recently, the phase field method (PFM) (Chen 2002) started to attract broad attention. In the materials science community, PFM is interpreted as the combination of the Cahn–Hilliard diffusion equation and the time-dependent Ginzburg–Landau (TDGL) equation which were introduced in Section 10.3.2. The simultaneous solution of both equations is supposed to provide microstructural evolution process in terms of both composition and configuration characterized by order parameters. The key to the PFM is the interfacial energy, which is included explicitly in the Ginzburg–Landau type total free energy in Eq. (120), which is

10.4 Model Calculations

subject to a minimization with respect to field variables. Hence, unlike the situation in most of the previous work using a sharp interface model, an interface is not a special entity within the PFM to be described separately, but merely a spatial localization of field variables.

The local free energy density f_0 in Eq. (120) determines the final equilibrium state as well as the path toward it; therefore the description of f_0 plays a determining role in the PFM. In the conventional PFM, the local free energy density is given by a phenomenological Landau-type series expansion by which the first- or second-order character of the transformation is quite easily described. Furthermore, one may investigate multimode phase transitions which involve more than a single phase transition, such as phase separation with ordering, by an efficient parameterization.

All of these fruitful features of the Landau-type free energy are connected with the phenomenological nature of this approach. Also, it is indispensable to select various parameters, including materials constants, effectively in order to reproduce the transition behavior of interest. These selected parameters, however, leave some ambiguities in the calculated results, as will be pointed out later. Recently, hybridization of the CVM free energy within the PFM was attempted (Ohno and Mohri 2001a,b, 2002, 2003; Ohno 2004). The advantages of the hybrid model are addressed by two points. One is that the atomistic nature of the CVM free energy provides an opportunity to determine all the necessary quantities appearing in the PFM without ambiguities. This is the clue to the first-principles phase field method which will be discussed in Section 10.4.2.2. The other is there is the possibility of performing multiscale calculations from atomistic to microstructural scale. Below, description of the time evolution of the antiphase boundary associated with $L1_0$ ordering by the hybrid model is exemplified. For simplicity, the composition is fixed at 1:1 stoichiometry, and our main focus is placed on configurational kinetics described merely by TDGL equations.

As stated in Section 10.2.5, the CVM free energy for the $L1_0$ ordered phase is given symbolically as Eq. (91), and the number of independent variables can be reduced at a fixed composition of 50 at.% due to the symmetry of the $L1_0$ ordered phase and the free energy was written as Eq. (93).

$$f^{L1_0}[\{\xi_i\}] = f_1(\xi_1^\alpha, \xi_1^\beta, \xi_2^{\alpha\alpha}, \xi_2^{\alpha\beta}, \xi_2^{\beta\beta}, \xi_3^{\alpha\alpha\beta}, \xi_3^{\alpha\beta\beta}, \xi_4^{\alpha\alpha\beta\beta}). \tag{91}$$

$$f^{L1_0}[\{\xi_i\}] = f_2(\xi_1^\alpha, \xi_2^{\alpha\alpha}, \xi_2^{\alpha\alpha}, \xi_3^{\alpha\alpha\beta}, \xi_4^{\alpha\alpha\beta\beta}) \tag{93}$$

It should be noted that temperature T and effective interaction energies $\{v_i\}$ are omitted from the original free energy formula. The calculated $L1_0$-disorder phase diagram was demonstrated in Fig. 10.8. The transition temperature is 1.893 and the spinodal ordering temperature is 1.633 on the normalized temperature scale with respect to the nearest-neighbor effective pair interaction energy v_2.

As was demonstrated in Fig. 10.14, the $L1_0$ ordered phase is characterized by the layered stacking of alternative atomic planes of A and B atoms along one of $\langle 100 \rangle$ directions. For the phase diagram calculations, the distinction between the

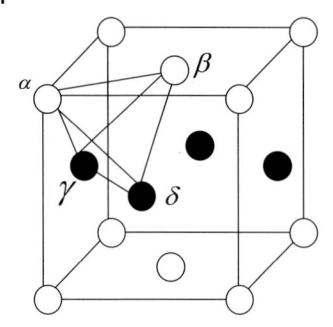

Fig. 10.19 The L1$_0$ ordered phase splits into four interpenetrating sublattices.

three layered directions need not be considered. However, for the microstructure calculation, it is essential to distinguish the difference in stacking directions, which results in the variants. The free energy given by Eq. (91) should be modified so that the four sublattices of the fcc lattice are taken into account. In fact, the fcc lattice can be decomposed into four interpenetrating simple cubic lattices, and the apices of a nearest-neighbor tetrahedron cluster termed α, β, γ, and δ in Fig. 10.19 each belong to a different simple cubic lattice. Hence, one can identify that the free energy is a function of 15 independent variables written as Eq. (178).

$$f^{L1_0} = f_3(\xi_1^\alpha, \xi_1^\beta, \xi_1^\gamma, \xi_1^\delta, \xi_2^{\alpha\beta}, \xi_2^{\alpha\gamma}, \xi_2^{\alpha\delta}, \xi_2^{\beta\gamma}, \xi_2^{\beta\delta}, \xi_2^{\gamma\delta}, \xi_3^{\alpha\beta\gamma}, \xi_3^{\alpha\beta\delta}, \xi_3^{\alpha\gamma\delta}, \xi_3^{\beta\gamma\delta}, \xi_4^{\alpha\beta\gamma\delta}) \tag{178}$$

Accordingly, three distinguishable LRO parameters are defined as in Eqs. (179–181).

$$\eta_{100} = \frac{1}{4}(\xi_1^\alpha + \xi_1^\beta - \xi_1^\gamma - \xi_1^\delta) \tag{179}$$

$$\eta_{010} = \frac{1}{4}(\xi_1^\alpha + \xi_1^\gamma - \xi_1^\beta - \xi_1^\delta) \tag{180}$$

$$\eta_{001} = \frac{1}{4}(\xi_1^\alpha + \xi_1^\delta - \xi_1^\beta - \xi_1^\gamma) \tag{181}$$

Again, by imposing the symmetry of the L1$_0$ ordered phase at a fixed 1:1 stoichiometric composition, the free energy can be simplified by Eq. (182), in which the number of independent variables is reduced from 15 in Eq. (178) to ten. Three of these independent variables, η_1, η_2, and η_3, are LRO parameters as defined by Eqs. (179–181). The rest of the correlation functions are regarded as SRO parameters.

$$f^{L1_0} = f_4(\eta_1, \eta_2, \eta_3, \xi_2^{\alpha\beta}, \xi_2^{\alpha\gamma}, \xi_2^{\alpha\delta}, \xi_3^{\alpha\beta\gamma}, \xi_3^{\alpha\beta\delta}, \xi_3^{\alpha\gamma\delta}, \xi_4^{\alpha\beta\gamma\delta}) \tag{182}$$

By substituting Eq. (182) into the TDGL equation (135), one can derive the time evolution behavior of the LRO parameters through Eq. (183).

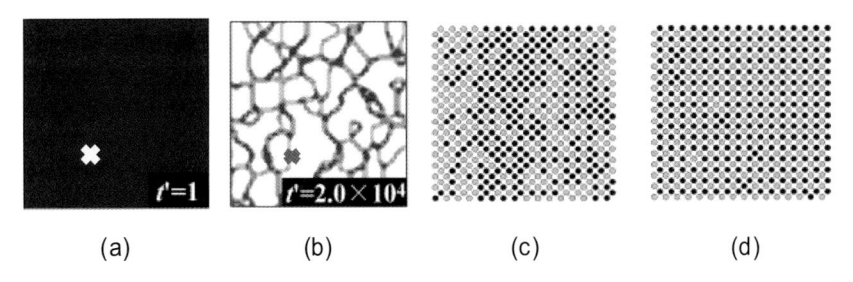

Fig. 10.20 Microstructural evolution from (a) a disordered state to (b) an ordered state. Magnification ×. (c) and (d) indicate the atomic configuration of the {001} plane at a fixed point of the microstructure (×) corresponding to (a) and (b) respectively (Ohno 2004; Mohri 2005; Mohri et al. 2005a,b, 2006).

$$\frac{\partial \eta_i}{\partial t} = -L_i \left(\frac{\partial f_4 \lfloor \{\eta_j, \xi_i\} \rfloor}{\partial \eta_i} - \kappa_i \nabla^2 \eta_i \right) \tag{183}$$

The numerical calculation of the TDGL equation was carried out by transforming it to the difference equation. Each spatially discretized point in the difference equation consists of an fcc lattice of which free energy is described by f_4. The LRO and SRO parameters at each point differ from one another; so interfaces are formed. Among such interfaces, we consider an antiphase boundary (APB) caused by the heterogeneity of the LRO, which can be observed by transmission electron microscopy (TEM) to be associated with an order–disorder transition.

The microstructural evolution of a system at 50 at.% is shown in Fig. 10.20(a,b); it is initially in the disordered phase (Fig. 20a) and is quenched to a temperature $T = 1.6$ (see Fig. 10.8), at which the L1$_0$ ordered phase is stabilized (Fig. 10.20b). The gray level in the contrast is proportional to the square of the LRO parameter which simulates the dark-field TEM image. Hence, the dark region indicates disordered and APB areas and bright regions suggest ordered domains. One can see that characteristic APBs are formed in the microstrucure and, importantly, one can find a triple point junction which is formed by the collision of the different variants of ordered domains. One can never reproduce the triple point junction if the free energy of f_2 in Eq. (93) is employed.

The advantage of the hybrid model is the fact that the atomistic information is derived simultaneously, as shown in Fig. 10.20(c,d); these configurations correspond to the fixed points in the micrographs indicated by the respective crosses in Fig. 10.20(a,b). Two edge directions correspond to the $\langle 100 \rangle$ direction. These are obtained based on the Williams algorithm (Williams 1976). Since the CVM free energy provides the correlation functions at each time step, the Williams procedure synthesizes the atomic configuration in $120 \times 120 \times 120$ fcc cubic lattice points. The error between the input and resultant correlation functions is less than 1%. One recognizes that ordering proceeds at the atomic level and the ini-

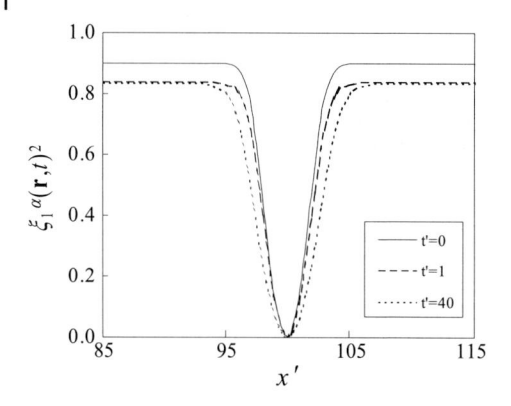

Fig. 10.21 Evolution of the square of the LRO parameter in the vicinity of an APB (Ohno and Mohri 2003; Ohno 2004).

tially disordered atomic arrangements (Fig. 10.20c) transform to the ordered arrangements (Fig. 10.20d) with some antisite atoms remaining.

Figure 10.21 shows the time evolution of the LRO in the vicinity of an APB of which the center position is located at $x' = 100$ on the normalized scale. It should be noted that across the APB, the LRO reverses the sign as was described in Fig. 10.11, while the vertical axis in Fig. 10.21 is the square of the LRO parameter. The profiles are calculated at three different times t', which is the normalized time, $t' = N \cdot v_{2,1} \cdot L \cdot t$. One can see that in the earlier period ($t' = 0$–1) LRO is reduced within each domain, and in the later period ($t' = 1$–40) the APB is widened. This is well known as the wetting phenomenon which is the characteristic of a first-order transition.

In this way, the hybrid model is capable of investigating multiscale phenomena covering the atomistic behavior within each domain (Fig. 10.20(c,d)) and interfacial phenomena (Fig. 10.21) as well as the microstructural evolution process (Fig. 10.20(a,b)). It is worth pointing out that the difference in the time constant associated with the spatial scales is also reproduced in the present model (Ohno and Mohri 2003; Ohno 2004).

10.4.2.2 Toward the First-Principles Phase Field Calculation

Despite various advantageous features, the hybrid model is basically qualitative in the sense that the length scale and time scale are not uniquely fixed in a self-consistent manner within the theoretical framework. In fact, the spatial scale is normalized as in Eq. (184), where the gradient energy coefficient κ_i remains unassigned although it has been amply demonstrated that the nearest-neighbor effective pair interaction energy $v_{2,1}$ can be determined by the first-principles calculation.

$$x' = x \cdot \sqrt{(N \cdot v_{2,1})/(2\kappa_i)} \qquad (184)$$

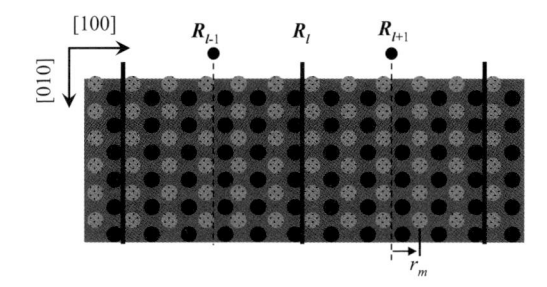

Fig. 10.22 Local and global coordinate system in [100] direction.

In order to assign the gradient energy coefficient within the hybrid model, one needs an atomistic model, which should be coherently connected to the continuous PFM.

First, we split the entire lattice into cells in a specified direction so that each cell contains the same number of lattice planes. Then, as shown in Fig. 10.22 (Ohno 2004), we represent the position of each atomic plane, p_n, by two coordinate systems, a global and a local one, through Eq. (185), where the global coordinate, R_l, indicates the position of each block, while the local coordinate, r_m, gives the position of an atomic plane within each cell.

$$p_n = R_l + r_m \tag{185}$$

The origin of the global coordinate is assigned arbitrarily, whereas for the local coordinate it is taken to be the center of each cell, so that Eq. (186) is satisfied.

$$\sum_m r_m = 0 \tag{186}$$

It should be noted that Fig. 10.22 indicates the (001) planes in the bcc lattice, but the procedure is the same in all directions (planes) for all structures.

We describe the free energy of the entire lattice as the sum of the free energy at each lattice plane, $f_{CVM}(p_n)$, by Eq. (187).

$$F = \sum_n f_{CVM}(p_n) \tag{187}$$

This indicates that both the internal energy and the entropy are defined at each lattice plane, for which the location of clusters such as the pair, triangle, tetrahedron, etc., are described consistently by the position of a lattice plane, p_n, and a lattice constant, a. By knowing that the CVM free energy is a function of the cluster probability or correlation function, Eq. (187) is rewritten as Eq. (188), where ϕ_s represents a configurational variable (cluster probability or correlation function) and $h(a)$ is a function of the lattice constant a. Such a description of the free energy is known as the inhomogeneous CVM (Kikuchi and Mohri 1997).

$$F = \sum_{l,m} f_{CVM}[\{\phi_s(R_l + r_m + h(a))\}] \tag{188}$$

By assuming that the spatial variation of ϕ_s is smooth enough for it to be expanded around the global coordinate, R_l, one derives the approximate expression in Eq. (189), where the higher-order terms beyond the third-order derivative are neglected.

$$\phi_s(R_l + r_m + h(a)) \cong \phi_s(R_l) + (r_m + h(a)) \cdot \nabla \phi_s(R_l)$$
$$+ \frac{1}{2}(r_m + h(a))^2 \cdot \nabla^2 \phi_s(R_l) \tag{189}$$

The substitution of Eq. (189) into Eq. (188) yields the symbolical free energy expression in Eq. (190).

$$F \cong \sum_{l,m} f_{CVM}(r_m + h(a), \{\phi_s(R_l)\}, \{\nabla \phi_s(R_l)\}, \{\nabla^2 \phi_s(R_l)\}) \tag{190}$$

If we expand the above free energy around the homogeneous state in which all the spatial derivatives of configurational variables vanish and the free energy is written only as $\{\phi_s\}$, the resultant expression is given by Eq. (191), where the zero indicates that the derivative is evaluated around the homogeneous state.

$$F \cong \sum_{l} \left\{ \sum_{m} f_{CVM}(\{\phi_s(R_l)\}) + \sum_{m,s} \left.\frac{\partial f_{CVM}}{\partial(\nabla \phi_s)}\right|_0 (\nabla \phi_s) + \sum_{m,s} \left.\frac{\partial f_{CVM}}{\partial(\nabla^2 \phi_s)}\right|_0 (\nabla^2 \phi_s) \right.$$
$$\left. + \frac{1}{2} \sum_{m,s,s'} \left.\frac{\partial^2 f_{CVM}}{\partial(\nabla \phi_s)\partial(\nabla \phi_{s'})}\right|_0 (\nabla \phi_s)(\nabla \phi_{s'}) \right\} \tag{191}$$

In this expansion, term(s) higher than second order are neglected. Equation (191) can be represented concisely as Eq. (192), where $C_{i,s}(r_m, a, \{J\})$ is easily identified as a coefficient term in Eq. (191) and depends on the local coordinate r_m, the lattice constant a, and the atomic configuration specified by $\{J\}$ on the cluster s. It is shown (Ohno 2004) that these coefficient terms can be determined by the symmetry of the lattice and the orientation of the crystallographic plane of concern. Note that in order to avoid complication, $\cong$ in Eq. (191) is replaced by an equality sign in Eq. (192).

$$F = \sum_l \left\{ \sum_m f_{CVM}(\{\phi_s(R_l)\}) + \sum_{s,m} C_{1,s}(r_m, a, \{J\}) \cdot (\nabla \phi_s) \right.$$
$$\left. + \sum_{m,s} C_{2,s}(r_m, a, \{J\}) \cdot (\nabla^2 \phi_s) + \sum_{m,s} C_{3,s}(r_m, a, \{J\})(\nabla \phi_s)^2 \right\} \quad (192)$$

By transforming the sum with respect to the global coordinate into an integral, one obtains Eq. (193), where L is the length of each cell.

$$F = \frac{1}{L} \int \left\{ \sum_m f_{CVM}(\{\phi_s(R_l)\}) + \sum_{s,m} C_{1,s}(r_m, a, \{J\}) \cdot (\nabla \phi_s) \right.$$
$$\left. + \sum_{m,s} C_{2,s}(r_m, a, \{J\}) \cdot (\nabla^2 \phi_s) + \sum_{m,s} C_{3,s}(r_m, a, \{J\})(\nabla \phi_s)^2 \right\} dx$$
$$(193)$$

One can readily prove that $\sum_m C_{1,s}(r_m, a, \{J\})$ vanishes, and partial integration, under the conventional assumption that the gradient of ϕ_s vanishes at a surface, yields Eq. (194), where N' is the number of the lattice plane in each cell and $\kappa_{s,s'}$ is given as Eq. (195). Here, $V_{s,s'}$ is called the generalized V-matrix (Ohno 2004), which is an extension of the V-matrix in Eq. (25) and contains the information about the atomic arrangements in a specified cluster.

$$F = \frac{1}{L} \int \left[N' \cdot f_{CVM}[\{\phi_s\}] + \sum_{s,s'} \kappa_{s,s'} (\nabla \phi_s)(\nabla \phi_{s'}) \right] \cdot dx \quad (194)$$

$$\kappa_{s,s'} = N \cdot k_B \cdot T \cdot \sum_J V_{s,s'}(r_m, a, \{J\}) \quad (195)$$

Mathematically, in the course of summation with respect to J, dependences of r_m are tacitly eliminated from Eq. (195), and therefore the gradient energy coefficient is finally given as Eq. (196).

$$\kappa_{s,s'} = \kappa_{s,s'}(T, \{\phi_s\}, a) \quad (196)$$

We see that the mathematical form of Eq. (194) is similar to a Ginzburg–Landau type of free energy expression with all the pertinent information at the atomistic scale condensed in the generalized gradient coefficient, which is not a constant as has been assumed in the conventional PFM calculations, but depends on temperature and SRO, $\{\phi_s\}$. It should be noted that the present derivation is a generalization of Kikuchi and Cahn's pioneering work (Kikuchi and Cahn 1962) on B2 ordered alloys. There are, however, a number of differences between the two derivations. Since detailed discussion is beyond the scope of the present chapter; the interested reader should consult the original articles. By stressing that the (gener-

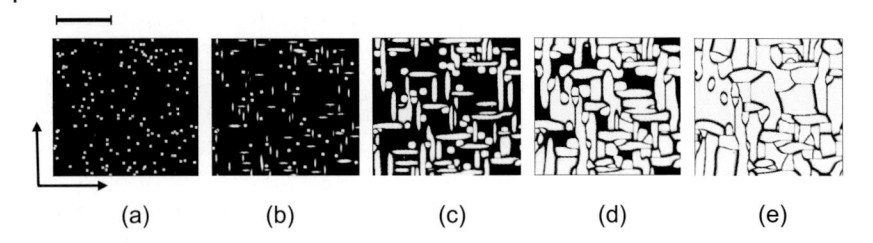

Fig. 10.23 Time evolution of an antiphase boundary for Fe–Pd at 50 at.% at 880 K (Mohri et al. 2005a,b; Mohri et al. 2006). Time t': (a) 0.01; (b) 0.83; (c) 3.59; (d) 7.04; (e) 16.70. The bar above (a) represents a length of 100 nm, and the arrows suggest the [100] and [011] directions.

alized) gradient energy coefficient is now given by all the atomistic information derived from the CVM, we go one step forward to calculate the microstructural evolution from first principles.

Figure 10.23(a–e) shows the time evolution of an APB when the 50 at.% Fe–Pd system is quenched from a disordered phase region to 880 K, at which the $L1_0$ ordered phase is stabilized as one can see in the phase diagram in Fig. 10.18(b). It is observed that an absolute length scale is assigned in the micrograph and the crystallographic orientation is also fixed uniquely. It should be emphasized that these calculations are performed based only on the two atomic numbers of Fe (26) and Pd (46). Rather strong anisotropy of the APB may be caused by neglect of the second nearest-neighbor pair interaction energy. This is one of the remaining subjects to be settled in future calculations by introducing the tetrahedron–octahedron approximation of the CVM (Sanchez and de Fontaine 1978; Mohri et al. 1985a,b).

For a fully consistent first-principles calculation of phase transitions, two remaining subjects should be addressed. One is to assign an absolute time scale. In fact, in Fig. 10.23, the time scale is normalized with respect to a relaxation constant. The other is more rigorous coarse graining operation. For these subjects, efficient incorporation of fluctuations into the theory may be inevitable.

10.5
Future Scope and Outlook

The theoretical basis in the present chapter is the cluster variation method, which is one of the most efficient tools for investigating alloy phase equilibria. Although the CVM is an approximate theory, it is assured that by choosing a bigger basic cluster one can approach an exact solution. One hopes that this account has provided readers with a rigorous framework for the thermodynamics for alloys, as well as practical techniques for investigating alloy problems through the sound theory of CVM.

It has been demonstrated that by combining the CVM with electronic structure total energy calculations, one may perform self-consistent first-principles calculation of phase equilibria. The main procedure of the first-principles calculations within the CVM is nearly established, and applications to versatile systems are a matter of numerical technique. In view of the recent development of high-performance computers, the author believes that the CVM has been further penetrated with the distribution of user-friendly program codes. The main task one has to tackle in the future is the extension to kinetics, i.e., development of the path probability method. As was demonstrated in Section 10.2.6, the applicability of the PPM has been limited to spin kinetics and no serious efforts have been directed to the vacancy mechanism. Upon introduction of a vacancy, one faces the difficulty of enumerating and identifying microscopic paths for configurational evolution/devolution processes. In order to circumvent this problem, it is desirable to develop a program code which formulates the path probability function for a specified set of crystal symmetry and a level of approximation prior to the numerical computation of simultaneous equations. Such an attempt has been initiated by author's group. This should pave the way to combining the PPM with the electronic structure calculations to perform a first-principles kinetics investigation which is expected to resolve various diffusion problems involved in ordered compounds.

Recent development of the phase field method is quite remarkable and various microstructural evolution processes have been analyzed efficiently by visualization of solutions. It was demonstrated that the CVM can even be combined with PFM to achieve multiscale calculations. Some preliminary calculations introduced in the present chapter are precursors to the future development of fully satisfactory multiscale calculations. For this, a major stumbling block is the coarse graining operation. One recipe introduced in Section 10.4.2.2 is an efficient way to determine the spatial scale and crystallographic orientation from atomistic information in a unique manner, but the time scale is still left as a subject for the future. It is believed that the PPM will play an essential role in determining the time scale by combining consistently with the PFM. In view of the recent trend of multiscale and multiphysics calculations on tera- and peta-flop computers, spatio-time coarse graining is quite a challenging topic.

One of the remarkable trends recently in the materials science community is the discovery and development of low-dimensional materials such as thin films, nanowires and nanoparticles. In fact, many discussions are provided by various authors in this book. A low-dimensional material is a thermodynamic system in which the surface predominates. Since the CVM was developed originally for a bulk system in which symmetry is preserved all over the crystal, how to deal with the broken symmetry at a surface is a central concern in the theoretical studies. Unfortunately, CVM-based studies on a low-dimensional material are far behind Monte Carlo and molecular dynamics simulations. It may be recalled, however, that the CVM is a unique theoretical tool which is able to perform a synthetic study involving a phase diagram, atomistic kinetics, and microstructural dynamics. The development of an efficient ingredient to extend the CVM to a system in which numerous kinds of symmetries including non-Bravais lattice symmetries coexist is urged. For this, a generalization of the continuous displacement cluster variation method (CDCVM) which was introduced in Section 10.4.1.3 has been readdressed.

Finally, it is emphasized that a long-lived theory is soundly based. This is the case with the CVM. Quite often, however, a sound theory is found to be too rigorous and not easily applicable to practical problems. Fortunately, the hierarchic structure of CVM assures a broad range of applications. The accuracy of the solution depends merely on the basic cluster, which can be further enlarged on tera- and peta-flop computers. The author believes that CVM-based alloy thermodynamics will be one of the important components of modern alloy physics and will be a stimulant to solid-state physics.

Appendix

CALPHAD Free Energy

In the course of this chapter, the main focus was placed on the atomistic description of alloy phase stability, phase equilibria, and phase transitions. Naturally, the final goal of such an approach is calculations from first principles. Although much effort has been made in recent studies to generalize the scheme so that versatile classes of materials can be dealt with, the applicability of the first-principles calculation is mostly limited to binary or ternary alloys with highly symmetric structures. The application of the first-principles scheme to practical multicomponent alloy systems with an arbitrary symmetry is not very realistic at the present stage.

One of the most powerful methods which can be applied to study thermodynamic properties of practical alloy systems is the CALPHAD method. CALPHAD is the abbreviation for *Calculation of Phase Diagram* and this method has been developed over nearly 40 years of international cooperation. The method is based on the Bragg–Williams description of the entropy and free energy; hence the mathematical treatment is not difficult. The lack of physical exactness originating from the primitive Bragg–Williams approximation is circumvented by the efficient parameterization of the free energy in which the precision of the experimental input data is the critical issue. An introduction to CALPHAD free energy is attempted in this appendix. Among various reference books, the author refers to Turchi's concise summary (Turchi 2003) on this topic.

Within the regular solution model, the free energy, $^\Phi G$, of a multicomponent solution of a particular phase Φ can be given by Eq. (A1), where $^\Phi G^0$ is the free energy due to the mechanical mixing of pure components written as in Eq. (A2), in which I designates a component, M the total number of components, and

c_I and ${}^{\Phi}G_I^0$ are the concentration and free energy, respectively, of a pure component i.

$$ {}^{\Phi}G = {}^{\Phi}G^0 + {}^{\Phi}G_{mix}^{ideal} + {}^{\Phi}G_{mix}^{XS} \tag{A1} $$

$$ {}^{\Phi}G^0 = \sum_{I=1}^{M} c_I \cdot {}^{\Phi}G_I^0 \tag{A2} $$

${}^{\Phi}G_{mix}^{ideal}$ originates from the configurational entropy contribution, which according to the regular solution model is given by Eq. (A3), which is nothing but the Bragg–Williams approximation given by Eq. (56) for a binary system.

$$ {}^{\Phi}G_{mix}^{ideal} = R \cdot T \cdot \sum_{I} c_I \cdot \ln c_I \tag{A3} $$

Finally, ${}^{\Phi}G_{mix}^{XS}$ is the excess free energy necessary to incorporate nonideal contributions. The simplest form is given by Eq. (A4), where $\Omega_{I,J}$ is termed an interaction parameter in thermochemistry.

$$ {}^{\Phi}G_{mix}^{XS} = \sum_{I} \sum_{J>I} \Omega_{I,J} c_I \cdot c_J \tag{A4} $$

In order to further improve the description of the extra energy, one may generalize Eq. (A4) in the series expansion of Eq. (A5), where $L_{i,j}^K$ is a kth-order binary interaction parameter between species I and J which is expressed as a polynomial in temperature T.

$$ {}^{\Phi}G_{mix}^{XS} = \sum_{I} \sum_{J>I} c_I \cdot c_J \cdot \sum_{k} {}^{\Phi}L_{I,J}^k (c_I - c_J)^k \tag{A5} $$

This particular expression is known as the Redlich–Kister expansion. It is readily realized that Eq. (A4) is a special case of Eq. (A5) in which the expansion is truncated to $k = 0$.

Two main features of the CALPHAD method are addressed. One is that the free energy expression of pure elements, ${}^{\Phi}G_I^0$, in Eq. (A2) is given by the particular form of Eq. (A6), where H_m^{SER} is the enthalpy at the standard element reference state of the substance in its most stable state at 298.15 °C, a, b, c, and d_i are model (materials) constants, and data have been compiled by Dinsdale (1991) in the Scientific Group Thermodata Europe (SGTE).

$$ G_m - H_m^{SER} = a + b \cdot T + c \cdot T \cdot \ln T + \sum_{i} d_i \cdot T_i \tag{A6} $$

It should be noted that this expression is adopted not only for a pure element but also for a stoichiometric compound. This is the starting building block for more complicated alloy systems and in order to secure accuracy the compiled data have often been updated.

The second feature of the CALPHAD method is that the Gibbs free energy for a multicomponent solution phase is described on the basis of a sublattice model. Essential points of the model are described below by taking as an example the two-sublattice model for an A–B binary alloy system.

Within the two-sublattice model, a phase consists of two sublattices. The two components enter both sublattices, which is denoted as $(A, B)_p(A, B)_q$ where subscripts p and q are the numbers of sites in each sublattice. The free energy, G, of a solution phase is then given by Eq. (A7), which is in the same form as $^\Phi G$ in Eq. (A1).

$$G = G^0 + G_{mix}^{ideal} + G_{mix}^{XS} \tag{A7}$$

Like $^\Phi G^0$, G^0 describes the contribution from the mechanical mixing; however, due to the introduction of the sublattices, four kinds of pure components are mixed. Two of them are pure elements A and B and the other two are the stoichiometric compounds, $A_p B_q$ and $A_q B_p$. Then, G^0 is written as Eq. (A8), where y indicates the concentration (site fraction) and the subscripts and superscripts denote the species and sublattice, respectively: y_A^I, for instance, is the concentration of species A in the Ith sublattice.

$$G^0 = y_A^I \cdot y_A^{II} \cdot G_{A;A}^0 + y_A^I \cdot y_B^{II} \cdot G_{A;B}^0 + y_B^I \cdot y_A^{II} \cdot G_{B;A}^0 + y_B^I \cdot y_B^{II} \cdot G_{B;B}^0 \tag{A8}$$

$G_{I:J}^0$ is the Gibbs energy of the compound $I_p J_q$ expressed by Eq. (A6). $G_{I:I}^0$ is equivalent to $^\Phi G_I^0$ in Eq. (A2), and $G_{I:J}^0$ ($G_{J:I}^0$) is the new contribution due to the introduction of the sublattices. One realizes that the prefactor c_I in Eq. (A2) is modified in Eq. (A8) by the product of the compositions of each species in the sublattices of concern.

G_{mix}^{ideal} is the mixing entropy term based on the Bragg–Williams expression given by Eq. (A9), and this corresponds to $^\Phi G_{mix}^{ideal}$ in Eq. (A3).

$$\begin{aligned}G_{mix}^{ideal} = & \, p \cdot RT(y_A^I \cdot \ln y_A^I + y_B^I \cdot \ln y_B^I) \\ & + q \cdot RT(y_A^{II} \cdot \ln y_A^{II} + y_B^{II} \cdot \ln y_B^{II})\end{aligned} \tag{A9}$$

One should note that the mixing entropy is evaluated in each sublattice and the net contribution is given as the weighted sum of p and q.

Finally, G_{mix}^{XS} is the excess Gibbs energy of mixing due to nonideal contributions. This is again the extension of Eq. (A5) and is given by Eq. (A10), where $L_{A,B;*}^k$ ($L_{*;A,B}^k$) is the kth-order interaction parameter between the components A and B in the first (second) sublattice.

$$G_{mix}^{xs} = y_A^I \cdot y_B^I \left(y_A^{II} \cdot \sum_{k=0} L_{A,B;A}^k \cdot (y_A^I - y_B^I)^k + y_B^{II} \cdot \sum_{k=0} L_{A,B;B}^k \cdot (y_A^I - y_B^I)^k \right)$$

$$+ y_A^{II} \cdot y_B^{II} \left(y_A^I \cdot \sum_{k=0} L_{A;A,B}^k \cdot (y_A^{II} - y_B^{II})^k + y_B^I \cdot \sum_{k=0} L_{B;A,B}^k \cdot (y_A^{II} - y_B^{II})^k \right)$$

(A10)

In this notation, a semicolon separates components occupying different sublattices, and a comma separates interacting components in the same sublattice. These equations can be generalized for phases with multicomponents and multi-sublattices, and they reduce to a random substitutional model when there is only one sublattice.

The variables participating in minimizing the free energy are basically only concentrations (site fractions), which simplifies the mathematical procedures. This is the most significant advantage of the CALPHAD method. Physically, however, the description of the free energy through Eqs. (A7–A10) does not take the atomic correlations – not only inter-sublattices but also intra-sublattices – into account. Such a deficiency is circumvented by the efficient parameterization of the interaction parameters used for fitting the experimental data.

In view of the complexity of practical alloys, the CALPHAD method is quite useful and recently (Kaufman et al. 2001; Turchi et al. 2005, 2007; Tokunaga et al. 2006) the combination of the first-principles electronic structure calculations with the CALPHAD type of description of the free energy has been actively developed.

Acknowledgments

The author is grateful to his coworkers, Professor Y. Chen of the University of Tokyo and Dr. M. Ohno of Hokkaido University, for their sincere contributions to the collaboration. Without their hard work, the CVM-based studies of first-principles calculations would not have been carried out at the level of detail presented here. In particular, Dr. Ohno permitted the author to cite some unpublished work from his Ph.D. thesis.

References

Allen, S.M., Cahn, J.W. (1972), *Acta Metal.* **20**, 423.

Allen, S.M., Cahn, J.W. (1979), *Acta Metal.* **27**, 1085.

Bacon, G.E., Crangle, J. (1963), *Proc. Royal. Soc. London* **272**, 387.

Barker, J.A. (1953), *Proc. Roy. Soc.* **A216**, 45.

Becker, J.D., Sanchez, J.M., Tien, J.K. (1990), *Proc. MRS Boston*,

Bellashchenko, K.D., Samolyuk, G.D., Vaks, V.G. (1999a), *J. Phys. Condens. Matter* **11**, 10 567.

Bellashchenko, K.D., Dobretsov, V.Yu., Pankratov, I.R., Samolyuk, G.D., Vaks,

V.G. (1999b), *J. Phys. Condens. Matter* **11**, 10 593.

Bernard, J., Zunger, A. (1990), *Phys. Rev.* **B42**, 9622.

Bragg, W.L., Williams, E.J. (1934), *Proc. Roy. Soc.* **A145**, 69.

Cahn, J.W. (1961), *Acta Metal.* **9**, 795.

Cahn, J.W., Hilliard, J.E. (1958), *J. Chem. Phys* **28**, 258.

Chen, L.-Q. (2002), *Annu. Rev. Mat. Res.* **32**, 113 and references therein.

Chen, Y., Atago, T., Mohri, T. (2002a), *J. Phys. Condens. Matter* **14**, 1903.

Chen, Y., Iwata, S., Mohri, T. (2002b), *CALPHAD* **26**, 583.

Chen, Y., Iwata, S., Mohri, T. (2005), *Proc. Int. Conf. Solid–Solid Phase Transformations in Inorganic Materials 2005*, Vol. 2, TMS, Warrendale, Pa, U.S.A., p. 663.

Connolly, J.W., Williams, A.R. (1983), *Phys. Rev.* **B27**, 5169.

de Fontaine, D. (1975), *Acta Metal.* **23**, 553.

de Fontaine, D. (1979), *Solid State Physics* **34**, 73.

de Fontaine, D. (1994), *Solid State Physics* **47**, 33.

Dinsdale, A. (1991), *CALPHAD* **15**, 317.

Dobretsov, V.Yu., Pankratov, I.R., Vaks, V.G. (2004), *JETP Lett.* **80**, 602.

Dobretsov, V.Yu., Pankratov, I.R., Stroev, A.Yu., Vaks, V.G. (2005), *Proc. Int. Conf. Solid–Solid Phase Transformations in Inorganic Materials 2005*, Vol. 1, TMS, Warrendale, Pa, U.S.A., p. 169.

Ducastelle, F. (1991), *Order and Phase Stability in Alloys (Cohesion and Structure*, Vol. 3), ed. de Boer, F.R., Pettifor, D.G., North-Holland, Amsterdam.

Fan, D., Chen, L.Q. (1997), *Acta Mater.* **45**, 3297.

Ferreira, L., Wei, S., Zunger, A. (1989), *Phys. Rev.* **B40**, 3197.

Ferreira, L., Ozolins, V., Zunger, A. (1999), *Phys. Rev.* **B60**, 1687.

Glauber, R.J. (1953), *J. Math. Phys.* **4**, 294.

Jansen, H.J.F., Freeman, A.J. (1984), *Phys. Rev.* **B30**, 561.

Kanamori, J., Kakehashi, Y. (1977), *J. Phys., Paris* **38**, C7-274.

Kaufman, L., Turchi, P.E.A., Huang, W., Liu, Z.-K. (2001), *CALPHAD* **25**, 419.

Kawasaki, K. (1966), *Phys. Rev.* **145**, 224.

Khachaturyan, A.G. (1983), *Theory of Structural Transformations in Solids*, John Wiley & Sons, New York.

Khromov, K.Yu., Pankratov, I.R., Vaks, V.G. (2005), *Proc. Int. Conf. Solid–Solid Phase Transformations in Inorganic Materials 2005*, Vol. 1, TMS, Warrendale, Pa, U.S.A., p. 217.

Kikuchi, R. (1951), *Phys. Rev.* **81**, 998.

Kikuchi, R. (1966), *Prog. Theor. Phys. Suppl.* **35**, 1.

Kikuchi, R. (1974), *J. Chem. Phys.* **60**, 1071.

Kikuchi, R. (1998), *J. Phase. Equilibria* **19**, 412–421.

Kikuchi, R., Beldjenna, A. (1992), *Physica* **A182**, 617.

Kikuchi, R., Cahn, J.W. (1962), *J. Phys. Chem. Solids* **23**, 137.

Kikuchi, R., Masuda-Jindo, K. (1999), *Comp. Mater. Sci.* **14**, 295.

Kikuchi, R., Mohri, T. (1997), *Cluster Variation Method*, Morikita Syuppan Co., Tokyo (in Japanese).

Kikuchi, R., Mohri, T., Fultz, B. (1992), *Mat. Res. Soc. Symp. Proc.* **205**, 387.

Laks, D., Wei, S., Zunger, A. (1992), *Phys. Rev. Lett.* **69**, 3766.

Lu, Z., Wei, S., Zunger, A., Frotapessoa, S., Ferreira, L. (1991), *Phys. Rev.* **B44**, 512.

Lu, Z., Klein, B., Zunger, A. (1995), *Modelling Simul. Mater. Sci. Eng.* **3**, 753.

Massalski, T.B. (ed.) (1986), *Binary Phase Diagrams*, ASM, Metals Park, OH.

Mbaye, A., Ferreria, L., Zunger, A. (1987), *Phys. Rev. Lett.* **58**, 49.

Mohri, T. (1990), *Acta Metal. Mater.* **38**, 2455.

Mohri, T. (1992), *Structural and Phase Stability of Alloys*, ed. J.L. Moran-Lopez, Plenum Press, New York, p. 87.

Mohri, T. (1993), *Interatomic Potential and Structural Stability*, ed. K. Terakura, Springer-Verlag, Heidelberg, p. 168.

Mohri, T. (1994a), *Solid–Solid Phase Transformations*, ed. W.C. Johnson. The Minerals, Metals, & Materials Society, Warrendale, PA, p. 54.

Mohri, T. (1994b), *Statics and Dynamics of Alloy Phase Transformations*, ed. P.E.A. Turchi, Plenum Press, New York, p. 665.

Mohri, T. (1994c), *Prog. Theoret. Phys. Suppl.* **115**, 147.

Mohri, T. (1996), *Stability of Materials*, ed. A. Gonis, Plenum Press, New York, p. 205.

Mohri, T. (1997), *Properties of Complex Inorganic Solids*, ed. A. Gonis, Plenum Press, New York, p. 83.

Mohri, T. (1999), *Z. Metallkunde* **B90**, 71.

Mohri, T. (2000a), *Modelling Simul. Mater. Sci. Eng.* **8**, 239.

Mohri, T. (2000b), *Properties of Complex Inorganic Solids 2*, ed. A. Meike, Kluwer Academic/Plenum Publishers, New York, p. 123.

Mohri, T. (2002), *Zairyo Systemgaku (Materials Systems)*, Asakura Shoten Co., Tokyo, (in Japanese).

Mohri, T. (2005), *The Science of Complex Alloy Phases*, ed. T.B. Massalski, P.E.A Turchi, TMS, Warrendale, Pa, U.S.A., p. 109.

Mohri, T., Chen, Y. (2002), *Mater. Trans.* **43**, 2104.

Mohri, T., Chen, Y. (2004a), *Mater. Trans.* **45**, 1478.

Mohri, T., Chen, Y. (2004b), *J. Alloys Compounds* **383**, 23.

Mohri, T., Chen, Y. (2004c), *J. Japan Inst. Metals*. **63**, 966 (in Japanese); the English version is under preparation.

Mohri, T., Ikegami, T. (1993a), *Defect and Diffusion Forum* **95–98**, 119.

Mohri, T., Ikegami, T. (1993b), *Diffusion in Ordered Alloys*, ed. B. Fultz, The Minerals, Metals, and Materials Society, Warrendale, PA, p. 79.

Mohri, T., Miyagishima, S. (1998), *Mater. Trans., JIM* **39**, 154.

Mohri, T., Sanchez, J.M., de Fontaine, D. (1985a), *Acta Metal.* **33**, 1463.

Mohri, T., Sanchez, J.M., de Fontaine, D. (1985b), *Acta Metal.* **33**, 1711.

Mohri, T., Nakamura, K., Ito, T. (1991), *J. Appl. Phys.* **70**, 1320.

Mohri, T., Takizawa, S., Terakura, K. (1993), *J. Phys. Condens. Matter* **5**, 1473.

Mohri, T., Nakahara, T., Takizawa, S., Suzuki, T. (1995), *J. Alloys Compounds* **220**, 1.

Mohri, T., Oh, C.-S., Takizawa, S., Suzuki, T. (1996a), *Intermetallics* **4**, S3.

Mohri, T., Ichikawa, Y., Nakahara, T., Suzuki, T. (1996b), *Theory and Applications of the Cluster Variation and Path Probability Methods*, ed. J.L. Moran-Lopez, Plenum Press, New York, p. 37.

Mohri, T., Ichikawa, Y., Suzuki, T. (1997), *J. Alloys Compounds* **247**, 98.

Mohri, T., Ohno, M., Chen, Y. (2005a), *Materials Science Forum* **475–479**, 3075.

Mohri, T., Ohno, M., Chen, Y. (2005b), *Proc. Int. Conf. Solid–Solid Phase Transformations in Inorganic Materials 2005*, Vol. 2, TMS, Warrendale, Pa, U.S.A., p. 633.

Mohri, T., Ohno, M., Chen, Y. (2006), *J. Phase Equil. Diffusion* **27**, 47.

Morruzi, V., Janak, J.F., Schwarz, K. (1988), *Phys. Rev.* **B37**, 790.

Nakamura, K., Mohri, T. (1993), *Modelling Simul. Mater. Sci. Eng.* **1**, 143.

Oh, C.-S., Mohri, T., Lee, D.N. (1994), *Mater. Trans., JIM* **35**, 445.

Ohnuma, I., Ishida, K. (2004), Tohoku University, private communications.

Ohno, M. (2004), Ph.D. Dissertation, Graduate School of Engineering, Hokkaido University.

Ohno, M., Mohri, T. (2001a), *Mater. Sci. Eng.* **A312**, 50.

Ohno, M., Mohri, T. (2001b), *Mater. Trans.* **42**, 2033.

Ohno, M., Mohri, T. (2002), *Mater. Trans.* **43**, 2189.

Ohno, M., Mohri, T. (2003), *Phil. Mag.* **83**, 315.

Ohno, M., Mohri, T. (2007), *Mater. Trans.* **47**, 2718. *Proc. 15th Int. Symp. on Transport Phenomena*, p. 420.

Ozolins, V., Wolverton, C., Zunger, A. (1998), *Phys. Rev.* **B57**, 6427.

Pankratov, I.R., Vaks, V.G. (2001), *J. Phys. Condens. Matter* **13**, 6031.

Perdew, J.P., Chevary, J.A., Vosko, S.H., Jackson, K.A., Pederson, M.R., Singh, D.J., Fiolhais, C. (1992), *Phys. Rev.* **B46**, 6671.

Richards, M., Cahn, J.W. (1971), *Acta Metal.* **19**, 1263.

Sanchez, J.M., de Fontaine, D. (1978), *Phys. Rev.* **B17**, 2926.

Sanchez, J.M., de Fontaine, D. (1981), in *Structure and Bonding in Crystals*, ed. M. O'Keeffeeb, A. Navrotsky, Vol. II, Academic Press, New York, p. 117.

Sanchez, J.M., Ducastelle, F., Gratias, D. (1984), *Physica (Utrecht)* **128A**, 33.

Santai, M., Wang, L.G., Zunger, A. (2003), *Phys. Rev. Lett.* **90**, 045 502-1.

Sluiter, M.H.F., Esfarjani, K., Kawazoe, Y. (1995), *Phys. Rev. Lett.* **75**, 3142.

Tokunaga, T., Ohtani, H., Hascbc, M. (2006), *CALPHAD* **30**, 201.

Turchi, P.E.A. (2003), in *Phase Diagrams as a Tool for Advanced Materials Design*, APDIC World Round Robin Seminar Text, ed. T. Mohri, Sapporo, p. 3.

Turchi, P.E.A., Drchal, V., Kudrnovsky, J., Colinet, C., Kaufman, L., Liu, Z.-K. (2005), *Phys. Rev.* **B71**, 094 206-1.

Turchi, P.E.A., Kaufman, L., Liu, Z.-K. (2006), *CALPHAD* **30**, 70.

Turchi, P.E.A., Abrikosov, I.A., Burton, B., Fries, S.G., Grimvall, G., Kaufman, L., Korzhavyi, P., Manga, V.R., Ohono, M., Pisch, A., Scott, A., Zhang, W. (2007), *CALPHAD*, in the press.

Uzawa, H., Mohri, T. (2001a), *Mater. Trans.* **42**, 422.

Uzawa, H., Mohri, T. (2001b), *Mater. Trans.* **42**, 1866.

Uzawa, H., Mohri, T. (2002), *Mater. Trans.* **43**, 2185.

Vaks, V.G., Samolyuk, G.D. (1999), *J. Exper. Theor. Phys.* **88**, 89.

Wei, S., Mbaye, A., Ferreria, L., Zunger, A. (1987), *Phys. Rev.* **B36**, 4163.

Wei, S., Ferreira, L., Zunger, A. (1990), *Phys. Rev.* **B41**, 8240.

Wei, S., Ferreira, L., Zunger, A. (1992), *Phys. Rev.* **B45**, 2533.

Wei, S., Zunger, A. (1993), *Phys. Rev.* **B48**, 6111.

Williams, R.O. (1976), *Report No. ORNL-5140*, Oak Ridge National Laboratory, Oak Ridge, TN.

Wolverton, C., Zunger, A. (1997), *Comp. Mater. Sci.* **8**, 107.

Wolverton, C., Zunger, A., Lu, Z. (1994), *Phys. Rev.* **B49**, 16 058.

Zarkevich, N.A., Johnson, D.D. (2004), *Phys. Rev. Lett.* **92**, 255 702-1.

Zunger, A., Wei, S., Mbaye, A., Ferreria, L. (1988), *Acta. Metal.* **36**, 2239.

11
Ab-Initio Methods and Applications

Stefan Müller, Walter Wolf, and Raimund Podloucky

11.1
Introduction

The idea of an ab-initio approach is to calculate properties of (solid) matter without the need for any empirical information. For this purpose two basic requirements have to be fulfilled: (a) the interaction of atoms is described by the fundamental laws of physics, and (b) the models and approximations one is forced to make are reasonably accurate. Concerning point (a), equations are needed which comprise the quantized interaction of electrons. These can be formulated traditionally by the Schrödinger equation involving many-body wavefunctions. As mentioned in Section 11.2.1 such equations are generally not applicable to solids, because of the large number of interacting particles. On the other hand, the electron density would be much better suited provided one finds a manageable formulation for the many-body part of the total energy of the system being a functional of the density. For practical reasons but still being exact, density functional theory is recast in terms of single-particle wavefunctions. Thus, equations are derived which look like simplified Schrödinger equations but, of course, the many-body complications are not lost. They are just reformulated in such a way that requirement (b) can be fulfilled, namely that reasonable approximations can be introduced giving useful results in a reasonable timeframe – provided the system is not too large. In that sense, a true ab-initio concept is established. Of course, approximations do not work well in all cases. One should, however, not speak of failures of density functional theory itself (as is rather common), since the shortcomings are due to the approximations to exchange–correlation interactions introduced into the numerical solutions.

A variety of numerical techniques and codes are available. Very briefly, some of the most important methods are discussed.

Because of their striking success, applications of density functional theory (DFT) became very popular and penetrated the field of materials science. In this chapter we will shortly describe DFT methods and give examples of very recent investigations in which important materials properties were derived from DFT concepts and calculations. This includes the ab-initio determination of vibrational properties, which provides, apart from the vibrational spectrum itself, the most important temperature-dependent thermodynamic properties as long as the temperature is well below the melting point. DFT applications – strictly valid only for $T = 0$ K – can thus be made temperature-dependent. Furthermore, elastic properties can be derived in a quite straightforward way from the calculation of the total energy or stress tensor as a function of suitable distortions.

Further exploitation of the total energy might lead to the derivation of structural and phase stabilities, for which energy differences between suitable reference systems have to be taken. This concept can be carried further for calculating defect properties involving a grand-canonical treatment based on total energies of supercells. Furthermore, migration of defects and impurities can be evaluated quantitatively in terms of temperature-dependent diffusion coefficients within transition state theory, making use of transition state barrier energies and vibrational spectra. More complex planar faults can also be analyzed and, as an application, impurity effects on grain boundary cohesion and intergranular fracture as obtained from total energy calculations are dicussed.

Going beyond consideration of total energies of long-range ordered solids only with fixed stoichiometry (apart from studying small variations as for the defect and impurity case), a much more general approach is pursued by the cluster expansion (see Section 11.3.5). There, DFT is used to provide a database of energies for a series of atomic configurations (so-called clusters). Once a convergent expansion in terms of these clusters is found, a wide area of materials problems can be studied quantitatively, such as the formation of precipitates as a function of temperature and time. To this end, the DFT database is used in Monte Carlo numerical simulations for probing the probability of site occupations.

As is shown in this contribution, ab-initio methods have reached a mature state and are capable of operating in a robust manner within the context of other advanced simulation techniques. This has widened the range of applicability to an unprecedently large variety of materials problems. In the present chapter, the topics mentioned will be dealt with by discussing very recent examples and publications, which – in most cases – are based on the work of the authors' groups. The selection of examples is representative of similar studies currently being done within the computational materials science community.

11.2
Theoretical Background

11.2.1
Density Functional Theory

The aim is to describe the interactions between electrons and atoms in a solid system on the most fundamental level, namely by the basic principles of quantum physics. Restricting the considerations to time-independent cases (i.e., on stationary ground states described by the time-independent Hamilton operator $\hat{H}$) one has to solve the time-independent Schrödinger eigenvalue equation (1), labeling the states by the (set of) quantum numbers i related to suitable boundary conditions, including also spin degrees of freedom.

$$\hat{H}\Psi_i = E_i\Psi_i \tag{1}$$

For our purpose and for simplicity, we consider only nonrelativistic systems for which the number of particles is conserved. Then, the spin part of the wavefunctions can be separated from the space-dependent functions. Furthermore, no temperature-driven excitations are considered, which means that all the following considerations are strictly valid only for $T = 0$ K. This, however, for most of the applications of DFT is not a severe restriction, because of the thermal properties of the (fermionic) electron gas.

The wavefunction Ψ_i is a function of $4N$ (spin and space) variables for the system of N electrons, providing the energy eigenvalues E_i. Because electrons are fermions, the wavefunction must be antisymmetric concerning the exchange of indistinguishable particles α and β, i.e., when the sets of coordinates for particle α and β are exchanged [Eq. (2)].

$$\Psi(\ldots \mathbf{x}_\alpha \ldots \mathbf{x}_\beta \ldots) = -\Psi(\ldots \mathbf{x}_\beta \ldots \mathbf{x}_\alpha \ldots) \tag{2}$$

The variable $\mathbf{x}_\alpha = (\mathbf{r}, \uparrow)_\alpha$ comprises space and spin coordinates for electron α. A direct consequence of Eq. (2) is the so-called exchange interaction as manifested by Pauli's exclusion principle. This purely quantum statistical interaction is accompanied by the so-called correlation which stems from the repulsive Coulomb repulsion $1/|\vec{r}_\alpha - \vec{r}_\beta|$ between electronic point charges, as it occurs in the Hamilton operator. Both, these quantum many-body effects are summarized under the name exchange–correlation. If one could solve Eq. (1) with acceptable precision as it is, all the further considerations would be obsolete. However, in practice direct many-body wavefunction solutions to the Schrödinger equation (1) are restricted to rather small molecules and are in general not applicable for solids with their $\approx 10^{23}$ electrons.

The vast majority of ab-initio applications for deriving ground-state properties of solids is based on DFT as derived by Hohenberg and Kohn (1964) and Kohn and Sham (1965). In the fundamental work of Hohenberg and Kohn the central role of the wavefunction is replaced by the electron density, which is assumed to determine uniquely the total energy of the ground state $E[\rho]$ of the electrons. Given the so-called external potential v_{ext} as generated by the nuclei of the system, the total energy is decomposed [Eq. (3)]

$$E[\rho] = \int v_{ext}\rho \, d\mathbf{x} + F[\rho] \tag{3}$$

The functional $F[\rho]$ is universal because it depends only on the electron density and not on the external potential. Overcoming all the problems of the mathematical existence of this postulated functional it can be shown that for reasonable electron densities ρ, such an ansatz exists. The functional F is a sum of the kinetic energy of the electrons T, the classical Coulomb interaction between electronic charge densities E_{coul}, and the remaining term E_{xc} which comprises all the complications of the quantum many-body interactions [Eq. (4)].

$$F[\rho] = T[\rho] + E_{coul}[\rho] + E_{xc}[\rho] \tag{4}$$

This very general formulation in combination with the variational principle results in a very elegant relationship involving only the electron density. It is, however, not (yet) suitable for reliable calculations because it is not accurate enough. A major step toward practical applications of DFT was made by Kohn and Sham (1965) by introducing a scheme of auxiliary wavefunctions which have the properties of single-particle wavefunctions (or orbitals). For that purpose, the kinetic energy functional T is split into a single-particle term T_s and the remainder which is supposed to be included in the many-body exchange–correlation functional. Although on one hand this functional is not known, on the other hand the rearrangement of terms presumably leads to cancellation of large errors when approximations are introduced.

The only true purpose of the Kohn–Sham orbitals ϕ_i is to build up the true ground-state electron density by summing over a suitable number of states [Eq. (5)], defining the total number of electrons N in the system.

$$\sum_i \phi_i^*(\mathbf{r})\phi_i(\mathbf{r}) = \rho(\mathbf{r}), \quad \int \rho(\mathbf{r})\,d\mathbf{r} = N \tag{5}$$

From this concept of Kohn–Sham orbitals a stationary Schrödinger-like equation of the form of Eq. (6) for the ith orbital is derived. The Hamilton operator is built up as in Eq. (7).

$$\hat{H}\phi_i = \epsilon_i \phi_i \tag{6}$$

$$\hat{H} := -\frac{\hbar^2}{2m}\Delta + v_{eff}(\mathbf{x}), \quad v_{eff} = v_{ext} + v_{coul} + v_{xc} \tag{7}$$

The effective potential v_{eff} is the sum of the external potential v_{ext}, the classical Coulomb interactions v_{coul}, and a term v_{xc} which contains all the many-body or exchange–correlation interactions. Like the functional $F[\rho]$, the potential $v_{xc}[\rho]$ (being the functional derivative $\delta E_{xc}/\delta \rho$) is universal and depends only on the electron density ρ. The potentials generally depend on the spin state (in the most simple form by spin polarization of the electronic states into majority and minority spin components), which enables treatment of systems with spontaneous magnetic ordering. The basic and most fundamental results of an application of DFT is the ground-state electron density ρ and its corresponding total energy $E[\rho]$. Other important quantities are derived from the total energy being dependent on physical parameters. For example, the force acting on an atom with position $\mathbf{r_i}$ is determined by the derivative $\partial E/\partial \mathbf{r_i}$.

The quality of the results of actual calculations for real systems then depends on the quality of the approximation one has to make for $v_{xc}[\rho]$. The first and surprisingly successful approximations were based on the properties of the homogeneous electron gas. Locally, the value of true electron density was assumed to be the density of the interacting homogeneous electron gas for which the exchange–correlation energy can be determined. Consequently, the exchange–correlation functional is constructed as in Eq. (8), in which ε_{xc} is the (local) exchange–correlation energy per particle of the homogeneous electron gas of density ρ. Obviously, this ansatz is called the local density approximation (LDA) (e.g., Vosko et al. 1980).

$$E_{xc}^{LDA}[\rho] = \int \rho(\mathbf{r})\varepsilon_{xc}(\rho)\,d\mathbf{r} \tag{8}$$

Applying the functional derivatives due to ρ, the LDA version of the Kohn–Sham equation is derived as Eq. (9).

$$\left(-\frac{\hbar^2}{2m}\Delta + v(\mathbf{r}) + \int \frac{\rho(\mathbf{r'})}{|\mathbf{r}-\mathbf{r'}|}\,d\mathbf{r'} + v_{xc}^{LDA}(\mathbf{r})\right)\phi_i = \varepsilon_i\phi_i \tag{9}$$

A variety of different LDAs exist which basically give rather similar results.

As is now well known, LDAs lead to serious errors for systems with valence electrons of a more localized nature. For example, the ground-state crystal structure and magnetic state for Fe are wrongly predicted by LDA calculations. Since the beginning of the 1990s extensive work has gone into the systematic improvement of LDAs, resulting in the concepts of the (semi-local) gradient approximations, including the gradient of the electron density $\vec{\nabla}\rho$ (e.g., Perdew and Wang 1992). The potential v_{xc}^{LDA} in the Kohn–Sham equations has then to be replaced by the corresponding potential v_{xc}^{GGA} of the so-called generalized gradient approximation (GGA). Similarly to the development of LDA potentials, a variety of GGA formulations exist, which might lead to significantly different results when treating rather inhomogenous systems, e.g., calculating adsorption energies and migration barriers of molecules on surfaces. Fortunately, for most of the bulk studies, the different GGA potentials yield the same results, which in general are in better agreement with experimental facts, as demonstrated for Fe in Fig. 11.1.

Figure 11.1 clearly shows that LDA (right panel) would predict the wrong ground-state structure (namely fcc) with an equilibrium volume which is far too small. On the other hand, GGA yields the correct bcc ground state and a reasonable volume when compared to experiment.

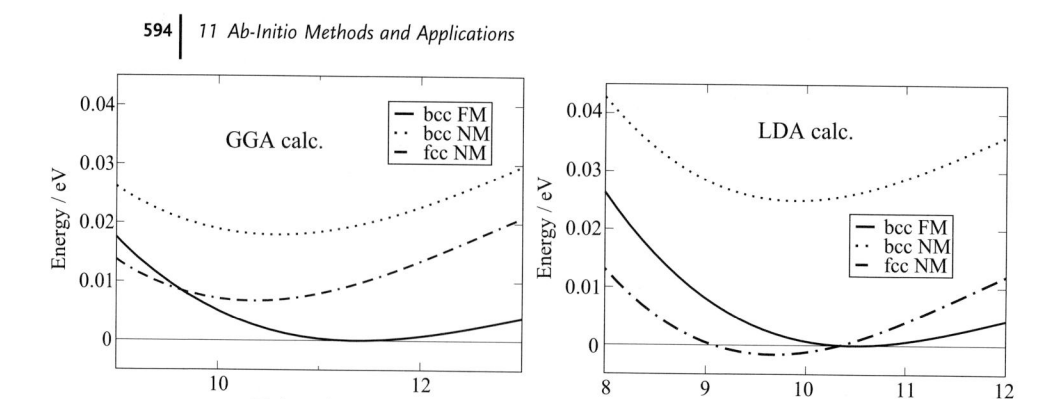

Fig. 11.1 Ab-initio total energies for Fe vs. atomic volume (in Å^3). The energy zero is defined by the minimum of energy of the ferromagnetic calculation for the bcc structure. NM: not spin-polarized, FM: ferromagnetic calculation. Left panel: GGA results, right panel: LDA results. Calculations were made by the FLAPW code flair (as discussed in Section 11.2.2).

Although applying GGA achieves considerable improvement in many cases in which LDA is just not sufficiently accurate, for some specific systems the many-body interactions need a more sophisticated treatment (for example, for highly correlated systems with strongly localized electronic states). At present, strong efforts are being made for an improved handling of the many-body interactions by partially going back to wavefunction-like concepts beyond the Kohn–Sham description. A very recent survey of some elaborate methods is given in a study on bulk MnO (Franchini et al. 2005). These highly theoretical considerations may have a strong impact on the energetics and hence also on the physical properties of alloys and metals, for example when Mn is involved (Chen et al. 2005b).

11.2.2
Computational Methods

The goal of performing ab-initio calculations is to work with a computational machinery which – without any empirical parameters – provides quantitatively useful numbers. This can be achieved by making use of density functional theory and proper numerical/technical implementations of sufficient methodolgical and numerical accuracy. As discussed in Section 11.2.1, within the Kohn–Sham formulation of density functional theory Schrödinger-like equations have to be solved for single-particle wavefunctions $\psi_{\mathbf{k}}^{\nu}(\mathbf{x})$. If periodic boundary conditions are chosen – which is the standard in solid state physics – the so-called Bloch states are labeled in terms of the wave vector $\mathbf{k}$ of the first Brillouin zone and the band index ν. Bloch states fulfill the boundary condition given by Eq. (10) under a translation by a real lattice vector $\mathbf{R}$.

$$\psi_{\mathbf{k}}^{\nu}(\mathbf{x}+\mathbf{R}) = e^{i\mathbf{kR}}\psi_{\mathbf{k}}^{\nu}(\mathbf{x}) \tag{10}$$

As is common for numerical applications, the wavefunctions are composed of linear combinations of suitable basis functions $\Phi_{\mathbf{k},i}$ with the coefficients $c_{\mathbf{k},i}^{\nu}$ [Eq. (11)].

$$\psi_{\mathbf{k}}^{\nu}(\mathbf{x}) = \sum_{i}^{N} c_{\mathbf{k},i}^{\nu} \Phi_{\mathbf{k},i}(\mathbf{x}) \tag{11}$$

The summation runs over a sufficiently large number of basis functions N so that numerical convergency is reached for all important physical properties, such as the electron density and the total energy. The total energy functional formed by the translationally invariant Hamiltonian $\hat{H}_{\mathbf{k}}$ describing all the interactions between atoms and electrons in the unit cell, Eq. (12),

$$\langle \psi_{\mathbf{k}}^{\nu}(\mathbf{x}) | \hat{H}_{\mathbf{k}} | \psi_{\mathbf{k}}^{\nu}(\mathbf{x}) \rangle \tag{12}$$

is minimized with respect to the variational parameters $c_{\mathbf{k},i}^{\nu}$ under the normalization condition Eq. (13).

$$\langle \psi_{\mathbf{k}}^{\nu}(\mathbf{x}) | \psi_{\mathbf{k}}^{\nu}(\mathbf{x}) \rangle = 1 \tag{13}$$

Following this variational procedure, the matrix eigenvalue equation (14) is derived, in which $H_{ij} = \langle \Phi_{\mathbf{k},i} | \hat{H} | \Phi_{\mathbf{k},j} \rangle$ is the matrix element of the Hamiltonian and $S_{ij} = \langle \Phi_{\mathbf{k},i} | \Phi_{\mathbf{k},j} \rangle$ denotes the overlap matrix element formed by the basis functions $\Phi_{\mathbf{k},j}$.

$$\sum_{j} H_{ij} c_{\mathbf{k},j}^{\nu} = \epsilon_{\mathbf{k}}^{\nu} \sum_{j} S_{ij} c_{\mathbf{k},j}^{\nu} \tag{14}$$

For nonorthogonal basis functions the matrix S is different from the unity matrix. The energies $\epsilon_{\mathbf{k}}^{\nu}$ are the eigenvalues and the coefficients $c_{\mathbf{k},j}^{\nu}$ are the components of the eigenvectors, which have to be calculated. Depending on the size of the system (i.e., the number of atoms per unit cell) and the chosen numerical methods, the dimensions of the matrices H_{ij} and S_{ij} could be quite large (e.g., of the order of 10^4 for about 100 atoms in a plane-wave based approach; see later). This means that, for many applications, most of the computational effort goes into the diagonalization of such large matrices.

Once Eq. (14) has been solved, the corresponding electron density is constructed by summing over the squares of the Kohn–Sham orbitals. Because for building the Hamiltonian the density must already be known (in order to construct the potentials), a self-consistency problem has to be solved. Achievement of self-consistency could be rather cumbersome, in particular for large systems with many electronic and geometric degrees of freedom. Therefore, elaborate schemes of mixing old and new densities and optimizing geometries have to be applied in the computer codes.

The most accurate and successful numerical methods are able to solve the Kohn–Sham equations without any further modeling restrictions such as assumptions about the geometrical shape of electron density and potential. In the following, we will give a very short description of pseudopotential methods (PP), full-potential linearized augmented plane-wave (FLAPW) applications, and full-potential linearized muffin-tin orbital (FLMTO) approaches. These methods are widely used in the community. For a more detailed insight the reader is referred by the book Martin (2004). Of course, the fundamental many-body electron–electron interactions have to be approximated, as described in Section (11.2.1), but these approximations should be independent of the numerical technique applied.

A natural choice of basis functions for describing Bloch functions are plane waves $e^{i(k+G_j)r}$ which are labeled by the reciprocal lattice vectors G_j. This kind of strategy is pursued in the PP approaches. The basic problem of a pure plane wave approach is the rapid change of the real potential and wavefunction near the nuclei. Any plain Fourier series of these quantities will not converge. Therefore, ways had to be invented to overcome this problem, which is done by "pseudolizing" the potential: not the bare Coulomb potential of the atomic nuclei but a potential screened by the inner-shell electronic states is introduced. Furthermore, following the argument that for the bonding only the wavefunction in the region between the atoms needs to be accurate, pseudoptential schemes and packages were developed. The latest development consists in working with the so-called projector augmented wave potentials (Blöchl 1994), by which the correct nodal structure of the wavefunctions is reintroduced but still most of the efficiency of a plane-wave basis is maintained. There is no longer any limitation to treat transition metal elements and localized states within this refined pseudopotential treatment. One of the most powerful modern pseudopotential packages is the Vienna ab-initio simulation package (VASP),[1] by which a variety of ground-state properties, including full relaxation of atomic positions and the cell shape, can be derived and also ab-initio molecular dynamics features can be utilized.

Close to an atomic nucleus, linear combinations of atomic-like functions centered at the nuclear position r_i are a more natural choice for basis functions, such as those given in Eq. (15) with radial functions R_l and spherical harmonics Y_{lm} labeled by the orbital quantum numbers l, m.

$$\Phi_{at}^{k}(x) = \sum_{l} \sum_{m} A_{lm}^{k} R_{l}(|r - r_i|, \uparrow) Y_{lm}(\widehat{r - r_i}) \tag{15}$$

The coefficients A_{lm} are undetermined as yet. Therefore, a combination of such atomic-like functions with plane waves in regions between the nuclei might

---

1) http://cms.mpi.univie.ac.at/vasp/

be reasonable basis functions, as utilized by the FLAPW method. The key point is the division of space into (a) nonoverlapping spheres (so-called atomic or muffin-tin spheres) centered at the nuclei and (b) an interstitial region in between the spheres. In both regions, the natural choice of basis functions is made, as described above. In order to construct suitable basis functions for the whole space it is required that plane waves and atomic-like functions match at the sphere boundaries, which is achieved by an appropriate definition of the cofficients A_{lm}. The matching condition mediates the Bloch property of the plane wave onto the atomic-like function; the overall basis function then has the correct translational symmetry. By this construction, the plane-wave basis has been augmented by the functions inside the spheres. In the late 1970s, the historical assumptions of spherically symmetric charge densities and potentials being constant in the interstitial region were overcome when the full-potential FLAPW approach was presented (Wimmer et al. 1981). The FLAPW method is considered to be one of the most precise concepts in the field. However, a price for computational costs has to be paid, in comparison to PP methods: because of the matching and the more complicated potential matrix elements inside the spheres, the FLAPW method is less fast than a pseudopotential approach such as VASP. As examples for existing FLAPW codes we mention the WIEN2k code[2] and the flair code,[3] both are based on the mentioned developments at the end of the 1970s.

The Bloch property of the above two approaches was mediated by plane waves involving the reciprocal lattice. However, Bloch functions might also be constructed in another way by summing over the real lattice **R**, in terms of linear combinations of atomic-like function ϕ_{lm} centered at the atom with position R [Eq. (16)].

$$\psi_{\mathbf{k}}^{v}(\mathbf{x}) = \sum_{l}\sum_{m} c_{\mathbf{k},lm}^{v} \sum_{\mathbf{R}} e^{i\mathbf{k}\mathbf{R}} \phi_{lm}(\mathbf{r} - \mathbf{R}, \uparrow) \tag{16}$$

The problem consists in the proper treatment of the tails which overlap and therefore comprise the bonding properties. Historically, such a concept was applied in a variety of approaches, in particular for space filling (and therefore overlapping) atomic spheres in terms of the so-called muffin-tin orbital method. A suitable treatment of the tails of these functions (apart from other technical details) originated in the derivation of the FLMTO method. (Methfessel 1988; Methfessel et al. 1989). An FLMTO code is available at the address given.[4]

The three concepts described above solve the Kohn–Sham equations (i.e., DFT in its – at present – most useful formulation) in a quantitatively precise way. There are numerous other approaches which also make use of approximations to the potential and density for describing systems with a large number of atoms.

---

2) http://www.wien2k.at/
3) http://www.uwm.edu/~weinert/flair.html
4) http://www.fplo.de

11.2.3
Elastic Properties

Elastic constants and elastic moduli are fundamental materials data providing detailed information on the mechanical properties of materials. The knowledge of these data may enable prediction of mechanical behavior in many different situations. Single-crystal elastic constants can be computed reliably from first principles with an accuracy that rivals experimental methods. For polycrystalline materials important quantities such as Young's moduli and shear moduli are then accessible from computed single-crystal elastic constants by suitable averaging schemes. Compared to the rather demanding experiments the ab-initio evaluation of elastic data can be automated and is extremely fast.

The calculation of elastic data from first principles is based on the general linear relationship connecting stress to strain as provided by the generalized Hooke's law (Einstein notation), Eq. (17).

$$\sigma_{mn} = C_{mnpr} \epsilon_{pr} \tag{17}$$

Herein σ_{mn} denotes the stress tensor and ϵ_{pr} the strain tensor, and the elements of the fourth-order tensor C_{mnpr} are the so-called elastic constants. Whereas σ_{mn} and ϵ_{pr} are symmetric and therefore have six independent elements, the number of 81 elastic constants is reduced by symmetry arguments to a total of 21. The elastic energy density U, which is defined as the total energy per volume, is obtained from the stress tensor (force per unit area) by integration of Hooke's law to give Eq. (18).

$$U = \frac{E}{V} = \frac{1}{2} C_{mnpr} \epsilon_{mn} \epsilon_{pr} \tag{18}$$

Introducing the convenient matrix-vector notation, where the six independent elements of stress and strain are represented as vectors (denoted here as Σ_i and ε_j where i and j run from 1 to 6 in the sequence xx, yy, zz, yz, xz, xy) and the fourth-order tensor C_{mnpr} is rewritten as a 6×6 matrix c_{ij}, these basic relationships can be simplified to Eqs. (19) (Voigt notation).

$$\Sigma_i = c_{ij} \varepsilon_j \quad U = \frac{E}{V} = \frac{1}{2} c_{ij} \varepsilon_i \varepsilon_j \tag{19}$$

Taking into account additional symmetry arguments imposed by the crystal lattice, the number of elastic constants decreases further. The total number of independent elastic constants for all crystal systems is summarized in Table 11.1.

There are two direct ways of computing single-crystal elastic constants from ab-initio methods: they may be called the energy–strain approach and the stress–strain approach. The energy–strain approach is based on the computed total

Table 11.1 The number of independent elastic constants for different crystal systems and point groups (from Le Page and Saxe 2001).

Crystal system (point group)	Number of independent elastic constants
Triclinic	21
Monoclinic	13
Orthorhombic	9
Tetragonal (4, $\bar{4}$, 4/m)	7
Tetragonal (422, 4mm, $\bar{4}$2/m, 4/mmm)	6
Hexagonal and rhombohedral (3, $\bar{3}$)	7
Hexagonal and rhombohedral (32, 3m, $\bar{3}$2/m)	6
Hexagonal (6, $\bar{6}$, 6/m, 622, 6mm, $\bar{6}$2m, 6/mmm)	5
Cubic	3

energies of appropriately selected strained states of the crystal. The crystal is strained in order to extract corresponding stiffness values preserving as much symmetry as possible. For each strain type several magnitudes of strain are applied and the corresponding total energies are computed with an ab-initio approach. The stiffness is then derived from the curvature of the energy–strain parabola by means of least-squares fits making use of Eq. (18). Some of the imposed strains may be related to a single elastic constant while others are described by a linear combination of elastic constants, from which the elastic constant tensor is finally evaluated. The stress–strain approach, on the other hand, relies on the ability of the ab-initio approach to directly calculate the stress tensor, which is by no means fulfilled by all available ab-initio methods. Once the stress tensor can be computed from first principles, the elastic constants matrix can be derived directly from the generalized Hooke's law of Eq. (17). Whereas within the energy–strain approach several magnitudes of strain need to be evaluated to probe a particular strain type from a parabolic fit to the energies, within the stress–strain approach just one evaluation is in principle sufficient to obtain the same information. However, to ensure high accuracy three strain values have been applied for all systems calculated here. Both approaches have been implemented in a symmetry-general form;[5] the underlying concepts are discussed in detail by Page and Saxe (2001) for the energy–strain approach and by Le Page and Saxe (2002) for the stress–strain approach. As a further alternative, elastic constants can be derived by specialized ab-initio codes within the density functional perturbation theory (see Hamann et al. 2005).

[5] MedeA MT, Materials Design Inc., Angel Fire (NM), USA, http://www.materialsdesign.com

Table 11.2 Elastic constants c_{ij} (in GPa) of metals and intermetallic compounds from experiment and ab-initio calculations.

	Ti			Al			TiAl		NiAl	
	exp.[a]	exp.[b]	calc.[c]	exp.[d]	exp.[e]	calc.	exp.[f]	calc.	exp.[g]	calc.
c_{11}	162.4	171.6	176.2	108.2	114.30	110.2	186	193.8	204.6	202.9
c_{12}	92.0	86.6	90.5	61.3	61.92	54.8	72	68.5	135.4	140.3
c_{13}	69.4	72.6	69.6	–	–	–	74	85.1	–	–
c_{33}	180.7	190.6	188.2	–	–	–	176	170.8	–	–
c_{44}	46.7	41.1	39.6	28.5	31.62	30.4	101	111.5	116.8	112.6
c_{66}	–	–	–	–	–	–	77	46.6	–	–

[a] Allard 1969.
[b] Ikehata et al. 2004.
[c] X.-Q. Chen, unpublished.
[d] Huntington 1958.
[e] Kamm and Alers 1964.
[f] He et al. 1995.
[g] Davenport et al. 1999.

By macroscopic averaging, elastic moduli of polycrystalline materials can also be derived from the computed single-crystal elastic constants. There are several averaging procedures available to derive the elastic moduli of a quasi-isotropic polycrystalline material from its single-crystal elastic constants. The averaging runs over all possible orientations of the crystal, and there is a well-defined lower and upper limit for the elastic moduli. Based on the averaging procedures, the ab-initio treatment for single crystals can be extended to polycrystalline samples.

In order to demonstrate the capability of calculating elastic properties from first principles without relying on any empirical input data, experimental and computed single-crystal elastic constants for simple metals and intermetallic compounds are listed in Table 11.2. The selected materials are of strategic importance for high-temperature, lightweight applications. In the case of TiAl it is noted that the experimental sample has a stoichiometry of $Ti_{44}Al_{56}$ whereas the calculation was performed for exact stoichiometry. Typically, the computed data are well within the experimental error bars. As further examples, computed and experimental elastic data of the rather complex minerals quartz and schorl are compared in Table 11.3. Quartz is used technologically in clocks because of its piezoelectric properties, and in addition plays a key role in semiconductor devices as gate oxide. Its elastic properties are therefore well known. Schorl has a complex tourmaline structure of stoichiometry $NaFe_3B_3Al_6Si_6O_{27}(OH)_3F$ with 53 atoms per unit cell. Summarizing, computed elastic constants are very reliable and are typically within the errors of experimental determinations.

Table 11.3 Elastic constants c_{ij} (in GPa) of quartz and schorl from experiment and ab-initio calculations.

	Quartz		Schorl	
	exp.[a]	calc.[b]	exp.[c]	calc.[d]
c_{11}	86.5	81.5	306	322
c_{12}	6.7	2.0	109	109
c_{13}	11.7	9.6	53	56
c_{14}	−18.0	−19.7	−8	−16
c_{33}	105.7	97.9	174	182
c_{44}	58.2	63.3	65	62

[a] Every and McCurdy 1992.
[b] Le Page et al. 2002.
[c] Helme and King 1978.
[d] P. Saxe, unpublished.

11.2.4
Vibrational Properties

A detailed knowledge of lattice vibrations is critical for the understanding and quantitative prediction of a wide variety of physical properties of solids. The interpretation of optical, infrared, Raman, and synchrotron spectroscopic methods as well as neutron scattering experiments rely on an accurate theoretical approach to lattice dynamics. The fundamental thermodynamic functions of internal and free energy, entropy, heat capacity as well as nonlinear properties such as thermal expansion and heat conduction are to a considerable extent determined by the vibrations of the constituent atoms in the lattice. Further phenomena such as electrical resistivity of metals, superconductivity, and the temperature dependence of optical spectra are governed by electron–phonon coupling. Fortunately, the quantum theory of lattice dynamics is well developed and has proven to be one of the most successful theories of solid-state physics and therewith provides an impressive reassurance of the validity and applicability of the concepts of quantum theory in general.

From a historical point of view, the fundamental theory of lattice dynamics was developed in the 1930s and was reviewed in the book by Born and Huang (1954). In those early days, the main interest focused on general properties of the dynamical matrix, e.g., its symmetry and analytical properties, but less emphasis was put on its underlying physical origin based on the interactions of the constituent ions and electrons. The first approaches to theoretically determine phonon dispersion relationships were based on lattice-dynamical models with adjustable parameters to be fitted to experimentally measured frequencies, e.g., from inelastic neutron scattering experiments, and were mainly used as a means to interpolate between experimental data points. These model potentials either did not account for electronic polarization at all (force-constant model, Born–von Kármán model, Born–Mayer and Lennard-Jones potentials (Born and Huang 1954), or polarization effects were approximated by electric dipoles generated by relative displacements of the shells of valence electrons relative to the ion cores (shell models – Dick and Overhauser 1958; Cochran 1959; Schröder 1966). For highly anisotropic electron distributions as found in covalently bonded crystals, models relying on angular forces (Keating models – Keating 1966) or bond charge models (Martin 1969; Weber 1977) have been developed. The predictive capabilities of these methods turned out to be quite limited. Although experimental phonon dispersions were accurately reproduced close to the data used for fitting, predictions along lower-symmetry directions frequently failed due to the lack of a rigorous description of the interactions among constituent atoms. The first attempts to investigate the connections between the electronic structure and the dynamical properties were made as late as in the early 1970s and were based on dielectric matrices or the electron polarizability (Pick et al. 1970; De Cicco and Johnson 1969).

The advent of DFT (Hohenberg and Kohn 1964; Kohn and Sham 1965) and the progress of numerical methods for solving quantum physical equations together with the emergence of more and more powerful computers made it feasible to describe accurately the interatomic interactions in crystals and molecules based on quantum mechanics. With methods as described in the previous sections it is nowadays possible to calculate a large variety of properties without relying on input data from experiments. In fact, ab-initio methods revolutionized the practical approaches to lattice dynamics, and accurate theoretical phonon dispersion curves can be calculated completely independently of any experimental knowledge.

Three different techniques for ab-initio evaluation of vibrational properties have been developed, namely (a) direct methods based on total energy changes or forces calculated for atoms displaced from their equilibrium position, (b) analytical calculation of force constants based on a perturbative expansion around the equilibrium geometry, and (c) Fourier transform of the atomic velocity autocorrelation function obtained from a molecular dynamics trajectory (Rahman 1964). The third option usually suffers from difficulties in reaching equilibrium within reasonable simulation times, although several technical tricks have been developed to address these problems (e.g., the Nosé-Hoover thermostat (Martyna et al. 1992) and multiple signal classification (MUSIC) algorithm (Marple 1987; Kohanoff et al. 1992; Kohanoff 1994)). The second option makes use of the fact that the lattice distortion associated with a phonon is a static perturbation acting on the electrons, causing a linear response of the electron density which determines the energy variation up to third order ($2n + 1$ theorem: Gonze and Vigneron 1989). The linear variation of the electron density is determined in terms of the linear response of the Kohn–Sham orbitals by applying first-order density functional perturbation theory (Baroni et al. 1987; Giannozzi et al. 1991). This method is quite accurate, efficient, and elegant, but it requires highly specialized ab-initio computer codes and considerable implementation efforts. For a review of the formalism and implementation of density functional perturbation theory and its application see Baroni et al. (2001).

Direct methods, option (a), require the evaluation of total energy and forces for the equilibrium geometry as well as of several distorted geometries from which the force constant matrix can be assembled. Several different techniques are subsumed under the title direct methods: in the early 1980s, the first attempts at ab-initio lattice dynamics simulation were based on frozen phonon total energy calculations to derive accurate phonon frequencies for specific high-symmetry points in the Brillouin zone (e.g., Yin and Cohen 1982). Phonon dispersion curves along specific high-symmetry directions in reciprocal space were determined by the method of interplanar force constants (Kunc and Martin 1982), where planes perpendicular to these directions are displaced within an elongated supercell. Since these one-dimensional force constants between high-symmetry planes are linear combinations of the three-dimensional interatomic force constants, the phonon dispersion can be derived, in principle, for general directions (Wei and Chou 1992, 1994). The most general direct approach to lattice dynamics is based on the ab-initio evaluation of forces on all atoms, produced by a set of finite displacements of a few atoms within an otherwise perfect crystal. The perfect crystal environment has to be sufficiently large to ensure that interactions of the perturbation with all its translational symmetry equivalent copies are small, which usually requires construction of suitable supercells. The techniques for selecting suitable supercells and atomic displacements, assembling force constant matrices from the calculated forces, and calculating phonon dispersion relations via Fourier transform are described by Kresse et al. (1995), Frank et al. (1995), and Parlinski et al. (1997). Phonon dispersion curves determined from direct approaches are very accurate, provided the studied crystal area (supercell) is larger than the interaction range of all constituent atoms. The main advantages of direct methods consist in the fact that no specialized ab-initio codes are required: any code capable of deriving reasonably accurate forces can be applied. The necessity to consider supercells in the case of a simple crystal structure with small unit cell extensions is one of the main drawbacks with regard to the linear response methods.

All applications presented later in this chapter were obtained by the general direct method as developed by Kresse et al. (1995), Frank et al. (1995), and Parlinski et al. (1997). Therefore, the method is briefly outlined in more detail. The ground-state energy E of a crystal as a function of atom positions $\mathbf{R}(n, m)$ (where n and m denote the unit cell and μ and ν the atom index) can be written as a Taylor expansion in terms of atomic displacements $\mathbf{u}$ around the equilibrium positions [Eq. (20)].

$$E(\ldots, \mathbf{R}(n,\mu), \ldots, \mathbf{R}(m,\nu), \ldots)$$
$$= E_0 + \frac{1}{2} \sum_{n,\mu,m,\nu} \Phi(n,\mu;m,\nu)\mathbf{u}(n,\mu)\mathbf{u}(m,\nu) + O(\mathbf{u}^3) \qquad (20)$$

E_0 is the equilibrium energy. Terms linear in **u** are absent due to the condition that at equilibrium all forces on the atoms vanish. The elements of the atomic force constant matrix Φ are defined by Eq. (21), with the gradients taken at the minimum energy configuration, for which all first-order derivatives vanish.

$$\Phi_{i,j}(n,\mu;m,v) = \left[\frac{\partial^2 E}{\partial R_i(n,\mu)\partial R_j(m,v)}\right]_0 \qquad (21)$$

The frequently used harmonic approximation consists in retaining in Eq. (20) only terms up to quadratic order in the displacements.

Within the harmonic approximation the classical equations of motion for each atom are given by Eq. (22), where M_μ is the atom mass of atom μ.

$$M_\mu \mathbf{u}(n,\mu) = \sum_{n,\mu,m,v} \Phi(n,\mu;m,v)\mathbf{u}(m,v) \qquad (22)$$

Solutions are required to exhibit Bloch waveform because of translational invariance. The **k** vectors are chosen to fulfill Born-von Kármán periodic boundary conditions [Eq. (23)].

$$\ddot{\mathbf{u}}(n,\mu) = \frac{1}{M_\mu} e(\mathbf{k}) e^{i\mathbf{k}R(n,\mu) - i\omega t} \qquad (23)$$

The equation of motion can thus be cast into the simple form of Eq. (24), where for each mode j the phonon frequencies $\omega^2(\mathbf{k}, j)$ are the eigenvalues and the polarization vectors $e(\mathbf{k}, j)$ are the eigenvectors of the dynamical matrix $D(\mathbf{k})$, which has been introduced as the discrete Fourier transform in Eq. (25).

$$D(\mathbf{k})e(\mathbf{k}, j) = \omega^2(\mathbf{k}, j)e(\mathbf{k}, j) \qquad (24)$$

$$D(\mathbf{k};\mu,v) = \frac{1}{\sqrt{M_\mu M_v}} \sum_m \Phi(0,\mu;m'v)e^{2\pi i\mathbf{k}[R(0,\mu)-R(m,v)]} \qquad (25)$$

The summation m runs over all atoms of the crystal. The forces $\mathbf{F}(n,\mu)$ on all atoms generated by the displacement $\mathbf{u}(m,v)$ of atom v can be written as Eq. (26), which relates the generated forces to the force constant matrices and atomic displacements. This is the central relationship used in the direct method.

$$F_i(n,\mu) = \sum_{m,v,j} \Phi_{i,j}(n,\mu;m,v)u_j(m,v) \qquad (26)$$

The complete quantum mechanical description for a system of ions and electrons requires additional approximations. Since the nuclear mass is much larger than that of an electron, it is reasonable to consider the nuclei in their equilibrium position while dealing with the electronic motion. In mathematical terms this so-called adiabatic (or Born–Oppenheimer) approximation yields a separation of the general Schrödinger equation into one for the motion of nuclei and one for the electrons. The nuclear motion is determined by the potential field generated by the average motion of the electrons and the corresponding nuclear Schrödinger equation yields formally the same results as the classical theory. The above equations remain valid, however, with the exact effective potential. The purely electronic Schrödinger equation can be solved within density functional theory, as discussed in previous sections.

In general, the ab-initio calculations for the direct approach are performed for a supercell constructed as a multiplication of the original cell large enough to ensure that forces arising from a displaced atom diminish inside the supercell. In a first step all atom positions of the unperturbed supercell are optimized. In principle, any computational method providing atomic forces is suitable for the direct approach, but usually first-principles methods are employed. After the initial total energy minimization process all forces should vanish, and the resulting structure is the reference and starting point for all further steps. In order to obtain complete information on all force constants it is necessary to displace each nonequivalent atom of the supercell along three nonequivalent directions, and to calculate for each of these perturbed supercells the forces on all other atoms generated by this displacement. The number of displacements may be considerably reduced by making use of symmetry relationships. For instance, for a cubic site symmetry, a single displacement along a single fourfold axis is adequate. If the site point group is tetragonal, one displacement along the fourfold symmetry axis and one perpendicular to it are sufficient. From this set of displacements and all the resulting forces, the force constant matrix can be evaluated from a multidimensional fit to Eq. (26). From the force constants the supercell dynamical matrix is obtained by discrete Fourier transform. Finally, the frequencies $\omega^2(\mathbf{k}, j)$ and polarization functions $e(\mathbf{k}, j)$ of phonon modes j are obtained by diagonalization of the supercell dynamical matrix for each wave vector $\mathbf{k}$.

Vibrational properties are conveniently expressed by phonon dispersion curves along a selected high-symmetry path through the Brillouin zone and as the phonon density of states (DOS), i.e., the frequency distribution of all normal modes. The contribution of lattice vibrations to thermodynamic functions such as internal energy, free energy, entropy, heat capacity, and thermal displacements are derived from the integrated phonon density of states $g(\omega)$ as a function of frequency ω. From the phonon DOS the phonon free energy F_{ph} as a function of temperature can be derived from Eqs. (27), where the phonon internal energy is given by Eq. (28), and phonon entropy is evaluated as in Eq. (29).

$$F_{ph}(T) = U_{ph}(T) - TS_{ph}(T) \tag{27}$$

$$U_{ph}(T) = \frac{1}{2}\int_0^\infty g(\omega)\hbar\omega \coth\left(\frac{\hbar\omega}{2k_BT}\right) d\omega \tag{28}$$

$$S_{ph}(T) = k_B \int_0^\infty g(\omega)\left(\frac{\hbar\omega}{2k_BT}\left[\coth\left(\frac{\hbar\omega}{2k_BT}\right) - 1\right] - \ln\left[1 - \exp\left(-\frac{\hbar\omega}{k_BT}\right)\right]\right) d\omega. \tag{29}$$

Ab-initio calculations have been largely viewed as being restricted to 0 K, or more precisely, as being condemned to operate without the concept of temperature at all. Based on Eqs. (27–29), this major constraint has been lifted, and routine methods to evaluate vibrational properties and temperature-dependent thermodynamic functions from first principles have become available.[6] A few applications of this will be presented in the following sections. Another contribution from thermodynamics has not yet been taken into account: configuration space and configurational entropy. This important part, in particular for alloy physics, will also be addressed in the later section on multiscale modeling (Section 11.3.5).

11.3
Applications

11.3.1
Structural and Phase Stability

The aim is to find the thermodynamical ground state of an ordered solid. The phase with the lowest energy is searched for. Restricting the considerations to $T = 0$ K (or sufficiently low temperatures) all the information needed is included in the total energy U_{DFT} of the DFT calculation. This quantity depends on the actual atomic distribution, i.e., on the crystal structure and the volume. Therefore, energies of different structures can be compared and the equilibrium volume found. An important elastic ground-state property is the bulk modulus B, which is derived from the second derivative of the volume-dependent DFT energy. It can also be expressed by a linear combination of elastic constants; see Section 11.2.3. It should be noted that the bulk modulus may be also denoted by the symbol K, as for example in Chapter 4. Phase stabilities are usually compared in energies of formation $\Delta H(S, V_0)$ in which the energy of the pure phases serve as a reference. For example, the formation energy (or enthalpy of formation at zero pressure) of the compound $A_n B_m(S, V_0)$ with crystal structure C at its equilibrium volume is defined by Eq. (30).

$$\Delta H(C, V_0) = U_{DFT}(A_n B_m(C, V_0)) - [n U_{DFT}(A) + m U_{DFT}(B)] \qquad (30)$$

The reference states for the constituent atoms A and B are in their respective ground states, corresponding to the experimental conditions. For example, for the fabrication of crystals of intermetallic compounds the pure metals have to be melted and put together for reaction.

As an example of the structural and phase stability the Laves phase compounds XCr_2 (X = Ti, Zr, Hf) (Chen et al. 2005a) are discussed. Three different crystal structures are considered for the compound. The reference energies are calculated for Zr metal in its hexagonal ground state, and for Cr in its bcc magnetic ground state, although Cr in the compound is nonmagnetic. Figure 11.2 shows

---

[6] Materials Design Inc., MedeA Phonon, Angel Fire (NM), USA, http://www.materialsdesign.com

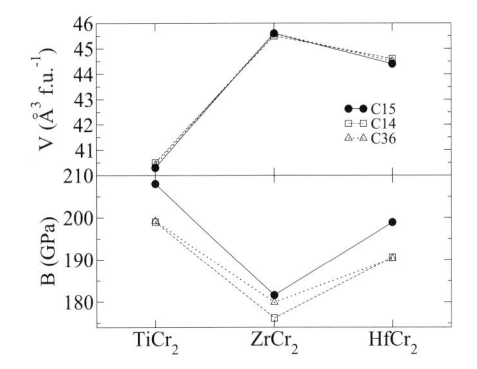

Fig. 11.2 DFT equilibrium volumes V_0 and bulk moduli B for the C14, C15, and C36 structures of $TiCr_2$, $ZrCr_2$, and $HfCr_2$.

the trend of volumes and bulk moduli B revealing the largest B values for the C15 structures, which are the stable ground-state phases according to the calculated energies, and also according to experiment. One realizes the exceptional nature of the $ZrCr_2$ compound for which the volumes for all three structures are the largest but the bulk moduli the smallest when compared to $TiCr_2$ and $HfCr_2$. Only two experimental results for B exist, namely for $TiCr_2$ in the C14 structure (159 GPa; DFT value is 199 GPa), and for $ZrCr_2$ in the C15 structure (162 GPa; DFT value is 182 GPa). The difference between experiment and calculation is due to the smaller calculated volumes (see Table 11.4). Concerning the volumes, the differences are attributed to the temperature dependence of the lattice parameters. This holds in particular for the C14 phase of $TiCr_2$, which only exists at high temperatures.

The calculated equilibrium enthalpies of formation with respect to the equilibrium bulk phases of the pure constituents are defined as the difference between the corresponding total energies U_{DFT} calculation according to Eq. (30), and are listed in Table 11.5 and illustrated in Fig. 11.3. The C15 structure is always the most stable one in accordance with the experimental findings up to rather high temperatures. The energy differences between the different structures are rather uniform with C36 being less stable than C15 by about 5 kJ mol^{-1}, and C14 being even less stable than C36 by about 2 to 3 kJ mol^{-1}.

Currently, three experimental values for the enthalpy of formation of $TiCr_2$ are published, showing considerable discrepancies. The experimental values seem to be rather questionable. For $HfCr_2$ an estimate of -5.4 kJ mol^{-1} at 298 K is known. However, for both $ZrCr_2$ and $HfCr_2$ no measured values are available. As recently demonstrated for $TiMn_2$ (Chen et al. 2003), for which calorimetric experimental data were measured also, a high-quality DFT approach yields reliable results for the enthalpy of formation, and can be used as a predictive tool. Furthermore, the DFT result of -141.2 kJ mol^{-1} for $ScAl_2$ (Chen et al. 2004a) is in very good agreement with a very recent experimental value of -141.6 kJ mol^{-1} obtained by reaction calorimetry (Cacciamini et al. 1999). Therefore, the DFT en-

Table 11.4 DFT and experimental structural parameters for $TiCr_2$, $ZrCr_2$, and $HfCr_2$ with C14, C15, and C36 crystal structures. Lattice parameters a and c are in Å, equilibrium volumes V_0 per formula unit in Å^3, and bulk moduli B in GPa.

	Structure	a	c	V_0	B	Method
$TiCr_2$	C15	6.857		40.3	208	DFT
		6.910		41.2		expt.
		6.993		42.7		expt.
	C14	4.885	7.830	40.5	199	DFT
		4.932	8.005	42.2		expt.
		4.900	7.927	41.2		expt.
					159	expt.
	C36	4.869	15.748	40.4	199	DFT
		4.932	16.001	42.1		expt.
$ZrCr_2$	C15	7.145		45.6	182	DFT
		7.204		46.7		expt.
		7.207		46.8		expt.
					162	expt.
	C14	5.094	8.103	45.5	176	DFT
		5.106	8.292	46.8		expt.
	C36	5.065	16.409	45.6	180	DFT
		5.100	16.611	46.7		expt.
$HfCr_2$	C15	7.079		44.4	199	DFT
		7.140		45.4		expt.
	C14	5.046	8.095	44.6	191	DFT
		5.067	8.237	45.9		expt.
	C36	5.022	16.301	44.5	190	DFT
		5.064	16.470	45.7		expt.

Table 11.5 DFT enthalpies of formation ΔH in kJ mol^{-1} as compared to the available experimental data and values obtained by the application of Miedema's model for the C14, C15, and C36 structures of $TiCr_2$, $ZrCr_2$, and $HfCr_2$.

	C14	C15	C36	Miedema	Expt.
$TiCr_2$	−30.6	−35.9	−33.0	−30.3	−27.9; −9.9^2; 0.3
$ZrCr_2$	−9.0	−14.5	−12.1	−51.6	−
		−14.6			
$HfCr_2$	−29.6	−34.9	−32.5	−29.2	−

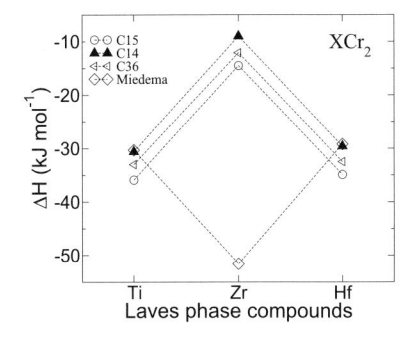

Fig. 11.3 DFT enthalpies of formation for TiCr$_2$, ZrCr$_2$ and HfCr$_2$ and values derived by Miedema's model.

thalpies of formation for the XCr$_2$ compounds in Table 11.5 can be considered to be reliable. In addition, the enthalpy of formation for ZrCr$_2$ of -14.5 kJ mol^{-1} agrees well with the value of -14.6 kJ mol^{-1} obtained by another DFT application (Krčmar and Fu 2003).

Figure 11.3 shows the trend of DFT enthalpies of formation, with ZrCr$_2$ being less stable by a factor of about two compared to TiCr$_2$ and HfCr$_2$. Also shown is the result obtained from Miedema's semiempirical model, which is totally unreliable. Miedema's model fails very badly for ZrCr$_2$, reversing the trend of the DFT results. Similar failures were also found for ZrCo$_2$, ZrPd, and ZrPd$_3$.

Utilizing a simple model, the variation of the average electron density as expressed by the number of valence electrons per atom is correlated to the cohesive properties of compounds which have a comparable type of chemical bonding. For the XCr$_2$ compounds the number of valence electrons are $n_X = 4$ and $n_{Cr} = 6$ which – by application of the relation $c = (n_X + 2n_{Cr})/3V_0 = 16/3V_0$ – for the C15 structures results in values of $c = 0.132, 0.117, 0.120$ e Å^{-3} for TiCr$_2$, ZrCr$_2$, and HfCr$_2$, respectively, for the equlibrium volumes V_0 as listed in Table 11.4. Clearly, this results in the correct trend compared to the formation energies, due to the opposite trend of the equilibrium volumes.

So far, nonmagnetic compounds have been discussed. For deriving enthalpies of formation for magnetic compounds the concept is the same (Chen et al. 2005b). However it should be realized that the gain of bonding energy due to the formation and ordering of magnetic moments (i.e., due to spin polarization and orientations of moments) could be substantial. Even the structural stability could be changed, as is the case for ZrMn$_2$ and ZrFe$_2$ (Chen et al. 2005b). Therefore, contributions to the total energy due to magnetism could strongly affect thermodynamic properties and -consequently- the finding of the ground state.

Vibrational properties are of particular importance for temperature-dependent phase and structural stabilities. In the following, some corresponding examples for the Laves phase compounds ZrCr$_2$ and ZrMn$_2$ (Chen et al. 2005a) are pre-

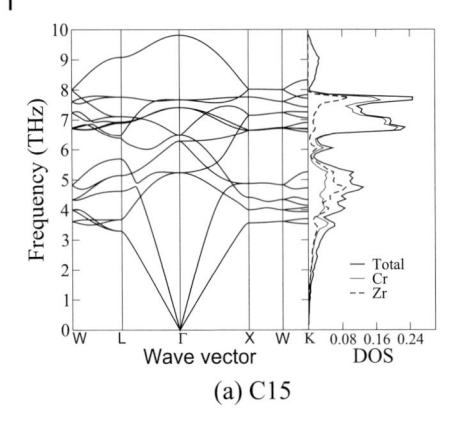

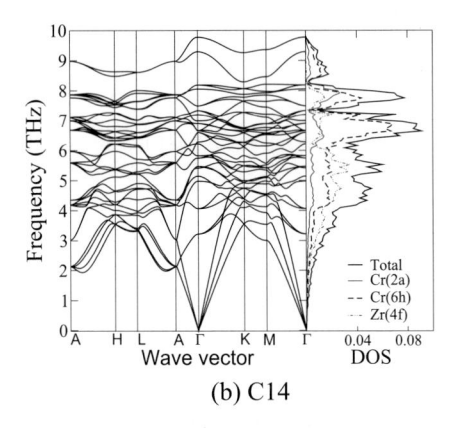

Fig. 11.4 Ab-initio phonon dispersion relations and density of states for ZrCr$_2$ in the C15 (a) and C14 (b) structure.

sented. All these results are derived by applying the concept presented in Section 11.2.4. Figure 11.4 shows the phonon dispersion relationships and the density of states (DOS) for ZrCr$_2$ calculated at the equilibrium volumes of the C14 and C15 structures. There are 18 (for C15) and 36 (for C14) phonon branches due to the number of atoms in the unit cell (6 and 12 for C15 and C14, respectively). Because of the higher symmetry, the features of the DOS for the C15 structure are much sharper. A further significant difference between the dispersions of the two structures is seen in the frequency range between 2 and 3.5 THz. There, no modes exist for the C15 structure, in contrast to C14 for which highly dispersive bands are seen in particular around the point A, and close to A a soft mode arises.

From the phonon DOS the phonon free energy F_{ph} can be derived from Eqs. (31) combining the phonon free energy $U_{ph}(T)$ and phonon entropy $S_{ph}(T)$.

$$F_{ph}(T) = U_{ph}(T) - TS_{ph}(T) \tag{31}$$

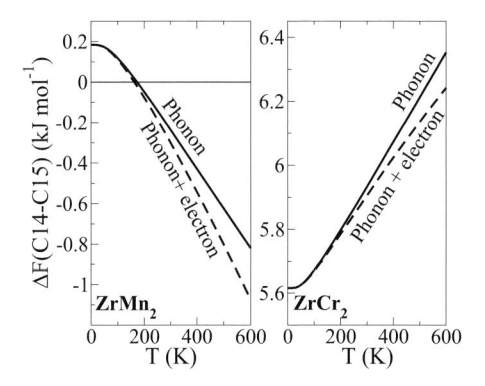

Fig. 11.5 Differences in total free energies $\Delta F = F_{C14} - F_{C15}$. Results with (solid line) and without (broken line) temperature-dependent electronic contributions (Chen et al. 2004b). Left panel: ZrMn$_2$; right panel: ZrCr$_2$. Positive values: C15 is the stable structure; negative values: C14 is the stable structure.

The difference in total free energies $\Delta F = F_{C14} - F_{C15}$ corresponding to the C14 and C15 structures of ZrCr$_2$ and ZrMn$_2$ are shown in Fig. 11.5. The energy scales in the two panels are different because of the very different structural energies: for ZrCr$_2$ the energy U_{DFT} for the C15 structure is much lower than for C14 whereas for ZrMn$_2$ the corresponding energies are close to being degenerate. Furthermore, the temperature dependence of the structural free energy difference (Chen et al. 2004b) for ZrMn$_2$ is totally opposite to ZrCr$_2$. Although for ZrMn$_2$ at very low temperatures C15 is stable, the C14 phase is stabilized above a transition temperature of 160 K due to the phonon free energy difference. The structural stability of ZrCr$_2$ is quite different: at very low temperature C15 is already much more stable than is the case for the Mn compound. Figure 11.6 shows the

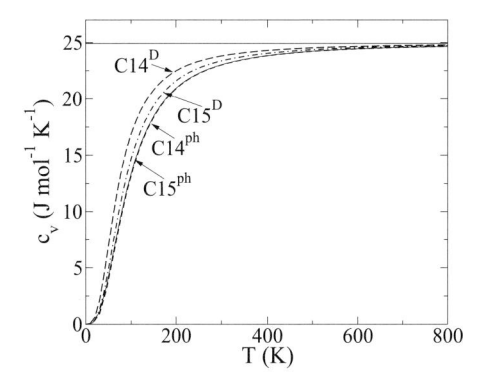

Fig. 11.6 Lattice specific heat for C14 and C15 structures of ZrCr$_2$. Full phonon calculation is denoted by ph; Debye results are marked by D.

specific heat for the C14 and C15 structures of ZrCr$_2$. Whereas for the full phonon calculation the curves nearly coincide, for the results based on the Debye approximation a significant splitting between C14 and C15 is observed with all the Debye results being larger than the full phonon results. The largest differences occur at about 200 K.

11.3.2
Point Defects

Here, point defects will be discussed for binary compounds in terms of supercells. The basic concepts can be easily extended to more complicated cases. The corresponding defect energies are derived from total energies of suitable supercells. The supercells containing the defect should be sufficiently large in order to avoid defect–defect interactions. Therefore, such supercells could be large and the calculations costly, depending on the type of material and defects under study. The advantage of the supercell concept is that all the advantages of a standard DFT calculation are available (such as accuracy, freedom of choice of crystal symmetry, geometry relaxations, spin polarization). True point defects might be modeled ideally by one defect atom within a matrix of an infinite number of host atoms. In that case, other techniques based on Green functions have to be applied which, however, are less generally applicable than a supercell approach. The Green function approach could be also costly, depending on the actual system and the accuracy required.

Returning to the treatment of point defects by supercells, within a sufficiently large unit cell for the perfectly ordered $A_m B_n$ defects are introduced by replacing one of the atom types A or B by a defect species X, resulting in the compositions $A_{m-1} B_n X$ or $A_m B_{n-1} X$. Species X might be a third atomic species, a vacancy, or an A, B atom sitting at a "wrong" lattice site (a so-called antisite defect). Another important type of defect is an X atom at an interstitial position. Considering as an example the case of $A_{m-1} B_n X$, a defect formation energy can be formulated in terms of the differences in DFT total energies by Eq. (32), including suitable reference states for the pure phases. This type of formation energy is a formation energy for a ternary compound.

$$\Delta E_1(X) = E_{DFT}(A_{m-1} B_n X) - [m E_{DFT}(A) + n E_{DFT}(B) + E_{DFT}(X)] \tag{32}$$

Quite often, one wants to know the energy for the formation of a defect X in the perfect compound of composition $A_m B_n$. Therefore, one might express another type of formation energy by Eq. (33).

$$\Delta E_2(X) = E_{DFT}(A_{m-1} B_n X) - [E_{DFT}(A_m B_n) - E_{DFT}(A) + E_{DFT}(X)] \tag{33}$$

Immediately the question arises about the physical state of the pure phases of A and X: should $E_{DFT}(A)$ and $E_{DFT}(X)$ be calculated for the solid phase, or the gas phase, depending on the history of how A is removed from the compound and is replaced by X? Clearly, such a concept is not very useful. The way out consists in constructing a thermodynamical model with suitable particle reservoirs described by chemical potentials for the species A and B, as derived from suitable supercell calculations. The formulation of a concept based on statistical mechanics for a compound is more complicated than for a one-atom type of system by the condition that the overall stoichiometry has to be preserved, which means that various kinds of defects have to coexist. Therefore, one has to work with a grandcanonical ensemble for which the number of particles can be adjusted by chemical potentials.

For low concentrations of substitutional defects, statistical models can be formulated (Meyer and Fähnle 1999; Rasamny et al. 2001) which make use of the supercell energies ϵ_X^λ for a defect species X at lattice site λ. For example, ϵ_A^B denotes the supercell energy of an antisite defect for which an A atom is sitting on a site of the B sublattice. The supercell energy for a vacancy, e.g., at the A sublattice is denoted by ϵ_{vac}^A. For noninteracting defects and sufficiently low concentrations of defects the energies ϵ_X^λ are independent of the concentration. The defect systems should also be at their ground states. However, this assumption can be lifted and more general expressions can be formulated including pressure–volume terms (Meyer and Fähnle 1999; Rasamny et al. 2001).

By minimization of the Gibbs free energy with respect to the number of point defects, expressions for the concentrations of point defects are derived. For example, for a vacancy at an A site one derives Eq. (34), in which M represents the total number of available sites, and M^A the available sites of the A sublattice.

$$c_{vac}^A = \frac{M^A}{M} \frac{e^{-\beta(\epsilon_{vac}^A + \mu_A)}}{1 + e^{-\beta(\epsilon_{vac}^A + \mu_A)} + e^{-\beta(\epsilon_B^A + \mu_A - \mu_B)}} \quad (34)$$

In a similar way, the concentration for an antisite defect is expressed by Eq. (35).

$$c_B^A = \frac{M^A}{M} \frac{e^{-\beta(\epsilon_B^A + \mu_A - \mu_B)}}{1 + e^{-\beta(\epsilon_B^A + \mu_A)} + e^{-\beta(\epsilon_B^A + \mu_A - \mu_B)}} \quad (35)$$

The concentrations for the A-rich defects (B vacancy and A antisite) are formulated correspondingly. The so-far unknown chemical potentials obey the condition expressed by Eq. (36).

$$G = \mu_A N_A + \mu_B N_B \quad (36)$$

The general composition of the compound including the defects in a suitable way is given by $N_A/N_B = x/(1-x)$ whereby N_i is the total number of atomic species, $i = \{A, B\}$, and x varies in a small range around the stoichiometric composition. Considering the conditions discussed and including the supercell energy ϵ_0 for the perfect compound, the chemical potentials can be calculated from a set of coupled equations. As visualized by Eqs. (34) and (35), the defects are intermixed in order to maintain the overall composition, which is kept fixed. Of course, only small deviations of the overall composition from the perfect stoichiometry are reasonable, which are typically in the order of 1–2%. In general, the chemical potentials depend on pressure, temperature, and concentration x. For example, the energy cost for the formation of an A vacancy for a given concentration x amounts to $\Delta H_{vac}^A = \epsilon_{vac}^A + x\mu_A$, in which the chemical potential $\mu_A(x)$ represents the energy for adding an A atom to the system.

The thermodynamic model presented contains some simplifications because at high temperatures more complicated defect combinations may contribute significantly to the configurational entropy. Furthermore, vibrational effects are not included.

As an example, calculated results for point defects in the compound $ZrMn_2$ for two crystal structures are presented, as published very recently (Chen et al. 2005b). Vacancies V(Zr) at the Zr and V(Mn) at the Mn sites, and antisite defects Zr^{Mn} for Zr occupying a Mn site and Mn^{Zr} for Mn occupying a Zr site were considered. Because of symmetry this results in four possible point defects for the high-symmetry cubic C15 structure, and in six different point defects for the hexagonal C14 structure due to crystallographically inequivalent Mn sites. Further complications arise because the Mn atoms carry magnetic moments, which depend on their local environement. For the DFT calculations, however, this is easy to handle: the total energy is minimized with respect to spin polarization.

The enthalpies of formation for Zr_xMn_{1-x} compounds derived from the total energies of the supercells are summarized in Fig. 11.7. For both structures the formation of antisites is clearly preferred when compared to vacancies, and the Mn^{Zr} antisite defects are the most favorable ones. For the C14 compound the supercell formation energy for a Mn^{Zr} defect is so low that it even falls below the broken line in Fig. 11.7 marking the boundary of thermodynamic stability with respect to the perfect compound $ZrMn_2$ and pure Mn. This means that the ordered compound with the supercell structure for the Mn^{Zr} defect is energetically more stable than a mixture of pure Mn and $ZrMn_2$ phases, which seems not realistic. This result indicates the possible existence of other ordered Zr–Mn compounds being richer in Mn, and it explains a broadening of the homogeneity range of the $ZrMn_2$ phase toward Mn-rich compositions. Up to now, the only experimentally prepared solid compound of the Zr–Mn system is the Laves phase $ZrMn_2$. Therefore, thermodynamical experiments on the Mn-rich part of the phase diagram are encouraged.

Table 11.6 lists the defect formation energies at 1500 K for the stoichiometric composition. Antisite defects are clearly preferred, as is also demonstrated by Table 11.7 showing the temperature dependence of defect concentrations.

According to Fig. 11.8, with increasing temperature the minimum of the free energy of C14 $ZrMn_2$ no longer occurs at the stoichiometric composition but is shifted toward Mn richer compositions. This implies that the congruent melting point also may not occur at the stoichiometric composition but at the Mn-rich side. In addition, it is expected that Mn antisites may produce a broad nonstoichiometric range toward the Mn-rich side, in agreement with the experimental phase diagram. It is well known that $ZrMn_2$ samples produced by the standard

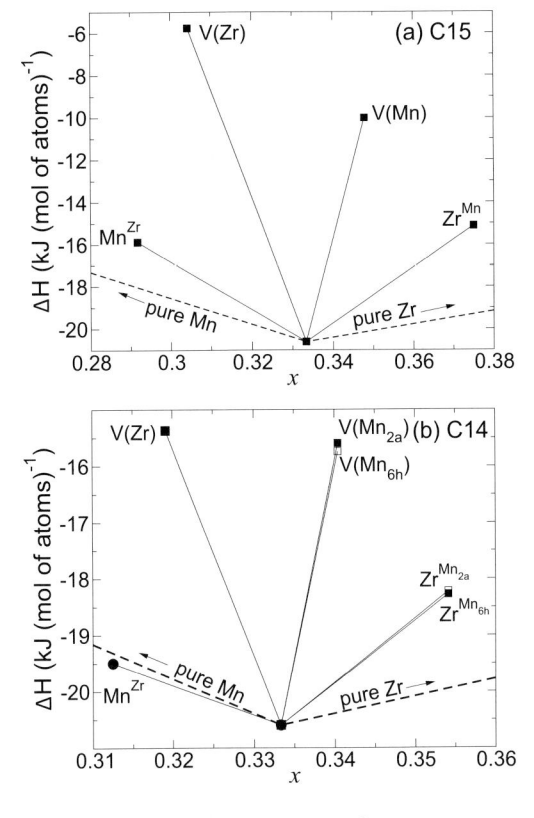

Fig. 11.7 DFT enthalpies of formation for Zr_xMn_{1-x} versus the concentration x of Zr atoms for the C15 structure (upper panel) and the C14 structure (lower panel). V(Zr): vacancy at Zr site, V(Mn): vacancy at Mn (2a) and (6h) sites, Zr^{Mn}: Zr at Mn (2a) and (6h) sites, Mn^{Zr}: Mn at Zr site.

melting techniques are off-stoichiometric with a slightly enriched Mn concentration.

Finally, the change in magnetic moments upon defect formation was investigated. A very striking effect is observed for the Mn antisite Mn^{Zr} defect, which

Table 11.6 Defect formation energies for the C14 and C15 structure of ferromagnetic $ZrMn_2$ at 1500 K.

	V(Mn)	V(Zr)	Mn^{Zr}	Zr^{Mn}
C15	2.64	3.74	1.19	1.19
C14	2.53	2.96	0.89	0.89

Table 11.7 Temperature-dependent defect concentrations for ferromagnetic and stoichiometric ZrMn$_2$.

T (K)	C15			C14		
	V(Mn)	V(Zr)	anti	V(Mn)	V(Zr)	anti
500	2.6E-27	1.8E-38	1.0E-12	2.7E-28	1.4E-30	1.1E-09
1000	5.1E-14	1.4E-19	1.0E-06	1.6E-13	1.2E-15	3.3E-05
1500	3.7E-09	2.6E-13	1.0E-04	3.0E-09	1.1E-10	1.0E-03

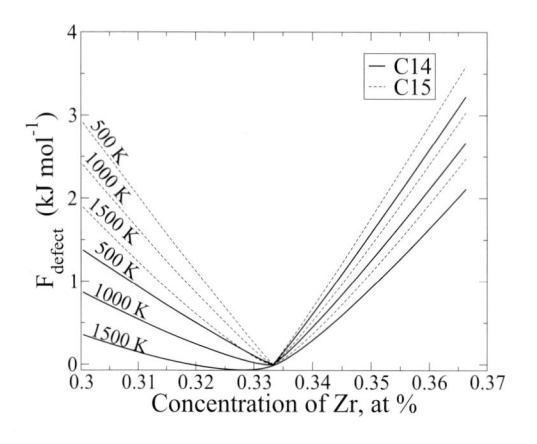

Fig. 11.8 Temperature-dependent defect free energy for Zr$_x$Mn$_{1-x}$ for the ferromagnetic C14 (solid lines) and C15 (broken lines) phases.

is also the dominant defect type at Mn-richer compositions: the local moment of the Mn atom increases from about 1 μ_B for the defect-free compound to 3.2 μ_B for the Mn^{Zr} defect, for both the C14 and C15 structures. This massive increase in the magnetic moments is due to the change in local coordination: compared to the defect-free compound the Mn defect atom at a Zr site has significantly increased nearest-neighbor distances to the surrounding Mn atoms, which causes stronger localization of the Mn states, and therefore the local magnetic moment is increased. Experiments should be able to detect the predicted large magnetic moments of Mn^{Zr} which – in reality – might appear as disordered local moments.

11.3.3
Diffusion Processes

Solid-state diffusion processes are behind the scene of a large variety of materials properties and processes such as segregation, phase transformation, environmental embrittlement, and corrosion (see Chapter 5). Despite this fundamental importance, reliable experimental mass diffusion data are rather scarce in some areas. For instance, the diffusion coefficients for hydrogen in metals such as titanium (Papazoglou and Heppworth 1968; Brauer et al. 1981) and aluminum (Young and Scully 1998) are scattered over many orders of magnitude and diffusion data for many important systems are not reported at all. Ab-initio calculations provide an alternative, reliable route to derive temperature-dependent diffusion coefficients quantitatively and, in addition, are capable of providing insight into the diffusion processes at an atomistic level. In this section the strategies and procedures to derive diffusion coefficients quantitatively from first principles are reviewed and, as an example of an application, results for diffusion processes in the lithium-aluminum alloy system are presented (Müller et al. 2007).

As a first step it is necessary to figure out possible diffusion mechanisms that might be responsible for observed mass diffusion processes. This requires the identification of stable configurations in the first place and possible rearrangements of atoms to reach adjacent stable configurations. In the case of impurity diffusion, for instance, the site preference of the impurity (interstitial or different substitutional sites) need to be established by total energy calculations and suitable thermodynamic models (along the lines discussed in Section 11.3.2). Once stable sites are identified, possible diffusion paths between these sites are figured out and energy profiles along these diffusion paths are calculated. Technically speaking, the so-called minimum energy path is required. The energy maximum along this path, i.e., the saddle point energy, defines the transition state that is of central importance for our further considerations in the framework of harmonic transition state theory (see Chapter 5). Of course, it might be necessary to investigate several possible diffusion paths and each diffusion path may involve multiple metastable sites and transition states. Those paths with lower-lying saddle point energies will dominate the diffusion process.

A number of ab-initio based methods have been developed to follow reaction paths (in chemistry) or diffusion paths in materials physics, and to identify the saddle points (see Henkelman et al. (2000) for a review). Some methods focus on the initial state only and explore reaction or diffusion possibilities either by following stepwise the path of slowest ascent, or by evaluating normal modes of a harmonic approximation to the local potential energy surface and following each of these modes until a saddle point is found. However, slowest ascent paths may not lead to saddle points and the latter approach requires evaluation and diagonalization of second derivatives that either may not be available or may quickly become computationally impracticable for larger systems. Other approaches take into account both the initial and the final configurations and try to find the minimum energy path between them. The most straightforward approach is to drag the migrating atom step by step along the diffusion path by linear interpolation between initial and final states. For each step this degree of freedom is kept fixed and all other degrees of freedom are optimized (molecular statics). This approach may work in simple cases but tends to fail for more complex energy hypersurfaces. More generally applicable algorithms for finding minimum energy paths between two points are the so-called chain-of-states methods. There, a chain of images is generated between the initial and final states. All the images are connected to trace out the diffusion path and are simultaneously optimized in a concerted manner. Whereas the initial images are located on a line connecting initial and final configurations, after simultaneous optimization of all images they lie close to the minimum energy path. Several variants of these methods have been developed; Henkelman et al. (2000) may serve as a review in this field. In a recent study of diffusion in the lithium–aluminum alloy system (Müller et al. 2007) the nudged elastic band method (NEB) (Jónsson et al. 1998) was applied, a variant of the chain-of-state methods including springs between the images to ensure that the images keep their distance. In addition, a minimization of the elastic band between images is carried out by projecting the perpendicular component of the spring force and the parallel component of the true force along the minimum energy path. This procedure ensures that the component perpendicular to the spring force does not drag images off the minimum energy path, and that the component parallel to the true force in the direction of the path does not slide images down the minimum energy path. The NEB method has been applied successfully for a wide range of applications such as diffusion at surfaces, sputter deposition, dissociative adsorption of molecules on surfaces, premelting of metal clusters, cross-slip of screw dislocations, and atomic exchange processes at semiconductor surfaces (see Jónsson et al. (1998) and references therein).

Here we show results of the NEB method for the identification of diffusion paths in the Li–Al alloy system (Müller et al. 2007). The calculations were performed for $2 \times 2 \times 2$ supercells containing about 32 atoms to determine vacancy and antisite formation energies, site preferences, and migration profiles of Li and Al atoms in $LiAl_3$ and aluminum. As an example, Fig. 11.9 shows the NEB results for the minimum energy path of an Al atom migrating into an adjacent Li vacancy within a supercell of stoichiometry Li_7Al_{24}. The NEB identified the minimum energy path, the saddle point energy, and the transition state geometry. Thus, the energy barrier to be overcome by the Al atom amounts to about 65 kJ mol^{-1} (0.67 eV) and the difference between initial and final states is about 32.5 kJ mol^{-1} (0.34 eV). These are the most important ingredients for the quantitative prediction of diffusion coefficients. Furthermore, the optimized structures of initial and final states and the transition state are required for the evaluation of temperature-dependent thermodynamic functions, which is the second prerequisite for applying harmonic transition state theory.

The quantitative description of diffusion is based on Eyring's concept of the activated complex (Eyring 1935) as applied by Wert and Zener (1949) to impurity diffusion in solids. Eyring's theory of the activated complex assumes that the reaction proceeds over a well-defined transition state, the so-called activated complex, which is in thermodynamic equilibrium with its surroundings. Furthermore, one translational degree of freedom along the diffusion path replaces the particular vibrational motion of the initial state leading to the transition state. The reaction rate is then given by (a) the ratio between the partition functions of the system in the transition state and the initial state including the Boltzmann factor $\exp[-\Delta E/(k_B T)]$, which involves the activation energy ΔE, and (b) the

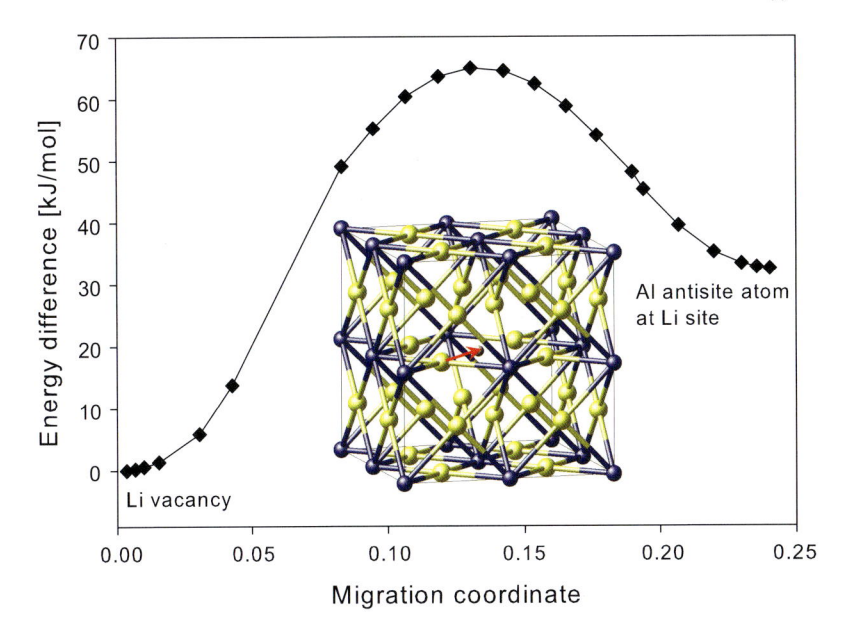

Fig. 11.9 Energy profile for the minimum energy path for migration of an Al atom into an adjacent Li vacancy in LiAl$_3$ as obtained by the nudged elastic band method. The diamonds represent the total energies of the optimized images. The supercell used for the calculation is shown: blue spheres represent Li and yellow spheres Al atoms; the small sphere in the center symbolizes the Li vacancy. The movement of the Al atom into the Li vacancy is shown by a red arrow (Müller et al. 2007).

mean velocity of the reactants crossing the transition state. The first term is directly obtained from ab-initio total energy and phonon calculations. The second term is taken from the temperature-dependent velocity distribution of classical particles at a temperature T.

Applying this approach to the diffusion of an atom in a solid, Eq. (37) is obtained for the temperature-dependent diffusion coefficient.

$$D(T) = n\beta d^2 \nu \exp{-\frac{\Delta G(T)}{k_B T}} \tag{37}$$

Herein, n is the number of nearest-neighbor stable sites, β is the probability that a jump to a nearest-neighbor site leads forward in the direction of diffusion, and d is the length of the jump projected onto the direction of diffusion. The frequency ν corresponds to the mode of the stable state which leads toward the transition state and ΔG is the effective free energy difference between transition state (TS) and initial state (denoted by 0) including electronic energies E_{el} obtained directly from ab-initio calculations and temperature-dependent vibrational energies $E_{vib}(T)$ and entropies $S_{vib}(T)$ from phonon calculations [Eq. (38)].

$$\Delta G(T) = E_{el}^{TS} - E_{el}^0 + E_{vib}^{TS}(T) - E_{vib}^0(T) - T[S_{vib}^{TS}(T) - S_{vib}^0(T)] \tag{38}$$

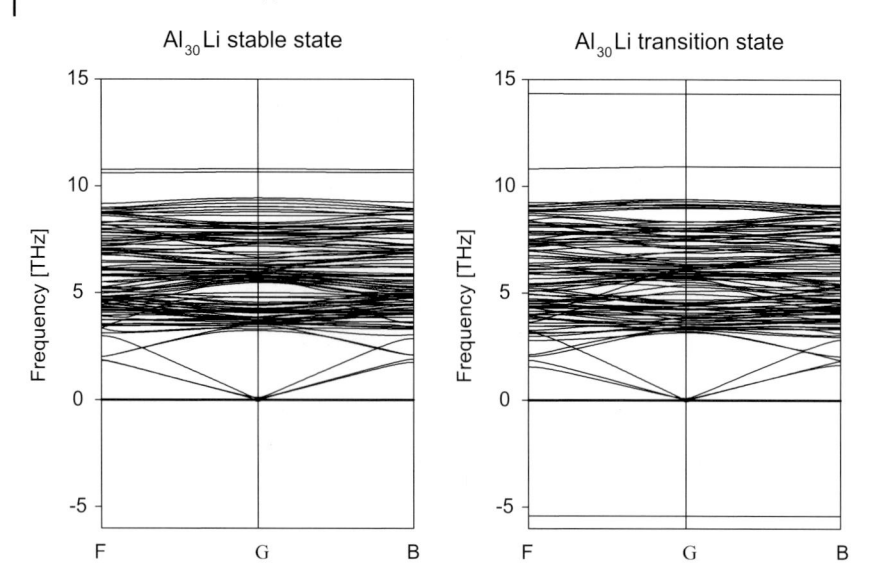

Fig. 11.10 Computed phonon dispersions for supercells $Al_{30}Li$ modeling vacancy-assisted diffusion of Li impurities in an aluminum matrix. Left panel: initial state with a Li atom substituting for an Al atom adjacent to an Al vacancy. Right panel: transition state of the Li atom migrating into the adjacent vacancy site. The imaginary frequency corresponds to the motion of the Li atom across the barrier and is visualized by a negative frequency (Müller et al. 2007).

As an example, the phonon dispersions of the initial state and transition state for a Li impurity migrating into an adjacent Al vacancy in aluminum within a supercell model of 32 atoms are shown in Fig. 11.10 (Müller et al. 2007). This represents the most probable model for diffusion of Li impurities in Al, i.e., a vacancy-assisted diffusion process. The phonons have been calculated by the direct method as described in Section 11.2.4. The Li atom in the initial state, substituting for an Al atom adjacent to an Al vacancy, is associated with three optical modes of high frequency (due to its smaller mass as compared to Al). Two of the modes are almost dispersionless and well separated from the other modes at 10.80 and 10.65 THz and the third one is located just above the Al modes between 9.2 and 9.4 THz and shows some dispersion. Inspection of the associated vibrations in real space reveals that the highest modes at 10.80 THz and 10.65 THz correspond to two different vibrations perpendicular to the diffusion path. The lowest-lying Li mode between 9.2 and 9.4 THz represents a vibration along the diffusion path leading into the transition state. The attempt frequency v is therefore associated with this particular phonon mode. It is noted that all phonon frequencies of the initial state are positive, indicating a true minimum on the energy hypersurface. At the transition state, the highest Li-related frequency is shifted toward higher frequencies (about 14 THz). This mode is a vibration of

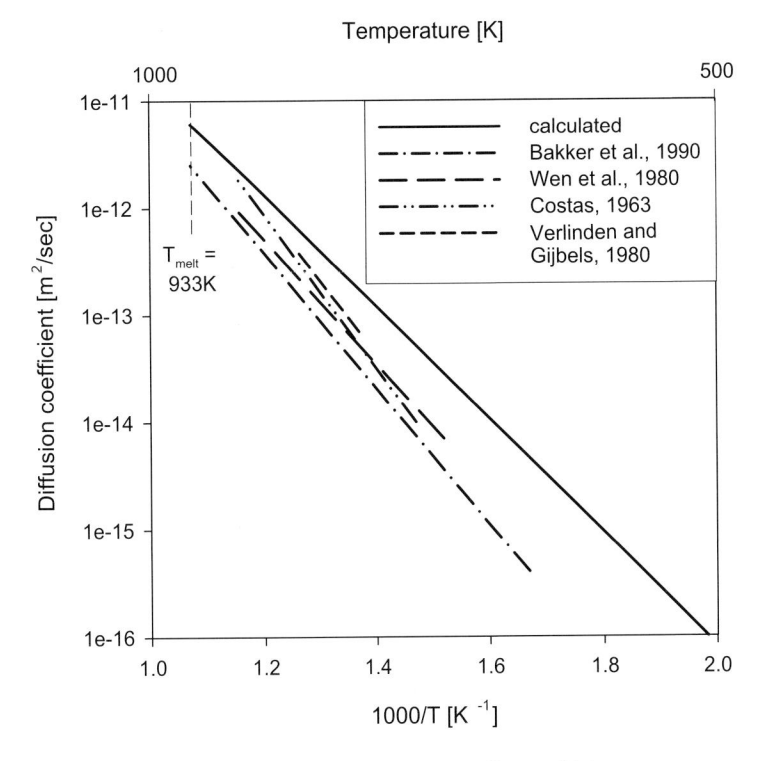

Fig. 11.11 Computed and experimental diffusion coefficient of lithium impurity atoms in aluminum (Müller et al. 2007).

the Li atom perpendicular to the diffusion path. By definition, a transition state is characterized by the occurrence of one negative eigenvalue in the dynamical matrix. This corresponds to a motion of the Li atom along the diffusion path. This mode exhibits an imaginary frequency that is represented in Fig. 11.10 by a negative frequency of about −5.4 THz. By sampling phonon frequencies throughout the whole Brillouin zone the phonon density of states is computed, which makes it possible to derive the temperature-dependent difference in free energy between the transition state and the initial state required for Eq. (37) of transition state theory.

The temperature-dependent diffusion coefficient as computed from Eq. (37) is shown in Fig. 11.11 and compared to available experimental data. Experimental diffusion data from Refs. (Bakker et al. 1990; Wen et al. 1980; Costas 1963; Verlinden and Gijbels 1980) typically agree somewhat better than within an order of magnitude, and the computed diffusion coefficient constitutes an upper envelope to these measured data. That the calculation may provide an upper limit for the diffusivity is consistent with the fact that in the theoretical model calculations ideal crystals are considered, and neither trapping at impurity and defect sites nor grain boundaries that may decrease the diffusivity are included in the present model. Whereas at higher temperatures the model is quite consistent with experimental determinations, at lower temperatures the experiments indicate an increasingly lower diffusivity as compared to the ab-initio results. The thermal energy at lower temperatures might not be sufficient to escape defect and impurity trapping.

The ab-initio approach for solid-state diffusion paradigmatically discussed here for the Li–Al system has so far been applied successfully to diffusion of impurities in various metals, both for interstitial impurities such as hydrogen, deuterium, tritium, oxygen, carbon, and helium, as well as for self-diffusion processes and vacancy-assisted diffusion of substitutional impurities (Wimmer et al. 2007).

11.3.4
Impurity Effects on Grain Boundary Cohesion

Ab-initio modeling has been used to investigate point defects in solids but planar defects such as grain boundaries or antiphase boundaries are also within its scope. As an application in this field we summarize here investigations on the effects of impurity elements He, Li, S, H, C, Zr, P, Fe, Mn, Nb, Cr, and B and some combinations thereof on the cohesion of a nickel $\Sigma 5\{001\}$ twist grain boundary. These calculations were able to explain metallurgical effects (e.g., why boron strengthens nickel grain boundaries) and the mechanisms of irradiation embrittlement (e.g., how boron transmutation results in grain boundary embrittlement), and address the influence of impurities on the environmentally assisted cracking resistance of nickel-base alloys (for a more detailed discussion see Young et al. (2003)).

For the quantitative assessment of the effect of impurities on the grain boundary cohesive strength, a Griffith-type fracture criterion [Eq. (39)] was used (Griffith 1920).

$$\Delta E_{\text{Griffith}} = (E_{\text{surf}} - E_{\text{surf}}^{\text{impurity}}) - (E_{\text{GB}} - E_{\text{GB}}^{\text{impurity}}) \tag{39}$$

The Griffith energy E_{Griffith} is the energy required to cleave the grain boundary (GB) creating two surfaces (surf). A positive value of the change in Griffith energy as expressed by Eq. (39) indicates weakening of the grain boundary due to the presence of impurity atoms. A negative value indicates grain boundary strengthening by the impurity. The Griffith criterion can also be expressed in terms of the critical energy release rate of a crack, given by Eq. (40), where σ is the stress, a is half the crack length, and E is the Young's modulus.

$$\text{Fracture Energy per Area} = \frac{\pi \sigma^2 a}{E} \tag{40}$$

The computational procedure to determine the fracture energy in the presence of a given impurity is as follows: (1) construction and optimization of the pure nickel grain boundary, (2) determination of the site preference of the impurity by optimization of corresponding models, (3) cleavage of the grain boundary, and (4) structure optimization of the free surfaces. For the pure nickel Σ5{001} twist grain boundary a theoretical fracture energy of 3.33 J m^{-2} is computed, which is considerably smaller than the energy of 4.30 J m^{-2} required for cleaving the nickel single crystal along an (001) plane. The determination of the site preference yielded hydrogen, boron, and carbon as interstitial impurities, while helium, phosphorus, sulfur, chromium, iron, manganese, and niobium are substitutional impurities.

The main results of the calculations are summarized in Fig. 11.12. The upper panel compares the tendency of the impurity element to segregate from the bulk to the grain boundary or to the free surface while the lower panel shows the effect of the impurities on the grain boundary fracture energy. The impurity elements are ranked in the sequence He, Li, S, H, C, Zr, P, Fe, Mn, Nb, Cr, and B in order of most embrittling to most strengthening. Helium is strongly embrittling ($\Delta E_{\text{Griffith}} = 1.07$ J m^{-2}), phosphorus has little effect ($\Delta E_{\text{Griffith}} = -0.05$ J m^{-2}) and boron has to offer considerable strengthening of the grain boundary ($\Delta E_{\text{Griffith}} = -0.54$ J m^{-2}). These ab-initio results are consistent with experimental observations since He, S, and H are, for instance, known embrittling agents whereas boron is applied as a strengthener in nickel-base alloys. Boron is embedded in the center of the grain boundary in a highly symmetric eight-fold coordination to Ni atoms and the analysis of the electronic structure reveals s-to-p electron transfer and hybridization of B p and Ni d electrons. For instance, in the case of carbon the s electrons are lower than the p electrons and s-to-p electron transfer and subsequent hybridization with Ni d electrons is inhibited. The strengthening of nickel-base alloys by boron can therefore be understood in terms of the electronic structure.

In addition to single impurities, selected impurity pairs H–Li, H–B, H–C, H–P, and H–S have been investigated, and the results are included in Fig. 11.12. Essentially no strong attraction between H and S, P, B, and C atoms is found in bulk, grain boundary, and surface environments. Since there is little interaction for the concentration tackled by the model, linear superposition of single impurity effects is a good approximation for the combined impurity effects. The results from ab-initio modeling are again consistent with the experimental observation that S and H atoms as well as H and P atoms embrittle grain boundaries, but that phosphorus alone has little effect on intergranular fracture. In addition to the intrinsic effects of an impurity on hydrogen embrittlement, effects on the hydrogen uptake have been studied. The experimental indication that phosphorus may promote hydrogen entry by poisoning the hydrogen recombination reaction $2H_{\text{ads}} \rightarrow H_2(g)$ is consistent with the calculations, since a weak interaction energy of about 8 kJ mol^{-1} between H and P atoms at the free surface is computed. This weak bonding suggests that an impurity enhancing the stability of adsorbed H atoms at the free surface can act as a hydrogen recombination poison. For a detailed discussion on implications of the study in the field of metallurgy, irradiation embrittlement, and cracking resistance of nickel-base alloys, see Young et al. (2003).

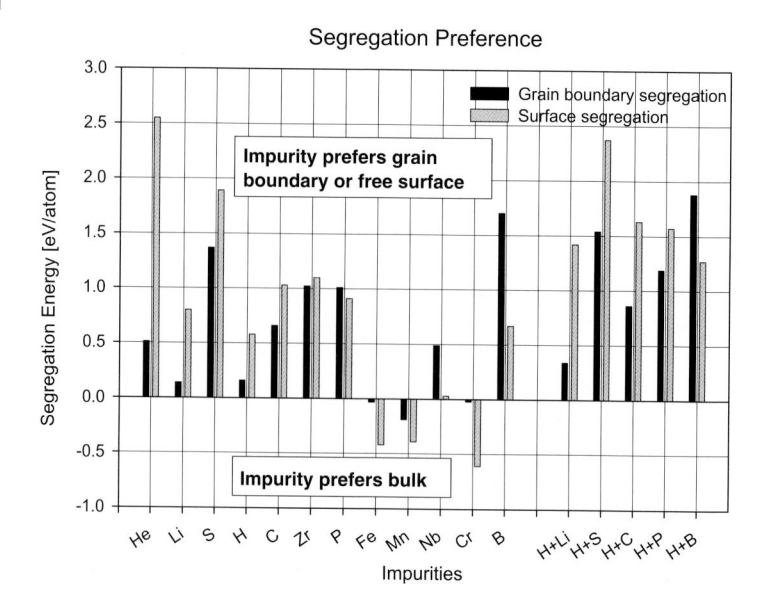

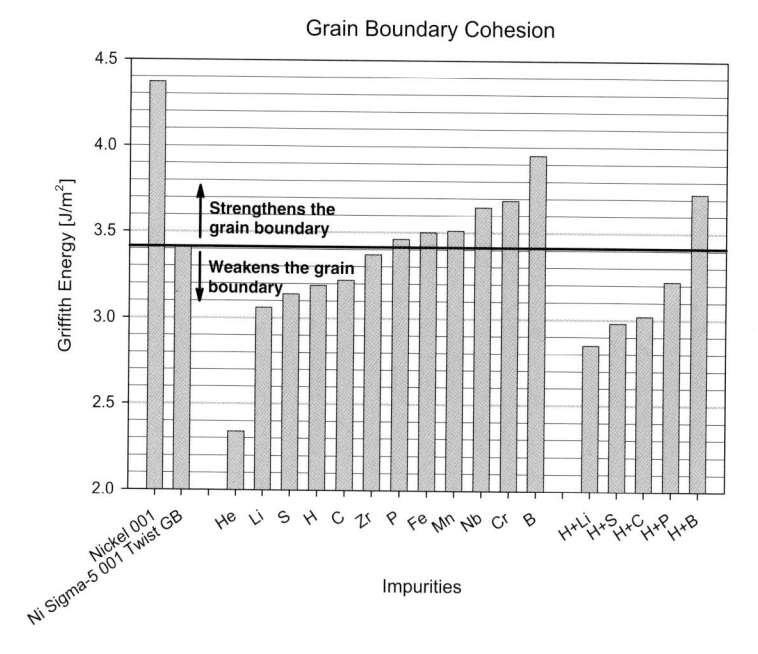

Fig. 11.12 Comparison of the tendency of impurity elements and pairs of impurities to segregate to the free surface or to the Ni Σ5{001} twist grain boundary (upper panel). Lower panel: Comparison of the effect of impurity elements and pairs of impurities on the cohesive strength of a Ni Σ5{001} twist grain boundary (Young et al. 2003).

11.3.5
Toward Multiscale Modeling: Cluster Expansion

Although DFT-based methods as described in the earlier sections allow for a detailed prediction of structural, vibrational, magnetic, and electronic properties of ordered compounds, standard DFT methods can be applied for only a rather small number of atoms per unit cell. The largest number to be feasible for ab-initio DFT calculations is a few hundreds of atoms in the unit cell. In this section, it will be demonstrated how the limitation to this rather small number of atoms can be overcome without losing the accuracy of DFT calculations, namely by combining DFT results with the concept of the cluster expansion (CE) (Sanchez et al. 1984), by which the atomic and mesoscopic length scales are joined together. For an understanding of processes such as decomposition reactions (i.e., formation of precipitates) it is essential that the energetics can be separated into elastic (mechanical) and chemical parts, which requires the treatment of the elastic energy in the framework of a direction- and concentration-dependent elasticity theory, which will be discussed first. The extension to finite temperatures is then realized by Monte Carlo (MC) simulations. Besides modeling thermodynamic properties, the use of the CE Hamiltonian in a kinetic MC approach allows the treatment of the alloy problem on a time scale of hours or even days (Müller 2003).

The starting point for modeling the elastic energy is the ideal, flat interface connecting crystals consisting of only A- or B atoms, as sketched in Fig. 11.13: the two "blocks" of material A and B possess a common in-plane lattice constant a_p perpendicular to the chosen direction $\hat{G}$. Along this direction both parts are allowed to relax geometrically. This situation is comparable to a pseudomorphic, epitaxial system without dislocations at the interface.

From the discussion above, it becomes clear that the epitaxial strain energy $\Delta E_A^{epi}(a, \hat{G})$ is the energy needed to deform the elemental solid A epitaxially (or biaxially) to the common lattice constant a_p in the two directions orthogonal to $\hat{G}$ while relaxing along $\hat{G}$. Forbidding relaxations along the direction $\hat{G}$ corresponds to a hydrostatic deformation. The corresponding energy is called the hydrostatic deformation energy $\Delta E_A^{hydro}(a)$ for the lattice parameter a. The relationship

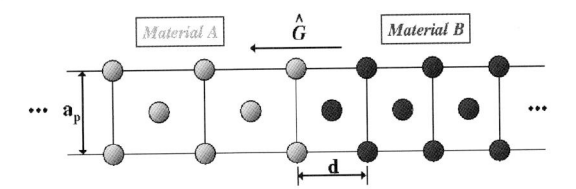

Fig. 11.13 Ideal interface perpendicular to direction $\hat{G}$ formed by two pieces of material A and B with a common in-plane lattice constant a_p. In the case of an epitaxial (biaxial) deformation both sides are allowed to relax along $\hat{G}$. The corresponding energy is called the epitaxial strain energy of the element A or B, respectively.

$\Delta E_A^{hydro}(a) > \Delta E_A^{epi}(a, \hat{G})$ is obvious, because it is always easier to deform a material epitaxially (biaxially) than hydrostatically (triaxially). The ratio of these two energies (Eq. (41) for $0 < q \leq 1$) defines the epitaxial softening function (Wood and Zunger 1988; Zunger 1994a)

$$q(a, \hat{G}) = \frac{\Delta E_A^{epi}(a, \hat{G})}{\Delta E_A^{hydro}(a)} \tag{41}$$

Small values of $q(a, \hat{G})$ indicate elastically soft directions $\hat{G}$. As an example, Fig. 11.4 shows the calculated softening functions, $q(a, \hat{G})$, for the fcc elements Pt and Rh along the three fundamental crystal directions. The vertical lines mark the DFT-derived equilibrium lattice parameters of Pt and Rh. For these values the harmonic elasticity theory must be valid, i.e., the directions have to follow a well-defined sequence with respect to their elastic softness. Indeed, in harmonic elasticity theory, q depends on only the direction $\hat{G}$, but *not* on the chosen lattice constant a_p (Bottomley and Fons 1996; Laks et al. 1992; Zunger 1994a), as in Eq. (42) with the cubic bulk modulus $B = \frac{1}{3}(C_{11} + 2C_{12})$ and the anisotropy parameter $\Delta = C_{44} - \frac{1}{2}(C_{11} - C_{12})$.

$$q_{harm}(\hat{G}) = 1 - \frac{B}{C_{11} + \Delta \gamma_{harm}(\hat{G})} \tag{42}$$

As described in Section 11.2.3 the elastic constants C_{11}, C_{12}, and C_{44} can be obtained easily from DFT calculations, and consequently Δ and B, too. The geometric function γ_{harm} depends on the polar angle Θ and the azimuth angle Φ of the direction vector $\hat{G}$ [Eq. (43)].

$$\gamma_{harm}(\Phi, \Theta) = \sin^2(2\Theta) + \sin^4(\Theta) \sin^2(2\Phi)$$
$$= \frac{4}{3}\sqrt{4\pi} \left[K_0(\Phi, \Theta) - \frac{2}{\sqrt{21}} K_4(\Phi, \Theta) \right] \tag{43}$$

Here, the parameter K_l is the cubic harmonic for the angular momentum l. However, this definition fails to describe the observed crossing of individual curves as found for Pt and Rh (Fig. 11.14) for values different from the equilibrium lattice parameter. This means that modeling strain in the framework of the harmonic theory is not sufficient for metals. Obviously, the crystallographic order of elastic softness can change as a function of the lattice parameter. For example, for Rh a decrease of 1% (Fig. 11.14) from the equilibrium lattice constant leads to a change in hierarchy with respect to the q values: now, the [110] is elastically harder than the [111] direction, while at equilibrium the opposite is true. This clearly indicates that for a description of strain effects in metals, not only the direction dependence of strain (anisotropic strain effects), but also the dependence of strain on the lattice parameter (anharmonic strain effects) must be taken into account (Ozoliņš et al. 1998a,b). This is achieved by introducing a geometric function γ, which depends on the lattice parameter a, according to Eq. (44).

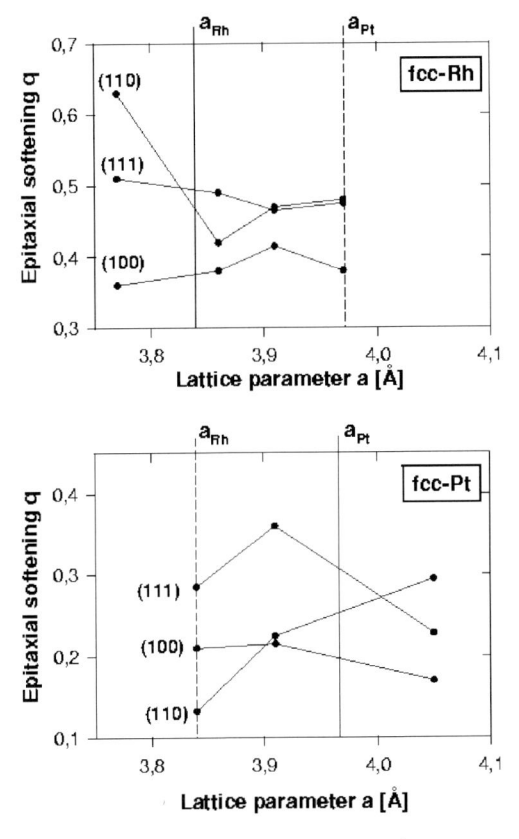

Fig. 11.14 Epitaxial softening function $q(a, \hat{G})$ Eq. (41), for Pt and Rh calculated by DFT. The vertical lines mark the calculated equilibrium lattice constants. Lines are to guide the eye.

$$\gamma(a, \hat{G}) = \gamma_{harm}(\hat{G}) + \sum_{l=0}^{l_{max}} b_l(a) K_l(\hat{G}) \qquad (44)$$

This relationship also includes now higher-order cubic harmonics as necessary to go beyond the harmonic approximation (for more details see, e.g., Ozoliņš et al. (1998a)). Then, Eq. (42) can be reformulated as Eq. (45).

$$q(a, \hat{G}) = 1 - \frac{B}{C_{11} + \Delta\gamma(a, \hat{G})} \qquad (45)$$

With the function $q(a, \hat{G})$ as derived from DFT calculations (Fig. 11.13), the quantity $\gamma(a, \hat{G})$ is defined by Eq. (45) and, in turn, the coefficients $b_l(a)$ result from applying Eq. (44). The determination of the coefficients $b_l(a)$ permits a generalization of the calculated epitaxial energies $\Delta E_A^{epi}(a, \hat{G})$ for a discrete set of directions to a function of arbitrary directions $\hat{G}$.

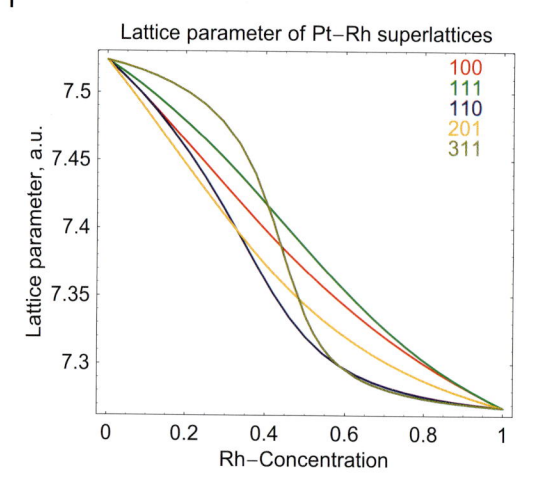

Fig. 11.15 Relationship between lattice parameter and concentration for Pt–Rh superlattices along five different directions.

The next step is parameterization of the equilibrium constituent (or coherency) strain energy $\Delta E_{CS}^{eq}(x, \hat{G})$ which is defined as the strain energy required to maintain coherency between a "piece" of material A and a "piece" of material B along an interface with orientation $\hat{G}$. Thus, a superlattice $A_n B_n$ along a certain direction $\hat{G}$ with $n \to \infty$ is constructed. In practice, the calculated elemental epitaxial energies are used to derive the constituent strain energy, which is determined by the equilibrium value of the composition-weighted sum of the epitaxial energies for A and B atoms [Eq. (46)].

$$\Delta E_{CS}^{eq}(x, \hat{G}) = \min_{a_p} [x \Delta E_A^{epi}(a_p, \hat{G}) + (1 - x) \Delta E_B^{epi}(a_p, \hat{G})] \tag{46}$$

The parameter $a_p(x)$ is the lattice parameter which minimizes ΔE_{CS}^{eq} at each concentration x. The minimization according to Eq. (46) allows a direct relationship between a_p and the average superlattice concentration x, as shown in Fig. 11.15. In this figure it is noticeable that, dependent on the direction, deviations from Vegard's law clearly occur. Figure 11.16 presents the resulting equilibrium constituent strain as a function of concentration and direction. As discussed above, the hierarchy of the elastic softness depends on the concentration of the superlattice.

The equilibrium constituent strain can be illustrated by a three-dimensional parameterization in terms of a sum of cubic harmonics allowing for a prediction of the constituent strain energy for any arbitrary direction, as shown in Fig. 11.17 for Pt–Rh superlattices corresponding to three different Rh concentrations. Thereby, the distance from the surface to the center of the cube represents the size of the strain energy in this crystallographic direction. Therefore, if the deviation from an ideal sphere represents a depletion as observed for $Pt_{0.75}Rh_{0.25}$ along the square face diagonal (or along the [110] direction), then this direction is elas-

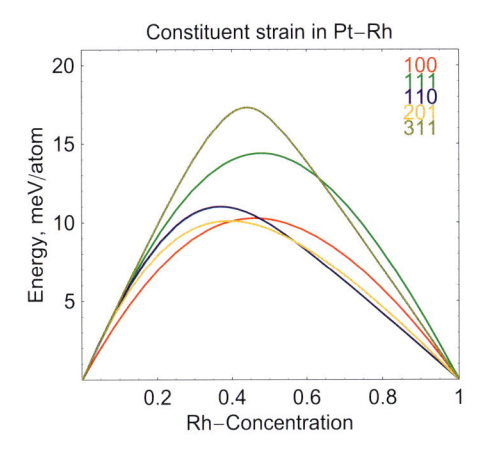

Fig. 11.16 Constituent strain energies ΔE_{CS}^{eq}, Eq. (46), for Pt–Rh superlattices vs. composition for five different directions. All energy differences are given with respect to the ideal undistorted fcc phases of Pt and Rh.

tically soft, while an elongated shape – such as occurs for the [111] direction – represents an elastically hard direction. In summary, Fig. 11.17 demonstrates the necessity to consider the concentration and direction dependence of strain in metal alloys.

A very efficient tool to bridge the gap between atomic and mesoscopic length scales is the concept of the cluster expansion (CE). The basic idea of CE (Sanchez et al. 1984) is to express the geometrically relaxed energy, $E(\sigma)$, for arbitrary lattice configurations σ on an underlying lattice as a linear expansion of energies which are characteristic for geometrical figures (or clusters) such as biatoms, triatoms, etc. (Fig. 11.18). For practical reasons, the "alloy problem" is transformed to an Ising model. To each atom i of the alloy $A_{1-x}B_x$ a spin value is assigned in such a way that $S_i = -1$ (i represents an A atom), and $S_i = +1$, (i represents B). Then, the energy of each configuration can be expressed by an Ising expansion, Eq. (47).

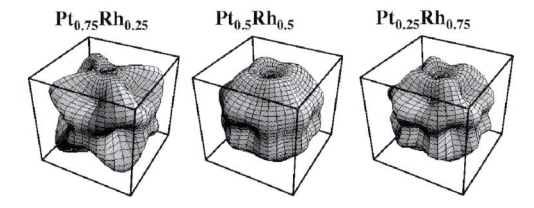

Fig. 11.17 Parameterized three-dimensional presentation of the constituent strain energy ΔE_{CS}^{eq} [Eq. (46)] for the Pt–Rh system for compositions of 25%, 50%, and 75% Rh. The distance from the surface to the center of the cube represents the size of the strain energy.

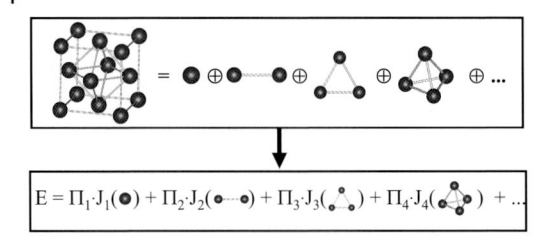

$$E = \Pi_1 \cdot J_1(\bullet) + \Pi_2 \cdot J_2(\bullet\text{---}\bullet) + \Pi_3 \cdot J_3(\triangle) + \Pi_4 \cdot J_4(\triangle\!\!\!\triangle) + \ldots$$

Fig. 11.18 The concept of cluster expansions. The crystal structure is decomposed into characteristic geometrical figures (or clusters), as shown for the fcc lattice. The energy of any configuration can then be written as a linear combination of the characteristic energies J_f of the figures.

$$E(\sigma) = J_0 + \sum_i J_i S_i(\sigma) + \sum_{j<i} J_{ij} S_i(\sigma) S_j(\sigma) + \sum_{k<j<i} J_{ijk} S_i(\sigma) S_j(\sigma) S_k(\sigma) + \cdots \tag{47}$$

The first two terms on the right define the energy of the random alloy (with zero mutual interactions), the third term contains all pair interactions, the fourth all three-body interactions, and so on. This equation can be made more compact by introducing a correlation function $\overline{\Pi}_F$ for each class of symmetry-equivalent geometrical figures F (Zunger 1994b), as in Eq. (48).

$$\overline{\Pi}_F(\sigma) = \frac{1}{ND_F} \sum_f S_{i_1}(\sigma) S_{i_2}(\sigma) \ldots S_{i_m}(\sigma) \tag{48}$$

The symbol D_F denotes the number of geometrical figures of class F per site of the lattice. The index f sums over all the ND_F geometrical figures in class F, whereas m denotes the number of sites of geometrical figure f. In this way, Eq. (47) transforms to Eq. (49) (Laks et al. 1992).

$$E(\sigma) = N \sum_F D_F \overline{\Pi}_F(\sigma) J_F \tag{49}$$

The key point now is *how* the interactions J_F can be determined. In principle, the direct inversion method of Conolly and Williams (1983) can be applied. In this procedure, a set (about 15–40) of N_σ mostly simple ordered compounds with typically no more than 2–16 basis atoms is chosen, for which the energies E_{DFT} of the geometrically relaxed structures are derived from DFT calculations. Then, the N_σ energies are used to construct the N_F interactions J_F by solving Eq. (50) with respect to the N_F values of the unknown cluster interactions J_F.

$$\sum^{N_\sigma} \left| E_{DFT}(\sigma) - N \sum^{N_F} D_F J_F \overline{\Pi}_F(\sigma) \right|^2 = \text{Minimum} \tag{50}$$

Clearly, for finding solutions the relationship $N_F \leq N_\sigma$ must hold. Obviously, there are two remaining problems which have to be solved, namely (a) how to select the set of input structures which describes the energetics of the system and (b) how to select relevant geometrical figures for which the cluster interactions will be constructed. As recently shown (Hart et al. 2005), the second problem can be handled by applying a genetic algorithm for the selection of geometrical figures. Typically, this algorithm can select a set of figures from all figures with up to, say, six sites (six-body interactions) for given maximum distances between the atomic sites. This is called the "inner loop" (for details, see, e.g., Hart et al. 2005). Regarding the first problem, one efficient tool with which to find structures with important "structure information" for the determination of the interactions is a ground-state search (Ferreira et al. 1991) in the early stage of the construction: for a "starting set" of about 15–30 DFT energies of arbitrary input structures with typically no more than 2–20 basis atoms, a fit by CE is made. The resulting interactions J_F are then used to predict the energy of *all possible structures* with, e.g., up to 16 atoms per unit cell. (This is indeed a very reasonable restriction, since most known stable structures of binary metal alloys have fewer than 16 atoms per unit cell.) Such an analysis based on Eq. (49) needs only a few minutes on a standard PC. At the end, the corresponding CE formation enthalpies ΔH_f of all the structures are plotted versus composition, and a ground-state line is constructed. Hereby, $\Delta H_f(\sigma)$ is defined as the energy gain or loss per atom with respect to the bulk constituents at their equilibrium lattice constants. For an intermetallic compound $A_m B_n$ (consisting of m A and n B atoms per unit cell) ΔH_f is given by Eq. (51), with a_A and a_B being the equilibrium lattice constants of the elements A and B in their bulk, and $E_A^{tot}(a_A)$ and $E_B^{tot}(a_B)$ the respective total energies.

$$\Delta H_f(A_m B_n) = \frac{1}{m+n} [E^{tot}(A_m B_n) - m \cdot E_A^{tot}(a_A) - n \cdot E_B^{tot}(a_B)] \quad (51)$$

The result of such a ground-state search is shown schematically in Fig. 11.19. An individual structure σ only contributes to the ground-state line if the energy average of the stable structures at the next higher and lower concentrations is energetically less favorable than the formation enthalpy of σ. More precisely, for three structures α, σ, and β with $x(\alpha) < x(\sigma) < x(\beta)$ which are the lowest in energy for their individual concentrations, the structure σ has to fulfill the thermodynamical stability condition of Eq. (52) to be the ground state for $x(\sigma)$.

$$\Delta H_f(\sigma) < \frac{x(\sigma) - x(\beta)}{x(\alpha) - x(\beta)} \Delta H_f(\alpha) + \frac{x(\sigma) - x(\alpha)}{x(\beta) - x(\alpha)} \Delta H_f(\beta) \quad (52)$$

If Eq. (52) holds, then a mixture of the phases α and β would be higher in energy than structure σ. With the ground-state line constructed, one checks for all structures which lie on or very close ("excited states") to it, if they have already been considered as input structures for the CE. If not, their DFT energy is calcu-

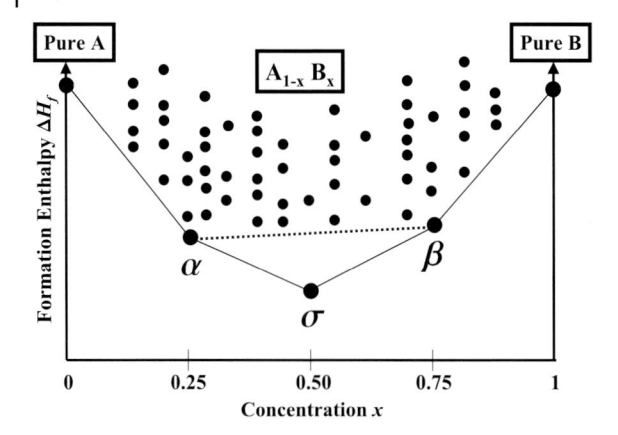

Fig. 11.19 Schematic groundstate diagram of a binary alloy $A_{1-x}B_x$ (Müller 2003). The groundstate line was constructed from 60 energies of geometrically relaxed structures (given by dots) by means of Eq. (52). Besides the pure elemental crystal the groundstate line is formed by the three structures α, σ, and β for the concentrations $x = 0.25, 0.50$, and 0.75, respectively. If the energy of σ would be above the dashed tie line between α and β, a mixture of α and β would be more stable than σ.

lated and added to the input set of structures, while input structures with ΔH values far above the ground-state line are removed from the input structure list. The procedure is repeated until the ground-state line becomes stable. This is called the "outer loop." An example for a ground-state line is discussed in Section 11.3.6.

Although the concept discussed allows for a prediction of formation enthalpies of ordered compounds and the random alloy case, it fails to predict the stability for long-range coherent systems on a mesoscopic scale, such as coherent nanoclusters in metal alloys or long-period superlattices: for a given superlattice $A_n B_n$, Eq. (49) predicts the asymptotic formation enthalpy $\Delta H_f = 0$ for $n \to \infty$. As first shown by Laks et al. (1992), this is an intrinsic fault of any finite CE and easy to understand: if one considers an atom A of an $A_n B_n$ superlattice "far" away from the A/B interface (i.e., all geometrical figures f connect the A atom exclusively to other A atoms), then according to the finite CE, atom A is in a bulk environment and consequently $\Delta H_f = 0$. However, the formation enthalpy of such an infinite superlattice should then be given by the remaining elastic energy, because in the limit $n \to \infty$ the superlattice formation enthalpy depends only on its strained constituents, and not on the interface properties. The problem can be solved (Laks et al. 1992; Zunger 1994b) by transforming a group of interactions to the reciprocal space and treating the constituent strain term explicitly as discussed at the beginning of this section in refs. (Wood and Zunger 1988; Zunger 1994a). The transformation is easiest for pair interactions. For that case, the Fourier transform of real-space pair interactions, $J_{pair}(\mathbf{k})$ and the structure factor $S(\mathbf{k}, \sigma)$ are introduced [Eqs. (53) and (54)].

$$J_{pair}(\mathbf{k}) = \sum_j J_{pair}(\mathbf{R}_i - \mathbf{R}_j) \exp(-i\mathbf{k}\mathbf{R}_j) \quad (53)$$

$$S(\mathbf{k}, \sigma) = \sum_j S_j \exp(-i\mathbf{k}\mathbf{R}_j) \quad (54)$$

Then the formation enthalpies for any arbitrary atomically relaxed configuration σ is given by Eq. (55) (Zunger 1994b).

$$\Delta H_{CE}(\sigma) = \sum_{\mathbf{k}} J_{pair}(\mathbf{k}) |S(\mathbf{k}, \sigma)|^2 + \sum_F D_F J_F \overline{\Pi}_F(\sigma) + \Delta E_{CS}(\sigma) \quad (55)$$

This relationship – as introduced by Zunger and coworkers (Laks et al. 1992; Zunger 1994b) – is the so-called mixed-space cluster expansion (MSCE). The first term includes all the geometrical pair figures in **k**-space. The second term represents many-body interactions and runs over symmetrically inequivalent clusters consisting of three or more lattice sites. It also includes J_0 and J_1 from Eq. (47). The symbol D_F again represents the number of equivalent clusters per lattice site, whereas $\overline{\Pi}_F(\sigma)$ denotes the structure-dependent geometrical coefficients of Eq. (48). The last term represents the constituent strain energy of σ, $\Delta E_{CS}(\sigma)$, and can be derived by expanding the equilibrium constituent strain energy [Eq. (46)], $\Delta E_{CS}^{eq}(x, \hat{k})$, as in Eq. (56), (Laks et al. 1992; Wolverton et al. 2000), utilizing Eq. (57).

$$\Delta E_{CS}(\sigma) = \sum_{\mathbf{k}} J_{CS}(x, \hat{k}) |S(\mathbf{k}, \sigma)|^2 \quad (56)$$

$$J_{CS}(x, \hat{k}) = \frac{\Delta E_{CS}^{eq}(x, \hat{k})}{4x(1-x)} \quad (57)$$

In this relationship, J_{CS} contains the correct long-period superlattice limit, namely the constituent strain energy.

Analogously to the real-space expansion in Eq. (49), the formation enthalpies $\Delta H_f(A_m B_n)$ of ordered compounds are needed as an input for the determination the coefficients $\{J_{pair}(\mathbf{k})\}$ and $\{J_f\}$ of the CE. The pair and multi-body interactions result from the fit to the N_σ formation enthalpies $\{\Delta H_f\}$, minimizing the root mean square (rms) error [Eq. (58)] (Zunger 1994b).

$$\Delta_{rms}^2 = \frac{1}{N_\sigma} \sum_\sigma w_\sigma [\Delta H_{CE}(\sigma) - \Delta H_{LDA}(\sigma)]^2 + \frac{t}{\alpha} \sum_{\mathbf{k}} J_{pair}(\mathbf{k})[-\nabla_{\mathbf{k}}^2]^{\lambda/2} J_{pair}(\mathbf{k}) \quad (58)$$

In this equation, λ and t are free parameters and α is a normalization constant (Laks et al. 1992). The second term on the right, i.e., the **k**-space smoothness criterion, automatically selects essential short-range pair interactions. After Eq. (58)

Table 11.8 Formation enthalpies of the Al–Zn system derived from DFT calculations ("direct") and cluster expansion ("CE") (compounds are sorted by superlattice direction and composition).[a]

Stoichiometry	x_{Al}	Formation enthalpies [meV atom^{-1}]					
		[100]	[110]	[111]	[201]	[311]	others
Zn	0.0						fcc
direct:							0.0
CE:							0.6
$AlZn_3$	0.25		Y3*	V3	$DO_{22}b$	W3	$L1_2$
direct:			24.8	4.3	13.3	9.0	5.3
CE:			24.2	3.7	12.5	11.0	10.2
$AlZn_2$	0.333	β2	γ2	α2			
direct:		14.9	17.7	2.0			
CE:		18.3	21.3	3.3			
Al_2Zn_3	0.40	Z7*					
direct:		7.0					
CE:		7.2					
$AlZn$	0.50	$L1_0$		$L1_1$			
direct:		23.5		7.4			
CE:		26.9		9.9			
Al_2Zn_2	0.50	Z2	Y2	V2	CH(40)	W2	
direct:		9.0	24.6	4.8	24.8	18.6	
CE:		9.6	26.8	7.3	28.8	22.3	
Al_3Zn_3	0.50	Z6*	Y6*	V6			
direct:		6.2	18.8	2.8			
CE:		5.7	16.4	3.2			
Al_4Zn_4	0.50			V8*			SQS8*
direct:				1.8			18.4
CE:				1.7			20.9
Al_3Zn_2	0.60	Z5*					
direct:		6.0					
CE:		6.2					
Al_2Zn	0.667	β1	γ1	α1			
direct:		17.4	32.6	15.9			
CE:		20.4	33.7	16.7			
Al_3Zn	0.75	Z1	Y1*	V1	$DO_{22}a$	W1*	$L1_2$
direct:		10.2	26.0	14.6	27.6	23.0	35.1
CE:		14.1	27.2	12.7	31.6	25.8	34.9

Table 11.8 *(continued)*

Stoichiometry	x_{Al}	Formation enthalpies [meV atom^{-1}]					
		[100]	[110]	[111]	[201]	[311]	others
Al_7Zn	0.875						D7
direct:							10.7
CE:							12.4
Al	1.00						fcc
direct:							0.0
CE:							−0.6

*Enthalpies not used in the CE fit, but represent predictions.
[a] Average error (standard deviation of the CE enthalpies of 25 input structures) = 1.5 meV; average prediction error (standard deviation of all 9 predicted structures) = 2.2 meV; maximum error (largest deviation between the CE and DFT value of the structures considered; Müller et al. 1999) = 4.1 meV.

has been applied to the pair interactions, a genetic algorithm is applied to select a set of figures allowing for a quantitative modeling of the alloy's energetics. As an example, Table 11.8 lists the $N_\sigma = 25$ input formation enthalpies ΔH_f ("direct") and the MSCE enthalpies ("CE") for Al–Zn. All enthalpies with an asterisk denote structures not used in the fit, i.e., they present "real" predictions. The table gives the stoichiometry of all input structures as well as information on whether the structures comprise a superlattice of Al and Zn layers along a certain direction. For example, V3 is defined by an $AlZn_3$ superlattice in the [111] direction for which there are always three Zn(111) layers followed by one Al(111) layer. The average error of fitting $N_\sigma = 25$ enthalpies is 1.5 meV whereas the resulting prediction error for the nine structures not used for the fit is 2.2 meV. Figure 11.20 shows the resulting pair and multibody interactions for the Al–Zn system. Clearly, pair interactions converge rapidly, and therefore 15–20 pair interactions are sufficient.

For finite temperature studies, Eq. (55) can be used in MC simulations. For studying thermodynamic properties we applied a simple Metropolis algorithm (Metropolis et al. 1953) allowing for mutual flipping of pairs of A and B atoms in an *arbitrary* distance. The aim is to reach the equilibrium configuration as fast as possible. In addition to the temperature dependence of the free energy, MC simulations can be used to calculate coherent phase boundaries in the phase diagram. Following the fluctuation–response theorem (Toda et al. 1983), the specific heat c_v at a certain temperature can be determined from the observation that c_v is proportional to the equilibrium fluctuations of the energy, $\langle E^2 \rangle - \langle E \rangle^2$. Since the energy exhibits a point of inflection for a second-order phase transition at the transition temperature T_{trans}, its response function $c_v = (\partial E / \partial T)_v$ has a

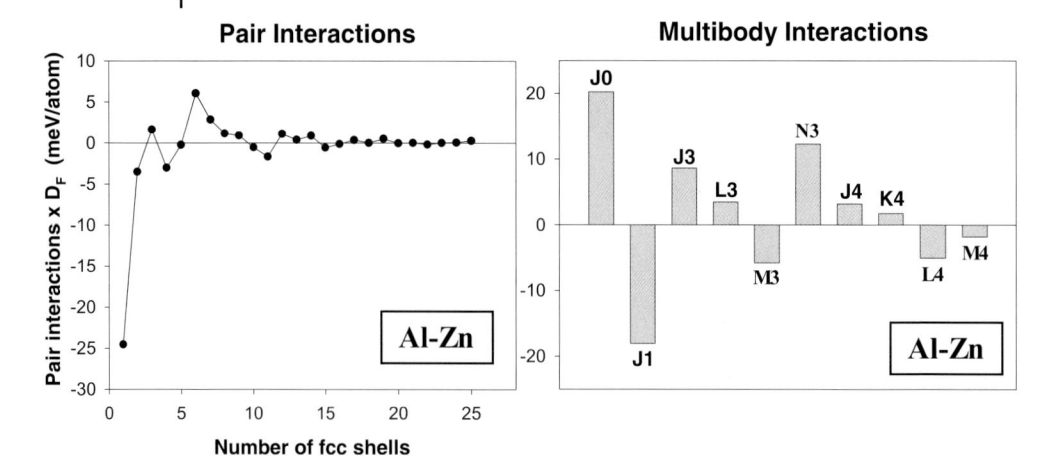

Fig. 11.20 Pair and multibody effective cluster interactions for Al–Zn deduced from the fit of Table 11.8. The symbols characterize individual multibody interactions: "3" denotes three- and "4" four-body interactions. In general, the distance between atoms increases with the letter, i.e., for example "J" means only nearest neighbors, "K" nearest neighbors and one second-nearest neighbor, and so on.

maximum at T_{trans} (Fig. 11.21a). Although a phase transition is strictly defined for an *infinite* system, one usually also speaks about a phase transition of a *finite* system, given by the maximum of c_v at the transition temperature, as illustrated in Fig. 11.21(a). If the MC simulations are applied for different concentrations x,

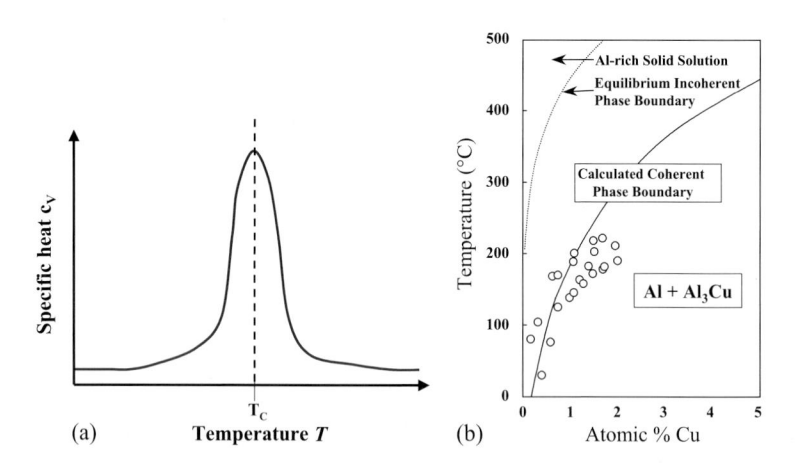

Fig. 11.21 (a) Schematic plot of the specific heat vs. temperature near a second-order phase-transition: c_v has a maximum at T_{trans}; (b) calculated coherent phase boundary for Al-rich Al–Cu (Müller et al. 1999) and comparison to experimental data (Murray 1985) (open circles).

the resulting temperatures T_{trans} can be used to construct the coherent phase boundary of a system, as displayed in Fig. 11.21(b) for the Al-rich side of the Al–Cu phase diagram (Müller et al. 1999). The open circles are measured values (Murray 1985). A small piece of the incoherent phase boundary is also shown. However, this boundary cannot be calculated by the method presented because it is restricted to *coherent* problems.

Besides the problem of bridging the length scales, many materials properties require simulation times reaching from fractions of a second to weeks. One important example is the decomposition of an alloy into its constituents by precipitation. Precipitates represent an important part of the microstructure of many alloy systems (see Chapter 7). Hereby, the dynamic evolution of precipitates takes place on a time scale of several hours, days, or even months. The CE Hamiltonian can help to solve this second scaling problem, too, by using the effective interactions in kinetic Monte-Carlo (KMC) simulations, which constitute one of the most successful approaches to describe diffusion, growth, and microstructure evolution in alloy systems (Jacobsen et al. 1995). The combination of MSCE and KMC simulations can be applied to simulate the aging of coherent precipitates in binary alloy systems. This decomposition reaction is sketched in Fig. 11.22 by a simplified two-dimensional presentation: a quenched solid solution (left frame) is aged at a given temperature. During this aging process islands are formed (right frame) which may show a characteristic size and shape distribution (it is assumed that islands are formed by black B atoms in an A-rich $A_{1-x}B_x$ alloy). The question is whether the distribution of these islands as a function of aging time can be calculated from first principles.

Since the main focus is on the distribution of the islands as a function of temperature and aging time while the path of an individual B atom through the crystal is less interesting in this case, it is assumed that the islands are formed by nearest-neighbor site flips only and not by continuous atomic motion. The activation barrier for the exchange process can be expressed in terms of the temperature-dependent diffusion coefficient $D(T)$. In order to calculate $D(T)$ by a first-principles approach, it is assumed that the exchange of atoms is given by a vacancy-controlled diffusion. Therefore, in a first step, activation barriers must be calculated as a function of the structural environment. In the case of precipitation in which the alloy contains only a tiny amount (typically 1–5%) of the precipitat-

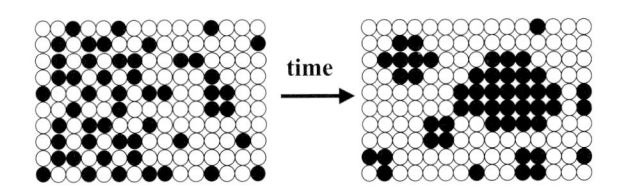

Fig. 11.22 Schematic crystal plane of an $A_{1-x}B_x$ alloy with characteristic islands formed by B (black) atoms during the aging process.

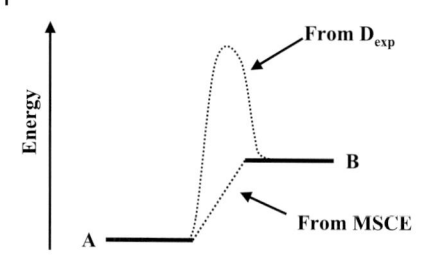

Fig. 11.23 Basic assumption in our simulations (Müller 2003): while the energy difference between two neighboring atoms can be derived easily from the MSCE, an average and temperature-dependent activation barrier is calculated from experimental temperature-dependent diffusion data.

ing element, one often restricts the calculation of activation barriers to the case of the dilute limit (atom B in an A crystal) and the structural environment at the interface between solid solution and precipitate. Although such activation barriers can – in principle – be used directly in KMC programs, they do not allow for a consideration of the temperature dependence as well as a transformation to real time scales. For this purpose, the complete phonon spectra for the relaxed structure corresponding to the vacancy formation, the migration, and the final configuration have to be calculated. This might be used in the framework of a transition state theory to predict the temperature-dependent diffusion constant of the system, $D(T)$ (see Chapter 5 and 11.3.3). Following classical diffusion theory the exchange frequency is proportional to the square of the atomic distance divided by the diffusion constant and the number of possible "jump directions" (e.g., six in a simple cubic lattice). If an exchange process between two certain neighbouring atoms has already been chosen, then, consequently, the frequency $1/\tau_0$ for a chosen exchange process as a function of temperature T is connected to $D(T)$ by Eq. (59), with a_{nn} being the average nearest-neighbor distance between atoms. Now, one can easily transform the KMC steps to real time.

$$\tau_0(T) = \frac{a_{nn}^2}{D_{exp}(T)} \tag{59}$$

The strength of the MSCE to control a huge configuration space can now be utilized to calculate the energy difference for *all possible* exchange processes, even if there are millions of them. This makes it possible to force atoms to move and to calculate the time which corresponds to this individual exchange process. The more unlikely an exchange process the longer is the corresponding time for this process. The concept is related to the "residence time algorithm" (Bortz et al. 1975) as discussed in Chapter 12 for nearest-neighbor exchange processes only (Müller et al. 2002). An accepted spin–flip demands a recalculation of $S(\mathbf{k}, \sigma)$ in Eq. (55). However, as shown by Lu et al. (1994), the MSCE method helps to avoid the necessity of recalculating $S(\mathbf{k},\sigma)$ after each atomic movement by directly calculating the *change* in $J_{pair}(\mathbf{k})|S(\mathbf{k},\sigma)|^2$ for each movement in real space (Lu et al. 1994). In the applied algorithm, a single KMC step is now no longer a constant real-time unit, but depends on the corresponding probability W_{tot}. A single kinetic MC step corresponds indeed to only a single exchange of one B atom with one A atom and *not* to one trial-flip for each B atom. Since the "flip channel" i is always chosen randomly and usually a large number of B atoms (typically 10^3–10^5) is considered to describe real aging processes, the probability that the same B atom is chosen in step i – when chosen already in step $(i-1)$ – is extremely small. Therefore, due to the large system size it is not necessary to forbid certain exchanges between A and B atoms, i.e., we do not have to give up the restriction that the algorithm should be based on the Markovian process.

11.3.6
Search for Ground-State Structures

In general, to find the thermodynamic ground-state structure for a solid compound at low temperatures seems to be straightforward: start from a given structure, relax all atomic positions and the cell shape, and finally select the structure with the lowest total energy. Although this procedure can be made to work (e.g., by a suitable implementation in a computer code such as VASP minimizing forces and relaxing the components of the stress tensor), it does not guarantee that the deepest energy minimum can be found. Because of symmetry, fixed numbers of atoms and large energy barriers, certain paths in structural configuration space could just be forbidden or prohibited, and – consequently – the search for the true ground-state structure is restricted to a certain class of structures. However, strategies have been developed which could at least substantially enlarge the configuration one is searching through. We will discuss such a procedure for binary compounds.

The search for the ground state of binary compounds by the standard procedure and diagrammatic approaches suffers largely from the requirement that the small group of crystallographic configurations considered must include the solution. No room is left for "surprises", but the set of possible ground states is restricted to the "usual suspects." The MSCE as introduced in Section 11.3.5 partly permits us to circumvent this problem by restricting the search to compounds with a limited number of basis atoms and calculating the formation enthalpy of *all possible configurations* via the MSCE Hamiltonian. In particular for metal alloys, where most of the observed long-range ordered ground states do not contain more than 16 basis atoms, such an ansatz appears to be a very reasonable approach. One should keep in mind that this restriction still allows for *more than about 130 000 structures*, a number which could never be treated by direct DFT calculations for all compounds. The calculation of all these formation enthalpies corresponding to the geometrically fully relaxed structures by applying the MSCE Hamiltonian can be done within about an hour on a high-performance PC.

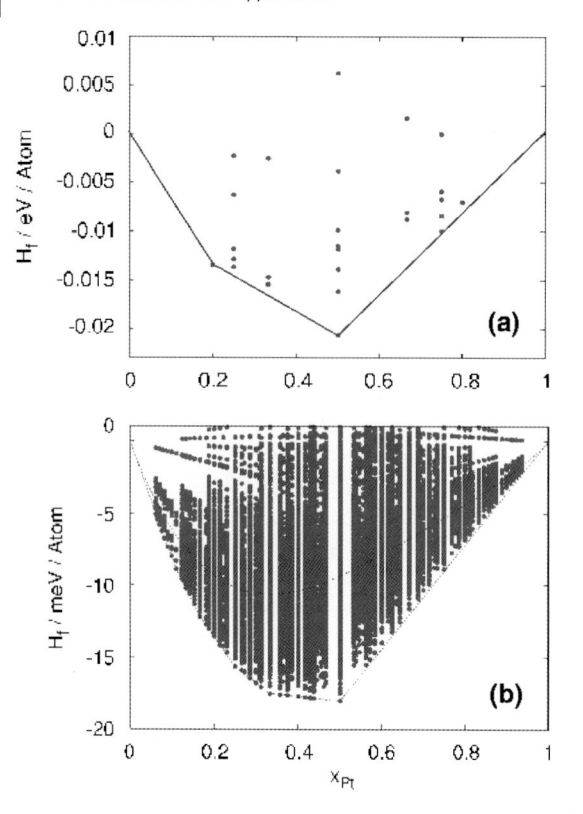

Fig. 11.24 Ground-state diagram for $T = 0$ K for the Pt–Rh system. (a) Enthalpies directly calculated from DFT calcualtions; (b) MSCE results considering all possible compounds with up to 16 basis atoms. The effective interactions were constructed from the DFT energies as shown in (a). The broken line represents the energetics of the random alloy.

As an example, Fig. 11.24 compares the ground-state diagram for the fcc-based Pt–Rh alloy system constructed (a) by simply plotting the formation enthalpies as derived from DFT calculations and (b) by using these enthalpies to construct effective interactions in order to consider all compounds with up to 16 basis atoms. Each point in the figure represents the formation enthalpy of a given structure. In case (a), the constructed ground-state line (for technical details, see Section 11.3.5) consists of two long-range ordered ground-state structures, namely the so-called $D1_a$ and CH structures (names correspond to the *Strukturbericht*). These two intermetallic compounds are displayed in Fig. 11.25: both structures are superlattices along [201] with the stacking sequence $PtRh_4$ ($D1_a$) and Pt_2Rh_2 (CH). If the ground-state search is now extended to all compounds with up to 16 basis atoms, a number of new ground states is predicted for the Rh-rich, but not for the Pt-rich side. Without going into detail, it should be emphasized that the consideration of more than 130 000 compounds leads to a drastically different

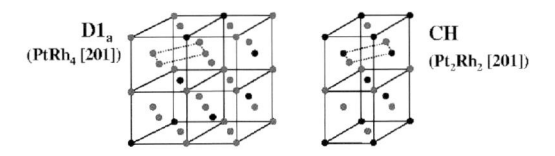

Fig. 11.25 The two ground-state structures as predicted directly by DFT enthalpies: $D1_a$ (PtRh$_4$ superlattice along [201]) and CH (Pt$_2$Rh$_2$ superlattice along [201]). As can be seen from the MSCE-based ground-state diagram (Fig. 11.24b) the extension of the configuration space to 16 basis atoms identifies $D1_a$ no longer as the ground state.

ground-state line. In particular, the $D1_a$ structure, which was predicted to be stable (Fig. 11.25a) is *not* the ground state any longer. Instead, no long-range ordered structure is found to be a ground state for a concentration of 20%. This result makes it clear that predictions with respect to the stability and formation of intermetallic compounds are only trustworthy if they are based on a huge configuration space. Besides the energetics of ordered compounds, the MSCE is also able to predict the enthalpy of the random alloy as a function of concentration in a quite straightforward manner. The corresponding curve is shown in Fig. 11.24(b) by a broken line, demonstrating that the energy of the random alloy clearly lies above the constructed ground-state line. One should, however, keep in mind that such a ground-state diagram describes the situation for $T = 0$ K (or suffiently low temperatures). Considering the fact that all energies displayed in Fig. 11.24 are within a range of 20 meV atom^{-1} it is not surprising that for the Pt–Rh system no intermetallic compounds are observed at finite temperatures. Nevertheless, such a ground-state diagram can provide important hints about what kind of substitutional short-range order is to be expected, because the short-range order in the solid solution is strongly influenced by the corresponding $T = 0$ K ground-state structure. Finally, it should be mentioned that – as discussed in Section 11.3.5 – the application of the MSCE Hamiltonian in MC simulations allows for calculations of mixing enthalpies for any arbitrary temperature and concentration. As recently shown (Kolb et al. 2006), the mixing enthalpies of Pt–Rh are negative *for all temperatures*, indicating that the assumed miscibility gap in the phase diagram of Pt–Rh for low temperatures does not exist. This result is supported by very recent experimental studies (Steiner et al. 2005).

11.3.7
Ordering and Decomposition Phenomena in Binary Alloys

As discussed in Section 11.3.5, the combination of the MSCE with MC simulations allows for a parameter-free description of materials systems on a mesoscopic scale for any arbitrary temperature. In the case of metal alloys this includes two important and characteristic properties which cannot be treated by DFT directly, namely short-range order and precipitation. This section will present examples of both phenomena. Regarding ordering, so far no definition has been intro-

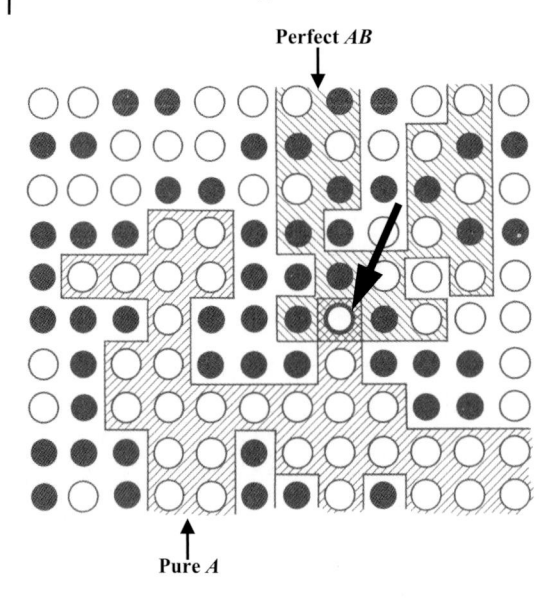

Fig. 11.26 The dilemma in describing ordering (see Ziman 1979): does the atom marked by a large arrow belong to "a cluster of pure A atoms," or to a "region of perfect AB order"?

duced of *how* to quantify the distribution of the different types of atoms over the lattice. Even more importantly a predictive model for the most probable distribution is missing. Ziman (1979) illustrated the difficulty of handling ordered zones in a disordered matrix by the help of Fig. 11.26: for the given configuration, one cannot decide whether the atom marked by an arrow belongs to a "cluster of pure A atoms" or to a "region of perfect AB order." By applying percolation theory Ziman demonstrated that almost every A atom belongs to an infinite cluster of A atoms. Paradoxically, if one is looking for ordered domains (Fig. 11.26), then almost every atom belongs to an infinite domain with perfect AB ordering. In this situation, statistical concepts (Muto and Tagaki 1955; Guttman 1956; Ziman 1979) may help: for a system consisting of N sites each surrounded by M neighbors, the probability of a bond being of AB-type is given by Eq. (60), with N_{AB} being the total number of AB-type bonds.

$$P_{AB} = \lim_{N \to \infty} \left(\frac{N_{AB}}{\frac{1}{2}MN} \right) \qquad (60)$$

The denominator gives the total number of bonds in the system. It is now easy to introduce a *nearest-neighbor correlation parameter* Γ_{AB}. Assuming that each site of the system is independently occupied by an A or B atom with probability x_A or x_B ($x_A + x_B = 1$), P_{AB} would be $2x_A x_B$. Then, Γ_{AB} is defined as the difference in Eq. (61).

$$\Gamma_{AB} = \frac{1}{2} P_{AB} - x_A x_B \tag{61}$$

Dividing Γ_{AB} by $-x_A x_B$ leads to the well-known Warren–Cowley short-range order (SRO) parameter [Eq. (62)] (Cowley 1950); see also Chapter 7.

$$\alpha_j = 1 - \frac{P_{AB}^j}{2 x_A x_B} \tag{62}$$

Here, α_j is already extended to arbitrary neighbor distances j. The sign of α_j indicates whether atoms in a given distance j prefer AB ordering ($\alpha_j < 0$) or clustering ($\alpha_j > 0$). The SRO parameter is normalized so that $-1 \leq \alpha_j \leq +1$; $\alpha_j = 0$ for all j stands for a perfect random alloy, i.e., an alloy without any atomic correlations. Since the parameters α_j (being the Fourier coefficients of the diffuse scattering intensity) can be determined from diffuse X-ray and neutron diffraction experiments (Sparks and Borie 1966; Krivoglaz 1996; Schönfeld 1999), a *quantitative* comparison between calculation and measurement is possible. For this purpose, Eq. (62) is reformulated as Eq. (63), where $P_{lmn}^{A(B)}$ is the conditional probability that – given an A atom at the origin – there is a B atom at site (lmn). For comparison with experiment, the so-called "shells" lmn are introduced which are defined by the distance between A and B atoms in terms of half lattice parameters, $\left(l\frac{a}{2}, m\frac{a}{2}, n\frac{a}{2}\right)$. For example, for an fcc lattice the nearest-neighbor (NN) distance would be described by the shell (110), the second-neighbor (NNN) distance by (200) and so on. As already mentioned, the sign of α indicates whether atoms in a given shell prefer to order ($\alpha < 0$) or to cluster ($\alpha > 0$). The SRO parameter may be written in terms of the CE pair correlations as Eq. (64) (Ozoliņš et al. 1998a),

$$\alpha_{lmn}(x) = 1 - \frac{P_{lmn}^{A(B)}}{x} \tag{63}$$

$$\alpha_{lmn}(x) = \frac{\langle \overline{\Pi}_{lmn} \rangle - q^2}{1 - q^2} \tag{64}$$

where $q = 2x - 1$ and $\langle \overline{\Pi}_{lmn} \rangle$ denotes the pair correlation function (according to Eq. (48)) for shell (lmn). In diffraction experiments the diffuse scattering due to SRO is proportional to the lattice Fourier transform of $\alpha_{lmn}(x)$ (Sparks and Borie 1966; Krivoglaz 1996).

$$\alpha(x, \mathbf{k}) = \sum_{lmn}^{n_R} \alpha_{lmn}(x) e^{i \cdot \mathbf{k} \cdot \mathbf{R}_{lmn}} \tag{65}$$

In this relationship, n_R is the number of real-space shells used in the transform. Equation (64) in combination with Eq. (65) allows the comparison of both the

Table 11.9 Warren–Cowley SRO parameter α_{lmn} as derived from combined CE and MC calculations, in comparison with experimental results (Reinhard et al. 1990) for $Cu_{0.689}Zn_{0.311}$ at $T = 473$ K.

(lmn)	α_{lmn}^{exp}	α_{lmn}^{theo}	(lmn)	α_{lmn}^{exp}	α_{lmn}^{theo}	(lmn)	α_{lmn}^{exp}	α_{lmn}^{theo}
000	1.0831	1.0000	400	0.0296	0.0279	510	−0.0107	−0.0186
110	−0.1373	−0.1689	330	−0.0134	−0.0211	521	−0.0019	−0.0092
200	0.1490	0.1863	411	0.0141	0.0306	440	−0.0050	−0.0104
211	0.0196	0.0196	420	0.0050	0.0825	433	0.0038	0.0092
220	0.0358	0.0883	332	−0.0005	−0.0050	530	−0.0066	−0.0057
310	−0.0404	−0.0453	422	−0.0050	−0.0050	442	−0.0084	−0.0145
222	−0.0077	0.0371	431	0.0068	0.0148	600	0.0130	0.0017
321	−0.0036	−0.0132						

measured and calculated diffuse diffraction patterns (in reciprocal space) and the SRO parameters (in real space).

For a quantitative comparison of the computed SRO with experiment, the Warren–Cowley SRO parameters α_{lmn} are calculated [Eq. (64)] for the first 22 shells of an $Cu_{0.689}Zn_{0.311}$ alloy at $T = 473$ K and compared to data from neutron scattering experiments (Reinhard et al. 1990) (see also Chapter 12), shown in Table 11.9. Considering the fact that the experimental error of α_{000} amounts to as much as 8% (since $\alpha_{000} = 1.000$ by definition), the predicted and measured values agree very well. As can be seen, α_{110} is negative, indicating that Zn atoms prefer Cu atoms as nearest neighbors. Furthermore, all α_{2n00} are positive, while all $\alpha_{(2n-1)10}$ are negative.

In experiments, the SRO parameters, Table 11.9, were used to construct a real-space image of the alloy by using them as input for an inverse MC approach in order to obtain characteristic interactions (Reinhard et al. 1990). In the case of α-brass, real-space images are of special interest for the following reason: as apparent from Table 11.9, all SRO parameters described by $(lmn) = (2n; 0; 0)$ are positive, while all SRO parameters described by $(2n − 1; 1; 0)$ are negative. This property should lead to characteristic chains of Zn atoms along the [100] direction. Figure 11.27 compares the real-space structure deduced from experiment (Reinhard et al. 1990) and from "CE plus MC" calculations (Zn atoms are displayed as dark atoms). In both cases, chains of Zn atoms are visible along [100], indicating that short-range order is essential for a quantitatively correct description of the physical properties of the disordered solid solution of α-brass.

Quenching a solid solution of a metal alloy deep into the two-phase region of the phase diagram, followed by sample aging, leads to the formation of characteristic nanoclusters, the nuclei of the precipitating phase, which define an important part of the microstructure of many alloy systems (see Chapter 7). The early stage of these reactions typically involves the formation of coherent precipitates that adopt the crystallographic lattice of the alloy from which they emerge (Guin-

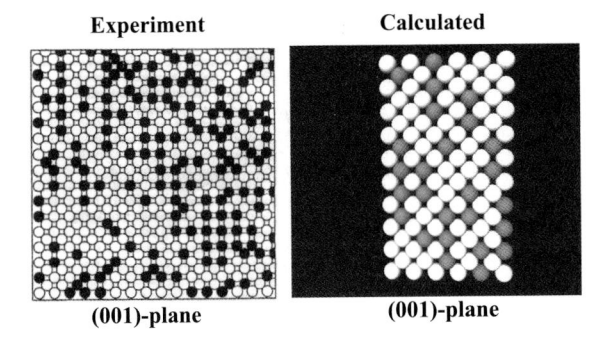

Fig. 11.27 Visualization of a (001) plane of α-brass (cut through the crystal) for $T = 473$ K. While the picture on the left results from a model crystal based on diffuse neutron scattering experiments (Reinhard et al. 1990), the right-hand picture is the result of MC simulations using ΔH_{CE}. In both cases, chains of Zn atoms (dark atoms) along [100] can be seen, indicating that SRO is present, and therefore the observed solid solution cannot be described as a random alloy.

ier 1959). Coherent precipitates have practical relevance, as they impede dislocation motion, and thus lead to "precipitation hardening" in many alloys (Guinier 1959; Khachaturyan 1983; Cohen 1986). Therefore, their size versus shape distribution as a function of temperature and aging time is of particular interest. Despite their importance, precipitate microstructures were amenable for the first time just a few years ago (Wolverton 1999) to first-principles theories, since their description requires "unit cells" containing 10^3–10^6 atoms or more. The concept presented in Section 11.3.5 gives access to such huge model systems.

The shape of precipitates is controlled by two competing energies (Khachaturyan 1983): interfacial or chemical energy E_{chem} and strain energy E_{CS}. While the former leads to a compact shape, the latter leads to a flattening along the elastically soft direction of the precipitate. The MSCE Hamiltonian allows for a separation into these two characteristic energy parts: as already mentioned in Section 11.3.5, the first two terms of Eq. (55) include information about the strength and importance of the indvidual pair and multibody interactions and therefore represent the chemical energy of the system. The last term, however, reflects the elastic properties of the alloy. This separation is used to analyze the ratio between chemical energy E_{chem} and strain energy E_{CS} as a function of the precipitate size. As examples, Al-rich fcc-based Al–Li and Al–Zn alloys are discussed. Their precipitate distributions for a given aging time and temperature are shown in the upper part of Fig. 11.28, where only the Li and Zn atoms are displayed. The MC cell used for the simulations consists of $56 \times 56 \times 56 = 175\,616$ atoms in order to achieve sufficient statistics.

In the case of Al–Li, precipitates of the size considered up to 5 nm never flatten, but always possess a spherical shape, in excellent agreement with experimental results (Lendvai et al. 1989; Sato and Kamino 1991). This behavior becomes understandable by analyzing E_{chem}/E_{CS} for different precipitate sizes as shown in

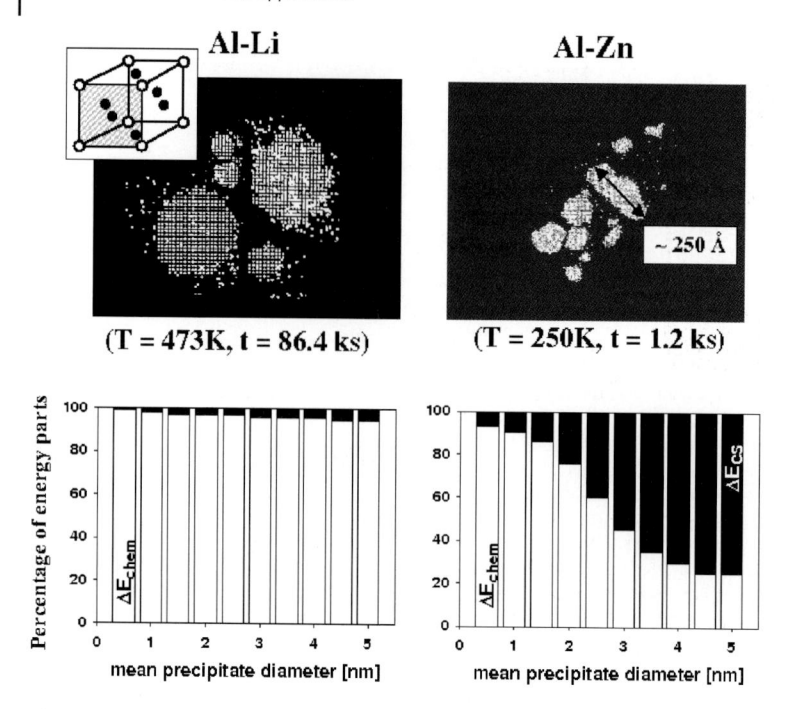

Fig. 11.28 Size and shape distributions of precipitates in Al-rich fcc-based Al–Li and Al–Zn alloys (no Al atoms are shown) and their corresponding percentages of strain and chemical energy as functions of the precipitate size. Precipitates of Al–Li form the $L1_2$ structure as shown in the top left-hand corner.

the lower part of Fig. 11.28 for a sample containing 9.7% Li atoms: for all sizes the chemical energy E_{chem} (white bars) clearly predominates over the strain energy E_{CS} (black bars). Following experimental results, it must be expected that this dominance of E_{chem} over E_{CS} holds even for precipitates about 50 nm in diameter, because they still possess a spherical shape. The Li atoms seem to form a simple cubic lattice. This is due to the fact that the precipitates themselves show an Al_3Li stoichiometry forming the $L1_2$ structure sketched at the top left-hand corner of Fig. 11.28: while all the corners of the unit cell are occupied by Li atoms, all the faces are occupied by Al atoms. Since only the Li atoms are displayed in the real-space image, they form a simple cubic lattice. This observation also makes clear that there are practically no antiphase boundaries within the Al_3Li precipitates which would demand the occupation of Al sites by Li atoms.

Contrarily to Al–Li, for Al–Zn a strong size dependence of the precipitate shape is found: Zn precipitates below 2.5 nm are more spherical, i.e., chemically dominated, while larger precipitates become more and more ellipsoidal (strain-dominated) (Müller et al. 2000). It is interesting to note that for low temperatures a third component becomes important, namely the anisotropy of the chemical

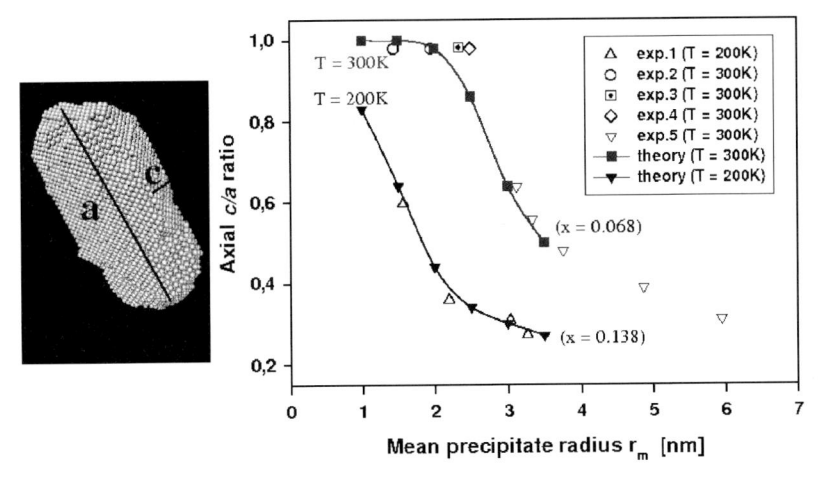

Fig. 11.29 Shape (c/a) vs. size relationship of Zn precipitates for two different temperatures. The lines denote the results from calculations (Müller et al. 2001), the open points are taken from different experimental studies: expts. 1–5 correspond to previous reports (Laslaz and Guyot 1977; Bubeck et al. 1985; Fumeron et al. 1980; Deguercy et al. 1982; Gerold et al. 1987, respectively).

part which controls quantitatively to what extent the precipitates facet at low temperatures.

Another remarkable feature of the coherent Zn precipitates is the fact that their short axis is always along the [111] (and symmetrically equivalent) directions. Indeed, at a first glance, this appears a bit of a surprise, because most fcc elements are elastically soft along the [100] direction, and consequently hard along [111]. However, fcc Zn is an unusual phase. While Zn is stable in the hcp structure, fcc Zn shows an instability when deformed rhombohedrally along [111] (Müller et al. 1999). As a consequence, fcc Zn precipitates flatten along this direction (Müller et al. 2000). This feature allows the definition of a c/a ratio, and therefore a quantitative measure for the description of the precipitate shape as used in many experimental studies and shown schematically on the left-hand side of Fig. 11.29: while a represents the long axis of the ellipsoid (perpendicular to [111]), c is its thickness (parallel to [111]). The size is given by the radius of the associated sphere having the same volume as the corresponding precipitate. Figure 11.29 compares the experimental size–shape relationship for two different aging times and concentrations (Laslaz and Guyot 1977; Fumeron et al. 1980; Deguercy et al. 1982; Bubeck et al. 1985; Gerold et al. 1987) with those predicted by the calculations. For both temperatures, the agreement is excellent, i.e., the calculational approach allows for a *quantitative* prediction of the size versus shape versus temperature relationship of coherent precipitates (Müller et al. 2001).

The examples above make it clear that the combination of DFT with MSCE and MC is probably one of the most successful approaches to studying binary alloy properties without any empirical parameters. At the moment, the limitation of the presented access is given by the underlying lattice which, for example, does not permit study of melting processes. Regarding ordering and decomposition phenomena in the solid phase, the method allows for a quantitative prediction of experimental data. The method presented is by no means restricted to ordering phenomena of a bulk phase, but can also be applied to, e.g., surface problems, such as adsorbate systems (Drautz et al. 2003), or surface segregation (Wieckhorst et al. 2004).

11.4
Outlook

The intention of the present chapter has been to give a flavor of density functional theory methods and some recent applications for studying properties of materials, in particular of intermetallic compounds. The central quantity we have focused on was the total energy of a system of atoms of a periodic solid. One should, however, keep in mind that for each evaluation of the total energy the complete electronic structure is available, because the Kohn–Sham orbitals (according to Eq. (10)) defining the electron density must always be calculated. The electronic structure can be used to analyze the bonding, the magnetic structure, or – going further – to calculate matrix elements and more intricate properties such as susceptibilities, optical spectra, and so on.

As demonstrated, for the actual calculation of "hard numbers" the approximations to the many-body electron–electron interactions (i.e., to the exchange–correlation functional) must be reasonably accurate, which – concerning ground-state energies – in many cases they are. For example, to derive enthalpies of formation for intermetallic compounds with a reasonably large number of atoms per unit cell (about 100 or fewer, to be conservative) the calculated enthalpies of formation (as discussed in Section 11.3.1) are at least as accurate as reliable experiments by calorimetric methods. Usually, the calculation is much faster and cheaper provided a reliable and fast computational package – as mentioned in Section 11.2.2 – is at hand. Standard density functional theory does not include the concept of temperature and is therefore applicable at $T = 0$ K only or, pragmatically speaking, at low temperatures. However, the most important temperature effects (at least for temperatures well below the melting point) can now be included by the determination of the vibrational states and derived quantities. As discussed in Section 11.2.4, within the harmonic approximation the phonon spectra may be derived from density functional theory potentials by calculating suitable force fields. Thus, consistent temperature-dependent structural phase stabilities are derived.

A first natural step beyond strict stoichiometries of particular ordered compounds are intrinsic defects, such as vacancies and antisites. Based on supercell calculations, as elaborated in Section 11.3.2 for binary compounds, chemical potentials can be derived which are then included into a statistical mechanics machinery in order to obtain temperature- and concentration-dependent properties of point-like defects. The migration of defects and impurities is dicussed in Section 11.3.3 and temperature-dependent diffusion coefficients are derived within transition state theory based on ab-initio total energy and phonon calculations. Extended defects such as grain boundaries are also within the scope, which is demonstrated in Section 11.3.4 by analyzing impurity effects on grain boundary cohesion and intergranular fracture.

A rather challenging task is to study structural phase stabilities over a wide range of compositions. For binary alloys, a very successful strategy is at hand in terms of the cluster expansion, described in Section 11.3.5. The basic concept is to structurally decompose the solid into small building blocks – clusters of atoms – considered to be glued together by cluster interactions. Provided these interactions are known and a finite cluster expansion reaches convergence, then a large variety of different structures can be built with numerically well defined energies. In fact, the cluster interactions can be derived from a suitably large database of supercell calculations performed in the framework of density functional theory. Such a strategy could assist to find the ground-state phase diagram as a function of composition, as discussed in Section 11.3.6. Once the cluster interactions are known, one could go far beyond the ground-state properties by making use of Monte Carlo techniques. Systems of millions of atoms might be treated without giving up the accuracy and predictive power of ab-initio methods. Thus, the most probable atomic distribution for a given temperature and concentration could be found, enabling for instance the study of precipitation formation, as discussed in Section 11.3.7. System sizes have already reached such large dimensions that they are tractable by continuum theory approaches. Therefore, with the atomistic-based strategy of the cluster expansion in combination with density functional theory approaches, length and time scales can be bridged, reaching from nanoscale atomic bonds to mesoscale microstructures and from picosecond atomic movements to realistic time scales of microstructural evolution. We are convinced that this development will be one of the hot topics of the future, both for the development of advanced simulation techniques and for applications of density functional theory based methods in alloy physics.

Acknowledgements

Work supported by the Austrian Science Fund FWF Project No. 16957.

References

Allard, S. (Ed.), **1969**. *Metals, Thermal and Mechanical Data*. Pergamon Press, New York, Vol. 16 of *International Tables of Selected Constants*.

Bakker, H., Bonzel, H., Bruff, C., Dayananda, M., Gust, W., Horváth, J., Kaur, I., Kidson, G., LeClaire, A., Mehrer, H., Murch, G., Neumann, G., Stolica, N., Stolwijk, N.

(Eds.), **1990**. Landolt–Börnstein. Vol. 26. Springer, Berlin.

Baroni, S., Giannozzi, P., Testa, A., **1987**. Phys. Rev. Lett. 58, 1861.

Baroni, S., de Gironcoli, S., dal Corso, A., Giannozzi, P., **2001**. Rev. Mod. Phys. 73, 515.

Blöchl, P., **1994**. Phys. Rev. B 50, 17 953.

Born, M., Huang, K., **1954**. *Dynamical Theory of Crystal Lattices*. Oxford University Press, Oxford.

Bortz, A. B., Kalos, M. H., Lebowitz, J. L., **1975**. J. Comp. Phys. 17, 10.

Bottomley, D. J., Fons, P., **1996**. J. Cryst. Growth 160, 406.

Brauer, E., Doerr, R., Gruner, R., Rauch, F., **1981**. Corros. Sci. 21, 449.

Bubeck, E., Gerold, V., Kostorz, G., **1985**. Cryst. Res. Technol. 20, 97.

Cacciamini, G., Riani, P., Borzone, G., Parodi, N., Saccone, A., Ferror, R., Pisch, A., Schmid-Fetzer, R., **1999**. Intermetallics 7, 101.

Chen, X.-Q., Witusiewicz, V. T., Podloucky, R., Rogl, P., Sommer, F., **2003**. Acta Mater. 51, 1239.

Chen, X.-Q., Wolf, W., Podloucky, R., Rogl, P., **2004a**. Intermetallics 12, 59.

Chen, X.-Q., Wolf, W., Podloucky, R., Rogl, P., Marsman, M., **2004b**. EuroPhys. Lett. 67, 807.

Chen, X.-Q., Wolf, W., Podloucky, R., Rogl, P., **2005a**. Phys. Rev. B 71, 174 101.

Chen, X.-Q., Wolf, W., Podloucky, R., Rogl, P., Marsman, M., **2005b**. Phys. Rev. B 72, 054 440.

Cochran, W., **1959**. Phys. Rev. Lett. 2, 495.

Cohen, J. B., **1986**. *Solid State Physics*. Vol. 39. Academic Press, New York, p. 131.

Conolly, J. W. D., Williams, A. R., **1983**. Phys. Rev. B 27, 5169.

Costas, L. P., **1963**. US Atomic Energy Commission Report DP-813.

Cowley, J. M., **1950**. J. Appl. Phys. 21, 24.

Davenport, T., Zhou, L., Trivisonno, J., **1999**. Phys. Rev. B 59, 3421.

De Cicco, P. D., Johnson, F. A., **1969**. Proc. R. Soc. London, Ser. A 310, 111.

Deguercy, J., Denanot, M. F., Fumeron, M., Guillot, J. P., Caisso, J., **1982**. Acta Metall. 30, 1921.

Dick, B. G., Overhauser, A. W., **1958**. Phys. Rev. 112, 90.

Drautz, R., Singer, R., Fähnle, M., **2003**. Phys. Rev. B 67, 035 418.

Every, A. G., McCurdy, A. K. (Eds.), **1992**. Landolt–Börnstein. Vol. 29. Springer, Berlin.

Eyring, H., **1935**. J. Chem. Phys. 3, 107.

Ferreira, L. G., Wei, S.-H., Zunger, A., **1991**. J. Supercomp. Appl. 5, 34.

Franchini, C., Bayer, V., Podloucky, R., Paier, J., Kresse, G., **2005**. Phys. Rev. B 72, 045 132.

Frank, W., Elsässer, C., Fähnle, M., **1995**. Phys. Rev. Lett. 74, 1791.

Fumeron, M., Guillot, J. P., Dauger, A., Caisso, J., **1980**. Scripta Metall. 14, 189.

Gerold, V., Siebke, W., Tempus, G., **1987**. Phys. Stat. Sol. (a) 104, 213.

Giannozzi, P., de Gironcoli, S., Pavone, P., Baroni, S., **1991**. Phys. Rev. B 43, 7231.

Gonze, X., Vigneron, J. P., **1989**. Phys. Rev. B 39, 13 120.

Griffith, A. A., **1920**. Phil. Trans. Roy. Soc. (London) A 221, 163.

Guinier, A., **1959**. *Solid State Physics*. Vol. 9. Academic Press, New York, p. 293.

Guttman, L., **1956**. *Solid State Physics*. Vol. 3. Academic Press, New York, p. 145.

Hamann, D. R., Wu, X., Rabe, K. M., Vanderbilt, D., **2005**. Phys. Rev. B 71, 035 117.

Hart, G. L. W., Blum, V., Walorski, M. J., Zunger, A., **2005**. Nat. Mater. 4, 391.

He, Y., Schwarz, R. B., Migliori, A., Whang, S. H., **1995**. J. Mater. Res. 10, 1187.

Helme, B. G., King, P. J., **1978**. J. Mater. Sci. 13, 1487.

Henkelman, G., Jóhannesson, G., Jónsson, H., **2000**. In: *Progress on Theoretical Chemistry and Physics*, edited by S. Schwartz, Kluwer Academic Publishers, New York.

Hohenberg, P., Kohn, W., **1964**. Phys. Rev. 136, B864.

Huntington, H., **1958**. *Solid State Physics*. Vol. 7. Academic Press, New York, p. 213.

Ikehata, H., Nagasako, N., Furuta, T., Fukumoto, A., Miwa, K., Saito, T., **2004**. Phys. Rev. B 70, 174113.

Jacobsen, J., Jacobsen, K. W., Norskov, J. K., Stoltze, P., **1995**. Phys. Rev. Lett. 74, 2295.

Jónsson, H., Mills, G., Jacobsen, K. W., **1998**. *Classical and Quantum Dynamics in Condensed Phase Simulations*. World Scientific, Singapore, p. 385.

Kamm, G. N., Alers, G. A., **1964**. J. Appl. Phys. 35, 327.

Keating, P. N., **1966**. Phys. Rev. 145, 637.

Khachaturyan, A. G., **1983**. *Theory of Structural Transformations in Solids*. John Wiley, New York.

Kohanoff, J., **1994**. *Comput. Mater. Sci.* 2, 221.

Kohanoff, J., Andreoni, W., Parrinello, M., **1992**. *Phys. Rev. B 46*, 4371.

Kohn, W., Sham, L. J., **1965**. *Phys. Rev. 140*, A1133.

Kolb, B., Müller, S., Botts, D. B., Hart, G. L. W., **2006**. *Phys. Rev. B 74*, 144 206.

Krčmar, M., Fu, C. L., **2003**. *Phys. Rev. B 68*, 134 110.

Kresse, G., Furthmüller, J., Hafner, J., **1995**. *Europhys. Lett. 32*, 729.

Krivoglaz, M. A., **1996**. *X-Ray and Neutron Diffraction in Nonideal Crystals*. Springer, Berlin.

Kunc, K., Martin, R. M., **1982**. *Phys. Rev. Lett. 48*, 406.

Laks, D. B., Ferreira, L. G., Froyen, S., Zunger, A., **1992**. *Phys. Rev. B 46*, 12 587.

Laslaz, G., Guyot, P., **1977**. *Acta Metall. 25*, 277.

Lendvai, J., Gudladt, H.-J., Wunderlich, W., Gerold, V., **1989**. *Z. Metallkde. 80*, 310.

Le Page, Y., Saxe, P., **2001**. *Phys. Rev. B 63*, 174 103.

Le Page, Y., Saxe, P., **2002**. *Phys. Rev. B 65*, 104 104.

Le Page, Y., Saxe, P., Rodgers, J. R., **2002**. *Phys. Stat. Sol. (b) 229*, 1155.

Lu, Z. W., Laks, D. B., Wei, S. H., Zunger, A., **1994**. *Phys. Rev. B 50*, 6642.

Marple, S. L., **1987**. *Digital Spectral Analysis with Applications*. Prentice-Hall, Englewood Cliffs, NJ.

Martin, R. M., **1969**. *Phys. Rev. 186*, 871.

Martin, R. M., **2004**. *Electronic Structure: Basic Theory and Practical Methods*. Cambridge University Press.

Martyna, G. J., Klein, M. L., Tuckerman, M., **1992**. *J. Chem. Phys. 97*, 2635.

Methfessel, M., **1988**. *Phys. Rev. B 38*, 1537.

Methfessel, M., Rodriguez, C. O., Andersen, O. K., **1989**. *Phys. Rev. B 40*, 2009.

Metropolis, N., Rosenbluth, A. W., Rosenbluth, M. N., Teller, A. H., Teller, E., **1953**. *J. Chem Phys. 21*, 1087.

Meyer, B., Fähnle, M., **1999**. *Phys. Rev. B 59*, 6072.

Müller, S., **2003**. *J. Phys.: Cond. Matter 15*, R1429.

Müller, S., Wang, L.-W., Zunger, A., Wolverton, C., **1999**. *Phys. Rev. B 60*, 16 448.

Müller, S., Wang, L.-W., Zunger, A., **2002**. *Model. Simul. Mater. Sci. Eng. 10*, 131.

Müller, S., Wolverton, C., Wang, L.-W., Zunger, A., **2000**. *Acta Mater. 48*, 4007.

Müller, S., Wolverton, C., Wang, L.-W., Zunger, A., **2001**. *Europhys. Lett. 55*, 33.

Müller, S., Wolf, W., Podloucky, R., **2007**. Unpublished.

Murray, J. L., **1985**. *Int. Met. Rev. 30*, 211.

Muto, T., Tagaki, Y., **1955**. *Solid State Physics*. Vol. 1. Academic Press, New York, p. 193.

Ozoliņš, V., Wolverton, C., Zunger, A., **1998a**. *Phys. Rev. B 57*, 4816.

Ozoliņš, V., Wolverton, C., Zunger, A., **1998b**. *Phys. Rev. B 57*, 6427.

Papazoglou, T. P., Heppworth, M. T., **1968**. *Trans. TMS-AIME 242*, 682.

Parlinski, K., Li, Z. Q., Kawazoe, Y., **1997**. *Phys. Rev. Lett. 78*, 4063.

Perdew, J. P., Wang, Y., **1992**. *Phys. Rev. B 45*, 13 244.

Pick, R. M., Cohen, M. H., Martin, R. M., **1970**. *Phys. Rev. B 1*, 910.

Rahman, A., **1964**. *Phys. Rev. 136*, A405.

Rasamny, M., Weinert, M., Fernando, G. W., Watson, R. E., **2001**. *Phys. Rev. B 64*, 144 107.

Reinhard, L., Schönfeld, B., Kostorz, G., Bührer, W., **1990**. *Phys. Rev. B 41*, 1727.

Sanchez, J. M., Ducastelle, F., Gratias, D., **1984**. *Physica A 128*, 334.

Sato, T., Kamio, A., **1991**. *Mater. Sci. Eng. A 146*, 161.

Schönfeld, B., **1999**. *Prog. Mater. Sci. 44*, 435.

Schröder, U., **1966**. *Solid State Commun. 4*, 347.

Sparks, C. J., Borie, B., **1966**. In: *Local Arrangements Studied by X-ray Diffraction*, edited by Cohen, J. B., and Hilliard, J. E., Met. Soc. Conf. 36, 5. Gordon and Breach, New York.

Steiner, C., Schönfeld, B., Portmann, M. J., Kompatscher, M., Kostorz, G., Mazuelas, A., Metzger, T., Kohlbrecher, J., Deme, B., **2005**. *Phys. Rev. B 71*, 104 204.

Toda, M., Rubo, R., Saito, N., **1983**. *Statistical Physics I*. Springer, Berlin.

Verlinden, J., Gijbels, R., **1980**. *Adv. Mass Spectrom. 8A*, 485.

Vosko, S. H., Wilk, L., Nusair, M., **1980**. *Can J. Phys. 58*, 1200.

Weber, W., **1977**. *Phys. Rev. B 15*, 4789.

Wei, S., Chou, M. Y., **1992**. *Phys. Rev. Lett. 69*, 2799.

Wei, S., Chou, M. Y., **1994**. *Phys. Rev. B 50*, 2221.

Wen, C. J., Weppner, W., Boukamp, B. A., Huggins, R. A., **1980**. *Metall. Trans. B 11*, 131.

Wert, C., Zener, C., **1949**. *Phys. Rev. 76*, 1169.

Wieckhorst, O., Müller, S., Hammer, L., Heinz, K., **2004**. *Phys. Rev. Lett. 92*, 195 503.

Wimmer, E., Krakauer, H., Weinert, M., Freeman, A., **1981**. *Phys. Rev. B 24*, 864.

Wimmer, E., Wolf, W., Sticht, J., Saxe, P., Geller, C. B., Najafabadi, R., Young, G. A., **2007**. Unpublished.

Wolverton, C., **1999**. *Phil. Mag. Lett. 79*, 683.

Wolverton, C., Ozoliņš, V., Zunger, A., **2000**. *J. Phys.: Condens. Matter 12*, 2749.

Wood, D. M., Zunger, A., **1988**. *Phys. Rev. Lett. 61*, 1501.

Yin, M. T., Cohen, M. L., **1982**. *Phys. Rev. B 26*, 3259.

Young, G. A., Najafabadi, R., Strohmayer, W., Baldrey, D. G., Hamm, W. L., Harris, J., Sticht, J., Wimmer, E., **2003**. In: *Eleventh International Conference on Environmental Degradation of Materials in Nuclear Power Systems – Water Reactors.* American Nuclear Society, p. 758.

Young, G. A., Scully, J. R., **1998**. *Acta Mater. 46*, 6337.

Ziman, J. M., **1979**. *Models of Disorder.* Cambridge University Press, Cambridge.

Zunger, A., **1994a**. In: Hurle, T. (Ed.), *Handbook of Crystal Growth.* Vol. 63. Elsevier, Amsterdam, p. 99, and references therein.

Zunger, A., **1994b**. In: Turchi, P. E. A., Gonis, A. (Eds.), NATO ASI on: *Statics and Dynamics of Alloy Phase Transformations.* Plenum Press, New York, p. 361.

12
Simulation Techniques

Ferdinand Haider, Rafal Kozubski, and T.A. Abinandanan

12.1
Introduction

Availability of computers opened completely new directions for the development of computational methods in all fields of science, but in particular in physics and in materials science. Using quantum mechanical models and refined computation methods, one is able to predict the electron structure of complicated alloys and thus the total binding energy of a given structure and of defects therein. Additionally, it is possible to compute forces between atoms, if there are deviations from equilibrium. Nevertheless until now it has been difficult to predict high-temperature stability, dynamics, and kinetics in an alloy.

In this chapter we present simulation techniques on a somewhat coarser scale, that is, on an atomistic or even continuum level. In molecular dynamics simulations (presented in Section 12.2) the ensemble of atoms forming the material is treated as a set of interacting classical particles. The interactions are often derived from quantum mechanical considerations, but are taken as given in the form of so-called phenomenological potentials. In this framework the dynamics of atoms (e.g., lattice vibrations) is correctly treated, but neglects any deviations from the Born–Oppenheimer approximation, that is, any influence of the atomic positions on the electronic system.

A further level of coarse graining is taken in Monte Carlo simulations (presented in Section 12.3 of the chapter). Here the atoms are fixed to rigid lattice positions, and the microstructure evolves via atomic jumps onto neighboring sites. The microscopic degrees of freedom are taken into account only as a heat bath, i.e., as a given temperature or thermal energy of the atoms.

Section 12.4 describes phase field models, a continuum technique which has become very popular during the last few years. Here, instead of considering discrete atoms, only fields of parameters describing the microstructure of an alloy are studied. Examples are the local concentration and local order parameter, but also crystallinity, orientation, or others. To follow the spatial and/or temporal evolution of a sample, one has to make assumptions on the underlying thermodynamics and mobility of the system.

The methods described here show limitations in the range of times accessible to the simulation as well as the maximum system size, and thus also limit the applicability of the respective method to "real world" problems. These limitations get less strict in the sequence of methods – while molecular dynamics imposes the strictest limits on both time and size, in Monte Carlo methods studies of the long-term evolution of an alloy are feasible, and in phase field models even complicated phenomena like solidification, sintering, or recrystallization can be studied.

The hierarchy of methods described in this chapter is far from being complete: on the fine scale, methods which couple electronic and structural degrees of freedom like the Car–Parrinello method (Car and Parrinello 1985) are missing, while on the continuum scale, the very powerful finite element (FEM) techniques for the study of external or internal fields like stress, temperature, or others are missing.

12.2
Molecular Dynamics Simulations

Molecular dynamics was originally developed in the 1950s to understand the properties of hard-sphere systems [Alder and Wainright 1957, 1959]. The first studies using continuous potentials were published in the 1960s, where problems of radiation damage were studied. Today it is a very common technique in computer simulation of materials – a literature database search using the keywords "molecular dynamics" and "alloys" gave 1080 hits, and for the period of January 2005–July 2005 still nearly 70. It is applied to all kinds of problems (see below), examples being radiation damage, properties of molten alloys, defect properties, and many more. General introductions can be found in many papers (e.g., Ercolessi 1997) and books (Rappaport 1995; Allen and Tildesley 1987).

12.2.1
Basic Ideas

Molecular dynamics is a simulation technique for studying the motion of a set of particles (in most cases single atoms) on a classical level. This may sound contradictory, since, as is well known, the behavior of atoms and their interaction are governed by quantum mechanics. Nevertheless, a simple estimate shows that, even though the interaction of atoms via their valence electrons is in principle a quantum mechanical problem, thanks to the Born principle one may to a good approximation decouple the changes in the electronic system (which occurs on a time scale of femtoseconds) from those of the atoms, occurring on a time scale of picoseconds. The basic explanation for that decoupling is based on the large mass ratio of electrons to atoms, roughly speaking 10^{-5}. The characteristic times of each process scale like the square root of the respective masses, so are different by a factor of 10^2–10^3.

The elementary idea of molecular dynamics is to integrate numerically the classical equations of motion (Newton's equations) for a large set of atoms, as in Eq. (1), where x_i denotes the location of atom i, m_i its mass. The force $f(x_i)$ acting on atom i is the sum of all the forces exerted by other atoms on the atom under consideration and leads to an acceleration of the atom.

$$m_i \ddot{x}_i = f(x_i) \tag{1}$$

After integration of the set of equations (1), one obtains a trajectory $x_i(t)$ in space and time for all atoms (which naturally includes also the time derivatives, i.e., the momenta p_i or velocities $\dot{x}_i$ for all atoms; thus it is a complete description in a 6N-dimensional phase space for a system containing N atoms). Of course, the integration includes any microscopic motion of an atom, i.e., all lattice vibrations and relaxations. To realize this, a time step for integration must be short compared to any atomic motion in the lattice – that is, in the order of femtoseconds.

In contrast to Monte Carlo methods, which introduce an element of stochasticity to simulate a heat bath in the broadest sense, MD is (at least in principle) a completely deterministic simulation technique. Nevertheless, since a simulation yields the evolution of the system in phase space, it can be used to compute any kind of thermodynamic averages of observables. Besides this, MD methods can be (and are) also used, but perhaps not mainly, for simulations of nonequilibrium processes such as, for example, irradiation effects, plastic deformation, fracture, etc.

Molecular dynamics is a powerful technique, but of course shows certain limitations. The first is a restriction of the method itself – treating atoms as classical particles requires that any quantum mechanical effects can be neglected. This holds in most cases, but can be a problem, for example, for light atoms like hydrogen, where tunneling can play a role in diffusion properties. The next problem (discussed in more detail below) is whether the forces give a realistic description of the physics of the system. The conventional approach – using a given interaction force or potential between the atoms – is restricted to cases where the electronic structure of the solid under consideration does not change during the process studied. This can be violated for cases where, for example, during a structural phase transition the electronic structure changes. An extreme case would be the transition of graphite to diamond, with a transition from sp^2 bonds to sp^3 bonds.

More technical restrictions are limits in space and time. The typical system usually consists of several thousand atoms (up to millions) and, more severely, since a time step is typically in the order of femtoseconds, the time span which can be treated in MD is limited to nanoseconds (up to microseconds). Restrictions in space (that is, finite size effects) can be partly overcome by using periodic boundary conditions, if the problem under consideration is appropriate. To overcome the limitations in time, there have been proposals for the use of "accelerated algorithms," which try to bridge the potentially very long time between "important" events (such as, for example, lattice vibrations until a diffusive jump occurs) either by an artificially increased temperature or by a lowered barrier (Voter 1997; Sorensen and Voter 2000; Voter et al. 2002).

12.2.2
Atomic Interaction, Potential Models

In order to perform the MD integration, the most important step is to set up a model for the interaction forces. These in most cases are derivatives of the potential energy with respect to atomic displacements, as expressed by Eq. (2), where $f(x_i)$ denotes the force on the atom at position x_i, $V(\{x\})$ the potential energy of the system with atoms at positions $\{x\} = (x_1, x_2, \ldots, x_3)$, and ∇_{x_i} the derivative with respect to atomic position x_i. The potential energy can be obtained with different degrees of sophistication, i.e., with more or less physical justification.

$$f(x_i) = -\nabla_{x_i} V(\{x\}) \tag{2}$$

12.2.2.1 Pairwise Interaction

The simplest choice for the potential is to assume pairwise interactions [Eq. (3)].

$$V(\{x\}) = \frac{1}{2} \sum_i \sum_{j>i} \phi(|x_i - x_j|) \tag{3}$$

An example of a simple pair potential is the so-called Lennard-Jones (LJ) potential (Lennard-Jones 1931), [Eq. (4)], which gives a reasonably realistic description for the interaction in solid noble gases; otherwise (in most cases of interest) it is a model potential, suitable for qualitative but not for quantitative studies:

$$V_{LJ}(r) = V_0 \left(\left(\frac{r_0}{r}\right)^{12} - 2\left(\frac{r_0}{r}\right)^6 \right) \tag{4}$$

The minimum of the potential is at $r = r_0$, and the depth is given by V_0. It is attractive for large distances and strongly repulsive at short distances. For practical reasons, a truncated LJ potential is often used, which is artificially forced to zero beyond a certain distance R_c [Eq. (5)].

$$V(r) = \begin{cases} V_{LJ}(r) - V_{LJ}(R_c) & \text{if } r \leq R_c \\ 0 & \text{else} \end{cases} \tag{5}$$

One particular problem of pair potentials (which makes them rather unsuitable in most real cases) is that they fulfill the Cauchy relation for elastic constants, meaning that in Voigt notation the elastic constants $c_{12} = c_{44}$, while in real alloys c_{44} is considerably larger than c_{12}.

Additionally, other physical quantities are only poorly modeled within the Lennard-Jones model or other pairwise interactions. For this reason in the 1980s many-body potentials were developed for alloys.

12.2.2.2 Many-Body Potentials, the EAM Method

Most of the many-body potentials for alloys in use today are tailored following a similar scheme: the interaction between a solute atom and its surroundings is assumed to be given by the density of neighbors (or, more precisely, by the electronic density acting due to the neighbors at the location of the atom). Qualitatively, using arguments from a tight binding approximation (Finnis and Sinclair 1984), one expects a dependence of the binding energy attributed to atom i, ϵ_i such as is given by Eq. (6), where $\langle i|H|j \rangle$ denotes the overlap integral between electronic states from atom i to its neighbors, and Z_i is its coordination number. This implies that the interaction energy should rather follow a square-root dependence of coordination instead of a linear dependence found in pair interaction.

$$\epsilon_i = \sqrt{\sum_j \langle i|H|j \rangle^2} \approx \sqrt{Z_i} \tag{6}$$

In more general schemes, this square-root dependence has been liberated in favor of an adjustable function, which led to the so-called embedded atom method (EAM) potentials. The total potential energy of the crystal [Eq. (7)] is made up of two parts, one being a pair potential, the other the so-called embedding part (Daw and Baskes 1984; Foiles et al. 1986; Baskes 1987, Voter and Chem 1987).

$$V = \frac{1}{2} \sum_{i,j=1}^{N} \Phi(r_{ij}) + \sum_{i=1}^{N} U(\rho_i) \tag{7}$$

The variable ρ_i denotes the electron density at the position of atom i, given in Eq. (8) again as the sum over the neighboring atoms.

$$\rho_i = \sum_{j=1}^{N} \Delta \rho_j(r_{ij}) \tag{8}$$

The embedding function U describes the contribution of this density to the total potential energy. In practice, these three functions are fitted to experimental data or to data obtained in ab-initio calculations, using suitable parameterizations such as combinations of polynomials and exponential functions.

For binary alloys, one needs in total seven functions: three pair potentials Φ_{XY} for AA, BB, and AB pairs, two density functions $\Delta \rho_X$ for A and B atoms, and two

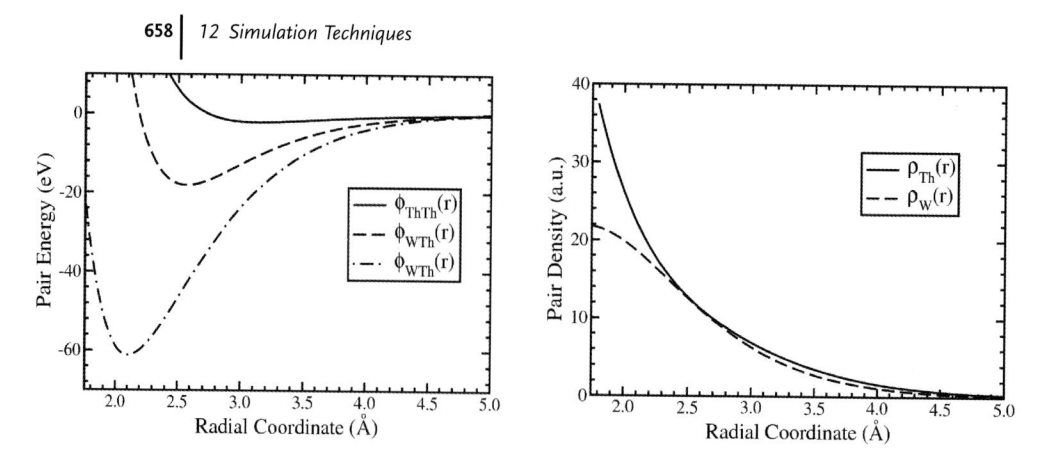

Fig. 12.1 EAM potential for tungsten and thorium (Eberhard 2006).

embedding functions U_X. In other words, one needs the full set of three functions for both atomic species and additionally only one (pair) function, describing the A–B interaction. A recent example derived from a fit to ab-initio data is shown in Fig. 12.1.

It must be emphasized that this type of potential contains *no* angular-dependent terms, so will be suitable only for cases where bonds are more or less undirected. To some extent this holds for metals, but certainly not for ionic or covalently bound materials. Therefore there are other approaches, such as the Tersoff or the Stillinger–Weber potential for semiconductors. But also for 4d and 5d transition metals or even more pronouncedly for f metals (e.g., the rare earth elements) problems can arise.

Another problem arising in the use of EAM and related potentials is the question of transferability, that is, whether it is really justified to use the same single element parts of a potential in different alloys. Due to an electronic transfer there is the possibility that even this part changes with alloying a second element, so that the simple pair potential ansatz for the A–B interaction is too simple.

A completely different approach, which does not refer to empirical potentials, is that introduced by Car and Parinello (Car and Parinello 1985), where the motion of the atoms is still treated as a classical problem, but where the forces are obtained directly by solving (in an adiabatic approximation) the electronic structure. This method of course avoids many of the problems related to empirical potentials, but at the cost of a much higher numerical effort, which then drastically limits the size and time scale of problems which can be treated.

Another approach, which also refers to ab-initio data, the so-called force matching method, (Ercolessi and Adams 1994) uses a somewhat similar idea: forces computed by ab-initio methods are parameterized for different atomic configurations and then used in MD. In some senses, this method can be seen as an "off-line version" of the Car–Parrinello method, but of course it is difficult to predict all the relevant configurations appearing in the course of a simulation.

12.2.3
Practical Considerations

Before one applies a potential from literature or other sources, one should verify its suitability for the problem under consideration. Usually, different measured quantities are reproduced with different precision. Quantities often used are: lattice constant, cohesive energy, bulk and other elastic moduli, stacking fault energy, vacancy formation energy, and others. The fitting of these quantities acts on different parts of the potential; elastic coefficients, for example, are essentially sensitive to small deviations from equilibrium positions, while vacancy properties reflect the long-distance tail of the potential.

12.2.4
Different Thermodynamic Ensembles: Thermostats, Barostats

In discussing the integration of Newton's equations of motion above, it was assumed implicitly that we are interested in a microcanonical ensemble, that is, a situation where total energy, volume, and number of particles are conserved (therefore it is often also called an NVE ensemble). For a typical experimental setup, it is usually much easier to control temperature (instead of energy) and pressure (instead of volume), and so to work in an NPT ensemble. This can be achieved by changing dynamically the average kinetic energy of the system (thus adjusting the temperature) or the system size (to adjust pressure) (Andersen 1980). An even more refined scheme was introduced by Parrinello and Rahman (1980, 1981), who allowed not only for changes in the system size but also for changes in the box shape. This can be important if phase transformations with a change in symmetry occur during the simulation.

Temperature in terms of microscopic quantities of the simulation is given through the equipartition formula, which states that in equilibrium the average of the kinetic energy $K(t)$ [Eq. (9)]

$$K(t) = \frac{1}{2}\sum_{i=1}^{N} m_i |v_i(t)|^2 \tag{9}$$

is equally distributed over the $3N$ degrees of freedom, each carrying a fraction $\frac{1}{2}k_B T$ [Eq. (10)].

$$\overline{K} = \frac{3}{2}Nk_B T \tag{10}$$

The simplest implementations of a thermostat just rescale the particle velocities periodically in order to keep temperature constant. This method, despite its appealing simplicity, suffers (at least theoretically) from a severe drawback: it does not respect the fluctuation–dissipation theorem [Eq. (11)], which relates the fluctuations of one thermodynamic observable (here: the temperature) with a corresponding response function (here specific heat).

$$\langle (\Delta T)^2 \rangle = \frac{k_B T^2}{c_V} \tag{11}$$

Better algorithms have been introduced as the Nosé–Hoover thermostat and variants (Hoover 1985). In the first case, the system is coupled to an external heat bath, which (via one extra degree of freedom) by itself follows an evolution equation similar to the evolution of particle trajectories.

Similar methods were introduced to keep pressure constant (Nosé and Klein 1983; Nosé 1984). Pressure follows from the virial function, as in Eq. (12), where F_i^{tot} denotes the total force acting on atom i. Its average is given by Eq. (13).

$$W(r_1, \ldots, r_N) = \sum_{i=1}^{N} r_i \cdot F_i^{tot} \tag{12}$$

$$\langle W \rangle = -3 N k_B T \tag{13}$$

The total force acting on particle i can be decomposed into an internal force (from the other atoms) and an external force (due to pressure) [Eq. (14)].

$$F_i^{tot} = F_i + F_i^{ext} \tag{14}$$

Similarly, the virial can be decomposed into an internal and an external contribution, where the external part can be evaluated by Eq. (15).

$$\langle W^{ext} \rangle = -3 p V \tag{15}$$

Solving Eqs. (13) and (15), one gets Eq. (16).

$$pV = N k_B T + \frac{1}{3} \sum_{i=1}^{N} r_i \cdot F_i \tag{16}$$

Again, the simplest idea, periodically readjusting the system size to keep pressure constant, violates the fluctuation–dissipation theorem (here relating the volume fluctuations ΔV with the isothermal compressibility κ).

$$\langle (\Delta V)^2 \rangle = -k_B T V_0 \kappa \tag{17}$$

More appropriate is the method of Andersen (1980), which introduces the box size as an extra degree of freedom and integrates this in order to keep pressure constant. The Parrinello–Rahman scheme (Parrinello and Rahman 1980, 1981) in principle works similarly, but uses six additional degrees of freedom, the sizes and angles of the simulation box. The particle coordinates are then integrated in coordinates scaled in terms of the box coordinates. The extra degrees of freedom all need a virtual mass (used in the Newton equation of motion), which in principle can be chosen arbitrarily but in a real application can cause problems by introducing oscillations or slow relaxation toward equilibrium.

12.2.5
Implementation of MD Algorithms

The essential part of any MD code is the time integration algorithm. For a time step Δt new positions, velocities, and forces have to be calculated. The simplest (but worst) way to do the integration would be to write Eq. (18).

$$x_i(t + \Delta t) = x_i(t) + v_i(t) \cdot \Delta t + \frac{F_i(t)}{2m} \Delta t^2 \tag{18}$$

Such a simple forward integration finite difference scheme is known to be numerically unstable and would require extremely short time steps to give reasonable accuracy.

Two popular and more efficient methods often applied in MD codes are the Verlet algorithm (Verlet 1967) and different kinds of predictor–corrector methods. The Verlet algorithm uses a forward and a backward Taylor expansion of positions, Eqs. (19) and (20), which can be added to give Eq. (21).

$$x_i(t + \Delta t) = x_i(t) + v_i(t) \cdot \Delta t + \frac{F_i(t)}{2m} \Delta t^2 + \frac{1}{6} b(t) \Delta t^3 + 0(\Delta t^4) \tag{19}$$

$$x_i(t - \Delta t) = x_i(t) - v_i(t) \cdot \Delta t + \frac{F_i(t)}{2m} \Delta t^2 - \frac{1}{6} b(t) \Delta t^3 + 0(\Delta t^4) \tag{20}$$

$$x_i(t + \Delta t) = 2x_i(t) - x_i(t - \Delta t) + \frac{F_i(t)}{m} \Delta t^2 + 0(\Delta t^4) \tag{21}$$

The truncation error is of the order of Δt^4, so it is called a fourth-order method. A variant is the velocity Verlet method (Swope et al. 1982) with the time steps described by Eqs. (22–25).

$$x_i(t + \Delta t) = x_i(t) + v_i(t) \cdot \Delta t + \frac{1}{2} a_i(t) \Delta t^2 \tag{22}$$

$$v_i\left(t + \frac{\Delta t}{2}\right) = v_i(t) + \frac{1}{2} a_i(t) \Delta t \tag{23}$$

$$a_i(t + \Delta t) = \frac{F_i(t + \Delta t)}{m} \tag{24}$$

$$v_i(t + \Delta t) = v_i\left(t + \frac{\Delta t}{2}\right) + \frac{1}{2} a_i(t + \Delta t) \Delta t \tag{25}$$

The second algorithm, the predictor–corrector method, also called Gear algorithm (Gear 1966, 1971), consists of three steps: from a Taylor expansion similar to that above, in the first step new positions, velocities, and accelerations are predicted. With these predicted quantities one evaluates the forces and computes the difference in acceleration between predicted and correct values. These corrections are used to correct positions, velocities, etc. The Gear scheme can be used to quite a high order (often to order six), but the efficiency is not necessarily better than that of the Verlet method.

12.2.6
Practical Aspects: Time Steps

As the first step in an MD simulation one has to choose a suitable atomic interaction potential which is known to fulfill the conditions discussed above. Additionally, one has to select a thermodynamic ensemble (e.g., NPT) and an algorithm to control the thermodynamic variables.

The next step is to set up initial atomic positions which define the problem, e.g., atoms in a perfect lattice or atoms at a grain boundary. In order to define a starting temperature, initial velocities (taken from a Maxwell distribution with zero average, corresponding to the temperature) have to be assigned randomly to the atoms.

Rather crucial for the efficiency and correctness of an MD-simulation is the proper choice of the integration time step. On the one hand, it has to be chosen as large as possible in order to increase the computation speed; on the other hand, it must not be too large, so as to avoid instabilities of the integration algorithm, and, more importantly, to avoid spurious overshooting of the particle motion across, e.g., local potential minima. The time step must be short compared to vibrational times in the solid, this being of the order of 10^{-12}–10^{-13} s. A fairly pragmatic but efficient way to choose an adequate time step is to start with a value like, say, 1 fs, then testing, in a short run in a small simulation box under constant energy, whether energy is conserved. If not, one has to reduce the time step; otherwise, one can try to enlarge it.

12.2.7
Evaluation of Data: Use of Correlation Functions

The first and sometimes most important results of an MD simulation are the atomic positions after a simulation run. Often it is useful not to take the instantaneous positions as obtained for a certain endpoint but to relax the positions to local minima of the potential energy ("quenching"). There are different ways to do this: either particle velocities are abruptly decreased to zero if the potential energy would increase during the next step, or individual particle velocities are set to zero if for that special atom the potential energy would increase. Another way is to use different minimization techniques (such as steepest descent or conjugate gradients) to find a local minimum of the potential energy.

Here good visualization techniques can be important, perhaps supported by an overlay of other quantities, such as local energy or coordination, onto the atomic positions.

Related to the inspection of atomic positions is the study of displacements and relaxations, where the difference between a reference and the actual position is computed.

Statistical averages over the whole simulation cell which are easily obtained as functions of simulation time are the average total, kinetic, and potential energy, total cell volume (in an NPT ensemble), or total pressure (in NVT). These quantities make it possible to monitor changes and equilibration.

At equilibrium one can perform time averages over the trajectories of single atoms or sets of atoms to obtain mean values of observables. The fluctuations of such observables again can be used to determine quantities like elastic moduli or specific heat by applying the fluctuation–dissipation theorem.

A further useful but somewhat more involved evaluation of simulation data is the computation of correlation functions in space and time. Real-space correlations are, for example, the density–density correlation $g(\mathbf{r})$, which gives the probability of finding an atom at a distance $\mathbf{r}$ away from another atom [Eq. (26)], or its spherical average. The density–density correlation function is especially useful for the study of liquid and amorphous structures, where it decays for large r. Partial correlation functions, taken only between one species of atoms, can also be of interest.

$$g(\mathbf{r}) = \frac{1}{N} \left\langle \sum_{i=1}^{N} \sum_{j=1}^{N} \delta(\mathbf{r} - \mathbf{r}_{ij}) \right\rangle \tag{26}$$

Comparison with scattering experiments is possible by the use of reciprocal space correlation functions like the (static) structure factor, Eq. (27) where $\rho(\mathbf{k})$ [Eq. (28)] is the spatial Fourier transform of the atomic distribution. It is easy to show that the structure factor is the Fourier transform of the density–density correlation function. The static structure factor is the quantity measured in an elastic scattering experiment from a diffraction pattern.

$$S(\mathbf{k}) = \frac{1}{N} \langle \rho(\mathbf{k})\rho(-\mathbf{k}) \rangle \tag{27}$$

$$\rho(\mathbf{k}) = \sum_{i=1}^{N} \exp(i\mathbf{k} \cdot \mathbf{r}_i) \tag{28}$$

The dynamic structure factor, Eq. (29) contains information about vibrational excitations and is measured experimentally in inelastic scattering (compare Chapter 13.1). The positions of peaks in the $(\mathbf{k}, \omega)$-plane show the dispersion of an excitation like a phonon.

$$\hat{S}(\mathbf{k},\omega) = \frac{1}{N}\int \langle \rho(\mathbf{k},t)\rho(-\mathbf{k},0)\rangle \exp(-i\omega t)\, dt \tag{29}$$

A final important correlation function is the velocity autocorrelation function, Eq. (30), which is directly proportional to the phonon density of states.

$$g(\omega) = \frac{2}{\pi}\int \frac{\langle v_i(t)v_i(0)\rangle}{\langle v_i(0)v_i(0)\rangle} \cos(\omega t)\, dt \tag{30}$$

12.2.8
Applications to Alloys, Alloy Dynamics, and Alloy Kinetics

MD methods allow to study static and dynamic properties of alloys, but are rather limited in the study of kinetic processes in alloys, since the total time accessible to MD simulations is limited to microseconds at most, while (diffusion-controlled) kinetic processes in alloys occur rather on time scales of seconds, minutes, or hours. Prominent examples are the study of defects like interfaces, especially grain boundaries (as shown in Fig. 12.2; see also Sutton and Ballu 1995), or of vacancies (Fig. 12.3). Figure 12.2 shows an MD computation of grain boundary energies for symmetrical tilt and twin boundaries. Here the MD approach essentially serves as an efficient method to relax the structure into a stable configuration. The example shown in Fig. 12.3 used finite temperature calculations to measure not only the internal energy but also the entropy by applying the method of overlapping distributions (Bennett 1976) (see also below).

As described above, phonon densities of states can be obtained by using velocity autocorrelation functions. An alternative way to determine phonon dispersion relationships is to measure the dynamic force matrix and to compute the phonon spectra in a harmonic approximation. An example is shown in Fig. 12.4.

Other important values which need to be determined are diffusion properties. In principle, MD simulations offer the possibility of measuring diffusion coefficients in alloys directly by following the mean square displacements [Eq. (31)], but since the time scale in MD simulation is limited to microseconds at best, this possibility is limited to either high temperature (and of course to liquids) or to situations with increased diffusivities, for example short-circuit diffusion in grain boundaries (compare Chapter 5).

$$\langle x^2 \rangle(t) = \frac{1}{N}\sum_{i=1}^{N}(x_i(t) - x_i(0))^2 = Dt, \tag{31}$$

Studying diffusional phase transformations in solids is more or less impossible since the kinetics of such transformations are too slow. More promising are studies of nondiffusional phase transitions such as martensitic transformations (Schneider and Stoll 1978).

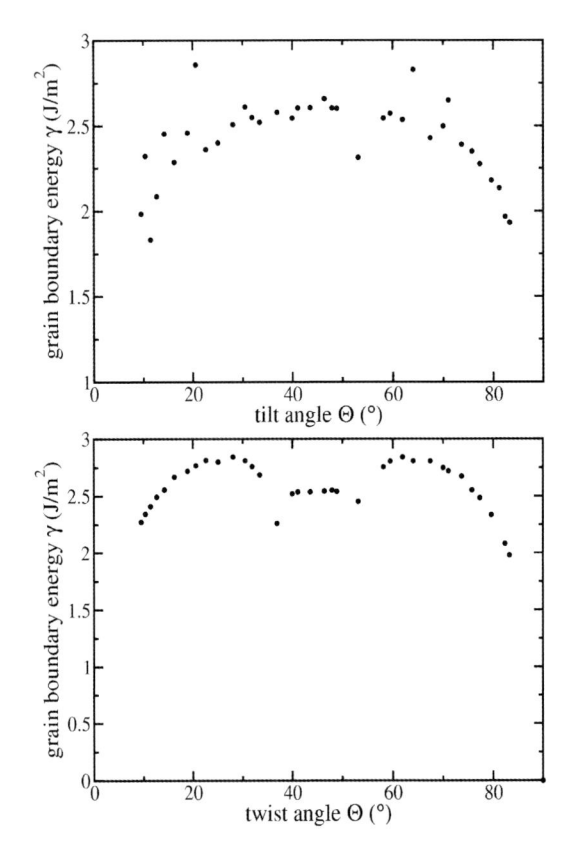

Fig. 12.2 Energy of symmetric tilt (above) and twist (below) grain boundaries in tungsten (Eberhard 2006).

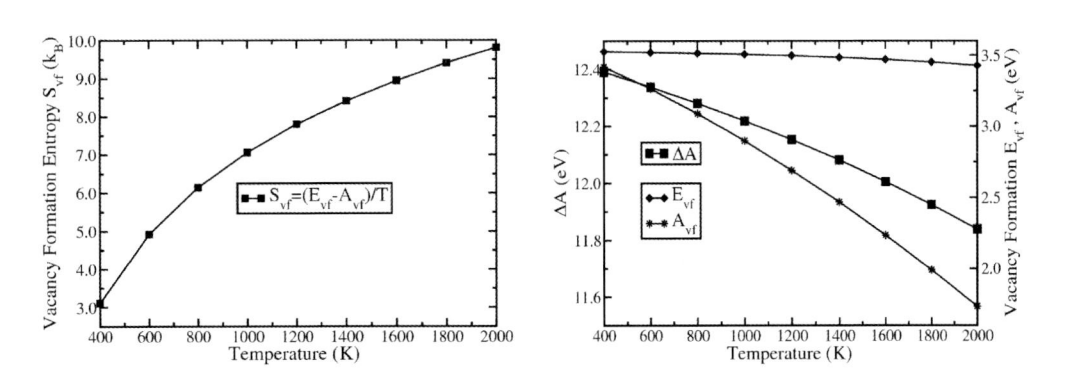

Fig. 12.3 Left: vacancy formation entropy and right: energy and free energy in tungsten (Eberhard 2006).

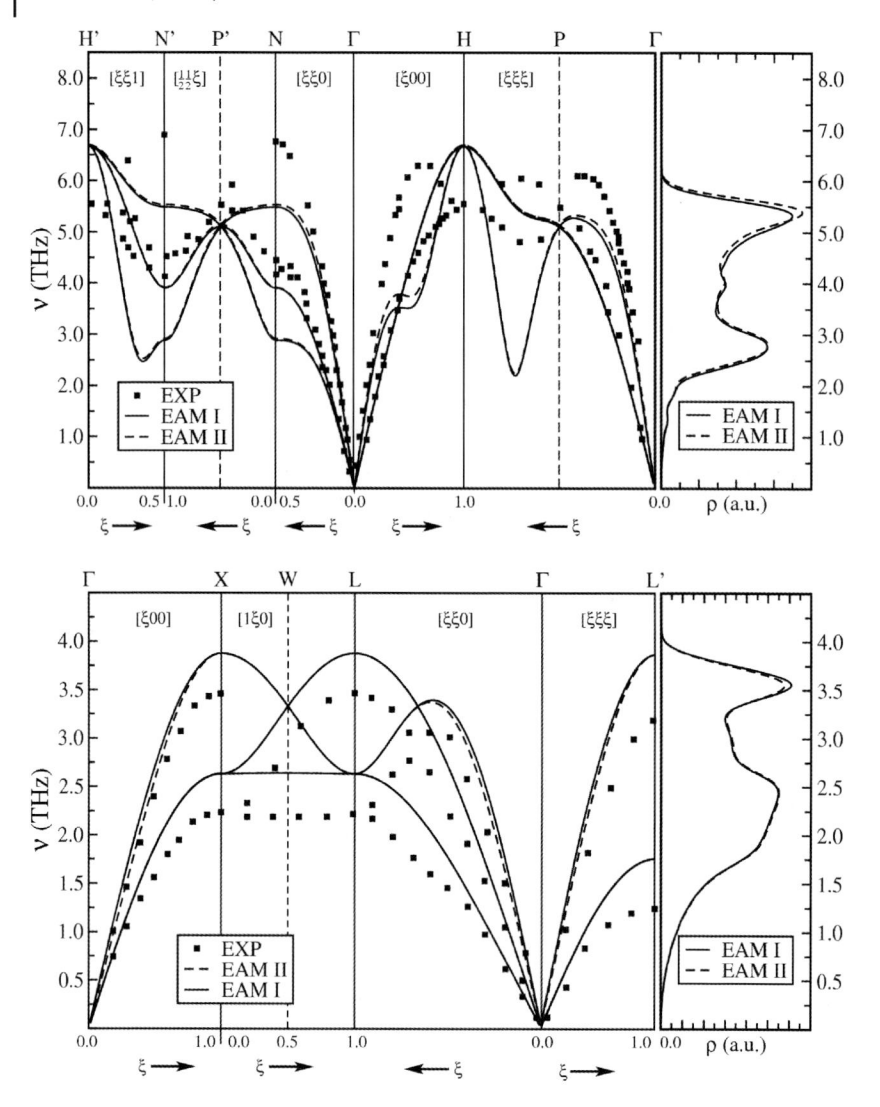

Fig. 12.4 Phonon dispersion curves for tungsten (upper panel) and thorium (lower panel), computed using a force matrix derived from MD simulations (Eberhard 2006). The right-hand inset shows the phonon density of states.

Nevertheless it is at least possible to study single-jump processes, the jump path, and the jump energetics as well as other quantities relevant for diffusion, such as vacancy formation enthalpies and entropies, solute–vacancy binding energies, etc. Figure 12.5 shows another example in tungsten, the energy profile of an atom displaced towards a vacancy along two different jump paths, one along a $\langle 111 \rangle$ and the other along a $\langle 100 \rangle$ direction.

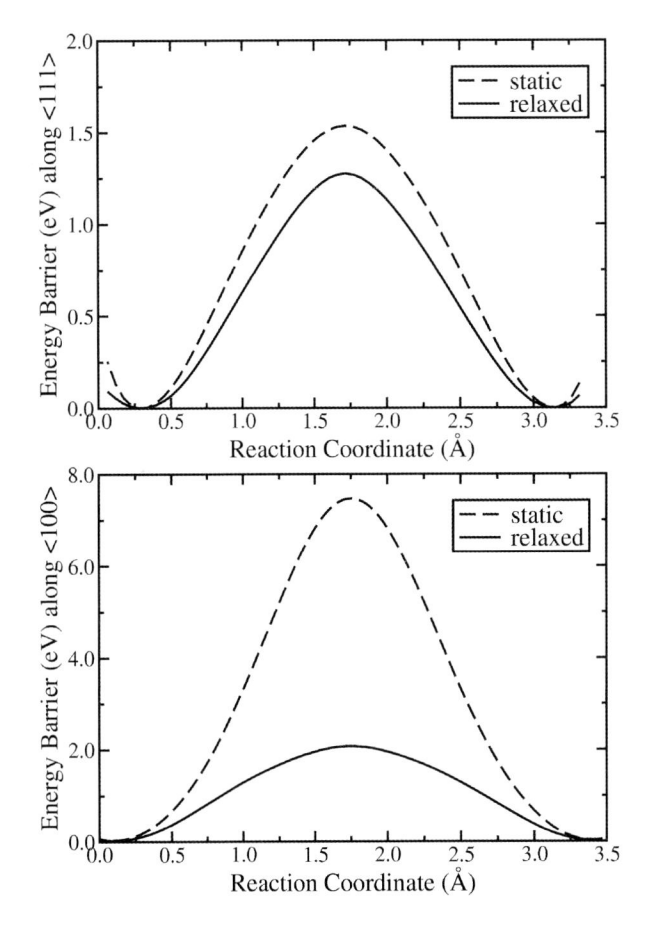

Fig. 12.5 Energy along the jump path of a tungsten self-diffusional jump along a $\langle 111 \rangle$ (upper panel) and a $\langle 100 \rangle$ direction (lower panel) (Eberhard 2006).

12.3
Monte Carlo Simulations

12.3.1
Foundations of Stochastic Processes – Markov Chains and the Master Equation

Statistical physics dealing with macroscopic systems, often composed of a number of components comparable to Avogadro's number, appeared an important beneficiary of new computer facilities. Each one of such systems may be found in one of its *macro*scopic states determined by particular *micro*scopic states $\{\sigma\}$ of all the components and classically represented by points in a $6N$-dimensional phase space (N is the number of system components). The macroscopic states characterized by macroscopic parameters (observables A) such as energy, volume, degree of chemical order, magnetization, etc. are usually highly degenerate with respect to the microscopic ones $\{\sigma\}$. The values of observables A measured under particular conditions (temperature, pressure, external field, etc.) are identified with corresponding averages $\langle A \rangle$ over all microscopic states $\{\sigma\}$ in which the system may be found under these conditions.

12.3.2
The Idea of Sampling

The central problem of statistical physics (statistical thermodynamics) is a calculation of the values of $\langle A \rangle$. The averaging is performed over an appropriate ensemble of macroscopic systems representing the related microscopic states. An ensemble is characterized by a so-called density $\rho(\sigma)$ defined in a way that $P(\sigma)$ [Eq. (32)] is a probability that a system in the ensemble is in the microscopic state σ (see, e.g., Huang 1963). The principle achievement of the founders of statistical physics was the derivation of formulae for the densities $\rho_{eq}(\sigma)$ corresponding to ensembles of systems in the equilibrium state.

$$P(\sigma) = \frac{\rho(\sigma)}{\sum_{\sigma} \rho(\sigma)} \tag{32}$$

A complete description of the system thermodynamics is derivable from the sum Z [Eq. (33)], called a partition function.

$$Z = \sum_{\sigma} \rho_{eq}(\sigma) \tag{33}$$

The types of ensembles and the rules concerning their choice when solving particular problems are discussed in text books on statistical thermodynamics (e.g., Huang 1963). Here, we will only summarize the problem by listing the ensembles and the corresponding densities ρ_{eq} in Table 12.1.

In particular, in the case of a microcanonical ensemble of isolated systems in equilibrium, all microstates σ with the energy E may appear with uniform probability.

If $A(\sigma)$ denotes the value of the observable A in the microscopic state σ, then Eq. (34) applies, where the sum covers all possible microscopic states σ (whose number is most often extremely large or even infinite) and its strict calculation is usually unfeasible.[1] The basic idea is to approximate the complete sum (34)

1) In an isobaric–isothermal ensemble

$$\langle A \rangle = \frac{\int_0^\infty dV [\sum_\sigma A(\sigma, V) \times \rho(\sigma, V)]}{\int_0^\infty dV [\sum_\sigma \rho(\sigma, V)]}$$

Table 12.1

Type of ensemble	Usage	Density function ρ_{eq}[a]
Microcanonical ensemble	isolated systems with fixed energy E	$\delta_{H(\sigma_i), E}$
Canonical ensemble	systems with fixed volume V and number of particles N studied at fixed temperature T determined by thermal bath	$\exp\left[-\dfrac{H(\sigma_i)}{k_B T}\right]$
Isothermal–isobaric ensemble	systems with fixed number of particles N studied at fixed pressure P and temperature T determined by thermal bath	$\exp\left[-\dfrac{H(\sigma_i) + PV}{k_B T}\right]$
Grandcanonical ensemble	Opened systems with fixed volume V studied at fixed temperature T determined by thermal bath	$\exp\left[-\dfrac{H(\sigma_i) + \sum_k \mu_k(\sigma_i) N_k}{k_B T}\right]$, N is the number of particles, μ is chemical potential

[a] H denotes the Hamiltonian of the system. $H(\sigma_i)$ used in Monte Carlo simulations *does not* contain kinetic energy terms.

by a partial one performed over some subset $\{\sigma_i, i = 1, \ldots, M\}$ of the microscopic states σ [Eq. (35)].

$$\langle A \rangle = \sum_\sigma [P(\sigma) \times A(\sigma)] = \frac{1}{Z} \sum_\sigma [\rho(\sigma) \times A(\sigma)] \tag{34}$$

$$\langle A \rangle \approx \frac{\sum_{i=1}^{M} [\rho(\sigma_i) \times A(\sigma_i)]}{\sum_{k=1}^{M} \rho(\sigma_k)} \tag{35}$$

It may be demonstrated that the efficient way to select states σ_i – called sampling – is a random choice realized numerically by means of the random number generators (see, e.g., Binder and Heermann 1997; Landau and Binder 2000 for numerous references).

Two variants of sampling may be applied:
- *Simple sampling*, where the random choice of the states σ_i runs according to a uniform distribution. The drawback of such an option is a possibility that many σ_i states may correspond to the low value of the density ρ, making the approximation [Eq. (35)] poor.

- *Importance sampling.* The procedure is essentially improved when the so-called importance sampling is applied. In this case the states σ_i are chosen randomly with a nonuniform distribution $\Pi(\sigma)$. In such a case, the average $\langle A \rangle$ is given by Eq. (36) (Heermann 1986), which is simplified dramatically by assuming Eq. (37), which leads to Eq. (38)

$$\langle A \rangle \approx \frac{\sum_{i=1}^{M}[p(\sigma_i) \times A(\sigma_i) \times \Pi^{-1}(\sigma_i)]}{\sum_{k=1}^{M}[p(\sigma_k) \times \Pi^{-1}(\sigma_k)]} \qquad (36)$$

$$\Pi(\sigma) = P(\sigma) = \frac{p(\sigma)}{\sum_{\sigma} p(\sigma)} \qquad (37)$$

$$\langle A \rangle = \frac{1}{M} \sum_{i=1}^{M} A(\sigma_i) \qquad (38)$$

12.3.3
Markov Chains as a Tool for Importance Sampling

The solution of a problem of generating a set of the system states showing the distribution [Eq. (37)] was proposed by Metropolis et al. (1953), who indicated that such a set may be constructed by generating a Markov chain (see, e.g., van Kampen 1987) with appropriate transition rates $W(\sigma_i \to \sigma_j)$ (i.e., transition probabilities per time unit) between the states $\{\sigma_i\}$.

If states of the systems in an ensemble change due to Markov processes, the time evolution of the probability distribution $P(\sigma)$ is given by a master equation, Eq. (39) (compare Chapter 5).

$$\frac{dP(\sigma_i)}{dt} = -\sum_{j} W(\sigma_i \to \sigma_j) \times P(\sigma_i) + \sum_{j} W(\sigma_j \to \sigma_i) \times P(\sigma_j) \qquad (39)$$

The evolution leads to a stationary distribution $P_{st}(\sigma)$ (for which $\frac{dP_{st}(\sigma_i)}{dt} = 0$),[2] provided that Eq. (40) holds.

$$\sum_{j} W(\sigma_i \to \sigma_j) \times P_{st}(\sigma_i) = \sum_{j} W(\sigma_j \to \sigma_i) \times P_{st}(\sigma_j) \qquad (40)$$

2) $P_{st}(\sigma)$ defines a stationary state, which differs from the equilibrium one by possible appearance of systematic flows.

$\dfrac{d\rho_{eq}}{dt} = 0$ must obviously hold for the *equilibrium* distribution $P_{eq}(\sigma)$, for which the sufficient, though not necessary, condition is that the transition frequencies W satisfy the so-called detailed balance of Eq. (41).

$$\frac{W(\sigma_i \to \sigma_j)}{W(\sigma_j \to \sigma_i)} = \frac{P_{eq}(\sigma_j)}{P_{eq}(\sigma_i)} = \frac{\rho_{eq}(\sigma_j)}{\rho_{eq}(\sigma_i)} \tag{41}$$

It can be proven (see, e.g., Binder 1976) that the detailed balance [Eq. (41)] guarantees a convergence of a Markov chain to $P_{eq}(\sigma)$.

The goal to generate a set of states $\{\sigma_i\}$ with the distribution of Eq. (37) is thus attainable by the realization of a Markov chain of the system microstates σ with the transition frequencies fulfilling the condition in Eq. (41).

This condition constitutes the basis for MC methods developed for particular problems of statistical physics solvable by considering appropriate ensembles determining the densities $\rho(\sigma)$ (Table 12.1).[3]

12.3.4
General Applicability

The Monte Carlo method may be basically applied to diverse kinds of problems in statistical physics:

12.3.4.1 Simulation and Characterization of System Properties in Thermodynamic Equilibrium

The procedure starts from a system in some (arbitrary) initial state. Subsequently, an evolution of the system is simulated as a Markov chain of microscopic states σ_i with the transition frequencies $W(\sigma_i \to \sigma_j)$ obeying detailed balance [Eq. (41)] corresponding to the particular conditions of the equilibrium state in question. The algorithm applied must enable the evolution of some macroscopic parameter of the system (for example its energy) to be followed, so that it is possible to observe the approach of equilibrium (microscopic states are in dynamical equilibrium with the distribution $P_{eq}(\sigma)$).[4] Once this stage is attained, the microscopic states σ_i of the system appearing at particular time moments may be randomly sampled and used in the averaging procedure [Eq. (38)] (effectively, time averaging is done).

Among a number of tasks solvable in this way, the determination of thermodynamic potentials from which all fundamental properties of macroscopic systems may be derived is of basic importance.

---

[3] In order to accelerate the saturation of a Markov chain, various modifications of the sampling procedure have been proposed: e.g., so called reweighting schemes (see, e.g., Binder and Heermann 1997; Bhattacharya and Sethna 1998) for "multicanonical sampling").

[4] More accurately, the most slowly relaxing macroscopic parameter should be monitored.

The procedure consists entirely or partially of importance sampling as described in Section 12.3.2, which may be combined with integration of standard thermodynamic equations. Let us focus attention on the evaluation of Helmholtz free energy F of a system, given by Eq. (42), where U denotes internal energy, T stands for absolute temperature, and S denotes entropy.

$$F = U - T \cdot S \qquad (42)$$

The internal energy U is conventionally evaluated according to Eq. (38) applied to system energies $E(\sigma_i)$ sampled from the generated Markov chain of microstates σ_i. As explained earlier, the sampling is performed on that part of the Markov chain which corresponds to thermal equilibrium with thermal fluctuations, i.e., on its finally quasi-constant part.

The entropy S may be evaluated either indirectly by means of other observables determined directly in MC (such as specific heat C, magnetization M, etc.), or by direct averaging, as proposed by Ma (1981).

The first technique involves standard thermodynamic equations; for example, the entropy may by found by performing an integration (Heermann 1986) to obtain Eq. (43) ($\beta = 1/k_B T$), which, however, requires that (a) the reference value of entropy $S(T')$ is known (which is often possible at either very low or very high temperature) and (b) U is evaluated along the whole path of integration.

$$S(T) = S(T') + \frac{U}{T} - \beta^{-1} \int_{\beta'}^{\beta} U \, d\beta \qquad (43)$$

Ma (1981) postulated a direct evaluation of S from Eq. (44), where $P(\sigma_i)$ is the probability of the occurrence of the microstate σ_i.

$$S = -k_B \sum_i P(\sigma_i) \times \ln[P(\sigma_i)] \qquad (44)$$

In practice, the determination of $P(\sigma_i)$ is highly nontrivial due to the low frequency of the occurrence of particular microstates with respect to their total number.

When discussing the Monte Carlo techniques for free energy determination one should mention a method proposed by Bennett (1976) which is commonly applied and often called the "overlapping distribution method". It is shown that given two ensembles (two systems or phases) defined in the same configurational space and described by Hamiltonians H_0 and H_1, respectively, one can find a ratio Z_1/Z_0 of their partition functions [Eq. (33)] by analyzing the distributions of these two ensembles over the configurational space. The distributions are generated by Monte Carlo simulation of a specific Markov chain over the configurational space, each step of which means a switch between H_0 and H_1 phases. Evaluation of Z_1/Z_0 means the direct evaluation of a difference between the corresponding free energies F_1 and F_0.

12.3.4.2 Simulation of Relaxation Processes Toward Equilibrium

The master equation (39) basically can be regarded as describing relaxation and migration processes (see e.g., Kozubski 1997 for a concise review and references). The Monte Carlo method becomes, therefore, a natural tool for simulating such processes. Instead of sampling the microscopic states σ_i after the saturation of the particular Markov chain, one now simulates and observes an ensemble of independent parallel Markov chains, and performs the averaging of the observable of interest over all of them at particular consecutive time moments also before saturation. In this way the time evolution (relaxation) $\langle A \rangle (t)$ of the observable is obtained. Since, for a unique choice of transition rates, the condition of detailed balance is not sufficient, the treatment of such problems is sophisticated, involving a number of problems including, e.g., the relationship between the computer and real-time scales (see Section 12.3.6). A choice of transition rates following a well-based microscopic model is therefore necessary. In general, the relaxation path toward equilibrium can depend sensitively on the particular physical model for the transition rates.

12.3.4.3 Simulation of Nonequilibrium Processes and Transport Phenomena

Nonequilibrium processes constitute a wide group of phenomena, very often of fundamental interest: diverse growth processes are typical examples. Such processes, consisting of effective transitions between microscopic states, constitute a natural subject for studies by means of MC methods (see Landau and Binder 2000 for an extended review). The nonequilibrium character of the phenomenon means that a detailed balance no longer must be obeyed by the transition frequencies which are to model particular microscopic reactions involved in the process. MC simulation of crystal growth according to the Kossel model (Kossel 1927) is a common example. In this model three atomistic-scale processes – deposition, evaporation and diffusion – compete. A particular frequency (rate) is modeled and attributed to each process and its selective influence on the overall effect (e.g., crystal growth rate) may then be studied by means of MC.

Transport phenomena may be studied by means of MC in both stationary and nonstationary states of the systems, simply by monitoring the process of transport during the simulated Markov chain.

The most common example is MC simulation of various kinds of diffusion and the determination of particular parameters, such as the diffusion constant or correlation factor. A basic review of the related methodology and topics examined has been given by Murch (1984).

The standard method (see, e.g., a recent study of B2-ordered FeAl by Weinkamer et al. (1998) consists of monitoring mean-square displacement $R^2_{A(V)}(t)$ of a tracer atom (A)/vacancy (V) as a function of MC time. The vacancy/tracer diffusion constant $D_{V(A)}$ is then calculated by means of Eq. (45) (Zhao et al. 1996).

$$D_{V(A)} = \lim_{t \to \infty} \left[\frac{1}{6N_{V(A)}} \frac{\partial}{\partial t} \left(\sum_{V(A)} R^2_{V(A)}(t) \right) \right] \tag{45}$$

Similarly to the case of relaxations, the analysis makes no sense without a proper definition of the time scale.

Monitoring of $R^2_{A(V)}(t)$ also leads to the estimation of the correlation factors for atoms A (Einstein 1905) by Eq. (46), where n_A denotes the number of jumps of length a performed by the A atom – as considered, e.g., by Murch (1982).

$$f_A = \frac{\langle R^2_A(n_A) \rangle}{a^2 \times n_A} \qquad (46)$$

The procedure may basically be implemented with various particular conditions (concentration/chemical potential gradient, external stress, etc.), which allows different types of diffusion to be simulated.

12.3.5
Limitations: Finite-Size Effects and Boundary Conditions

Thermodynamic laws in statistical physics correspond to the thermodynamic limit $N \to \infty$, where N denotes the number of particles in the system. It is obvious that when simulating a system numerically, one always operates with a finite value of N. This generates so-called finite size effects, resulting in a variety of problems in the thermodynamic interpretation of simulation results. The "simplest" task aiming in the elimination of the parasitic influence of the sample limits on the simulated effects is usually realized by the application of periodic boundary conditions consisting of the consideration of the opposite sample boundaries (e.g., the very first and the very last crystallographic planes if a crystalline system is in question) as neighboring parts. The well-known two-dimensional analogue of this idea is a transformation of a plane into a torus. Despite compensating the boundary (surface) effects, periodic boundary conditions cannot remove the limitation caused by finite size in the consideration of any distance dependencies. This applies, for example, to correlation lengths ξ (which cannot exceed the sample size) and results in characteristic blunting (rounding) of singularities marking continuous and discontinuous phase transitions. The effect may be described by specific finite-size scaling rules derivable for particular relationships (Binder and Heermann 1997). An example are effects concerning the relaxation times τ and the transition temperatures T_t for simulated discontinuous "order–disorder" transitions: both quantities scale with characteristic critical exponents according to the sample size L.

Instead of diverging to infinity at the temperature of phase transition, the relaxation time increases up to a maximum value as in Eq. (47).

$$\tau_{max} \propto L^z \qquad (47)$$

In the case of the transition temperature, we observe, in turn, a shift from the thermodynamic value described by Eq. (48).

$$T_t(L) - T_t(L \to \infty) \propto L^{-\lambda} \tag{48}$$

Another unwanted effect of periodic boundary conditions can be the stabiliziation (or suppression) of certain superstructures matching (or not matching) the periodicity of the lattice at the boundaries.

Finally, it should also be mentioned that in the case of MC simulatons on rigid lattices (crystalline phases, lattice gas) the finite size of the simulated sample limits the possible reduction of the concentration of system components. The effect is important, e.g., when modeling vacancy-controlled processes in alloys.

12.3.6
Numerical Implementation of MC

Most basically, a simulation algorithm aims at generating microscopic states σ_i of a system (e.g., particular atomic or spin configurations) corresponding to a particular macroscopic state (characterized by macroscopic parameters: temperature, pressure, degree of order, etc.). Conceptually, this is achieved by generating a Markov chain of microscopic states and the related particular ideas consist of a choice of a formula for the transition frequency $W(\sigma_i \to \sigma_j)$ fulfilling the detailed balance.

12.3.6.1 Classical Realization of Markov Chains

Let the system energy be equal to E_{ini} in the microstate σ_i and to E_{fin} in the microstate σ_j. Then after Metropolis et al. (1953) one writes Eq. (49), where $\Delta E = E_{fin} - E_{ini}$, k_B and T denote the Boltzmann constant and temperature, respectively, and τ is a time scale constant.

$$W(\sigma_i \to \sigma_j) = \begin{cases} \tau^{-1} \times \exp\left[-\dfrac{\Delta E}{k_B T}\right], & \Delta E > 0 \\ \tau^{-1}, & \Delta E < 0 \end{cases}$$

$$\equiv \min\left\{\tau^{-1}, \tau^{-1} \times \exp\left[-\dfrac{\Delta E}{k_B T}\right]\right\} \tag{49}$$

Although the Metropolis transition frequencies (49) obviously fulfill the detailed balance (in a canonical ensemble) and a corresponding Markov chain definitely converges at the equilibrium distribution of the microstates $\{\sigma_i\}$, in some particular cases their use may be disadvantageous at high temperatures where, due to the transition probabilities approaching the value of 1, the system being off-equilibrium keeps oscillating between different microscopic states, which makes the simulated process not perfectly ergodic (Landau and Binder 2000).

An alternative concept for the transition frequencies $W(\sigma_i \to \sigma_j)$ goes back to Glauber (1963); Eq. (50) for $W(\sigma_i \to \sigma_j)$ is proposed.

$$W(\sigma_i \to \sigma_j) = (\tau)^{-1} \times \frac{\exp\left[-\frac{E_{fin}}{k_BT}\right]}{\exp\left[-\frac{E_{ini}}{k_BT}\right] + \exp\left[-\frac{E_{fin}}{k_BT}\right]}$$

$$= (\tau)^{-1} \times \frac{\exp\left[-\frac{\Delta E}{k_BT}\right]}{1 + \exp\left[-\frac{\Delta E}{k_BT}\right]} \tag{50}$$

Fulfillment of detailed balance follows directly from Eq. (50). It is also immediately found that at high temperatures both $W(\sigma_i \to \sigma_j)$ and $W(\sigma_j \to \sigma_i)$ tend to the level of $1/(2\tau)$.

The simulation algorithms involving the above expressions for $W(\sigma_i \to \sigma_j)$ usually work in the following cycles.

- The system is in some microscopic state σ_i.
- Another microscopic state $\sigma_j \neq \sigma_i$ is chosen at random from the set $\{\sigma_i\}$.
- Transition $\sigma_i \to \sigma_j$ is executed *or suppressed* according to the probability $\tau \times W(\sigma_i \to \sigma_j)$.

The above scheme yields a probabilistic rationale of Eq. (50). The probability of an event "the system either transforms from σ_i to σ_j or remains in σ_i" is in this particular case equal to 1. On the other hand, it must be equal to a sum of the two corresponding probabilities. Equation (50) is clearly consistent with these assumptions.

The time-scale constant τ is usually kept fixed over all the Markov chain; sometimes it is related to temperature.

The drawback of the procedure is that, especially at low temperatures, a number of MC steps are lost (the transition is suppressed). The drawback becomes especially troublesome when simulating phenomena where $\Delta E = E_{fin} - E_{ini}$ is always positive (e.g., in the case of kinetic Monte Carlo simulations of diffusion or relaxation processes involving activation barriers – see Section 12.3.4).

12.3.6.2 "Residence Time" Algorithm

The idea stemmed from Young and Elcock (1966) and Bortz et al. (1975), then was applied and developed in many papers on the basis of the method proposed by Gillespie (1976) for the simulation of the kinetics of coupled chemical reactions. The basic idea is that, instead of making a number of unsuccessful attempts to perform a transition, one computes the (average) time for the system to stay in its microstate i, then performs one of the possible transitions with a choice which respects the proper weights.

The following steps are used.

- All possible transitions $\sigma_i \to \sigma_j$ between the microstates of the system are listed and numbered and their corresponding

Metropolis frequencies [Eq. (49)] $W_k = W(\sigma_i \to \sigma_j) = \tau^{-1} \times \exp\{-[E(\sigma_j) - E(\sigma_i)]/k_B T\}$ are determined. This appears feasible in many cases when particular systems and particular transition mechanisms are considered (see, e.g., Bortz et al. 1975).
- Two random numbers R_1 and R_2 between 0 and 1 are generated.
- The transition k *is chosen*, for which Eq. (51) holds.

$$\sum_{i=1}^{k-1} \frac{W_i}{\sum_j W_j} < R_1 \leq \sum_{i=1}^{k} \frac{W_i}{\sum_j W_j} \qquad (51)$$

- Time is incremented by Δt [Eq. (52)], where Δt has an obvious meaning of a "residence time" of the system in the initial microscopic state.

$$\Delta t = -\frac{\ln R_2}{\sum_j W_j} \qquad (52)$$

It was shown (Novotny 1995; Athenes et al. 1997) that the standard Metropolis-type algorithms are equivalent to the "residence-time" one in the sense that the average residence time $\langle \Delta t \rangle$ (52) equals the time resulting from the cumulation of unsuccessful "Metropolis" transition attempts (see Appendix).

The classical "residence-time" algorithm has been generalized towards reducing the effect of back and forth transitions (Athenes et al. 1997) – in particular application to vacancy diffusion.

12.3.6.3 The Problem of Time Scales
The interpretation of results of MC simulations in terms of natural physical phenomena requires that the MC time – i.e., the sequence of simulation steps, is related to real time. A solution of the problem depends on the particular algorithm applied.

Metropolis-type algorithms
- In each MC step a number N of *possible* transitions $\sigma_i \to \sigma_j$ has to be determined.
- Each MC step is associated with time increment $\Delta t = \tau/N$.

As thoroughly analyzed by Athenes (Athenes 1997), the realization of the above idea requires that the value of N and its possible variation during the simulated process is examined carefully.

If applied to different situations (e.g., different temperatures) there is not a unique translation from MC simulation to real time.

"Residence time" algorithm As mentioned before, the time increment Δt is weighted by a sum of frequencies of possible transitions [Eq. (52)].

Among a number of proposed justifications (see, e.g., Nowotny 1995; Athenes et al. 1997) the one proposed by Athenes (1997) seems to be the simplest.

Let the system be in the microstate σ_i at $t = 0$ and let $P(t)$ be a probability that it still remains in this microstate at $t > 0$ (t is thus the residence time). If $\{W_i\}$ denotes the frequencies of all possible transitions starting from σ_i the differential equation (53) holds, with Eq. (54) as the obvious solution, which is the probability distribution for residence times with the normalization factor $C = \sum_i W_i$.

$$\frac{d}{dt} P(t) = -\left(\sum_i W_i\right) \times P(t) \tag{53}$$

$$P(t) = C \times \exp\left[-\left(\sum_i W_i\right) \times t\right] \tag{54}$$

It is clear that the application of Eq. (52) means a numerical realization of the evaluation of the residence time Δt according to the probability given by Eq. (53).

From Eq. (54) we find Eq. (55) gives the value most often assigned to the residence time in practical applications.

$$\langle t \rangle = \frac{1}{\sum_i W_i} \tag{55}$$

12.3.7
Applications to Alloys

12.3.7.1 General Assumptions

We will focus our attention on the configurational thermodynamics of alloys: the search for equilibrium configurations and investigation of relaxation kinetics (features of the system evolution toward the equilibrium configuration). Most of the information will concern MC simulations of canonical ensembles. A recent account of the achievements and trends has been given by Schweika (1998).

Particular meanings will now be attributed to the general notions introduced in the preceding sections.

The microstates σ_i correspond to atomic configurations, from which one can easily deduce macroscopic parameters such as short- and long-range order parameters, internal energy, and precipitate size and distribution.

The transitions between microstates consist of elementary steps of atomic migration. When searching for equilibrium configurations only, direct exchanges of atoms can be considered. In the case of kinetics in alloys, we mainly consider atomic jumps to nearest-neighbor (NN) vacancies (Fig. 12.6), which have been described in detail in Chapter 5. It is, however, worth noticing that as vacancies are

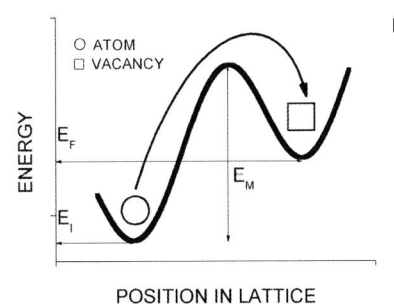

Fig. 12.6 Atomic jump energetics.

most often treated as additional alloy components, in both cases one has to do with actual exchanges between particles (atoms and vacancies).

The corresponding MC simulations are run by applying either of the algorithms discussed in the preceding sections [Eqs. (49–52)]. As pioneering works, the papers of Flinn and McManus (1961) for Metropolis-type and of Young and Elcock (1966) for the "residence time" algorithms are quoted. There are, however, two alternative ways in which particular energies are substituted for E_{fin}: in classical simulations E_{fin} is identified with the energy E_F (Fig. 12.6) – the final energy of the system after the atomic jump is executed. More appropriate from the point of view of transition-state theory is an alternative assumption, where the saddle-point energy E_M is substituted for E_{fin}.[5]

The second option is more suitable for simulating kinetics and, therefore, the corresponding method is called kinetic Monte Carlo (KMC) (see, e.g., Landau and Binder 2000). However, no strict proof for the latter has been given as yet, and to the contrary, when applied to specific problems of kinetics both standard MC and KMC yield analogous results (Oramus et al. 2001a,b).

12.3.7.2 Physical Model of an Alloy

Realization of the general task requires that a particular model of an alloy is considered. The absolute majority of MC simulations of alloys have up to now been done with the Ising model (see Chapter 10), although, due to the development of more or less rigorous ab-initio calculations considerably increasing in materials science during recent years, progress beyond this approximation has been achieved.

The classical Ising model was proposed to describe a magnetic system as a lattice of pairwise interacting spins, which were assumed to flip without changing positions in the lattice. Translation of the problem of this "classical" Ising system with temperature-dependent magnetization M and domain structure into the problem of an alloy with, e.g., a temperature-dependent degree of chemical long-

---

[5] In the case of Metropolis-type algorithms [Eqs. (49), (50)], a transition from the state i to the state f across the saddle point (SP) is realized by dividing the process into two, $i \to SP$ and $SP \to f$. It can easily be shown that a detailed balance then holds for the whole transition $i \to f$.

range order η and a network of antiphase domains immediately suggests the correspondence: spins-up $\leftrightarrow$ A-atoms; spins-down $\leftrightarrow$ B-atoms; $M \leftrightarrow \eta$, where, however, contrary to the "concentrations" of up- and down-spins in the ferromagnet, atomic concentrations in an alloy must be conserved. The bridge between the two topics is mostly due to Kawasaki (1966), who proposed a formalism assuming the conservation of magnetization to consider critical spin diffusion in a ferromagnet.

Application of the Ising model to an alloy and its implementation with MC simulation algorithms require that the energy of a configuration can be described as the sum of pair interactions e_{ij} between the component atoms.

In particular, energy increments/decrements ΔE attributed to the exchanges of atomic positions (including vacancies) and controlling the values of the exchange frequencies [Eqs. (49), (50)] should be known. As shown in Chapter 10, in any case ΔE depends on the atomic pair interactions e_{ij} via their linear combinations $v_{ij} \propto 2e_{ij} - e_{ii} - e_{jj}$ $(i, j = A, B)$ (the proportionality coefficient depends on particular definitions of other parameters) sometimes called "ordering energies" or just "pair-interaction energies," which is highly misleading. Therefore, for practical use in classical MC (not in KMC!) it is enough to know only the v_{ij} parameters. While for both "i" and "j" being atoms, v_{ij} may be evaluated experimentally (e.g., using chemical short-range order parameters deduced from diffuse scattering data – see Section 12.3.9.2 and Chapter 13.1), in the case of vacancy–atom or vacancy–vacancy interactions experimental access to the parameters is much more difficult. Very often pair interactions with and between vacancies are neglected (put equal to zero) but then, in addition to the v_{ij} parameters, so-called "asymmetry energies" $w_{ij} \propto e_{ii} - e_{jj}$ parameterize the ΔE. Evaluation of the latter most often consists of speculations supported by theoretical calculations. For an example of such an approach see Oramus et al. (2001a).

Another source for the values of pair-interaction energies to be used in MC simulations are more precise simulation methods such as ab-initio or molecular statics/dynamics (see, e.g., Kim 1991).

As long as a system is described by means of an Ising Hamiltonian, ab-initio calculations may be addressed in such a way that the appropriate cohesion and defect-formation energies expressed within the cluster variation method (CVM) in terms of pair interactions are assumed to be equal to the values resulting from the quantum electron theory. An example in this field may be found in Kozłowski et al. (2005). It should, however, be made clear that more interesting results following from the combination of quasi- or strictly ab-initio calculations with MC simulation of alloys are obtained when going beyond the Ising model and by involving many-body potentials. This topic is discussed briefly in Section 12.3.10.

In the case of KMC, except for pair interactions determining the energies of atoms in lattice sites, the values of saddle-point energies E_M (Fig. 12.6) are needed. In many papers arbitrary fixed values of E_M are either assigned to particular atoms (neglecting the influence of local configurations), or, as for example in Pitsch and Gahn (1993); Athenes et al. (1996), E_M is calculated by assuming that pair interaction energies $e^{(M)}{}_{ij}$ bonding the jumping atoms on the saddle point are proportional to the corresponding e_{ij} parameters related to atoms occupying lattice sites. In such a way, the reliability of the simulations with an atomic jump model involving saddle-point energy E_M is somewhat weakened due to the difficulty in the evaluation of E_M.

Experimentally, saddle-point energies of atoms migrating in a crystal are deducible, e.g., from the analysis of phonon spectra (Schober et al. 1992) or diffusion/relaxation data, provided activation energies for vacancy formation are known. The problem is, however, most often solvable by resorting to ab-initio or quasi-ab-initio calculations of the alloy energetics and by running molecular statics for atoms drawn along the path of an elementary jump (see, e.g., Oramus et al. 2001b; Schweiger et al. 2001). Saddle-point energies evaluated in this way were applied in KMC simulations of ordering kinetics in Ni_3Al (Oramus et al. 2001b).

12.3.8
Practical Aspects

In most of the related studies a crystal of an alloy with particular symmetry is simulated. The size of the sample depends on the computer power available; nowadays, the simulations usually involve a range of at least 10^5 atoms. A large simulation cell minimizes finite size effect and, in particular, allows treatment of realistically low concentrations (especially that of vacancies). In the case of bulk properties being in question, periodic boundary conditions are applied in all x, y, and z directions (see the previous section and, e.g., Binder and Heermann 1997). A study of surface or general boundary effects requires that periodic boundary conditions are partially or totally removed.

The following ways exist to implement an MC algorithm with atomic jumps in a crystalline material.

(a) *Pair exchange*: Metropolis-type simulation of atoms moving by direct exchange – applied when only equilibrium states are studied
- random choice of two atoms of different chemical nature
- execution or rejection of their exchange according to probabilities given by Eq. (49) or (50).
- increment of time by τ/N, where N is the total number of atoms.

(b) *Vacancy-atom exchange*: Metropolis-type simulation of atoms moving with vacancy mechanism
- introduction of vacancies to the system. Most often just one vacancy is introduced by random removal of an atom. To maintain stoichiometry strictly, it can be necessary to remove more than one atom.
- random choice of an atom in the first coordination shell of the vacancy
- execution or rejection of its exchange with the vacancy – according to probabilities given by Eq. (49) or (50)

- repetition of the above steps for each vacancy in the system
- increment of time by τ/Z, where Z is the first-shell coordination number (for explanation see the Appendix).

(c) *Residence time algorithm*: As already mentioned, Metropolis-type algorithms appear inefficient in this case because numerous atomic jumps are rejected. Problems with vacancy traps appear. The present-day common solution is the "residence time" algorithm.
- Metropolis transition frequencies W_i ($i = 1, \ldots, Z$) [Eq. (49)] are calculated for all potential exchanges of a vacancy with atoms in the first coordination shell.
- The exchange chosen randomly according to Eq. (45) is executed.
- Time is incremented either by $\Delta t = -\dfrac{\ln R}{\sum_j W_j}$ (by generating a random fraction R each time), or by an average $\langle \Delta t \rangle = \dfrac{1}{\sum_i W_i}$. Note that the increments are by definition proportional to the reciprocal attempt frequency τ. This frequency is usually taken as constant, but some authors have proposed taking into account its temperature dependence (Bozzolo et al. 2002).

In the Appendix, the question of "time scale" is addressed again, now with definite reference to atomic jumps. A very consistent explanation of the problem was given by Athenes et al. (1997).

12.3.9
Review of Current Applications in Studies of Alloys

Monte Carlo methods have nowadays become one of the most important and powerful tools in modeling of alloys. A review of databases shows that over 1200 related papers have been published during the past decade (over 400 during the past three years) and the investigations cover all topics in the field. The classical interest in the search for ground-state configurations and, consequently, phase diagram calculations (Landau and Binder 2000) has been extended to time evolutions including diffusion phenomena and the kinetics of diverse structural transformations. Most recent MC studies of alloys concern the hottest topics, such as the properties of nanostructured materials (see, e.g., Guofeng-Wang et al. 2004a,b; Canzian et al. 2004; Kozłowski et al. 2005) and hydrogen storage in metals (see, e.g., Bhatia et al. 2004; Kamakoti and Sholl 2005).

Below, some of the less routine MC approaches to the alloy physics problems are briefly scanned.

12.3.9.1 Computation of Phase Diagrams using Grandcanonical Ensemble

Calculation of phase diagrams for binary alloys is an important task in order to determine the solubility limits, supersaturation, existence range of ordered phases, etc., but also to test the validity of potential parameters. While in principle it is possible to calculate phase diagrams using standard MC techniques in the canonical ensemble, precision and computational speed are higher if one uses instead a grand canonical ensemble (Binder et al. 1981). Practically speaking, this means that instead of applying an evolution of microstates via AB exchanges of vacancy diffusion (both conserving total concentration), we transform A atoms into B and vice versa, again according to MC rules, but now with an additional term in the transition probability. This additional term, (56), corresponds to an external applied field in the case of the magnetic Ising model. Here, $\Delta\mu$ is the difference in chemical potential of species A and B, and ΔN_B is the change of the number of B atoms during the MC step.

$$\Delta\mu \times \Delta N_B \qquad (56)$$

Running the simulation with such an algorithm (with $\Delta\mu$ given) after equilibration leads to an equilibrium concentration. Varying $\Delta\mu$ now yields a curve $c_B(\Delta\mu)$ with jumps at phase boundaries, and, after repetition at different temperatures, one can construct the complete phase diagram.

A complication of this procedure is due to metastability – often there is a certain difference in c_B values depending on whether $\Delta\mu$ was increased or decreased during the calculation. One solution is to take an average of both values, or to put "seeds" of the second phase into the simulation box.

12.3.9.2 Reverse and Inverse Monte Carlo Methods: from Experimental SRO Parameters to Atomic Interaction Energies

While standard MC simulations are used for the generation of equilibrium states (configurations) or possibly for the study of kinetics (relaxation toward the equilibrium state) on the basis of the system energetics and external conditions (temperature, pressure), the method may also be used to determine atomistic energy parameters on the basis of the equilibrium state (for a review see, e.g., Schweika 1998).

The first step is the so-called reverse MC method-going back to Gehlen and Cohen (1965), who fitted correlation functions (short-range order (SRO) parameters) to atomic configuration, so that the experimentally observed pattern of diffuse scattered intensity is reproduced. The method is thus a specific technique for numerical fitting rather than a simulation of a physical phenomenon. The algorithm constructs a Markov chain of configuration changes $\sigma_i \to \sigma_f$ (due to atomic displacements) according to the transition frequencies [Eq. (39)] given by McGreevy and Pusztai (1988) as in Eq. (57), where χ is defined by Eq. (58), $\alpha_k^{(theor)}$ and $\alpha_k^{(ex)}$ are theoretical and experimental values of SRO parameters, respectively, and ε_k denotes the corresponding standard deviation.

$$W = \exp\left[-\frac{\chi_f^2 - \chi_i^2}{2}\right] \tag{57}$$

$$\chi^2 = \sum_k \frac{(\alpha_k^{(theor)} - \alpha_k^{(ex)})^2}{\varepsilon_k} \tag{58}$$

As discussed by Schweika, in order to make the procedure converge to the definite solution, one has to perform a number of runs starting from different initial conditions.

The method is very efficient as a fitting tool and is also used by one of the present authors in the calculations of Laplace transformations of the MC-simulated "order–order" relaxation isotherms (Oramus et al. 2001a). Several successful applications in the studies of atomic correlation in amorphous phases have recently been reported (see, e.g., Hanada et al. 2004).

The inverse Monte Carlo method is an application of the above fitting procedure of the values of pair-interaction energies to the equilibrium set $\{\alpha_k\}$ of SRO parameters obtained from experiment. The original idea is due to Gerold and Kern (1987). If the crystal configuration (described by $\{\alpha_k\}$) corresponds to thermal equilibrium, the value of configurational energy U is dynamically constant – i.e., despite permanent fluctuations due to atomic migration its average value does not change.

An exchange of an A–B pair of atoms brings about the change of U expressed in Eq. (59), where ΔN_i and v_i denote the resulting change in the number of A–A pairs and the effective pair-interaction energy (actually, the so-called ordering energy $v \propto 2 \times e_{AB} - e_{AA} - e_{BB}$, see Section 12.3.7.2) in the ith coordination shell, respectively.

$$\Delta U = \sum_i \Delta N_i v_i \tag{59}$$

The probability of the A–B exchange is assumed to obey the Glauber formula [Eq. (60)].

$$p = \frac{\exp\left[-\frac{\Delta U}{k_B T}\right]}{1 + \exp\left[-\frac{\Delta U}{k_B T}\right]} \tag{60}$$

As $\langle \Delta U \rangle = $ const, $\langle \Delta N_i \rangle$ should be equal to zero for each coordination shell; therefore Eq. (61) applies.

$$\langle N_i \rangle = \sum_{exchanges} p \times \Delta N_i = 0 \tag{61}$$

The following algorithm realizes the idea.
(1) A crystal configuration with the equilibrium $\{\alpha_k\}$ is generated.
(2) An A–B exchange is *virtually* executed in a randomly selected pair of atoms and the resulting values of the ΔN_i are calculated. Note: the actual configuration of the crystal *does not change*.
(3) Point (2) is repeated many times.
(4) The evaluated set of ΔN_i numbers are substituted into Eq. (60) and combined with Eq. (59) to form a set of equations for pair interaction energies v_i.

In the original paper Gerold and Kern tested the method on several binary intermetallic systems. Among recent applications, studies of Ag–Al (Yu et al. 2004) and Fe–Pd (Mehaddene 2005) systems may be found.

The effective pair-interaction energy parameters v_i may be evaluated in a more straightforward way involving standard MC simulation – as proposed by Livet (1987). Here, an equilibrium atomic configuration at temperature T (above the "order–disorder" transition) is generated using the standard Metropolis algorithm with some starting set $\{v_i^{(0)}\}$ and the values of $\{\alpha_k^{(calc)}\}$ are calculated. The values of $v_i^{(0)}$ may be determined within the classical Clapp–Moss model (Clapp and Moss 1966). An iterative procedure is then applied, where subsequent corrections Δv_i to the pair-interaction parameters are calculated until the agreement between the calculated and experimental SRO parameters $\{\alpha_k\}$ is achieved.

12.3.10
Going beyond the Ising Model and Rigid-Lattice Simulations

Numerous papers published during past decades show that Monte Carlo methods of alloy modeling may easily operate in continuum space (i.e., beyond the rigid lattice approximation), the domain classically attributed to molecular dynamics. The corresponding studies concern liquid phases (see, e.g., Landau and Binder 2000), amorphous systems or surface phenomena (Dereli et al. 1989; Bhattacharya and Sethna 1998; Vauth et al. 2003; Rojas 2004). The amorphous distribution of atoms is either approximated (lattice gas-like approach (Vauth et al. 2003)), or modeled in real continuum (Dereli et al. 1989). The results concerning the formation of amorphous atomic configurations simulated by means of molecular dynamics and off-lattice MC were compared by Bhattacharya et al. (1998).

On the other hand, off-lattice MC simulations may be run for crystalline systems in order to model lattice strains and relaxations. Pioneering results in this field date from the 1990s (see, e.g., Lee 1991; Fratzl and Penrose 1995). An interesting study in this field has recently been performed on Al–Cu–Mg alloys by Mason et al. (Mason et al. 2004). The paper is recommended as a very good reference regarding the combination of MC concepts with the methodology typical of the continuous space treatment.

Another area of development of MC techniques in alloy physics is an implementation of the method with rigorous (or partially rigorous) physical models of interatomic interactions. This means going beyond the traditional Ising atomic pair interactions and accounting for many-body potentials. Literature from the field since the mid-1990s offers a number of interesting studies covering a wide range of research from the application of EAM potentials (Kim 1991; Rojas 2004) up to the involvement of rigorous ab-initio calculations (de Gironcoli and Gianozzi 1991; Keshari and Ishikawa 1994; Jellinek et al. 1998; Dorfman et al. 2001; Bozzolo et al. 2002; Wang et al. 2004; Vogtenhuber et al. 2005). In the latter case, however, rather small clusters of atoms have been simulated up to now.

12.3.11
Monte Carlo Simulations in View of other Techniques of Alloy Modeling

The adequacy of the Monte Carlo technique as a tool for solving problems in equilibrium and nonequilibrium statistical thermodynamics has been definitely proven and verified, mostly due to the extensive research output of Kurt Binder. The method may, therefore, be considered as parallel or complementary to other modeling techniques, which allows for interesting comparative studies.

Maugis et al. (2005) have recently studied precipitation in a model binary B2-ordered system by means of MC simulations, cluster dynamics, and a model based on classical laws of nucleation, growth, and coarsening. It was concluded that classical laws are in good agreement with simulations as long as the regular solution approximation remains adequate for the system. On the other hand, good agreement between MC and cluster dynamics results has been achieved when precisely describing the energetics of the smallest clusters.

Another example of a comparative study is found in the previously mentioned investigation of "order–order" kinetics in binary intermetallics (Oramus et al. 2001a, 2003). Here, the MC technique appeared the most efficient when simulating entire relaxations, while the correctness of basic assumptions (e.g., the vacancy mechanism of atomic migration) and the reliability of detailed findings (particular sequences of atomic jumps) were verified by molecular dynamics.

12.4
Phase Field Models

12.4.1
Introduction

Microstructures in materials may be thought of as bulk features (grains, domains, phases) that are separated by interfaces (grain boundaries, antiphase boundaries, interphase interfaces). In this picture, microstructural evolution is a process that comes about through the migration of these interfaces. Phase field models are essentially models that allow us to study the energy and kinetics of migration of interfaces. These models have proved to be very useful for understanding microstructural evolution in a wide variety of settings, and phase transformations in particular. Since these models are computationally intensive, their increasing popularity is also driven by the availability of cheaper and more powerful computers.

We begin this section with a tutorial introduction to the Cahn–Hilliard model (Cahn and Hilliard 1958) of a diffuse interface. The arguments that lead to the development of this model are firmly rooted in the physics of inhomogeneous alloys, and allow us to identify the salient – but generic – features of other more complex phase field models. We then introduce the Cahn–Allen model (Allen and Cahn 1979) for an antiphase boundary (APB) between two ordered domains. These two models are prototypical in that, together, they cover many of the phenomena of interest in materials science, and generalizations to other more complicated phenomena (such as solidification and grain growth) are often based on these two basic models. We will end this section with a brief description of features that need to be incorporated into phase field models for studying interesting effects such as anisotropy in interfacial energy.

The review by Chen (2002), and some parts of the reviews by Fratzl et al. (1999), Binder and Fratzl (2001), and Thornton et al. (2003) cover the major developments in phase field models and their applications.

For this section, we will confine our attention to condensed systems under isothermal conditions. From a thermodynamic point of view, then, microstructural evolution – by which we mean the evolution of the field variable $c(\mathbf{r}, t)$ – that the system undergoes must be such that it always decreases the system's free energy.

12.4.2
Cahn–Hilliard Model

We start with a description of the model of Cahn and Hilliard (1958), culminating in the Cahn–Hilliard equation (Cahn 1961). We will present some of the key results on interfacial energy and width. We will also discuss its numerical solution, and results from computer simulations.

Following Cahn and Hilliard, we consider a compositionally inhomogeneous binary A–B alloy with a local composition $c(\mathbf{r})$, and a (constant) density N_V of atoms (molecules) per unit volume.

12.4.2.1 Energetics

Defining f as the free energy per atom, the total free energy of the system, F, is expressed as an integral of f over the system volume in Eq. (62), where N_V is the number of atoms (molecules) per unit volume, assumed to be a constant.

$$F = N_v \int_V f \, dV \tag{62}$$

When the composition field $c(\mathbf{r})$ is non-uniform, the local, coarse-grained free energy f per atom is dependent not only on the local composition c, but also on the local spatial derivatives of the composition. Thus, we may expand f in a Taylor series about $f_o(c)$, the free energy per atom of a solution of uniform composition c [Eq. (63)], where L_i, κ_{ij}^A and κ_{ij}^B are the Taylor series coefficients, and are material property tensors, and we have defined $p_i = \partial c/\partial x_i$, and $q_{ij} = \partial^2 c/\partial x_i \partial x_j$.

$$f(c, p_i, q_{ij}, \ldots) = f_o(c) + L_i p_i + \frac{1}{2}\kappa_{ij}^A p_i p_j + \kappa_{ij}^B q_{ij} + \cdots \tag{63}$$

We have adopted the Einstein convention of summation over repeated indices. The Taylor series coefficients are formally defined by Eqs. (64–66).

$$L_i = \frac{\partial f}{\partial p_i} \tag{64}$$

$$\kappa_{ij}^A = \frac{\partial^2 f}{\partial p_i \partial p_j} \tag{65}$$

$$\kappa_{ij}^B = \frac{\partial f}{\partial q_{ij}} \tag{66}$$

For centrosymmetric crystals and amorphous solids, $L_i = 0$; also, since κ_{ij}^A and κ_{ij}^B are symmetric, second-rank tensors, they are isotropic for cubic and amorphous materials; i.e., $\kappa_{ij}^A = \kappa^A \delta_{ij}$ and $\kappa_{ij}^B = \kappa^B \delta_{ij}$, where δ_{ij} is the isotropic second-rank unity tensor (Kronecker symbol). Assuming only small composition gradients, and negligible higher spatial derivatives, we retain only those terms with up to second-rank tensors in Eq. (63). We first apply the divergence theorem to the term with κ_{ij}^B in Eq. (67), where $\hat{n}$ is the unit normal vector to the surface S.

$$\int_S \kappa_{ij}^B p_j n_i \, dS = \int_V \frac{\partial \kappa_{ij}^B}{\partial x_i} p_j \, dV + \int_V \kappa_{ij}^B q_{ij} \, dV \tag{67}$$

Since we are interested in the bulk behavior in an infinitely extended system, the composition gradient p_i may be assumed to be zero everywhere on the surface at infinity. Applying the chain rule $\partial \kappa_{ij}^B/\partial x_i = (\partial \kappa_{ij}^B/\partial c)(\partial c/\partial x_i)$, and rearranging, we obtain Eq. (68).

$$\int_V \kappa_{ij}^B q_{ij} \, dV = -\int_V \frac{\partial \kappa_{ij}^B}{\partial c} p_i p_j \, dV \tag{68}$$

Using this equation in Eq. (63) yields Eq. (69) for the free energy of the system in Eq. (62) (from now on, we neglect the higher-order terms in the Taylor series expansion in Eq. (63)), where $\kappa_{ij} = (\kappa_{ij}^A/2) - (\partial \kappa_{ij}^B/\partial c)$.

$$F = N_v \int_V [f_0(c) + \kappa_{ij} p_i p_j] \, dV \qquad (69)$$

For cubic crystals and amorphous solids, $\kappa_{ij} = \kappa \delta_{ij}$ is an isotropic tensor, and Eq. (69) reduces to Eq. (70).

$$F = N_v \int_V [f_0(c) + \kappa (\nabla c)^2] \, dV \qquad (70)$$

In most phase field models, κ is assumed to be a constant, and yields an isotropic interfacial energy.

12.4.2.2 Interfacial Energy and Width

In phase separating systems at low temperatures, the phase diagram exhibits a miscibility gap (Fig. 12.7; compare also Chapter 7). The plot of the free energy $f_0(c)$ versus composition for a temperature T below the critical temperature T_c exhibits a double well as shown in Fig. 12.8; any alloy whose composition lies inside the miscibility gap at T would, on equilibration, have a planar interface separating two phases of composition c_α and c_β. This interface would also be compositionally diffuse (as shown in Fig. 12.9) and hence have a finite width, within which the composition would change continuously from c_α to c_β. The interface finite width can be thought of as determined by a competition between two energy terms: if it is too wide, the system would have a large region of intermediate composition with higher $f_0(c)$. On the other hand, if the width is too narrow, the system would have a high energy due to the gradient energy term in Eq. (69). Thus, the system chooses that composition profile for which the functional F in Eq. (69) is minimized. This exercise yields the governing equation for $c^*(x)$, the equilibrium composition profile, Eq. (71), where $\Delta f(c) = f_0(c) - (1-c)\mu_A^e - c\mu_B^e$, and

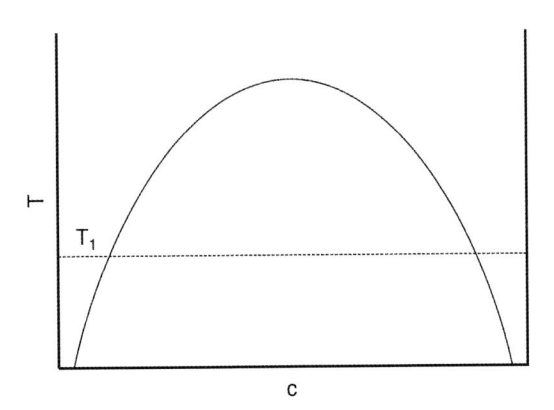

Fig. 12.7 Schematic phase diagram of a binary A–B phase separating system exhibiting a miscibility gap at low temperatures.

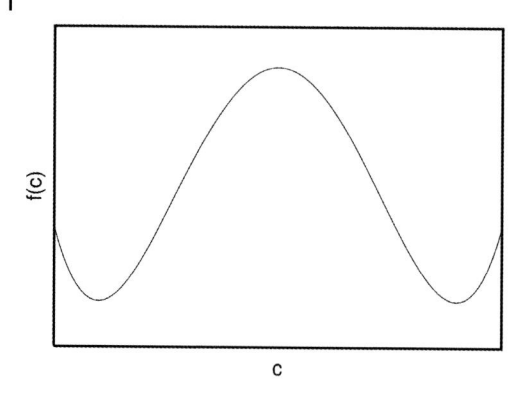

Fig. 12.8 Schematic depicting the dependence of bulk free energy density f_o on composition for the binary system depicted in Fig. 12.7 at a low temperature such as T_1.

μ_A^e and μ_B^e are the chemical potentials of species A and B, respectively, evaluated in a homogeneous alloy of composition c_α (or, equivalently, c_β).

$$\Delta f(c^*(s)) = \kappa \left(\frac{\partial c^*}{\partial s}\right)^2 \tag{71}$$

If we approximate the function $f_o(c)$ to the form of the double well potential in Eq. (72), where A_c sets the height of the barrier between the two energy wells in Fig. 12.8, it can be shown that the equilibrium composition profile is given by Eq. (73), where $a = (c_\beta - c_\alpha)\sqrt{A_c/\kappa}$. From this profile, the interfacial free energy can be shown to be $\sigma = (1/3)N_V(c_\beta - c_\alpha)^3\sqrt{A_c\kappa}$ and the interfacial width $w = 4(c_\beta - c_\alpha)^{-1}\sqrt{(\kappa/A_c)}$.

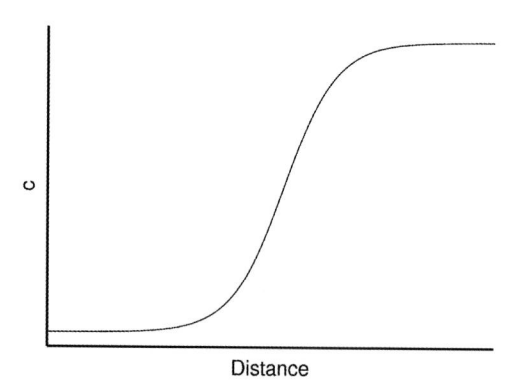

Fig. 12.9 Schematic of the equilibrium composition profile across a compositionally diffuse interface between two equilibrium phases of compositions c_α (to the left) and c_β (to the right).

$$f_o(c) = A_c(c - c_\alpha)^2(c_\beta - c)^2 \tag{72}$$

$$c = c_\alpha + (c_\beta - c_\alpha)\frac{e^{as}}{1 + e^{as}} \tag{73}$$

12.4.2.3 Dynamics

The evolution of the (inhomogeneous) composition field $c(\mathbf{r}, t)$ is modeled by writing J, the net flux of B atoms, as proportional to the gradient in chemical potential (to be defined below), as in Eq. (74), where M is a mobility tensor of second rank.

$$J = -M\nabla\mu \tag{74}$$

Using this form for the diffusional flux in the continuity equation, we obtain the equation that governs the dynamics of the composition field, Eq. (75).

$$\frac{\partial c}{\partial t} = \nabla \cdot M\nabla\mu \tag{75}$$

The (generalized) chemical potential, μ, is defined by Eq. (76).

$$\mu = \frac{\delta(F/N_v)}{\delta c} = \left[\frac{\partial f_o}{\partial c} - 2\kappa\nabla^2 c\right] \tag{76}$$

For cubic crystals and amorphous solids, the mobility tensor is isotropic: $M_{ij} = M\delta_{ij}$. Assuming M to be independent of concentration, Eqs. (75) and (76) yield Eq. (77).

$$\frac{\partial c}{\partial t} = M\nabla^2\left[\frac{\partial f_o}{\partial c} - 2\kappa\nabla^2 c\right] \tag{77}$$

In the literature, this equation (originally derived by Cahn 1961) is referred to as the Cahn–Hilliard equation. It is a nonlinear partial differential equation, since it contains terms such as $c^2, c^3, \ldots$, that enter through $\partial f_o/\partial c$ in Eq. (77).

Starting with an initial microstructure, given by an initial condition $c(\mathbf{r}, t = 0)$, solving Eq. (77) allows us to obtain $c(\mathbf{r}, t)$, and hence the microstructure, at any later time; since the Cahn–Hilliard equation cannot be solved analytically, microstructural evolution can be studied only by solving Eq. (77) numerically.

12.4.3
Numerical Implementation

In computer simulations of microstructure, the amount of available memory constrains us to study only a finite system; in order to avoid any spurious effects that may arise from the surfaces of the finite system, it is customary to employ periodic boundary conditions. The early computer simulations employed a simple, finite difference technique, with an explicit Euler technique for time-stepping. For example, for a two-dimensional (2D) system (generalization to 3D is straightforward), with an $N \times N$ square grid of size $\Delta x = \Delta y$, discretization of Eq. (77) yields Eq. (78), where $\mu_{p,q}^t = \mu(p\Delta x, q\Delta y, t)$ is given by Eq. (79).

$$\frac{c_{p,q}^{t+\Delta t} - c_{p,q}^t}{\Delta t} = M \left[\frac{\mu_{p+1,q}^t + \mu_{p-1,q}^t + \mu_{p,q+1}^t + \mu_{p,q-1}^t - 4\mu_{p,q}^t}{(\Delta x)^2} \right] \tag{78}$$

$$\mu_{p,q}^t = \left(\frac{\partial f_o}{\partial c}\right)_{p,q} - \left[\frac{c_{p+1,q}^t + c_{p-1,q}^t + c_{p,q+1}^t + c_{p,q-1}^t - 4c_{p,q}^t}{(\Delta x)^2} \right] \tag{79}$$

While such a discretization is simple to formulate, it has problems, the most important being the limit on the size of time step Δt that one can use (see Zhu et al. (1999) for details).

Most of the modern, large-scale studies are based on Fourier spectral techniques, one of which, referred to as a semi-implicit Fourier spectral method, developed by Zhu et al. (1999), is described below.

We start with the Fourier transform of the governing equation.

Since the composition field is described on N points on a regular grid of spacing Δx, the Fourier transform is also described on a regular grid in k-space, with a spacing of $2\pi/N\Delta x$. Using a tilde ($\sim$) on a real-space variable to denote its Fourier transform, the Fourier-transformed governing equation in k-space has the form of Eq. (80), where $\tilde{g}$ is the Fourier transform of $g(c) = \partial f_o/\partial c$.

$$\frac{d\tilde{c}}{dt} = -Mk^2 \tilde{g}(k,t) - 2M\kappa k^4 \tilde{c}(k,t) \tag{80}$$

Discretizing this equation with backward difference for time, we get Eq. (81), which can be rearranged to yield Eq. (82).

$$\frac{d\tilde{c}}{dt} = -Mk^2 \tilde{g}(k + \Delta t) - 2M\kappa k^4 \tilde{c}(k, t + \Delta t) \tag{81}$$

$$\tilde{c}(k, t + \Delta t) = \frac{\tilde{c}(k,t) - M\kappa k^2 \tilde{g}(k, t + \Delta t)}{1 + 2M\kappa k^4 \Delta t} \tag{82}$$

Note that the right-hand side of Eq. (82) has the term $\tilde{g}$, which needs to be evaluated at time $t + \Delta t$; since this function in real space $g(c) = \partial f_o/\partial c$ has nonlinear terms in c, it is difficult to evaluate its Fourier transform at the end of the time step. Thus, we approximate it using $\tilde{g}(k, t + \Delta t) \approx \tilde{g}(k, t)$ to get a semi-implicit formulation. It is this formulation that most modern studies use; it is applicable for bulk systems that are modeled using periodic boundary conditions. We note that Zhu et al. (1999) have also shown how this method can be extended to study systems in which the atomic mobility M is a position-dependent variable.

Since the governing equations are all partial differential equations, other techniques are also available; we note in particular the finite element method (Canuto et al. 1988) and the control volume method (Patankar 1980; Ferziger and Peric 2002).

For a given number of grid points N and a given total number of time steps, we would like to use in our simulations the largest possible values of Δx and Δt. However, since the interface is compositionally diffuse, and has a finite width, we have to ensure a sufficiently large number of grid points at the interface to resolve the composition profile there. Typically, we use four to six grid points to resolve the interface; i.e., $w = n\Delta x$, with $4 \leq n \leq 6$.

Thus, for a new phase field model, one has first to compute, as accurately as possible, the equilibrium profile of the field variable across a planar interface, and estimate its width w. In the more elaborate, large-scale simulations, one then ensures that w spans at least four grid points.

12.4.4
Application: Spinodal Decomposition

Figure 12.10 shows the evolution of the composition field during spinodal decomposition in a symmetric alloy $c = 0.5$ quenched to a low temperature at which it finds itself well inside the miscibility gap in Fig. 12.7. Since the free energy $f_o(c)$ exhibits negative curvature, this alloy is unstable with respect to small composition fluctuations, which are introduced into the simulations through a small random noise at every grid point.

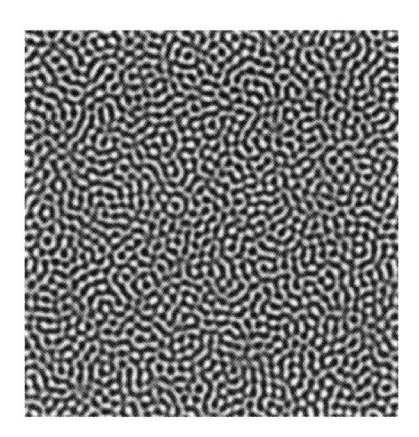

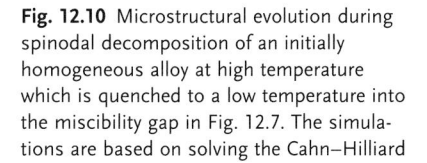

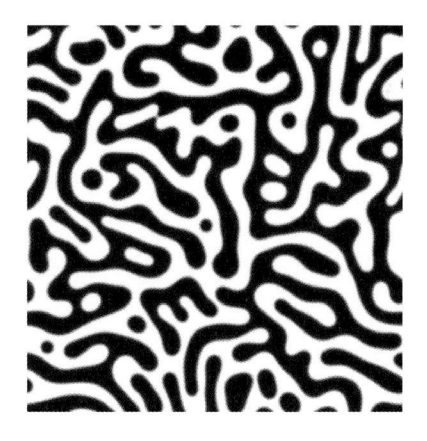

Fig. 12.10 Microstructural evolution during spinodal decomposition of an initially homogeneous alloy at high temperature which is quenched to a low temperature into the miscibility gap in Fig. 12.7. The simulations are based on solving the Cahn–Hilliard equation [Eq. (75)]. The dark and light shades represent A-rich and B-rich phases, respectively. The average composition c_0 remains conserved during the simulations. The figure on the left is for an earlier time than the right-hand figure.

However, we must mention that the Cahn–Hilliard model, as developed here, is incapable of studying early stages of phase separation in alloys outside the spinodal. In these alloys (say, A-rich alloys), phase separation requires nucleation of B-rich islands; since the process of nucleation increases the system free energy, and the Cahn–Hilliard equation ensures a strictly decreasing F, nucleation events cannot be studied. A possible way out of this problem is to use thermally induced fluctuations (a noise term) in the equation (such an equation is termed the Cahn–Hilliard–Cook equation (Cook 1970)) at every time step (see Binder and Fratzl (2001) for further details on the formulation). Simulation of nucleation through this technique, however, is time-consuming. Moreover, in many cases one is interested in microstructural evolution during growth and, more importantly, during coarsening (Ostwald ripening). In these situations, one simply introduces viable nuclei of the B-rich phase into the system at the beginning of the simulations.

12.4.5
Cahn–Allen Model

Some alloys (for example, Cu–Zn) exhibit a disorder-to-order transition; at high temperatures, the alloy is in the form of a random solid solution which, at low temperatures, becomes ordered. For example, a bcc alloy, with an A2 structure, may undergo this transition to exhibit a B2 structure, in which atoms of species A and B occupy, preferentially, the corner and body-center sites, respectively (or vice versa). If the A2 structure is thought to be composed of a sum of two interpenetrating simple cubic lattices (referred to as α- and β-sublattices), the B2 structure is obtained when the two sublattices are differentially occupied by species B; in the parent A2 structure, species B occupies the two sublattices randomly.

Clearly, an alloy in the B2 structure possesses a long-range order, but is disordered in the A2 structure at high temperatures. A parameter η that quantifies this long-range order may be defined in many ways. A popular choice for η for describing the B2 structure is by Eq. (83), where x_B^α (or x_B^β) is the fraction of α (or β) sublattice sites occupied by B atoms.

$$\eta = (x_B^\beta - x_B^\alpha) \tag{83}$$

From this definition, configurations with $(+\eta)$ and $(-\eta)$ are equivalent. However, a region in the alloy that has B atoms on the β sublattice (i.e., $\eta > 0$) and another that has them on the α sublattice (with $\eta < 0$) would find themselves separated by an "antiphase boundary" (APB).

The order parameter as defined above is not in all cases just a scalar quantity; it can be a more dimensional vectorial parameter for structures made of more than two sublattices, such as the L1$_2$ structure. In contrast to the central quantity of the Cahn–Hilliard theory, the concentration, the order parameter is a nonconserved quantity, i.e., its integral value over the whole sample is not constant during the temporal evolution.

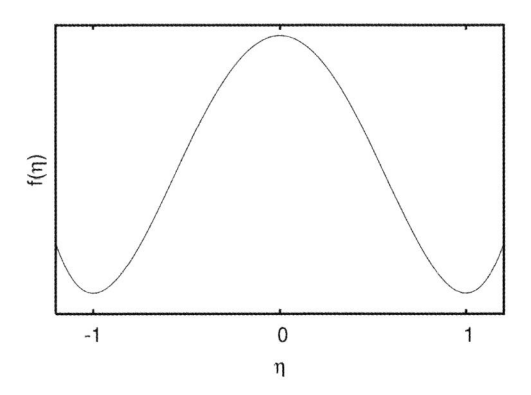

Fig. 12.11 Schematic of the dependence of the bulk free energy density on the order parameter η in the Cahn–Allen model.

The model of Allen and Cahn (1979) is for the energy and dynamics of such an APB, which may be seen as a region in which the order parameter varies continuously (from $+\eta$ to $-\eta$) over a region of finite width. The arguments used for developing the Cahn–Allen model, therefore, are the same as those for the Cahn–Hilliard model.

Following the same steps as the Cahn–Hilliard model, the total free energy may be written as Eq. (84), where, $f_\eta(\eta)$ is the (bulk) free energy density of a homogeneous alloy with uniform η; $f_\eta(\eta)$ is shown schematically in Fig. 12.11.

$$F = \int_V [f_\eta(\eta) + \kappa_\eta (\nabla \eta)^2] \, dV \tag{84}$$

12.4.5.1 Kinetics

Since the order parameter is not conserved, Cahn and Allen postulated Eq. (85) for the dynamics, where L is a (positive) relaxation coefficient. This postulate implies that, locally, the order parameter would evolve in response to the local driving force $\delta F / \delta \eta$.

$$\frac{\partial \eta}{\partial t} = -L \frac{\delta (F/N_v)}{\delta \eta} \tag{85}$$

Substituting Eq. (84) into this equation, we get the Cahn–Allen equation (which, in physics literature, goes by the name "time-dependent Ginzburg–Landau (TDGL) equation") [Eq. (86)].

$$\frac{\partial \eta}{\partial t} = -L \left[\frac{\partial f}{\partial \eta} - 2\kappa \nabla^2 \eta \right] \tag{86}$$

Since this equation is similar in form to (but simpler than) the Cahn–Hilliard equation, the derivation of the semi-implicit Fourier spectral procedure for nu-

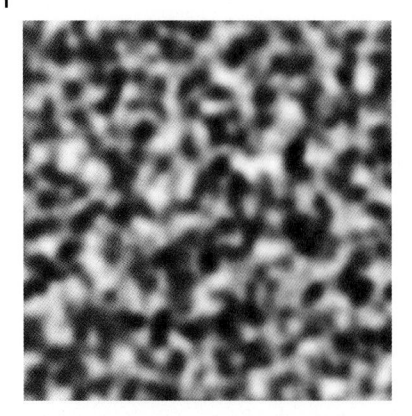

Fig. 12.12 Microstructural evolution in an initially disordered ($\eta = 0$) alloy quenched to a low temperature (below the disorder-to-order, or critical, temperature T_c). The dark and light shades represent regions with $\eta < 0$ and $\eta > 0$, respectively. The simulation is based on solving the Cahn–Allen equation (86). In contrast to Fig. 12.10, the average value of η in these figures is not conserved during the simulation. The left-hand figure is for an earlier time than the right-hand figure.

merically solving this equation is not presented here. Figure 12.12 shows the evolution of ordered domains during the ordering of an initially disordered alloy, simulated by solving the Cahn–Allen equation, above.

12.4.6
Generalized Phase Field Models

12.4.6.1 Key Features of Phase Field Models

Our discussion of the Cahn–Hilliard and Cahn–Allen models now allows us to list the key features of almost all phase field models used for studying microstructural evolution in materials.

First, they use a continuum mathematical description in which one or more position-dependent field variables are used for specifying a system's configuration (microstructure). Each configuration is associated with an appropriate energy; this energy is written as a functional of the field variables. In particular, this functional has two contributions: a bulk free energy density, and a gradient energy density. The dynamics of the field variables are governed by a set of partial differential equations: the Cahn–Hilliard equation for conserved fields, and the Cahn–Allen equation for nonconserved fields.

As with the other simulation techniques in this chapter, phase field models (a) provide a mathematical description of an instantaneous configuration of the system, (b) describe an appropriate energy (in the case of isothermal, isobaric systems, it would be the Gibbs free energy) of a configuration, and (c) describe a method by which the system evolves from one configuration to the next.

In this section, let us consider such generalized phase field models and how they have been applied in studies of interesting phenomena. The most important step is in the choice of an appropriate set of field variables to describe a microstructure. This choice then drives the remaining steps.

12.4.6.2 Precipitation of an Ordered Phase

A simple generalization is a model for studying precipitation of an ordered phase. Since the matrix and precipitate phases have different compositions, $c(\mathbf{r}, t)$ is a relevant field variable; further, the precipitate is also an ordered phase, so $\eta(\mathbf{r}, t)$ is also a relevant field variable, yielding Eq. (87), with the bulk free energy density $f(c, \eta)$ obtained from (bulk) thermodynamics of the alloy system. It must yield the two-phase equilibrium between the matrix (characterized by $(c, \eta) = (c_\alpha, 0)$) and the precipitate (with $(c, \eta) = (c_\beta, \eta^*)$) phases. Microstructural evolution is governed by a set of two coupled equations; one is the Cahn–Hilliard equation for composition, and the other a Cahn–Allen equation for the order parameter.

$$F = \int_V [f(c, \eta) + \kappa_c (\nabla c)^2 + \kappa_\eta (\nabla \eta)^2] \, dV \qquad (87)$$

It is now possible to generalize this model further to take into account more than one order parameter. Additional η values are needed when one considers ordered phases that have several (more than two) variants. For example, in Ni–Al-based superalloys, the matrix is disordered, with an fcc structure, and the precipitate is ordered, with the L1$_2$ structure. Considering the fcc lattice as a combination of four interpenetrating simple cubic (sub)lattices, the L1$_2$ structure is obtained when one of them is preferentially occupied by B atoms, and the other three by A atoms; in the perfectly ordered state, the alloy composition would be A_3B. An (imperfectly) ordered alloy with L1$_2$ may be described using three independent concentration waves, whose amplitudes can be associated with three order parameters (η_1, η_2, η_3) (Wang et al. 1998).

The four variants of the ordered structure that can be formed from the same disordered fcc lattice are then characterized by $(\eta, \eta, -\eta)$, $(\eta, -\eta, \eta)$, $(-\eta, \eta, \eta)$, and $(-\eta, -\eta, -\eta)$. With $f(c, \eta_1, \eta_2, \eta_3)$ as the bulk free energy density, the total free energy would be given by Eq. (88).

$$F = N_v \int_V \left[f(c, \eta_1, \eta_2, \eta_3) + \kappa_c (\nabla c)^2 + \sum_{p=1}^{3} \kappa_p (\nabla \eta_p)^2 \right] dV \qquad (88)$$

The expression for the bulk free energy density $f(c, \eta_1, \eta_2, \eta_3)$ must obey certain symmetry properties; see Wang et al. (1998) for details. The evolution equations now consist of a Cahn–Hilliard equation for composition, and a set of three Cahn–Allen equations, one for each order parameter η_p.

Using such a model, Wang et al. (1998) have studied precipitation of γ' precipitates from the γ matrix in Ni–Al alloys. Similar models have been used by Wen et al. (1999) in their study of precipitation of an orthorhombic phase from a hexagonal phase in Ti–Al–Nb alloys.

12.4.6.3 Grain Growth in Polycrystals

A polycrystalline material has grains separated by grain boundaries. At sufficiently high temperatures, the grain boundaries migrate due to their curvature, leading to grain coarsening (often referred to as grain growth). In order to study this process, Fan and Chen (1997) used a model in which each grain (say, the p-th grain) is associated with an "orientational order parameter" η_p such that $\eta_i = 1$ in the region occupied by the ith grain, and η_i is zero everywhere outside it. In this model, a grain boundary between the pth and qth grains is a region in which η_p and η_q exhibit sharp gradients. Further, for a polycrystal with n grains, this model requires the use of n orientational order parameters. In effect, this model can be thought of as the continuum equivalent of the n-state Potts model (Potts 1952).

Since the grain boundary can migrate, thus transforming regions from one grain to another, the order parameter is clearly not conserved; Fan and Chen derived a set of Cahn–Allen equations to describe the dynamics of the order parameters.

To see how this model works, here is the free energy functional [Eq. (89)].

$$F = \int_V \left[f(\eta_p) + \sum_{p=1}^{n} \kappa_p (\nabla \eta_p)^2 \right] dV \qquad (89)$$

The bulk free energy is formulated in such a way that it has n degenerate minima (wells of equal depth) at locations $(1, 0, 0, \ldots, 0), (0, 1, 0, \ldots, 0), \ldots (0, 0, 0, \ldots, 1)$, For a system with only two grains, only η_1, and η_2 are sufficient to describe the microstructure; for such a system $f(\eta_1, \eta_2)$, illustrated in Fig. 12.13, would have two absolute minima at (η_1, η_2) values of $(1, 0)$ and $(0, 1)$; further, since these two

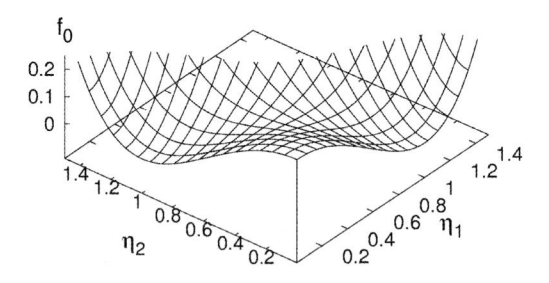

Fig. 12.13 Schematic of the dependence of the bulk free energy density on two order parameters (which represent two arbitrarily different grain orientations) in the Fan–Chen model. The free energy surface has two absolute minima at $(\eta_1, \eta_2) = (1, 0)$ and $(0, 1)$. The free energy values at the two minima are the same.

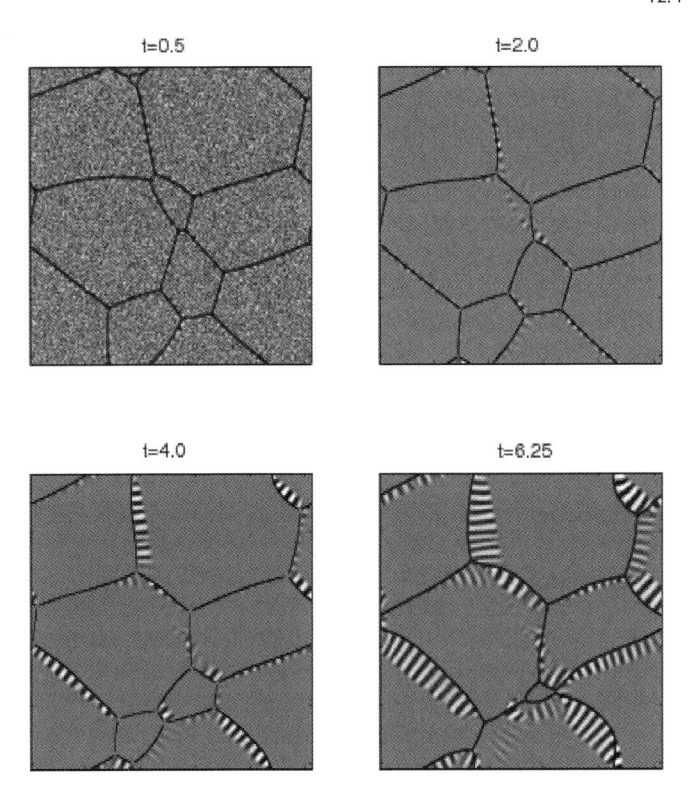

Fig. 12.14 Microstructural evolution during spinodal decomposition in a polycrystalline alloy. The simulation is based on a phase field model that combines the Cahn–Hilliard model for a phase separating alloy with the Fan–Chen model for a polycrystal. This particular system is characterized by large values of atomic mobility M and grain boundary mobility L. The scaled simulation time is shown above each microstructure (from Ramanarayan 2004).

configurations (which represent two different grains) are equivalent to each other, the free energy values at these two minima would also be equal.

The Fan–Chen model was combined with the Cahn–Hilliard model in a recent study (Ramanarayan 2004) of grain boundary effects on spinodal decomposition. A sample result from this work is shown in Fig. 12.14. In this example, migrating grain boundaries act as "transformation fronts" at which spinodal decomposition takes place (due to fast diffusion at the grain boundary); the resulting microstructure exhibits alternating lamellae of (incompletely transformed) A-rich and B-rich regions lying normal to the (migrating) grain boundary. This example shows clearly that phase field models can be used in simulations of phenomena that could lead to highly complex microstructures.

12.4.6.4 Solidification

As a final example, let us consider a phase field model which is in wide use in solidification studies. For simplicity, we consider solidification in a pure metal. The microstructure is described in terms of a phase field variable η that takes a value of zero in the liquid and unity in the solid; the solid–liquid interface is a region where η exhibits a large (but continuous) gradient. This continuum description has several advantages over the sharp interface models that were in wide use earlier: with a phase field model, there is no need to track the location of the interface at every time step. Thus, it is easy to study complex shapes and topological changes (such as a merger of two dendritic arms).

Solidification of a region from the liquid state to the solid state is signified by a rise in the value of η in and near the interface region, accompanied by an evolution of latent heat which has to be transported away. Thus, one uses a coupled set of equations governing the evolution of two field variables: the phase field η and the temperature field T; see Langer (1986) and Kobayashi (1993) for details.

Since the phase field variable η has been introduced into the model largely for mathematical convenience, early studies were devoted to establishing that this model does lead to results that are consistent with those from classical sharp-interface theories. They established, for example, that it reproduces the Gibbs–Thomson correction to the melting point due to a curved interface. We refer the reader to a recent review by Karma (2001) and to the text by Davis (2001) for further details.

12.4.7
Other Topics

The popularity of phase field techniques is due to the ease with which many important effects can be incorporated into the formalism. A recent review by Chen (2002) discusses a wide variety of phenomena where these models have been used to study microstructural evolution. In addition to the ones we have covered, these phenomena include ferroelectric transformations, martensitic transformations, phase transformations in thin films, dislocation dynamics, and crack propagation. In this section, we restrict ourselves to a brief discussion of developments in incorporating into phase field models (a) an interfacial energy anisotropy and (b) elastic stress effects.

12.4.7.1 Anisotropy in Interfacial Energy

In many crystalline systems, the interfacial energy σ is anisotropic. Since the gradient energy coefficient tensor κ_{ij} is of second rank, systems with lower symmetries (tetragonal, orthorhombic, etc.) can be studied in a limited way using anisotropic versions of κ_{ij}; however, for cubic systems, this tensor is isotropic. Thus, systems with anisotropic σ (particularly those with cubic anisotropy) require other strategies. The most popular strategy (due to Kobayashi 1993) is to use a scalar (or isotropic) value of κ which depends on the interface orientation (identified with the direction of the local gradient in order parameter). A second method is to extend the original Cahn–Hilliard formalism to include higher-order terms in the Taylor series expansion of the local free energy f in Eq. (63). This idea, alluded to by Langer (1986) and Taylor and Cahn (1998), was pursued by Abinandanan and Haider (2001), who worked out all the fourth-order terms necessary for describing cubic anisotropy in σ, and used one of them to study the finite size dependence of particle shapes. A third possibility is to work with multiple-order parameter models, such as those required for describing the $L1_2$ structure (Braun et al. 1998).

12.4.7.2 Elastic Strain Energy

In many solid systems with multiple phases (such as the superalloys which have a Ni-rich fcc solid solution as the matrix and the $L1_2$-ordered intermetallic solution Ni_3Al as the precipitate phase), the interface is coherent. In such systems, a lattice parameter mismatch between the two phases introduces elastic strains everywhere, which have been shown to have a strong influence on microstructural evolution. For example, in the Ni–Al system, particles of the precipitate phase change their shape from sphere to a cuboid and, still later, to plates. These particles also become aligned along the $\langle 100 \rangle$ directions of the Ni-rich matrix (Khachaturyan 1983; Johnson 1999).

These elastic stress effects have been incorporated into the phase field models by using the simplest approximation, in which the total free energy of the system F_t is a simple sum of the elastic strain energy of the system F_{el} and the chemical free energy F_{ch} [Eq. (90)].

$$F_t = F_{ch} + F_{el} \tag{90}$$

The chemical part is given in the usual form of a functional, such as Eq. (70). The rest of the formalism (for example, the definition of chemical potential in Eq. (76)) remains the same, except that F in these equations is replaced by F_t. The elastic stress effects work in the following way in such models: the order parameter determines the lattice parameter (or, effectively, the misfit strain), which, in turn, results in a stress field. Thus, even though F_{el} is introduced as an additive term in the above equation, it has an implicit dependence on the order parameter distribution and hence the microstructure; it is through this coupling that it is able to exert an influence back on the microstructure and its evolution.

Diffusional equilibration (relaxation) of the composition field is far slower than that of the elastic (mechanical) equilibration of the stress field. Thus, for a given configuration (at the beginning of the time-stepping procedure), we may compute the elastic stress field, and use it in solving for the evolution equation for the order parameter fields at the end of the time step. This implies that incorporating elastic stress effects into the phase field formalism requires, first, a description of how the misfit strain depends on the order parameter(s) and, second, an additional module that computes the elastic stress field for a given distribution of misfit strains. For the second part, Fourier transform formalisms exist (Khachaturyan 1983; Mura 1982), making it easy to include this module in the usual (Fourier transform-based) solution of the evolution equations.

12.5
Outlook

Further progress in the development of computational techniques, but of course also of computer technology, will make it possible to study larger systems on longer time scales. Nevertheless, no drastic progress can be expected here, but it is necessary to combine the respective methods into multiscale integrated codes. The somewhat futuristic expectation is to generate effective potentials from ab-initio codes, derive short-term (and local) development using molecular statics or dynamics methods, and finally use Monte Carlo, phase field, or finite element codes to compute the large-scale, long-term evolution of an alloy (see, e.g., Yip 2005; Lu 2005). On the other hand, reduced dimensions as in nanotechnology and in thin-film devices get into length scales which can be realistically modeled with current techniques.

Appendix

Let us assume that a Metropolis-type MC algorithm is used. In such a case the probability p_i of the particular ith atom from the first coordination shell of the vacancy executing the jump is given by Eq. (A1), where the meaning of w_i (atomic jump frequency) is analogous to $W(\sigma_i \rightarrow \sigma_j)$ in Eqs. (49) and (50).

$$p_i = \frac{1}{Z} \times \tau \times w_i \tag{A1}$$

The factor $1/Z$ accounts for the initial random choice among the Z nearest neighbors of the vacancy. The parameter τ has in fact the meaning of a reciprocal jump-attempt frequency discussed in Chapter 5.

The corresponding probability for the event that the ith atom is selected, but the jump is rejected is given by Eq. (A2).

$$\bar{p}_i = \frac{1}{Z} \times (1 - \tau \times w_i) = \frac{1}{Z} - p_i \tag{A2}$$

The probability p_i, referred to as the "a posteriori" probability (Athenes et al. 1997), defines the effective time scale constant $\tau' = \tau/Z$ of the simulated process. Accordingly, after each Metropolis-type MC step time has to be incremented by Δt [Eq. (A3)]

$$\Delta t = \tau/Z \tag{A3}$$

This becomes still more clear if one calculates a probability for the event that the ith atom jumps to the vacancy after n subsequent *unsuccessful* attempts preceding this jump. The probability of the event is given by Eq. (A4), and the total time increment to be attributed to the event by Eq. (A5).

$$p_{n,1} = \left[\sum_i \bar{p}_i\right]^n \times p_i \tag{A4}$$

$$\Delta t_{n,1} = \frac{(n+1) \times \tau}{Z} \tag{A5}$$

It is obvious that $\Delta t_{n,1}$ has a meaning of the residence time for the ith atom on the initial lattice site.

The average residence time for this atom is, therefore, given by Eq. (A6), which is perfectly consistent with the considerations presented in Section 12.3.6.3 [Eqs. (54) and (55)].

$$\overline{\Delta t_{n,1}} = \frac{\tau}{Z} \times \sum_{n=1}^{\infty} (n+1) \times p_{n,1} = \frac{1}{\sum_i w_i} \tag{A6}$$

References

Abinandanan, T. A., Haider, F. (2001), *Phil. Mag. A*, **81**, 2457.

Allen, M. P., Tildesley, D. J. (1987), *Computer Simulation of Liquids*. Oxford University Press, Oxford.

Allen, S. M., Cahn, J. W. (1979), *Acta Metall.* **27**, 1085.

Alder, B. J., Wainright, T. E. (1957), *J. Chem. Phys.* **27**, 1208.

Alder, B. J., Wainright, T. E. (1959), *J. Chem. Phys.* **31**, 459.

Andersen, H. C. (1980), *J. Chem. Phys.*, **72**, 2384.

Athenes, M. (1997), Thesis, Université Paris 6.

Athenes, M., Bellon, P., Martin, G., Haider, F. (1996), *Acta Mater.* **44**, 4739.

Athenes, M., Bellon, P., Martin, G. (1997), *Phil. Mag. A* **76**, 565.

Baskes, M. I. (1987), *Phys. Rev. Lett.* **59**, 2666.

Bennett, C. H. (1976), *J. Comput. Phys.* **22**, 245.

Bhatia, B., Xinjun-Luo, Sholl, C. A., Sholl, D. S. (2004), *J. Phys.: Condens. Matter* **16**, 8891.

Bhattacharya, K. K., Sethna, J. P. (1998), *Phys. Rev. E* **57**, 2553.

Binder, K. (1976), in: *Phase Transitions and Critical Phenomena*, Vol. 5b, Domb, C., Green, M. S. (Eds.), Academic Press, New York, p. 1.

Binder, K., Heermann, D. W. (1997), *Monte Carlo Simulation in Statistical Physics. An Introduction*, Springer, Berlin, p. 5.

Binder, K., Lebowitz, J. L., Phani, M. K., Kalos, M. H. (1981), *Acta Metall.* **29**, 1655.

Binder, K., Fratzl, P. (2001), in: *Phase Transformations in Materials*, Kostorz, G. (Ed.), Wiley-VCH, Weinheim, Chapter 6.

Bortz, A. B., Kalos, M. H., Lebowitz, L. J. (1975), *J. Comput. Phys.* **17**, 10.

Bozzolo, G. H., Khalil, J., Noebe, R. D. (2002), *Comput. Mater. Sci.* **24**, 457.

Braun, R. J., Cahm, J. W., NcFadden, G. B., Rushmeier, H. E., Wheeler, D. D. (1998). *Acta Mater.* **46**, 1.

Cahn, J. W. (1961), *Acta Metall.* **9**, 795.

Cahn, J. W., Hilliard, J. E. (1958), *J. Chem. Phys.* **28**, 258.

Canuto, C., Hussaini, M. Y., Quarteroni, A., Zang, T. A. (1988), *Spectral Methods in Fluid Dynamics*, Springer, Berlin.

Canzian, A., Mosca, H. O., Bozzolo, G. (2004), *Surf. Rev. Lett.* **11**, 235.

Car, R., Parrinello, M. (1985), *Phys. Rev. Lett.* **55**, 2471.

Chen, L.-Q. (2002), *Annu. Rev. Mater. Res.* **32**, 113.

Clapp, P. C., Moss, S. C. (1966), *Phys. Rev.* **142**, 418.

Cook, H. E. (1970), *Acta Metall.* **18**, 297.

Davis, S. H. (2001), *Theory of Solidification*, Cambridge University Press, Cambridge.

Daw, M. S., Baskes, M. I. (1984), *Phys. Rev. B*, **29**, 6443.

de Gironcoli, S., Giannozzi, P. (1991), *Phys. Rev. Lett.* **66**, 2116.

Dereli, G., Yalabik, M. C., Ellialtioglu, S. (1989), *Phys. Scr.* **40**, 117.

Dorfman, S., Mundim, K. C., Liubich, V., Fuks, D. (2001), *J. Appl. Phys.* **90**, 705.

Eberhard, B. (2006), Ph.D. Thesis, University of Augsburg.

Einstein, A. (1905), *Annln. Phys.* **17**, 549.

Ercolessi, F., Adams, J. B. (1994), *Europhys. Lett.* **26**, 583.

Fan, D., Chen, L.-Q. (1997), *Acta Mater.* **45**, 1115.

Ferziger, J. H., Peric, M. (2002), *Computational Methods for Fluid Dynamics*, 3rd edn., Springer, Berlin.

Finnis, M. W., Sinclair, J. E. (1984), *Phil. Mag. A*, **50**, 45.

Flinn, P. A., McManus, G. M. (1961), *Phys. Rev.* **124**, 54.

Foiles, S. M., Baskes, M. I., Daw, M. S. (1986), *Phys. Rev. B*, **33**, 7983.

Fratzl, P., Penrose, O. (1995), *Acta Metall. Mater.* **43**, 2921.

Fratzl, P., Penrose, O., Lebowitz, J. L. (1999), *J. Stat. Phys.* **95**, 1429.

Gear, C. W. (1966). The numerical integration of ordinary differential equations of various orders. Report ANL 7126, Argonne National Laboratory.

Gear, C. W. (1971). *Numerical Initial Value Problems in Ordinary Differential Equations*. Prentice-Hall, Englewood Cliffs, NJ.

Gehlen, P. C., Cohen, J. B. (1965), *Phys. Rev.* **139**, 844.

Gerold, V., Kern, J. (1987), *Acta Metall.* **35**, 393.

Gillespie, D. T. (1976), *J. Comput. Phys.* **22**, 403.

Glauber, R. J. (1963), *J. Math. Phys.* **4**, 294.

Guofeng-Wang, Van-Hove, M. A., Ross, P. N., Baskes, M. I. (2004), *J. Chem. Phys.* **121**, 5410.

Guofeng-Wang, Van-Hove, M. A., Ross, P. N., Baskes, M. I. (2004), in: *Nanoparticles and Nanowire Building Blocks Synthesis, Processing, Characterization and Theory*, Glembocki, O. J., Hunt, C. E. (Eds.), Materials Research Society, Warrendale, p. 89.

Hanada, T., Hirotsu, Y., Ohkubo, T. (2004), *Mater. Trans.* **45**, 1194.

Heermann, D. W. (1986), *Computer Simulation Methods in Theoretical Physics*, Springer Verlag, Berlin, p. 74.

Hoover, W. G. (1985), *Phys. Rev. A*, **31**, 1695.

Huang, K. (1963), *Statistical Mechanics*, Wiley & Sons, New York, p. 139.

Jellinek, J., Srinivas, S., Fantucci, P. (1998), *Chem. Phys. Lett.* **288**, 705.

Johnson, W. C. (1999), in: *Lectures in the Theory of Phase Transformations*, 2nd edition, Aaronson, H. I. (Ed.), TMS, Warrendale.

Kamakoti, P., Sholl, D. S. (2005), *Phys. Rev. B* **71**, 1 4301.

Karma, A. (2001), Phase field methods, in: *Encyclopedia of Materials Science and Technology*, Buschow, K. H. J., Cahm, R. W., Flemings, N. C., Ilschmer, B. (Eds.), Elsevier, Oxford.

Kawasaki, K. (1966), *Phys. Rev.* **145**, 224.

Keshari, V., Ishikawa, Y. (1994), *Chem. Phys. Lett.* **218**, 406.

Khachaturyan, A. G. *Theory of Structural Trnasformations*, Wiley and Sons, New York.

Kim, S. M. (1991), *J. Mater. Res.* **6**, 1455.

Kobayashi, R. (1993), *Physica D* **63**, 410.

Kossel, W. (1927), *Nachr. Akad. Wiss. Göttingen, Math. Phys. Kl.*, 135.

Kozlowski, M., Kozubski, R., Pierron-Bohnes, V., Pfeiler, W. (2005), *Comput. Mater. Sci.* **33**, 287.

Kozubski, R. (1997), *Prog. Mater. Sci.* **41**, 1.

Landau, D. P., Binder, K. (2000), *A Guide to Monte Carlo Simulations in Statistical*

Physics, Cambridge University Press, Cambridge, p. 48.

Langer, J. S. in: *Directions in Condensed Matter Physics* (1986), Grinstein, G., Mazenko, G. (Eds.), World Scientific, Singapore.

Lee, J. K. (1991), *Metall. Trans. A*, **22**, 1197.

Lennard-Jones, J. E. (1931), *Proc. Camb. Phil. Soc.* **27**, 469.

Livet, F. (1987), *Acta Metall.* **35**, 2915.

Lu, G., Kaxiras, E. (2005), in: *Handbook of Theoretical and Computational Nanotechnology*, Vol. X, Rieth, M., Schommers, W. (Eds.), American Scientific Publishers, Los Angeles, pp. 1–33.

Ma, S. K. (1981), *J. Stat. Phys.* **26**, 221.

Mason, D. R., Rudd, R. E., Sutton, A. P. (2004), *J. Phys.: Condens. Matt.* **16**, S2679.

Maugis, P., Soisson, F., Lae, L. (2005), *Defect and Diffusion Forum* **237–240**, 671.

McGreevy, R. L., Pusztai, L. (1988), *Mol. Sim.* **1**, 359.

Mehaddene, T. (2005), *J. Phys.: Condensed Matt.* **17**, 485.

Metropolis, N., Rosenbluth, A. W., Rosenbluth, N. N., Teller, A. H., Teller, E. (1953), *J. Chem. Phys.* **21**, 1087.

Mura, T. (1982), *Micromechanics of Defects in Solids*, Martinus Nijhoff, The Hague.

Murch, G. E. (1982), *Philos. Mag.* **45**, 941.

Murch, G. E. (1984), in: *Diffusion in Crystalline Solids*, Murch, G. E., Nowick, A. S. (Eds.), Academic Press, Orlando, p. 379.

Nosé, S. (1984), *Mol. Phys.* **52**, 255.

Nosé, S., Klein, M. L. (1983), *Mol. Phys.* **50**, 1055.

Novotny, M. A. (1995), *Comput. Phys.* **9**, 46.

Oramus, P., Kozubski, R., Pierron-Bohnes, V., Cadeville, M. C., Pfeiler, W. (2001a), *Phys. Rev. B* **63**, 174 109.

Oramus, P., Kozubski, R., Pierron-Bohnes, V., Cadeville, M. C., Massobrio, C., Pfeiler, W. (2001b), *Defect and Diffusion Forum* **194–199**, 453.

Oramus, P., Massobrio, C., Kozłowski, M., Kozubski, R., Pierron-Bohnes, V., Cadeville, M. C., Pfeiler, W. (2003), *Comput. Mater. Sci.* **27**, 186.

Parrinello, M., Rahman, A. (1980), *Phys. Rev. Lett.* **45**, 1196.

Parrinello, M., Rahman, A. (1981), *J. Appl. Phys.* **52**, 7182.

Patankar, S. V. (1980), *Numerical Heat Transfer and Fluid Flow*, Taylor and Francis London.

Pitsch, W., Gahn, U. (1993), *Mater. Technol. Steel Res.* **64**, 484.

Potts, R. B. (1952), *Proc. Cambridge Philos. Soc.* **48**, 106.

Ramanarayan, H. (2004), Grain boundary effects on spinodal decomposition, Ph.D. Thesis, Indian Institute of Science, Bangalore.

Rojas, M. I. (2004), *Surf. Sci.* **569**, 76.

Schneider, T., Stoll, E. (1978), *Phys. Rev. B*, **17**, 1302.

Schober, H. R., Petry, W., Trampenau, J. (1992), *J. Phys.: Condens. Matter* **4**, 9321; **5**, 993.

Schweiger, H., Podloucky, R., Püschl, W., Pfeiler, W. (2001), *Mater. Res. Soc. Symp. Proc.* **646**, N5.11.1-6.

Schweika, W. (1998), *Disordered Alloys: Diffuse Scattering and Monte Carlo Simulations*, Springer Verlag, Berlin, p. 1.

Sorensen, M. R., Voter, A. F. (2000), *J. Chem. Phys.* **112**, 9599.

Sutton, A. P., Ballu, R. W. (1995), *Interfaces in Crystalline Materials*, Oxford Science Publishing, Oxford.

Swope, W. C., Andersen, H. C., Berens, P. H., Wilson, K. R. (1982), *J. Chem. Phys.* **76**, 637.

Taylor, J. E., Cahn, J. W. (1998), *Physica D* **112**, 381.

Thornton, K., Agren, J., Voorhees, P. W. (2003), *Acta Mater.* **51**, 5675.

van Kampen, N. G. (1987), *Stochastic Processes in Physics and Chemistry*, North-Holland, Amsterdam, p. 76.

Vauth, S., Streng, C., Mayr, S. G., Samwer, K. (2003), *Phys. Rev. B* **68**, 205 425.

Verlet, L. (1967), *Phys. Rev.* **159**, 98 (1967), *Phys. Rev.* **165**, 201.

Vogtenhuber, D., Houserova, J., Wolf, W., Podloucky, R., Pfeiler, W., Püschl, W. (2005), *Mater. Res. Soc. Symp. Proc.* **842**, S5.28.1.

Voter, A. F., Chen, S. P. (1987), *Mater. Res. Soc. Symp. Proc.* **82**, 175.

Voter, A. F., Montalenti, F., Germann, T. C. (2002), *Annu. Rev. Mater. Res.*, **32**, 321.

Voter, A. F. (1997), *Phys. Rev. Lett.*, **78**, 3908.

Wang, S., Mitchell, S. J., Rikvold, P. A. (2004), *Comput. Mater. Sci.* **29**, 145.

Wang, Y., Banerjee, D., Su, C. C., Khachaturyan, A. G. (1998), *Acta Mater.* **46**, 2983.

Weinkamer, R., Fratzl, P., Sepiol, B., Vogl, G. (1998), *Phys. Rev. B* **58**, 3082.

Wen, Y. H., Wang, Y., Chen, L.-Q. (1999), *Acta Mater.* **47**, 4375.

Yip, S. (2005) (Ed.), *Handbook of Materials Modeling*, Springer, Berlin.

Young, W. M., Elcock, E. W. (1966), *Proc. Phys. Soc.* **89**, 735.

Yu, S. Y., Schönfeld, B., Heinrich, H., Kostorz, G. (2004), *Prog. Mater. Sci.* **49**, 561.

Zhao, L., Najafabadi, R., Srolovitz, D. J. (1996), *Acta Mater.* **44**, 2737.

Zhu, J. Z., Chen, L.-Q., Shen, J., Tikare, V. (1999), *Phys. Rev. E* **60**, 3564.

13
High-Resolution Experimental Methods

13.1
High-Resolution Scattering Methods and Time-Resolved Diffraction
Bogdan Sepiol and Karl F. Ludwig

13.1.1
Introduction: Theoretical Concepts, X-Ray, and Neutron Scattering Methods

Detailed information about the microstructure is without doubt the most important and fundamental information necessary to start further analysis of any alloy. Generally, one can recognize a few levels for microstructure analysis: (1) the way the atoms of an alloy are arranged on a theoretical mean structure; (2) all the kinds of displacements from the mean structure; and (3) the dynamical behavior of constituent atoms including their vibrational properties and the motion of atoms between lattice sites. Unfortunately, investigation of any of these levels for a specific alloy is an extremely time-consuming and difficult task. Up to now, only very few alloys have been explored in all or in most facets. Usually, only selected features of the investigated system are studied, and analogies to similar known structures are used to make conclusions about those properties which are not accessible to direct exploration. A deep knowledge of some selected model systems therefore appears even more important.

A detailed characterization of the structure of the system on an atomic scale is the basis for a thorough understanding of diffusion in the solid state, where it is necessary to be familiar with both the collective (phonons) and single-particle dynamics (usually called diffusion). In most cases this requires the combination of different experimental techniques. X-ray and thermal neutron scattering are standard techniques for the investigation of the structural properties of alloys. Information about the mean structure is contained in the Bragg peaks and for this kind of study electron scattering is also a valuable technique. Displacements of atoms from the mean structure are, however, better detected quantitatively by X-ray and thermal neutron scattering. These displacements result in a scattering intensity between Bragg peaks and are caused by static atomic displacements and by short-range order phenomena. The corresponding approaches are generally called diffuse scattering methods (for reviews see, e.g., Schweika 1998; Schönfeld

Alloy Physics: A Comprehensive Reference. Edited by Wolfgang Pfeiler
Copyright © 2007 WILEY-VCH Verlag GmbH & Co. KGaA, Weinheim
ISBN: 978-3-527-31321-1

13 High-Resolution Experimental Methods

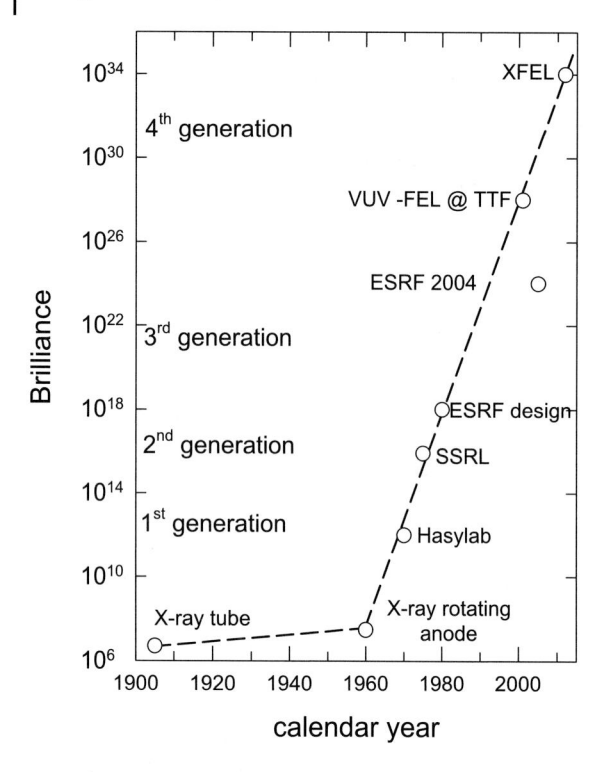

Fig. 13.1 Development of the brilliance of X-ray sources, i.e., the number of photons emitted in unit time by a unit source area in a unit solid angle in a relative bandwidth of 10^{-3} for a given energy.

1999). Because of the low cross-section for scattering, the extensive development of X-ray diffuse scattering methods has relied on the rapid growth in available X-ray brilliance at successive generations of synchrotron sources (Fig. 13.1).

The dynamical properties of hard and soft condensed matter are of considerable importance from a fundamental point of view, but are in addition essential for the functionality of nanoscale devices. The wavelengths of X-rays and neutrons in principle allows them to probe dynamics at much shorter length scales than is possible with visible light scattering (see Fig. 13.2), though X-ray techniques in this area have only become feasible with the advent of synchrotron radiation sources. The role of dynamical properties becomes increasingly relevant with decreasing size of the structural units, and even novel dynamical phenomena are expected in nanostructures. The performance of nanoscale devices will be strongly influenced by finite-size effects and the dimensionality of the system. The microscopic origin of these effects is far from being understood yet. A thorough understanding opens the way to tailor the development of future functional nanoscale systems. Since the properties of low-dimensional structures are signifi-

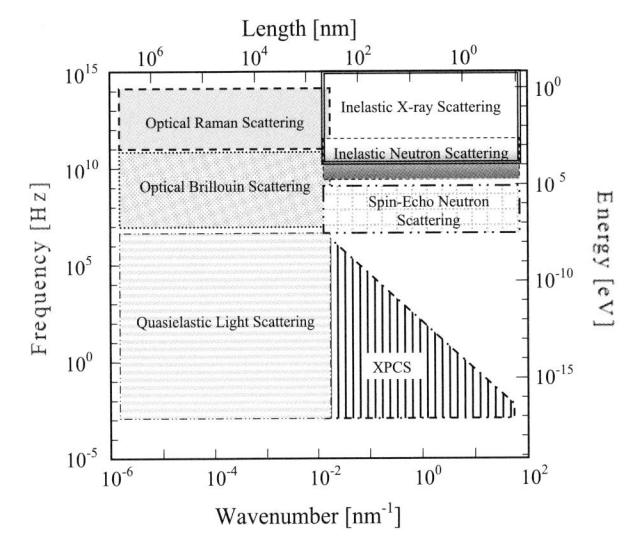

Fig. 13.2 Frequency/energy and spatial/wavenumber domains of inelastic scattering probes of dynamics in materials.

cantly different from those of corresponding bulk materials, new methods have to be developed for the experimental characterization and the theoretical modeling. An efficient way to achieve this is to use the brilliant X-rays from modern synchrotron radiation sources or to use neutron scattering methods to study the dynamical properties, usually under ultrahigh vacuum conditions. Nuclear resonant scattering of synchrotron radiation and neutron scattering are powerful methods for studying vibrational properties, diffusion and growth, and magnetic processes especially due to their high spatial and temporal resolutions. A distinct advantage of the technique of nuclear resonant scattering is that it is isotope-specific. Compared to other methods, the signal is essentially free from contributions from surrounding materials. Moreover, probe layers can be selectively deposited to study the magnetic and dynamical properties with atomic resolution.

In the mid-1990s, only very sparse information on diffusion and dynamics in pure metals and in intermetallic phases was available. Diffusion is a necessary precondition for most microstructural processes such as, for example, the nucleation of new phases, coarsening, and recrystallization, with a wide use in current technology, e.g., for surface hardening, changing deformation behavior by nucleation, diffusion doping, or sintering. Intermetallics further attracted attention as suitable materials for high-temperature applications (Sauthof 1995; Liu et al. 2002). Knowledge of the diffusion behavior of intermetallics is, therefore, of fundamental importance for basic materials science as well as for their use in technological applications.

13.1.2
Magnetic Scattering

Neutron- and X-ray scattering techniques continue to play a central role in elucidating the magnetic structures of alloys, whether in bulk, thin films, or nanostructures. The spin of neutrons makes them a sensitive tool for the study of local magnetic moments in a material. Since the original work of Shull and collaborators beginning in the late 1940s (Shull and Smart 1949), elastic neutron scattering has most often been the technique of choice for determining magnetic structure and has successfully elucidated even complex ordering behavior, such as helices, helifans, cycloids, and multi-wavelength structures. Inelastic neutron scattering also has a long history of success in examining magnetic fluctuations and excitations. In the more recent decades, neutron capabilities have continued to evolve with the widespread application of polarized neutron scattering.

In general, the interaction of the neutron spin with magnetic moments in materials does not distinguish between contributions to the local magnetic density from electron orbital and from spin angular momentum. Nor does it distinguish between contributions of different chemical species. Moreover, some materials have sufficiently high neutron absorption cross-sections to make magnetic scattering studies difficult. The nonresonant elastic magnetic X-ray scattering amplitude for linearly polarized photons is below that of elastic charge scattering by a factor of $E_\gamma/m_e c^2$. However, the high brilliance of synchrotron X-ray sources has allowed magnetic X-ray scattering to develop in the past 20 years into a powerful alternative to neutron scattering, one that is able to overcome some of the limitations of magnetic neutron scattering. Making use of the natural polarization of synchrotron radiation, magnetic X-ray scattering can often separate the spin and orbital contributions to the magnetic moment by a polarization analysis of the diffracted beam. Moreover, the ability to tune the photon energy at these sources to utilize resonant enhancement effects now routinely provides opportunities to examine chemically specific magnetic structure. Thus, magnetic X-ray scattering has developed into a strong complementary tool to magnetic neutron scattering. The literature on magnetic neutron- and X-ray scattering studies of alloys is so large that we do not attempt to survey it here. Instead, we discuss the current technical capabilities of each approach and some recent examples of their use.

13.1.2.1 Magnetic Neutron Scattering

There are a number of good review sources discussing the capabilities of magnetic neutron scattering (Bacon 1975; Lovesey 1986; Sköld and Price 1987; Williams 1988; Balcar and Lovesey 1989; Squires 1997; Shirane et al. 2002). The differential cross-section for elastic scattering of neutrons can be written as Eq. (1), where the sum is over atoms l, s_i and s_f are the initial and final neutron spin directions.

$$\frac{d\sigma}{d\Omega}(\mathbf{q}) \propto \left| \sum_l e^{i\mathbf{q}\cdot\mathbf{r}_l} U_l^{s_i s_f} \right|^2 \tag{1}$$

If, for simplicity, we neglect scattering from the nuclear magnetic moment, Eq. (2) can be written for $U_l^{S_iS_j}$, where b_l is the nuclear scattering amplitude, $p = \gamma g e^2/2mc^2$ the magnetic scattering amplitude, γ the neutron gyromagnetic ratio, $\mu_{\perp l}$ the component of the local magnetic moment that is perpendicular to **q**, and σ the Pauli spin matrix of the neutrons.

$$U_l^{S_iS_j} = b_l - p\mu_{\perp l} \cdot \sigma \tag{2}$$

We can therefore write the scattering cross-section as a sum of nuclear and magnetic unit cell form factors as in Eq. (3), where the sum is over all unit cells and F_N and F_M are nuclear and magnetic scattering structure factors for the unit cell.

$$\frac{d\sigma}{d\Omega}(\mathbf{q}) \propto \left|\sum_i (F_{iN} + F_{iM\perp})e^{i\mathbf{q}\cdot\mathbf{r}_i}\right|^2 \tag{3}$$

We first discuss unpolarized magnetic neutron scattering. In this case the neutron spins are randomly oriented so that the average of the cross-product between nuclear and magnetic scattering vanishes. The nuclear and magnetic scattering amplitudes then add incoherently. If the neutron beam is unpolarized and the sample has a uniaxial magnetic moment orientation $\hat{\mathbf{m}}$, it is useful to write the elastic magnetic scattering cross-section as the square of the Fourier transform of the component of the magnetic form factor that is perpendicular to **q**, $\mu_\perp(\mathbf{q})$ [Eq. (4)].

$$\left(\frac{d\sigma}{d\Omega}\right)_{mag}(\mathbf{q}) \propto \left(\frac{\gamma e^2}{2mc^2}\right)^2 (1 - (\hat{\mathbf{q}}\cdot\hat{\mathbf{m}})^2)|\mu_\perp(\mathbf{q})|^2 \tag{4}$$

This term must be separated from the nuclear scattering if both terms contribute at the reciprocal space point of interest. With different degrees of success this can be done by varying the direction or amplitude of an applied magnetic field, or by increasing the sample temperature above the Curie or Néel temperature.

By measuring a set of magnetic intensities at Bragg peaks in a single crystal or a polycrystalline sample, a set of magnetic form factor amplitudes can be determined and the magnetic structure can often be solved in real space. Results of relatively early magnetic structure determinations were compiled by Olés et al. (1976) and by Cox (1972). While polarized magnetic neutron scattering has grown considerably in popularity relative to unpolarized studies, the high flux of unpolarized neutrons available continues to make them attractive for experiments with low scattering cross-sections. For example, diffuse scattering can be examined to learn about short-range magnetic order in amorphous or paramagnetic materials in the wide-angle scattering regime (see, for instance, the study of amorphous Gd–Si alloys by Chumakov et al. 2005) or about magnetic inhomogeneities on larger scales in the small-angle scattering regime (see, for instance, the study of ferromagnetic amorphous Fe–Zr by Calderón et al. 2005).

Inelastic and quasielastic unpolarized magnetic neutron scattering are also powerful methods for the study of time-dependent magnetic phenomena. Using a triple-axis geometry, they essentially measure the dynamic structure factor [Eq. (5)].

$$S(\mathbf{q}, \omega) = \frac{1}{2\pi} \int_{-\infty}^{\infty} dt e^{i\omega t} \sum_{l} e^{i\mathbf{q} \cdot \mathbf{r}_l} \langle \mu_{0\perp}(0) \mu_{l\perp}(t) \rangle \tag{5}$$

The energy resolution ΔE of neutron spectrometers is usually greater than 10 μeV (corresponding to time scales shorter than 10^{-10} s) and more typically 0.1–0.5 meV. Quasielastic scattering refers to the examination of magnetic fluctuations occurring over a broad range of energy scales. Although it has traditionally dealt with relatively small energy scales, comparable to the resolution of the spectrometer, magnetic fluctuations with broad spectra can give considerable weight even at energies of several milli-electronvolts. As an example, Gaulin et al. (2002) have used quasielastic unpolarized magnetic neutron scattering to examine the energy scale and temperature dependence of incommensurate spin fluctuations in the heavy fermion superconductor UNi_2Al_3. Another interesting recent example is a study of the nature of magnetic waves in Zn–Mg–Tb quasicrystal alloys by Sato et al. (2006).

Inelastic scattering from excitations with definite energies is usually distinct from quasielastic scattering. Inelastic unpolarized magnetic neutron scattering has been perhaps the greatest source of information about the dynamics of spin waves in materials, both zero-point fluctuations and thermal excitations. A nice recent example is the work of Szuszkiewicz et al. (2006), measuring spin waves in MnTe and modeling their dispersion relations to obtain exchange parameters in the material.

One of the difficulties that can be encountered in interpreting the observed elastic neutron scattering is separating the magnetic scattering from the nuclear scattering. The use of polarized neutrons can resolve the situation, however, and polarized magnetic neutron scattering experiments have grown rapidly in number during the past 25 years (for a comprehensive discussion see Williams 1988). Polarized neutron beams can be created by making use of interference between magnetic and nuclear Bragg scattering from ferromagnetic crystals (such as the Heusler alloy Cu_2MnAl), by reflection from magnetized neutron mirrors ("supermirrors" in which the critical angles of reflection are different for different spin directions) or by using selective filters such as ^{149}Sm or polarized ^3He nuclei.

The term $U_l^{S_iS_f}$ in the neutron scattering cross-section changes sign when the neutron spin is reversed. Thus, one method of isolating the magnetic scattering component of the signal is to reverse the neutron spin using a spin-flipping coil – a "flipper" – that changes the spin by an angle of π. A polarizing monochromator crystal first produces a beam of polarized neutrons and a magnetic field at the sample orientates the sample magnetic moment. A flipper coil between the monochromator and the sample can then be selectively turned on to use the Larmor precession of spins in a magnetic field to reverse the spin directions. The intensity of a Bragg peak is measured with the flipper alternately turned off and on. The ratio of these intensities in a centrosymmetric crystal is given by Eq. (6), where the last equality assumes that $F_M \ll F_N$.

$$R = \frac{I_{on}}{I_{off}} = \left(\frac{F_N + F_{M\perp}}{F_N - F_{M\perp}}\right)^2 \approx 1 + 4\frac{F_{M\perp}}{F_N} \tag{6}$$

This greatly enhances the relative value of magnetic signal over the $(F_M/F_N)^2$ that is expected in general for unpolarized neutron scattering. The technique has been widely used for examining magnetization densities using Bragg peaks in single crystals, for studying short-range magnetic order in amorphous materials and binary alloys, and for the small-angle scattering examination of magnetic inhomogeneities on nanometer length scales (see, for example, the separation of chemical and magnetic structure due to precipitation in an amorphous Ni–P alloy by Tatchev et al. 2005). The study of larger (micron)-scale magnetic inhomogeneities can be performed with polarized super ultra-small angle neutron scattering (SUSANS), using multiple-bounce channel cut crystals to limit the beam divergence and to measure precisely the scattering angle (Wagh et al. 2005).

More detailed information can be obtained from polarized magnetic neutron scattering by using a triple-axis spectrometer with a spin flipper and polarizer in the scattered beam as well as in the incident beam. The term $U_I^{S_i S_f}$ in the neutron-scattering cross-section gives the probability that a neutron will change its spin direction on scattering. In analyzing this probability, it is convenient to designate directions relative to the neutron polarization in a right-handed cartesian coordinate system: (ξ, η, ζ) with the initial polarization in the ζ direction (Moon et al. 1969). Then the terms for spin-flip and non-spin-flip scattering become those in Eqs. (7).

$$U_{no-flip} = b \pm p\mu_{\perp\zeta}$$
$$U_{flip} = -p(\mu_{\perp\xi} \pm i\mu_{\perp\eta}) \tag{7}$$

We see that the nuclear scattering factor b does not contribute to the spin-flip scattering. In the case where **q** is parallel to the direction of neutron polarization, all magnetic scattering is in the spin-flip channel, since $\mu_{\perp\zeta} = 0$. If the monochromator and analyzer crystals are set so that both reflect the same spin orientation, then setting both flippers either on or off lets the scattered beam component pass that has not flipped. Alternatively, setting one flipper off and the other on lets that component of the scattered beam pass that has its spin flipped. By separately measuring spin-flip and non-spin-flip scattering with the neutron polarization alternately parallel and perpendicular to **q**, the individual components of magnetic moment can be examined. When measuring the magnetic form factor at Bragg peaks, this technique largely avoids the difficulties that occur in unpolarized magnetic neutron scattering when separation of the nuclear and magnetic scattering is attempted, even in cases of relatively weak magnetic scattering (see, for example, the examination of ferrimagnetism in Fe–B metallic glasses by Cowlam et al. 2005).

Because of increasing interest in thin-film magnetic structures, in particular in giant magnetoresistance (GMR) materials, polarized neutron reflection (PNR) has grown significantly more popular since the mid-1990s. When either incident polarization is used alone or both incident and transmitted polarization are employed, fits of the reflection curves give considerable information about magnetic layer orientation, thickness, and magnetic interface roughness. For example, Singh et al. (2005) have used PNR to examine the magnetic moment of Ni in Ni/Cu multilayer structures. In addition to investigating the average vertical magnetic structure with the specular reflectivity, the diffuse scattering at low reflected angles gives information about the lateral magnetic structure, such as that due to magnetic domains. A review of some of the recent investigations using this approach is given by Felcher and te Velthuis (2001).

Polarized neutron beams can also be used for neutron depolarization (ND) studies (for an early review, see Endoh et al. 1992). This technique utilizes the depolarization of an initially polarized beam of neutrons when it transits through a ferromagnetic sample. In its most sophisticated form, the entire 3×3 matrix is measured relating incident polarization directions to transmitted ones and is interpreted in terms of domain size and shape. As an example, Sakarya et al. (2005) used the technique to examine domains in the ferromagnetic superconductor UGe_2, part of an unusual class of materials in which ferromagnetism and superconductivity apparently coexist.

A very different use of neutron spin is made in neutron spin echo (NSE) spectroscopy: see Mezei (2003). In NSE the spins of polarized neutrons precess in a magnetic field both before and after scattering. For a given field strength and length, the amount of precession for each neutron depends on its velocity. A π spin flipper is placed just before the sample so that the second precession exactly compensates for the first precession, if the exit velocity is equal to the incident velocity. If the scattering is inelastic so that the outgoing neutron velocity is slightly changed, then the resulting beam is slightly depolarized. The depolarization measurement determines the change in neutron velocity on scattering. A key attribute of the technique is that only the change in velocity of each neutron is important, so that the overall distribution of energies in the incident neutron beam can be much wider than the energy resolution of the experiment. The NSE technique gives access to time scales much longer than with traditional quasielastic scattering – on order of 100 ns. It has provided unique information about slow relaxation in spin glasses (see, for example, Pappas et al. 2003).

A variation of NSE is spin echo small-angle neutron scattering (SESANS), which uses the spin echo principle to perform elastic small-angle scattering. It utilizes the decrease in polarization resulting from the change in neutron precession along the altered path of the scattered beam through the second magnetic field. Thus, just as NSE has an effective energy resolution narrower than the energy width of the incident neutron beam, SESANS has an effective angular resolution smaller than the divergence of the neutron beam. Length scales up to a micron are accessible. Although the technique was originally developed for nonmagnetic applications, it also possesses the capability to examine long length scale magnetic inhomogeneities (Grigoriev et al. 2006). In this case, the magnetic scattering can itself act as a flipper, allowing the separation of magnetic and nuclear scattering.

13.1.2.2 Magnetic X-Ray Scattering

Discussions about the possibilities for magnetic X-ray scattering from materials were started in 1970 by Platzman and Tzoar (Platzman and Tzoar 1970) with the first elastic scattering studies by de Bergevin and Brunel (1972), the first magnetic Compton scattering experiments by Sakai and Ono (1976), and the first resonant scattering studies by Namikawa et al. (1985). Due to the need for brighter sources, the field began to develop rapidly only with the growing availability of synchrotron X-ray sources in the early 1980s. Since then the field has prospered as new sources and approaches continue to be developed. A number of excellent reviews on magnetic X-ray scattering are now available (Lovesay and Collins 1996; Cooper and Stirling 1999; Laundy 1999; Altarelli 2001; Lebech 2002; Beaurepaire et al. 2006).

As discussed in more detail below, magnetic X-ray scattering offers a number of complementary capabilities to magnetic neutron scattering. Using polarization analysis, the orbital and spin contributions to the form factor of the magnetic moment can be separated. The magnetic contributions of individual chemical species can be investigated with resonance techniques. The high collimation of synchrotron X-ray sources permits the precise study of magnetic periods in complex structures and of long-wavelength critical fluctuations near phase transitions. Meanwhile the high brilliance of these sources also allows the study of magnetism at surfaces, in thin films, and in nanoparticles. Further, the brilliance of synchrotron sources facilitates the development of microdiffraction tools. In addition, there are important elements such as Gd and Sn which have high thermal neutron absorption cross-sections, complicating their study by neutron scattering. It should be noted that "elastic" X-ray scattering integrates over a much wider range of energies than do most neutron scattering measurements. It therefore usually probes time scales of 10^{-15} s while neutron scattering usually examines magnetic structure on time scales approximately three orders of magnitude longer. In most cases, this is not believed to cause any observable differences but it should be kept in mind. Finally, magnetic Compton scattering measures the spin momentum density, a quantity of fundamental interest which is not accessible to neutron scattering.

The cross-section for nonresonant elastic X-ray scattering from a material can be written as Eq. (8) (Blume 1985), where the sums are over electron positions r_j, p is the electron momentum, and a_1, a_2 and a_3 polarization-dependent vectors given by Eqs. (9), in which $\hat{k}$, $\hat{k}'$ are unit vectors in the directions of the incident and scattered photons respectively, and $\hat{\varepsilon}$, $\hat{\varepsilon}'$ are their respective polarization vectors.

$$\frac{d\sigma}{d\Omega}(\mathbf{q}) = \left(\frac{e^2}{mc^2}\right) \left\{ \left\langle \psi \left| \sum_j e^{i\mathbf{q}\cdot\mathbf{r}_j} \right| \psi \right\rangle \mathbf{a}_1 - i\frac{E_\gamma}{mc^2} \left[\left\langle \psi \left| \sum_j e^{i\mathbf{q}\cdot\mathbf{r}_j} \mathbf{s}_j \cdot \mathbf{a}_2 \right| \psi \right\rangle \right. \right.$$
$$\left. \left. + \left\langle \psi \left| \sum_j e^{i\mathbf{q}\cdot\mathbf{r}_j} \frac{\mathbf{q} \times \mathbf{p}}{\hbar^2 q^2} \cdot \mathbf{a}_3 \right| \psi \right\rangle \right] \right\}^2 \tag{8}$$

$\mathbf{a}_1 = \hat{\varepsilon} \cdot \hat{\varepsilon}'$

$\mathbf{a}_2 = \hat{\varepsilon}' \times \hat{\varepsilon} + (\hat{\mathbf{k}}' \times \hat{\varepsilon}')(\hat{\mathbf{k}}' \cdot \hat{\varepsilon}) - (\hat{\mathbf{k}} \times \hat{\varepsilon})(\hat{\mathbf{k}} \cdot \hat{\varepsilon}') - (\hat{\mathbf{k}}' \times \hat{\varepsilon}') \times (\hat{\mathbf{k}} \times \hat{\varepsilon})$ (9)

$\mathbf{a}_3 = \hat{\varepsilon}' \times \hat{\varepsilon}$

The first term in Eq. (8) describes the Thomson charge scattering while the others describe "magnetic" scattering from the electron spin (second term) and orbital angular momentum (third term). The magnetic scattering amplitudes are lower than that of the charge scattering by a factor $E_\gamma/m_e c^2$ of approximately 10^{-4} at typical X-ray energies. They are also imaginary, so that they do not interfere with the real part of the charge scattering but only with the typically smaller imaginary part (not shown in Eq. (9)), if linearly polarized X-rays are used.

When using linearly polarized X-rays the different polarization dependences of the second and third terms allow the determination of the individual contributions of spin and orbital motion to the total magnetic moment (Blume and Gibbs 1988). In particular, the matrix elements linking the polarization components perpendicular to the plane of scattering (σ-polarization) with the components in the plane of scattering (π-polarization) have nonzero terms in both diagonal and non-diagonal elements (Eqs. (10), where the scattering vector $\mathbf{q}$ is in the z-direction and the x direction is perpendicular to the y–z plane of scattering).

$M_{\sigma\sigma} = S_x \sin 2\theta$

$M_{\sigma\pi} = 2 \sin^2 \theta [(S_y + L_y) \cos \theta + S_z \sin \theta]$ (10)

$M_{\pi\pi} = \sin 2\theta [2L_x \sin^2 \theta + S_x]$

Analysis of the polarization of the scattered beam is usually performed using an analyzer crystal diffracting at 90°. Since the scattering is nonresonant, the energy of the X-ray beam can typically be chosen to ensure this condition.

In cases where the symmetry of the magnetic moment is the same as the charge symmetry (as in a ferromagnet) and when linearly polarized X-rays are used, it is possible, but often difficult, to separate the nonresonant magnetic scattering signal from the much larger charge scattering by reversing the direction of an applied magnetic field or by analyzing changes in polarization. However, nonresonant magnetic X-ray scattering has been most often used in cases where there are specific Bragg reflections primarily due to the magnetic structure. In the early work of Gibbs et al. (1988), this approach was used to identify orbital and spin contributions in Ho, which has a large magnetic moment ($10.6\mu_B$) and a spiral antiferromagnetic structure. Despite the small cross-section for nonresonant magnetic X-ray scattering, the increasing availability of high-brilliance beams from third-generation synchrotron sources has broadened its range of use. In one interesting application, microfocused nonresonant magnetic scattering was used to image spin density wave domains in Cr on the micron scale (Evans et al. 2002). In another case, nonresonant magnetic X-ray scattering has been used to investigate the differences between bulk and near-surface ordering in $NdCu_2$ alloys (Schneidewind et al. 2001).

The use of relatively high-energy X-rays (40–150 keV) offers a particular set of possibilities for nonresonant magnetic X-ray scattering studies. The relatively small penetration depths (on the order of microns) of typical hard 5–30 keV X-rays for most magnetic materials means that, in contrast to magnetic neutron scattering, only the near-surface regions of bulk samples are examined. While this can be an advantage for those studies interested in the near-surface region, it also means that the measurements may be sensitive to surface preparation and quality, a disadvantage if the primary subject of interest is the bulk alloy behavior. Using high-energy X-rays, experiments can be performed in transmission geometry, helping to ensure that the scattering originates from a bulk phenomenon. In addition, in this limit, the contribution of orbital scattering to the magnetic scattering cross-section becomes negligible and thus only the spin component perpendicular to the scattering plane is measured. By comparing magnetic neutron and high-energy X-ray magnetic scattering, the orbital and spin structure factors can be separated without using polarization measurements, as has been shown by Strempfer et al. (2000) in the case of Cr.

An alternative to using the small signals produced by the nonresonant magnetic scattering with linear polarization is to use circularly polarized X-rays. These can be obtained using a synchrotron beam slightly outside the orbital plane, with a phase plate or an appropriate insertion device such as an elliptical multipole wiggler. Circularly polarized X-rays produce interference between the real part of the charge scattering and the magnetic scattering, significantly enhancing the contribution of the magnetic scattering. Although the magnetic scattering then occurs at the same place in reciprocal space as the charge scattering, it is possible to moderate the influence of charge scattering. This can be done, for instance by diffracting near the horizontal direction at a synchrotron to take advantage of the suppression of charge scattering by polarization, by examining changes in the observed scattering under reversal of photon helicity, or by reversal of an applied magnetic field. By utilizing a white beam and an energy-dispersive detector, this approach has allowed the very efficient measurement of the magnetic form factor in ferromagnets such as Fe (Laundy et al. 1991; Collins et al. 1992; Ito and Hirano 1997), $HoFe_2$ (Collins et al. 1993), and Ni (Laundy et al. 1998). Other experimenters have used a monochromated beam with circular polarization to determine the temperature dependences of the spin and orbital moments in a ferromagnetic Sm–Gd–Al alloy (Taylor et al. 2002). An alternative approach to mixing the charge and magnetic scattering to increase the magnetic scattering intensity would be to examine noncentrosymmetric structures, since the structure factor there is complex. However, this approach has apparently not been attempted yet.

In addition to the terms shown in Eq. (8), there enters a term $\mathbf{A} \cdot \mathbf{p}$ from perturbation theory which gives rise to photoelectric absorption in its first order. Here $\mathbf{A}$ is the vector potential of the electromagnetic field. Absorption spectroscopy, particularly magnetic circular and linear dichroism spectroscopy, has become a very powerful tool for measuring the spin-dependent density of states (see the review by Wende 2004). However, we confine ourselves here to a discussion of scattering techniques. The second-order $\mathbf{A} \cdot \mathbf{p}$ term contribution to the elastic scattering amplitude [expression (11)]

$$\sum_o \frac{\langle i|\mathbf{A}_{\text{out}} \cdot \mathbf{p}|o\rangle\langle o|\mathbf{A}_{\text{in}} \cdot \mathbf{p}|i\rangle}{E_i - E_o - E_\gamma - i\Gamma/2} \tag{11}$$

becomes dominant in resonance conditions when the incident photon energy is comparable to the difference in energy between the ground state and a virtual excited state. Here the state of the electron system is $|i\rangle$ and the sum is over virtual excited states $|o\rangle$. The cases of interest are virtual excitations between a core and an unfilled valence shell. Although the resonant elastic scattering term above does not explicitly include a dependence on magnetic moment, it is sensitive to magnetic order through differences in cross-section that depend on the electron spin states. In the simplest cases, this can arise from the exchange splitting of the valence states probed. Resonant magnetic scattering can often be observed in cases where its observation by nonresonant techniques would have been difficult or impossible. In addition, using photon energies near the absorption edges of different elements, the magnetic contribution of those elements can be probed individually, a unique capability. Moreover, relatively small samples can typically be studied using resonance enhancement. This has facilitated the study of, for instance, samples under high pressure. As an example Kernavanois et al. (2005) have examined the resonant magnetic scattering from $Ce(Co_xFe_{1-x})_2$ at pressures up to nearly 10 kbar.

Resonant scattering is large near absorption edges and is particularly sensitive to magnetic order in the partially unfilled 3d shell states in the 3d transition metals, 4f states in lanthanides and 5f states in actinides. In the electric dipole approximation the change in electron orbital momentum is ± 1 while the higher-order electric quadrupole term gives a change of ± 2. Thus the strongest resonant contributions (coming from the electric dipole term) would be expected to occur near 3d transition metal $L_{2,3}$ edges (2p–3d resonance), lanthanide $M_{4,5}$ edges (3d–4f resonance), and actinide $N_{4,5}$ edges (4d–5f resonance). However, most of these edges are in the soft X-ray regime. The short absorption length of soft X-rays in air necessitates the use of specialized vacuum spectrometers for such experiments. Resonant diffraction experiments with hard X-rays must instead utilize the transition metal K-edge (accessing primarily the 1s–4p resonance), lanthanide L-edges (accessing primarily the 2p–5d resonance) and actinide M-edges (accessing primarily the 3d–5f resonance). In the case of the transition metals and the lanthanides, hybridization of the probed valence states with the polarized

3d or 4f states is believed to play the dominant role in creating a magnetic resonance effect.

Resonant magnetic scattering experiments near K-edges are not strongly sensitive to the partially empty valence 3d states and hence resonance effects there are usually relatively modest, particularly if no polarization analysis is used, as seen in the early study of ferromagnetic Ni by Namikawa et al. (1985). Although relative signal rates are not large, polarization analysis can show significant enhancements of $\sigma\pi$ scattering near K-edges. This has been utilized in the study of the spin wave state in Cr (Mannix et al. 2001a), though its widest use has been in the study of magnetic, charge, and orbital ordering in transition metal oxides (see the review by Hill 2001). Interestingly, large resonances have been observed at the K-edges of nonmagnetic elements in U antiferromagnets (Mannix et al. 2001b). Though a detailed understanding is still controversial, the phenomenon opens new possibilities for K-edge resonant experiments.

Diffraction near lanthanide L- and actinide M-edges often offers the possibility of larger resonant enhancements in the hard X-ray regime. In the first observations of resonant scattering by Gibbs et al. (1988), a strong resonance was seen at a Ho L-edge. The exact nature of the enhancement appears to be complex (van Veenendaal et al. 1997). Even larger resonant effects are typically found near the $M_{4,5}$ edges of actinides where a 3d electron can make a virtual transition into a partially unoccupied polarized 5f shell. The nature of the resonance scattering complicates quantitative calculation of the magnetic moments themselves. However, hard X-ray resonant scattering at L- and M-edges has been widely used to examine specific chemical aspects of magnetic ordering, including the nature of magnetic phase transitions in bulk alloys (Pengra et al. 1994; Vigliante et al. 1998; Hupfeld et al. 2000; Lidstrom et al. 2000; Goff et al. 2001; Stunault et al. 2001, 2002, 2004; Okuyama et al. 2004; Good et al. 2005; Kuzishita et al. 2005), the microdomain structure and anisotropy in thin films (Beutier et al. 2004), complex behavior in multilayers (Nelson 1999; Jaouen 2002; Marrows et al. 2005), magnetic order in heavy-fermion thin films (Jourdan et al. 2005), and magnetism at surfaces (Ferrer et al. 1997).

As discussed above, the strongest magnetic resonances are in the soft X-ray regime, at the relevant transition metal L-edges, the lanthanide M-edges, and the actinide N-edges. The huge enhancements in this regime (by a factor of up to 10^7 at actinide N-edges) allow the magnetism of surfaces and extremely thin films to be studied. Taking advantage of these opportunities requires working completely in a vacuum environment to avoid X-ray absorption in the air; this considerably complicates experiments. Because of the low scattering wavenumbers accessible, resonant soft X-ray magnetic scattering studies have often focused on magnetic structure on nanometer length scales via scattering or on surface/interface roughness via reflectivity. As with hard X-ray magnetic scattering, polarization analysis of the scattered photons can help separate charge and magnetic scattering. Also, circularly polarized X-rays can be used to produce interference between the charge and magnetic scattering. The reversal of helicity can then be used to help isolate the interference term between charge and magnetic scattering. However, for the case of reflectivity at least, clearly interpreting the observed scattering in terms of magnetic and atomic structure can be tricky (Osgood et al. 1999; Kortright and Kim 2000). Recent studies include that of Dudzik et al. (2000) examining striped magnetic domains using the Fe L_3-edge in FePd. There, varying the helicity of the circularly polarized X-rays allowed the investigators to determine the structure of closure domains. A number of experiments have meanwhile been examining Co/Pt multilayers whose structure and magnetic interactions can be tuned over a significant range (Kortright et al. 2001; Hellwig et al. 2003; Pierce et al. 2003). Other investigations have studied stripe domains in Co/Pt multilayers using a transmission geometry (Kortright et al. 2001), the magnetism of coupled layers in spin-valve and giant magnetoresistance materials (Hase et al. 2000; Mirone et al. 2000), magnetic heterogeneity in a recording media alloy (Kortright et al. 2003), magnetic correlations in dense nanoparticle assemblies (Kortright et al. 2005), separation of chemical and magnetic interfacial roughness in CoFe thin films (Freeland et al. 1999; Kelly et al. 2002), thickness dependence of the magnetic behavior of ultrathin films (Weschke et al. 2005), and the magnetization of buried interfaces in Co films (MacKay et al. 1996).

The enhanced scattering from soft X-ray resonance conditions has been used to observe coherent magnetic scattering. Yakhou et al. (2001) were able to follow domain wall fluctuations in UAs and Chesnel et al. (2002) could examine transverse strip domain morphology in FePd wires. As discussed further in the section on coherent scattering, Pierce et al. (2003) were able to use coherent magnetic scattering to determine the extent of microscopic return point memory in Pt/Co multilayers. Even the imaging of magnetic domains in a Co/Pt multilayer using lensless holography has been demonstrated by Eisebitt et al. (2004).

Inelastic magnetic X-ray scattering is another method being applied. While extensive use has been made of X-ray emission spectroscopy to examine magnetic structure, we will confine ourselves here to true inelastic scattering processes; most of these experiments use magnetic Compton scattering. Compton scattering uses energy transfers that are large compared to valence electron binding energies so that electron motion during the interaction is negligible – this is essentially the impulse approximation. Therefore magnetic Compton scattering is sensitive only to the spin of the electrons and experiments using circularly polarized X-rays can measure the projection of the magnetic Compton momentum profile in a given direction [Eq. (12)].

$$J_{mag}(p_z) = \iint [n_\uparrow(\mathbf{p}) - n_\downarrow(\mathbf{p})] \, dp_x \, dp_y \tag{12}$$

This is a unique capability, as the magnetic momentum density, which is not accessible to neutron scattering, can often be compared directly with theoretical predictions. The cross-section for spin magnetic Compton scattering is given by Eq. (13), where P_c is the degree of circular polarization.

$$\left(\frac{d^2\sigma}{d\Omega\, dE_s}\right)_{mag} = \left(\frac{e^2}{mc^2}\right)^2 \left(\frac{E_s}{E_i}\right) \left(\frac{1}{2\hbar cq}\right) (\cos\theta - 1) P_c \hat{\sigma} \cdot (\mathbf{k_i} \cos\theta + \mathbf{k_s}) J_{mag}(p_z)$$
(13)

The subscript "i" refers to the incident photon and the subscript "s" to the scattered photon. This contribution must be separated from the charge Compton scattering by reversing the direction of polarization or by reversing the direction of an applied field. Typically, the contribution of the magnetic term to the overall Compton scattering cross-section is 1% or less. Therefore, high photon statistics are required. In order to maximize the signal and momentum resolution, experiments are typically performed at relatively high X-ray energies (>40 keV) and in a backscattering geometry.

Sakai and Ono (1976) performed the first experiments in the field (using a radioactive source) and Sakai (1996) has reviewed the development of the method. As an example of more recent work utilizing synchrotron radiation, Cooper et al. (1993) have measured the spin-resolved Compton profile of $HoFe_2$. The differing lineshapes expected for the Ho and Fe momentum distributions were used to analyze the significantly different behaviors of the spin moments of the atoms as a function of temperature. In another study, Lawson et al. (1997) were able to take advantage of the insensitivity of magnetic Compton scattering to orbital moments to measure the spin moment on uranium. The spin moment is antiparallel to the orbital moment in this case, resulting in only a small net moment, which complicated previous studies by other techniques. In transition metals, a particular advantage of magnetic Compton scattering compared to elastic scattering is its sensitivity to delocalized valence electron spin moments. It has been used, for instance, in studies of Ni which are able to challenge current theoretical understanding (Dixon et al. 1998). In a ferromagnetic Gd–Y alloy, Duffy et al. (2000) have examined induced spin polarization, while others have studied the boron contribution to the spin moment in amorphous Fe–B alloys (Taylor et al. 2001), the spin momentum densities in Fe (Montano et al. 2000), in Pd–Co alloys (Taylor et al. 2002a), in invar (Fe_3Pt) (Srajer et al. 1999; Taylor et al. 2002b), in a Ce–Fe–Ru alloy (Ahuja et al. 2002), in the Ni_2MnSn Heusler alloy (Deb et al. 2001), in Fe_3Si and Fe_3Al alloys (Zukowski et al. 2000), and in $CrRh_3B_2$ (Sakurai et al. 2003), spin transitions in a Ce–Fe–Ru alloy (Sharma et al. 2005), and the metamagnetic transition in UCoAl (Tsutsui et al. 2005).

13.1.3
Spectroscopy

In this section, spectroscopic methods are discussed which provide some of the most complete dynamic structural information available. Section 13.1.3.1 is devoted to coherent X-ray scattering; Section 13.1.3.2 deals with studies of phonon-excitation spectra by X-ray scattering. Both methods have been employed to an increasing extent in recent years. Section 13.1.3.3 is about methods which have been present for significantly longer in the scientific community, though they

13.1.3.1 Coherent Time-Resolved X-Ray Scattering

The continuing increase in brilliance of synchrotron X-ray sources has allowed the development of "coherent" X-ray scattering as a technique for examining dynamics and kinetics in metallic alloys. Traditional X-ray scattering uses photon beams that have effective coherence lengths significantly smaller than the size of the beam itself. That is, the X-ray wave vector **k** varies among different incident photons so that diffraction from points on the sample that are separated beyond some coherence length (longitudinal and transverse) smaller than the beam size does not interfere at the detector, but rather adds incoherently. The availability of high-brilliance hard X-ray beams from third-generation synchrotron sources has opened the possibility of making small X-ray beams (typically of order 10 μm in diameter) that exhibit significant coherence throughout the scattering volume. Here we call the use of such beams in scattering experiments "coherent" X-ray scattering, though traditional X-ray diffraction experiments also rely on a more limited coherence to create interference between scattering from objects that are not too far apart spatially (i.e., within the limits set by the traditional "resolution function"). The use of coherent X-ray beams has also been termed X-ray intensity fluctuation spectroscopy (XIFS) or X-ray photon correlation spectroscopy (XPCS). While currently available hard X-ray beams do not exhibit the level of coherence available from optical lasers, they nonetheless enable some of the techniques traditionally used in the optical regime to be used now in the hard X-ray regime. In particular, the use of "coherent" X-rays offers the potential to examine dynamics, kinetics and microscopic reversibility on length scales of 10^{-1}–10^3 nm in metallic alloys.

It has now been just over a decade since the first coherent hard X-ray scattering experiments were performed (Sutton et al. 1991) and there have been continued technical and conceptual advances in the technique. However, the limitations on coherent hard X-ray flux available have restricted the application of coherent X-ray scattering to a limited number of situations in which the coherent scattered intensity is sufficiently high for its evolution to be monitored accurately on the time scales of interest. Thus, while there have been some coherent X-ray studies of metallic alloys, most experiments using the technique have so far focused on macromolecular materials that have high effective scattering power (a large number of atoms within each macromolecule and low absorption) and relatively long time constants.

On a typical undulator beamline at a third-generation synchrotron source, the effective size of the electron beam serving as the X-ray source is on the order of 1000 μm × 20 μm ($H \times V$) FWHM. The flux and coherence available for a scattering experiment depend on the source and beamline optics. Typical fluxes through a 10 μm pinhole located $R \sim 50$ m from the source are on the order of 10^{10} photons s^{-1} using a crystal monochromator. This geometry gives a transverse coherence length in each direction of $\xi_t \sim \lambda R/(2s)$, where s is the appropriate source size and λ is the X-ray wavelength. Typical transverse coherence areas are then on the order of 5 μm × 500 μm ($H \times V$) FWHM and coherent fractions are 20–30% in beams of 10 μm diameter.

The longitudinal coherence length $\xi_l \sim \lambda^2/2\Delta\lambda$, where $\Delta\lambda$ is the beam's spread in wavelength, is also an important parameter. For a typical crystal monochromator, the wavelength spread $\Delta\lambda/\lambda$ is about 10^{-5}–10^{-4} so that ξ_l is a few microns. Thus, if photon pathlengths for scattering processes from different atoms differ by more than a few microns, the scattering contributions will add incoherently at the detector. This is not usually a problem for small-angle scattering experiments, because pathlength differences are small. In that case, the energy resolution can actually be relaxed significantly from the crystal monochromator value, for example by using a multilayer monochromator. For wide-angle scattering in a symmetric reflection geometry at incident angle θ, however, the pathlength difference will typically be as large as $2l_{abs} \sin^2 \theta$, where l_{abs} is the X-ray absorption length in the material. Thus for typical wide-angle scattering, l_{abs} should be less than about 10 μm.

In order to work in the near field of the pinhole of diameter d defining the beam, the sample must be placed a distance L_1 from the pinhole with $L_1 \ll d^2/\lambda$. In order to measure the scattering as a function of momentum transfer $\hbar\mathbf{q}$, the scattered photons must be detected in the far field so that the distance between sample and detector L_2 must be $L_2 \gg h^2/\lambda$, where h is a projected beam size on the sample. Coherent scattering from a material without long-range order gives a stochastic speckle pattern with typical speckle sizes at the detector of about 30 μm.

To collect scattering signals efficiently with minimal background, direct-illumination deep-depletion layer CCD X-ray area detectors are in many ways ideal if the time scales of interest are sufficiently slow. When count rates are low and it is desirable to do so, data can be analyzed in a photon counting mode using a droplet algorithm (Livet et al. 2000) to identify individual photons and even to energy-discriminate among scattered photons with a typical resolution of about 190 eV. In this mode, correlations are usually calculated post facto. Alternatively, at the fastest time scales, fast avalanche photodiode point detectors have been used in conjunction with digital autocorrelators to give 50 ns temporal resolution (Sikharulidze et al. 2002).

13.1.3.1.1 Homodyne X-Ray Studies of Equilibrium Fluctuation Dynamics

Since the speckle pattern in a coherent scattering experiment is a function of the position of each scatterer in the sample, it changes with time according to Eq. (14) if the scatterers are moving.

$$I(\mathbf{q},t) = \left|\sum_i f_i e^{i\mathbf{q}\cdot\mathbf{r}_i(t)}\right|^2 = \rho(\mathbf{q},t)\rho^*(\mathbf{q},t) \tag{14}$$

Here f_i is the scattering factor for the ith atom, $r_i(t)$ is its position at time t, and $\rho(\mathbf{q},t)$ is (in the case of hard X-rays) the Fourier transform of the electron density.

This changing intensity is the basis of the widely used dynamic light scattering (DLS) technique, performed with optical lasers. In homodyne optical DLS, the dynamics of macromolecules or clusters on the micron size scale is examined by measuring the decay of the intensity autocorrelation function of the speckle pattern. This gives the second-order correlation function Eq. (15).

$$g_2(\mathbf{q},t) = \frac{\langle I(\mathbf{q},t')I(\mathbf{q},t'+t)\rangle_{t'}}{\langle I(\mathbf{q},t')\rangle_{t'}^2} = \frac{\langle \rho(\mathbf{q},t')\rho^*(\mathbf{q},t')\rho(\mathbf{q},t'+t)\rho^*(\mathbf{q},t'+t)\rangle_{t'}}{\langle \rho(\mathbf{q},t')\rho^*(\mathbf{q},t')\rangle_{t'}^2} \tag{15}$$

While $g_2(\mathbf{q},t)$ is directly accessible from the intensity autocorrelation function, theory usually calculates the first-order correlation function $g_1(\mathbf{q},t) = \langle \rho(\mathbf{q},t')\rho(\mathbf{q},t'+t)\rangle_{t'}/\langle I(\mathbf{q},t')\rangle_{t'}$. If fluctuations are Gaussian, however, $g_2(\mathbf{q},t)$ is related to $g_1(\mathbf{q},t)$, the first-order correlation function, by Eq. (16).

$$g_1(\mathbf{q},t) = 1 + |g_2(\mathbf{q},t)|^2 \tag{16}$$

For simple translational diffusion, the correlation functions exhibit an exponential decay $\exp(-q^2 Dt)$, where D is the diffusion constant. More generally, the decay of the normalized autocorrelation functions reveals how fluctuations grow and decay on length scales $2\pi/q$.

The advent of coherent X-ray scattering allows DLS to be extended to the X-ray regime so that dynamics in opaque materials can, at least in theory, be examined down to atomic length scales. As discussed above, practical limitations restrict the technique's applicability to cases in which the clusters of scattering atoms have a large scattering cross-section and the time scales are not too fast. This is considered in a quantitative manner in an estimate [Eq. (17)] due to G.B. Stephenson (2003) of the typical count rate N^{SP} that can be expected per speckle; σ_{el} is the atomic cross-section for elastic scattering, t is the effective sample thickness, A is the cross-section area of the incident X-ray beam with N_{in} coherent incident photons, and the normalized incoherent structure factor is $S(q)$.

$$N^{SP} = \frac{\lambda^2 \sigma_{el} S(q)}{2\pi At} N_{in} \tag{17}$$

In many cases, it is the contrast between regions that causes the observed scattering; then σ_{el} must be replaced by an appropriately weighted difference in scattering cross-sections between the regions. For typical coherent scattering experiments using samples containing mostly medium-Z elements (e.g. Cu), we can estimate that $N^{SP} \sim 1 \times 10^{-13} S(q) N_{inc}$. A typical incident flux is about 10^{10} photons s^{-1}. If we take as a minimum requirement that $N_{inc} \sim 10^{-2}$ per correlation time in order to perform correlation spectroscopy, then we would need a normalized structure factor $S(q)$ on the order of 10 in order to measure correlation times on the order of 1 s. Of course, this is a very approximate calculation, but it can clearly be seen that for studies of subsecond dynamics, sizeable clusters of atoms scattering collectively are required.

Consistently with this conclusion, X-ray fluctuation spectroscopy has been applied to examine the dynamics of relatively large groups of atoms, such as colloidal nanoparticles, latex spheres in solution, and block copolymer micelles. There have been fewer coherent X-ray studies of equilibrium dynamics near phase transitions in metallic alloys. In a relatively early experiment, Brauer et al. (1995) examined the critical fluctuations near the B2–D0$_3$ order–disorder transition in Fe$_3$Al. More recently, Sutton's group at McGill has revisited the issue, bringing significant improvements in beam intensity and stability since the earlier work (Sutton 2005). These studies have been able to measure carefully the fluctuation dynamics and compare it to theoretical expectations for the first time in an alloy. The experiments rely on the relatively large scattering very near the critical point due to the large fluctuation length scales (over 500 nm in these investigations), and on the relatively slow dynamics in the alloy at this point (tens of seconds). Francoual et al. (2003) have also examined the dynamics of phason fluctuations in an icosohedral AlPdMn alloy, again comparing the fluctuation dynamics with theory for the first time.

In summary, measurements of equilibrium fluctuation dynamics with coherent X-ray scattering have developed considerably in the past decade, but intensity considerations remain quite limiting. Potentially this limitation can be relaxed using heterodyne correlation spectroscopy, as discussed next.

13.1.3.1.2 Heterodyne X-Ray Studies of Equilibrium Fluctuation Dynamics

The first heterodyne hard X-ray studies of fluctuation dynamics have been reported recently (Gutt et al. 2003; Livet et al. 2006; Madsen et al. 2004). Heterodyne studies use a scattered reference beam to interfere with the scattered sample beam. If we again denote the scattering amplitudes from the reference and sample objects as ρ_r and ρ_s respectively, then the measured intensity is given by Eq. (18).

$$I = |\rho_r + \rho_s|^2 = I_r + 2\,\mathrm{Re}(\rho_r \rho_s^*) + I_s \tag{18}$$

The autocorrelation of the total intensity is then obtained from Eq. (19), where I_r and I_s now refer to averaged quantities.

$$\langle I(\mathbf{q}, t')I(\mathbf{q}, t'+t)\rangle_{t'} = (I_r + I_s)^2 + 2I_s I_r\,\mathrm{Re}[g_1(\mathbf{q}, t)] + I_s^2 g_2(\mathbf{q}, t) \tag{19}$$

If $I_r \gg I_s$, then the second term will be much larger than the third term, i.e., the heterodyne signal will be much larger than the available homodyne signal. There are other advantages to the heterodyne approach as well. The heterodyne technique is less sensitive to large, slow-moving contaminants (e.g., dust) and it yields $g_1(\mathbf{q}, t)$ more directly.

There are, however, important constraints on the implementation of the heterodyne approach. Heterodyning requires that the scattering from the reference and sample beams interferes at the detector. This is possible only if pathlengths for photons scattered by the reference and sample differ by less than the longitudinal coherence length (typically on the order of a micron for a hard X-ray beam, as discussed above). Moreover, if the reference is not tightly coupled mechanically to the sample, relative vibrational motions can cause severe problems.

An additional consideration is that the random noise on the reference signal I_r sets a limit to the increased sensitivity practically possible. In optical heterodyning, ratios I_r/I_s at least as high as 10^4 have been used (Earnshaw 1997), though smaller values are more common. For a fully coherent beam (not realized in current sources), Eq. (18) suggests that, for a given coherence time, the shot noise on the reference signal at each pixel will be smaller than the fluctuation-caused changes to the second term if $N_s > N_r^{1/2}$, where the N represent average photon numbers from the two scatterers. However, in making the autocorrelation of Eq. (19), many coherence times will typically be averaged over. If we average over N_c coherence times, we might expect the shot noise to be sufficiently small if $N_c N_s > (N_c N_r)^{1/2}$, i.e., $N_s > (N_s/N_c)^{1/2}$. This conjecture, however, remains to be verified.

In recent hard X-ray heterodyne spectroscopy studies of capillary waves on liquid surfaces by Gutt et al. (2003), ratios of reference to sample signal of 1–3:1 are cited. The reference signal in this work is believed to be the component of the incident beam that is totally reflected from the surface, not a separate scattering object. In the work of Livet, Sutton, and collaborators (Livet et al. 2006), scattering from an aerosol served as the reference for the study of coherent scattering from a filled polymer.

Effective heterodyning has also been used in "lensless" soft X-ray holography (McNulty et al. 1992; Howells et al. 2001; Eisebitt et al. 2003). Here several approaches have been investigated to create a reference beam. These include the use of a small auxiliary hole in a mask to give a pinhole diffraction reference signal and the use of small-angle scattering reference signals from a static aerogel.

Thus hard X-ray heterodyne fluctuation spectroscopy, which offers the potential of higher sensitivity than homodyne techniques, has now been demonstrated but its development remains in its infancy.

13.1.3.1.3 Studies of Critical Fluctuations with Microbeams

Mocuta et al. (2005) have developed a very different approach to the study of fluctuations near a critical point. They focused an X-ray beam to a micron length scale in the critical region of Fe_3Al and observed the temporal variation in scattered intensity as individual fluctuations appeared and disappeared. They were able to compare their results carefully with theory and to identify the crossover from noncritical to critical dynamics behavior.

13.1.3.1.4 Coherent X-Ray Studies of the Kinetics of Nonequilibrium Systems

Coherent X-ray scattering can also provide unique information about the *kinetics* of phase transitions, though theoretical interpretation is a key issue. In the case of a material evolving toward equilibrium, the structure is no longer stationary and the autocorrelation function defined above in Eq. (15) is no longer meaningful. An alternative approach is to define the two-time correlation function, Eq. (20), where the average represents an ensemble average often estimated from the X-ray data by smearing out the speckle pattern.

$$C(\mathbf{q}, t_1, t_2) = \left(\frac{I(\mathbf{q}, t_1) - \langle I(\mathbf{q}, t_1)\rangle}{\langle I(\mathbf{q}, t_1)\rangle}\right)\left(\frac{I(\mathbf{q}, t_2) - \langle I(\mathbf{q}, t_2)\rangle}{\langle I(\mathbf{q}, t_2)\rangle}\right) \quad (20)$$

Of course, the two-time correlation function is again an experimental quantity which is not accessible from traditional "incoherent" scattering experiments.

In the case of late stage coarsening in ordering and phase separating alloys, Langevin theory and simulations (Brown et al. 1997, 1999) have predicted how $C(\mathbf{q}, t_1, t_2)$ decays as a function of wave vector $\mathbf{q}$ and difference in observation times $\Delta t = t_2 - t_1$. They predict that the speckle pattern from coarsening alloys is quite long-lived, with correlation decay times comparable to the mean time $t_m = (t_1 + t_2)/2$ since the ordering began. They also predict a scaling relationship between decay time τ and mean ordering time t_m with scaling variable $x \equiv q^2 t$.

The general theory and simulation predictions of the behavior of $C(\mathbf{q}, t_1, t_2)$ have been borne out in coherent small-angle X-ray scattering (SAXS) studies of phase separation (Malik et al. 1998; Livet et al. 2001) and wide-angle X-ray-scattering (WAXS) studies of ordering (Fluerasu et al. 2005; Ludwig et al. 2005). While the experiments confirmed some of the theoretical predictions about coarsening kinetics, there were also features of the coherent X-ray scattering that were unexpected. For example, Fig. 13.3(a) shows the normalized two-time correlation functions at three different mean times t_m during the domain coarsening of a $Cu_{0.79}Pd_{0.21}$ long-period superlattice (LPS) alloy (Ludwig et al. 2005). The width of the peaks gives the correlation decay time, which is linear in mean ordering time, as predicted by theory (Fig. 13.3b). However, the dimensionless ratio between them is different than expected. Langevin theory, Langevin simulations (Brown et al. 1999), and Ising model simulations (Ludwig et al. 2005) on a variety of ordering models all give ratios of 1.3–1.4:1 between correlation decay time and mean time of coarsening. Experiments on Cu–Pd give ratios of about 0.5:1 and independent experiments on Cu–Au alloys give slightly higher values (Fluerasu et al. 2005), but still significantly lower than 1.3:1. The cause and significance of the discrepancy are still unclear.

In addition, the studies on Cu–Pd and Cu–Au alloys (Fluerasu et al. 2005; Ludwig et al. 2005) find small superlattice peak shifts during ordering that would correspond to average lattice changes of approximately 0.1–0.2%. At least in the Cu–Pd case, the superlattice peaks shift while the individual speckles making up the peak do *not*. Since a uniform change in lattice parameter during ordering would shift the speckles, we conclude that the peak shift is due to some inhomogeneous process – perhaps the elimination of strain at antiphase domain boundaries as suggested by Fluerasu et al. (2005).

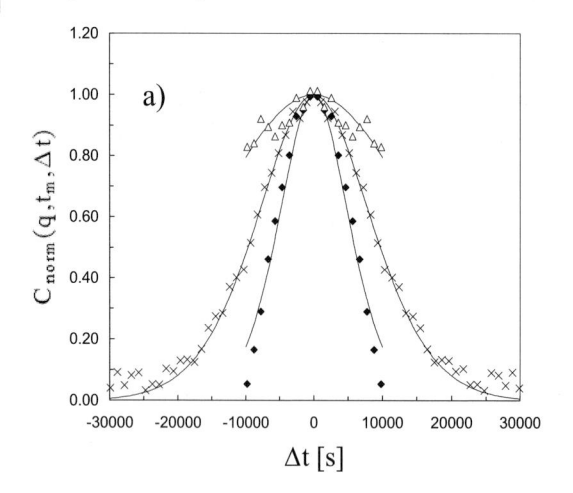

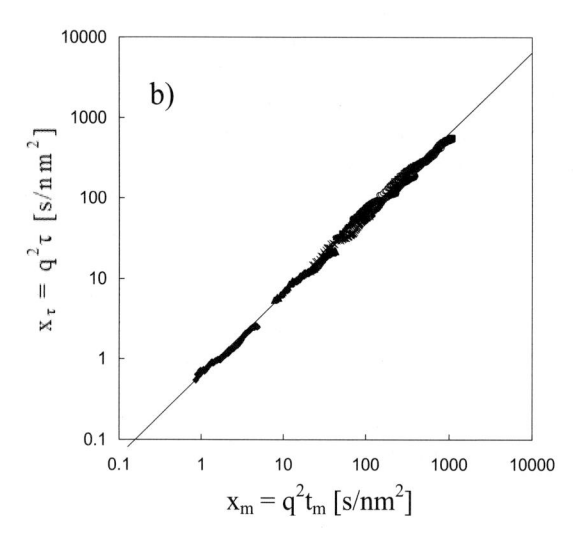

Fig. 13.3 Behavior of two-time correlation functions during ordering of $Cu_{0.79}Pd_{0.21}$ alloy: a) normalized two-time correlation function for mean ordering times t_m of (from bottom to top) 5676 s, 17 286 s, and 30 440 s; b) scaling of reduced two-time correlation function decay time τ with reduced mean time. The two are proportional, as predicted by theory.

An alternative approach to using the two-time correlation function of Eq. (20) for examining late-stage coarsening kinetics with coherent X-ray scattering is to use a detrending fluctuation analysis instead (Stadler et al. 2003). This allows one to examine the speckle intensity fluctuations about a varying mean. Stadler et al. (2003, 2004) used this approach to distinguish between competing coarsening mechanisms in Al–Ag and Al–Zn alloys.

In summary, the coherent scattering study of phase transition kinetics is still young. It is clear that much more information is available from coherent studies than from traditional X-ray scattering. However, theoretical developments are also needed to interpret the extra information obtained from coherent scattering in more general cases.

13.1.3.1.5 Coherent X-Ray Studies of Microscopic Reversibility

In first-order phase transitions, the development and growth of the product phase inside the parent phase is often heterogeneous, with nucleation and growth pathways determined by local defects and stresses. This raises the important question of to what extent such processes are microscopically reversible. Since the coherent scattering speckle pattern is sensitive to the microscopic structure of the material, variations in the speckle pattern allow differences between microscopic states to be measured quantitatively. The first such measurement has been performed by Pierce et al. (2003) using coherent magnetic soft X-ray scattering to examine the magnetic domain evolution around the hysteresis loop of Co–Pt multilayer films. For a given magnetic field the magnetic domain structure and hence the speckle pattern are largely static. As the magnetic field is varied around either a minor hysteresis loop or the full, major hysteresis loop, the speckle pattern is monitored. The microscopic correlation between the domain structure of the sample at two fields $\mathbf{B}_1$ and $\mathbf{B}_2$ is calculated with the speckle cross-correlation function, basically a sum over the detector pixels of the product of suitably normalized intensities $\sum_\mathbf{q} I(\mathbf{q}, \mathbf{B}_1) I(\mathbf{q}, \mathbf{B}_2)$. At low fields, the microscopic memory loss grows exponentially with field. For smooth films, without strong pinning sites, saturation completely erased the microscopic memory while for rough films significant microscopic memory remained after going around the hysteresis loop.

Though the technique has only recently been demonstrated, the use of coherent X-ray scattering has clearly shown its potential to follow microscopic reversibility and memory quantitatively.

13.1.3.2 Phonon Excitations

Studies of collective motion of atoms in condensed matter, better known as phonon or vibrational dynamics measurement, is traditionally the domain of neutron spectroscopy. Neutrons as probing particles are particularly suitable, due to:
- a large penetration depth because of the sufficiently low neutron–nucleus scattering cross-section
- the energy of neutrons being comparable to the energies of collective excitations such as phonons
- neutron momentum making it possible to probe the whole dispersion scheme out to several Å^{-1}, in contrast to inelastic light scattering techniques such as Brillouin and Raman scattering which can only determine acoustic and optic modes at very small momentum transfers.

Whereas there are a number of good reviews discussing phonon investigations with neutron spectroscopy (Krivoglaz 1969; Squires 1997; Lovesey 1986; Bee 1988; Scherm and Fåk 2006), here only new types of spectroscopy that have been developing rapidly during the last few years will be discussed.

It was pointed out in textbooks (e.g., Ashcroft and Mermin 1988) a long time ago that X-rays can in principle be utilized to determine the phonon spectra. However, it was stressed that it will be an enormously difficult experimental challenge caused by the extremely high-energy resolution of an X-ray instrument necessary to resolve excitations resulting from the collective atom motions. Since that time, however, enormous technical improvements have enabled this kind of measurement. Moreover, measurements of vibrational density of states (VDOS) spectra are nowadays possible with two powerful and partially complementary high-resolution techniques, namely by inelastic X-ray scattering and by nuclear inelastic scattering. Both techniques will be described in the following sections.

13.1.3.2.1 Inelastic X-Ray Scattering

Considering X-ray photons with a wavelength of about 1 Å (with an equivalent photon energy of about 12 keV) compared to the vibrational excitations in condensed matter, which are in the meV range, requires a relative energy resolution of at least $\Delta E/E \cong 10^{-7}$. The use of photons is, however, sometimes significantly advantageous compared to the use of neutrons. This can be particularly important in the studies of disordered media: it is not possible to study acoustic excitations propagating with the speed of sound v_s using a probe particle with a speed v smaller than v_s. This limitation is not relevant in neutron spectroscopy studies of crystalline samples because acoustic excitations can be studied in high-order Brillouin zones. The situation is very different in the studies of phonons in disordered systems such as liquids, glasses, and gases. In crystals, the mesoscopic space–time domain corresponding to a momentum–energy region of 0.01–10 nm^{-1} and 0.1–20 meV is traditionally studied by inelastic neutron scattering (INS) (Lovesey 1986; Squires 1997). The neutron technique has been used successfully to investigate the dynamics in disordered systems, at momentum transfer typically larger than 10 nm^{-1} (Suck et al. 1992; Buchenau et al. 1996). In the mesoscopic scale neutrons *cannot* easily probe the acoustic branch as soon as the speed of sound v_s is higher in the measured disordered material. Considering that the typical values of v_s in glasses and liquids are either comparable or considerably larger than 1500 ms^{-1}, it can be explained why, in disordered systems, a comprehensive experimental picture of the high-frequency collective dynamics is still missing[1] (Sette et al. 1998).

To a certain extent another advantage of the inelastic X-ray scattering (IXS) technique is even more important. This is the fact that very small beam sizes of

---

1) Only in fluids of heavy atoms and low-density gases is the speed of sound v_s lower than about 1000 ms^{-1} and only in these media can the acoustic dynamics be investigated effectively via neutron spectroscopy.

the order of a few tens of microns can be obtained currently at third-generation synchrotron sources, allowing studies of systems available only in small quantities down to only a few thousand μm^3. Another consequence of the very narrow beam is the possibility of investigating samples in extreme conditions, especially at very high pressures up to 100 GPa and at high temperatures.

These differences with respect to inelastic neutron scattering motivated the development of the very high-resolution inelastic X-ray scattering technique. The pioneering experiments performed in 1986 (Burkel et al. 1986; Pattison et al. 1986; Suortti et al. 1986) triggered rapid evolution of the IXS technique. To date there are two instruments operational at the ESRF, one at APS and one at Spring-8, and several more under construction.

The optical layout of the instrument is based on the triple-axis principle; it consists of the high-energy resolution monochromator, the sample goniometer, and the crystal analyzer. Details of the instrumentation can be found in Burkel (2000), Ruocco and Sette (2003), and Sette and Krisch (2006). Due to the backscattering geometry the beamline is fairly long in order to acquire a sufficient beam offset between the incident photon beam from the X-ray source and the focused, high energy-resolution beam at the sample position.

The general applicability of the method for the study of elementary solids can be estimated easily (Bosak and Krisch 2005). The overall scattered intensity is proportional to $nsf^2(\langle Q \rangle) \exp(-s/\lambda)$, where n is the concentration of scatterers, s is the sample thickness, $f(\langle Q \rangle)$ the atomic form factor and λ is the X-ray attenuation length $\lambda = 1/\mu$. The optimum signal is attained if $s = \lambda$; on the other hand s must not exceed the focal depth of field of the spectrometer. For typical VDOS setups the depth of field is limited to about 3 mm.

As a consequence, the X-VDOS can be determined with an appropriate accuracy for essentially all elemental solids. One of the potential applications of this novel technique is the VDOS determination of samples submitted to high pressures. The most commonly used high-pressure device is the diamond anvil cell (DAC). For elements heavier than scandium the signal level should be sufficiently high to allow measurements in a DAC. This is of particular interest for geophysical studies, where the determination of the VDOS in combination with low Q measurements allows the determination of average sound and average longitudinal speed and the derivation of the shear velocity without precise knowledge of the equation of state. With respect to INS, the amount of material needed is three to five orders less and anomalous absorption of atoms such as B, Cd, or G, or an anomalously high cross-section as in H, is not present. The validity of the approach has been checked very recently by IXS measurement on diamond and on MgO and by successful comparison with ab-initio and thermodynamical results (Bosak and Krisch 2005).

IXS measurements in solids are not numerous but due to the novelty of the technique other recent applications should be mentioned. The first is the measurement of the lattice dynamics of molybdenum in the DAC cell, where complete phonon dispersion curves were collected at high pressure and compared with ab-initio calculations (Farber et al. 2006). The second contribution deals with the dispersion of acoustic and optical surface phonon modes at the $2H$ polytype of $NbSe_2$ by IXS under grazing incidence conditions, demonstrating surface sensitivity of the method by selective studies of either surface or bulk lattice dynamics in a single experiment (Murphy et al. 2005).

13.1.3.2.2 Nuclear Inelastic Scattering

As was stressed in Section 13.1.3.2 for scattering of visible light, momentum transfer from phonons covers only a minute part of the Brillouin zone. For scattering of X-rays, energy changes through phonon interaction are orders of magnitude smaller than X-ray energies and therefore extreme resolution is demanded to resolve the tiny changes. For recoilless nuclear resonance absorption, the extreme energy resolution of the Mössbauer effect goes without saying. There was an insurmountable intensity problem earlier, however, due to the width of the phonon line being greater by orders of magnitude and therefore the amplitude being tiny compared to the Mössbauer line. The breakthrough came through synchrotron radiation. Instead of measuring scattering from electronic shells, one can use resonant scattering from nuclei where increases in scattering amplitude of several orders of magnitude are possible. Therefore, the counting rate in nuclear inelastic absorption experiments can be quite high using the present instrumentation with the third generation of synchrotron sources, even from extremely small samples. The high brilliance of synchrotron radiation permits the tuning of the incoming monochromatized X-ray energy, so that it exactly matches the nuclear excitation energy and can be resonantly absorbed at the "Mössbauer level" of a suitable nucleus, usually the 14.4 keV nuclear level of ^{57}Fe.

If a phonon is now created or absorbed by the photon, the following nuclear resonance works only when, after (Minkiewicz et al. 1967) the phonon interaction, the X-ray energy just matches the energy of the nuclear transition. Therefore, by tuning the X-ray energy, the phonon spectrum of a solid can be scanned. The incoherent scattering (re-emission after absorption at the nuclear level) of the X-rays reproduces the vibrational phonon density of states (VDOS). The accompanying prompt electronic scattering is separated using the pulse structure of synchrotron radiation and the relatively long lifetime of nuclear excitation. The incident radiation is monochromatized down to an energy width of about 1 meV.

The method is termed nuclear inelastic scattering (NIS). A distinguishing feature of nuclear inelastic scattering compared to other phonon-sensitive techniques is its unique isotope selectivity: no nuclear level other than the first excited level of ^{57}Fe will match the 14.4 keV incoming X-ray energy. Nuclear inelastic scattering therefore selects no species other than iron atoms.

The development of the method is described in a number of publications (see, e.g., Seto et al. 1995; Sturhahn et al. 1995; Chumakov et al. 1995); a review of some of the recent investigations using this method is given by Chumakov and Sturhahn (1999) and by Burkel (2000). For the theoretical aspects of the method, refer to Sturhahn and Kohn (1999).

Since NIS is a resonant method, the X-ray scattering amplitude is several orders of magnitude higher than the Thomson scattering from the electrons. Count rates in inelastic experiments are usually quite high and even the vibrational density of states from *submonolayer* thin samples can be measured. Despite the short time since the first experiments, the method has been applied to a variety of systems ranging from crystals to disordered and amorphous materials as well as from multilayers to monolayers and nanocrystalline islands. The feasibility of the method has been demonstrated first by measurements of the phonon density of states of α-iron foil as a function of temperature (Chumakov et al. 1996a) or pressure (Lübbers et al. 2000). From the huge variety of systems studied since the mid-1990s, only a small selection can be presented here – for instance, studies of VDOS in oriented hcp iron (Giefers et al. 2002); phonon projected density of states in $FeBO_3$ (Kohn et al. 1998); of VDOS in magnetite (Handke et al. 2005), in iron alloys such as Fe_3Al (Fultz et al. 1998), in Pt_3Fe (Yue et al. 2002), in icosahedral quasicrystals (Brand et al. 1999), and in nanocrystalline iron (Fultz et al. 1997); and of protein dynamics in myoglobin (Achterhold et al. 2002) and in thin iron films in grazing incidence geometry (Röhlsberger et al. 1999). Finally one should mention that phonon densities can be measured using not only the ^{57}Fe Mössbauer isotope but also ^{161}Dy (Chumakov et al. 2001), ^{119}Sn (Barla et al. 2000), and other isotopes, or even in samples containing no resonant isotope in a modified setup where photons scattered inelastically in the nonresonant sample are analyzed by a foil containing the resonant isotope in front of the detector (Chumakov et al. 1996b). The last technique especially seems to be very promising if the next, much more powerful, generation of synchrotron sources start running.

13.1.3.3 Quasielastic Scattering: Diffusion

Diffusion in solids is investigated typically on a macroscopic scale following the interpenetration of two atomic species across an interface related to the chemical diffusion coefficient. The self-diffusion coefficient can be measured, on the other hand, by labeling atoms with radioactive isotopes. This technique, also known as the tracer or macroscopic method of diffusion investigation (see, e.g., Mehrer 1990, 1996), deduces atomic events from macroscopic effects, such as, for instance, the concentration gradient of diffusing isotopes. Diffusion can be studied also at the microscopic level, i.e., by revealing the jump vector and the jump rate of the atoms. The study of the microscopic jump diffusion mechanism is the topic of this section. Different scattering methods can be used to resolve the atomic dynamics on various time scales: Mössbauer spectroscopy, quasielastic neutron scattering and neutron spin–echo, nuclear resonant scattering, quasielastic scattering of helium atoms, or X-ray photon correlation spectroscopy.

Whereas quasielastic methods like Mössbauer spectroscopy and neutron scattering both work in the energy domain, some new methods work in the time domain, i.e., interference between different frequencies is observed as a function of time. X-rays from the synchrotron source monochromatized in two steps are transmitted through the sample, exciting the resonant nuclei. The delayed, re-emitted radiation is mapped as a function of elapsed time, constituting the time spectrum of the nuclei in the sample. The method, called nuclear resonance scattering of synchrotron radiation, is a daughter of Mössbauer spectroscopy. Relative methods working in the time domain are neutron spin–echo spectroscopy and X-ray photon correlation spectroscopy. One should note that considerable proportions of the dynamics investigations were performed in recent years with synchrotron radiation, which turns out to be an exceptionally versatile tool for diffusion studies.

In order to verify which diffusion mechanism operates in a specific material it is necessary to develop a mathematical model of the diffusion process. This is done from knowledge of the system, or it is simply an educated guess. The mathematical model is constructed as a system of differential equations. Usually it can be solved analytically and yields the function describing the atomic motion. The free parameters in the model are adjusted in such a way that the best possible agreement with the experiment is achieved.

The simplistic picture of atoms jumping on the lattice can be formulated precisely using the van Hove (1954) correlation function $G(\mathbf{r}, t)$[2]. If we know the function $G(\mathbf{r}, t)$, we have full information about the diffusing system. The van Hove correlation function (or pair correlation function) has the following classical meaning.

> $G(\mathbf{r}, t) \, d^3\mathbf{r}$ is the probability of finding a nucleus in the volume element $\mathbf{r} \ldots \mathbf{r} + d^3\mathbf{r}$ for a given time t, if *this or another nucleus* has been at the position $\mathbf{r} = 0$ at time $t = 0$. Its self-part $G_S(\mathbf{r}, t) \, d^3\mathbf{r}$ is the probability of finding a nucleus in the volume element $\mathbf{r} \ldots \mathbf{r} + d^3\mathbf{r}$ for a given time t if *the same particle* was at the origin $\mathbf{r} = 0$ at $t = 0$, averaged over all the starting positions of the nucleus.

Neglecting the discrete structure of a real material, i.e., for a continuous medium, the diffusion of particles must obey Fick's equation. The solution of this equation neglects the discontinuous nature of real materials. In a real material and at not too elevated temperatures, diffusion can be considered as jump diffusion as in the model of Chudley and Elliott (1961). This model, although originally developed for liquids, found more applications in solid-state diffusion with the following assumptions (cf. Chapter 5, Section 5.2).

- All lattice sites are equivalent, i.e., we have a Bravais (translation-invariant) lattice.
- The diffusion is decoupled from lattice vibrations.
- Only jumps to nearest-neighbor sites are allowed; l_i is an ith vector connecting these sites.
- The jump time is negligible compared with the residence time on the lattice site τ.
- Successive jumps are uncorrelated, i.e., the jump direction of the following jump is independent of the preceding one. In

---

2) $\mathbf{r}$ denotes here displacements of atoms (in our case jump vectors) and not positions.

other words, the present state of the system is determined only by the past state at a particular time (Markov process = uncorrelated diffusion).

If we denote a jump rate of the particle from one site to any nearest-neighbor site by τ^{-1} and the number of neighboring sites (coordination number) by N, then the probability of finding this particle at position **r** at time $t + \Delta t$ is given by Eq. (21), i.e., two probability terms are included: the first term describes particles that were at the neighboring site i at a distance $\mathbf{l}_i$ from **r** and jump into **r** during the time interval Δt; the second term particles that stay at the site **r**.

$$G_S(\mathbf{r}, t + \Delta t) = \sum_{j=1}^{N} \frac{1}{N\tau} G_S(\mathbf{r} + \mathbf{l}_j, t)\Delta t + G_S(\mathbf{r}, t)\left(1 - \frac{1}{\tau}\Delta t\right) \quad (21)$$

In the limit of infinitesimally small Δt the master equation, Eq. (22), is obtained.

$$\frac{\partial}{\partial t} G_S(\mathbf{r}, t + \Delta t) = \frac{1}{N\tau} \sum_{j=1}^{N} \{G_S(\mathbf{r} + \mathbf{l}_j, t) - G_S(\mathbf{r}, t)\}. \quad (22)$$

Equation (22) is usually solved by a Fourier transform in space leading to the differential equation for the so-called intermediate scattering function $I_S(\mathbf{k}, t)$.[3]

$$\frac{\partial}{\partial t} I_S(\mathbf{k}, t) = -\frac{1}{N\tau} \sum_{j=1}^{N} [1 - \exp(-i\mathbf{k} \cdot \mathbf{l}_j)] I_S(\mathbf{k}, t) \quad (23)$$

The solution of Eq. (23) with the boundary condition corresponding to $G_S(\mathbf{r}, 0) = \delta(\mathbf{r}) \Rightarrow I_S(\mathbf{k}, 0) = 1$ is an exponential function of time, Eq. (24).

$$I_S(\mathbf{k}, t) = I_S(\mathbf{k}, 0) \exp[-\gamma(\mathbf{k})t/\tau], \quad \gamma(\mathbf{k}) = N^{-1} \sum_{j=1}^{N} [1 - \exp(-i\mathbf{k} \cdot \mathbf{l}_j)] \quad (24)$$

$I_S(\mathbf{k}, t)$ transformed into the energy domain by a Fourier transform in time yields the scattering function $S(\mathbf{k}, \omega)$ [Eq. (25)], which is a Lorentzian function and can be measured experimentally by Mössbauer spectroscopy or quasielastic neutron scattering.

$$S(\mathbf{k}, \omega) \propto \frac{\Gamma(\mathbf{k})/2}{(\Gamma(\mathbf{k})/2)^2 + (\hbar\omega)^2}, \quad \Gamma(\mathbf{k}) = (2\hbar/\tau)\gamma(\mathbf{k}) \quad (25)$$

---

[3] We use the same terminology for all methods. Thus for Mössbauer spectroscopy and for quasielastic neutron scattering we use **k** for the wave vector, $I_S(\mathbf{k}, t)$ for the intermediate scattering function, and $S(\mathbf{k}, \omega)$ for the scattering function.

The linewidth of this Lorentzian scattering function $\Gamma(\mathbf{k})$ depends on $\mathbf{k}\cdot\mathbf{l}$ and hence on the relative orientations of the radiation and the crystalline structure. This model can describe not only nearest-neighbor jumps; it can be extended straightforwardly for jumps into further neighbors' shells by Eq. (26).

$$\gamma(\mathbf{k}) = 1 - \sum_i W_i \left(\frac{1}{N^i} \sum_j^{N^i} \exp(-i\mathbf{k}\cdot\mathbf{l}_j^i) \right) \quad (26)$$

Here W_i is the probability of a jump to a coordination shell i and N^i denotes the number of lattice sites in the coordination shell i. The sum over probabilities W_i must be normalized to unity.

We can calculate the line broadening explicitly using Eq. (26) for different lattices; e.g., for nearest-neighbor jumps on a cubic bcc lattice with a lattice constant a, the coordinate number $N^1 = 8$ and k_x, k_y, and k_z the components of the outgoing wave vector $\mathbf{k}$ referred to the crystal axes. The eight possible jump vectors $\mathbf{l}_i$ are: $\pm(a/2, a/2, a/2)$, $\pm(a/2, -a/2, a/2)$, $\pm(-a/2, a/2, a/2)$ and $\pm(-a/2, -a/2, a/2)$. The function $\gamma(\mathbf{k})$ calculated due to Eq.(26) reads as Eq. (27).

$$\gamma(\mathbf{k}) = 1 - \cos\left(k_x \frac{a}{2}\right) \cos\left(k_y \frac{a}{2}\right) \cos\left(k_z \frac{a}{2}\right) \quad (27)$$

In the same way "structure factors" $\gamma(\mathbf{k})$ can be calculated for other Bravais lattices. For example, diffusion on a bcc lattice is realized by iron atoms in Fe–Al disordered alloys. Diffusional line broadening $\Gamma(\mathbf{k})$ in disordered bcc Fe–Al is shown in Fig. 13.4. In this case we can conclude immediately that Fe atoms jump only via the nearest-neighbor distance and that all lattice sites are equivalent due to disorder. A comparison with the theoretical line broadening confirms this simplest microscopic diffusion model.

One should note that for calculating the function $I_S(\mathbf{k}, t)$ or $\Gamma(\mathbf{k})$, only the jump rate τ^{-1} and the explicit directions of atomic jumps $\mathbf{l}_i$ must be applied. However, in order to derive from Eq. (26) a function that can be used for calculating nuclear resonant scattering (NRS) spectra, the further transformation discussed in Section 13.1.3.3.2 is required.

Intermetallic alloys are compounds of metals whose crystal structures are different from those of the constituents and do not form "simple" solid solutions, i.e., a Bravais lattice with sites occupied *statistically* by atoms of the alloy constituents. Intermetallic alloys exhibit instead an additional type of atomic order consisting of superlattices occupied by only one species (cf. Chapter 2). Intermetallics form because the strength of bonding between unlike atoms is stronger than that between like ones. This leads to particular crystal structures with a more or less ordered distribution of atoms (Ferrero and Saccone 1991). For stoichiometric compositions and at low temperatures they are well ordered and their degree of order can be controlled by temperature and/or chemical composition. Due to

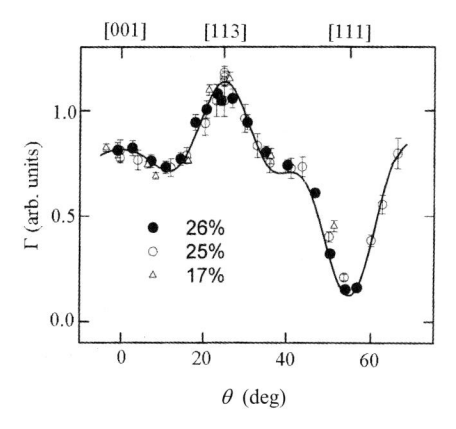

Fig. 13.4 Angular dependence of the line broadening of disordered $Fe_{1-x}Al_x$, $x = 0.26, 0.25$, and 0.17 (Feldwisch 1996): θ is the angle from [001] direction in the $(1\bar{1}0)$ plane; the solid line was calculated from Eq. (27) for nearest-neighbor jumps in a bcc lattice.

their advantageous mechanical behavior (yield stress anomaly; cf. Chapter 6) and their corrosion resistance, they are of high technological interest for high-temperature applications (Sauthof 1995). It is energetically unfavorable to produce antistructure defects by atoms jumping to "wrong" lattice sites in these structures. This complicates the diffusion mechanisms, such as jumps of a given atom species leading to wrong sites may be markedly less probable than jumps to sites on their own sublattice, i.e., there are different jump rates between sublattices (cf. Chapter 5). In this case it is not sufficient to set up only one master equation such as Eq. (22) for the Bravais lattice. For a superstructure with m sublattices one needs m master equations (a jump matrix). The theory of Chudley and Elliott (1961) was extended by Rowe et al. (1971) for hydrogen diffusion on interstitial sites of a non-Bravais lattice for investigations by quasielastic neutron scattering. The Rowe theory assumes equal occupation of different sublattices (sites with different "local symmetry" in the unit cell), and attempts to extend the theory for systems with differently occupied sublattices have been undertaken (Kutner and Sosnowska 1977). Randl et al. (1994) examined the theory in detail for diffusion in non-Bravais structures.

An atom in a non-Bravais lattice can have energetically less and more favored lattice sites (some jumps can be less probable), i.e., we will have different jump rates $1/\tau_{ij}$, which is the jump rate from a site of symmetry i to any nearest-neighbor site of symmetry j. The jump rates between nearest-neighbor sites are constrained by the detailed balance, which demands that the number of atoms jumping in a time unit from one sublattice into another must be equal to the number of reverse jumps according to Eq. (28), where c_i is the probability of the occupation of the ith sublattice ($\sum c_i = 1$) and each site of the ith sublattice is surrounded by n_{ij} sites of the jth sublattice.

$$\frac{c_i}{n_{ij}\tau_{ij}} = \frac{c_j}{n_{ji}\tau_{ji}} \tag{28}$$

The solution of the set of master equations is obtained by calculation of the eigenvalues and eigenvectors of the appropriate jump matrix (Randl et al. 1994). The resulting scattering function is a sum of m Lorentzian functions, each with the particular width Γ_p and weighting factor w_p.

If the jump mechanism and jump frequencies are known, we can in any case calculate the macroscopic diffusion coefficient and compare it with the tracer diffusion result. The diffusion coefficient in a non-Bravais lattice with m nonequivalent sublattices can be calculated as a sum of partial diffusion coefficients between sublattices [Eq. (29)].

$$D = \frac{1}{6} \sum_{i,j} (\mathbf{l}_i - \mathbf{l}_j)^2 \tau_{ij}^{-1} c_i \tag{29}$$

13.1.3.3.1 Quasielastic Methods: Mössbauer Spectroscopy and Neutron Scattering

Our interest in diffusion studies concentrates on methods which can be used to resolve details of the atomic jump process, i.e., features which are not directly accessible to tracer diffusion studies. Nowadays there are three different methods derived from nuclear physics with this property:
- quasielastic Mössbauer spectroscopy (QMS)
- quasielastic neutron scattering (QNS)
- nuclear resonant scattering of synchrotron radiation (NRS).

For neutron scattering $\mathbf{k}$ is the scattering vector and $\hbar\omega$ the energy transfer. As $\hbar\omega$ is very small the scattering process is nearly elastic or "quasielastic;" that is why the method is called quasielastic neutron scattering (QNS). In analogy we speak of quasielastic Mössbauer spectroscopy (QMS) because the method is based on the same theory as QNS, $\hbar\omega$ again being very small.

Due to the relatively slow diffusive motion in solids a very good energy resolution is required in order to be capable of measuring this slow motion. On the other hand, diffusive jumps are over very short distances and, hence, large momentum transfers are required. Diffusion processes or, more precisely, jumps of atoms are by their nature a random phenomenon, i.e., their influence on the measured system can be treated as a perturbance and can lead to a deterioration of the spectrum. In particular, a broadening of the spectral line in the energy domain due to diffusion jumps can be observed. Importantly, this destructive action of diffusion can be investigated as a function of the wave vector. This enables differentiation between various microscopic diffusion mechanisms.

Quasielastic neutron scattering is a method promoted in the late 1950s by Brockhouse (1959) and was first applied to diffusion jumps in metals in the 1980s, whereas quasielastic Mössbauer spectroscopy started in the late 1960s

(Knauer and Mullen 1968). The phenomenon of the emission or absorption of a γ-ray photon without energy loss due to recoil of the nucleus and without thermal broadening is known as the Mössbauer effect. For early reviews, see Wertheim (1964), Flinn (1980), Bauminger et al. (1986), and Nienhaus and Parak (1994). A γ-quantum emitted by a nucleus decaying from an excited state to the ground state is absorbed resonantly at the ground level of a second nucleus of the same isotope. It was Rudolf Mössbauer's discovery that, for nuclei placed in *solids* at low temperature there is, however, a finite probability that emission or absorption of the γ-ray photon will take place without absorption or emission of a phonon. This means that the solid will be in the same internal state before and after the event, so that, in effect, the recoil is taken up by the crystal as a whole and not by an individual atom. This makes the recoil energy immeasurably small. The result is a spectroscopy having the resolution of the lifetime uncertainty of the excited nuclear state, which is usually in the 10^{-9} eV range. In practice, the energy is scanned by repeatedly moving a radioactive source toward and away from an absorber. Due to the corresponding Doppler shift, the energy of the γ-rays arriving at the absorber is varied, causing a varying count rate at the detector behind the absorber.

For ^{57}Fe, by far the most frequently used isotope in Mössbauer spectroscopy, the energy difference between the excited and ground states, i.e., the energy of the γ-quantum, is 14.4 keV, corresponding to a length of its wave vector k of 7.3 Å^{-1}. A very much simplified picture can be helpful in understanding this phenomenon: at low temperature the energy width of the emitted radiation is determined by the mean lifetime τ_0 of the excited Mössbauer level. If at sufficiently high temperatures, however, the mean residence time τ of diffusing atoms between two successive jumps is of the same order of magnitude as, or even smaller than the mean lifetime τ_0 of the Mössbauer level, each emitting Mössbauer atom on average changes its position during the emission process. Thus the wave train emitted by a diffusing atom is "cut" into several shorter wave trains, which leads to a greater width via the Heisenberg uncertainty principle. As these wave trains are emitted by one and the same nucleus, they are coherent. The interference between wave trains emitted by the same nucleus depends on the relative orientations of the jump vector and the direction of the wave vectors (cf. Eq. (26) and Fig. 13.4). Therefore, in certain crystal directions the linewidth is small, while in others it is very large. In this picture the broadening will be greater, the stronger the fragmentation of the wave train. A simplified, semiclassical explanation of the diffusional line broadening is shown in Fig. 13.5.

An analogous argument is possible for QNS where the wave train of neutrons scattered by a diffusing atom is cut into several wave trains. It will be noticed that in QNS the calculated scattering function $S(\mathbf{k}, \omega)$ in Eq. (25) has to be convoluted with the energy resolution of the given experimental setup. The high-energy resolution of the QNS technique is achieved by the application of backscattering spectrometers, which are similar to triple-axis spectrometers with the difference that the scattering angle 2ϑ at the monochromator as well as at the analyzer is

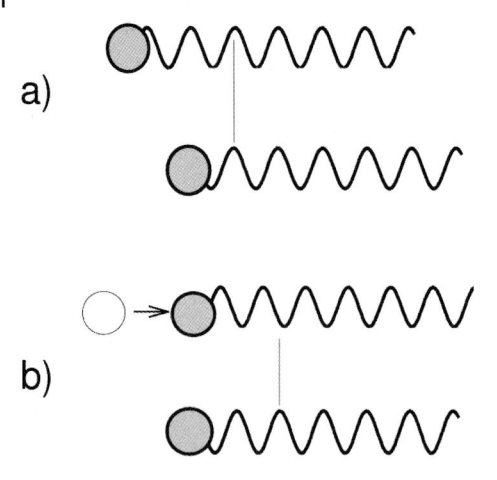

Fig. 13.5 In a nuclear resonant experiment, the jumping atom itself represents the source and therefore conserves its phase relative to the emitted wave during the jump. This is why a) the phase matching with emitted waves from other nuclei is established by the instantaneous excitation by the X-ray pulse. b) Phase matching is generally lost after the jump process. This leads to an accelerated decay of the intensity emitted in the forward direction (NRS) and to the linewidth broadening (QMS, QNS), respectively (Kaisermayr 2001).

fixed at nearly 180°. On the one hand, this restricts the range of possible energy transfers, which can no longer be obtained by changing the angular position of the monochromator and analyzer. On the other hand, however, it significantly increases the resolution in energy (Hempelmann 2000).

QMS and QNS have proven to be appropriate tools for diffusion studies for different classes of alloys QMS has been useful in the studies of iron diffusion in iron aluminides and silicides. QNS, on the other hand, can be applied for diffusion investigations in alloys containing nickel, cobalt, and titanium,[4] provided the diffusivity is sufficiently fast, because the energy resolution of backscattering spectrometers IN10 or IN16 at the high-flux reactor ILL Grenoble is lower by a factor of approximately 100 than the QMS energy resolution (Hempelmann 2000).

Very recently a new neutron spin-echo (NSE) technique was applied (Kaisermayr et al. 2001a), where the energy resolution close to that of ^{57}Fe QMS is possible. NSE has been successfully applied to the study of the dynamics in amorphous systems (proteins, polymers, glassy dynamics, etc.) but it was shown that it is also suitable for the investigation of diffusion on lattices giving direct access

---

4) We are discussing diffusion studies of *metallic* species in intermetallic alloys here. Studies of hydrogen diffusion in organic materials and of hydrogen interstitial diffusion in metals provide, however, the greater part of QNS applications.

to the jump mechanism. NSE is a Fourier method and is sensitive to the time-dependent correlation function yielding directly the intermediate scattering function $I(\mathbf{Q}, t)$, Eq. (24). NSE combines the high-energy resolution from QNS with the high intensity of a beam which is only moderately monochromatic. In NSE the velocity change of neutrons after scattering by a sample is measured by comparing the Larmor precession in known magnetic fields before and after the scattering. This comparison is made for each neutron individually, thus the resolution of the velocity change can be much better than that corresponding to the width of the incident beam.

For a review of the QMS method in diffusion studies see, e.g., Vogl (1990, 1996) and Vogl and Sepiol (2005); for QNS see Springer (1998) and Hempelmann (2000) and references therein.

13.1.3.3.2 Nuclear Resonant Scattering of Synchrotron Radiation

The theoretical principles of nuclear resonant scattering (NRS) of Mössbauer radiation were worked out just after the discovery of the Mössbauer effect but until Ruby (1974) suggested that synchrotron radiation (SR) could be used for a resonant excitation of nuclei, it was not considered for applications. Only 11 years later Gerdau et al. (1985) observed the resonant effect with X-rays for the first time. The next step was the experimental observation of the time structure of the scattered SR in the forward direction, which, primarily, opened the vast field of hyperfine structure for investigation. Finally, Smirnov and Kohn (1995) (also Kohn and Smirnov 1998) proposed a theory of NRS in the presence of diffusive motion of nuclei.

The NRS technique combines the properties of a small and intense synchrotron beam with the spatial and energetic resolution of conventional Mössbauer spectroscopy and can be regarded as its time-based analogue (Mössbauer spectroscopy in the time domain). The first experiment showing the applicability of SR to diffusion studies was performed at the synchrotron radiation source ESRF in Grenoble (Sepiol et al. 1996). NRS allows the determination of the diffusion mechanism on an atomic scale in space and time; see Vogl and Sepiol (1999a) for a systematic discussion of theoretical and experimental aspects of NRS.

The principle of NRS is based on the unique sharpness of the nuclear levels (on the order of 10^{-9} eV) if the scattering process is slow. The enormous brilliance of SR produced by an undulator specially designed for the energy required to excite the resonant level of a Mössbauer atom (14.4. keV for the ^{57}Fe isotope) allows the extraction of a very narrow band in the range of few meV from the energy spectrum. This is achieved with so-called nested monochromators that are especially designed for NRS and which have been the object of intensive work of X-ray optics groups over the last decade. To collect photon counts with good time resolution an avalanche photodiode is usually used as a detector. ^{57}Fe isotope nuclei are excited by synchrotron radiation tuned to the resonance energy. In turn these excited states decay with their characteristic lifetime modulated by coherence effects. Due to the pulsed time structure of synchrotron radiation, the discrimination between electronic (fast) and nuclear (slow) scattering is quite simple.

The most important difference between QMS, QNS, and NRS is that NRS works in the time domain, whereas the first two operate in the energy domain. The principal idea of NRS is in the strict sense very similar to that of both quasielastic methods: the extreme coherence of the synchrotron radiation in the forward direction after nuclear resonance absorption in the sample is destroyed by diffusion, which leads to a faster decay of the scattered intensity with respect to an undisturbed scattering process. From this "diffusionally accelerated" decay, details of the diffusion process can be derived due to the angular dependence of the decay as a function of the X-ray wave vector. Differences between NRS and QMS/QNS techniques consist in the mathematical handling of the intermediate scattering function $I_S(\mathbf{k}, t)$ [Eq. (24)] obtained from the explicit jump diffusion model, which must be, however, be transformed twice to provide an experimentally measurable quantity. These two transformations, $I_S(\mathbf{k}, t) \Rightarrow \varphi(\mathbf{k}, \omega) \Rightarrow I_{FS}(\mathbf{k}, t)$, are set out in Eqs. (30).

$$I_{FS}(\mathbf{k}, t) \propto \left| \int_{-\infty}^{+\infty} d\omega \exp(-i\omega t) \exp\left[-\frac{L}{4\tau_0} \varphi(\mathbf{k}, \omega) \right] \right|^2 ,$$

$$\varphi(\mathbf{k}, \omega) = \int_0^\infty dt \exp\left[i\omega t - \frac{\Gamma_0}{2\hbar} \right] I_S(\mathbf{k}, t).$$

(30)

$I_{FS}(\mathbf{k}, t)$ is the intensity of forward scattered radiation as a function of time and the help function $\varphi(\mathbf{k}, \omega)$ is called the universal resonance function calculated by a time-to-energy Laplace transformation of the intermediate scattering function. Here L is the effective sample thickness, which is proportional to the temperature-dependent Debye–Waller factor and resonant isotope ^{57}Fe content. In the thin-sample approximation Eqs. (30) can be solved analytically, e.g., for the Bravais lattice with the linewidth $\Gamma(\mathbf{k})$ from Eq. (25), which gives Eq. (31), which means that the logarithm of the decay rate is proportional to the width of the diffusional broadening $\Gamma(\mathbf{k})$ as measured in classical QMS.

$$I_{FS}(\mathbf{k}, t) \propto \frac{L^2}{4\tau_0} \exp\left[-\frac{t}{\hbar}(\Gamma_0 + \Gamma(\mathbf{k})) \right]$$

(31)

One should note that Eq. (31) is correct only in the thin-sample approximation, due to the much more important role of the effective sample thickness in NRS than in QMS or QNS. Usually Eqs. (30) can be calculated only numerically, but if the sample is enriched in ^{57}Fe isotope, i.e., thick for X-rays, one can obtain very high count rates and much shorter measuring times than the quasielastic methods from Section 13.1.3.3.1.

While the first studies of diffusion with SR were performed in bulk materials, the field where the unique properties of SR are truly exploited is the application of NRS to surface studies. These measurements are carried out in grazing-incidence condition, i.e., the technique of NRS and grazing-incidence reflection are combined for studying the structure, the hyperfine parameters, and the dynamics of thin films (Röhlsberger 1999). It is known that NRS in grazing-incidence geometry provides depth selectivity, so it can also be used for investigation of diffusion phenomena on the surface or in near-surface regions of metallic films containing iron (or other Mössbauer isotopes). Measurements in the grazing-incidence geometry can be performed only if: (a) the beam divergence is very low, (b) the energetic resolution is in the milli-electronvolt range, and both conditions are excellently fulfilled by the current X-rays sources.

Two features make grazing-incidence NRS useful for studies of surface dynamics: (a) the scattering intensity is proportional to the *square* of the number of resonant nuclei (i.e., the effective thickness) L^2 [cf. Eq. (31)]; (b) the thickness scales as $1/\alpha$, α being the very small angle of incidence (usually a few milliradians). The factor of $1/\alpha$ increasing the effective layer thickness allows measurements of very thin layers or even sub-monolayers containing resonant atoms (Sladecek et al. 2004).

13.1.3.3.3 Pure Metals and Dilute Alloys

The first investigations of the diffusion mechanism were restricted to simple Bravais systems. Such a system is for instance bcc β-titanium, where self-diffusion of titanium atoms belongs among the fastest self-diffusion processes in metals and, moreover, titanium is an excellent incoherent scatterer. The question of the elementary diffusion jump was solved definitely by the use of QNS (Vogl et al. 1989) and it was not very surprising that the dominant diffusion mechanism was a jump into a nearest-neighbor (NN) vacancy. However, no other diffusion mechanism was found which could explain anomalous self-diffusion in β-Ti or in other bcc metals such as β-Zr or β-Hf. Anomalous effects in bcc metals are not only extraordinarily fast self-diffusion but also curvature of Arrhenius plots. These effects were finally explained by phonons. At elevated temperatures the phonon spectrum of titanium contains some very soft phonon modes corresponding to large vibration amplitudes "pushing" Ti atoms to vacant NN sites and in this way considerably lowering the migration energies. Thus the high diffusivity in β-titanium could be explained.

Diffusion of dilute ^{57}Fe atoms in fcc single crystals of Al and Cu has been investigated by QMS (Mantl et al. 1983; Steinmetz et al. 1986). In this case the situation is not so simple as in β-titanium because iron atoms are impurities in the host lattice, attracting vacancies via lattice strain. Extended theory of impurity-atom diffusion via bound vacancies elaborated by Le Claire (1978) could explain broadening of the linewidth measured in different directions versus Al and Cu single crystals.[5] Diffusion coefficients obtained from Eq. (29) agree well with the tracer diffusion data supporting the validity of the method.

---

5) Samples in these measurements were sources of radiation resulting from the decay of ^{57}Co as a precursor of ^{57}Fe. This method, called emission Mössbauer spectroscopy, uses the same theoretical description by Eqs. (25) and (26).

13.1.3.3.4 Ordered Alloys

B2 alloys (CsCl structure) B2-structures are the simplest intermetallic alloy structures; nevertheless they are extremely interesting because NN jumps in these lattices always lead to antistructure sites and therefore should be energetically disfavored. The diffusion mechanism in B2 lattices was a subject of vigorous discussion in the 1990s (Fähnle et al. 1999; Wolff et al. 1999; Frank et al. 2001; Bester et al. 2002). The most intriguing question for a well-ordered B2 structure is whether diffusing atoms jump to NN sites, or whether they perform jumps to e.g., second- or third-NN sites belonging to their own sublattice. Quasielastic measurements on selected intermetallic alloys contributed substantially to the understanding of atomic jumps in B2 lattices.

$Fe_{1-x}Al_x$ alloys crystallize in the B2 structure within a wide homogeneity range on the iron-rich side. The data obtained by measuring iron diffusion in well-ordered FeAl by QMS (Vogl and Sepiol 1994; Feldwisch et al. 1995) and by NRS (Vogl et al. 1998) methods could be fitted only with a superposition of the functions from Eq. (26) corresponding to [100] and [110] jumps in the ratio of 1.9($\pm$0.1):1. This is quite unobvious result for the Fe atoms, indicating a priority of effective jumps to third-NN sites over second-NN sites. $Fe_{65}Al_{35}$, which is still B2 ordered but – due to the composition – exhibits a considerable amount of disorder, provided quite different spectra due to an increased amount of excess iron atoms. The spectra of this alloy clearly consist of two Lorentzian lines (Feldwisch et al. 1995), which is a certain sign that the lattice sites visited by the iron atom do not belong to a Bravais lattice. Finally for the composition close to Fe_3Al (see Fig. 13.4), diffusivity of iron can be very well fitted by simple jumps on the bcc lattice due to Eq. (27). An atomistic diffusion model based on these results is as follows: starting near the ideal B2 composition, FeAl iron atoms diffuse via NN jumps to antistructure sites with a remarkably *short* residence time on the aluminum sublattice. With increasing iron content the residence time of iron atoms on the aluminum sublattice increases and in "disordered" Fe_3Al iron diffuses on a Bravais lattice.

A Monte Carlo simulation with vacancy exchange exclusively with NN atoms, but also with an interaction between the vacancy and the atoms, fitted experimental results perfectly (Weinkamer et al. 1999). The interpretation is as follows: the vacancy migrates through the lattice, destroying and restoring the lattice order. It interacts particularly with defects just created by its own movement, which causes different future paths of the vacancy to be energetically unequal. In the particular FeAl alloy, the vacancy prefers such sequences of NN jumps, which result in effective jumps of iron atoms to third-nearest-neighbor sites ([110] jumps). Recent defect-structure simulations (Bester et al. 2002) confirm the NN jump as the most reasonable elementary diffusion jump in B2 FeAl.

The QNS method was applied successfully to the study of Ni diffusion in NiGa (Kaisermayr et al. 2000), and of Co in CoGa (Kaisermayr et al. 2001b). In both intermetallic alloys only NN jumps were found and the next-nearest-neighbor jumps of Ni suggested by Donaldson and Rowlings (1976) could not be found.

The conclusion is that transition metal (Ni, Co) atoms in NiGa and CoGa jump via antistructure sites.

$D0_3$ alloys (Fe_3Si structure) The $D0_3$ superstructure of Fe_3Si is more complicated than the B2 superstructure. The cubic supercell can be divided into four sublattices, i.e., three iron sublattices (α_1, α_2, and γ) and one silicon (β) sublattice. The lattice constant of these sublattices of the supercell is twice as large as the lattice constant of the small bcc cells. Mössbauer studies were the first to detect that despite the high degree of order in Fe_3Si, iron diffusion in this compound is very fast (Sepiol and Vogl 1993; Vogl and Sepiol 1999b). The most surprising result was that iron diffusion becomes *slower* with *decreasing* order, i.e., below 25 at.% silicon content. This was later confirmed by tracer studies (Gude and Mehrer 1997) and later on by studies with NRS (Sepiol et al. 1996, 1998). The spectra measured by QMS and NRS turn out to be simpler. In contrast to the situation in B2 structures, an iron atom in Fe_3Si can diffuse exclusively between iron sublattices via NN jumps without creating antistructure defects. The diffusion mechanism turns out to be a rather conventional one: diffusion in Fe_3Si is fast due to a large concentration of vacancies (Kümmerle et al. 1995) and relatively soft phonon modes (Randl et al. 1995) favoring low migration energies, and occurs via NN jumps. Off-stoichiometric Fe–Si alloys contain many fewer vacancies and thus diffusion is slower. These experimental results have been confirmed basically by the recent ab-initio simulations of Dennler and Hafner (2006).

Another compound with a $D0_3$ structure is Ni_3Sb. Nickel diffusion in Ni_3Sb, which contains many vacancies, is the fastest ever found in an intermetallic alloy. Results obtained with the QNS method (Vogl et al. 1996a) fitted well under the assumption that nickel atoms jump between NN sites, while other jump models give noticeably worse results.

Fast diffusion of Sn was investigated in the Cu_3Sn sample (Thiess et al. 2003) with the NRS method, supporting the nearest-neighbor jumps model as well.

B8 alloys (NiSb structure) Diffusion of nickel in NiSb (Vogl et al. 1993) and of iron in FeSb (Sladecek et al. 2001) were performed by QNS and QMS methods, respectively. The B8 structure of NiSb and FeSb is rather open and it is not surprising that the atomic jumps are performed via interstitial sites.

13.1.3.3.5 Amorphous Materials

Diffusion in amorphous metals and alloys (metallic glasses), has been a subject of lively discussion since the mid-1990s (Faupel et al. 2003). Generally, glasses are textbook examples of dense random packing systems. The metastability of these systems is their most interesting aspect, particularly due to their relevance in living organisms, but also due to their increasing technical importance (Hilzinger et al,. 1999; Buchanan 2002). Their diffusional behavior is responsible for the metastability, giving rise to various rearrangement processes at moderately elevated temperatures. Obviously, understanding the diffusion processes in glasses is of fundamental importance and, moreover, plays a significant role in their production. The classical tools for measurements of diffusion and dynamics in glasses were neutron spectroscopy (Springer 1972, 1998) and photon correlation spectroscopy (Pusey 1991). Optical methods are, however, limited to studies of nonopaque media by very low wave vector transfers. In recent times it has not been possible to study long-range atomic transport in metallic glasses because of their strong tendency to crystallize when heated through the glass transition. Only recently has discovery of multicomponent bulk metallic glasses exhibiting a much greater resistance to crystallization (Inoue et al. 1990; Peker and Johnson 1993) triggered intensive investigations of their dynamical behavior. Tracer methods of diffusion investigations since that time have been feasible (Knorr et al. 1999). By means of broadband inelastic neutron scattering the vibrational properties and relaxational motion in different metallic glasses could be studied (Meyer et al. 1996b, 1999; Meyer 2002). The energy resolution is, however, always the "Achilles heel" of neutron scattering methods. Some limited access to the much higher-energy resolution, of the order of nano-electronvolts, provides application of the Mössbauer effect, but in practice this is limited to only one Mössbauer isotope which can be reasonably measured, i.e., the ^{57}Fe isotope. Fast relaxation in a metastable metallic melt has been studied with this technique (Meyer 1996a). Numerous Mössbauer spectroscopy studies were performed on amorphous or nanocrystalline soft- and hard-magnetic alloys; see, e.g., Miglierini and Grenèche 1999, 2003; Stankov et al. 2005). Even more frequent have been studies of dynamics in organic glasses, e.g., by Lichtenegger (1999) and Parak (2003a,b).

In order to elucidate the role of high-resolution scattering methods and time-resolved diffraction in the studies of amorphous materials, a few illustrative techniques will be presented. Most of these techniques are brand new and until now only feasibility tests have been performed to present their possibilities but, in our opinion, their potential for development justifies their presentation in this review.

The main feature distinguishing amorphous from crystalline materials is the lack of long-range order. This difference, which is sometimes regarded as negligible, is, however, the origin of delicate and demanding experimental problems. A very basic approach to the calculation of the structure factor will be presented below.

Description of atomic movements by the self-correlation function $G_S(\mathbf{r}, t)$ is actually only a special case (usually conditional on the measuring method) of the more general description by the pair correlation function $G(\mathbf{r}, t)$ (see the beginning of Section 13.1.3.3). An expression for *coherent* scattering function on Bravais lattices has been given by Ross and Wilson (1978); this problem is also discussed by Springer (1972, 1998). One can split the pair-correlation function (Kaisermayr 2001; Kaisermayr et al. 2001c) into a time-dependent part $G'(\mathbf{r}, t)$ and a static part, as in Eq. (32), where $\mathbf{r}_i$ is the ith lattice site, N is the total number of lattice sites, and c is the concentration of the scattering atoms on the Bravais lattice.

$$G(\mathbf{r}, t) = G'(\mathbf{r}, t) + c \sum_{i=0}^{N} \delta(\mathbf{r} - \mathbf{r}_i) \tag{32}$$

Inserting this equation into Eq. (22) and solving by Fourier transformation with the boundary condition $G(\mathbf{r}, 0) = (1 - c)\delta(\mathbf{r}) + c\sum_{i=0}^{N}\delta(\mathbf{r} - \mathbf{r}_i)$, the coherent scattering function is obtained [Eq. (33)].

$$S(\mathbf{Q}, \omega) = c(1-c)\frac{\frac{1}{2}\Gamma(\mathbf{Q})}{\left(\frac{1}{2}\Gamma(\mathbf{Q})\right)^2 - (\hbar\omega)^2} + cN\delta(\omega)\delta(\mathbf{Q} - \mathbf{G}) \tag{33}$$

Equation (33) describes coherent scattering on Bravais lattices (Kaisermayr et al. 2001c). A quasielastic term with a Lorentzian lineshape is, apart from a factor $c(1 - c)$, identical with the scattering function calculated from the self-correlation function $G_S(\mathbf{r}, t)$ for incoherent scattering, Eq. (25). A purely elastic term describes scattering in Bragg directions, where $\mathbf{Q} = \mathbf{G}$. A quasielastic term describes isotropic diffuse scattering (note the prefactor $c(1 - c)$ characteristic of the Laue diffuse scattering (Schönfeld 1999)). Note that Eq. (33) is derived for an unrealistic lattice occupied by one type of scattering atoms only. Derivation for the case of a non-Bravais lattice with one or more scattering atoms is also possible and was obtained for the first time by Kaisermayr (2001) and Kaisermayr et al. (2001c). The general conclusions are similar to those for Bravais lattices – no quasielastic broadening apart from a negligible contribution from diffuse scattering can be expected in the Bragg reflections, irrespective of whether these are fundamental or superstructure peaks. If the lattice is occupied by more than one scattering element, the different quasielastic parts are obtained by simple summation of all the elementary contributions. This prediction was proven experimentally by measuring diffusion in the B2 ordered intermetallic phase $Co_{60}Ga_{40}$ using time-domain interferometry of synchrotron radiation (Kaisermayr et al. 2001d).

From Eq. (33) and from equivalent derivation for non-Bravais lattices (Kaisermayr et al. 2001c) the following conclusions can be drawn.

- The coherent scattering function is elastic for scattering in fundamental and in superstructure Bragg directions.
- In the regions between the reciprocal lattice points, quasielastic diffuse scattering can be observed, i.e., scattering due to lattice disorder (Laue diffuse scattering). The scattering function $S(\mathbf{Q}, \omega)$ is calculated in the same way as the scattering function for the incoherent scattering, i.e., it is calculated from the self-correlation function equations (23) and (25).
- The maximum intensity of the quasielastic component is in the superstructure lattice direction of the non-Bravais lattice; however, it is *not* possible to measure quasielastic broadening in these positions since the diffuse intensity will be completely hidden under the elastic Bragg line.

- Observation of the diffuse scattering is difficult due to the very low intensities, and thus large detectors are necessary. Higher intensities are expected by scattering on samples without lattice structure, i.e., on glassy samples. Such a glassy sample was measured in the first time-domain interferometry experiment of Baron et al. (1997).

It is well known that the energy resolution of X-rays is limited due to the laws of electronic Bragg diffraction (see, e.g., Chumakov et al. 1997; Toellner et al. 1997). In order to improve the resolution into the nano-electronvolt range, elastic nuclear resonant scattering has to be applied (see the Section 13.1.3.3.2). It is self-evident that an ideal instrument for inelastic scattering with nano-electronvolt energy resolution would be a resonant triple-axis spectrometer (Burkel 2000). Re-emission from the nuclear state excited by the synchrotron radiation pulse defines the energy width and a nuclear absorber as analyzer is used to study the behavior of the sample containing no resonant isotopes. A pilot experiment of this kind was performed by Baron (1997) and is called time-domain interferometry. A reference interference pattern is temporally modulated by the scattering from the sample, providing a quantum beats pattern variable as a function of the sample temperature and the momentum transfer. The deconvolution of the beat pattern determines directly the scattering function $S(\mathbf{q},t)$ and thus the full information about dynamics in the sample. The feasibility test was performed on the glass-forming liquid glycerol covering the time scale from about 15 ns to about 200 ns corresponding to 5–50 neV on the energy scale. An extension of the method to investigate the dynamics via quasielastic scattering of X-rays on amorphous solid systems actually has only the problem of the beam intensity, which will be increased significantly in the future due to application of free-electron lasers.

With increasing intensity of the beamlines another method, SR-based perturbed angular correlation (SRPAC) (Baron et al. 1996), has now become accessible for applications in glassy states. SRPAC is a scattering variant of time-differential perturbed angular correlation, a method using special radioactive isotopes with a γ–γ cascade applied to characterize the probe's atom lattice location via precession of probe nuclei, which is proportional to the internal nuclear hyperfine field interactions, see, e.g., Butz (1996).

In SRPAC the intermediate nuclear level is not excited via a cascade originating from the decay of a radioactive parent, but from the ground state during resonant excitation of a ^{57}Fe resonant atom by SR. Directional selection and timing are obtained in SRPAC by the direction and the timing of the incident SR short pulse of radiation. The intermediate nuclear level split by magnetic dipole or electric quadrupole interaction allows one to investigate the hyperfine interactions and rotational dynamics of the nuclear probe atoms.

One of the first experiments with the SRPAC method was performed on the rotational dynamics above the glass transition up to the liquid state of the molecular glass former dibutyl phthalate (DBP) with ferrocene molecules as probes (Sergueev et al. 2006). In the regime of slow relaxation, the damping of the hyperfine beats can also be observed by nuclear forward scattering (see Section 13.1.3.3.2) on ^{57}Fe-enriched ferrocene molecules. At higher temperatures, however, in the regime of fast relaxation, only SRPAC spectra can be measured which approach the natural decay in a characteristic way. The probe molecules reproduce the dynamics of the glass former and enable extraction of the pure translational relaxation rate (Asthalter et al. 2006).

The potential of SRPAC relies on the fact that single-nucleus scattering depends neither on recoil-free emission and absorption, nor on translational motion. Therefore, SRPAC allows one to continue Mössbauer investigations of hyperfine interactions and rotational dynamics into regions where the Lamb–Mössbauer factor vanishes, i.e., in viscous glasses and liquids. It is expected that SRPAC using the ^{57}Fe resonance can cover a dynamic range of relaxation times of five orders of magnitude, from ∼10 ps up to 1 µs (Blachowski and Ruebenbauer 2004).

Brilliance of third-generation synchrotron sources at the present time is still insufficient for the measurement of diffusion in metallic amorphous materials. It is, however, important to take this method into account due to its prospective development if the next generation of synchrotron sources become available, especially the free-electron laser sources with their brilliance higher by up to ten orders of magnitude (cf. Section 13.1.4.1).

13.1.4
Time-Resolved Scattering

High temporal resolution is of increasing experimental interest in order to study the kinetics of nonequilibrium alloys or to investigate equilibrium dynamics. Real-time investigations have natural advantages over post-facto studies. In real-time studies, there is a complete temporal record, ensuring that no kinetic features of the evolution are missed. There is no concern about sample changes in the period between the end of processing and the post-facto measurement. This is of particular concern if the sample evolution occurs at a high temperature where relaxation can be relatively rapid. Finally, in building a quantitative kinetic record of the sample evolution for a given set of experimental parameters, real-time studies do not suffer from run-to-run processing variations that can occur if the kinetic evolution is instead reconstructed through post-facto measurements made on samples from different runs.

Because of the complexity of many growth and processing geometries, it is often difficult to implement real-space probes (e.g., electron microscopy or atomic force microscopy) for real-time study. Instead, scattering techniques are naturally applicable and high time resolution studies are therefore of high interest. The definition of what constitutes "high" resolution in this context is open to interpretation. The important time scale in a given material is, of course, set by the physical process of interest. In an alloy undergoing atomic ordering, phase separation, or crystallization, diffusion is necessary and sets the fundamental time scale. Since atomic diffusivities are thermally activated, this fundamental time scale can easily vary between times of the nanosecond order near the melting point to times greater than the age of the universe at very low temperatures! On the other hand, if nonequilibrium lattice vibrations are of interest, then the appropriate time scale is femtoseconds. Given the widely disparate time scales relevant to different materials, here we will take a relatively broad view of what constitutes "high resolution" in the time domain.

The current state-of-the-art in traditional scattering studies of metallic alloys is discussed first below, including technical possibilities and the type of alloy science being performed.

13.1.4.1 Technical Capabilities

At least three things are necessary in order to perform time-resolved studies successfully: a source of sufficient intensity, an appropriate physical stimulation to initiate the desired alloy response, and appropriate detectors. We begin by briefly discussing source technology, since that has been the most rapid to develop. While time-resolved neutron scattering studies continue to be important, the continuing increase in the X-ray intensity available from synchrotron sources has been a primary factor in facilitating the development of time-resolved studies since the mid-1980s. The typical X-ray intensity available in the hard X-ray regime at a third-generation is of the order of 10^{14} photons s^{-1}. The scattering cross-section plays a pivotal role in determining the time resolution accessible with reasonable counting statistics. The diffuse scattering from an amorphous or disordered alloy, or from an alloy surface, is sufficiently weak for times shorter than a millisecond not to be usually accessible – indeed, many such studies more naturally operate in the second regime. On the other hand, the evolution of an intense Bragg peak can be monitored on the nanosecond or shorter time scale. However, most time-resolved studies of metallic alloy evolution have been on the slower time scales.

Due to difficulties in temporally resolving X-ray scattering patterns on fast time scales, as well as the desire to synchronize experiments with a "pump" process starting a transformation, a pulsed source-structure is most often employed for the highest time resolution experiments. The bunch structure of electrons/positrons in storage rings is one available source of short X-ray pulses. Typical X-ray pulses from a single electron/positron bunch in a third-generation storage ring have a duration of ~100 ps. In cases where it is desirable to extract pulses of shorter duration, it has recently been shown that the X-ray beam intensity can be modulated on the subpicosecond time scale using acoustic pulses induced by a fast optical laser pulse in a transmitting crystal (DeCamp et al. 2001). New fourth-generation sources are already producing intense hard X-ray pulses in the femtosecond regime. The subpicosecond pulse source (SPPS) at the Stanford Linear Accelerator (SLAC) in Stanford, California, uses chirped electron bunches, energy-dispersive magnetic chicanes and undulators to produce 100 fs X-ray pulses. Toward 2010, it is expected that 100 fs coherent hard X-ray pulses with ~10^{12} photons per pulse will become available from the X-ray free-electron lasers (XFELs) being constructed at the Linac Coherent Light Source (LCLS) project in

California, the European XFEL Project at the Deutches Electronen-Synchrotron (DESY) in Hamburg, and at the SPring-8 XFEL project in Japan. At this point, few studies of alloys have been contemplated on such fast time scales. Most studies on semiconductors so far have focused on phonon behavior and energy relaxation from excited electrons to phonons.

Laser-induced plasmas are an alternative to accelerator-based pulse sources and are also undergoing rapid development. Because these sources emit isotropically, their intrinsic brightness is lower than that of current accelerator-based sources. However, with focusing, plasma X-ray sources have now achieved 10^4 photons per 100 fs pulse. This is sufficient for studies of Bragg peak changes on these time scales and these X-ray beams have proven to be a valuable tool for ultrafast kinetics studies of phonon dynamics in semiconductors (Bargheer et al. 2004).

In order to observe the scattering evolution from a material in a meaningful manner, it is also usually necessary that the entire illuminated volume is evolving uniformly. This requires that the state of the sample be changed suddenly on a time scale shorter than that of the physical processes of interest. For the fastest processes, such as nonequilibrium laser-induced melting, the only pump available is a fast laser pulse. Slower processes, however, can be induced by shock loading, sudden changes in electric or magnetic field, or sudden changes in temperature or pressure. Typically metallic alloy phases are not particularly sensitive to pressure (unlike the situation for soft condensed matter), so inducing phase transitions by pressure jumps is usually not feasible. Very fast upward jumps in temperature are possible with laser heating. Relatively rapid drops in temperature are also possible if only a surface layer is heated so that the heat can diffuse back through the colder part of the sample. For macroscopic samples, limited thermal conductivity can increase cooling times to milliseconds and beyond. Perhaps the fastest controllable temperature changes in macroscopic samples are achieved with resistive heating of thin ribbons (Brauer et al. 1990).

A final requirement for successful time-resolved scattering studies is the availability of appropriate detectors. At the slowest end, CCD-based area detectors have traditionally had read times of the order of 1 s, though faster read times are possible. It has been shown that in the high-energy X-ray study of diffuse scattering from an alloy, an area detector is a very powerful tool for quickly recording the scattering over a significant portion of a plane (Reichert et al. 2005). For the future, this offers attractive options for the time-resolved study of diffuse scattering evolution. For faster times, silicon avalanche photodiode detectors with nanosecond resolution are becoming more widely available – these facilitate the use of pump–probe studies, though this technique has not yet been widely applied to the study of metallic alloys. Streak cameras, in which the recording medium is physically moved rapidly behind an aperture during exposure, give the highest time resolution accessible and are now capable of subpicosecond resolution.

13.1.4.2 Time-Resolved Studies – Examples

Time-resolved scattering has been used to examine a number of different phenomena in metallic alloys, but studies of diffusive phase transition kinetics have been particularly widespread because of their technological importance, significant theoretical interest, and (sometimes) accessible time scales. In examining "phase ordering" processes of atomic ordering or phase separation, a distinction is usually made between the "early-stage" transformation kinetics, during which the material transforms locally to the new phase, and the "late-stage" transformation kinetics, during which domains of the ordered phase coarsen to reduce their interfacial energy. The late-stage coarsening process is typically much slower than the early-stage processes and has been studied widely. In this case, theory and simulation predict that the average domain size grows with time as $t^{1/n}$ where $n = 2$ for a nonconserved order parameter (e.g., atomic ordering) and $n = 3$ for a conserved order parameter (e.g., phase separation). Theory also predicts that the evolving structure factor obeys a dynamic scaling with $S(q,t) = t^{d/n} F(qt^{1/n})$, where $F(x)$ is a scaling function. These general predictions agree well in some cases with scattering studies of late-stage coarsening kinetics (Shannon et al. 1992), but other alloys appear to show more complex behavior (Mazumder et al. 1999); it is not yet clear whether this is due to elastic interactions or other factors.

Fewer experiments have examined in detail the early-stage kinetics. The development of order can proceed by nucleation and growth or by a spinodal (continuous) process. Following a quench from the disordered phase into the ordered phase, fluctuations grow. If the disordered phase is quenched into an unstable region of a phase diagram, below a second-order transition line or below a mean-field instability (spinodal) line, then some fluctuations are unstable. If the disordered phase is instead quenched into a metastable region of the phase diagram, below a first-order transition line, then most fluctuations will equilibrate to a new metastable level. Statistically rare fluctuations will eventually nucleate the ordered phase. In some cases, there is a mean-field spinodal, or instability, buried below a first-order transition. In such cases, we would expect a crossover in ordering kinetics behavior for disordered samples rapidly quenched to temperatures above and below the instability. Ludwig et al. (1988) and Tanaka et al. (1994) have examined the equilibration of fluctuations and its relationship to this crossover in the case of ordering in the classic Cu–Au alloys. In this case, the initial kinetics is relaxational in nature, even for quenches to temperatures below the mean-field instability point. The simplest theory of ordering or decomposition kinetics in the unstable regime is the linear theory of Cahn, Hilliard, Khachaturian, and Cook (Cahn and Hillard 1959; Khachaturian 1968; Cook et al. 1969) (cf. Chapters 5, 7, and 10). It neglects coupling between fluctuations at different wave vectors $\mathbf{q}$ and predicts that at each wave vector the structure factor should either relax exponentially (if fluctuations at this wave vector are stable) or grow exponentially (if fluctuations at this wave vector are unstable). Many of the time-resolved scattering studies of early-stage kinetics in alloys have sought to examine to what extent the simple linear theory could describe fluctuation equilibration in the disordered phase and the growth of fluctuations during the early stages of a spinodal phase transformation. It is now generally believed that nonlinear theories are necessary to describe early-stage fluctuation growth quantitatively. Among the clearest experiments testing nonlinear theory are the time-resolved X-ray studies of early-stage spinodal decomposition in Al–Zn alloys by Mainville et al. (1997) and Hoyt et al. (1989). For ordering, Livet et al. (2002) have carefully examined the fluctuation equilibration process experimentally above the critical point, as well as the ordering process just below.

Although most experiments have focused on the development of order from disorder, there have also been interesting studies of order relaxation following temperature changes within the ordered phase (Park et al. 1992; Park 1996) and of the dissolution of order (Okuda et al. 1997).

While much of the time-resolved work in the field has focused on very fundamental comparisons with statistical mechanics models employing simple pair interactions or Ginzburg–Landau free energies, most alloys exhibit more complex ordering kinetics. Gao and Fultz (1993) have identified the formation of a clear precursor phase during ordering in Fe–Al alloys. In other cases, time-resolved studies have been able to examine the kinetics in the presence of complex phase behavior, with competing phase separation and ordering tendencies, as in Fe–Al (Allen et al. 1992), or the presence of long-period superlattice (LPS) phases in $Cu_{0.79}Pd_{0.21}$ (Wang et al. 2005). In the latter case, the satellite peaks associated with the modulated LPS structure are observed to grow faster initially than the superlattice peaks associated with the short-range order (Fig. 13.6a). This is apparently due to the presence of an instability to modulated order below the first-order transition and can be reproduced in Monte Carlo simulations of the ordering process (Fig. 13.6b) using effective pair potentials derived from diffuse scattering data, as discussed below in Section 13.1.5.2.

The agreement with the real-time X-ray scattering data suggests that the real-space kinetics of the Monte Carlo simulation (see Fig. 13.7) is reproducing the actual growth kinetics reasonably well. Figure 13.7 shows the simulated development of the modulated order in real space. Here the lattice has been broken into a set of unit cells, and the dominant sublattice on which order exists in each unit cell has been identified. A single plane of unit cells has been selected for display. The four possible sublattice orderings on the fcc lattice are denoted by the four non-black shades; those unit cells without a clear sublattice ordering are denoted by black. Antiphase boundaries (APBs) between degenerate ordered regions can be either "conservative" or "nonconservative." A conservative APB separates two ordered regions that differ from each other by a translation vector parallel to the plane of the APB interface, while a nonconservative APB separates two ordered regions that differ from each by a translation vector with a component perpendicular to the APB interface. Because the nearest-neighbor atomic environment is not changed by the presence of conservative APBs, these are typically lower in energy than nonconservative APBs, and they are the equilibrium APB's in the 1D LPS structure of $Cu_{0.79}Pd_{0.21}$. Careful examination of the first frame shows that, at 10 MCS (Monte Carlo steps), local ordered regions are already forming, but there are many nonconservative antiphase relationships between neighboring ordered regions. This is consistent with the initial growth primarily of the satellite peaks in the X-ray data. The second frame (30 MCS) shows that, with passing time, the conservative APBs have propagated at the expense of the neighboring

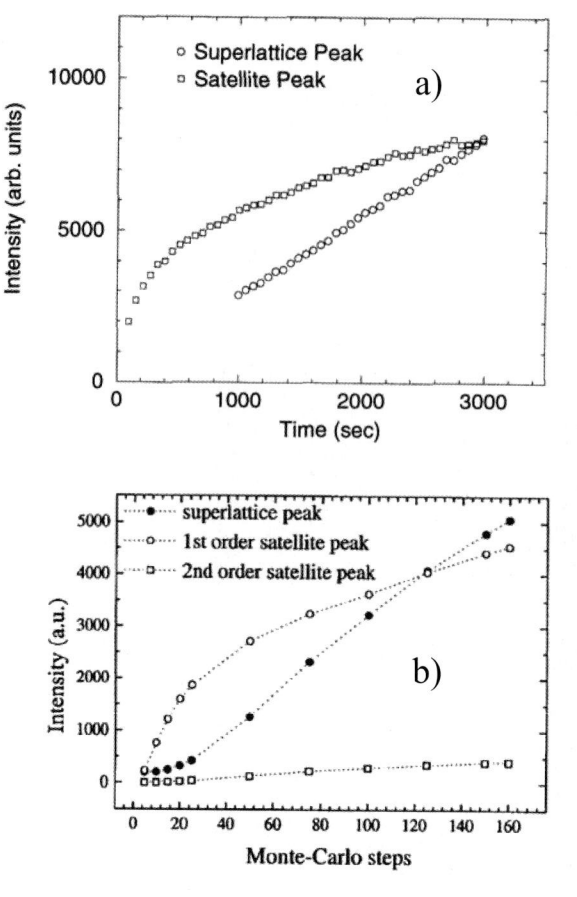

Fig. 13.6 Kinetics of ordering in a $Cu_{0.79}Pd_{0.21}$ alloy: a) the growth of X-ray peak intensities for the satellite (due to modulated order) and superlattice peaks following a quench to 723 K. The satellite peak initially grows faster than the superlattice peak. b) Results from Monte Carlo simulations of the ordering kinetics using effective pair interactions derived from diffuse X-ray scattering.

regions having nonconservative antiphase relationships. This produces elongated domains, typical of the true 1D LPS structure. By the third frame (150 MCS), we can clearly see small 1-d LPS regions of different variants, and the same variant, abutting each other. The kinetics is now slowing considerably. The fourth frame (600 MCS) largely shows the coarsening effects between different 1D LPS regions. A particularly interesting boundary occurs slightly to the left and below the center. Here two modulated structures confront each other; they have a horizontal modulation wave vector and a conservative boundary at their interface.

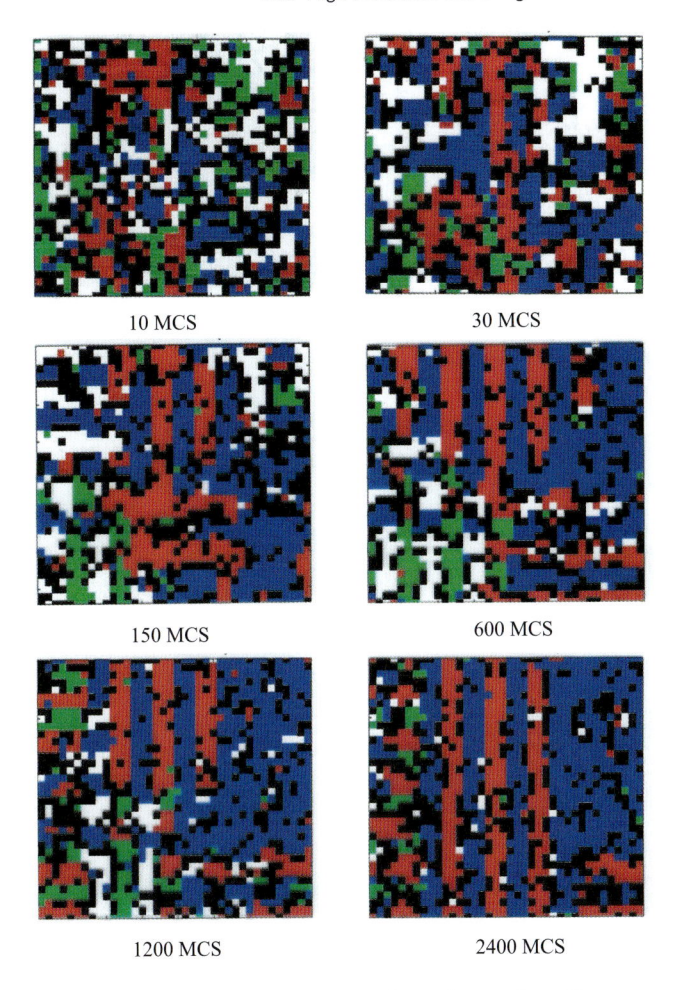

Fig. 13.7 The evolution of order in one plane of atoms from Monte Carlo simulations of ordering in 1D LPS $Cu_{0.79}Pd_{0.21}$ alloy. Times are in Monte Carlo steps and each cell has been colored to represent on which of the four possible sublattices local ordering has occurred.

While most studies of ordering and phase separation kinetics have focused on bulk phenomena, Reichert et al. (1997, 2001a) have used surface-sensitive time-resolved X-ray scattering to examine how ordering kinetics differs at the surface. Among other effects, they have found distinct anisotropies in the rate of ordering perpendicular to the surface and parallel to it.

The processes of crystal and quasicrystal formation from metallic glasses have also been widely studied with time-resolved X-ray scattering. Studies have particularly focused, during crystallization from the glass, on transient phase formation

(Sutton et al. 1989; Brauer et al. 1992; Bruning et al. 1996) and nanocrystallization (Antonowicz 2005). Crystallization from the liquid has been studied less, probably because the nucleation occurs in a very heterogeneous manner and subsequent local growth is rapid. In an interesting experiment, however, Yasuda et al. (2004) were able to examine dendritic growth during uniaxial solidification and observe the detachment of fragmented dendrite arms (cf. Chapter 3).

Though not as well studied as atomic ordering and phase separation, diffusionless transformations in alloys are also of great interest and time-resolved studies can offer unique insights. For example, Babu et al. (2005) were able to examine lattice parameter fluctuations in austenite and how they evolved during the transformation to bainite. Abe et al. (1994) carefully examined the kinetics of the martensitic transition in In–Tl alloys and could make detailed comparisons with theory. The studies of transition kinetics between the cubic and the orthorhombic phases in Ni_3Sb after sudden changes in temperature by Svensson et al. (1997) elucidate well the continuity between the concepts of athermal and isothermal martensitic transformations.

One of the great strengths of the time-resolved scattering technique is its wide applicability. Thus, in addition to its use for more fundamental studies of alloy kinetics, it has been increasingly employed as a unique tool for investigating applied materials problems. Thus, Meneghini et al. (2003) have used time-resolved scattering to examine microstructure evolution in Co–Cu giant magneto-resistance (GMR) alloys, Chen et al. (2003) examined the evolution of a nickel aluminide bond coat at high temperatures, Elmer et al. (2004) tracked phase formation during the welding of steel, Choo et al. (2004) followed the strain evolution during tensile loading and creep in a TiAl–W alloy, and Wada and Tabira (2005) examined the structural evolution of gas-storage alloys during chemical reactions.

13.1.5
Diffuse Scattering from Disordered Alloys

Many alloys of interest are disordered at the atomic scale, either being amorphous or being crystalline without long-range chemical order. In either case, the local structure of the material is best studied through diffuse scattering. X-ray and neutron diffuse scattering studies of alloy structures with X-ray and neutrons have a long history, but the development of synchrotron sources has enabled the full development of techniques which were previously quite difficult. The broad energy spectrum of synchrotron sources has enabled anomalous X-ray scattering to be used to separate different contributions to diffuse scattering. Meanwhile the high collimation of synchrotron beams has facilitated the further development of small-angle X-ray scattering (SAXS) and surface-sensitive scattering to study long-range chemical fluctuations and near-surface structure, respectively. Finally, the use of high-energy synchrotron X-rays has enabled the measurement of diffuse scattering over entire surfaces at a single shot using position-sensitive detectors. Schweika (1998) has written a general review of the field.

13.1.5.1 Metallic Glasses and Liquids

The elastic scattering intensity from a single-component amorphous material can be simply written as Eq. (34), where the sum is over the N atoms, b_i is the scattering factor (or scattering length in the case of neutrons), and $\mathbf{r}_{ij} = \mathbf{r}_i - \mathbf{r}_j$.

$$I(\mathbf{q}) \propto \sum_{i,j} b_i b_j^* e^{i\mathbf{q}\cdot\mathbf{r}_{ij}} = |b|^2 \sum_{i,j} e^{i\mathbf{q}\cdot\mathbf{r}_{ij}} = N|b|^2 S(\mathbf{q}) \qquad (34)$$

For an isotropic material, the structure factor $S(\mathbf{q})$ depends only on the magnitude of $\mathbf{q}$. The quantity of interest is usually the real-space pair correlation function $g(r)$, which can be obtained from the experimental $S(q)$ by transformation [Eq. (35)].

$$g(r) \equiv \frac{1}{4\pi r^2 N} \sum_{i \neq j} \delta(r - r_{ij}) = \frac{2}{\pi} \int_0^\infty q(S(q) - 1)\sin(qr)\,dq \qquad (35)$$

The quantity $4\pi r^2 g(r)\,dr$ gives the average number of atoms around a central atom at a distance between r and $r + dr$. Thus, for instance, $g(r)$ yields the average number and distance of nearest neighbors. Because of the averaging over orientation and individual atoms taking place in $g(r)$, some researchers have instead used the scattering data as input to inverse Monte Carlo simulations, which then yield real-space models of the material structure (cf. Chapter 12). This has been particularly useful when examining the intermediate range order in alloys, i.e., order on length scales of about 1–2 nm. The pair distribution function is less sensitive to such structures than is $S(q)$, particularly in the relatively low-q region. While an inverse Monte Carlo simulation result is not unique, the procedure does allow input of other physical information, such as steric constraints and average density, as well as data from other experimental techniques, such as extended X-ray absorption fine structure (EXAFS). For example, Sheng et al. (2006) have recently used this approach to examine short-range and medium-range order in a Ni–P metallic glass. Also helpful in discerning structural features is the collection of high-q data from high-energy X-ray or neutron scattering. In performing the transform to obtain $g(r)$, this ensures that peak widths are not broadened by the truncation of the integral and better allows individual coordination shells to be identified (Billinge 2004).

In alloys with N chemical components, where each component has a different average scattering factor (or length), the scattering intensity can be written as the sum of $N(N+1)/2$ partial structure factors associated with each kind of pair [Eq. (36)].

$$I(q) \approx \sum_{\alpha,\beta} \sum_{i \in \alpha, j \in \beta} b_\alpha b_\beta^* e^{i\mathbf{q}\cdot\mathbf{r}_{ij}} = \sum_{\alpha\beta} N_\alpha b_\alpha b_\beta^* S_{\alpha\beta}(q) \qquad (36)$$

If the individual partial structure factors have been determined, they can be transformed to yield the chemically specific partial distribution functions $g_{\alpha\beta}(r)$ [Eq. (37)].

$$g_{\alpha\beta}(r) \equiv \frac{1}{4\pi r^2 N} \sum_{i \in \alpha, j \in \beta} \delta(r - r_{ij}) = \frac{2}{\pi} \int_0^\infty q(S_{\alpha\beta}(q) - 1) \sin(qr) \, dq \qquad (37)$$

Obviously these are more revealing than the total pair correlation function in a metallic glass alloy. To determine the individual partial structure factors, the scattering factors f_α can be varied (Keating 1963), either by isomorphous substitution, by isotopic substitution to vary the neutron scattering lengths in the case of neutron scattering, or by utilizing the "anomalous," or resonant, change in the X-ray-scattering factor near the absorption edges of the elements as in Eq. (38), where $f_0(q)$ is the atomic form factor, the Fourier transform of the real-space electron distribution of an atom, and f' and f'' are the real and imaginary anomalous scattering factors, respectively.

$$b_\alpha(q, E) = f_{0\alpha}(q) + f'_\alpha(E) + i f''_\alpha(E) \qquad (38)$$

In the case of isomorphous substitution there is always concern that the amorphous material prepared with an isomorphous replacement will not have an identical local structure to the original; therefore the methods of isotopic substitution or resonant X-ray scattering are more often preferred. Unless isotopes are available to change the sign of one or more of the components' scattering lengths, however, the resulting equations are usually rather ill-conditioned, making accurate determination of the $S_{\alpha\beta}(q)$ difficult. Again, one way to try to overcome this problem is to use inverse Monte Carlo simulations to fit several independent sets of scattering data while also incorporating known physical constraints.

An alternative to taking $N(N+1)/2$ independent data sets to determine all the partial structure factors fully is to change only the scattering factor of a single chemical component and measure the difference, so that all correlations not involving that atomic species cancel. This is differential anomalous scattering (Fuoss et al. 1981); it is particularly convenient using two X-ray photon energies just below the absorption edge of one atomic species we designate A. Then we may often neglect the change in scattering that occurs for the other atomic species and Eq. (39) applies, where we assume that the scattering factor is real for simplicity.

$$\Delta I(q) = I(q, E_1) - I(q, E_2) = \Delta f'_A \sum_\beta f_\beta S_{A\beta} \qquad (39)$$

This can be normalized and transformed to give a "differential distribution function," which is sensitive only to the environment around that species of atom. Recently this approach has been used, for instance, to examine short- and medium-range order in complex Zr–Cu–Ni–Ta–Al amorphous alloys (Hufnagel and Brennan 2003).

13.1.5.2 Diffuse Scattering from Disordered Crystalline Alloys

Disordered crystalline alloys exhibit diffuse scattering because of the nonperiodic arrangement of individual atomic species on lattice sites, the displacement of atoms from their ideal lattice positions, and thermal vibrations. Thus the scattered intensity is given by Eq. (40), where $\mathbf{R}_{ij} = \mathbf{R}_i - \mathbf{R}_j$ is the difference in ideal lattice position of two atoms i and j in the crystal and δ_i, δ_j are the displacements of these atoms from their ideal positions.

$$I(\mathbf{q}) \approx \sum_{i,j} b_i b_j^* e^{i\mathbf{q}\cdot\mathbf{r}_{ij}} = \sum_{i,j} b_i b_j^* e^{i\mathbf{q}\cdot\mathbf{R}_{ij}} e^{i\mathbf{q}\cdot(\delta_i - \delta_j)} \tag{40}$$

In the case of a binary alloy we can write Eq. (41) (Borie and Sparks 1971), where c_A and c_B are the chemical compositions, and $p_{ij}^{\alpha\beta}$ is the (conditional) probability of finding an atom of species β on site j if there is an atom of species α on site i.

$$\begin{aligned}\langle b_i b_j^* e^{i\mathbf{q}\cdot(\delta_i - \delta_j)} \rangle \\ = c_A p_{ij}^{AA} b_A b_A^* \langle e^{i\mathbf{q}\cdot(\delta_i^A - \delta_j^A)} \rangle + c_A p_{ij}^{BA} b_B b_A^* \langle e^{i\mathbf{q}\cdot(\delta_i^B - \delta_j^A)} \rangle \\ + c_B p_{ij}^{AB} b_A b_B^* \langle e^{i\mathbf{q}\cdot(\delta_i^A - \delta_j^B)} \rangle + c_B p_{ij}^{BB} b_B b_B^* \langle e^{i\mathbf{q}\cdot(\delta_i^B - \delta_j^B)} \rangle \end{aligned} \tag{41}$$

The exponentials with the displacements can also be expanded in a power series, Eq. (42), where $\delta_{ij} \equiv \delta_i$.

$$\langle e^{i\mathbf{q}\cdot\delta_{ij}} \rangle = 1 + i \langle \mathbf{q}\cdot\delta_{ij} \rangle - \frac{\langle |\mathbf{q}\cdot\delta_{ij}|^2 \rangle}{2} - \cdots \tag{42}$$

The total intensity can then be written as the sum of a fundamental intensity, which is independent of the local order, a term dependent only on the short-range order (SRO), a first-order displacement term (1D), and higher-order displacement terms (hD), as in Eq. (43), where the first term on the left-hand side is given by Eq. (43), using thermal Debye–Waller factors as in Eq. (44).

$$I(q) = I_{fund}(q) + I_{SRO}(q) + I_{1D} + I_{hD}(q),$$

$$I_{fund}(q) = |c_A b_A e^{-M_A q^2} + c_B b_B e^{-M_B q^2}|^2 \sum_{ij} e^{i\mathbf{q}\cdot\mathbf{R}_{ij}} \tag{43}$$

$$e^{-Mq^2} = |\langle e^{i\mathbf{q}\cdot\delta} \rangle| = e^{-\langle (\mathbf{q}\cdot\delta)^2 \rangle/2} \tag{44}$$

We introduce the Warren–Cowley SRO parameter $\alpha_{lmn} \equiv 1 - \langle p_{ij}^{AB} \rangle / c_B$, where the indices l, m, n refer to the relative positions of two atoms in lattice units. In a cubic crystal, the SRO intensity is given by Eq. (45), where h_1, h_2, h_3 are the indices of the scattering wave vector in reciprocal lattice units and the exponential term represents an effective Debye–Waller factor between neighbors separated by l, m, n lattice units.

$$I_{SRO}(q) = Nc_A c_B |b_A - b_B|^2 \sum_{l,m,n} \alpha_{lmn} e^{-2M\Phi_{lmn} q^2} \cos[\pi(h_1 l + h_2 m + h_3 n)] \quad (45)$$

The different factors entering the diffuse scattering, SRO, atomic displacements and thermal vibrations, can be separated by using their different symmetries in reciprocal space and by varying the scattering factors of the atomic species. To vary the scattering factors, isotopic substitution in neutron scattering or resonant X-ray scattering can be used. Isomorphous substitution is also possible, but, as is the case with diffuse scattering studies of amorphous materials, it can always suffer the problem that the structure of separate alloy samples prepared using isomorphous elements may not be identical. A relatively recent review of the subject, within the context of using resonant X-ray scattering to help separate the different components, has been written by Ice and Sparks (1999).

Most diffuse scattering studies examining SRO and displacements have traditionally used a point detector to measure the diffuse scattering throughout a symmetry-chosen volume of reciprocal space. However, using relatively high X-ray or neutron energies in conjunction with an area detector, a relatively flat plane of diffuse scattering data can be recorded in a single exposure (Reichert et al. 2001b). This considerably facilitates the study of local alloy structure as a function of temperature and/or concentration.

With the information on SRO and displacements available from diffuse scattering, much can be learned about the nature of interactions in the alloy. As some recent examples, local atomic arrangements in a Fe–Ni invar alloy have been correlated with the material's anomalous thermal properties (Robertson et al. 1999), and Mezger et al. (2006) have been able to determine effective interactions in Ni–Pd alloys. Wang et al. (2006) examined the diffuse scattering from classic LPS Cu–Pd alloys. These exhibit diffuse peaks at satellite positions associated with the modulated order, rather than at the superlattice sites themselves (Figs. 13.8a, 13.9a). From the diffuse scattering data, they derived effective pair potentials which exhibit Friedel oscillations (Fig. 13.8b), presumably due to Fermi-surface nesting in the $\langle 110 \rangle$ directions (Sato and Toth 1961). The experimental effective pair potentials were then used in Monte Carlo simulations to reproduce the temperature dependence of the diffuse scattering fine structure in the disordered phase (Fig. 13.9b) and the details of the ordering kinetics, as discussed above in Section 13.1.4.2.

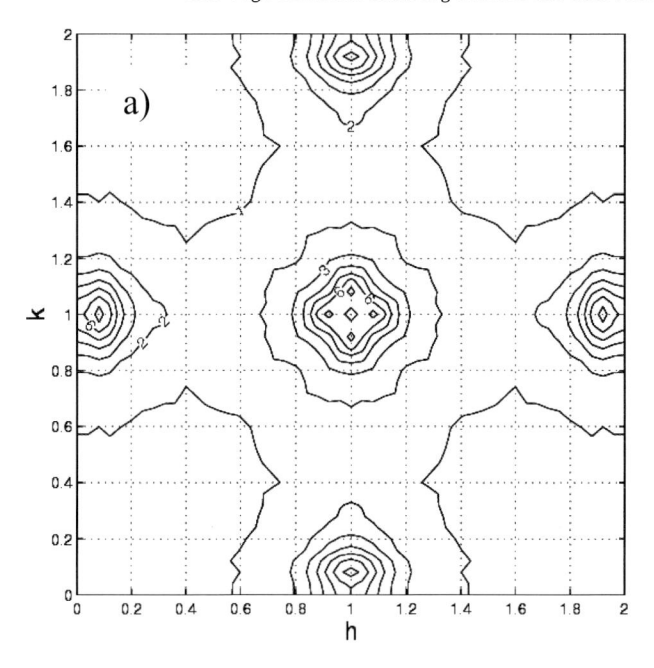

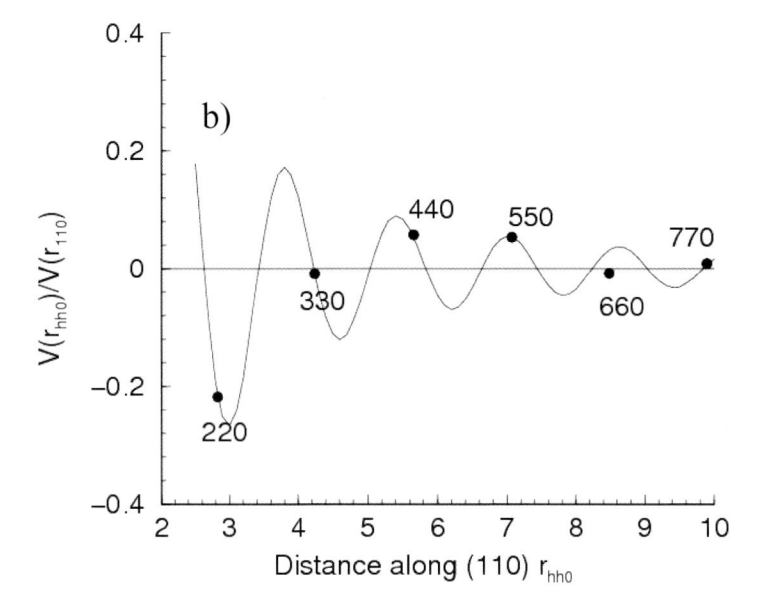

Fig. 13.8 a) Short-range order diffuse X-ray scattering in disordered $Cu_{0.79}Pd_{0.21}$ alloy; b) effective pair interaction between atoms derived from the diffuse scattering; the pair potential shows Friedel oscillations.

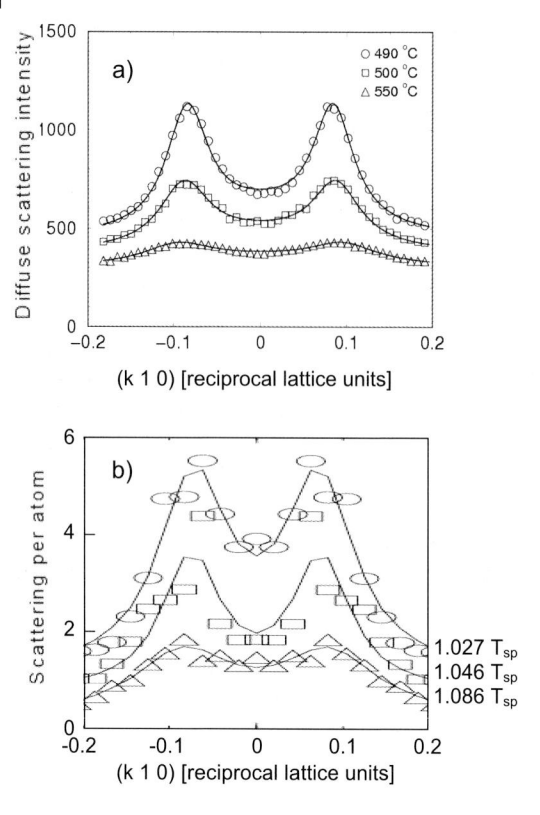

Fig. 13.9 a) Temperature dependence of observed diffuse scattering fine structure in disordered $Cu_{0.79}Pd_{0.21}$ alloy; b) results of Monte Carlo simulations of the diffuse scattering fine structure as a function of temperature above the spinodal ordering point T_{sp}.

13.1.6
Surface Scattering – Atomic Segregation and Ordering near Surfaces

While X-rays have traditionally been considered a probe of the bulk structure of a material due to their relatively long penetration depth (typically microns), the high brightness of synchrotron sources has facilitated the development of surface-sensitive X-ray scattering. This uses glancing incidence/exit angles comparable to the critical angle of total reflection (of order 1°), or the weak X-ray scattering from surfaces at specific reciprocal space positions to achieve its surface sensitivity. Most of the early applications of the technique focused on surface structure in ultrahigh vacuum (UHV); an overview can be found in Robinson and Tweet (1992). The variety of surface structural information that can be obtained from surface-sensitive X-ray diffraction is remarkable. In the case of surface reconstruction (or any process yielding a surface layer of different structure than the bulk) the atomic structure of the surface can be determined with two-dimensional crystallography, in analogy to traditional three-dimensional crystallographic methods. Two-dimensional structures, such as a surface reconstruction or the simple termination of a bulk crystal, have Fourier transforms that fall off slowly in the normal direction. Thus, instead of Bragg peak spots, surface structure gives rise to truncation rods of diffracted intensity normal to the surface. These crystal truncation rods are also very sensitive to the structure normal to the surface, particularly at the anti-Bragg points, halfway between the normal Bragg reciprocal lattice points perpendicular to the surface. At these anti-Bragg points, alternating crystal layers give scattering amplitudes with alternating signs, so that the scattering there is very sensitive to the surface structure. The rods can be analyzed to determine the depth of the surface structure, surface roughness, surface chemical segregation, vertical relaxation of outer surface layers, and registry of the surface structure with the underlying crystal.

Of particular interest for alloy physics is preferential atomic segregation or ordering at a surface (Dosch and Reichert 2000). This can become quite complex, with ordering and segregation apparently competing in many cases, as has been observed in Ni–Al alloys (Drautz et al. 2001). The nature of critical phenomena near surfaces, and how it relates to the bulk phase transition, can also be elucidated, as seen, for instance, in the case of V_2H (Trenkler et al. 1998).

13.1.7
Scattering from Quasicrystals

X-ray and neutron scattering have been crucial for elucidating the structure of quasicrystals. Kelton (1993) conducted an older review, while Steurer (2003, 2004) has reviewed much of the more recent structural work (cf. Chapter 2). In general, the diffraction pattern from a perfect quasicrystal consists of a mathematically dense set of Bragg spots. Often it is useful to reference the observed three-dimensional reciprocal space structure to a Bravais lattice in a higher-dimensional space. As in conventional three-dimensional crystallography, the intensity of Bragg spots can then be used in an effort to determine the atomic structure of the quasicrystal. Due to the large number of reflections from a quasicrystal, however, this is difficult. In practice, of course, only a finite set of peaks will be sufficiently intense to be observed in a given diffraction experiment. Indeed, in real quasicrystals, the diffuse scattering between the most intense Bragg peaks can often be dominated by long-wavelength phason modes rather than the weaker Bragg peaks from the perfect quasicrystal (de Boissieu and Francoual 2005). Chemical and local structural disorder may also be significant. Given these difficulties in determining the long-range order of perfect quasicrystals, much more information has so far been gained about the nature of short-range order.

13.1.8
Outlook

Development of new experimental techniques has always revitalized old scientific disciplines, pushing their evolution and growth in an unexpected way. Over the past few decades, the same has happened with X-rays. In the 1970s, following decades of steady progress in the theoretical understanding of the interaction of X-rays with matter, a technique over a century old (1895: W.C. Röntgen) received an enormous impulse, thanks to the development of a new radiation source. It was realized that the synchrotron radiation emitted from electrons circulating in storage rings is potentially a much more intense and versatile source of X-rays than the widely used X-ray tubes. The period since the mid-1970's has been a time of continuous development of this kind of X-ray source, culminating to date in the so-called third-generation synchrotron sources as indicated in Fig. 13.1. The next generation of sources, based on a free-electron laser (FEL), is already operating and will culminate with the hard X-ray FELs expected to be turned on near 2010. The unprecedented increase in photon flux from synchrotron sources was certainly the main reason for application of X-rays in an enormously broad range of materials and life sciences, and it revolutionized the whole of X-ray physics. In particular, nanoscience benefits significantly from the new brilliant sources through development of spectroscopic methods revealing fundamental dynamical processes in nanoscale materials with unprecedented spatial and temporal resolution. The very low emittance defined by the size and the angular divergence of the electron beam circulating in the rings contributes to the high coherence of the new sources, which, as a consequence, has permitted development of time correlation techniques previously feasible with visible-light lasers only. New experimental techniques based on inelastic X-ray and nuclear resonant scattering reveal fundamental dynamical processes in nanoscale materials. Within a decade the technique of using X-rays for the study of vibrational excitations achieved amazing results and opened a qualitatively "new window" of scientific research.

It is clear that in some applications X-ray methods are and will in the future continue to be complementary to other techniques, especially to neutron scattering. One should take into account, however, that neutron spectroscopy seems to be in a relatively mature phase of its scientific development. Over three decades, the maximum intensity of the available neutron sources has not significantly increased. Moreover, more and more techniques previously reserved for neutrons can now be performed with X-rays, e.g., magnon spectroscopy, phonon spectroscopy, and the study of magnetic switching and magneto-optics.

In sum, we are confident that X-ray scattering techniques will dominate the alloy research landscape during many decades of the 21st century. Since a fundamental goal of materials research is to understand and successfully predict the evolution of materials during thermal or other processing, the kinetics and dynamics of alloys is likely to continue to grow in interest. Ideally, the continued development of computational speed will increasingly facilitate the comparison of first-principles predictions with experiment. Looking ahead, we cannot say in detail what new science will develop if a new generation of radiation sources with still 10^9 times higher brilliance appears, but it will be certainly very exciting.

Acknowledgments

We thank Dr. C. Nelson for her comments on magnetic X-ray scattering. The work on this project at Boston University was partially supported by NSF DMR-0508630 and at the University of Vienna by the Austrian bm:bwk GZ 45.529/2-VI/B/7a/2002, the FWF P17775, and the EU project DYNASYNC NMP4-CT-2003-001516.

References

Abe, H., Ishibashi, M., Ohshima, K., Suzuki, T., Wuttig, M., Kakurai, K. (1994), *Phys. Rev. B* **50**, 9020.

Achterhold, K., Sturhahn, W., Alp, E. E., Parak, F. G. (2002), *Hyperfine Interact.* **141–142**, 3.

Ahuja, B., Ramesh, T., Sharma, B., Chaddah, P., Roy, S., Kakutani, Y., Koizumi, A., Hiraoka, N., Toutani, M., Sakai, N., Sakurai, Y., Itou, M. (2002), *Phys. Rev. B* **66**, 012 411.

Allen, S. M., Park, B., Stephenson, G. B., Ludwig Jr., K. F. (1992), *Proceedings of the TMS Spring 1992 Meeting* ASM, Pittsburgh.

Altarelli, M. (2001), *J. Magnetism Magnetic Mater.* **233**, 1.

Antonowicz, J. (2005), *J. Non-Cryst. Sol.* **351**, 2383.

Ashcroft, N. W., Mermin, N. D. (1988), *Solid State Physics*, Saunders College, Philadelphia.

Asthalter, T., Sergueev, I., van Bürck, U., Dinnebier, R. (2006), *J. Phys. Chem. Solids* **66**, 2271.

Babu, S., Specht, E., David, S., Karapetrova, E., Zschack, P., Peet, M., Bhadeshia, H. (2005), *Met. Mater. Trans. A* **36**, 3281.

Bacon, G. (1975), *Neutron Diffraction*, Clarendon Press, Oxford.

Balcar, E., Lovesey, S. W. (1989), *Theory of Magnetic Neutron and Photon Scattering*, Oxford University Press, New York.

Bargheer, M., Zhavoronkov, N., Gritsai, Y., Woo, J. C., Kim, D. S., Woerner, M., Elsaesser, T. (2004), *Science* **306**, 1771.

Barla, A., Rüffer, R., Chumakov, A. I., Metge, J., Plessel, J., Abd-Elmeguid, M. M. (2000), *Phys. Rev. B* **61**, R14 881.

Baron, A. Q. R., Chumakov, A. I., Rüffer, R., Grünsteudel, H., Grünsteudel, H. F., Leupold, O. (1996), *Europhys. Lett.* **34**, 331.

Baron, A. Q. R., Franz, H., Meyer, A., Rüffer, R., Chumakov, A. I., Burkel, E., Petry, W. (1997), *Phys. Rev. Lett.* **79**, 2823.

Bauminger, E. R., Novik, I. (1986), in *Mössbauer Spectroscopy*, Ed. Dickson, D. P. E., Berry. F. J., Cambridge University Press, Cambridge, p. 219.

Beaurepaire, E., Bulou, H., Scheurer, F., Kappler, J.-P. (2006), *Magnetism: A Synchrotron Radiation Approach*, Springer, Berlin.

Bee, M. (1988), *Quasielastic Neutron Scattering: Principles and Applications in Solid State Chemistry, Biology and Materials Science*, Hilger, Bristol.

Bester, G., Meyer, B., Fähnle, M., Fu, C. L. (2002), *Mater. Sci. Engng. A* **323**, 487.

Beutier, G., Marty, A., Chesnel, K., Belakhovsky, M., Toussaint, J. C., Gilles, B., van der Laan, G., Collins, S., Dudzik, E. (2004), *Physica A* **345**, 143.

Billinge, S. J. L. (2004), *Zeit. Krist.* **219**, 117.

Blachowski, A., Ruebenbauer, K. (2004), *Synchrotron Radiation in Natural Science* **3**, 2.

Blume, M. (1985), *J. Appl. Phys.* **57**, 3615.

Blume, M., Gibbs, D. (1988), *Phys. Rev. B* **37**, 1779.

Borie, B., Sparks, C. J. (1971), *Acta Cryst. A* **27**, 198.

Bosak, A., Krisch, M. (2005), *Phys. Rev. B* **72**, 224 305.

Brand, R. A., Coddens, G., Chumakov, A. I., Calvayrac, Y. (1999), *Phys. Rev. B* **59**, R14 145.

Brauer, S., Ryan, D. H., Strom-Olsen, J., Sutton, M., Stephenson, G. B. (1990), *Rev. Sci. Inst.* **61**, 2214.

Brauer, S., Strom-Olsen, J. O., Sutton, M., Yang, Y. S., Zaluska, A., Stephenson, G. B., Koster, U. (1992), *Phys. Rev. B* **45**, 7704.

Brauer, S., Stephenson, B., Sutton, M., Brüning, R., Dufresne, E., Mochrie, S. G. J., Grübel, G., Als-Nielsen, J., Abernathy, D. L. (1995), *Phys. Rev. Lett.* **74**, 2010.

Brockhouse, N. (1959), *Phys. Rev. Lett.* **2**, 287.

Brown, G., Rikvold, P. A., Sutton, M., Grant, M. (1997), *Phys. Rev. E* **56**, 6601.

Brown, G., Rikvold, P. A., Sutton, M., Grant, M. (1999), *Phys. Rev. B* **60**, 5151.

Brüning, R., Zaluska, A., Sutton, M., Strom-Olsen, J. O., Brauer, S. (1996), *J. Non-Cryst. Sol.* **207**, 540.

Buchanan, O. (2002), *MRS Bull.* **27**, 850.

Buchenau, U., Pecharroman, C., Zorn, R., Frick, B. (1996), *Phys. Rev. Lett.* **77**, 659.

Burkel, E. (2000), *Rep. Prog. Phys.* **63**, 171.

Burkel, E., Illini, Th., Peisl, J., Dorner, B. (1986), *HASYLAB Annual Report*, DESY, Hamburg, p. 337.

Butz, T. (1996), *Z. Naturforsch. A* **41**, 396.

Cahn, J., Hilliard, J. (1959), *J. Chem. Phys.* **31**, 688.

Chen, M. W., Ott, R., Hufnagel, T. C., Wright, P. K., Hemker, K. J. (2003), *Surf. Coat. Tech.* **163**, 25.

Chesnel, K., Belakhovsky, M., Livet, F., Collins, S. P., van der Laan, G., Chesi, S., Attane, J., Marty, A. (2002), *Phys. Rev. B* **66**, 172 404.

Choo, H., Seo, D., Beddoes, J., Bourke, M., Brown, D. (2004), *Appl. Phys. Lett.* **85**, 4654.

Chudley, C. T., Elliott, R. J. (1961), *Proc. Phys. Soc.* **77**, 353.

Chumakov, A. I., Sturhahn, W. (1999), *Hyperfine Interact.* **123/124**, 781.

Chumakov, A. I., Rüffer, R., Grünsteudel, H., Grünsteudel, H. F., Grübel, G., Metge, J., Leupold, O., Goodwin, H. A. (1995), *Europhys. Lett.* **30**, 427.

Chumakov, A. I., Rüffer, R., Grünsteudel, H., Grünsteudel, H. F. (1996a), *Phys. Rev. B* **54**, 9596.

Chumakov, A. I., Baron, A. Q. R., Rüffer, R., Grünsteudel, H., Grünsteudel, H. F., Meyer, A. (1996b), *Phys. Rev. Lett.* **76**, 4258.

Chumakov, A. I., Rüffer, R., Baron, A. Q. R., Metge, J., Grünsteudel, H., Grünsteudel, H. F. (1997), *SPIE* **3152**, 262.

Chumakov, A. I., Rüffer, R., Leupold, O., Barla, A., Thiess, H., Gil, J. M., Alberto, H. V., Vilão, R. C., Ayres de Campos, N, Kohn, V. G., Gerken, M., Lucht, M. (2001), *Phys. Rev. B* **63**, 172 301.

Chumakov, N. K., Tugushev, V. V., Gudenko, S. V., Nikolaeva, O. A., Lazukov, V. N., Goncharenko, I. N., Alekseev, P. A. (2005), *JETP Lett.* **81**, 292.

Collins, S. P., Laundy, D., Rollason, A. (1992), *Phil. Mag. B* **65**, 37.

Collins, S. P., Laundy, D., Guo, G. (1993), *J. Phys. Cond. Matt.* **5**, L637.

Cook, H., de Fontaine, D., Hilliard, J. (1969), *Acta Met.* **17**, 765.

Cooper, M. J., Stirling, W. G. (1999), *Rad. Phys. Chem.* **56**, 85.

Cooper, M., Zukowski, E., Timms, D., Armstrong, R., Itoh, F., Tanaka, Y., Ito, M., Kawata, H., Bateson, R. (1993), *Phys. Rev. Lett.* **71**, 1095.

Cowlam, N., Hanwell, M., Wildes, A., Jenner, A. (2005), *J. Phys. Cond. Matt.* **17**, 3585.

Cox, D. 1972, *IEEE Trans. Magn.* **MAG-8**, 161.

Deb, A., Hiraoka, N., Itou, M., Sakurai, Y., Onodera, M., Sakai, N. (2001), *Phys. Rev. B* **63** 205 115.

De Bergevin, F., Brunel, M. (1972), *Phys. Lett.* **A39**, 141.

de Boissieu, M., Francoual, S. (2005), *Zeit. Krist.* **220**, 1043.

DeCamp, M. F., Reis, D. A., Bucksbaum, P. H., Adams, B., Caraher, J. M., Clarke, R., Conover, C. W. S., Dufresne, E. M., Merlin, R., Stoica, V., Wahlstrand, J. K. (2001), *Nature* **413**, 825.

Dennler, S., Hafner, J. (2006), *Phys. Rev. B* **73**, 174 303.

Dixon, M., Duffy, J., Gardelis, S., McCarthy, J., Cooper, M., Dugdale, S., Jarlborg, T.,

Timms, D. (1998), *J. Phys. Cond. Mater.* **10**, 2759.

Donaldson, A. T., Rowlings, R. D. (1976), *Acta Met.* **24**, 285.

Dosch, H., Reichert, H. (2000), *Acta Mat.* **48**, 4387.

Drautz, R., Reichert, H., Fahnle, M., Dosch, H., Sanchez, J. (2001), *Phys. Rev. Lett.* **87**, 236 102.

Dudzik, E., Dhesi, S., Durr, H., Collins, S., Roper, M., van der Laan, G., Chesnel, K., Belakhovsky, M., Marty, A., Samson, Y. (2000), *Phys. Rev. B* **62**, 5779.

Duffy, J., Dugdale, S., McCarthy, J., Alam, M., Cooper, M., Palmer, S., Jarlborg, T. (2000), *Phys. Rev. B* **61**, 14331.

Earnshaw, J. C. (1997), *Appl. Optics* **36**, 7583 (1997).

Eisebitt, S., Lorgen, M., Eberhardt, W., Luning, J., Stohr, J., Terrner, C., Hellwig, O., Fullerton, E., Denbeaux, G. (2003), *Phys. Rev. B* **68**, 104419.

Eisebitt, S., Luning, J., Schlotter, W., Lorgen, M., Hellwig, O., Eberhardt, W., Stohr, J. (2004), *Nature* **432**, 885.

Elmer, J. W., Palmer, T. A., Babu, S. S., Zhang, W., DebRoy, T. (2004), *Welding J.* **83**, 244S.

Endoh, Y., Itoh, S., Watanabe, T., Mitsuda, S. (1992), *Physica B* **180**, 34.

Evans, P., Isaacs, E., Aeppli, G., Cai, Z., Lai, B. (2002), *Science* **295**, 1042.

Fähnle, M., Mayer, J., Meyer, B. (1999), *Intermetallics* **7**, 315.

Farber, D. L., Krisch, M., Antonangeli, D., Beraud, A., Badro, J., Occelli, F., Orlikowski, D., (2006), *Phys. Rev. Lett.* **96**, 115 502.

Faupel, F., Frank, W., Macht, M.-P., Mehrer, H., Naundorf, V., Rätzke, K., Schober, H. R., Sharma, S. K., Teilcher, K. (2003), *Rev. Mod. Phys.* **75**, 237.

Felcher, G. P., te Velthuis, S. G. E. (2001), *Appl. Surf. Sci.* **182**, 209.

Feldwisch, R. (1996), PhD Thesis, Universität Wien.

Feldwisch, R., Sepiol, B., Vogl, G. (1995), *Acta Met. Mater.* **43**, 2033.

Ferrer, S., Alvarez, J., Lindgren, E., Torrelles, X., Fajjardo, P., Boscherim, F. (1997), *Phys. Rev. B* **56**, 9848.

Ferrero, R., Saccone, A. (1991), in *Materials Science and Technology: A Comprehensive Treatment*, Vol. 1, Ed. Haasen, P., Weinheim, VCH.

Flinn, P. A. (1980), in *Applications of Mössbauer Spectroscopy*, Vol. II, Ed. Cohen, R. L., Academic Press, New York, p. 393.

Fluerasu, A., Sutton, M., Dufresne, E. (2005), *Phys. Rev. Lett.* **94**, 055 501.

Francoual, S., Livet, F., de Boissieu, M., Yakhou, F., Bley, F., Létoublon, A., Caudron, R., Gastaldi, J. (2003), *Phys. Rev. Lett.* **91**, 225 501.

Frank, St., Divinski, S. V., Södervall, U., Herzig, Chr. (2001), *Acta Mater.* **49**, 1399.

Freeland, J., Bussmann, K., Udzerda, Y., Kao, C.-C. (1999), *Phys. Rev. B* **60**, R9923.

Fultz, B., Ahn, C. C., Alp, E. E., Sturhahn, W., Toellner, T. S. (1997), *Phys. Rev. Lett.* **79**, 937.

Fultz, B., Stephens, T. A., Sturhahn, W., Toellner, T. S., Alp, E. E. (1998), *Phys. Rev. Lett.* **80**, 3304.

Fuoss, P. H., Eisenberger, P., Warburton, W. K., Bienenstock, A. (1981), *Phys. Rev. Lett.* **46**, 1537.

Gao, Z. Q., Fultz, B. (1993), *Phil. Mag. B* **67**, 787.

Gaulin, B. D., Mao, M., Wiebe, C. R., Qiu, Y., Shapiro, M., Broholm, C., Lee, S.-H., Garrett, J. D. (2002), *Phys. Rev. B* **66**, 174520.

Gerdau, E., Rüffer, R., Winkler, H., Tolksdorf, W., Klages, C. P., Hannon, J. P. (1985), *Phys. Rev. Lett.* **54**, 835.

Gibbs, D., Harshman, D. R., Isaacs, E. D., McWhan, D. B., Mills, D., Vettier, C. (1988), *Phys. Rev. Lett.* **61**, 1241.

Giefers, H., Lübbers, R., Rupprecht, K., Wortmann, G., Alfè, D., Chumakov, A. I. (2002), *High Press. Res.* **22**, 501.

Goff, J., Sarthour, R., McMorrow, D., Yakhou, F., Vigliante, A., Gibbs, D., Ward, R., Wells, M. (2001), *J. Magn. Magn. Mat.* **226**, 1113.

Good, W., Kim, J., Goldman, A., Wermeille, D., Canfield, P., Cunningham, C., Islam, Z., Lang, J., Srajer, G., Fisher, I. (2005), *Phys. Rev. B* **71**, 224 427.

Grigoriev, S., Kraan, W., Rekveldt, M., Kruglov, T., Bouwman, W. (2006), *J. Appl. Cryst.* **39**, 252.

Gude, A., Mehrer, H. (1997), *Phil. Mag. A* **76**, 1.

Gutt, C., Ghaderi, T., Chamard, V., Madsen, A., Seydel, T., Tolan, M., Sprung, M.,

Grubel, G., Sinha, S. K. (2003), *Phys. Rev. Lett.* **91**, 076 104.

Handke, B., Kozłowski, A., Parlinski, K., Przewoznik, J., Slezak, T., Chumakov, A. I., Niesen, L., Kakol, Z., Korecki, J. (2005), *Phys. Rev. B* **71**, 144 301.

Hase, T., Pape, I., Tanner, B., Durr, H., van der Laan, G., Vaures, A., Petroff, F. M. (2000), *Phys. Rev. B* **61**, R3792.

Haskel, D., Srajer, G., Lang, J., Pollmann, J., Nelson, C., Jiang, J., Bader, S. (2001), *Phys. Rev. Lett.* **87** 207 201.

Hellwig, O., Denbeaux, G., Kortright, J., Fullerton, E. (2003), *Physica B* **336**, 136.

Hempelmann, R. (2000), *Quasielastic Neutron Scattering and Solid State Diffusion*, Oxford Science, Oxford.

Hill, J. (2001), *Synch. Rad. News* **14**, 21.

Hilzinger, R., Reichert, K. (1999), in *Proceedings of the International Conference on High Frequency Magnetic Materials*, Gorham Intertech, Santa Clara, CA.

Howells, M., Jacobsen, C., Marchesini, S., Miller, S., Spence, J., Weirstall, U. (2001), *Nucl. Instrum. Meth. Phys. Res. A* **467**, 864

Hoyt, J., Clark, B., de Fontaine, D., Simon, J. P., Lyon, O. (1989), *Acta Met.* **37**, 1597 and 1611.

Hufnagel, T. C., Brennan, S. (2003), *Phys. Rev. B* **67**, 014 203.

Hupfeld, D., Schweika, W., Strempfer, J., Mattenberger, K., McIntyre, G., Bruckel, T. (2000), *Europhys. Lett.* **49**, 92.

Ice, G. E., Sparks, C. J. (1999), *Annu. Rev. Mater. Sci.* **29**, 25.

Inoue, A., Zhang, T., Masumoto, T. (1990), *Mater. Trans., JIM* **30**, 965.

Ito, M., Hirano, K. (1997), *J. Phys. Cond. Matt.* **9**, L613.

Jaouen, N., Tonnerre, J., Raoux, D. Bontempi, E., Ortega, L., Muenzenberg, M., Felsch, W., Rogalev, A., Durr, H., Dudzik E., van der Laan, G., Maruyama, H., Suzuki, M. (2002), *Phys. Rev. B* **66**, 134 420.

Jourdan, N., Zakharov, A., Hiess, A., Charlton, T., Berhoeft, N., Mannix, D. (2005), *Euro. Phys. J. B* **48**, 445.

Kaisermayr, M. (2001), PhD Thesis, Universität Wien.

Kaisermayr, M., Combet, J., Ipser, H., Schicketanz, H., Sepiol, B., Vogl, G. (2000), *Phys. Rev. B* **61**, 12038.

Kaisermayr, M., Pappas, C., Sepiol, B., Vogl, G. (2001a), *Phys. Rev. Lett.* **87**, 175 901.

Kaisermayr, M., Combet, J., Ipser, H., Schicketanz, H., Sepiol, B., Vogl, G. (2001b), *Phys. Rev. B* **63**, 054 303.

Kaisermayr, M., Sepiol, B., Vogl, G. (2001c), *Physica B* **301**, 115.

Kaisermayr, M., Sepiol, B., Thiess, H., Vogl, G., Alp, E. E., Sturhahn, W. (2001d), *Eur. Phys. J. B* **20**, 335.

Keating, D. T. (1963), *J. Appl. Phys.* **34**, 923.

Kelly IV, J., Barnes, B., Flack, F., Lagally, D., Savage, D., Friesen, M., Lagally, M. (2002), *J. Appl. Phys.* **91**, 9978.

Kelton, K. F. (1993), *Int. Mater. Rev.* **38**, 105.

Kernavanois, N., Deen., P., Paolasini, L., Braithwaite, D. (2005), *Rev. Sci. Inst.* **76**, 083 909.

Khachaturian, A. (1968) *Soviet Physics: Solid State* **9**, 2040.

Knauer, R. C., Mullen, J. G. (1968), *Phys. Rev.* **174**, 711.

Knorr, K., Macht, M.-P., Freitag, K., Mehrer, H. (1999), *J. Non-Cryst. Solids* **250**, 669.

Kohn, V. G., Smirnov, G. V. (1998), *Phys. Rev. B* **57**, 5788.

Kohn, V. G., Chumakov, A. I., Rüffer, R. (1998), *Phys. Rev. B* **58**, 8437.

Kortright, J. B., Kim, S.-K. (2000), *Phys. Rev. B* **62**, 12216.

Kortright, J. B., Kim, S.-K., Denbeaux, G. P., Zeltzer, G., Takano, K., Fullerton, E. E. (2001), *Phys. Rev. B* **64**, 092 401.

Kortright, J. B., Jiang, J. S., Bader, S. D., Hellwig, O., Marguiles, D., Fullerton, E. (2003), *Nucl. Inst. Meth. Phys. Res. B* **199**, 301.

Kortright, J. B., Hellwig, O., Chesnel, K., Sun, S., Fullerton, E. (2005), *Phys. Rev. B* **71**, 012 402.

Krivoglaz, M. A. (1969), *Theory of X-Ray and Thermal-Neutron Scattering by Real Crystals*, Plenum Press, New York.

Kümmerle, E. A., Badura, K., Sepiol, B., Mehrer, H., Schaefer, H.-E. (1995), *Phys. Rev. B* **52**, R6947.

Kutner, R., Sosnowska, I. (1977), *J. Phys. Chem. Soc.* **38**, 741.

Kuzushita, K., Ishii, K., Ohwada, K., Murakammi, Y., Kaneko, K., Metoki, N., Lander, G., Ikeda, S., Haga, Y., Onuki, Y. (2005), *Physica B* **359**, 1045.

Laundy, D. (1999), *Rad. Phys. Chem.* **56**, 151.

Laundy, D., Collins, S. P., Rollason, A. (1991), *J. Phys. Cond. Matt.* **3**, 369.

Laundy, D., Brown, S., Cooper, M. Bowyer, D., Thompson, P., Paul, D., Stirling, W. (1998), *J. Synch. Rad.* **5**, 1235.

Lawson, P., Cooper, M., Dixon, M., Timms, D., Zukowski, E., Itoh, F., Sakurai, H. (1997), *Phys. Rev. B* **56**, 3239.

Lebech, B. (2002), *Physica B* **318**, 251.

Le Claire, A. D. (1978), *J. Nucl. Mater.* **69–70**, 70.

Lichtenegger, H., Doster, W., Kleinert, T., Birk, A., Sepiol, B., Vogl, G. (1999), *Biophys J.* **76**, 414.

Lidstrom, E., Mannix, D., Hiess, A., Rebizant, J., Wastin, F., Lander, G., Marri, I., Carra, P., Vettier, C., Longield, M. (2000), *Phys. Rev. B* **61**, 1375.

Liu, C. T., Whang, S. H., Pope, D. P., Yamaguchi, M., Vehoff, H. (2002), *Structural and Functional Intermetallics*, Elsevier, Amsterdam.

Livet, F., Bley, F., Mainville, J., Sutton, M., Caudron, R., Mochrie, S. G. J., Geissler, E., Dolino, G., Abernathy, D. L., Grübel, G. (2000), *Nucl. Instrum. Methods Phys. Res. A* **451**, 596.

Livet, F., Bley, F., Caudron, R., Geissler, E., Abernathy, D., Detlefs, C., Grubel, G., Sutton, M. (2001), *Phys. Rev. E* **63**, 036 108.

Livet, F., Bley, F., Simon, J. P., Caudron, R., Mainville, J., Sutton, M., Lebolloc'h, D. (2002), *Phys. Rev. B* **66**, 134 108.

Livet, F., Bley, F., Ehrburger-Dolle, F., Morfin, I. Geissler, E., Sutton, M., (2006), *J. Synch. Rad.* **13**, 453.

Lovesey, S. W. (1972, 1986), *The Theory of Neutron Scattering from Condensed Matter*, Vols. I and II, Oxford University Press, Oxford.

Lovesey, S. W., Collins, S. P. (1996), *X-Ray Scattering and Absorption by Magnetic Materials*, Oxford University Press, Oxford.

Lübbers, R., Grünsteudel, H. F., Chumakov, A. I., Wortmann, G. (2000), *Science* **287**, 1250.

Ludwig, K. F., Stephenson, G. B., Jordan-Sweet, J., Mainville, J., Yang, Y., Sutton, M. (1988), *Phys. Rev. Lett.* **61**, 1859.

Ludwig, K. F., Livet, F., Bley, F., Simon, J.-P., Caudron, R., Le Bolloc'h, D., Moussaid, A. (2005), *Phys. Rev. B* **72**, 144 201.

Lumma, D., Lurio, L. B., Borthwick, M. A., Falus, P., Mochrie, S. G. J. (2000), *Phys. Rev. E* **62** 8258.

Lurio, L. B., Lumma, D., Sandy, A. R., Borthwick, M. A., Falus, P., Mochrie, S. G. J., Pelletier, J. F., Sutton, M., Regan, L., Malik, A., Stephenson, G. B. (2000), *Phys. Rev. Lett.* **84**, 785.

MacKay, J., Teichert, C., Savage, D., Lagally, M. (1996), *Phys. Rev. Lett.* **77**, 3925.

Madsen, A., Seydel, T., Sprung, M., Gutt, C., Tolan, M., Grubel, G. (2004), *Phys. Rev. Lett.* **92**, 096 104

Mainville, J., Yang, Y. S., Elder, K. R., Sutton, M., Ludwig Jr., K. F., Stephenson, G. B. (1997), *Phys. Rev. Lett.* **78**, 2787.

Malik, A., Sandy, A. R., Lurio, L. B., Stephenson, G. B., Mochrie, S. G. J., McNulty, I., Sutton, M. (1998), *Phys. Rev. Lett.* **81**, 5832.

Mannix, D., de Camargo, P., Giles, C., de Oliveira, A., Yokaichiya, F., Vettier, C. (2001a), *Eur. Phys. J. B* **20**, 19.

Mannix, D., Stunault, A., Bernhoeft, N., Paolasini, L., Lander, G., Vettier, C., de Bergevin, F., Kaczorowski, D., Czopnik, A. (2001b), *Phys. Rev. Lett.* **86**, 4128.

Mantl, S., Petry, W., Schroeder, K., Vogl, G. (1983), *Phys. Rev. B* **27**, 5313.

Marrows, C., Steadman, P., Hampson, A., Michez, L., Hickey, B., Telling, N., Arena, D., Dvorak, J., Langridge, S. (2005), *Phys. Rev. B* **72**, 024 421.

Mazumder, S., Sen, D., Batra, I. S., Tewari, R., Dey, G. K., Banerjee, S., Sequeria, A., Amenitsch, H., Bernstorff, S. (1999), *Phys. Rev. B* **60**, 822.

McNulty, I., Kirz, J., Jacobsen, C., Anderson, E. H., Howells, M. R., Kern, D. P. (1992), *Science* **256**, 1009.

Mehrer, H. (Ed.) (1990), *Diffusion in Metals and Alloys*, Landolt-Börnstein, Numerical Data and Functional Relationships in Science and Technology, New Series Vol. III/26, Springer, Berlin.

Mehrer, H. (1996), *Mater. Trans. JIM* **37**, 1259.

Meneghini, C., Prieto, A. G., Fdez-Gubieda, M. L., Mobilio, S. (2003), *J. Magn. Magn. Mater.* **262**, 92.

Meyer, A. (2002), *Phys. Rev. B* **66**, 134 205.
Meyer, A., Franz, H., Sepiol, B., Wuttke, J., Petry, W. (1996a), *Europhys. Lett.* **36**, 379.
Meyer, A., Wuttke, J., Petry, P., Peker, A., Bormann, R., Coddens, G., Kranich, L., Randl, O. G., Schober, H. (1996b), *Phys. Rev. B* **53**, 12 107.
Meyer, A., Busch, R., Schober, H. (1999), *Phys. Rev. Lett.* **83**, 5027.
Mezei, F., Pappa, C., Gutberlet, T. (Eds.) (2003), *Neutron Spin Echo Spectroscopy: Basics, Trends and Applications*, Springer, New York.
Mezger, M., Reichert, H., Ramsteiner, I., Udyansky, A., Shchyglo, O., Bugaev, V., Dosch, H., Honkimaki, V. (2006), *Phys. Rev. B* **73**, 184 206.
Miglierini, M., Grenèche, J.-M. (1999), *Hyperfine Interact.* **122**, 121.
Miglierini, M., Grenèche, J.-M. (2003), *J. Phys. Cond. Matt.* **15**, 5637.
Minkiewicz, V. J., Shirane, G., Nathans, R. (1967), *Phys. Rev.* **162**, 528.
Mirone, A., Sacchi, M., Dudzik, E., Durr, H., van der Laan, G., Vaures, A., Petroff, F. (2000), *J. Magn. Magn. Mat.* **218**, 137.
Mocuta, C., Reichert, H., Mecke, K., Dosch, H., Drakopoulos, M. (2005), *Science* **308**, 1287.
Montano, P., Ruett, U., Beno, M., Jennings, G., Kimball, C. (2000), *J. Phys. Chem. Sol.* **61**, 353.
Moon, R., Riste, T., Koehler, W. (1969), *Phys. Rev.* **181**, 920.
Murphy, B. M., Requardt, H., Stettner, J., Serrano, J., Krisch, M., Müller, M., Press, W. (2005), *Phys. Rev. Lett.* **95**, 256 104.
Namikawa, K., Ando, M., Nakajima, T., Kawata, H. (1985), *J. Phys. Soc. Jap.* **54**, 4099.
Nelson, C. S. (1999), Ph.D. Thesis, Northwestern University.
Nienhaus, G. U., Parak, F. (1994), *Hyperfine Interact.* **90**, 243.
Okuda, H., Tanaka, M., Osamura, K., Arnemiva, Y. (1997), *J. Appl. Cryst.* **30**, 592.
Okuyama, D., Matsumura, T., Murakami, Y., Wakabayashi, Y., Sawa, H., Li, D. (2004), *Physica B* **345**, 63.
Oleś, A., Kajzar, F., Kucab, M., Sikora, W. (1976), *Magnetic Structures Determined by Neutron Diffraction*, Polska Akademia Nauk, Warsaw.

Osgood III, R., Sinha, S., Freeland, J., Idzerda, Y., Bader, S. (1999), *J. Magn. Magn. Mat.* **199**, 698.
Pappas, C., Mezei, F., Ehlers, G., Manuel, P., Campbell, I. (2003), *Phys. Rev. B* **68**, 054 431.
Parak, F. G. (2003a), *Cur. Opin. Struct. Biol.* **13**, 552.
Parak, F. G. (2003b), *Rep. Prog. Phys.* **66**, 103.
Park, B. (1996), *Jap. J. Appl. Phys. Part 2-Letters* **35**, L1287.
Park, B., Stephenson, G. B., Ludwig Jr, K. F., Allen, S. M. (1992), *Mater. Res. Soc. Symp. Proc.* **205**, 119.
Pattison, P., Suortti, P., Weyrich, W. (1986), *J. Appl. Cryst.* **19**, 353.
Peker, A., Johnson, W. L. (1993), *Appl. Phys. Lett.* **63**, 2342.
Pengra, D., Thoft, N., Wulff, M., Feidenhans, R., Bohr, J. (1994), *J. Phys. Cond. Matt.* **6**, 2409.
Pierce, M. S., Moore, R. G., Sorensen, L. B., Kevan, S. D., Hellwig, O., Fullerton, E., Kortright, J. B. (2003), *Phys. Rev. Lett.* **90**, 175 502.
Platzman, P. M., Tzoar, N. (1970), *Phys. Rev. B* **2**, 3556.
Pusey, P. N. (1991), in *Liquids, Freezing and Glas Transition*, Part II, Les Houches Session LI NATO ASI, Eds. Hansen, J. P., Levsque, D., Zinn-Justin, J., North-Holland, Amsterdam, p. 841.
Randl, O. G., Sepiol, B., Vogl, G., Feldwisch, R., Schroeder, K. (1994), *Phys. Rev. B* **49**, 8768.
Randl, O. G., Vogl, G., Petry, W., Hennion, B., Sepiol, B., Nembach, K. (1995), *J. Phys.: Cond. Matt.* **7**, 5983.
Reichert, H., Eng, P. J., Dosch, H., Robinson, I. K. (1997), *Phys. Rev. Lett.* **78**, 3475.
Reichert, H., Bugaev, V., Shchyglo, O., Schops, A., Sikula, Y., Dosch, H. (2001b), *Phys. Rev. Lett.* **87**, 236 105.
Reichert, H., Dosch, H., Eng, P. J., Robinson, I. K. (2001a), *Europhys. Lett.* **53**, 570.
Reichert, H., Schöps, A., Ramsteiner, I. B., Bugaev, V. N., Shchyglo, O., Udyansky, A., Dosch, H., Drautz, R., Honkimäki, V. (2005), *Phys. Rev. Lett.* **95**, 235 703.
Robertson, J. L., Ice, G. E., Sparks, C. J., Jiang, X., Zschack, P., Bley, F., Lefebvre, S., Bessierre, M. (1999), *Phys. Rev. Lett.* **82**, 2911.

Robinson, I., Tweet, D. (1992), *Rep. Prog. Phys.* **55**, 599.

Röhlsberger, R. (1999), in *Nuclear Resonant Scattering of Synchrotron Radiation*, Eds. Langouche, G., de Waard, H., *Hyperfine Interact.* **123/124**, 301.

Röhlsberger, R., Sturhahn, W., Toellner, T. S., Quast, K. W., Hession, P., Hu, M., Sutter, J., Alp, E. E. (1999), *J. Appl. Phys.* **86**, 584.

Ross, D. K., Wilson, D. L. T. (1978), in *Neutron Inelastic Scattering*, 383, IAEA, Vienna.

Rowe, J. W., Sköld, K., Flotow, H. E., Rusch, J. J. (1971), *J. Phys. Chem. Solids* **32**, 41.

Ruby, S. L. (1974), *J. Phys.* **35**, C6-209.

Ruocco, G., Sette, F. (2003), in *Inelastic X-Ray Scattering: A New Spectroscopy Tool to Investigate the Atomic Dynamics in Condensed Matter*, Ed. Mobilio, S., Vlaic, G., Conference Proceedings, Vol. 82, Synchrotron Radiation: Fundamentals, Methodologies and Applications, p. 623, Italian Physical Society.

Sakai, N. (1996), *J. Appl. Cryst.* **29**, 81.

Sakai, N., Ono, K. (1976), *Phys. Rev. Lett.* **37**, 351.

Sakarya, S., van Dijk, N., Brück, E. (2005), *Phys. Rev. B* **71**, 174 417.

Sakurai, Y., Itou, M., Tamura, J., Nanao, S., Thamizhavel, A., Inada, Y., Galatanu, A., Yamamoto, E., Onuki, Y. (2003), *J. Phys.– Cond. Matt.* **15**, S2183.

Sato, H., Toth, R. S. (1961), *Phys. Rev.* **124**, 1833.

Sato, T., Takakura, H., Tsai, A. P., Shibata, K. (2006), *Phys. Rev. B* **73**, 054 417.

Sauthof, G. (1995), *Intermetallics*, VCH, Weinheim.

Scherm, R., Fåk, B. (2006), in *Neutron and X-Ray Spectroscopy*, Eds. Hipert, F., Geissler, E., Hodeau, J. L., Lelievre-Berna, E., Regnard, J.-R., Springer, Dordrecht, p. 361. Also: Currat, R. p. 383; Cywinski, R., p. 427, Eccleston, R., p. 457.

Schneidewind, A., Loewenhaupt, M., Hiess, A., Kramp, S., Reif, T., Neubeck, W., Vettier, C. (2001), *J. Magn. Magn. Mat.* **233**, 113.

Schönfeld, B. (1999), *Prog. Mater. Sci.* **44**, 435.

Schweika, W. (1998), *Disordered Alloys: Diffuse Scattering and Monte Carlo Simulations*, Springer, Berlin.

Sepiol, B., Vogl, G. (1993), *Phys. Rev. Lett.* **71**, 731.

Sepiol, B., Meyer, A., Vogl, G., Rüffer, R., Chumakov, A. I., Baron, A. Q. R. (1996), *Phys. Rev. Lett.* **76**, 3220.

Sepiol, B., Meyer, A., Vogl, G., Franz, H., Rüffer, R. (1998), *Phys. Rev. B* **57**, 10 433.

Sergueev, I., van Bürck, U., Chumakov, A. I., Asthalter, T., Smirnov, G. V., Franz, H., Rüffer, R., Petry, W. (2006), *Phys. Rev. B* **73**, 024 203.

Seto, M., Yoda, Y., Kikuta, S., Zhang, X. W., Ando, M. (1995), *Phys. Rev. Lett.* **74**, 3828.

Sette, F., Krisch, M. (2006), in *Neutron and X-ray Spectroscopy*, Eds.: Hipert, F., Geissler, E., Hodeau, J. L., Lelievre-Berna E., Regnard, J.-R., Springer, Dordrecht, p. 169.

Sette, F., Krisch, M. H., Masciovecchio, C., Ruocco, G., Monaco, G. (1998), *Science* **280**, 1550.

Seydel, T., Madsen, A., Tolan, M., Grubel, G., Press, W. (2001), *Phys. Rev. B* **63**, 073 409.

Shannon, R. F., Nagler, S. E., Harkless, C. R., Nicklow, R. M. (1992), *Phys. Rev. Lett.* **46**, 40.

Sharma, B. K., Purvia, V., Ahuja, B. L., Sharma, M., Chaddah, P., Ro, S. B., Kakutani, Y., Koizumi, A., Nagao, T., Omura, A., Kawai, T., Sakai, N. (2005), *Phys. Rev. B* **72**, 132 405.

Sheng, H., Luo, W., Alamgir, F., Bai, J., Ma, E. (2006), *Nature* **439**, 419.

Shirane, G., Shapiro, S., Tranquada, J. (2002), *Neutron Scattering with a Triple-Axis Spectrometer*, Cambridge University Press, Cambridge.

Shull, C. G., Smart, J. S. (1949), *Phys. Rev.* **76**, 1256.

Sikharulidze, I., Dolbnya, I., Fera, A., Madsen, A., Ostrovskii, B., de Jeu, W. (2002), *Phys. Rev. Lett.* **88**, 115 503.

Singh, S., Basu, S., Poswal, A., Gupta, M. (2005), *Solid State Comm.* **136**, 400.

Sköld, K., Price, D. (Eds.) (1987), *Neutron Scattering*, Parts A, B, C, Academic Press, Orlando.

Sladecek, M., Miglierini, M., Sepiol, B., Ipser, H., Schicketanz, H., Vogl, G. (2001), *Defect and Diffusion Forum* **194–199**, 369.

Sladecek, M., Sepiol, B., Korecki, J., Slezak, T., Rüffer, R., Kmiec, D., Vogl, G. (2004), *Surf. Sci.* **372**, 566.

Smirnov, G. V., Kohn, V. G. (1995), *Phys. Rev. B* **52**, 3356.

Springer, T. (1972), *Quasielastic Neutron Scattering for the Investigation of Diffusive Motions in Solids and Liquids*, Springer Tract in Modern Physics **64**, Springer, Berlin.

Springer, T. (1998), in *Diffusion in Condensed Matter*, Eds. Kärger, J., Heitjas, P., Haberlandt, R., Vieweg Verlag, Braunschweig/Wiesbaden, p. 59.

Squires, G. L. (1997), *Introduction to the Theory of Thermal Neutron Scattering*, Cambridge Dover Publication, New York.

Srajer, G., Yahnke, C., Haeffner, D., Mills, D., Assoufid, L., Harmon, B., Zuo, Z. (1999), *J. Phys. Cond. Matt.* **11**, L253.

Stadler, L.-M., Sepiol, B., Weinkamer, R., Hartmann, M., Fratzl, P., Kantelhardt, J. W., Zontone, F., Grübel, G., Vogl, G. (2003), *Phys. Rev. B* **68**, 180 101.

Stadler, L.-M., Sepiol, B., Kantelhardt, J. W., Zizak, I., Grubel, G., Vogl, G. (2004), *Phys. Rev. B* **69**, 224–301.

Stankov, S., Sepiol, B., Kaňuch, T., Scherjau, D., Würschum, R., Miglierini, M. (2005), *J. Phys. Cond. Matt.* **17**, 3183.

Steinmetz, K. H., Vogl, G., Petry, W., Schroeder, K. (1986), *Phys. Rev. B* **34**, 107.

Stephenson, G. B. (2003), private communication and in "LCLS: The first experiments", SLAC-R-611, available at www.slac.stanford.edu/cgi-wrap/getdoc/slac-r-611.pdf.

Steurer, W. (2003), *Zeit. Krist.* **219**, 391.

Steurer, W. (2004), *J. Non-Cryst. Sol.* **334**, 137.

Strempfer, J., Bruckel, T., Caliebe, W., Vernes, Ebert, H. Y., Prandl, W., Schneider, J. (2000), *Europ. Phys. J. B* **14**, 63.

Stunault, A., Bernhoeft, N., Vettier, C., Dumesnil, K., Dufour, C. (2001), *J. Magn. Magn. Mater.* **226**, 1116.

Stunault, A., Dumesnil, K., Dufour, C., Vettier, C., Bernhoeft, N. (2002), *Phys. Rev. B* **65**, 064 436.

Stunault, A., Vettier, C., Regnault, L., de Bergevvin, F., Paolasini, L., Henry, J. (2004), *Physica B* **345**, 74.

Sturhahn, W., Toellner, T. S., Alp, E. E., Zhang, X., Ando, M., Yoda, Y., Kikuta, S., Seto, M., Kimball, C. W., Dabrowski, B. (1995), *Phys. Rev. Lett.* **74**, 3832.

Sturhahn, W., Kohn, V. G. (1999), *Hyperfine Interact.* **123/124**, 367.

Suck, J.-B., Egelstaff, P. A., Robinson, R. A., Sivia, D. A., Taylor, A. D. (1992), *Europhys. Lett.* **19**, 207.

Suortti, P., Pattison, P., Weyrich, W. (1986), *J. Appl. Cryst.* **19**, 336.

Sutton, M. (2005), private communication.

Sutton, M., Mochrie, S. G. J., Greytak, T., Nagler, S. E., Berman, L. E., Held, G. A., Stephenson, G. B. (1991), *Nature* **352**, 608.

Sutton, M., Yang, Y. S., Mainville, J., Jordan-Sweet, J., Ludwig, Jr., K. F., Stephenson, G. B. (1989), *Phys. Rev. Lett.* **62**, 288.

Svensson, S., Vogl, G., Kaisermayr, M., Kvick, A. (1997), *Acta Mater.* **45**, 4205.

Szuszkiewicz, W., Dynowska, E., Witkowska, B., Hennion, B. (2006), *Phys. Rev. B* **73**, 104 403.

Tanaka,Y., Udo, K., Hisatsune, K., Yasuda, K. (1994), *Phil Mag. A* **69**, 925.

Tatchev, D., Hoell, A., Kranold, R., Armyanov, S. (2005), *Physica B* **369**, 8.

Taylor, J., Duffy, J., Bebb, A., Cooper, M., Dugdale, S., McCarthy, J., Timms, D., Greig, D., Xu, Y (2001), *Phys. Rev. B* **63**, 220 404.

Taylor, J., Duffy, J., Poulter, J., Bebb, A., Cooper, M., McCarthy, J., Timms, D., Staunton, J., Itoh, F., Sakurai, H., Ahuja, B. (2002a), *Phys. Rev. B* **65**, 024 442.

Taylor, J., Duffy, M., Bebb, A., McCarthy, J., Lees, M., Cooper, M., Timms, D. (2002b), *Phys. Rev. B* **65**, 224 408.

Thiess, H., Baron, A., Ishikawa, D., Ishikawa, T., Miwa, D., Sepiol, B., Tsutsui, S. (2003), *Phys Rev. B* **67**, 1 843 021.

Thurn-Albrecht, T., Steffen, W., Patkowski, A., Meier, G., Fischer, E. W., Grubel, G., Abernathy, D. L. (1996), *Phys. Rev. Lett.* **77**, 5437.

Thurn-Albrecht, T., Meier, G., Muller-Buschbaum, P., Patkowski, A., Steffan, W., Grubel, G., Abernathy, D. L., Diat, O., Winter, M., Koch, M. G., Reetz, M. T. (1999), *Phys. Rev. E* **59**, 642 (1999).

Toellner, T. S., Hu, M. Y., Sturhahn, W., Quast, K., Alp, E. E. (1997), *Appl. Phys. Lett.* **71**, 2112.

Tolan, M., Seydal, T., Madsen, A., Grubel, G., Press, W., Sinha, S. K. (2001), *Appl. Surf. Sci.* **182**, 236.

Tonnerre, J., Seve, L, Raoux, D., Soullie, G., Rodmaco, B., Wolfers, P. (1995), *Phys. Rev. Lett.* **75**, 740.

Trenkler, J., Chow, P. C., Wochner, P., Abe, H., Bassler, K. E., Paniago, R., Reichert, H., Scarfe, D., Metzger, T. H., Peisl, J., Bai, J., Moss, S. C. (1998), *Phys. Rev. Lett.* **81**, 2276.

Tsui, O. K. C., Mochrie, S. G. J. (1998), *Phys. Rev. E* **57** 2030.

Tsutsui, S., Sakurai, Y., Itou, M., Matsuda, T., Haga, Y., Onuki, Y. (2005), *Phys. B – Cond. Matt.* **359**, 1117.

Van Hove, L. (1954), *Phys. Rev.* **95**, 249.

Van Veenendaal, M., Goedkoop, J., Thole, B., (1997), *Phys. Rev. Lett.* **78**, 1162.

Vigliante, A. Christensen, M., Hill, J., Helgesen, G., Sorensen, S., McMorrow, D., Gibbs, D., Ward, R., Wells, M. (1998), *Phys. Rev. B* **57**, 5941.

Vogl, G. (1990), *Hyperfine Interact.* **53**, 197.

Vogl, G. (1996), in *Mössbauer Spectroscopy Applied to Magnetism and Material Science*, Eds. Long, G. J., Grandjean, F., Plenum Press.

Vogl, G., Sepiol, B. (1994), *Acta Metall. Mater.* **42**, 3175.

Vogl, G., Sepiol, B. (1999a), *Hyperfine Interact.* **123/124**, 595.

Vogl, G., Sepiol, B. (1999b), *Hyperfine Interact.* **123/124**, 595.

Vogl, G., Sepiol, B. (2005), in *Diffusion in Condensed Matter*, 2nd ed., Eds. Heitjans, P., Kärger, J., Springer, Berlin Heidelberg.

Vogl, G., Petry, W., Flottmann, Th., Heiming, A. (1989), *Phys. Rev. B* **39**, 5025.

Vogl, G., Randl, O. G., Petry, W., Hünecke, J. (1993), *J. Phys.: Cond. Matt.* **5**, 7215.

Vogl, G., Kaisermayr, M., Randl, O. G. (1996a), *J. Phys.: Cond. Matt.* **8**, 4727.

Vogl, G., Sepiol, B., Czichak, C., Rüffer, R., Weinkamer, R., Fratzl, P., Fähnle, M., Meyer, M., (1998), *Mater. Res. Soc. Symp. Proc.* **527**, 197.

Wada, M., Tabira, Y. (2005), *J. Jap. Inst. Met.* **69**, 206.

Wagh, A. G., Rakhecha, V. C., Stroble, M., Treimer, W. (2005), *J. Res. Nat. Inst. Stand. Tech.* **110**, 231.

Wang, X., Mainville, J., Ludwig, K. F., Flament, X., Caudron, R., Finel, A. (2005), *Phys. Rev. B* **72**, 024 215.

Wang, X., Ludwig, K. F., Flament, X., Caudron, R., Final, A. (2006), unpublished.

Weinkamer, R., Fratzl, P., Sepiol, B., Vogl, G. (1999), *Phys. Rev. B* **59**, 8622.

Wende, H. (2004), *Rep. Prog. Phys.* **67**, 2105.

Wertheim, G. K. (1964) *Mössbauer Effect: Principles and Applications*, Academic Press, New York.

Weschke, E., Ott, H., Schierle, E., Schussler-Langeheine, B., Vyalikh, D., Kaindle, G., Leiner, V., Ay, M., Schmitte, T., Zabel, H., Jensen, P. (2005), *Phys. Rev. Lett.* **93**, 157 204.

Williams, W. G. (1988), *Polarized Neutrons*, Clarendon Press, Oxford.

Wolff, J., Franz, M., Broska, A., Kerl, R., Weinhagen, M., Köhler, B., Brauer, M., Faupel, F., Hehenkamp, Th. (1999), *Intermetallics* **7**, 289.

Yakhou, F., Letoublon, A., Livet, F., de Boissieu, M., Bley, F. (2001), *J. Magn. Magn. Mater.* **233**, 119.

Yasuda, H., Ohnaka, I., Kawasaki, K., Suglyama, A., Ohmichi, T., Iwane, J., Umetani, K. (2004), *J. Cryst. Growth* **262**, 645.

Yue, A. F., Papandrew, A., Bogdanoff, P. D., Halevy, I., Lin, J. G.-W., Fultz, B., Sturhahn, W., Alp, E. E., Toellner, T. S. (2002), *Hyperfine Interact.* **141/142**, 249.

Zukowski, E., Andrejczuk, A., Dobrzynski, L., Kaprzyk, S., Cooper, M., Duffy, J., Timms, D. (2000), *J. Phys. Cond. Matt.* **12**, 7229.

13.2
High-Resolution Microscopy
Guido Schmitz and James M. Howe

In this chapter we describe techniques to investigate the micro- and nanostructure of metals and alloys. For the most part, the physics of alloys deals with defects – local deviations from the ideal structure or composition – since these defects are responsible for many application-relevant properties. Point defects, like vacancies and impurities, or interstitials and antisite defects, determine the kinetics of atomic transport. Structure, shape, and stability of interfaces and surfaces control the reaction mechanism of phase transformations and catalytic processes, respectively. Dislocations, their spatial distribution, and their interaction with precipitates determine plasticity, hardness, and toughness of structural materials. The size distribution and spatial arrangement of precipitates are decisive for pinning mechanisms in superconductors or ferromagnets in order to increase the critical current or coercivity, respectively. These examples emphasize the role of defects and justify the enormous experimental effort that is frequently undertaken to clarify the defect structure of materials.

To some extent, the analysis of defects is possible by integral methods such as scattering techniques or various types of spectroscopy, such as Mössbauer spectroscopy, positron annihilation, or nuclear magnetic resonance. Also, indirect methods such as the measurement of electric conductivity are quite helpful to determine the number density of known defects. However, in all these cases the interpretation of data requires an a-priori model based on a considerable amount of pre-knowledge about the nature and geometry of the respective defects. Thus, whenever the nature of defects needs to be identified, different types of defects overlap in their effect, or their absolute position within the specimen must be clarified, integral techniques are not sufficient. Instead, microscopic methods which directly image the defect arrangement or provide local chemical or spectroscopic information are the suitable choice.

As the number of major microscopy techniques is large and the possibilities of their variation even larger, we restrict this discussion to review three major branches of high-resolution microscopy, namely surface analysis by scanning probe microscopy (SPM), structural imaging by high-resolution transmission electron microscopy (HRTEM), and local chemical analysis by atom probe tomography (APT). As their common outstanding feature these techniques are distinguished by atomic resolution. Each of them represents the state-of-the-art in their respective fields.

The application of scanning probe microscopy in materials science is rapidly expanding, so that textbooks and comprehensive reviews are published regularly. Thus, we restrict our discussion to a general orientation on the various branches of the technique and their importance for materials science. Transmission electron microscopy is widely utilized among materials laboratories throughout the world. Since the physics of imaging and wave optics is already taught routinely in university courses, we focus our review on a dedicated variation of the method, the in-situ microscopy of structural transformations, to emphasize this important aspect of alloy physics. HRTEM is able to image single point defects (see Chapter 5) in materials in only special situations. In contrast, field ion microscopy (FIM) can be used to study vacancies and surface atoms and therefore extends the available resolution to atomic dimensions. When combined with a mass spectrometer, the field ion microscope becomes an atom probe (APFIM), easily capable of compositional analysis of local areas of samples. This technique is therefore highly complementary to HRTEM and is discussed in Section 13.2.3. The latest version of analytical FIM so-called atom probe tomography, is currently used by only a few groups throughout the world, so that we expect the reader to have less knowledge. Therefore, a somewhat more detailed description of the physical principles is presented before recent applications to the physics of alloys are discussed.

13.2.1
Surface Analysis by Scanning Probe Microscopy

Scanning probe microscopy (SPM) has become an absolutely essential tool for the analysis of surfaces down to the nanometer scale. The term SPM identifies a family of techniques that are based on the surprisingly simple principle of mechanical scanning. Binning and Rohrer (1982) were the first to demonstrate that the surface of a conductive material may be scanned at atomic resolution, if piezoelectric actuators are controlled by the tunnelling current between a tip-shaped probe and the sample. Plotting the height of the probe versus its lateral position, the topography of the surface could be imaged. The scanning tunneling microscope (STM) was born. Owing to its atomic resolution, the STM has been used extensively to study the growth mechanisms of metallic thin films. Various growth modes – layer by layer Frank–van der Merwe mode, Vollmer–Weber island nucleation, or the two-stage Stranski–Krastanov mode – could be easily distinguished from the topography of the growing film. Furthermore, the nucleation of first islands and even details of the behavior of individual adatoms, such as general surface diffusion or the influence of the Schwoebel barrier at the edges of terraces, could be investigated quantitatively. Other important applications are the study of atomic reconstruction and order of alloy or intermetallic surfaces (Göken et al. 2002). In addition to imaging, local spectroscopy is possible by varying the bias voltage. This scanning tunneling spectroscopy (STS) reveals information on filled and empty states of the electronic structure. The local density of states, including band gaps and surface states, can be measured.

Although it is not critical within the context of this book on alloy physics, a general disadvantage of STM is its restriction to conductive materials and surfaces. Shortly after the introduction of STM, it was shown that the required control of the probe–specimen distance could also be achieved by measuring the atomic interaction forces between the sample and probe (Binning et al. 1986). With this so-called atomic force microscopy (AFM), a standard resolution in the nanometer range is obtained in the so-called repulsive contact mode. Applying a more elaborate dynamic mode, even atomic resolution has been demonstrated (Giessible 1995) and the atomic surface structure of nonconductive materials could be imaged for the first time. The emergence of the AFM triggered the development of many different branches of scanning probe techniques, among them magnetic or electric force microscopy (MFM, EFM) (Hartmann 1999) or scanning near-field optical microscopy (SNOM). For the latter, the surface is excited by a local optical field a few tens of nanometers in diameter that is scanned across the surface by manipulating a suitable optical fiber. The variety of different techniques and modifications is so large that providing a thorough overview is difficult. The interested reader may find detailed descriptions of scanning probe techniques in textbooks (e.g., Maganov and Whangbo 1996; Meyer et al. 2003) or recent reviews (Meyer et al. 2004).

AFM and the scanning probe techniques in general do not need a vacuum environment. They are performed in many different atmospheres, even under liquids, so that some materials are easily studied that could not be investigated with conventional electron microscopy due to their lacking vacuum compatibility. Therefore, the number of applications to soft matter such as polymer films or biomaterials under their living conditions is increasing. In addition, AFM often requires minimal effort in specimen preparation so that this method is also advantageous in the analysis of traditional materials, provided that surface information is sufficient to solve the relevant problem. In the context of metallic materials, the possibility of measuring local mechanical properties is of particular interest, since alloys are still used predominantly as structural materials. By replacing the usual cantilever with a diamond nanoindenter that is driven by capacitive actuators, local measurements of hardness are possible with a spatial resolution of a few tens of nanometers. This way, even local stress fields at crack tips or the plastic properties of embedded particles can be investigated (Kempf et al. 1998).

Below, we briefly describe the functional principles of scanning tunneling and atomic force microscopy, albeit with a focus on the latter technique. The application to typical problems in the physics of alloys is demonstrated using several examples, including surface reconstruction of ordered intermetallics investigated by STM, the characterization of heterogeneous microstructures by AFM, the study of grain boundary grooving, local measurement of plasticity, and also the imaging of magnetic domains and the substructure of domain boundaries with MFM techniques.

13.2.1.1 Functional Principle of Scanning Tunneling and Atomic Force Microscopy

The various branches of scanning probe microscopy follow a common principle: a probe as sharp as possible scans a surface using very accurate mechanics. The respective interaction between the specimen and probe is measured and the signal is used in a feedback loop to control the height/distance between the surface and probe. An image of the surface topology is obtained by plotting the height position of the probe versus the scanned area.

13.2 High-Resolution Microscopy

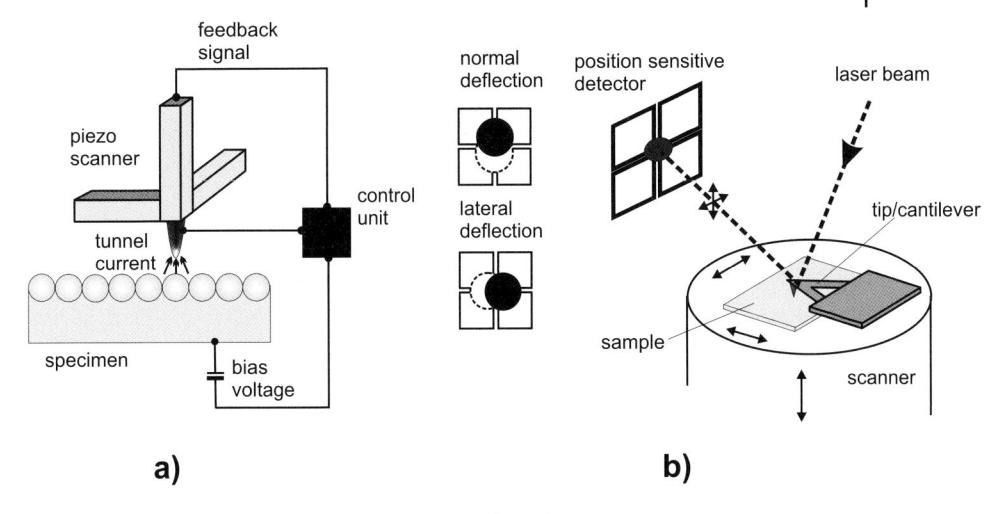

Fig. 13.10 Schematic representations of scanning probe techniques:
a) scanning tunneling microscope; b) atomic force microscope.

In Fig. 13.10(a), a schematic representation of the original scanning tunneling microscope is shown. A metallic tip produced by electrochemical or mechanical means is mounted on a piezoelectric drive, which allows to move the tip in 3D space with sub-ångstrom accuracy. If a bias potential of a few volts is supplied to the probe with respect to the conducting sample, a tunneling current develops as soon as the tip approaches the surface at distances below a few nanometers. According to quantum mechanics, the tunneling current is proportional to the density of electronic states of the specimen and probe, and proportional to the transfer matrix, which depends on the relative spacing. For practical purposes a simple exponential decay function (46) is often sufficient to estimate the current.

$$I(h) \propto \exp[-\kappa_{\mathit{eff}} \cdot h] \tag{46}$$

Since the decay constant κ_{eff} is of the order of a reciprocal ångstrom, the current depends very sensitively on the probe height h (i.e., the sample–probe distance). During the lateral scan, the height of the tip is controlled via the driving voltage of the respective piezoactuator so that the tunneling current is always maintained constant at a certain preset value. Plotting the required driving voltage versus the lateral position yields an image of the surface topology. If the sharpness of the probe tip is sufficient – in ideal cases only one final atom may protrude from the apex – even the atomic corrugation of the surface becomes visible, so that the lateral resolution amounts to about 2–3 Å. The accuracy of the height profile can be better than 0.1 Å. In the case of alloys, the local electronic density, and thus the tunneling current, depend on the atomic species. As a consequence, the image contains a superposition of local topology and chemical information, which can be used to identify individual atoms.

Tunneling currents can only be measured when conductive specimens are observed. In order to extend the principle of mechanical scanning to insulating materials, local atomic interaction forces, instead of tunneling currents, are measured in an AFM. Therefore, a fine, tip-shaped probe is attached to an elastic cantilever as illustrated in Fig. 13.10(b). In most cases, the cantilever deflection is measured by means of an optical beam, which is reflected onto a position-sensitive detector (Meyer and Amer 1988). Using a four-quadrant device, as indicated in Fig. 10(b), has the interesting advantage that deflection of the beam may be determined in two independent directions, so that interaction forces normal to the surface as well as lateral friction forces may be measured. Alternatively, capacitive sensors can be used to detect the deflection of the cantilever. State-of-the-art detectors are very precise – a shift of the probe by fractions of an Ångstrom is readily detected. Since cantilevers are produced by methods of microfabrication such as lithography, anisotropic etching, and ion beam sputtering, they may be formed in varying geometry and dimensions. Typical spring constants are in the range of 10^{-2}–5×10^2 N m^{-1}, so that using Hooke's law, interaction forces in the piconewton range may be measured.

Using modern techniques of microfabrication, the resonance frequency of cantilevers is adjusted in the range of 10–1000 kHz, which allows a typical scanning rate of about 500 Hz. If, for example, an area of 30×30 µm^2 should be scanned in 512×512 pixels, the total measuring time per frame amounts to 10 min, so that there is little possibility of increasing the field of view significantly. This is one of the important drawbacks of SPM in comparison to electron microscopy.

At first hand, any microscopy should provide an image of the real specimen. Thus, any modifications of the specimen structure during the imaging process should be carefully avoided. But, in addition to merely imaging, the scanning probe technique may be used to intentionally modify the local surface structure of the sample. Thus, the probe is applied as a unique nanotool. Pushing and positioning of individual atoms on a surface, like writing with "atomic ink" on a sheet of micrometer dimensions, is certainly the most exciting example performed with low-temperature STM, which allows the vision of data storage in an extreme packaging density. There are other examples of using the scanning probe as a kind of nanolaboratory. Since the AFM is a device for producing and measuring local forces, it is natural to suggest using this instrument for local mechanical testing. However, the usual cantilever geometry is not suited for this purpose because the maximum forces that can be supplied are not sufficient, and furthermore, because of the lever geometry, vertical movement is always accompanied by a slight lateral shift, which hinders an exact vertical indentation and thus the correct measurement of hardness. Finally, the conventional probe tips are just too fragile to be used as a nanoindenter. As a result of these features, an instrumental modification known as NI-AFM (nanoindenter AFM) has been successfully applied in local elasticity and hardness tests. Its principle is illustrated by Fig. 13.11. Instead of a cantilever spring, a capacitive transducer is used to operate a small pyramidal diamond indenter (Bhushan et al. 1996) in analogy to a measurement of conventional Vickers hardness. Forces ranging from 100 nN up to 10 mN

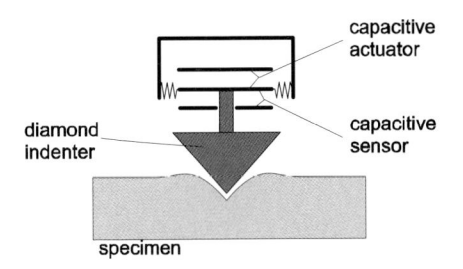

Fig. 13.11 A capacitive transducer for an NI-AFM.

are produced by the electrostatic field of a capacitor. The penetration depth is measured via a capacitive sensor, as indicated by the three electrodes in the schematic drawing. Before and after the indentation the diamond tip may be used to produce a topology image, just as with a standard cantilever AFM.

13.2.1.2 Modes of Measurement in AFM

To produce an atomic force image, two particular modes of operation are common practice, the so-called contact mode and the dynamic mode. For imaging in the contact mode, the cantilever deflection is used directly as a feedback signal. In other words, a preset value of interaction force is maintained during the scan, so that the image represents a surface of constant interaction force. Usually, the repulsive core interaction is probed. A typical preset for the force may be 100 pN. The important advantage of the contact mode is the ease of operation, i.e., complex electronics are not required. However, since the probe is in direct contact with the sample, the probability of specimen damage is relatively high. On the other hand, in direct contact with the surface lateral friction forces may be measured, which makes the AFM an important tool for tribology. A simple estimate demonstrates that atomic resolution is hardly achieved in contact mode. For real atomic resolution, a probe with a protruding area in the range of one atom would be required, which leads to a contact area of about $5 Å^2$. Given a force of 100 pN, a stress of 2 GPa would develop underneath the tip, which is close to the theoretical strength of either the specimen or probe material. Therefore, probes for the contact mode are usually much more blunted than this. If image details of apparent atomic resolution appear under these conditions, they are usually due to Moiré effects and do not provide real information at the ångstrom level. In practice, the lateral resolution of the contact mode ranges from 1 to 10 nm.

In the dynamic mode the tip is excited to periodic oscillations near the resonance frequency of the cantilever. During the major part of the oscillation period the probe is far (10–100 nm) from the surface, so that predominantly long-range forces, e.g., van der Waals, or magnetic and electrostatic interactions may be sensed. Since contact with the specimen is reduced to a short intermittent "tapping," the technique is also called the tapping mode. The danger of specimen damage is significantly reduced in this dynamic mode. Since periodically varying quantities can be measured very accurately with lock-in amplifiers, the dynamic mode is much more sensitive than the contact mode, and under optimum conditions even atomic resolution may be obtained. Usually the oscillation amplitude or phase with respect to the excitation at constant frequency is used as a feedback signal to control the piezodrive of the height position.

The oscillating cantilever may be described as a damped harmonic oscillator with an equation of motion given by Eq. (47), in which ω_0, h_0, δ_0 denote the resonance frequency, average position, and steady-state amplitude of the free oscillator without specimen interaction; γ is a damping factor that is related to the quality factor of the oscillator.

$$m\frac{\partial^2 h}{\partial t^2} + \gamma \frac{\partial h}{\partial t} + m\omega_0^2(h - h_0) = \delta_0 m\omega_0^2 \cos \omega t + \left(F_0 + \frac{\partial F}{\partial h}h\right) \tag{47}$$

The interaction with the specimen is taken into account by the term in brackets on the RHS of Eq. (47), which represents a linear approximation of the interaction force. Without any lengthy calculation, it is seen by shifting the force terms to the LHS of the equation that the constant force term will just modify the zero position of the oscillator, while the first derivative, the compliance of the interaction, will modify the resonance frequency as in Eq. (48).

$$m\tilde{\omega}_0^2 = m\omega_0^2 - \frac{\partial F}{\partial h} \tag{48}$$

Since the amplitude of the probe oscillation is given by Eq. (49) and the phase by Eq. (50), the shift of the resonance frequency will also modify the oscillation amplitude and the phase.

$$\delta = \frac{\delta_0 \tilde{\omega}_0^2}{\sqrt{(\omega^2 - \tilde{\omega}_0^2)^2 + \gamma^2 \omega^2}} \tag{49}$$

$$\alpha = \arctan \frac{\gamma \omega}{\omega^2 - \tilde{\omega}_0^2} \tag{50}$$

Frequency, amplitude, and phase are well measurable quantities that may be used to map the local variation of $\partial F/\partial h$. Furthermore, the amplitude δ and in particular the phase α depend on the damping factor. As a consequence, a phase image at constant frequency of oscillation will deliver information on the local damping factor, which depends on the local viscosity or internal friction of the specimen. Thus, imaging at constant phase angle may be also termed "viscosity microscopy." Such viscosity imaging has proven to be particularly useful in distinguishing soft rubber-like precipitates embedded in a thermoplastic matrix, a typical microstructure of high-performance polymer construction materials (Bar and Meyers 2004).

13.2.1.3 Cantilever Design for the AFM

In the early days of STM, obtaining sharp probe tips was an art performed in the laboratory, but this situation has changed. Nowadays, cantilevers for AFM are produced by state-of-the-art microfabrication using the latest lithography and etching techniques on Si wafers. They are usually purchased from specialized companies. For imaging in the contact mode, a triangular cantilever in combination with an Si_3N_4 tip, formed as a square pyramid, is often used (see Fig. 13.12a). Silicon nitride is much harder than Si and so reduces the risk of probe damage in direct contact with the sample. The curvature radius at the apex varies from 5 to 50 nm. Microscopy in the dynamic mode prefers the geometry shown in Fig. 13.12(b). The cantilever is formed as a rectangular beam equipped with a Si tip, which may be produced with sharper radii below 10 nm.

In the design of the cantilever geometry, one has to compromise between a low spring constant, which results in high force sensitivity, and a large resonance frequency, which is important for high scanning rates. The minimum force that can be measured is limited by thermal noise to F_{min}, as given by Eq. (51), where B, k_s, f_0, and Q denote the bandwidth (which means the pixel frequency of the measurement), spring constant, resonance frequency of the cantilever, and quality factor, respectively (Yasumura et al. 2000). The latter is defined as usual by $Q = 2\pi E_0/\Delta E$, the inverse of the relative energy dissipation per oscillation period.

$$F_{min} = \sqrt{\frac{4kTB}{2\pi}\frac{k_s}{f_0 Q}} \tag{51}$$

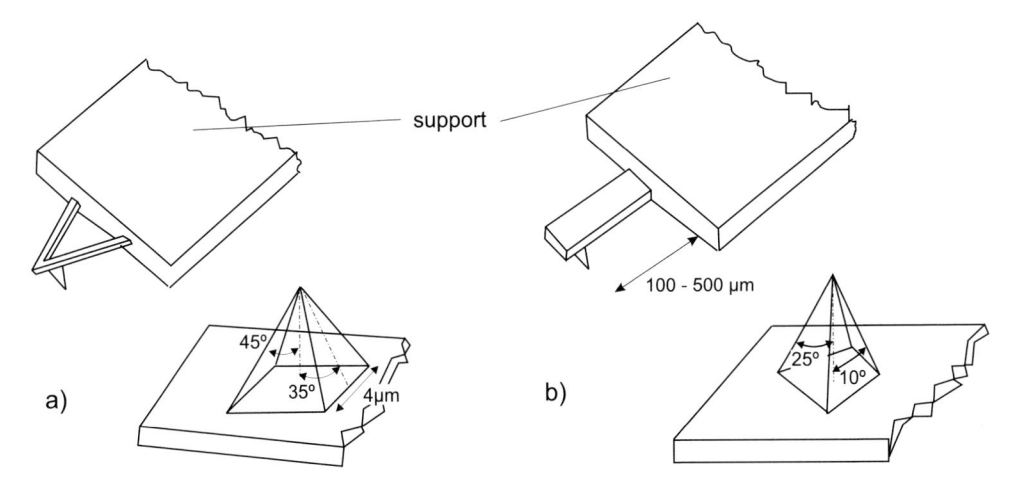

Fig. 13.12 Cantilever geometries: a) triangular cantilever for the contact mode equipped with an Si_3N_4 tip formed as a regular square pyramid; b) rectangular beam with a Si tip as preferred for work in the dynamic mode.

Thus, high force sensitivity requires a small spring constant and at the same time a high resonance frequency, two requirements which normally oppose each other. The only way to decrease the spring constant without lowering the frequency is by optimizing the shape of the cantilever in accordance with Eq. (51) (Frederix et al. 2004). The spring constant of a rectangular cantilever bar of length l, thickness t, and width w is proportional to wt^3/l^3, while the frequency is proportional to $\sqrt{k_s/\rho} = t/l^2$. Therefore, obtaining a high sensitivity requires minimizing the parameter combination wt^2/l. In other words, very thin and narrow cantilevers are advantageous to optimize sensitivity, which imposes a challenge to microfabrication. In addition, the reduction of the spring constant finds a further natural limit, if attractive forces should be measured. As soon as the increase in the interaction force exceeds the spring constant when approaching the surface, i.e., $\partial F/\partial h > k_S$, the height position becomes unstable and the tip is drawn into the sample surface. As a consequence, measurements close to the sample surface become impossible if the spring constant is too low. Following typical rules of design, usual spring constants range from 10^{-2} to 5×10^2 N m^{-1} and the resonance frequency from 10 to 1000 kHz.

In addition to the standard geometries discussed above, various modifications are in use for special purposes. For example, in tribology studies cantilevers may be equipped with the contact material in question; see, e.g., Fig. 13.13(a), which shows a glass sphere attached to a cantilever (Perry 2004), or the tip may be coated with a ferromagnetic layer to perform magnetic force microscopy (MFM), which probes the interaction with the magnetic stray field of the sample surface. By depositing a carbon–hydride supertip as an etching mask and subsequent ion beam sputtering, a single magnetic domain 50 nm in diameter may be formed at the tip apex as shown in Fig. 13.13(b) (Hartmann 1999).

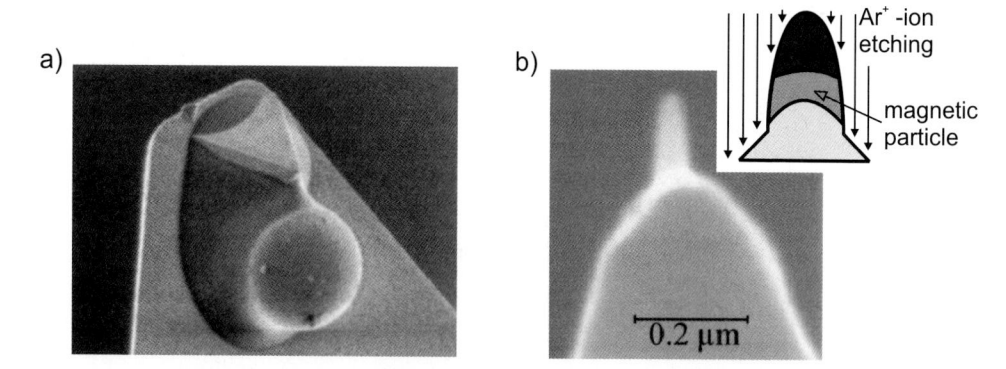

Fig. 13.13 Scanning electron micrographs of special cantilever tips:
a) sodium-borosilicate microsphere (5 μm radius) for tribology studies;
b) magnetic supertip deposited on top of a Si tip for high-resolution MFM. (After Perry 2004; Hartmann 1999.)

13.2.1.4 Exemplary Studies by Scanning Probe Microscopy

By discussing a few experimental studies, the various possibilities of applying scanning probe microscopy to the field of metals and alloys are illustrated. Considering STM, general AFM, and MFM as well, we try to document the broad spectrum of these techniques using a small set of examples.

13.2.1.4.1 Chemical Contrast by STM and Surface Ordering

It is clear that STM provides a topological image of the atomic structure. In addition, since the tunneling current depends on the density of electronic states of the tip and sample, one may expect some chemical contrast between different atomic species, provided that the electronic states are not too delocalized. However, this expectation is in conflict with the general electronic property of metals. If the corrugation seen by the STM is estimated with a calculation of the local density of states (LDOS), the predicted contrast is usually so low that it is hardly distinguishable from the background noise. As such, it was a surprise that Schmid et al. (1993) could nevertheless demonstrate chemical contrast at atomic resolution on dense-packed surfaces of PtNi alloys. Meanwhile, chemical contrast has been found also on other alloy systems such as PtNi, PtRh, PtCo, PtAu, and AgPd (Varga and Schmid 1999). As all these alloys contain at least one noble element known for catalytic activity, knowledge of surface structure, potential segregation, and chemical ordering may be quite important.

Figure 13.14 shows an example of an STM image of a $Pt_{25}Ni_{75}$ specimen. The image was obtained in a constant current mode. To obtain atomic resolution with

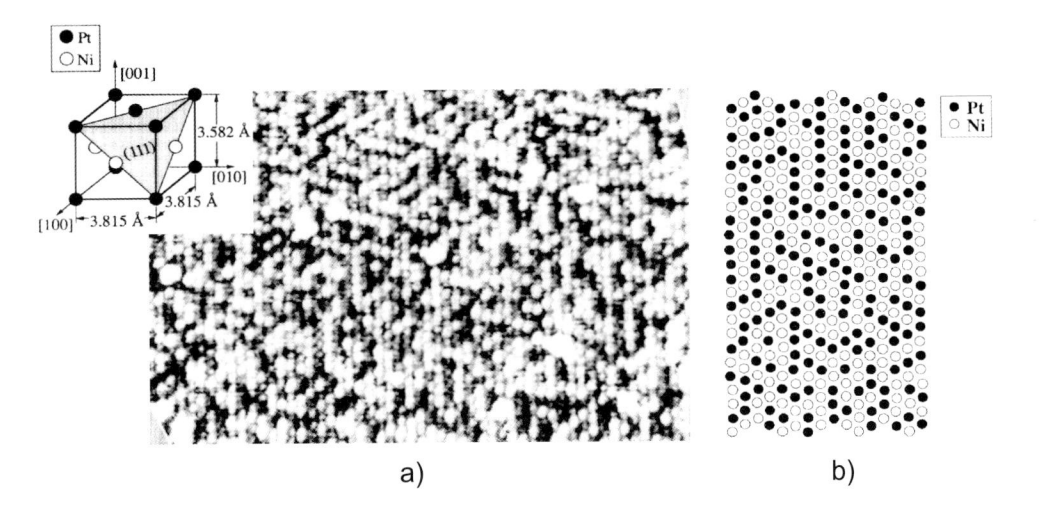

Fig. 13.14 a) STM image produced at a constant current of 16 nA; sample bias −0.5 mV. The image size represents an area of 125 × 100 Å^2; b) atomic arrangement at the surface of a Ni–37 at.% Pt alloy at 420 K as predicted by MC simulation. (After Schmid et al. 1993.)

the quality shown, the sample had to be cleaned under UHV conditions by several cycles of ion sputtering and annealing. It is easily seen without any quantitative contrast analysis that two kinds of atoms can be distinguished by bright and dark intensity. If "dark" and "bright" species are counted, the average composition of the surface layer is determined to have concentration of about 50 at.%. However, the relative amount of "dark" species increases with sputter cleaning time. Since it is known that Ni has a higher sputtering probability, this observation indicates that the "dark" species represent Pt, while the "bright" ones represent Ni atoms.

The clarity of the chemical contrast is remarkable. If image intensity is converted into corrugation, a height difference of 0.3 Å between the two species is found, which is about twice the difference in atomic radius of Ni and Pt, so that a simple geometric explanation of the contrast is not possible. Furthermore, also from the LDOS, a significantly smaller height contrast of only 0.03 Å is expected.

Even more puzzling, the better the cleaning of the specimen before investigation, the lower the probability of obtaining pronounced chemical contrast. Therefore, Varga and Schmid have suggested an indirect mechanism of adsorbate atoms. If an adsorbate atom, for example an oxygen or sulfur atom, is picked randomly by the probe tip, and this adsorbate has different chemical affinity to the different constituents of the alloy, a situation as sketched in Fig. 13.15 may appear. The adsorbate is slightly drawn toward the surface atom of the attractive species, establishing a kind of temporary chemical bond, so that the tunneling current increases due to the elevated density of states between the surface and the adsorbate.

The experiment discussed is also very interesting from the point of view of alloy physics, since short-range order on surfaces has been documented for the first time by direct imaging. Chemical (atomic) order is an important phenomenon of intermetallic compounds which determines plasticity and also functional properties such as conductivity or magnetization. In the bulk case, order may be completely described by the degree of long-range order and its fluctuations. At surfaces, however, several additional phenomena may be present. First, the composition may deviate due to segregation to the surface. Second, in intermetallic compounds different kinds of crystal planes can be distinguished in certain low-indexed stacking directions. What is the preferred terminating layer at the surface? Last, but not least, one may ask whether the order symmetry of the bulk is

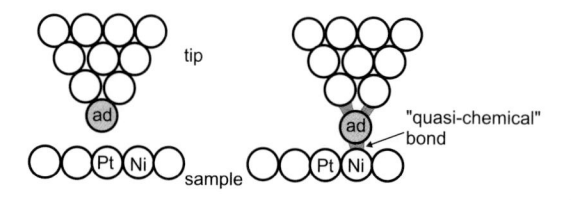

Fig. 13.15 Principle of contrast formation by chemically selective adsorbates.

disturbed or reconstructed at the surface. The experimental result in Fig. 13.14(a) can answer these kinds of questions, if the detected surface arrangement is compared to Monte Carlo (MC) simulations, e.g., based on embedded atom potentials, as has been pointed out by Schmid and coworkers.

In contrast to a random arrangement of both species, the STM image in Fig. 13.14 reveals a row-like alignment of like atoms, while, according to the bulk phase diagram, the $Pt_{25}Ni_{75}$ alloy should be a disordered solid solution. In addition, a Pt content of about 50 at.% is seen in the surface layer, in striking contrast to the mean composition of the sample. All these facts are well understood, if it is assumed that the material close to the surface becomes enriched in Pt to about 37 at.% by selective sputtering. According to the result of MC simulations shown in Fig. 13.14(b), the surface layer will be further enriched under these circumstances to about 50 at.% Pt by segregation. At this composition a bulk alloy would be ordered in an $L1_0$ superlattice, of which the (111) net planes are characterized by rows of like atoms along $\langle 011 \rangle$ directions (see the inset of Fig. 13.14a). As is clear from the experimental image and MC simulation, the terminating plane prefers this ordered state of aligning like atoms in nearest-neighbor chains. However, since the $L1_0$ structure is tetragonal, long-range order in the surface would lead to considerable strain with respect to the cubic bulk material below. To avoid the related stress, Nature prefers splitting the ordered surface structure into small domains, so that strain vanishes on a coarser scale. In consequence, the surface plane appears to be short-range ordered, as is evident from the STM image.

13.2.1.4.2 Microstructure Characterization and Surface Topology by AFM

Atomic resolution of the STM has been an important prerequisite in the previous example. Usually, atomic resolution in AFM is achieved only with considerable effort, in vacuum and with special surface preparation. Therefore, apart from a few exceptions, the main use of AFM is for studies on a somewhat coarser scale, say 1–10 nm. In this regime, AFM has the important advantage of rather simple specimen preparation and operation of the instrument, since the investigation may be performed under the ambient atmosphere. In this regard, the AFM is even replacing optical microscopy for surface metallography, since it offers much better resolution.

The application of AFM in metallography has been checked in a detailed study (Durst and Göken 2001) by comparison with results from transmission electron microscopy (TEM) and conventional scanning electron microscopy (SEM). The decomposed microstructure of a superalloy ("Waspalloy") was characterized in terms of the volume fraction, average size, and size distribution of the precipitates. This specific microstructure had the advantage of a bimodal size distribution of the γ' precipitates with 16 nm and about 110 nm mean diameter. Therefore particle contrast and resolution could be determined in different size regimes. Similarly to established practice in SEM work, the surface of the sample was mechanically polished with 0.25 μm diamond paste and electrochemically etched to achieve suitable particle contrast by selective etching. In Fig. 13.16(a),

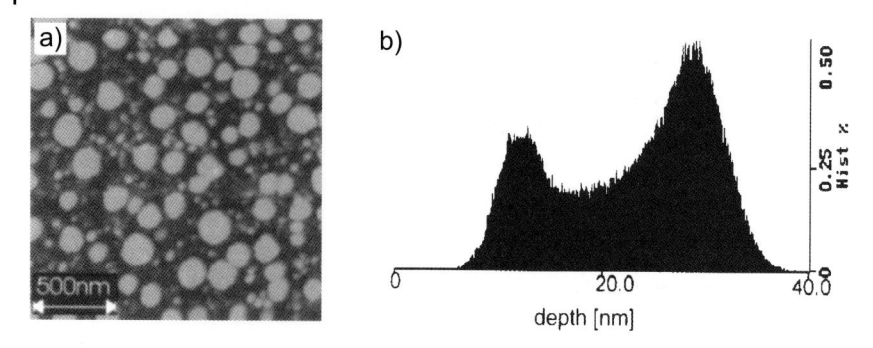

Fig. 13.16 Microstructure of "Waspalloy" with γ' particles in a disordered matrix as observed by contact-mode AFM: a) topology micrograph; b) histogram of pixel intensities corresponding to surface heights. (After Durst and Göken 2001.)

an AFM image of the bimodal particle microstructure is shown which was produced in contact mode. Both particle fractions are clearly imaged with very similar brightness. Therefore, only two maxima appear in the histogram of the height profile (Fig. 13.16b), corresponding to the matrix and precipitate phase, respectively. By correct choice of a threshold between them, precipitates are reliably distinguished from the matrix independently of their size. This ease of contrast interpretation is an important advantage of AFM, when micrographs need to be analyzed automatically.

From the spacing between the two maxima in the histogram, the effect of selective etching is determined quantitatively. In the example shown, γ' precipitates tend to protrude from the matrix level by about 16 nm. Quantitative comparison of the size distributions obtained by AFM and both electron microscopy techniques revealed that the size scale of the AFM is systematically shifted by a certain amount. In comparison to electron microscopy, the average size of the small and the large particles appeared enlarged by the same amount. The schematic drawing in Fig. 13.17 illustrates how this discrepancy is understood as an artifact introduced by the finite radius of the probe tip. However, having identified this situation, the particle size measured by AFM can be corrected if the tip radius and height difference between protruding particles and matrix are known. By geometric considerations, the erroneous overestimate x of the particle radius is given in Eq. (52) (for definitions of the symbols, see Fig. 13.17).

$$R = \frac{x^2 + h^2}{2h} \qquad (52)$$

The height difference h for a particular measurement can easily be determined from a histogram like that presented in Fig. 13.16(b), while the unknown radius of the probe tip must be calibrated by measurements of known microstructures or on special samples containing sharp vertical edges. The authors (Durst and

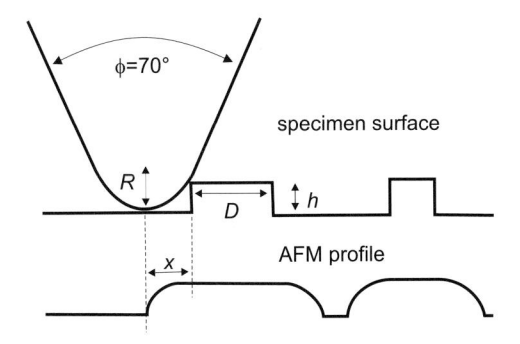

Fig. 13.17 Schematic diagram of particle contrast formation with AFM. Particles protrude by a height difference h with respect to the matrix level. The ideal height profile (top) appears experimentally broadened (bottom) at a scale defined approximately by the tip radius R. (After Durst and Göken 2001.)

Göken 2001) conclude that AFM is a well recommended alternative to optical microscopy. As the main advantages, they point to a much better spatial resolution than is achieved with optical microscopy or even SEM, the uncomplicated specimen preparation in comparison to TEM, and the clear image contrast allowing an easy numerical evaluation of the information. AFM is also used for the routine microstructural characterization of technical polymers, which often consist of a two-phase microstructure of thermoplastic matrix and soft rubber-like particles to improve the toughness. In this case, the tapping mode in combination with a phase image delivers optimal contrast due to the different viscoelastic properties of the polymer materials (Bar and Meyers 2004).

In order to characterize heterogeneous microstructures as in the previous example, an artificial height contrast has to be produced by specimen preparation. Thus, the surface sensitivity of SPM is used only as an indirect means to obtain information on volume properties. There are many other cases in which the surface plays a direct role in the physical mechanism of interest, and surface information is desired. An illustrative example is the study of thermal grain boundary grooving, which allows the determination of surface diffusion coefficients and of grain boundary energies as well. At the intersection of a grain boundary with the free surface, a groove develops during annealing because thermal equilibrium requires that the tensions of the free surface and the grain boundary compensate each other. A general theory that allows the quantitative evaluation of the grooving profile was developed by Mullins (1957). Therefore, grain boundary grooving has already been investigated in the past by means of optical microscopy on cross-sections of the surface.

However, low-indexed grain misorientations are distinguished by particularly small grain boundary energies. The resulting shallow grooves are hardly detectable by optical microscopy, so that AFM provides an important improvement in this area. In Fig. 13.18(a), a contact-mode AFM image of a surface region is

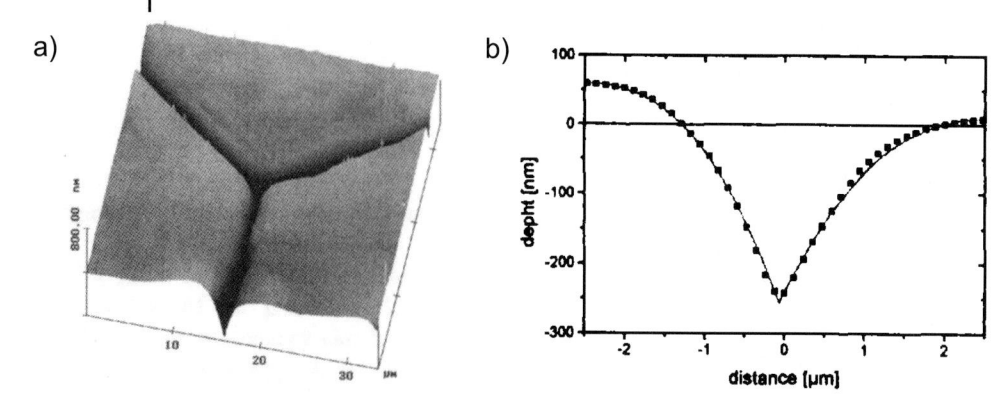

Fig. 13.18 a) AFM at the surface of a Cu polycrystal, showing a region with a triple-junction of three grains and the related surface grooves. b) Height profile measured by AFM across a grain boundary groove of Ni. The solid line represents a solution of Mullins' model. (After Weber et al. 2000.)

shown, where three Cu grains meet in a triple line (Weber et al. 2000). All three grain boundaries are marked by well-developed grooves. An experimental height profile across such a grain boundary is presented in Fig. 13.18(b). The trace of the profile is well described by Mullins' theory. By quantitative evaluation of the height profiles, Weber and coworkers could determine the energy of various low-sigma grain boundaries (even $\Sigma 3$) and the temperature dependence of the surface diffusion coefficients for Ni and Cu polycrystals. The improved accuracy of the AFM method has been utilized also to investigate in detail deviations from Mullins' theory that are due to the anisotropy of the surface energy (Rabkin et al. 2000).

As a final example of modern metallography by AFM methods, the measurement of local hardness by nanoindentation is presented. For this, the special transducer of the NI-AFM technique is used, as discussed previously (see Section 13.2.1.1). Since the transducer may also be used to produce an image of the topology, a heterogeneous microstructure, suitably etched, may be imaged before and after the hardness test. In this way, the location of the indent can be controlled very precisely, even to an accuracy of a few nanometers. In Fig. 13.19, an NI-AFM image is presented from a study by Göken and coworkers (Göken and Kempf 1999). The heterogeneous microstructure of a Nimonic superalloy is shown that is characterized by ordered cube-shaped γ' particles (bright) embedded in disordered matrix (dark). Due to the small size of the diamond indenter, hardness measurements could be performed on precipitate and matrix phases independently. Four indents distributed on the matrix and precipitates are seen as dark triangles, about 100 nm in size. If the maximum load is reduced to about 100 µN, the size of the indents may even be reduced to about 20 nm. The indents located on the matrix phase seem to be slightly larger than those on the γ'

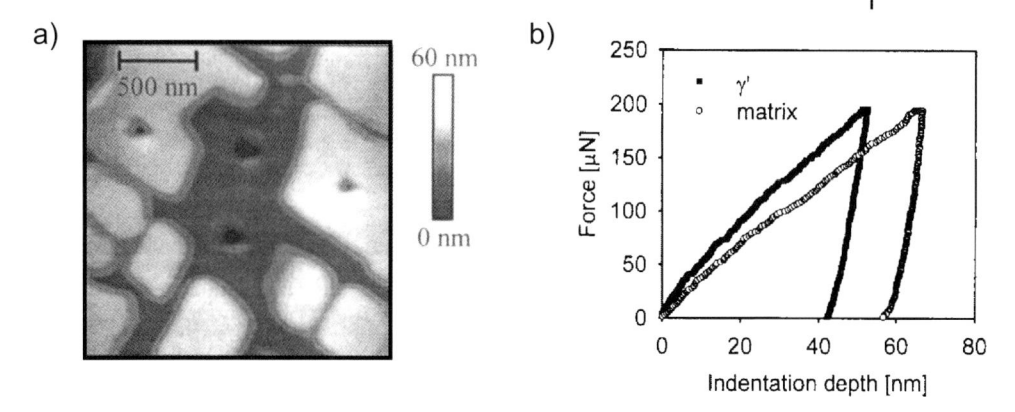

Fig. 13.19 a) NI-AFM image of a heterogeneous Nimonic superalloy structure. Indents produced by the nanohardness tester with a maximum force of 500 µN are evidenced by dark triangles on the matrix (dark) and the precipitate phase (bright). b) Local stress–strain curves of the matrix and precipitate phase. (After Göken and Kempf 1999.)

phase, which indicates that the intermetallic compound has a greater hardness. However, on the short length scale, the boundary of the indents is not well defined, which makes quantitative evaluation of the indent's cross-section – usually used for hardness measurements – difficult. In order to overcome this difficulty, use of the unique feature of a local force measurement is suggested. Thus, stress–strain curves are determined separately for each phase, as presented in Fig. 13.19(b). These curves provide the desired information on irreversible plastic deformation and confirm the greater hardness of the intermetallic precipitates. They yield further additional information on local elasticity.

13.2.1.4.3 Imaging of Nanomagnets by Magnetic Force Microscopy

Using special probes, the AFM principle allows to study a broad range of particular interaction forces. Here, magnetic force microscopy (MFM) is discussed as a variant which is especially important for the study of metals and alloys. In order to investigate magnetism on a microscopic scale, a single-domain nanoparticle is attached to the probe tip as described previously in Section 13.2.1.3. Usually, the dynamic mode with a comparatively large specimen–tip distance is used to avoid artifacts, which may be induced by the tip field. Frequency-shift images reflect the interaction of the tip with the stray field of the sample as discussed in Section 13.2.1.2. The MFM technique dates back to 1987 (Martin and Wickramasinghe 1987; Saenz et al. 1987). Important breakthroughs are the resolution of the internal fine structure of domain boundaries (Göddenhenrich et al. 1990) and the imaging of individual flux lines in superconducting materials (Moser et al. 1995; see Chapter 14, Section 14.6), which were made possible by attaching nanoscale ferromagnetic probes to the tip (see Fig. 13.13). Nowadays, AFM is used routinely in industry for checking the domain structure and quality of data storage devices such as surfaces of hard disk coatings or giant magneto-resistance (GMR) recording heads (see Chapter 14, Section 14.4). Its widespread use is particularly due to the fact that no special specimen preparation is required and that the technique is hardly influenced by surface contamination.

Scientifically, MFM is widely used to investigate artificial magnetic nanostructures, which currently attract considerable interest due to their potential in ultrahigh-density storage and spin electronics (see Chapter 14, Section 14.4). An experimental example (Zhu and Grütter 2004) is shown in Fig. 13.20. A regular array of single domain particles made of Permalloy (NiFe) was produced by lithography. Each particle has an elongated shape, $240 \times 90 \times 10$ nm^3 in size, so that only two well-defined magnetizations parallel to the long axis are allowed by shape-anisotropy. If an external field is supplied, the magnetization may be switched between these two states. In the micrographs, each particle appears as

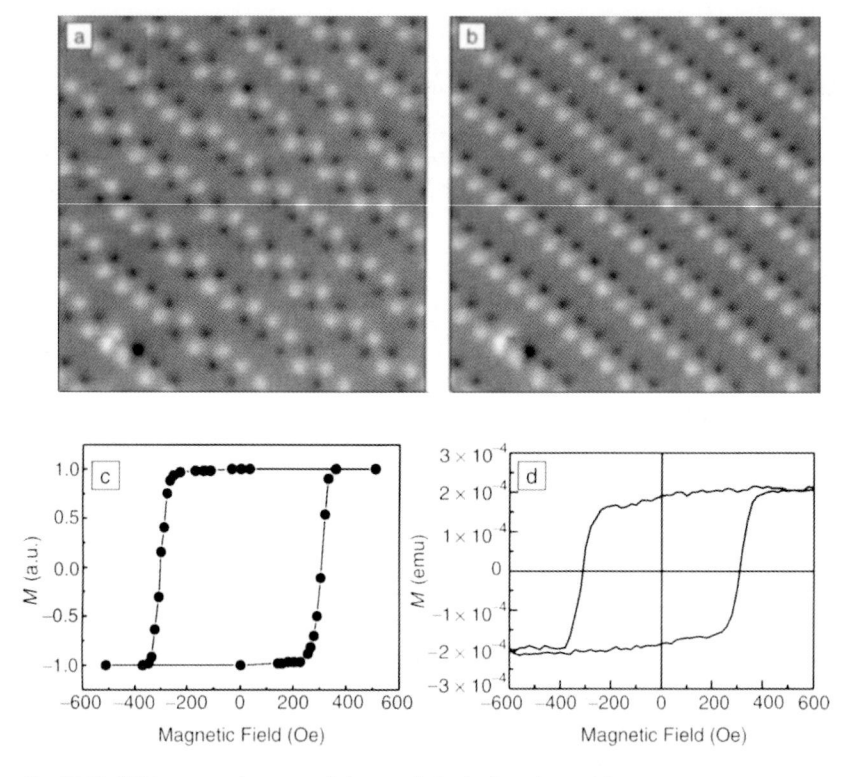

Fig. 13.20 AFM at a regular array of elongated single domain particles: a) remanent state after applying a field of 304 Oe along the long axis of the particles; b) after applying 510 Oe. c) Ensemble hysteresis curve determined as the average of the microscopically observed individual states. d) Hysteresis curve measured by macroscopic magnetometry. (After Zhu and Grütter 2004.)

a pair of black and white dots, which is due to the positive and negative interactions with the stray field at both ends of the magnetic dipoles. Thus, the direction black-to-white indicates the individual magnetic state of a particle. The possibility of switching the elements is seen by a comparison of Figs. 13.20(a) and (b). If similar pictures are obtained after applying different magnetic fields, a complete hysteresis loop may be calculated by counting the number of individual "up" and "down" states, as shown in Fig. 13.20(c), which compares reasonably well with a hysteresis curve obtained by macroscopic magnetometry, shown in Fig. 13.20(d).

In the present example, the spacing of the particles was chosen to avoid their interaction. The interaction between particles has been studied with chain-like patterns having a smaller spacing. The observed process may be described as propagation of information. In addition, the switching of individual elements could be investigated with AFM with the application of external fields. An intermediate vortex state was observed (Zhu and Grütter 2004).

13.2.2
High-Resolution Transmission Electron Microscopy and Related Techniques

High-resolution transmission electron microscopy (HRTEM) provides unique capabilities for determining the atomic structure, composition, and behavior of defects in alloys at or near the atomic level (Spence 1988; Williams and Carter 1996; Fultz and Howe 2002). Such information is critical to understanding the atomic structures of dislocations (see Chapters 3 and 6), or the mechanisms of phase transformations (see Chapters 5, 7, and 8), for example (Amelinckx and van Dyck 1992; Howe 1999). In this section, we describe the technique of HRTEM, particularly as related to in-situ dynamic studies, and then show how the technique can be used to understand the behavior of materials at the atomic level. From the examples, we see that in-situ HRTEM allows direct observation of the dynamic behavior of atomic processes, limited mainly by the time resolution of the image recording devices (~ 0.03 s) and the averaging that occurs through the specimen along the projection direction. The latest generations of HRTEMs can obtain image resolutions of less than 0.1 nm, so that the image resolution is adequate for most materials problems of interest (Batson et al. 2002; Pennycook et al. 2006). In fact, HRTEM is such a powerful tool for the characterization of materials that some microstructural features are defined in terms of their visibility in HRTEM images.

Chemical analysis is not discussed in this section, but current HRTEMs have the ability to form electron probes smaller than 0.1 nm, so that chemical analyses can be performed near this level of spatial resolution using either energy-dispersive X-ray spectroscopy (EDXS) (Joy et al. 1986) or electron energy-loss spectroscopy (EELS) (Egerton 1996). This is because the high-energy electrons (usually with 100–400 keV of kinetic energy) in an HRTEM cause electronic excitations of the atoms in the specimen. The important spectroscopic techniques of EDXS and EELS make use of these excitations by incorporating suitable detectors into the HRTEM. In EDXS, an X-ray spectrum is collected from small regions of the specimen illuminated with a focused electron probe, using a solid-state detector. Characteristic X-rays of each element are used to determine the concentrations of the different elements present in the specimen. In EELS, a magnetic prism is used to separate the electrons according to their energy losses after having passed through the specimen. Energy loss mechanisms such as plasmon excitations and core-electron excitations cause distinct features in EELS. These can be used to quantify the elements present, to provide information about atomic bonding, and for a variety of other useful phenomena. The reader is referred to the following references for more information on these techniques: Joy et al. (1986); Egerton (1996); Williams and Carter (1996).

In scanning transmission electron microscopy (STEM), a focused beam of electrons (typically less than 0.5 nm in diameter) is scanned in a television-style raster pattern across the specimen, as in a scanning electron microscope (SEM) (Keyse et al. 1998). In synchronization with the raster scan, products resulting from the interaction of the electron beam with the specimen are collected, such as emitted X-rays, or secondary or backscattered electrons, to form images. Electrons that pass through the specimen can also be detected to form images that are similar to usual HRTEM images. An annular detector can be used to collect the scattered transmitted electrons and this leads to so-called Z-contrast imaging (Browning et al. 1997). The STEM mode of operation is particularly useful for spectroscopic analysis, since it permits the acquisition of a chemical map of the sample with subnanometer spatial resolution under optimum conditions. For example, an image of the distribution of Fe in a sample can be made if one were to record in synchronization with the raster pattern, either the emission from the sample of Fe K_α X-rays (with the EDXS spectrometer), or transmitted electrons with energy losses greater than that of the Fe L-edge (with the EELS spectrometer).

A fully equipped HRTEM has the capability to record the variations in image intensity across the specimen using mass thickness or diffraction contrast techniques, to reveal the atomic structure of materials using so-called high-resolution (phase-contrast) imaging or Z-contrast (incoherent) imaging, to obtain electron diffraction patterns from small areas of the specimen using a selected-area aperture or a focused electron probe, and to perform EELS and EDXS measurements with a small probe. Additional lenses can be installed in conjunction with an EELS spectrometer to create an energy filter, enabling one to form energy-filtered TEM images (EFTEM). These images allow one to map the chemical composition of a specimen with subnanometer spatial resolution. A block diagram of such an HRTEM is shown in Fig. 13.21 (Fultz and Howe 2002).

In addition to the six main techniques of: (a) conventional imaging, (b) phase-contrast imaging, (c) Z-contrast imaging, (d) electron diffraction, (e) EDXS, and (f) EELS, it is possible to perform many other analyses in a HRTEM. For example, when electrons pass through a magnetic specimen, they are deflected slightly by Lorentz forces, which change direction across a magnetic domain wall. In a method known as Lorentz microscopy, special adjustments of lens currents per-

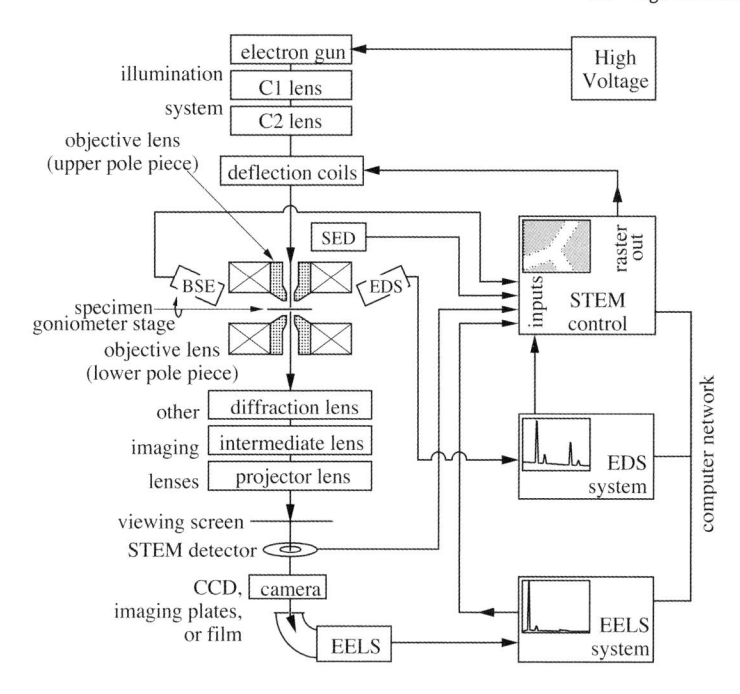

Fig. 13.21 Block diagram of a typical TEM with STEM capability (from Fultz and Howe 2002).

mit imaging of these domain walls. Using in-situ microscopy, phase transformations and microstructural changes in a specimen can be observed directly as the specimen is heated, cooled, or deformed in the microscope, using special specimen stages. Differential pumping can be used to allow the introduction of gases into the microscope column surrounding a thin foil, making it possible to follow chemical reactions in a HRTEM in situ. Many of these techniques can be performed at spatial resolutions of a few tenths of a nanometer. The possibilities are almost endless, and that is why HRTEM is an indispensable tool in materials research (Williams and Carter 1996). Since it is not possible to cover all of these techniques, this section focuses on the theory and practice of HRTEM imaging and in-situ HRTEM techniques.

13.2.2.1 Principles of Image Formation in and Practical Aspects of High-Resolution Transmission Electron Microscopy

13.2.2.1.1 Principles of Image Formation

HRTEM is phase contrast imaging (Spence 1988). In phase contrast imaging, two or more beams are allowed to interfere to form an image. With two or a row of beams contributing, a fringe pattern is obtained, while three or more non-collinear beams can produce a lattice image. The spacings which one wants to resolve in the image must be within the resolution limits of the instrument; i.e., the lens system must preserve the coherence of the image-forming beams. This section gives a general outline of phase contrast imaging. Practical aspects of HRTEM imaging and interpretation of HRTEM images are discussed in Section 13.2.2.1.2.

The definitive feature of HRTEM is perhaps the two-dimensional nature of the information it provides. While a specimen may be tilted in conventional TEM to obtain three-dimensional information, one is generally limited with HRTEM to observations of projected atomic structures along low-index zone axes with planar spacings that are within the resolution limits of the microscope. For this reason, one would like to interpret a HRTEM image directly as a simple map of the projected structure (crystal potential). However, such an interpretation is only possible for a very narrow range of instrumental parameters and specimen thicknesses. This situation is well described theoretically and so provides the starting point for the following discussion (Cowley 1975; Spence 1988).

A useful approximation in the description of the interaction of the electron beam with the specimen is found in the assumption of phase changes only, that is, a phase object. An incident wavefunction ψ_o incurs a phase change proportional to the crystal potential expressed by Eq. (53), where σ is the electron interaction parameter, t is the specimen thickness, and $\phi_p(x, y)$ is the projected specimen potential in the x–y plane, normal to the electron beam.

$$\psi = \psi_o \exp[-i\sigma t \phi_p(x, y)] \tag{53}$$

By additionally assuming a weak interaction, one obtains the weak phase object (WPO) approximation, Eq. (54).

$$\psi \approx [1 - i\sigma t \phi_p(x, y)] \tag{54}$$

Application of the WPO approximation requires that Eq. (55) holds.

$$t \ll 1/\sigma \phi_p(x, y) \tag{55}$$

Hence, the required thickness decreases with increasing atomic number, though it is less than 10 nm for most materials of interest. The above form of the exit wavefunction has a straightforward physical interpretation. The dominant transmitted wave is approximated by ψ_o (unit amplitude) while to this is added a relatively weak scattered wave of amplitude $\sigma t \phi_p(x, y)$ and phase $-\pi/2$ relative to the unscattered portion, as it is purely complex. For such thicknesses and assuming an ideal Scherzer lens, i.e., one that imposes a phase change of $-\pi/2$ on all diffracted beams, a linear relationship between image intensity $I(x, y)$ and projected crystal potential is obtained [Eq. (56)].

$$I(x, y) = 1 - 2\sigma t \phi_p(x, y) \tag{56}$$

The effect of the objective lens must then be considered. In an ideal lens, diffracted beams undergo a phase shift that is strictly a function of their angle from the optic axis. Assuming the transmitted beam is aligned along the optic axis, this angle is $2\theta_B$, twice the Bragg angle of the respective spatial frequency. Allowing for spherical aberration and a lens defocus Δf, the phase change is given by Eq. (57),

$$\chi = \pi \lambda g^2 [1/2\lambda^2 C_s g^2 - \Delta f] \tag{57}$$

where λ is the electron wavelength, g is the magnitude of the reciprocal lattice vector, and C_s is the spherical aberration coefficient of the objective lens. By definition, Δf is the distance from the specimen exit plane to the lens object plane. The defocus is taken as negative for an underfocused lens (obtained by reducing the lens current).

The effect of the objective lens is then to multiply each diffracted beam by the phase factor $\exp[i\chi(g)]$. Under the WPO approximation, the transmitted beam ($\chi = 0$) and the real components of the diffracted beams interfere to form the image, i.e., a diffracted beam's contribution is proportional to $\sin \chi$. For negative $\sin \chi$, the diffracted beams interfere destructively with the transmitted beams, producing "black" atom images, while a positive $\sin \chi$ leads to "white" atoms.

It is usual to plot a linear contrast transfer function (CTF) to describe the lens action, that is, a plot of $\sin \chi$ versus g. For a WPO, the largest number of beams will contribute to the image by maximizing the portion of the CTF where $\sin \chi = -1$ or close to it. Requiring that $\chi \leq -1/\sqrt{2}$ leads to the Scherzer defocus value, Eq. (58) (Scherzer 1949).

$$\Delta f_{Sch} = -1.2(C_s \lambda)^{1/2} \tag{58}$$

The first zero crossover gives the Scherzer resolution limit [Eq. (59)].

$$d_{Sch} = 0.7 C_s^{1/4} \lambda^{3/4} \tag{59}$$

This is the highest resolution at which one may hope to interpret an HRTEM image directly. Decreasing the defocus value leads to higher-order passbands at Δf_n given by Eq. (60), which may be employed to resolve finer detail, $n = 0$ giving the Scherzer passband.

$$\Delta f_n = [1/2 C_s \lambda (8n + 3)]^{1/2} \tag{60}$$

Since higher passbands exclude some lower-angle scattering, their direct interpretation is not possible, particularly in the case of defect imaging, for which diffuse scattering possesses important image information. In addition, electronic instabilities effectively damp higher frequencies, limiting the number of useful passbands. This damping limit is usually called the information resolution limit.

As seen above, proper selection of defocus is employed to optimize resolution. While working at the microscope, one uses the minimum contrast condition as a reference point. At this defocus setting, diffracted beams are as close to a $-\pi/2$ total phase shift as possible, minimizing interference with the transmitted beam: In other words, $\sin \chi$ is close to zero over a maximized range. This range is given by Eq. (61), and is about one-third the Scherzer defocus value.

$$\Delta f_{min} = -0.44(C_s \lambda)^{1/2} \qquad (61)$$

The above concepts are commonplace in discussions of HRTEM results even though the WPO approximation is often invalid because it is quite difficult to produce a good specimen only a few nanometers thick. However, they provide an acceptable starting point for selecting proper imaging conditions and for image interpretation. Proper evaluation of defect atomic positions requires concurrent image simulations using hypothetical models until good matching with experimental images is achieved.

13.2.2.1.2 Practical Aspects of HRTEM

In order to interpret HRTEM images, one needs to know how various specimen and microscope parameters affect image contrast. There are many factors that can potentially contribute to HRTEM image contrast and it is not possible to discuss all of them here. Instead, we examine how some of the more important parameters that tend to be less easy to control and/or quantify affect image contrast. Particular attention is paid to practical guidelines that can be used to obtain as much information about defect structures as possible without expending too much effort (Fultz and Howe 2002).

Below are listed most of the specimen and microscope parameters that must be determined or considered when interpreting HRTEM images:
- sample thickness
- objective lens focus
- beam coherence
- beam convergence
- objective aperture
- crystal orientation
- beam tilt
- interface geometry
- surface effects
- thin-foil relaxation
- exposure time
- specimen drift
- projection problem
- beam damage.

In terms of the microscope, one must optimize and quantify as many of the microscope parameters as possible, just as one would do for any HRTEM investigation. Several microscope parameters such as the accelerating voltage, spherical aberration coefficient of the objective lens, and Gaussian spread of focus are provided by the manufacturer and can be directly input into image simulation programs. Other parameters that are fairly straightforward to determine include the objective lens defocus, which can be determined by taking a fast Fourier transform (FFT) of the amorphous edge of a specimen, the semi-angle of beam convergence, which can be determined directly from the diameter of the disks in the diffraction pattern, and the objective aperture radius (if used) and position, which can also be determined by using a double-exposure to photograph the objective aperture on the diffraction pattern. Although it is possible to quantify the degree of astigmatism in an HRTEM image, one ideally tries to eliminate any astigmatism, since it cannot be incorporated into most image simulation programs (O'Keefe 1984).

One of the most difficult microscope parameters to quantify is the amount of beam tilt present during imaging. It has been demonstrated that for most defect structures, alignment of the electron beam to within one milliradian of the optic axis or better is necessary for reliable interpretation of the atomic structure, particularly in ordered crystals with large unit cells (Smith et al. 1983, 1985). While the effect of beam tilt can be readily included in image calculations for comparison with experimental images, it is more desirable to minimize this effect before imaging. One technique for obtaining the required accuracy in alignment is to apply equal and opposite tilts to the incident beam and then compare the resulting images of the amorphous region at the foil edge. The tilts can be adjusted in the image mode until the same image appearance is produced for opposite tilts. A similar result can be achieved by taking FFTs of the two images and then comparing their diffraction patterns, which should be identical. It is now possible to perform this operation online using an iterative computer program, so that beam tilt and astigmatism can be corrected to the required levels of accuracy by the computer after an initial manual alignment.

Suitable specimens are essential; preferably flat, clean, gradually decreasing to zero thickness at an amorphous edge. Alignment of the specimen along a zone axis is critical. This can be particularly challenging with bent specimens, as often occurs for metals. Normally, Kikuchi bands or convergent-beam patterns are employed for precise tilt adjustment; however, any portion of specimen displaying such effects is too thick for HRTEM imaging. The thin edge is often slightly bent so one must use a selected area aperture, tilting the specimen until the intensity of diffracted spots appears to be balanced about the transmitted beam.

13.2.2.2 In-Situ Hot-Stage High-Resolution Transmission Electron Microscopy

In-situ HRTEM provides unique capabilities for quantifying the dynamic behavior of materials at the atomic level (Sinclair et al. 1988; Howe 1999; Gai and Boyes 2003). This section provides a brief description of particular requirements for performing in-situ hot-stage HRTEM studies, used to induce phase transformations in materials.

The specimen and microscope requirements for in-situ hot-stage HRTEM imaging are not different from those of static HRTEM, except that one must have a heating holder and some method of recording and analyzing dynamic images (Butler and Hale 1981; Sinclair et al. 1988). At present, most HRTEMs are equipped with charge-coupled device (CCD) or TV-rate cameras that are fiber-optically coupled to the HRTEM for recording high-resolution images. The simplest and least expensive method for recording in-situ images is to send the output from the TV-rate camera into a standard videocassette recorder (VCR). A reasonably good VHS format VCR and videotapes can be used to store hours of in-situ data for low cost. The disadvantage of this method is that the images often need to be converted to digital form for image processing and/or analysis and the TV camera response is not linear over a large range. Recording images digitally using a CCD camera eliminates these problems, but computer memory and acquisition rates can become limiting factors. TV-rate cameras generally acquire 30 frames s^{-1}, so that the time resolution of the videorecording is $\sim$0.03 s. Faster video systems are available, but the signal-to-noise in the images then starts to become an issue. Specimen drift is usually not a significant problem when images are being acquired at TV rates, but substantial drift makes subsequent batch processing and computer comparison of digital images difficult. It is possible to install drift compensators on the microscope that alleviate this problem, but these are not readily available commercially.

The most versatile specimen holder for materials science applications is a double-tilt hot stage, and several manufacturers offer such holders (Williams and Carter 1996). A typical water-cooled holder can achieve a temperature of 800–1000 °C with a full range of tilt. As in static HRTEM, the area of interest ideally needs to be in a zone-axis orientation in order to interpret the atomic structure in the images. Knowledge of the temperature of the specimen in the area of interest as well as the heating/cooling rate when the current is changed are critical for quantitative studies and can vary from the hot-stage thermocouple readout depending on the sample conductivity, how well it is in contact with the holder, etc. In most cases, experience has shown that the sample temperature is usually within about 20 °C of the thermocouple readout for most holders of this type.

Just about any type of phenomenon that can be induced by temperature changes within the limits of the holder can be examined by in-situ hot-stage HRTEM. (The same applies to in-situ cooling experiments using liquid-nitrogen or liquid-helium holders, although holder vibration due to bubbling can be a consideration when using cooling liquids.) These phenomena include order/disorder reactions (Howe et al. 2006), grain-boundary motion (Merkle et al. 2002), melting and freezing of materials (Sasaki and Sasa 1991), precipitation-dissolution (Howe and Benson 1995), interfacial reaction (Sinclair 1994), crystallization (Sinclair et al. 1993), twin and martensitic motion (Howe 1993), oxidation/reduction (Kang and Eyring 1991), dislocation motion (Sinclair et al. 1988), faceting/roughening transitions (Gabrisch et al. 2001), nanoparticle reactions (Lee and Mori 2005; see Chapter 9), coarsening (Iijima 1985), quasimelting (Marks 1994), catalytic reaction (Gai and Boyes 2003), etc.

13.2.2.3 Examples of HRTEM Studies of Dislocation and Interphase Boundaries

13.2.2.3.1 Disclinations in Mechanically Milled Fe Powder

Mechanical milling is a technique for producing metallic alloys with ultrafine grain sizes by severe plastic deformation (cf. Chapter 8, Section 8.2). These nanocrystalline alloys have unique mechanical properties, such as hardnesses and yield strengths that are several times higher than those of conventional alloys, that make them attractive for a variety of applications (Meyers et al. 2006). The mechanisms by which materials deform during mechanical milling to produce ultrafine grains are not known, although it is thought that turbulent shear processes requiring crystal rotation are operative and that disclinations contribute to this process (Romanov and Vladimirov 1992; Seefeldt 2001). Disclinations in crystalline materials can alternatively be described in terms of individual line defects in the atomic structure called dislocations (see Chapter 6). It is possible to image the atomic structures of defects such as dislocations and disclinations in crystalline solids using HRTEM (Murayama et al. 2002).

Figure 13.22(a) shows an HRTEM image of mechanically milled, nanocrystalline Fe powder. The image is approximately 20.5 nm wide. The grain that occupies most of the figure is in a $\langle 111 \rangle$ orientation. The hexagonal arrangement of columns of Fe atoms in this orientation are visible as white spots, as verified by image simulations. White lines were drawn periodically along the three sets of edge-on $\{110\}$ planes in the grain, and these are shown superimposed on the HRTEM image in Fig. 13.22(b). Two sets of $\{110\}$ planes running vertically in the figure are straight (or nearly so), but the set that is approximately horizontal in the figure bends considerably. Dark lines were drawn on all of the nearly horizontal, bent $\{110\}$ planes in Fig. 13.22(b) to indicate their positions accurately. The black and white lines in Fig. 13.22(b) were removed from the HRTEM image and are shown separately in Figure 13.22(c) for clarity. The discussion that follows emphasizes the bent $\{110\}$ planes shown in Fig. 13.22(c), but it should be remembered that these planes are derived from the actual atomic positions visible in Fig. 13.22(a).

The arrows in Fig. 13.22(c) mark two wedge-shaped regions that together form a partial disclination dipole. The wedge-shaped regions are approximately 3.5 nm apart. Each of the wedge-shaped regions contains a number of terminating $\{110\}$ planes, which are individual dislocations with a Burgers vector, which is the displacement vector that describes the magnitude and direction of their strain field, of $\mathbf{b} = a/2\langle 111 \rangle$. The partial dislocation dipoles in Fig. 13.22(c) appear to be wedge disclinations with a Frank vector $\mathbf{w}$, which is the rotational vector which describes the distortional power of the disclination, which is parallel to the defect line (in this case, the $\langle 111 \rangle$ viewing direction). Unfortunately, it is not possible to determine exactly the wedge and twist components of the partial disclinations in Fig. 13.22(c), because an HRTEM image reveals atomic displacements only perpendicular to the electron beam direction, and there may be displacements parallel to the beam that are not visible. However, this image demonstrates that it is possible to observe directly the individual dislocations that constitute partial disclination

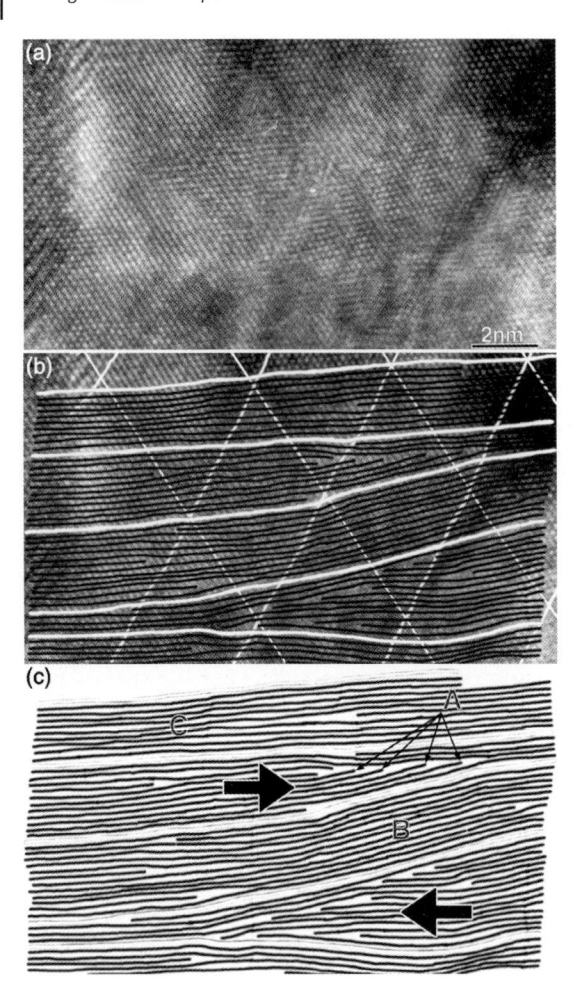

Fig. 13.22 a) An experimental HRTEM image of mechanically milled, nanocrystalline Fe powder taken near Scherzer defocus. The hexagonal arrangement of white spots in the image corresponds to columns of Fe atoms in a $\langle 111 \rangle$ orientation. b) White lines shown superimposed periodically on the three sets of {110} planes in a) to highlight the distortion of the nearly horizontal set of {110} planes. Black lines were also superimposed on this set of planes to indicate their position clearly. c) The nearly horizontal black and white lines in b) removed from the HRTEM image so that they are more clearly visible. The various labels are explained in the text (from Murayama et al. 2002).

dipoles in metals at the atomic level, even in mechanically milled powders that have undergone severe plastic deformation.

The set of terminating {110} planes that constitute the individual partial disclinations, such as the ones labeled A in Fig. 13.22(c), can also be considered terminating tilt grain boundaries (Hirth and Löthe 1968). Compared to complete tilt grain boundaries, terminating tilt grain boundaries contain missing dislocations

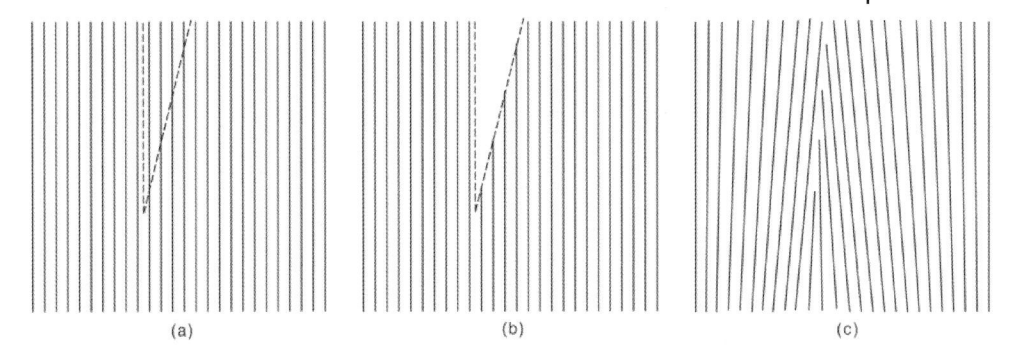

Fig. 13.23 The elastic distortion associated with a partial wedge disclination (or terminating tilt grain boundary). a) One set of planes in a perfect crystal, such as the perfect {110} planes seen in Fig. 13.22(b). b) A wedge-shaped piece of material is removed from the crystal. c) The new surfaces are allowed to close, to fill the wedge. The resulting crystal contains a terminating tilt grain boundary, i.e., a terminating array of edge dislocations, and considerable elastic distortion, which increases the elastic strain energy of the remaining solid (from Murayama et al. 2002).

and these are replaced by rotational elastic deformation in the crystal (Michler et al. 1998). Such a configuration can be interpreted as a wedge of material added to or removed from an ideal crystal, as illustrated in Fig. 13.23, and this is evident from the wedge shape of the bent white lines in Fig. 13.22(b,c). The crystal rotation produced by the partial disclinations (or terminating tilt grain boundaries) in Fig. 13.22 is also evident. For example, the {110} planes located between the two partial disclinations (labeled B in Fig. 13.22c) are rotated approximately 9° relative to the {110} planes located outside the dipole (such as the nearly horizontal planes labeled C in Fig. 13.22c). This observation provides direct confirmation that crystals can rotate and thereby undergo turbulent behavior during severe plastic deformation by the action of partial disclinations.

The terminating dark lines in Fig. 13.22(c) reveal the arrangement of individual dislocations that constitute the partial disclinations associated with the dipole in this figure, as well as other, isolated dislocations in the metal. The generation and interaction of partial wedge disclinations allows reorientation of crystal volumes only several nanometers in size. This mode of deformation on such a fine scale probably' facilitates the fragmentation and reorientation process of metal grains undergoing severe plastic deformation (Romanov and Vladimirov 1992; Seefeldt 2001), leading to an ultrafine grain size. Thus, partial disclination defects such as those in Fig. 13.22 can contribute to both the deformation response and the strengthening of metals. The generation of partial dislocation defects provides an alternative mechanism to grain boundary sliding, which has been suggested to allow rotation of nanosize crystals during mechanical milling (Schiotz et al. 1998). It is not possible from the HRTEM image to determine exactly how the partial disclination dipoles in Fig. 13.22(c) formed, although the dislocations probably nucleated at pre-existing defects such as grain boundaries or cell walls in the metal and rearranged into the terminating arrays in Fig. 13.22.

13.2.2.3.2 Interphase Boundaries in Metal Alloys

Interfaces that separate two crystals that differ in composition, Bravais lattice, or both, are referred to as interphase boundaries (Wolf and Yip 1992; Sutton and Balluffi 1995; Howe 1997). Understanding the details of how the atoms arrange at such interfaces and cross from one phase to form the other are essential to understanding the nature of phase transformations in materials (Wayman et al. 1994; Aaronson 2002). Interphase boundaries are often divided into three classes based on the degree of atomic matching or coherency across the interface (Howe et al. 2000; see Chapter 7), as:

- coherent interfaces, where there is complete continuity of atomic planes across the interface between two phases
- partly coherent interfaces, where the mismatch between two crystal structures across the interface is accommodated by periodic misfit dislocations, i.e., strain localization, in the interface, and
- incoherent interfaces, where atomic matching is sufficiently poor for there to be no correspondence of atom planes across the interface or localization of misfits into dislocation defects.

In addition to this classification, which is based on the degree of atom matching and/or misfit localization, an interphase boundary may be sharp or diffuse, depending on whether the composition or structural changes occur abruptly across a single plane, or smoothly over several or more planes (Cahn and Hilliard 1958; Sutton and Balluffi 1995; Howe 1997). The following examples illustrate how HRTEM has been a valuable technique for determining the dynamic behavior and mechanisms of motion of both diffuse and abrupt interphase boundaries displaying different degrees of coherency, as listed above.

13.2.2.3.3 Diffuse Interface in Cu–Au

Figure 13.24(a) shows an HRTEM image of a diffuse interphase boundary between the long-range ordered AuCu-I phase and disordered α phase in a Au–41 at.% Cu alloy sample taken as a frame from a videotape at approximately 305 °C during in-situ heating in the HRTEM (Howe et al. 2006). The position where an intensity profile 25 pixels wide was taken across the interphase boundary is indicated by the white rectangle in the figure. The corresponding intensity profile is shown in Fig. 13.24(b). In this profile, the ends of the diffuse interphase boundary on the ordered and disordered sides are indicated by the lines labeled O and D, respectively, and the same locations are also indicated in Fig. 13.24(a). Detailed HRTEM image simulations (e.g., inset in Fig. 13.15a) show that the Au-rich (001) planes in the ordered AuCu-I phase appear as bright lines in the image in Fig. 13.24(a) and corresponding high-intensity peaks on the left-hand side of the intensity profile in Fig. 13.24(b). The Cu-rich planes in between are barely visible

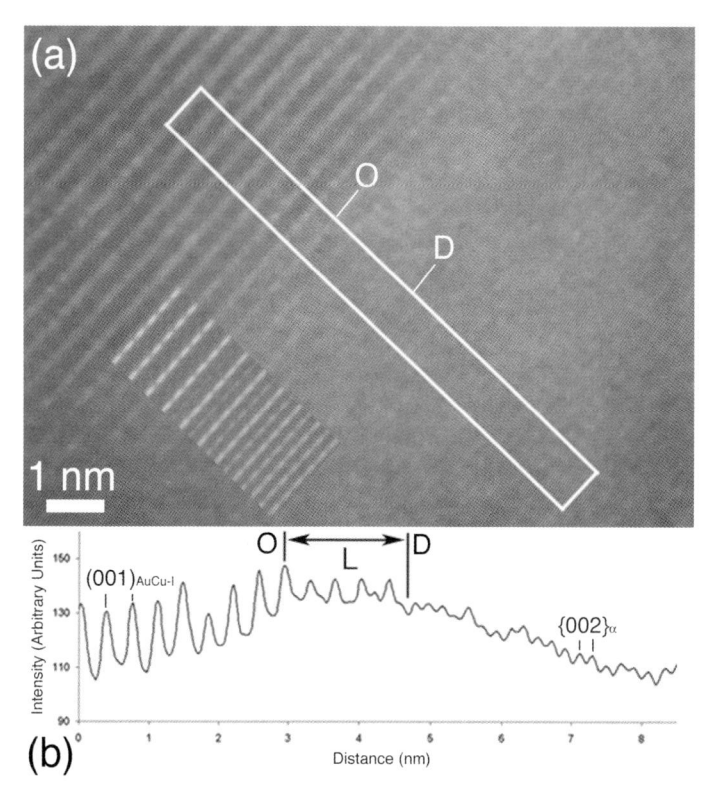

Fig. 13.24 (a) HRTEM image of an order–disorder interphase boundary at 305 °C with a simulated image of the interphase boundary for the same sample and microscope conditions shown superimposed on the experimental image. The Au-rich (001) planes in the AuCu-I phase appear bright on the left-hand side of the interphase boundary, as do the {002} planes that are spaced approximately half as far apart in the α phase on the right-hand side; (b) an intensity profile 25 pixels wide taken across the interphase boundary over the region indicated by a white box in (a). The positions O and D, which indicate the ordered and disordered sides of the diffuse interphase boundary, respectively, are indicated in both (a) and (b) (from Howe et al. 2006).

because they appear dark in the HRTEM image in Fig. 13.24(a) under these experimental conditions. The lower intensity peaks spaced only half this distance apart to the right of position D correspond to the {002} planes in the disordered α phase. In the region between O and D, the intensities change from left to right, leading to the interphase boundary thickness labeled L in Fig. 13.24(b).

The positions O and D in Fig. 13.24 were determined for each frame (time interval of 0.03 s) over a period of 60 s and these are plotted in Fig. 13.25(a). The data in Fig. 13.25(a) show several important features. First, it is evident that the disordered side of the diffuse interphase boundary (top graph) generally moves more frequently and often over larger distances than the ordered side of the interphase boundary (bottom graph) with time. The shortest distance present in the

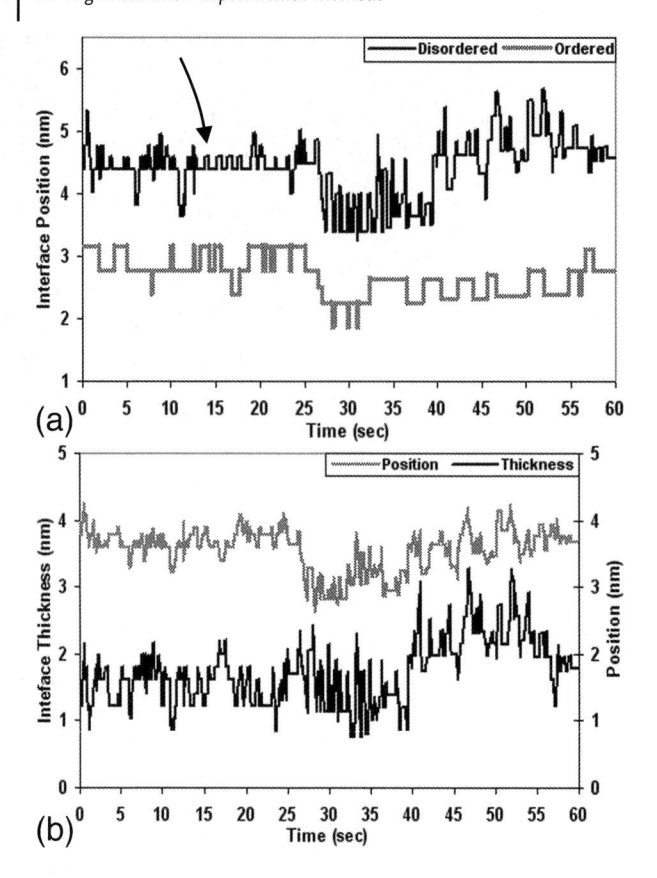

Fig. 13.25 (a) Graphs showing the positions of the disordered (top) and ordered (bottom) sides of the diffuse interphase boundary in Fig. 13.24(a) over a period of 60 s; (b) corresponding graphs showing the interphase boundary mean position (top) and thickness (bottom) over the same period. The interphase boundary mean positions in (b) fluctuate around a value of approximately 3.5 nm, which is roughly the midpoint between the two intensity profiles in Fig. 13.244(b) (from Howe et al. 2006).

top graph (arrow) is the spacing of one {002} plane in the disordered α phase. Following the disordered side of the interphase boundary with time, it is evident that the interphase boundary fluctuates over a distance of one to as many as six {002} α planes. In contrast, using the {002} plane spacing as a reference, it is apparent that the ordered side of the interphase boundary in the lower graph typically moves twice this distance, which is the spacing between the Au-rich (001) planes in the AuCu-I phase. In addition, it fluctuates with a frequency that is several times less than that of the disordered side. Thus, the position of the disordered side of the interphase boundary rapidly fluctuates between a number of adjacent {002} α planes with time, while the ordered side fluctuates more slowly and only between Au-rich planes in the ordered AuCu-I phase.

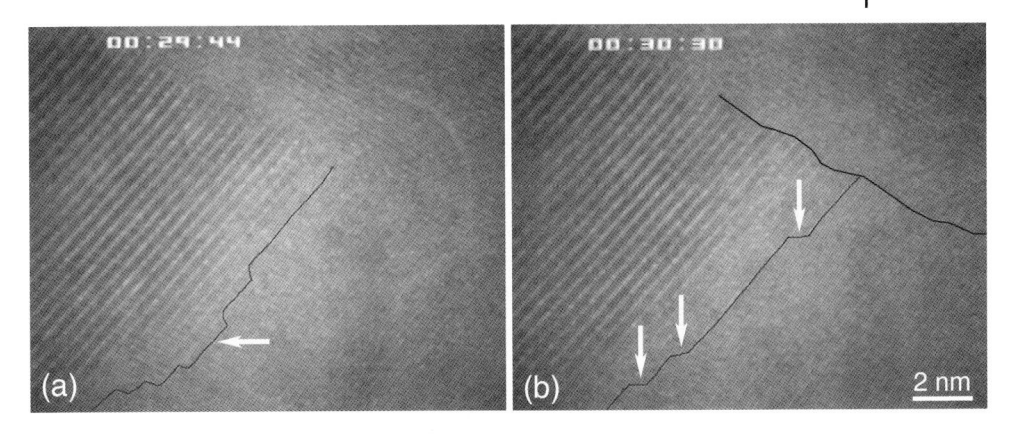

Fig. 13.26 HRTEM images showing mechanisms of interface movement. Fluctuations nucleate homogeneously on the interface and expand by steps, one (001) AuCu-I plane high, to form (a) a pillbox shape (white arrow) and (b) a series of steps (see arrows). The interface profiles are outlined.

The different interphase boundary behaviors observed in Fig. 13.25(a) are probably due to a combination of several factors occurring at the interphase boundary, namely: (a) the tetragonality of the AuCu-I structure, which leads to an energetically stable (001) plane that tends to keep the ordered side of the interphase boundary parallel to this plane; (b) a lower average diffusivity in the AuCu-I phase as compared to the disordered phase at 305 °C; and related to this, (c) a longer atomic jump distance (0.357 nm) for diffusion perpendicular to the ordered (001) planes in the AuCu-I ordered phase; and (d) the fact that the interphase boundary must move this same distance between Au-rich planes on the ordered side (Schweika et al. 2004; Butrymowicz et al. 1974). A combination of these effects leads to a significantly higher activation energy for movement of the ordered side of the interphase boundary, causing the experimentally observed interphase boundary behavior in Fig. 13.25(a). In fact, if one performs an approximate calculation for the average diffusivities and corresponding jump frequencies of the atoms in the ordered and disordered phases, respectively, using the equation $\Gamma = 6D/\lambda^2$, where Γ is the jump frequency (s^{-1}), D is the diffusivity at 305 °C (m^2 s^{-1}) and λ is the jump distance (m), one finds that the jump frequency in the disordered phase is approximately 3×10^5 s^{-1}, as compared to 3×10^4 s^{-1} in the ordered phase, consistently with the considerations mentioned above. However, these frequencies are approximately four orders of magnitude greater than the actual frequencies of interphase boundary movement recorded for the ordered and disordered sides of the interphase boundary, indicating that collective volumes, rather than individual atom jumps, are involved in the process. In other words, movement of the interface involves the formation of critical-size fluctuations involving many atoms (Howe and Benson 1995).

An outline of the ordered side of the interface was drawn for a number of individual frames, based on the positions of the bright (001) planes. Figure 13.26 shows two different interface profiles that were typically found, reflecting the mechanisms by which interface movement (or fluctuation) occurred (Howe et al. 2006). Figure 13.26(a) shows a situation where a perturbation in the ordered phase appeared to nucleate homogeneously on the interface (arrow) away from any other features. This perturbation expanded into the disordered phase and laterally along the order–disorder interface, propagating it forward and resulting in the interface shape outlined in Fig. 13.26(a). The image in Fig. 13.26(b) shows a second case, where a fluctuation in the ordered phase nucleated near a boundary between two different ordered domains. This ordered portion then propagated along the order–disorder interface as a series of steps, one (001) high (indicated by arrows in Fig. 13.26b), again resulting in net movement forward. In either case, the interface then receded by the opposite movement of such fluctuations (or steps). In most of the frames examined in the video, it was found that the ordered phase advanced into the disordered phase at more than one region along the interface, with nucleation at the domain boundary constituting only a small fraction of the fluctuations. Hence, this intersection did not appear to have a large effect on the overall interface behavior.

Based on Fig. 13.26, it is clear that the interface does not fluctuate by single-atom jumps, but by a collective mechanism involving the nucleation and spreading of critical-size (001) steps parallel to the interface plane. This is similar to the behavior of highly faceted interfaces in other diffusional transformations such as precipitation and crystallization (Howe et al. 1996) as shown below, and different from the behavior of fluctuations at solid/liquid interfaces (Hoyt et al. 2003), which are not constrained to a certain height, or energy minimum perpendicular to the interface (e.g., one (001) AuCu-I plane spacing). It also demonstrates that the anisotropy in the interfacial energy of the AuCu-I phase, specifically the (001) interphase boundary energy, is sufficient at this temperature to influence the morphology and behavior of fluctuations at the diffuse interface. Such behavior is consistent with previous observations of interfaces in other similar systems (Loiseau et al. 1994; Chatterjee et al. 2004).

Since HRTEM images such as those in Figs. 13.24 and 13.26 are projections of the three-dimensional interface structure, the interface fluctuations seen in projection also occur through the specimen thickness along the viewing direction, so that the profile in Fig. 13.26(a) for example, most probably represents a set of somewhat faceted, concentric pillboxes in three dimensions. It is also quite likely that fluctuations in the interface through the specimen thickness are similar, if not identical, to what is observed in the plane of projection, since the interface plane is (001). The minimum size perturbation (or fluctuation) observed at the interface was one (001) plane spacing high (~0.37 nm) and about three to four times this wide (1.2–1.6 nm), so that the critical fluctuation volume is on the order of 0.6 nm^3, or about 40 atoms, assuming a circular pillbox shape, or slightly larger if a square faceted shape is assumed. Note that this is about the minimum volume of specimen that one would expect to be observable in a specimen 20 nm thick.

The interphase boundary thickness and mean position were determined from the data in Fig. 13.25(a) and these are plotted in Fig. 13.25(b). The mean interphase boundary position was determined as the midpoint between positions O and D. The average frequency of interphase boundary movement is 1.2 s^{-1}, with a maximum frequency of 7.7 s^{-1} and a minimum frequency of 0.4 s^{-1}. This temporal variation reflects mainly that of the disordered side, which fluctuates at a higher frequency. Also note that the mean interphase boundary position moves up more often than down in Fig. 13.25(b), again reflecting the character of the disordered side of the interphase boundary, which tends to move into the disordered phase, expanding the interphase boundary. Thus, the interphase boundary fluctuations are not symmetric about the mean position, but favor the disordered side. The average velocities of the ordered and disordered sides of the interphase boundary were also determined over 60 s using the data in Fig. 13.25(a), and were found to be 0.3 and 1.3 nm s^{-1}, respectively.

The interphase boundary thickness is shown in the bottom graph in Fig. 13.25(b). The average thickness is approximately 1.7 nm, or the equivalent of seven {002} planes (or about four unit cells) in the disordered α phase, although the thickness averages slightly more than 2.0 nm during the latter 20 s of the video sequence, where the disordered side of the interphase boundary moves a few planes further away from the ordered side, and slightly less than 1.7 nm earlier on. The maximum and minimum interphase boundary thicknesses observed were 3.3 and 0.8 nm, respectively, which represent variations in thickness on the order of 100%. The average interphase boundary thickness measured from Fig. 13.25(b) is comparable to recent studies on strain-induced incomplete wetting above the critical temperature T_c, where the alloy becomes disordered, at AuCu-I (001) surfaces (Schweika et al. 2004), and is also similar to previous calculations for a Au–Cu alloy (Kikuchi and Cahn 1979) and recent atomistic calculations performed on Al–Li and Al–Ag alloys slightly below T_c, using embedded atom potentials and Monte Carlo (MC) methods (Sluiter and Kawazoe 1996; Asta and Hoyt 2000). In these systems, which are similar to the alloy in the Au–Cu case, the calculations indicate that the diffuse interphase boundary should extend over four to six unit cells (i.e., $L \sim 1.6$–2.4 nm). In the Au–Cu system, the strain-induced incomplete wetting phenomenon leads to order in AuCu (001) surfaces that extends five to seven planes (i.e., $L \sim 1.0$–1.4 nm) into the crystal at temperatures much greater than T_c.

13.2.2.3.4 Partly Coherent Interfaces in Al–Cu

Recent in-situ hot-stage HRTEM studies of precipitate plates in Al–Cu–Mg–Ag and Al–Ag alloys performed both parallel and perpendicular to the plate faces, and comparison of these studies with prior HRTEM and conventional in-situ hot-stage TEM investigations, have clearly established the terrace–ledge–kink (TLK) mechanism as the primary atomic mechanism involved in growth and dis-

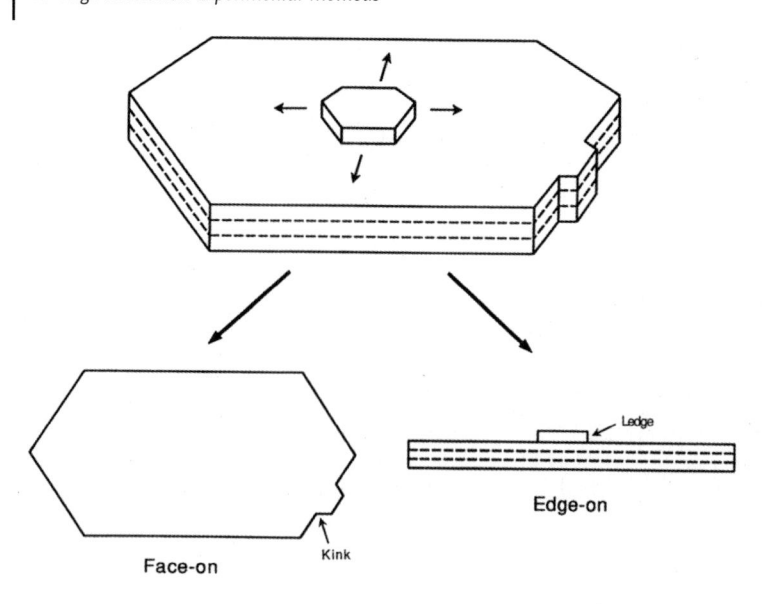

Fig. 13.27 Perspective view (top), face-on (bottom left), and edge-on (bottom right) views of a precipitate plate growing by a terrace–ledge–kink mechanism (from Howe et al. 1996).

solution of faceted precipitates in metal alloys (Laird and Aaronson 1969; Howe et al. 1987; Howe 1997). This process is illustrated schematically in Fig. 13.27, which shows a perspective view of a precipitate plate growing by ledges that nucleate on the habit plane and propagate out to the edge, where they stack one above the other. Figure 13.27 shows two additional views, one perpendicular to the habit plane of the plate, i.e., face-on, and the other parallel to the habit plane, i.e., edge-on. It is clear from the two lower illustrations, that when the plate is viewed edge-on parallel to the facets, it is possible to observe the atomic structure and dynamics of individual ledge motion and also the motion of the ledges stacked at the plate edges by in-situ HRTEM. When the edge of the plate is viewed in the face-on orientation, the electron beam is parallel to the ledges and this makes it possible to observe kinks that form on and propagate along the ledges. By combining information from these two orientations, it is possible to obtain a three-dimensional description of the atomic mechanisms of interfacial motion, as illustrated by the following data for $\{111\}$ θ plates in an Al–Cu–Mg–Ag alloy (Howe and Benson 1995; Benson and Howe 1997).

Structural and kinetic analyses in an edge-on orientation The orientation relationship of the $\{111\}$ θ phase with the matrix is $(\bar{1}10)_\theta \| (111)_\alpha$, $[110]_\theta \| [101]_\alpha$, and $[001]_\theta \| [1\bar{1}1]_\alpha$, which is a low-energy orientation relationship for θ phases designated as a Vaughan II orientation relationship. Figure 13.28(a) shows an HRTEM image of a ledge on the face of a θ plate viewed edge-on along a $[001]_\theta \| [1\bar{2}1]_\alpha$ di-

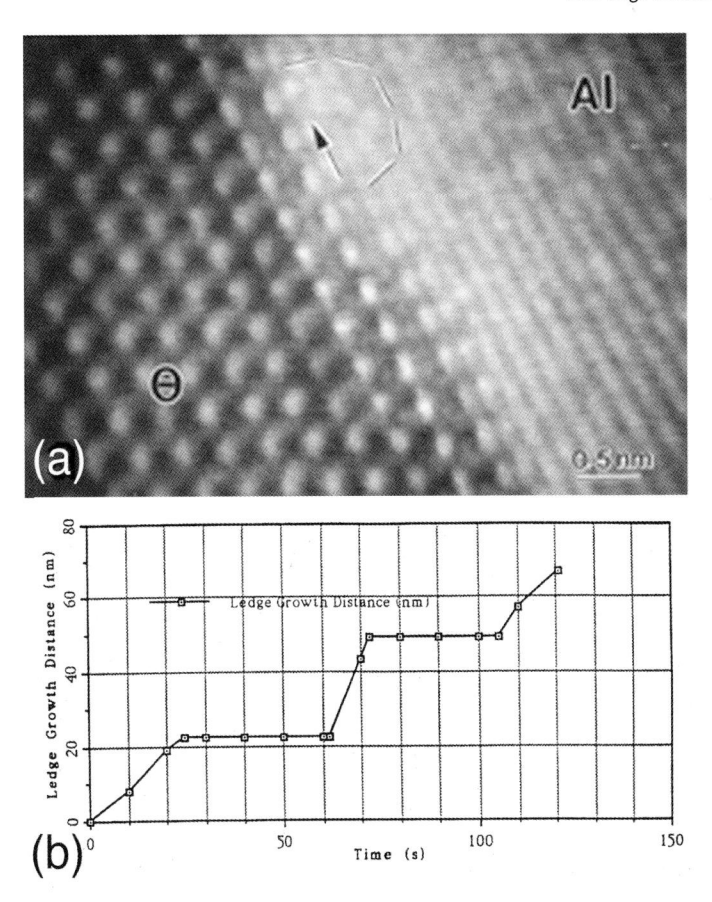

Fig. 13.28 (a) HRTEM image of a single ledge on a θ plate during growth at about 220 °C, and (b) graph of growth distance versus time for the ledge in (a) (from Benson and Howe 1997).

rection, as in the lower-right illustration in Fig. 13.27. The ledge is approximately two $\{111\}_\alpha$ matrix planes high, or half of a unit cell of the θ structure (0.424 nm). This was the smallest ledge size that was observed on the faces of the θ plates and higher ledges were often observed (Benson and Howe 1997). At about 220 °C the ledge was observed to oscillate several times per second over a distance of about two unit cells of the θ phase along the precipitate face while moving slowly across the face toward the precipitate edge in the direction indicated by an arrow. In-situ experiments that were performed perpendicular to the plate face indicate that the oscillatory motion is due to the formation and annihilation of kinks along the ledge, as demonstrated in the next section. The videocassette recording also revealed direct experimental evidence of enhanced atomic motion in the matrix just ahead of the ledge; this leads to slight blurring in the photograph, which is visible in the enclosed area in Fig. 13.28(a). It is important to note that the precipitate structure only one unit cell behind the ledge appears completely transformed, indicating that the structural and compositional changes which are necessary for diffusional growth occur simultaneously within a few atomic distances of the ledge. Although the ledge appeared to move smoothly across the precipitate face over short periods of time, it displayed start–stop behavior when viewed over longer times, as shown in Fig. 13.28(b). Such periodic lack of mobility during the migration of ledges has been observed previously and attributed to a lack of sites for atomic attachment along the ledges as they align along low-energy matrix directions, in this case $\langle 121 \rangle_\alpha$ (Garg et al. 1993).

Structural and kinetic analyses in a face-on orientation Figure 13.29 shows an HRTEM image taken at the edge of a θ plate viewed face-on along a $[\bar{1}10]_\theta \| [111]_\alpha$ direction, as in the lower-left illustration in Fig. 13.27, during an in-situ hot-stage experiment at about 275 °C. The prominent rectangular pattern of white spots outlined in Fig. 13.28 with dimensions of 0.244×0.429 nm^2 relates directly to positions of Cu atoms in the θ structure, as determined by HRTEM image simulation. During the in-situ HRTEM experiments, the θ plate was observed to grow by the nucleation of double kinks half a unit cell high (0.429 nm) along the $(110)_\theta \| (101)_\alpha$ plate edge, which propagated along the edge until they reached the intersecting facet. The arrow in Fig. 13.29 indicates the end of one such kink. The smallest kinks were one-half of the θ unit cell in height (one rectangular pattern of white spots about 0.429 nm long) but sometimes two or three kinks nucleated and/or dissolved in rapid succession in an oscillatory manner about an average position, similarly to the behavior described for the ledge in Fig. 13.28. It is important to note that the kink in Fig. 13.20 is well defined to within two or three half unit cells of the {111} θ structure. Although the image in Fig. 13.29 was taken at the edge of the θ plate where several ledges may be stacked vertically parallel to the electron beam direction, when this perspective

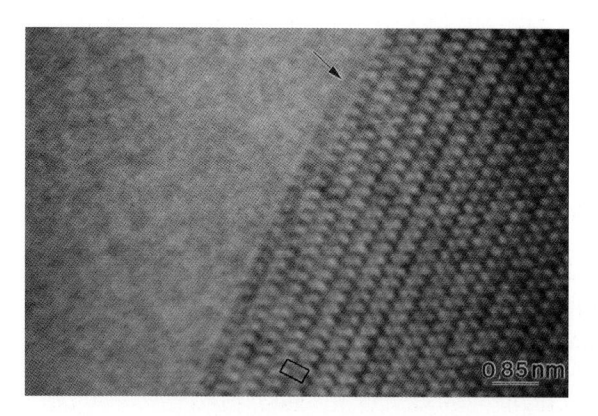

Fig. 13.29 An isolated kink (indicated by the arrow) traveling along the $[110]_\theta$ facet of a {111} θ precipitate plate during growth at about 275 °C (from Howe and Benson 1995).

is combined with the one in Fig. 13.28, it is possible to conclude that the phase transformation is occurring at kinks in ledges on the θ plates and that the transformation is completed within a volume as small as about one unit cell of the θ phase along the $[\bar{1}10]_\theta$ and $[110]_\theta$ directions. This volume contains about four atoms of Cu and eight atoms of Al. Thus, performing in-situ HRTEM allows observation of the atomic mechanisms of the transformation (the TLK mechanism) as well as the dynamics of transformation. It is also possible to study the dynamics of the kinks at the edges of the plates, as described in detail elsewhere (Howe and Benson 1995; Benson and Howe 1997).

13.2.2.3.5 Incoherent Interfaces in Ti–Al

In contrast to the case of coherent and partly coherent interfaces described above, the atomic mechanisms by which solid–solid incoherent interfaces move are not well understood (Howe et al. 2000; Aaronson 2002; Massalski et al. 2006). This is true for both high-angle grain boundaries (Sutton and Balluffi 1995; Schonfelder et al. 1997; Merkle 2002) and interphase boundaries. During dissolution or at static interfaces when nucleation at the interface is not required for motion, local fluctuations can occur due to thermal effects. In-situ HRTEM studies on grain boundaries have revealed such reversible fluctuations (Merkle 2002).

Recent in-situ heating HRTEM experiments performed on massive transformation interfaces in a TiAl alloy have shown evidence of both continuous and stepwise motion during growth at the atomic level, depending on the orientation relationship and interface plane (Howe et al. 2002). These two types of growth behavior were previously observed at much lower levels of resolution during cinematographic studies of the massive transformation in Cu–Ga alloys (Kittl and Massalski 1967) and during in-situ heating TEM experiments of massive interfaces in Ag–Al (Perepezko and Massalski 1975) and Cu–Zn alloys (Bäro and Gleiter 1974). One problem associated with all of the previous, lower-resolution in-situ studies is that the actual structures of the massive transformation interfaces were not known, but assumed to be incoherent. This assumption has been challenged repeatedly in the literature, particularly since many massive products display definite crystallographic orientation relationships with the parent phase (Aaronson 2002). The following results were obtained from frame-by-frame analysis of in-situ HRTEM data obtained on a massive transformation interface in a Ti–Al alloy growing by a continuous mechanism, i.e., by atom attachment continuously along the interface rather than at well-defined crystallographic facets or steps in the interface (Raffler and Howe 2006). The results reveal new phenomena associated with continuous growth at solid–solid interfaces and are compared with existing experimental and atomistic simulation results on grain and interphase boundaries at the same level of spatial resolution.

Figure 13.30 shows an HRTEM image of an α_2/γ_m incoherent interface obtained in a Ti–46.54 at.% Al alloy that was solutionized at 1360 °C for 1200 s and quenched with water or ice–brine to nucleate grains of the γ_m (massive) phase in the α (parent) phase, after the specimen had been heated for about 5400 s (1.5 h) at 575 °C in situ in the TEM (Howe et al. 2002). The interface is

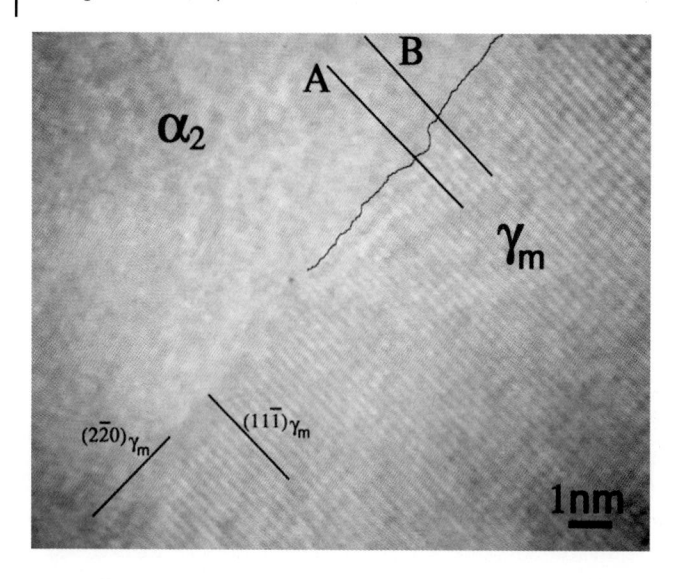

Fig. 13.30 HRTEM image of an α_2/γ_m incoherent interface obtained after the specimen had been heated for about 5400 s (1.5 h) at 575 °C in situ in the TEM. The $(2\bar{2}0)\gamma_m$ interface plane and the $(11\bar{1})\gamma_m$ planes in the γ_m phase are indicated. Lines A and B show the positions of two $(11\bar{1})\gamma_m$ planes that were used to obtain Fig. 13.34 and a typical trace of the α_2/γ_m interface is outlined in the top half of the image (from Raffler and Howe 2006).

edge-on and nearly parallel to a $(2\bar{2}0)\gamma_m$ plane, or roughly perpendicular to the $(11\bar{1})\gamma_m$ planes that are clearly visible and indicated in the γ_m phase in the image. The orientation relationship between the two phases across this interface was approximately: $[112]\gamma_m\|[122]\alpha_2$ with about 1.4° tilt between the two zone axes; and $(11\bar{1})\gamma_m\|(\bar{2}23)\alpha_2$, and the viewing direction is approximately $[112]\gamma_m\|[122]\alpha_2$ in Fig. 13.30. In the video recording, the interface appeared to advance into the α_2 phase by continuous overall motion, without evidence of the start–stop growth behavior which is often observed during motion of coherent and partly coherent interfaces between solid phases with low-index orientations and interface planes by steps or ledges, as shown above for θ phases. The interface also appeared to undulate slightly as it advanced, rather than remaining planar during growth. These features were examined in detail in the subsequent frame-by-frame analyses over a period of 50 s (or 1500 frames).

A typical trace of the α_2/γ_m interface that was analyzed in this study is indicated by a line in the top-center part of Fig. 13.30. Figure 13.31(a) shows the actual positions of 10 such traces obtained from each frame of the video recording during the first 0.3 s. In Fig. 13.31(b), the traces were separated by a distance of 0.52 nm, starting with the original frame on the right, to better reveal their characteristics. The positions of various advances and recessions in the traces of the interface are indicated by arrows in Fig. 13.31(b). Initial examination of the traces gives the

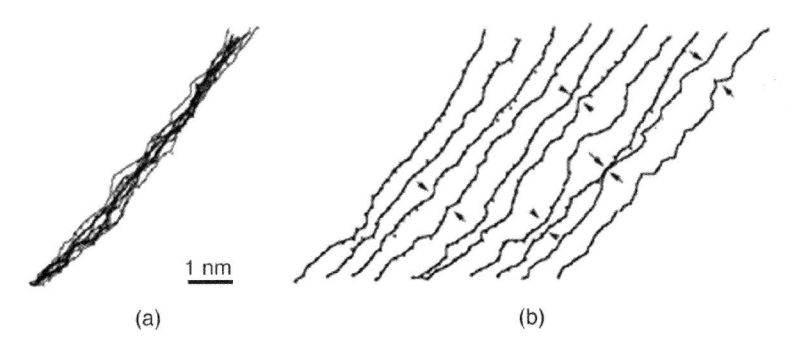

Fig. 13.31 (a) The positions of 10 traces obtained from each frame of the video recording during the first 0.3 s. (b) The same traces separated by a distance of 0.52 nm, starting with the original frame on the right, to better reveal their characteristics. The positions of various advances and recessions in the traces of the interface are indicated by arrows (from Raffler and Howe 2006).

impression that motion of the interface is highly irregular, although every trace shows at least one location where the interface advances into the α_2 phase. Further examination of the arrows shows that when the interface advances in one location in a frame, it subsequently recedes in the same location in the next frame. This pattern of advance and recession leads to an oscillatory behavior in the interface position as a function of time, as illustrated using only the $(11\bar{1})$ planes in the γ_m phase in Fig. 13.32(a). (Note that the regularity of the $(11\bar{1})$ planes shown in Fig. 13.32(a) is only for purposes of illustration and not meant to imply that the motion of these planes at the interface is so regular.) In addition to the oscillatory behavior caused by the advances and recessions, movement of the interface appeared to occur by the spreading of certain advances along the interface. This behavior is illustrated in Fig. 13.32(b). Both of these features at the interface appeared to involve the cooperative growth and dissolution among groups of individual $(11\bar{1})$ plane edges in the γ_m phase at the interface.

The distances between advancements in the traces of the interface were measured in each frame and found to display characteristic spacings of approximately 1.15 and 1.61 nm. This feature is illustrated in Fig. 13.33(a), where the distances between the advances are indicated by solid dots. Figure 13.33(b) shows the same traces with broken lines drawn tangentially to the advances and recessions along the interface traces and a darker, parallel solid line drawn halfway between these. The distances between the broken lines varied from 0.27 to 0.58 nm, with an average value of 0.41 nm. If the amplitudes and distances of the advances/ recessions at the interface are treated as a wave-like fluctuation, the fluctuation has an amplitude and fundamental wavelength of approximately 0.21 nm and 1.15 nm, respectively. It is interesting to note that the second characteristic wavelength associated with the fluctuation of 1.61 nm is approximately $\sqrt{2}$ times the fundamental wavelength.

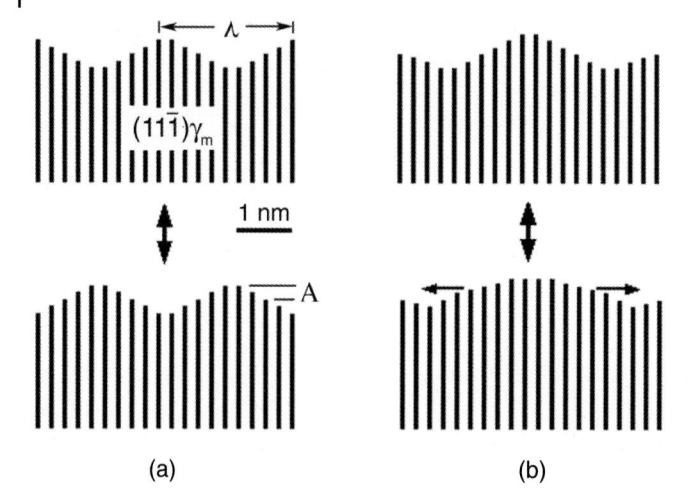

Fig. 13.32 Illustrations of: (a) the oscillatory behavior of the interface as a function of time, and (b) movement of the interface by the spreading of an advancement in the directions indicated by the horizontal arrows. The amplitude (A) and wavelength (λ) of the fluctuations are indicated in (a) (from Raffler and Howe 2006).

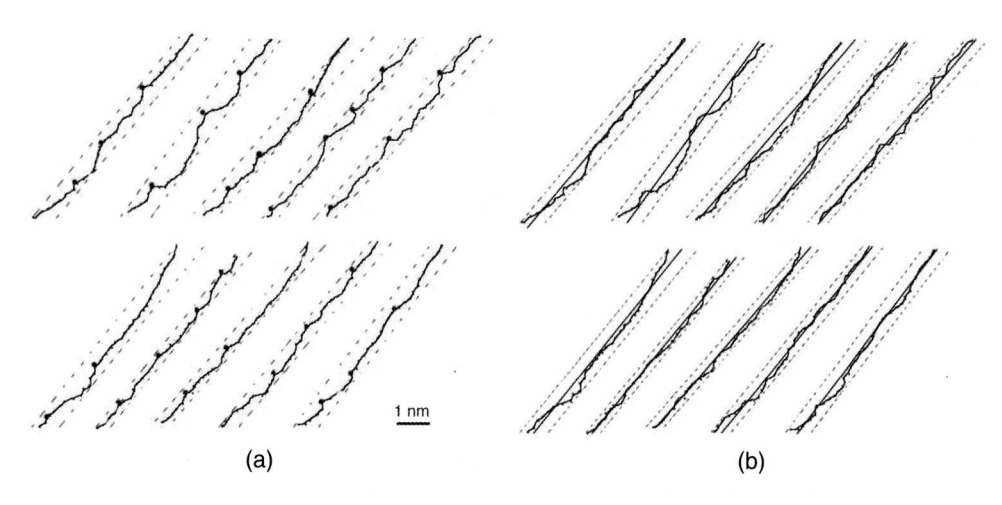

Fig. 13.33 (a) The same traces as in Fig. 13.20, with the distances between advances indicated by solid dots. These were used to determine the wavelength of the fluctuations. (b) The same traces with broken lines drawn tangentially to the advances and recessions along the interface traces and a darker, parallel solid line drawn halfway between these. These were used to determine the amplitude of the fluctuations (from Raffler and Howe 2006).

The interface dynamics was analyzed over a longer period of time by similarly plotting the interface traces from video frames obtained every 5 s, over the duration of the video sequence (50 s). The same interface characteristics were observed over this longer time interval as for the previous short one, indicating that the interface behavior is the same over all time scales within the time resolution of the experiment (i.e., 0.03 s). In the 50 s segment analyzed, the distances between the advances in the interface were found to be approximately 1.19 and 1.69 nm and the distances between the advances and recessions of the fluctuation varied from 0.32 to 0.53 nm. These values yield an average amplitude and fundamental wavelength for the interface fluctuation of approximately 0.21 nm and 1.19 nm, respectively, with the higher wavelength of 1.69 nm again being approximately $\sqrt{2}$ times the fundamental wavelength. The same analyses performed at intermediate time intervals yielded similar amplitudes and wavelengths, indicating that they are characteristic of the interface over all time scales of observation from 0.3 to 50 s.

The average velocity of the interface based on the initial and final positions of the interface traces over the 50 s of observation was 0.023 nm s^{-1}. In order to understand further the instantaneous velocity/behavior of the interface, the edges of individual $(11\bar{1})\gamma_m$ planes were followed frame-by-frame in two locations at the interface, indicated by lines parallel to the planes labeled A and B in Fig. 13.30. Figure 13.34(a,b) shows the resulting plots of these $(11\bar{1})\gamma_m$ plane edges every 30th frame over 1500 frames, or a time of 50 s. These plots clearly illustrate the somewhat irregular but definite oscillatory behavior of the interface, i.e., its wave-like fluctuations versus time. In addition, it is apparent that the average velocity of the interface is constant, as indicated by the straight lines that were fitted and superimposed on the plots in Fig. 13.34(a,b), although the instantaneous position of the interface oscillates about the average position with time at these locations. The equations for the lines are also given in the plots. This behavior was typical of all such $(11\bar{1})\gamma_m$ planes at the interface and reflects the behaviors illustrated in Fig. 13.31. It is important to note that although the average interface velocity was 0.023 nm s^{-1}, the instantaneous velocity at any given time could be more than an order of magnitude higher, as evidenced by the slopes of the oscillations in Fig. 13.34.

The results from the frame-by-frame analyses above indicate that this massive transformation interface, which moves forward continuously as a planar interface overall, rather than by an obvious ledge mechanism, displays quasiperiodic fluctuations along its length that can be characterized by a wave-like function with an amplitude of approximately 0.21 nm (roughly the spacing of the $(11\bar{1})\gamma_m$ planes) and a fundamental wavelength of approximately 1.15 nm. In addition, a second wavelength of 1.6 nm, or approximately $\sqrt{2}$ times the fundamental wavelength (1.15 nm × $\sqrt{2}$ = 1.6 nm) is also present at the interface. This oscillatory interface behavior, which accompanies the interface as it propagates forward with constant average velocity, is superimposed on the average position of the interface versus time. The oscillations are not regular, but they display a characteristic am-

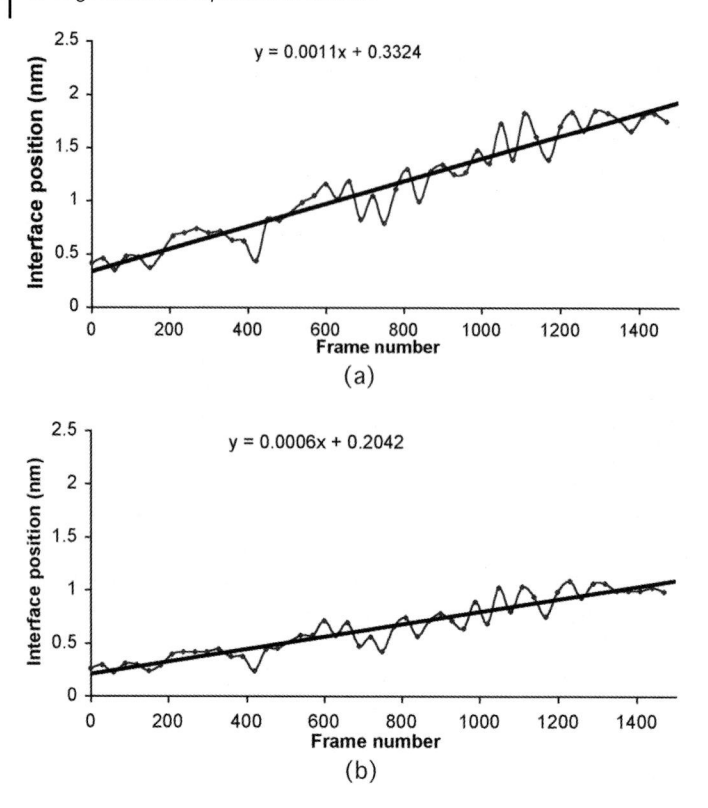

Fig. 13.34 Frame-by-frame plots of the $(11\bar{1})\gamma_m$ plane edge positions at locations A and B in Fig. 13.30, with data points indicated every 30th frame over the duration of 1500 frames (50 s). The equations for the straight lines fitted to the data points in the figure are also given in the plots (from Raffler and Howe 2006).

plitude and wavelength. There was no obvious connection between either the amplitude or the wavelength at the interface with any structural features in the α_2 phase, such as a characteristic ledge spacing or height, although this aspect deserves further examination. It would also be interesting to compare the behavior of this interface at other boundary orientations obtained in the in-situ HRTEM experiments and this work is currently in progress.

The creation of advances at the interface and spreading of some of these advances along the interface indicates that the interface must be overcoming some barrier to migration, rather than simply moving forward uniformly everywhere. As mentioned at the start of this section, the presence of local barriers to interface migration typically leads to oscillatory behavior at interfaces in diffusional transformations. The present interface is different from the previous Al–Cu–Mg–Ag diffusional interface in that it is a (relatively) planar, incoherent interface between two crystals with a high-index orientation relationship and does not involve long-range diffusion, as opposed to being a partly coherent interface between two crystals with a low-index orientation relationship and cusp-oriented interfaces that grow by a terrace–ledge–kink mechanism, involving long-range diffusion. In this regard, it is more like a general, high-angle grain boundary migrating under a driving force (Sutton and Balluffi 1995; Schonfelder et al. 1997). Unlike the case of highly faceted cusp-oriented interfaces, it appears that the lack of well-defined structure at this interface allows it to undergo regular wave-like spatial fluctuations in position with time, some of which grow to a critical size and then spread along the interface, moving it forward. The size of the critical fluctuations for this interface appears to be just greater than approximately 0.21 nm high and 1.69 nm wide, since fluctuations larger than these are not commonly observed. In this regard, its motion is similar to that of a planar interface growing by nucleation and spreading of two-dimensional nuclei (Cahn et al. 1964; Benson and Howe 1997; Howe 1997), except that the nuclei are not well-defined crystallographically and their location is a stochastic process.

The dynamic characteristics of the present interface are remarkably similar to those observed for the massive transformation in Cu–Ga alloys obtained at a much lower spatial resolution (i.e., at the micrometer scale), but at a higher time resolution (at 64 frames s^{-1}) and at much greater velocities (up to 1 mm s^{-1}) using cinematographic studies (Kittl and Massalski 1967). In fact, if one compares the interface traces in Figs. 13.31 and 13.33 with those in Figs. 8 and 9 in Kittl and Massalski (1967), the traces are essentially identical, indicating scaling of certain interface behaviors in these massive transformations over great variations in length, velocity, and time. The benefit of the Ti–Al case is that the much higher spatial resolution reveals the previously unknown atomic details of interfacial structure associated with the dynamic behavior. The oscillatory character of the interface and the constant velocity (i.e., the linear displacement versus time in Fig. 13.34) also compare reasonably well with results from a recent kinetic MC study of a massive transformation interface between fcc and bcc crystals (see Figs. 3 and 5 in Bos et al. (2004), for example), although in this study the low-index close-packed planes in the two crystals were parallel, so that the interface is cusp-oriented and two-dimensional bcc nuclei are clearly present at the interface at low driving forces. It is also interesting to note that some oscillations are present in molecular dynamics simulations of grain-boundary migration in planar, high-angle twist grain boundaries (see Fig. 5 in Schonfelder et al. (1997), for example).

13.2.3
Local Analysis by Atom Probe Tomography

Whereas the previous sections on microscopy focused on surface and structural characterization, we now address a technique dedicated to local chemical analysis. Scientists who are planning chemical micro- or nanoanalysis of materials have nowadays a choice among a vast number of different techniques. These are based either on inelastic interactions of electrons and ions with matter (e.g., energy-dispersive X-ray spectroscopy (EDXS), electron energy loss spectroscopy (EELS),

Rutherford backscattering (RBS)) or on direct mass spectrometry (e.g., secondary ion mass spectrometry (SIMS)). Atom probe tomography (APT) represents a branch of the latter group. Since this technique is particularly suited to the investigation of metals and alloys and it has the unique capability of producing 3D maps with single-atom sensitivity, it is worth choosing this method for a detailed description.

Atom probe tomography applies the principle of field ion microscopy (FIM) and represents the latest progress in this area. Thus, the method dates back to the pioneering work of E. W. Müller, an ingenious physicist, who invented the field ion microscope in 1951 (Müller 1951) after years of experiments with electron emission microscopy that turned out to be resolution-limited by the finite Fermi energy of the electrons. With field ion microscopy, he could already demonstrate atomic resolution images of tungsten surfaces in 1957 (Müller 1957). He and his coworkers were also the first to combine FIM with time-of-flight mass spectrometry in 1965, creating in this way a unique tool for quantitative chemical analysis in the nanometer range, which attracts particular attention today in the era of nanotechnology (Müller et al. 1968). Although being able to achieve image magnifications in the order of 10^6 and thereby of atomic resolution, an FIM is a surprisingly simple instrument if compared to the complex electron optics of modern electron microscopes, since not a single imaging lens is needed. Owing to its simple projective geometry the instrument does not suffer from all the tedious stability problems to which an electron microscopist is accustomed. However, only recent progress in detector technology, of which the important milestones were the introduction of microchannel plates, CCD cameras, fast charge-to-digital converters, and picosecond time measurement, made available single-ion detectors with sufficient spatial resolution and quite high detection frequencies. Using this modern equipment the early atom probe of Müller advanced to a modern instrument that allows the almost complete, three-dimensional reconstruction of the spatial arrangement of a few million of atoms contained in a volume of typically 10^5 nm^3.

Meanwhile various instrumental concepts of 3D atom probes have been designed and put into operation. The first instrument functioning well was the so-called position-sensitive atom probe (PoSAP) described in 1988 by Cerezo et al. (1988). An improved instrument with the capability to process multiple events is the tomographic atom probe (TAP) presented somewhat later (1993) by Blavette and coworkers (Deconihout et al. 1993). Since this instrument was delivered to locations throughout the world it has become popular and has lent its name to the general term of the method: "atom probe tomography." Nowadays technical development is in continuous progress and new instruments are introduced regularly. The most recent achievement is the so called local electrode atom probe (LEAP) (Kelly et al. 2004). This instrument moderates the serious restrictions in specimen geometry, namely that a needle of high aspect ratio is needed, by placing a micrometer-sized electrode in front of an array of microtips. Also, the total number of atoms analyzed per measurement, and thus the size of the reconstructed volume, increase by one to two orders of magnitude.

This section can provide only a rather short overview of the latest atom probe methods. First, the physical principles of the measurement tools are described and the algorithms for 3D reconstruction are introduced. Then, the accuracy of the method and its limitations are discussed on the basis of simulated measurements. Finally, in Section 3.2.3.7, a few selected case studies are presented to demonstrate the application of the method to the physics of alloys. Readers who are interested in further details are referred to recent textbooks (Miller et al. 1996; Miller 2000) on the method and also to basic literature on field ion microscopy (Tsong 1990).

13.2.3.1 The Functional Principle of Atom Probe Tomography

All FIM techniques make use of the fact that electrical fields are concentrated at tips of sharp curvature. With moderate voltages supplied, enormous field strengths in the range of some 10 V nm^{-1} are easily obtained at the apex of nanometer-sized tips, whereas fields of such a magnitude would never be obtained with macroscopic geometries. So, a typical field ion microscope consists of an ultrahigh vacuum chamber with a specimen stage holding the sample tip, a high-voltage supply, and a viewing screen with the capability to image single ion impacts; see Fig. 13.35. A positive potential is supplied to the metallic specimen while the entrance face of the screen is kept at ground. In order to reduce thermal energies, a cryostat is required to cool the tip to about 20 K. The field at the tip

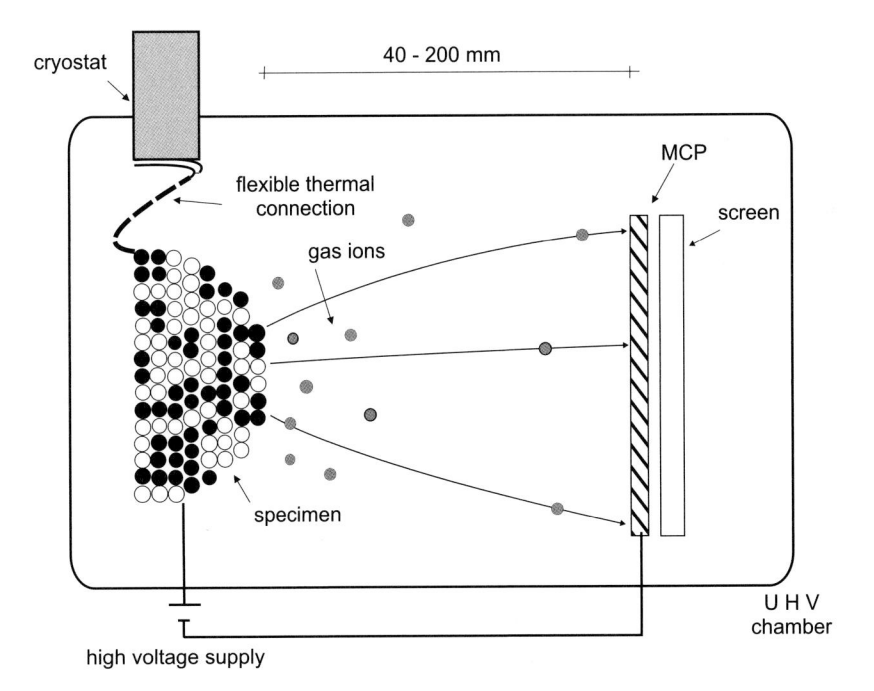

Fig. 13.35 Principle of a field ion microscope.

surface is controlled by the voltage supplied, according to Eq. (62), in which V and β denote the voltage supplied to the tip and a shape factor.

$$E_{surf} = \frac{V}{\beta R} \qquad (62)$$

For electrodes of ideal spherical symmetry this shape factor would be exactly equal to 1. But in reality the shaft part of the needle-shaped specimens leads to a field modification that is reflected by a β factor in the range 7–10 depending on the specimen's geometry. With increasing distance from the surface, the field decays logarithmically, which means that the dominant part of the field already drops in the first few millimeters.

In order to produce a field ion micrograph of the tip surface, an imaging gas, usually He, Ne, or a mixture of both, is introduced into the vacuum chamber. The gas atoms are polarized within the inhomogeneous field close to the apex and drawn toward the tip. The energetic situation for the shell electrons of a gas atom is sketched in Fig. 13.36. If a positive voltage is supplied to the tip, the electronic states of the gas atom become elevated with respect to the Fermi energy of the metallic sample. Provided the field strength is sufficient and the atom is at a suitable distance, a finite probability exists that an electron tunnels from the atomic state into the band structure of the metallic specimen. After having lost an elec-

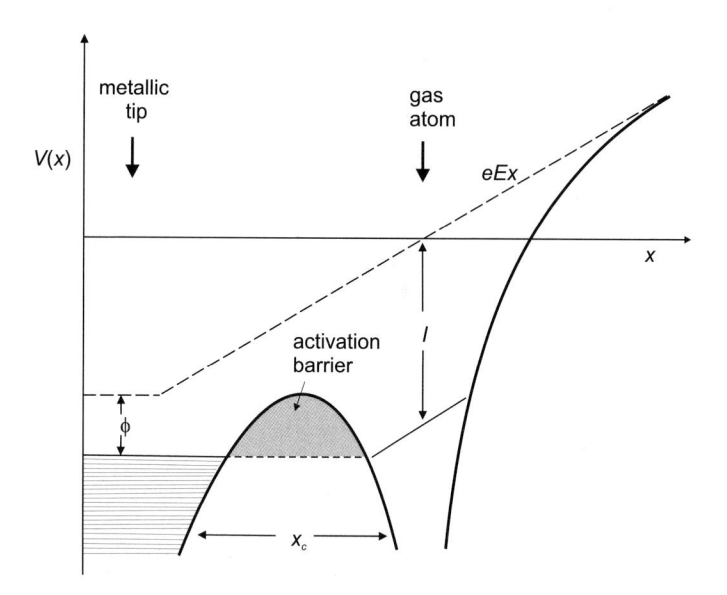

Fig. 13.36 Electronic states of the metallic sample and an image gas atom close to the tip. Atomic states are elevated by the field E with respect to the metallic states. ϕ and I denote the work function of the metal and the ionization energy of the gas atom, respectively.

tron, the positively charged gas atoms are accelerated toward the imaging screen. There, by means of a multichannel plate and a phosphorus anode, each ion impact produces a visible light flash via the principle of secondary electron multiplication.

The potential well for the electron transfer by tunneling is sensitively controlled by the local field strength. As a consequence, the ionization rate of the gas atoms is a function of (a) the tip voltage and (b) the surface topography. Because of the discrete atomic structure of the surface, the local field varies periodically along the surface with a wavelength in correlation with the interatomic distance. In particular, the edges of the atomic terraces are protruding features, and thus regions of elevated field strength and pronounced ionization rate.

The ion trajectory is determined by the shape of the electric field. Since the thermal energy of the ions is rather low compared to the energy gain within the field, the ions are practically starting at rest. As a consequence, there is a one-to-one correspondence between the location of ionization at the tip surface and the imaged position on the screen. Regions of high ionization rate at the specimen surface are represented by locations of high brightness on the screen. As a consequence, protruding edges of lattice planes, the atomic terraces, are imaged as bright concentric rings surrounding low-indexed poles in the case of crystalline lattice structures. This is illustrated by Fig. 13.37, which presents a comparison of a field ion micrograph and a corresponding ball model of the imaged structure.

To achieve a clear field ion micrograph the so-called best imaging field must be established at the tip surface. To a first approximation this field strength is characteristic of the imaging gas (e.g., 30 V nm^{-1} and 41 V nm^{-1} for Ne and He, respectively). If the field strength exceeds this imaging level, a further process is

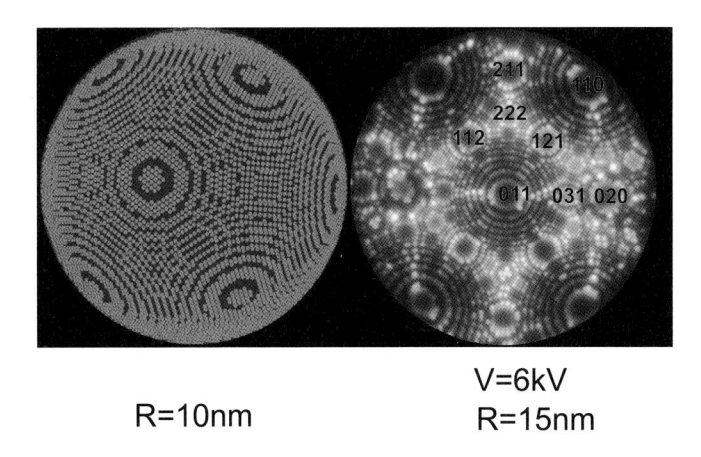

Fig. 13.37 Ball model of a spherical apex (left) in comparison to an experimental field ion micrograph. Protruding atoms are represented by bright dots. The image structure comprising concentric rings is a natural consequence of the crystalline packaging and the tip-shaped geometry. (By courtesy of V. Vovk, University of Münster.)

initiated, so-called field evaporation. Reaching a critical evaporation field characteristic of the material under investigation, the specimen atoms themselves are ionized, field desorbed and accelerated toward the screen, similarly to the previously discussed imaging gas atoms. This field evaporation mechanism represents the basis of the local chemical analysis by an atom probe. The evaporation is triggered by high-voltage pulses of a few nanoseconds' duration. With a suitable control of the pulse amplitude the evaporation rate can be properly controlled (e.g., to about 0.01 atoms per pulse), so that the probability of multiple events becomes quite low. Thus, atoms may be counted and evaluated one by one.

In principle, such an analytical instrument, the 3D atom probe, consists of the same components as the FIM shown in Fig. 13.35. Only the screen has to be replaced by a position-sensitive detector system of single-ion sensitivity, and the high-voltage supply must allow the required voltage pulsing. Using the voltage pulse as the start signal for an accurate time measurement and the detector signal for stopping, the time-of-flight (ToF) of the ions and thus their mass and chemical identity can be measured individually. The flight path between tip and detector amounts to about 0.5 m, leading to typical flight times in the microsecond range. During a measurement, up to several million detector events, the respective flight times, and the hitting positions are recorded. From these raw data the original spatial arrangement of the different atomic species is reconstructed subsequently to the measurement by efficient numerical methods which will be discussed in more detail below.

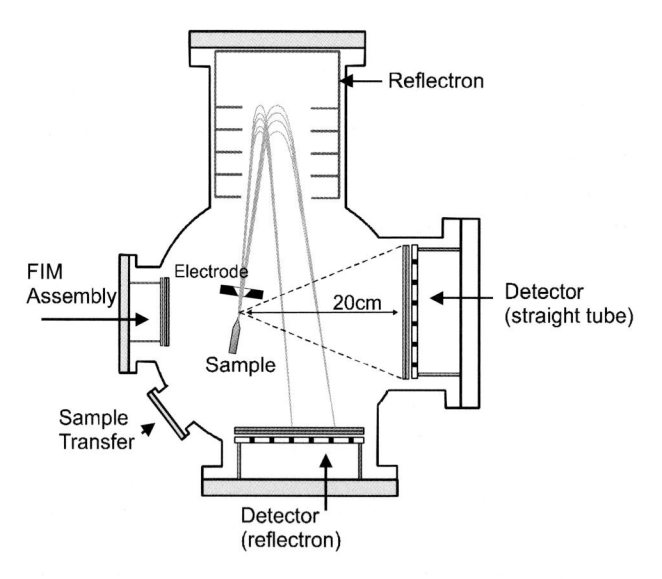

Fig. 13.38 Design of a modern tomographic atom probe. The dedicated chamber layout makes it possible to switch between different detector geometries: the straight flight tube allowing a short flight path to optimize the open angle of the instrument; and the reflectron arrangement yielding an improved mass resolution.

Figure 13.38 is a schematic drawing of a modern conventional 3D atom probe, for illustration. In contrast to the concepts introduced above, two modifications should be noticed.

- Instead of a straight flight tube, a reflectron geometry is used with an ion mirror which leads to parabola-shaped ion trajectories. This geometry compensates for fluctuations in the kinetic energy of individual ions. A faster ion will penetrate deeper into the mirror field and so will have a longer flight path. If length and voltage of the reflectron are adjusted correctly, ions of identical mass but leaving the tip with slightly different velocities will hit the detector after identical flight times. In this way, the mass resolution is significantly improved.
- The evaporation pulse is supplied as a negative pulse to an extraction electrode in front of the tip, which allows shorter and better defined pulse shapes.

13.2.3.2 Two-Dimensional Single-Ion Detector Systems

The rapid development of atom probe tomography in recent years was only made possible by the remarkable evolution of spatially resolving detector systems with single-ion sensitivity. In the last two decades several detector concepts have been proposed and put into operation. Systems that are currently in use will be discussed briefly.

All available detector concepts are based on a stack of two to three "multichannel plates" (MCPs). Such an MCP is an amplifier working according to the principle of secondary electron multiplication. It is made of thousands of small glass tubes, about 25 µm in diameter and 1 mm in length, which are packed in parallel alignment to form a plate of about 1 mm thickness (see Fig. 13.39). The front and back sides of the plate are coated by thin metallic films serving as electrodes to supply a voltage in the region of 1 kV. Usually the tubes are tilted by about 10° to the plate normal to achieve a higher detection probability. An ion hitting the inner wall of such a glass tube will induce a few secondary electrons, which are accelerated further by the supplied field. On their way toward the back of the plate, they hit the glass wall several times and produce further secondary electrons. This cascade process will finally yield a cloud of about 10^4 electrons. If a stack of two or three MCPs is used instead of a single one, the individual amplification factors will be multiplied, so that a single ion hitting the front with sufficient energy will produce finally a cloud of about 10^8 elementary charges, which is sufficient for further electronic evaluation. An MCP is a fast device. With optimized electronic circuitry the rise time of single ion pulses is in the region of 100 ps. The spatial resolution of the MCP is controlled by the dimension of the glass tubes.

For mere imaging, as for conventional FIM, it is sufficient to place a phosphor coated screen behind the exit face of the MCP, so that each electron cloud produces a short light flash. To obtain quantitative positional information, either

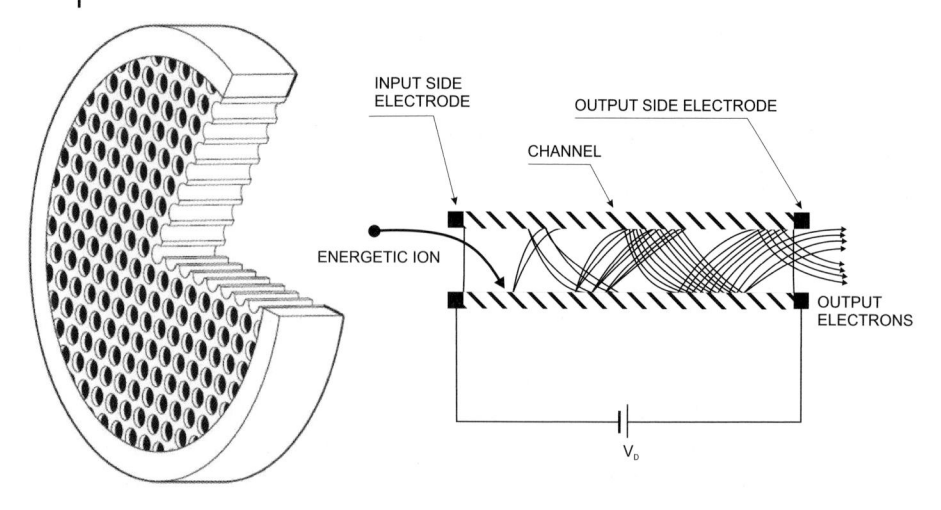

Fig. 13.39 Funcitonal principle of an multichannel plate (MCP), an arrangement of thousands of secondary electron multipliers working in parallel. Each of them is made of a tiny glass tube, 25 µm in diameter.

the optical information of the screen can be recorded by a CCD camera or the electron clouds need to be measured directly by suitable electronics. For the latter case, a multiple anode array is placed behind the MCP and connected to sensitive preamplifiers and fast converters to transform analogous charge or time information into digital data. Various concepts may be distinguished by their different layouts of the anode array and the complexity of the electronics. Historically, the first system to function well was the so called position-sensitive atom probe (PoSAP) build around a "wedge and strip anode" (Cerezo et al. 1988). The name of this anode type is self-explanatory regarding the layout sketched in Fig. 13.40(a). The geometry is designed in such a way that the relative fractions of

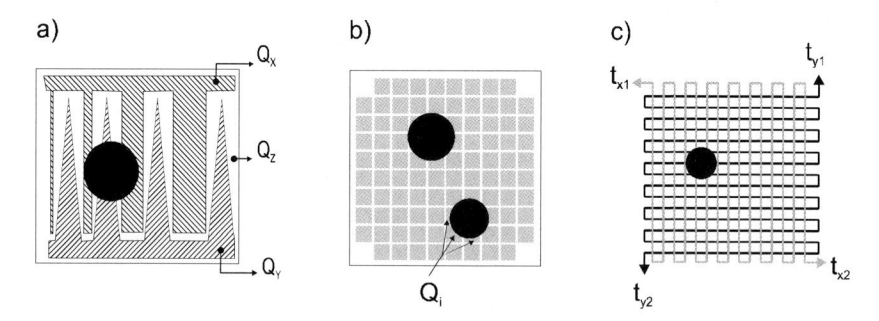

Fig. 13.40 Anode layouts for the readout of the positional information of a channel plate: a) wedge and strip detector; b) TAP detector; c) delay line detector. (Instead of the double-wire Lecher lines, only a single wire is shown to illustrate the principle.)

the total charge measured on the three electrodes Q_X, Q_Y, and Q_Z vary with the position of the electron cloud. For the layout shown, the position may be calculated in a straightforward manner by applying Eqs. (63). (Here and in the following, uppercase letters represent detector coordinates while positions on the specimen are described by lowercase letters.)

$$X \propto \frac{Q_X}{Q_X + Q_Y + Q_Z}; \quad Y \propto \frac{Q_Y}{Q_Y + Q_Z}. \tag{63}$$

Since only three independent anodes are used, the required electronics is reasonably simple. However, the layout has the important drawback that the total area of each electrode and consequently the respective capacities are quite large so that the drain of charges after the impact takes a considerable time. If two ions hit the detector within this time gap, they cannot be separated. To reduce the risk of such double events, the operator is forced to perform the evaporation of the tip very slowly.

To circumvent this problem, the so-called tomographic atom probe (TAP) detector was designed shortly afterward as a square-anode array of 96 independent electrodes (see Fig. 13.40b). Choosing the correct distance between the MCP and anode, the size of the electron cloud is adjusted so that an individual electron cloud hits at least three or four electrodes and its central position may be determined by weighting the charge fractions on the respective electrodes. Since many other unused electrodes are available, besides those that are hit by a first event, most multiple events can be separated. Due to this parallel design the possible analysis rate of the TAP detector becomes much higher than with the earlier PoSAP, but of course this is at the cost of much more elaborate electronic equipment. Nevertheless, there is still a certain probability that the charge clouds of neighboring events may overlap and cannot be resolved.

To improve the spatial resolution, systems have been built which combine anode arrays of low resolution with a gated CCD camera that localizes individual light flashes on the screen, with high accuracy. However, since the gating of a CCD camera is rather slow, the flight time and in particular the sequence of impacts hitting the detector after a pulse cannot be determined from the optical information of the camera alone. The CCD data have to be complemented by the measurement of hitting times on an auxiliary anode array and correlated suitably. To realize this concept, in the so-called optical PoSAP (Cerezo et al. 1994) the information of the screen is divided by a beam splitter. In this way, part of the image intensity can be used to determine the impact flashes on an auxiliary anode array using the scintillation principle; another part is evaluated by the CCD camera as sketched in Fig. 13.41. From the figure it becomes obvious that rather complex equipment is required. As an alternative, optically transparent electrodes with phosphor coatings may be used (Deconihout et al. 1998). Nevertheless, in both cases the data rate is limited by the rather slow scanning rates of the CCD camera so that the optical principle is not competitive – at least at present.

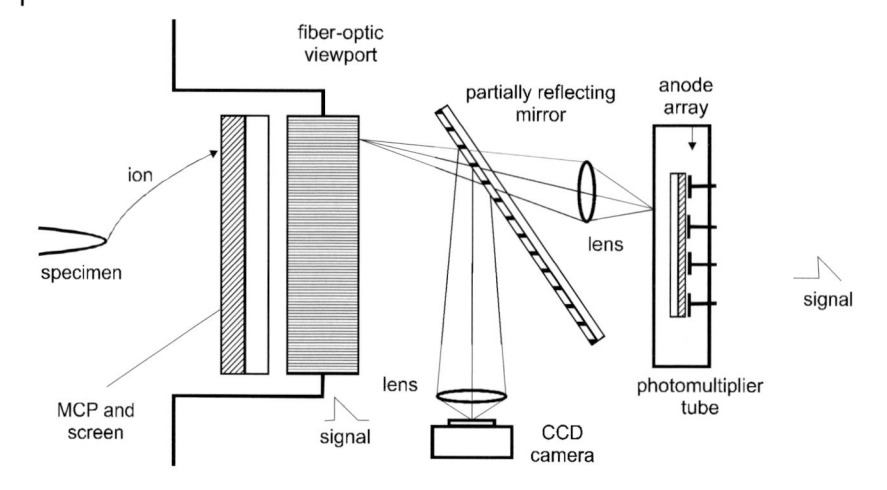

Fig. 13.41 Principle of the optical PoSAP detector. The combination of a CCD camera and a resistive anode array allows high positional accuracy and at the same time multievent detection.

Most recent instruments are equipped with a detector applying the delay line principle which was proposed first in 1987 (Keller et al. 1987). Instead of flat electrode areas, two independent double-wire spirals are used which are wound along the X and Y axes of the detector used to determine the X and Y position, as indicated schematically in Fig. 13.40(c). Each double wire represents a Lecher line, on which the pulse signal propagates with the velocity of light. Both ends of the Lecher line are connected to a fast multichannel time-to-digital converter (TDC) with subnanosecond resolution. If the ion hits at the center of the detector, the pulse signals will propagate symmetrically to both ends of the Lecher line and will reach the TDC at exactly the same time, whereas for an asymmetric impact position the two time signals will differ considerably. In Fig. 13.42 the sum and the difference of time signals are presented as they have been collected for many independent impacts on a circular detector of 120 mm diameter and spiral spacing of the anode of 1 mm. The sum of the two time signals represents an instrumental constant and corresponds approximately to the propagating time from one end of the line to the other. A time interval of 150 ns can be measured well. The time difference between the two signals is proportional to the position according to Eq. (64), where the calibration parameter v_p denotes the propagation velocity along the spiral axis, which amounts to about 0.4 mm ns^{-1}.

$$X = v_p(t_X^{(l)} - t_X^{(r)}) \tag{64}$$

An analogous equation holds for the Y direction. Since a fast time-to-digital converter is already required for ToF mass spectrometry, delay line detectors are a very economical solution. Furthermore, they allow very high data rates, so that evaporation pulses may be applied with frequencies of up to 200 kHz.

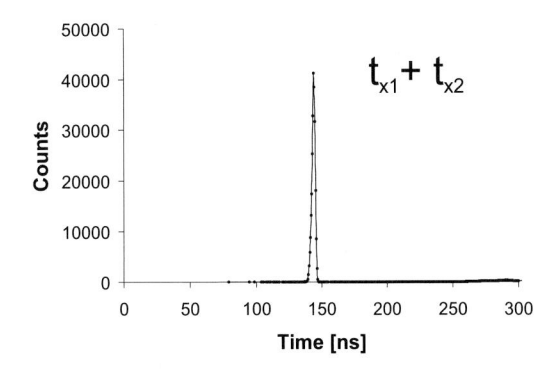

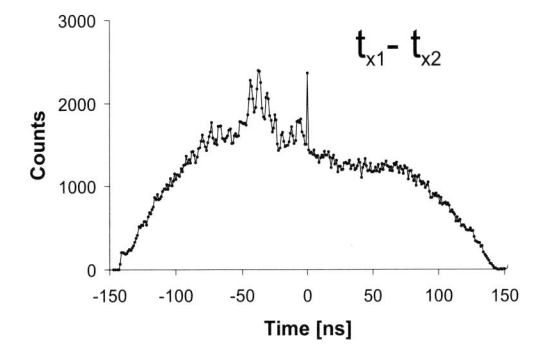

Fig. 13.42 Evaluation of time signals of the delay line. The sum (top) of the two signals is a constant which may be used to relate time signals correctly to individual events. The difference is a direct measure for the position. Due to a circular channel plate, the frequency distribution resembles a sinusoidal. (By courtesy of P. Stender, University of Münster.)

13.2.3.3 Ion Trajectories and Image Magnification

In order to understand the properties of field ion micrographs and the quality of the volume reconstruction, the ion trajectories must be discussed in more detail. An idealized specimen will have axial symmetry, so that the potential and the field may be considered in a two-dimensional space with r and z, the coordinates perpendicular and parallel to the rotational axis of symmetry, respectively. For convenience, we split the expression for the electrical potential Φ into the absolute tip voltage V and a spatial distribution function φ so that Eq. (65) holds, and the equations of motion can be written in accordance to classical mechanics as Eq. (66), where n and m denote the charge state and the mass of the ion, respectively.

$$\Phi(r,z) = V \cdot \varphi(r,z) \tag{65}$$

$$\frac{d^2 r}{dt^2} = -\frac{neV}{m}\frac{\partial \varphi}{\partial r}; \quad \frac{d^2 z}{dt^2} = -\frac{neV}{m}\frac{\partial \varphi}{\partial z} \tag{66}$$

Without any further calculation it is seen that the acceleration in both coordinate directions depends on mass, charge state, and voltage in the same way. Thus, having the ion initially at rest, the trajectory becomes independent of all these variables. All the different species will follow the same trajectory; only the required flight time is characteristic of the given charge state and mass. This has the important consequence that the fundamental imaging relationship between tip and detector is universal for all species and that, besides minor modifications due to a slight difference in initial position, imaging gas and specimen atoms will follow exactly the same trajectory. Therefore, the position-sensitive detector system may be calibrated by means of field ion micrographs.

A quite reasonable geometric representation of a well-prepared tip may be given by a truncated cone closed by a hemispherical cap, as indicated in Fig. 13.43. In addition, field lines and model trajectories are shown. Since the ions start at rest, their motion follows the field lines initially. This means that they leave the spherical apex in a radial direction. Later, they deviate from the curvature of the field lines because of the forces of inertia and the trajectory becomes almost straight. While the initial position of the ion at the spherical cap is described with the polar angle ϑ, the imaged position at the detector may be characterized by a smaller angle ϑ' (see Fig. 13.43). In order to calculate the initial position on the tip's surface from the image, only the function between the two angles must be known. This function varies in detail from tip to tip as the individual

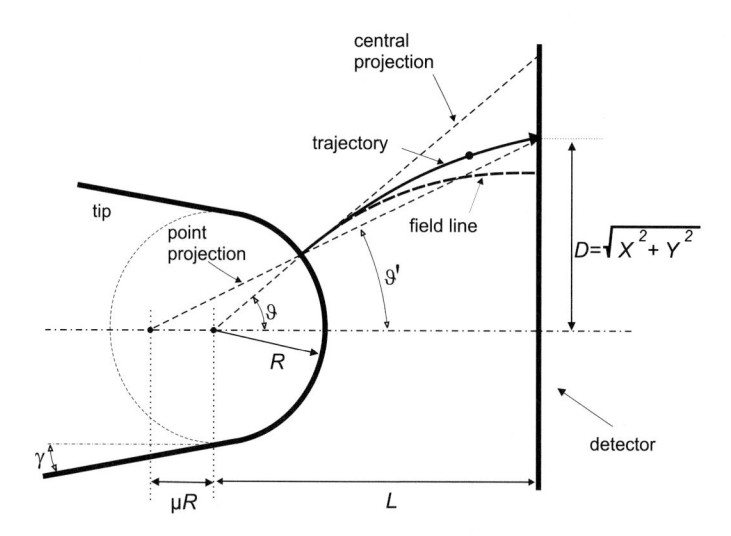

Fig. 13.43 Electrical field and trajectory of an evaporated atom. The impact position and initial location at the tip surface are related by a point projection. The center of projection is shifted relative to the center of the spherical cap by μR. As an alternative the ratio between the polar angle ϑ and the projection angle ϑ' may be used for evaluation.

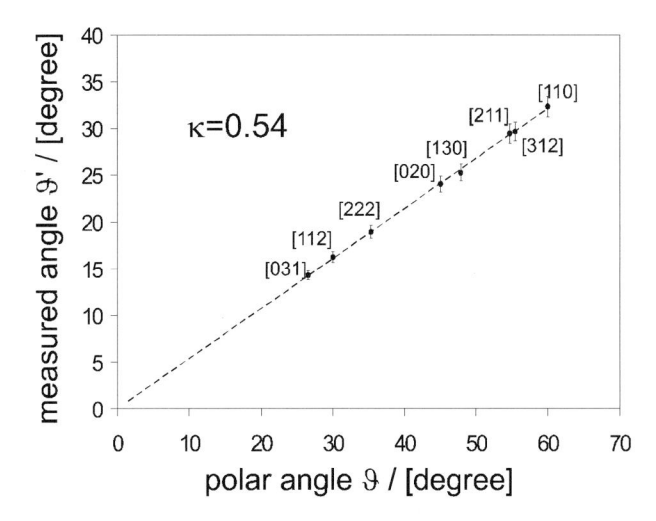

Fig. 13.44 Relationship between measured angle ϑ' and polar angle ϑ as determined for electropolished tungsten tips. The polar angle has been determined from crystallography. Tip axis aligned parallel to [011]. (By courtesy of P. Stender, University of Münster.)

shape does. However, from a practical point of view, it turns out that a simple proportionality holds, which is conveniently described by an imaging compression factor κ, defined by Eq. (67).

$$\kappa := \vartheta'/\vartheta \tag{67}$$

The compression factor can be calibrated by means of field ion micrographs of single crystalline specimens, in which low-indexed lattice directions are identified from symmetry considerations. See, for example, the indexed poles on the micrograph of Fig. 13.37. The crystallographic angle between different poles and thus the polar angle ϑ is known from crystallography, while the detection angle ϑ' is determined from the position of the pole on the FIM micrograph and the flight distance L between tip and detector. Typical data determined for an electropolished and field-developed tungsten tip are shown in Fig. 13.44. Obviously, the linear relationship of Eq. (67) is well fulfilled. For this individual specimen a compression factor of $\kappa = 0.54$ is determined.

It is easy to quantify the magnification of the analytical microscope on the basis of the geometric model. The polar distance on the image, $D = L \cdot \sin \vartheta' \approx L \cdot \vartheta'$, has to be compared to the distance at the hemispherical cap, $d \approx R \cdot \vartheta$, where R denotes the current radius of the apex. Thus, the magnification is given by Eq. (68).

$$M := \frac{D}{d} = \frac{L \cdot \kappa}{R} \tag{68}$$

Recalling that a typical tip radius amounts to about 30 nm and the distance between detector and tip may reach 0.5 m, a magnification of 10^7 is easily obtained.

In Eq. (68) an interesting detail is noteworthy. The magnification of the microscope itself depends on the tip radius of the specimen. In other words, the specimen represents the essential lens of the microscope. Therefore, atom probe tomography will only function in a reliable manner if specimens are prepared carefully and can be well reproduced in shape. Furthermore, the tip radius increases during the measurement, as the specimen field-evaporates continuously. As a consequence, the magnification decreases during the measurement. In order to reconstruct the spatial arrangement of the atoms after the measurement, the development of the radius must be recorded or estimated in a suitable way. The introduced geometric model of a tip neglects the roughness of the surface on the atomic scale. In reality, edges of the atomic terraces and a faceting of the spherical surface – low-indexed surface orientations are emphasized in size due to anisotropy of the evaporation probability – will lead to slight modifications of the trajectories, which will be discussed later.

13.2.3.4 Tomographic Reconstruction

The atomic reconstruction of the investigated volume can be performed with surprisingly simple algorithms, which are outlined in this section. To reduce the mathematical effort, we assume that the specimen axis is aligned perpendicular to the detector plane, which is a suitable approximation for the majority of experimental work. The outlined scheme is based on the work of Bas et al. (1995). The general case, taking into account a relative tilt between tip and detector, can be treated in an analogous way. The interested reader may find appropriate formulae in Miller (2000) and Al-Kassab et al. (2003).

The evaluation of data is conveniently subdivided into three steps: 1) the specific mass m/n is calculated from the ToF, 2) the lateral position at the tip surface is calculated from the impact position at the detector, and 3) the depth scale along the symmetry axis of the specimen is determined from the data sequence. In the following, we express the impact position at the detector by Cartesian coordinates X and Y (see also Fig. 13.43 for further geometric parameters).

Since the field lines are concentrated at the tip apex, the ions have already taken up the major fraction of their kinetic energy during the very first millimeter of their trajectory. Later the motion is almost straight and uniform, so that from conservation of energy the specific mass is calculated to sufficient approximation by Eq. (69).

$$\frac{m}{n} = \frac{2t_{\text{ToF}}^2 e(V_{\text{tip}} + V_{\text{pulse}})}{L^2 + X^2 + Y^2} \tag{69}$$

Using the image compression factor introduced earlier and the appropriate geometric relation for the detection angle [Eq. (70)],

$$\tan \vartheta' = \sqrt{X^2 + Y^2}/L, \tag{70}$$

the Cartesian coordinates of the position at the tip's surface are determined by Eqs. (71).

$$x = \frac{X}{D} R \sin \vartheta = \frac{X}{D} R \sin(\vartheta'/\kappa)$$

$$y = \frac{Y}{D} R \sin(\vartheta'/\kappa) \tag{71}$$

$$\tilde{z} = R(1 - \cos(\vartheta'/\kappa))$$

In the last equation the axial coordinate $\tilde{z}$ has been marked by a tilde to express that at this step the axial position is only a preliminary one. It is given relative to the position of the tip front z_0, but as this reference point shifts during the measurement, we have to correct the depth position in the final evaluation step: With each evaporated atom the specimen is eroded by one atomic volume. Thus, the number of atoms already detected represents a natural depth scale. To establish this scale, the actual image magnification, which relates the sensitive area of the detector to the investigated area at the apex, needs to be taken into account. Expressing all this in a differential equation, we obtain Eq. (72), where Ω and N denote the average volume per atom and the number of detected atoms, respectively. The factor p takes into account the fact that the detector has a limited probability for detection ($p \approx 0.5$), and the magnification M is calculated by means of Eq. (68).

$$dz_0 = \frac{\Omega}{p A_{measured}} dN = \frac{\Omega \cdot M^2}{p A_{detector}} dN \tag{72}$$

Applying Eq. (72), the total shift of the tip front relative to its initial position at the beginning of the measurement is found by integration and the final z coordinate results from summing the two contributions: $z = z_0 + \tilde{z}$.

In order to evaluate Eqs. (71) or (72), the instantaneous tip radius R must be known. If the evaporation properties of the investigated material are reasonably homogeneous, this radius can be concluded from the total voltage (tip plus electrode voltage) that is required to obtain a given evaporation rate. The evaporation field strength at this rate is known as a material parameter so, by inversion of Eq. (62) we can get Eq. (73) to determine the actual tip radius.

$$R = \frac{U_{tot}}{\beta \cdot E_{evap}} \tag{73}$$

However, this method is only feasible if the evaporation field strength E_{evap} stays constant during the measurement. In heterogeneous specimens, e.g., thin-film multilayers, this prerequisite is usually not fulfilled. In this case, the evolution of the radius must be estimated under the assumption of a constant cone angle of the tip. Having measured the initial radius by transmission electron micros-

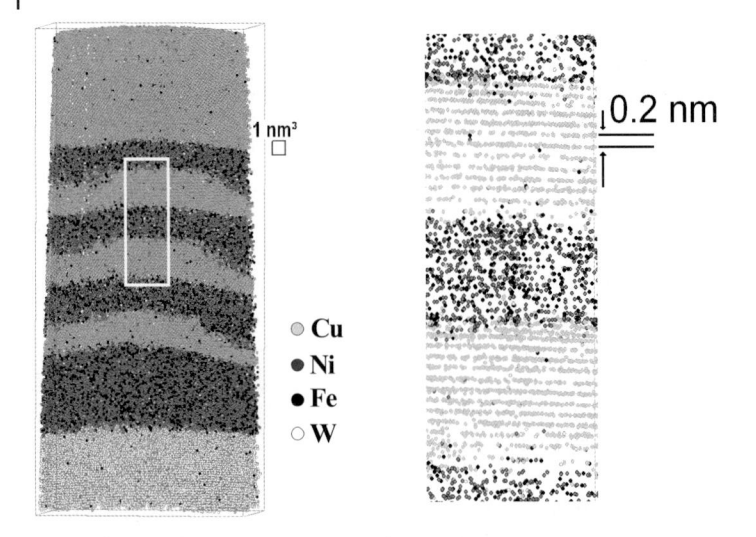

Fig. 13.45 Reconstruction of the atomic arrangement of a Cu/Py (Py = $Ni_{79}Fe_{21}$, Permalloy) multilayer deposited onto a tungsten substrate. The atomic positions within the marked box are plotted in higher magnification on the right. (After Ene et al. 2005).

copy or estimated it from the tip voltage when the measurement starts, the increase of radius can be calculated based on simple geometry (see Fig. 13.43) with Eq. (74).

$$dR = \frac{\sin \gamma}{1 - \sin \gamma} dz \tag{74}$$

A typical reconstruction of experimental data, calculated by the method outlined, is shown in Fig. 13.45. In this case the volume was taken from a Cu/NiFe multilayer specimen. In most reported measurements the analyzed region corresponds to a rectangular prism of typically $15 \times 15 \times 100$ nm^3. The position of each detected atom is marked by a gray-scale-coded dot.

After the spatial distribution of the atoms has been reconstructed, various averages, composition profiles, 2D compositional maps, and isoconcentration surfaces may be derived by sorting and counting the different species in suitable subvolumes. As shown in the detail of Fig. 13.45 (right-hand side), the reconstruction often reveals low-index lattice planes that are orientated almost perpendicular to the tip axis. If the required parameters – either the evaporation field strength and field compression factor or the initial radius and shaft angle – are chosen correctly, the spacing between the resolved planes indeed corresponds to the spacing of the lattice structure under consideration. Fourier methods are being used in order to detect such periodicities inside the reconstructed volumes.

13.2.3.5 Accuracy of the Reconstruction

As already discussed, the fundamental imaging law relies on a good description of the shape of the specimen, which needs to be produced by continuous field evaporation prior to the measurement. This process is also called the "field development" of the specimen. As soon as a steady state of tip development is reached, the tip geometry represents a surface of constant evaporation rate. For a homogeneous material, the tip is then bounded by an isosurface of the field strength. If this process of development is performed too fast or insufficiently, a variety of artifacts may be produced. Particularly dangerous is the partial fracture of a specimen before or during the measurement; such fracture usually produces surface topologies that are unsuitable for a reliable analysis. Furthermore, the depth scale will be erroneously calibrated due to the intermediate loss of material.

Even if the experimentalist obeys all the rules of good experimental practice, the positioning of the atoms within the reconstruction is never perfect, at least as long as the evaluation scheme outlined in Section 13.2.3.4 is used. It is based on two important assumptions: (a) the tip apex may be represented by a perfect sphere; (b) the critical field strength required to evaporate the atoms is homogeneous along the surface. The first assumption is only true as long as the atomic-scale roughness of the surface can be neglected. The second assumption is no more valid when alloys or intermetallic phases are investigated that comprise several atomic species of different evaporation probabilities. Figure 13.46 represents the situation when only a few atoms of a lattice plane are left, still sticking to the next plane. In this case, the electrical field that controls the trajectory of the atom next in the evaporation line will definitely deviate from the idealized field of the spherical surface due to the local disturbance of the remaining protruding atoms. Thus, to calculate the trajectories exactly, the structure which should be determined from just these trajectories needs to be known in advance. In other words, we are faced with an implicit conundrum: This is one of the problems to

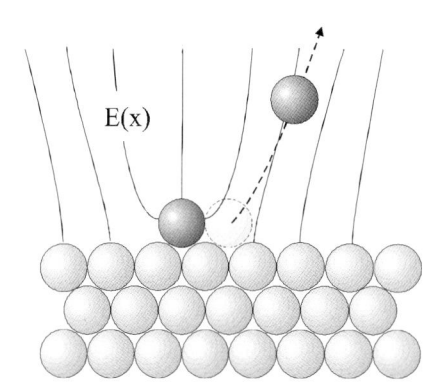

Fig. 13.46 Schematic field distribution on a low-indexed lattice plane just before it is completely evaporated. The last-but-one atom is accelerated by a field which is significantly deformed by the last atom.

be solved in the future. Meanwhile, for a sound evaluation of atom probe data, the only promising strategy is to simulate the evaporation sequence of hypothetical specimen structures and to compare the simulated reconstructions with those of real experiments. This procedure is quite analogous to the usual practice in high-resolution electron microscopy, where experimental images are compared to those simulated for hypothetical structures until a good match is found.

For that task, Vurpillot et al. (2000, 2001) derived a simulation scheme that allowed evaluation of the spatial accuracy of the tomography and the influence of heterogeneous evaporation properties on theoretical grounds. As the simplest geometrical model, which still reflects the microscopic features on the atomic scale, they suggest constructing the apex from a simple cubic arrangement of Wigner–Seitz cells. In a first step, the electrical field between the model tip and the detector is calculated by solving Poisson's equation numerically for the electrical potential in vacuum by means of a finite element method. The electrical field strength at the positions of the surface atoms is determined, and the position of highest field strength selected. In a second step, the atom, that means the Wigner–Seitz cell at this specific position, is removed from the apex model and the field is recalculated for the new configuration. In a final step, the trajectory of the selected ion is calculated between its initial position at the apex and the detector, based on classical mechanics and considering the acceleration of the ion within the electrical field. In the case of alloys or heterogeneous systems, the different evaporation probabilities of the atomic species must be taken into account. This means that, before the position of highest field is selected in the first step, the electrical field is scaled artificially by a factor varying from atom to atom in order to reflect the respective evaporation probability.

By repeating this scheme recursively to predict the impact positions of several thousand atoms, an atom probe measurement can be simulated quite realistically. General features of experimental data are well reproduced, although an artificially short flight distance between tip and detector and rather small specimens of about 20 nm radius had to be used to limit the computational effort. In Fig. 13.47 the impact positions of several thousands of atoms on the detector are shown, as calculated from the simulated evaporation of a few lattice planes. In contrast to the naïve expectation, the spatial density of the events on the detector is by no means homogeneous. Instead, lines of significant redistribution of the atoms are seen, which are related to low-index zones of the crystal structure. Obviously the redistribution is related to the special situation sketched in Fig. 13.46. The last few atoms sticking on a flat, low-index surface are affected by severe field distortions, and therefore their trajectory is disturbed significantly in comparison to a simple point projection.

These subtle redistributions of the atoms limit the spatial resolution of the atom probe method and indicate to the experimentalist that specimens should be tilted during the measurement in such a way that low-indexed poles are avoided inside the measured volume. Deviations of the atom trajectories by the local surface topology are the main limiting factor for the instrument's resolution. This is made particularly clear by the positional shift of atom positions in an

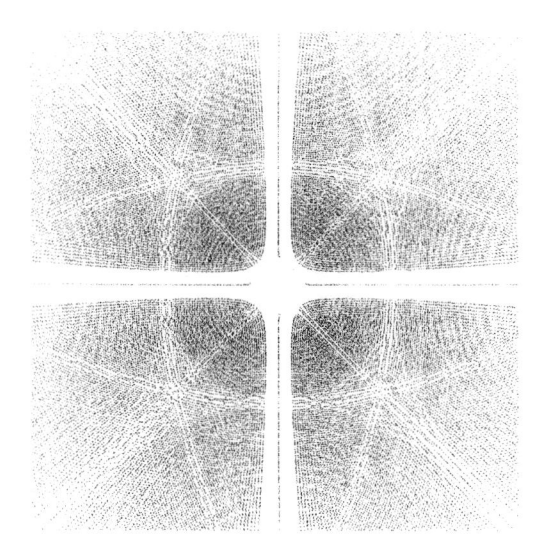

Fig. 13.47 Simulated impact positions on the detector area, during evaporation of a few lattice planes of a simple cubic model alloy. Tip axis aligned along [001]. (After Vurpillot et al. 2000.)

alloy compared to that of a pure specimen, as represented in Fig. 13.48. Because of the different evaporation probabilities of the two species, the local fields for the alloy are distorted differently than for a pure specimen. Thus, the reconstructed positions of the atoms are different for both model tips. However, as can be seen from Fig. 13.48, most positions agree to within about one lattice constant. Only a

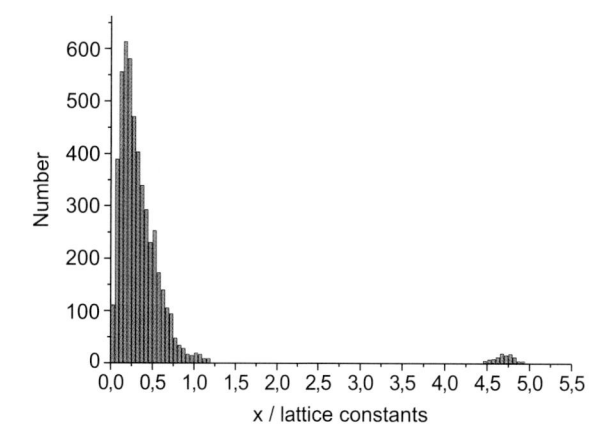

Fig. 13.48 Shift of reconstructed atom positions of an AB model alloy with respect to their ideal position in a homogeneous specimen (x in units of lattice constant). (After Vurpillot et al. 2000.)

minor fraction of atoms originating from zone line positions are shifted by much larger amounts, up to five lattice constants. In this way, the simulation indicates the lateral accuracy of the tomography to be slightly better than one lattice constant, as long as poles and zone lines are avoided.

13.2.3.6 Specimen Preparation

Since atom probe tomography requires a dedicated sample geometry, possible preparation methods should be discussed briefly. Sharp, needle-shaped specimens are used for any field ion microscopy technique. Traditionally, these needles have been produced by electropolishing of thin metallic wires under in-situ control by means of optical microscopy. But at present thin films and other complex nanostructures, for which no conventional wire is usually available, are becoming more and more important. Thus, the art of specimen preparation is under rapid development. For many recent studies on thin-film materials, layer systems have been deposited onto tungsten tips, which serve as a substrate. To achieve an optimum shape, the freshly prepared tungsten substrates are field-developed before the deposition. Since considerable stress is induced by the electrical field during the measurement, many investigations are hindered by insufficient mechanical stability of the interface between the substrate tip and coating. Thus, this interface needs special care. Sometimes an interlayer, often chromium, is used as an adhesion aid; ion-beam cleaning of the substrates immediately before deposition has also proven to be advantageous (Schleiwies and Schmitz 2002). In Fig. 13.49(a), a typical thin-film specimen produced in this way, an Al/Cu/Al trilayer deposited on tungsten, is presented. In this geometry, the main analysis direction is aligned normal to the interfaces, so that these specimens are especially suited to investigation of reactive diffusion by local depth profiles.

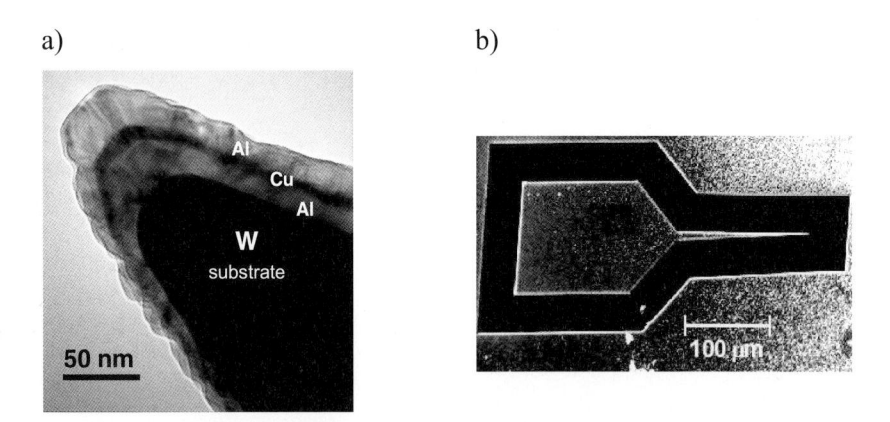

Fig. 13.49 (a) Example of a Al/Cu/Al tri-layer on a tungsten substrate tip, deposited by ion beam sputtering technique (By courtesy of C. Ene, Göttingen). (b) Scanning electron micrograph of a layer specimen prepared by electron beam lithography (by courtesy of J. Schleiwies, Göttingen).

However, when interpreting the analytical results, one has to keep in mind that the films are deposited onto a curved surface. Usually the curvature induces a rather small grain size, so that the microstructure is not directly comparable to that of thin films deposited onto conventional planar wafer substrates (Lang and Schmitz 2003).

If this variation in microstructure cannot be tolerated, for example because the properties of technical devices are to be determined, tips have to be cut by lithography (Hono et al. 1993) or focused ion-beam techniques. A planar layer system is first coated with a suitable photoresist, exposed to electron beam lithography, and developed chemically to obtain a suitable etching mask. A typical sample is shown in Fig. 13.49(b). The tip is attached to a "handle" about 100 µm in size. The wedge-shaped needles pointing to the right-hand side taper to a maximum thickness of 100 nm at the apex. After etching by means of an ion beam, the tips are removed from the substrate and glued to a supporting wire. In this specimen geometry the interesting interfaces are aligned parallel to the main direction of analysis; thus the method is well suited to investigation of interfacial roughness or pinholes in multilayers of small periodicity.

As an alternative approach, Larson (2001) described deposition of multilayers on the planar top surface of rectangular Si posts having a typical cross-section 10×10 µm^2 square. Since this size is already much larger than the film thicknesses typically investigated, the growth conditions on top of the posts should correspond to those on larger planar substrates. To allow the atom probe measurement, the posts with their coated top surfaces need to be further sharpened in a final preparation step. Focused ion beams have been used successfully in this case. Since energetic Ga beams of 15–30 keV are applied, the ion beam introduces considerable irradiation damage, which leads for example to ballistic mixing (cf. Chapter 8) close to the surface of the needle. Therefore one has to be careful to avoid artifacts. Usually this means that the atom probe measurement should include sufficiently large-volume regions located deep below the damaged surface region.

13.2.3.7 Examples of Studies by Atom Probe Tomography

With the following examples, the application of atom probe tomography to essential problems of alloy physics will be documented. The topics selected here cannot be a complete survey over all the activities in the field, and the choice is certainly biased by the individual interests of the authors. Nevertheless, we try to include a fairly broad range of phenomena and to highlight the remarkable strength of the method in terms of a real three-dimensional analysis with outstanding spatial resolution. The historical development should also be reflected somehow.

13.2.3.7.1 Decomposition in Supersaturated Alloys

In order to understand the early stages of decomposition, two fundamental mechanisms are distinguished from a theoretical point of view. In the limit of weak supersaturation, nuclei with a solute concentration close to thermodynamic equilibrium need to be formed by statistical fluctuations in order to initialize the decomposition. With increasing solute content, the activation barrier to nucleation decreases, until it vanishes at a critical composition, called the spinodal. Within the spinodal composition regime, decomposition starts spontaneously. The mathematical model of this process dates back to as early as 1958 (Cahn and Hilliard 1958). First, the fundamental evolution equation was solved in a linear approximation, which makes it possible to represent the composition field by a superposition of compositional waves. Waves within a certain range of wavelengths will increase in amplitude by uphill diffusion, which is a consequence of the thermodynamic driving forces of the alloy system. The wavelength of fastest growth will dominate the microstructure, so that periodic structures develop. (For spinodal decomposition, see also Chapter 7.)

Many experimental studies have been focused on the kinetics of spinodal decomposition. Especially, for systems of long-range interaction such as polymer blends, the Cahn–Hilliard theory was well confirmed by the experimental data (Binder 2001). But in the case of binary alloys, it has been a long-term issue to prove the predicted continuous increase in compositional amplitudes by direct local analysis. If initial heterogeneities are too small to be resolved by the microscopic analysis, an intermediate average composition is determined. During coarsening, an increasing fraction of heterogeneities will overcome the limit of resolution. Thus, there is a real probability that an experimentally observed growth of concentration amplitudes is just a side effect of limited resolution of the instrument and coarsening of the microstructure. Interestingly, one had to wait for the development of atom probe tomography to make possible a sufficiently detailed analysis of morphology and local composition for a direct comparison with the theory of spinodal decomposition to be made.

A complete and very detailed study in this field has been presented by Hyde, Miller, and coworkers on supersaturated FeCr alloys (Miller et al. 1995). Although this study was done with the early PoSAP instrument and therefore suffered from some experimental limitations, it is nevertheless worth reporting here, since this work also documents one of the very early direct comparisons of atom probe data sets to Monte Carlo (MC) simulations. Such a comparison is particularly advantageous, because both methods work with data sets of discrete atoms, and their data represent roughly the same number, of atoms say about 10^6. Meanwhile, this strategy has become common practice for the interpretation of results by atom probe tomography. In the study discussed here, the comparison to simulated atomic structures was also used to estimate the influence of experimental limitations, such as finite resolution and low data rate of the PoSAP, which leads to unbalanced detection probabilities for both species (Cr 30%, Fe 50%). For the MC simulation, a Metropolitan algorithm based on pairwise interaction and the swapping of randomly selected pairs of nearest neighbors was used. Besides the simulations, the experimental results were also interpreted by numerical solutions of the nonlinear Cahn–Hilliard–Cook theory of spinodal decomposition.

Different FeCr alloys were prepared, solution-treated, and isothermally annealed at a temperature of 773 K. The following discussion is restricted to the alloy with a composition close to the middle of the miscibility gap (Fe–45 at.% Cr),

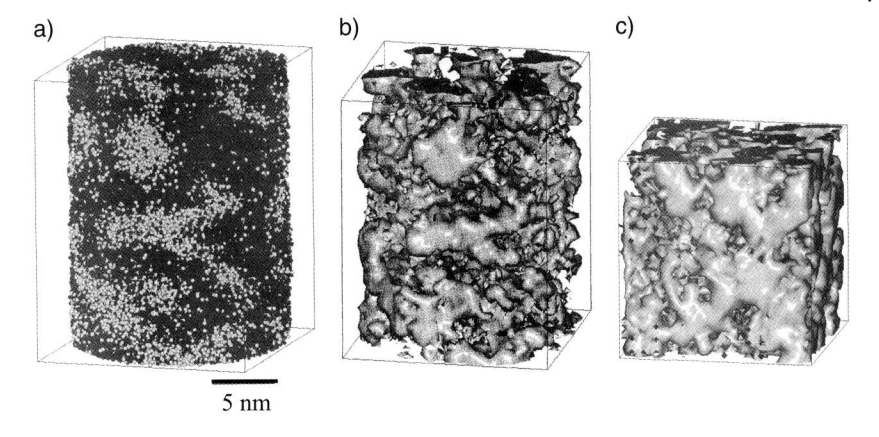

Fig. 13.50 Atom probe tomography of Fe–45 at.% Cr after 500 h of annealing at 773 K (a, b) in comparison with a corresponding microstructure simulated by a Monte Carlo algorithm based on nearest-neighbor swapping (c). For differences between (a) and (b), see the text. (After Miller et al. 1995.)

which is distinguished by the highest thermodynamic driving force. In Fig. 13.50(a), a tomographic reconstruction of the atomic arrangement is shown as determined from a specimen annealed for 500 h. The decomposition into Fe- and Cr-rich regions is obvious. Isoconcentration surfaces are very helpful in working out the interconnectivity of the 3D morphology, which is typical of a process of decomposition into almost balanced volume fractions of the two phases. To determine local compositions, a regular array of grid points is spanned throughout the investigated volume.

Around each grid point, a cube-shaped subvolume is defined, and the composition attributed to this point is calculated by counting all the different atomic species within the respective subvolume. Having determined a 3D composition field in this way, 1D profiles, 2D composition maps, and also isoconcentration surfaces are easily determined by data reduction. Since the statistical accuracy of a concentration value scales with the reciprocal square root of the number N of evaluated atoms [Eq. (75)],

$$\Delta c = \sqrt{\frac{c(1-c)}{N}} \tag{75}$$

when choosing the subvolume size one needs to compromise between low statistical scattering and high spatial resolution. Usually, subvolumes have a dimension of one to two nanometers, while the grid spacing amounts to only a few ångstroms. Thus, the composition field represents a moving average. Isoconcentration surfaces calculated in this way from the experimental data are presented in Fig. 13.50(b), and those determined from comparative MC data in Fig. 13.50(c).

In both cases identical evaluation methods have been used. In a qualitative sense, both representations indicate comparable microstructures that are characterized by interconnected Cr-rich domains, the size of which varies over a considerable range of length scales.

In order to check the validity of the established theories, Hyde et al. (1995) performed a very detailed quantitative evaluation of parameters that are suitable to characterize the microstructure. The size and interspacing of the Cr-rich domains were determined by the autocorrelation technique. For the small analysis volumes of conventional 1D atom probes, the statistics of autocorrelation functions have been notoriously bad. This situation changed significantly with the introduction of the 3D versions of the instrument. Statistical scattering becomes significantly reduced due to the increased number of atoms involved. Now radial autocorrelation functions R_k [Eq. (71)] may be determined by evaluating the atoms in spherical shells around arbitrarily chosen central atoms of the measured data set.

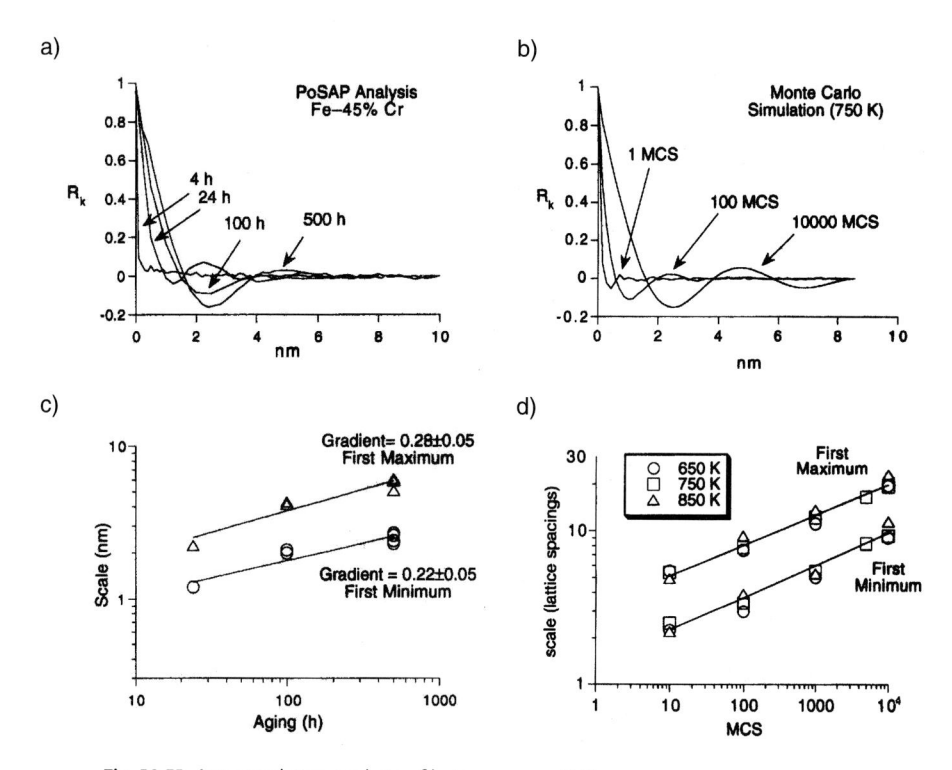

Fig. 13.51 Autocorrelation analysis of heterogeneous FeCr microstructures as measured by atom probe tomography (a,c) and simulated with a Monte Carlo technique. The correlation functions indicate the average domain size (first minimum) and the typical spacing between Cr-rich regions (first maximum) after various annealing times. The respective kinetics are presented in the lower row (c,d). (After Hyde et al. 1995.)

$$R_k = \frac{1}{\sigma^2} \sum_{r}^{r_{max}-k} (C_{(r)} - \bar{C})(C_{(r+k)} - \bar{C}) \tag{76}$$

As shown in Fig. 13.51 these radial autocorrelation functions develop clearly distinguished first minima and maxima that are interpreted as the average size and spacing of Cr-rich domains, respectively. The growth of these parameters with annealing time demonstrates the coarsening of the structure. Double logarithmic plots (Fig. 13.51c,d) prove a power law behavior in the experiment and the MC simulation as well. With an exponent of 0.22 ± 0.05, the experimental kinetics agrees very well with the MC simulation, for which the exponent amounts to 0.21. The same autocorrelation analysis was done for microstructures calculated by the nonlinear continuum theory. Surprisingly, a growth exponent of 0.33 is found, which resembles conventional coarsening kinetics. This is in remarkable disagreement with the experimental finding.

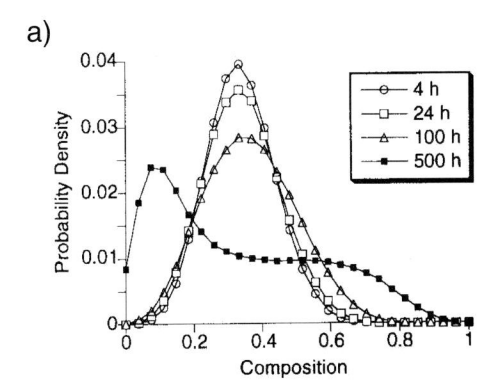

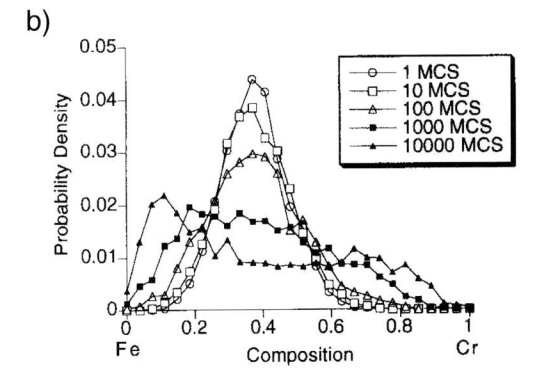

Fig. 13.52 Frequency of composition distribution as determined from (a) experimental analysis of Fe–45 at.% Cr alloy during annealing at 773 K; (b) simulated structures. Pronounced decomposition is reflected by the appearance of two distinct maxima. (After Hyde et al. 1995.)

As has already been pointed out, a critical test for the spinodal mechanism would be the observation of growing composition amplitudes. To analyze these amplitudes, the frequency distribution of the compositions found for the subvolumes mentioned have been plotted in Fig. 13.52 for the experimental and MC-simulated data sets. Beginning with the homogeneous alloy represented by a unique probability maximum at the average composition, the decomposition leads to double-peaked frequency distributions in later stages; see, for example,

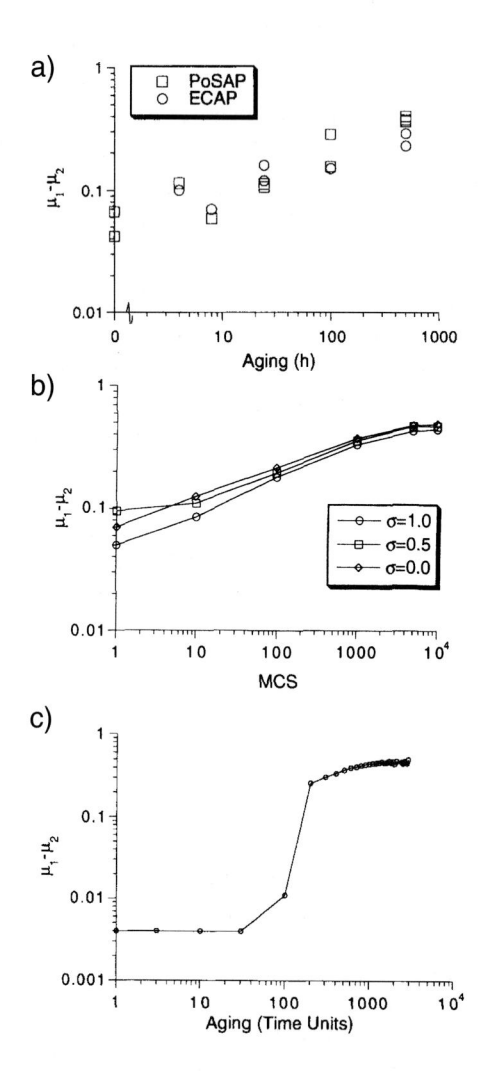

Fig. 13.53 Growth of concentration amplitude during spinodal decomposition as determined (a) from the atom probe analysis; (b) by Monte Carlo simulation; (c) by the nonlinear Cahn–Hilliard continuum model. (After Hyde et al. 1995.)

the very clear curve after 500 h of annealing in the experimental data (Fig. 13.52a). Following an earlier suggestion by Langer, Hyde approximates these frequency distributions by two overlapping Gaussian peaks that represent the two phases of the heterogeneous microstructure (Hyde et al. 1995). The centers of these peaks mark two compositions μ_1 and μ_2, of which the difference quantifies the amplitude of compositional segregation. Despite considerable experimental fluctuations, the compositional amplitude of the atom probe data sets increase significantly during annealing, as does the amplitude of the MC simulated data similarly (Fig. 13.53a,b). By contrast, the conventional Cahn–Hilliard theory yields a very different prediction. In this latter case, the concentration amplitude is expected to increase in a step-like manner after a well-defined incubation period (see Fig. 13.53c). The deficits of the widely accepted Cahn–Hilliard continuum model in reproducing the experimentally determined behavior are quite remarkable. The authors of the study suggest that the obvious discrepancy may be due to interfacial transport, which is completely neglected in the Cahn–Hilliard theory.

13.2.3.7.2 Nucleation of the First Product Phase

Owing to the technological trend toward miniaturization, the very early stages of reactive diffusion at thin-film interfaces shifted into the focus of research. Frequently it has been argued that the thermodynamic driving force to form the first reaction product is usually so high that the critical thickness of nucleation ranges down to the size of a lattice constant, or even smaller. Consequently, nucleation should not be a rate-controlling step at all. However, the first evidence that this is not true came from calorimetric studies of reactive diffusion in metallic thin films. Although only one product phase forms, double-peaked heat releases were observed during the reactions in several binary systems (Bergmann et al. 2001; Roy and Sen 1992; Michaelsen et al. 1997). This experimental finding was interpreted by Coffey et al. (1989) as a two-stage mechanism. In the first stage, nuclei form at the initial interface and quickly grow in lateral directions, whereas the second heat release is attributed to parabolic thickness growth by volume diffusion. The process of nucleation is still quite unclear. Several mechanisms have been proposed to explain the apparent reduction in driving force in the presence of a sharp composition gradient (Desré and Yavari 1990; Gusak 1990; Hodaj and Desré 1996). As a common feature, they predict a critical composition gradient that must be established by interdiffusion before nucleation of the first intermetallic compound becomes possible. However an experimental verification for this interpretation has been missing.

A recent nanoanalytical study (Vovk et al. 2004) has been aimed to shed some light on the early nucleation stages. The reaction couple Al/Co was chosen as a particularly clear model system, since a double-peaked heat release has been reported in the literature, and furthermore the first product phase Al_9Co_2 is an exact stoichiometric compound with an existance range smaller than 1 at%. The latter fact is expected to increase the influence of a composition gradient. For the experiments, bilayers of Co and Al, each 20–30 nm in thickness, were deposited

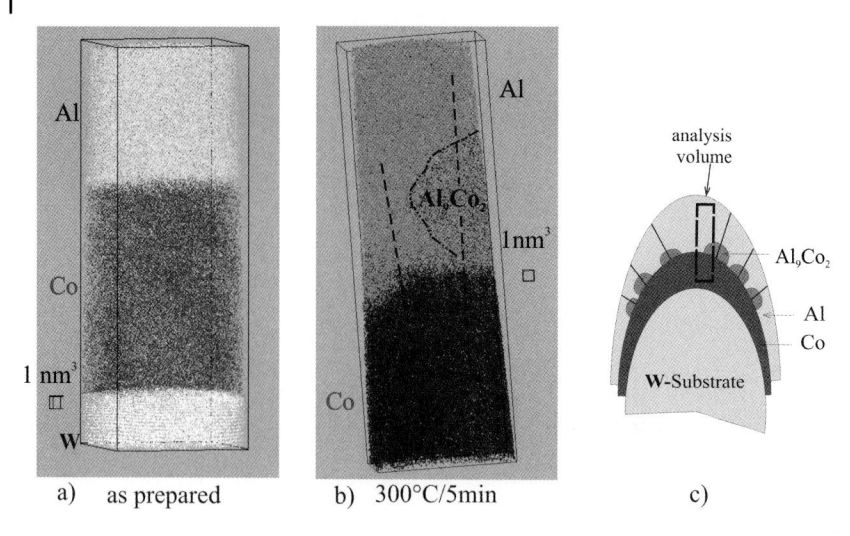

Fig. 13.54 (a), (b) Atomic reconstructions of the Al/Co interface (a) in the as-prepared state and (b) after annealing at 300 °C for 5 min. Positions of individual atoms are marked by gray-coded dots. (c) Sketch of specimen geometry. (After Vovk et al. 2004.)

on tungsten substrate tips. In Fig. 13.54 two examples of 3D reconstructions of the Co/Al interface are presented. Although the layers are deposited on curved surfaces, the initial interface appears practically flat, since the radius of curvature is still significantly larger than the width of the rectangular prism analyzed. This flat interface is preserved for short annealings, so that the earliest stages may be characterized by one-dimensional composition profiles determined normal to the interface as shown in Fig. 13.55. Due to the outstanding resolution, minor chemical modifications at the interface become noticeable. After a heat treatment of 5 min at 300 °C, the zone of interfacial mixing at the interface has broadened from about 1 nm in the as-prepared state to 3.5 nm, indicating a significant mixing of the components. However, the composition profile in this annealing state is well fitted by an error function. It therefore clearly results that only interdiffusion is observed, instead of the formation of a new intermetallic reaction product, which would appear as a plateau of almost constant composition.

Nucleation of a new phase is first observed in some of the measurements after 5 min annealing at 300 °C. In these cases, globular particles are detected at the interface toward the Al side (see Fig. 13.54b). The fact that these particles appear only in some of the measurements after identical annealing conditions emphasizes the statistical nature of a nucleation process. Furthermore, from the volume reconstructions it becomes clear that nucleation takes place at heterogeneous sites at the interface as sketched in Fig. 13.54(c). A composition profile across the newly formed phase (right-hand broken line of Fig. 13.54b) identifies the product as Al_9Co_2 (see Fig. 13.55), while a profile determined across the remaining interface (left-hand broken line in Fig. 13.54b) confirms the interdiffusion at a

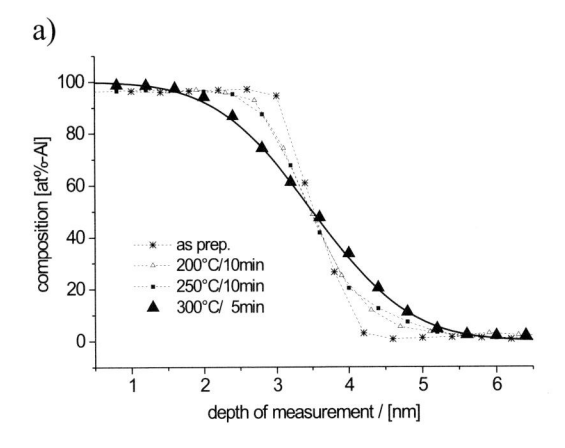

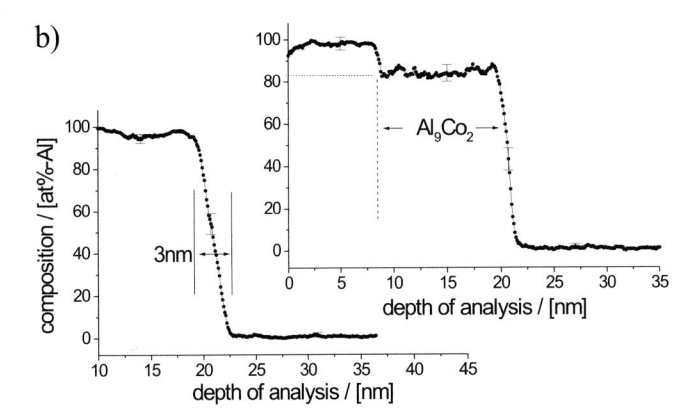

Fig. 13.55 (a) Composition profiles determined perpendicular to the initial Al/Co interface after several annealing stages. (b) Composition profiles determined along the left- and right-hand broken lines in Fig. 13.54(b), respectively. (After Vovk et al. 2004.)

depth of 3–4 nm as described previously. In this way, atom probe tomography yields a clear demonstration that there is significant interdiffusion before nucleation of the product. According to the experimental analysis, the critical diffusion depth before nucleation starts amounts to 3.5 nm in the case of Al/Co at 300 °C.

If the theoretical nucleation thickness of the intermetallic product is estimated by the usual balance between volume driving force and interfacial energy, a value of $d = 2\sigma/g_v = 0.2$ nm is predicted (Pasichnyy et al. 2005). In view of this value it is very surprising that the product Al_9Co_2 is formed only after the intermixed zone has already reached a minimum thickness of 3.5 nm. However, by a quantitative argument (Pasichnyy et al. 2005), it could be demonstrated that this behavior is very consistent with a polymorphic nucleation mechanism. This mechanism assumes that the nucleus of the new phase is produced by transforming

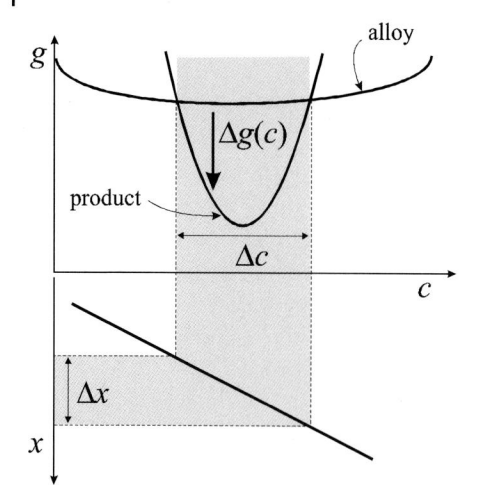

Fig. 13.56 Schematic Gibbs free energy curves (top) and idealized linear composition profile (bottom). Polymorphic transformation leads to reduction of energy only within the gray shaded composition range, which is established only in a thin layer within the interdiffusion zone. x is the position of the nucleus within the diffusion zone.

the lattice structure into that of the product phase without modifying the local composition. Since any nucleus must have a minimum size in order to overcome the nucleation barrier, the ideal stoichiometric composition can only be achieved in the center of the nucleus, while in the boundary region of the nucleus the composition must deviate due to the existing concentration gradient. This situation is sketched in Fig. 13.56, giving evidence that the composition range may be directly related to a depth range. In other words, nucleation is only probable within a thin-layer fraction of the total diffusion zone and the thickness of this layer shrinks with increasing concentration gradient. In consequence, high concentration gradients will prevent nucleation.

For the quantitative calculation, the free energy required to form a nucleus was calculated by integrating the difference between the two Gibbs free energies (see Fig. 13.56) over the volume of the nucleus. For clarity, the nucleus may be approximated by a cube, so that the rather simple formula in Eq. (77) is obtained, in which $2R$, x_c, and σ are the width of the cube, the position of the center of the nucleus within the diffusion zone, and the specific interfacial energy, respectively.

$$\Delta G(R, x_c) = 24R^2 \cdot \sigma + 4R^2 \int_{x_c-R}^{x_c+R} \Delta g(c(x))\, dx \tag{77}$$

Numerical results of Eq. (77) are presented in Fig. 13.57 for three different widths of the diffusion zone: 3.0, 3.5, and 4.0 nm. In these plots, the size of the nucleus is expressed by N, the number of atoms contained. The concentration c_x repre-

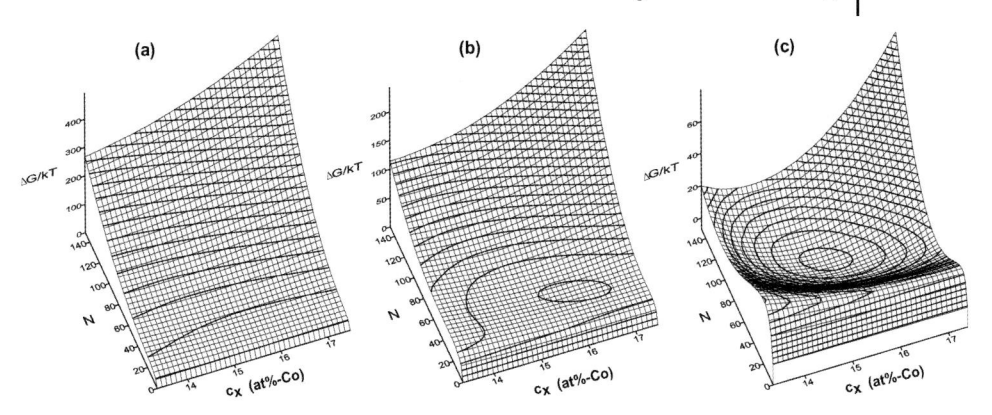

Fig. 13.57 Surface $\Delta G(N, c_x)$ for the polymorphic transformation of a cubic volume into the Al_9Co_2 phase inside diffusion fields of width: (a) 3.0 nm, (b) 3.5 nm, (c) 4.0 nm. Thermodynamic functions were considered for a temperature of 573 K. (After Pasichnyy et al. 2005.)

sents the composition at the center of nucleus. For a diffusion width of 3.0 nm, the Gibbs free energy to form a nucleus still increases monotonically with its size. Thus, nucleation is forbidden. At a width of 3.5 nm the situation has already changed slightly. A weak local minimum appears in the energy landscape. For a width of 4.0 nm, this minimum has become more pronounced and its magnitude becomes negative, which means that a particle of the product phase may now be formed with a gain of energy. A nucleation barrier of 25 kT is determined from the energy landscape, which is quite a realistic value to obtain a reasonable nucleation rate. If the critical diffusion width is defined as that thickness for which a particle of the new phase becomes stable, this critical width must be slightly greater than 3.5 nm, in remarkably good agreement to the experimental observation by atom probe tomography. The same calculation was also performed for other nucleation mechanisms suggested. For example, the calculation for the so-called transversal mode (Desré and Yavari 1990) yielded a critical diffusion width smaller than 1.0 nm, in obvious contradiction to the measurements. In consequence, we can conclude that the atom probe analysis yielded convincing evidence for an interdiffusion process taking place before nucleation of the first product, although the zone of mixing ranges down to only a few nanometers. In this way it was possible to determine the critical diffusion depth with an accuracy better than 1 nm, which is important to distinguish between different nucleation modes. The transversal mode, which has been discussed for more than a decade, can be excluded for the Al/Co reaction couple.

13.2.3.7.3 Diffusion in Nanocrystalline Thin Films

In nanocrystalline matter with grain sizes down to about 10 nm, the volume fraction attributed to grain boundaries (GBs) can easily exceed 50%. Since there is vast experimental evidence that the diffusion along GBs is much faster than bulk diffusion (Kaur 1989), atomic transport in nanocrystalline thin-film materials is expected to be accelerated significantly by the presence of GBs. With further decreasing grain size, an additional necessary topological feature of the boundary arrangement – the so-called triple line – may affect atomic transport. Along a triple line, three GBs meet forming a one-dimensional defect. The structure of a triple junction is expected to differ considerably from that of an ordinary GB. Presumably, it will be much more disordered, so that an even faster transport rate can be assumed. However, up to now only one measurement of triple line diffusion by analytical electron microscopy has been reported (Bokstein et al. 2001). The authors interpret their results in terms of a much faster diffusion rate along the triple line. The difficulty of the measurement stems from the small effective cross-section of the defect, which requires chemical analysis of the highest possible resolution. In the 2D projection of a TEM sample the triple line is always overlapped by grain volume. So, the measured level of segregation is much reduced.

In recent atom probe experiments (Schmitz et al. 2006), Au (15 nm in thickness) and Cu layers (25 nm in thickness) were deposited on top of tungsten tips. In order to slow down the intermixing of the soluble metals Cu and Au, a thin Co barrier (6 nm thick) was inserted in between. Tips were annealed in a UHV furnace at 295 °C for 30 min. Due to the strong substrate curvature, thin films deposited on substrate tips tend to be very fine-grained, with grain sizes down to 5 nm (Lang and Schmitz 2003). Thus, these specimens are ideal candidates in which to observe the transport along topological singularities of the GB arrangement. A typical field ion micrograph of the upper Cu layer is shown in Fig. 13.58(a). Discontinuities in the structure of concentric bright rings mark the presence of grain boundaries. A few of them are marked by broken lines. Obviously, a polycrystalline structure with a grain size of about 15 nm has formed. Triple junctions are also seen that are aligned approximately parallel to the tip axis, so that they can be analyzed in the main measurement direction; "T" marks a particularly clear one.

By selecting such areas for analysis, the local concentration field around the junction could be measured. The geometry of the three GBs and an example of a 2D composition map is shown in Fig. 13.58(b). Since the atom probe delivers real three-dimensional data, 2D composition maps can be determined in any arbitrary direction subsequently to the measurement. In the example shown, the map is aligned perpendicular to a triple line. It is seen quite clearly that the line is locally enriched in Au and also that the three GBs are distinguished by a measurable Au content, though at a somewhat lower level than inside the junction. The measured concentration fields around the triple line were evaluated by means of the approximate solution of Klinger et al. (1997), which represents an extension of the established Fisher model of GB diffusion (see Chapter 5). In the case of the triple line, the atomic transport may be understood by a three-level cascade process instead of only two levels in the case of ordinary GB diffusion. First, the material is transported along the triple line. Second, leakage into the re-

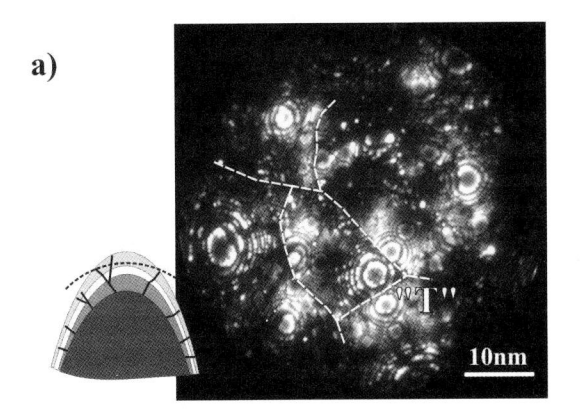

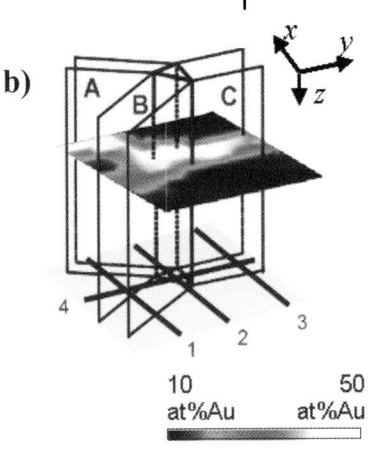

Fig. 13.58 (a) Field ion micrograph of a sputter-deposited Cu layer. Grain boundaries are marked by broken lines, a triple junction by "T". (b) Two-dimensional composition map in gray scale representation determined as a cross-section through a triple line joining three Cu grains with their boundaries A, B, and C. (After Schmitz et al. 2006.)

lated three boundaries takes place, and finally, atoms are drained from the GB into the bulk grain volume. Quantitatively the concentration field is described by Eq. (78).

$$c(x,y,z,t) = c_0 \cdot \exp\left(-\frac{\sqrt{3} \cdot \sqrt[4]{D_{GB}\delta} \cdot \sqrt[8]{4D_B/\pi t}}{\sqrt{D_{TJ}s}} \cdot z\right)$$
$$\times \exp\left(-\frac{\sqrt[4]{4D_B/\pi t}}{\sqrt{D_{GB} \cdot \delta}} \cdot y\right) \times \left[1 - \mathrm{erf}\left(\frac{x}{2\sqrt{D_B t}}\right)\right] \quad (78)$$

In this equation the x, y, and z coordinates are directed perpendicular to a representative GB, along a grain boundary perpendicular to the triple line, and along the triple line, respectively (see the axes in Fig. 13.58b). The meaning of the indices of the three diffusion constants D_i is self-explanatory. The effective thickness δ of the GB may be estimated as about 1 nm and the effective cross-section s of the triple junction is approximated by $s = \delta^2$. Conveniently, the right-hand side of Eq. (78) separates into three independent terms on the space coordinates.

Since the Au content within the grain volume falls short of the noise level of the atom probe method, the authors used the known bulk diffusion coefficient to interpret their data. Evaluating the compositional slope along the y axes (see Fig. 13.59a), the GB coefficient D_{GB} can be determined using the second term on the RHS of Eq. (78). Then, by means of the first term, the slope along the z axis (see Fig. 13.59b) yields the diffusion coefficient within the triple junction. From the measurements at Cu/Au films the following numerical results were obtained:

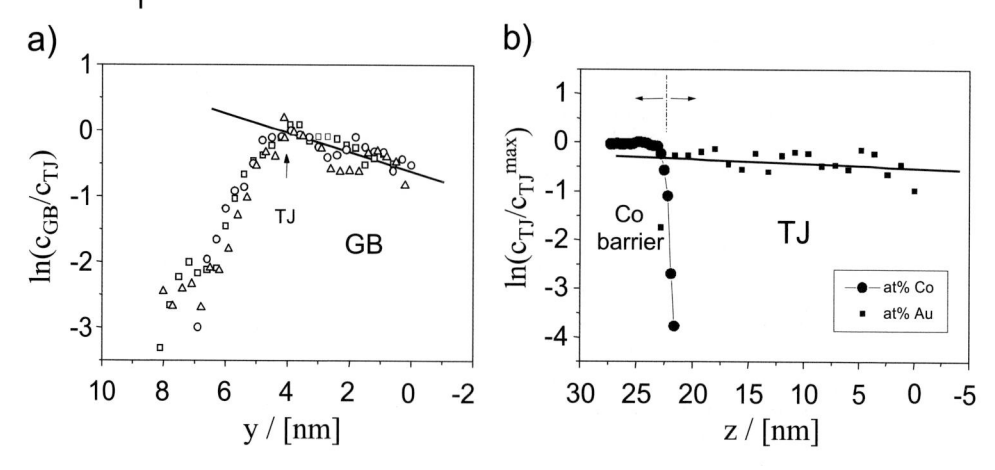

Fig. 13.59 Au diffusion in Cu at 295 °C: Normalized concentration profiles determined along grain boundary (a) and triple junction (b) as determined by atom probe tomography. (After Schmitz et al. 2006.) For details see the text.

$$D_B = 2.1 \times 10^{-23} \text{ m}^2 \text{ s}^{-1}$$

$$D_{GB} = 1.7 \times 10^{-17} \text{ m}^2 \text{ s}^{-1}$$

$$D_{TJ} = 9.6 \times 10^{-14} \text{ m}^2 \text{ s}^{-1}$$

Thus, at the diffusion temperature of 295 °C, the diffusion coefficient of the triple line is indeed about *5600 times* larger than the GB diffusion coefficient. Actually, atom probe tomography is the only method allowing such detailed measurements in complex nanocrystalline microstructures.

13.2.3.7.4 Thermal Stability of GMR Sensor Layers

In recent years reading heads based on the giant magnetoresistance effect (GMR) have made possible a dramatic increase in magnetic recording density (see Chapter 14, Section 14.4). Since the period of the required multilayers ranges down to a few nanometers, spatially resolved analysis is a challenge for atom probe tomography, too. Co/Cu and Cu/Ni$_{79}$Fe$_{21}$ are two of the metallic systems most often used. The soft magnetic alloy Ni$_{79}$Fe$_{21}$ (Permalloy, Py) is especially suited for position or orientation sensors, since the effect of hysteresis is very low. For many potential applications, such as angular sensors in motor vehicles, the thermal stability of the device is an important issue. It is known that the Cu/Py system is much more sensitive to thermal load than Cu/Co. With the former, the GMR amplitude is already degenerating at temperatures of 150 °C, while for the latter the amplitude remains stable up to 400 °C (Hecker et al. 2002). Different mechanisms have been proposed to be responsible for GMR degradation: van Loyen et al. (2000) argue that at least two effects should contribute, namely grain boundary diffusion and inter- or demixing at the interface. Hecker et al. (2002) conclude

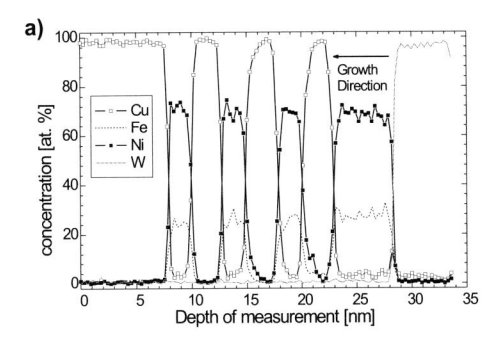

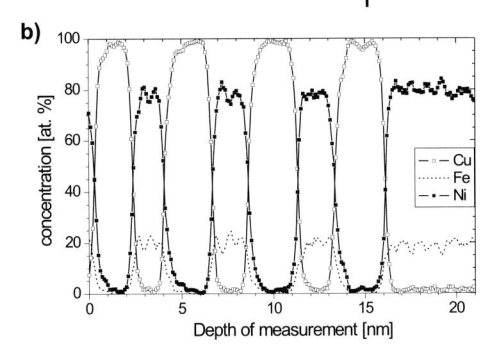

Fig. 13.60 Composition profiles of a Cu/NiFe multilayer determined perpendicular to the layer interfaces. (a) as-prepared state, (b) state after 20 min of annealing at 350 °C. (After Ene et al. 2005.)

that the alloying tendency of Ni and Cu above 250 °C controls the decay of the GMR in the Py systems.

In view of potential technical applications it is important to identify the mechanisms of thermal reaction. In an experimental study (Ene et al. 2005) $Cu_{2.5\ nm}$/$Py_{2.5\ nm}$ multilayers were deposited onto substrate tips and annealed in a UHV furnace. A typical volume reconstruction of an as-prepared state is shown in Fig. 13.45, which we have already discussed above. The depth resolution is sufficient to distinguish individual (111) planes of Cu, which are used to calibrate the length scale of the reconstruction. Concentration profiles determined along the tip axis direction are presented in Fig. 13.60 for the as-prepared state (a) and the 350 °C/ 20 min annealed state (b). The resolution of analysis, which reproduces sharp transitions at the interfaces from almost 100–0 at.% Cu, is noteworthy. It would be not possible to achieve such a selectivity with analytical TEM. If the profiles of the as-prepared state are compared to those after the annealing at 350 °C, no difference is seen at first sight. In particular, the integrity of the thin films is preserved and no grain boundary effects are detected. This is all the more striking, since the magnetoresistivity has already vanished completely at lower temperatures.

However, due to the outstanding sensitivity of TAP, it is possible to determine even the smallest modifications of the interfacial chemistry. In this case, the slope of the concentration profile at the interfaces was used as a characteristic parameter (for a definition, see the insert in Fig. 13.61a). Clearly, during the annealing the slope decreases, starting with its initial maximum value of 1.5 nm^{-1} (Fig. 13.61a). At the temperature of GMR breakdown (250 °C) it has already fallen to 1 nm^{-1}. Such a variation is not negligible, because in view of the small layer thickness a considerable zone at the interface has alloyed, which is obviously already sufficient to degenerate the GMR effect. This observation by TAP has been found to be in complete agreement with measurements of electrical resistivity (Hecker et al. 2002), which had shown that the base resistance of the multilayer,

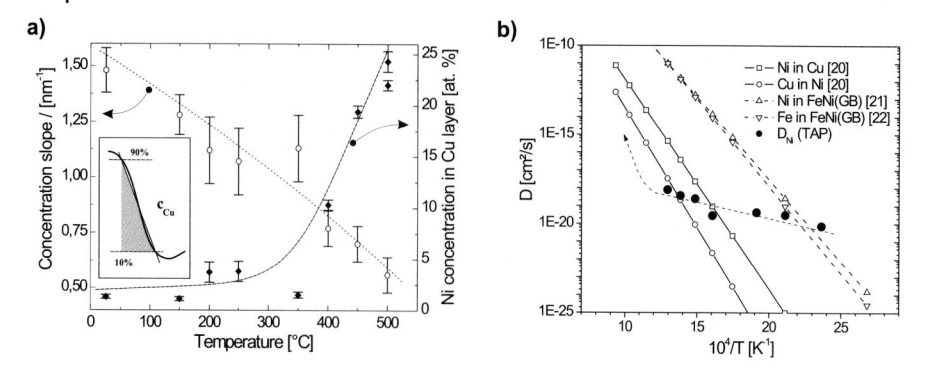

Fig. 13.61 (a) Temperature dependence of the concentration slope at the interfaces (left-hand scale) and the Ni content of the Cu layers (right-hand scale), both after 20 min of isochronal annealing. The broken lines are merely guidelines. (b) Arrhenius representation of the interdiffusion coefficients determined. (After Ene et al. 2005.)

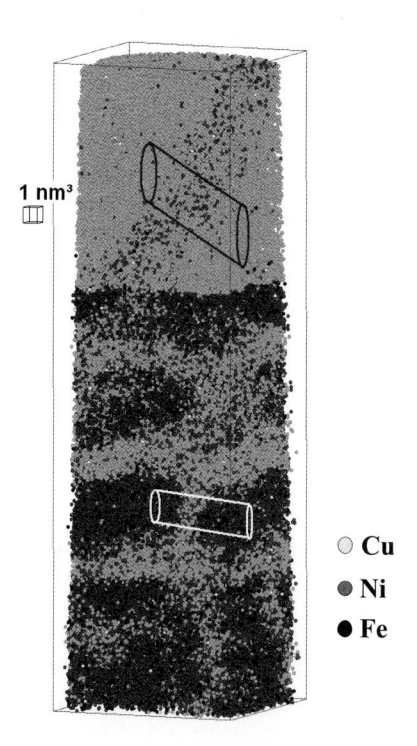

Fig. 13.62 Grain boundary diffusion in nanocrystalline Ni/Py multilayers after 30 min of annealing at 400 °C. (After Ene et al. 2005.)

indicating the complete alloying of the layers, increases significantly only at temperatures higher than 250 °C, at which the magnetoresistance has already vanished. In addition, the decrease in the interfacial slope has also been analyzed in terms of a volume diffusion coefficient. Measurements at different temperatures allowed an Arrhenius plot presented in Fig. 13.61(b), where it is compared with literature diffusion coefficients for Cu, Ni, and GB diffusion in Py. The activation energy is remarkably small, even smaller than that of GB diffusion. This suggests that the observed short-range diffusion at the interfaces is induced by nonequilibrium point defects.

It must be pointed out that the thermal reaction described so far is not influenced by neighboring GBs; only volume diffusion effects are observed, though on a very short length scale. With the help of FIM images like that shown in Fig. 13.58(a), GBs may be selected for the analysis and investigated for segregation which would point at GB transport. Indeed, significant GB transport could be observed as shown in Fig. 13.62, but only at temperatures above 300 °C. Ni diffuses into the grain boundaries of Cu, and in turn Cu into the grain boundaries of Py. After annealing for 30 min at 400 °C, the Ni segregation level in Cu amounts to 25 at.% and that of Cu in Py to about 55 at.%. While the former is far too low to introduce ferromagnetic shortage across the Cu spacer layer, the latter would be sufficient to fragment the ferromagnetic layers into isolated domains. However, at realistic temperatures leading to the degeneration of the amplitude of the GMR effect, no GB segregation was detected at all. Therefore, based on the TAP measurements, GB transport could be ruled out as a potential mechanism.

13.2.4
Future Development and Outlook

The purpose of this chapter has been to demonstrate how high-resolution microscopic imaging and analysis are used to understand the atomic structure and dynamics of defects in materials. Particular emphasis has been placed on interphase boundaries and surfaces, since these are important to an understanding of solid-state reactions and – utilizing these reactions – to development of the structure and properties of materials. With the methods presented it is possible to image, analyze, and even manipulate individual atoms, and furthermore to study their cooperative motion even in situ. Having reviewed so many impressing experimental examples, one may ask whether there is still room for any further development at all. However, a closer look at the severe restrictions with which each of the three methods is currently faced will help us to define important starting points for essential improvements.

In the case of atom probe tomography, the volume investigated and thus the number of identified atoms is so small, let us say 10^6, that statistics limits the accuracy of any composition measurement in general to about 0.1 at.%. The microstructure of materials is often complex, with at least one characteristic length scale in the micrometer range. The probability of finding a certain defect on this length scale within the analysis volume of $10 \times 10 \times 100$ nm^3 is negligible. Last but not least, all the examples presented dealt with metallic specimens, while modern applications often require functional materials such as ceramics or semiconductors. How can the method, which is a classical tool of physical metallurgy, overcome the barrier toward other classes of materials?

A possible way to achieve a larger analyzed volume would be to elevate the voltage supplied to the tip. The radius of the tip scales with the voltage and so does the lateral dimension of the analyzed volume. However, this strategy finds its natural limit for practical reasons, since there is no way to produce a voltage pulse sufficiently short and with appropriate frequencies for amplitudes larger than 5 kV. Another possibility is to increase the aperture angle of the instrument, which can be achieved by reducing the flight distances. This in turn requires a very accurate time-of-flight measurement. Nevertheless, equipped with modern time-to-digital converters, very recent tomographic atom probes have been put into operation with an aperture angle of up to $\pm 35°$. This leads to a lateral size range of analysis of 40–80 nm, as reported at relevant conferences (Deconihout et al. 2006; Stender et al. 2006). Together with the volume, the total number of atoms to be evaluated increases dramatically. A typical data set will then consist of several million atoms. To measure such a large number of events in a reasonable period of time, pulse frequencies need to be increased significantly, which is difficult to achieve for the high voltage needed in a conventional atom probe.

A very intelligent method of circumventing this technical problem is the so-called scanning atom probe, which had already been suggested in 1994 by Nishikawa and Kimoto (1994). By placing a tiny extraction electrode of some 10 μm bore size close to the tip (see Fig. 13.63), the voltage required to obtain a given evaporation field strength is reduced by a factor 2–4. In consequence much lower voltage pulses, which can be produced with repetition rates higher than 100 kHz, are sufficient. After related technical problems were solved, the concept has recently been put into operation. Meanwhile, these instruments are functioning well and are even available commercially (Kelly et al. 2004). Their efficiency of analysis is impressive. A data rate higher than 10 000 atoms s^{-1} is obtainable and data sets with more than 10^8 atoms have been achieved routinely. Besides the large aperture angle and high data rate, this most recent branch of atom probe tomography has an important advantage regarding specimen preparation. As indicated by Fig. 13.63, instead of a single tip made from a supporting wire, an array of microtips may be used. Since the requirement for a large aspect ratio is considerably relaxed by the electrode geometry, these microtips may be produced conveniently by sputtering through a suitable mask. Very recently, InAs nanowires, grown naturally by using nanopatterned Au catalysts, could be analyzed by a tomographic atom probe equipped with such a microelectrode (Perea et al. 2006).

Currently several groups are experimenting with pulsed laser irradiation to extend APT to nonmetallic materials with a conductivity below 10^{-2} Ω^{-1} cm^{-1}, which has been the important limit for high-voltage pulsed evaporation (Melmed

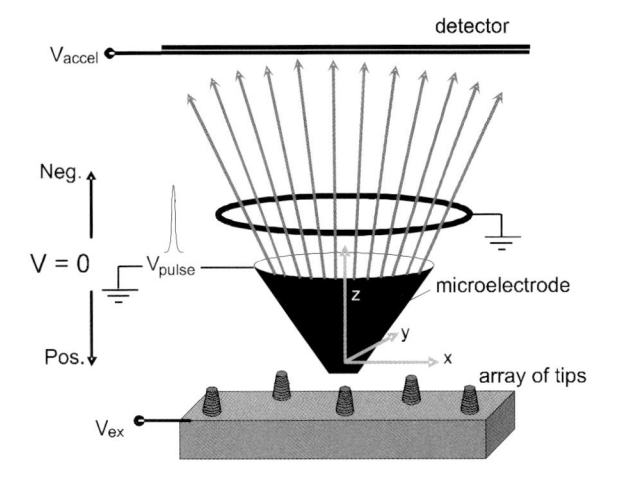

Fig. 13.63 Principle of a scanning atom probe: an extraction electrode with an open diameter of about 10 µm is placed close to a microtip to concentrate the electric field. By post-acceleration between the ring-shaped electrode and the detector entrance, mass resolution is improved. (see Kelly et al. 2004).

et al. 1981). The idea dates back to early work by Tsong and coworkers (Kellogg and Tsong 1980). Presumably, it has been due to the primitive technique of early laser systems that this idea has not been explored to its full extent and has been finding a renaissance only recently after the emergence of the femtosecond laser technique. Owing to the very short pulse width of these lasers, the evaporation may even be assisted by a direct field produced by the light wave instead of the mere thermal pulsing postulated in the early laser work. Evidence for this has been collected by the atom probe group in Rouen (Gault et al. 2005; Vella et al. 2006). Other authors opposed this interpretation (Cerezo et al. 2006). Be that as it is, at least several very recent reports have documented that very "difficult" materials like oxide ceramics can be analyzed reliably by laser-assisted atom probe tomography (Oberdorfer et al. 2006; Thompson et al. 2006; Vurpillot et al. 2006). Indeed, this signals an important breakthrough, which will certainly help to bring atom probe analysis toward a broader acceptance in materials science.

The purpose of the section on HRTEM was to demonstrate how static and in-situ HRTEM are used to understand the atomic structure and dynamics of defects in materials. It was seen that in-situ HRTEM is able to reveal the projected atomic structure of defects and their mechanisms of motion involving collective atom processes, but not individual atoms or jumps.[1] Since most dynamic pro-

---

[1] Of course, HRTEM is able to resolve individual atoms and their motion on surfaces in projection.

cesses at interphase boundaries appear to involve collective atom motion, e.g., fluctuations at order–disorder interfaces and growth of precipitate interfaces by a TLK mechanism, and HRTEM is generally able to resolve these features, this is not a great limitation. Hence, in-situ HRTEM is able to provide a wealth of information about detailed atomic processes that occur in materials. This situation will only continue to improve, as rapid progress is being made in increasing the spatial, temporal, and chemical resolution of HRTEMs, the capability of imaging devices, storage and processing, and precise computer control of the microscope.

More generally in the field of electron microscopy, recent developments include the introduction of dedicated corrector optics, in order to eliminate the spherical and chromatic aberrations of the objective lens, and thereby to improve the resolution of the instrument significantly. This allows straightforward interpretation of high-resolution images. In analytical electron microscopy, both the introduction of monochromators to narrow the energy spread of the electron beam, and that of in-column energy filters, have been important recent developments. With conventional optics, the spatial resolution of energy-filtered images is limited to about 1 nm owing to the nature of inelastic scattering and the chromatic aberration of the objective lens. Thus, the combination of chromatic aberration correction and energy filtering is an appealing approach to the ultimate goal of chemical mapping with atomic resolution in a TEM, while monochromation and spherical aberration correction can provide atomic-level imaging and chemistry in an STEM.

Another important development is taking place in electron tomography. By applying stereographic methods and powerful image reconstruction, the former restriction to two-dimensional image projections, from which electron microscopy has suffered in the past, can be overcome. Tomography is already an established practice in the characterization of complex biological molecules with electron microscopy. However, inorganic materials are crystalline in most cases, so that strong diffraction contrast effects are present, which complicate easy application of stereographic methods. Overcoming this problem by a combination of Z-contrast or energy filtering imaging and stereography is quite promising and excellent examples are beginning to appear in the literature. It is reasonable to expect that it will be possible to image the 3D structures of materials with atomic resolution in the near future.

In scanning probe microscopy, a remarkable breakthrough is not so much expected in terms of improving the ultimate resolution limit. Instead, important progress is currently being made in the development of new dedicated nanoprobes which realize the idea of a "lab on top of the tip" in many different fields, such as electrochemistry, magnetism, thermometry, or tribology. Continuously improved methods of microfabrication will produce even smaller cantilevers and also faster piezoscanner systems, which allow higher resonance frequencies and much faster scanning rates. In this way, in-situ measurements with the AFM may become a suggested option to study time-dependent processes in biological objects or nanostructured materials, similarly to what we have discussed for HRTEM in this section.

Acknowledgments

The authors would like to dedicate this chapter to Professor H. I. Aaronson, in recognition of his constant enthusiasm and prodding to understand the atomic mechanisms of phase transformations. Furthermore, the authors are grateful to the many students and colleagues whose work is referenced herein. This research was supported by the National Science Foundation under Grants DMR-9908855 and DMR-0554792 (JMH).

References

Aaronson, H.I. (2002), *Metall. Mater. Trans.* **48A**, 2285.

Al-Kassab, T., Wollenberger, H., Schmitz, G., Kirchheim, R. (2003), in: *High Resolution Imaging and Spectrometry of Materials*, Ernst, T., Rühle, M. (Eds.), Springer, Berlin, p. 290.

Amelinckx, S., van Dyck, D. (1992), in: *Electron Diffraction Techniques*, Vol. 2, Cowley, J.M. (Ed.), International Union of Crystallography, Oxford University Press, Oxford.

Asta, M., Hoyt, J.J. (2000), *Acta Mater.* **48**, 1089.

Bar, G.K., Meyers, G.F. (2004), *MRS Bulletin* **29**, 464.

Bäro, G., Gleiter, H. (1974), *Acta Metall.* **22**, 141.

Bas, P., Bostel, A., Deconihout, B., Blavette, D. (1995), *Appl. Surf. Sci.* **87/88**, 298.

Batson, P.E., Delby, N., Krivanek, O.L. (2002), *Nature* **418**, 617.

Benson, W.E., Howe, J.M. (1997), *Phil. Mag. A* **75**, 1641.

Bergmann, C., Emeric, E., Clugnet, G., Gas, P. (2001), *Def. Diff. Forum*, **194–199**, 1533.

Bhushan, B., Kulkarni, A.V., Bonin, W., Wyrobek, J.T. (1996), *Phil. Mag. A* **74**, 1117.

Binder, K. (2001), in: *Phase Transformations in Materials*, Kostorz, G. (Ed.), Wiley-VCH, Weinheim.

Binning, G., Quate, C.F., Gerber, Ch. (1986), *Phys. Rev. Lett.* **56**, 930.

Binning, G., Rohrer, H. (1982), *Helv. Phys. Acta* **55**, 726.

Bokstein, B., Ivanov, V., Oreshina, O., Pteline, A., Peteline, S. (2001), *Mater. Sci. Eng. A* **302**, 151.

Bos, C., Sommer, F., Mittemeijer, E.J. (2004), *Acta Mater.* **52**, 3545.

Browning, N.D., Wallis, D.J., Nellist, P.D., Pennycook, S.J. (1997), *Micron* **28**, 334.

Butler, E.P., Hale, K.F. (1981), *Dynamic Experiments in the Electron Microscope*, North-Holland, New York.

Butrymowicz, D.B., Manning, J.R., Read, M.E. (1974), *J. Phys. Chem. Ref. Data* **3**, 527.

Cahn, J.W., Hilliard, J.E. (1958), *J. Chem. Phys.* **28**, 258.

Cahn, J.W., Hillig, W.B., Sears, G.W. (1964), *Acta Metall.* **12**, 1421.

Cerezo, A., Godfrey, T.J., Smith, G.D.W. (1988), *Rev. Sci. Instrum.* **59**, 862.

Cerezo, A., Godfrey, T.J., Hyde, J.M., Sijbrandij, S.J., Smith, G.D.W. (1994), *Appl. Surf. Sci.* **76/77**, 374.

Cerezo, A., Smith, G.D.W., Clifton, P.H. (2006), *Appl. Phys. Lett.* **88**, 154 103.

Chatterjee, K., Howe, J.M., Johnson, W.C., Murayama, M. (2004), *Acta Mater.* **52**, 2923.

Coffey, K.R., Clevenger, L.A., Barmak, K., Rudman, D.A., Thompson, C.V., (1989), *Appl. Phys. Lett.* **55**, 852.

Cowley, J.M. (1975), *Diffraction Physics*, 2nd edn., North-Holland, Amsterdam.

Deconihout, B., Bostel, A., Menand, A., Sarrau, J.M., Bouet, M., Chambreland, S., Blavette, D. (1993), *Appl. Surf. Sci.* **67**, 444.

Deconihout, B., Renaud, L., Da Costa, G., Bouet, M., Bostel, A., Bavette, D. (1998), *Ultramicroscopy* **73**, 253.

Deconihout, B., Vurpillot, F., Gault, B., Gilbert, M., Vella, A., Bostel, A., Da Costa, G., Menand, A., Blavette, D. (2006), *Proc. IFES*, IEEE Conference Publication Management Group, p. 55.

Desré, P.J., Yavari, R. (1990), *Phys. Rev. Lett.* **64**, 1533.

Durst, K., Göken, M. (2001). *Prakt. Metallogr.* **38**, 197.

Egerton, R.F. (1996), *Electron Energy-Loss Spectroscopy in the Electron Microscop*, 2nd edn., Plenum Press, New York.

Ene, C.B., Schmitz, G., Kirchheim, R., Hütten, A. (2005), *Acta Mater.* **53**, 3383.

Frederix, P.L.T.M., Hoogenboom, B.W., Fotiadis, D., Müller, D. J., Engel, A. (2004), *MRS Bulletin* **29**, 449.

Fultz, B.T., Howe, J.M. (2002), *Transmission Electron Microscopy and Diffractometry of Materials*, 2nd edn., Springer-Verlag, Berlin.

Gabrisch, H., Kjeldgaard, L., Johnson, E., Dahmen, U. (2001), *Acta Mater.* **49**, 4259.

Gai, P.L., Boyes, E.D. (2003), *Electron Microscopy in Heterogeneous Catalysis*, Institute of Physics Publishing, Bristol.

Garg, A., Chang, Y.-C., Howe, J.M. (1993), *Acta Metall. Mater.* **41**, 235.

Gault, B., Vurpillot, F., Bostel, A., Menand, A., Deconihout, B. (2005), *Appl. Phys. Lett.* **86**, 094 101.

Giessible, F.J. (1995), *Science* **267**, 1451.

Göddenhenrich, T., Lemke, H., Hartmann, U., Heiden, C. (1990), *Appl. Phys. Lett.* **56**, 2578.

Göken, M. (2002), in: *Intermetallic Compounds*, Vol. 3. *Principles and Practice*, Westbrook, J.H., Fleischer, R.L. (Eds.), John Wiley & Sons.

Göken, M., Kempf, M. (1999), *Acta mater.* **47**, 1043.

Gusak, A.M. (1990), *Ukr. Phys. J.* **35**, 725.

Hartmann, U. (1999), *Annu. Rev. Mater. Sci.* **29**, 53.

Hecker, M., Tietjen, D., Wendrock, J., Schneider, C.M., Cramer, N., Malinski, L. (2002), *J. Magn Magn. Mater.* **247**, 62.

Hirth, J.P., Lothe, J. (1968), *Theory of Dislocations*, McGraw-Hill, New York.

Hodaj, F., Desré, P.J. (1996), *Acta Mater.* **44**, 4485.

Hono, K., Hasegawa, N., Okano, R., Fujimori, H., Sakurai, T. (1993), *Appl. Surf. Sci.* **67**, 407.

Howe, J.M. (1993), in: *Proc. Inter. Conf. Martensitic Transformations '92 (ICOMAT-92)*, Wayman, C.M., Perkins, J. (Eds.), Monterey Institute of Advanced Studies, Carmel, p. 185.

Howe, J.M. (1997), *Interfaces in Materials: Atomic Structure, Thermodynamics and Kinetics of Solid–Vapor, Solid–Liquid and Solid–Solid Interfaces*, John Wiley, New York.

Howe, J.M. (1999), in: *International School of Electron Microscopy, 8th Course: Impact of Electron and Scanning Probe Microscopy on Materials Research*, Rickerby, D.G. et al. (Eds.), Kluwer Academic, The Netherlands.

Howe, J.M., Aaronson, H.I., Hirth, J.P. (2000), *Acta Mater.* **48**, 3977.

Howe, J.M., Benson, W.E. (1995), *Interface Sci.* **2**, 347.

Howe, J.M., Dahmen, U., Gronsky, R. (1987), *Phil. Mag A* **56**, 31.

Howe, J.M., Benson, W.E., Garg, A., Chang, Y.-C. (1996), in: *Advances in Physical Metallurgy – 94*, Gordon and Breach Science Publishers, Amsterdam, p. 277.

Howe, J.M., Reynolds, Jr., W.T., Vasudevan, V.K. (2002), *Metall. Mater. Trans.* **33A**, 2391.

Howe, J.M., Gautam, A.R.S., Chatterjee, K., Phillipp, F. (2006), *Acta Mater.*, in press.

Hoyt, J.J., Asta, M., Karma, A. (2003), *Mater. Sci. Eng.* **R41**, 121.

Hyde, J.M., Miller, M.K., Hetherington, M.G., Cerezo, A., Smith, G.D.W., Elliott, C.M. (1995), *Acta Metall. Mater.* **43**, 3403.

Iijima, S. (1985), *J. Electron Micros.* **34**, 249.

Joy, D.C., Romig, A.D., Goldstein, J.I. (Eds.) (1986), *Principles of Analytical Electron Microscopy*, Plenum Press, New York.

Kang, Z., Eyring, L. (1991), *Metall. Trans.* **22A**, 1323.

Kaur, I., Gust, W., Kozma, L. (1989), *Handbook of Grain and Interphase Boundary Diffusion Data*, Ziegler Press, Stuttgart.

Keller, H., Klinghöfer, G., Kankelheit, E., (1987), *Nucl. Instr. Meth.* **A258**, 221.

Kellogg, G., Tsong, T. (1980), *J. Appl. Phys.* **51**, 1184.

Kelly, T.F., Gribb, T.T., Olson, J.D., Martens, R.L., Shepard, J.D., Wiener, S.A., Kunicki, T.C., Ulfig, R.M., Lenz, D.R., Strennen, E.M., Oltman, E., Bunton, J.H., Strait, D.R. (2004), *J. Microsc. Microanal.* **10**, 373.

Kempf, M., Göken, M., Vehoff, H. (1998), *Appl. Phys. A* **66**, 843.

Keyse, R.J., Garratt-Reed, A.J., Goodhew, P.J., Lorimer, G.W. (1998), *Introduction to Scanning Transmission Electron Microscopy*, BIOS Scientific Publishers, New York.

Kikuchi, R., Cahn, J.W. (1979), *Acta Metall.* **27**, 1337.

Kittl, J.E., Massalski, T.B. (1967), *Acta Mater.* **15**, 161.

Klinger, L.M., Levin, L.A., Petelin, A.L. (1997), *Def. Diff. Forum* **143–147**, 1523.
Laird, C., Aaronson, H.I. (1969), *Acta Metall.* **17**, 505.
Lang, C., Schmitz, G. (2003), *Mater. Sci. Eng. A* **353**, 119.
Larson, D. (2001), *Microsc. Microanal.* **7**, 24.
Lee, J.-G., Mori, H. (2005), *J. Mater. Res.* **20**, 1708.
Loiseau, A., Ricolleau, C., Potez, L., Ducastelle, F. (1994), in: *Proc. Intl. Conf. Solid–Solid Phase Transformations*, Johnson, W.C., Howe, J.M., Laughlin, D.E., Soffa, W.A. (Eds.), The Metals, Minerals and Materials Society, Warrendale, p. 385.
Maganov, S.N., Whangbo, M.-H., *Surface Analysis with STM and AFM*, VCH, Weinheim.
Marks, L.D. (1994), *Rep. Prog.* (1996) *Phys.* **57**, 603.
Martin, Y., Wickramasinghe, H.K. (1987), *Appl. Phys. Lett.* **50**, 1455.
Massalski, T.B., Soffa, W.A., Laughlin, D.E. (2006), *Metall. Mater. Trans.* **37A**, 825.
Melmed, A.J., Martinika, M., Girvin, S.M., Sakurai, T., Kuk, Y. (1981), *Appl. Phys. Lett.* **39**, 416.
Merkle, K.L., Thompson, L.J., Phillipp, F. (2002), *Phys. Rev. Lett.* **88**, 225 501-1.
Meyer, G., Amer, N.M. (1988), *Appl. Phys. Lett.* **53**, 1045.
Meyer, E., Hug, H.J., Bennewitz, R. (2003), *Scanning Probe Microscopy: The Lab on a Tip*, Springer-Verlag, Berlin.
Meyer, E., Jarvis, S.P., Spencer, N.D. (2004), *MRS Bulletin* **29**, 443 and subsequent articles therein.
Meyers, M.A., Mishra, A., Benson, D.J. (2006), *Prog. Mater. Sci.* **51**, 421.
Michaelsen, C., Barmak, K., Weihs, T.P. (1997), *J. Phys. D*, **30**, 3167.
Michler, J., von Kaenel, Y., Steigler, J., Blank, E. (1998), *J. Appl. Phys.* **83**, 187.
Miller, M.K. (2000), *Atom Probe Tomography*, Kluwer Academic, New York.
Miller, M.K., Hyde, J.M., Hetherington, M.G., Cerezo, A., Smith, G.D.W., Elliott, C.M. (1995), *Acta Metall. Mater.* **43**, 3385.
Miller, M.K., Cerezo, A., Hetherington, M.G., Smith, G.D.W. (1996), *Atom Probe Field Ion Microscopy*, Oxford Science Publications, Oxford.
Moser, A., Hug, H.J., Parashikov, I., Stiefel, B., Fritz, O. (1995), *Phys. Rev. Lett.* **74**, 1847.
Müller, E.W. (1951), *Z. Physik* **131**, 136.
Müller, E.W. (1957), *J. Appl. Phys.* **28**, 1.
Müller, E.W., Panitz, J.A., McLane, S.B. (1968), *Rev. Sci. Instrum.* **39**, 83.
Mullins, W.W. (1957), *J. Appl. Phys.* **28**, 333.
Murayama, M., Howe, J.M., Hikada, H., Takaki, S. (2002), *Science*, **295**, 2433.
Nishikawa, O., Kimoto, M. (1994), *Appl. Surf. Sci.* **76/77**, 424.
Oberdorfer, C., Stender, P., Reinke, C., Schmitz, G. (2006), *J. Microsc. Microanal.* in press.
O'Keefe, M.A. (1984), in: *Proc. 3rd Pfeffercorn Conf. Electron Optical Systems*, Hren, J.J., Lenz, F.A., Munro, E., Sewell, P.B., Bhatt, S.A. (Eds.), Scanning Electron Microscopy, Inc., Illinois, p. 209.
Pasichnyy, M.O., Schmitz, G., Gusak, A.M., Vovk, V. (2005), *Phys. Rev. B* **72**, 014 118.
Pennycook, S.J., Varela, M., Hetherington, C.J.D., Kirkland, A.I. (2006), *Mater. Res. Soc. Bull.* **31**, 36.
Perea, D.E., Allen, J.E., May, S.J., Wessels, B.W., Seidman, D.N., Lauhon, L.J. (2006), *Nano Letters* **6**, 181.
Perepezko, J.H., Massalski, T.B. (1975), *Acta Metall.* **23**, 621.
Perry, S.S. (2004), *MRS Bulletin* **29**, 481.
Rabkin, E., Klinger, L., Semenov, V. (2000), *Acta Mater.* **48**, 1533.
Raffler, N., Howe, J.M. (2006), *Metall. Mater. Trans.* **37A**, 873.
Romanov, A.E., Vladimirov, V.I. (1992), in: *Dislocations in Solids*, Vol. 9, *Dislocations and Disclinations*, Nabarro, F.R.N. (Ed.), North-Holland, Amsterdam, p. 101.
Roy, R., Sen, S.K. (1992), *J. Mater. Sci.*, **27**, 6098.
Saenz, J.J., Garcia, N., Grütter, P., Meyer, E., Heinzelmann, H., Wiesendanger, R., Rosenthaler, L., Hidber, H.R., Güntherod, H.-J. (1987), *J. Appl. Phys.* **62**, 4293.
Sasaki, K., Saka, H. (1991), *Phil. Mag. A* **63**, 1207.
Scherzer, O. (1949), *J. Appl. Phys.* **20**, 20.
Schiotz, J., Di Tolla, F.D., Jacobsen, K.W. (1998), *Nature* **391**, 561.
Schleiwies, J., Schmitz, G. (2002), *Mater. Sci. Eng.* **A327**, 94.
Schmid, M., Stadler, H., Varga, P. (1993), *Phys. Rev. Lett.* **70**, 1441.
Schmitz, G., Ene, C., Lang, C., Vovk, V. (2006), *Adv. Sci. Technol.* **46**, 126.

Schonfelder, B., Wolf, D., Philpot, S.R., Furtkamp, M. (1997), *Interface Sci.* **5**, 245.

Schweika, W., Reichert, H., Babik, W., Klein, O., Engemann, S. (2004), *Phys. Rev. B* **70**, 041 401R.

Seefeldt, M. (2001), *Rev. Adv. Mater. Sci.* **2**, 44.

Sinclair, R. (1994), *Mater. Res. Soc. Bull.* **XIX**, 26.

Sinclair, R., Yamashita, T., Parker, M.A., Kim, K.B., Holloway, K., Schwartzman, A.F. (1988), *Acta Crystallogr.* **A44**, 965.

Sinclair, R., Morgiel, J., Kirtikar, A.S., Wu, I.-W., Chiang, A. (1993), *Ultramicroscopy* **51**, 41.

Sluiter, M., Kawazoe, Y. (1996), *Phys. Rev. B* **54**, 1 0381.

Smith, D.J., Saxton, W.O., O'Keefe, M.A., Wood, G.J., Stobbs, W.M. (1983), *Ultramicroscopy* **11**, 263.

Smith, D.J., Bursill, L.A., Wood, G.J. (1985), *Ultramicroscopy* **16**, 19.

Spence, J.C.H. (1988), *Experimental High-Resolution Electron Microscopy*, 2nd edn., Oxford University Press, Oxford.

Stender, P., Oberdorfer, C., Artmeier, M., Pelka, P., Spaleck, F., Schmitz, G. (2006), *Ultramicroscopy* in press.

Sutton, A.D., Balluffi, R.W. (1995), *Interfaces in Crystalline Materials*, Clarendon Press, Oxford.

Thompson, K., Bunton, J.H., Kelly, T.F., Larson, D. (2006), *J. Vac. Sci. Technol. B* **24**, 421.

Tsong, T.T. (1990), *Atom-Probe Field Ion Microscopy*, Cambridge University Press, Cambridge.

van Loyen, L., Elefant, D., Tietjen, D., Schneider, C.M., Hecker, M., Thomas, J., (2000), *J. Appl. Phys.* **87**, 4852.

Varga, P., Schmid, M. (1999), *Appl. Surf. Sci.* **141**, 287.

Vella, A., Vurpillot, F., Gault, B., Menand, A., Deconihout, B. (2006), *Phys. Rev. B* **73**, 165 416.

Vovk, V., Schmitz, G., Kirchheim, R., (2004), *Phys. Rev. B* **69**, 104 102.

Vurpillot, F., Bostel, A., Cadel, E., Bavette, D. (2000), *Ultramicroscopy* **84**, 213.

Vurpillot, F., Bostel, A., Bavette, D. (2001), *Ultramicroscopy* **89**, 137.

Vurpillot, F., Gault, B., Vella, A., Bouet, M., Deconihout, B. (2006), *APL* **88**, 094 105.

Wayman, C.M., Aaronson, H.I., Hirth, J.P., Rath, B.B. (Eds.) (1994), *Proc. Pacific Rim Conf. Role of Shear and Diffusion in the Formation of Plate-Shaped Transformation Products*, *Metall. Mater. Trans.* **25A**, 1781.

Weber, T., Marx, M., Göken, M., Vehoff, H. (2000), in: *Proc. Euromat '99*, Vol. 4, Jouffrey, B. (Ed.), Wiley-VCH Weinheim, p. 208.

Williams, D.B., Carter, C.B. (1996), *Transmission Electron Microscopy: A Textbook for Materials Science*, Plenum Press, New York.

Wolf, D., Yip, S. (Eds.) (1992), *Materials Interfaces: Atomic-Level Structure and Properties*, Chapman & Hall, London.

Yasumura, K.Y., Stowe, T.D., Chow, E.M., Pfafman, T., Kenny, T.W., Stipe, B.C., Rugar, D. (2000), *J. Microelectromech. Systems* **9**, 117.

Zhu, X., Grütter, P. (2004), *MRS Bulletin* **29**, 457.

14
Materials and Process Design

14.1
Soft and Hard Magnets
Roland Grössinger

14.1.1
What do "Soft" and "Hard" Magnetic Mean?

A simple classification of ferromagnetic materials can be made on the basis of their coercivity. Soft magnetic materials are characterized by a small area of the hysteresis loop and a low coercivity ($H_c < 1000$ A m^{-1}) whereas hard magnetic materials (permanent magnets) show a high coercivity ($H_c > 30$ kA m^{-1}). Figure 14.1 shows typical hysteresis loops of soft and hard magnetic materials.

Applications for soft magnetic materials include electromagnets, motors, transformers and relays, and electromagnetic shielding. Applications for hard magnetic materials are relays stepper motors, generators, etc. Semihard materials are used for data storage applications.

The demand for saving energy in electrical engineering and the continuing increase in operating frequencies in electric circuits require a constant improvement of the quality of soft and hard magnetic materials. In recent years, in addition to the standard materials (soft: Fe–Si, soft magnetic ferrites), amorphous and

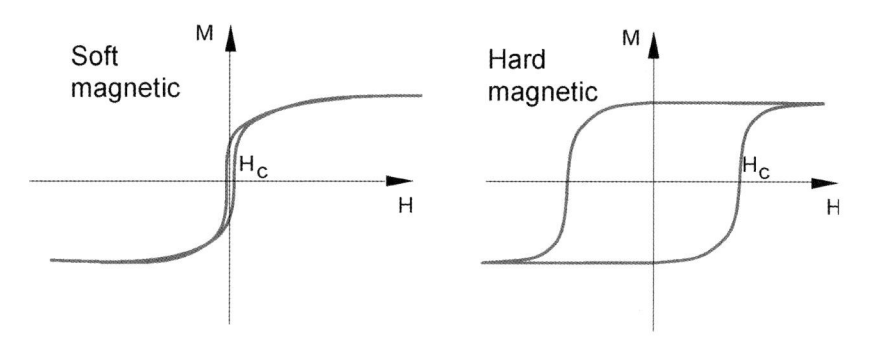

Fig. 14.1 Hysteresis loops of ferromagnetic soft and hard magnetic materials.

Alloy Physics: A Comprehensive Reference. Edited by Wolfgang Pfeiler
Copyright © 2007 WILEY-VCH Verlag GmbH & Co. KGaA, Weinheim
ISBN: 978-3-527-31321-1

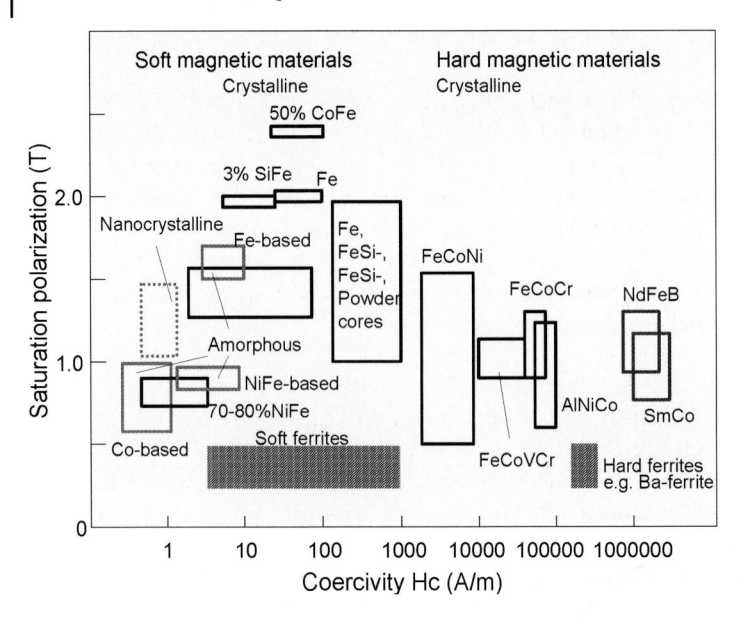

Fig. 14.2 Overview of soft and hard magnetic alloys: saturation polarization achieved as a function of the coercivity. Note that the coercivities of the different materials range more than seven orders of magnitude.

nanocrystalline ferromagnetic ribbons and wires have been considered very promising for this purpose (see, e.g., Luborsky 1983; Vazquez 1994; Herzer 1997). Besides the standard hard magnetic materials (Alnico, ferrites) nowadays rare-earth-based materials (Sm–Co, Nd–Fe–B) but also nanocrystalline hard

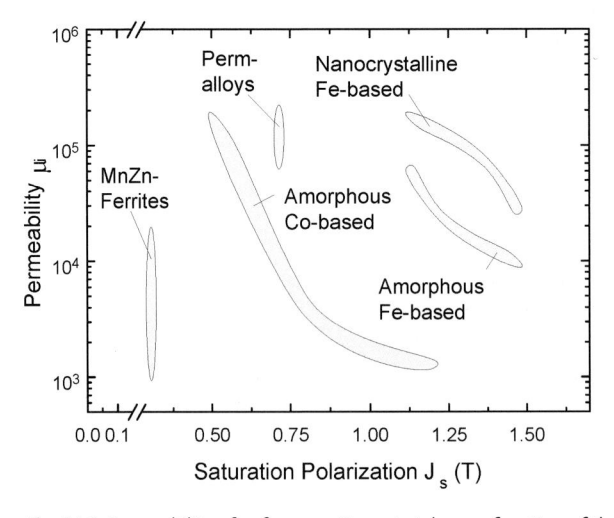

Fig. 14.3 Permeability of soft magnetic materials as a function of the saturation polarization.

magnetic materials are used (for a survey see Schultz and Müller 1998; Grössinger et al. 2003, 2004).

As an overview of the magnetic properties of soft and hard magnetic alloys, Fig. 14.2 shows the saturation polarization achieved as a function of the coercivity. Note that the coercivity of the different materials ranges over more than seven orders of magnitude. Figure 14.3 shows the permeability of soft magnetic materials as a function of the saturation polarization.

14.1.1.1 Intrinsic Properties Determining the Hysteresis Loop (Anisotropy, Magnetostriction)

14.1.1.1.1 Anisotropy

The magnetic properties of a ferromagnetic material are strongly influenced by various contributions of anisotropy. The magnetic anisotropy is the dependence of the internal energy ($dF = \vec{H}d\vec{M}$) on the direction of the spontaneous magnetization. This energy term is called the magnetic anisotropy energy. Different kinds of anisotropy exist.

- Shape anisotropy is macroscopically important because it describes the macroscopic stray field: $H_S = -D \cdot M_S$ (D = demagnetizing factor); it also plays a role inside a crystalline material where the local stray field acts between the grains influencing the coercivity. In the case of magnetic powders the local shape anisotropy of the grains has to be considered. The shape anisotropy also determines the working points in all kinds of magnet circuits.
- The origin of the crystalline anisotropy is the spin–orbit coupling. The coupling is given by the lowest energy condition of the direction of the spin moment and the field produced by the orbital contribution of the electron wavefunction. For a crystalline hexagonal system such as cobalt, the free energy can be expressed by expanding F in a series of powers of $\sin^2 \theta$ [Eq. (1)].

$$F = K_0 + K_1 \sin^2 \theta + K_2 \sin^4 \theta + \cdots \qquad (1)$$

- For hard magnetic materials especially, this is the most important contribution. For cubic crystals such as iron and nickel it is more convenient to express F in terms of the direction cosines of the internal magnetization with respect to the three cube edges. Thus the expression is given by Eq. (2).

$$F = K_0 + K_1(\cos^2 \alpha_1 \cos^2 \alpha_2 + \cos^2 \alpha_2 \cos^2 \alpha_3 \\ + \cos^2 \alpha_3 \cos^2 \alpha_1) + K_2 \cos^2 \alpha_1 \cos^2 \alpha_2 \cos^2 \alpha_3 + \cdots \qquad (2)$$

- Induced anisotropy is important for stress- or field-annealed soft magnetic materials; in this case K_u depends on the microstructure but also on the elastic constants (Kraus and Duhaj 1990; Herzer 1994). It influences over the magneto-elastic energy $3/2\lambda.\sigma$ the shape of the hysteresis loop, which is important for sensor applications.

14.1.1.1.2 Magnetostriction

Generally magnetostriction is either the change in the shape or the change in the volume of a magnetically ordered material when a magnetic field is applied. Therefore this is a general effect when a material becomes magnetized. It appears in all solids, in metals as well as in insulators. Due to Hooke's law the deformation suffered is proportional to the sample size, say ℓ, and therefore it is convenient to express the deformation by the relative variation $\Delta = \Delta\ell/\ell$, which is called the *linear* magnetostriction. In a crystal $\Delta\ell$ is doubly anisotropic: it depends on the crystallographic direction of measurement and on the direction along which the magnetization, **M**, is oriented by the applied magnetic field, **H**. Because the saturation magnetostriction λ_S depends on the crystalline directions of $\Delta\ell$ and **M** (β and α, respectively) we have to specify the magnetostriction as $\Delta(\alpha,\beta)$. But also, in a polycrystalline material formed by small randomly distributed crystallites, λ measured along **H**, called the parallel magnetostriction $\lambda_\parallel$, can be different from that measured along the perpendicular direction, $\lambda_\perp$. It can happen that $\lambda_\perp \approx \lambda_\parallel/2$ (isotropic, no volume change) but usually this is not the case and then the volume of the solid changes by the amount $\Delta V/V = \omega = \lambda_\parallel + 2\lambda_\perp$, called the *volume* magnetostriction. The *volume* magnetostriction is of interest from a scientific point of view; it allows general conclusions about the origin of the magnetostriction. Usually the volume magnetostriction is about 10–100 times smaller than the shape magnetostriction.

On the other hand, the shape (or form of the unit cell) of the material is modified, and a measure of this effect is the difference $\lambda_t = \lambda_\parallel - \lambda_\perp$, called the shape magnetostriction. The shape magnetostriction is determined by the magnetization process and is of technical relevance for sensor and actuator applications.

14.1.1.2 Extrinsic Properties – Microstructure

A technically relevant magnetic material generally consists of more than one phase. The favorable properties of a technically usable material are based on the existence of interplay between different phases. Therefore the microstructure is important. It is determined by: grain size and grain shape; grain boundaries; and phases with different physical (magnetic) properties. It depends on the production process.

For investigating the microstructure, one can use optical methods or, better still, scanning electron microscopy (SEM), transmission electron microscopy (TEM), or atomic force microscopy (AFM); compare Chapter 13, Section 13.2. The different crystallographic phases are characterized by diffraction experiments which can also be applied locally.

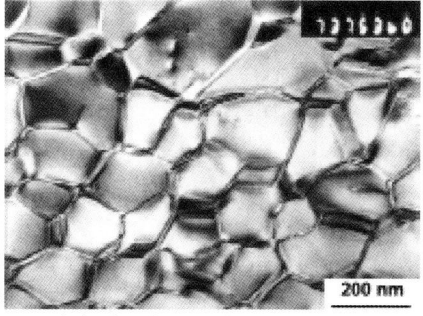

complete development   incomplete development

Fig. 14.4 Typical microstructure of a 2/17-type magnet taken by TEM. Left: completely developed lamellar structure; right: incompletely developed microstructure (Matthias et al. 2002).

In soft magnetic materials coercivity, permeability, and magnetic losses are determined by the alloy microstructure. These properties are therefore called "extrinsic."

In hard magnetic materials the microstructure determines the coercivity and the magnetization process. A high coercivity is achievable only with a certain microstructure. In modern Sm–Co magnets the pinning of narrow domain walls occurs in phases with different crystal structures and consequently different magnetic properties (especially anisotropy) (Kronmüller et al. 1984). As an example, Fig. 14.4 shows the microstructure of a 2/17-type permanent magnet (Matthias et al. 2002): the typical lamellar structure consists of 2/17-type material inside and 1/5 $Sm(Co,Cu)_5$ material between the grains (see Section 14.1.3.3).

14.1.2
Soft Magnetic Materials

The ranges of permeability and saturation polarization of most soft magnetic alloys are shown in Fig. 14.3. The most important conventional and new soft magnetic materials are as follows: Fe–Ni (e.g., Permalloy, Mumetal), Fe–Si, Fe–Al (Sendust, i.e., FeSiAl), Fe–Co, soft ferrites (Mn–Zn ferrites etc.), amorphous alloys (Fe- or Co-based), and nanocrystalline alloys.

A soft magnetic material which is well suited for technical applications should exhibit the following properties.

- *High saturation magnetization at room temperature*: The saturation magnetization determines the necessary volume in transformers for the transmission of a certain power. The higher the saturation magnetization of the material, the lower the necessary volume. At room temperature Fe has the

highest saturation magnetization, but it has to be very pure in order to achieve good soft magnetic properties.
- *High Curie temperature*: Therefore mainly materials with Fe, Co or Ni (3d elements) are used for technical applications.
- *Low coercivity*: a low coercivity is generally accompanied by a high permeability (these properties are reciprocal).
- *High permeability*: A high permeability is of great importance, especially for shielding applications.
- *Low conductivity*: This is important for ac applications.
- *Low losses*: The losses are caused by the hysteresis losses and the eddy current losses (normal and anomalous).

These properties (coercivity, permeability, losses) are correlated with the magnetization process, where the interaction between domain walls and the microstructure (grain size, phases, grain boundaries, local stress centers, etc.) are of importance.

The introduction of various rapid quenching techniques (such as planar flow casting, atomic vaporization, and others) allowed the production of new types of materials. By rapid quenching three types of interesting soft magnetic materials can be produced:
- crystalline materials like Fe–Si and Fe–Al–Si (Sendust alloy)
- amorphous materials
- nanocrystalline materials.

Rapid solidification from molten material leads to a new state which can be amorphous, nanocrystalline, or just crystalline. Applying a high quenching rate to a molten alloy can result in a material which presents generally new and interesting physical properties. Such materials are said to be "rapidly solidified."

Amorphous metallic ribbons can be fabricated by bringing a stream of molten alloy in contact with a rapidly moving substrate surface. The most common substrate surfaces described in the literature are the insides of drums or metallic wheels, the outsides of wheels, between twin rollers, and on belts.

Very often the so-called planar flow-casting technique is used, in which the nozzle is situated very close to the moving substrate surface (at a distance around 0.3 mm). The rapidly quenched material obtained is a thin (about 20–50 µm) and usually narrow (up to 10 mm) ribbon (see Fig. 14.5). In industrial production lines, much wider ribbons are possible.

Annealing of amorphous alloys is a very important technique for reducing local stresses, achieving chemical homogenization and improvement of magnetic properties (adjustment of the shape of the hysteresis loop) of the samples. Annealing can cause stress relaxation, controlled development of induced anisotropy, adjustment of a well-defined domain structure, controlled microstructural changes, and nanocrystallization.

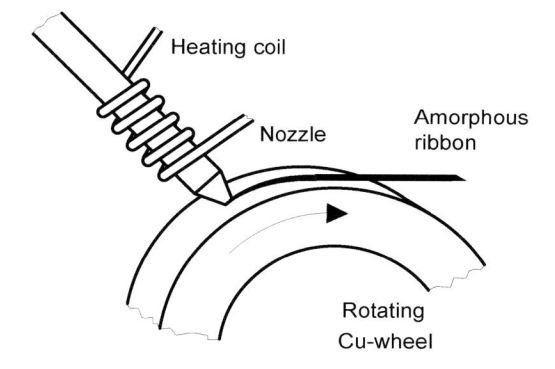

Fig. 14.5 Melt spinning technique.

14.1.2.1 Pure Fe and Fe–Si

Iron and iron–silicon alloys are the most important soft magnetic materials in use today. Improvements in magnetic properties have been achieved primarily by minimizing chemical impurities and controlling crystal orientation. The term "pure" iron generally refers to iron of 99.9% purity. Nonmetallic impurity elements such as C, O, S, or N which enter the lattice interstitially reduce the soft magnetic properties. A maximum permeability for commercial iron of 10 000–20 000 can be increased to 100 000 and more by annealing in H_2 and in vacuum. This additionally causes a decrease in the coercivity H_C; Table 14.1 gives a short summary of the effect of purification of Fe. Small additions of C or N (of the order of 100 ppm and more) cause a drastic increase in the coercivity. For special applications, cores produced from Fe powder are used. This is favorable because of the low losses at frequencies up to 1 MHz.

The so-called silicon steel had already been introduced by the beginning of the 20th century and became very successful for electric transformers. The magnetic properties of the Fe–Si system can be summarized as follows: the saturation magnetization, the Curie temperature, and the magnetocrystalline anisotropy decrease due to Si substitution, whereas the components of the magnetostriction

Table 14.1 Room-temperature values of different treated "pure" Fe materials.

Material	$\mu_0 M_S$ [T]	H_C [A m^{-1}]	Permeability (max.)
Vacuum-melted Fe	2.15	25	20 000
Electrolytic Fe	2.15	16	up to 40 000
Pure Fe (H_2-treated)	2.16	4	up to 100 000

constants λ_{100} and λ_{111} behave different. However, close to 6 wt.% Si λ_{100} approaches zero. This composition, on the other hand, is not favorable because at 6% Si the material becomes too brittle for technical applications. The electrical resistivity increases strongly due to Si substitution, which is very important, especially for ac applications.

For transformer applications all kinds of losses have to be considered: for example, hysteresis losses, eddy current losses, and anomalous eddy current losses. Whereas hysteresis losses are independent of frequency, eddy current losses and anomalous eddy current losses which are due to domain wall movements are frequency-dependent.

The core losses can be affected by: composition (silicon content), impurities, grain orientation (texture), applied stress, grain size, thickness, and surface conditions. As early as 1934 (Goss 1934) it was shown that the development of a texture leads to a reduction of core losses.

In transformer applications the magnetic flux lies predominantly in the length of the laminations. It is therefore desirable to enhance the permeability in this direction. This can be achieved by various hot- and cold-rolling stages which lead to textured sheets. Good results were achieved with grain-oriented silicon–steel, with the [001] direction in the length of the lamination. The [001]-type crystal directions are the easy directions of magnetization which lead to the greatest permeability. Grain orientation of silicon–iron can be performed by a rolling procedure as shown in Fig. 14.6.

A very detailed description of the properties of silicon–iron can be found in Chin and Wernick (1980). Besides Fe–Si, Fe–Al and Fe–Al–Si soft magnetic alloys are also of technical importance. These materials are characterized by a high electrical resistivity, high hardness, high permeability, and low losses. Famous in this respect is the so-called Sendust family of Fe–Al–Si alloys (4–7%Al, 7–13%Si). This composition is defined by the lowest magnetostriction value as well as a low crystalline anisotropy in the ternary phase diagram.

14.1.2.2 Ni–Fe Alloys

Special Ni–Fe alloys (trade name: Permalloy), defined by a zero or low magnetostriction and zero or low magnetic anisotropy, are also known under the name

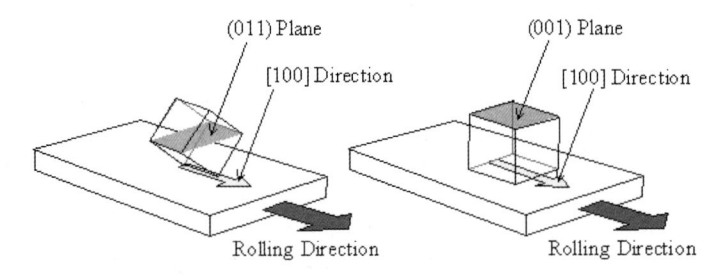

Fig. 14.6 Two possible ways to produce grain orientation in Fe–Si by a rolling procedure.

"mu-metal," which is produced by a careful heat treatment and minor additions of Cu and Cr. These compounds exist over a wide range of compositions, from 30–80 wt.% Ni with varying magnetic properties. We distinguish between alloys with a high Ni content, which are characterized by a high permeability and a Ni content of about 50 wt.%, which exhibit a high saturation; and those with a low Ni content, which have a rather high electrical resistivity.

Ni–Fe alloys with low anisotropy and low magnetostriction have extremely high permeability, up to 300 000, and an intrinsic coercivity as low as 0.4 A m^{-1}. The high permeability makes mu-metal very effective as a screening material against static or low-frequency magnetic fields, which cannot be attenuated by any other method. Mu-metal requires special heat treatment – annealing in a hydrogen atmosphere, which increases the magnetic permeability about 40-fold. The annealing alters the crystal structure, aligning the grains and removing some impurities, especially carbon. Mechanical treatment may disrupt the grain alignment, leading to a drop in permeability in the affected areas, which can be restored by repeating the hydrogen annealing step.

14.1.2.3 Soft Magnetic Ferrites

A ferrite is a ferrimagnetic oxidic compound which contains magnetically ordered iron and which is derived from magnetite (Fe_3O_4 more exactly $Fe^{2+}O.Fe_2^{3+}O_3$) by substituting divalent metal ions for Fe^{2+}. These materials exhibit generally good soft magnetic properties combined with an excellent high-frequency behavior due to the fact that they are insulators. Trivalent metal ions substitute for Fe^{3+} and ions with other valences (1+, 4+, 5+) also can be incorporated by considering charge compensation. Mn, Fe, and Co are generally the most common divalent metals, but Cu, Zn, Mg and others are also used. Ferrites exhibit a close-packed cubic spinel ($MgAl_2O_4$) structure, where the divalent ions replace Mg and the trivalent ions replace Al. The origin of magnetism in ferrites is due to (a) unpaired 3d electrons, (b) superexchange between adjacent metal ions, and (c) nonequivalence in number of A and B sites. Since the common ferrite ions (Mn^{2+}, Fe^{2+}, Ni^{2+}, Co^{2+}) have more than five 3d electrons, the magnetic moments are aligned antiparallel between A and B sites. This results in a ferrimagnetic type of order where, due to the occupation number, the B sites dominate. Most of the ferrites of the formula MFe_2O_4 are inverse spinel structures ($Fe^{3+}\uparrow$ [$M^{2+}\downarrow Fe^{3+}\uparrow$]$O_4$) and the moment of the compound is given by that of M^{2+}.

An advantage of ferrites is their high electrical resistivity due to the fact that the material is an insulator. A large saturation magnetization is desired but, due to the ferrimagnetic structure, the total magnetization is generally low in soft magnetic ferrites. Zn substitution causes an increase in the saturation magnetization (especially at low temperatures), but a decrease in ordering temperature.

For soft magnetic applications the following properties are important.
- *High permeability*: This depends on spin rotation in the domains, which are defined by intrinsic parameters ($\mu_R = 1 + 2\pi M_S^2/K$; μ_R = rotational part of permeability), but also on wall displacements, which are dominated by micro-

Table 14.2 Magnetic properties at $T = 20$ °C of ferromagnetic spinels.

Ferrite	$J_S = \mu_0 M_S$	Ion moment [μ_B]	Curie temperature T_C [°C]	Lattice parameter [nm]	Density [10^3 kg m^{-3}]
MnFe$_2$O$_4$	0.50	5.0	300	0.850	5.00
FeFe$_2$O$_4$	0.60	4.1	585	0.839	5.24
CoFe$_2$O$_4$	0.53	3.7	520	0.838	5.29
NiFe$_2$O$_4$	0.34	2.3	585	0.834	5.38
CuFe$_2$O$_4$	0.17	1.3	455	0.8445	5.42
MgFe$_2$O$_4$	0.14	1.1	440	0.836	4.52
Li$_{0.5}$Fe$_{2.5}$O$_4$	0.39	2.6	670	0.833	4.75
γ-Fe$_2$O$_3$	0.52	2.3	575	0.834	4.89
ZnFe$_2$O$_4$	0	0	–	0.840	5.40
CdFe$_2$O$_4$	0	0	–	0.873	5.76

structural parameters ($\mu_W = 1 + 0.75\pi M_S^2 D/\gamma$; μ_W = wall displacement part of permeability). The best candidates for high permeability are Mn and Ni ferrites.

- Low coercivity: This needs the lowest value of magnetocrystalline anisotropy and a low magnetostriction.

Magnetic properties at $T = 20$ °C of ferromagnetic spinels are summarized in Tables 14.2 and 14.3.

Note that the first-order anisotropy constant K_1 of Mn–Zn ferrite is exceptional low (Table 14.3). Therefore Mn–Zn ferrites are considered as high-permeability materials. The most important commercial soft ferrites are: Mn–Zn and Ni–Zn

Table 14.3 Typical values of the magnetocrystalline anisotropy coefficient and saturation magnetostriction at room temperature for various ferrites.

Ferrite	K_1 [10^3 J m^{-3}]	λ_S [$\times 10^6$]
MnFe$_2$O$_4$	−4	−5
FeFe$_2$O$_4$	−12	40
CoFe$_2$O$_4$	200	−110
NiFe$_2$O$_4$	−7	−26
MgFe$_2$O$_4$	−4	−6
γ-Fe$_2$O$_3$	−5	–
Mn$_{0.62}$Zn$_{0.41}$Fe$_{1.97}$O$_4$	0.2	–

Table 14.4 Properties of Mn–Zn and Ni–Zn ferrites.

Property	Mn–Zn	Ni–Zn
Initial permeability μ_i	500–20000	10–2000
Saturation polarization J_S [T]	0.3–0.5	0.1–0.36
Curie temperature T_C [°C]	100–250	100–500
Coercivity H_C [A m^{-1}]	4–100	16–1600
Resistivity ρ [Ω m]	0.02–20	10–10^7
Density [10^3 kg m^{-3}]	4.6–4.8	4.8–4.9
Total power loss [W m^{-3}]	50–200	very low

ferrites – see Table 14.4. Mn–Zn ferrites are favorable for applications up to 1 MHz, whereas Ni–Zn ferrites are suited to higher frequencies. The range of properties depends very much on preparation conditions and material parameters such as composition, grain size, porosity, etc.

14.1.2.4 Amorphous Materials

The production of amorphous 3d-metal (Fe,Co,Ni)-based materials yields a great variety of compounds with all kinds of magnetic properties. Amorphous materials are of interest here because the magnetic properties can be changed systematically by substituting different elements without changing any kind of "structure." It should be mentioned that at the beginning it was very surprising that magnetic order is possible without the periodicity of a crystal structure. This problem was solved by realizing that the short-range order of the atoms causes a local density of state which is very similar to that of crystalline materials. The technical breakthrough was achieved by the invention of the so-called melt-spinning process (e.g., the single roller technique; cf. Fig. 14.5), which allows the easy and fast production of large amounts of thin ribbons (typically between 10 and 50 µm thick) of amorphous materials. For magnetic applications two interesting families of amorphous 3d-metal-based materials exist:
- 3d metal–metalloid (metalloid = B, Si, C, etc.)
- early transition element–late transition element. Here the Fe–Zr and Co–Zr systems are famous – both systems show an unusual magnetization behavior which can be explained to be due to invar-type behavior (see Section 14.2).

Great progress in developing new soft magnetic materials has been achieved since the mid-1980s (see, e.g., Pfeifer and Radeloff 1980; Luborsky 1983). The amorphous alloys are produced with the general formula (3d metal)$_{80}$(metalloid)$_{20}$ (3d metal = Fe, Co, Ni; metalloid = B, Si, Al, C, etc.). The metalloid is necessary for lowering the melting point of the alloy, thus stabilizing the amorphous state. Amorphous materials are generally obtained as thin ribbons (thickness 20–50 μm) by the single roller technique (Hasegawa 1982; Luborsky 1983). The absence of grains and grain boundaries leads to excellent soft magnetic properties (see, e.g., Egami et al. 1974). The easy substitution over a broad concentration range makes these materials interesting from a basic point of view, but the excellent magnetic properties are also interesting for many applications (Boll and Hinz 1985). The most promising systems are here Fe–Co-based amorphous alloys. Therefore many studies were performed on these materials, describing the basic magnetic properties such as the concentration and temperature dependence of the magnetization, the Curie temperatures (O'Handley et al. 1976; O'Handley and Bordreaux 1978), the crystallization behavior (Köster 1984), the magnetostriction in these systems (O'Handley et al. 1976, 1977; Grössinger 1990), the induced anisotropy (Vazquez et al. 1977), etc.

14.1.2.5 Nanocrystalline Materials

Nanocrystalline materials form the second group of new soft magnetic compounds. Amorphous ribbons with the approximate composition $Fe_{73.5}Cu_1Nb_3Si_{13.5}B_9$ are heat-treated (550 °C for 1 h), thus achieving a nanocrystalline state. The composition is mainly $Fe_{73.5}(Si,B)_{22.5}$ where the addition of Cu and Nb hinders the growth of the grains, thus leading to crystals of about 10–15 nm (Köster et al. 1991). This material was developed by Hitachi Metals and is on the market under the name Finemet (Yoshizawa et al. 1988). The material consists in the nanocrystalline state of small α-Fe–Si grains which are embedded in an amorphous matrix. An important factor is that the ordering temperature T_{ca} of the amorphous matrix is lower (about 300 °C) than that of the Fe–Si T_{cc} (about 550 °C). The excellent soft magnetic properties occur because the grain size is smaller than the magnetic exchange length. This leads to an averaging of the magnetic anisotropy of the crystalline Fe–Si and consequently to a low value of the effective anisotropy (Herzer 1990). A similar reduction occurs for the magnetostriction, where the positive contribution of the amorphous matrix is reduced by the negative contribution of the Fe–Si (Herzer 1995; Sato Turtelli et al. 2000).

The magnetic behavior is strongly determined by the ratio of crystalline to amorphous phases. This becomes especially important at elevated temperatures (above T_{ca}) where the coupling between the grains is reduced, which leads to a magnetic hardening (Grössinger et al. 1995; Hernando et al. 1994).

The second nanocrystalline system of importance is based on a Fe–Zr–Cu–B amorphous alloy which also transforms into the nanocrystalline state after a heat treatment at about 600 °C for 1 h (Suzuki et al. 1994; Dahlgren at al. 1996a; Estevez Rams et al. 1996). In this material nanocrystals of α-Fe are embedded in an amorphous matrix. In both cases the exchange coupling between the nanocrystals leads to excellent soft magnetic properties at room temperature. The Fe–Zr–Cu–B system has the technical disadvantage of difficult production – it needs a protective gas because of the highly reactive Zr. However, it exhibits a higher saturation magnetization in the nanocrystalline state and from an experimental point of view it is a simpler two-phase system. As mentioned already, α-Fe appears there as nanocrystals. This is much easier to detect – e.g., by Mössbauer spectroscopy. Additionally, the ordering temperature of the amorphous phase in this case lies between 50 and 100 °C which is technically disadvantageous.

14.1.3
Hard Magnetic Materials

The development of hard magnetic materials started with the discovery of Fe–C steels. During the 20th century the technical performance was dramatically improved. A key number for permanent magnets is the energy density (described in kJ m^{-3}) stored at the optimum working point of a magnetic material. The stored energy density $(B.H)_{max}$ could be improved from less than 40 kJ m^{-3} in about 1900 to more than 400 kJ m^{-3} which is available nowadays. Figure 14.7 shows the improvement of $(B.H)_{max}$ within the last 100 years. The dramatic enlargement of the coercivity became possible by the discovery of materials, such as the rare-earth 3d intermetallics, with a high intrinsic magnetocrystalline anisotropy.

One has to distinguish between the intrinsic and extrinsic properties which are important for a high quality permanent magnet. The intrinsic properties combined with the actual microstructure (grain size and shapes, phases, structures, etc.) determine the extrinsic properties.

The following intrinsic properties are important for permanent magnets.
- *High saturation magnetization at room temperature*: Therefore 3d metals such as Fe or Co are used. A high saturation magnetization is also necessary to achieve a high-energy product.

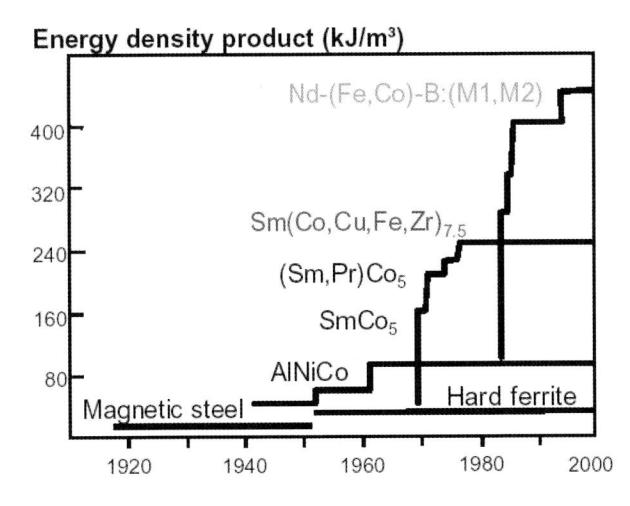

Fig. 14.7 Development of energy density achieved over the last 100 years of commercial permanent magnets.

- *High Curie temperature, T_C*: Here again the 3d metals are necessary, because the direct exchange between the 3d electrons causes a high ordering temperature. A high T_C and consequently a high operating temperature are necessary for all industrial applications.
- *High anisotropy*: The use of 4f metals is favorable because they generally exhibit an orbital moment **L** which causes a high magnetocrystalline anisotropy. The anisotropy can be described either by the so-called anisotropy field H_A or by the anisotropy constants $K_1, K_2, K_3 \ldots$. The anisotropy constants generally determine the nucleation but also the pinning field, which are important parameters for the mobility of the domains and consequently for the magnetization process. The condition of a high uniaxial anisotropy is necessary for achieving a high coercivity.

The following extrinsic properties are important for permanent magnets.
- *High-energy product* $(B.H)_{max}$, which scales with B_r^2: Therefore a high saturation magnetization determines together with the degree of alignment the maximum possible stored energy. For isotropic, uniaxial material $B_r = B_S/2$, and therefore $(B.H)_{max}$ in an isotropic material is principally much smaller! Generally, a large energy density is necessary for application with a high degree of miniaturization. Polycrystalline, isotropic hard magnetic materials are important for applications where a magnetic code is to be written.
- *High coercivity*: Besides the magnetocrystalline anisotropy, a well-adjusted microstructure determines mainly the coercivity – see Fig. 14.4.
- *Thermal stability*: This is given by the temperature dependence of the magnetization of all phases involved.
- *Mechanical properties*: Here again the microstrucure plays a leading role.
- *Corrosion stability*: This depends on the chemical composition of the phases involved as well as on the microstructure.

Figure 14.8 shows the demagnetizing curve B as a function of applied field H in the second quadrant for different hard magnetic materials. It also demonstrates how the working point is found. The arrow indicates a working point, which is given by the geometry of the sample (the slope is the reciprocal of the demagnetizing factor) (Dahlgren 1998).

Nowadays there are four important industrial permanent magnet families:
- AlNiCo
- hard magnetic ferrites

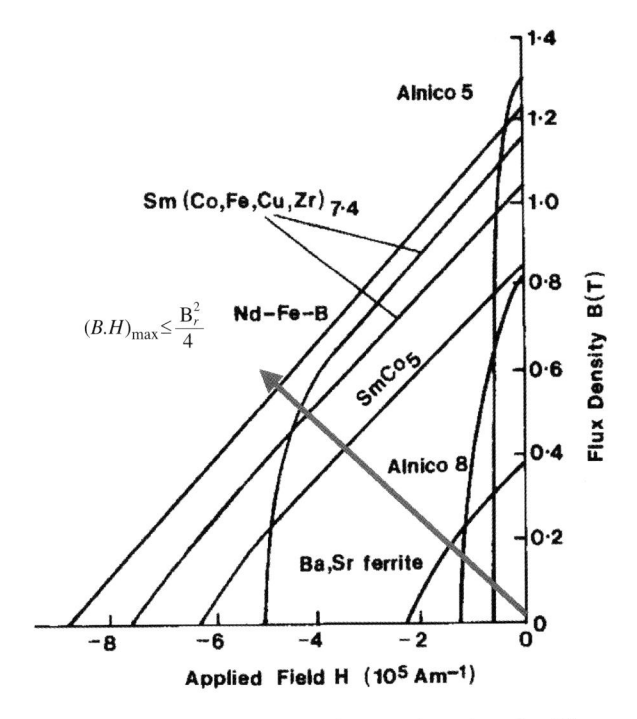

Fig. 14.8 Demagnetizing curve in the second quadrant for different permanent magnets.

- Sm–Co magnets: 1/5-based materials and 2/17-based compounds which are significantly different in their magnetization behavior
- Nd–Fe–B magnets.

Table 14.5 Comparison of permanent magnet materials.

Material	Cost index [%]	$(B.H)_{max}$ [MGOe]	Coercivity $_iH_C$ [kOe]	Max. operating temp. [°C]	Machinability
Nd–Fe–B (sintered)	65	45	up to 30	130	fair
Nd–Fe–B (bonded)[a]	50	10	up to 11	100	good
Sm–Co (sintered)	100	30	up to 50	300	difficult
Sm–Co (bonded)*	85	12	up to 10	120	fair
Alnico	30	10	up to 2	550	difficult
Hard ferrite	5	4	up to 3	300	fair
Flexible*	2	2	up to 3	100	excellent

[a] Bonded or flexible magnets are plastic-bonded magnets, which are useful for industrial applications which do not need the highest magnetic properties but do need every kind of shape.

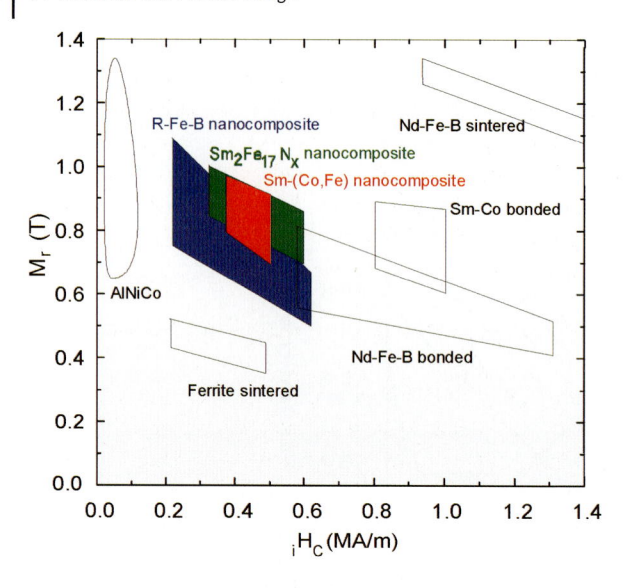

Fig. 14.9 Remanence versus coercivity field for different permanent magnet materials (Dahlgren 1998).

The properties of industrial permanent magnets are shown in Table 14.5. The relative prices are normalized to that of a sintered Sm–Co magnet (100%) (see the first entry). The last entry shows that all kind of bonded materials are superior with respect to machinability. However, these materials exhibit a lower working temperature and naturally a lower energy density.

In Fig. 14.9 the remanence M_r of various permanent magnet materials is plotted versus the coercivity. One can see very clearly that the sintered Nd–Fe–B-based materials exhibit the highest values of remanence and coercivity at room temperature.

14.1.3.1 AlNiCo

Alnico magnets are made up of an alloy of iron, aluminum, nickel, and cobalt with small amounts of other elements added to enhance the properties of the magnet. Alnico magnets have a high corrosion resistance, a good temperature stability (see Table 14.5), and good shock resistance, but are easily demagnetized due to the rather low coercivity field. Alnico magnets are produced by two typical methods, casting or sintering. Sintering offers superior mechanical characteristics, whereas casting delivers higher-energy products (up to 5.5 MGOe) and allows the design of intricate shapes. Anisotropic Alnico, which provides a preferred direction of magnetic orientation, can also be produced. Its coercivity is generally based on the shape anisotropy of the Fe–Co rich phase.

14.1.3.2 Ferrites

As a result of their favorable performance/price ratio, ceramic anisotropic ferrite magnets are located in the center of the permanent magnet market, where requirements with respect to performance and/or allowable magnet volume are demanding but not extreme. The main application for high quality ferrite magnets is as segments in various DC motors for the automotive industry. The manufacture of ceramic ferrite magnets consists of fine milling of the prefired material followed by several pressing and sintering steps which have been optimized in the last 50 years.

The intrinsic properties correspond to the M-type crystal structure (PG_3/mmc) crystal structure and, notably, from the five distinct Fe sublattices. Two intrinsic properties are crucial: saturation magnetization (J_s) for B_r and anisotropy field strength (H_A) for H_{cJ}. The five Fe sublattices are coupled by superexchange, allowing only parallel (up) or antiparallel (down) orientation. Their mutual orientation is given by the Gorter model (Went et al. 1952; Stuijts et al. 1955) 2a (up), 4f1 (down), 12k (up), 4f2 (down), 2b (up). Taking into account that the magnetic moment for Fe^{3+} amounts to $5\mu_B$, the total moment per mole of $AFe_{12}O_{19}$ at 0 K amounts to $20\mu_B$, in agreement with the observed saturation magnetization for the pure compound (J_s). The temperature dependence of J_s is shown in Fig. 14.10, implying $J_{s(300 K)} = 478$ mT. It is remarkable that the J_s–T curve is almost linear in a broad T region ($dJ_s/dT = -0.9$ mT K^{-1}). Mössbauer analysis revealed that the latter stems from the 12k sublattice (Stuijts et al. 1959). The magnetization is strongly bound to the hexagonal c axis. The anisotropy field derives from K_1 and J_s: $H_A = 2K_1/J_s$. The temperature dependence of K_1 is analogous to that of J_s, but its increase at decreasing temperature is somewhat less, resulting in a

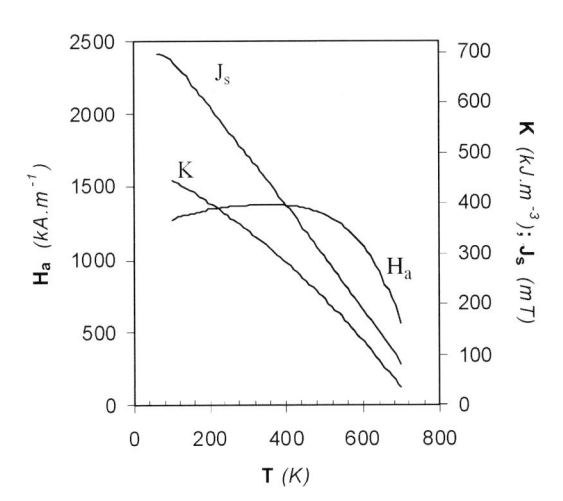

Fig. 14.10 Temperature dependence of saturation magnetization J_s, anisotropy constant K, and the anisotropy field $H_a = H_A$ for $BaFe_{12}O_{19}$.

flat H_A–T curve, having a maximum around 300 K (Fig. 14.10). There is not yet a clearcut model for the magnetocrystalline anisotropy. The contribution of dipole–dipole interaction has been calculated to be relatively small. So, the spin–orbit coupling of the Fe^{3+} ions must play the main role, in spite of the fact that (free) Fe^{3+} has no orbital momentum. Mostly, the contribution to the overall spin–orbit coupling is associated with the 2b site, but the joint 12k sites also play a significant role (Stuijts et al. 1959; Lotgering 1974; Kools 1994).

The hard magnetic properties of M-type ferrites originate in the magnetocrystalline anisotropy of Fe^{3+} in the magnetoplumbite structure (space group $P6_3/mmc$). M-type ferrites were the first permanent magnets where the coercivity was determined by the magnetocrystalline anisotropy.

The invention of La- and Co-substituted hard magnetic ferrites with improved properties renewed the interest in these materials (Kools et al. 2002).

14.1.3.3 Sm–Co

Since 1966 a new family of magnet materials has been evolving that is known as the "rare-earth permanent magnets" (REPMs). For the application of REPMs, a sufficiently high Curie temperature, a high magnetization, and a high magnetocrystalline anisotropy are important. The first two of these properties are provided mainly by the sublattice of the 3d element, whereas the rare-earth sublattice is mainly responsible for the last. The magnetic anisotropy of the rare-earth sublattice is a single-ion crystal field induced anisotropy. A strong magnetic coupling between the rare-earth sublattice is required in order to extend the anisotropy to the whole lattice and to preserve this anisotropy at elevated temperatures. The so-called Sm–Co magnets are multiphase metallurgical systems with complex mi-

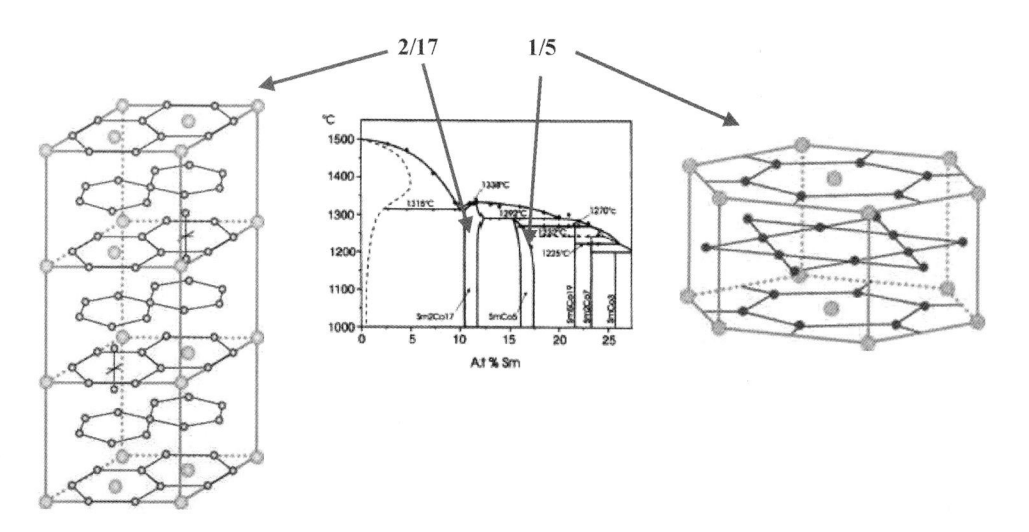

Fig. 14.11 Phase diagram of the Sm–Co system together with the unit cells of the 1/5 and 2/17 structures.

crostructures and they always contain more than two elements. The crystalline structures of the phases from 1:3 to 2:17 stoichiometry are closely related.

Samarium–cobalt-based rare-earth magnet materials have a much higher magnetic strength (energy product) than AlNiCo or ceramic ferrite materials. Introduced to the market in the 1970s, samarium–cobalt magnets continue to be used today. They are divided into two main groups: $SmCo_5$ and Sm_2Co_{17} (commonly referred to as 1/5 and 2/17). Figure 14.11 shows the Sm–Co phase diagram and the hexagonal structures of the 1/5 and the 2/17 unit cells.

In order to achieve the optimum microstructure a well-designed heat treatment is very important. In recent years special Sm–Co based magnets which are stable at high temperatures and can be used up to 500 °C were developed (Liu et al. 1999).

14.1.3.4 Nd–Fe–B

Nd–Fe–B-based magnets were invented about 20 years ago (Sagawa et al. 1984; Croat et al. 1984). The excellent hard magnetic properties are based on the high saturation polarization of the tetragonal $Nd_2Fe_{14}B$ compound together with the high anisotropy field at room temperature (H_A about 70 kOe). Figure 14.12(a) shows the tetragonal structure of this material, and Fig. 14.12(b) the complex phase diagram of Nd–Fe–B. This third generation of rare-earth magnets contains the most powerful and advanced commercialized permanent magnets today. Since they are made from neodymium, one of the most plentiful rare-earth elements, and inexpensive iron, Nd–Fe–B magnets offer the best value in cost and performance. Rare-earth iron–boron magnets became commercially available in the mid-1980s and have been growing increasingly popular ever since.

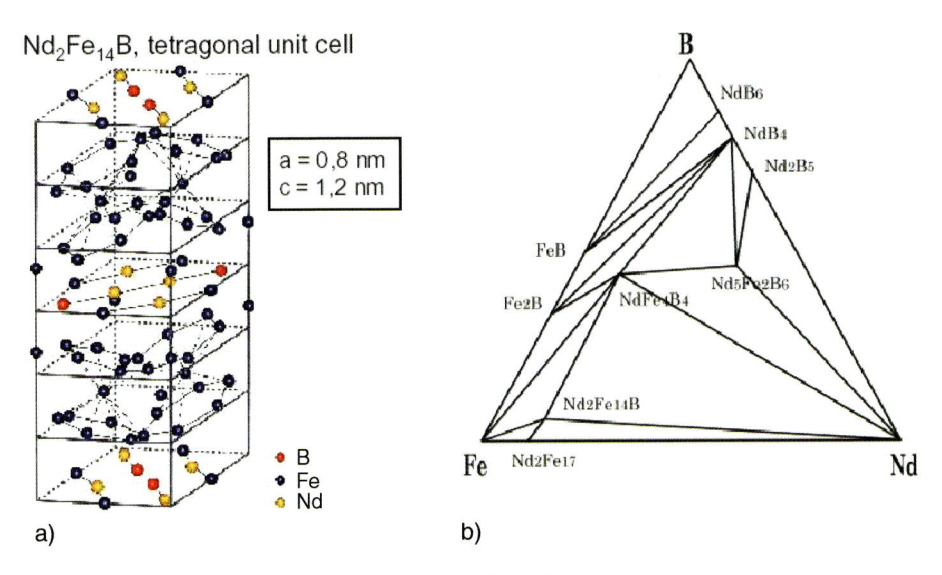

Fig. 14.12 (a) Tetragonal structure of $Nd_2Fe_{14}B$; (b) phase diagram of Nd–Fe–B.

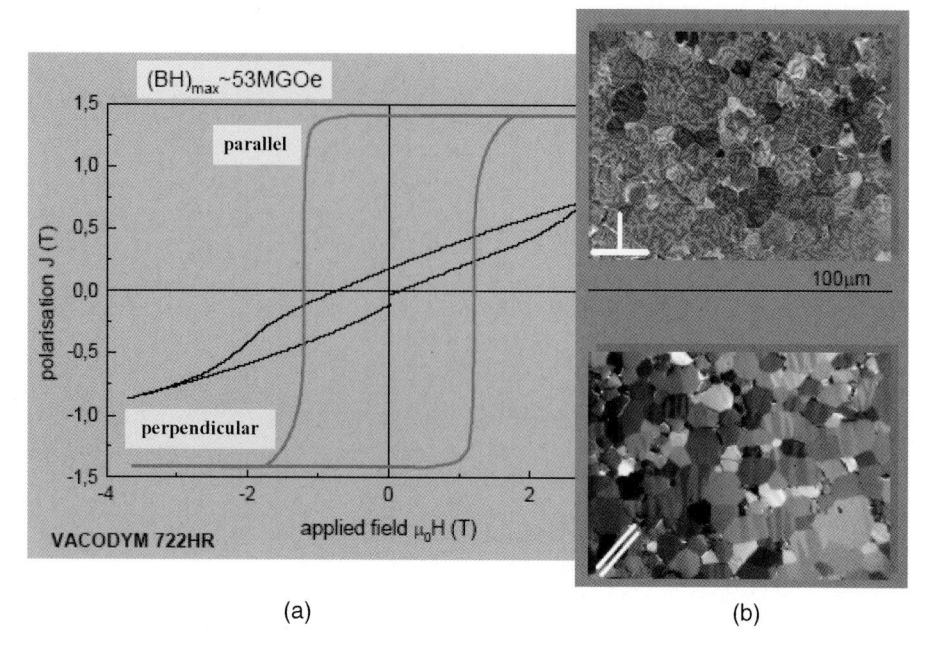

Fig. 14.13 (a) Hysteresis loop measured parallel (grey) and perpendicular (black) to the preferential orientation (Rodewald et al. 2002); (b) microstructure of a world record Nd–Fe–B-based permanent magnet (Gutfleisch 2001).

Nd–Fe–B magnets are available in both sintered and bonded forms. Sintered Nd–Fe–B offers the highest magnetic properties (28–50 MGOe), whereas bonded Nd–Fe–B offers lower energy products for the advantage of easy machinability. Although bonded magnets do not possess magnetic properties as advanced as those of sintered magnets, they can be made in shapes and sizes that are difficult to achieve with sintering. A variety of coatings can be applied to the magnets' surface to overcome the principle drawback of neodymium-based magnets, namely their tendency to corrode easily.

In order to demonstrate the excellent properties of this kind of material, Fig. 14.13 shows the hysteresis loop as well as the microstructure of a world record Nd–Fe–B-based permanent magnet with $(B.H)_{max} = 56$ MGOe (Rodewald et al. 2002).

14.1.3.5 Nanocrystalline Materials

Some years ago nanocrystalline hard magnetic materials were invented and studied (Coehoorn et al. 1988; Jha et al. 1989; Liou et al. 2004). For grains below approximately 30 nm the grain size becomes comparable to the magnetic exchange length. This leads to a remanence enhancement, which consequently causes a higher stored energy $(B.H)_{max}$. The remanence enhancement was first observed in Nd–Fe–B–Si (Keem et al. 1987) and also later in many other compounds. It was shown that nanocrystalline hard magnetic materials exhibit magnetic properties which are different from those of microcrystalline magnetic materials (Dahlgren at al. 1996b, 1997a,b). For applications the most interesting feature is that, due to the enhanced remanence, the energy product is increased even for isotropic grains with a uniaxial easy axis. Here the ratio M_r/M_s is higher than the expected 0.5 for noninteracting uniaxial grains according to Stoner and Wohlfarth (1948). Additionally, they exhibit a single-phase type of magnetic hysteresis loop instead of the expected two-phase type of hysteresis loop that is measured for microcrystalline two-phase magnets. Also, the shape of the recoil curves is characteristic of so-called spring type magnets (McCormick et al. 1996).

In Fig. 14.9, the remanence as a function of the coercivity of different Sm–Co- and Nd–Fe–B-based materials is compared. The special position of the nanocrystalline hard magnetic materials is obvious. By varying the actual composition as well as the heat treatment, the renamence and the coercivity of these materials can be changed dramatically. A comparison of the actual microstructure with the macroscopic data (hysteresis loop, magnetization, anisotropy) can be used as input for any kind of modeling (Hauser and Grössinger 2001).

Most investigations of nanomagnets were performed on RE–Fe–B-based samples. These isotropic nanocrystalline materials can be classified as three different compositions (Dahlgren 1997b).

- *Low RE content*: In this case nanocrystalline $RE_2Fe_{14}B$ grains are mixed with nanocrystalline soft magnetic α-Fe grains, or if the amount of B is high enough, with soft magnetic Fe_3B. Here, an even higher remanence enhancement than for stochiometric RE–Fe–B occurs, due to polarization of the α-Fe with a higher magnetic moment – these are called nanocomposite magnets.
- *Stoichiometric RE–Fe–B*: There the pure exchange coupling between the nanocrystalline $RE_2Fe_{14}B$ grains can be studied.
- *High RE content*: The nanocrystalline $RE_2Fe_{14}B$ grains are isolated from an R-rich phase, which at room temperature is nonmagnetic. This material is technically irrelevant.

Davies and his group (Davies et al. 1993), as shown in Fig. 14.14, carefully investigated the effect of varying the Nd content on the magnetic properties of Nd–Fe–B.

The development of new production methods combined with the discovery of the nanocrystalline state opened new horizons for magnetic materials research, and revolutionized the world of permanent magnets. Since then, many hard magnetic systems starting from Nd–Fe–B but also including the structure types 1/12, 2/17, 3/29, 1/5 etc. have been made with nanosize microstructures and with large values of coercivity (Pinkerton and Winterton 1989; Ding et al., 1993; Kuhrt et al. 1993; Shen et al. 1993). Besides melt spinning (Dahlgren et al. 1997b) and splat cooling, other rapid solidification techniques have been used, such as vapor depo-

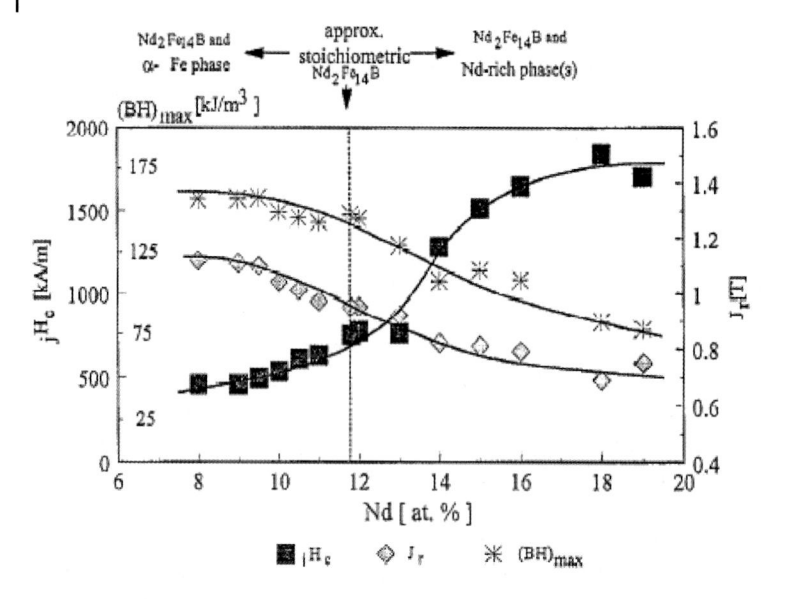

Fig. 14.14 Dependence of the coercivity $_jH_c$, the remanence J_r and the energy product $(B.H)_{max}$ on the Nd content in a Nd–Fe–B alloy. (after Davies et al. 1993.)

sition (Cadieu 1992, 1995; Sellmyer 1992), atomization (Narasimhan et al. 1986) and mechanical alloying (McCormick et al. 1998; Schultz 1990) for the fabrication of nanophase magnets. Recently, sputtering techniques have also been used to prepare nanosize Sm–Co alloys (Lambeth 1996; Sellmyer 1996) and CoPt and FePt alloys for high-density recording media (Liu et al. 1998; Starroyiannis et al. 1998).

Mechanical alloying is another method for producing nanocrystalline materials. This method is always followed by a heat treatment (Ding et al. 1994). In order to produce nanocrystalline materials in larger amounts, which are industrially of interest, severe plastic deformation (SPD) was suggested (Giguère et al. 2002).

14.1.3.6 Industrial Nanocrystalline Hard Magnetic Materials

Whereas soft magnetic nanocrystalline materials are produced widely and used by industry, the situation for hard magnetic nanocrystalline materials is different. Only Magnequench really offers commercial powder in which nanocrystalline Nd–Fe–B is used. MQP™ powders are based upon Magnequench's patented RE–Fe–B alloy compositions, which are rapidly solidified from the molten state at extremely high cooling rates, on the order of 10^6 s^{-1}, by melt spinning. This rapid solidification results in a material which has an extremely fine (typically 30–50 nm) metallurgical grain structure. Because the resulting grain size is smaller than the critical size for a single magnetic domain, these materials are magnetically isotropic. Further, in contrast to the fine, anisotropic powders that are used to manufacture sintered RE–Fe–B magnets, MQP™ powder is relatively stable against oxidation-induced demagnetization. These characteristics make MQP™ powders ideally suited for the production of bonded permanent magnets (see: http://www.magnequench.com/).

14.1.4
Outlook

The newest developments concerning hard magnetic materials are application-dominated. Nd–Fe–B-based bulk magnets with high coercivity but without the expensive Dy or Tb are a target of research. Magnets based on Sm–Co which are stable at high temperatures are also under development. Another trend goes in the direction of producing high-quality and stable hard magnetic films, mainly for micromachines, as well as high-density recording applications. Here the study (experimental as well as by modeling) of dynamic magnetization process is a challenging task.

In soft magnetic materials amorphous micro- and nanowires are being studied because they are of interest for sensor applications. This is related to their magnetic behavior, characterized by a square hysteresis loop. This phenomenon is very useful for a large number of technological applications. The study of highly ordered arrays of magnetic wires with diameters typically in the range of tens to a hundred nanometers is also a topic of growing interest for storage applications. The ordering by self-assembly which occurs between the nanowires during their production, together with the magnetic nature of nanowires, gives rise to outstanding cooperative properties different from bulk and even from thin-film systems; this is of fundamental and technological interest.

References

Boll, R., Hinz, G. (1985), *Techn. Messen* **52**(5), 189.

Cadieu, F.J. (1992), *Physics of Thin Films* **16**, Academic Press, San Diego.

Cadieu, F.J. (1995), *Int. Mater. Rev.* **40**, 137.

Chin, G.Y., Wernick, J.H. (1980), *Ferromagnetic Materials*, Vol. 2, Ed. E.P. Wohlfarth, North-Holland Publ. Amsterdam, p. 55.

Coehoorn, R., de Mooij, D.B., Duchateau, J.P.W.B., Buschow, K.H.J. (1988), *J. Phys.* **C8**, 669.

Croat, J.J., Herst, J.F., Lee, R.W., Pinkerton, F.E. (1984), *J. Appl. Phys.* **55**, 2078.

Dahlgren, M. (1998), Thesis, T.U. Vienna.

Dahlgren, M., Grössinger, R., Hernando, A., Holzer, D., Knobel, M., Tiberto, P. (1996a), *J. Magn. Magn. Mater.* **160**, 247.

Dahlgren, M., Kou, X.C., Grössinger, R., Wecker, J. (1996b), *Proc. 9th Internat. Symp. Magnetic Anisotropy and Coercivity*; Ed. F.P. Missel, V. Villas-Boas, H.R. Rechenberg, F.J.G. Landgraf, World Scientific, Sao Paulo, p. 307.

Dahlgren, M., Grössinger, R., de Morais, E., Gama, S., Mendoza, G., Liu, J.F., Davies, H.A. (1997a), *IEEE Trans. Magn.* **MAG-33**, 3895.

Dahlgren, M., Kou, X.C., Grössinger, R., Liu, J.F., Ahmad, L., Davies, H.A., Yamada, K. (1997b), *IEEE Trans. Magn.* **MAG-33**, 2366.

Davies, H.A., Manaf, A., Leonowicz, M., Zhang, P.Z., Dobson, S.J., Buckley, R.A. (1993), *Proc. 1st Internat. Conf. Nanostructured Materials*, Cancun (1992); (1993) *Nanostruct. Mater.* **2**, 197.

Ding, J., McCormick, P.G., Street, R. (1993), *J. Alloys Compounds* **191**, 197.

Ding, J., Liu, Y., McCormick, P.G., Street, R. (1994), *J. Appl. Phys.* **75**, 1032.

Egami, T., Flanders, P.J., Graham, C.D. (1974), *AIP Conf. Proc.* **24**, 697.

Estevez Rams, E., Fidler, J., Dahlgren, M., Grössinger, R., Knobel, M., Tiberto, P., Allia, P., Vinai, F. (1996), *J. Appl. Phys. D* **29**, 848.

Giguère, A., Hai, N.H., Dempsey, N.M., Givord, D. (2002), *J. Magn. Magn. Mater.* 242–245, 581.

Goss, N.P. (1934), U.S.Patent 1 965 559.

Grössinger, R. (1990), *Anal. Fisica Ser. B* **86**, 135.

Grössinger, R., Sato, R. (2003), *Hard Magnetic Properties of Rapidly Solidified Alloys and Spring Magnets*, Ed. P. Tiberto, F. Vinai, Research Signpost India, p. 183.

Grössinger, R., Holzer, D., Kussbach, C., Sinnecker, J.P., Sato Turtelli, R., Dahlgren, M. (1995), *Proc. IV Internat. Workshop on Non-Crystalline Solids*, Ed. Vazquez, M., Hernando, A., World Scientific, Singapore, p. 480.

Grössinger, R., Sato Turtelli, R. (2004), *Workshop Metallurgy and Magnetism*, Ed.: Kawalla, R., Institut für Metallformung; Deutsche Bibliothek, Freiberg.

Gutfleisch, O. (2001), *Hartmagnete*, Heraeus Summer School, TU Dresden, September 10.

Hasegawa, R. (1982), *Glassy Metals: Magnetic, Chemical and Structural Properties*, CRC Press, Boca Ratou.

Hauser, H., Grössinger, R. (2001), *J. Magn. Magn. Mater.* **226–230**, 1254.

Hernando, A., Kulik, T. (1994), *Phys. Rev. B* **49**(10), 7064.

Herzer, G. (1995), *Proc. IV Internat. Workshop on Non-Crystalline Solids*, Ed. Vazquez, M., Hernando, A., World Scientific, Singapore, p. 449.

Herzer, G. (1990), *IEEE Trans. Magn.* **MAG-26**, 1397.

Herzer, G. (1994), *IEEE Trans. Magn.* **30**, 4800.

Herzer, G. (1997), *Handbook of Magnetic Materials*, Ed. Buschow, K.H.J., Elsevier, Amsterdam, p. 415.

Jha, A., Davies, H.A., Buckley, R.A. (1989), *J. Magn. Magn. Mater.* **80**, 109.

Keem, J.E., Clement, G.B., Kadin, A.M., McCallum, P.G. (1987) *Proc. Conf. ASM Materials Week* **87**, Ed. Salsgiver, J.A., American Society for Metals, p. 87.

Kools, F. (1994), *Kirk-Othmer Encyclopedia of Chemical Technology*, Vol. 10, 4th edn.

Kools, F., Morel, A., Grössinger, R., Le Breton, J.M., Tenaud, P. (2002), *J. Magn. Magn. Maert.* **242–245**, 1270.

Köster, U. (1984), *Z. Metallkunde* **75**(9), 691.

Köster, U., Schünemann, U., Blanke-Bewersdorff, M., Brauer, S., Sutton, M., Stephenson, G.B. (1991), *Mater. Sci. Engng.* **A133**, 611.

Kraus, L., Duhaj, P. (1990), *J. Magn. Magn. Mater.* **83**, 337.

Kronmüller, H., Durst, K.D., Ervens, W., Fernengel, W. (1984), *IEEE Trans. Magn.* **MAG-20**, 1569.

Kuhrt, C., Schnitzke, K., Schultz, L. (1993), *J. Appl. Phys.* **73**, 6026.

Lambeth, D. (1996), *Magn. Hysteresis in Novel Magnetic Materials*, Ed.: Hadjipanayis, G.C., NATO ASI Series No. 338, p. 767.

Liou, Y. (2004), *Proc. 18th Int. Workshop on High Performance Magnets and their Applications*, Vol. 1, HPMA France, p. 28.

Liu, J.F., Zhang, Y., Hadjipanayis, G.C. (1999), *J. Magn. Magn. Mater.* **202**, 69.

Liu, J.P., Liu, Y., Luo, C.P., Shan, Z.S., Sellmyer, D.J. (1998), *Appl. Phys. Lett.* **72**, 483.

Lotgering, F. (1974), *J. Phys. Chem. Solids* **35**, 1633.

Luborsky, F.E. (1983) *Amorphous Metallic Alloys*, Ed. Luborsky, F.E., Butterworth, London, p. 360.

Matthias, T., Zehetner, G., Fidler, J., Scholz, W., Schrefl, T., Schobinger, D., Martinek, G. (2002), *J. Magn. Magn. Mat.* **242–245**, 1353.

McCormick, P.G., Ding, J., Feutrill, E.H., Street, R. (1996), *J. Magn. Magn. Mater.* **157–158**, 7.

McCormick, P.G., Miao, W.F., Smith, P.A.I., Ding, J., Street, R. (1998), *J. Appl. Phys.* **83**, 6256.

Narasimhan, K., Willman, C., Dulis, E.J. (1986), US Patent 4 588 439.

O'Handley, R.C. (1977), *Solid State Commun*, **21**, 1119.

O'Handley, R.C., Boudreaux, D.S. (1978), *Phys. Stat. Sol. a* **45**, 607.

O'Handley, R.C., Hasegawa, R., Ray, R., Chew, C.P. (1976), *Appl. Phys. Lett.* **29**, 330.

Pfeifer, F., Radeloff, C. (1980), *J. Magn. Magn. Mater.* **19**, 190.

Pinkerton, F.E., Van Wingerden, D.J. (1989), *IEEE Trans. Magn.* **MAG-25**, 3306.

Rodewald, W., Wall, B., Katter, M., Uestuener, K. (2002), *IEEE Trans. Magn.* **MAG-38**, 2955.

Sagawa, M., Fujimura, S., Togawa, M., Yamamoto, H., Matsuura, Y. (1984), *J. Appl. Phys.* **55**, 2083.

Sato Turtelli, R., Duong, V. H., Grössinger, R., Schwetz, M., Ferrara, E., Pillmayr, N. (2000), *IEEE Trans. Magn.* **MAG-36**, 508.

Schultz, L. (1990), *Science and Technology of Nanostructured Materials*, Ed. Prinz, G., Hajipanayis, G.C., NATO ASI Series No. 259, p. 583.

Schultz, L., Müller, K.H. (Eds.) (1998), *Proc. 15th Internat. Workshop Rare Earth Magnets and their Applications*, Vol. 1, Werkstoff Informationsgesellschaft, Deutschland.

Sellmyer, D.J. (1992), *J. Alloys Compounds* **181**, 397.

Sellmyer, D.J. (1996), *Magnetic Hysteresis of Novel Magnetic Materials*, Ed. Hadjipanayis, G.C., NATO ASI Series No. 338, p. 419.

Shen, B.G., Kong, L.S., Wang, F.W., Cao, L. (1993), *Appl. Phys. Lett.* **63**, 2288.

Starroyiannis, S., Panagiotopoulos, I., Niarchos, D., Christodoulides, J., Zhang, Y., Hadjipanayis, G.C. (1998), *Appl. Phys. Lett.* **73**, 3453.

Stoner, E.C., Wohlfarth, E.P. (1948), *Phil. Trans. Roy. Soc. A* **240**, 5299.

Stuijts, A. (1955), *Philips Tech. Rev.* **16(7)**, 205.

Stuijts, A. (1959), US Patent 2 900 344.

Suzuki, K., Makino, A., Inoue, A., Masumoto, T. (1994), *IEEE Trans. Magn.* **MAG-30**, 4776.

Vazquez, M., Ascasibar, E., Hernando, A., Nielson, O.V. (1977), *J. Magn. Magn. Mater.* **66**, 37.

Vazquez, M., Gomez-Polo, C., Chen, D.X., Hernando, A. (1994), *IEEE Trans. Magn.* **MAG-30**, 907.

Went, J. (1952), *Philips Tech. Rev.* **13(7)**, 194.

Yoshizawa, Y., Oguma, S., Yamauchi, K. (1988), *J. Appl. Phys.* **64**, 6044.

14.2
Invar Alloys
Peter Mohn

14.2.1
Introduction and General Remarks

By the name "Invar alloys" one nowadays understands a class of magnetic materials, which show a pronounced departure in their thermal expansion from the normal Grüneisen behavior (for normal metals the linear thermal expansion coefficient α at room temperature is of the order of $(10-20) \times 10^{-6}$ K^{-1}). The name "Invar" was coined by the swiss physicist Charles Édouard Guillaume who, in his search for a cheaper material for the meter standard to replace Pt–Ir, discovered that face-centered cubic (fcc) alloys of iron (Fe) and nickel (Ni) with a composition of about $Fe_{65}Ni_{35}$ exhibit almost vanishing thermal expansion in a broad temperature range around room temperature (Guillaume 1897).

Figure 14.15 shows this concentration dependence, whereas Fig. 14.16 depicts schematically the temperature dependence of the thermal expansion coefficient α for $Fe_{65}Ni_{35}$. Because of these outstanding properties the material was given the name Invar, attributed to its temperature-invariant volume. Guillaume's discovery

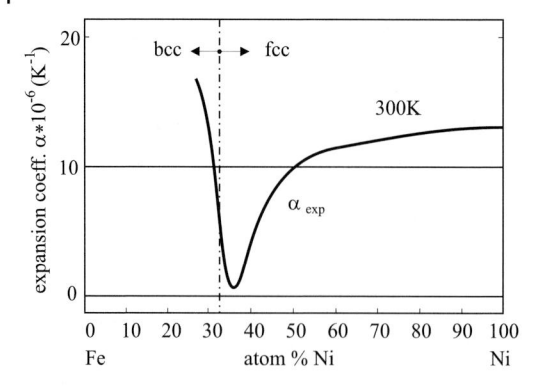

Fig. 14.15 Concentration dependence of the linear coefficient of thermal expansion α as a function of the alloy concentration for the Invar alloy Fe–Ni. (After Wassermann 1990).

immediately found widespread application in the construction of calibrated, high-precision mechanical instruments.

Applications are also found in electronic devices such as shadow masks in TV and computer screens, and resonant cavities of microwave and laser instruments. In addition, Invar alloys are used as structural components ranging from liquid gas containers to crude oil vessels and as core wires of long-distance power cables. Due to its importance for technological applications many different Invar alloys have been developed, among them many ternary and quaternary alloys. A material with a thermal expansion coefficient (TEC) of almost zero ($<10^{-7}$ K^{-1}), the so-called *super Invar* consists of 32 wt.% Fe–4 wt.% Ni–Co. A material which, besides its small TEC, also shows resistance to chemical corrosion, is stainless

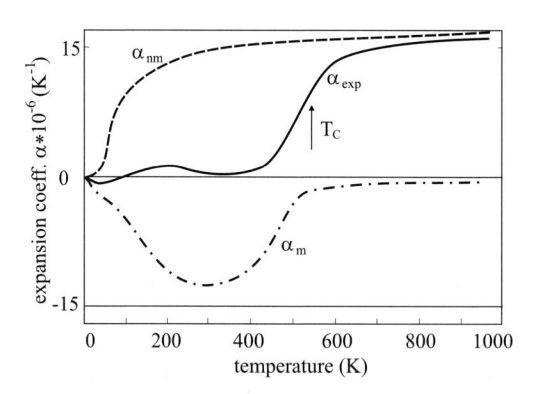

Fig. 14.16 Schematic temperature dependence of the thermal expansion coefficient (full line) in Fe$_{65}$Ni$_{35}$ which consists of a non-magnetic part α_{nm} (phonons) and a magnetic contribution α_m (magnetostriction). (After Wassermann 1990).

Invar, 54 wt.% Fe–9 wt.% Co–Cr. As structural materials exhibiting a tensile stress of >1 GPa, the work-hardening alloy Fe–Ni–Mo–C and the precipitation-hardening alloy Fe–NI–Co–Ti are in use. A further important alloy consists of 36 wt.% Fe–12 wt.% Ni–Cr, the so-called Elinvar, which shows temperature-invariant elastic properties and for decades was used as a material for springs in mechanical watches. It should be mentioned that for the discovery of Elinvar, C. E. Guillaume was awarded with the physics Nobel prize in 1920. Since all Invar materials are ferro- or anti-ferromagnetic (e.g., Cr–Fe–Mn) and the Invar-related properties occur only below the Curie/Neel temperature, it became evident very early that magnetism is inevitably entangled with the effect. These assumptions were also the basis of the earliest attempts to explain the Invar behavior. Most prominent among these early models is the 2γ state model (Weiss 1963). There it was argued that in fcc Fe–Ni Invar there should exist two magnetically ordered localized states, a ferromagnetic one with a larger volume γ_1 and an anti-ferromagnetic one with a smaller volume γ_2, where the latter can become thermally excited and thus compensate for the thermal expansion caused by the phonons. At first modern electronic structure theory seemed to confirm this main idea, when first-principles calculations of ferromagnetic γ-iron and ordered Fe_3Ni showed the existence of two magnetic states, a low-spin low-volume state and a high-spin high-volume state (Williams et al. 1982; Moruzzi 1990). Close to the Invar concentration the binding energy curves in these systems were found to consist of two distinct branches with a small energy difference between their respective minima, which has been viewed as a success of the 2γ-state model (Wassermann 1990) although the calculations grossly overestimate the volume decrease as a function of concentration. On the other hand, there are several experimental observations that contradict the 2γ-state model. For example, the phase transition from the high-spin to the low-spin state as a function of concentration should be first order (Akai et al. 1993; Abrikosov et al. 1995), while experimentally it is not (see, e.g., the specific heat results of Bendick and Pepperhof 1979). In addition, the two-minimum shape of the binding energy curve should lead to some discontinuities in the pressure dependence of several physical properties which, however, have not been seen experimentally. It must be said clearly that there is no experimental evidence whatsoever for the existence of two separate magnetic phases. Also, the earlier theoretical results (Williams et al. 1982; Moruzzi 1990), which originally seemed to support the 2γ-state model, turned out to be an artifact of the constraints applied during the calculations. Unfortunately and most probably because of its simple setup, the 2γ-state model is still employed to explain experimental results. Although it gives a phenomenological explanation for the macroscopic properties of Invar alloys, it does not make any direct connection with electronic structure theory.

The present microscopic understanding is based on the assumption of an interplay between local magnetic moments, which gradually become disordered with increasing temperature. These local moments couple with the conduction electron system, which itself becomes polarized and forms an itinerant component of the total magnetic moment, being responsible for the strong magnetovolume coupling. The success of rather complex types of interactions like this has been demonstrated by van Schilfgaarde et al. (1999), who in their calculations also allowed for noncollinear magnetic order. They find that at large volumes (corresponding to low temperature) the ground state is indeed given by parallel (ferromagnetic) spin alignment. When they reduce the volume the spins gradually depart from parallel alignment so that the spin directions become increasingly disordered. Since spin disorder is normally caused by finite temperatures, their result shows that, as far as the magnetic part of the problem is concerned, increased temperature leads to a smaller equilibrium volume. For Invar alloys this increasing disorder leads to a decreasing volume which again compensates the vibrational thermal expansion in the desired way.

14.2.2
Spontaneous Volume Magnetostriction

The central property which distinguishes Invar alloys from normal magnetic alloys is the anomalously large spontaneous volume magnetostriction ω_{s0}. Figure 14.17 shows this behavior schematically. Below the magnetic ordering temperature the fractional volume change departs strongly from the usual "nonmagnetic" behavior of the lattice ω_{lat} (broken line).

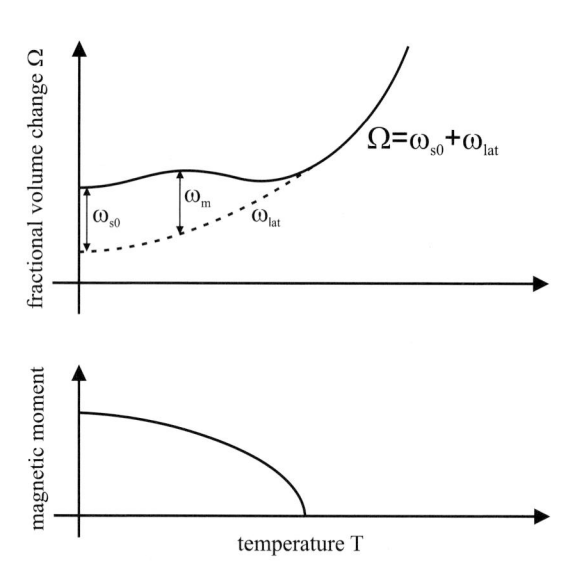

Fig. 14.17 Schematic temperature dependence of the volume change. The difference between the experimental volume and the hypothetical nonmagnetic volume (broken line) ($T = 0$ K) defines the spontaneous volume magnetostriction ω_{s0}. For temperatures above T_c the normal thermal expansion behavior due to the anharmonicity of the lattice vibrations (phonons) is regained (Grüneisen behavior).

The difference between the actual volume and a hypothetical ω_{lat} defines the spontaneous volume magnetostriction ω_{s0} at $T = 0$ K. For temperatures above T_c the normal thermal expansion behavior due to the anharmonicity of the lattice vibrations (phonons) is regained (Grüneisen behavior). It is exactly this volume anomaly which in Invar systems compensates for the thermal expansion due to the phonons. Any theory which attempts to explain the Invar effect has thus to account for the magnetic part of the volume expansion ω_m.

14.2.3
The Modeling of Invar Properties

As already stated earlier (Wassermann 1990), the name Invar is actually somewhat misleading, since in Invar alloys one observes a multitude of related anomalies. According to Wassermann, a better name would be "moment–volume instabilities in 3d-element-rich systems". In his review (Wassermann 1990) he also gives a list of Invar-related properties, which is partly repeated here (Table 14.6).

An analytic treatment of the magneto–volume effects within the Landau theory of phase transitions becomes relatively easy and straightforward if one describes the magnetic and mechanical properties in the lowest order by assuming the polynomial representation of Eq. (3), where the coefficients are related to physically easily accessible quantities. E_0 is an (arbitrary) energy zero point, which can also be taken as zero.

$$E(M, V) = E_0 + \frac{A}{2}M^2 + \frac{B}{4}M^4 + \beta V + \gamma V^2 + \delta M^2 V \tag{3}$$

The equilibrium magnetic moment M_0 and volume V_0 at $T = 0$ K are given by Eqs. (4) and (5).

Table 14.6 Invar-related properties.

Property	Variable
Thermal expansion	$(\Delta l/l)(T)$, $\alpha(T)$
Lattice constant	$a(T)$
Spontaneous volume magnetostriction	$\omega_s(T)$, $\omega_s(T=0) = \omega_{s0}$
Forced volume magnetostriction	$(d\omega/dH)(T)$
High-field susceptibility	$\chi_{hf}(T)$
Pressure dependence of the magnetization	$-(dM/dP)_{T,H}$
Pressure dependence of the Curie temperature	$-(dT_c/dP)$
Young's and bulk modulus	$\mathcal{E}(T)$, $\mathcal{B}(T)$
Elastic constants	$c_L(T)$, $c_{44}(T)$, $c'(T)$
Magnetization	$M(T)$

$$M_0^2 = -\frac{A + 2\delta V_0}{B} \tag{4}$$

$$V_0 = -\frac{\beta + \delta M_0^2}{2\gamma} \tag{5}$$

The coefficients A and B are related to the susceptibility χ_0 at $T = 0$ K via Eqs. (6) and (7).

$$A = -\frac{1}{2\chi_0} - 2\delta V_0 \tag{6}$$

$$B = \frac{1}{2\chi_0 M_0^2} \tag{7}$$

γ describes the harmonic part of the binding energy, whereas β, which depends on the isotropic bulk modulus $\mathscr{B}_0$, is responsible for the anharmonicity of the binding energy [Eq. (8)].

$$\mathscr{B}_0 = -2\gamma V_0 = \beta - \delta M_0^2 \tag{8}$$

Finally, and most importantly for the Invar effect, is the magneto–volume coupling constant δ, which is related to the critical pressure for the disappearance of magnetism P_c. Thus all quantities are related to ground-state properties of the respective alloy.

The next and crucial step is to introduce effects of finite temperature. As long as we are only interested in the magnetic part of the thermal expansion, we can restrict ourselves to magnetic thermal excitations and add the phonon contribution according to the Grüneisen relationship, which assumes that the thermal expansion due to the phonons is proportional to the Debye law for the specific heat.

An elegant way to introduce effects of finite temperature is to assume thermally induced fluctuations of the magnetic moment. One replaces the bulk magnetization M^{2n} by a new quantity, which also contains spin fluctuations $\langle m_\parallel^2 \rangle$ and $\langle m_\perp^2 \rangle$ parallel and perpendicular to the direction of $\vec{M}$. The idea behind the inclusion of magnetic fluctuations is that they, in an elegant way, resemble the effects of collective excitations. It should be noted that in our formulations magnetic fluctuations are treated classically, allowing only for a high-temperature description of the magnetic properties, which however is appropriate for the present problem. A detailed derivation is given elsewhere (Mohn 2005). Including these magnetic fluctuations, the free energy expansion now reads as Eq. (9).

$$\begin{aligned}\Delta F = E_0 &+ \frac{A}{2}(M^2 + \langle m_\parallel^2 \rangle + 2\langle m_\perp^2 \rangle) \\ &+ \frac{B}{4}\begin{pmatrix} M^4 + M^2(6\langle m_\parallel^2 \rangle + 4\langle m_\perp^2 \rangle) \\ +8\langle m_\perp^2 \rangle^2 + 3\langle m_\parallel^2 \rangle^2 + 4\langle m_\parallel^2 \rangle\langle m_\perp^2 \rangle \end{pmatrix} \\ &+ \beta V + \gamma V^2 + \delta V(M^2 + \langle m_\parallel^2 \rangle + 2\langle m_\perp^2 \rangle) \end{aligned} \tag{9}$$

The equations of state for the magnetic field H and the pressure P become Eqs. (10) and (11).

$$H = AM + BM^3 + \frac{B}{2}M(6\langle m_\parallel^2 \rangle + 4\langle m_\perp^2 \rangle) + 2\delta VM - P \tag{10}$$

$$= \beta + 2\gamma V + \delta(M^2 + \langle m_\parallel^2 \rangle + 2\langle m_\perp^2 \rangle) \tag{11}$$

At the Curie temperature T_c the bulk magnetization M vanishes and the parallel and perpendicular magnetic fluctuations $\langle m_\parallel^2 \rangle$ and $\langle m_\perp^2 \rangle$ become equal. In these conditions, the mean-square of the fluctuating magnetic moment $\langle m_c^2 \rangle$ at T_c is given by Eq. (12).

$$\langle m_c^2 \rangle = \frac{\gamma A - \delta(\beta - P)}{3\delta^2 - 5\gamma B} \tag{12}$$

Since numerical evaluations found that to a good approximation $\langle m_{\perp,\parallel}^2 \rangle$ varies linearly with temperature and that $\langle m_\parallel^2 \rangle$ and $\langle m_\perp^2 \rangle$ are essentially equal for an isotropic system, one can simplify Eq. (9) considerably by assuming Eq. (13) to hold.

$$\langle m_{\perp,\parallel}^2 \rangle = \langle m_c^2 \rangle (P=0) \frac{T}{T_c(P=0)}$$

$$= \langle m_c^2 \rangle (P) \frac{T}{T_c(P)} \tag{13}$$

We are now able to derive a number of important quantities for the description of Invar alloys [Eqs. (14–23)].

- The magnetic contribution α_m to the thermal expansion coefficient:

$$\alpha_m = \frac{1}{3V_0}\frac{dV}{dT} = \frac{1}{3T_c}\frac{2B\delta\langle m_c^2\rangle}{A\delta - B(\beta + P)} \tag{14}$$

- The critical pressure P_c for the disappearance of magnetism:

$$P_c = \frac{\gamma A - \beta\delta}{\delta} \tag{15}$$

- The pressure dependence of the Curie temperature, which follows from the assumed pressure independence of $\langle m_c^2 \rangle / T_c$ [Eq. (13)]:

$$\frac{dT_c}{dP} = -\frac{T_c}{P_c} \tag{16}$$

- The high-field susceptibility χ_{hf} taken at constant pressure:

$$\chi_{hf} = \chi_P = \chi_V \cdot \eta \tag{17}$$

$$\frac{1}{\chi_P} = -\frac{2\delta}{\gamma} P_c \left[1 - \frac{T}{T_c}\right] \tag{18}$$

$$\frac{1}{\chi_V} = 2BM^2 \tag{19}$$

$$\eta = \left[1 - \frac{\delta^2}{B\gamma}\right]^{-1} \tag{20}$$

χ_V is the susceptibility at constant volume, and η is a temperature-independent magneto–mechanical enhancement factor. It has been shown that for, e.g., the Fe–Ni system, η becomes anomalously large in the vicinity of the Invar compositions (Shimizu 1981).

- The forced magnetostriction h:

$$h = \frac{d\omega}{dH} = \frac{1}{V}\frac{dV}{dH} = \frac{1}{V}\frac{dV}{dM}\frac{dM}{dH}\bigg|_P$$

$$= \frac{d\omega}{dM}\frac{dM}{dH}\bigg|_P = -\frac{\delta M}{\gamma V}\chi_P \tag{21}$$

According to Eq. (18) for $T = P = H = 0$ Eq. (21) reduces to Eq. (22)

$$h = \frac{M_0}{2V_0}\frac{1}{P_c} \tag{22}$$

- The spontaneous volume magnetostriction ω_{s0} as described in Fig. 14.17 is then given by Eq. (23).

$$\omega_{s0} = \omega_m(T=0) - \omega_{lat}(T=0) = -\frac{\delta M_0^2}{\mathscr{B}_0} \tag{23}$$

One finds that the figure of merit, ω_{s0}, depends directly on the magneto–volume coupling constant δ and the equilibrium magnetic moment M_0. Since also the bulk modulus $\mathscr{B}_0$ enters Eq. (23), it turns out that the desired large values for ω_{s0} require lattices with a reasonably small bulk modulus.

It should be noted that all quantities depend essentially on ground-state properties and that a few physically intuitive parameters are sufficient to allow a phenomenological description of the magneto–volume effects.

14.2.4
A Microscopic Model

When calculating ω_{s0} according to Eq. (23), it has to be assumed that the nonmagnetic reference volume is really a fictitious nonmagnetic state. Nonmagnetic in this context means that there are no magnetic moments present. It is of course well accepted that the paramagnetic state (the state above the Curie temperature) in general can be seen as a state where the magnetic moments are completely disordered, rather than a state where all magnetic moments have vanished. A microscopic modeling of the Invar properties must therefore take into account this magnetically disordered state. An elegant way to simulate this magnetic state is provided by DLM (disordered local moment) calculations, which make it possible to model magnetic disorder on the basis of a CPA (coherent potential approxima-

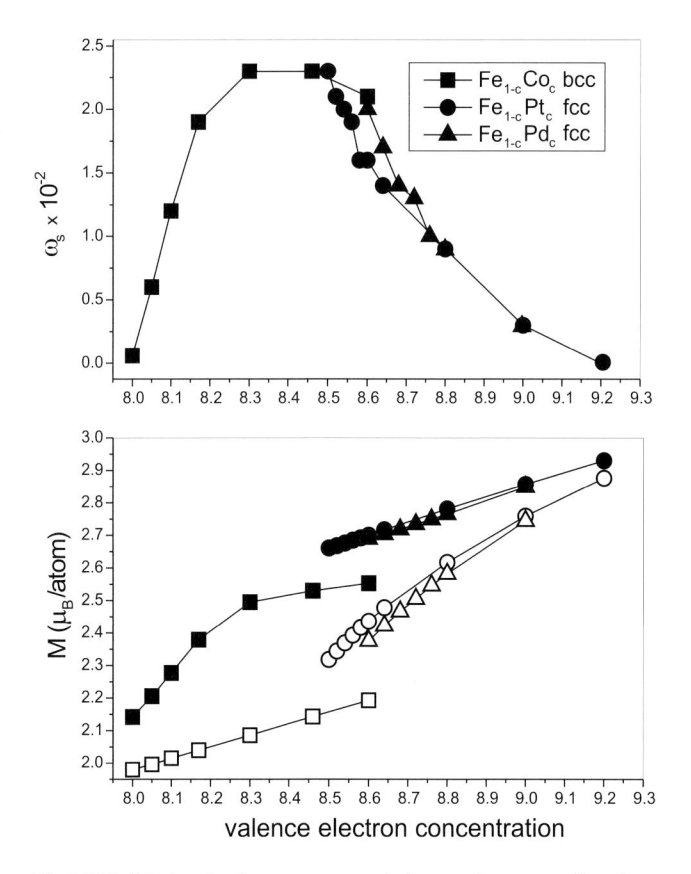

Fig. 14.18 Calculated volume magnetostriction ω_{s0} (upper panel) and local atomic moments of Fe alloys (lower panel). For the local moments full symbols refer to the ferromagnetic and open symbols to the disordered local moment state, respectively.

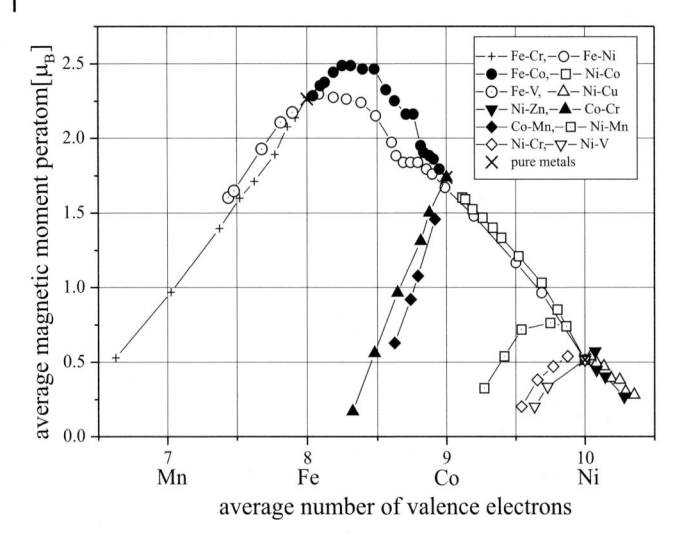

Fig. 14.19 Slater–Pauling curve for ferromagnetic transition metal alloys.

tion) calculation (Cyrot 1970; Hubbard 1981; Gyorffy et al. 1985; Khmelevskyi et al. 2003). The results from such a calculation are shown in the upper panel of Fig. 14.18, where the spontaneous volume magnetostriction ω_{s0} is plotted as a function of the average valence electron number of the binary alloys Fe–Co, Fe–Pt, and Fe–Pd. It is immediately seen that for all alloys there exists a common maximum for ω_{s0} around 8.4 electrons/atom. The lower panel of Fig. 14.18 depicts the respective ferromagnetic (FM) ground-state moments and the DLMs (magnetic moments in the fully disordered state).

The maximum in ω_{s0} correlates with the maximum in the difference between the FM moment and the DLM, which resembles the result given in Eq. (23). This behavior is also in agreement with the Slater–Pauling curve (Fig. 14.19), where one plots the magnetic moment of an alloy as a function of the valence electron concentration. The general behavior again shows a maximum around 8.4 electrons/atom, where the magnetic behavior changes from weakly to strongly ferromagnetic.

One can thus conclude that Invar behavior always occurs if an alloy is on the verge of this change in the magnetic behavior and at the same time exhibits a large enough magnetic ground-state moment. In this sense Invar is not a singular feature of some alloys, but rather a general property, which, if all prerequisites are fulfilled, leads to the observed temperature independence of the thermal expansion coefficient.

14.2.5
Outlook

Although it has been on going for more than a century, Invar research is still an highly active field. In particular, the modern preparation methods allow production of tailor-made properties and applications, e.g., as temperature-driven actuators in micromachinery. A rather new field are the anti-Invar systems, which exhibit an exceptionally large thermal expansion coefficient, an effect which often occurs together with antiferromagnetic order (e.g., in fcc Fe).

References

Abrikosov, I. A., Eriksson, O., Söderlind, P., Skriver, H. L., Johansson, B. (1995), *Phys. Rev. B* **51**, 1058.

Akai, H., Dederichs, P. H., (1993), *Phys. Rev. B* **47**, 8747.

Bendick, W., Pepperhoff, W. (1979), *J. Phys. F: Metal Physics* **9**, 2185.

Cyrot, M. (1970), *Phys. Rev. Lett.* **25**, 871.

Guillaume, C. E. (1897), *C.R. Acad. Sci.* **125**, 235.

Gyorffy, B. L., Pindor, A. J., Staunton, J., Stocks, M., Winter, H. (1985), *J. Phys. F: Met. Phys.* **15**, 1337.

Hubbard, J. (1981), in *Electron Correlation and Magnetism in Narrow Band Systems*, Ed. T. Moriya, Springer, New York.

Khmelevskyi, S., Turek, I., Mohn, P. (2003), *Phys. Rev. Lett.* **91**, 037 201.

Mohn, P., *Magnetism in the Solid State* (2005), Springer Series of Solid State Science, Vol. 133, Springer, New York.

Moruzzi, V. L. (1990), *Phys. Rev. B* **41**, 6939.

Shimizu, M. (1981), *Rep. Prog. Phys.* **44**, 329.

van Schilfgaarde, M., Abrikosov, I. A., Johansson, B. (1999), *Nature* **400**, 46; Mohn, P. (1999), ibid. **400**, 18.

Wassermann, E. F. (1990), in *Ferromagnetic Materials*, Eds. K. H. Buschow, E. P. Wohlfarth, Vol. 5, North-Holland, Amsterdam, p. 237.

Weiss, R. J. (1963), *Proc. R. Soc. London* **82**, 281.

Williams, A. R., Moruzzi, V. L., Gelatt, C. D. Jr., Kübler, J., Schwarz, K. (1982), *J. Appl. Phys.* **53**, 2019.

14.3
Magnetic Media
Laurent Ranno

14.3.1
Data Storage

14.3.1.1 Information Storage

In our technological age, a huge amount of information needs to be stored in a digital manner. The trend is exponential and shows no signs of calming down. This is the driving force in the field of data storage and makes it a fast-evolving industry. Regarding microelectronics, Moore's law is the well-known self-replicating trend which predicts that the number of transistors per surface area will double every 18 months. In the last decade the growth of hard disk drive (HDD) bit density has been faster than that of microelectronics and has reached 100% bit density increase per year. A density of 100 gigabits/in^2 (Gb in^{-2}), originally thought to be a dream value, has been reached in laboratories in 2002 and is already in commercial HDD. The new target is 1 terabit/in^2 (Tb in^{-2}; 1 in = 2.54 cm). Data storage is a large industry and many technologies compete. The life cycle of technologies is only a few years long and innovation is constant. Before discussing the field of high density magnetic recording, let us detail some of the criteria that govern storage technologies.

Data The quantity of bits to store is a first parameter. A microprocessor will deal with 32 or 64 bits at a time and a few megabits (Mb) will be stored in cache memory. One hour of music (i.e., a CD-ROM) will be equivalent to a few 100 megabytes (MB) while one hour of video (i.e., a DVD-ROM) will need a few gigabytes of storage. The typical capacity of a personal computer hard disk drive has now reached 100 GB . A small archiving system will deal with terabytes (10^{12} bytes) and a large archiving system will store petabytes (10^{15} bytes). The total capacity of hard disk drives produced per year is already a few exabytes (10^{18} bytes).

Data access Not only do data need to be stored but they must also be accessed. Sequential access (e.g., tapes) and random access (e.g., random access memory, RAM) are two possibilities, the choice depending on the use. Sequential access is suitable for archiving or to watch a video, whereas accessing files on a hard disk needs random access, as any bit can be required at any moment. The access time and the access rate also matter. The timescale can span orders of magnitude from minutes to access a tape to nanoseconds to read an individual RAM cell, and the required data flow can reach gigabytes per second.

Volatility Data storage should show a certain lack of volatility. However, nonvolatility will be designed taking into consideration the relevant time scale. This can span orders of magnitude, from years in the archiving business down to microseconds in computing.

14.3.1.2 Competing Physical Effects

Bistable state A digital memory requires two well-defined and stable states to store a bit. Many physical effects offer such configurations. Bits can be stored using two physical levels by mechanically engraving the storage medium: bumps on a CD-ROM or DVD-ROM, tracks on old-fashioned LPs (analogue storage). Optical properties of the media can be modified to create reflecting/nonreflecting bits (recordable (CD-R) and erasable (CD-RW)) or bits with opposite magneto-optical properties (magneto-optical storage). The data can also be materialized by the presence or absence of an electrical charge as in semiconductor memories (SRAM, DRAM, flash), the presence or absence of electrical polarization like ferroelectric RAM (FeRAM), the crystallized or amorphous nature of an alloy as in phase change media (PCRAM memory) discussed in Section 14.5, and of course the direction of a magnetic moment (magnetic tapes, hard disk media, magnetic RAM), which is the subject of this section of Chapter 14.

Reading technique Depending on the way data have been stored and on data density (bit size), the appropriate reading technique is required. Mechanical bits (LPs) were read by using a mechanical contact; current CD-ROMs are read by using a focused infrared laser reflected from bumps in polycarbonate (bump height is a quarter of the wavelength in polycarbonate) so that the laser beam undergoes destructive interference when reflected from the edge of a bump. Optical bits are also read by measuring a reflected laser beam. Electrically stored bits can be read because they control (open or closed) the channel of a transistor, by monitoring the discharge of a capacitor or by sensing the electrical resistance of the bit (phase change RAM). Magnetic bits are read by sensing the stray magnetic field above them, or more directly by monitoring changes of the polarization of a laser (magneto-optical effects) or by measuring their resistance (magnetoresistive effect). Unexpected physical effects can be used in innovative systems such as the thermal dissipation change used in IBM's Millipede program.

Characteristics of magnetic recording A ferromagnetic material can possess a well-defined axis (the easy axis) along which its magnetic moments will align preferentially. The two possible directions will define two stable states. When the easy axis is well defined (large remanence of the magnetization along the easy axis in the absence of an applied field) and if the field necessary to reverse the magnetic moment (the coercive field) is large enough, the material will be called a permanent magnet and can be used to store information. The minimum size of a stable magnetic particle is less than 10 nm (depending on the material), which translates directly into high-density storage, and the intrinsic switching time of a particle magnetization is in the nanosecond range, which leads to short data read and write times.

14.3.1.3 Magnetic Storage

14.3.1.3.1 Aspects of magnetism for storage

Magnetization Ferromagnetic materials carry a spontaneous magnetic moment m (in A m^2). Normalized by the sample volume it is called the magnetization **M** (in A m^{-1}), which is also written as $\mathbf{J} = \mu_0 \mathbf{M}$ (in T). The value of magnetization depends on the nature of the magnetic elements. When an atom carries a magnetic moment, partially filled electronic shells must exist. Two series of elements with partially filled shells are at the origin of magnetism in most magnetic materials: 3d transition metals and 4f rare earths. Possessing magnetic atoms is not the only condition for observing ferromagnetism. Interaction between magnetic moments must exist and parallel coupling must dominate. Antiparallel coupling will induce antiferromagnetic order resulting in no net magnetic moment, i.e., no magnetization. Only a few elements are ferromagnetic at room temperature (3d transition metals Ni, Co, and Fe), iron being the most magnetic of them ($\mu_0 M = 2.2$ T). Of course the list of ferromagnetic materials is much longer and numerous alloys and compounds are ferromagnetic (Table 14.7). In ferromagnetic materials, magnetization is reduced to 0.5 T for oxides (oxygen does not carry any magnetic moment) and may increase to 2.4–2.5 T for FeCo alloy or Fe nitride. The magnetic state of the elements does not always give clues about the magnetic state of a compound. Cr is antiferromagnetic but CrO$_2$ is a ferromagnetic oxide used in tape recording. FeMn and NiO are antiferromagnetic. In

Table 14.7 Curie temperature and magnetization of selected materials.

Material	Fe	Co	Ni	$Fe_{50}Pt_{50}$
Magnetization $\mu_0 M$ [T] at 300 K	2.16	1.72	0.61	1.43
Curie temperature [K]	1043	1394	631	750
Material	$Fe_{50}Co_{50}$	$BaFe_{12}O_{19}$	$Co_{68}Cr_{20}Pt_{12}$	$Ni_{80}Fe_{20}$
Magnetization $\mu_0 M$ [T] at 300 K	2.5	0.48	0.63	1.04
Curie temperature [K]	1253	723		595

recording, high magnetization is not a necessary requirement. When stray fields are being read, the signal is proportional to magnetization, but for magneto-optical media, magneto-optical properties of the material will matter and they do not scale as magnetization. In both cases large magnetizations induce large demagnetizing effects, which have to be addressed for long-term bit stability.

Curie temperature Spontaneous magnetization M_s is the value of magnetization in a magnetic domain in the absence of an applied field. It is the value which quantifies the magnetic order in zero field. M_s decreases with increasing temperature and disappears at the Curie temperature T_c (Fig. 14.20). T_c characterizes the phase transition between long-range magnetic order and disorder, i.e., the competition between the ferromagnetic exchange energy, characterized by the exchange constant A (in pJ m^{-1}) and thermal energy kT. The exchange energy represents the cost of non-uniform magnetization, $E_{exchange}$ [Eq. (24)]

$$E_{exchange} = \int_V A \left(\nabla \frac{M_i}{M_s} \right)^2 dV \quad \text{and} \quad i = x, y, z \tag{24}$$

T_c should be well above room temperature (i.e., >450–500 K) for most applications, since not only magnetization but anisotropy and coercive field also go to zero at T_c. The nonmagnetic elements (Ta, Cr, B) which are added to some alloys to control their microstructure can cause a decrease of both T_c and magnetization.

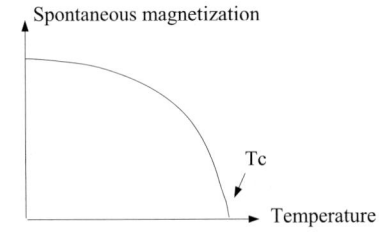

Fig. 14.20 Temperature dependence of spontaneous magnetization.

Anisotropy – magnetic easy axis To be considered as a permanent magnet with two stable antiparallel configurations, a magnetically easy axis should be developed in the material. The required anisotropy has several possible origins: shape anisotropy, crystalline anisotropy, or induced anisotropy. An elongated shape favors the alignment of magnetization along the long axis e.g., needle-like particles used for low-density storage (tapes). Tetragonal or hexagonal (e.g., cobalt) crystallographic structures can induce a strong easy axis along the c axis. Such a strong magnetocrystalline anisotropy cannot exist in a cubic structure, which is too symmetrical. Deposition or thermal annealing under magnetic field may slightly order a disordered alloy and induce an easy axis ($Ni_{80}Fe_{20}$). When an easy axis exists, the uniaxial anisotropy is characterized by the anisotropy constant K (in J m^{-3}) according to Eq. (25), θ being the angle between the magnetization and the easy axis.

$$E_{anisotropy} = K \sin^2 \theta \tag{25}$$

$K > 0$ leads to an easy axis and $K < 0$ to a hard axis (easy plane), which is not relevant for storage media.

Shape – demagnetizing effects When a particle or a sample is uniformly magnetized, the exchange energy is a minimum. However, the stray magnetic field around the sample represents a large magnetic energy stored in space. To evaluate the stray field, a mathematical analogy with electrostatics can be used, where magnetism is replaced by surface pseudo-charges (density $\sigma = \mathbf{M}.\mathbf{n}$) and volume pseudo-charges (density $\rho = -\mathrm{div}\,\mathbf{M}$). The energy balance for a large sample usually gives the stable state of a ferromagnetic sample as the magnetic configuration with the minimum stray field, i.e., the demagnetized state. To get zero net magnetic moment without paying too much exchange energy, magnetic domains with opposite magnetizations are created. Between domains, the region where the magnetization rotates is called a domain wall. The width δ and the energy γ of a domain wall are mainly determined by the exchange energy A, which favors smooth transitions, and the anisotropy energy K, which favors sharp transitions [Eqs. (26)].

$$\delta(\mathrm{m}) = \pi\sqrt{\frac{A}{K}} \quad \text{and} \quad \gamma(\mathrm{J\ m}^{-3}) = 4\sqrt{AK} \tag{26}$$

Depending on the domain wall geometry, both prefactors in Eqs. (26) may vary (here both prefactors, π and 4, correspond to a 180° Bloch wall). A typical value for A is 10 pJ m^{-3}. For cobalt ($K = 0.5$ MJ m^{-3}) it gives rise to domain walls of width 14 nm and wall energy 9 mJ m^{-2}.

A direct consequence of demagnetizing effects is that at zero field, the measured magnetization is usually lower than the spontaneous magnetization (Fig. 14.21). It is called M_r, the remanent magnetization, and this is the magnetization which is sensed by the read head. The field needed to reverse half the

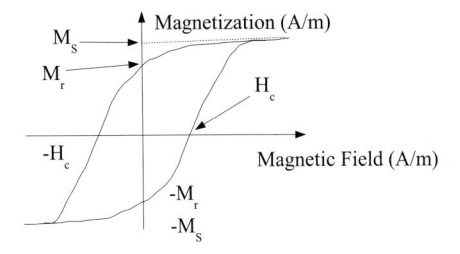

Fig. 14.21 Magnetization as a function of magnetic field for a ferromagnet.

magnetization (i.e., to reach zero net magnetization) is the coercive field H_c (in A m^{-1}).

The art of permanent magnet design is to prevent demagnetizing fields from demagnetizing the magnet. Limiting domain nucleation or pinning domain walls to impede domain wall propagation are possible strategies.

Magnetic states As far as recording is concerned, only two magnetic states per particle are required. Once an easy axis has been established, the particle size should be reduced below a critical size: the single domain size. This critical size represents the limit below which a multidomain magnetic configuration becomes unfavorable. Since the domain wall energy scales as a surface and the demagnetizing energy scales as a volume, below the critical size the cost of introducing a domain wall will not be balanced by the gain in demagnetizing energy. For a spherical particle (radius r), assuming that introducing one domain wall (cost $\gamma\pi r^2$) reduces the demagnetizing energy by 50% (gain $\frac{1}{12}\mu_0 M^2 V$), one gets the critical radius r_c for a spherical particle from Eq. (27) (compare Table 14.8)

$$r_c = \frac{9\gamma}{\mu_0 M^2} \tag{27}$$

Magnetization reversal Good recording media must possess properties quite similar to those of a permanent magnet. However, in most media, information is to be erased and rewritten. This property implies that an applied magnetic field larger than the coercive field must be produced by the write head. This limits the upper value of the coercive field of recording media. The coercive field of high-density media has recently reached $\mu_0 H_c = 0.5$ T.

Table 14.8 Single domain radii for selected ferromagnetic materials.

Material	Fe	Ni	Co	Nd$_2$Fe$_{14}$B	SmCo$_5$
Single domain radius r_c [nm]	6	14	35	107	828

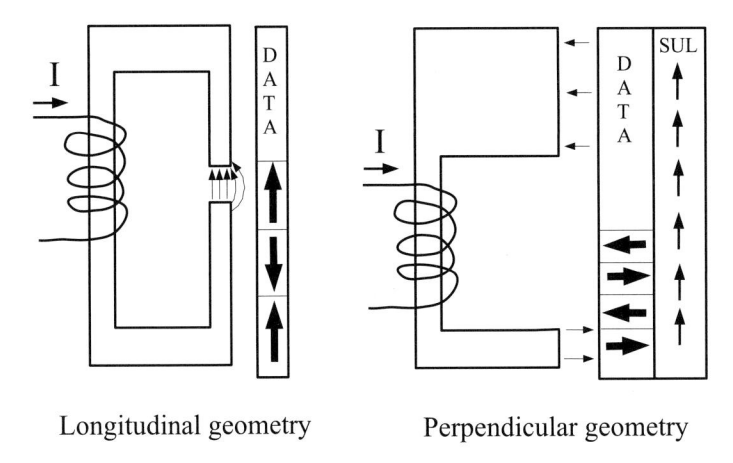

Fig. 14.22 Write head geometry in the case of longitudinal and perpendicular recordings.

Write head The principle of the write head is that of an electromagnet, which includes a soft magnetic circuit, a gap, and windings (Fig. 14.22). For longitudinal recording the applied field must be parallel to the medium surface which is next to the gap. For perpendicular recording, the field must be perpendicular to this surface. The **B** flux is applied through a narrow gap to the medium. Since induction lines are closed loops, the flux is guided through a soft underlayer (SUL) and comes back to the head through a second gap, which must be wide enough not to overwrite written bits. **B** flux is conserved, so a wide gap means a smaller field. In perpendicular recording, the magnetic bits are effectively in the gap of the write head, which allows larger fields to be applied.

Reversal dynamics The dynamics of magnetic moment reversal is governed by Larmor precession. In an applied field the magnetic moment precesses around the field and aligns with it. The Larmor precession frequency is of the order of 1 GHz in a 1 T field. More quantitatively, the dynamics of magnetization follows the Landau–Lifshitz equation [Eq. (28)], which includes a precession term and a damping term.

$$\frac{d\mathbf{M}}{dt} = \gamma \mathbf{M} \times \mathbf{H}_{\text{eff}} + \alpha \mathbf{M} \times (\mathbf{M} \times \mathbf{H}_{\text{eff}}) \tag{28}$$

In Eq. (28), γ is the gyromagnetic ratio, α is the damping constant, and $\mathbf{H}_{\text{eff}}$ is the effective field, given by Eq. (29).

$$\mathbf{H}_{\text{eff}} = -\frac{1}{\mu_0} \frac{\partial E}{\partial \mathbf{M}} \tag{29}$$

The effective field includes contributions from the applied field (Zeeman energy), the demagnetizing field (shape anisotropy), and magnetocrystalline and exchange energies. The demagnetizing field is not a local field, which makes micromagnetic calculations quite elaborate for objects too large to be considered as a single magnetic moment (macrospin approximation).

Subnanosecond reversals of submicronic magnetic elements have been obtained experimentally. The reversal mechanism for larger particles involves domain wall propagation, which is a much slower mechanism (an upper limit for domain wall velocity is the velocity of sound in the material).

Magnetostriction When magnetization rotates, due to the spin–orbit coupling, which couples spin moments and orbital moments (i.e., the electron distribution), the material experiences distortions. $\lambda_\parallel$ and $\lambda_\perp$ are defined as the magnetostriction coefficients parallel and perpendicular to the field direction. High magnetostriction induces stress and can lead to mechanical failure. Repetitive magnetic reversals will require materials with low magnetostrictive coefficients.

14.3.1.3.2 Read Head

One option for reading information stored as a magnetic bit is to sense the stray magnetic field. Any ferromagnetic material will create a magnetic field (called the demagnetizing field within the material and the stray field outside the material). An electrostatic analogy may be used. Magnetic pseudo-charges are created when the magnetization has a component perpendicular to a surface or when the divergence of magnetization is not zero. In longitudinal recording (Fig. 14.23), magnetization lies in the plane of the tape or disk and no surface pseudo-charges exist, but pseudo-charges will be present in the transition region between two antiparallel magnetizations (non-uniform **M**).

In principle the read head will then act as a magnetic field sensor. One of two physical effects are exploited: induction or magnetoresistance.

Induction When a relative movement between the head and the medium exists, an inductive head may be used. The signal is the induced voltage in a pick-up coil, given by Eq. (30).

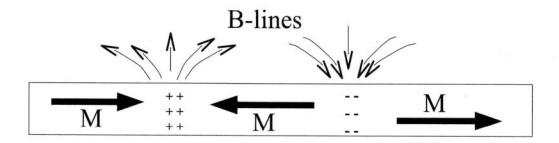

$$V = -\frac{d\mathbf{B}.\mathbf{S}}{dt} \tag{30}$$

Fig. 14.23 Stray magnetic fields above a longitudinal recording medium.

The main limitation of this sensing technique is miniaturization: the inductive signal scales as the surface of the pick-up coil.

Magnetoresistance   For high-density media (HDD), a magnetoresistive (MR) effect is preferred. The MR element can be down-sized, keeping its resistance constant. The relative head–medium velocity has no longer a direct influence on the signal amplitude. Nevertheless high velocity is still required in order to keep a fast access time.

The reading resolution must be adapted to the bit size, i.e., thin film technology has to be used. The fly height of the sensing element matters also. For tapes, a clean and controlled environment is not possible. The preferred situation is to have a mechanical contact between the head and the medium (which limits velocity and requires a lubrication layer). In a HDD, the environment is controlled (no dust particles), and the head can fly over the disk without permanent mechanical contact, which results in faster movements (10000 rpm and $r = 3$ cm gives 30 m s^{-1}).

Magneto-optical effect   Magneto-optical effects may also be used to read magnetic bits. A ferromagnetic material does not obey time-reversal symmetry. Its dielectric tensor has nonzero off diagonal elements which are the source of magneto-optical effects. When a linearly polarized laser beam is reflected perpendicularly to the magnetic surface (polar Kerr geometry), the plane of polarization of the reflected beam rotates. The rotation angle θ_K is $Q.M_z$ with Q the Verdet constant and M_z the perpendicular component of magnetization. Q depends on the material and on the wavelength. Shorter wavelengths will allow smaller bits to be read. However, since magneto-optical (M-O) properties strongly depend on wavelength, a change of laser requires a change in materials. Commercial M-O disks use lasers similar to those used for CDs and DVDs. So the trend is to go from infrared (780 nm) to red (650 nm) and now blue lasers (405 nm).

The application of the polar Kerr effect to recording necessitates the use of magnetic media with perpendicular anisotropy, and the large magnetization criterion will be replaced by a large Kerr angle criterion. More precisely the figure of merit is the intensity of the rotated polarization, which includes reflectivity and Kerr angle (a large Kerr angle with small reflectivity gives a poor signal/noise ratio).

Stability: superparamagnetism   A reduction in the size of magnetic bits requires a decrease in the size of crystallographic grains. For uniaxial materials, the anisotropy energy varies as $K.V \sin^2 \theta$, V being the volume of the grain (Fig. 14.24). Keeping the anisotropy of the crystal constant means that the energy barrier $K.V$ which separates the two stable states ($\theta = 0$ and $\theta = \pi$) becomes smaller.

When a ferromagnetic particle is small enough, it may behave as a paramagnetic particle, i.e., as if it had no permanent magnetic moment. This behavior is called superparamagnetism. Experimentally it is observed when the characteristic time for magnetic moment measurement is longer than the characteristic time

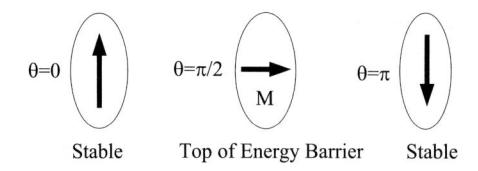

Fig. 14.24 Energy stability as a function of the magnetization orientation of a single domain particle.

Table 14.9 Critical grain diameter for a ten-year stability versus thermal reversal at 300 K.

Material	Fe	Co	PtFe	Ni
Critical diameter [nm]	22	10	4	50

for thermally induced magnetic reversal. For recording it means that the information is lost before it is read. Superparamagnetism occurs when the thermal energy kT is sufficiently large compared to $K.V$ for the magnetization reversal probability to be non-negligible in the absence of any applied field. The characteristic time τ for a reversal follows an Arrhenius law, where $1/\tau_0$ is the attempt frequency (10^9–10^{10} Hz) as in Eq. (31).

$$\tau = \tau_0 e^{KV/kT} \tag{31}$$

The critical grain size for recording is calculated using a ten-year limit for temperature-induced reversal of a grain's magnetization at 300 K (Table 14.9). There are typically 100 grains per magnetic bit, so that the reversal of a few grains does not destroy the information. In such conditions, the minimum volume of a grain is given by $60kT = K.V$.

Media noise The write head does not control the position of the written magnetic bits with respect to the crystallographic grains. The transition width between two bits will be at least as large as the grain size (Fig. 14.25). A distribution in the ori-

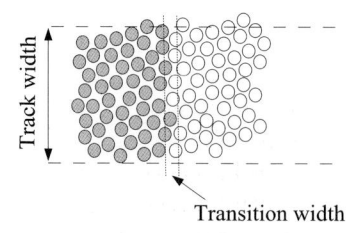

Fig. 14.25 Media noise associated to bit transition width.

entation of the anisotropy axes will also contribute to media noise. On the one hand, narrow transition regions are required to have well-defined magnetic bits and clear signals to read. On the other hand, sharp transitions lead to large volume pseudo-charges and thus to large demagnetizing fields. If one assumes that magnetization in the transition region can be written as Eq. (32) with a the transition width, then the demagnetizing field H_d created by the magnetic pseudo-charges associated with this non-uniform magnetization is given by Eq. (33), t being the thickness of the longitudinal medium. Since the demagnetizing field must be smaller than the coercive field, the transition width has to be kept larger than $M_r t/H_c$. Small $M_r t$ and large H_c are required to achieve high-density recording.

$$M(x) = M_r arctan\left(\frac{x}{a}\right) \tag{32}$$

$$H_d = -\frac{2M_r t}{\pi a} \frac{\frac{x}{a}}{1+\frac{x^2}{a^2}} \tag{33}$$

14.3.2
Magnetic Recording Media

14.3.2.1 Particulate Media

A magnetic storage medium consists of ferromagnetic particles. The position of the magnetic bit transition is not predetermined (the write head does not aim at magnetic grains), so neighboring grains should be magnetically decoupled to be able to locate the bit transition anywhere. The first magnetic media used were magnetic powders screen-printed onto a substrate. This method is still in use for low-density applications (floppy disks, audio and video tapes). The particle size is submicronic. The origin of the particle anisotropy gives two families of particles. For materials with small magnetocrystalline anisotropy, shape anisotropy is required and particles are needle-like (e.g., CrO_2, γ-Fe_2O_3, Fe). For systems with large magnetocrystalline anisotropy, particles can be spherical (e.g., $BaFe_{12}O_{19}$ hexagonal ferrites). Particulate media can be screen-printed onto large areas and onto flexible media (tapes). However, the volume fraction of magnetic materials is limited to about 60%. To improve this fraction, continuous film materials have been developed.

Floppy disks and audio tapes have been around for many years (1.44 MB 3.5 in floppy IBM 1987). Recent improvements have lead to 100–250 MB floppy disks and high-density digital tapes. The typical characteristics of present day tapes are as follows. A polymer binder 100 nm thick (charged with elongated iron-based particles 60 nm long) is coated onto a polyethylene substrate (5–10 μm thick). A prototype 1 TB tape makes it possible to write bits 76 nm long and to reach 1 Gb in^{-2} storage density. Positioning the head on a floppy medium is less accurate than onto a HDD, which leads to wider tracks. However, the linear character of a tape means that multiple tracks can be written and read in parallel.

14.3.2.2 Continuous Media – Film Media

Particulate media (particles, screen-printed to a tape) allow low density recording. Higher density requires the use of continuous media (films). Crystallographic grains must be magnetically decoupled, which, in a continuous medium, will involve a nonmagnetic intergranular phase. The size of the magnetic bits should be as close as possible to that of the superparamagnetic limit, and therefore a granular structure with a narrow grain size distribution is required. Grains are roughly equiaxed, so the origin of anisotropy must be magnetocrystalline and alloys based on hexagonal cobalt are widely used. Control of the crystallographic texture improves the remanence and thus the signal. For HDD media, the present magnetic bit density has reached 100 Gb in^{-2}, with the following characteristics (Fujitsu prototye 2002). The size of one magnetic bit is roughly 30 nm along the track (7.5×10^5 bits per inch: 750 kbpi) and the track is 180 nm wide (1.42×10^5 tracks per inch: 142 ktpi). The medium grain size is less than 10 nm and has to be as uniform as possible. To achieve stable bits and narrow transition widths, $\mu_0 H_c$ reaches 0.4 T and $M_r t$ is 0.37 emu cm^{-2} (1 emu = 10^{-3} A m^2).

14.3.2.2.1 Cobalt-Based Alloys

Structure Glass disks or Ni–P-coated Al–Mg (4 at.%) alloy substrates are used for HDD (the glass disks are harder and stiffer). A thick layer (10 μm) of amorphous Ni–P is electrolessly plated onto the Al–Mg disk. Ni–P can be polished mirror-like (roughness 5 nm) and a surface texture can be added. Ni grains are superparamagnetic in the Ni–P layer so they do not contribute any magnetic signal. The magnetic storage layer will be isolated from the substrate by an underlayer which induces crystallographic texture and a seed layer which controls the granular size distribution. Its top surface is protected by depositing a mechanically hard capping layer (diamond-like carbon or dense CN_x) and finally a lubricating layer is added.

Storage layer The magnetic storage layer is a CoCr-based alloy. Cr segregates, leaving magnetic Co-rich grains, separated by a Cr-rich, nonmagnetic, intergranular phase which decouples the Co grains. The Co grains must retain the hcp crystal structure, which is at the origin of the magnetocrystalline anisotropy and the large coercive fields. Ta or B addition will favor Cr segregation to the grain boundaries. However Cr addition to Co grains reduces the magnetocrystalline anisotropy. The role of Pt addition is then to increase Co anisotropy and such compositions are used to reach coercivities of 0.5 T. A typical composition is Co containing 15 at.% Cr, 10 at.% Pt, and a lower percentage of Ta or B. The CoCr alloy undergoes a transition from a highly anisotropic hcp to a low anisotropic fcc symmetry if too much Pt is substituted.

Underlayer A bcc-Cr underlayer develops a (110) texture, which induces cobalt (10–10) texture (*c* axis slightly out-of-plane). To get the Co *c* axis in-plane, Cr(002) textured layers are used. NiAl B2 structure grows with (112) texture. When a quaternary CoCrPtTa storage layer is being used, a better lattice match and crystallographic texture may require the use of Cr alloys (CrV, CrTi) as underlayers. Extra in-plane texture can be induced by substrate scratching. Laser texturing of the substrate can induce circumferential anisotropy, while leaving the substrate sufficiently smooth.

14.3.2.2.2 Laminated and Antiferromagnetically Coupled Media

Media grain size should be reduced to keep the transition width narrow and to get a large number n of grains per magnetic bit (media noise scales as $1/\sqrt{n}$). However thermal stability requires new strategies to be used since reducing grain size while keeping the properties similar induces superparamagnetic behavior. To reduce grain size while maintaining stability, new media have been proposed. The idea is to keep low $M_r t$, which is the figure of merit for media, by using two antiferromagnetically coupled magnetic layers. One example is Co/Ru/Co. When the ruthenium spacer is 0.6–0.7 nm thick, antiparallel coupling between the ferromagnetic layers is induced by the oscillating spin polarization of conduction electrons in the nonmagnetic (Ru) spacer (RKKY mechanism). The Co/Ru/Co multilayer is crystallographically coherent. When both magnetic layers have the same thickness, the structure has no net magnetic moment; it is then called a synthetic or artificial antiferromagnet. The strength of the antiparallel coupling can be tuned by controlling the Ru layer. So the total thickness of a grain can be larger than that of single layer media, pushing back the superparamagnetic limit, keeping a low effective $M_r t$. Double and triple magnetic layers create laminated media. Noise measurements show that some structures behave as if one grain were replaced by two or three grains, reducing the noise ($1/\sqrt{n}$ law).

14.3.2.2.3 New Materials

Further improvements may be made by shifting to $L1_0$ materials (FePd, FePt, CoPt etc.) or rare earth–transition metal intermetallics. The $L1_0$ structure is a natural multilayer. Pt and Pd have large spin–orbit coefficients which translate into large magnetocrystalline anisotropies. The $L1_0$ structure is also quite interesting because of the tunable character of the atomic order. It is possible to reduce the anisotropy and thus to suppress the out-of-plane easy axis, by reverting to the disordered fcc alloy (A1 phase).

Intermetallics relevant for recording are compounds with a well-defined easy c axis. The presence of rare earth atoms induces very large magnetocrystalline anisotropies. Processing rare earth elements and dealing with huge coercive fields are two key points to progress in this direction.

14.3.2.3 Perpendicular Recording

14.3.2.3.1 Magneto-Optical Media

The magneto-optical Kerr effect is exploited to read so-called magneto-optical (M-O) media. Since it is sensitive to the optical properties, the absolute value of magnetization is less relevant. However, the stability of magnetic bits makes demagnetizing effects important still. Since polar geometry is used, perpendicular magnetization is required.

Writing Writing on the medium requires a perpendicular applied field. A large coil is used to create a spatially homogeneous field and the submicronic bits are written by reducing the coercive field in a controlled manner, i.e., by local heating using the focused laser at higher power than for reading. Materials with perpendicular anisotropy and a large temperature dependence of coercive field are the best choices. Heating above the Curie temperature was the initial strategy in the 1960s and MnBi the first alloy (Kerr rotation 2°).

Ferrimagnetism Intermetallic alloys such as TbFeCo have a ferrimagnetic configuration, i.e., the transition metal (TM) moments are antiparallel to the rare earth (RE) ones. The temperature dependences of both sublattice magnetizations are quite different and when the composition is close to $RE_{0.25}TM_{0.75}$, the two sublattice magnetic moments compensate each other at room temperature (RT). The net magnetization becomes zero at the compensation temperature and the coercive field diverges since Zeeman energy no longer exists. When these alloys are heated above RT (i.e., above the compensation temperature), the coercive field decreases rapidly, allowing for writing. RE–TM alloys have a larger Kerr rotation for infrared wavelengths and were the first commercialized systems. Perpendicular anisotropy can be developed in amorphous layers, and thermal cycling does not degrade the material. However, terbium is easily oxidized and capping of the layer is necessary. Since Kerr rotation is only 0.2°, capping layers include an anti-reflection layer and internal reflections to eliminate the nonrotated component of light and to enhance the rotated component of the reflected beam.

Multilayers such as Co/Pt or Co/Pd are also applicable as M-O media. Perpendicular anisotropies can be dominant when the individual layers are kept thin (a few atomic planes). Moreover, large Kerr effects have been measured in these multilayers at shorter wavelengths (relevant for blue laser diode) and T_c can be tuned when alloying Co with Ni to get a strong temperature dependence of H_c.

The lack of an industry standard for magneto-optical disks and the development of large-capacity Flash and HDD transportable media make M-O recording a niche market. The ultimate density is limited by the diffraction-limited focusing, i.e., $d = \lambda/2NA$, NA being the numerical aperture of the optical system, and by the angular resolution power $d = 1.6\lambda/D$, with D the lens diameter. Super-resolution strategies have been developed using amplification layers but they may not be enough to warrant the survival of this technology.

14.3.2.3.2 High Density Perpendicular Recording
To improve the stability of HDD media and to allow for further improvements in the storage density, a transition from longitudinal to perpendicular recording (magnetic moments are perpendicular to the magnetic film) is now taking place. Commercial products have been announced and have been available since the end of 2005. The main change that perpendicular recording brings is the modification of the bit aspect ratio. Magnetic grains become column-like. Furthermore, when the density is increased, the magnetic bits become more elongated, i.e., the

demagnetizing field does not increase. The new geometry requires the design of a new write head because the applied field is perpendicular to the surface. The storage multilayer requires also a new design which includes a soft underlayer (SUL) below the recording layer to guide the B flux (Fig. 14.22).

As far as materials are concerned, out-of-plane anisotropy has to be induced. The distribution of easy axes can be better defined compared to longitudinal (in-plane) anisotropy, which is quasi-isotropic in-plane since the substrates are intrinsically isotropic in-plane. The recording medium is effectively in the gap of the write head, so it experiences the maximum field from the head. Therefore, larger applied fields are available, films with larger coercive fields will be writable, and thus larger magnetocrystalline anisotropies may be used.

The first perpendicular recording layers are based on CoCrPt alloy with out-of-plane c axis. SiO_x may be used as a segregation agent to obtain a narrow grain size distribution. The crystallographic texture of the recording layer has to be developed on top of the SUL (200 nm thick) covered with a nonmagnetic interlayer so that both magnetic layers are decoupled. The SUL should be thick enough to guide the B flux. A material with high magnetization is preferred and it must maintain soft properties (H_c smaller than 10 A m^{-1}) with a controlled magnetic state (induced radial or longitudinal anisotropy) and remain domain-free to eliminate noise in the read signal. To eliminate domains, the SUL can be biased by an antiferromagnetic layer such as $Ir_{20}Mn_{80}$ or a hard layer such as SmCo. Controlled anisotropy may require a specific heat treatment under a magnetic field to release deposition stress and induce a well-defined anisotropy direction.

Soft underlayers previously studied include $Fe_{19}Ni_{81}$, amorphous $Co_{91}Nb_7Zr_2$, FeCoB, $Fe_{95}Ta_3N_2$, $Co_{92}Ta_3Zr_5$, FeAlSi, etc.; the nonmagnetic interlayer can be 2–5 nm of Ag, MgO, or Cr, depending on miscibility and texture.

To increase bit density, while reducing the superparamagnetic limit, higher-anisotropy materials are being studied. $Fe_{50}Pt_{50}$ ordered alloy ($L1_0$ phase), TbFeCo and Co/Pd multilayers are being investigated as layers with a high perpendicular anisotropy. These are the same materials as the ones studied to push back the superparamagnetic limit in longitudinal recording. Their easy c axis is now perpendicular to the plane.

This will require drastic changes from the well-established CoCr alloys. Keeping the coercive field low enough to allow writing may not be possible, so strategies using heating to decrease the coercive field during the writing process are currently being studied.

14.3.3
Outlook

Perpendicular recording is being implemented and will be improved in the years to come toward the dream density of 1 Tb in^{-2}. In 2005, prototypes by Hitachi (230 Gb in^{-2}) and Seagate (245 Gb in^{-2}) have already been presented.

The next improvement may be heat-assisted writing to allow for the limitation of the maximum field a head can create (2.4 T). Reducing the coercive field while writing is already a strategy used in magneto-optical recording.

The next revolution will be the so-called "patterned media." Using continuous films, one magnetic bit contains roughly a hundred crystallographic grains, the volume of each of them being close to the superparamagnetic limit. Moreover these grains are randomly distributed on the surface. Patterned media rely on one bit-one grain storage. Each bit is then better defined, there is no longer a transition region between bits, and the noise associated with this transition, but the bit positions should be organized on the disk.

To pattern a disk, several techniques are being developed in laboratories: 5–10 nm magnetic particles formed using a chemical route can self-organize on the surface, with the help of pre-patterned features such as grooves to maintain long-range order. Nanoimprinting of a polymer, similar to CD-ROM imprinting, allows the production of sub-100 nm features on large surfaces, the master press being produced through conventional electron beam lithography. Patterning of discrete tracks 140 nm wide has been demonstrated on a 95 mm disk by Komag in 2005. Direct electron lithography is a process too slow to be compatible with mass production. Since magnetic storage is a small market compared to microelectronics, the need for lithography implies that microelectronic's technological roadmap (Moore's law) produces the tools necessary for patterning at the relevant nanometric scale. This technological bottleneck may limit the growth of high-density storage to Moore's law while it was faster than the microelectronics trend in recent years.

Recording materials require specific properties which can be finely tuned by alloying. The most relevant property is the magnetocrystalline anisotropy, which requires anisotropic crystallographic structures, i.e., compounds or ordered alloys. At the same time, the nanostructure at a nanometer scale must be controlled over wide areas and modern media involve multilayers comprising nucleation layers, texture-controlling layers, magnetic hard and soft layers, mechanically hard layers, and lubricating layers. Each of these layers plays an important role and has to be optimized.

Further Reading

Cebollada, A., Farrow, R.F.C., Toney, M.F. (2002), *Magnetic Nanostructures* (Ed.: H.S. Nalwa), American Scientific, Los Angeles.

Hadjipanayis, G.C. (Ed.) (2001), *Magnetic Storage Systems Beyond 2000*, NATO Science Series, Kluwer Academic Publishers, Dordrecht.

Mansuripur, M. (1995), *The Physical Principles of Magneto-optical Recording*, Cambridge University Press, Cambridge, UK.

O'Handley, R.C. (2000), *Modern Magnetic Materials, Principles and Applications*, Wiley-Interscience, New York.

Up-to-date reports about high-density recording can be found in the proceedings of the yearly Intermag conference, usually published in *IEEE Trans. Magn.*

14.4
Spin Electronics (Spintronics)
Laurent Ranno

The transport properties of metallic and semiconducting materials have been studied and used for a long time. Present electronics is based on the transport of electrical charges in metallic conductors and semiconducting channels and makes use of the influence of electrical field on charge drift and charge density. For quite a long time the spin of electrons or holes was not considered when designing an electrical transport device, which can be understood since in nonmagnetic conductors, the properties of electrons do not depend on spin. This, however, is no longer valid when considering ferromagnetic conducting materials. Using the spin as an extra parameter or degree of freedom widens the field of electronics and is called spin electronics or spintronics. This is a fast-developing field, which was created at the end of the 1980s when giant magnetoresistance (GMR) was discovered and has already given fruitful applications in the field of sensing. Magnetoresistive sensors have rapidly overcome inductive sensors in the field of high-density recording, and prototypes have appeared in the field of data storage: magnetic RAM (MRAM).

14.4.1
Electrical Transport in Conductors

14.4.1.1 Conventional Transport
The two main parameters used to define the electrical properties of conductors are the density of carriers and the mobility of these carriers.

Carrier density When the Fermi level intersects bands, a finite density of states exists (Fig. 14.26). The material is then a metal and, in general, its density of states depends only weakly on temperature. When the Fermi level lies in a bandgap, the density of carriers is zero at low temperature (intrinsic semiconductor) and then it increases exponentially when the temperature increases, which creates thermally excited carriers. Addition of impurities (i.e., controlled doping) may create energy levels close to the gap edges and may strongly enhance the carrier density. Small carrier densities may be modified by electrical field (depletion layers); they gave birth to the field-effect transistor (FET).

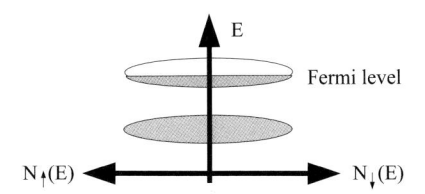

Fig. 14.26 Band structure of a nonmagnetic metal.

Mobility A high density of carriers is not enough to induce low electrical resistivity ρ (i.e., high electrical conductivity $\sigma = 1/\rho$). Mobility of the carriers may vary by several orders of magnitude. The relevant scattering mechanisms which control mobility have to be considered. The quality of the crystalline state is a major parameter: the better the crystal (topological) order, the larger the conductivity σ. A perfectly ordered crystal would have infinite conductivity, i.e., the electron wavefunctions could be described as plane Bloch waves. Several kinds of imperfections exist in real materials. First of all, quantum mechanics prevents atoms from being motionless (i.e., on a perfectly periodic lattice) and experimentally accessible temperatures are always above 0 K. Crystal defects (granular structure, atomic disorder in alloys, dislocations, point defects, dopants) do not depend on temperature and contribute a constant conductivity. Lattice vibrations (phonons) can also be considered as defects but have a strong temperature dependence – they make the main scattering contribution in pure metals at room temperature. In addition, several minor scattering mechanisms can also take place, e.g., electron–electron scattering, electron–magnon scattering, etc.

Conductivity In the relaxation–time approximation, the different scattering mechanisms are assumed to be independent. Each of them is characterized by a scattering time τ_i. Being independent, the probabilities of collisions (scattering) are simply additive. This means that the inverse scattering rates add: $1/\tau = 1/\tau_1 + 1/\tau_2$. Since resistivity is proportional to $1/\tau$, this finally leads to Matthiessen's rule $\rho(T) = \rho_1 + \rho_2$. For example, resistivity can usually be considered to be the sum of a residual resistivity due to static defects (impurities, structural defects), which is the low-temperature limit of $\rho(T)$ and a thermal resistivity due to phonons. For a conductor, with carrier density n, carrier charge q, and carrier effective mass m^*, the Drude formula relates the conductivity to these intrinsic parameters through the characteristic scattering time τ [Eq. (34)]. The mean free path λ is related to the characteristic scattering time τ through the Fermi velocity v_F. The carrier drift velocity $\mathbf{v}$ (which defines the current $\mathbf{j}$) is proportional to the applied electrical field, the mobility μ being the proportionality factor [Eq. (35)]

$$\sigma = \frac{nq^2\tau}{m^*} \quad \text{and} \quad \lambda = \frac{v_F}{\tau} \tag{34}$$

$$\mathbf{v} = \mu \mathbf{E} \quad \text{and} \quad \mu = \frac{\tau q}{m^*} \tag{35}$$

The Drude relaxation time τ is 2×10^{-14} s for a very good metal such as silver and decreases to 1.4×10^{-15} s for a ferromagnetic metal such as iron. To improve the model, the real Fermi surface has to be taken into account and the k dependence of the scattering times should be included: realistic calculations of resistivities are quite elaborate!

Characteristic values of resistivity at 300 K (Table 14.10) are 1 $\mu\Omega$ cm for a good metallic element and 10 $\mu\Omega$ cm for a magnetic element. Pure element resistivities

Table 14.10 Resistivity at room temperature of a few metallic conductors.

Metallic material	Ag	Co	$Ni_{80}Fe_{20}$	Pt	Ta	Bi	Fe + x%Si	Steel
Resistivity [$\mu\Omega$ cm] (300 K)	1.6	5.8 (a axis) 10.3 (c axis)	15	10.4	13.1	156	10 + 7.x	75

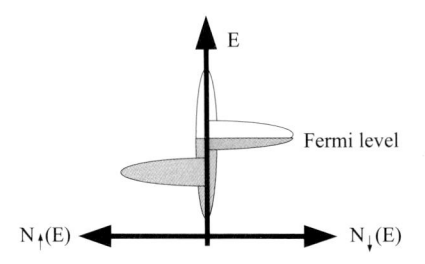

Fig. 14.27 Band structure of a strong ferromagnetic metal.

span two orders of magnitude. Two key parameters are the electron symmetry in the conduction band and the density of electrons. For example, Bi is a semi-metal with a small carrier density. The larger resistivity of magnetic elements can be interpreted by taking into account s–d scattering. In strong ferromagnetic metals (Ni, Co) the 3d↑ band is full and E_F lies in the 3d↓ band (Fe is a weak ferromagnet; the Fermi level passes through both 3d bands). Please note that strong/weak ferromagnetism refers to 3d band filling and does not correlate with properties such as magnetization, magnetocrystalline anisotropy, or coercive field. In transition metals a partially filled 4s band also exists at the Fermi level (Fig. 14.27). The s bands are not intrinsically spin-polarized and only acquire some polarization through hybridization with 3d bands. The 4s electrons have smaller effective masses (are lighter) compared to heavier 3d electrons, which are more localized. 4s-like electrons carry most of the current. The direct consequence for mean free paths is that s↓–d↓ scattering occurs whereas s↑ electrons do not experience such events since there are no d↑ states at the Fermi level.

The maximum value of resistivity for a metal is of the order of 200 $\mu\Omega$ cm. It corresponds to the limit where the electron mean free path becomes similar to the interatomic distance (Ioffe–Regel criterion).

14.4.1.2 Role of Disorder

In alloys, atomic disorder plays a leading role. When impurities are introduced in a normal metal, Nordheim's rule states that the increase in resistivity depends linearly on the concentration of impurities and the Linde–Norbury rule adds that the increase varies as the valence electron difference between the impurity and the host matrix: $\Delta\rho \propto c.\Delta Z^2$. For example, iron resistivity (10 $\mu\Omega$ cm at 300 K) increases by 7 µΩ cm for each 1% of Si added. Si addition is used to reduce eddy currents in Fe-core transformers without diluting iron magnetization. A direct consequence of disorder is that the temperature dependence of resistivity is weak and it can be used to monitor atomic ordering during annealing (cf. Chapter 5). Disordered alloys will have larger resistivities than pure elements and, as active spintronics materials, it is mainly alloys with small ΔZ which are considered, such as FeNi or FeCo.

14.4.1.3 Transport in Magnetic Conductors

The main difference in a ferromagnetic conductor is the spin dependent density of states. The densities of spin-up and spin-down carriers are different, which allows for a new parameter P: the spin polarization of the conduction band. It is now necessary to define two carrier populations: the majority carrier and minority carrier populations. The majority carriers do not necessarily have their spin parallel to the material magnetization – it depends on the filling of bands at the Fermi level.

Traditional transition metals and alloys have spin polarization of the order of 40–50% (Table 14.11). Since magnetoresistive effects are directly related to spin polarization, a search for higher polarization is a long-term trend. Ferromagnetic half-metallicity, i.e., the existence of a gap at E_F for one spin population while the other spin carriers are metallic, has been predicted to exist in several compounds. These are crystals with no s bands at E_F and narrow bands to allow for gaps. Some conducting oxides (CrO_2, $La_{0.7}Sr_{0.3}MnO_3$, Fe_3O_4, etc.) and some compounds with the semi-Heusler (C1b) structure (NiMnSb, PtMnSb, etc.), CoS_2, and some chalcogenides have fully spin-polarized conduction bands according to ab-initio calculations. Experimentally, spin polarizations in excess of 80% have been measured at low temperatures but 3d metals and alloys are still the best spin-polarized materials at room temperature.

Spin–flip Spin–flip scattering allows both carrier populations to interact. It is a rare event compared to non-spin–flip mechanisms, so both populations are well defined and a two-fluid picture is quite commonly used, especially at low temperature. Nonbalanced spin populations can be induced by polarized light absorption or when an electric current passes through an interface between two materials

Table 14.11 Typical spin polarizations of ferromagnetic metals.

Material	Ni	Co	$Ni_{80}Fe_{20}$	Fe	$Co_{50}Fe_{50}$	NiMnSb, CrO_2, $La_{0.7}Sr_{0.3}MnO_3$
Spin polarization	20–30%	35–45%	32–48%	40–44%	50%	≫ 50% at low temperature

Table 14.12 Characteristic spin diffusion lengths.

Material	Cu	Ni$_{80}$Fe$_{20}$	Co
Spin diffusion length [nm]	350 (300 K)	4 (4 K)	59 (77 K) 38 (300 K)

with different spin polarizations. Equilibrium of both populations is restored by the relevant spin–flip scattering mechanisms. The spin–flip characteristic time is τ_{sf}. The length scale to restore spin equilibrium is the spin diffusion length l_{sf} [Eq. (36) and Table 14.12]

$$l_{sf} = \sqrt{\frac{v_F \tau_{sf} \lambda}{3}} \tag{36}$$

Since a two-fluid picture is relevant, one introduces $\alpha = \sigma\!\uparrow/\sigma\!\downarrow = \rho\!\downarrow/\rho\!\uparrow$, the resistivity asymmetry coefficient; $\alpha > 1$ for transition metals because $s\!\downarrow\!-\!d\!\downarrow$ scattering impedes $\sigma\!\downarrow$. For Co, α is 2 to 3, for FeNi it increases up to $\alpha = 9$. At high temperature a $\rho\!\downarrow\!\uparrow$ spin mixing term can be introduced to take spin–flip into account.

Interfaces In nanostructures, interfaces between materials with different spin polarizations can be treated as spin-dependent interfacial resistances. This interfacial resistance is of the order of 1×10^{-15} Ω m^2. Its value changes by a factor 5 depending on spin in the case of the FeNi/Cu interface. Spin-dependent transport through an interface between two different spin-polarized materials creates a spin accumulation zone. To generalize and solve Ohm's law in a magnetic nanostructure, use of the spin-dependent chemical potential is made. The spin-dependent currents are proportional to the spin-dependent chemical potential gradients.

14.4.2
Magnetoresistance

Application of a magnetic field on a metal modifies the electron distribution in the material and leads to several magnetoresistive (MR) effects. Some MR effects are volume effects (cyclotron and anisotropic magnetoresistance) and have been known for a long time. Since the fabrication of nanostructured materials has been mastered, especially when in multilayer form, new effects related to spin transport through interfaces have been discovered (giant magnetoresistance and tunnel magnetoresistance, spin injection and spin torque effects).

14.4.2.1 Cyclotron Magnetoresistance

In any metal, even nonmagnetic ones, an applied magnetic field will disturb the electron trajectories. Longitudinal and transverse cyclotron magnetoresistances can be defined. The effect is small and it corresponds to an increase of resistance under the field. It is generally proportional to the square of the field. Most metals follow Kohler's law, i.e., $\Delta\rho/\rho = f(B/\rho)$, with f being a material-dependent function which is close to a quadratic law. The lower the temperature (smaller ρ) and the higher the field, the higher is the cyclotron MR. The order of magnitude of the effect is 0.1% in a 1 T field. In fact, this MR at high field can be shown to be very sensitive to the electronic orbits on the Fermi surface (open orbits or closed orbits). By using single-crystalline metals, and varying the amplitude and the direction of the applied field compared to the crystallographic axes, Shubnikov–De Haas oscillations of the resistivity can be observed. The period of these oscillations is proportional to $1/B$ and gives insight into details of the Fermi surface.

14.4.2.2 Anisotropic Magnetoresistance (AMR)

AMR is also a volume effect. In ferromagnetic metals, s and d electrons are present in the conduction band. If the asymmetry of the d band is large, s–d scattering depends on the configuration between magnetization M and the carrier wave vector k. Two resistivities can be defined: $\rho_\parallel$, where the current and the magnetization are parallel, and $\rho_\perp$ where the current and the magnetization are perpendicular (Fig. 14.28). The angular dependence of AMR can be written as Eq. (37).

$$\rho = \rho_\perp + (\rho_\parallel - \rho_\perp) \cos^2(k, M) \tag{37}$$

The AMR can reach a few percent in nickel and nickel alloys ($Ni_{80}Fe_{20}$ 2% at room temperature, 20% at low temperature). $(\rho_\parallel - \rho_\perp)$ is positive for transition metal systems. This kind of MR was the first one used in sensors; for example, it is used in compasses (resolution milliTesla for a DC field, nanoTesla at higher frequencies). It should be noted that a SQUID sensor resolution is better but it requires a cryogenic environment since it is based on a superconducting device.

14.4.2.3 Giant MR (GMR) and Tunnel MR (TMR)

GMR and TMR are MR effects due to interfaces. The basic requirements are two ferromagnetic layers which are magnetically decoupled in order to be able to obtain parallel and antiparallel magnetization configurations (Fig. 14.29). The MR effect will be observed when comparing the resistance of the two configurations. In order to decouple two ferromagnetic electrodes, one needs a nonmagnetic

Fig. 14.28 AMR configurations to define the parallel $\rho_\parallel$ and perpendicular $\rho_\perp$ resistivities.

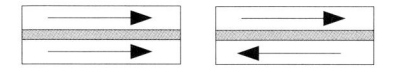

Fig. 14.29 Parallel and antiparallel magnetic configurations in GMR and TMR junctions.

spacer. If the spacer is a metal (e.g., Cu), then the GMR effect may be observed; if the spacer is an insulator (Al_2O_3, MgO), then a TMR may be measured.

GMR The mechanism for GMR is the spin-dependent scattering experienced by electrons in ferromagnetic layers. A large difference in spin resistivity should exist. In a given layer spin ↑ electrons can be majority electrons or minority electrons, depending on the magnetization configuration. If spin ↑ electrons experience low scattering in both electrodes, then they will short-circuit the spin ↓ current. In this case the GMR junction resistance will be low. In the antiparallel magnetization configuration, both spin currents will experience low scattering in one electrode and large scattering in the other one, giving a large net resistance to the junction.

Two electrical configurations exist. If the current is in the plane of the trilayer, we have the CIP (current-in-plane) GMR. If the current passes perpendicular to the interfaces it is the CPP-GMR (current-perpendicular-to-plane) (Fig. 14.30). In CIP the relevant length scale for the spacer thickness is the electron mean free path. Electrons should visit both ferromagnetic electrodes. The typical thickness of the spacer is a few nanometers. In CPP geometry, the spacer thickness can be larger (but small compared to the spin diffusion length in the spacer, which could be as large as 100 nm).

TMR The TMR mechanism is different. Only current perpendicular to the spacer is possible (otherwise there would be two independent currents on both sides of the insulating barrier). Conduction occurs by a tunneling effect through the barrier. It depends on the densities of states at E_F on both sides. Tunneling electrons maintain their spin so the tunnel probability involves the filled states in the starting electrode and the empty states with the same spin in the final electrode. A simple model from Jullière [Eq. (38)] relates the TMR value to the spin polarizations P_i of both electrodes.

$$TMR = \frac{R_{antiparallel} - R_{parallel}}{R_{parallel}} = \frac{2P_1 P_2}{1 - P_1 P_2} \qquad (38)$$

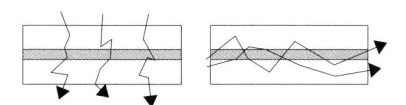

Fig. 14.30 Electrical configurations for GMR: current-perpendicular-to-plane (CPP) and current-in-plane (CIP).

TMR values as large as 50% can be obtained using an amorphous alumina barrier. Julliére's model has to be completed to really understand the tunnel junction. In particular, the evanescent wavefunction in the barrier has to be taken into account. Depending on their symmetry, wavefunctions can have different penetration depths in the barrier, leading to spin filtering. A full quantum mechanical treatment of the structure is then required. Recently, crystallized MgO barriers have been shown theoretically and then experimentally to spin-filter one spin population very efficiently. MgO barrier implementation has led to TMR ratios greater than 200%.

The relevant length scale for the junction is the thickness of the barrier (1–2 nm). The electrodes play the role of resistances in series and their thickness is usually kept as thin as a few nanometers. Since only the spin polarization at the two barrier interfaces is relevant, high spin polarization CoFe is usually inserted on both sides of the barrier.

14.4.2.4 Magnetic Field Sensors

One main advantage of sensors based on a magnetoelectrical effect (Hall effect or magnetoresistance) is the possibility of reducing the size of the sensor, keeping the amplitude of the effect. Unlike inductive sensors the response of which is proportional to the size of the coil and to the time derivative of the field, magnetoelectrical sensors can work in static mode.

Hall sensors require high-quality semiconducting layers involving epitaxial deposition and controlled doping. Moreover, four electrical contacts are required. MR sensors can be deposited in a polycrystalline state onto Si wafers using the sputtering technique. Two current leads are enough to read the signal.

To include a TMR or a GMR junction in a sensor, several criteria must be met. The magnetic configuration of the junction has to be single-valued and reproducible. One of the magnetic layers, called the reference layer, is chosen so that its magnetization is fixed. Only the magnetization of the free layer will be allowed to rotate and such a junction is called a spin valve.

Reference layer Blocking of the reference layer can be achieved by using a high-coercivity material such as Co, but a preferred situation is to pin this layer to an antiferromagnetic (AF) layer. Antiferromagnetism only responds to very high magnetic fields. The magnetic moment of the surface of the AF layer acts as a bias field on the reference layer. The reference layer magnetization loop will be shifted by the bias exchange field and the coercive field may be enhanced by the AF layer. One magnetization direction becomes more favorable (Fig. 14.31).

Direct exchange coupling between the pinned layer and the AF layer is obtained by sequentially depositing a NiFe ferromagnetic electrode and an AF layer (FeMn, IrMn, NiO, PtMn) or an artificial AF layer (Co/Ru/Co). Parameters from which to choose the AF layer include its Néel temperature (i.e., its transition temperature) and the blocking temperature, i.e., the temperature above which the exchange bias field disappears. The blocking temperature depends on the AF thickness. When the AF is metallic, it also gives a non-MR contribution to the

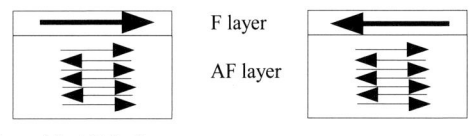

Favorable AF-F alignment   Unfavorable AF-F alignment

Fig. 14.31 Microscopic origin of the exchange bias between a ferromagnetic layer and an antiferromagnetic pinning layer.

junction (short-circuit (CIP) or serial resistance (CPP)), so its thickness must be kept small ($\ll 50$ nm). A deposition under a field or an adequate thermal treatment under a magnetic field will pin the reference layer in the chosen direction. Such field processing may be carried out before the free-layer deposition.

Free layer   The second ferromagnetic layer is the sensing electrode or free layer. A soft material such as $Fe_{20}Ni_{80}$ is usually chosen. In the absence of field its magnetization direction must be defined. A uniaxial anisotropy may be induced by applying a magnetic field or by patterning the free layer (shape anisotropy), or it may be induced during growth. To get linear sensing, biasing of the junction may be required. In effect, GMR and TMR depend as $\cos\theta$ on the position of the free layer magnetization. Linear sensing requires that the work position must be around $\theta = 90°$. Either orthogonal anisotropies are created during the deposition annealing process or external biasing must be used to rotate the free layer $90°$ away from the reference layer. Special care must be taken not to modify the bottom-layer anisotropy when inducing anisotropy in the second layer. This usually defines the deposition sequence.

Shields   In the case of high-density sensing, the GMR or TMR structure must be inserted between two soft magnetic shields which screen stray magnetic fields coming from regions away from the region to be read (Fig. 14.32). These shields are based on Permalloy ($Fe_{20}Ni_{80}$) or Sendust ($Fe_{84}Si_{10}Al_5$). Shields can be laminated with oxide layers (2–4 nm SiO_2, for example) to reduce eddy currents since hard disk drive sensing is much faster than 100 MHz. One shield of the sensor acts also as a shield for the write head which is deposited next to the read head.

GMR and TMR structures are more complicated than AMR structures. Because of their larger responses (GMR $\approx 20\%$, TMR $> 50\%$ compared to AMR $\approx 2\%$)

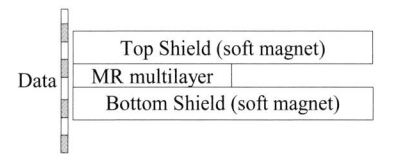

Fig. 14.32 Two magnetic shields determine the high spatial resolution of the read head.

they have been implemented in high-density recording read heads for hard disk drives. TMR effect is even larger $\approx$ 200% using MgO barriers instead of alumina barrier. When taking into account the full structure (buffer layer, capping layer, pinning layer, current leads, etc.) the MR ratio is much lower than that of the best trilayers because of the contribution of the non-MR layers.

An example of a TMR sensor structure is: Ta (3 nm)/ $Ir_{26}Mn_{74}$ (25 nm)/ $Co_{82}Fe_{18}$ (4 nm) reference layer/ Al_2O_3 (1 nm)/ $Co_{82}Fe_{18}$ (3 nm)/ $Ni_{80}Fe_{20}$ (7 nm) free layer/ Ta (9 nm).

14.4.2.5 Magnetic RAM

Once uniaxial anisotropies have been induced and an electrode has been pinned, GMR and TMR junctions possess two equilibrium positions with well-defined resistances (high state and low state). This is the basis for magnetic RAM (MRAM). Such a memory element can be inserted into a matrix architecture similar to that of semiconducting RAM. A word-line and a bit-line address each memory cell, which is enough to read the magnetic state (parallel or antiparallel). A transistor controls the opening of each line. To write a bit, the first generation of MRAM used two extra lines. Two current pulses create two magnetic field pulses around the lines. Only at the cross-point between the lines would the total applied magnetic field be sufficient to reverse the free-layer magnetization (Fig. 14.33).

Advantages for the development of MRAM are the nonvolatile character of magnetic configurations (unlike electrical leakage of DRAM), the intrinsic fast write time (magnetization reversal of a nanomagnet is faster than 1 ns, faster than flash memory), the infinite cyclability of GMR and TMR, and the possible downscaling of resistance keeping its value and its MR ratio. Magnetism is also less sensitive to radiation, especially ionizing radiation, compared to semiconducting (electrostatic) memories. MRAM has been called the universal memory. However, prototype capacities are still in the megabyte range, so for the moment MRAM does not compete with multi-gigabyte memories.

A first commercial MRAM was proposed by Freescale in June 2006. It is a 4 Mb (= 512 kB) unit with one tunnel junction and one transistor per memory cell. A 180 nm CMOS (complementary metal-oxide-semiconductor) process has been

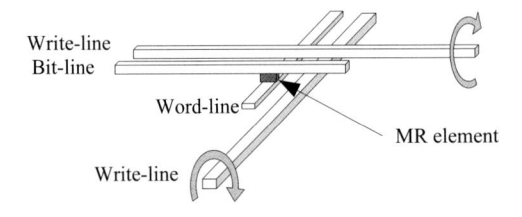

Fig. 14.33 MRAM memory cell. Two lines make it possible to measure the junction resistance state (high or low); two lines make it possible to send current pulses creating magnetic field pulses larger than the free layer's coercive field.

used and the read and write cycles are 35 ns long. The price is similar to that of DRAM or flash memories, so the price/bit is larger by a factor of 1000 than that of gigabyte semiconducting memories. To improve the cell selectivity, next-generation MRAM concepts include thermally assisted writing where a current is passed through the selected junction during writing to reduce the free layer's coercive field. It has also been proposed that the spin torque mechanism be used. A large spin-polarized current pulse has been shown to be able to reverse the magnetization of a small element using a spin torque. No write lines would be necessary any more, improving the cell selectivity and the form factor of one cell.

14.4.3
Outlook

Spin electronics is a very active research field due to the application potential of spin-dependent transport effects. Electron mean free path is at the nanometer scale, so the development of controlled nanostructures is required in order to enhance these spin-dependent effects. The mechanism details are not fully understood and new effects such as spin injection and spin torque are still between theoretical proposals and experimental validation. The improvement in the fabrication of structures, the use and discovery of new materials (fully spin-polarized metals, room-temperature ferromagnetic semiconductors) lead to the conclusion that spin electronics has a great future both in research and in application laboratories.

Further Reading

Rossiter, P.L. (1987), *The Electrical Resistivity of Metals and Alloys*, Cambridge University Press, Cambridge, UK.

Ziese, M. (Ed.) (2001), *Spin Electronics*, Lecture Notes in Physics Series, Springer, Berlin.

For an introduction to quantitative treatment of GMR structures:

Valet, T., Fert, A. (1993), *Phys. Rev. B*, **48**, 7099.

14.5
Phase-Change Media
Takeo Ohta

14.5.1
Electrically and Optically Induced Writing and Erasing Processes

Great advances have been made in memory devices such as magnetic tapes, floppy disks, hard disk drives (HDDs), and semiconductor memories. Hard disks have high data transfer rates but generally do not have read-only functionality like

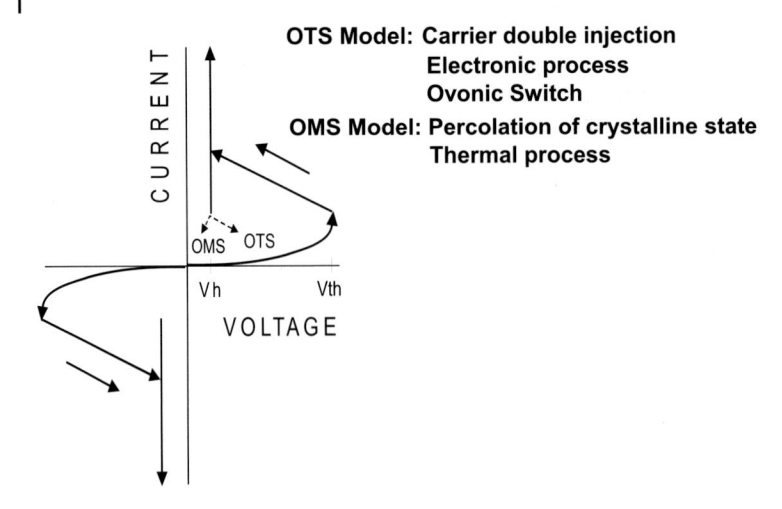

Fig. 14.34 Reversible electric current–voltage switching (I–V) model for an amorphous semiconductor switching device. OTS (Ovonic threshold switch): switches back from the low-resistance (crystalline) state to the high-resistance (amorphous) state below V_h. OVS (Ovonic memory switch): the resistance state remains after the current is removed and depends on the amount of current passed through the device.

that found in optical disks. With the increasing use of multimedia applications, phase-change rewritable optical disks are becoming more popular due to their flexibility of use and compatibility with pre-programmed read-only optical disks.

Ovshinsky (1968) discovered the switching and memory effect in amorphous semiconductors (the "Ovonic" effect). Ovonic[1] is a name derived from the Ovshinsky effect, which is the underlying characteristic of a broad class of disordered and amorphous materials (Evans et al. 1970). It was named after Stanford R. Ovshinsky (born 1922), an American inventor. His inventions opened a new field of physics and materials research and, in 1971, laser- recorded rewritable phase-change memory devices were announced (Feinleib et al. 1971).

I–V characteristics of two types of electronic switching phenomena of chalcogenide materials are shown in Fig. 14.34. When a voltage pulse is applied at a level exceeding a threshold value (V_{th}) (determined by the material and its thickness) to an Ovonic threshold switch (OTS) in the amorphous state, the device instantly switches to a low-resistive state by a phase transformation from an amorphous to a crystalline state, and the current flow increases dramatically. When the voltage is decreased to a value less than the holding voltage (V_h), the device returns to its high-resistive, amorphous state as indicated by the arrow in

---

1) The *American Heritage Dictionary of the English Language*: Ovonic: adj, of or relating to a device whose operation is based on the Ovshinsky effect; Ovshinsky effect: n, The effect by which a specific glassy thin film switches from a nonconductor to a semiconductor upon application of minimum voltage.

Fig. 14.34. The basic disordered structure is maintained in the material in both states when appropriate chalcogenide alloy compositions are used to display the Ovonic threshold switching phenomena.

An Ovonic memory switch (OMS) can be made using different compositions and bonding configurations. This switch also displays the threshold switching phenomenon of the OTS but, at specific current levels, a low-resistive state remains after the applied pulse is terminated. The decrease in resistance results from atomic restructuring into a crystalline configuration. The desired amorphous or crystalline structure is obtained by varying the pulse amplitude and duration. The low-resistive crystalline state can be induced by applying a pulse with moderate amplitude and long duration (20–50 ns) (Lai and Lowrey 2001). The amorphous state is programmed using a pulse with a large amplitude and short duration (20 ns). The ability to retain the desired resistive state after the current has been removed allows the OMS to be used as nonvolatile memory (NVM).

Whereas state changes in the OTS and OMS were induced with voltage pulses, the structure of the chalcogenide material in phase-change optical data storage devices is altered using a laser pulse. Changes in the atomic configuration state for optical memory and electronic memory (OMS) are shown in Fig. 14.35. Figure 14.35(a) illustrates atomic order (crystalline) and disorder (amorphous) phase-change states; Fig. 14.35(b) gives the temperature dependence of the enthalpy of the crystalline and amorphous disordered states. A short, high-intensity

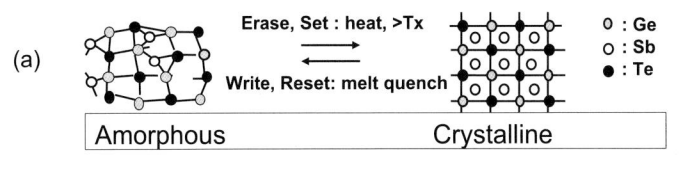

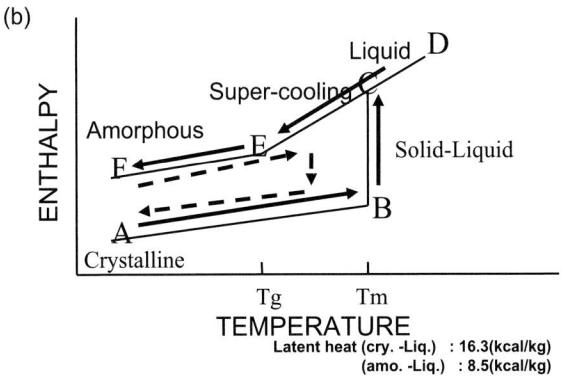

Fig. 14.35 Model of phase-change memory. (a) Schematic representation of the change between amorphous and crystalline atomic configuration; (b) enthalpy change between the crystalline and amorphous states.

laser pulse increases the temperature of the crystalline film past the melting temperature T_m, starting at state A and passing through states B, C, and D (solid arrows in Fig. 14.35b). When the laser pulse is removed, the material cools rapidly without time for atomic rearrangement and ends up in the disordered amorphous state, F. A longer, low-intensity laser pulse applied to an amorphous film, F, will raise the temperature above the glass transition temperature, T_g (E), and crystallize the material (dotted arrows).

When the chalcogenide alloy is switched from the amorphous to the crystalline structure, the electronic resistance is similar to that in a metallic state and it is around 1000 times lower than in the amorphous state. The optical characteristics change as the optical absorption edge of the material shifts to a longer wavelength. This is accompanied by a change in the complex refractive index $N = n + ik$ (n is the refractive index and k is the extinction coefficient), which leads to a difference in reflectivity. The laser records low-reflective amorphous data bits in the high-reflective crystalline background. The data bits can be recrystallized (erased) for rewritable data storage (Ohta et al. 1989a).

In the differential scanning calorimeter (DSC) scan for $Ge_2Sb_2Te_5$ (Fig. 14.36) an exothermic crystallization transition is visible at 130 °C and the endothermic melting transition occurs at 610 °C (Yamada et al. 1991). The phase diagram of the pseudo-binary $GeTe$–Sb_2Te_3 system is shown in Fig. 14.37 (Abrikosov and Danilova-Dobroyakova 1965; Yamada et al. 1987; Suzuki et al. 1988; Ohta et al. 1989b). The melting temperatures for all the compounds in this system are around 600 °C and the initial crystalline structure for all is face-centered cubic (fcc). The optical constants and conductivities of the two structures are shown in

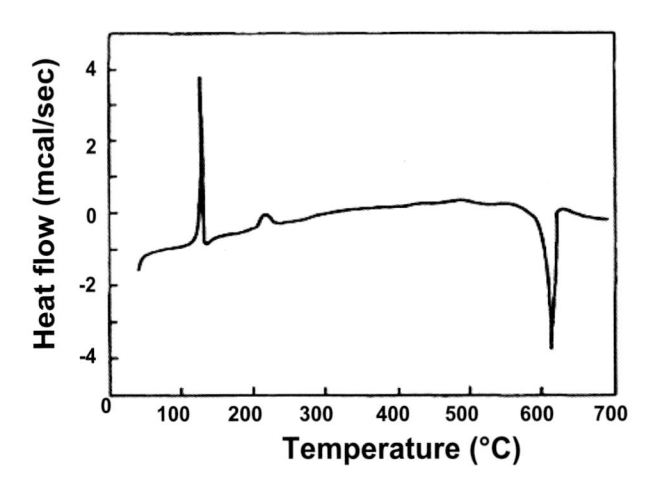

Fig. 14.36 DSC results for $Ge_2Sb_2Te_5$ at a heating rate of 10 °C min^{-1}. Taken from Yamada et al. (1991). Latent heat of crystallization at 130 °C = 85. kcal kg^{-1}; latent heat of crystalline–liquid transition at 610 °C = 16.3 kcal kg^{-1}.

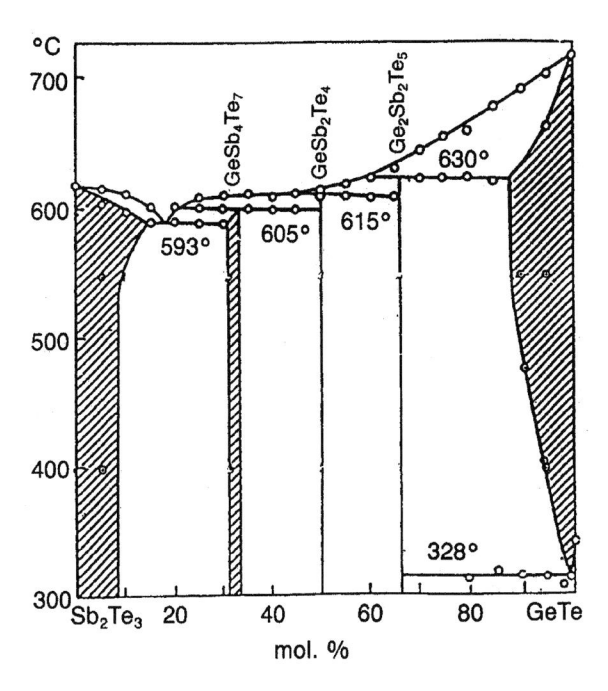

Fig. 14.37 Phase diagram for the GeTe–Sb$_2$Te$_3$ system (Abrikosov and Danilova-Dobroyakova 1965).

Table 14.13 Optical constants and change in resistance of Ge$_2$Sb$_2$Te$_5$ phase-change material.

	Amorphous	Crystalline
Refractive index n	small, $n_a = 4.9$	large, $n_c = 5.7$
Extinction coefficient k	small, $k_a = 1.4$	large, $k_c = 3.4$
Electrical resistance R [Ω]	high ($1E+6$)	low ($5E+3$)
Chemical and physical characteristics	remarkably different	

Table 14.13. Another phase-change optical disk material of the AgInSbTe system has been proposed for CD-RWs (rewritable compact disks) (Iwasaki 1997).

14.5.2
Phase-Change Dynamic Model

As storage densities increase, higher data transfer rates are desired. The crystallization speed of the two structural changes in a material (amorphous to crystalline and crystalline to amorphous) determines the rate at which data can be over-

(a) Nucleation type: GeTe-Sb$_2$Te$_3$

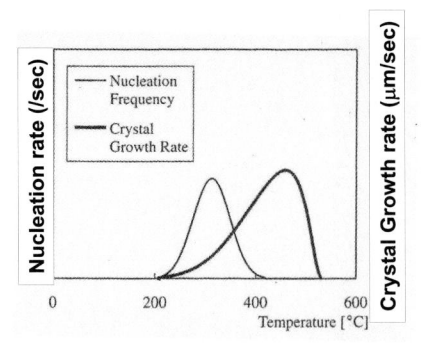

(b) Growth type: Sb$_{70}$Te$_{30}$

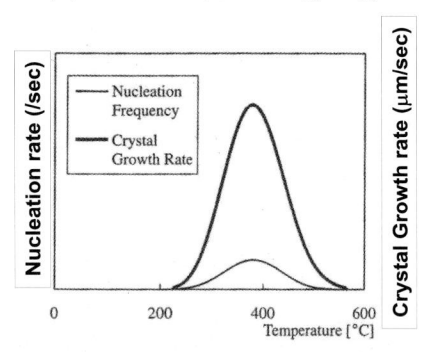

Fig. 14.38 Nucleation and growth rates obtained by TEM (Nishi et al. 2002) for two-material systems with different crystallization mechanisms: (a) GeTe–Sb$_2$Te$_3$; (b) Sb$_{70}$Te$_{30}$.

written and thus the speed of data transfer. There are two mechanisms for crystallization in phase-change materials. Materials such as those of the GeTe–Sb$_2$Te$_3$ pseudo-binary system undergo crystallization through a nucleation-dominated process, where crystalline growth proceeds from crystalline nuclei. Compositions based on the Sb$_{70}$Te$_{30}$ eutectic crystallize through a growth-dominated process, where crystalline growth continues from existing crystalline areas.

The nucleation and growth rates of both material systems are shown in Fig. 14.38 (Nishi et al. 2002). The GeTe–Sb$_2$Te$_3$ system (Fig. 14.38a) has a high nucleation rate peaking at a temperature lower than that of the growth rate. The nucleation and growth rates in this system are about the same. The Sb$_{70}$Te$_{30}$ system (Fig. 14.38b) has a very low nucleation rate while the growth rate is much greater. The peaks of both of these rates occur around the same temperature, just below 400 °C.

The speed of crystallization and consequently the overwrite speed can be altered through changes in the structure of the material layers and through changes in the laser modulation, or the write strategy. Recording simulations are important in optimizing the desired effect (Nishi et al. 2002). The Kolmogorov–Johnson–Mehl–Avrami (KJMA) equation [Eq. (39)] (see Chapter 7) is often used to calculate the crystalline fraction in thermal modeling (Avrami 1939, 1940, 1941; Senkader and Wright 2004; Wright at al. 2004):

$$\chi = 1 - \exp(-k(T)t^n) \qquad (39)$$

Here χ is the transformed crystal fraction, $k(T)$ the rate constant, n the Avrami exponent relating to growth dimensions, t the time, and T the temperature. A schematic plot of a KJMA calculation of the crystalline fraction as a function of time is shown in Fig. 14.39(a).

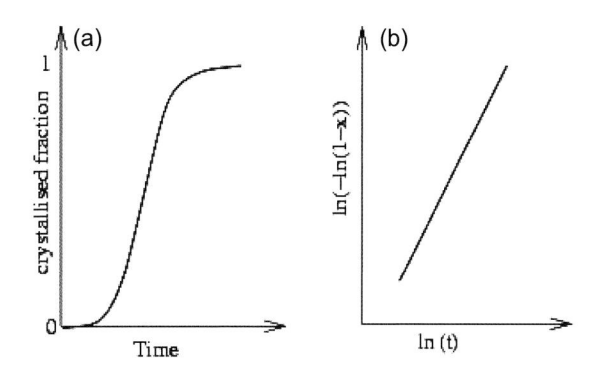

Fig. 14.39 Crystallization analysis obtained with the KJMA equation [Eq. (40)]: (a) crystallized fraction as calculated with KJMA Eq. (40); (b) KJMA plot yielding the Avramic constant.

Equation (39) can be rearranged into Eq. (40), which is useful for determining the Avrami exponent n and the rate constant k (Avrami plot, Fig. 39a).

$$\ln(-\ln(1-\chi)) = n\ln(t) + \ln k \tag{40}$$

The slope of the line in Fig. 14.39(b) yields the Avrami constant, n, and the rate constant, k, is obtained from the intercept.

Avrami coefficient, n, and rate constant, $k(T)$, have been obtained by reflectivity measurement on nucleation-type $Ge_2Sb_2Te_5$ phase-change film during isothermal heating using a heater stage. With $R(t)$ measuring the reflectance at time t of heating at temperature T, R_{cry} the reflectance (crystalline state), and R_{am} the reflectance (amorphous state), Eq. (39) can be written as the fraction of reflectance change [Eq. (41)].

$$\chi(t) = (R(t) - R_{am})/(R_{cry} - R_{am}) \tag{41}$$

The measured results are shown in Fig. 14.40(a,b). The plot of Fig. 14.40(a) includes the incubation time of the initial nucleation process, whereas in the plot of Fig. 14.40(b) the incubation time is subtracted as $t' = t - t_{inc}$. The Avrami constant of $Ge_2Sb_2Te_5$ film including incubation time results as $n = 5.8$. The value obtained from the plot neglecting the incubation time is $n = 2.5$, which corresponds to a nearly three-dimensional crystalline growth.

Introducing in Eq. (39) a nucleation factor $I(t)$ and growth factor $U(t)$ (as there is three-dimensional volume growth of $(4\pi/(3R^3))$, an extended KJMA equation, Eq. (42), results (Davies 1976, Wright et al. 2004).

$$\chi(t) = 1 - \exp\left\{-4\pi/3 \int_0^{r_2} I(t) R^3(t)\, dt\right\} \tag{42}$$

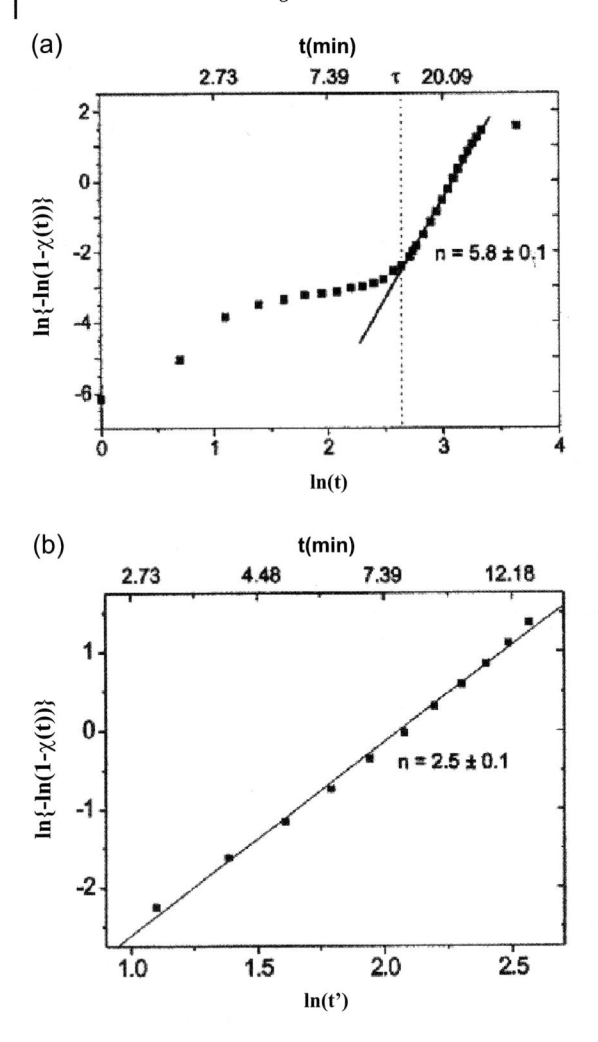

Fig. 14.40 Measurement of the Avrami coefficient n with a KJMA plot: (a) KJMA plot including the incubation time: $t_{inc} = 14.3$ min; $n = 5.8$; (b) KJMA plot with the incubation time subtracted: $t' = t - t_{inc}$; $n = 2.5$.

Equation (42) can be used to calculate the nucleation and growth rates for $Ge_2Sb_2Te_5$. The temperature dependence of the nucleation rate and the growth rate are shown in Fig. 14.41(a,b), respectively (Nishi et al. 2002). The results correlate well with the experimental data for nucleation rate and the crystalline growth rate in Fig. 14.38(a). This correspondence gives evidence that the extended KJMA equation can be used in thermal modeling of amorphous mark formation. An example of such a calculation is shown in Fig. 14.42 (Nishi et al. 2002). The laser recording strategy (laser modulation pattern) is shown in Fig. 14.42(a). Here the laser pulse width used is $t_w = 17.1$ ns, the linear disk velocity $LTV = 8.2$

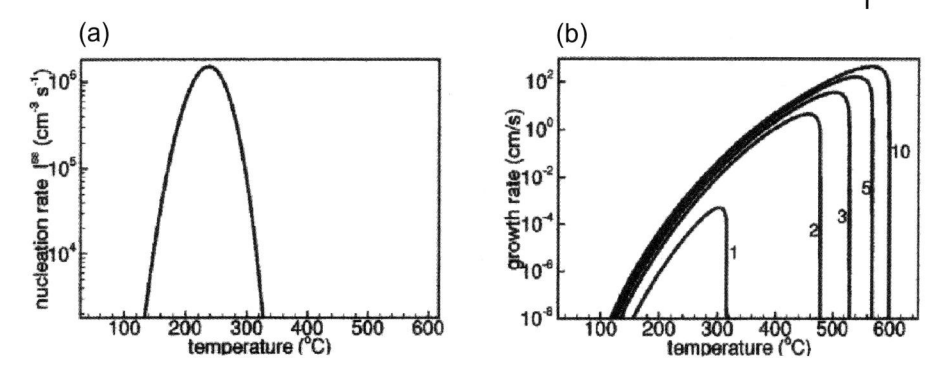

Fig. 14.41 (a) Nucleation rate $I(T)$ and (b) growth rate $U(T)$ of Ge$_2$Sb$_2$Te$_5$ as calculated by the extended KJMA equation [Eq. (42)]. The parameters on the curves indicate the radius (nm) of the clusters.

m s^{-1}, the laser wavelength $\lambda = 650$ nm, and the numerical aperture of the lens $NA = 0.6$. The formation of an amorphous mark for the nucleation-type material GeSbTe is shown in Fig. 42(b) for $t = 0$–$4.5t_w$ with a step $0.5t_w$ as calculated with the above formalism. This modeling shows that the trailing edge of the amorphous mark is influenced by thermal diffusion. This indicates how the write strategy can be modified to control mark formation and position. Thermal modeling

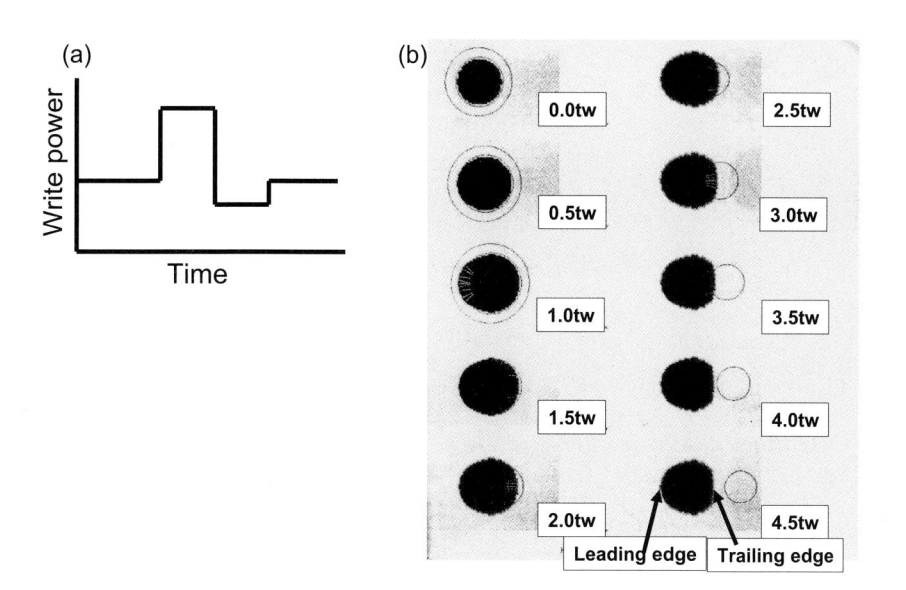

Fig. 14.42 Simulation of amorphous mark formation for GeSbTe as calculated by the extended KJMA model: (a) laser write strategy (laser modulation pattern) to write a 3T mark (T = tw = 1.7 lns, pulse width); (b) amorphous mark formation for times $t = 0$–$4.5t_w$.

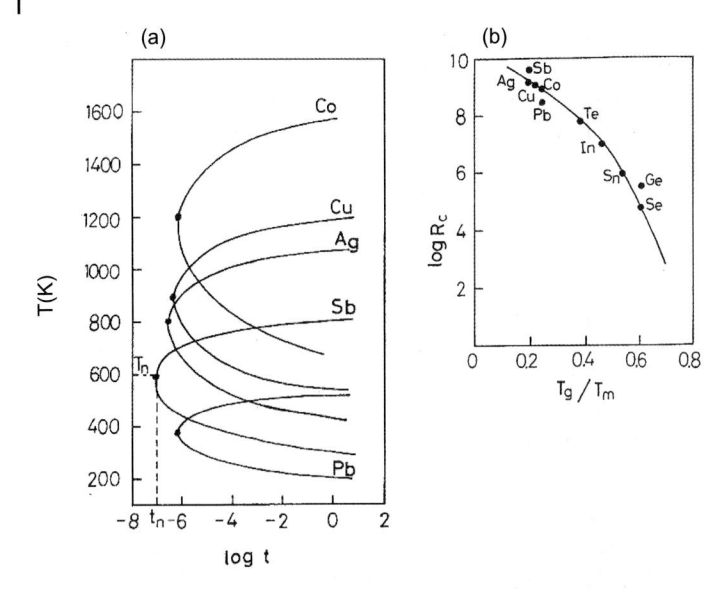

Fig. 14.43 Increasing crystallization of phase-change materials by addition of elements (Jiang and Okuda 1991): (a) time–temperature transformation curve; (b) variation of critical cooling rate with reduced temperature, T_g/T_m.

results of a growth-type material, $Ag_x In_y(Sb_{70}Te_{30})_z$, show that the mark edge shrinks much less in these materials (Nishi et al. 2002).

The speed of crystallization of growth-type materials can be affected by the addition of transition and main group metals. A temperature–time–transformation (T–T–T) plot for various additional elements is shown in Fig. 14.43(a) (Jiang and Okuda 1991). The curves represent a crystallized volume fraction of 10^{-6}. The point on the curves indicated with a full circle (T_n, t_n) corresponds to the shortest time, t_n, for crystallization (fraction of 10^{-6}) occurring at temperature T_n.

The critical cooling rate, R_c, can be calculated by Eq. (43) using t_n and T_n and the melting temperature T_m.

$$R_c = (T_m - T_n)/t_n \tag{43}$$

The relationship between the R_c and T_g/T_m for some elements is shown in Fig. 14.43(b). The elements at the top of the curve, Sb, Ag, and Cu, are predicted to contribute to fast crystallization (Okuda et al. 1992). Table 14.14 shows the critical parameters related to $\log R_c$ for these elements. From Table 14.14, the critical cooling rate of Sb can be determined as $R_c = 4.1 \times 10^9$ K s^{-1}, a value close to the measured critical cooling rate of GeSbTe, $R_c = 3.4 \times 10^9$ K s^{-1} (Ohta et al. 1989a).

Phase-change optical disks are manufactured with a four-layer structure on a polycarbonate substrate (bottom dielectric layer $\sim$ 100 nm; phase-change

Table 14.14 Estimated values of T_g and T_g/T_m and calculated values of T_n, t_n and log R_c of certain elements.

Element	T_m [K]	T_g [K]	T_g/T_m	T_n [K]	t_n [s]	$\log_{10} R_c$ [K s^{-1}]
Sb	895	182	0.20	600	7.41E–8	9.61
Ag	1235	250	0.20	800	2.98E–7	9.16
Cu	1357	298	0.22	900	3.38E–7	9.13
Co	1768	445	0.25	1200	6.54E–7	8.94
Pb	601	152	0.25	375	7.07E–7	8.50
Te	722	285	0.39	500	5.75E–6	7.54
Ge	1212	750	0.62	945	8.96E–4	5.47

layer ~ 20 nm; upper dielectric layer ~ 20 nm; metal reflection layer ~ 100 nm). The dielectric layers serve two purposes in controlling the temperature of the phase-change material. First, they offer thermal protection by the dissipation of heat. Second, they are used in the optical design of the layer structure to optimize laser absorption efficiency (contrast between the amorphous and crystalline reflectance). The metal reflection layer is necessary for the quenching process involved in creating amorphous marks. The optical, mechanical, and thermal characteristics of optical disk layers are listed in Table 14.15.

Early in their development, phase-change rewritable disks had limited overwrite cycle performance. Amorphous marks are formed by laser heating above the melting temperature, 600 °C. Erasing of marks through crystallization requires temperatures of at least near 400 °C. When optical disks were first being developed in the 1980s, the cycle life was projected to be 10–100 cycles. But there are essential differences between the phase-change optical disk, with its thin layered structure, and the usual melt solidification metallurgy process. The dimensions on a disk are localized: time in nanoseconds and space in nanometers. Diffusion of materials and segregation of elements are localized to nanoscale dimensions. Crystallization and melt-quenching on these scales are very different from common metallurgical processes which are recorded in hours and tons. A breakthrough in cycle performance was made by looking closely to the two main cycle degradation mechanisms, first the increase in noise and second the signal pulse dropout with write–erase cycle time.

The first degradation mechanism was analyzed as the grain growth of the protection dielectric layers, bottom and upper layers. The original dielectric layer material was a ZnS dielectric layer; after an overwrite cycle the grain size increased and therefore the noise level also increased. A new dielectric protection layer was found in a mixture of ZnS and SiO_2 with a composition of $(ZnS)_{80}(SiO_2)_{20}$. A comparison of the grain structures of ZnS and ZnS–SiO_2 films is shown in Fig. 14.44. The grains of ZnS–SiO_2 after annealing at 700 °C for 5 min were very small, ~2 nm, while the grains of ZnS became rather large (about 50 nm) and

Table 14.15 The optical, mechanical, and thermal properties of materials.

Material	Refractive index (l = 830 nm)	Density [kg m^{-3}]	Young's modulus [N m^{-2}]	Poisson's ratio	Specific heat	Thermal conductivity [W m^{-1} K^{-1}]	Coeff. of linear expansion [K^{-1}]
GeTe–Sb$_2$Te$_3$–Sb (amorphous)	$4.9 + 1.4i$	6150	5.49×10^{10}	0.33	0.209×10^3	0.581	1.1×10^{-5}
GeTe–Sb$_2$Te$_3$–Sb (crystalline), 2:1:0.5 mol. ratio	$5.7 + 3.4i$	6150	5.49×10^{10}	0.33	0.209×10^3	0.581	1.1×10^{-5}
ZnS–SiO$_2$, 4:1 mol. ratio	2.0	3650	7.81×10^{10}	0.2	0.563×10^3	0.657	7.4×10^{-6}
SiO$_2$	1.46	2202	7.81×10^{10}	0.2	0.753×10^3	0.313	5.5×10^{-7}
Al alloy	$2.2 + 7.5i$	2750	7.03×10^{10}	0.345	0.892×10^3	0.215×10^3	2.2×10^{-5}
Polycarbonate	1.58	1200	2.26×10^9	0.3	0.126×10^2	0.223	7.0×10^{-5}

Latent heat: crystal–liquid: 0.682×10^5 J kg^{-1}; amorphous–liquid: 0.356×10^5 J kg^{-1}.

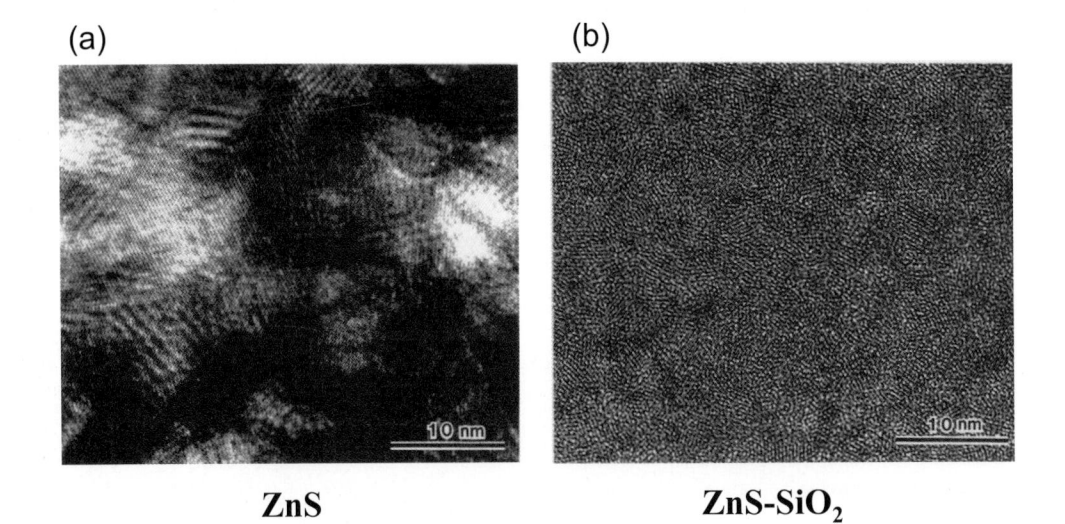

Fig. 14.44 TEM of grain structure of ZnS and ZnS–SiO$_2$ layers after annealing for 5 min at 700 °C (Ohta et al. 1990).

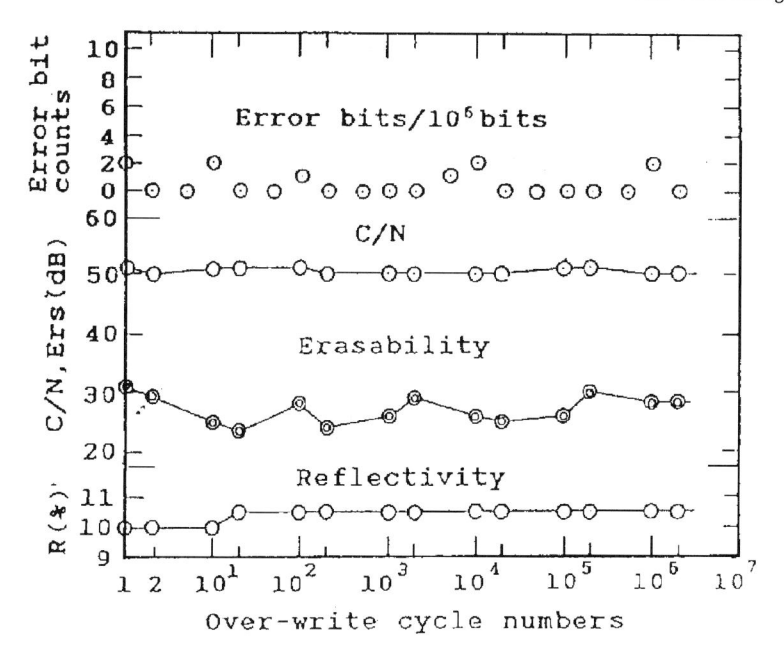

Fig. 14.45 The effect of an additional SiO$_2$ protection layer: 2×10^6 overwwrite cycles of phase-change optical disk (Ohta et al. 1989b).

were still growing to 100 nm and more with longer annealing time (Inoue et al. 1992). Grain growth in ZnS–SiO$_2$ was not observed after annealing at 700 °C for 5 min and the cycle characteristics were improved for more than 1000 cycles.

The second mechanism of overwrite cycle life degradation concerns the decrease in the absorbed laser pulse during write–erase cycles. This is related to the change in layer thickness along the track. The reason for this is a displacement of the phase-change layer element by deformation of the disk layers, which is due to thermal expansion and contraction during the recording and erasing process. This deformation can be reduced by adding a layer which has a small thermal expansion coefficient. The additional layer is inserted between the phase-change and the upper dielectric layers. The thermal expansion coefficient of SiO$_2$ (5.5×10^{-7} K^{-1}) is less by a factor of more than 10 than that of ZnS–SiO$_2$ (6.1×10^{-6} K^{-1}). The addition of an SiO$_2$ layer extended the overwrite cycle life from about 10^3 to over 2×10^6 cycles, as shown in Fig. 14.45 (Ohta et al. 1989b).

14.5.3
Alternative Functions

Phase-change optical CDs and DVDs are widely used for data storage; they include the next-generation products which are increasing their share of the mar-

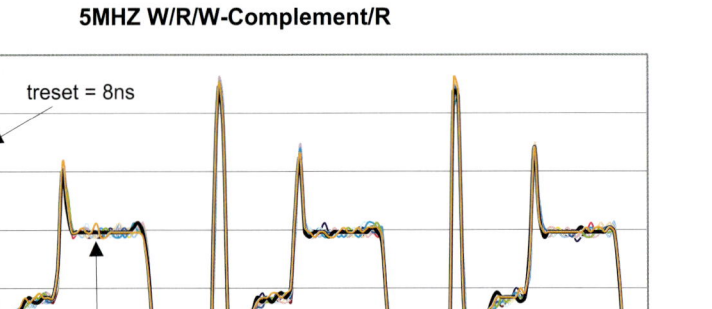

Fig. 14.46 Operation of Ovonic unified memory (OUM): oscilloscope traces of the voltage drop across an OUM cell during repetitive cycling (5 MHz programming). Read pulse voltage $V = 0.2$ V (Gill et al. 2002).

ket, namely Blu-ray Disc (BD) with a capacity of 50 GB and high-density DVD (HD-DVD), capacity 36 GB. Phase-change materials are also used more and more as electrical memory in applications such as for nonvolatile memory of flash memories and other memories, e.g., SRAM (static random access memory) and DRAM (dynamic random access memory).

To use phase-change materials as electrical memory, a simple device structure is preferred (Lai and Lowrey 2001). Operation characteristics of so-called Ovonic Unified Memory (OUM) are shown in Fig. 14.46 (Gill et al. 2002). A set pulse of $V = 0.6$ V, $t = 85$ ns, is used to change the material from high resistance (85 kΩ) to low resistance (2 kΩ). A reset pulse of $V = 0.8$ V, $t = 8$ ns, changes the material back to high resistance. The power consumption is calculated to be 2.6 pJ, which is much less than the power consumption of 0.8 nJ of phase-change material in an optical disk.

In the Ovonic memory switching (OMS) process, the current flow heats the high-resistance amorphous material sufficiently to crystallize it, giving a low-resistance state. A short, high-intensity pulse melts the crystalline filament, which then rapidly cools back to the amorphous state. These devices will cycle more than 10^{13} times. The electrical memory has a much longer cycle life than the optical memory because of the fixed operation of the phase-change material. Movement of atoms is minimal in electrical switching, while the dynamic scanning of the laser in optical memory induces large movement, and thus potential for failure in the material.

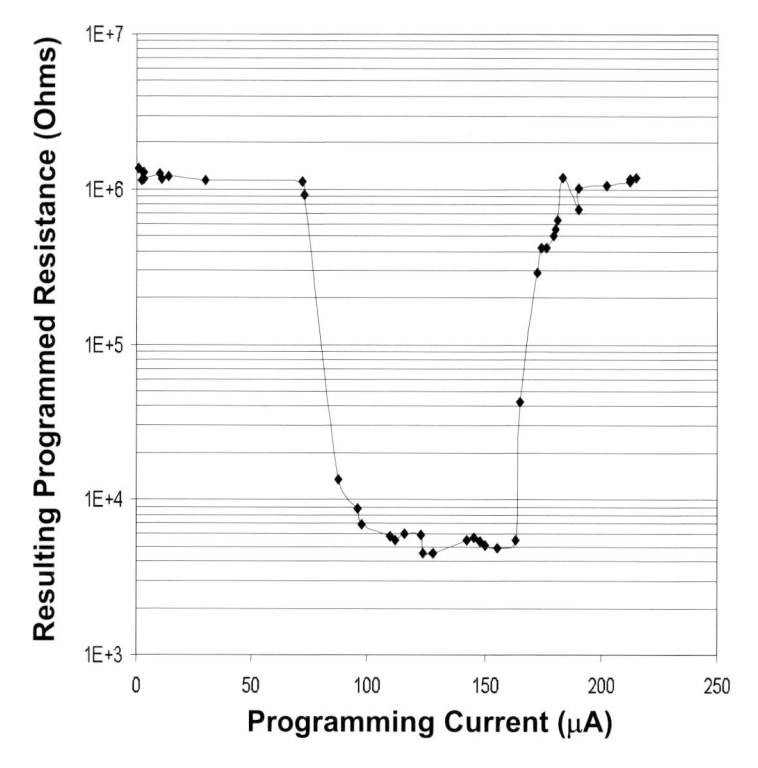

Fig. 14.47 Resistance characteristics of phase-change RAM (PRAM). Set pulse: 80 ns; reset pulse: 20 ns.

An example of a change in resistance by a supplied current pulse for phase-change RAM (PRAM) is shown in Fig. 14.47. When the pulse amplitude is lower than the set current, the device does not change state and the resistance remains constant. At the set current, the resistance drops to a low-resistance state.

What is the speed limit of the phase change between the amorphous and crystalline states? Applications requiring high-speed switching include fiber communications and high-resolution laser processing. Lasers having pulse widths on the order of pico- and femtoseconds have enabled new studies of materials to be carried out (Miura et al. 1997). The recording speed of the phase-change optical memory has been increased from 10 Mb s^{-1} to 140 Mb s^{-1} (Kato et al. 2001). Now femtosecond laser pulses will provide an opportunity to develop materials and devices for even higher-speed phase-change optical memory.

Due to thermal diffusion, the phase-change material heated by a laser is wider than the laser spot. The size and position of a mark are determined not only by the diffraction-limited spot size ($\lambda/2NA$), but also by the pulse duration and thermal characteristics of the disk. The response of phase-change memory to femtosecond laser pulses was measured (Morilla et al. 1997; Ohta et al. 2001). Marks formed with 60 ns and 120 fs pulses are shown in Figure 14.48. The main differ-

(a) Recrystallization **(b)** No recrystallization

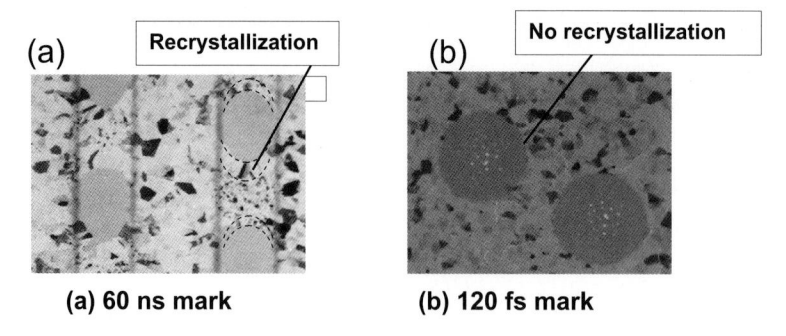

(a) 60 ns mark   (b) 120 fs mark

Fig. 14.48 Comparison by TEM of marks made with (a) nanosecond and (b) femtosecond pulses (Wang et al. 2003): (a) conventional laser recording: $\lambda = 780$ nm, $NA = 0.5$, $t = 60$ ns; (b) femtosecond pulse laser recording: $\lambda = 780$ nm, $NA = 0.95$, $t = 120$ fs.

ence between the conventional 60 ns pulse and the 120 fs pulse is the recrystallized band around the amorphous mark. The longer pulse, Fig. 14.48(a), shows a large-grain crystalline band around the mark, while the mark made with the short pulse, Fig. 14.48(b), does not. The data rate of phase-change optical memory recorded with femtosecond laser pulses is expected to be more than 1 Tb s^{-1}.

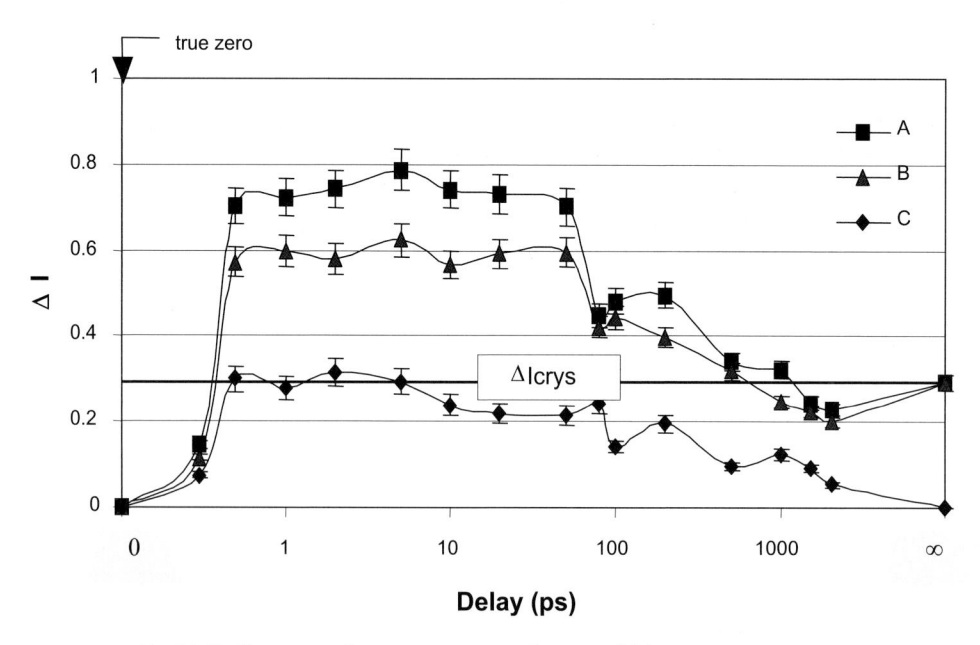

Fig. 14.49 Change in reflective intensity as a function of delay time after the femto laser pulse (t) exposure: mean (pulse width of 120 fs) influence = 30 mJ cm^{-2}, $\Delta I = I(t) - I_a$. (I_a is the reflection intensity of as-deposit amorphous state).

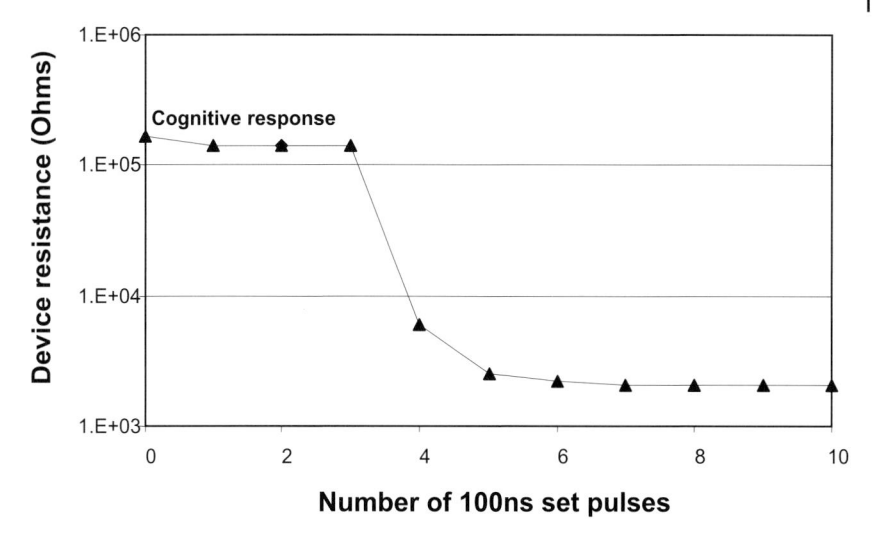

Fig. 14.50 Behavior of an Ovonic electrical cognitive device (Ovshinsky 2003; Ovshinsky and Pashmakov 2004): change in resistance versus number of set pulses.

T. C. Chong and his group investigated crystallization using a femtosecond laser (Wang et al. 2003). The reflectivity change measured by time-resolved microscopy after a 130 s pulse is shown in Fig. 14.49. The results show that the reflectivity continues to change long after the end of the pulse.

Ovshinsky recently proposed another use for phase-change materials: cognitive switching (Ovshinsky 2003; Ovshinsky and Pashmakov 2004). The behavior of an Ovonic electrical cognitive device is shown in Fig. 14.50. No change is obtained until a certain number of pulses have been applied, after which the device switches to a low-resistance state. The device "remembers" previous pulses applied and only switches once the threshold pulse number is reached. A comparison of Ovonic switching memory and Ovonic cognitive response is shown in Fig.

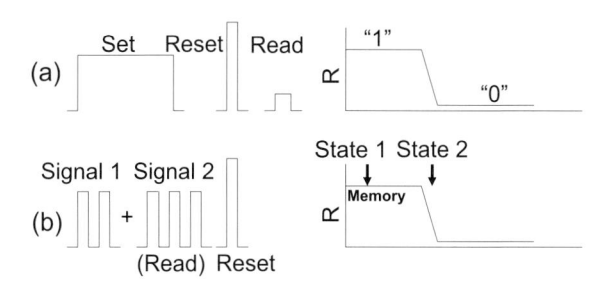

Fig. 14.51 Comparison of an Ovonic memory device and an Ovonic cognitive device: (a) PRAM (OUM): binary memory; (b) cognitive function: processing and memory.

14.51. A possible explanation of the cognitive function is that nucleation and grain growth begin during the application of the initial pulses. At the threshold pulse, percolation occurs and the device switches to low resistance. This provides, in a single device, functionality similar to that of a neurosynaptic cell in the brain.

14.5.4
Outlook

Phase changes have been a well-known phenomenon for a long time (see Chapters 1 and 7). Since Ovshinsky found the effective functions of the memory effect named the Ovonic effect, it took around 20 years to promote the phase-change rewritable optical disks such as CD-RWs, DVD-RAMs and BDs, HD-DVDs and so on by proposal of new materials and the development in device structure. Laser marks on the disks are getting smaller and smaller, and lie today at about 100 nm in the nanometer range. Research in the field of phase-change electronic memory (PRAM) is very exciting at the moment with respect to the development of nonvolatile memory: the flash memory market, for example, is more than doubling every year. PRAM is a candidate for a higher density of data storage than flash memory and also has DRAM speed overwrite characteristics; it is therefore a candidate for replacing DRAM.

The main pathway of scientific research and technical development in the field of phase-change media will involve:
- gaining a more detailed understanding of nucleation and growth parameters during the phase change
- increasing the phase-change speed limit
- exploring the threshold switching mechanisms
- searching for corresponding phase changes which do not need crystallographic long-range order.

In any case, further research in the field of materials physics with respect to phase-change mechanisms and size limitations will be necessary to secure an important place in the nanostructure device market for phase-change media in the future.

References

Abrikosov, N. Kh., Danilova-Dobroyakova, G. T. (1965), *InorganicMaterials*, **1**, 187.
Avrami, M. (1939), *J. Chem. Phys.*, **7**, 1103.
Avrami, M. (1940), *J. Chem. Phys.*, **8**, 212.
Avrami, M. (1941), *J. Chem. Phys.*, **9**, 177.
Davies, H.A. (1976), *Phys. Chem. Glass.*, **17**, 159.
Evans, E.J., Helbers, J.H., Ovshinsky, S.R. (1970), *J. Non-Cryst. Solids*, **2**, 334.
Feinleib, J., deNeufville, J., Moss, C., Ovshinsky, S.R. (1971), *Appl. Phys. Lett.*, **18**, 254.
Gill, M., Lowrey, T., Park, J. (2002), *Solid State Circuits Conference, 2002. Digest of Technical Papers. ISSCC. 2002 IEEE International*, p. 202.
Inoue, K., Furukawa, S., Yoshioka, K., Kawahara, K., Ohta, T. (1992), *Proc. ASME*, **2**, 593.

Iwasaki, H. (1997), *Proc. SPIE*, **3109**, 12.
Jiang, F., Okuda, M. (1991), *Jpn. J. Appl. Phys.*, **30**, 97.
Kato, T., Hirata, H., Inoue, H., Shingai, H., Utsunomiya, H. (2001), *Tech. Digest, ISOM 2001*, **Fr-K-01**, 200.
Lai, S., Lowrey, T. (2001), IEEE *IEDM Technical Digest*, p. 803.
Miura, K., Qiu, J., Inoue, H., Mitsuyu, T., Hirao, K. (1997), *Appl. Phys. Lett.*, **71**, 3329.
Morilla, M.C., Solis, J., Afonso, C.N. (1997), *Jpn. J. Appl. Phys.*, **36**, 1015.
Nishi, Y., Kando, H., Terao, M. (2002), *Jpn. J. Apply. Phys.*, **41**, 631.
Ohta, T., Inoue, K., Furukawa, S., Akiyama, T., Uchida, M., Nakamura, S. (1989a), *Electro. & Comun. Technical Research Meeting Rep.* **CPM 89-84**, 41.
Ohta, T., Uchida, M., Yoshioka, K., Inoue, K., Akiyama, T., Furukawa, S., Kotera, K., Nakamura, S. (1989b), *Proc. SPIE*, **1078**, 27.
Ohta, T., Inoue, K., Furukawa, S., Yoshioka, K., Uchida, M., Nakamura, S. (1990), *Electro. & Comun. Technical Research Meeting Rep.*, **PM 90-35**, 43.
Ohta, T., Yamada, N., Yamamoto, H., Mitsuyu, T., Kozaki, T., Qiu, J., Hirao, K. (2001), *Mat. Res. Soc. Symp. Proc.*, **674**, V1.1.1.
Okuda, M., Naito, H., Matsushita, T. (1992), *Jpn. J. Appl. Phys.*, **28**, 466.
Ovshinsky, S.R. (1968), *Phys. Rev. Lett.*, **21**, 1450.
Ovshinsky, S.R. (2003), *Tech. Digest ISOM 2003*, **Tu-A-02**, p. 2.
Ovshinsky, S.R., Pashmakov, B. (2004), *Mater. Res. Soc. Symp. Proc.*, **803**, 49.
Senkader, S., Wright, C.D. (2004), *J. Appl. Phys.*, **95**, 504.
Suzuki, M., Doi, I., Nishimura, K., Morimoto, I., Mori, K. (1988), *Proc. Optical Memory Symp. '88*, JJAP, p. 41.
Wang, Q., Shi, L., Huang, S., Chong, T. (2003), *Mater. Res. Soc., Symp. Proc.*, **803**, 239.
Weidenhof, V., Friedrich, I., Ziegler, S., Wuttig, M. (2001), *J. Appl. Phys.*, **89**, 3168.
Wright, C.D., Aziz, M.M., Armand, M., Senkader, S., Yu, W. (2004), *Proc. EPCOS*, Unaxis, p. 1.
Yamada, N., Ohno, E., Akahira, N., Nishiuchi, K., Nagata, K., Takao, M. (1987), *Proc. Int. Symp. on Optical Memory, Jpn. J. Appl. Phys.*, p. 61.
Yamada, N., Ohno, E., Nishiuchi, K., Akahira, N. (1991), *J. Appl. Phys.*, **69**, 2849.

14.6
Superconductors
Harald W. Weber

14.6.1
Fundamentals

The discovery of superconductivity by Heike Kamerlingh Onnes (in 1911) was prompted by two major developments at the beginning of the 20th century: firstly the successful liquefaction of the "last gas," i.e., helium, by Onnes himself in 1908, which opened a new field of research at very low temperatures (4.2 K and below); and secondly, the quest, mainly by solid-state theorists, to understand the temperature dependence of the electrical resistivity, especially at temperatures close to absolute zero. Consequently the metal that could be prepared with the smallest amount of impurities at that time, Hg, was chosen by Onnes for his investigations of the resistivity as a function of temperature for $T \to 0$. The result, i.e., the complete disappearance of dc resistivity at a certain (low) temperature, the transition temperature T_c, set the stage for a fascinating new field of research that has been keeping theorists and experimentalists busy for nearly 100 years now. The final breakthrough onto the market in the electric power industry, medical diagnostics, metrology, and electronics ("key technology of the 21st century") is expected in the very near future.

It soon turned out that ideal conductivity was accompanied by perfect diamagnetism ($\chi = -1$, the "Meissner effect"), i.e., the complete expulsion of magnetic flux from the interior of the superconductor. These two unique properties of the solid state are linked, as already indicated by the first phenomenological theory of the electrodynamics of superconductivity in the early 1930s by the London brothers, who showed that "supercurrents" flowing in a thin surface layer (of dimension λ, the magnetic penetration depth, ~100 nm) were able to shield the interior completely from flux penetration. Inserting typical numbers, we find current *densities*, J_c, of these shielding currents of the order of 10^{11} A m^{-2}, but – because the current flow is restricted to these thin surface layers – *currents* of only around 70 A (e.g., through a "massive" wire with a diameter of 1 mm). Another issue refers to the stability of the superconducting phase. It was found very early on that *both* the ideal conductivity and the perfect diamagnetism could be suppressed at temperatures well in the superconducting regime ($T < T_c$) either by a certain current, the critical current I_c, or by a certain magnetic field, the critical field H_c, which are again linked through a simple relation, $I_c = H_c/2\pi r$ (Silsbee), i.e., a phase transition back into the normal conducting state can be induced at any temperature in the superconducting phase (Fig. 14.52a). The temperature dependence of these critical parameters follows a parabolic law.

To find a suitable explanation for the occurrence of superconductivity in metals, alloys, and compounds proved to be one of the most demanding challenges for the community of solid-state theorists and took nearly 50 years in the end. The fundamental idea put forward by Bardeen, Cooper, and Schrieffer ("BCS theory," in 1957) is based on a mechanism that can be summarized as follows. At sufficiently low temperatures, a quantum mechanical exchange interaction mediated by the lattice vibrations (phonons) becomes strong enough to establish an attrac-

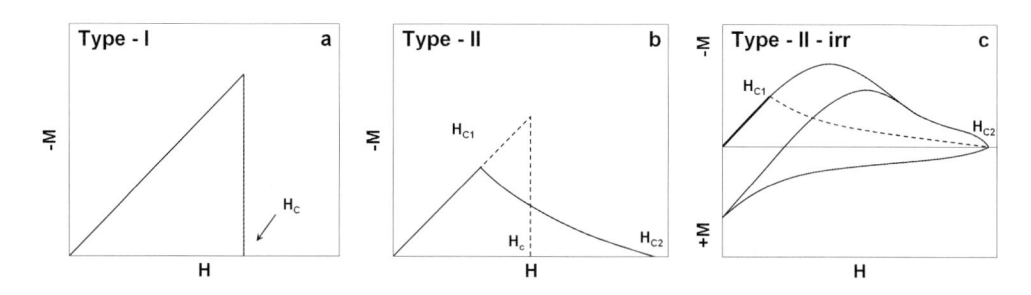

Fig. 14.52 Magnetization curves of superconductors: (a) type I superconductor; (b) type II superconductor; (c) type II superconductor with flux pinning.

tive interaction between two valence electrons (with opposite wave vector and spin) which overcomes the (partly strongly reduced) Coulomb repulsion between these charge carriers in the lattice and leads to a definite correlation between them (Cooper pair formation) over a certain distance (BCS coherence length ξ_0, 10–100 nm), which can be understood as the spatial extension of this new particle (Cooper pair). In view of this "huge" size on the scale of the crystal lattice ($\sim$0.1 nm), these new charge carriers (responsible for the current transport in the superconducting state) are strongly correlated and strongly overlapping. Moreover, due to the spin selection condition, the Cooper pair has a total spin of 0, and therefore escapes the limitations set up by the Fermi–Dirac statistics and the Pauli exclusion principle, thus rendering the usual solutions for the density of states at the Fermi level unstable and opening an energy gap $\Delta(T)$ at the Fermi energy E_F. Moreover, all Cooper pairs are allowed to attain the same quantum mechanical state and can be described by a single wavefunction with an amplitude that corresponds to the density of Cooper pairs in the metal, and with a well-defined phase, thus making superconductivity a "macroscopic quantum phenomenon" or a "phase coherent state" with wide consequences. Firstly, being in the same quantum mechanical state, individual Cooper pairs cannot interact with the crystal lattice or any of its defects, i.e., scattering processes by impurities are impeded and the resistance drops to zero. Secondly, this mechanism prevails only so long as any energy transferred to the superconductor does not exceed the "binding energy" of the Cooper pairs, i.e., the strength of the attractive interaction provided by the electron–phonon mechanism ("condensation energy", $\mu_0 H_c^2/2$ per unit volume). In other words, if this energy is exceeded, e.g., by transport currents or external fields, the Cooper pairs break up and the normal conducting state is re-established (even at temperatures well below T_c), thus explaining the existence of critical parameters (I_c, J_c, H_c) introduced above (J_c is actually called the "depairing" critical current density $J_{c,dep}$). Thirdly, the electron–phonon coupling is directly reflected by the magnitude of the energy gap $\Delta(T)$ at the Fermi energy E_F (a few milli-electronvolts at typical Fermi energies of a few electronvolts) and its influence on the density of states in the superconducting phase $N_s(E)$. $\Delta(T)$ is of course correlated with the condensation energy (e.g., at $T = 0$, $N(E_F)\Delta(T)^2/2$, where $N(E_F)$ denotes the density of states at the Fermi energy) and is assessed best by tunnel experiments between two thin superconducting films separated by an insulating layer (Giaever et al. in 1961). An outstanding result related to the theoretical analysis of such tunnel contacts should be mentioned here, i.e., the prediction of tunneling Cooper pairs by Brian Josephson in 1962 (the "Josephson effect"), i.e., the flow of phase-coherent supercurrents across such barriers. This paved the way for the development of one of the most important fields for applications of superconductivity (SQUIDs, superconducting quantum interference devices), which are based on a phase modulation of the macroscopic phase-coherent wavefunction by the vector potential.

All of these fundamental predictions (and later extensions of the basic ideas) were subsequently confirmed by experiment and therefore provide us with a solid knowledge of the microscopic mechanisms leading to superconductivity. However, gaining a detailed understanding of the behavior of superconductors in magnetic fields was a slow process. It was soon shown that the simple picture of the Meissner effect dominating the magnetic phase diagram could not be proven by experiment, because magnetization curves demonstrated that magnetic flux had to be able to penetrate the superconductor in some way under some circumstances. Using Landau's theory of second-order phase transitions as a basis, Ginzburg and Landau (in 1950) introduced a phenomenological theory for local variations of the "order parameter" $\psi(r)$ (whose square turned out to be the "local" density of Cooper pairs) and the "local field" within the superconductor (GL equations), which led to a classification of their magnetic properties according to the relative size of these characteristic variation lengths, i.e., the spatial variation distance of the Cooper pair density, ξ_{GL}, the GL coherence length, and the variation distance of the local field, λ, the magnetic field penetration depth. Their ratio, $\kappa = \lambda/\xi_{GL}$, the GL parameter, proved decisive for the energy provided by phase boundaries between the normal and superconducting phase in the material, which is positive for $\kappa < 1/\sqrt{2}$ and negative for $\kappa > 1/\sqrt{2}$. Accordingly, type-I superconductors, i.e., those with $\kappa < 1/\sqrt{2}$, will generally not allow such phase boundaries to be formed (because they would enhance the total free energy of the system) and complete flux expulsion determines their magnetic phase diagram. On the other hand, type II superconductors, i.e., those with $\kappa > 1/\sqrt{2}$, may take advantage of the negative phase boundary energy to reduce their total free energy. This situation was investigated in detail by the theorist A. A. Abrikosov (in 1957), who found "spectacular" solutions for the current and field distributions in these materials, as follows. At a certain critical field (Fig. 14.52b), the lower critical field H_{c1}, the formation of field-containing areas becomes energetically favorable for the first time, i.e., the loss of condensation energy by the formation of these field-containing areas is compensated by the energy gain provided by the negative phase boundary energy. The field penetrates in the form of so-called "flux lines," i.e., normal conducting spots, surrounded by circulating supercurrents ("vortices"), which build up the local field and form a regular hexagonal lattice throughout the superconductor. Due to its special spatial configuration, each of the flux lines carries exactly one magnetic flux quantum, $\phi_0 = h/2e = 2.067 \times 10^{-15}$ Wb. With increasing field, the lattice spacing of the flux line lattice becomes smaller, the normal conducting cores finally overlap at a certain critical field, the upper critical field H_{c2}, where a second-order phase transition into the normal conducting state occurs. This field, $H_{c2} = H_c \kappa \sqrt{2} = \phi_0/2\pi\mu_0\xi^2_{GL}$, can obviously become very much larger than the thermodynamic critical field H_c, depending on the magnitude of κ or ξ, thus extending the stability range of superconductivity in this so-called "mixed state" to very high magnetic fields. The final step was achieved by Gor'kov (in 1958 and 1959), who showed that the order parameter $\psi(r)$ of the phenomenological GL theory was proportional to the gap parameter $\Delta(r)$ of BCS theory, thus linking GL theory to the microscopic theory of superconductivity, the proportionality being governed by the so-called Gor'kov function $\chi(\alpha)$, where $\alpha = 0.882\xi_0/\ell$ is the "impurity" parameter, ξ_0 denotes the

BCS coherence length at $T = 0$ in the "clean limit," and ℓ the mean free electron path. Since ξ_{GL} is proportional to ℓ, we immediately find that small GL parameters κ (i.e., type I superconductivity) will always occur when $\ell \to \infty$ (i.e., mostly in pure metals), whereas type II superconductivity prevails for $\ell \to 0$, i.e., in alloys and compounds.

It took several years for the community to realize the real impact of this theory complex alternative: set of theories, for two reasons. The most important was that Abrikosov's paper was published in the same year as the BCS theory, which attracted worldwide attention. Furthermore, magnetization curves such as those schematically depicted in Fig. 14.52b had not been confirmed by experiment but showed a rather broad and hysteretic behavior (Fig. 14.52c) which, at the time, was analyzed (again) in terms of phenomenological models, e.g., the Bean model, where a "material parameter," the critical current density, J_c, was introduced to explain the hysteretic magnetization and the loss-free current transport in magnetic fields. Indeed, according to Abrikosov's picture, loss-free current transport cannot exist in the mixed state, since the Lorentz forces set up by the field and the circulating supercurrents would immediately start to drive the flux lines (with their "normal conducting" cores) through the superconductor and thus create an electric field E, which would lead to energy dissipation $W = |E| |J|$, unless this motion could be prevented in some way. This subject represents the most important issue for applications of superconductors in magnetic fields and is commonly referred to as the "flux pinning problem." It can be considered on a global basis, i.e., by taking the volume-averaged action of such "pinning forces" over the entire volume of the superconductor into account (the critical state equation: $F_L = J_c x B = -F_P$), or by modeling the individual pinning forces, f_p, between a single flux line and a certain defect in the superconductor matrix and then trying to sum them appropriately. The principle of flux pinning is easily explained. Since the normal conducting flux line core represents a volume where condensation energy is lost compared to the undisturbed superconducting state, this energy loss can be avoided by placing the flux line core, e.g., onto a normal conducting precipitate of appropriate size (i.e., with a diameter of $\sim 2\xi_{GL}$), which would make it energetically favorable for the flux line to remain at this particular position, thus establishing the pinning force f_p of this defect. In practice, the optimization of defect structures suitable for flux pinning turned out to be one of the most difficult tasks and it is still ongoing today, even in the "best known technical" superconductors.

To sum up this section, we have learned that tailoring superconductors appropriately is most important for applications. Firstly, we need to adjust the electron mean free path ℓ in such a way that type II superconductivity and, hence, high upper critical fields are achieved. Secondly, we need to establish an optimized network of metallurgical defects in the superconductor, each on the scale of the coherence length, i.e., of a few to a few ten nanometers, in order to take advantage of the enhanced field range for loss-free current transport. The limit for the latter task will always be set by the intrinsic superconducting properties of the material, i.e., the condition $J_c \leq J_{c,dep}$.

14.6.2
Superconducting Materials

Beginning from Hg with its transition temperature of 4.15 K, the highest known T_c values developed slowly but steadily (Fig. 14.53), initially among the metallic elements, then among alloys and compounds. They reached quite a stable plateau in the class of so-called A15 superconductors slightly above 20 K, the boiling temperature of liquid hydrogen, in the early 1970s. It came as a big and unexpected surprise, therefore, when in 2001 Akimitsu's group reported on superconductivity in the "very simple" compound MgB$_2$ with a T_c of nearly 40 K ("How could we miss it?"). In the meantime, several classes of material with highly interesting and exciting properties in the superconducting state had been discovered, such as the "heavy fermion" superconductors (e.g., CeCu$_2$Si$_2$, UBe$_{13}$, UPt$_3$), the ruthenate superconductors (Sr$_2$RuO$_4$), the "magnetic superconductors" (e.g., the quaternary boron carbide compounds of the type RNi$_2$B$_2$C, where R stands for a rare earth element), the organic superconductors, or even the fullerenes, i.e., (alkali metal)-doped C$_{60}$ molecules, but none of them were able to participate in the

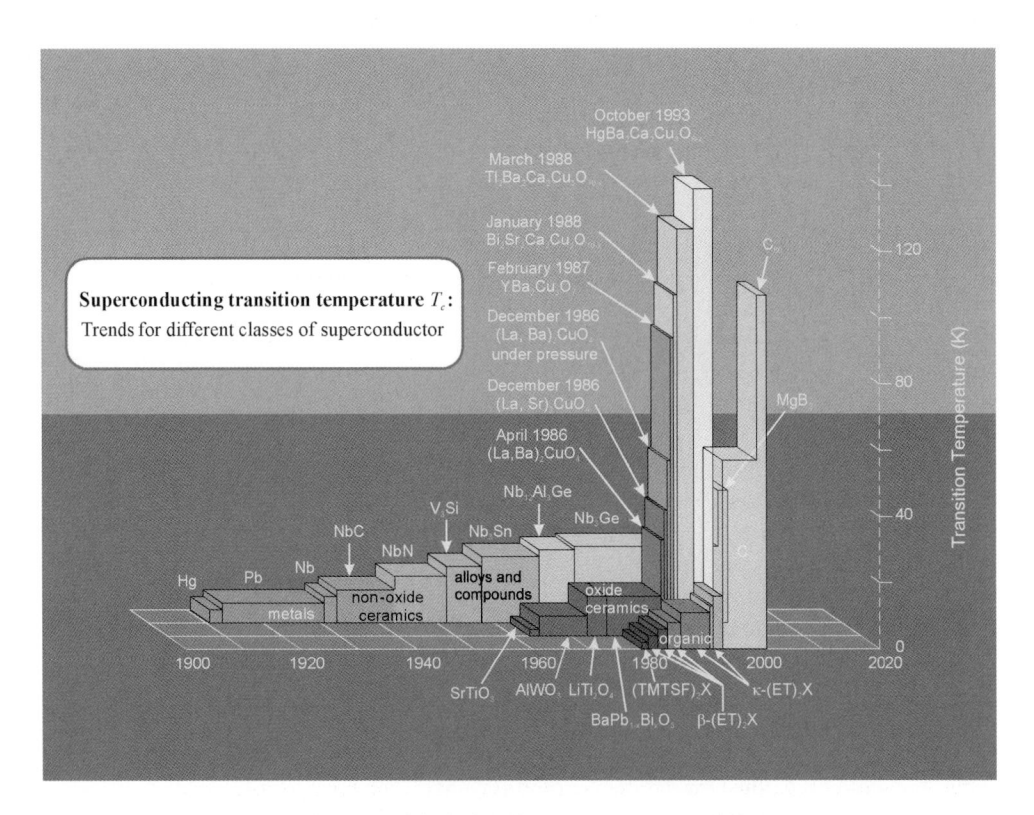

Fig. 14.53 Development of the highest known transition temperatures since the discovery of superconductivity in 1911.

"race" for the highest known T_c, maybe with the exception of charge-doped pure C_{60}, which disappeared again in the context of the fraud scandal around G. Schön, however.

The most outstanding event during the nearly 100 years of superconductor research was certainly the discovery of "high-temperature superconductivity" (HTS) by Bednorz and Müller in 1986. In the course of research on ceramic oxides (some of which had been known to become superconducting at temperatures around 10 K), they started to investigate cuprate-based compounds and found evidence for superconductivity above 30 K in $(La,Ba)_2CuO_4$, a discovery that led not only to a "gold rush" atmosphere among scientists, but also attracted the attention of politicians and the public in general. The highest known T_c values rocketed to temperatures well above the boiling temperature of liquid nitrogen (77 K), thus revolutionizing the issue of cryogenics for applications. Among all the cuprates, $YBa_2Cu_3O_{7-\delta}$ (YBCO) and the Bi-based compounds, with transition temperatures of ~ 93 and up to ~ 110 K, respectively, play the dominant role.

Today, nearly 20 years after its discovery, HTS is still far from being fully understood. Although we know that Cooper pairs are again formed (in most cases, however, with a different pairing symmetry), the pairing mechanism itself is still under intensive discussion and major modifications to the original BCS concepts developed for metals will certainly be required. Concerning their physical properties, the cuprates offer an enormous variety of challenges for physicists, crystallographers, metallurgists, and engineers. Due to their unit cell structure with relatively small a,b plane-, but "very large" c axis-dimensions (typically 0.4 versus 3.5 nm), and due to the periodic formation of CuO_2 layers along the c axis, extraordinary properties occur in the normal as well as in the superconducting state, the most prominent being the high anisotropy between the basal planes and the c axis. Since the Cooper pairs are mainly formed within the CuO_2 planes, which – in some extreme cases – may be separated by several layers of "nonsuperconducting" material, superconductivity itself becomes two-dimensional (2D), rather than three-dimensional (3D) as is the case in conventional superconductors. As a consequence, the standard flux line lattice structure in the mixed state of these extreme type II superconductors becomes "interrupted" if the field is applied along the c direction, and breaks up into a loosely connected array of flux line sections ("pancakes") which are extremely difficult to pin by metallurgical defects, especially at the high temperatures where the superconductor is supposed to operate. This is because of the constant supply of thermal energy $k_B T$, which is about 20 times higher at 77 K than at 4.2 K, and counteracts the pinning energy. Moreover, the coherence lengths are extremely small (just slightly larger than the unit cell dimensions) and highly anisotropic (with factors of 7 up to a few 100 between the basal plane and the c axis), thus rendering all other mixed state parameters (including the upper critical fields) anisotropic in the same way. As a consequence, sintered ceramics (i.e., the original material form of the cuprates) are completely unsuitable for carrying loss-free supercurrents of reasonable magnitude, since the current would have to cross a huge number of grain boundaries between grains of different orientations and thus of hugely different superconducting properties. Complex texturing techniques are therefore required as a way out, 2D texturing (i.e., the alignment of the grains' c axis) being sufficient for the Bi-based materials, whereas 3D texturing (i.e., the additional alignment of the a,b planes) is needed for YBCO. Returning to the flux pinning issue, the small coherence length makes nanoengineering of the defect structures mandatory, since very small defects ("point defects") can only act collectively and usually lead to very small pinning forces (at least at high temperatures), whereas large defects (i.e., those that often occur during processing) are far too large for efficient pinning. Excellent examples of such efficient defects include fast-neutron-induced collision cascades (with a diameter of $\sim$6 nm, almost exactly 2ξ in YBCO at 77 K) or the recently reported addition of nanoparticles of a second phase to YBCO bulk materials before processing.

In summary, a few thousand metals, alloys, compounds, ceramics, or organic substances are known today to become superconducting at temperatures between a few microkelvins and $\sim$135 K. Among this enormous variety – not only with regard to their material form, but also concerning their spectrum of physical properties in the normal and superconducting state – fewer than ten are suitable for applications ("technical superconductors") either in the high-current sector or for thin-film applications, an alarmingly small number that meet the four basic application requirements: high T_c, high H_{c2}, high J_c, and mechanical stability against thermal cycling as well as against deformation.

14.6.3
Technical Superconductors

Imagine you have identified a material that fulfills all the four requirements just mentioned at the end Section 14.6.2, and further, that you have managed to produce it in the form of a wire with a length of few hundred meters and that you have successfully wound a superconducting magnet coil. From the critical material parameters and your design of the coil, you would certainly expect to be able to operate the magnet safely, e.g., at 4.2 K and at a load current of e.g., 120 A, to produce a magnetic field of, e.g., 8 T. However, this was not the case initially, i.e., the magnet "quenched" (turned normal conducting) at currents that were lower than the "short sample" critical currents by a factor of 5–10. This effect, called "degradation," represented the last obstacle to high-current applications and was only solved by a careful analysis of the operating conditions within a typical "densely packed" wire configuration in a coil. It turns out that heat may be produced locally (either by less efficient exposure to the cooling agent or by mechanical energy transferred by vibrations or Lorentz forces), thus locally increasing the temperature of this coil segment, which in turn results in a decrease of the pinning forces and the initiation of flux movement. In this way a "vicious circle" is initiated, since flux movement again increases the local temperature, further reducing pinning there, and so on, until the normal state is reached at that spot. Due to the sudden appearance of the full normal-state resistance under full current load, an "enormous" amount of heat is generated, which quickly spreads through the rest of the coil and turns it to entirely normal conducting. This can be avoided by "stabilizing" the superconducting material by surrounding it with a good conductor (Cu), in order to provide an alternative path for the current flow (with low dissipation) and thus to enable the superconductor to return to its normal operating conditions. This basic idea has been refined over the past decades and has led to the concept of "multifilamentary wires," where hundreds of tiny superconducting wires with diameters in the micrometer range are embedded in a Cu matrix (Fig. 14.54a) and arranged according to a sophisticated scheme that allows transposure and twisting of the entire wire ensemble with a twist pitch of a few centimeters, in order to keep the wires decoupled under moderate field sweep rates. This concept has led to advanced high-tech processes, by which wires are commercially produced in kilometer lengths and assembled into strands, cables, and finally suitable conductors, depending on the desired application (Fig. 14.54b).

At present, the materials for high-current applications are dominated by two metallic low-temperature superconductors, the alloy Nb–Ti (with T_c values between 8.5 and ~10 K, depending on composition) and the compound Nb_3Sn ($T_c \approx 18$ K).

The "workhorse" for superconducting magnets is certainly Nb–Ti, which is produced in quantities of roughly 1000 tons per year, mainly for the magnets of magnetic resonance imaging (MRI) devices. The success of Nb–Ti is due to its combination of excellent strength and ductility with high current-carrying capac-

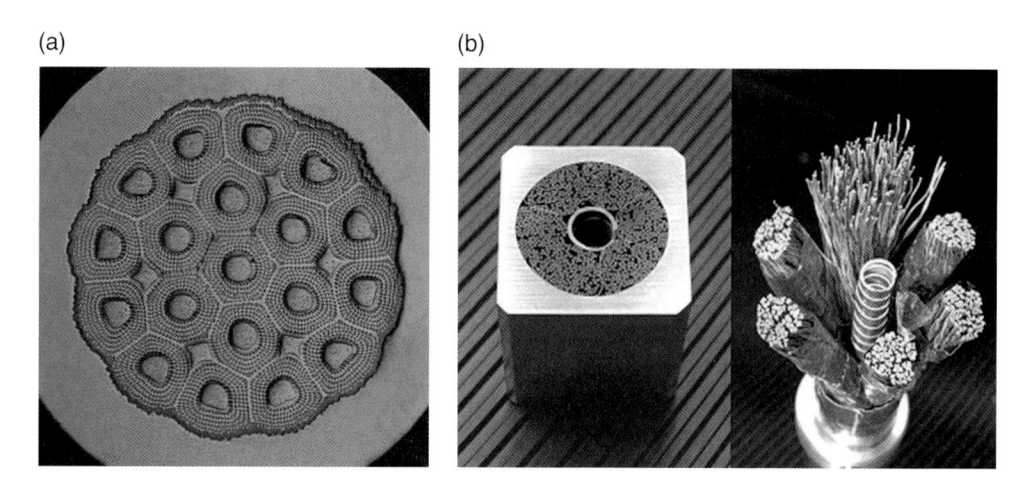

Fig. 14.54 (a) Multifilamentary superconductor: Nb_3Sn strand made by the internal tin process (19 × 219 filaments, single Ta barrier, Cu/non-Cu = 1:1, wire diameter: 0.81 mm, $J_c > 10^9$ A m^{-2} at 12 T and 4.2 K); (b) Nb_3Sn conductors manufactured for the toroidal field model coil (TFMC) of ITER for operation at a current of 80 kA and a field of 9.7 T (4.5 K). 720 Nb_3Sn strands are arranged in a stainless steel jacket (cable diameter: 37.5 mm).

ity at magnetic fields sufficient for most applications (i.e., ~8 T at 4.2 K and ~10 T at 1.9 K). Extensive work on this alloy system has demonstrated that both the transition temperature and the upper critical field are optimal near ~40 at.% Ti, a composition in which – according to the phase diagram – a significant fraction of the normal conducting α-Ti phase is formed as a precipitate under suitable processing conditions and acts as pinning centers together with the dislocation cell structure and other crystalline defects. After nearly 40 years of development, significant progress has only recently been achieved by optimizing all processing steps toward the defect structure on a nanometer scale at the final wire diameter. Critical current densities of $(4-4.6) \times 10^9$ A m^{-2} (4.2 K, 5 T) have been reported; room for further improvements can still be expected.

Unfortunately, the intrinsic material parameters limit the range of applicability to magnetic fields of ~8 T at 4.2 K. Higher fields can only be achieved by turning to the intermetallic compound Nb_3Sn, which offers much less handling comfort for the magnet assembly due to its brittleness (small strain tolerance of about 0.2–0.4%). Consequently, a "wind and react" concept has to be employed for the magnet assembly in most cases, i.e., the entire wire and cable production and the winding of the magnet (or magnet sections) have to be done with a "precursor" material. The "final product" is then subjected to a heat treatment procedure involving fairly elevated temperatures (600–700 °C), which are needed to form the superconducting A15 phase. Unfortunately, these temperatures are far too high for one essential component of the magnet, i.e., the conductor insulation, which usually consists of glass-fiber-reinforced plastics (usually epoxies). Therefore, the windings have to be carefully separated (without bending them too much) to provide room for the application of the glass fiber tapes (Fig. 14.55). After reassembly, the entire winding pack is then subjected to a "vacuum impregnation" process in which the epoxy is introduced and allowed to form the plastic compound at a certain curing temperature (~120 °C), thus providing the winding pack with the required electrical insulation as well as with sufficient strength against shear and compression loads exerted by the Lorentz forces.

In order to actually produce Nb_3Sn multifilamentary wires, a variety of techniques have been devised, among them the "bronze process" and the "internal tin process", which will be outlined roughly here. In the bronze process, Nb rods are placed in a bronze (Cu–Sn alloy) cylinder and arranged there in the desired configuration. After extrusion and repeated drawing and annealing steps, the Cu stabilizer and diffusion barrier materials are added, and a second billet configuration is assembled, again subjected to similar mechanical and thermal cycles, and finally twisted. The decisive heat treatment step will then allow Sn to diffuse from the bronze into the Nb rods and to react there to the desired A15 phase. The internal tin process relies on a different way of supplying Sn to the superconductor. In this case, Nb rods are again arranged in the desired form, but now in a hollow Cu cylinder, forming the initial billet which is extruded. Insertion of a Sn rod in the central hole is followed by drawing and annealing cycles, in much the same way as described above. In all cases, great care must be taken to achieve the best possible degree of reaction, i.e., to ensure that most

Fig. 14.55 Production step of the TFMC: application of the winding insulation (white glass-fiber tape, arrow) after the reaction heat treatment.

of the Nb volume is indeed transformed into Nb_3Sn by sufficient Sn uptake. Further improvements of the performance have been achieved by adding small amounts of Ti or Ta to the Nb rods, which in the end act as additional scattering centers for the charge carriers in the A15 phase, enhancing H_{c2} and thus the critical current densities at 4.2 K. Depending on design details and application requirements (stabilization, ac losses), J_c values of $(1–4) \times 10^9$ A m^{-2} (4.2 K, 12 T) over lengths of several kilometers, and of $\sim 10^8$ A m^{-2} even at 19 T (and 4.2 K), have been reported.

Turning now to the "emerging" conductors, comparatively rapid progress could be achieved in the fabrication of long conductors based on the cuprate HTS $Bi_2Sr_2Ca_2Cu_3O_{10+\delta}$ (BiSCCO-2223, $T_c \sim 110$ K). As mentioned in Section 14.6.2, texturing is required to align the c axis orientation of the grains. The route to success is based on the "powder-in-tube" (PIT) technique employing silver as the tube material. The principle is as follows. BiSCCO powder or rods are inserted in a silver tube (or several such tubes are arranged in a suitable form in a silver block for the production of multifilamentary conductors) and subjected to similar cold working (pressing, drawing, rolling) steps to those mentioned above for the metallic conductors, until a tape with typical dimensions of 3 mm × 0.3 mm is obtained. The tape is wound onto a drum and then subjected to a heat treatment cycle to form the superconducting phase. The temperature window for this process is extremely narrow (around 835 °C) and also depends on the sheath material (i.e., whether or not alloyed silver is used to increase the mechanical strength of the conductor). Great care has to be taken to optimize the density of the superconductor within the filaments as well as to achieve the highest possible volume fraction of the 2223 phase. At present, commercial processes are in place for the production of tapes in kilometer lengths offering critical currents in the range of 70 kA (at 77 K in self-field). The problem with this material lies in its very low current-carrying capability in external magnetic fields at this temperature (because of weak flux pinning), i.e., J_c drops to zero at around 400 mT at 77 K. This imposes severe limitations on its application range, e.g., to power cables or current leads, where the magnetic field at the conductor remains small. Of course, these restrictions no longer apply if lower temperatures are considered, because flux pinning drastically increases at temperatures below ∼40 K. Interestingly enough, recent development rather favors the sister compound Bi-2212 ($T_c \sim 85$ K) for applications at temperatures between 4.2 and 20 K, e.g., as insert coils for enhancing the ultimate magnetic field produced by such a "hybrid" magnet.

Returning to applications at temperatures around 77 K, higher magnetic fields can be achieved only by employing YBCO. In this case, however, the intrinsic physical properties of this material require "full" texturing, i.e., the additional alignment of the basal planes, a task that can be solved only by highly sophisticated techniques ("coated conductors," Fig. 14.56) as follows. The RABiTS technique (*Rolling-Assisted Biaxially Textured Substrate*) is based on texture formation by cold-working Ni or Ni alloy tapes leading (after a recrystallization heat treatment) to a biaxial grain structure with typical grain misorientation angles of the order of 10°. As the next step, a buffer layer (e.g., yttrium oxide, yttrium-stabilized zirconia (YSZ), ceria, or a combination of these) is deposited, which grows epitaxially on top of the substrate and serves as a protection layer against interdiffusion between the substrate metal and the superconductor (thickness: several 100 nm). The YBCO film is deposited next (by any suitable deposition technique), the alignment (being transferred to the YBCO grains via the buffer layer) is of the order of 5–6°, the thickness of the superconductor around 1.5 μm. The second technique, IBAD (*Ion Beam Assisted Deposition*), starts from an untextured metallic substrate (either of the same materials as above or stainless steel) and induces an in-plane texture of the buffer layer (initially YSZ) during its deposition by an assisting ion beam. The degree of texture is again of the

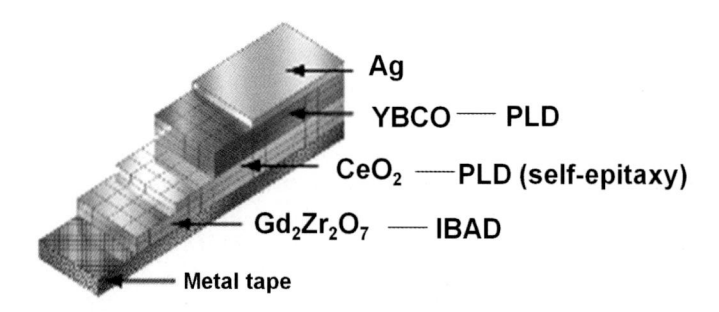

Fig. 14.56 Layout of a coated conductor.

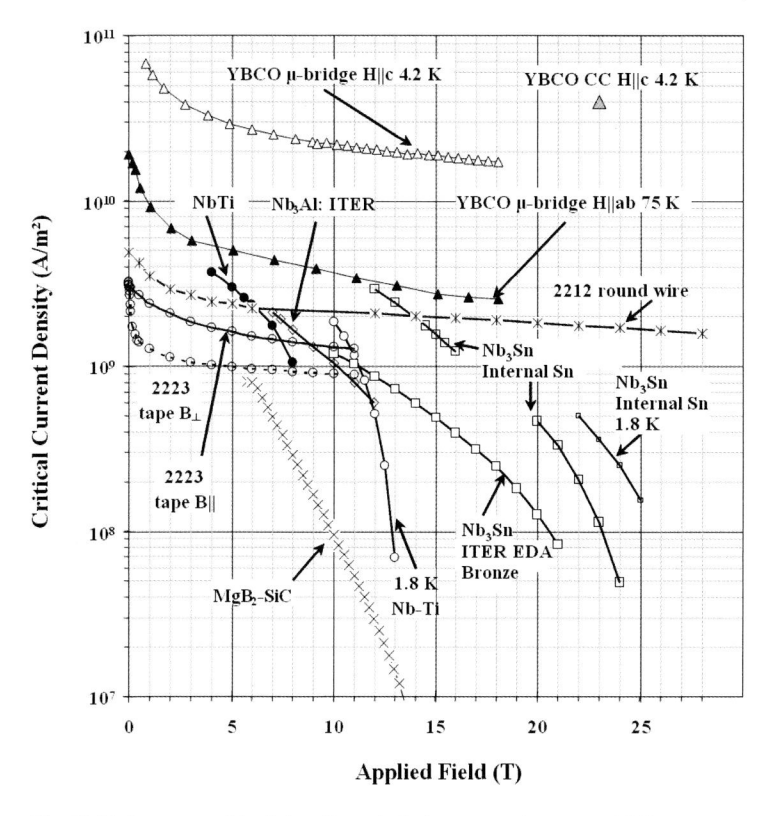

Fig. 14.57 Summary of J_c–B data for technical superconductors at 4.2 K (unless otherwise stated).

order of 10° or less, and the deposition of the superconductor proceeds as above. In this way, coated conductors with critical current densities exceeding 10^{10} A m^{-2} at 77 K and self-field have been achieved and maintained at a significant level in fields up to several tesla due to the much more favorable flux pinning conditions in YBCO at high temperatures. The present challenge consists mainly in developing these techniques for long length production and for devising suitable conductor architectures for high current transport.

The "new" superconductor MgB$_2$ has a certain application potential. It is competing with both the metallic and the ceramic superconductors, mainly because of its very high upper critical fields and the very low cost of the materials, but only at temperatures below ~28 K. It has already been developed in the form of multifilamentary wires (again based on the PIT technique) and can be produced in lengths of several hundred meters, but a significant improvement in its current-carrying capability is clearly necessary. A summary of the present situation, illustrated by a J_c–B diagram, is depicted in Fig. 14.57.

To conclude this section, a few remarks on materials for small-scale applications seem to be appropriate. The "workhorse" here is the elemental superconductor Nb ($T_c = 9.2$ K), which is used practically exclusively for all SQUID magnetometers and electronic devices, especially because of its robustness against thermal cycling and its moisture tolerance. NbN ($T_c = 16$ K) is certainly a candidate material for SQUID applications (because of the reduced cooling requirements compared to Nb, especially for mobile magnetometers), but also for integrated circuit technology. The same holds for YBCO thin films on single-crystalline substrates or bicrystals, as long as the advantage of the high operating temperature (extremely favorable cooling requirements) is not counterbalanced by the unavoidable enhancement of the noise level at these temperatures.

14.6.4
Applications

In 1915, Onnes predicted that applications of superconductivity were "not far away," an error that can obviously be made even by one of the greatest and most ingenious scientists. The short review of the development of superconductivity presented in the previous sections should, however, enable the reader to appreciate the numerous problems and obstacles provided by the intrinsic physics of the superconducting state as well as by technological issues that had to be solved in order to pave the way to success. Indeed, the development of superconducting magnets revolutionized the measuring techniques in solid-state, nuclear, low-temperature physics and chemistry laboratories worldwide, as did the successful development of SQUID magnetometers based on the Josephson effect. Huge scientific projects in high-energy physics, such as the accelerators at the Fermi National Laboratory (Batavia, IL, USA: Tevatron) or at CERN (Geneva, Switzerland: Large Hadron Collider), could not have been built without superconducting magnets. The same holds for the largest international project to be undertaken in the decade beginning from 2006, i.e., the construction of the nuclear fusion device ITER (at Cadarache, France) which is designed to demonstrate the feasibility of nuclear fusion for electric power generation.

Whereas all of these applications are somehow related to a scientific/technological environment, the largest and commercially most successful application of superconductivity has occurred in medicine, where superconducting magnets of unique uniformity and homogeneity form the basis for magnetic resonance imaging, a novel diagnostic tool with unprecedented resolution, especially in their latest versions employing 7 T magnets. In the same discipline, Josephson junctions arranged in the form of multiarray sensors have opened a new field of research and diagnostics, "biomagnetism," i.e., a new way of detecting (without any contacts) certain functions of our body and thus complementing the traditional electrocardiogram or encephalogram techniques.

Most of these applications do not really suffer severely from the requirement of cooling the superconductor with liquid helium. However, a much broader range of applications could easily be imagined, if liquid nitrogen cooling or the much more readily available cooling capacity of cryocoolers at temperatures in that range could be employed. Indeed, first power cables based on the high-temperature superconductor Bi-2223 are already being installed, YBCO thin-film filters for cellphone base stations have undergone successful field tests, fault current limiters for the protection of the power grid, motors, and transformers are in their final development stages, etc., etc. At the same time, the programmes for optimization of materials mentioned in Section 14.6.3 progress rapidly, thus making the prediction of a real breakthrough into the market in the near future much less risky than a few years ago.

Further Reading

Cardwell, David A., Ginley, David S. (Eds.) (2003), *Handbook of Superconducting Materials*, Institute of Physics Publishing, Bristol.

Cooley, L.D., Ghosh, A.K., Scanlan, R.M. (2005), *Supercond. Sci. Technol.* **18**, R51–R65.

Komarek, Peter (1995), *Hochstromanwendung der Supraleitung*, Teubner, Stuttgart.

Seeber, Bernd (Ed.) (1998), *Handbook of Applied Superconductivity*, Institute of Physics Publishing, Bristol.

Tinkham, Michael (1996), *Introduction to Superconductivity*, McGraw Hill, New York.

Index

a
ab-initio approach 589
accelerated algorithms 656
activated-state rate theory 378, *also see* transition state theory
activation barriers 637, 638
activation energy 284
activation energy of self-diffusion 310, *also see* self diffusion activation energy
activation entropy 284
activation volume 284
activity coefficient 236, 261
actuator, piezoelectric 777
ad-atoms 270
ad-vacancy 270
additional elements 930
adiabatic approximation 120, 132
advective motion 260
AFM 778, 781, 785, 786, 788
  – cantilever 778, 781
  – capacitive transducer 778
  – contact mode 779
  – dynamic mode 779
  – measurement of size distribution 786
  – metallography by 785, 788
  – nano-indenter (NI-AFM) 778, 788
  – nanoindentation 788
  – nanolaboratory 778
  – nanotool 778
  – tapping mode 779
aging time, precipitates 637
Ag–Co 444, 478
Ag–Cu 444, 478, 481
Ag–Fe 444, 449
Ag–Ni 444, 449
Al_3Li 384
Allen–Cahn equation 556
allotriomorph 396
  – grain boundary 396
AlNiCo 874, 875, 876

aluminum 620
  – attempt frequency 620
  – diffusion coefficient of lithium impurity atoms in aluminum 621
  – diffusion of Li impurities in 620
  – measured and calculated 621
  – phonon dispersions of the transition state 620
  – temperature-dependent diffusion coefficient 621
  – trapping 621
  – vacancy-assisted diffusion 620
alloys 19
alloys under irradiation 439, 440, 443
  – ballistic jumps 443
  – elementary effects 440
  – primary recoil spectrum 440
  – recombination 443
  – sink 443
Al–Cu system 636
  – specific heat 636
Al–Cu–Mg–Ag alloy 807
Al–Li alloys, precipitation 645 ff
  – size and shape distributions of precipitates 646
  – strain and chemical energy as functions of the precipitate size 646
Al–Sc 431
Al–Zn alloys, precipitation 444, 645 ff
  – size and shape distributions of precipitates 646
  – strain and chemical energy as functions of the precipitate size 646
Al–Zn system 634 ff
  – effective cluster interactions 636
  – formation enthalpies 634
  – ground state structure 634
  – input structures for CE 634, 635
  – pair and multibody interactions 635
Ammann lines 51 ff, 52

amorphization 445 ff, 483
– displacement cascades 445
amorphous nanoparticles 498
amorphous magnets 861, 865, 866
amplification factor 474
anelastic relaxation 184
anharmonic(ity) 123, 142, 148–154, 164, 369, 418
anisotropic, anisotropy 123, 131, 135, 161, 863, 874
anisotropic magnetoresistance 915 ff
anisotropy, magnetic 899, 920
– induced anisotropy 899, 919, 920
– magnetocrystalline anisotropy 899
– shape anisotropy 899, 919
annihilation of dislocations 320
annihilation (of phonons) 140, 157 ff
anomalous scattering 758
anti-Invar 895
anti-site pairs 248
antiferromagnetic state 564, 570
antiphase boundary (APB) 237, 332, 557, 753, see also APB
antiphase domain boundaries 557, 727
antisite atoms (antisites) 174, 175, 359
antisite disorder equation 203
antistructure atoms 175
antistructure bridge (ASB) mechanism 249
antistructure defect 175, also see antisite atoms
APB 576, 580
aperiodic crystals 48 ff
approximants 51, 54
APT 831, 832, 833, 836, 839
– accuracy of reconstruction 833
– lattice plane 832
– reconstruction 832
– specimen preparation 836
– statistical accuracy 839
– tip radius 831
Arrhenius law 178
Arrhenius plot 180, 184, 191, 247, 284
– differential dilatometry 180
– negative curvature 247
– positron annihilation 184
– upward curvature at high temperature 191
Ashby maps 453
athermal martensite 409
athermal temperature 288
atom configurations 348
atom jump 178, 217
– phenomenological description 178
– statistical mechanics 217

atom probe tomography (APT) 817, 819, also see APT
atomic correlations 525
atomic diffusivity 494
– interdiffusion 494
– self-diffusion 494
atomic force microscopy (AFM) 776, also see AFM
atomic interaction energy 525, 535
atomic ordering 359
atomic relocation distance 469
atomic resolution 818
atomic structure imaging 791
Au–47.5 at.%Cd 409
AuCu-alloy 802
AuCu-I phase 802, 804
austenite 409
autocatalytic formation of Martensite 407
autocorrelation analysis 841
avalanche photodiode 723, 741

b

B19 (MgCd) structure 207
B2 (CsCl) structure 207, 210, 239, 240, 243, 246, 249, 251
– antistructure bridge (ASB) mechanism 249
– atom jump types 239
– correlation effects 240
– NNN jumps vs. triple-defect mechanism 251
– point defect concentrations 210
– PPM treatment of diffusion 243
– six-jump cycle 246
B2 alloy 744
B2 structure 239
B8 alloys NiSb 745
backscattering spectrometer 739
Bain distortion 413
Bain strain 412
ballistic diffusion coefficient 461
ballistic mixing 462, 465
ballistic relocation 456
band energy 122 ff
band structures 27, 28, 29, 31
Barker's formula 544
barostats 659
basic cluster 541
$BaTiO_3$ 406
beam tilt alignment 797
beam-assisted deposition 447
Bergman cluster 56
$(B.H)_{max}$ 873, 874
binary alloy 347

binding energy   120 ff
binomial expansion   348
biomagnetism   953
BiSCCO-2223   949
Bloch states   594
Boltzmann factor   361, 378, 618
Boltzmann–Arrhenius dependence   495
Boltzmann–Matano analysis   257
bond stiffness versus bond length   368
bonding   19, 22, 25, 26, 33, 34, 35, 36, 119
   – covalent   22, 25, 26, 34, 36
   – directed   36
   – exohedral   33
   – ionic   22, 26
   – metallic   26
   – multicenter   34, 35
   – skeletal   33
bonds   32, 33, 35, 36
   – covalent   32, 33
   – four-center   35
   – six-center   35
   – three-center   35, 36
   – two-center   33, 35, 36
   – two-electron   33, 36
Born principle   654
Born–Oppenheimer approximation   217
Bose-Einstein distribution   141, 144, 152 ff, 158
boundary conditions of MC simulation   674
Bragg–Williams approximation   199, 204, 361, 435, 525, 537, 583
Bragg–Williams description   582
brass β 37
brass γ 27, 29, 37, 42
brilliance   708
bulk modulus   123, 128 ff, 150 ff, 606
Burgers vector   281

c

c/a ratio   562
Cahn–Allen equation   695
Cahn–Allen model   694
Cahn–Hilliard diffusion equation   264, 526, 556, 572, 691
Cahn–Hilliard free energy   405
Cahn–Hilliard model   402, 552, 687
Cahn–Hilliard–Cook equation   694
CALPHAD   544, 582
canonical ensemble   669
cantilever   782
   – spring constant   782
capillary waves   726
carbon   446
carbon steel   406

Car–Parrinello method   658
cascade sizes   441
Cauchy relationship   127
CCD detectors   723
cell periodicity   71–74
central atomic forces   566
chalcogenide materials   922
charge transfer   368
chemical bonding   19, 20, 25, 26, 32, 34, 35, 37
   – covalent   20
   – ionic   20
   – metallic   20
chemical contrast   784
chemical disordering   441
chemical force   313
chemical potential   235
chemical preferences   381
chemical spinodal   405
chemical stability   546
Clapp lattice instability model   417
classical elasticity theory   125 ff, 166
classical nucleation theory   427, 432
climb   282, 341
close-packed structure   119, 148 ff, 154
cluster approximation   364, 365
cluster concentration   527
cluster dynamics (CD)   426, 427
cluster expansion method (CEM)   121 ff, 534, 564, 565, 625 ff
   – anharmonic strain effects   626
   – anisotropic strain effects   626
   – bridging the length scales   625
   – bridging the time scales   637
   – chemical energy   625
   – cluster interactions   630
   – constituent strain energy   628
   – construction of   630
   – elastic energy   625
   – epitaxial softening function   626
   – epitaxial strain energy   625, 627
   – geometrical figures   630
   – hydrostatic deformation energy   625
   – input structures   635
   – Ising model   629
   – superlattice concentration   628
   – Vegard's law   628
cluster functionals   121
cluster probability   527, 543, 578
cluster, subcritical fluctuation   386
cluster variation method (CVM)   365, 525, 536, 538
coagulation   265, *also see* particle coalescence
CoAl   198, 214

coarse graining operation   526
coarsening mechanism   265, 728
coated conductors   950
Co–Cu   451, 478
Coercivity   861, 862, 874
CoGa   181, 214, 216, 744
coherence length   722
coherent phase boundaries   635
coherent precipitation   383, 425
coherent scattering   726
coherent spinodal   405
cohesion   119 ff, 128
cohesive energy   119, 187, 499, 535
columnar structure   101–102
columnar-equiaxed transition   107–110
combined reactions   375
common tangent construction   353, 547
compactness   119
competing dynamics   475
compliant substrate   166 ff
composite model   326
composition   371
 – nonequilibrium   371
compositional patterning   442, 469, 480
compounds   19, 21, 22, 23, 25, 27, 29, 31, 32, 33, 34, 36, 37, 40, 43
 – covalent   32
 – Grimm–Sommerfeld   33
 – intermetallic   19, 22, 25, 29, 32, 34, 36, 37, 42, 43, 44
 – ionic   32
 – organic   25
 – polar intermetallic   32, 36
 – valence   21, 22, 23, 27, 31, 32
compressibility   128
compression   127, 132
computation of phase diagrams   683
concentration fluctuations   400, 404
 – strain energy   404
conductivity, see resistivity   866, 912 ff
configurational enthalpy   363
configurational entropy   363, 537, 583
configurational fluctuation   549
configurational thermodynamics   527
configurational variable   527
conjugate variable   545
conservation of solute   352
conservative field variables   555
constituent strain energy   629
constitutional defects   209
constitutional supercooling   373
continuity equation   225, 256, 555
continuous displacement cluster variation method (CDCVM)   569
continuous solid solubility   351

continuous solubility   351
convection   68, 82, 110–113
convective flow   480
coordination   20, 22, 34, 36, 37, 39
 – environments   39
 – numbers   20, 22, 34, 39
 – polyhedra   37
coordination number   534, 543
coordination shell   736
CoPt, $CoPt_3$ 123, 131, 161 ff, 164 ff, 244
correlation correction factor   539
correlation effects in diffusion   228, 229, 230, 232, 240
 – correlation factor   229
 – correlation factors for various lattice structures   230
 – impurity diffusion   232
 – ordered structure   240
correlation functions   528, 533, 543, 662
correlations   363
 – short-range   363
corrosion   449, 874
Cottrell–Stokes law   293, 325
coverings   54
CPA (coherent potential approximation)   893, 894
creation (of phonons)   140, 157 ff
critical cluster   388
critical cooling rate   930
critical current   940
critical current density   943
critical gradient concept   843
critical pressure   891
critical radius   385, 386
critical temperature   361, 364, 373, 380
cross-slip   305, 332
crystal rotation   801
crystal structure   19 ff
crystal-to-amorphous   452
crystalline anisotropy   863
cubic lattice (crystal, structure)   119, 127 ff, 135 ff, 146, 159 ff
Curie temperature   866, 870, 871, 874, 891, 898
Curie/Neel temperature   887
Cu–Al   407
Cu–Al–Ni   410
Cu–Co   384
Cu–Mo   444
Cu–W   449
Cu–Zn   407
Cu–Zn system   643
 – short-range order   644
 – Warren–Cowley parameters   644
Cu–Zn–Al   410

CVM 572, 580
cyclic response 454
cyclotron magnetoresistance 915 ff
CVM-PFM hybrid model 573, 577

d

D-sorbitol 483
$D0_3$ alloys Fe_3Si 745
$D0_{19}$ (Ni_3Sn) structure 207
$D0_3$ (Fe_3Al) structure 207
DAC (diamond anvil cell) 731
Darken equation 260, 261
Debye approximation 612
Debye frequency 145, 153, 569
Debye temperature 145 ff, 153, 569
Debye wave vector 153
Debye's model 144 ff, 149
Debye–Gruneisen model 569
Debye–Scherrer rings 519
decagonal quasicrystals 53
decomposition of alloys 837
defect annihilation 377
defect concentrations 209, 210
  – B2 ordered alloys 210
  – $L1_0$ ordered alloys 209
defects in metals and alloys 63, 173
deformation 124 ff, 129 ff, 154, 167
degradation mechanism 931
delay line detector 826
demagnetizing curve 874, 875
demagnetizing field 902
dendrite kinetics 78–82
dendrite periodicity 83–85
dendrites 74
  – intensive treatment 74–86
density (ensemble) 668
density functional theory 590 ff, 591
  – effective potential 592
  – exchange–correlation 591
  – generalized gradient approximation (GGA) 593
  – ground-state electron density 592
  – Kohn–Sham orbitals 592
  – local density approximation (LDA) 593
  – total energy of the ground state 591
density of states 27, 28, 29, 36
detailed balance 235, 425, 450, 457, 671, 737
detrending fluctuation analysis 728
diagonalize free energy matrix 549
diamond 446
diamond anvil cell 731
diatomic linear chain 136

differential anomalous scattering 758
differential dilatometry 179
differential thermal analysis 495
diffracted beam 795
diffuse scattering 549, 643, 683, 756
diffusion 198, 217, 259, 263, 265, 266, 268, 270, 371, 616 ff, 673, 847
  – diffusion barrier 617
  – diffusion mechanisms 617
  – diffusion paths 617
  – dislocation-core 266
  – grain-boundary 268
  – intermetallic compounds 198
  – nanocrystalline materials 847
  – nonreciprocal 259
  – pipe 266
  – processes at an atomistic level 617
  – short-circuit 265
  – suppressed 371
  – surface 270
  – uphill 263
diffusion coefficient 227, 236, 237, 259, 261, 495, 619, 621, 638
  – chemical 259
  – interdiffusion 259, 261
  – intrinsic 259
  – random walk 227
  – thermodynamic factor 236
  – tracer 237
diffusion couple 257
diffusion equation 256, 264, 402, 554
diffusion of solute atoms 291
diffusion potentials 433, 435, 436
diffusion theory of thermal activation 221
diffusion, ab-initio methods 617 ff
  – activated complex Eyring 618
  – attempt frequency 620
  – chain-of-states method 618
  – diffusion coefficient 619
  – diffusion path 618
  – minimum energy paths 617
  – molecular statics 617
  – nudged elastic band method 618
  – phonon dispersions of the transition state 620
  – slowest ascent 617
  – the transition state 618
  – transition state theory 617
  – Wert and Zener 618
diffusion-controlled glide 291
diffusionless transformation 406
diffusivity 550
disclination 799
  – dipole 800
  – partial 799

dislocation 95, 125, 166 ff, 175, 266, 281, 449, 799
- climb 309
- diffusion pathway 266
- dipole 799
- exhaustion 320
- line tension 286
- multiplication 317
- patterning 447
- sources 318
- source/sink of point defects 175
dislocation–dislocation interaction 321
dislocation–solute interaction 285
disorder-to-order transition 694
disordered phase 533, 548
disordering 444
displacement cascades 441, 468, 479
- cascade sizes 479
- disordered zones 479
displacive transformation 412
dissipative systems 443
dissolution of ordered precipitates 444
divacancy 174
divergence theorem 553
DLM (disordered local moment) 893
domain wall 899
Doppler shift 739
double well potential 558
DQCs (decagonal quasicrystals) 53
driven alloys 438
droplet algorithm 723
Drude formula 912
3DTAP 437, see also three-dimensional tomographic atom probe
Dulong and Petit (Law of) 143 ff, 153
dynamic light-scattering (DLS) 724
dynamic recovery 329
dynamic strain aging 291, 331
dynamic structure factor 663
dynamical equilibrium phase diagram 450
- control parameters 450
dynamical equilibrium phase diagram 471, 479
dynamical matrix 134 ff, 138, 163
dynamical phase transitions 453
dynamical sources of entropy 363
dynamical structure factor 158
dynamical theories of thermal activation 221

e

EAM (embedded atom method) 657
effective anisotropy 872
effective cluster interaction energy 534, 564, 565, 570
effective cluster interactions 636
effective diffusion coefficient 480
effective free energy 447, 465
effective Hamiltonian 457
effective interactions 477, 483
effective pair interaction (EPI) energies 204
effective potentials 121, 165
effective stress 283
effective surface energy 477
effective temperature 462, 470, 476, 483
EFTEM (energy filtered TEM) 792
eigenvalue, eigenvector, eigenstate 135, 137, 141, 149, 163, 549
elastic constant (modulus) 125 ff, 128 ff, 146, 161, 166 ff
elastic modulus 521
elastic properties 598 ff
- elastic constants from ab-initio methods 598
- elastic constants of Ti, Al, TiAl, NiAl 600
- elastic contants of Quartz and Schorl 601
- elastic energy density 598
- elastic moduli 600
- Hooke's law 598
- number of elastic constants for crystal systems and point groups 599
- polycrystalline material 600
- strain tensor 598
- stress tensor 598
- Voigt notation 598
elastic strain 393
elastic strain energy 701
elastic waves 146
elasticity 124 ff, 130, 161, 166 ff
electrical resistometry 182, 238
- kinetics of ordering 238
- point defects 182
electron 20, 21, 24, 27, 28, 29, 33, 34, 362
- concentrations 20, 21, 24, 27, 28, 29
- counts 27, 33, 34
- exchange and correlation 362
electron energy-loss spectroscopy (EELS) 791
electron localizability indicator 34
electron localization function 34
electron tomography 856
electron-to-atom ratio 367
electronegativity 20, 21, 23, 24, 27, 32, 367
electronic memory 923
electronic structure calculation 418
electronic total energy 569
electron–phonon interactions 370

Index | 961

ELF (electron localization function) 34, 36
ELI (electron localization indicator) 34
Elinvar 887
elongation, elastic 127
embedded atom method (EAM) 657
embedding function 657
embryo, subscritical fluctuation 386
energy 349, 362
    – configurational 349
    – exchange and correlation 362
energy density 873, 874
energy-dispersive X-ray spectroscopy (EDXS) 791
energy-filtered TEM 792
enthalpy of formation 606, 607
    – calorimetry 607
    – DFT enthalpy Miedema's model 608
    – magnetic compounds 609
entropy 349, 350, 363, 366, 536, 672
    – and CVM 536
    – atom vibrations 363
    – configurational 349
    – dynamical 363
    – electronic excitations 363
    – logarithmic singularity 350
    – magnetic spins 363
    – phonon 366
environment 37
    – atomic 37
epitaxial softening function 626
equation of motion 130, 133 ff, 137, 146
equiaxed dendrite 77, 81–82
equiaxed structure 101–103
equilibrium constituent strain energy 628, 633
equilibrium state of order 380
Euler–Lagrange equation 402
eutectic spacing 93
eutectic structures 90–93
eutectic temperature suppression 375, 496
eutectoid phase diagram 358
evaporation rates 430
excess free energy 583
exchange bias, see exchange field 919
exchange field 918
exhaustion hardening 322
exhaustion of dislocations 317
experimental investigation of phonons 156 ff
explicit Euler technique 692
external forcing 438
external strain 126 ff, 134
extraction electrode 823
extrinsic dynamics 438
extrusions 335

f

faceted interface 63
far-from-equilibrium transition 552
fast variable 464
    – adiabatic elimination 464
fatigue 333, 454
fcc-equivalent 564
Fe 865, 866, 867
Fe–Si 867
Fe–Zr–Cu–B 872
Fe_3Pd ($L1_2$) 564
Fe–Al 182, 448, 452, 744
FeCr alloy 838
femtosecond laser 855
Fe–Ni 409
Fe–Ni $L1_0$ 570
FePd 244
FePd ($L1_0$) 564
FePd, $FePd_3$ 136, 142 ff, 162 ff
$FePd_3$ ($L1_2$) 564
FePt 244
ferrimagnetism 908
Ferrites 866, 870
ferromagnetic state 564, 570
Fibonacci sequence 50 ff
Fick's laws 256
field evaporation 822
field ion microscope (FIM) 818, 819, see also FIM
field variables 552
FIM 821, 829
    – best imaging field 821
    – magnification 829
Finemet 872
finite-range ballistic jumps 476
finite-size effects 674
finite-size scaling 674
first-nearest-neighbors (nearest neighbour atoms) 348
first-order transition 549
first-principles calculation 576
Fisher model 848
FLAPW (full-potential linear augmented plane wave) method 561
flash memory 934
fluctuation kinetics 550
fluctuation–dissipation 475
fluctuations 384, 429, 433, 442, 459, 474, 479, 546
    – gas of clusters 429
    – source 433
    – subcritical 433
flux lines 942
flux pinning 943

fluxes 445
– chemical 445
– point defects 445
focused ion beam 837
Fokker–Planck equation 265, 389
force constants (matrix) 133 ff, 137, 140, 164
force matching method 658
forced atomic mixing 441
forced magnetostriction 892
forcing intensity 482
forest dislocations 321
forest mechanism 292, 325
forging 373
formation energy of a compound 606
formation enthalpy of a compound 631
Fourier spectral method 692
Fourier transformation, prediction of fluctuations 549
Fowler–Guggenheim correction factor 364
fracture 335, 338
frame-by-frame analyses 815
free energy 354, 362, 381, 542, 554
– minimize 354, 362
– surface 381
– versus composition 354
free energy function 362
free-electron laser 750, 764
free-electron-gas 26
freezing 373
– dendritic 373
Frenkel defect 173
Frenkel pairs 440
full-potential linearized augmented plane-wave (FLAPW) 596
full-potential linearized muffin-tin orbital (FLMTO) approach 596

g

GB (grain boundary) diffusion 848
Gear algorithm 662
(generalized) chemical potential 691
generalized gradient approximation (GGA) 561
(generalized) gradient energy coefficient 580
generalized V-matrix 579
genetic algorithm 459
genetic algorithm for the selection of geometrical figures 631
geometrical condition 528, 532, 533
giant magneto resistance (GMR) 850
giant magnetoresistance 915 ff
Gibbs coefficient 69
Ginzburg–Landau equations 254
Ginzburg–Landau free energy 552
Ginzburg–Landau type 554
Ginzburg–Landau type of free energy 572, 579
glass 374
glass transition 522
glass transition temperature 499
Glauber 675
glide of dislocations 282
global stability analysis 473
GMR (giant magnetoresistance) 919 ff
GMR sensors 850
– thermal stability 850
golden mean 50
Gorter model 877
gradient energy coefficient 402, 554, 558
grain boundary 98–100, 384, 395
– low-angle 384
– surface energy 395
grain boundary (as diffusion pathway) 268
grain boundary diffusion 852
grain boundary energies 664
grain boundary grooving 787
grain boundary, impurity effects 622 ff
– ab-initio modelling 622
– boron strengthener 623
– cleavage energy 623
– cleavage of 623
– cracking resistance 623
– critical energy release rate of a crack 622
– embrittling agents 623
– grain boundary embrittlement 622
– Griffith criterion 622
– Griffith energy 622
– hydrogen embrittlement 623
– hydrogen recombination poison 623
– hydrogen uptake 623
– theoretical fracture energy 623
grain growth 698
Grain structure 101–103
– important sternum 77
grandcanonical ensemble 669
graphite 446
grazing-incidence reflection 743
Green's function of diffusion equation 225
ground state search 639 ff
ground-state 525
ground-state analysis 561
ground-state diagram 640
– configuration space 641
– random alloy 641
ground-state structures 639
– united-space cluster expansion 639

groundstate diagram of a binary alloy   632
group velocity   142
growth of precipitates   265
Gruneisen parameter   151
Grüneisen behavior   885
Gummelt decagons   54

h
habit plane   407
half-metallicity   914
Hall–Petch law   337
hard disk drive (HDD)   895, 896, 920, 921
hard magnetic materials   861, 862, 863, 873
hardening   316
hardening stages   323
hardness measurement   788
  – local   788
harmonic approximation   132 ff, 137, 139 ff, 147
  – beyond ...   149 ff
harmonic oscillator   780
heat capacity   143, 145, 150 ff
heat flow   152 ff
Helmholtz free energy   349, 542, 568, 672
heterodyne X-ray studies   725
heterogeneous nucleation   383
heterophase fluctuations   432
$HfCr_2$: phase stability   607 ff
    – bulk modulus   607
    – enthalpy of formation   608
    – structural parameters   608
hierarchy of energies   545
high-density DVD (HD-DVD)   14
high resolution electron microscopy   519
high-energy ball milling   438, 448, 480, 483
high-energy electron irradiation   467
high-field susceptibility   892
high-resolution transmission electron microscopy (HRTEM)   791
Hohenberg and Kohn (1964)   591
homodyne X-ray studies   723
homogeneous nucleation   383
homogeneous state   578
homogeneous system   552
Hooke's law   126, 133, 369
HRTEM   791, 793, 796, 797
    – hot-stage   797
    – image resolution   791
    – in-situ   797
    – microscope parameters   796
    – practical aspects   796
    – principles of image formation   793
Hume-Rothery phases   20, 27, 33, 53
Hume–Rothery rules   367

hybrid model CVM-PFM   573, 577
hydrogen interstitial   193
hypercrystals   48 ff, 51
hysteresis curve   790, 791
hysteresis loop   861, 863

i
icosahedral quasicrystals   55
imaging compression factor   829
imaging gas   820
impingement rates   430
importance sampling   670
impurity diffusion   232
    – five-frequency model   232
in-situ microscopy   793
incubation time   391
indomethacin   483
induced anisotropy   864
industrial permanent magnets   874
inelastic neutron scattering   731
inelastic X-ray scattering   730
infinite-temperature dynamics   443
information resolution limit   795
inhomogeneity   552
inhomogeneous CVM   578
interaction forces   656
interaction parameters   504, 535
interatomic force constants   368
interchange energy   349
α_2/γ_m interface   812
interface   384, 802, 806, 807, 810, 811, 813, 815, 851
    – chemical sharpness   851
    – coherent   384, 802
    – dynamics   815
    – incoherent   384, 802, 811
    – kink   810
    – migration   802 ff
    – order-disorder   806
    – oscillatory motion   815
    – partly coherent   802, 807
    – semicoherent   384
    – velocity   815
    – wave-like fluctuation   813
Interface (Solid-Liquid)   63–64
    – detailed sternum   63–70
Interface curvature   70
interface fluctuation   806
Interface kinetics   65
Interface stability   71
interfacial energy   384, 432, 496, 689
interfacial free energy   429
interfacial width   689
intermediate scattering function   735

intermetallic alloys 20, 37, 330, 332, 736
intermetallics 736, 907 ff
internal stress 282, 321
interphase boundaries 802
interphase boundary 803, 807
– diffuse 803
– position 807
– thickness 807
interstitial solute atom 178, 195
– migration 178
– sites 195
interstitialcy mechanism 229, 230
– correlation effects 230
interstitials 449
intrinsic dynamics 438
intrinsic stability 544
Invar 885
inverse Kirkendall effect 445, 461, 463, 465
inverse Monte Carlo simulation 683, 757
inversion method of Conolly and Williams 630
ion trajectory 827
ion-beam mixing 477
irradiation 438, 444, 460, 478
– electron 444
irradiation-induced precipitation 463
irradiation-induced segregation 463
Ising model 679
isobaric–isothermal ensemble 668
isomorphous substitution 758
isothermal–isobaric ensemble 669
isothermal martensite 408
isothermal section 359
isotopic substitution 758
isotropic magnetic materials 864, 874, 881
isotropic system (Ginzburg–Landau approach) 553
IXS (inelastic X-ray scattering) 730

j

Jackson parameter 65
Jagodzinki symbol 40
Jagodzinski–Wyckoff notation 40
jogs 309, 310, 311
Josephson effect 941
jump diffusion 734
jump frequency 379

k

k-space formulation 549
Kagome nets 44
Kear–Wilsdorf lock 332
Kerr effect 903, 907 ff
Kinchin–Pease formula 440

kinetic master equation 379
kinetic Monte-Carlo simulations 247, 424, 456, 458, 637, 679
– aging of coherent precipitates 637
– diffusion 637
– growth 637
– microstructure evolution 637
– precipitation 637
kinetic path 380
kinetic processes 371
– deviations from equilibrium 371
kinetics, approach to equilibrium 371
kinetics of concentration fields 436, 460
kink-pair mechanism 293
Kirkendall effect 259
KMC see kinetic Monte Carlo 430
Kohn and Sham (1965) 591
Kolmogorov–Johnson–Mehl–Avrami (KJMA) growth equation 397, 926
Kurdjumov–Sachs relationship 409

l

$L1_0$ 565
$L1_0$ (CuAu–I) structure 209, 244
– diffusion properties 244
– point defect concentrations 209
$L1_0$ (CuAu) structure 207
$L1_0$ ordered phase 542, 571, 572, 573, 574
$L1_0$ structure 562
$L1_0$-disorder phase boundary 570
$L1_1$ ordered phase 564, 565, 567
$L1_2$ (Cu_3Au) structure 207, 242
– jump types 242
$L1_2$ structure 562
lactose 483
$La_2Zr_2O_7$ 446
Lagrange multiplier 542
laminated media 907
Landau-type series expansion 573
Langevin equation 474, 475
laser-assisted atom probe tomography 854
latent heat (freezing) 372
lattice gas 426
lattice phase field 437
lattice specific heat 611
lattice vibration effect 569
Laves phases 20, 21, 37, 43, 44, 46, 606
ledge motion 809
length scales 441, 468
– displacement cascades 441
– self-organization during external forcing 468
Lennard-Jones (LJ) potential 534, 656
Lever rule 68, 351

Lewis theory   34
Li–Al alloy   618
– diffusion paths   618
– energy barrier   618
– LiAl$_3$ 618
limited vacancy mobility   228
limits in MD   655
linear magnetostriction   864
linear stability analysis   463, 472
liquid phase   372
– undercooled liquid   372
liquidus (liquidus line)   356
lithography   837
local atomic displacement   569
local chemical analysis   817
local density approximation   362
locking–unlocking mechanism   295, 297
logarithmic singularity   350
long-period superlattice   727
long-range order (LRO)   197, 206, 240, 264, 359, 364
– Bragg–Williams theory   204
– correlation effects   240
– parameter   206, 360, 547
– point defects   197
– spinodal ordering   264
longitudinal recording   902
Lorentz microscopy   792
losses   865, 866, 868
– core losses   868
– current   866
– magnetic   865
LRO parameters   574
LSW (Lifshitz–Slyozov–Wagner) theory   265
Lyapunov function   462

m

machinability   875
Mackay cluster   56
Macro-segregation   110–113
MAGNEQUENCH   882
magnetic bit   897, 902, 904 ff, 910
magnetic Compton scattering   720
magnetic fluctuations   890
magnetic force microscopy (MFM)   782, 789
magnetic ground state   564
magnetic neutron scattering   710
magnetic RAM (MRAM)   896, 920
magnetic recording   896 ff
magnetic resonance imaging   953
magnetic storage   897, 906, 910
magnetic X-ray scattering   715
magnetization   897 ff

– remanent magnetization   899
– spontaneous magnetization   898
magneto-optics   897, 903, 907 ff
magnetocrystalline anisotropy   906
magnetoplumbite structure   878
magnetoresistance   903, 915 ff
– anisotropic-AMR   915
– cyclotron   915
– giant-GMR   915
– tunnel-TMR   915
Magnetostriction   864
magneto–volume coupling   887, 889, 890
magnons   363
many-body interactions   525
many-body potentials   657
Markov chains   667, 671, 675
Markov process   735
martensite   406, 413, 415
– characteristics   406
– crystallography   413
– models   415
martensite finish temperature   408
martensite midrib   407
martensite plate   407
martensite start temperature   408
martensitic transformation   406, 407
massive transformation   411, 811
master equation   253, 430, 667, 735
master equation method   253
Matano plane   258
mean curvature   560
mean free path   913, 917
mean-field approximations   253
– crystal energetics   199
– kinetic models   253
mean-square displacement   673
mechanical activation   483
mechanical milling   799
mechanical stability   546
media noise   904
melting temperature depression   505
memory device   921
Mendeleev number   24
Meso-segregation   110–113
metallic glass   86, 375, 745, 757
metallic radius   367
metastable states   382, 549
method of inequality   536
method of overlapping distributions   664
Metropolis-type algorithms   675, 677
MgB$_2$ 944, 952
micro-segregation   85
microbeams   726
microcanonical ensemble   659, 668, 669

microscopic master equation 456
microscopic methods 774
microstructure 347, 372, 373, 864–866, 873, 880
    – dendritic 372
    – evolution 347
    – of magnetic materials 864–866, 873
migration profiles 618
milling intensity 448, 452
mixed-space cluster expansion 633
mixing (forced mixing) 481
    – superdiffusive 481
Molecular dynamics (MD) 221, 266, 441, 455, 654
    – atom jump 221
    – special diffusion paths 266
moment–volume instabilities 889
Monte Carlo (MC) simulation 247, 419, 458, 635, 667, 744, 838
    – bridging the time scales 637
    – coherent phase boundaries 635
    – compared to other descriptions of kinetics 255
    – correlated jump cycles 247
    – phase transition 636
    – specific heat 635, 636
Morse potential 569
Mössbauer effect 244, 738
Mössbauer spectroscopy 244, 251
    – quasielastic 251, 738
motion, oscillatory of ledges 809
multibody interaction energy 561
multichannel plate 823
multichannel plate (MCP) 824
multifilamentary wires 947
multiple anode array 824
multiply twinned structure 520
multiscale calculation 526, 560
multiscale modeling 625 ff
    – bridging the length scales 625
multivalue function 546
Mumetal 865
mushy zone (or mush) 110

n

8-N rule 32, 33
n-point cluster 531
n-point correlation function 531
n-state Potts model 698
n^c-fold symmetry 49
nanoclusters 644
nanocrystalline alloys 336, 866, 872, 880
nanohardness 789
nanomagnets 789

nanostructured materials 419
nanostructures 480
natural iteration method 542
Nb 952
Nb_3Sn 406, 948
Nb–Ti 947
Nd-Fe-B 879
near-equilibrium transition 552
nearest-neighbor pair interaction 525, 536
nearest-neighbor pair probability 527, 533
neutron depolarization 714
neutron source 367
    – pulsed 367
    – reactor 367
neutron spin echo 714, 740
neutron weight 367
    – correction 367
Newton–Raphson procedure 542
Newton's equations of motion 655
Ni–Al 182, 197, 214, 444, 469
    – diffusion 197
Ni_2Al_3 216
Ni_3Al 197, 208
    – diffusion 197
Ni–Fe 868
Ni_2Ga_3 216
Ni_3Ga_4 216
Ni_4Mo 444
Ni_3Sb 197, 216
    – diffusion 197
nickel Σ5{001} twist grain boundary 622 ff
    – boron strengthens 622
    – grain Boundary Cohesion 624
    – grain boundary segregation 624
    – impurity elements He, Li, S, H, C, Zr, P, Fe, Mn, Nb, Cr, and B 622
    – segregation Preference 624
    – surface segregation 624
NiGa 214, 744
Ni–Si 444
Ni–Zr 448, 452
noise from fluctuations 442
non-equilibrium 371
    – compositions 371
    – phase fractions 371
nonconservative field variable 555, 556
noncrystallographic symmetry 49
nonpolarized calculations 562
nonresonant magnetic scattering 717
non-volatile memory (NVM) 923, 924
normalization condition 532, 542
Nosé–Hoover thermostat 660

NPT ensemble (number of particles, pressure, temperature) 659
nuclear collisions 440
nuclear inelastic scattering 732
nuclear resonant scattering 741
nuclear waste 446
nucleation 264, 373, 382, 384, 387, 391, 392, 395, 396
– and growth rate 926
– coherent 383
– Gibbs droplet 387
– grain boundary 396
– heterogeneous 395
– heterogeneous, incoherent 384
– homogeneous 395
– homogeneous, coherent 384
– incoherent 383
– kinetics 373
– rate 387
– time-dependent 391, 392
nucleation, growth, and coarsening 436
nucleation of product phase 843
nucleation rate 388
– steady state 388
nucleation theory 386, 466
– irradiation 466
nucleation time lag 391
nucleus 386, 393
– strain energy 393
nudged elastic band method 222
– atom jump 222
– dislocation core diffusion 267
null cluster 536

o

occupation operator 528
octahedron cluster 541
octet rule 25, 26
Olson–Cohen model 416
Onsager coefficients 436, 461
Onsager's transport coefficients 435
operating temperature (permanent magnets) 875
optical disk 922
optical memory 923
order parameter 360, 547, 552
order–disorder 206, 452
order–disorder interphase boundary 803
order–disorder transition 206, 478, 575
ordered alloys 197, 238
– atom migration 238
– point defects 197
ordered arrangement 531
ordered phase 533

ordering 350, 752
ordering transformation 359
organic compounds 448, 483
Orowan law 282, 315, 317
oscillator 368
– frequency 368
oscillatory regimes 453
overcounting problem 364
overlapping distribution method 672
overwrite cycle performance 931
overwrite speed 926
Ovonic effect 922
Ovonic Electrical Cognitive device 937
Ovonic memory switch (OMS) 923, 934
Ovonic threshold switch (OTS) 922
Ovonic Unified Memory (OUM) 934
Ovshinsky effect 922

p

packing 40
– closest 40
packings 39, 45
– close 39, 45
– closest 39, 45
pair and multibody interactions 635
pair approximation 363, 378
pair correlation function 530, 734, 757
pair exchange 681
pair interaction model 187, 204
– Bragg–Williams approximation 204
– vacancy formation 187
pair potentials 349
pair probabilities 378, 527, 531
pair variables 362, 364
pair-correlation function 746
pair-interaction energies 680
pairwise interaction 656
parallel space 49
partial diffusion coefficients 738
partial dislocations 416
partial structure factor 758
particle coalescence 478
particulate media 905
partition function 219, 348, 365, 668
– oscillator 366
path probability function 550, 551
path probability method (PPM) 243, 255, 264, 526, 549
– B2 diffusion 243
– compared to other descriptions of kinetics 255
– spinodal ordering 264
patterning 477
patterning of chemical order 478

Pauli's exclusion principle 591
Peierls potential 293
Peierls-type friction forces 293
peierls-type mechanisms 298
Penrose tilings 51 ff, 58
peritectic (phase diagram) 356
  – melting 357
peritectoid 358
Permalloy (NiFe) 790, 865, 868
Permeability 862, 865, 866
perpendicular recording 907 ff, 908 ff
perpendicular space 49
persistent slip bands (PSBs) 333
pharmaceutical compounds 483
ω-phase 412
phase boundary 355
phase change 499
  – crystalline-to-amorphous 499
phase change optical disk 925
phase change RAM (PRAM) 935
phase diagram 66–88, 95, 347, 355–359, 878, 879
  – binary alloy 357
  – continuous solid solubility 355
  – eutectic 356
  – peritectic 356
  – pseudo-binary 359
  – ternary 358
  – unmixing 355
phase equilibrium 347
phase field 437
phase object imaging 794
phase separation 263, 531, 755
phase transformation 347, 373, 377, 382
  – nucleation and growth 373, 382
  – second-order 377
phase transition 497
  – crystalline-to-liquid 497
phase-field modeling (PFM) 82–114, 419, 459, 460, 526, 576, 686
phases (intermetallic compounds) 19
phason 51
phason walls 341
phonon density of states 141, 366, 664
phonon dispersion 603
phonon dispersions, ab-initio 610
phonon DOS 367
phonon electronic structure calculation 367
  – phonon 367
phonon entropy 365–368
  – electronegativity 368
  – trends 367
phonon free energy 366
  – phonon 366

phonon occupancy 366
phonon spectrum 732
phonon–phonon interactions 370
phonons 604
  – Born–Oppenheimer approximation 604
  – Bornūvon Kármán periodic boundary condition 604
  – dynamical matrix 604
  – entropy 605
  – equations of motion 604
  – force constant matrix 604
  – free energy 605
  – harmonic approximation 604
  – internal energy 605
  – phonon density of states 605
phonos, calculation of 602
  – ab-initio methods 602
  – bond charge models 602
  – density functional perturbation theory 602
  – direct methods 603 ff
  – force-constant model 602
  – frozen phonon 603
  – interplanar force constants 603
  – Keating models 602
  – linear response 602
  – model potentials 602
  – molecular dynamics trajectory 602
  – phonon dispersion 603
  – shell models 602
  – velocity autocorrelation function 602
piezoelectric drive (SPM) 777
planar flow casting 866
Planck distribution 366
plastic deformation 447, 480
point approximation 362, 541
point correlation function 529
point defect supersaturations 443, 449
point defects 94, 173, 174, 176–178, 180, 182, 195, 197, 204, 612
  – Bragg–Williams model 204
  – chemical potentials 613
  – defect formation energies 612, 615
  – elastic interaction 195
  – enthalpy of formation 182
  – enthalpy of migration 182
  – equilibrium concentration 176, 180
  – free energy of association 177
  – free energy of binding 177
  – free energy of formation 176
  – free energy of migration 178
  – Gibbs free energy 613

- grandcanonical ensemble   613
- in ordered alloys   174, 197
- in pure metals   174
- mobility   178
- non-equilibrium production   177
- noninteracting defects   613
- supercell model   612
- vacancy antisite   612
point probability   527
point-defect complexes   176
polarization, spin   913 ff
polarized neutron reflection   714
polarized neutron scattering   710
polymorphic nucleation   845
Portevin–le Châtellier (PLC) effect   291
position-sensitive atom probe (PoSAP)   824
positron annihilation   183
potential (Lennard-Jones type)   534
precipitate   384
- coherent   384
- incoherent   384
precipitate plate   807
precipitates   469
- coherent   644
- disordering   469
- dissolution   469
precipitates, shape   645
- chemical energy   645
- strain energy   645
precipitates, size   645
precipitation   383, 424, 468, 641, 697
- irradiation-induced   468
precipitation hardening   645
predictor–corrector method   662
projector augmented wave potentials   596
pseudopotential methods   596
pseudostable state   382
PtNi alloy   783
Pt–Rh system   626 ff
- constituent strain energies   629
- effective cluster interactions   640
- Epitaxial softening function for Pt and Rh   627
- ground-state diagram   640
- ground-state structures   641
- mixing enthalpies   641
- softening functions for Pt and Rh   626
pyrochlore compound   446

q
QCs   49 ff
quantum size effect   491
quasi-molten state   500

quasi-steady states   443
quasicrystals   48 ff, 339, 763
quasielastic Mössbauer spectroscopy   738
quasielastic neutron scattering   738
quasiharmonic   369
quasiharmonic approximation   569
quasilattice   51
quasiparticle interactions   363
- high temperature   363
quasiperiodic structures   50

r
radiation-induced segregation   445
radii   20, 22, 23
- atomic   23
- orbital   23
- pseudopotential   22
radius   20, 44
- ratio   44
- ratios   20
random solid solution   348, 531, 535, 536
random walk   222, 227
- definition of diffusion coefficient   227
rapid heating   371
rapid Solidification   86–90
rate theory   432
reactive diffusion   843
read head   902, 919
real-space pair interactions   632
recording media   905
recording tape   896
recovery   316
Redlich–Kister expansion   583
refinement   103–107
reflectron geometry   822, 823
relative stability   544
remanence   876
remanence enhancement   880, 881
replacement collision sequences   441
residence time algorithm   247, 458, 638, 676, 678, 682
residual resistometry   182, 238
- kinetics of ordering   238
- point defects   182
resistivity   182, 238, 249
resonant magnetic scattering   718
restoring force   546
reverse Monte Carlo   683
Richardson pair   480
rough interface   64
rule   25, 26, 32, 33
- 8 – N   32, 33
- octet   25, 26
- Wade's   33

S

saddle-point energies 680, 681
saddle points 382
sampling 668
Samson cluster 57
SAXS 727
scaling approach 479
scanning atom probe 854
scanning probe microscopy (SPM) 774–776, see also SPM
scanning transmission electron microscopy (STEM) 792
scanning tunneling microscope (STM) 775, 776, see also STM
scattering function 735
Scheil law 68
Scherzer defocus 795
Scherzer resolution limit 795
Schottky defect 175
Schottky product equation 203
Schrödinger equation 591
Scientific Group Thermodata Europe (SGTE) 583
segregation 784
– chemical 67–69
segregation coefficient 67, 266
segregation limit 544
self diffusion activation energy 190
self-diffusion coefficient 310
self-healing 482
self-interstitial atoms 173, 193–195
   – enthalpy of formation 194
   – enthalpy of migration 195
   – entropy of formation 194
   – equilibrium concentration 194
   – in pure metals 193
   – volume of formation 194
self-interstitials 443
self-organization 468, 469
Sendust 865, 866
sensors 902, 916, 918
sequential reactions 377
severe plastic deformation 799
shape anisotropy 863
sharp interface model 573
shear 412
   – invariant 412
shearing 449
shield, magnetic 919
Shoji–Nishiyama relations 416
short-circuit diffusion 265
short-range ballistic jumps 476
short-range order 364, 378, 641
   – as nearest-neighbor correlation parameter Γ 642
   – diffuse scattering 643
   – pair correlations 643
   – Warren–Cowley short-range order parameter 643, 644
short-range order (SRO) parameters 547, 759
simple sampling 669
simulation of evaporation sequence 834
single crystal 94
single ion detector 823
single roller technique 871, 872
sink for defects 454
six-frequency model 242
six-jump cycle (6JC) 246
Slater–Pauling curve 894
Sm-Co 875, 878
small-angle neutron scattering 714
small-angle X-ray-scattering 727
Snoek relaxation 196
soft magnetic ferrites 869
soft magnetic materials 861, 862, 865
soft mode 417
soft mode transition 417
soft phonon modes 745
soft underlayer (SUL) 901, 909
solid solubility 518
solid solution 347
solid-state amorphization 375
solidification 63, 700
solidus (solidus line) 355
solubility 351, 354, 374, 467
   – irradiation 467
solutal boundary layer 67, 68
solute atoms 285
solute atoms in metals 195
solute partitioning 375
sound propagation, velocity 130
spatial scale 362
spectrometer 367
   – chopper 367
   – triple axis 367
spherical aberration of objective lens (TEM) 795, 797
spin diffusion lengths 915, 917
spin disorder 888
spin flipping kinetics 552
spin flipping probability 550
spin operator 528
spin polarizations 917 ff
spin-polarized 562
spinel 869

spinodal composition 838
spinodal decomposition 263, 365, 399, 402, 546, 693, 753, 838
 – Cahn and Hilliard 402
spinodal disordering 549
spinodal instability 473
spinodal ordering 264, 549, 762
spinodal phase transformation 752
SPM 774, 775
spontaneous volume magnetostriction 888, 892
SQUID 952
SRO parameters 574, 579
SRPAC 748
$SrTiO_3$ 446
stabilization 411
stainless Invar 886, 887
2γ state model 887
static random access memory (DRAM) 934
statistical kinetics 377
statistical mechanic 349
steady state 443
Stirling approximation 349, 364
STM 775, 783
 – chemical contrast 783
stoichiometric 881
strain energy 521
strain hardening 335
stress instabilities 289, 331
stress–strain curve 789
 – local 789
structural 19, 21, 29
 – motif 21
 – motifs 19
 – patterns 29
structural defects 216
structural instability 500
structural vacancies 216
structure 20, 21
 – maps 21
 – types 20
structure factor 632
structure of $Nd_2Fe_{14}B$ 879
subcluster 541
sublattice 173, 360, 527, 574
sublattice model 584
substitutional disorder 175
super Invar 886
super ultra-small angle neutron scattering 713
super-abundant vacancies 193
superconducting magnets 952
superconductors 939, 944
 – application

superparamagnetism 903 ff
supersaturated phase 383
surface 784
surface effect 491
surface energy 385, 496
surface ordering 783, 785
surface scattering 762
surface topology 785
SUSANS 713
symmetry 33, 36, 40
synchrotron radiation 708

t

T_0 line (locus of intersections of solid and liquid free energy curves) 375
tapes 905
Taylor Law 293, 321
TDGL equation 559, 574
technical superconductors 946
TEM 856, *see also* transmission electron microscopy
 – corrector optics 856
temperature-dependent phase and structural stabilities 609 ff
terrace–ledge–kink (TLK) mechanism 807, 808
tetragonal structure 123, 127, 135 ff, 164
tetragonal distortion 571
tetragonality 563
tetrahedron approximation 541–543
tetrahedron atomic interaction energy 536
tetrahedron–octahedron (TO) approximation 541, 572, 580
texture 868
thermal activation 178, 217, 283, 551
thermal conductivity 132, 149, 151 ff
thermal expansion 150, 369, 885
thermal expansion coefficient 151, 886
thermal expansion of a crystal 132, 147, 149, 150 ff
thermal gravimetric analysis 495
thermal stability 546
thermally activated mechanisms 283
Thermo-Calc 502
thermodynamic equilibrium 347
thermodynamic model 428, 429
thermostats 659
thin film 166 ff
Thomson scattering 716
three-dimensional tomographic atom probe 426
Ti-Al alloy 208, 244, 811

$TiCr_2$: phase stability 607 ff
– bulk modulus 607
– enthalpy of formation 608
– structural parameters 608
tie-lines 359
tight binding formalism 122
tilings 54
tilt grain boundary 800
time domain (Mössbauer spectroscopy) 733
time of flight spectroscopy 822
time scales 442, 677
time step for integration (MD) 655
time-dependent Ginzburg–Landau (TDGL) equation 254, 526, 554, 556, 572, 695
time–temperature–transformation 373
time–temperature–transformation (TTT) diagram 373
TMR (tunnel magnetoresistance) 919 ff
TO (tetrahedron-octahedron) approximation 543
tomographic atom probe 825
tomographic Volume reconstruction 830
total energy 119–23, 126, 131, 150, 153, 161 ff
tracer diffusion 733
train wheel 482
transferability 658
transformation 411
– displacive 411
– massive 411
transition kinetics 552
transition rate 457
transition state search 617
– chain-of-states methods 618
– identify the saddle points 617
– minimum energy paths 617
– molecular statics 617
– normal modes 617
– nudged elastic band method 618
– slowest ascent 617
transition state theory 217
transition temperature 197, 549, 939
transmission electron microscopy (TEM) 492, also see TEM
– dark field 480
trehalose 483
tribochemistry 481
triple defect (TD) mechanism 214, 250
triple junction 848
– transport 848
triple point junction 575
Tsai-cluster 57

TTT-diagram 89, 373
tunnel magnetoresistance (TMR) 915 ff
tunneling current 777, 783
twins 97
two-dimensional square lattice 533
two-time correlation function 727

u
umklapp phonon process 150, 153
undercooling 66, 69, 373
– capillary undercooling 70
– kinetic undercooling 66
*under*saturated solid solutions 463
unmixing 350, 355
– chemical 355
unpolarized neutron scattering 713

v
V-matrix 532
V_3Si 406
vacancy 173, 187–191, 193, 213, 309, 310, 335, 378, 449
– abundance in intermetallic compounds 213
– enthalpy of formation 188
– enthalpy of migration 191
– entropy of formation 189
– equilibrium concentration 190
– in pure metals 187
– interaction with hydrogen 193
– volume of formation 188
vacancy mechanism 217, 230
– correlation effects 230
vacancy wind 259
vacancy-atom exchange 681
vacancy-controlled diffusion 637
vacancy-ordered phases 216
vacancy–solute complexes 192
Van Hove singularity 142
velocity of sound propagation 130
Verlet algorithm 661
vibrational density of states (VDOS) 730
vibrational entropy 147 ff, 168
vibrational free energy 569
vibrational properties 601 ff
– relevance of 601
vibrational stabilization of alloys 147
vibrations (of lattice, atoms) (see Lattice vibrations)
videorecording 798
viscosity imaging 780
Voice law 329
Voigt's indices 125

Volterra process 281, 339
volume magnetostriction 864

w
Wade's rule 33
WAXS 727
weak phase object approximation (WPO) 794
wear 448
– Bielby layer 448
– mechanically mixed layers 448
– third body 448
wetting phenomenon 576
wide angle tomographic atom probe 854
wide-angle X-ray-scattering 727
work of formation 385, *also see* formation energy
write head 901, 919

x
X-ray absorption fine structure spectroscopy 496
X-ray intensity fluctuation spectroscopy 722
X-ray photon correlation spectroscopy 722
XAFS 496
XIFS 722
XPCS 722

y
YBCO 945, 950
yield stress anomalies 330
Young's modulus 128 ff, 167

z
Zeldovich constant 435
Zeldovich factor 389
Zener relaxation 196
zero-point energy 569
Zintl phases 32
$ZrCr_2$: phase stability 607 ff
– bulk modulus 607
– enthalpy of formation 608
– phonon density of states 610
– phonon dispersion 610
– specific heat 611
– structural parameters 608
– total free energy 611
$ZrMn_2$ 611
– total free energy 611
$ZrMn_2$: point defects 614 ff
– defect concentration 614
– defect formation energy 614
– defect free energy 614
– enthalpy of formation 614
– magnetic moments 616
ZrO_2 406

Related Titles

Herlach, D. M. (Ed.)

Solidification and Crystallization

322 pages with 204 figures and 20 tables
2004
Hardcover
ISBN-13: 978-3-527-31011-1
ISBN-10: 3-527-31011-8

Jackson, K. A.

Kinetic Processes
Crystal Growth, Diffusion, and Phase Transitions in Materials

424 pages with 279 figures and 11 tables
2004
Hardcover
ISBN-13: 978-3-527-30694-7
ISBN-10: 3-527-30694-3

Kostorz, G. (Ed.)

Phase Transformations in Materials

724 pages with 392 figures and 33 tables
2001
Hardcover
ISBN-13: 978-3-527-30256-7
ISBN-10: 3-527-30256-5

Sauthoff, G.

Intermetallics

165 pages with 33 figures and 3 tables
1995
Hardcover
ISBN-13: 978-3-527-29320-9
ISBN-10: 3-527-29320-5